INTERMEDIATE ALGEBRA

Connecting Concepts through Applications

MARK CLARK
PALOMAR COLLEGE

CYNTHIA ANFINSON
PALOMAR COLLEGE

BROOKS/COLE
CENGAGE Learning

Australia • Brazil • Japan • Korea • Mexico • Singapore • Spain • United Kingdom • United States

BROOKS/COLE
CENGAGE Learning

Intermediate Algebra: Connecting Concepts through Applications
Mark Clark & Cynthia Anfinson

Publisher: Charlie Van Wagner

Developmental Editor: Don Gecewicz, Carolyn Crockett, Rita Lombard

Assistant Editor: Stefanie Beeck

Editorial Assistant: Jennifer Cordoba

Media Editor: Heleny Wong

Marketing Manager: Gordon Lee

Marketing Assistant: Shannon Myers

Marketing Communications Manager: Darlene Macanan

Content Project Manager: Cheryll Linthicum

Art Director: Vernon T. Boes

Print Buyer: Karen Hunt

Rights Acquisitions Specialist: Roberta Broyer

Rights Acquisitions Specialist: Don Schlotman

Production Service: TBC Project Management, Dusty Friedman

Text Designer: Diane Beasley

Art Editor: Leslie Lahr, Lisa Torri

Photo Researcher: Bill Smith Group

Text Researcher: Isabel Alves

Copy Editor: Barbara Willette

Illustrator: Matrix Art Services, Jade Myers; Lori Heckelman

Cover Designer: Lawrence Didona

Cover Image: Masterfile Royalty Free

Compositor: MPS Limited, a Macmillan Company

© 2012 Brooks/Cole, Cengage Learning

ALL RIGHTS RESERVED. No part of this work covered by the copyright herein may be reproduced, transmitted, stored, or used in any form or by any means graphic, electronic, or mechanical, including but not limited to photocopying, recording, scanning, digitizing, taping, Web distribution, information networks, or information storage and retrieval systems, except as permitted under Section 107 or 108 of the 1976 United States Copyright Act, without the prior written permission of the publisher.

> For product information and technology assistance contact us at
> **Cengage Learning Customer & Sales Support, 1-800-354-9706.**
> For permission to use material from this text or product,
> submit all requests online at **www.cengage.com/permissions.**
> Further permissions questions can be e-mailed to
> **permissionrequest@cengage.com.**

Library of Congress Control Number: 2010935511

Student Edition:

ISBN-13: 978-0-534-49636-4
ISBN-10: 0-534-49636-9

Loose-leaf Edition:

ISBN-13: 978-1-111-56912-9
ISBN-10: 1-111-56912-6

Brooks/Cole
20 Davis Drive
Belmont, CA 94002-3098
USA

Cengage Learning is a leading provider of customized learning solutions with office locations around the globe, including Singapore, the United Kingdom, Australia, Mexico, Brazil, and Japan. Locate your local office at **www.cengage.com/global**.

Cengage Learning products are represented in Canada by Nelson Education, Ltd.

To learn more about Brooks/Cole, visit **www.cengage.com/brookscole**

Purchase any of our products at your local college store or at our preferred online store **www.cengagebrain.com**.

To my wife Christine for her love and support
throughout our lives together.
And to our children Will and Rosemary.
MC

To my husband Fred and son Sean, thank you
for your love and support.
CA

Extra thanks go to Jim and Mary Eninger
for opening up their family cabin for me
to spend many a day and night writing.
Also to Dedad and Mimi, who built the
cabin in the 1940's.
MC

About the Authors

MARK CLARK graduated from California State University, Long Beach, with a Bachelor's and Master's in Mathematics. He is a full-time Associate Professor at Palomar College and has taught there for the past 13 years. Through this work, he is committed to teaching his students through applications and using technology to help them both understand the mathematics in context and to communicate their results clearly. Intermediate algebra is one of his favorite courses to teach, and he continues to teach several sections of this course each year.

CYNTHIA (CINDY) ANFINSON graduated from UC San Diego's Revelle College in 1985, summa cum laude, with a Bachelor of Arts Degree in Mathematics and is a member of Phi Beta Kappa. She went to graduate school at Cornell University under the Army Science and Technology Graduate Fellowship. She graduated from Cornell in 1989 with a Master of Science Degree in Applied Mathematics. She is currently an Associate Professor of Mathematics at Palomar College and has been teaching there since 1995. Cindy Anfinson was a finalist in Palomar College's 2002 Distinguished Faculty Award.

About The Cover

Each cover image tells a story. Our goal with this cover was to represent how people interact and *connect* with technology in their daily lives. We selected this cover as it illustrates the fundamental idea of the Clark/Anfinson series—*connecting* concepts to applications and rote mathematics to the real world since the skills and concepts in this series have their foundation in applications from the world around us.

Brief Contents

CHAPTER 1 LINEAR FUNCTIONS 1

CHAPTER 2 SYSTEMS OF LINEAR EQUATIONS AND INEQUALITIES 129

CHAPTER 3 EXPONENTS, POLYNOMIALS AND FUNCTIONS 221

CHAPTER 4 QUADRATIC FUNCTIONS 291

CHAPTER 5 EXPONENTIAL FUNCTIONS 413

CHAPTER 6 LOGARITHMIC FUNCTIONS 483

CHAPTER 7 RATIONAL FUNCTIONS 551

CHAPTER 8 RADICAL FUNCTIONS 611

CHAPTER 9 CONIC SECTIONS, SEQUENCES, AND SERIES 683

APPENDIX A BASIC ALGEBRA REVIEW A-1

APPENDIX B MATRICES B-1

APPENDIX C USING THE GRAPHING CALCULATOR C-1

APPENDIX D ANSWERS TO PRACTICE PROBLEMS D-1

APPENDIX E ANSWERS TO SELECTED EXERCISES E-1

Contents

1 Linear Functions 1

- **1.1** Solving Linear Equations 2
- **1.2** Using Data to Create Scatterplots 15
  Using Data to Create Scatterplots • Graphical Models • Intercepts, Domain, and Range
- **1.3** Fundamentals of Graphing and Slope 35
  Introduction to Graphing Equations • Slope-Intercept Form of a Line • The Meaning of Slope in an Application • Graphing Lines Using Slope and Intercept
- **1.4** Intercepts and Graphing 54
  The General Form of Lines • Finding Intercepts and Their Meaning • Graphing Lines Using Intercepts • Horizontal and Vertical Lines
- **1.5** Finding Equations of Lines 64
  Finding Equations of Lines • Parallel and Perpendicular Lines • Interpreting the Characteristics of a Line: A Review
- **1.6** Finding Linear Models 80
  Using a Calculator to Create Scatterplots • Finding Linear Models
- **1.7** Functions and Function Notation 93
  Relations and Functions • Function Notation • Writing Models in Function Notation • Domain and Range of Functions

Chapter 1 Summary 112
Chapter 1 Review Exercises 120
Chapter 1 Test 123
Chapter 1 Projects 125

2 Systems of Linear Equations and Inequalities 129

- **2.1** Systems of Linear Equations 130
  Definition of Systems • Graphical and Numerical Solutions • Types of Systems
- **2.2** Solving Systems of Equations Using the Substitution Method 143
  Substitution Method • Consistent and Inconsistent Systems
- **2.3** Solving Systems of Equations Using the Elimination Method 155

2.4 Solving Linear Inequalities   163
Introduction to Inequalities • Solving Inequalities • Systems as Inequalities • Solving Inequalities Numerically and Graphically

2.5 Absolute Value Equations and Inequalities   175
Absolute Value Equations • Absolute Value Inequalities Involving "Less Than" or "Less Than or Equal To" • Absolute Value Inequalities Involving "Greater Than" or "Greater Than or Equal To"

2.6 Solving Systems of Linear Inequalities   189
Graphing Linear Inequalities with Two Variables • Solving Systems of Linear Inequalities

Chapter 2   Summary   203
Chapter 2   Review Exercises   209
Chapter 2   Test   212
Chapter 2   Projects   214
Equation-Solving Toolbox Chapters 1 - 2   216
Cumulative Review Chapters 1 - 2   217

3 Exponents, Polynomials, and Functions   221

3.1 Rules for Exponents   222
Rules for Exponents • Negative Exponents and Zero as an Exponent • Rational Exponents

3.2 Combining Functions   234
The Terminology of Polynomials • Degree • Adding and Subtracting Functions • Multiplying and Dividing Functions

3.3 Composing Functions   251

3.4 Factoring Polynomials   260
Factoring Out the GCF • Factoring by Grouping • Factoring Using the AC Method • Factoring Using Trial and Error • Prime Polynomials

3.5 Special Factoring Techniques   272
Perfect Square Trinomials • Difference of Squares • Difference and Sum of Cubes • Multistep Factorizations • Trinomials in Quadratic Form • Choosing a Factoring Method

Chapter 3   Summary   281
Chapter 3   Review Exercises   286
Chapter 3   Test   288
Chapter 3   Projects   290

4 Quadratic Functions   291

4.1 Quadratic Functions and Parabolas   292
Introduction to Quadratics and Identifying the Vertex • Identifying a Quadratic Function • Recognizing Graphs of Quadratic Functions and Indentifying the Vertex

4.2 Graphing Quadratics in Vertex Form 304
Vertex Form • Graphing Quadratics in Vertex Form • Domain and Range

4.3 Finding Quadratic Models 321
Finding Quadratic Models • Domain and Range

4.4 Solving Quadratic Equations by the Square Root Property and Completing the Square 335
Solving from Vertex Form • Completing the Square • Converting to Vertex Form • Graphing from Vertex Form with *x*-Intercepts

4.5 Solving Equations by Factoring 349
Solving by Factoring • Finding an Equation from the Graph

4.6 Solving Quadratic Equations by Using the Quadratic Formula 362
Solving by the Quadratic Formula • Determining Which Algebraic Method to Use When Solving a Quadratic Equation • Solving Systems of Equations with Quadratics

4.7 Graphing Quadratics from Standard Form 375
Graphing from Standard Form • Graphing Quadratic Inequalities in Two Variables

Chapter 4   Summary   389
Chapter 4   Review Exercises   398
Chapter 4   Test   402
Chapter 4   Projects   404
Equation-Solving Toolbox Chapters 3 - 4   407
Cumulative Review Chapters 1 - 4   408

5 Exponential Functions 413

5.1 Exponential Functions: Patterns of Growth and Decay 414
Exploring Exponential Growth and Decay • Recognizing Exponential Patterns

5.2 Solving Equations Using Exponent Rules 431
Recap of the Rules for Exponents • Solving Exponential Equations by Inspection • Solve Power Equations

5.3 Graphing Exponential Functions 439
Exploring Graphs of Exponentials • Domain and Range of Exponential Functions • Exponentials of the Form $f(x) = a \cdot b^x + c$

5.4 Finding Exponential Models 451
Finding Exponential Equations • Finding Exponential Models • Domain and Range for Exponential Models

5.5 Exponential Growth and Decay Rates and Compounding Interest 462
Exponential Growth and Decay Rates • Compounding Interest

Chapter 5   Summary   472
Chapter 5   Review Exercises   476
Chapter 5   Test   478
Chapter 5   Projects   480

x Contents

6 Logarithmic Functions   483

6.1 Functions and Their Inverses   484
Introduction to Inverse Functions • One-to-One Functions

6.2 Logarithmic Functions   496
Definition of Logarithms • Properties of Logarithms • Change of Base Formula • Inverses • Solving Logarithmic Equations

6.3 Graphing Logarithmic Functions   505
Graphing Logarithmic Functions • Domain and Range of Logarithmic Functions

6.4 Properties of Logarithms   512

6.5 Solving Exponential Equations   517
Solving Exponential Equations • Compounding Interest

6.6 Solving Logarithmic Equations   528
Applications of Logarithms • Solving Other Logarithmic Equations

Chapter 6   Summary   536
Chapter 6   Review Exercises   541
Chapter 6   Test   543
Chapter 6   Projects   544
Equation-Solving Toolbox Chapters 5 - 6   545
Cumulative Review Chapters 1 - 6   547

7 Rational Functions   551

7.1 Rational Functions and Variation   552
Rational Functions • Direct and Inverse Variation • Domain of a Rational Function

7.2 Simplifying Rational Expressions   567
Simplifying Rational Expressions • Long Division of Polynomials • Synthetic Division

7.3 Multiplying and Dividing Rational Expressions   577
Multiplying Rational Expressions • Dividing Rational Expressions

7.4 Adding and Subtracting Rational Expressions   581
Least Common Denominator • Adding Rational Expressions • Subtracting Rational Expressions • Simplifying Complex Fractions

7.5 Solving Rational Equations   592

Chapter 7   Summary   601
Chapter 7   Review Exercises   606
Chapter 7   Test   608
Chapter 7   Projects   609

Contents xi

8 Radical Functions 611

8.1 Radical Functions 612
Radical Functions That Model Data • Domain and Range of Radical Functions • Graphing Radical Functions

8.2 Simplifying, Adding, and Subtracting Radicals 629
Square Roots and Higher Roots • Simplifying Radicals • Adding and Subtracting Radicals

8.3 Multiplying and Dividing Radicals 635
Multiplying Radicals • Dividing Radicals and Rationalizing the Denominator • Conjugates

8.4 Solving Radical Equations 644
Solving Radical Equations • Solving Radical Equations Involving More than One Square Root • Solving Radical Equations Involving Higher-Order Roots

8.5 Complex Numbers 655
Definition of Imaginary and Complex Numbers • Operations with Complex Numbers • Solving Equations with Complex Solutions

Chapter 8 Summary 664
Chapter 8 Review Exercises 669
Chapter 8 Test 671
Chapter 8 Projects 673
Equation-Solving Toolbox Chapers 7 - 8 676
Cumulative Review Chapters 1 - 8 678

9 Conic Sections, Sequences, and Series 683

9.1 Parabolas and Circles 684
Introduction to Conic Sections • Revisiting Parabolas • A Geometric Approach to Parabolas • Circles

9.2 Ellipses and Hyperbolas 703
Ellipses • Hyperbolas • Recognizing the Equations for Conic Sections

9.3 Arithmetic Sequences 717
Introduction to Sequences • Graphing Sequences • Arithmetic Sequences

9.4 Geometric Sequences 729

9.5 Series 738
Introduction to Series • Arithmetic Series • Geometric Series

Chapter 9 Summary 748
Chapter 9 Review Exercises 755
Chapter 9 Test 758
Chapter 9 Projects 759
Cumulative Review Chapters 1 - 9 761

A Basic Algebra Review   A-1

Number Systems • Rectangular Coordinate System • Operations with Integers • Operations with Rational Numbers • Order of Operations • Basic Solving Techniques • Scientific Notation • Interval Notation

B Matrices   B-1

Solving Systems of Three Equations • Matrices • Matrix Row Reduction • Solving Systems with Matrices • Solving Systems of Three Equations to Model Quadratics

C Using the Graphing Calculator   C-1

Basic Keys and Calculations • Long Calculations • Converting Decimals to Fractions • Absolute Values • Complex Number Calculations • Entering an Equation • Using the Table Feature • Setting the Window • Graphing a Function • Tracing a Graph • Graphing a Scatterplot • Graphing an Inequality with Shading • Error Messages • Additional Features • Zooming to an Appropriate Window • Zero, Minimum, Maximum, and Intersect Features • Regression

D Answers to Practice Problems   D-1

E Answers to Selected Exercises   E-1

Index I-1

Applications Index

Business and Industry
Airlines, 237–239
Amusement parks, 381–382
Appliances, 196–197
Auto companies, 241, 242, 249, 359
Auto detailing, 173
Auto stereos, 248
Autos, used, 78
Backhoes, 62
Beer companies, 92
Bicycle companies, 4–6, 47–48, 190–191, 195–196, 200, 372
Business machines, 248
Business promotions, 13, 29, 385, 737
Business survival, 14, 51, 62
Cabinets, 173
Candy, 10, 89, 728
CDs, 14, 112–113, 115, 301, 347, 494
Cell phones, 328–330, 333–334, 372
Charities, 66, 71–72, 563, 590, 598
Christmas lights, 115–116
Clothing, 29–30, 93
Computers, 249, 256, 372, 382, 386
Cycling gloves, 332
Delivery services, 51, 92
Digital cameras, 241–242, 385–386
Electronics, 319, 494
Fruit, 108, 404
Golf clubs, 14
Hardware stores, 25–27, 30, 248, 386
Heaters, 624, 653
Jewelry, 200
Manufacturing, 140–141
Oil companies, 90
Online stores, 474–475
Open source code, 452–454, 455
Orange juice, 252–253
Pest companies, 13, 47
Pet stores, 113–114, 118
Photography, 12, 352–353
Pies, 29
Planes, 353
Pottery, 141, 151
Publishing companies, 11, 248, 257, 494
Record companies, 112–113, 115, 301, 347, 494
Restaurants, 48, 56, 78, 354
Salary structure, 11, 12, 51, 62, 142, 147–148, 153, 161, 174, 427, 564, 734, 737, 746, 747
Shoe companies, 85–87, 248, 372
Snow cones, 12–13
Sunglasses, 386
Supercomputers, 526
Tablecloths, 151, 173
Tire companies, 89
Tooling machine companies, 329
Toy companies, 78–79, 256, 494
T-shirts, 11, 14, 153, 302
Vacation time, 256–257
Water bottles, 332
Window cleaning, 13
Work rate, 597

Education
Athletes, 152, 167–169
Enrollment, 141, 152, 169, 245
Expenses, 51
Expenses/enrollment, 239, 360
Faculty, 172
Graduates/degrees, 141, 142, 151–152
Math class enrollment, 10–11, 55–56
Public libraries, 325–327, 365
Student-teacher ratio, 152, 558–559
Teacher salaries, 32
Tutoring, 78

Electronics
Electric current, 372, 564
Ohm's Law, 599
Wind energy, 78

Entertainment
Cable TV systems, 301–302, 323, 336–337, 373
Movie theaters, 353–354
Radio station play time, 112, 536
TV violence beliefs, 206
TV watching, 185, 312–313, 337

Farm Management
Egg production, 91
Farms, 109
Meat consumption, 89, 133–134, 145
Milk consumption, 92, 245
Weed growth rate, 455–456, 526

Finance
Auto depreciation, 65–66, 71, 117, 725, 737, 753
Business machinery depreciation, 738
Compound interest, 467, 469, 471, 475–476, 523–525
Hedge funds, 51
Investments, 148–149, 153, 197–198, 202, 318, 347
Loan sharks, 417, 435
Savings, 728, 746–747
Subprime loans, 246
Total income, 742, 744

Geometry
Area of a window, 238, 247
Box construction, 359
Circle circumference, 564
Fencing, 360
Framing, 360

Home Management
Candy consumption, 108
Cell phone plans, 12, 72, 173, 214
Clothing costs, 75, 79, 109
Computer owners, 333
Drywall repair costs, 235
Flooring costs, 125–126
Fruit/vegetable consumption, 245, 319, 494
Home improvement costs, 125–126, 161, 235
Internet use, 91–92, 333, 373, 458, 459
Spending, 598
Water consumption, 744

Medicine and Health
Alzheimer's disease, 519–521
Birth weight, 185
Birthrate, 173, 174
Body temperature, 188
Bypass surgery, 75, 79
Calories needed, 202, 556
Children's growth rates, 614–615
Diabetes incidence/costs, 78, 146–147
Disease spread, 437, 737
Health care costs, 92–93, 245
Health insurance costs, 251–252
Heart failure incidence, 318
Height, 188, 625–626
Homicide rate, 11, 16–18, 21, 24–25, 55–56, 494
IQ, 188
Life expectancy, 31
Medical costs, 77–78
Medical office costs, 6
Medicare, 62, 109, 559, 594–595
Medicine half-life, 737
Mortality rates, 31, 119, 152, 319, 739–740
Number of hospitals, 75–76, 79
Prescription drugs, 246–247
Sexual activity percentage, 62
Smoking rate, 47–48, 132
Swine flu, 739–740
Tumor growth, 544
Venereal disease incidence, 72–73, 91
Weight, 75, 79, 185
West Nile virus, 119

Miscellaneous

Beer head decay, 469–470, 527
Fibonacci sequence, 759
Fundraisers, 360
Lead paint use, 333, 372
Mixtures, 156–157, 160–161
Monthly temperature, 301, 321–323, 332, 387
Newspaper circulation, 142
Recycling, 78
Rumor spread rate, 416–417, 437

Politics, Government, and Military

GDP, 465
Housing, 598
Lottery revenue, 237–238
Marriage rate, 174
Missile production, 458–459
Population, immigrants, 249
Population, non-U.S., 28, 172, 465, 466, 469, 527
Population, U.S., 13, 31, 78, 240–241, 249–250, 426, 494, 599
Population, U.S. states, 18, 21, 25, 31, 62, 75, 77, 79, 80–82, 83–85, 101–103, 108, 465, 494
Population, world, 420, 469, 526
Population/national debt, 240–241
Population/pollution, 253
Poverty level, 317–318, 333, 347
Prison inmates, 246
Social Security benefits, 108
State expenditures, 598–599
Tax revenue, 91
Traffic accidents, 652
Unemployment, 365

Recreation and Hobbies

Car collection, 752–753
Concert tickets, 728
Cyclists, 108
Exercise, 372, 738
Fairs, 161, 257
Golf, 52, 161
Hobby stores, 67, 72
Parties, 564
Prize money, 28, 87–88
Snowmobiles, 255–256
Stadium attendance, 302–303
Surfing, 78
Tennis, 28, 248
Vacations, 12, 593

Science and Engineering

Air pressure, 11, 48, 56, 486
Amphitheater design, 742
Ant colony growth rate, 463–464
Bacterial growth, 418–420, 425–426, 434, 437, 449, 472
Bridge design, 701–702, 716
Butterfly allometry, 613–614, 625, 653–654
Comet orbits, 759
Cooling rate, 480
Cup stacking, 728, 747
Distance to the horizon, 653, 667
Drainage rate, 348, 459
Drop time, 624–625, 653
Earthquake magnitude/intensity, 529–530, 534
Encryption, 290
Fill rate, 597
Fire spread, 426
Flashlight design, 695–696
Force/acceleration, 557, 564–565
Gas volume/pressure, 564, 565
Gravity, 564
Humpback whales, 470, 526
Light decay, 422–423, 556–557, 564, 590, 599, 609
Lizard allometry, 625
Machine part tolerance, 182, 188
Mammal allometry, 625, 653
Period of a pendulum, 645–646, 652
PH, 530–531, 534
Planetary distances, 654, 709, 715
Planetary orbits, 709
Planetary periods, 673–674
Population decay, 475
Projectile path, 302, 318, 332, 333–334, 347–348, 358–359, 373, 386–387, 405, 737
Rabbit population growth, 480
Radioactive half-life, 420–422, 427–428, 449, 459, 737
Resource loss, 459
River depth, 257, 459
Satellite dish design, 696, 701
Sea otter population growth, 464–465, 522
Solar flare cycle, 391–392
Sound intensity, 534
Sound speed, 646–647, 653
Table design, 715
Turn radius, 673
Wall design, 728, 747
White-tailed deer population, 469, 526, 734
Wind energy, 78

Travel

Airline tickets, 318–319
Auto rental, 161
Auto repair, 141, 151, 555
Bus rental, 553–554
Gas consumption, 178, 386
Gas mileage and speed, 178
Gas prices, 11
Hotel costs, 564
Miles driven, 91, 179, 186
Taxi cost, 3, 11
Tour companies, 153, 563, 598
Trip costs, 153, 563, 590, 598
Truck rental, 2–3, 12, 37–39, 47, 55, 62, 131–132, 166–167, 173–174, 485–486

Preface

In teaching the Intermediate Algebra course, we wanted to help our students apply traditional mathematical skills in real-world contexts. So our quest was to find the right combination of algebra skill development and the use of those skills within real world applications. We taught from "traditional" math texts and supplemented the applications; this resulted in students resenting the "extra work." Later, we taught from an applications- and exploration-based text and supplemented skills problems; this met with the same response. We required a text with great applications that students could relate to, *as well as* skills that would lay a strong foundation for their future math classes and lives. This text is our attempt to address that challenge and is the result of many years of using this approach in our classrooms, refined with the help of many of you who shared your thoughtful feedback with us throughout our extensive development process.

Our *application-driven* approach is designed to engage students while they master algebraic concepts, critical thinking, and communication skills. Our goal is simple: to present mathematics as concepts in action rather than as a series of techniques to memorize; and to have students *understand and connect* with mathematics—*what it means in real-world contexts*—while developing a solid foundation in algebra.

What follows are the tools we developed to help your students master algebra . . . .

Discovery-Based Approach

Worked Examples with Practice Problems

Examples range from basic skills and techniques to realistic applications with manageable data sets to intriguing applications that call for critical thinking about phenomena and daily activities. In-text practice problems help students to gain a sense of a concept through an example and then to practice what was taught in the preceding discussion.

Worked examples and exercises require students to answer by explaining the meaning of basic concepts such as slope, intercept, and vertex in context. The benefit of requiring verbal explanations is that the student learns to communicate mathematical concepts and applications in a precise way that will carry over to other disciplines and coursework such as psychology, chemistry, and business.

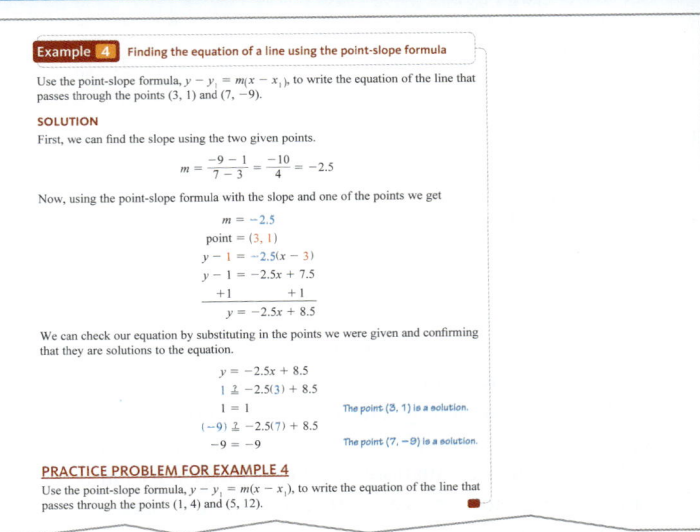

Example 4 Finding the equation of a line using the point-slope formula

Use the point-slope formula, $y - y_1 = m(x - x_1)$, to write the equation of the line that passes through the points (3, 1) and (7, −9).

SOLUTION
First, we can find the slope using the two given points.

$$m = \frac{-9 - 1}{7 - 3} = \frac{-10}{4} = -2.5$$

Now, using the point-slope formula with the slope and one of the points we get

$$m = -2.5$$
$$\text{point} = (3, 1)$$
$$y - 1 = -2.5(x - 3)$$
$$y - 1 = -2.5x + 7.5$$
$$+1 \quad\quad +1$$
$$y = -2.5x + 8.5$$

We can check our equation by substituting in the points we were given and confirming that they are solutions to the equation.

$$y = -2.5x + 8.5$$
$$1 \stackrel{?}{=} -2.5(3) + 8.5$$
$$1 = 1 \quad\quad \text{The point (3, 1) is a solution.}$$
$$(-9) \stackrel{?}{=} -2.5(7) + 8.5$$
$$-9 = -9 \quad\quad \text{The point (7, −9) is a solution.}$$

PRACTICE PROBLEM FOR EXAMPLE 4
Use the point-slope formula, $y - y_1 = m(x - x_1)$, to write the equation of the line that passes through the points (1, 4) and (5, 12).

■ **CONCEPT INVESTIGATION**
Is that a reasonable solution?

In each part, choose the value that seems the most reasonable for the given situation. Explain why the other given value(s) do not make sense in that situation.

1. If P is the population of the United States in millions of people, which of the following is a reasonable value for P?
   a. $P = -120$
   b. $P = 300$
   c. $P = 5,248,000,000$
2. If H is the height of an airplane's flight path in feet, which of the following is a reasonable value for H?
   a. $H = -2000$
   b. $H = 3,500,000$
   c. $H = 25,000$
3. If P is the annual profit in dollars of a new flower shop the first year it opens, which of the following is a reasonable value for P?
   a. $P = -40,000$
   b. $P = 50,000$
   c. $P = 3,000,000$

Concept Investigations

Concept Investigations are directed-discovery activities. They are ideal as group work during class, as part of a lecture, or as individual assignments to investigate concepts further. Inserted at key points within the chapter, each Concept Investigation helps students explore and generalize patterns and relationships such as the graphical and algebraic representations of the functions being studied.

Visualization

Graphing
Graphing is demonstrated with a step by step approach, using ideas such as symmetry to simplify the graphing process. A strong focus on graphing by hand as well as reading graphs is carried throughout the book to help students build these skills.

Hand-drawn Graphs
A hand-drawn style graph is often used when illustrating graphing examples and answers to exercises. The hand-drawn style helps students visualize what their work should look like.

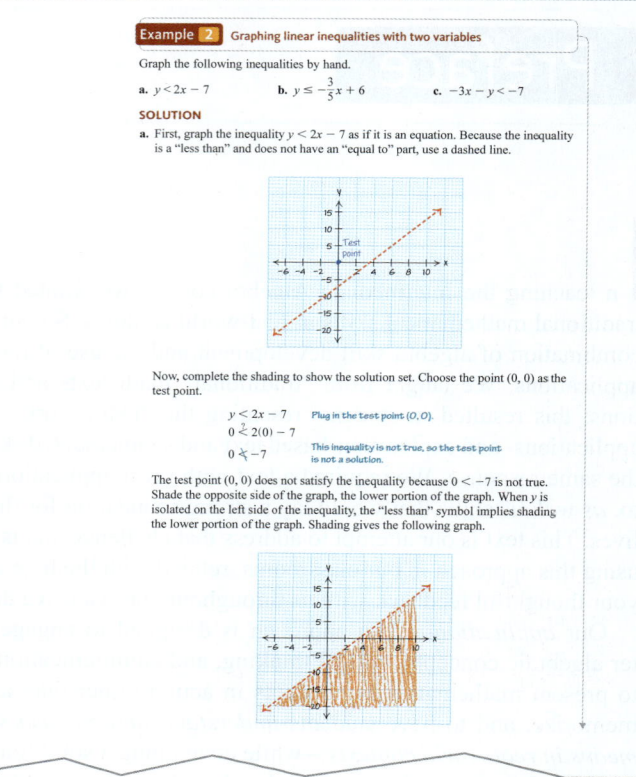

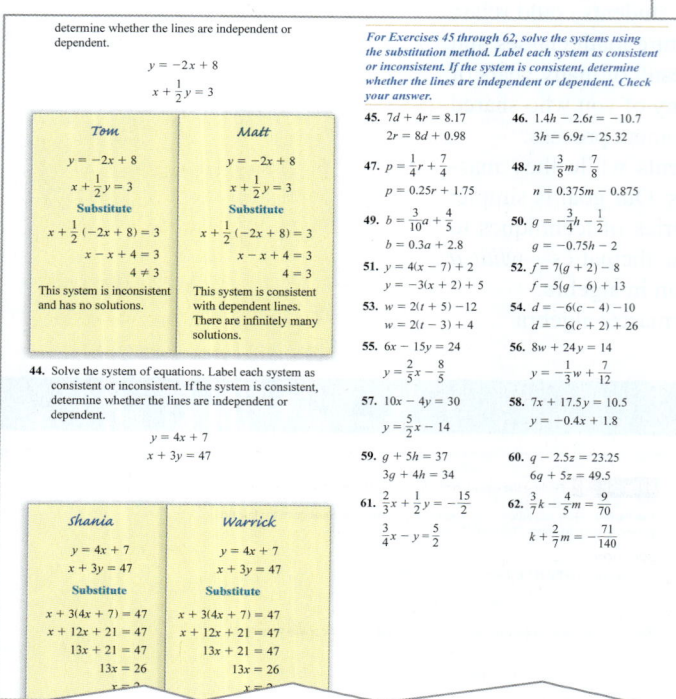

Integrated "Student Work"
Clearly identifiable examples of "student work" appear throughout the text in Examples and Exercises. These boxes present examples of correct and incorrect student work. Students are asked to identify and fix common student errors.

The "Eyeball Best-Fit" Approach
Linear, quadratic, and exponential functions are analyzed through modeling data using an eyeball best-fit approach. These models investigate questions in the context of real-life situations. Creating models by hand leads students to analyze more carefully the parts of each function. This reinforces solving techniques and makes a better connection to the real-life data and how the data effects the attributes of the function's graph. Graphing calculators are used to plot data and to check the fit of each model.

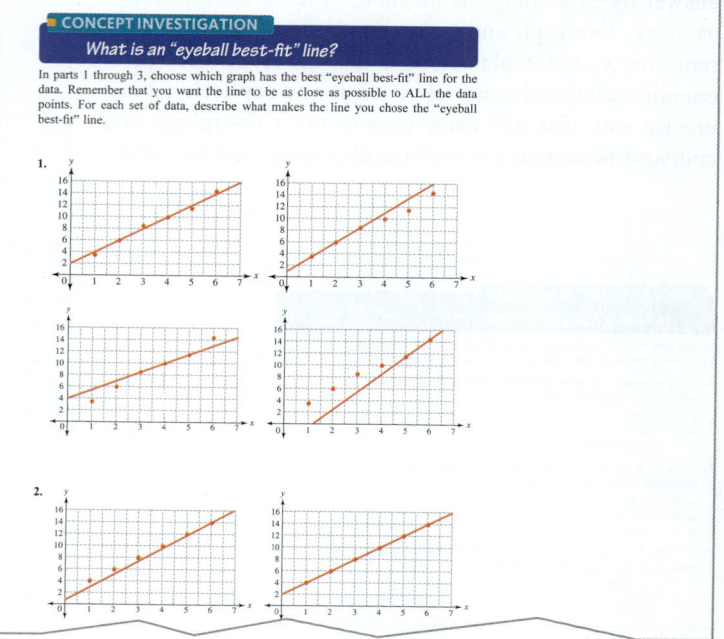

Margin Notes

The margin contains four kinds of notes written to help the student with specific types of information:

1. *Skill Connections* provide a just-in-time review of skills covered in previous sections of the text or courses, reinforcing student skill sets.
2. *Connecting the Concepts* reinforce a concept by showing relationships across sections and courses.
3. The specific vocabulary of mathematics and applications are helpfully defined and reinforced through margin notes called *What's That Mean?*
4. *Using Your TI-Graphing Calculator* offers just-in-time calculator tips for students. Additional calculator help is available in the *Using the Graphing Calculator* Appendix.

Exercise Sets

The exercise sets include a balance of both applications- and skill-based problems developed with a clear level of progression in terms of level of difficulty. Most exercise sets begin with a few warm-up problems before focusing on applications and additional skills practice. A combination of graphical, numerical, and algebraic skill problems are included throughout the book to help students see mathematics from several different perspectives. End-of-book answers are written in full sentences to underscore the emphasis on student communication skills.

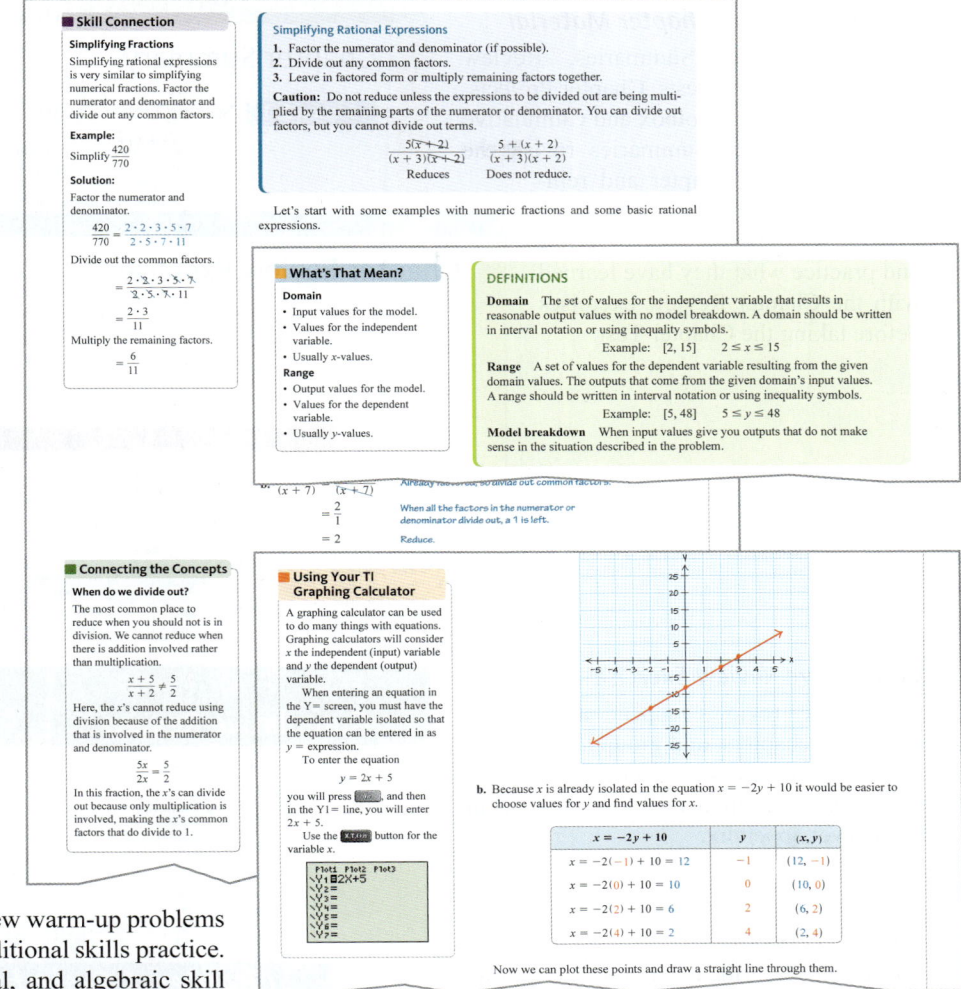

Making Connections

Application problems require a complete-sentence answer, encouraging students to consider the reasonableness of their solution. As a result, model breakdown is discussed and is used as a way of teaching critical thinking about the reasonableness of answers in some of the examples and problems throughout each chapter.

Learning with Technology

Appropriate Use of the Calculator

Graphing calculators are used in this text to help students understand mathematical concepts and to work with real-world data. Graphing calculators are used to create scatterplots of real data and to check the reasonableness of algebraic models for the data. The calculator is also used to check solutions both graphically and numerically and to do some numerical calculations. However, students are also required to solve problems, graph, and do other algebraic skills by hand.

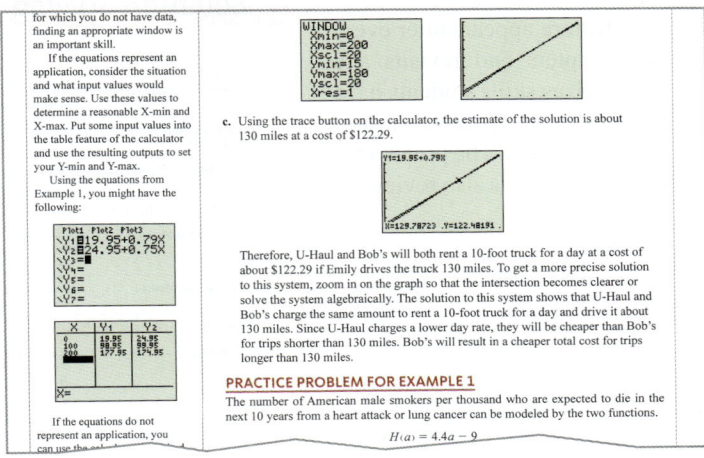

xviii Preface

Extensive End-of-Chapter Material
Includes Chapter Summaries, Review Exercises, Chapter Tests, Chapter Projects, Equation Solving Toolbox and Cumulative Reviews. Chapter Summaries revisit the big ideas of the chapter and reinforce them with *new* worked out examples. Students can also review and practice what they have learned with the Chapter Review exercises before taking the Chapter Test.

Cumulative Review Exercises
Cumulative reviews appear after every two chapters, and group together the major topics across chapters. Answers to all the exercises are available to students in the answer appendix.

Chapter Projects
To enhance critical thinking, end-of-chapter projects can be assigned either individually or as group work. Instructors can choose which projects best suit the focus of their class and give their students the chance to show how well they can tie together the concepts they have learned in that chapter. Some of these projects include on-line research or activities that students must perform to analyze data and make conclusions.

Equation-Solving Toolbox
This feature appears after every two chapters and revisits and summarizes skills students have used to solve equations in preceding chapters. The process for finding models is also reviewed.

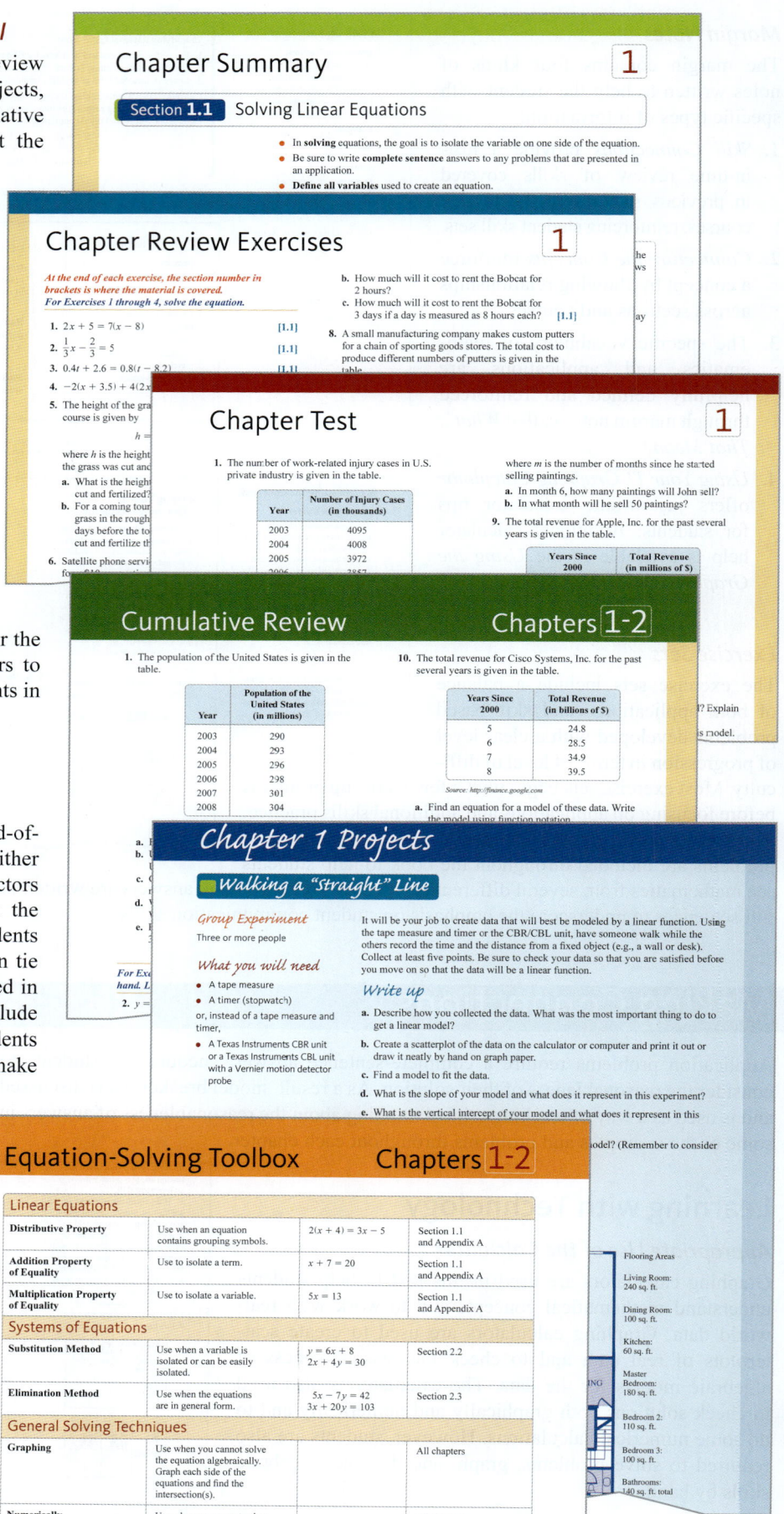

Chapter-by-Chapter Overview

Chapter 1, Linear Functions: Some of the mathematical skills covered in this chapter are a review of the topics covered in a beginning algebra course. The chapter is used to help students adjust to the higher expectations of thinking about problems in the context of real-world applications. Starting in the first section, students are asked to explain the meaning of answers in the context of the application and analyze the reasonableness of their answers. The concepts of linear equations and their graphs are covered thoroughly. All of the skills related to linear equations are pulled together in a section focused on finding linear models from real-life data. The chapter concludes with an introduction to functions and function notation. The theme of functions is then carried throughout the text.

Chapter 2, Systems of Linear Equations and Inequalities: This chapter focuses on working with systems of linear equations. Graphing lines is reinforced when students solve systems graphically. Both the substitution and elimination method of solving systems are presented. Linear and absolute value inequalities are discussed leading up to solving systems of linear inequalities.

Chapter 3, Exponents, Polynomials, and Functions: This chapter covers some basic skills needed as a foundation for other chapters in the book, as well as continues the development of the concept of functions. The rules for exponents support work with combining functions and factoring techniques. Combining functions is introduced from an applications approach and reinforced with basic skills practice. Composing functions is covered in a separate section, making the section optional. Factoring techniques conclude the chapter in preparation for solving quadratics and working with rational expressions.

Chapter 4, Quadratic Functions: In this chapter the concepts of quadratic functions are discussed. The chapter starts with the basic features of quadratics and then focuses on the vertex form of a quadratic. This gives many students a new way to look at quadratics and their graphs. Graphing is introduced early and carried throughout the chapter to help students build graphing skills. Solving quadratics using the square root property, completing the square, factoring, and the quadratic formula are covered in the next sections of the chapter. Finally, graphing from standard form reinforces the characteristics of parabolas, and many of the connections between the vertex and standard form of a quadratic function.

Chapter 5, Exponential Functions: This chapter starts with an introduction to exponential functions by means of a pattern-finding approach. This helps students explore the characteristics of exponential growth and decay by building a table and looking for patterns. This pattern-finding approach may be different for many students, but it helps to reinforce the basic idea of repeated multiplication and exponents. The rules for exponents are reviewed and rational exponents are used to solve power equations. Graphing exponential functions and finding their domain and range is discussed to prepare students to find exponential models. This chapter is wrapped up with a look at exponential growth and decay rates as well as other applications of exponentials. Compounding interest and the number e are introduced in the final section to be revisited again in the later sections of the next chapter.

Chapter 6, Logarithmic Functions: This chapter begins with inverse functions and their applications. Logarithms are introduced with some of the basic rules for logarithms including the change-of-base formula. This introduction gives the students a chance to absorb the idea of this new function before doing too much work with logs. The graphs of logarithms and their domain and range are discussed, followed by the properties of logarithms. The Chapter concludes with two sections on solving exponential and logarithmic equations. Applications of logarithms such as pH in chemistry and calculating the magnitude of earthquakes are covered and compounding interest problems are also re-visited.

Chapter 7, Rational Functions: This chapter presents rational functions. Applications of rational functions are introduced in the first section, including inverse variation and problems derived from division of two simpler functions. The domain of rational functions is covered, along with some of the characteristics of graphs for rational functions. The next sections focus on the basic operations with rational expressions including long division and synthetic division. The chapter concludes with solving rational equations including applications. These equations include some work that require solving quadratic equations to reinforce the techniques taught in Chapter 4.

Chapter 8, Radical Functions: Applications of radical functions open this chapter. Basic graphs of radical functions are covered, as well as domain and range. Simplifying radical functions and performing basic operations with radicals is covered in the next sections. The basic operations are followed by solving radical equations including those with radicals on both sides of the equal sign. Again, some radical equations that result in quadratic equations reinforce the skills taught in Chapter 4. This chapter concludes with a section on complex numbers and complex solutions to quadratic equations.

Chapter 9, Conics, Sequences, and Series: Chapter 9 covers the topics of conic sections, sequences, and series. The first two sections look at the equations and graphs of the common conic sections including circles, parabolas, ellipses, and hyperbolas. The next two sections introduce arithmetic and geometric sequences. The final section of this chapter includes an introduction to arithmetic and geometric series.

Appendices:

Appendix A	Basic Algebra Review is available with additional prealgebra and beginning algebra topics. Many of these topics are covered throughout the main chapters, but each section of this appendix provides instruction and a small exercise set to help students who need a focused review.
Appendix B	This appendix covers matrices and systems of three equations and three variables, it is included for those schools that wish to extend the material covered in chapters 2 and 4.
Appendix C	Using the Graphing Calculator appendix covers the basic steps of how to use a Texas Instruments 83/84 graphing calculator. Key by key instructions including screen shots are provided for each feature used in the book. Common error messages and how to fix them are also discussed. Additional features of the calculator not discussed in the main chapters are also provided for more advanced users.
Appendix D	Answers to Practice Problems
Appendix E	Answers to Selected Exercises

For the Instructor:

Annotated Instructor's Edition

The *Annotated Instructor's Edition* provides the complete student text with answers next to each respective exercise. The annotated instructor's edition includes **new classroom examples** with answers to use in lecture that parallel each example in the text. These could also be used in class for additional practice problems for students. **Teaching tips** are also embedded to help with pacing as well as key points that instructors may want to point out to their students.

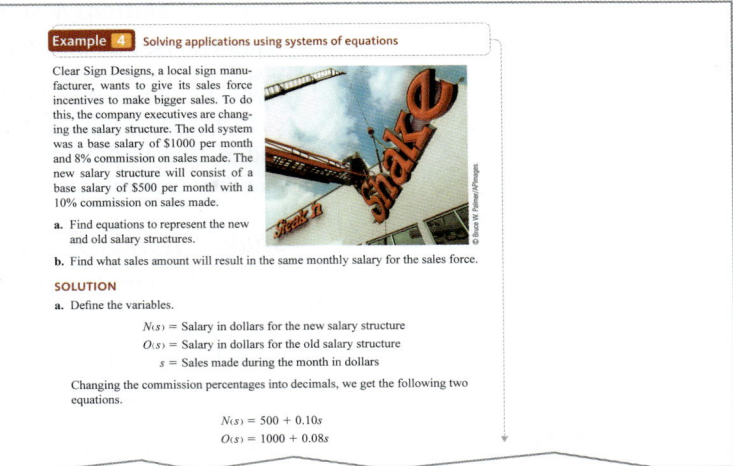

Complete Solutions Manual

Gina Hayes, *Palomar College*; **Karen Mifflin,** *Palomar College*

The *Complete Solutions Manual* provides worked-out solutions to all of the problems in the text.

Instructor's Resource Binder

Maria H. Andersen, *Muskegon Community College*

Each section of the main text is discussed in uniquely designed *Teaching Guides* containing instruction tips, examples, activities, worksheets, overheads, assessments, and solutions to all worksheets and activities.

PowerLecture with ExamView®

This CD-ROM provides the instructor with dynamic media tools for teaching. Create, deliver, and customize tests (both print and online) in minutes with ExamView® Computerized Testing Featuring Algorithmic Equations. Microsoft® PowerPoint® lecture slides, figures and additional new examples from the annotated instructor's edition are also included on this CD-ROM.

Solution Builder

This online instructor database offers complete worked solutions to all exercises in the text, allowing you to create customized, secure solutions printouts (in PDF format) matched exactly to the problems you assign in class.
Visit **www.cengage.com/solutionbuilder**

Enhanced WebAssign with eBook

Exclusively from Cengage Learning, Enhanced WebAssign®, used by over one million students at more than 1,100 institutions, allows you to assign, collect, grade, and record homework assignments via the Web. This proven and reliable homework system includes thousands of algorithmically generated homework problems, links to relevant textbook sections, video examples, problem-specific tutorials, and more.

The authors worked very closely with their media team, writing problems with Enhanced WebAssign in mind and selecting exercises that offer a balance of applications and skill development. They class tested these along with other instructors this past year. Additionally, the "Master It" questions focus primarily on modeling exercises that contain real-life data. They break the question into multiple parts reflecting the step-by-step process that is taught in the text. Instructors can choose to assign these or use the tutorial version. These problems support the conceptual applied approach of the textbook and force students to confirm their understanding as they work through each step of the problem.

In addition, diagnostic quizzing for each chapter identifies concepts that students still need to master, and directs them to the appropriate review material. Students will appreciate the interactive eBook, which offers search, highlighting, and note-taking functionality, as well as links to multimedia resources. All of this is available to students with Enhanced WebAssign.

For the Student:

Student Solutions Manual (0534496415)
Gina Hayes, *Palomar College*; **Karen Mifflin**, *Palomar College*

Contains fully worked-out solutions to all of the odd-numbered end-of section exercises as well as the complete worked-out solutions to all of the exercises included at the end of each chapter in the text, giving students a way to check their answers and ensure that they took the correct steps to arrive at an answer.

Student Workbook (1111568898)
Maria H. Andersen, *Muskegon Community College*

Get a head-start. The *Student Workbook* contains all of the assessments, activities, and worksheets from the *Instructor's Resource Binder* for classroom discussions, in-class activities, and group work.

Enhanced WebAssign with eBook (0538738103 One Term Access Card)
Exclusively from Cengage Learning, Enhanced WebAssign®, used by over one million of your fellow students at more than 1,100 institutions, allows your instructor to assign, collect, grade, and record homework assignments via the web. This proven and reliable homework system includes thousands of algorithmically generated homework problems and offers many tools to help you including links to relevant textbook sections, video examples and problem-specific tutorials. You also have access to an interactive eBook with search, highlighting, and note-taking functionality as well as links to multimedia resources.

In addition, diagnostic quizzing for each chapter identifies concepts that you may still need to master, and directs you to the appropriate review material. All of this is available to you with Enhanced WebAssign.

Accuracy and Development Process

Periodically during the authoring process, there were several phases of development where our manuscript was sent out to be reviewed by fellow college mathematics instructors either via a traditional "paper" review and/or via in-person focus groups. The feedback we received, very often including very thoughtful markups of the manuscript, related to accuracy, pacing, ordering of topics, accessibility and reading level, integration of technology, and completeness of coverage. At each stage, we analyzed and incorporated this feedback into the manuscript.

The manuscript also benefited from student and instructor feedback from the many class tests conducted with the 3 different printed Class Test versions. Bearing in mind the developmental math student and our applications-first approach, we took great pains to ensure the reading level was appropriate, working closely with a developmental reading and writing instructor to confirm this.

While revising the manuscript, a 10-person advisory board—made up of instructors who had seen the manuscript in different iterations—was specifically tasked to consider content queries and, design and art questions.

The final manuscript was delivered to Brooks/Cole-Cengage Learning for production after being through this thorough review, development and revision process. Once in

Photo credits clockwise from the top: istockphoto.com/DOUGBERRY; Image © seanelliottphotography 2010. Used under license from Shutterstock.com; © Ian Leonard/Alamy; Image © ZTS. Used under license from Shutterstock.com; Courtesy of White Loop Ltd; © Istockphoto.com/Chris Schmidt; © Istockphoto.com/clu; © Istockphoto/Carmen Martínez Banús

production, the manuscript was cycled through a process by specialized team members: production editor, copyeditor, accuracy reviewer, designer, proofreader, art editor, artists, and compositor. Each team member pays special attention to accuracy and completeness. As each phase of the cycle was completed, the manuscript was sent to us to verify any suggested changes or corrections. By the time this manuscript was published, it had been through the many phases of the production process and at least 10 pair of highly trained eyes verified its accuracy and completeness. Please be assured that accuracy was our primary goal and we take great pride in having partnered with many people to deliver an accurate, engaging and meaningful teaching tool for your intermediate algebra classes.

Reviewers

Authoring a textbook is a huge undertaking and we are very grateful to so many colleagues who assisted us throughout the many stages of this text's development. It was a painstaking process made possible by your willingness to collaborate with us and share your thoughtful feedback and comments. We thank you most sincerely for your many efforts on our behalf. You all gave of your time and expertise most generously: reviewing *multiple* rounds of manuscript, preparing for and attending our detailed development focus groups, class testing the manuscript in its many iterations and class testing and reviewing the online homework problems we created. We couldn't have done this without you and your students and the text and its accompanying ancillary program are a testimony to that.

Advisory Board Members

Scott Barnett, *Henry Ford Community College*
Maryann Firpo, *Seattle Central Community College*
Kevin Fox, *Shasta College*
Brian Karasek, *South Mountain Community College*
Krystyna Karminska, *Thomas Nelson Community College*
Amy Keith, *University of Alaska, Fairbanks*
Brianne Lodholtz, *Grand Rapids Community College*
Marilyn Platt, *Gaston College*
Janice Roy, *Montcalm Community College*
James Vallade, *Monroe County Community College*

Reviewers

Tim Allen, *Delta College*
Alex Ambriosa, *Hillsborough Community College, Brandon*
Ken Anderson, *Chemeketa Community College*
Frank Appiah, *Maysville Community & Technical College*
Farshad Barman, *Portland Community College*
Scott Barnett, *Henry Ford Community College*
Michelle Beerman, *Pasco-Hernando Community College*
Nadia Benakli, *New York City College of Technology*
David Bendler, *Arkansas State University*
Joel Berman, *Valencia Community College*
Rebecca Berthiaume, *Edison Community College*
Scott Berthiaume, *Edison Community College*
Nina Bohrod, *Anoka-Ramsey Community College*
Jennifer Borrello, *Grand Rapids Community College*
Ron Breitfelder, *Manatee Community College*
Reynaldo Casiple, *Florida Community College*
James Chesla, *Grand Rapids Community College*
Lisa Christman, *University of Central Arkansas*
Astrida Cirulis, *Concordia University Chicago*
Elizabeth Cunningham, *Santa Barbara City College*
Amy Cupit, *Copiah-Lincoln Community College*

Robert Diaz, *Fullerton College*
William Dickinson, *Grand Valley State University*
Susan Dimick, *Spokane Community College*
Randall Dorman, *Cochise College*
Maryann Faller, *Adirondack Community College*
Maryann Firpo, *Seattle Central Community College*
Thomas Fitzkee, *Francis Marion University*
Nancy Forrest, *Grand Rapids Community College*
Kevin Fox, *Shasta College*
Marcia Frobish, *Grand Valley State University*
John Golden, *Grand Valley State University*
Gail Gonyo, *Adirondack Community College*
James Gray, *Tacoma Community College*
Joseph Haberfeld, *Columbia College Chicago*
Mahshid Hassani, *Hillsborough Community College*
Stephanie Haynes, *Davis & Elkins College*
Julie Hess, *Grand Rapids Community College*
Kalynda Holton, *Tallahassee Community College*
Sharon Hudson, *Gulf Coast Community College*
Daniel Jordan, *Columbia College Chicago*
Brian Karasek, *South Mountain Community College*
Krystyna Karminska, *Thomas Nelson Community College*
Ryan Kasha, *Valencia Community College*
Fred Katirae, *Montgomery College*
Amy Keith, *University of Alaska, Fairbanks*
Tom Kelley, *Henry Ford Community College*
Dennis Kimzey, *Rogue Community College*
Ivy Langford, *Collin County Community College*
Richard Leedy, *Polk State College*
Mary Legner, *Riverside City College*
Andrea Levy, *Seattle Central Community College*
Brianne Lodholtz, *Grand Rapids Community College*
David Maina, *Columbia College Chicago*
Jane Mays, *Grand Valley State University*
Tim Mezernich, *Chemeketa Community College*
Pam Miller, *Phoenix College*
Christopher Milner, *Clark College*
Jeff Morford, *Henry Ford Community College*
Brian Moudry, *Davis & Elkins College*
Ellen Musen, *Brookdale Community College*
Douglas Nerig, *Ivy Tech Community College*
Coleen Neroda, *Austin Community College*
Ken Nickels, *Black Hawk College*
Marilyn Platt, *Gaston College*
Sandra Poinsett, *College of Southern Maryland*
Linda Reist, *Macomb Community College*
David Reynolds, *Treasure Valley Community College*
Patricia Rhodes, *Treasure Valley Community College*
Janice Roy, *Montcalm Community College*
Lily Sainsbury, *Brevard Community College, Cocoa*
Sailakshmi Srinivasan, *Grand Valley State University*
Jeannette Tyson, *Valencia Community College*
James Vallade, *Monroe County Community College*
Matt Williamson, *Hillsborough Community College, South Shore*

Deborah Wolfson, *Suffolk County Community College*
Carol Zavarella, *Hillsborough Community College, Ybor*

Developmental Focus Group Participants

Khadija Ahmed, *Monroe Community College*
Elaine Alhand, *Ivy Tech Community College*
Ken Anderson, *Chemeketa Community College*
Katherine Jankoviak Anderson, *Schoolcraft College*
Margaret Balachowski, *Everett Community College*
Farshad Barman, *Portland Community College*
Lois Bearden, *Schoolcraft College*
Clairessa Bender, *Henry Ford Community College*
Kathy Burgis, *Lansing Community College*
Caroline Castel, *Delta College*
Mariana Coanda, *Broward College*
Greg Cripe, *Spokane Falls Community College*
Alex Cushnier, *Henry Ford Community College*
Susan Dimick, *Spokane Community College*
Julie Fisher, *Austin Community College*
Will Freeman, *Portland Community College*
Irie Glajar, *Austin Community College*
Deede Furman, *Austin Community College*
James Gray, *Tacoma Community College*
Mahshid Hassani, *Hillsborough Community College*
Farah Hojjaty, *North Central Texas College*
Susan Hord, *Austin Community College*
Liz Hylton, *Clatsop Community College*
Eric Kean, *Western Washington University*
Beth Kelch, *Delta College*
Dennis Kimzey, *Rogue Community College*
Becca Kimzey
Mike Kirby, *Tidewater Community College*
Patricia Kopf, *Kellogg Community College*
Kay Kriewald, *Laredo Community College*
Riki Kucheck, *Orange Coast College*
Joanne Lauterbur, *Baker College*
Ivy Langford, *Collin County Community College*
Melanie Ledwig, *Victoria College*
Richard Leedy, *Polk State College*
Brianne Lodholtz, *Grand Rapids Community College*
Babette Lowe, *Victoria College*
Vinnie Maltese, *Monroe Community College*
Tim Mezernich, *Chemeketa Community College*
Michael McCoy, *Schoolcraft College*
Charlotte Newsome, *Tidewater Community College*
Maria Miles, *Mount Hood Community College*
Karen Mifflin, *Palomar College*
Pam Miller, *Phoenix College*
Christopher Milner, *Clark College*
Carol McKilip, *Southwestern Oregon Community College*
Vivian Martinez, *Coastal State College*
Anna Maria Mendiola, *Laredo Community College*
Joyce Nemeth, *Broward College*
Katrina Nichols, *Delta College*
Joanna Oberthur, *Ivy Tech Community College*

Ceci Oldmixon, *Victoria College*
Mary B. Oliver, *Baker College*
Diana Pagel, *Victoria College*
Joanne Peeples, *El Paso Community College*
Sandra Poinsett, *College of Southern Maryland*
Bob Quigley, *Austin Community College*
Jayanthy Ramakrishnan, *Lansing Community College*
Patricia Rhodes, *Treasure Valley Community College*
Billie Shannon, *Southwestern Oregon Community College*
Pam Tindel, *Tyler Junior College*
James Vallade, *Monroe Community College*
Thomas Wells, *Delta College*
Charles Wickman, *Everett Community College*
Eric Wiesenauer, *Delta College*
Andy Villines, *Bellevue Community College*
Beverly Vredevelt, *Spokane Falls Community College*

Class Testers

Tim Allen, *Delta College*
Scott Barnett, *Henry Ford Community College*
Jennifer Borrello, *Grand Rapids Community College*
Damon Ellingston, *Seattle Central Community College*
Maryann Firpo, *Seattle Central Community College*
Marcia Frobish, *Grand Valley State University*
Rathi Kanthimathi, *Seattle Central Community College*
Mary Stinnett, *Umqua Community College*
Randy Trip, *Jefferson Community College*
Thomas Wells, *Delta College*
Michael White, *Jefferson Community College*
Eric Wiesenauer, *Delta College*
Paul Yu, *Grand Valley State University*

Acknowledgments

We would like to thank Karen Mifflin and Gina Hayes for their many helpful suggestions for improving the text. We are also grateful to Dennis Kimzey and Brianne Lodholtz for helping with the accuracy checking for this text. We also thank the editorial, production, and marketing staffs of Brooks/Cole: Charlie Van Wagner, Carolyn Crockett, Rita Lombard, Stefanie Beeck, Jennifer Cordoba, Cheryll Linthicum, Gordon Lee, Mandee Eckersley, Heleny Wong, and Darlene Macanan for all of their help and support during the development and production of this edition. Thanks also to Vernon Boes, Leslie Lahr, and Lisa Torri for their work on the design and art program, and to Barbara Willette and Marian Selig for their copyediting and proofreading expertise. We especially want to thank Don Gecewicz, who did an excellent job of ensuring the accuracy and readability of this edition, mentoring us through this process. Our gratitude also goes to Cheryll Linthicum and Dusty Freidman who had an amazing amount of patience with us throughout production. We truly appreciate all the hard work and efforts of the entire team.

Mark Clark
Cynthia Anfinson

To the Student

Welcome to a course on Intermediate Algebra. This text was designed to help you learn the algebra skills and concepts you need in future math, science and other classes you may take. Driving these skills and concepts is a foundation in applications from the world around us. Much of the data and applications throughout the book are taken from current information in media and from the Web.

We hope that you will find the real life applications and data to be interesting and helpful in understanding the material. An emphasis is placed on thinking critically about the results you find throughout the book and communicating these results clearly. The *answers in the back* of the book will help demonstrate this level of thinking and communication. We believe improving these skills will benefit you in many of your classes and in life.

There are several features of the book that are meant to help you review and prepare for exams. Throughout the sections are *definition boxes* and *margin features* that point out key skills or vocabulary that will be important to your success in mastering the skills and concepts. We encourage you to use the *end of chapter reviews* and *chapter tests* as an opportunity to try a variety of problems from the chapter and test your understanding of the concepts and skills being covered. An *Equation-Solving Toolbox* is provided after every two chapters. These toolboxes are a quick summary of different solving techniques and modeling processes that have been covered so far in the text. They can also refer you back to the sections where those techniques were discussed so you can review them further.

There is an appendix in the back of the book with some *elementary algebra review* as well as help with using a graphing calculator. We hope these will be a good reference if you find yourself in need of a little more help. Remember that your instructor and fellow classmates are some of the best resources you have to help you be successful in this course and in using this text.

Good luck in your Intermediate Algebra course!

Sincerely,
Mark Clark
Cynthia Anfinson

Linear Functions

- **1.1** Solving Linear Equations
- **1.2** Using Data to Create Scatterplots
- **1.3** Fundamentals of Graphing and Slope
- **1.4** Intercepts and Graphing
- **1.5** Finding Equations of Lines
- **1.6** Finding Linear Models
- **1.7** Functions and Function Notation

The U.S. Census Bureau found that the average number of square feet of floor area in new one-family houses in 1980 was 1740 square feet. In 2003, the average number of square feet of floor area increased to 2330 square feet. In this chapter, we will discuss how to use linear models to analyze trends in real-life data. One of the chapter projects will ask you to investigate the costs associated with installing new flooring in a home.

1.1 Solving Linear Equations

LEARNING OBJECTIVES

- Solve linear equations.
- Write complete answers to application problems.
- Determine whether a solution is reasonable.
- Solve literal equations.

Equations can be used to represent many things in life. One of the uses of algebra is to solve equations for an unknown quantity, or variable. In this section, you will learn how to solve linear equations for a missing variable and how to write a complete solution. A complete solution will include a sentence that gives the units and the meaning of the solution in that situation. Providing these details will demonstrate that you have a clear understanding of the solution.

Example 1 Solving applications and providing complete solutions

U-Haul charges $19.95 for the day and $0.79 per mile driven to rent a 10-foot truck. The total cost to rent a 10-foot truck for the day can be represented by the equation

$$U = 19.95 + 0.79m$$

where U is the total cost in dollars to rent a 10-foot truck from U-Haul for the day and m is the number of miles the truck is driven.

a. Determine how much it will cost you to rent a 10-foot truck from U-Haul and drive it 75 miles.

b. Determine the number of miles you can travel for a total cost of $175.00.

SOLUTION

a. Because the number of miles driven was given, replace the variable m in the equation with the number 75 and solve for the missing variable U as follows:

$$U = 19.95 + 0.79m$$
$$U = 19.95 + 0.79(75)$$
$$U = 19.95 + 59.25$$
$$U = 79.20$$

This answer indicates that renting a 10-foot truck from U-Haul for the day and driving it 75 miles would cost you $79.20.

b. Because the total cost of $175.00 is given in the statement, substitute 175.00 for the variable U and solve for the missing variable m.

$$U = 19.95 + 0.79m$$
$$175.00 = 19.95 + 0.79m$$
$$\underline{-19.95 \quad -19.95} \quad \text{Subtract 19.95 from both sides.}$$
$$155.05 = 0.79m$$
$$\frac{155.05}{0.79} = \frac{0.79m}{0.79} \quad \text{Divide both sides by 0.79.}$$
$$196.266 \approx m$$

■ Connecting the Concepts

What is the difference between the (=) symbol and the (≈) symbol?

In mathematics, we use these symbols and others to show a relationship between two quantities or between two expressions.

The equal sign (=) is used when two quantities or expressions are equal and exactly the same.

The approximation symbol (≈) is used to show that two quantities or expressions are approximately the same. The approximation symbol will be used whenever a quantity is rounded.

Since U-Haul would charge for a full mile for the 0.266, round down to 196 miles to stay within the budget of $175. Check this answer by substituting $m = 196$ to be sure U will equal $175.

$$U = 19.95 + 0.79(196)$$
$$U = 19.95 + 154.84$$
$$U = 174.79$$

This answer indicates that for a cost of $175.00, you can rent a 10-foot truck from U-Haul for a day and drive it 196 miles.

PRACTICE PROBLEM FOR EXAMPLE 1

While you are on spring break in Fort Lauderdale, Florida, the cost, C, in dollars of your taxi ride from the airport to your hotel can be represented by the equation

$$C = 2.10 + 2.40m$$

when the ride is m miles long.

a. What will an 8-mile taxi ride cost?

b. How many miles can you ride if you budget $35 for the taxi?

(*Note:* The answers to the Practice Problems are in Appendix D.)

Connecting the Concepts

How are we going to round?

In general, we will round values to at least one more decimal place than the given numbers in the problem.

In some applications, rounding will be determined by what makes sense in the situation.

When a specific rounding rule is stated in a problem, we will follow that rule.

Example 1 uses an algebraic equation to represent the cost of the rental in terms of the mileage.

Many problems in this book will investigate applications that involve money and business. Defining some business terms will help in understanding the problems and explaining the solutions. Three main concepts in business are **revenue, cost,** and **profit**.

A simple definition of revenue is the total amount of money that is brought into the business through sales. For example, if a pizza place sells 10 pizzas for $12 each, the revenue would be 10 pizzas · $12 per pizza = $120. Revenue is often calculated as price times the quantity sold. The revenue for a business can not be a negative number.

Cost is defined as the amount of money paid out for expenses. Expenses often are categorized in two ways: fixed costs and variable costs. The same pizza place would probably have fixed costs such as rent, utilities, and perhaps salaries. It would have variable costs of supplies and food ingredients per pizza made. The cost for the business would be the fixed costs and the variable costs added together.

The profit for a business is the revenue minus the cost. If this pizza place had a cost of $100 when making the 10 pizzas, they would have a profit of $120 − $100 = $20. Although a business cannot have a negative revenue, profit can be negative. When profit is negative, it is sometimes called a loss, but it is understood to be a negative profit.

A business is often interested in what its **break-even point** is. That is when the revenue from a product is the same as the cost. The break-even point is also when the profit is zero. The point when profit changes from negative to positive is important to a company that is considering a new product and wants to know how many should be produced or sold for the company to start making a profit.

These definitions should help you to understand some of the examples and exercises throughout this book.

Skill Connection

Solving equations

In solving an equation for a variable, the goal is to isolate the variable (get it by itself) on one side of the equation and simplify the other side. When you have more than one term on the same side of the equation where the variable is, start by undoing any addition or subtraction using the addition property of equality. Next you can use the multiplication property of equality to undo any multiplication and division using the opposite operation. For more of a review of solving equations see Appendix A.

DEFINITIONS

Revenue The amount of money brought into a business through sales. Revenue is often calculated as

revenue = price · quantity sold

Cost The amount of money spent by a business to create and/or sell a product. Cost usually includes both fixed costs and variable costs. Fixed costs are the

Continued

same each month or year, and variable costs change depending on the number of items produced and/or sold.

$$\text{cost} = \text{fixed cost} + \text{variable cost}$$

or

$$\text{cost} = \text{fixed cost} + \text{cost per item} \cdot \text{quantity sold}$$

Profit The amount of money left after all costs.

$$\text{profit} = \text{revenue} - \text{cost}$$

Break-even point A company breaks even when their revenue equals their cost or when their profit is zero.

$$\text{revenue} = \text{cost}$$
$$\text{profit} = 0$$

Example 2 Revenue cost and profit

A small bicycle company produces high-tech bikes for international race teams. The company has fixed costs of $5000 per month for rent, salary, and utilities. For every bike they produce, it costs them $755 in materials and other expenses related to that bike. The company can sell each bike for an average price of $1995, but it can produce a maximum of only 20 bikes per month.

a. Write an equation for the monthly cost of producing b bikes.

b. How much does it cost the bicycle company to produce 20 bikes in a month?

c. Write an equation for the monthly revenue from selling b bikes.

d. How much revenue will the bicycle company make if they sell 10 bikes in a month?

e. Write an equation for the monthly profit the company makes if they produce and sell b bikes. (You can assume that they will sell all the bikes they make.)

f. What is the profit of producing and selling 15 bikes in a month?

g. How many bikes does the company have to produce and sell in a month to make $15,000 profit?

h. How many bikes does the company have to produce and sell in a month to make $30,000 profit?

SOLUTION

a. First define the variables in the problem.

b = The number of bikes produced each month. (Remember that a maximum of 20 bikes can be produced each month.)

C = The monthly cost, in dollars, to produce b bikes.

Each bike cost $755 for materials and other expenses, so multiply b by 755, and then add on the fixed costs, to get the total monthly cost. This gives the following equation.

$$C = 755b + 5000$$

b. The number of bikes produced is given. Substitute 20 for b and simplify the right side to find C.

$$C = 755b + 5000$$
$$C = 755(20) + 5000$$
$$C = 20100$$

A monthly production of 20 bikes will result in a total monthly cost of $20,100.

c. First define the variables in the problem. Recall that b was already defined in part a.

b = The number of bikes produced each month.
R = The monthly revenue, in dollars, from selling b bikes.

The bicycle company can sell each bike for an average price of $1995, so the revenue can be calculated by using the equation

$$R = 1995b$$

d. The number of bikes is given, so substitute 10 for b.

$$R = 1995b$$
$$R = 1995(10)$$
$$R = 19950$$

The total monthly revenue from selling 10 bikes is $19,950.

e. Profit is calculated by taking the revenue and subtracting any business costs.

b = The number of bikes produced each month.
P = The monthly profit, in dollars, from producing and selling b bikes.

Use the equations for revenue and cost written earlier.

$$P = R - C$$
$$P = (1995b) - (755b + 5000) \quad \text{Substitute for } R \text{ and } C.$$

This profit equation can be simplified by distributing the negative and combining like terms.

$$P = (1995b) - (755b + 5000)$$
$$P = 1995b - 755b - 5000 \quad \text{Distribute the negative sign.}$$
$$P = 1240b - 5000 \quad \text{Combine like terms.}$$

f. The number of bikes is given, so substitute 15 for b.

$$P = 1240(15) - 5000$$
$$P = 13600$$

The monthly profit from producing and selling 15 bikes is $13,600.

g. The amount of profit desired is given. Substitute 15000 for P and solve for b.

$$P = 1240b - 5000$$
$$15000 = 1240b - 5000$$
$$\underline{+5000 \qquad\qquad +5000} \quad \text{Add 5000 to both sides.}$$
$$20000 = 1240b$$
$$\frac{20000}{1240} = \frac{1240b}{1240} \quad \text{Divide both sides by 1240.}$$
$$16.129 \approx b$$

Because there is a decimal answer, compare the profits for the number of bikes represented on both sides of this decimal.

$$P = 1240(16) - 5000$$
$$P = 14840$$

■ Skill Connection

Using the Distribution Rule in Subtraction

When subtracting an expression in parentheses, remember that subtraction can be represented as adding the opposite.

$$a - b = a + (-1)b$$

Since subtraction is defined as adding the opposite, we see that we use the distributive property first to distribute the factor of -1. This will ensure that we subtract all the terms of the second expression.

$$25x - (5x + 30)$$
$$= 25x + (-1)(5x + 30)$$
$$= 25x - 5x - 30$$
$$= 20x - 30$$

In most cases, we will do this process without writing the -1. We will say to distribute the negative sign, implying the -1 that is not seen in the original expression or equation.

$$P = 1240(17) - 5000$$
$$P = 16080$$

To make at least $15,000 profit for the month, the company will need to produce at least 17 bikes. Since producing 16 bikes would not quite make $15,000 profit, round up to the next whole number quantity of bikes.

h. The amount of profit desired is given. Substitute 30000 for P and solve for b.

$$P = 1240b - 5000$$
$$30000 = 1240b - 5000$$
$$\underline{+5000 \qquad\qquad +5000} \qquad \text{Add 5000 to both sides.}$$
$$35000 = 1240b$$
$$\frac{35000}{1240} = \frac{1240b}{1240} \qquad \text{Divide both sides by 1240.}$$
$$28.226 \approx b$$

Check this answer by substituting in 28.226 for b and confirming that it gives a profit of $30,000.

$$P = 1240(28.226) - 5000$$
$$P = 35000.24 - 5000$$
$$P = 30000.24$$

The answer was rounded, so the check is not exact, but it is close enough.

Again, there is a decimal answer, so round to the whole number of bikes that will produce the desired profit. That would give 29 bikes produced in a month. However, this level of production is not possible, since the problem stated that the company could produce a maximum of 20 bikes per month. Therefore, the correct answer is that the company cannot make $30,000 profit in a month with its current production capacity.

PRACTICE PROBLEM FOR EXAMPLE 2

A local chiropractor has a small office where she cares for patients. She has $8000 in fixed costs each month that cover her rent, basic salaries, equipment, and utilities. For each patient she sees, she has an average additional cost of about $15. The chiropractor charges her patients or their insurance company $80 for a visit.

a. Write an equation for the total monthly cost when v patient visits are done in a month.

b. What is the total monthly cost if the chiropractor has 100 patients visit during a month?

c. Write an equation for the monthly revenue when v patients visit a month.

d. Write an equation for the monthly profit the chiropractor makes if she has v patient visits in a month.

e. What is the monthly profit when 150 patients visit in a month?

f. How many patient visits does this chiropractor need to have in a month for her profit to be $5000.00?

Example 2 shows that you should check each answer to determine whether or not it is a reasonable answer. Many times, this is something that requires only some common sense; other times, a restriction that is stated in the problem should be considered.

In both of the previous examples, it is important to pay attention to the definition of each variable. The definitions of the variables help you to determine which variable value was given and which variable you want to solve for. In some questions, you will have to define your own variables. Use meaningful variable names to make it easy to remember what they represent. For example,

- t = time in years
- h = hours after 12 noon

- p = population of San Diego (in thousands)
- P = profit of IBM (in millions of dollars)
- S = Salary (in dollars per hour)

Units, or how the quantity is measured, are very important in communicating what a variable represents. The meaning of $P = 100$ is very different if profit for IBM is measured in dollars and not millions of dollars. The same for $S = 6.5$. If S represents your salary for your first job out of college, it would be great if S were measured in millions of dollars per year and not dollars per hour. Units can make a large difference in the meaning of a quantity. When defining variables, be sure to include units.

When solving an equation that represents something in an application, you should always check that the answer you find is a reasonable one for the situation. Use the following concept investigation to practice determining which answers might be reasonable and which would not make sense in the situation given.

CONCEPT INVESTIGATION

Is that a reasonable solution?

In each part, choose the value that seems the most reasonable for the given situation. Explain why the other given value(s) do not make sense in that situation.

1. If P is the population of the United States in millions of people, which of the following is a reasonable value for P?
   a. $P = -120$
   b. $P = 300$
   c. $P = 5,248,000,000$

2. If H is the height of an airplane's flight path in feet, which of the following is a reasonable value for H?
   a. $H = -2000$
   b. $H = 3,500,000$
   c. $H = 25,000$

3. If P is the annual profit in dollars of a new flower shop the first year it opens, which of the following is a reasonable value for P?
   a. $P = -40,000$
   b. $P = 50,000$
   c. $P = 3,000,000$

When solving equations that involve fractions, you could work with the fractions or eliminate the fractions at the beginning of the problem by multiplying both sides of the equation by the least common denominator. Finish solving the equation as you would any other equation.

Example 3 Solving equations

Solve the following equations.

a. $\dfrac{2}{3}x + \dfrac{5}{6} = 7$

b. $\dfrac{1}{4}(x + 5) = \dfrac{1}{2}x - 6$

c. $4t - 2(3.4t + 7) = 5t - 17.3$

SOLUTION

a. To eliminate the fractions, multiply both sides of the equation by the least common denominator 6 and then continue solving.

$$\frac{2}{3}x + \frac{5}{6} = 7$$

$$6\left(\frac{2}{3}x + \frac{5}{6}\right) = 6(7) \qquad \text{Multiply both sides by the least common denominator 6.}$$

$$\frac{6}{1}\left(\frac{2}{3}x\right) + \frac{6}{1}\left(\frac{5}{6}\right) = 6(7)$$

$$\frac{\cancel{6}^2}{1}\left(\frac{2}{\cancel{3}}x\right) + \frac{\cancel{6}}{1}\left(\frac{5}{\cancel{6}}\right) = 6(7) \qquad \text{Reduce to eliminate the fractions.}$$

$$4x + 5 = 42 \qquad \text{Finish solving.}$$

$$\underline{\quad -5 \quad -5\quad}$$

$$4x = 37$$

$$\frac{4x}{4} = \frac{37}{4}$$

$$x = \frac{37}{4}$$

$$\frac{2}{3}\left(\frac{37}{4}\right) + \frac{5}{6} \stackrel{?}{=} 7 \qquad \text{Check the answer.}$$

$$\frac{\cancel{2}}{3}\left(\frac{37}{\cancel{4}_2}\right) + \frac{5}{6} \stackrel{?}{=} 7$$

$$\frac{37}{6} + \frac{5}{6} \stackrel{?}{=} 7$$

$$\frac{42}{6} \stackrel{?}{=} 7$$

$$7 = 7 \qquad \text{The answer works.}$$

Therefore the answer is $x = \frac{37}{4} = 9\frac{1}{4} = 9.25$. This answer can be written as an improper fraction, a mixed number, or a decimal as shown.

b. For this equation, distribute the $\frac{1}{4}$ first and then eliminate the fractions by multiplying both sides of the equation by the least common denominator 4.

$$\frac{1}{4}(x + 5) = \frac{1}{2}x - 6$$

$$\frac{1}{4}x + \frac{1}{4}(5) = \frac{1}{2}x - 6 \qquad \text{Distribute the } \frac{1}{4}.$$

$$\frac{1}{4}x + \frac{5}{4} = \frac{1}{2}x - 6$$

$$4\left(\frac{1}{4}x + \frac{5}{4}\right) = 4\left(\frac{1}{2}x - 6\right) \qquad \text{Multiply both sides of the equation by the least common denominator 4.}$$

$$4\left(\frac{1}{4}x\right) + 4\left(\frac{5}{4}\right) = 4\left(\frac{1}{2}x\right) - 4(6) \qquad \text{Distribute the 4 through both sides.}$$

$$\cancel{4}\left(\frac{1}{\cancel{4}}x\right) + \cancel{4}\left(\frac{5}{\cancel{4}}\right) = \cancel{4}^2\left(\frac{1}{\cancel{2}}x\right) - 4(6) \qquad \text{Reduce and multiply.}$$

$$x + 5 = 2x - 24$$

$$\underline{-2x \qquad\quad -2x\quad} \qquad \text{Subtract 2x from both sides of the equation to get the variable terms together.}$$

$$-x + 5 = -24$$

$$\underline{\quad -5 \quad -5\quad} \qquad \text{Subtract 5 from both sides of the equation to get the variable term isolated.}$$

$$-x = -29$$

$$-1(-x) = -1(-29) \qquad \text{Multiply by } -1 \text{ to make } x \text{ positive.}$$

$$x = 29$$

$$\frac{1}{4}(29 + 5) \stackrel{?}{=} \frac{1}{2}(29) - 6 \quad \text{Check the answer.}$$

$$\frac{1}{4}(34) \stackrel{?}{=} \frac{29}{2} - \frac{12}{2}$$

$$\frac{17}{2} = \frac{17}{2} \quad \text{The answer works.}$$

c. To start solving, distribute the negative 2 through the parentheses.

$$4t - 2(3.4t + 7) = 5t - 17.3$$
$$4t - 6.8t - 14 = 5t - 17.3 \quad \text{Distribute the } -2.$$
$$-2.8t - 14 = 5t - 17.3 \quad \text{Combine like terms.}$$
$$\underline{+2.8t \qquad\qquad +2.8t} \quad \text{Add 2.8t to both sides of the equation}$$
$$-14 = 7.8t - 17.3 \quad \text{to get the variable terms together.}$$
$$\underline{+17.3 \qquad +17.3} \quad \text{Add 17.3 to both sides of the equation}$$
$$3.3 = 7.8t \quad \text{to isolate the variable term.}$$
$$\frac{3.3}{7.8} = \frac{7.8t}{7.8} \quad \text{Divide both sides of the equation by 7.8 to isolate } t.$$
$$0.423 \approx t$$

Check the answer.

$$4(0.423) - 2(3.4(0.423) + 7) \stackrel{?}{=} 5(0.423) - 17.3$$
$$1.692 - 2(8.4382) \stackrel{?}{=} 2.115 - 17.3$$
$$-15.1844 \approx -15.185 \quad \text{The answer works.}$$

Because of rounding, the two sides will not be exactly the same, but they are very close.

PRACTICE PROBLEM FOR EXAMPLE 3

Solve the following equations.

a. $\frac{2}{5}(x - 6) = \frac{1}{4}x + 15$

b. $3 + 2.5(3x + 4) = 1.75x - 8$

Equations that contain more than one variable are called **literal equations**. Formulas are literal equations that are used to express relationships among physical quantitites. We use literal equations and formulas in many areas of our lives. For example, $D = rt$ is a formula to calculate the distance traveled when you are given the rate at which you are traveling and for how long (time) you traveled that rate. This formula can be rearranged (solved) for one of the other variables to make it easier to use to find the rate if you know the distance and time or to find the time if you know the distance and rate. Solving literal equations and formulas for other variables uses the same steps as those used in solving other equations. One difference in solving literal equations, is that many of the calculations will not be able to be simplified until values for the variables are known.

Example 4 Solving literal equations for a different variable

Solve the following literal equations for the variable indicated.

a. $D = rt$ for r.

b. Distance in free fall: $D = \frac{1}{2}Gt^2$ for G.

c. $y = mx + b$ for m.

SOLUTION

a. Because r is being multiplied by t, divide both sides by t.

$$D = rt$$
$$\frac{D}{t} = \frac{rt}{t} \quad \text{Divide both sides by } t.$$
$$\frac{D}{t} = r$$

b. To solve for G, multiply both sides by 2 to undo the multiplication by $\frac{1}{2}$.

$$D = \frac{1}{2}Gt^2$$
$$2D = 2\left(\frac{1}{2}Gt^2\right) \quad \text{Multiply both sides by 2 to undo the } \frac{1}{2}.$$
$$2D = Gt^2$$
$$\frac{2D}{t^2} = \frac{Gt^2}{t^2} \quad \text{Divide both sides by } t \text{ squared.}$$
$$\frac{2D}{t^2} = G$$

c. To solve for m, isolate the term with m and then divide both sides by x.

$$y = mx + b \quad \text{Subtract } b \text{ from both sides to isolate the } mx \text{ term. Because } y \text{ and } b \text{ are not the same variable, we cannot subtract them but must leave the left side as } y - b.$$
$$\underline{-b \qquad\quad -b}$$
$$y - b = mx$$
$$\frac{y - b}{x} = \frac{mx}{x} \quad \text{Divide both sides by } x \text{ to isolate the } m.$$
$$\frac{y - b}{x} = m$$

PRACTICE PROBLEM FOR EXAMPLE 4

Solve the following literal equations for the variable indicated.

a. Velocity in free fall: $v = Gt$ for t.

b. Velocity: $v = v_0 + at$ for a. (Note: v_0 is the initial velocity.)

1.1 Exercises

For Exercises 1 through 10, solve each equation.

1. $2x + 10 = 40$
2. $3x + 14 = 35$
3. $-4t + 8 = -32$
4. $-7m + 20 = 48$
5. $2.5x + 7.5 = 32.5$
6. $3.4x - 8.2 = 15.6$
7. $20 = 5.2x - 0.8$
8. $45 = -3.6c + 189$
9. $0.05(x - 200) = 240$
10. $0.03(n - 500) = 108$

11. During the first day of training on the job, a new candy maker gets faster at making candies. The number of candies a new employee can produce during an hour can be represented by $C = 10h + 20$ candies, where h is the number of hours of training.

   a. Find the number of candies a new employee can produce in an hour after 1 hour of training.
   b. Find the number of candies a new employee can produce in an hour after 4 hours of training.
   c. How many hours of training must an employee receive before being able to produce 150 candies an hour?

12. The number of students who are enrolled in math classes at a local college can be represented by $E = -17w + 600$, where E represents the math class enrollment at the college w weeks after the start of the fall semester.

a. Find the total enrollment in math classes at the college at the beginning of the fall semester. (*Hint:* Because the semester is just starting, $w = 0$.)
b. During which week will the total enrollment be 430 students?
c. What will the total enrollment be in math classes after 8 weeks?

13. The number of homicides, N, of 15- to 19-year-olds in the United States t years after 1990 can be represented by the equation $N = -315.9t + 4809.8$.
    Source: Based on data from Statistical Abstract 2001.
    a. Find the number of homicides of 15- to 19-year-olds in the United States in 1992. (1992 is 2 years after 1990, so $t = 2$.)
    b. Find the number of homicides of 15- to 19-year-olds in the United States in 2002.
    c. In what year was the number of homicides 7337?

14. The gasoline prices in Southern California can increase very quickly during the summer months. The equation $p = 2.399 + 0.03w$ represents the gasoline prices p in dollars per gallon w weeks after the beginning of summer.
    a. What does gasoline cost after 5 weeks of summer?
    b. During what week of summer will gasoline cost $2.759 per gallon?

15. $P = 1.5t - 300$ represents the profit in dollars from selling t printed T-shirts.
    a. Find the profit if you sell 100 printed T-shirts.
    b. Find the profit if you sell 400 printed T-shirts.
    c. How many printed T-shirts must you sell to make $1000 in profit?

16. $P = 5.5b - 500.5$ represents the profit in dollars from selling b books.
    a. Find the profit if you sell 75 books.
    b. Find the profit if you sell 200 books.
    c. How many books must you sell to make $3600 in profit?

17. The total cost, C, in dollars for a taxi ride in New York City can be represented by the equation
    $$C = 2.50 + 2.0m$$
    when the trip is m miles long.
    a. Determine the cost for a 25-mile taxi ride in New York City.
    b. How many miles can you ride in a New York City taxi for $100?

18. A team of engineers is trying to pump out the air in a vacuum chamber to lower the pressure. They know that the following equation represents the pressure in the chamber:
    $$P = 35 - 0.07s$$
    where P is the pressure in pounds per square inch (psi) of the vacuum chamber and s is the time in seconds.

a. What will the pressure be after 150 seconds?
b. When will the pressure inside the chamber be 1 psi?

For Exercises 19 through 22, determine which given value seems the most reasonable for the given situation. Explain why the other given values do not make sense in that situation.

19. P is the population of Kentucky in thousands of people.
    a. $P = 3.5$ b. $P = 4200$ c. $P = -210$

20. R is the revenue in dollars from selling kettle corn at a two-day rodeo.
    a. $R = 20$ b. $R = -3000$ c. $R = 4500$

21. T is the temperature in degrees Fahrenheit at the South Pole.
    a. $T = -50$ b. $T = 75$ c. $T = 82$

22. S is a cook's monthly salary in dollars from working at White Castle Hamburgers.
    a. $S = 10.50$ b. $S = 1600$ c. $S = 28,000$

23. Salespeople often work for commissions on the sales that they make for the company. As a new salesperson at a local technology company, you are told that you will receive an 8% commission on all sales you make after the first $1000. Your pay can be represented by $p = 0.08(s - 1000)$ dollars, where s is the amount of sales you make in dollars.
    a. How much total pay will you earn from $2000 in sales?
    b. How much total pay will you earn from $50,000 in sales?
    c. If you need at least $500 per week to pay your bills, what sales do you have to make per week?

24. At a new job selling high-end clothing to women, you earn 6% commission on all sales you make after the first $500. Your pay can be represented by $p = 0.06(s - 500)$ dollars, where s is the amount of sales you make in dollars.
    a. How much total pay will you earn from $2000 in sales?
    b. How much total pay will you earn from $5000 in sales?
    c. If you need at least $450 per week to pay your bills, what sales do you have to make per week?

25. Budget charges $29.95 for the day and $0.55 per mile driven to rent a 10-foot moving truck.
   Source: Budget.com.
   a. Let B be the cost of renting a 10-foot moving truck from Budget for a day and driving the truck m miles. Write an equation for the cost of renting from Budget.
   b. How much would it cost to rent a 10-foot truck from Budget if you were to drive it 75 miles?
   c. How many miles could you drive the truck if you could pay only $100 for the rental?

26. A local cellular phone company has a pay-as-you-talk plan that costs $10 per month and $0.20 per minute you talk on the phone.
   a. Write an equation for the total monthly cost C of this plan if you talk for m minutes.
   b. Use your equation to determine the total monthly cost of this plan if you talk for 200 minutes.
   c. How many minutes did you talk on the phone if your bill for June was $37?

27. A salesperson is guaranteed $250 per week plus a 7% commission on all sales.
   a. Write an equation for your total pay per week, P, if you make s dollars of sales.
   b. What will your total pay be if you have sales of $2000?
   c. How many dollars of sales do you need to make to have a total weekly pay of $650?

28. A salesperson is guaranteed $300 per week plus a 5% commission on all sales:
   a. Write an equation for your total pay per week, P, if you make s dollars of sales.
   b. What will your total pay be if you have sales of $4000?
   c. How many dollars of sales do you need to make to have a total weekly pay of $750?

29. You are planning a trip to Las Vegas and want to calculate your expected costs for the trip. You found that you can take a tour bus trip for up to 7 days, and it will cost you $125 for the round trip. You figure that you can stay at a hotel and eat for about $100 per day.
   a. Write an equation for the total cost of this trip depending on the number of days you stay. (We will ignore the gambling budget.)
   b. How much will it cost for a 3-day trip?
   c. If you have $700 and want to gamble half of it, how many days can you stay in Las Vegas, assuming that you do not win any money?

30. Your family is planning a trip to Orlando, Florida, to visit the amusement parks. You want to budget for your expected costs. You find round trip flights for your four-person family that total $1000. You expect the hotel, food, and admissions to cost about $400 per day.
   a. Write an equation for the total cost of this trip depending on the number of days you stay.
   b. How much will it cost for a 5-day trip?
   c. If your family can afford to spend $3500 on this trip, how many days can you stay in Orlando?

31. A professional photographer has several costs involved in taking pictures at an event such as a wedding. Editing and printing proofs of the photos cost $5.29 each. The photographer also has to pay salaries of $400 for the day.
   a. Write an equation for the total cost to shoot a wedding depending on the number of proofs the photographer edits and prints.
   b. How much will it cost the photographer if she edits and prints 100 proofs?
   c. How many proofs can the photographer edit and print if the total cost cannot exceed a budget of $1250?

32. The photographer from Exercise 31 charges her clients a $7.50 fee for each proof she edits and prints plus a flat fee of $250 for the wedding.
   a. Write an equation for the total revenue for shooting the wedding depending on the number of proofs she edits and prints.
   b. How much will the photographer charge the client for a wedding that she edits and prints 100 proofs for?
   c. Write an equation for the profit made by the photographer depending on the number of proofs she edits and prints.
   d. How much profit will the photographer make on a wedding if she edits and prints 100 proofs?
   e. How many photos must the photographer edit and print to break even? (Breaking even means that profit = 0.)

33. A snow cone vendor on the Virginia Beach boardwalk has several costs of doing business. She pays salaries of $5500 per month and kiosk rental of $1500 per month. Each snow cone sold costs her 45 cents.
   a. Write an equation for the total cost of selling snow cones for a month depending on the number of snow cones sold.
   b. What will the monthly cost be if she sells 3000 snow cones in a month?
   c. How many snow cones can she sell if the total monthly cost cannot exceed a budget of $10,600?

34. The snow cone vendor from Exercise 33 charges $2.50 per snow cone.
    a. Write an equation for the total revenue from selling snow cones for a month, depending on the number of snow cones sold.
    b. How much revenue will the vendor make if she sells 3000 snow cones in a month?
    c. Write an equation for the profit made by the snow cone vendor depending on the number of snow cones sold.
    d. How much profit will the vendor make from selling 4500 snow cones?
    e. How many snow cones must the vendor sell in a month to break even? (Breaking even means that profit = 0.)

35. The Squeaky Clean Window Cleaning Company has several costs included in cleaning windows for a business. The materials and cleaning solutions cost about $1.50 per window. Insurance and salaries for the day will cost about $230.
    a. Write an equation for the total cost to clean windows for a day depending on the number of windows cleaned.
    b. How much will it cost if the company cleans 60 windows?
    c. How many windows can the company clean if the total cost cannot exceed a budget of $450?

36. The Squeaky Clean Window Cleaning Company from Exercise 35 charges companies $7 per window cleaned plus a travel charge of $50.
    a. Write an equation for the total revenue for cleaning windows at a business depending on the number of windows cleaned.
    b. How much will the Squeaky Clean Window Cleaning Company charge a business to clean 20 windows?
    c. Write an equation for the profit made by the Squeaky Clean Window Cleaning Company depending on the number of windows cleaned.
    d. How much profit will the company make from cleaning 40 windows for a business?
    e. How many windows must the company clean in a day to break even? (Breaking even means that profit = 0.)

For Exercises 37 and 38, compare the two students' work to determine which student did the work correctly. Explain what mistake the other student made.

37. A small publicity company will custom-label water bottles for your company or event. It costs the publicity company $75 to set up the label design and 55 cents for each bottle and custom label. Write an equation for the cost of each order depending on the number of bottles ordered.

Javier
C = Cost of each order in dollars
b = number of bottles ordered
$C = 75 + 55b$

Maria
C = Cost of each order in dollars
b = number of bottles ordered
$C = 75 + 0.55b$

38. The same publicity company as in Exercise 37 has the following revenue equation:

$$R = 0.95b$$

where R is the revenue in dollars from an order of b custom-labeled bottles of water. Write an equation for the profit the publicity company will earn from an order of b bottles.

Rosemary
P = Profit of each order in dollars
b = number of bottles ordered
$P = 0.95b - (75 + 0.55b)$
$P = 0.95b - 75 - 0.55b$
$P = -75 + 0.40b$

Will
P = Profit of each order in dollars
b = number of bottles ordered
$P = 0.95b - 75 + 0.55b$
$P = -75 + 1.50b$

39. Enviro-Safe Pest Management charges new clients $150 for an in-home inspection and initial treatment for ants. Monthly preplanned treatments cost $38.
    a. Write an equation for the total cost for pest management from Enviro-Safe Pest Management depending on the number of months a house is treated.
    b. If the house has an initial treatment and then is treated monthly for 1.5 more years, how much will Enviro-Safe charge?

40. The population of the United States during the 1990s can be estimated by the equation $P = 2.57t + 249.78$, where P is the population in millions t years since 1990.
    Source: Based on data from Statistical Abstract 2001.
    a. What was the population of the United States in 1993?
    b. In what year, was the population of the United States 270,000,000?
    c. In what year, did the population of the United States reach 300 million?

41. A small manufacturer of golf clubs is concerned about monthly costs. The workshop costs $23,250 per month to run in addition to the $145 in materials per set of irons produced.

    a. Write an equation for the monthly costs of this club manufacturer.
    b. What are the monthly costs for this company if they make 100 sets of irons?
    c. How many sets of irons does this manufacturer need to produce for their costs to be $20,000?
    d. If this company wants to break even making 100 sets of irons per month, what should they charge for each set? (To break even, the company needs the revenue to equal cost. Use the cost from part b and the fact that revenue can be calculated as the price times quantity.)

42. You are in charge of creating and purchasing T-shirts for a local summer camp. After calling a local silk-screening company, you find that to purchase 100 or more T-shirts, there will be a $150 setup fee and a $5 charge, per T-shirt.

    a. Write an equation for the total cost, C, of making t T-shirts.
    b. How much would 300 T-shirts cost?
    c. How many T-shirts can you purchase with a budget of $1500?
    d. If this camp wants to break even selling 300 T-shirts, what should they charge for each T-shirt? (To break even the camp needs the revenue to equal cost. Use the cost from part b and the fact that revenue can be calculated as the price times quantity.)

43. Rockon, a small-town rock band, wants to produce a CD before their next summer concert series. They have looked into a local recording studio and found that it will cost them $1500 to produce the master recording and then an additional $1.50 to make each CD up to 500.

    a. Write an equation for the total cost, C, in dollars, of producing n CDs.
    b. How much will it cost Rockon to make 250 CDs?
    c. If Rockon has $2000 to produce CDs, how many can they order?
    d. If Rockon has $3000 to produce CDs, how many can they order?

44. The percent P of companies that are still in business t years after the fifth year in operation can be represented by the equation

$$P = -3t + 50$$

(*Hint:* Be very careful with how the variables are defined.)

    a. What percentage of companies are still in business after 1 year in operation?
    b. What percentage of companies are still in business after 25 years in operation?
    c. After how many years are there only 35% of companies still in business?

For Exercises 45 through 66, solve each equation.

45. $5x + 60 = 2x + 90$

46. $6x + 20 = 9x + 5$

47. $\frac{2}{5}d + 6 = 14$

48. $\frac{3}{4}x - 17 = 20$

49. $\frac{1}{3}m + \frac{4}{3} = 4$

50. $\frac{1}{2}x + \frac{3}{2} = 5$

51. $-3x - 6 = 14 + 8x$

52. $5r - 9 = 18r + 2$

53. $\frac{5}{7}d - \frac{3}{10} = \frac{4}{7}d + 4$

54. $\frac{3}{8}p - \frac{4}{9} = \frac{5}{8}p + 7$

55. $1.25d - 3.4 = -2.3(5d + 4)$

56. $3.7m - 4.6 = -1.8(6m + 8)$

57. $3(c + 5) - 21 = 107$

58. $5k + 7 = 2(6k - 14) + 56$

59. $1.7d + 5.7 = 29.7 + 5d$

60. $2.1m + 3.4 = 7.2 - 9.4m$

61. $\frac{3}{7}(2z - 5) = \frac{4}{7}(-3z + 9)$

62. $\frac{2}{5}(3r - 8) = \frac{3}{5}(-4r + 6)$

63. $-3(2v + 9) - 3(3v - 7) = 4v + 6(2v - 8)$

64. $4(2x + 7) - 6(4x - 8) = 12x + 3(4x - 9)$

65. $-\frac{8}{9}(3t + 5) = \frac{2}{3}t - 12$

66. $-\frac{2}{7}(4x + 2) = \frac{3}{28}x - 15$

For Exercises 67 through 84, solve for the indicated variable.

67. Force: $F = ma$ for a.

68. Weight (newtons): $W = mg$ for m.

69. Impulse: $J = Ft$ for F.

70. $P = 10h$ for h.

71. Angular acceleration: $\omega = \omega_0 + \alpha t$ for α. (*Note:* ω is the Greek symbol "omega," and α is the Greek symbol "alpha.")

72. $y = mx + b$ for b.

73. Rotational kinetic energy (J): $K = \frac{1}{2}I\omega^2$ for I.

74. Elastic potential energy (J): $U = \frac{1}{2}kx^2$ for k.

75. Kinetic energy (J): $k = \frac{1}{2}mv^2$ for m.

76. $y = \frac{1}{2}xz^2$ for x.

77. $ax + by = c$ for y.

78. $2x - y = z$ for x.

79. $ax + 5 = y$ for x.

80. $4m + n = p$ for m.

81. $b = 2c + 3d$ for c.

82. $x = 3y + 5z$ for y.

83. $5x^2 + 3y = z$ for y.

84. $4a - 5b^2 = c$ for a.

85. If a digital thermometer gave the outside temperature as 73.4 degrees Fahrenheit, would you round to the nearest whole degree? Explain your reasoning.

86. If a digital thermometer gave your child's temperature as 100.3 degrees Fahrenheit would you round to the nearest whole degree? Explain your reasoning.

87. When calculating the discounted price of a TV, you get the result 236.5725. How would you round the result to find the discounted price? Explain your reasoning.

88. Using a cost equation, you find that 2200.8 pens can be produced with a budget of $500. How should you round the number of pens? Explain your reasoning.

89. Using a profit equation, you find that 312.25 cars needed to be washed to make a profit of $400 a week. How many cars should the company try to wash a week to make $400 profit? Explain your reasoning.

90. Give your own example of a real-life situation in which the math rounding rule does not apply. Explain why it does not apply.

1.2 Using Data to Create Scatterplots

LEARNING OBJECTIVES

- Create a graphical model of data by hand.
- Find a reasonable domain and range for a model.
- Identify and interpret intercepts in an application.
- Identify model breakdown.

Data are collected in many ways and places around us. Grocery stores collect data on how and what we buy in their stores. Scientists collect data during experiments to study different characteristics of the subject. Historians collect data to study trends in past events. The Centers for Disease Control and Prevention (CDC) collect data on cases of infectious diseases to determine when an outbreak has occurred or how dangerous a virus might be. Governments collect population data to predict future trends in public funds and needs for services.

After data have been collected, the observations have to be organized and presented in a useful form. Often, data are organized in a table.

Year	Number of Deaths
1994	3532
1995	3262
1996	2894
1997	2601
1998	2283
1999	2093

Source: Centers for Disease Control and Prevention.

This table alone does not give enough information about the situation to be of much help. The number of deaths in this table could represent many different things. Some possibilities may include the following:

- Number of deaths due to jaywalking
- Number of deaths caused by poisoning

- Number of deaths of bears living in Canada
- Number of homicides of 15- to 19-year-olds in the United States

These data were collected by the Centers for Disease Control and Prevention and actually do represent the number of homicides of 15- to 19-year-olds in the United States. We can see that as the years have gone by, the number of deaths has been decreasing.

Using Data to Create Scatterplots

Although a table is useful to display data, it sometimes is better to graph the data, giving a more visual picture of the situation. To create a graph, first define the variables involved and determine which variable depends on the other. In cases such as renting a U-Haul truck and driving it m miles for the day, it is clear that the cost U depends on how many miles the truck is driven. In many cases, it is not as clear which variable depends on the other. You may find that it is often easier to determine which variable is independent rather than dependent. In the number of deaths data given in the table, the following variables can be defined:

t = Year

N = Number of homicides of 15- to 19-year-olds in the United States

Since the year does not depend on the number of homicides committed, t is called the **independent variable** (input variable), thus making N the **dependent variable** (output variable). Someone might say that N depends on t. It is important to note that the number of homicides is not caused by what year it is; the number of homicides depends on what year you want to discuss.

Now that we know which variable depends on the other, we can build a **scatterplot**. The independent variable is usually placed on the horizontal axis, and the dependent variable is usually placed on the vertical axis. It is best to label each axis with at least the units for each variable being represented.

> **What's That Mean?**
>
> **Different names for data**
> The following words or phrases all mean the same thing.
>
> **Independent Variable**
> - input variable
> - input
> - domain value
> - usually x
>
> **Dependent Variable**
> - output variable
> - output
> - range value
> - usually y
>
> **Scatterplot**
> - scatter diagram
> - scattergram
> - statplot (Texas Instruments term)
>
> Some of these terms will be discussed in later sections.

Example 1 Creating a scatterplot

Create a scatterplot of the data given in the table.

Year	Number of Deaths
1994	3532
1995	3262
1996	2894
1997	2601
1998	2283
1999	2093

Source: Centers for Disease Control and Prevention.

SOLUTION

First we will use the variables as defined above.

t = Year

N = Number of homicides of 15- to 19-year-olds in the United States

Since t is the independent variable, it will be on the horizontal axis, and the dependent variable N will be on the vertical axis. The next thing to be decided is the **scale** (spacing) for each axis. Please note that the scale does not have to be the same on the horizontal and vertical axes. For any one axis, though, the scale must remain equal. The scale remaining equal on an axis means that every space of the same size represents the same number of units. For the horizontal axis here, representing the year, the scale can be 1 and start at about 1990; for the vertical axis, representing the number of deaths, the scale could be 200 starting at about 2000. This results in the following graph.

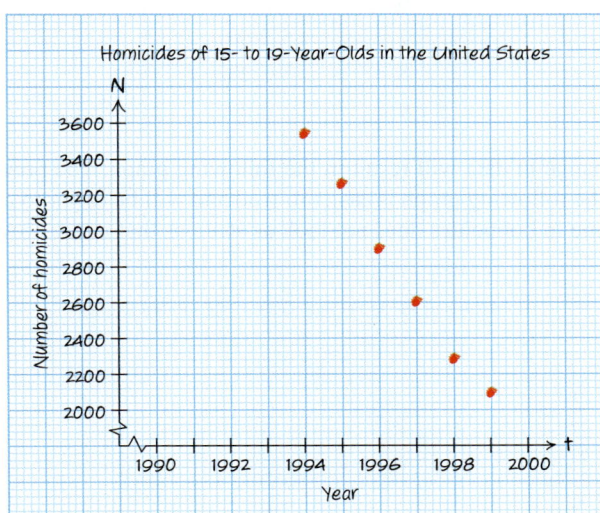

Notice that both axes in this case do not start at 0, so a zigzag pattern is placed at the beginning of each axis to indicate a jump in the numbering. Such a zigzag pattern is sometimes called a **break** in the axis. This is a valid way of making the graph easier to create, but it does distort how the relationship looks. With this scale, the decrease in homicides appears very steep. Using a different scale would cause this same decrease to appear much less drastic. By adjusting the definitions of the variables and the data, you can get an accurate graph that is not distorted. One option for such changes would be

t = Time in years since 1990

N = Number of homicides of 15- to 19-year-olds in the United States (in thousands)

Adjusting the data allows for smaller labels on the axes and often a more readable graph. The years have been represented with the year 1990 being year 0. You can choose any year you would like for the base year in a problem. This base year only gives you a place to start counting from; it does not determine the starting year of the data. Choosing a "nice" year such as 1990 or 1980 makes it easier to figure out what each number represents. For example, for a base year 1990, $t = 7$ represents 1997; for a base year 1980, 1997 would have been represented by $t = 17$.

Be careful that once you define your variables, you stay consistent with the values you use for each variable.

To adjust the years to time in years since 1990, subtract 1990 from each year. For example, 1994 – 1990 = 4. Because 1994 is 4 years since 1990, $t = 4$ would represent 1994.

To change the number of homicides data into thousands, multiply each number by

$$\frac{1 \text{ thousand}}{1000}$$

This is basically dividing each number by 1000 and adding the word *thousand* to the end of the number. For example,

$$3532 \cdot \frac{1 \text{ thousand}}{1000} = 3.532 \text{ thousand}$$

Changing the years and number of homicides will result in the following adjusted data table.

t	N
4	3.532
5	3.262
6	2.894
7	2.601
8	2.283
9	2.093

The above change in data will result in the following graph. The previous graph was reasonable, but the following graph is easier to read and will be easier to use in the future. The rate of decline in homicides does not look as steep as that in the previous graph.

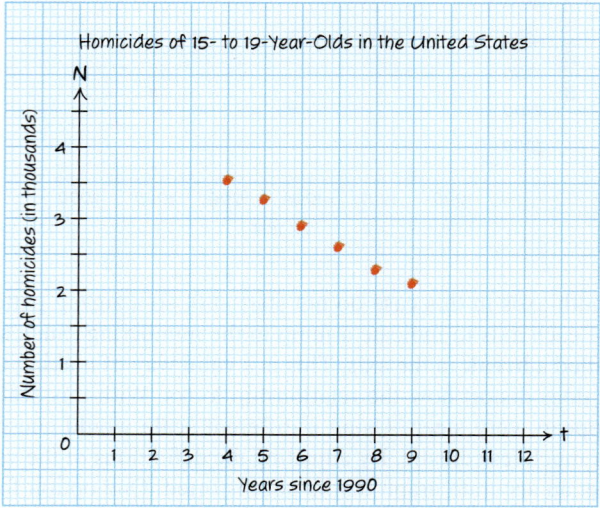

On this graph, you should notice that although the data for the years start at $t = 4$, we still must include 1, 2, and 3 on the horizontal axis. You should not skip numbers on the horizontal axis if you can avoid it.

PRACTICE PROBLEM FOR EXAMPLE 1

The state of Arizona's population is given in the table.

Year	Population of Arizona
2000	5,165,993
2001	5,295,929
2002	5,438,159
2003	5,577,784
2004	5,739,879
2005	5,939,292
2006	6,166,318*

*Estimate.
Source: U.S. Census Bureau.

a. Define variables for these data.

b. Adjust the data.

c. Create a scatterplot.

Graphical Models

The data shown thus far are considered to be **linearly related** because they generally fall along the path of a straight line. In the following concept investigation, we will draw an eyeball best-fit line through the data points on the scatterplots. This is best done with a small clear ruler. Choose a line that comes as close to all the plotted points as possible. The points that miss the line should be equally spread out above and below the line. This will allow for each point that the line misses to be balanced out by another point missed by the line.

CONCEPT INVESTIGATION

What is an "eyeball best-fit" line?

In parts 1 through 3, choose which graph has the best "eyeball best-fit" line for the data. Remember that you want the line to be as close as possible to ALL the data points. For each set of data, describe what makes the line you chose the "eyeball best-fit" line.

1.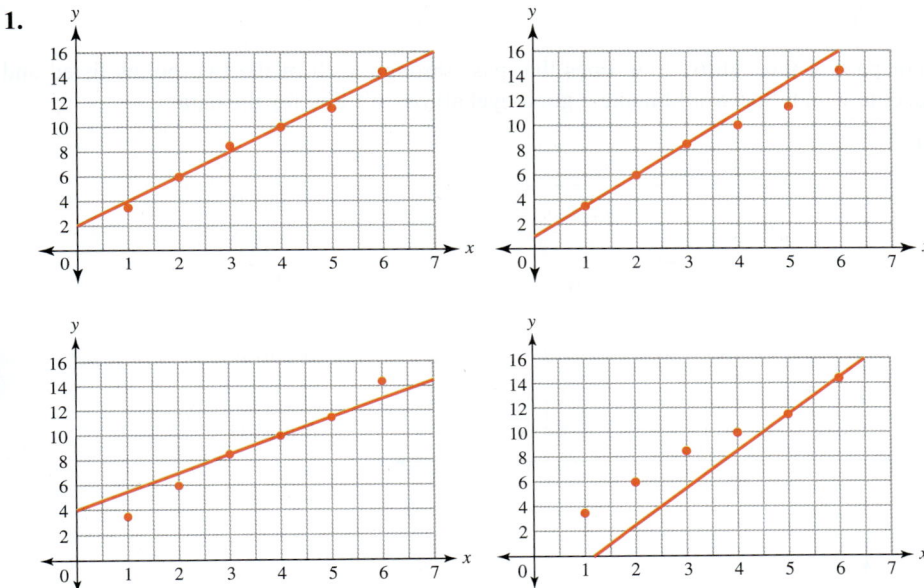

2.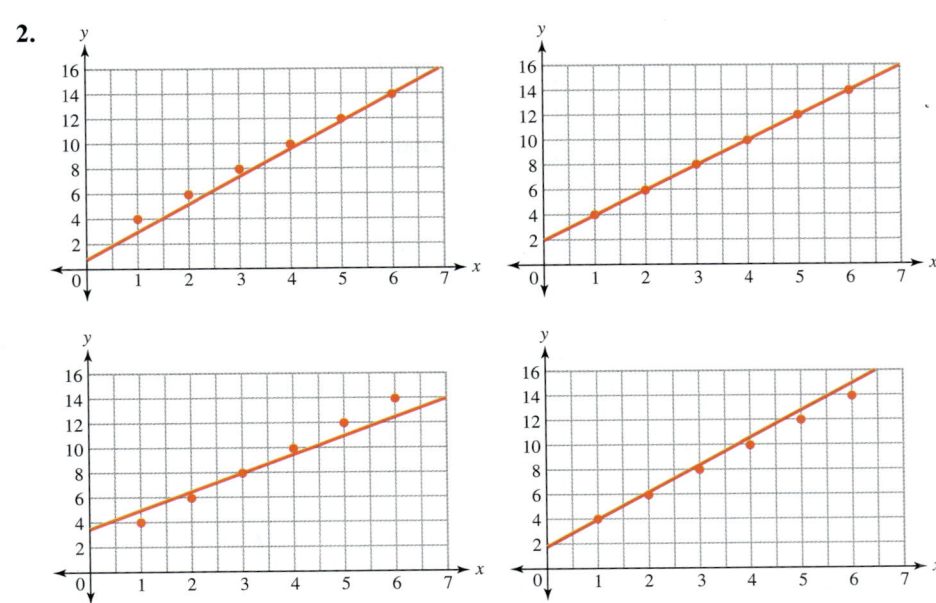

3.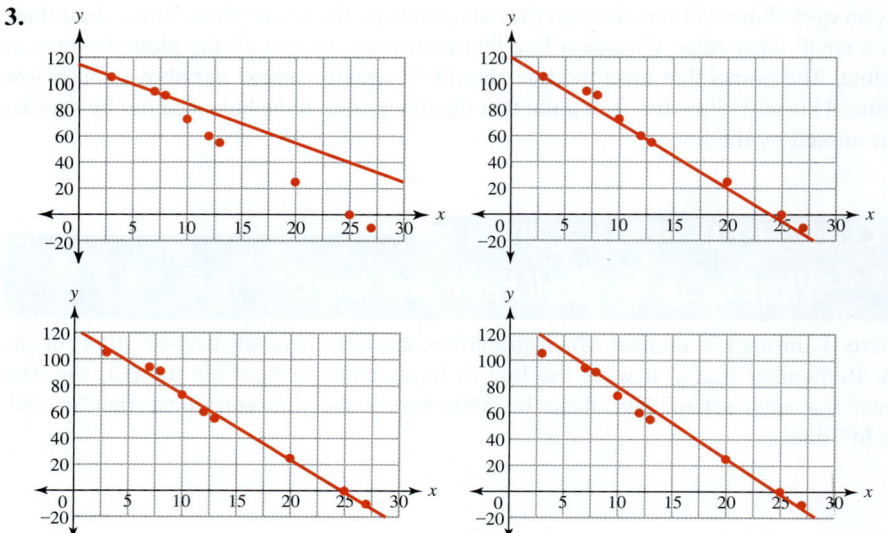

For parts 4 through 6, draw lines that pass through each of the two points listed and decide which line would make a good eyeball best-fit line for the data.

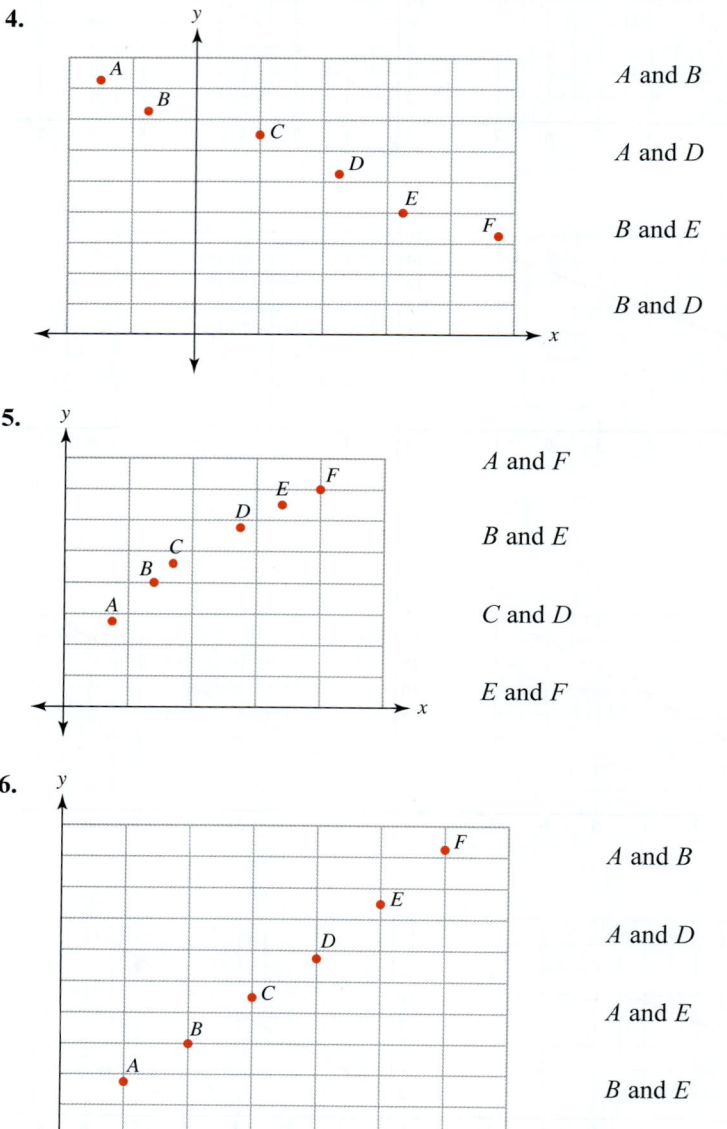

4.
A and B

A and D

B and E

B and D

5.
A and F

B and E

C and D

E and F

6.
A and B

A and D

A and E

B and E

Remember that the eyeball best-fit line will not necessarily hit any points on the graph but will be as close to ALL the points as possible. This line can be considered a **graphical model** of the data. A graphical model is used to gain additional information about the situation described by the data. Models may also be used to make predictions beyond the given data.

Example 2 Drawing eyeball best-fit lines

a. Using the scatterplot of the data from Example 1, draw an eyeball best-fit line through the data.

b. Using your eyeball best-fit line, make a prediction for the number of homicides of 15- to 19-year-olds in 1993.

SOLUTION

a. See the following graph.

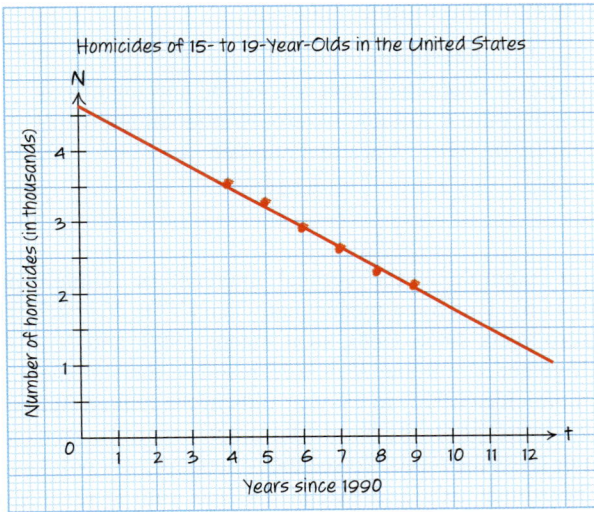

t	N
4	3.532
5	3.262
6	2.894
7	2.601
8	2.283
9	2.093

b. We are asked to estimate the number of homicides in 1993. The year 1993 is represented by $t = 3$, which is on the horizontal axis. Start at the 3 on the horizontal axis and look up to the line. The output value for the line when $t = 3$ is about 3.8 (slightly less than 4). Therefore, the graph indicates that in 1993, the number of homicides was approximately 3.8 thousand.

PRACTICE PROBLEM FOR EXAMPLE 2

Draw an eyeball best-fit line for the population of Arizona data you graphed in the Practice Problem for Example 1.

Intercepts, Domain, and Range

The points at which a graph crosses the axes are called **intercepts**. The **vertical intercept** or *y*-intercept is the point where the graph crosses the vertical axis. A vertical intercept will always have an input (*x*) value of zero. The **horizontal intercept** or *x*-intercept is the point where the graph crosses the horizontal axis. A horizontal intercept will always have an output (*y*) value of zero.

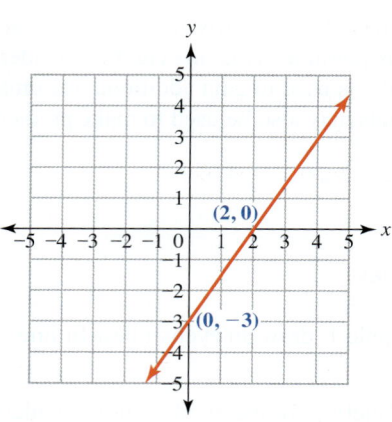 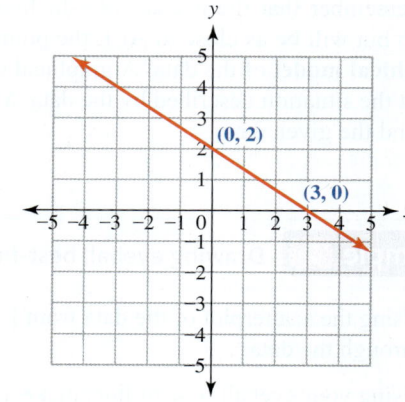

What's That Mean?

Different names for intercepts

The following words or phrases all mean the same thing.

Vertical Intercept
- y-intercept
- Can take on the name of the variable representing the vertical axis, such as N-intercept in Example 2.

Horizontal Intercept
- x-intercept
- Can take on the name of the variable representing the horizontal axis, such as t-intercept in Example 2.

DEFINITIONS

Vertical intercept   The point where the graph crosses the vertical axis. This will always occur when the input variable is zero. Vertical intercepts are written as an ordered pair $(0, k)$, where k is a real number.

$(0, 5)$ $(0, 2.3)$ $(0, -6)$

Horizontal intercept   The point where the graph crosses the horizontal axis. This will always occur when the output variable is zero. Horizontal intercepts are written as an ordered pair $(k, 0)$, where k is a real number.

$(2, 0)$ $(3.7, 0)$ $(-4, 0)$

In the homicides graph in Example 2, the vertical intercept or N-intercept is approximately at the point $(0, 4.75)$. This point means that when $t = 0$, $N = 4.75$. The $t = 0$ can be translated as zero years since 1990, so it means 1990, and $N = 4.75$ represents 4750 homicides. Together, $t = 0$ and $N = 4.75$ result in the statement: "In 1990, there were about 4750 homicides of 15- to 19-year-olds in the United States."

The horizontal intercept or t-intercept for the homicides graph cannot be seen, since the line does not extend to hit or cross the t-axis. To see the t-intercept, you will have to continue the values on the horizontal axis and extend the line until it reaches the t-axis. Determining intercepts is one reason to make the graph extend past the data given in the problem. If the graph were to reach the horizontal axis, N would be zero, meaning no homicides of 15- to 19-year-olds. Having no homicides will probably never happen, unfortunately, so the horizontal intercept would not make sense in this situation.

Example 3   Reading a graph

Use the graph to make the following estimates:

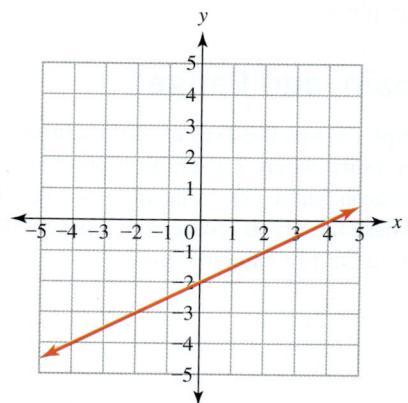

a. Estimate the vertical intercept.
b. Estimate the horizontal intercept.
c. Estimate the input value that makes the output of this graph equal -1.
d. Estimate the input value that makes the output of this graph equal -3.5.
e. Estimate the output value of this graph when the input value is -2.

SOLUTION

a. This graph crosses the vertical axis (y-axis) at about -2, so the vertical intercept is $(0, -2)$.
b. This graph crosses the horizontal axis (x-axis) at about 4, so the horizontal intercept is $(4, 0)$.
c. First locate the output -1 on the vertical axis and trace over to the line.

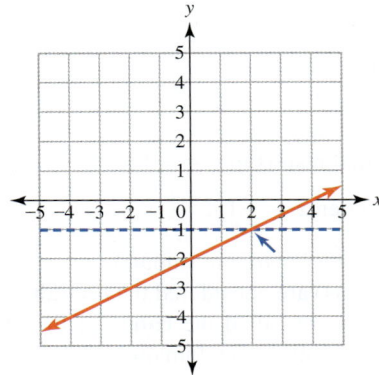

The line has an output of -1 when the input, x, is equal to 2.

d. We locate the output of $y = -3.5$ on the vertical axis and trace over to the line.

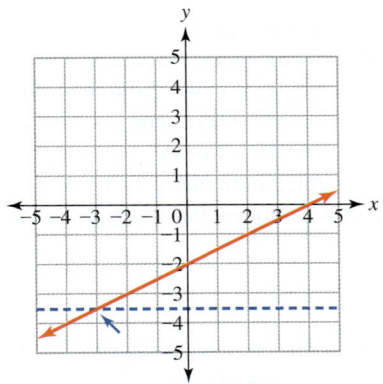

The line has an output of $y = -3.5$ when the input is $x = -3$.

e. We locate the input value of $x = -2$ on the horizontal axis and trace down to the line.

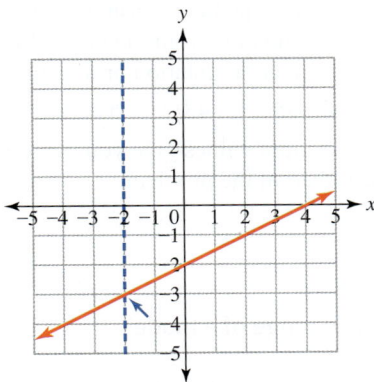

The line has an output of $y = -3$ when the input is $x = -2$.

PRACTICE PROBLEM FOR EXAMPLE 3

Use the graph to make the following estimates:

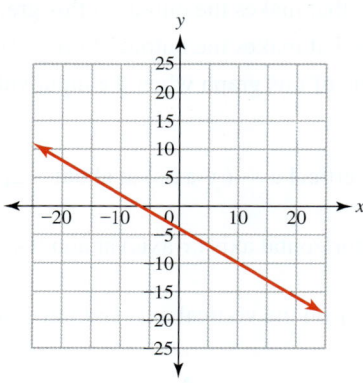

a. Estimate the y-intercept.

b. Estimate the x-intercept.

c. Estimate the value of x that results in $y = -10$.

d. Estimate the value of y when $x = -15$.

The values of the independent variable (inputs) that result in reasonable values for the dependent variable (outputs) are considered the **domain** of the model. The resulting outputs from a given domain are called the **range** of the model. Interval notation or inequalities will be used when stating the domain and range of a model. For a refresher on interval notation, see the review in Appendix A. In using an eyeball best-fit line drawn on a scatterplot, the domain and range will be estimated according to the graph and what you believe is reasonable. This means that every student's domain and range may be different but still equally correct.

Because a model is often meant to be used to extrapolate, that is, to predict a future or past value, the domain should extend beyond the data whenever reasonable. The main thing that you will want to avoid is model breakdown. **Model breakdown** occurs when a domain value results in an output that does not make sense in the situation or makes an equation undefined mathematically. An example of model breakdown in the number of homicides example would be a negative number of homicides. Another example of model breakdown would be a percentage of people or things that was over 100% or a negative percentage.

What's That Mean?

Domain
- Input values for the model.
- Values for the independent variable.
- Usually x-values.

Range
- Output values for the model.
- Values for the dependent variable.
- Usually y-values.

DEFINITIONS

Domain The set of values for the independent variable that results in reasonable output values with no model breakdown. A domain should be written in interval notation or using inequality symbols.

Example: [2, 15] $2 \leq x \leq 15$

Range A set of values for the dependent variable resulting from the given domain values. The outputs that come from the given domain's input values. A range should be written in interval notation or using inequality symbols.

Example: [5, 48] $5 \leq y \leq 48$

Model breakdown When input values give you outputs that do not make sense in the situation described in the problem.

Example 4 Domain and range of a model

Determine a reasonable domain and range for the homicide model found in Example 2.

SOLUTION

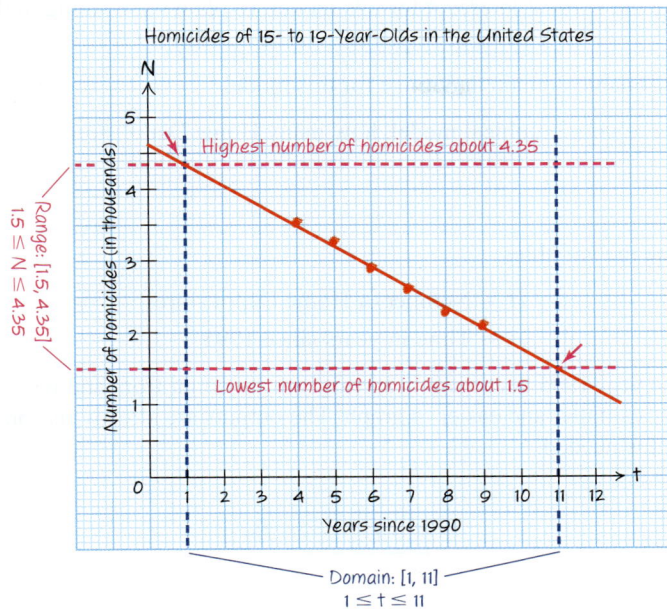

Because the model fits the data pretty well, we should be able to extend our domain beyond the data values. How far we spread out from the data is somewhat arbitrary. Because no obvious model breakdown shows up on the graph, such as negative numbers of homicides, we will be able to extend the input values somewhat in both directions. The data set starts in year 4 (1994) and goes to year 9 (1999), so in this situation, we could choose a domain of $1 \leq t \leq 11$.

Looking at the graph, we see that the lowest number of homicides predicted during the years within the domain is about 1.5 thousand in year 11 (2001). The highest number of homicides predicted during the years within the domain seems to be about 4.35 thousand in year 1 (1991). This gives us a range of $1.5 \leq N \leq 4.35$.

PRACTICE PROBLEM FOR EXAMPLE 4

Determine a reasonable domain and range for the graphical model you drew in the Practice Problem for Example 2.

Example 5 Interpreting intercepts, domain, and range

The number of Home Depot stores has been increasing at a very consistent pace. The bar graph gives the number of Home Depot stores for each year.

Source: Home Depot Annual Reports www.Homedepot.com

a. Create a scatterplot for these data and draw an eyeball best-fit line through the data.

b. Determine the vertical intercept for this model. Explain its meaning in this situation.

c. Find a reasonable domain and range for this model.

d. According to your graphical model, how many Home Depot stores were there in 2009?

SOLUTION

a. First define the variables.

$$H = \text{Number of Home Depot stores}$$
$$t = \text{Time in years since 1995}$$

We use the year 1995 because it is an easy year to count from and will make the input values smaller and easier to graph. The outputs are already reasonable, so no adjustment is needed.

t	H
4	930
5	1134
6	1333
7	1532
8	1807
9	1890
10	2042
11	2147

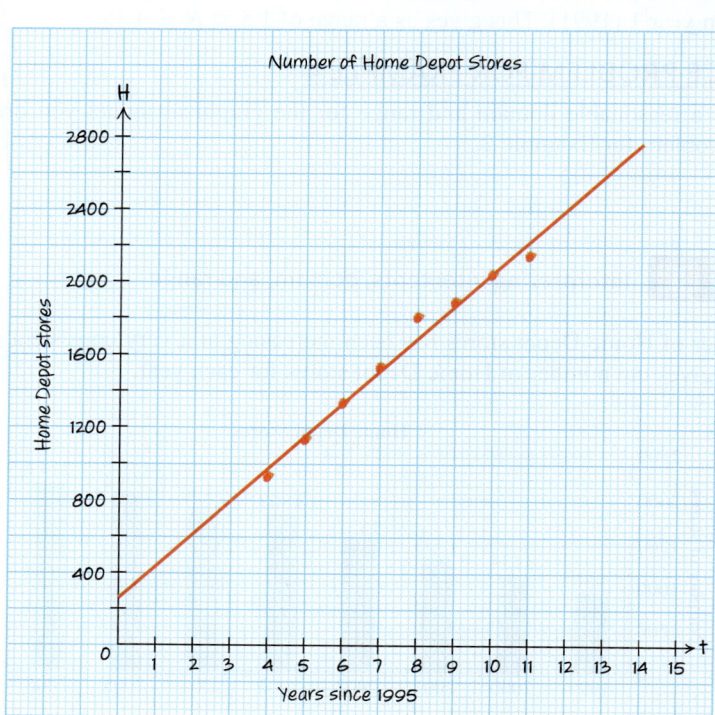

b. The vertical intercept for this model is when the line crosses the *H*-axis at approximately (0, 300), as shown in the graph. The 0 represents the year 1995, and the 300 represents number of Home Depot stores. Together, this means that in 1995, Home Depot had approximately 300 stores.

c. Because this model does a reasonable job of representing these data, we should be able to extend the domain beyond the given data. One possible domain for this model would be [2, 13] or $2 \leq t \leq 13$, which would represent the years 1997 to 2008. The range that results from that model would be the number of stores that the model predicts for Home Depot between the years 1997 and 2008.

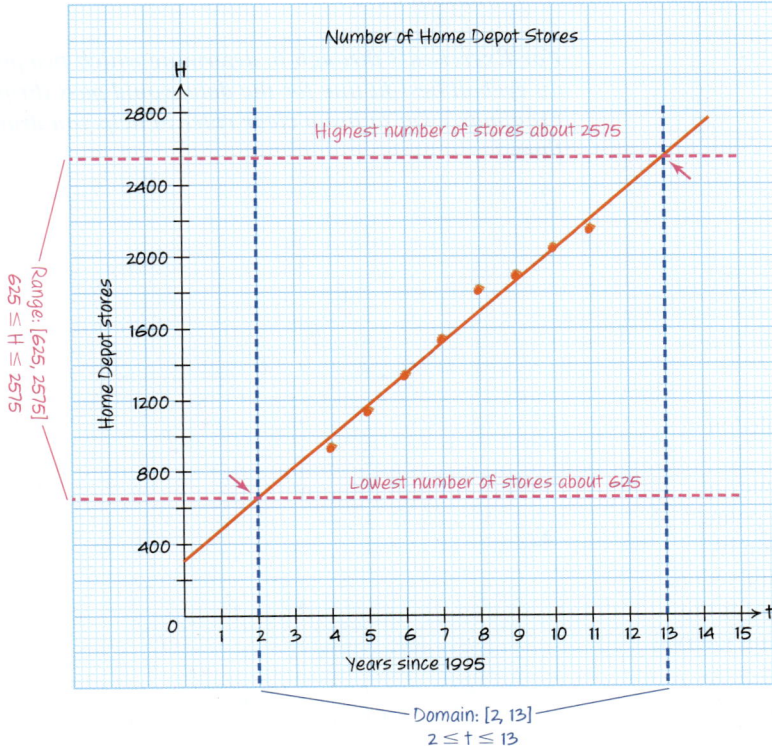

From the graph, we can estimate that the least number of stores during those years is about 625 and the most number of stores is about 2575. This results in a range of [625, 2575] or $625 \leq H \leq 2575$.

d. The year 2009 would be represented by $t = 14$, and according to the graph, the model is at approximately 2750 when $t = 14$. Thus, in 2009, there were approximately 2750 Home Depot stores.

Domain and range are essential parts of any model. They help the user of the model to know when it is appropriate to use that model and when it is not. In most applications, the domain is set by what values you think make the model reasonable. Therefore, domains will vary from one student to another. Always set your domain first. The range will then be found by using the model and the previously set domain.

Remember to set your domain by spreading out some from your input values and avoid model breakdown. Then use your model to look for the lowest and highest output values that your model produces within the set domain. This process will work for all the models we consider in this text.

1.2 Exercises

1. Use the given graphical model for the total prize money at Wimbledon to answer the following questions. (*Note:* Answers will vary.)

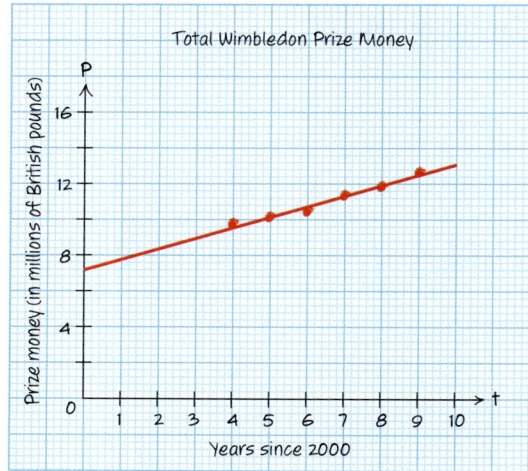

Source: www.wimbledon.org

   a. Estimate the total prize money at Wimbledon in 2002.
   b. Estimate in what year the total prize money at Wimbledon was about 9 million British pounds.
   c. Estimate the vertical intercept and explain its meaning in regard to prize money at Wimbledon.
   d. What are a reasonable domain and range for this graphical model? Write your answer using interval notation.

2. Use the given graphical model for the population of Russia to answer the following questions. (*Note:* Answers will vary.)

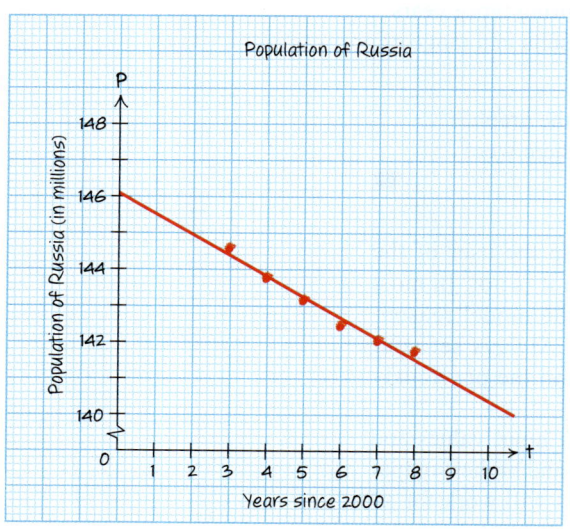

Source: www.gks.ru. Russia Federal State Statistics Service.

   a. Estimate the population of Russia in 2001.
   b. Estimate the year in which the population will be about 140 million people.
   c. Estimate the vertical intercept and explain its meaning in regard to the population of Russia.
   d. What are a reasonable domain and range for this graphical model? Write your answer using interval notation.

For Exercises 3 through 6, determine which two points an eyeball best-fit line for the data would pass through. (You may want to use a clear ruler to help you draw lines.)

3.

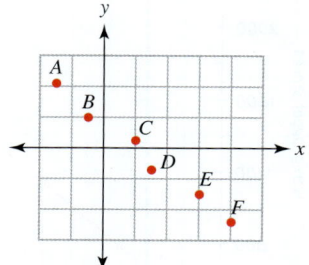

4.

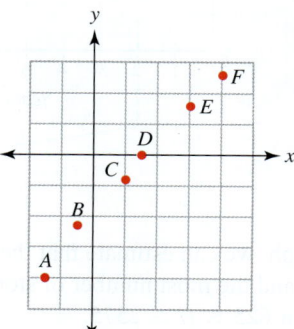

5.

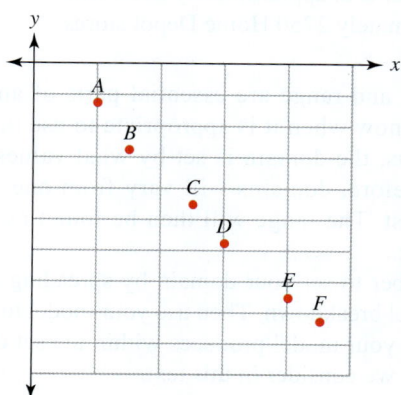

6.

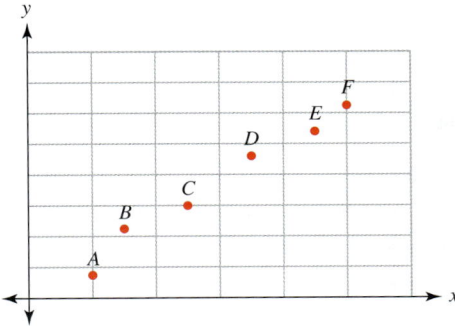

For Exercises 7 through 14, determine a scale that will allow the given data to fit on the graph without breaking the axis. Answers may vary.

7.

x
100
170
210
340

8.

x
25
55
100
135

9.

x
0.02
0.06
0.09
0.15

10.

x
0.005
0.002
0.001
0.0005

11.

x
−500
−300
−100
0

12.

x
−2005
−3040
−4000
−6500

13.

x
23,000
35,000
48,000
62,000

14.

x
5,000,000
7,000,000
11,000,000
20,000,000

For Exercises 15 through 18, the answers may vary.

15. The cost to produce chocolate-dipped key lime pie on a stick bars is given in the table.

Bars	Cost ($)
10	252.50
20	265.00
50	302.50
100	365.00
150	427.50

a. Define the variables for this problem. Identify which is the independent variable and which is the dependent variable. Adjust the data if needed.
b. Create a scatterplot and draw an eyeball best-fit line through the data.
c. Using your graphical model, estimate the cost to produce 200 bars.
d. What are a reasonable domain and range for your graphical model? Write your answer using inequalities.

16. Malcolm's Power Pens makes personalized pens for business promotions. The cost to produce a custom-printed metal pen is given in the table.

Pens	Cost ($)
100	90
200	125
1000	405
3000	1105
5000	1805

a. Define the variables for this problem. Identify which is the independent variable and which is the dependent variable. Adjust the data if needed.
b. Create a scatterplot and draw an eyeball best-fit line through the data.
c. Using your graphical model, estimate the cost to produce 500 pens.
d. What are a reasonable domain and range for your graphical model? Write your answer using inequalities.

17. Quiksilver, Inc., the makers of the Quiksilver and Roxy clothing lines, has had steadily increasing profits over the past several years. The profits for Quicksilver, Inc. are given in the bar graph.

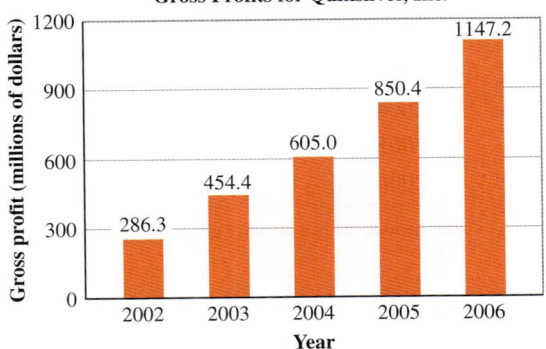

Source: CBS.Marketwatch.com

a. Define the variables for this problem. Identify which is the independent variable and which is the dependent variable. Adjust the data if needed.
b. Create a scatterplot and draw an eyeball best-fit line through the data.
c. Using your graphical model, estimate the profit for Quiksilver, Inc. in 2010.

d. What are a reasonable domain and range for your graphical model? Write your answer using inequalities.

18. Home Depot's net sales are given in the bar graph

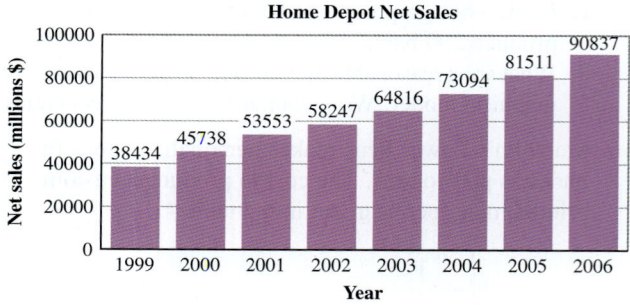

Source: Home Depot annual reports, www.homedepot.com

a. Define the variables for this problem. Identify which is the independent variable and which is the dependent variable. Adjust the data if needed.
b. Create a scatterplot and draw an eyeball best-fit line through the data.
c. Using your graphical model estimate the net sales for Home Depot in 2009.
d. What are a reasonable domain and range for your graphical model? Write your answer using inequalities.

For Exercises 19 through 22, use the given data for the incidence of hepatitis A to decide whether the given student work is correct or not. If the student work is not correct, give a correct answer. Answers may vary.

The incidence of acute viral hepatitis A in the United States is given in the table.

Year	Cases per 100,000 population
2001	3.7
2002	3.1
2003	2.6
2004	1.9
2005	1.5
2006	1.2

Source: www.cdc.gov. Health, United States, 2008.

19. Two students gave the following graphical models. Decide which model best fits the data. Explain your reasoning.

Ashley's model.
Let:
A = Number of cases per 100,000 of hepatitis A in the United States
t = Time in years since 2000

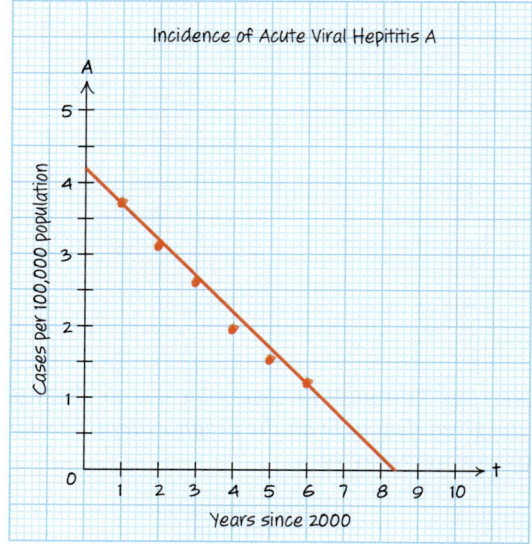

Maria's model.
Let:
A = Number of cases per 100,000 of hepatitis A in the United States
t = Time in years since 2000

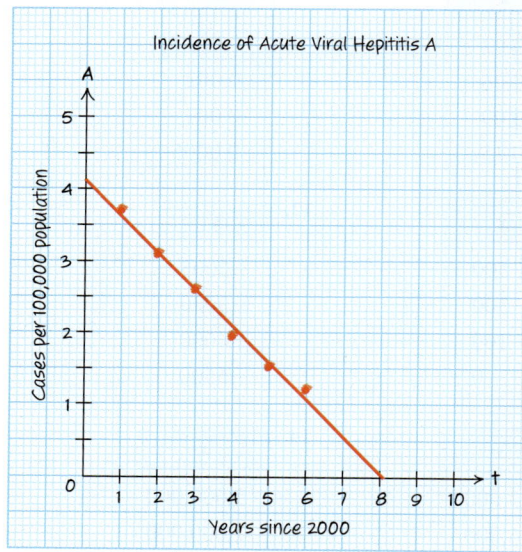

20. Using Maria's graphical model from Exercise 19, Maria estimated the vertical intercept to be (8.2, 0) and that it means there would be no cases of hepatitis A in about 2008. Is this correct? If not, explain what Maria did wrong and give the correct answers.

21. Using Maria's graphical model, explain why a domain of $0 \leq t \leq 10$ is not reasonable. Give a reasonable domain and range for this model.

22. Maria estimates that there was approximately 1 case of hepatitis A per 100,000 people in the United States in 2007. Is this correct? If not, give a correct answer and explain your reasoning.

In Exercises 23 through 28, answers may vary.

23. The population of the United States is given in the bar graph.

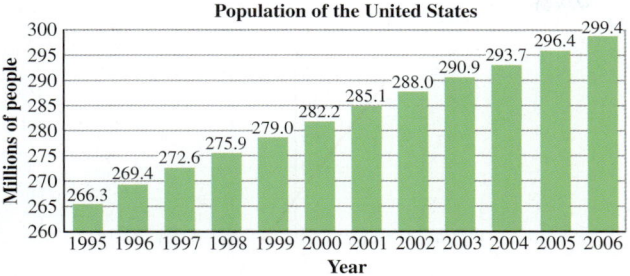

Source: U.S. Census Bureau.

a. Define the variables for this problem. Identify which is the independent variable and which is the dependent variable. Adjust the data if needed.
b. Create a scatterplot and draw an eyeball best-fit line through the data.
c. Using your graphical model, estimate the population of the United States in 2009.
d. What are a reasonable domain and range for your graphical model? Write your answer in interval notation.
e. What is the vertical intercept for your model?
f. What does the vertical intercept mean in terms of the population of the United States?

24. The population of Florida is given in the bar graph.

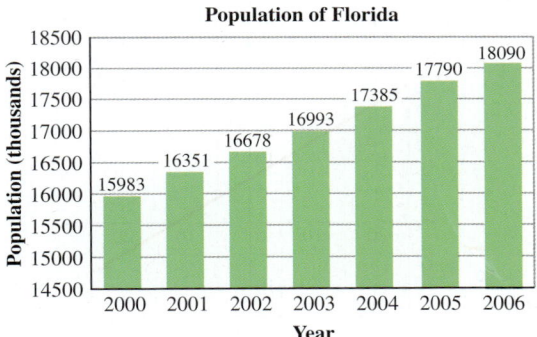

Source: U.S. Census Bureau.

a. Define the variables for this problem. Identify which is the independent variable and which is the dependent variable. Adjust the data if needed.
b. Create a scatterplot and draw an eyeball best-fit line through the data.
c. Using your graphical model, estimate the population of Florida in 2015.
d. What are a reasonable domain and range for your graphical model? Write your answer in interval notation.
e. What is the vertical intercept for your model?
f. Explain the meaning of the vertical intercept in terms of the population of Florida.

25. The number of deaths induced by illegal drugs of women in the United States is given in the table.

Year	2000	2001	2002	2003	2004
Deaths	6583	7452	9306	10297	11349

Source: Centers for Disease Control and Prevention.

a. Create a graphical model for these data. (Remember to define variables.)
b. Using your graphical model, estimate the number of drug-induced deaths of females in the United States in 2007.
c. Use your graphical model to estimate in what year the number of deaths was 4000.
d. What are a reasonable domain and range for your graphical model? Write your answer using inequalities.

26. The death rate (per 100,000 people) for heart disease in the United States for various years is given in the table.

Year	Death Rate
1999	266.5
2000	257.6
2001	247.8
2002	240.8
2003	232.3
2004	217.5

Source: Centers for Disease Control and Prevention.

a. Create a graphical model for these data.
b. According to your model, when will the death rate be approximately 200 deaths per 100,000 people?
c. What are a reasonable domain and range for your model?
d. What does your model predict the death rate to be in 1995?
e. What is the vertical intercept for your model?
f. What does the vertical intercept represent in regard to the death rate for heart disease?

27. The expected number of years someone will live is given in the table:

Age	Years
0 (at birth)	77.5
10	68.2
20	58.4
30	48.9
40	39.5
50	30.6
60	22.2

Source: Statistical Abstract of the United States, 2007, www.census.gov

a. Create a graphical model for these data.
b. According to your graphical model, what is the expected number of years a 45-year-old person would live?
c. What are a reasonable domain and range for your model?
d. What does your graphical model predict the number of years a 90-year-old person should expect to live?
e. What is the vertical intercept for your model?
f. What does the vertical intercept represent in this situation?

28. The amount teachers are paid per year is typically determined by their years of experience. The salaries for several different teachers at one school district are given in the table.

Years of Experience	Salary (dollars)
0	49,900
2	52,400
5	62,100
8	67,000
10	74,300
12	79,200

a. Create a graphical model for these data.
b. According to your graphical model, what is the expected salary if a teacher has 7 years of experience?
c. What are a reasonable domain and range for your model?
d. How many years of experience does a teacher need in order to earn $60,000?
e. What is the vertical intercept for your model?
f. What does the vertical intercept represent in this situation?

In Exercises 29 through 34, use the graph to make the estimates. Answers may vary.

29.

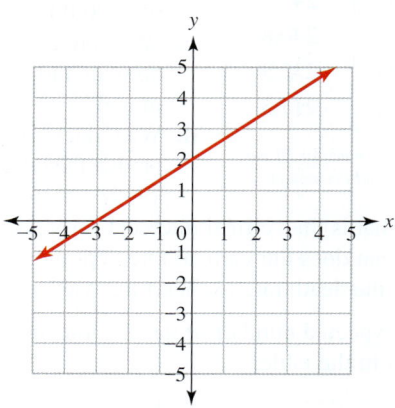

a. Estimate the y-intercept. Write it as a point (x, y).
b. Estimate the x-intercept. Write it as a point (x, y).
c. Estimate the value of x that results in $y = 4$.
d. Estimate the value of y when $x = 1$.

30.

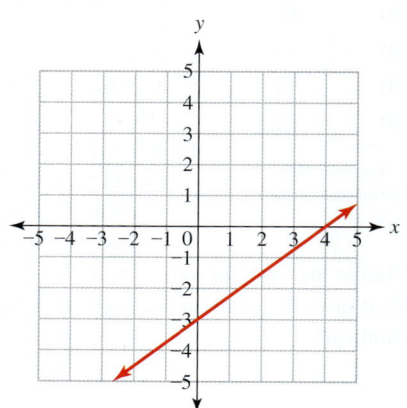

a. Estimate the y-intercept. Write it as a point (x, y).
b. Estimate the x-intercept. Write it as a point (x, y).
c. Estimate the value of x that results in $y = -2$.
d. Estimate the value of y when $x = -2$.

31.

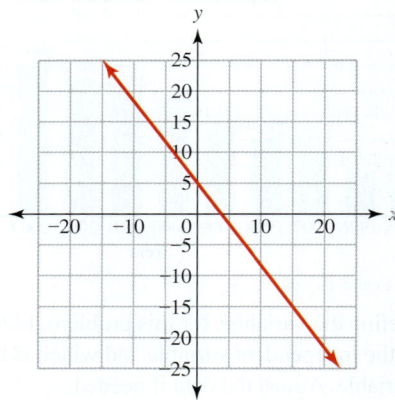

a. Estimate the vertical intercept. Write it as a point (x, y).
b. Estimate the horizontal intercept. Write it as a point (x, y).
c. Estimate the input value that makes the output of this graph equal 18.
d. Estimate the input value that makes the output of this graph equal -15.
e. Estimate the output value of this graph when the input value is 10.

32.

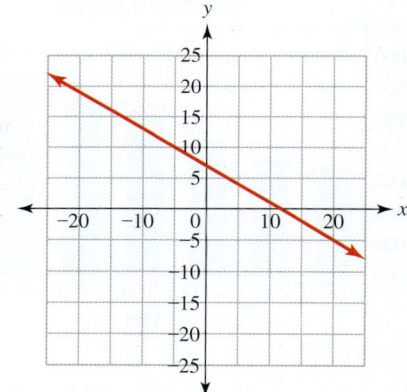

a. Estimate the vertical intercept. Write it as a point (x, y).
b. Estimate the horizontal intercept. Write it as a point (x, y).
c. Estimate the input value that makes the output of this graph equal 15.
d. Estimate the input value that makes the output of this graph equal -2.
e. Estimate the output value of this graph when the input value is -10.

33.

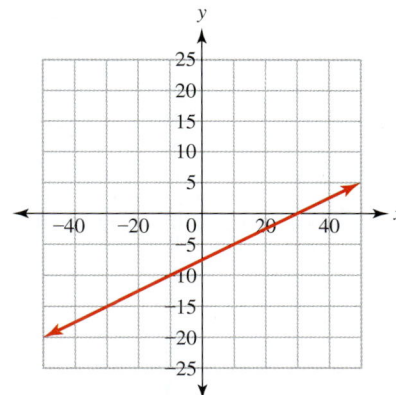

a. Estimate the y-intercept. Write it as a point (x, y).
b. Estimate the x-intercept. Write it as a point (x, y).
c. Estimate the value of x that results in y = −5.
d. Estimate the value of x that results in y = −10.
e. Estimate the value of y when x = 20.

34.

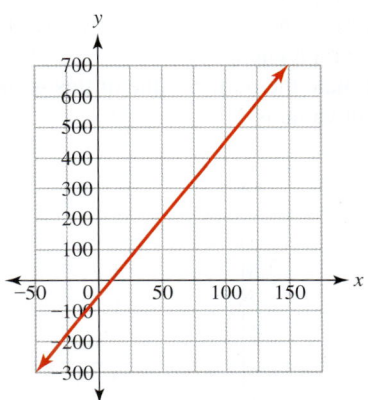

a. Estimate the x-intercept. Write it as a point (x, y).
b. Estimate the y-intercept. Write it as a point (x, y).
c. Estimate the value of x that results in y = 500.
d. Estimate the value of x that results in y = −200.
e. Estimate the value of y when x = 100.

In Exercises 35 and 36, decide whether the given student's answers are correct or not. If the answer is not correct, give the correct answer. Answers may vary.

35.

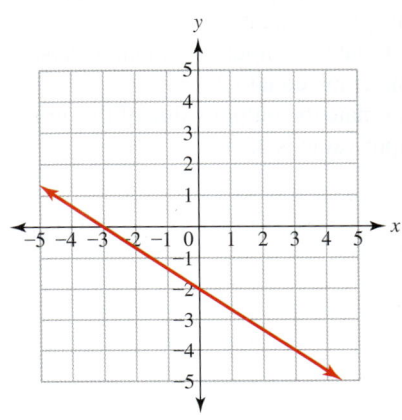

a. Estimate the vertical intercept. Write it as a point (x, y).

   Vertical intercept = (−2, 0)

b. Estimate the horizontal intercept. Write it as a point (x, y).

   Horizontal intercept = (−3, 0)

c. Estimate the input value that makes the output of this graph equal 1.

   $y = -2.5$

d. Estimate the input value that makes the output of this graph equal −1.

   $y = -1.5$

e. Estimate the output value of this graph when the input value is 3.

   $y = -4$

36.

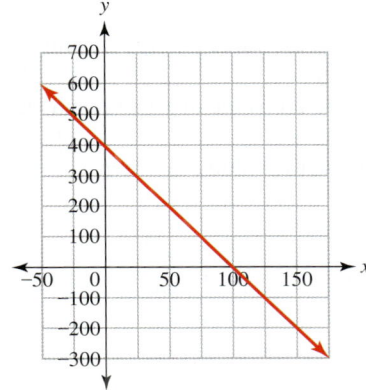

a. Estimate the x-intercept. Write it as a point (x, y).

   x-intercept = (100, 500)

b. Estimate the y-intercept. Write it as a point (x, y).

   y-intercept = (0, 0)

c. Estimate the value of x that results in y = 500.

   $x = -25$

d. Estimate the value of x that results in y = −200.

   $x = 100$

e. Estimate the value of y when x = 50.

   $y = 200$

In Exercises 37 through 40, use the graph to make the estimates. Answers may vary.

37.

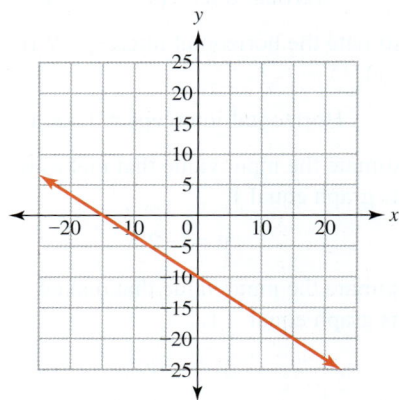

a. Estimate the x-intercept. Write it as a point (x, y).
b. Estimate the y-intercept. Write it as a point (x, y).
c. Estimate the value of x that results in $y = -20$.
d. Estimate the value of x that results in $y = -15$.
e. Estimate the value of y when $x = -10$.

38.

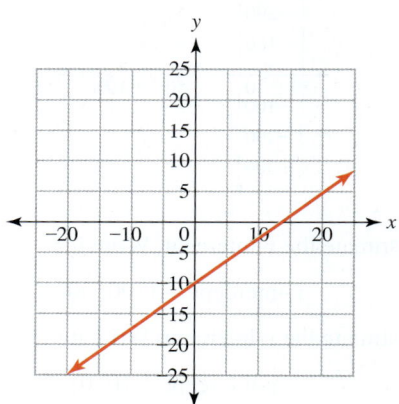

a. Estimate the x-intercept. Write it as a point (x, y).
b. Estimate the y-intercept. Write it as a point (x, y).
c. Estimate the value of x that results in $y = 8$.
d. Estimate the value of x that results in $y = -17$.
e. Estimate the value of y when $x = 20$.

39.

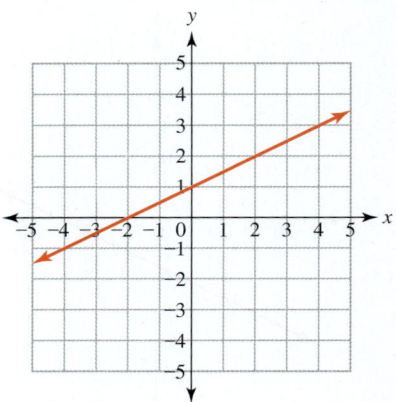

a. Estimate the vertical intercept. Write it as a point (x, y).
b. Estimate the horizontal intercept. Write it as a point (x, y).
c. Estimate the input value that makes the output of this graph equal 1.5.
d. Estimate the input value that makes the output of this graph equal -1.
e. Estimate the output value of this graph when the input value is 3.

40.

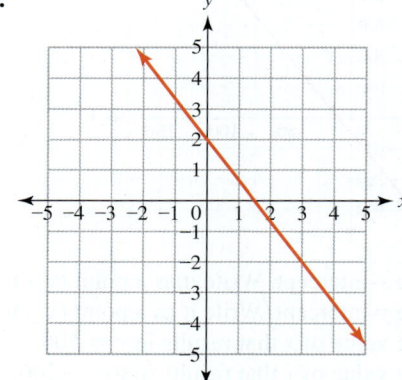

a. Estimate the vertical intercept. Write it as a point (x, y).
b. Estimate the horizontal intercept. Write it as a point (x, y).
c. Estimate the input value that makes the output of this graph equal 4.
d. Estimate the input value that makes the output of this graph equal -2.
e. Estimate the output value of this graph when the input value is 4.

1.3 Fundamentals of Graphing and Slope

LEARNING OBJECTIVES
- Graph equations by plotting points.
- Calculate the slope of a line.
- Interpret the meaning of slope.
- Graph lines using the slope-intercept form of a line.

Introduction to Graphing Equations

In Section 1.2, we created scatterplots of data by hand. We also drew an eyeball best-fit line on that scatterplot to give us an approximate graphical model for the data and situation we were considering. Although the graphical model is useful to estimate some values, it is often easier to find values for a model if we have an algebraic equation for it. In this section, we will investigate the different characteristics of a linear equation and its graph.

Example 1 Graphing equations by plotting points

Graph the equations by creating a table of values and plotting the points.

a. $y = 3x - 8$

b. $x = -2y + 10$

c. $y = x^2 + 3$

SOLUTION

a. Begin by finding ordered pairs that satisfy the equation. Because y is already isolated in this equation, it is easier to choose values for x and find the y-values that go with them.

x	$y = 3x - 8$	(x, y)
-2	$y = 3(-2) - 8 = -14$	$(-2, -14)$
0	$y = 3(0) - 8 = -8$	$(0, -8)$
2	$y = 3(2) - 8 = -2$	$(2, -2)$
3	$y = 3(3) - 8 = 1$	$(3, 1)$

Plot these points.

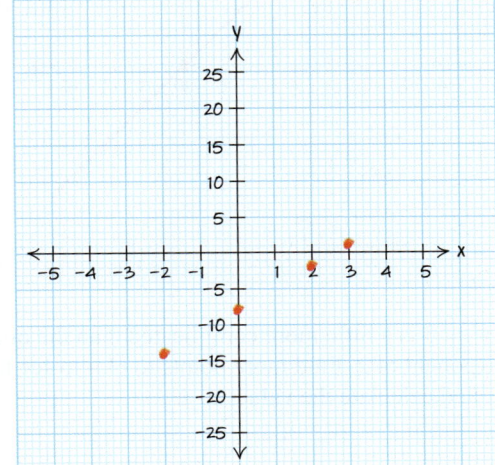

All of these points lie on a straight line. Draw a line through the points. Because this equation would have an infinite number of points that satisfy it, extend the line beyond the points drawn and put arrows on the ends of the line to indicate that the line continues in both directions infinitely. The line itself actually represents all of the possible combinations of *x* and *y* that satisfy this equation.

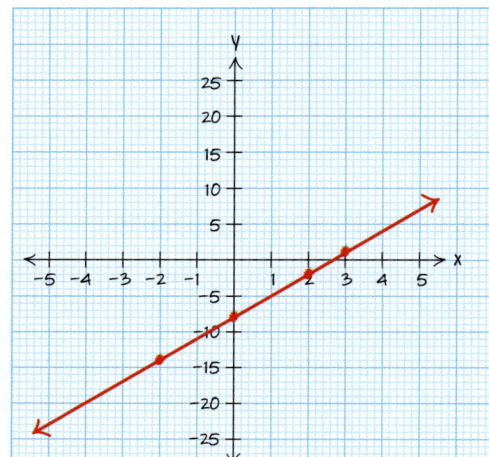

Using Your TI Graphing Calculator

A graphing calculator can be used to do many things with equations. Graphing calculators will consider *x* the independent (input) variable and *y* the dependent (output) variable.

When entering an equation in the Y= screen, you must have the dependent variable isolated so that the equation can be entered in as *y* = expression.

To enter the equation

$$y = 2x + 5$$

you will press [Y=], and then in the Y1= line, you will enter $2x + 5$.

Use the [X,T,θ,n] button for the variable *x*.

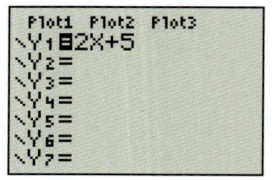

b. Because *x* is already isolated in the equation $x = -2y + 10$ it would be easier to choose values for *y* and find values for *x*.

$x = -2y + 10$	y	(x, y)
$x = -2(-1) + 10 = 12$	-1	$(12, -1)$
$x = -2(0) + 10 = 10$	0	$(10, 0)$
$x = -2(2) + 10 = 6$	2	$(6, 2)$
$x = -2(4) + 10 = 2$	4	$(2, 4)$

Now we can plot these points and draw a straight line through them.

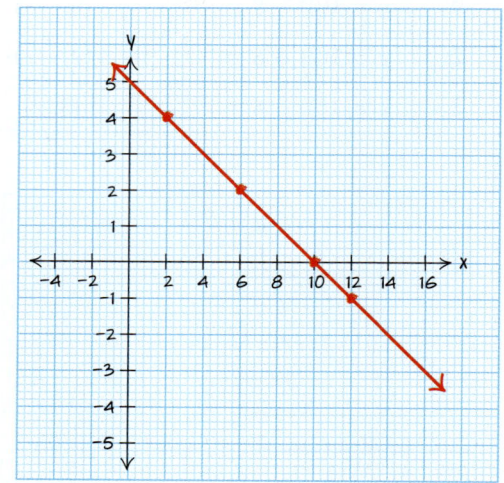

c. The equation $y = x^2 + 3$ is a little more complicated than the others because the variable *x* has an exponent of 2. Find more points to plot to see the graph's shape. Choose values of *x* and calculate the values of *y*.

x	$y = x^2 + 3$	(x, y)
-2	$y = (-2)^2 + 3 = 7$	$(-2, 7)$
-1	$y = (-1)^2 + 3 = 4$	$(-1, 4)$
0	$y = (0)^2 + 3 = 3$	$(0, 3)$
1	$y = (1)^2 + 3 = 4$	$(1, 4)$
2	$y = (2)^2 + 3 = 7$	$(2, 7)$
3	$y = (3)^2 + 3 = 12$	$(3, 12)$

Plot the points and draw a smooth curve through them.

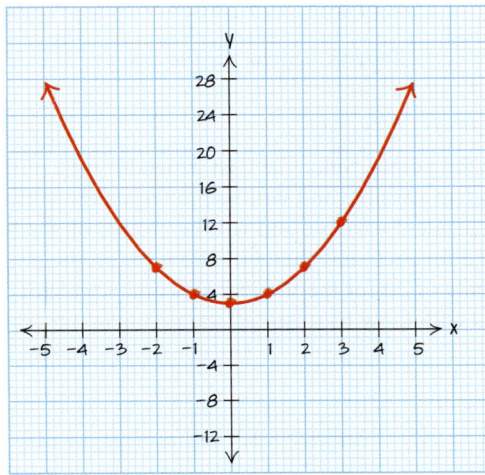

Using Your TI Graphing Calculator

Using the TABLE Feature: First you should enter the equation into the Y= screen. To set up the table, you first press 2nd WINDOW (TBLSET F2) to get the table setup menu.

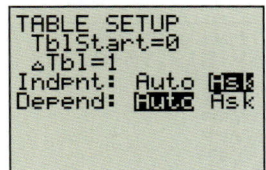

You should be sure the Indpnt and Depend settings are correct. Indpnt should be set to **Ask**. Depend should be set to **Auto**.

These settings will allow you to enter a value for the input variable. The calculator will automatically calculate the related output.

Now press 2nd GRAPH (TABLE F5) to get the table feature. Now you can enter the values of x that you want.

When the Indpnt setting is set to Auto, the table will automatically fill with input values based on the TblStart setting and the ΔTbl (change in table) setting. This is used when you want an incremental list.

For more on this feature, see the calculator guide in Appendix C.

This graph is more complicated than the lines in the first two parts of this example. By graphing more points, a better idea of the shape of the graph develops. In general, plotting three points will be enough for graphing a line, but five or more are needed to graph more complicated equations. This type of equation is addressed more in Chapter 4.

PRACTICE PROBLEM FOR EXAMPLE 1

Graph the equations by creating a table of values and plotting the points.

a. $y = 2x - 6$

b. $y = x^2 - 8$

Graphing equations by plotting points and then connecting them with a smooth line or curve can be used with most equations. This technique is the most basic method to graph an equation and can be used in applications as well as in basic algebra problems.

Example 2 Graphing equations by plotting points

In Section 1.1, we were given the equation $U = 19.95 + 0.79m$, where U is the cost in dollars to rent a 10-foot truck from U-Haul when it is driven m miles.

a. Create a table of points that satisfy this equation.

b. Create a graph for the equation using your points. Remember to label your graph with units.

Using Your TI Graphing Calculator

The table feature of the calculator can be used to calculate several values of an equation very quickly. First you must enter the equation you are working with into the Y= screen.

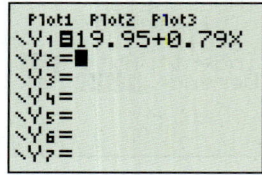

Now go to the table using **2nd** **GRAPH** and enter in the input values you want to evaluate the equation for.

SOLUTION

a. Since we are investigating an equation about renting a truck and driving it m miles, we should choose values of m that would make sense for miles driven. We will start with zero miles just to remind us of how much the rental costs even if we don't happen to go anywhere.

m	$U = 19.95 + 0.79m$	(m, U)
0	$U = 19.95 + 0.79(0) = 19.95$	$(0, 19.95)$
50	$U = 19.95 + 0.79(50) = 59.45$	$(50, 59.45)$
100	$U = 19.95 + 0.79(100) = 98.95$	$(100, 98.95)$
200	$U = 19.95 + 0.79(200) = 177.95$	$(200, 177.95)$
300	$U = 19.95 + 0.79(300) = 256.95$	$(300, 256.95)$

b. Now we can plot the points. Since the cost depends on the miles driven, the cost will be the dependent variable and will be plotted on the vertical axis. The miles driven is the independent variable and will be plotted on the horizontal axis. We will choose a scale for each axis that will allow for all the points to be plotted. We could choose a scale of 20 for the vertical axis and a scale of 25 for the horizontal axis. These scales will allow us to see all the points in our table as well as a little beyond them.

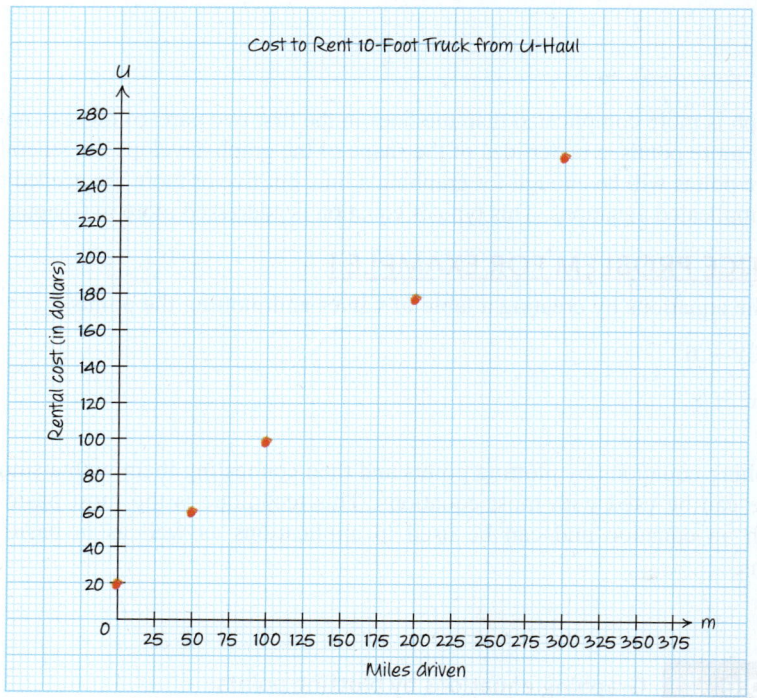

All the points lie on the same line. Draw that line and extend it farther out to represent more possible combinations of miles and costs that satisfy this equation.

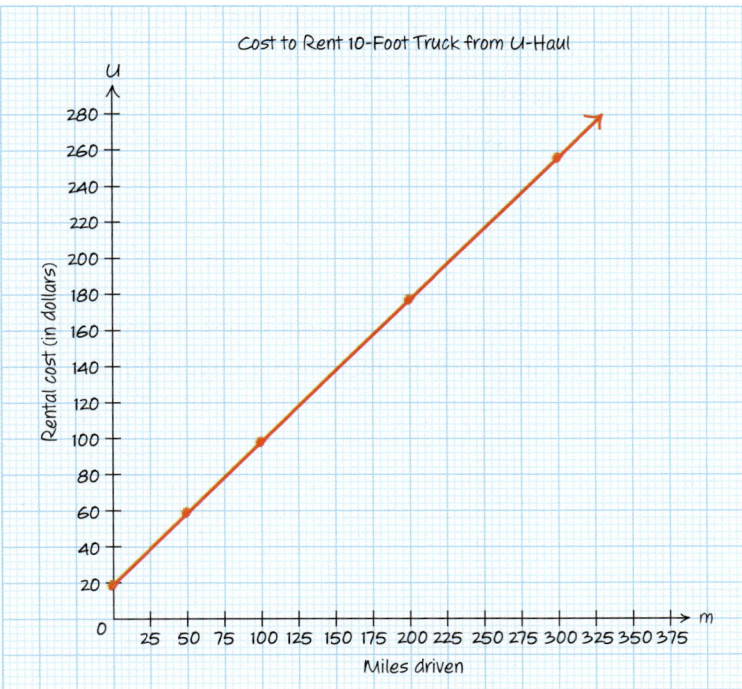

Slope-Intercept Form of a Line

Most equations can be graphed by finding points and plotting them. Graphing calculators use this technique to graph an equation that you enter. When you graph by hand, an understanding of the basic characteristics of the graph of the equation can make graphing it quicker and more accurate. In the concept investigation below, some of the characteristics of the graph of a line and how they are related to the equation of a line will be examined.

CONCEPT INVESTIGATION
What direction is that line going in?

Use your graphing calculator to examine the following.
Start by setting up your calculator by doing the following steps.

- Clear all equations from the Y= screen. (Press [Y=], [CLEAR].)
- Change the window to a standard window. (Press [ZOOM], [6] (ZStandard).)

Now your calculator is ready to graph equations. The Y= screen is where equations will be put into the calculator to graph them or evaluate them at input values. Several simple equations will be graphed to investigate how the graph of an equation for a line reacts to changes in the equation. (Note that your calculator uses y as the dependent (output) variable and x as the independent (input) variable.)

1. Graph the following equations that have positive coefficients on a standard window. Enter each equation in its own row (Y1, Y2, Y3, . . .).
   (*Note:* To enter an x, you use the [X,T,θ,n] button next to the [ALPHA] button.)
   a. $y = x$
   b. $y = 2x$
   c. $y = 5x$
   d. $y = 8x$

What's That Mean?

Coefficient

The number in front of a variable expression is the coefficient.
  For example:
$$-7x \qquad 5x^2$$
-7 is the coefficient for x.

5 is the coefficient of x^2.

  Remember that a variable that is by itself (x) has a coefficient of 1.

> **Using Your TI Graphing Calculator**
>
> In entering fractions in the calculator, it is often best to use parentheses.
>
> $$y = (1/5)x$$
>
> On many graphing calculators, parentheses are needed in almost all situations. In some calculators, when
>
> $$1/5x$$
>
> is entered, the calculator will interpret this as
>
> $$\frac{1}{5x}$$
>
> instead of
>
> $$\frac{1}{5}x$$
>
> To be sure the calculator does what you intend, using parentheses is a good idea.
>
> The TI-83 does not need parentheses in some situations, but in other situations, they are required. To keep confusion down, one option is to use parentheses around every fraction. Extra parentheses do not usually create a problem, but not having them where they are needed can cause miscalculations.

In your own words, describe what the coefficient (number in front) of x does to the graph. Remember to read graphs from left to right.

2. Now graph the following equations that have negative coefficients.

   a. $y = -x$
   b. $y = -2x$
   c. $y = -5x$
   d. $y = -8x$

   In your own words, describe what a negative coefficient of x does to the graph.

3. Graph the following equations with coefficients that are between zero and one.

   a. $y = x$
   b. $y = \frac{1}{5}x$
   c. $y = \frac{1}{2}x$
   d. $y = \frac{2}{3}x$
   e. $y = 0.9x$

   In your own words, describe what a coefficient of x between 0 and 1 does to the graph.

The three sets of graphs in the Concept Investigation demonstrate the **slope** of a line. Slope can be described in several ways and is basically the direction or steepness of the line. The graph of a line will have the same steepness (direction) over the entire graph. When considering slope or direction of a graph, always look from left to right. Here are some ways of thinking about slope of a line:

- The steepness of a line
- The direction in which a line is traveling (left to right)
- How fast something is changing

In mathematics, the slope of a line is calculated as a ratio (fraction) of the vertical change and horizontal change. This ratio must stay constant no matter where you are on the line. Change in mathematics is the difference between two quantities or variables and is calculated by subtraction. This concept results in some of the following ways to remember how to calculate the slope of a line; m is often the letter that is used to represent the slope of a line.

- $m = \dfrac{\text{rise}}{\text{run}}$ The vertical rise divided by the horizontal run.

- $m = \dfrac{\text{change in } y}{\text{change in } x} = \dfrac{y_2 - y_1}{x_2 - x_1}$ where (x_1, y_1) and (x_2, y_2) are two distinct points on the line.

All of these descriptions can be used to remember the idea of slope and how to calculate it. If you use one of the word descriptions, remember that change in mathematics is most often measured by using subtraction. When interpreting the slope of a model, remember that slope is a measurement of how fast something is changing.

In particular, the slope measures the increase or decrease in the output variable for a unit change in the input variable. This is often stated as the increase or decrease in y for a unit change in x. It is important to remember that all graphs in mathematics are read from left to right. Therefore, an increase in y will cause the line to go up from left to right and will result in a positive slope. A decrease in y will cause the line to go down from left to right and result in a negative slope.

DEFINITION
Slope

- The ratio of the vertical change and horizontal change of a line.
- The increase or decrease in y for a unit change in x.
- For a line going through the two distinct points (x_1, y_1) and (x_2, y_2),

$$\text{slope} = \frac{\text{rise}}{\text{run}} = \frac{\text{change in } y}{\text{change in } x} = \frac{y_2 - y_1}{x_2 - x_1}$$

- Slope represents the amount the output variable is changing for every unit change in the input variable.

What's That Mean?
Slope
- slope
- rate of change
- *m* is the slope in the slope-intercept form of a line

When asked for the slope of the line $y = 4x + 9$, be sure to give only the constant 4 as the slope. The variable *x* is NOT part of the slope.

Example 3 Finding slope from the graph of a line

Use the graph to estimate the slope of the line and determine whether the line is increasing or decreasing.

a.

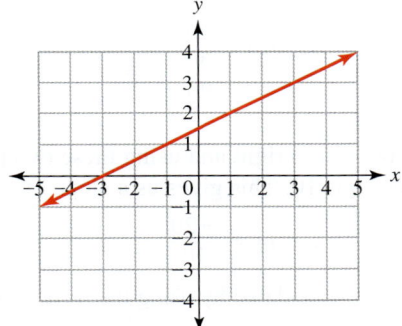

b.

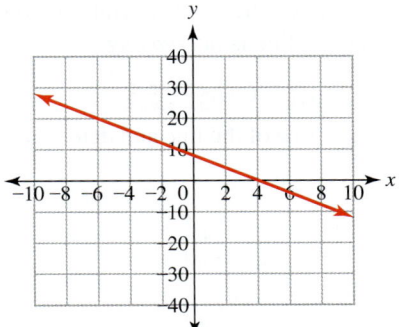

SOLUTION

a. On reading the graph, it appears that the points (1, 2) and (3, 3) lie on the line.

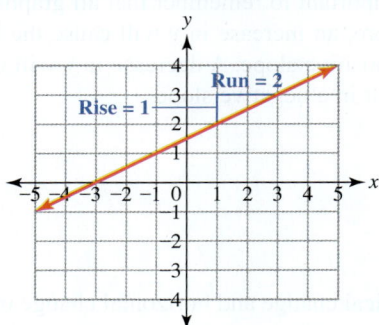

Reading the graph from left to right and using these two points, we can see that the rise is 1 when the run is 2. This gives us a slope of

$$m = \frac{1}{2}$$

If we pick another set of points on this line, we should get the same slope. Because this line is going up from left to right, it confirms that the slope should be positive. We say that this line is *increasing*.

b. On reading this graph, it appears that the points $(-6, 20)$ and $(4, 0)$ lie on the line.

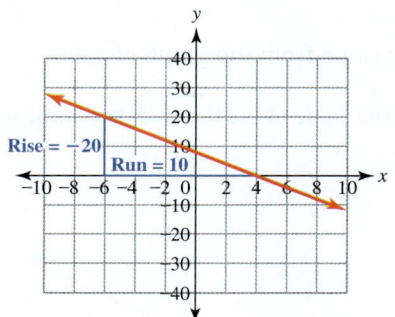

Reading the graph from left to right and using these two points, we can see that the rise is -20 when the run is 10. This gives us a slope of

$$m = \frac{-20}{10} = -2$$

This could also have been calculated by using the formula for slope, resulting in

$$m = \frac{20 - 0}{-6 - 4} = \frac{20}{-10} = -2$$

Because this line is going down from left to right, it confirms that the slope should be negative. We say that this line is *decreasing*.

PRACTICE PROBLEM FOR EXAMPLE 3

Use the graph to estimate the slope of the line and determine whether the line is increasing or decreasing.

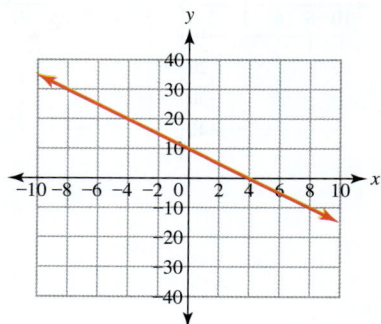

SECTION 1.3 Fundamentals of Graphing and Slope 43

Example 4 Finding slope from a table

Find the slope of the line passing through the points given in the table.

x	y
−3	29
0	20
2	14
11	−13
14	−22

SOLUTION

If we calculate the change in the values given in the table, we can find the slope.

x	y	Slope
−3	29	
0	20	$\dfrac{20 - 29}{0 - (-3)} = \dfrac{-9}{3} = -3$
2	14	$\dfrac{14 - 20}{2 - 0} = \dfrac{-6}{2} = -3$
11	−13	$\dfrac{-13 - 14}{11 - 2} = \dfrac{-27}{9} = -3$
14	−22	$\dfrac{-22 - (-13)}{14 - 11} = \dfrac{-9}{3} = -3$

All of the slope calculations are equal to -3. This confirms that these points would all lie on a line with slope -3.

PRACTICE PROBLEM FOR EXAMPLE 4

Find the slope of the line passing through the points given in the table.

x	y
0	140
5	100
10	60
15	20
20	−20

In Example 4, if the points in a table all lie on the same line, each pair of points will have the same slope. Therefore, calculate the slope by selecting *any* two points on the line. The points given in a table all lie on a line if the slopes remain constant between each pair of points. If the slope between two points is not the same as that of the other pairs of points, then it does not lie on the same line.

Example 5 Determining if given points lie on a line

Determine whether or not the points given in the table all lie on a line.

a.

x	y
1	−15
3	−5
5	5
7	15

Using Your TI Graphing Calculator

When calculating slope on the calculator, you can perform the calculations one piece at a time or all together as one calculation. One piece at a time:

$$\dfrac{14 - 20}{2 - 0} = \dfrac{-6}{2} = -3$$

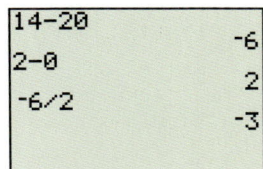

As one calculation:

$$\dfrac{14 - 20}{2 - 0} = -3$$

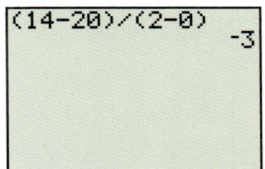

To do this as one calculation, you must add parentheses to both the numerator and the denominator of the fraction.

b.

x	y
−3	1
0	2
3	3
9	8

SOLUTION

a. To determine whether or not the points all lie on a line, calculate the slope between each pair of points and see if the slopes remain constant.

x	y	Slope
1	−15	
3	−5	$\dfrac{-5-(-15)}{3-1} = \dfrac{10}{2} = 5$
5	5	$\dfrac{5-(-5)}{5-3} = \dfrac{10}{2} = 5$
7	15	$\dfrac{15-5}{7-5} = \dfrac{10}{2} = 5$

All of the slopes are equal to 5, so these points lie on the same line. Confirm this by plotting the points and drawing a line.

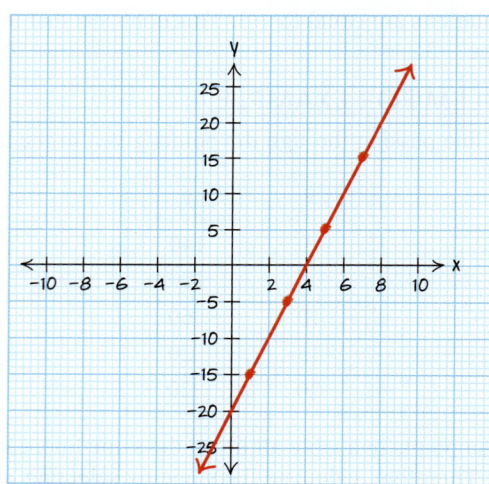

b. Check to see whether or not the slopes remain constant.

x	y	Slope
−3	1	
0	2	$\dfrac{2-1}{0-(-3)} = \dfrac{1}{3}$
3	3	$\dfrac{3-2}{3-1} = \dfrac{1}{3}$
9	8	$\dfrac{8-3}{9-3} = \dfrac{5}{6}$

The slopes do not remain constant, so these points do not all lie on the same line. This is confirmed by plotting the points and trying to draw a line through them.

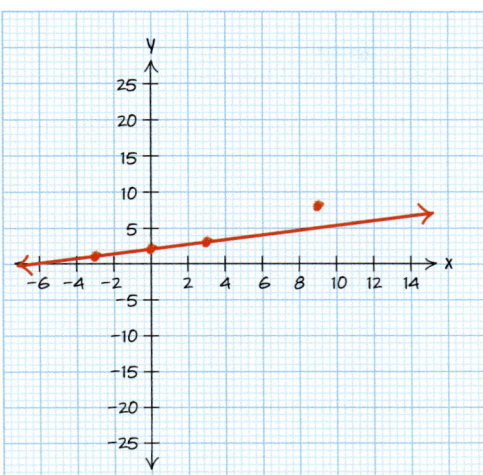

The last point does not lie on the same line as the first three.

PRACTICE PROBLEM FOR EXAMPLE 5

Determine whether or not the points given in the table all lie on a line.

a.

x	y
1	−5
4	1
7	7
10	13

b.

x	y
−5	8
0	6
5	4
10	7

■ CONCEPT INVESTIGATION

What's moving that graph?

On your calculator, start by clearing the equations from the Y= screen.

1. Graph the following equations on the same calculator screen.

   a. $y = x$

   b. $y = x + 1$

What's That Mean?

Constant Term

A constant term is a number, in an expression or equation, that is not being multiplied by a variable. The constant term in the equation $y = 4x + 9$ is 9. The number 4 is not a constant term, since it is being multiplied by x.

    c. $y = x + 2$

    d. $y = x + 3$

    e. $y = x + 7$

In your own words, describe what the constant term does to the graph.

2. Graph the following equations on the same calculator screen.

    a. $y = x$

    b. $y = x - 1$

    c. $y = x - 2$

    d. $y = x - 3$

    e. $y = x - 7$

In your own words, describe what a negative constant term does to the graph.

The constant term of a linear equation in this form represents the vertical intercept of the linear graph. In mathematics, b represents the vertical intercept. With the slope and vertical intercept, an equation of a line can be easily written. Every line can be described with two pieces of information: the slope of the line and a point on the line. Both of these are needed, since a slope does not tell you where the line is and a point does not tell you where to go.

An equation for a line can be written in many forms, but the **slope-intercept form** is the most common and most useful form. On a graphing calculator, the slope-intercept form is the easiest to use. The slope-intercept form of an equation of a line is represented by the equation $y = mx + b$. In this equation, m represents the slope of the line, and b represents the y-coordinate of the y-intercept (vertical intercept).

The variables x and y are the independent and dependent variables, respectively. Any equation that can be simplified into the form $y = mx + b$ is the equation of a line.

DEFINITIONS

Linear Equation An equation is linear if it has a constant rate of change (the slope is constant). That is, for every unit change in the input, the output has a constant amount of change.

Slope-Intercept Form of a Line

$$y = mx + b$$

Slope The increase or decrease in the output variable for a unit change in the input variable. In the slope-intercept form of a line, slope is represented by m.

Vertical Intercept The point where the line crosses the vertical axis. In the slope-intercept form of a line, the vertical intercept is $(0, b)$. This is often called the y-intercept.

Example 6 Determining slope and y-intercept of linear equations

Find the slope and y-intercept of the following lines.

    a. $y = 2x + 5$ b. $y = \frac{3}{2}x - \frac{7}{2}$

    c. $2x + 5y = 20$

SOLUTION

a. The equation $y = 2x + 5$ is in slope-intercept form, so the slope is 2, and the y-intercept is (0, 5).

b. The equation $y = \frac{3}{2}x - \frac{7}{2}$ is in slope-intercept form, so the slope is $\frac{3}{2}$, and the y-intercept is $\left(0, -\frac{7}{2}\right)$.

c. The equation $2x + 5y = 20$ should be put into slope-intercept form before we try to read the slope and intercept.

$$\begin{aligned} 2x + 5y &= 20 \\ -2x & -2x \qquad \text{Isolate } y. \\ \hline 5y &= -2x + 20 \\ \frac{5y}{5} &= \frac{-2x + 20}{5} \\ y &= -\frac{2}{5}x + 4 \end{aligned}$$

Therefore, the slope is $-\frac{2}{5}$, and the y-intercept is (0, 4).

PRACTICE PROBLEM FOR EXAMPLE 6

Find the slope and y-intercept of the following lines.

a. $y = -3x + 7$ b. $y = \frac{2}{7}x - 8$ c. $3x - 4y = 15$

The Meaning of Slope in an Application

Understanding the meaning of the slope of a line is an important part of understanding the role slope plays in a situation. The slope is the rate of change of the line and tells how much the output variable changes for a unit change in the input variable. The idea of a rate of change is easier to understand if you consider what it means in an application.

In Exercise 39 in Section 1.1, we found the equation $C = 38m + 150$ for the cost in dollars for pest management from Enviro-Safe Pest Management when m months of services are provided. This equation is in slope-intercept form, so the slope is the coefficient of the input variable.

$$38 = \frac{38}{1} = \frac{\$38}{1 \text{ month}}$$

This slope means that the cost for pest management is increasing by \$38 per month of service. Since the output variable represents cost and the slope is positive, the cost is increasing for each month.

In the U-Haul example, $U = 19.95 + 0.79m$, the slope was

$$0.79 = \frac{0.79}{1} = \frac{\$0.79}{1 \text{ mile}}$$

The cost for renting a 10-foot truck from U-Haul for the day increases by 0.79 dollar—that is, 79 cents—per mile driven.

If the variables are defined, writing this sentence is just a matter of putting the pieces together properly. Always remember that slope is the output change per input change.

Example 7 Finding and interpreting slope

Find the slope of the model and explain its meaning in the given situation.

a. Let $P = 115b - 4000$ be the profit in dollars Bicycles Galore makes from selling b bikes.

b. Let $S = -0.76a + 33.04$ be the percentage of women a years old who smoked cigarettes during their pregnancy.
Source: U.S. National Center for Health Statistics, National Vital Statistics Reports.

SOLUTION

a. The slope of the model $P = 115b - 4000$ is

$$115 = \frac{115}{1} = \frac{\$115}{1 \text{ bike}}$$

The output variable is P profit in dollars, and the input variable is b bikes sold. Therefore, the slope means: The profit for Bicycles Galore in increasing by $115 per bike sold.

b. The slope of the model $S = -0.76a + 33.04$ is

$$-0.76 = \frac{-0.76}{1} = \frac{-0.76 \text{ percentage points}}{1 \text{ year of age}}$$

The output variable is S percent of women who smoked cigarettes during their pregnancy, and the input variable is a years old. Therefore the slope means: The percent of women who smoke cigarettes during their pregnancy is decreasing by 0.76 percentage points per year of age. In other words, the older the women get, the fewer of them smoke during pregnancy.

Note that when a model is measuring a percentage, the slope will measure the number of percentage points that the output is changing, not the percentage itself.

PRACTICE PROBLEM FOR EXAMPLE 7

Find the slope of the model and explain its meaning in the given situation.

a. The pressure inside a vacuum chamber can be represented by $P = 35 - 0.07s$, where P is the pressure in pounds per square inch (psi) of the vacuum chamber after being pumped down for s seconds.

b. The cost for making tacos at a local street stand can be represented by $C = 0.55t + 140.00$, where C is the cost in dollars to make tacos at the local street stand when t tacos are made.

> ■ **Connecting the Concepts**
>
> **Percents or Percentage Points?**
>
> When working with percentages, you may encounter some confusion between a percent change and a percentage points change.
>
> If the percentage of people who smoke is currently 40% and it increases to 45%, that is a 5 percentage point increase.
> It is not a 5% increase. A 5% increase would be only to 42%—the 2 being 5% of 40.
>
> In working with how much a percentage is changing (slope), it will be interpreted as a percentage point increase or decrease.

Graphing Lines Using Slope and Intercept

Using the information gained from the slope-intercept form of a line, we can graph linear equations. Use the vertical intercept as a starting point on the line, then use the slope as the direction from left to right that you should follow to find additional points on the line. Remember that slope is the rise over the run that the line takes from one point to another. Once you have two or three points, you can draw the entire line by connecting these points and extending the line in both directions. In most cases, you will want to graph the line so that both the vertical and horizontal intercepts are shown.

Example 8

Sketch the graph of the following lines. Label the vertical intercept.

a. $y = \frac{2}{3}x - 2$

b. $y = 2x + 5$

c. $y = -3x + 2$

SOLUTION

a. The vertical intercept is $(0, -2)$, and the slope is $\frac{2}{3}$, so the rise is 2 when the run is 3. Working from left to right and using the vertical intercept and slope, we get the following graph.

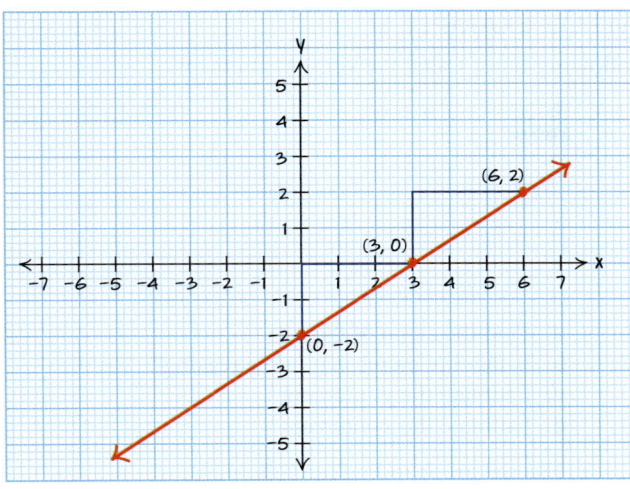

Starting at (0, −2) and going up 2 and over 3, we get to (3, 0). Doing this again gets us to (6, 2). Using these three points, we connect the points and extend the line in both directions.

b. The y-intercept is (0, 5), and the slope is $2 = \frac{2}{1}$, so the rise is 2 when the run is 1. Working from left to right and using the vertical intercept and slope, we get the following graph.

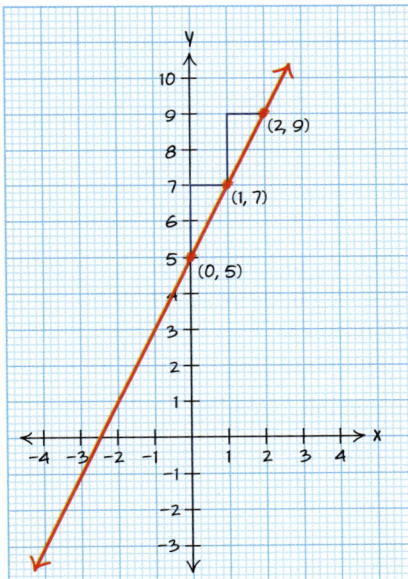

Starting at (0, 5) and going up 2 and over 1, we get to (1, 7). Doing this again gets us to (2, 9). Using these three points, we connect the points and extend the line in both directions.

c. The y-intercept is (0, 2), and the slope is $-3 = \frac{-3}{1}$, so the rise is -3 when the run is 1. Because the rise is negative, the graph will actually go downward from left to right. Using this information, we get the following graph.

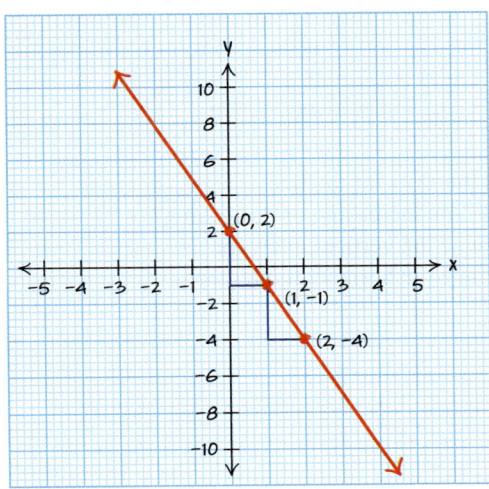

PRACTICE PROBLEM FOR EXAMPLE 8

Sketch the graph of the following lines. Label the vertical intercept.

a. $y = 4x - 5$
b. $y = -\dfrac{3}{2}x + 6$

1.3 Exercises

1. Use the graph to find the following.

   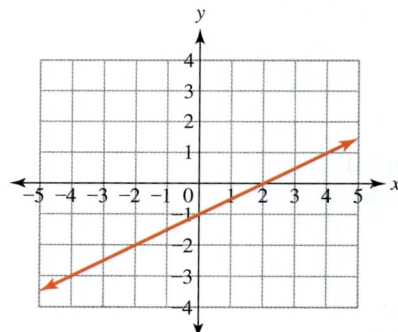

   a. Find the slope of the line.
   b. Is the line increasing or decreasing?
   c. Estimate the vertical intercept.
   d. Estimate the horizontal intercept.

2. Use the graph to find the following.

   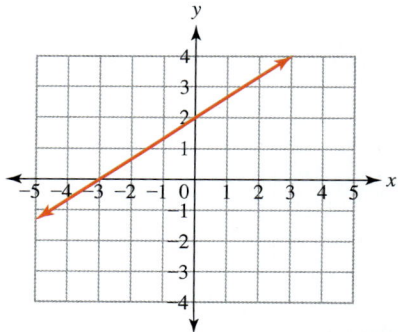

   a. Find the slope of the line.
   b. Is the line increasing or decreasing?
   c. Estimate the vertical intercept
   d. Estimate the horizontal intercept

3. Use the graph to find the following.

   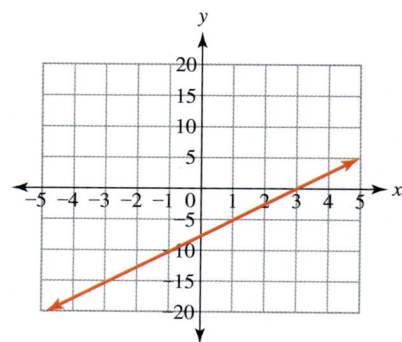

   a. Find the slope of the line.
   b. Is the line increasing or decreasing?
   c. Estimate the y-intercept
   d. Estimate the x-intercept

4. Use the graph to find the following.

   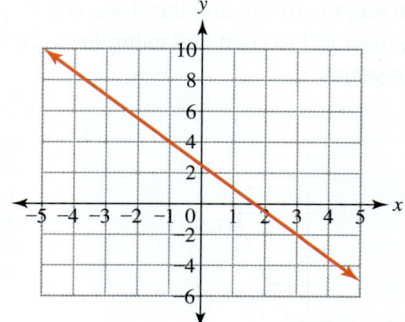

   a. Find the slope of the line.
   b. Is the line increasing or decreasing?
   c. Estimate the y-intercept
   d. Estimate the x-intercept

5. Use the graph to find the following.

   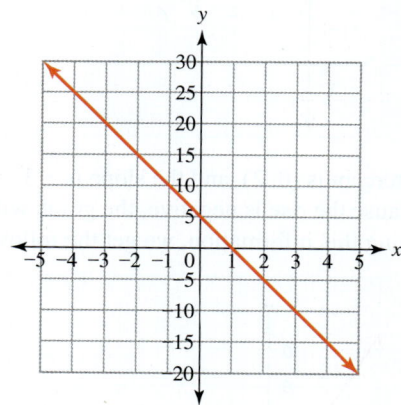

   a. Find the slope of the line.
   b. Is the line increasing or decreasing?
   c. Estimate the vertical intercept.
   d. Estimate the horizontal intercept.

6. Use the graph to find the following.

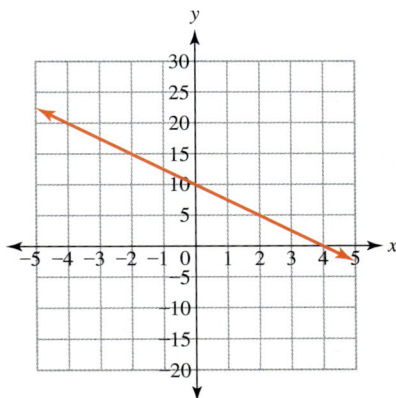

a. Find the slope of the line.
b. Is the line increasing or decreasing?
c. Estimate the vertical intercept.
d. Estimate the horizontal intercept.

For Exercises 7 through 22, graph the equations by plotting points. For linear equations, use at least three points. For more complicated equations, use at least five points.

7. $y = 2x + 3$
8. $y = 3x - 8$
9. $x = 5y + 4$
10. $x = -2y - 7$
11. $y = x^2 + 2$
12. $y = x^2 - 4$
13. $y = \frac{2}{3}x + 6$
14. $y = \frac{1}{4}x - 3$
15. $x = \frac{2}{3}y - 4$
16. $x = \frac{3}{4}y + 1$
17. $y = 0.5x - 3$
18. $y = -0.4x + 5$
19. $x = -1.5y + 7$
20. $x = -2.5y + 10$
21. $y = -2x^2 + 15$
22. $y = -1.5x^2 + 8$

23. In Section 1.1 exercises, we investigated the equation $B = 0.55m + 29.95$, which represented the cost in dollars for renting a 10-foot truck from Budget and driving it m miles.
    a. Create a table of points that satisfy this equation.
    b. Create a graph for the equation using your points. Remember to label your graph with units.

24. One type of investment that institutions and very wealthy people use is called a *hedge fund*. These funds typically give a high return on the money invested. That return comes at a price. A typical hedge fund charges 20% on the earnings of the fund. If someone invested 5 million dollars, their fees could be calculated as $F = 0.20e + 100,000$, where F represented the fee in dollars charged on the investment that had e dollars worth of earnings.
    Source: Equation based on information from Hundreds of Millions for Top Hedge Funders, by Bob Jamieson, abcnews.go.com, May 26, 2006.

    a. Create a table of points that satisfy this equation. (*Note:* Earnings for this type of investment can easily be $500,000 or more.)
    b. Create a graph for the equation using your points. Remember to label your graph with units.

25. A sales clerk at a high-end clothing store earns 4% commission on all of her sales in addition to her $100 base weekly salary.
    a. Write an equation for the sales clerk's salary per week when she sells s dollars of merchandise during the week.
    b. Create a table of points that satisfy the equation you found in part a.
    c. Create a graph for the equation using your points. Remember to label your graph with units.

26. A sales clerk at a computer store earns 3% commission on all of his sales in addition to his $200 base weekly salary.
    a. Write an equation for the sales clerk's salary per week when he sells s dollars of merchandise during the week.
    b. Create a table of points that satisfy the equation you found in part a.
    c. Create a graph for the equation using your points. Remember to label your graph with units.

27. Western Washington University charges each resident undergraduate student taking up to ten credits $137 per credit in tuition plus $208 in other fees.
    a. Write an equation for the total tuition and fees that Western Washington University charges its resident undergraduate students who take c credits.
    b. Create a table of points that satisfy the equation you found in part a.
    c. Create a graph for the equation using your points. Remember to label your graph with units.

28. A local specialty delivery service charges a fee of $5 for each delivery plus 25 cents for each mile over 10 miles.
    a. Write an equation for the cost in dollars for a delivery of m miles over 10 miles. (*Note:* A 15-mile delivery would be represented by $m = 5$.)
    b. Create a table of points that satisfy the equation you found in part a.
    c. Create a graph for the equation using your points. Remember to label your graph with units.

29. The percentage P of companies that are still in business t years after the fifth year in operation can be represented by the equation, $P = -3t + 50$.
    a. Create a table of points that satisfy this equation.
    b. Create a graph for the equation using your points. Remember to label your graph with units.

30. The number E of golf events on the Champions Tour t years after 2000 can be represented by the equation, $E = -2t + 39$.

Source: Model derived from data found in USA Today, May 25, 2005.

  a. Create a table of points that satisfy this equation.
  b. Create a graph for the equation using your points. Remember to label your graph with units.

For Exercises 31 through 36, find the slope of the line passing through the points given in the table.

31.

x	y
0	7
2	11
4	15
6	19

32.

x	y
-4	-18
0	-4
4	10
8	24

33.

x	y
-2	29
0	15
2	1
4	-13

34.

x	y
-3	14.25
0	9
3	3.75
6	-1.5

35.

x	y
2	-18
5	-6
8	6
11	18

36.

x	y
4	7
9	-5.5
14	-18
19	-30.5

For Exercises 37 through 40, determine whether the points given in the table all lie on a line.

37.

x	y
-2	-14
0	-6
2	2
4	10

38.

x	y
0	5
1	2
2	-1
3	-4

39.

x	y
-6	12
-3	10
0	8
3	5

40.

x	y
-8	-8
-2	-3.5
0	-2
10	7.5

For Exercises 41 through 50, determine the slope and y-intercept of the given linear equations.

41. $y = 3x + 5$

42. $y = 7x + 12$

43. $y = -4x + 8$

44. $y = -9x + 25$

45. $y = \frac{1}{2}x + 5$

46. $y = \frac{2}{3}x + 43$

47. $y = 0.4x - 7.2$

48. $y = 2.3x - 8.4$

49. $y = \frac{4}{3}x + \frac{7}{5}$

50. $y = -\frac{6}{5}x + \frac{4}{9}$

51. A student was asked for the slope and y-intercept of the equation $y = 4x - 9$. The student answered

$$\text{slope} = 4 \qquad y\text{-intercept} = (-9, 0)$$

Was the student correct? If not, give the correct answers.

52. A student was asked for the slope and y-intercept of the equation $y = -\frac{1}{3}x + \frac{4}{3}$. The student answered

$$\text{slope} = \frac{4}{3} \qquad y\text{-intercept} = \left(0, -\frac{1}{3}\right)$$

Was the student correct? If not, give the correct answers.

For Exercises 53 through 58, determine the slope and y-intercept of the given linear equations.

53. $2x + y = 20$

54. $5x + y = 11$

55. $4x - 2y = 20$

56. $6x - 3y = 18$

57. $3x + 5y = 12$

58. $-4x + 2y = 7$

59. A student was asked for the slope and y-intercept of the equation $3x + 2y = 6$. The student answered

$$\text{slope} = 3 \qquad y\text{-intercept} = (0, 6)$$

Was the student correct? If not give the correct answers.

60. A student was asked for the slope and y-intercept of the equation $7x - y = 4$. The student answered

$$\text{slope} = -7 \qquad y\text{-intercept} = (0, 4)$$

Was the student correct? If not give the correct answers.

61. Using the cost equation for Budget found in Exercise 23, give the slope of the equation. Explain its meaning in regard to the cost of renting a truck from Budget.

62. Using the equation for fees found in Exercise 24, give the slope of the equation and explain its meaning in regard to fees charged for investing in hedge funds.

63. Using the weekly salary equation found in Exercise 25, give the slope of the equation and explain its meaning in regard to the weekly salary.

64. Using the weekly salary equation found in Exercise 26, give the slope of the equation and explain its meaning in regard to the weekly salary.

65. Using the equation for the tuition and fees found in Exercise 27, give the slope of the equation and explain its meaning in this situation.

66. Using the equation for delivery costs found in Exercise 28, give the slope of the equation and explain its meaning in this situation.

67. Using the equation for the percentage of companies still in business found in Exercise 29, give the slope of the equation and explain its meaning in this situation.

68. Using the number of golf events equation found in Exercise 30, give the slope of the equation and explain its meaning in this situation.

For Exercises 69 through 76, sketch the graph on graph paper and label the vertical intercept. Round the values of the intercepts to two decimal places if needed.

69. $y = 2x - 7$
70. $y = 3x - 12$
71. $y = \dfrac{4}{5}x - 6$
72. $y = \dfrac{2}{3}x + 4$
73. $y = -2x + 5$
74. $y = -4x - 3$
75. $y = -\dfrac{7}{5}x + 11$
76. $y = -\dfrac{3}{4}x + 2$

For Exercises 77 through 80, describe what is wrong with the graph of the given equation.

77. $y = \dfrac{1}{2}x - 3$

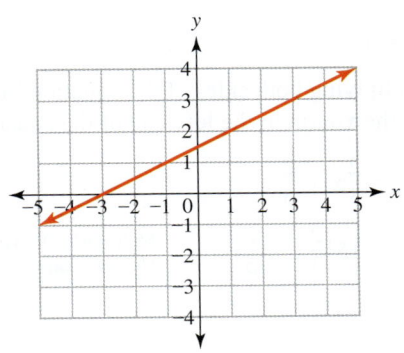

78. $y = 3x + 1$

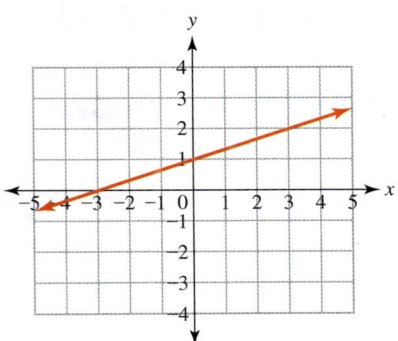

79. $y = -\dfrac{2}{3}x - 1$

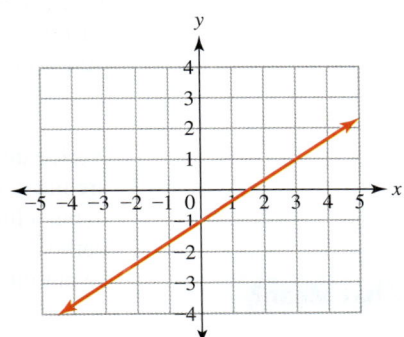

80. $y = \dfrac{1}{2}x + 2$

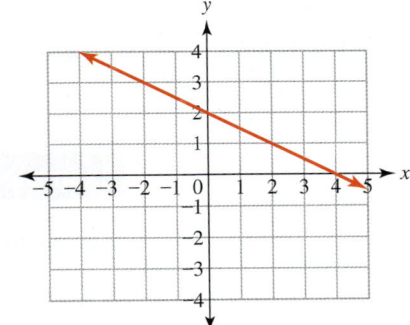

For Exercises 81 through 90, sketch the graph on graph paper and label the vertical intercept. Round the values of the intercepts to two decimal places if needed.

81. $y = 0.5x + 4$
82. $y = 0.75x - 8$
83. $y = -0.25x + 2$
84. $y = -0.3x + 6$
85. $y = -\dfrac{2}{5}x + \dfrac{3}{5}$
86. $y = -\dfrac{1}{4}x + \dfrac{9}{4}$
87. $y = \dfrac{1}{4}x - \dfrac{3}{4}$
88. $y = \dfrac{1}{8}x - \dfrac{5}{8}$
89. $y = -0.75x + 2.5$
90. $y = -0.4x + 3.6$

1.4 Intercepts and Graphing

LEARNING OBJECTIVES

- Identify the general form of a line.
- Find and interpret the intercepts of linear equations.
- Graph lines using the intercepts.
- Identify and graph horizontal and vertical lines.

The General Form of Lines

Linear equations can take another form, called the *general form*. The general form is written with both variables on one side of the equal sign and the constant term on the other side.

$$Ax + By = C$$

Notice that the variables x and y both have exponents of 1. The constants A, B, and C are integers, and A should be nonnegative. To have the constants be integers means that to put a linear equation into general form, you will remove any fractions by multiplying both sides of the equation by the least common denominator. All linear equations can be put into general form.

> **What's That Mean?**
>
> **Integer**
>
> Integers are the whole numbers and their opposites. So integers include all positive and negative whole numbers and zero.
>
> $\ldots -3, -2, -1, 0, 1, 2, 3, \ldots$

> **DEFINITION**
>
> **General Form of a Line**
>
> $$Ax + By = C$$
>
> where A, B, and C are integers and A is nonnegative.

Example 1 Rewriting equations in general form

Rewrite the following equations in general form.

a. $y = -4x + 12$

b. $y = \dfrac{2}{3}x + \dfrac{2}{5}$

SOLUTION

a. To put the equation into general form, bring the x-term to the left side of the equation.

$$y = -4x + 12$$
$$\underline{+4x \quad\quad +4x\phantom{{}+12}}$$
$$4x + y = 12$$

b. First, eliminate the fractions by multiplying both sides of the equation by the least common denominator and then get the x-term to the left side of the equation.

$$y = \dfrac{2}{3}x + \dfrac{2}{5}$$

$$15(y) = 15\left(\dfrac{2}{3}x + \dfrac{2}{5}\right) \quad\quad \text{Multiply by 15 to eliminate the fractions.}$$

$$15y = 10x + 6$$
$$\underline{-10x \quad -10x} \qquad \text{Move the }x\text{-term to the}$$
$$-10x + 15y = 6 \qquad \text{left side of the equation.}$$
$$-1(-10x + 15y) = -1(6) \qquad \text{Multiply by } -1 \text{ to make the}$$
$$10x - 15y = -6 \qquad \text{coefficient of }x\text{ positive.}$$

PRACTICE PROBLEM FOR EXAMPLE 1

Rewrite the following equations in general form.

a. $y = 2x + 10$ **b.** $y = \frac{1}{2}x - \frac{3}{7}$

Finding Intercepts and Their Meaning

Recall that the intercepts of a graph are the points where the graph crosses the vertical, y, and horizontal, x, axes.

Explaining the meaning of the intercepts of a line is an important part of understanding what role they play in an application. The intercepts of a line are points on the graph, so they represent combinations of input and output values that satisfy the equation.

Both parts of each intercept will have a meaning and should be explained in terms of the situation given in the problem. In the U-Haul example from Section 1.1, the vertical intercept, the U-intercept, is $(0, 19.95)$. This intercept means that if you rented a truck from U-Haul and drove it zero miles, it would cost $19.95 for the day. It may not make much sense to rent a truck and not drive it, but this is still the meaning of the vertical intercept. The horizontal intercept, m-intercept, is $(-33.81, 0)$. This means that if you rented a truck from U-Haul and drove it negative 33.81 miles, the truck would not cost you anything. This is definitely model breakdown, since you cannot drive a truck a negative number of miles. In the world of mathematics, both of these intercepts exist, but they may not make much sense in the given situation. Always explain what both numbers in the intercept mean in the given situation and remember to state if an intercept is a case of model breakdown.

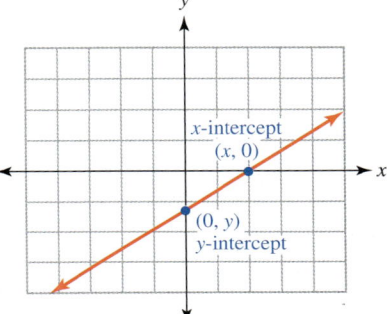

To find an intercept from an equation, set the "other" variable equal to zero and solve. What this means is that if you are looking for a y-intercept, you make $x = 0$ and solve for y. If you want an x-intercept, you make the $y = 0$ and solve for x. In an application problem such as the U-Haul equation, $U = 0.79m + 19.95$, to find the U-intercept, you make $m = 0$ and solve for U. To find the m-intercept, you make $U = 0$ and solve for m.

> **Finding the Intercepts of an Equation**
>
> Set the "other" variable equal to zero and solve.
>
> To find the y-intercept, set $x = 0$ and solve for y.
> To find the x-intercept, set $y = 0$ and solve for x.

Example 2 Finding and interpreting intercepts

Find the vertical and horizontal intercepts. Explain their meaning in the given situation.

a. The number of students who are enrolled in math classes at a local college can be represented by $E = -17w + 600$, where E represents the math class enrollment at the college w weeks after the start of the fall semester.

b. The number of homicides, N, of 15- to 19-year-olds in the United States t years after 1990 can be represented by the equation $N = -315.9t + 4809.8$.

SOLUTION

a. The equation $E = -17w + 600$ is in slope-intercept form so the vertical intercept, the E-intercept, is $(0, 600)$. We can also find this intercept by setting $w = 0$ and solving for E.

$$E = -17(0) + 600$$
$$E = 600$$

The vertical intercept means that during week zero, the beginning of the fall semester, 600 students are enrolled in math classes at the college. The horizontal intercept, the w-intercept, can be found by substituting zero for E and solving for w.

$$E = -17w + 600 \quad \text{Substitute zero for } E \text{ and solve for } w.$$
$$0 = -17w + 600$$
$$\underline{-600 \qquad\qquad -600} \quad \text{Subtract 600 from both sides.}$$
$$-600 = -17w$$
$$\frac{-600}{-17} = w \quad \text{Divide both sides by } -17.$$
$$35.29 \approx w$$

Therefore, the horizontal intercept is $(35.29, 0)$. This means that about 35 weeks after the start of the fall semester, there will be no students enrolled in math classes at the college. This is model breakdown, since a semester would not last 35 weeks and some students love math and would never drop their math class.

b. The equation $N = -315.9t + 4809.8$ is in slope-intercept form, so the vertical intercept is $(0, 4809.8)$. This means that in 1990, there were about 4810 homicides of 15- to 19-year-olds in the United States. In this case, the zero would be years after 1990, and zero years after 1990 would be 1990. Also, round the intercept values to something that is reasonable to the situation. A number of 4809.8 homicides has 4809 plus 0.8 of a homicide; 0.8 of a homicide does not make sense, so you should round the intercept to 4810. The horizontal intercept can be found by substituting zero for N and solving for t.

$$N = -315.9t + 4809.8 \quad \text{Substitute zero for } N \text{ and solve for } t.$$
$$0 = -315.9t + 4809.8$$
$$\underline{-4809.8 \qquad\qquad -4809.8} \quad \text{Subtract 4809.8 from both sides.}$$
$$-4809.8 = -315.9t$$
$$\frac{-4809.8}{-315.9} = t \quad \text{Divide both sides by } -315.9.$$
$$15.23 \approx t$$

Therefore, the horizontal intercept is $(15.23, 0)$. This means that in about 2005, there were no homicides of 15- to 19-year-olds in the United States. This must be model breakdown because there were definitely homicides of 15- to 19-year-olds in the United States in 2005. Note the 15.23 was rounded and represented the year, and the zero meant that there were no homicides.

PRACTICE PROBLEM FOR EXAMPLE 2

Find the vertical and horizontal intercepts. Explain their meaning in the given situation.

a. The pressure inside a vacuum chamber can be represented by $P = 35 - 0.07s$, where P is the pressure in pounds per square inch (psi) of the vacuum chamber after being pumped down for s seconds.

b. The cost for making tacos at a local stand can be represented by $C = 0.55t + 140.00$, where C is the cost in dollars to make tacos at the neighborhood stand when t tacos are made.

SECTION 1.4 Intercepts and Graphing

With a linear equation in the general form, solving for both intercepts is simple. It is easier to solve for x or y when the other variable is zero.

Example 3 Finding intercepts from the general form

Find the horizontal and vertical intercepts of the following equations.

a. $3x + 4y = 24$

b. $5x - 2y = 35$

SOLUTION

a. To find the horizontal intercept, the x-intercept, set $y = 0$ and solve for x.

$$3x + 4(0) = 24$$
$$3x = 24$$
$$x = 8$$

The horizontal intercept is $(8, 0)$. To find the vertical intercept, the y-intercept, set $x = 0$ and solve for y.

$$3(0) + 4y = 24$$
$$4y = 24$$
$$y = 6$$

The vertical intercept is $(0, 6)$.

b. To find the horizontal intercept, set $y = 0$ and solve for x.

$$5x - 2(0) = 35$$
$$5x = 35$$
$$x = 7$$

The horizontal intercept is $(7, 0)$. To find the vertical intercept, set $x = 0$ and solve for y.

$$5(0) - 2y = 35$$
$$-2y = 35$$
$$y = -\frac{35}{2} = -17.5$$

The vertical intercept is $(0, -17.5)$.

PRACTICE PROBLEM FOR EXAMPLE 3

Find the horizontal and vertical intercepts of $8x + 2y = 40$.

Graphing Lines Using Intercepts

Graphing lines that are in the general form by hand can be done most easily by using the intercepts. To graph a line, you have to plot only two points on the line, and the intercepts are two points that are easy to find.

Example 4 Graphing lines using intercepts

For each equation, find the intercepts and graph the line.

a. $2x - 5y = 20$

b. $5x + 3y = 40$

SOLUTION

a. First, find both intercepts.

$$2x - 5(0) = 20$$
$$2x = 20$$
$$x = 10$$
$$(10, 0)$$ Horizontal intercept.

To find the horizontal intercept, make $y = 0$ and solve for x.

$$2(0) - 5y = 20$$
$$-5y = 20$$
$$y = -4$$
$$(0, -4)$$ Vertical intercept.

To find the vertical intercept, make $x = 0$ and solve for y.

Plot these points and draw the line.

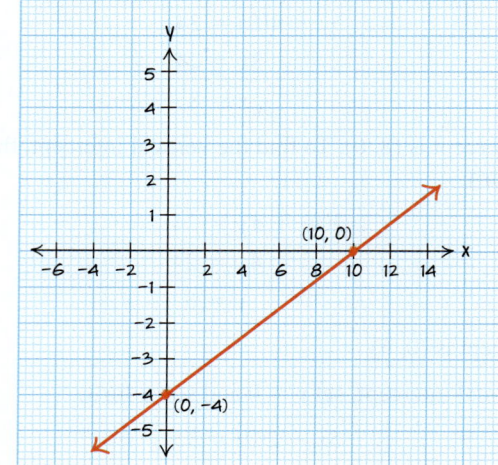

b. First, find both intercepts.

$$5x + 3(0) = 40$$
$$5x = 40$$
$$x = 8$$
$$(8, 0)$$ Horizontal intercept.

To find the horizontal intercept, make $y = 0$ and solve for x.

$$5(0) + 3y = 40$$
$$3y = 40$$
$$y = \frac{40}{3} \approx 13.3$$
$$\left(0, \frac{40}{3}\right)$$ Vertical intercept.

To find the vertical intercept, make $x = 0$ and solve for y.

Plot these points and draw the line.

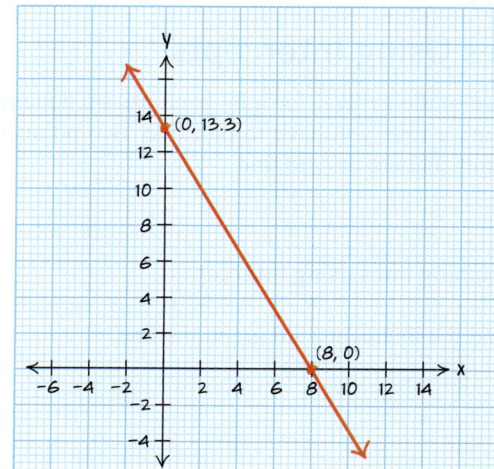

SECTION 1.4 Intercepts and Graphing

PRACTICE PROBLEM FOR EXAMPLE 4
Find the intercepts and graph the line $3x - 4y = 36$.

Horizontal and Vertical Lines

The general form of a line leads to two unique types of lines: vertical and horizontal lines. When either A or B in the general form has the value zero, we get the equation of a horizontal or vertical line.

When $A = 0$, in the general form, the x-term of the equation is eliminated.

$$Ax + By = C$$
$$0x + By = C$$
$$By = C$$

If you isolate the y, you get an equation of the form

$$y = \frac{C}{B} \quad \text{or} \quad y = k$$

where k is a constant. An equation of the form $y = k$ will result in a horizontal line. The following graph is the line $y = 3$.

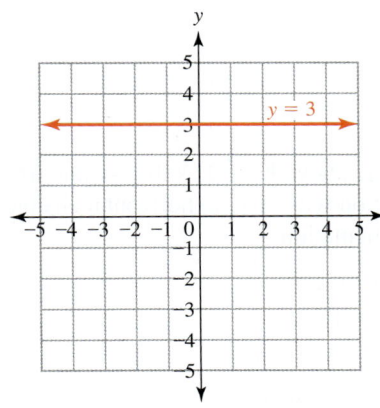

A horizontal line with the equation $y = 3$.

■ **Connecting the Concepts**

Why is division by zero undefined?

If you have a group of 15 people, can you divide them into zero groups? No.

$\dfrac{15}{0}$ is undefined.

Dividing by zero would be trying to put things into no piles. There has to be somewhere to put them. *Undefined* is the same as *does not exist* (DNE).

When you divide zero by something else, you are dividing nothing into piles, so each pile has nothing in it. Therefore, dividing zero by any real number is always zero.

A graphing calculator will say one of two things when you use zero in division.

Divide by zero:

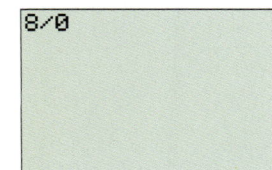

You get an error because it is undefined.

Divide into zero:

All of the points that lie on the horizontal line above will have a y-value of 3. If we look at any two points on this line, there will be no vertical change. If we pick the points $(-4, 3)$ and $(2, 3)$ and calculate the slope, we get

$$(-4, 3) \quad \text{and} \quad (2, 3)$$
$$m = \frac{3 - 3}{2 - (-4)}$$
$$m = \frac{0}{6}$$
$$m = 0$$

Because there will never be any vertical change, the slope will always be zero. If you consider the general form for this line and build a table, you can see that the value of x will not affect the value of y.

x	$0x + y = 3$	(x, y)
-2	$0(-2) + y = 3$	$(-2, 3)$
	$y = 3$	
0	$0(0) + y = 3$	$(0, 3)$
	$y = 3$	
15	$0(15) + y = 3$	$(15, 3)$
	$y = 3$	

The answer is zero.

When $B = 0$, in the general form, the y-term of the equation is eliminated.

$$Ax + By = C$$
$$Ax + 0y = C$$
$$Ax = C$$

If you isolate the x, you get an equation of the form

$$x = \frac{C}{A} \quad \text{or} \quad x = k$$

where k is a constant. An equation of the form $x = k$ will result in a vertical line. The following graph is the vertical line $x = 3$.

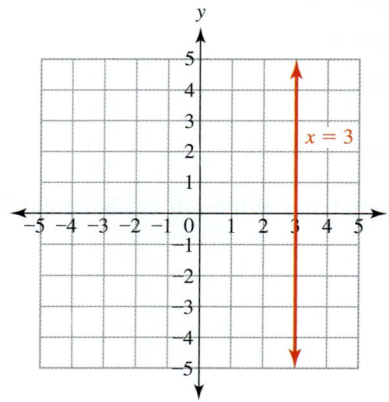

A vertical line with the equation $x = 3$.

All of the points that lie on the vertical line above have 3 as the x-value. If we pick any two points on this line, the slope calculation will have no horizontal change. Using the points (3, 5) and (3, 1) gives the slope calculation

$$(3, 5) \quad \text{and} \quad (3, 1)$$

$$m = \frac{5 - 1}{3 - 3}$$

$$m = \frac{4}{0} \quad \text{is undefined.}$$

The slope of this line is undefined, since any two points on the line will result in the denominator being zero.

If you consider the general form for this line and build a table, you can see that the value of y will not affect the value of x.

y	$x + 0y = 3$	(x, y)
-7	$x + 0(-7) = 3$ $x = 3$	$(3, -7)$
4	$x + 0(4) = 3$ $x = 3$	$(3, 4)$
18	$x + 0(18) = 3$ $x = 3$	$(3, 18)$

DEFINITIONS

Horizontal Line A horizontal line has an equation of the form $y = k$ and a slope $m = 0$.

Vertical Line A vertical line has an equation of the form $x = k$ and a slope m undefined.

SECTION 1.4 Intercepts and Graphing

Example 5 Graphing horizontal and vertical lines

Sketch the graph of the following lines.

a. $x = 2$

b. $y = -4$

SOLUTION

a. This equation is of the form $x = k$, so it will be a vertical line at $x = 2$.

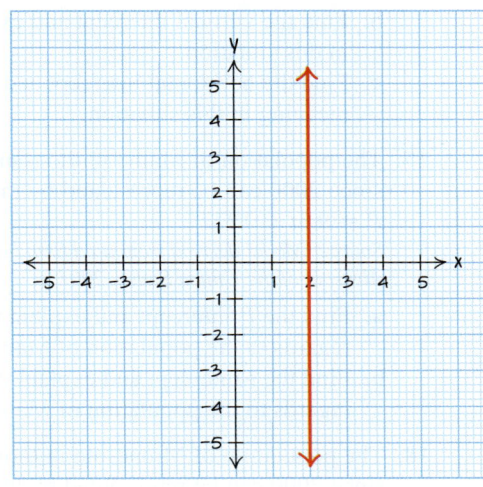

b. This equation is of the form $y = k$, so it will be a horizontal line at $y = -4$.

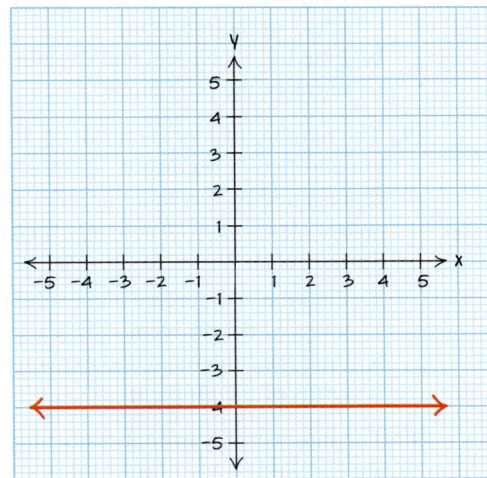

PRACTICE PROBLEM FOR EXAMPLE 5

Sketch the graph of the following lines.

a. $x = -3$

b. $y = 1.5$

1.4 Exercises

For Exercises 1 through 10, rewrite the equations in the general form of a line.

1. $y = 5x + 8$
2. $y = 7x + 20$
3. $y = -4x + 15$
4. $y = -3x + 9$
5. $y = \frac{2}{3}x - 8$
6. $y = \frac{3}{7}x - 4$
7. $y = \frac{1}{2}x + \frac{2}{5}$
8. $y = \frac{2}{9}x - \frac{7}{4}$
9. $y = -\frac{4}{5}x - \frac{1}{3}$
10. $y = -\frac{2}{7}x - \frac{7}{5}$

11. In Section 1.1 exercises, we investigated the equation $B = 0.55m + 29.95$, which represented the cost in dollars for renting a 10-foot truck from Budget and driving it m miles.
    a. Find the vertical intercept and explain its meaning in terms of renting a truck from Budget.
    b. Find the horizontal intercept and explain its meaning in terms of renting a truck from Budget.

12. The cost in dollars to rent a backhoe for h hours can be represented by the equation $C = 85h + 55$.
    a. Find the vertical intercept and explain its meaning in regard to renting a backhoe.
    b. Find the horizontal intercept and explain its meaning in regard to renting a backhoe.

13. The percentage of girls a years older than 10 years who are sexually active can be represented by the equation $P = 5a + 7$.
    a. Find the P-intercept and explain its meaning in terms of this situation.
    b. Find the a-intercept and explain its meaning in the terms of this situation.

14. The percentage P of companies that are still in business t years after the fifth year in operation can be represented by the equation $P = -3t + 50$.
    a. Find the P-intercept and interpret the intercept in terms of the information given in the problem.
    b. Find the t-intercept and interpret the intercept in terms of the information given in the problem.

15. A salesperson at a jewelry store earns 5% commission on all of her sales in addition to her $500 base monthly salary.
    a. Write an equation for the monthly salary that the salesperson earns if she makes s dollars in sales.
    b. Find the vertical intercept and explain its meaning in the terms of this situation.
    c. Find the horizontal intercept and explain its meaning in the terms of this situation.

16. A loan officer at a bank earns a $100 commission on each of his loans in addition to his $2000 base monthly salary.
    a. Write an equation for the monthly salary that the loan officer earns if he has l loans a month.
    b. Find the vertical intercept and explain its meaning in this situation.
    c. Find the horizontal intercept and explain its meaning in this situation.

17. The population of Maine in thousands t years since 2000 can be represented by the equation $P = 9.5t + 1277$.
    Source: Model derived from data found in the Statistical Abstract of the United States, 2007.
    a. Find the P-intercept and explain its meaning in regards to the population of Maine.
    b. Find the t-intercept and explain its meaning in regards to the population of Maine.

18. The population of Minnesota in thousands t years since 2000 can be represented by the equation $P = 41t + 4934$.
    Source: Model derived from data found in the Statistical Abstract of the United States, 2007.
    a. Find the P-intercept and explain its meaning in the terms of the problem.
    b. Find the t-intercept and explain its meaning in the terms of the problem.

19. The number of Florida residents in thousands who are enrolled in Medicare t years since 2000 can be represented by the equation $M = 44t + 2798$.
    Source: Model derived from data found in the Statistical Abstract of the United States, 2007.
    a. Find the M-intercept and explain its meaning in the terms of the problem.
    b. Find the t-intercept and explain its meaning in the terms of the problem.

20. The number of Vermont residents in thousands who are enrolled in Medicare t years since 2000 can be represented by the equation $M = 1.3t + 89$.
    Source: Model derived from data found in the Statistical Abstract of the United States, 2007.
    a. Find the M-intercept and explain its meaning in this situation.
    b. Find the t-intercept and explain its meaning in this situation.

SECTION 1.4 Intercepts and Graphing

For Exercises 21 through 34, find the intercepts and graph the line. Label the vertical and horizontal intercepts.

21. $2x + 4y = 8$
22. $4x + 2y = 20$
23. $3x - 5y = 15$
24. $5x - 2y = 20$
25. $4x + 6y = 30$
26. $10x + 8y = 16$
27. $-3x + 2y = 16$
28. $-4x + 3y = 18$
29. $-2x - 4y = -20$
30. $-7x - 5y = -70$
31. $x = 4$
32. $x = -3$
33. $y = 3$
34. $y = -6$

For Exercises 35 through 38, describe what is wrong with the graph of the given equation.

35. $2x + 3y = 6$

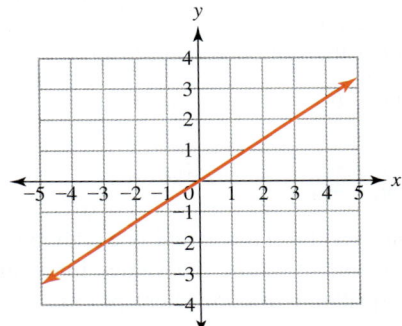

36. $x - 5y = 5$

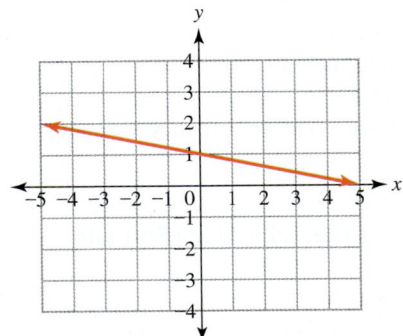

37. $x = -1$

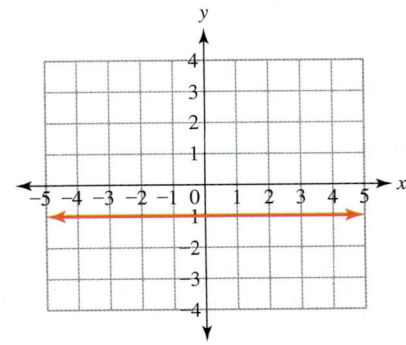

38. $y = 4$

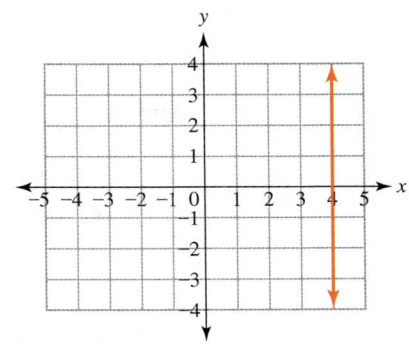

For Exercises 39 through 64, graph the lines using any method. Label the vertical and horizontal intercepts.

39. $y = \frac{3}{7}x - 6$
40. $y = \frac{4}{5}x - 10$
41. $y - 5 = 3(x - 4)$
42. $y + 4 = 2(x + 3)$
43. $y - 4 = \frac{1}{2}(x + 1)$
44. $y + 7 = \frac{2}{3}(x - 9)$
45. $y = 7$
46. $y = 2.5$
47. $x = 9$
48. $x = 3.5$
49. $y = -4.5$
50. $y = -6.5$
51. $x = -8$
52. $x = -15$
53. $y + 3x = 6\left(\frac{1}{2}x - 2\right)$
54. $2y - 4x = -2(2x + 7)$
55. $5y + x = 2y + 3(y - 10)$
56. $y + x = 2x - 8 + y$
57. $y + 2 = 2(x + 4) - 2x$
58. $y - 5 = 3x + y - 14$
59. $y = \frac{2}{15}x - 6$
60. $y = \frac{3}{11}x - 6$
61. $10x - 25y = 100$
62. $2x - 100y = 400$
63. $y = 0.001x - 20$
64. $y = -0.002x + 14$

1.5 Finding Equations of Lines

LEARNING OBJECTIVES

- Find the equation of a line using the slope-intercept form.
- Find the equation of a line using the point-slope form.
- Identify parallel and perpendicular lines.

Finding Equations of Lines

Now that we have learned the basic characteristics of a line and know how to find the slope of a line, we can use these skills to find the equation of any line we may want. Unless we are given the slope of the line we are looking for, we will use two points to find the slope and determine the equation of the line that we want. In previous sections, we found several equations using the written descriptions of the situation. In each of those exercises, the slope was given to you. In other applications, you may not be given the slope but instead may be given two points that you can use to calculate the slope.

Once you have the slope, you can use either the slope-intercept form

$$y = mx + b$$

or the **point-slope formula**

$$y - y_1 = m(x - x_1)$$

to find the equation. The point-slope formula uses the slope m and one additional point (x_1, y_1) to find the equation of a line. After you substitute the values for m and the additional point, you can isolate y to put the equation into slope-intercept form.

> **Steps to Find the Equation of a Line Using the Slope-Intercept Form** $y = mx + b$
> 1. Use any two points to calculate the slope.
> 2. Substitute in the slope and a point to find the value of b.
> 3. Write the equation in slope-intercept form.
> 4. Check the equation by plugging in the points to make sure they are solutions.

> **Steps to Find the Equation of a Line Using the Point-Slope Formula** $y - y_1 = m(x - x_1)$
> 1. Use any two points to calculate the slope.
> 2. Substitute in the slope and a point into the point-slope formula.
> 3. Write the equation in slope-intercept form.
> 4. Check the equation by plugging in the points to make sure they are solutions.

Example 1 Finding the equation of a line

Write the equation of the line that passes through the points (2, 5) and (4, 9).

SOLUTION

First, find the slope using the two given points.

$$m = \frac{9-5}{4-2} = \frac{4}{2} = 2$$

Now, using the slope and one of the points, substitute them into the slope-intercept form of a line and find b.

$$m = 2$$
$$\text{point} = (2, 5)$$
$$y = 2x + b$$
$$5 = 2(2) + b$$
$$5 = 4 + b$$
$$1 = b$$

Substitute the slope and the y-intercept into the slope-intercept form, and we will have the equation of the line.

$$y = 2x + 1$$

Check the equation by substituting in the points we were given and confirming that they are solutions to the equation.

$$y = 2x + 1$$
$$5 \stackrel{?}{=} 2(2) + 1$$
$$5 = 5 \qquad \text{The point (2, 5) is a solution.}$$
$$9 \stackrel{?}{=} 2(4) + 1$$
$$9 = 9 \qquad \text{The point (4, 9) is a solution.}$$

PRACTICE PROBLEM FOR EXAMPLE 1

Use the slope-intercept form, $y = mx + b$, to write the equation of the line that passes through the points (5, 4) and (20, 7).

Example 2 — Finding an equation for straight line depreciation

A business purchased a car in 2008 for $35,950. For tax purposes, the value of the car in 2011 was $20,550. If the business is using straight line depreciation, write the equation of the line that gives the value of the car based on the age of the car in years.

SOLUTION

First we define the variables that we will use. We are discussing the value of the car for tax purposes and its age in years, so we can define the variables as follows:

$a = $ Age of the car in years

$v = $ Value of the car, for tax purposes, in dollars

Now the question has given us two examples of the value of the car: In 2008, the car was new, or 0 years old, and worth $35,950; and in 2011, the car was 3 years old and worth $20,550. These two quantities represent the points

$$(0, 35950) \qquad (3, 20550)$$

Using these two points, we can calculate the slope of the equation.

$$m = \frac{y_2 - y_1}{x_2 - x_1}$$
$$m = \frac{20550 - 35950}{3 - 0}$$
$$m = \frac{-15400}{3}$$
$$m \approx -5133.33$$

What's That Mean?

Depreciation

Depreciation is a term used in accounting to describe how to calculate the way an asset like a car loses value over time.

Straight line depreciation means that the value decreases the same amount each year, and thus the value of the asset can be calculated using a linear equation.

The first point we were given is actually the vertical intercept, so we have the value of b. Therefore, our equation is

$$v = -5133.33a + 35950$$

We can check this equation by using both points in the equation and verifying that they are solutions.

$$v = -5133.33a + 35950$$
$$(0, 35950)$$
$$35950 \stackrel{?}{=} -5133.33(0) + 35950$$
$$35950 = 35950$$
$$(3, 20550)$$
$$20550 \stackrel{?}{=} -5133.33(3) + 35950$$
$$20550 \approx 20550.01 \qquad \text{Not exact because of rounding.}$$

Both given points satisfy the equation. Therefore, the equation is correct.

Example 3 — Finding an equation to estimate charitable donations

Compassion International is a charity that helps children all over the world who live in poverty. In 2002, donations were about $123.0 million, and in 2005, donations were about $215.8 million. Assume that Compassion International's donations are growing at a constant rate and write an equation to represent this situation.
Source: www.CharityNavigator.org

SOLUTION

Because we are told that the donations are growing at a constant rate, we know that we want to find a linear equation. We are given two examples of the donations in different years, and we can use them as our two points. First, we define the variables we want to use. Let

t = Years since 2000

D = Donations in millions of dollars that Compassion International receives during a year

With these definitions, we get the two points $(2, 123.0)$ and $(5, 215.8)$. Find the slope and the vertical intercept to write the equation.

$$m = \frac{215.8 - 123.0}{5 - 2} = \frac{92.8}{3} \approx 30.93 \qquad \text{Use the two points to find the slope. Round to one extra decimal place.}$$
$$D = 30.93t + b \qquad \text{Substitute the slope into } m.$$
$$123.0 = 30.93(2) + b \qquad \text{Use a given point to help find } b.$$
$$123.0 = 61.86 + b$$
$$\underline{-61.86 \quad -61.86}$$
$$61.14 = b$$
$$D = 30.93t + 61.14 \qquad \text{Substitute the slope and the vertical intercept into the final equation.}$$
$$123.0 \stackrel{?}{=} 30.93(2) + 61.14 \qquad \text{Check that both the given points satisfy the equation.}$$
$$123.0 = 123.0$$
$$215.8 \stackrel{?}{=} 30.93(5) + 61.14$$
$$215.8 \approx 215.79 \qquad \text{Both points check.}$$

Therefore, the equation $D = 30.93t + 61.14$ represents the donations D in millions of dollars received by Compassion International t years since 2000.

PRACTICE PROBLEM FOR EXAMPLE 3

The number of sports, cards hobby stores has been declining steadily. In 1995, there were 4500 stores, and in 2005, there were only 1500. Write an equation for the line that gives the number of sports, cards hobby stores given the number of years since 1995.
Source: Today's Local News, July 7, 2007.

Example 4 — Finding the equation of a line using the point-slope formula

Use the point-slope formula, $y - y_1 = m(x - x_1)$, to write the equation of the line that passes through the points $(3, 1)$ and $(7, -9)$.

SOLUTION

First, we can find the slope using the two given points.

$$m = \frac{-9 - 1}{7 - 3} = \frac{-10}{4} = -2.5$$

Now, using the point-slope formula with the slope and one of the points we get

$$m = -2.5$$
$$\text{point} = (3, 1)$$
$$y - 1 = -2.5(x - 3)$$
$$y - 1 = -2.5x + 7.5$$
$$\underline{+1 \qquad\qquad +1}$$
$$y = -2.5x + 8.5$$

We can check our equation by substituting in the points we were given and confirming that they are solutions to the equation.

$$y = -2.5x + 8.5$$
$$1 \stackrel{?}{=} -2.5(3) + 8.5$$
$$1 = 1 \qquad \text{The point (3, 1) is a solution.}$$
$$(-9) \stackrel{?}{=} -2.5(7) + 8.5$$
$$-9 = -9 \qquad \text{The point (7, -9) is a solution.}$$

PRACTICE PROBLEM FOR EXAMPLE 4

Use the point-slope formula, $y - y_1 = m(x - x_1)$, to write the equation of the line that passes through the points $(1, 4)$ and $(5, 12)$.

Example 5 — Writing the equation of a line

a. Write the equation of the line that passes through the points in the table.

x	y
2	7
4	15
6	23
8	31

b. Write the equation of the line shown in the graph.

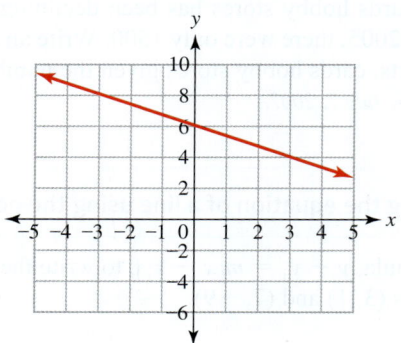

SOLUTION

a. First, find the slope of the line from the table. Use the slope and one of the points listed in the table in the point-slope formula to write the equation.

x	y	Slope
2	7	
4	15	$\frac{15-7}{4-2} = \frac{8}{2} = 4$
6	23	$\frac{23-15}{6-4} = 4$
8	31	$\frac{31-23}{8-6} = 4$

$$m = 4$$
$$\text{point} = (2, 7)$$
$$y - 7 = 4(x - 2)$$
$$y - 7 = 4x - 8$$
$$y = 4x - 1$$

Substitute the value of the slope and a point into the point-slope formula.

Write the equation in slope-intercept form.

Check this equation by substituting in a couple of points from the table into the equation.

$$y = 4x - 1$$
$$15 \stackrel{?}{=} 4(4) - 1$$
$$15 = 15 \qquad \text{The point } (4, 15) \text{ is a solution.}$$
$$23 \stackrel{?}{=} 4(6) - 1$$
$$23 = 23 \qquad \text{The point } (6, 23) \text{ is a solution.}$$

b. Using the graph, we can estimate that the points $(-3, 8)$ and $(3, 4)$ are on the line.

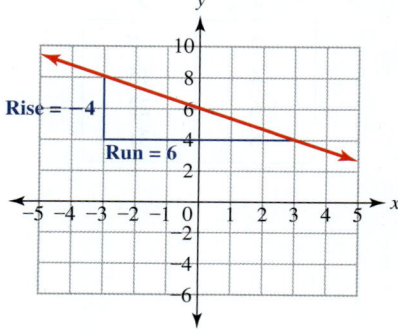

With a rise of −4 and a run of 6, the slope is

$$m = \frac{-4}{6} = -\frac{2}{3}$$

Using this slope and estimating the *y*-intercept to be (0, 6), we get the equation

$$y = -\frac{2}{3}x + 6$$

The graph is decreasing from left to right, so the slope should be negative, and the *y*-intercept seems to be (0, 6). The equation for the line should be correct.

PRACTICE PROBLEM FOR EXAMPLE 5

Write the equation of the line that passes through the points in the table.

x	y
1	6
4	1
7	−4
10	−9

Parallel and Perpendicular Lines

When looking at two lines, we notice that a couple of special relationships occur. We say that two lines are **parallel** if they never cross. Parallel lines must go in the same direction and have the same slope, or they would eventually cross.

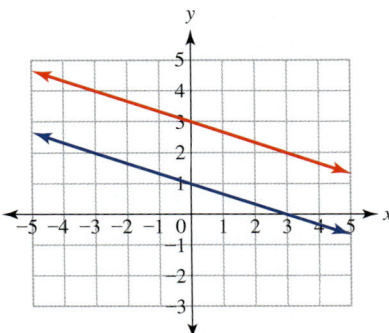

Two lines are said to be **perpendicular** if they cross at a right angle (90 degrees). For lines to meet at a right angle, their slopes must be opposite reciprocals of one another. For example, if the slope of one line is $m = \frac{2}{3}$, then the slope of a line perpendicular to it must be $m = -\frac{3}{2}$.

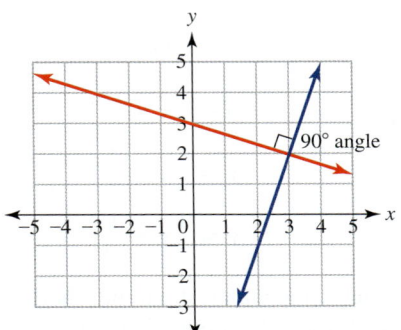

Using Your TI Graphing Calculator

When graphing equations on the calculator, you should be aware that the screen is a rectangle and will distort graphs by stretching them out along the horizontal axis. This will happen whenever the window is not set to the correct proportions that will account for the window's being wider than it is tall. The calculator has a ZOOM feature called ZSquare that will take any window and change it to the correct proportions to display a graph correctly.

The following two screens show the same pair of perpendicular lines (lines that cross at a 90-degree angle).

Standard window:

Xmin −10, Xmax 10
Ymin −10, Ymax 10

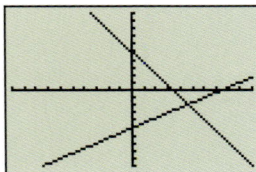

The standard window does not show these lines as perpendicular.

Window after ZSquare:

Xmin −15.161, Xmax 15.161
Ymin −10, Ymax 10

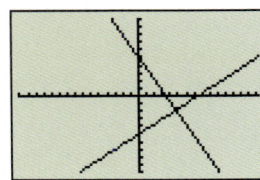

The square window does show these lines correctly as perpendicular.

> **DEFINITIONS**
>
> **Parallel Lines** Two different lines are parallel if their slopes are equal.
>
> $$m_1 = m_2$$
>
> **Perpendicular Lines** Two lines are perpendicular if their slopes are opposite reciprocals.
>
> $$m_1 = -\frac{1}{m_2}$$

We can use the relationship between the slopes of parallel lines and between the slopes of perpendicular lines to help us write equations.

Example 6 Equations of parallel and perpendicular lines

a. Write the equation of the line that goes through the point (7, 10) and is parallel to the line $y = 2x + 5$.

b. Write the equation of the line that goes through the point (2, 6) and is perpendicular to the line $-3x - y = 8$.

c. Write the equation of the line that goes through the point (5, 8) and is parallel to the x-axis.

SOLUTION

a. Because the line that we want is parallel to the line $y = 2x + 5$, it will have the same slope, $m = 2$. Using this slope and the given point, (7, 10), in the point-slope formula, we get

$$y - 10 = 2(x - 7)$$
$$y - 10 = 2x - 14$$
$$y = 2x - 4$$

b. Because the line that we want is perpendicular to the line $-3x - y = 8$, we want to know the slope of this line. To find the slope, put the line into slope-intercept form.

$$-3x - y = 8$$
$$-y = 3x + 8$$
$$y = -3x - 8$$

This line has a slope of $m = -3$. The line that we want will have the opposite reciprocal slope $m = \frac{1}{3}$. Using this slope and the given point, (2, 6), in the point-slope formula, we get

$$y - 6 = \frac{1}{3}(x - 2)$$
$$y - 6 = \frac{1}{3}x - \frac{2}{3}$$
$$y = \frac{1}{3}x - \frac{2}{3} + \frac{18}{3}$$
$$y = \frac{1}{3}x + \frac{16}{3}$$

c. Because the line is parallel to the x-axis, it will be a horizontal line and have a slope of zero. The line must go through the point (5, 8), so it will have the y-value 8. Therefore, we get the line $y = 8$.

PRACTICE PROBLEM FOR EXAMPLE 6

a. Write the equation of the line that goes through the point (4, 5) and is parallel to the line $y = 4x + 1$.

b. Write the equation of the line that goes through the point (4, −3) and is perpendicular to the line $y = -\frac{2}{5}x - 9$.

Interpreting the Characteristics of a Line: A Review

Example 7 Interpreting slope and intercepts

Using the value of the car equation found in Example 2, page 65, answer the following questions.

$$v = -5133.33a + 35950$$

a. What is the slope of the equation? What does it represent in regard to the value of the car?

b. What is the vertical intercept of the equation? What does it represent in regard to the value of the car?

c. What is the horizontal intercept of the equation? What does it represent in regard to the value of the car?

SOLUTION

a. The slope is

$$m = -5133.33$$
$$m = \frac{-5133.33}{1}$$
$$m = \frac{\$(-5133.33)}{1 \text{ year of age}}$$

The slope means that the value of the car for tax purposes is decreasing by $5133.33 per year of age.

b. The vertical intercept was given as (0, 35950). This means that when the car was new, zero years old, it was worth $35,950.

c. To find the horizontal intercept, make $v = 0$ and solve for a.

$$0 = -5133.33a + 35950$$
$$\underline{-35950 \qquad\qquad -35950}$$
$$-35950 = -5133.33a$$
$$\frac{-35950}{-5133.33} = \frac{-5133.33a}{-5133.33}$$
$$7.003 \approx a$$

The horizontal intercept is (7.003, 0). This means that a car that is 7 years old is worth nothing for tax purposes.

Example 8 Interpreting slope and intercepts

Using the donations equation $D = 30.93t + 61.14$ found in Example 3, page 66, answer the following questions.

a. What is the slope of the equation you found, and what does the slope mean in regard to donations received by Compassion International?

b. What is the vertical intercept of the equation you found, and what does that intercept mean in regard to donations received by Compassion International?

SOLUTION

a. The slope of this equation is 30.93. The donations received by Compassion International are increasing by about $30.93 million per year.

b. The vertical intercept of this equation is (0, 61.14). In 2000, Compassion International had about $61.14 million in donations.

PRACTICE PROBLEM FOR EXAMPLE 8

Using the application and equation $H = -300t + 4500$ found in Example 3 Practice Problem, page 67, answer the following questions.

a. What is the slope of the equation? What does the slope represent in regard to the number of sports-cards hobby stores?

b. What is the vertical intercept of the equation? What does that intercept represent in regard to the number of sports-cards hobby stores?

c. What is the horizontal intercept of the equation? What does that intercept represent in regard to the number of sports-cards hobby stores?

Example 9 Putting it all together in an application

A local cellular phone company has a pay-as-you-talk plan that costs $15 per month and $0.23 per minute that you talk on the phone.

a. Write an equation for the total monthly cost C of this plan if you talk for m minutes.

b. What is the slope of the equation you found in part a? What does it represent?

c. What is the vertical intercept of the equation you found in part a? What does it represent?

SOLUTION

a. We are looking for a monthly cost, so we are considering only one month at a time. Since the plan costs $15 for the month, we will take this cost and add to it the cost for the minutes used. It costs $0.23 per minute, so we will have a cost equation of

$$C = 0.23m + 15$$

where C is the total monthly cost in dollars when m minutes are used.

b. The slope of this equation is

$$m = 0.23 = \frac{0.23}{1} = \frac{\$0.23}{1 \text{ minute}}$$

This slope means that the total monthly cost will increase by $0.23 per minute used.

c. The equation is in slope-intercept form, so the vertical intercept is (0, 15). This means that if you do not use any minutes in a month, the plan will still cost you $15.

PRACTICE PROBLEM FOR EXAMPLE 9

In 2000, there were 709.5 thousand cases of chlamydia infection, a sexually transmitted disease, in the United States. In 2003, there were 877.5 thousand cases of chlamydia infection in the United States.

Source: Centers for Disease Control and Prevention, Health, United States, 2005.

■ **Skill Connection**

Creating Two Points to Find the Equation of the Line

In Example 9, we used the wording to determine the slope of the equation. Another way to find this equation would be to create two points and use them to calculate the slope and find the equation.

In this situation, the monthly cost would be $15 if you did not use the phone, so the point (0, 15) is on the line. If you use the phone for 1 minute, it will cost you an additional 23 cents, for a total cost of $15.23. Therefore, the point (1, 15.23) is also on the line.

Using these two points, you can calculate the slope and find the equation.

a. Assuming that the number of cases follows a linear trend, write an equation to represent this information.

b. What is the slope of the equation you found? What does it represent?

c. What is the vertical intercept of the equation you found? What does it represent?

1.5 Exercises

For Exercises 1 through 14, write the equation for the line shown in the given graph. Give your answer in slope-intercept form, $y = mx + b$. Use points that clearly cross the intersections of the graph paper.

1.

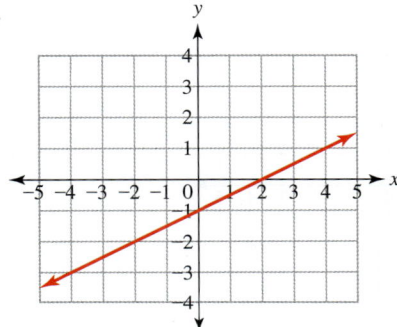

2.

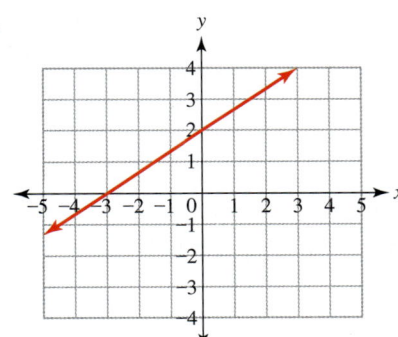

3.

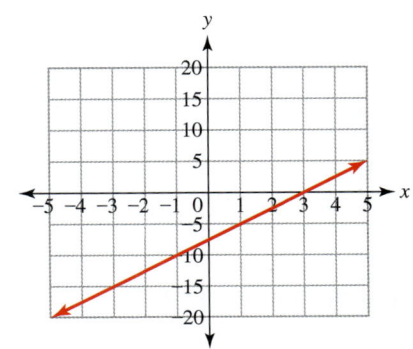

4.

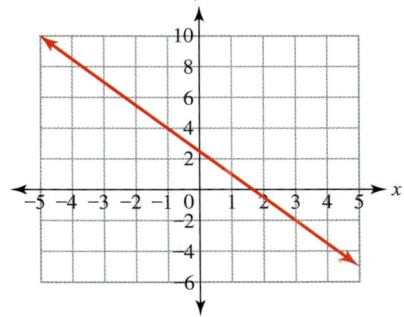

5.

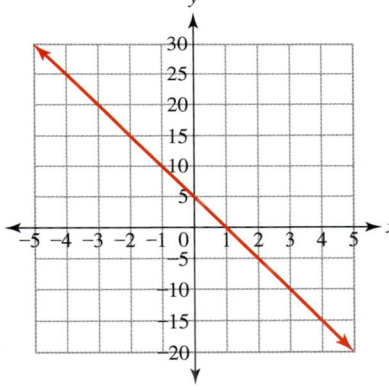

6.

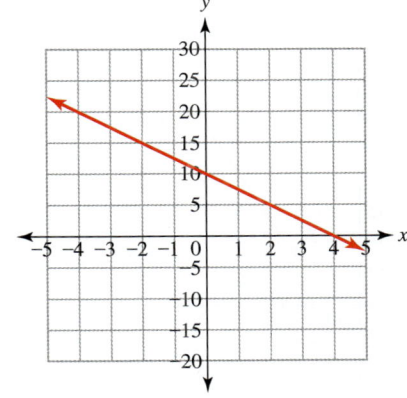

7.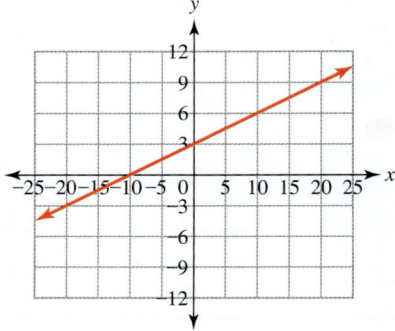

8.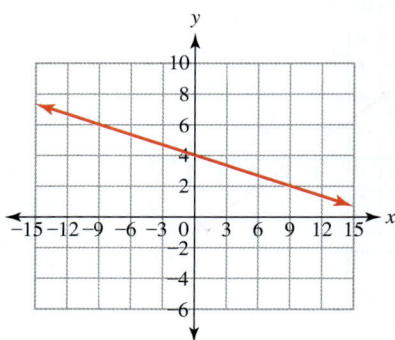

9.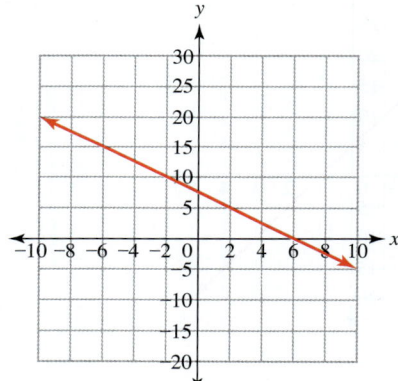

10.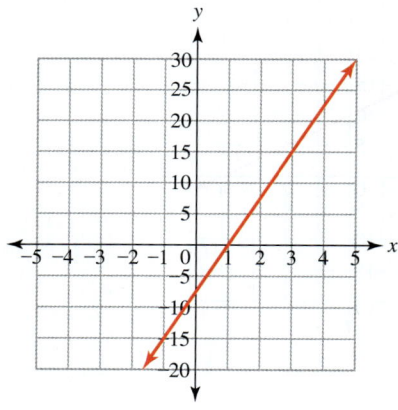

11.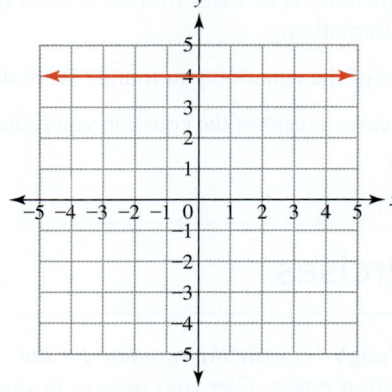

12.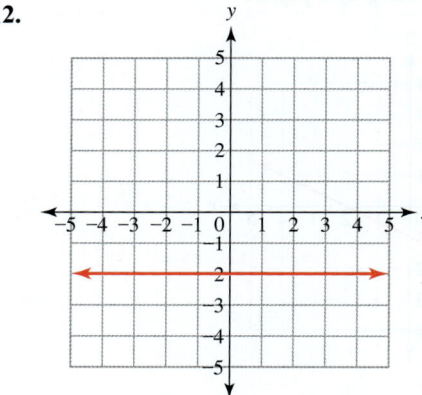

13.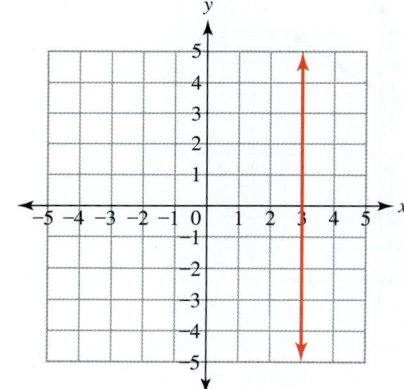

14.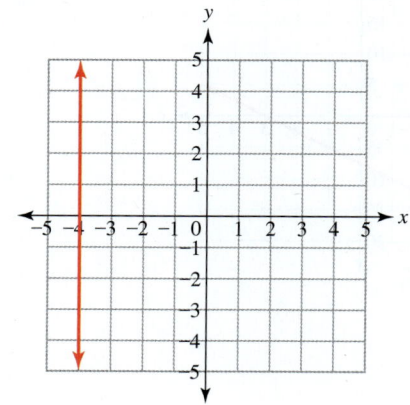

SECTION 1.5 Finding Equations of Lines

For Exercises 15 and 16, compare the two students' work to determine which student did the work correctly. Explain what mistake the other student made.

15. Write the equation for the line shown in the graph. Give your answer in slope-intercept form, $y = mx + b$. Use points that clearly cross the intersections of the graph paper.

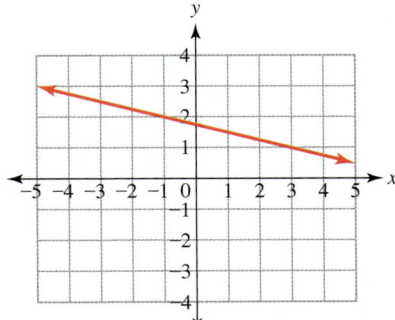

Cheryll

This is a vertical line where $x = -1$ for every point on the line. The equation is
$$x = -1$$

Lance

This is a vertical line at -1. The equation is
$$y = -1$$

Maritza

Pick two points and calculate the slope.

$(-1, 2)$ $(0, 1.5)$

$m = \dfrac{1.5 - 2}{0 - (-1)} = -0.5$

The point $(0, 1.5)$ is the y-intercept, so $b = 1.5$. The equation is
$$y = -0.5x + 1.5$$

Sherry

Pick two points and calculate the slope.

$(-1, 2)$ $(3, 1)$

$m = \dfrac{1 - 2}{3 - (-1)} = -\dfrac{1}{4} = -0.25$

Plug in a point to find b.

$y = -0.25x + b$
$2 = -0.25(-1) + b$
$2 = 0.25 + b$
$1.75 = b$

The equation is
$$y = -0.25x + 1.75$$

16. Write the equation for the line shown in the graph. Give your answer in slope-intercept form, $y = mx + b$. Use points that clearly cross the intersections of the graph paper.

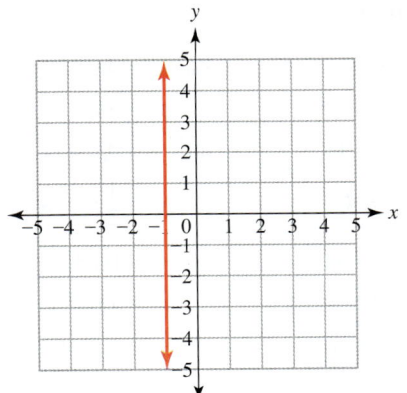

17. If 10 shirts cost $110 and 30 shirts cost $280, write a linear equation that gives the cost for n shirts.

18. If 15 sports jerseys cost $360 and 25 jerseys cost $500, write a linear equation that gives the cost for j jerseys.

19. The U.S. Census Bureau projects the populations of all states through 2030. They project the population of the state of Washington to be 6951 thousand in 2015 and 7996 thousand in 2025. Assuming that the population growth is linear, write an equation that gives the population of Washington t years since 2000.
Source: Statistical Abstract of the United States, 2007.

20. The U.S. Census Bureau projects the population of Nevada to be 2691 thousand in 2010 and 3863 thousand in 2025. Assuming that the population growth is linear, write an equation that gives the population of Nevada t years since 2000.
Source: Statistical Abstract of the United States, 2007.

21. According to Kristine Clark Ph.D., R.D., director of sports nutrition and assistant professor of nutritional sciences at Penn State University, a women's optimal weight is 100 pounds at 5 feet and 130 pounds at 5.5 feet. Use this information to write a linear model for a woman's optimal weight in pounds based on the number of inches her height is above 5 feet.
Source: Natural Health, March 2006.

22. According to Kristine Clark Ph.D., R.D., (see Exercise 21), the optimal weight is 106 pounds for a 5-foot man and 142 pounds for a 5.5-foot man. Use this information to write a linear model for a man's optimal weight in pounds based on the number of inches his height is above 5 feet.
Source: Natural Health, March 2006.

23. In 2001, 408 teenagers underwent gastric bypass surgery. In 2003, 771 teenagers underwent gastric bypass surgery. Assuming that the number of teens having gastric bypass surgery is growing at a constant rate, write an equation for the number of teenagers having gastric bypass surgery t years since 2000.
Source: Associated Press, March 2007.

24. The number of hospitals in the United States has been declining for many years. In 2001, there were 5801 hospitals; but in 2004, there were only 5759.

Assuming that this trend is linear, write an equation for the number of hospitals in the United States t years since 2000.
Source: Statistical Abstract of the United States, 2007.

For Exercises 25 through 38, write the equation for the line using the given information. Write the equation in slope-intercept form.

25. Write the equation of the line with slope $= -2$ and passing through the point $(3, 7)$.

26. Write the equation of the line with slope $= 3$ and passing through the point $(-4, -10)$.

27. Write the equation of the line with slope $= \frac{2}{3}$ and passing through the point $(12, 20)$.

28. Write the equation of the line with slope $= -\frac{5}{7}$ and passing through the point $(-2, 12)$.

29. Write the equation of the line that passes through the points $(1, 3)$ and $(4, 12)$.

30. Write the equation of the line that passes through the points $(2, 4)$ and $(7, 24)$.

31. Write the equation of the line that passes through the points $(7, 6)$ and $(21, -1)$.

32. Write the equation of the line that passes through the points $(-5, -2)$ and $(-3, -10)$.

33. Write the equation of the line that passes through the points $(-4, -5)$ and $(-1, 7)$.

34. Write the equation of the line that passes through the points $(-8, -3)$ and $(-12, -15)$.

35. Write the equation of the line that passes through the points $(7, -3)$ and $(7, 9)$.

36. Write the equation of the line that passes through the points $(4, 5)$ and $(4, 12)$.

37. Write the equation of the line that passes through the points $(2, 8)$ and $(4, 8)$.

38. Write the equation of the line that passes through the points $(-8, -15)$ and $(10, -15)$.

For Exercises 39 through 44, write the equation for the line passing through the points given in the table.

39.

x	y
0	7
2	11
4	15
6	19

40.

x	y
-4	-18
0	-4
4	10
8	24

41.

x	y
-2	29
-1	22
2	1
4	-13

42.

x	y
-3	14.25
-1	10.75
3	3.75
6	-1.5

43.

x	y
2	-18
5	-6
8	6
11	18

44.

x	y
4	7
9	-5.5
14	-18
19	-30.5

For Exercises 45 through 58, determine, without graphing, whether the two given lines are parallel, perpendicular, or neither. Check your answer by graphing both lines on a calculator or by hand.

45. $y = 3x + 5$
    $y = 3x - 7$

46. $y = 4x + 8$
    $y = 4x - 3$

47. $2x + 3y = 15$
    $y = \frac{3}{2}x + 4$

48. $y = 2x + 5$
    $y = \frac{1}{2}x + 8$

49. $2x - 5y = 40$
    $-4y = 10x + 10$

50. $y = 0.25x - 9$
    $x = 4y + 3$

51. $4x - 3y = 20$
    $12x - 9y = 30$

52. $y + 7 = 4(x + 3) - x$
    $y = -\frac{1}{3}x + 2$

53. $x = 8$
    $y = 9$
54. $x = -3$
    $x = 7$
55. $2x + 5y = 20$
    $10x - 4y = 20$
56. $3x + 2y = 8$
    $-6y = 9x + 12$
57. $5x + y = 7$
    $2y = 10x - 9$
58. $x = -10$
    $y = 3$

59. Write the equation of the line that goes through the point (2, 8) and is parallel to the line $y = 4x - 13$.

60. Write the equation of the line that goes through the point (4, 6) and is parallel to the line $y = -3x + 24$.

61. Write the equation of the line that goes through the point (−6, 8) and is parallel to the line $10x - 15y = -12$.

62. Write the equation of the line that goes through the point (−4, −9) and is parallel to the line $-x + 2y = -0.5$.

63. Write the equation of the line that goes through the point (1, 7) and is perpendicular to the line $y = 2x - 1$.

64. Write the equation of the line that goes through the point (6, 5) and is perpendicular to the line $y = 3x + 8$.

65. Write the equation of the line that goes through the point (5, 1) and is perpendicular to the line $4x + y = 5$.

66. Write the equation of the line that goes through the point (−4, −3) and is perpendicular to the line $6x + y = -11$.

67. Write the equation of the line that goes through the point (2, 3) and is perpendicular to the line $y = \frac{1}{5}x - 8$.

68. Write the equation of the line that goes through the point (−9, 3) and is perpendicular to the line $3x + 7y = -21$.

69. Write the equation of the line that goes through the point (4, 3) and is perpendicular to the line $y = 8$.

70. Write the equation of the line that goes through the point (3, 7) and is perpendicular to the line $x = -4$.

71. A student says that a line perpendicular to the line $y = \frac{2}{3}x + 5$ has slope $m = \frac{3}{2}$. Explain what is wrong with this statement. Give the correct slope.

72. A student says that a line perpendicular to the line $y = -5x + 12$ has slope $m = 5$. Explain what is wrong with this statement. Give the correct slope.

73. A student says that a line parallel to the line $2x + 5y = 14$ has slope $m = -\frac{1}{2}$. Explain what is wrong with this statement. Give the correct slope.

74. A student says that a line parallel to the line $6x - 3y = 10$ has slope $m = 6$. Explain what is wrong with this statement. Give the correct slope.

75. A student says that the line with slope = 4 passing through the point (2, 6) is $y = 4x + 6$. Explain what is wrong with this equation. Give the correct equation.

76. A student says that the line with slope = −3 passing through the point (7, 0) is $y = -3x + 7$. Explain what is wrong with this equation. Give the correct equation.

77. The population of Washington state can be estimated by the equation $P = 75.8t + 5906.2$, where P represents the population of Washington in thousands of people t years since 2000.
    Source: Model derived from data found in the Statistical Abstract of the United States, 2006.

    a. Find the slope of the equation and give its meaning in this situation.
    b. Find the vertical intercept for the equation and give its meaning in this situation.
    c. Find the horizontal intercept for the equation and give its meaning in this situation.

78. The population of New York state can be estimated by the equation $P = 62.6t + 19005.4$, where P represents the population of New York in thousands of people t years since 2000.
    Source: Model derived from data found in the Statistical Abstract of the United States, 2006.

    a. Find the slope of the equation and give its meaning in regard to the population of New York.
    b. Find the vertical intercept for the equation and give its meaning in regard to the population of New York.
    c. Find the horizontal intercept for the equation and give its meaning in regard to the population of New York.

79. The annual amount of public expenditure on medical research in the United States can be estimated by the equation $M = 4.02t + 25.8$, where M represents billions of dollars of public money spent on medical research t years since 2000.
    Source: Model derived from data found in the Statistical Abstract of the United States, 2006.

a. Find the slope of the equation and give its meaning in this example.
   b. Find the vertical intercept for the equation and give its meaning in this example.
   c. Find the horizontal intercept for the equation and give its meaning in this example.

80. The amount of wind energy produced in Alaska can be estimated by the equation $W = 0.225t + 0.45$, where W represents the megawatts of wind energy produced in Alaska t years since 2000.
   Source: American Wind Energy Association, AWEA.org.
   a. Find the slope of the equation and interpret what it means in the terms of the situation.
   b. Find the vertical intercept for the equation and interpret what it means in the terms of the situation.
   c. Find the horizontal intercept for the equation and interpret what it means in the terms of the situation.

81. The monthly profit for a small used car lot can be estimated by the equation $P = 500c - 6000$, where P represents the profit in dollars for the car lot when c cars are sold in a month.
   a. Find the slope of the equation and give its meaning in regard to the profit.
   b. Find the vertical intercept for the equation and give its meaning in regard to the profit.
   c. Find the horizontal intercept for the equation and give its meaning in regard to the profit.

82. The monthly profit for a coffee shop can be estimated by the equation $P = 0.75c - 3500$, where P represents the profit in dollars for the coffee shop when c customers visit in a month.
   a. Find the slope of the equation and give its meaning in regard to the profit.
   b. Find the vertical intercept for the equation and give its meaning in regard to the profit.
   c. Find the horizontal intercept for the equation and give its meaning in regard to the profit.

83. Dan gives surfing lessons over the summer and earns $30 for each 1 hour lesson given. His surfboards and supplies for the summer cost him $700.
   a. Write an equation for the profit in dollars that Dan makes from giving s surf lessons.
   b. Find the slope of the equation in part a and give its meaning in this situation.
   c. Find the vertical intercept of the equation and give its meaning in this situation.
   d. Find the horizontal intercept of the equation and give its meaning in this situation.

84. Janell tutors math students and earns $20 an hour for each session. She spends about $100 a month on transportation and other costs related to tutoring.
   a. Write an equation for the monthly profit in dollars that Janell makes from tutoring for h hours a month.
   b. Find the slope of the equation in part a and give its meaning in terms of her monthly profit.
   c. Find the vertical intercept of the equation and give its meaning in terms of her monthly profit.
   d. Find the horizontal intercept of the equation and give its meaning in terms of her monthly profit.

85. The Parent Teacher Association (PTA) at Mission Meadows Elementary School is starting a recycling program to help raise money for a new running track to be installed on the campus. For each pound of aluminum cans recycled, they earn $1.24. The PTA started their fund raising with $2000 donated from the parents.
   a. Write an equation for the total amount the PTA has raised for the track depending on how many pounds of aluminum cans they recycle.
   b. What is the slope of the equation you found? Give its meaning in this situation.
   c. Find the vertical intercept of the equation and give its meaning in this situation.
   d. Find the horizontal intercept of the equation and give its meaning in this situation.

86. Through the year 2050, the population of the United States is expected to grow by about 2.8 million people per year. In 2006, the population of the United States was estimated to be 298.4 million.
   Source: U.S. Census Bureau.
   a. Write an equation to represent the population of the United States t years since 2000.
   b. Use the equation, from part a to estimate the population of the United States in 2020.
   c. When will the population of the United States reach 400 million?
   d. What is the vertical intercept of the equation you found? Explain what it means using the terms of the problem?

87. The percentage of Americans who have been diagnosed with diabetes has been growing steadily over the years. In 2002, 4.8% of Americans had been diagnosed with diabetes. In 2004, 5.1% of Americans had been diagnosed with diabetes.
   Source: CDC Diabetes Program.
   a. Assuming that the percentage of Americans diagnosed with diabetes continues to grow at a constant rate, write an equation to represent this situation.
   b. Use the equation from part a to estimate the percentage of Americans who will have been diagnosed with diabetes in 2010.
   c. What is the slope of the equation you found. Explain what it means using the terms of the problem?

88. A toy manufacturer finds that if they produce 1000 toy cars an hour, 1% of the cars are defective.

If production is increased to 1500 toys an hour, 1.5% of the cars are defective.

   a. Assuming that the percentage of cars that are defective is linearly related to the number of cars produced an hour, write an equation for the percentage of cars that are defective if t toys are produced an hour.
   b. Use the equation from part a to find the percentage of cars that are defective if 2500 cars are produced an hour.
   c. What is the slope of the equation you wrote? What does it mean in regard to the percentage of cars that are defective?

89. Use the equation for the cost of shirts that you wrote in Exercise 17 on page 75 to answer the following questions.
   a. Use the equation to find how much 50 shirts will cost.
   b. What is the slope of the equation? What does it mean in regard to the cost of shirts?
   c. Find the vertical intercept of the equation and give its meaning in regard to the cost of shirts.
   d. Find the horizontal intercept of the equation and give its meaning in regard to the cost of shirts.

90. Use the equation for the cost of sports jerseys that you wrote in Exercise 18 on page 75 to answer the following questions.
   a. Use the equation to find how much 40 jerseys will cost.
   b. What is the slope of the equation? What does it mean in this situation?
   c. Find the vertical intercept of the equation and give its meaning in this situation.
   d. Find the horizontal intercept of the equation and give its meaning in this situation.

91. Use the equation for the population of Washington that you wrote in Exercise 19 on page 75 to answer the following questions.
   a. What does the equation predict the population of Washington will be in 2030?
   b. What is the slope of the equation? Interpret its meaning in terms of the problem.
   c. Find the vertical intercept of the equation and interpret its meaning in terms of the problem.
   d. Find the horizontal intercept of the equation and interpret its meaning in terms of the problem.

92. Use the equation for the population of Nevada that you wrote in Exercise 20 on page 75 to answer the following questions.
   a. What does the equation predict the population of Nevada will be in 2030?
   b. What is the slope of the equation? What does it mean in regard to the population of Nevada?
   c. Find the vertical intercept of the equation and give its meaning in regard to the population of Nevada.
   d. Find the horizontal intercept of the equation and give its meaning in regard to the population of Nevada.

93. Use the equation for the optimal weight of a woman that you wrote in Exercise 21 on page 75 to answer the following questions.
   a. What does the equation give as the optimal weight of a woman who is 6 feet tall?
   b. What is the slope of the equation? What does it mean in this situation?
   c. Find the vertical intercept of the equation and give its meaning in this situation.
   d. Find the horizontal intercept of the equation and give its meaning in this situation.

94. Use the equation for the optimal weight of a man that you wrote in Exercise 22 on page 75 to answer the following questions.
   a. What does the equation give as the optimal weight of a man who is 6 feet tall?
   b. What is the slope of the equation? Interpret its meaning in terms of the problem.
   c. Find the vertical intercept of the equation and interpret its meaning in terms of the problem.
   d. Find the horizontal intercept of the equation and interpret its meaning in terms of the problem.

95. Use the equation for the number of teenagers who had gastric bypass surgery that you wrote in Exercise 23 on page 75 to answer the following questions.
   a. What does the equation predict as the number of teenagers who had gastric bypass surgery in 2005?
   b. What is the slope of the equation? What does it mean in this situation?
   c. Find the vertical intercept of the equation and give its meaning in this situation.
   d. Find the horizontal intercept of the equation and give its meaning in this situation.

96. Use the equation for the number of hospitals in the United States that you wrote in Exercise 24 on page 75 to answer the following questions.
   a. What does the equation predict as the number of hospitals in the United States in 2010?
   b. What is the slope of the equation? What does it mean in regard to the number of hospitals?
   c. Find the vertical intercept of the equation and give its meaning in regard to the number of hospitals.
   d. Find the horizontal intercept of the equation and give its meaning in regard to the number of hospitals.

1.6 Finding Linear Models

LEARNING OBJECTIVES

- Use a graphing calculator to create a scatterplot of data.
- Find a linear model for data.
- Use a linear model to make predictions.

Creating scatterplots by hand can be very tedious, slow, and inaccurate. Using technology to help us create scatterplots and investigate models is a great help. In this section, we will discuss how to create scatterplots on a Texas Instruments graphing calculator. We will also write algebraic linear models using the data and these scatterplots. Algebraic models are equations that can represent the graphical model we learned about in Section 1.2. These models can be used more accurately and often more easily than a graphical model. In this text, the word *model* will refer to an algebraic model unless otherwise stated.

Using a Calculator to Create Scatterplots

After you have defined the variables and adjusted the data, you are ready to create a scatterplot. To create a scatterplot, or statplot, on the calculator, you will complete the following tasks.

1. Clear all equations in Y= screen.
2. Input the adjusted data into the calculator.
3. Set a window (tell the calculator where to look to see the data).
4. Set up the stat plot and graph it.

Example 1 Creating a scatterplot on the calculator

Create a scatterplot on your graphing calculator for the population data for Texas given in the chart.

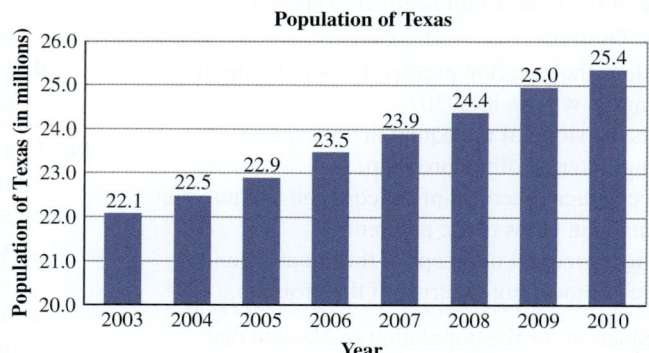

Source: Estimates and projections from the Texas State Data Center.

SOLUTION

First, define the variables as follows.

$$t = \text{Years since 2000}$$
$$P = \text{Population of Texas (in millions)}$$

The adjusted data are shown in the table.

t	Population of Texas (in millions)
3	22.1
4	22.5
5	22.9
6	23.5
7	23.9
8	24.4
9	25.0
10	25.4

To input these data into the calculator, push the STAT button and then  1 (EDIT) for the edit screen. You should get the following two screens.

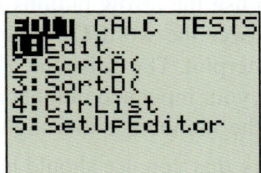 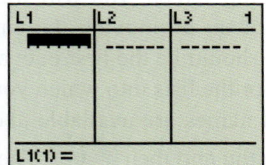

Using Your TI Graphing Calculator

When putting sets of data into the calculator, use the lists in the STAT menu. To get to the lists:

STAT , 1 (Edit)

To clear old data from lists:

Using the arrow buttons, move the cursor up to the title of the list (such as L1) and press CLEAR, ENTER.

Important note:
Do not try to clear a list using the DEL button. This will erase the entire list. If you do erase a list or if all of your lists are not available, go to

STAT , 5 (SetUpEditor),

ENTER , ENTER

When you return to the stat list screen, all the original lists will be back.

The second screen shows the first three lists (L1, L2, and L3) in which you can input data. You will see a black cursor in the first list ready for your first input value. Now enter your first input value and press ENTER. Repeat for all the input data and then use the ▶ button to move over to list 2 and do this again for the output data.

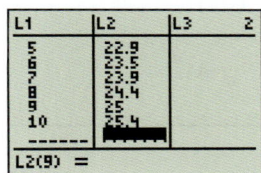

Now that the data have been entered, the window (the graphing area that the calculator is going to show you) needs to be set. Press the  button at the top of the calculator and enter Xmin, Xmax, Ymin, and Ymax values. The best way to choose these values is as follows:

- Xmin = A slightly smaller value than your smallest input
- Xmax = A slightly larger value than your largest input
- Ymin = A slightly smaller value than your smallest output
- Ymax = A slightly larger value than your largest output

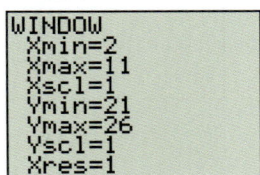

Although there are many windows that will show your data, you must be sure that the min and max you choose include your data values. Now that the window is fixed, set up the statplot and graph it. To set up the statplot, open the STAT PLOTS menu by pressing 2nd , Y= . You should get the following screen.

Using Your TI Graphing Calculator

Common Error Messages

ERR:DIM MISMATCH

This error happens when the numbers of data in your lists are not the same. Go back to the stat edit screen and check your data. This can also happen if your stat-plot is set up to use the wrong lists.

ERR: WINDOW RANGE

This error happens when your window is not set correctly. Usually, the error is that your Xmin is larger than your Xmax or Ymin is larger than your Ymax.

ERR: SYNTAX

This error occurs when you have entered something in the equation incorrectly. Use the goto option to see what the error is. This most often happens when a subtraction is used instead of the negative sign. The negative sign is to the left of the Enter button.

On the STAT PLOTS screen, you will want to select the plot you want to work with. You can either highlight the plot you want and press ENTER or press the number of the plot. By pressing 1, you will get the following plot setup screen.

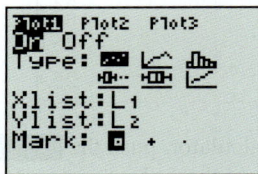

On the plot setup screen, you will want to use the arrow buttons to move around. First, turn the plot on by moving the cursor over the word **On** and pressing ENTER. The Type of plot should be the first one: a scatterplot. The Xlist and Ylist have to have the same names for the lists into which you put your input and output data, respectively. The standard list names are available above the numbers 1 thorough 6. To enter L1, use the 2nd button and then 1. The Mark that you use should be squares or plus signs. The dot is too hard to read. Once your screen is set like the one above, press the GRAPH button on the top right corner of the calculator, and you should get a scatterplot like the following.

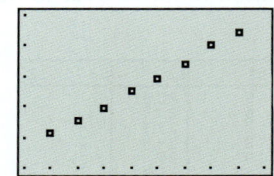

This entire process will get easier with practice and will be a vital part of working many problems throughout this textbook.

PRACTICE PROBLEM FOR EXAMPLE 1

Use a graphing calculator to create a scatterplot of the following set of data.

Year	Population of Arizona (in thousands)
2001	5,296
2002	5,439
2003	5,578
2004	5,740
2005	5,939
2006	6,239
2007	6,432
2008	6,623
2009	6,812
2010	7,000

Source: Arizona Department of Economic Security, Research Administration, Population Statistics Unit.

Finding Linear Models

Because we are creating scatterplots with the calculator, we cannot draw by hand an eyeball best-fit line on the calculator screen, so we want another method to create a model from our data. In Section 1.3, we learned about the slope-intercept form of a line and how to write the equation of a line. In perfect linear relationships, the data will have an equal amount of change between them. In real-life data, this change should still be somewhat equal but will typically have some variation in it. The number of homicides per year was decreasing by about 300 deaths per year. The population of Texas was increasing by about 0.4 million people per year. These amounts of change are somewhat consistent throughout the data. Although linearly related data will have a fairly consistent amount of change for each equal amount of change in inputs, look at the scatterplot to pick two points that will make a good eyeball best-fit line. We will combine the steps to find an equation of a line from Section 1.5 with the scatterplots to find a linear model and check its reasonableness.

Year	Number of Deaths
1994	3532
1995	3262
1996	2894
1997	2601
1998	2283
1999	2093

Source: Centers for Disease Control and Prevention.

Modeling Steps
1. Define the variables and adjust the data (if needed).
2. Create a scatterplot.
3. Select a model type.
4. **Linear Model:** Pick two points and find the equation of the line.
5. Write the equation of the model.
6. Check the model by graphing it with the scatterplot.

Year	Population of Texas (in millions)
2003	22.1
2004	22.5
2005	22.9
2006	23.5
2007	23.9
2008	24.4
2009	25.0
2010	25.4

Source: Estimates and projections from the Texas State Data Center.

Example 2 Finding a linear model

Find an equation for a model of the population of Texas data in Example 1.

SOLUTION

Step 1 Define the variables and adjust the data (if needed).

We will define the variables the same way as in example 1, so that we have:

t = Years since 2000

P = Population of Texas (in millions)

The adjusted data are shown below.

t	Population of Texas (in millions)
3	22.1
4	22.5
5	22.9
6	23.5
7	23.9
8	24.4
9	25.0
10	25.4

Source: Estimates and projections from the Texas State Data Center.

Step 2 Create a scatterplot.

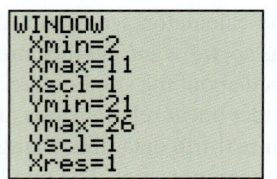 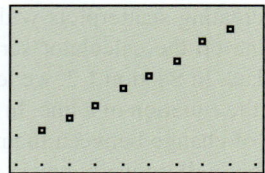

Step 3 Select a model type.
These data seem to be linearly related. Find a linear model.

Step 4 **Linear Model:** Pick two points and find the equation of the line.
Using a clear ruler, choose the second point and the last point because they seem to line up well with the rest of the data. First, calculate the slope.

$$(4, 22.5) \quad \text{and} \quad (10, 25.4)$$

$$m = \frac{25.4 - 22.5}{10 - 4}$$

$$m = \frac{2.9}{6}$$

$$m \approx 0.483$$

The data points given in the table have one decimal place so we should round to two or more decimal places. In this case, we are rounding to two additional decimal places.

Substitute the slope and one of the points into the point-slope formula and find the equation of the line. We could also use the slope-intercept form, $y = mx + b$, as we did in Example 1.

$$P - 22.5 = 0.483(t - 4)$$ *Substitute the slope and a point.*
$$P - 22.5 = 0.483t - 1.932$$
$$\underline{+22.5 \qquad\qquad +22.5}$$ *Isolate P to put the equation into slope-intercept form.*
$$P = 0.483t + 20.568$$

Step 5 Write the equation of the model.
The model is

$$P = 0.483t + 20.568$$

Step 6 Check the model by graphing it with the scatterplot on the calculator.
To check the model, enter the model into the Y= screen and graph it with the data. To do this, press the [Y=] button and enter the equation using the [X,T,θ,n] button for the input variable. After you enter the equation, press the [GRAPH] button to view the graph.

 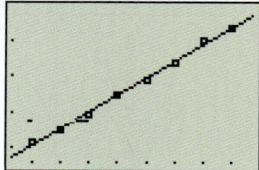

This equation gives us a line that comes close to all the points in the data and looks like a good fit. The points that are not on the model are well balanced on the top and bottom of the line. If the line did not fit well, we could adjust the slope or vertical intercept to try to get a better fit.

PRACTICE PROBLEM FOR EXAMPLE 2

Find a model for the population of Arizona data from Example 1 Practice Problem.

SECTION 1.6 Finding Linear Models

Year	Population of Arizona (in thousands)
2001	5,296
2002	5,439
2003	5,578
2004	5,740
2005	5,939
2006	6,239
2007	6,432
2008	6,623
2009	6,812
2010	7,000

Source: Arizona Department of Economic Security, Research Administration, Population Statistics Unit.

Once we have found a model, it can be used to extrapolate information about the situation. The characteristics of the line can be interpreted just as we did for equations in previous sections. We can also use the equation to find the domain and range of the model instead of estimating values from the graph. As with the graphical models we found in Section 1.2, we will determine a domain based on the given data, being careful to avoid model breakdown. The range will then be calculated using the model by substituting the inputs that will give us the lowest and highest output values based on the given domain.

Example 3 Finding and using a linear model

The following chart gives the revenues in the Asia Pacific countries for Nike, Inc. for various years.

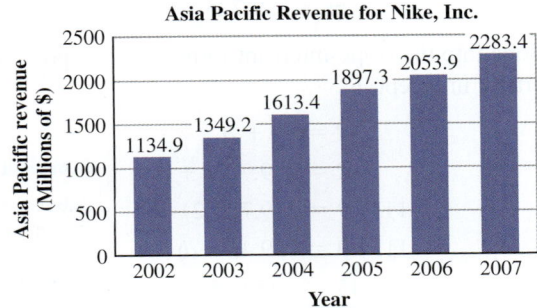

Source: Nike, Inc. Annual Reports, 2004 and 2007.

a. Find an equation for a model of these data.

b. Using your model, estimate the Asia Pacific revenues for Nike in 2010.

c. What is the slope of your model? What does it mean in regard to Nike's revenue?

d. Determine a reasonable domain and range for the model.

SOLUTION

a. **Step 1** Define the variables and adjust the data (if needed).

t = Years since 2000

R = Asia Pacific revenues for Nike, Inc. (in millions of dollars)

Adjust the data.

t	Asia Pacific Revenues (in millions of $)
2	1134.9
3	1349.2
4	1613.4
5	1897.3
6	2053.9
7	2283.4

Step 2 Create a scatterplot.

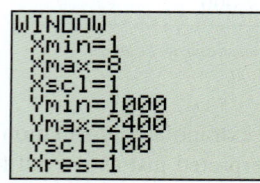 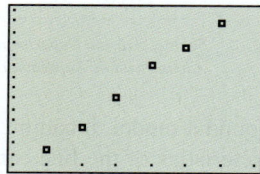

Step 3 Select a model type.
These data seem to be linearly related, so find a linear model.

Step 4 Linear Model: Pick two points and find the equation of the line.
Using a clear ruler, choose the first point and the next to last point because they seem to line up well with the rest of the data. First, calculate the slope.

$$(2, 1134.9) \quad \text{and} \quad (6, 2053.8)$$

$$m = \frac{2053.8 - 1134.9}{6 - 2}$$

$$m = \frac{918.9}{4}$$

$$m = 229.725$$

Substitute the slope into the slope-intercept form. Use one point to find the output value for the vertical intercept, b.

$$R = mt + b$$
$$R = 229.725t + b \quad \text{Substitute in the slope for } m.$$
$$1134.9 = 229.725(2) + b \quad \text{Substitute the point}$$
$$1134.9 = 459.45 + b \quad (2, 1134.9)$$
$$\underline{-459.45 \quad -459.45} \quad \text{Subtract 459.45 from both sides.}$$
$$675.45 = b$$

Step 5 Write the equation of the model.
Using the slope and b found in step 4, write the equation.

$$R = 229.725t + 675.45$$

Step 6 Check the model by graphing it with the scatterplot on the calculator.
To check the model, put the equation into the Y= screen and graph it with the scatterplot to see how it fits the data.

 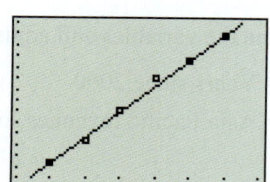

This equation gives a line that comes close to all the points in the data and looks like a good fit. The points that are not on the model are well balanced on the top and bottom of the line. If the line did not fit well, you could adjust the slope or vertical intercept to try to get a better fit.

b. The year 2010 is represented by $t = 10$, so substitute 10 for t and solve for R.

$$R = 229.725(10) + 675.45$$
$$R = 2972.7$$

In 2010, Nike, Inc. had about $2972.7 million revenue from the Asia Pacific countries.

c. The slope for this model is

$$m = \frac{229.725}{1} = \frac{\$229.725 \text{ million}}{1 \text{ year}}$$

The Asia Pacific revenues for Nike, Inc. are increasing by approximately $229.725 million per year.

d. The input values from the data go from 2 to 7. Expand beyond the data a few years in both directions. One possible domain of the model is [0, 10] or $0 \le t \le 10$. Now the range can be found by looking for the lowest and highest points on the model within that domain. On the graph, the lowest point within the domain occurs when $t = 0$. Since you now have an equation, you can calculate the output for this point.

$$R = 229.725(0) + 675.45$$
$$R = 675.45$$

The highest point within this domain occurs at $t = 10$. Using the equation, you get

$$R = 229.725(10) + 675.45$$
$$R = 2972.7$$

When these output values are used, the range is [675.45, 2972.7] or $675.45 \le R \le 2972.7$.

Many mistakes that occur in the modeling process can be avoided by checking your work along the way. When calculating the slope in Step 4, be sure to think about the slope you found and consider whether it agrees with the trend that you see in your scatterplot. If your data are decreasing from left to right, the slope should be negative. If the data are increasing from left to right, the slope should be positive.

PRACTICE PROBLEM FOR EXAMPLE 3

The following chart gives the total prize money given out at professional rodeo events in the United States.

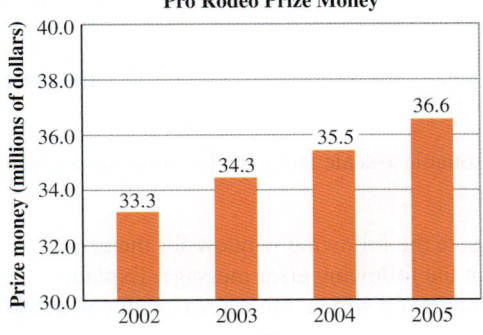

Source: Statistical Abstract of the United States, 2007.

88 CHAPTER 1 Linear Functions

a. Find an equation for a model of these data.

b. Using your model, estimate the total prize money given out in 2009.

c. What is the slope of your model? What does it mean in this situation?

d. Determine a reasonable domain and range for this model.

1.6 Exercises

For Exercises 1 through 8, choose the window that will best display the given data in a scatterplot.

1.

x	y
2	5
4	12
10	45
12	58

a. Xmin = 2, Xmax = 12
   Ymin = 5, Ymax = 58
b. Xmin = 1, Xmax = 13
   Ymin = 0, Ymax = 60
c. Xmin = −10, Xmax = 20
   Ymin = −10, Ymax = 80

2.

x	y
8	10
16	20
20	25
35	40

a. Xmin = 0, Xmax = 60
   Ymin = 0, Ymax = 80
b. Xmin = 8, Xmax = 35
   Ymin = 10, Ymax = 40
c. Xmin = 5, Xmax = 38
   Ymin = 5, Ymax = 45

3.

x	y
−20	1.2
−17	1.5
−12	1.7
−9	1.8

a. Xmin = −21, Xmax = −8
   Ymin = 0, Ymax = 5
b. Xmin = −30, Xmax = 10
   Ymin = 1, Ymax = 5
c. Xmin = −22, Xmax = −8
   Ymin = 1, Ymax = 2

4.

x	y
−42	−3.25
−36	−3.27
−22	−3.28
−12	−3.29

a. Xmin = −45, Xmax = −10
   Ymin = −3.3, Ymax = −3.24
b. Xmin = −10, Xmax = −45
   Ymin = −2, Ymax = −3
c. Xmin = −45, Xmax = −10
   Ymin = −4, Ymax = −3

5.

x	y
100	50
115	200
130	600
135	1000

a. Xmin = 50, Xmax = 200
   Ymin = 0, Ymax = 1050
b. Xmin = 90, Xmax = 145
   Ymin = 40, Ymax = 1010
c. Xmin = 95, Xmax = 140
   Ymin = 0, Ymax = 1050

6.

x	y
0	2002
1500	2010
2800	2015
8700	2018

a. Xmin = 0, Xmax = 8700
   Ymin = 1000, Ymax = 3000
b. Xmin = −20, Xmax = 8720
   Ymin = 1950, Ymax = 2050
c. Xmin = −500, Xmax = 9500
   Ymin = 2000, Ymax = 2020

7.

x	y
2.25	120
2.26	130
2.27	140
2.28	150

a. Xmin = 2.245, Xmax = 2.285
   Ymin = 115, Ymax = 155
b. Xmin = 2.2, Xmax = 2.33
   Ymin = 100, Ymax = 175
c. Xmin = 2, Xmax = 3
   Ymin = 90, Ymax = 200

8.

x	y
−1.1	−2000
−0.9	−3000
−0.5	−4000
−0.2	−5000

a. Xmin = 0, Xmax = −2
   Ymin = −1000, Ymax = −6000
b. Xmin = −1.5, Xmax = 0
   Ymin = −5500, Ymax = −1500
c. Xmin = −1.2, Xmax = −0.1
   Ymin = −5400, Ymax = −1600

For Exercises 9 through 16, the answers may vary.

9. Give a reasonable x-scale and y-scale for the data given in Exercise 1.

10. Give a reasonable x-scale and y-scale for the data given in Exercise 2.

11. Give a reasonable x-scale and y-scale for the data given in Exercise 3.

12. Give a reasonable x-scale and y-scale for the data given in Exercise 4.

13. A student gave the following window for the given data and got the following error message. Explain what the student did wrong. Give a reasonable window for the data.

x	y
−40	−15
−25	10
−20	20
−10	30

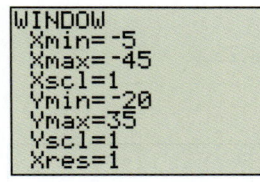 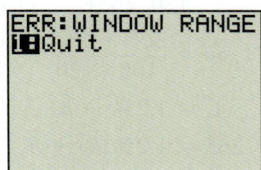

14. A student gave the following window for the given data and got a blank graph. Explain what the student did wrong. Give a reasonable window for the data.

x	y
100	4
300	6
500	8
700	10

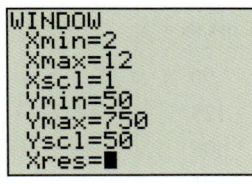 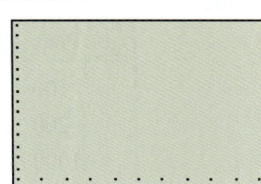

15. The revenue for Quick Tire Repair, Inc. is given in the table.

Year	Revenue*
2006	608
2007	611
2008	616
2009	620
2010	625

(in thousands of $)

a. Find an equation for a model of these data.
b. Give a reasonable domain and range for this model.
c. According to your model, when will Quick Tire Repair, Inc. revenue be $700 thousand?
d. What is the slope of your model? Explain its meaning in regard to the revenue.

16. The time it takes to make personalized chocolate bars is given in the table.

Chocolate Bars	Time (in hours)
20	6.25
50	12.0
100	22.25
250	52.0
500	102.5

a. Find an equation for a model of these data.
b. Give a reasonable domain and range for this model.
c. According to your model, how many chocolate bars can they produce in 75 hours?
d. What is the slope of your model? Explain its meaning in this situation.

For Exercises 17 through 20, graph the data and the equation on your calculator. Adjust m and/or b to get an eyeball best-fit. Your answers will vary.

17. $y = 2.5x + 7$

x	y
3	15.5
5	20.5
8	28.0
11	35.5

18. $y = -4x - 3.5$

x	y
−3	7
2	−13
6	−29
9	−41

19. $y = -1.25x + 20$

x	y
−6	24.5
−2	21.5
0	20.0
4	17.0

20. $y = 0.6x - 3$

x	y
−2	−4.4
0	−3.0
3	−0.9
8	2.6

For Exercises 21 and 22, the answers may vary.

21. The world's production of beef and pork is given in the following chart.

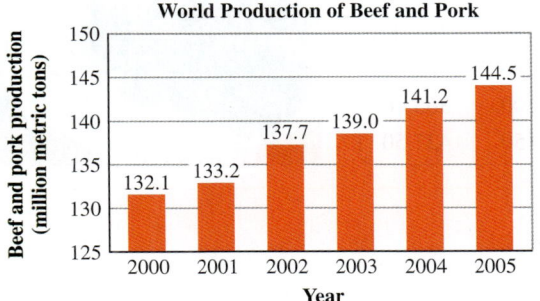

Source: Adapted from Mark R. Vogel, "Where's The Beef?," Found at http://www.seekingsources.com/cuts_of_beef.htm. Reprinted with Permission from Mark R. Vogel.

a. Find an equation for a model of these data.
b. According to your model, what was the world production of beef and pork in 2006?
c. When will the world production of beef and pork reach 175 million metric tons?
d. Give a reasonable domain and range for this model.
e. What is the slope of your model? Explain its meaning in the terms of the problem.

22. The gross profit for Exxon Mobil Corporation for various years is given here.

Year	Gross Profit (in billions of $)
2002	37
2003	47
2004	61
2005	75
2006	84

Source: http://finance.google.com

a. Find an equation for a model of these data.
b. Give a reasonable domain and range for this model.
c. According to your model, when will Exxon Mobil Corporation's gross profit be 100 billion dollars?
d. What is the slope of your model? Explain its meaning in regard to the gross profit.
e. According to your model, what was Exxon Mobil Corporation's gross profit in 2008?

For Exercises 23 and 24, decide whether the student below found a reasonable model for the data provided. If not, explain what the student did wrong. Answers may vary.

The cost to produce chocolate-dipped key lime pie on a stick bars is given in the table.

Bars	Cost ($)
10	252.50
20	265.00
50	302.50
100	365.00
150	427.50

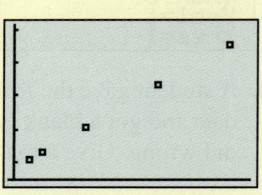

23.

Vivian

Let
k = Number of key lime pie on a stick bars produced
C = Cost in dollars to produce the bars
$(20, 265)$ $(100, 365)$
$m = \dfrac{100 - 20}{365 - 265} = \dfrac{80}{100} = 0.8$
$C = 0.8k + b$
$265 = 0.8(20) + b$
$265 = 16 + b$
$249 = b$
$C = 0.8k + 249$

24.

Chaya

Let
k = Number of key lime pie on a stick bars produced
C = Cost in dollars to produce the bars
$(20, 265)$ $(100, 365)$
$m = \dfrac{365 - 265}{100 - 20} = \dfrac{100}{80} = 1.25$
$C = 1.25k + b$
$265 = 1.25(20) + b$
$265 = 25 + b$
$\dfrac{265}{25} = 10.6 = b$
$C = 1.25k + 10.6$

For Exercises 25 and 26, decide whether the student found a reasonable model for the data provided. If not, explain what the student did wrong. Answers may vary.

The cost to produce a custom-printed metal pen is given in the table.

Pens	Cost ($)
100	90
200	125
1000	405
3000	1105
5000	1805

25.

Jim

Let
p = Number of custom-printed metal pens produced
C = Cost in dollars to produce the pens
$(100, 90)$ $(3000, 1105)$
$m = \dfrac{90 - 1105}{3000 - 100} = \dfrac{-1015}{2900} = -0.35$
$C = -0.35p + b$
$90 = -0.35(100) + b$
$90 = -35 + b$
$125 = b$
$C = -0.35p + 125$

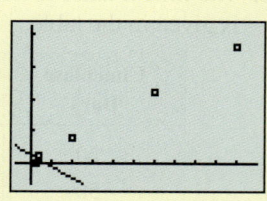

26.

Nolan

Let
p = Number of custom-printed metal pens produced
C = Cost in dollars to produce the pens
$(100, 90)$ $(3000, 1105)$

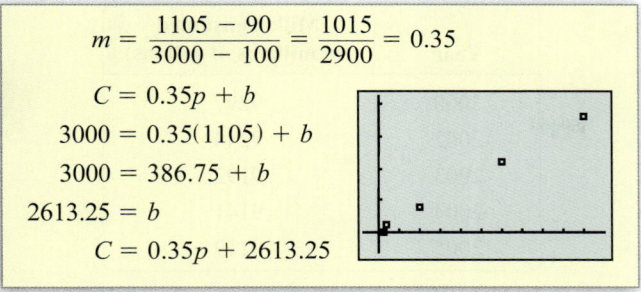

$$m = \frac{1105 - 90}{3000 - 100} = \frac{1015}{2900} = 0.35$$

$$C = 0.35p + b$$
$$3000 = 0.35(1105) + b$$
$$3000 = 386.75 + b$$
$$2613.25 = b$$
$$C = 0.35p + 2613.25$$

For Exercises 27 through 40, the answers may vary.

27. The number of reported cases of the sexually transmitted disease chlamydia reported for several years is given in the table.

Year	Reported Chlamydia Cases (in thousands)
2000	702
2001	783
2002	835
2003	877
2004	929

Source: U.S. Centers for Disease Control and Prevention.

a. Find an equation for a model of these data.
b. What is the slope of this model? Explain its meaning in this situation.
c. Estimate the number of chlamydia cases that were reported in 2006.

28. According to the North American Transportation Statistics Database, the total number of vehicle-miles driven on the road is given in the table.

Year	Vehicle Road Miles (in billions)
2000	2747
2001	2798
2002	2856
2003	2891
2004	2963
2005	2989

Source: North American Transportation Statistics Database.

a. Find an equation for a model of these data.
b. Give a reasonable domain and range for this model.
c. Estimate the total number of vehicle road miles in 2008.
d. What is the slope of your model? Explain its meaning in the terms of this problem.

29. The amount of eggs produced in the United States is given in the following chart.

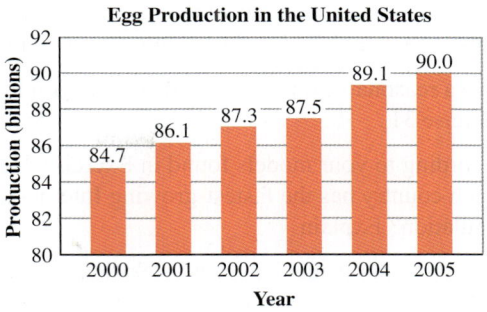

Source: Statistical Abstract of the United States, 2007.

a. Find an equation for a model of these data.
b. According to your model, how many eggs were produced in 2008?
c. Give a reasonable domain and range for this model.
d. What is the slope of your model? Explain its meaning in regard to egg production.

30. The total revenue for Costco Wholesale Corporation is given in the table.

Year	Total Revenue (in millions of $)
2002	37,863
2003	42,546
2004	48,110
2005	52,952
2006	60,151
2007	64,400

Source: http://finance.google.com

a. Find an equation for a model of these data.
b. What is the slope of this model? Explain its meaning in this situation.
c. Estimate the total revenue for Costco Wholesale Corporation in 2010.

While the Internet was growing in popularity in the late 1990's, many countries saw an increase in the number of Internet users. The number of European Internet users, in millions, is given in the following chart. Use these data for Exercises 31 through 34.

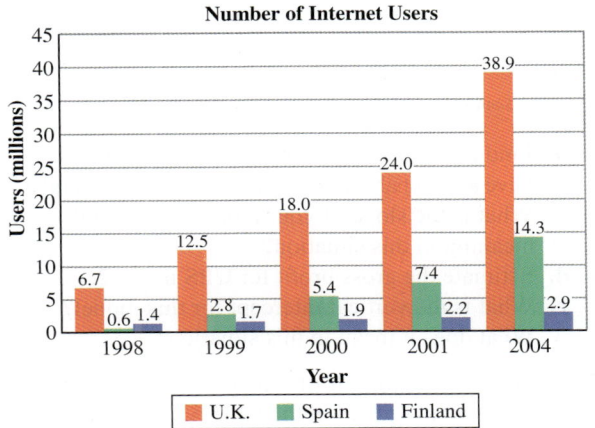

Source: NetStatistica.com and Internet worldstats.com

31. Find equations for models of the number of Internet users in the United Kingdom, Spain, and Finland.

32. Give a reasonable domain for the models you found in Exercise 31.

33. According to your models found in Exercise 31, which country has the fastest-growing Internet user population? Explain.

34. What is the horizontal intercept for Spain's Internet users model found in Exercise 31? Explain its meaning in regard to the number of Internet users.

35. The revenues for FedEx for several years are given in the table.

Year	Revenue (in millions of $)
2000	18,257
2001	19,629
2002	20,607
2003	22,487
2004	24,710
2005	29,363
2006	32,294
2007	35,214

    Source: FedEx.com

    a. Find an equation for a model of these data.
    b. Give a reasonable domain and range for this model.
    c. What is the slope of your model? Explain its meaning in the terms of this problem.
    d. Estimate the revenue for FedEx in 2010.

36. The gross profit for United Parcel Service, Inc. has been increasing steadily for the past several years. The gross profits for UPS are given in the table.

Years Since 2000	Gross Profit (in millions of $)
2	9,702
3	10,401
4	11,254
5	12,807
6	13,820

    Source: http://finance.google.com

    a. Find an equation for a model of these data.
    b. Give a reasonable domain and range for this model.
    c. What is the slope of your model? Explain its meaning in this situation?
    d. Estimate the gross profit for UPS in 2012.
    e. What is the vertical intercept for this model? What does it mean in this situation?

37. The number of gallons of milk consumed in the United States is given in the table.

Year	Milk Consumed (in millions of gallons)
2000	9297
2002	9244
2003	9192
2004	9141
2005	9119

    Source: USDA/Economic Research Service.

    a. Find an equation for a model of these data.
    b. Give a reasonable domain and range for this model.
    c. What is the vertical intercept for this model? Explain its meaning in regard to milk consumed.
    d. In what year did Americans drink 9000 million gallons of milk?
    e. What is the horizontal intercept for your model? Explain its meaning in regard to milk consumed.

38. The total operating expenses for Anheuser-Busch Companies, Inc. is given in the table.

Year	Total Operating Expenses (in millions of $)
2002	10,587
2003	10,947
2004	11,761
2005	12,549
2006	12,998

    Source: http://finance.google.com

    a. Find an equation for a model of these data.
    b. Give a reasonable domain and range for this model.
    c. Estimate the total operating expenses for Anheuser-Busch in 2009.
    d. What is the slope of your model? Explain its meaning in this situation.

39. The amount spent by individuals on health care expenses in the United States is given in the table.

Year	National Health Expenditures by Individuals (in billions of $)
2006	251
2007	266
2008	281*
2009	299*
2010	317*
2011	335*

    *Projected
    Source: Centers for Medicare and Medicaid Services.

    a. Find an equation for a model of these data.
    b. Give a reasonable domain and range for this model.

c. What is the slope of your model? Explain its meaning in the terms of the problem.
d. Estimate the amount spent by individual Americans on health care in 2015.
e. When will the amount spent be $500 billion?

40. Quiksilver, Inc., the makers of the Quiksilver and Roxy clothing lines, has had steadily increasing profits over the past several years. The profits for Quiksilver, Inc. are given in the chart.

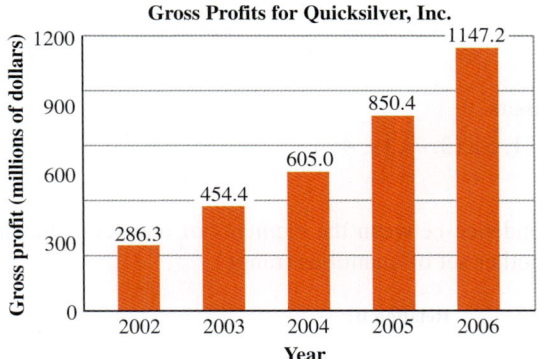

Source: CBS.Marketwatch.com

a. Find an equation for a model of these data.
b. Use your model to estimate the profit for Quiksilver, Inc. in 2009.
c. Give a reasonable domain and range for this model.
d. What is the slope for your model? Explain what this means in regard to Quiksilver's profit.

For Exercises 41 through 52, find an equation for a model of the given data. Answers may vary.

41.
x	y
4	4
7	22
10	40
12	52

42.
x	y
2	−6
5	4
9	9
14	24

43.
x	y
1	5
5	−7
8	−16
13	−31

44.
x	y
4	3
8	−24
20	−51
28	−69

45.
x	y
3	−4
9	−1.75
12	−1
21	1.75

46.
x	y
1	3.4
6	5.4
8	6.6
11	7.0

47.
x	y
−8	4.167
−5	3.500
3	2.167
7	1.333

48.
x	y
−4	9.45
1	5.45
5	2.45
11	−2.45

49.
x	y
10	492
14	552
19	627
27	747

50.
x	y
−9	370
12	625
24	775
35	915

51.
x	y
7	−3755
15	−4083
22	−4370
34	−4862

52.
x	y
2	5301
13	5587
34	6133
51	6575

1.7 Functions and Function Notation

LEARNING OBJECTIVES

- Identify a function.
- Apply the vertical line test for functions.
- Use function notation.
- Identify the domain and range of a function.

Relations and Functions

The way that things in life are related to one another is important to understand. In mathematics, we are also concerned with how different sets are related. These relationships can be as simple as the relationship between a person and his or her height, age, or weight. We could look at the day of the week and the number of work absences at a certain company. We might want to know the names of the brothers and sisters of each student in a class. Another relationship that we might consider is the number of credits each student is taking this semester. All of these represent what we, in mathematics, would call a **relation**. A relation is any connection between the elements of a set of input(s) (domain) and the elements of a set of output(s) (range). It is typically represented by a set of ordered pairs or by using an equation. The equation $y = 5x + 7$ represents a relation that relates the *x*-values with a corresponding *y*-value using arithmetic operations. Using this equation, we can calculate an infinite number of ordered pairs that represent the inputs and outputs of this relation.

> **DEFINITION**
>
> **Relation**   A set of ordered pairs.
>
> Relation A: $(1, 5), (3, 7), (9, 4), (-2, 4), (3, -1)$

A relation represents a correspondence between the elements of one set of quantities (domain) and the elements of another set of quantities (range).

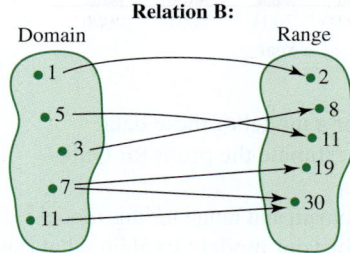

A special type of relation is one in which each input is paired with only one output. That is, when you put in one value, you only get out exactly one value. This type of relation is called a **function** in mathematics. In determining whether a relation is a function, it is important to consider whether each and every input has exactly one output associated with it. In the following example, we will determine which relations are functions and which are not.

> **DEFINITION**
>
> **Function**   A relation in which each input is related to only one output. For each input value in the domain, you must have one and only one output value in the range.

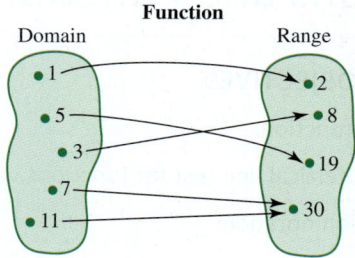

Example 1   Determining whether sets and word descriptions are functions

Determine whether the following descriptions of relations are functions or not. Explain your reasoning.

a. The set $S = \{(1, 3), (5, 7), (7, 9), (15, 17)\}$

b. The set $B = \{(2, 8), (2, 7), (3, 16), (4, 11)\}$

c.
Day	1	2	5	7
Height of Plant (in cm)	3	5	12	17

d.
Age of Student	7	8	7	6	6	9	10
Grade Level	2nd	3rd	3rd	1st	1st	4th	4th

e. The relationship between Monique's age, in days, and her height

f. The advertised prices of Sony 32-inch TVs in this Sunday's newspaper

SOLUTION

In each part, consider whether for each input value, there is exactly one output.

a. The set S is a function, since each input has exactly one output value.

b. The set B is not a function, since the input 2 has two different output values.

c. This table is a function because each day has one plant height associated with it.

d. In this table, the age of the student could be associated with more than one grade level. The 7-year-olds in this table go to either second or third grade, so this relation is not a function.

e. If we consider just one age, Monique will have only one height, so this is a function.

f. Sony 32-inch TVs would be advertised for several different prices, so this is a relation but not a function.

PRACTICE PROBLEM FOR EXAMPLE 1

Determine whether the following descriptions of relations are functions or not. Explain your reasoning.

a.
Units Produced	100	150	200	250	300	350	400
Total Cost (in $)	789	1565	2037	2589	3604	4568	5598

b.
First Name of Student	John	Mary	Mark	Fred	Juan	Karla	John
Number of Credits	10	12	9	15	16	21	4

c. The amount of profit that a company makes each month of a year

d. The population of California each year

In Example 1, functions can be represented by sets of data or words. If you consider the data sets from the last two sections, you will see that they all represent functions. Each input value had exactly one output value. Any set of data can be considered a function if it holds to the requirement that each input has only one output associated with it. Functions can also be represented as formulas or by using graphs. Most of the equations that we will work with in this book will be functions.

When we are given an equation to consider, it may seem harder to determine whether the equation represents a function. When given an equation, look for anything that is out of the ordinary, such as a $\pm$ symbol that might result in two answers for any one input. Most equations and sets of data that we will consider in this book will be functions, but it is best to have ways of recognizing when an equation is not a function. One way to get an idea whether an equation represents a function is to pick an arbitrary input and substitute it into the equation to see whether one or more outputs come out. If the equation gives only one output, it may be a function. We will consider whether there are any values that could be used as inputs that would result in more than one output. If we consider the linear equation $y = 2x + 5$, any x value that we substitute into this equation will give us only one output. For example, $x = 3$ results in

$$y = 2(3) + 5$$
$$y = 11$$

The input $x = 3$ results in the output $y = 11$. If we picked any other x-value, it would also result in only one output. Because each input results in only one output, the equation represents a function. An equation such as $x = y^2$ would not satisfy the condition that each input have only one output.

$$4 = y^2$$
$$4 = (-2)^2 \qquad 4 = (2)^2$$
$$y = -2 \qquad y = 2$$

In this example, we can see the input $x = 4$ results in two outputs: $y = 2$ and $y = -2$. Therefore, this equation is not a function.

Another way to determine whether or not an equation is a function is to look at its graph. Consider the following graph

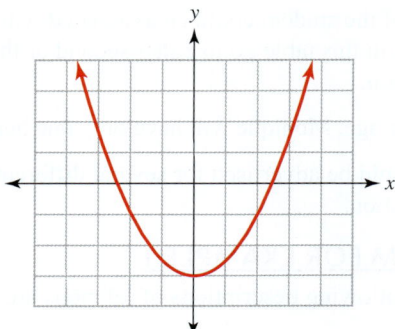

Every point that lies on this curve represents an input (x) and an output (y). Using this graph, we will determine whether each input (x) has exactly one output (y) associated with it. To determine whether an input (x) has only one output (y), we pick an x-value on the horizontal axis, draw an imaginary vertical line at that location, and see how many times the vertical line crosses the curve.

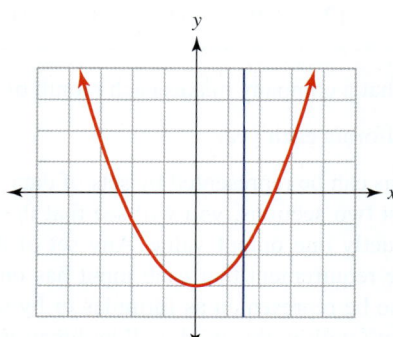

In this example, we chose the input $x = 1.5$ and drew a vertical line, which crosses the graph in only one place. If every vertical line that we draw crosses the graph only once, then we have the graph of a function. In this graph, all of the vertical lines cross the graph only once. This is the graph of a function.

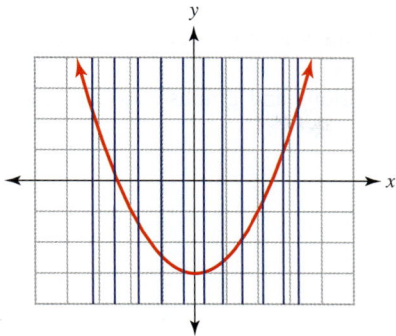

This process of testing vertical lines through a graph is called **the vertical line test.** This test can be used with most equations as long as we can graph them or put them in a graphing calculator and see its graph. When using the vertical line test on a calculator, we must be sure to get a viewing window that shows the overall characteristics of the graph, or we might incorrectly decide that the graph passes the vertical line test. Although the graphing calculator is a great tool, it can show us only what we ask it to. We can mislead ourselves if we are not thorough and careful.

> ### DEFINITION
>
> **The vertical line test for a function** If any vertical line intersects a graph in at most one point, the graph represents a function.

An example of the vertical line test proving that a curve is not a function is the following graph.

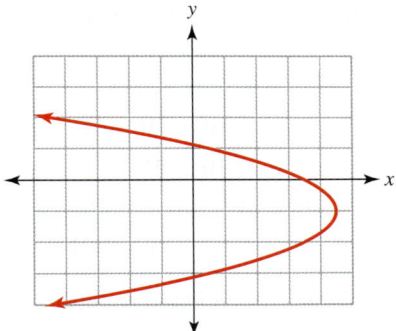

Drawing an arbitrary vertical line through the graph shows that it intersects the curve more than once. Therefore, this is not the graph of a function.

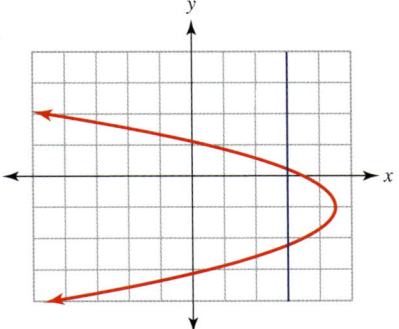

> **Example 2** Determining whether equations and graphs are functions

Consider the following equations and graphs and determine whether or not they are functions.

a. $P = 2.57t + 65$

b. $W = 2g^2 + 5g - 9$

c. $y = 2x \pm (6x - 9)$

d.

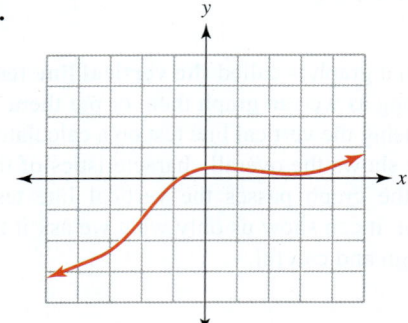

e.

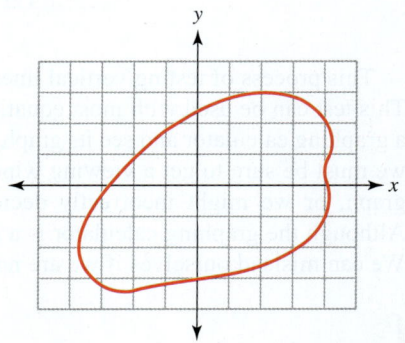

SOLUTION

a. This is a linear equation, and for each input t, there is a single output P. Therefore, this equation represents a function.

b. This equation is not linear, yet it still has only one output, W, associated with each input value, g. Therefore, it is a function.

c. This equation has a $\pm$ symbol, which means that almost all inputs will result in more than one output. For example, if $x = 5$,

$$y = 2(5) \pm (6(5) - 9) \quad \text{Substitute } x = 5.$$
$$y = 10 \pm (30 - 9)$$
$$y = 10 \pm 21$$

| $y = 10 + 21$ | $y = 10 - 21$ | The $\pm$ symbol means that we write two equations. |
| $y = 31$ | $y = -11$ | We get two results. |

This equation gives two outputs for an input, so this equation does not represent a function.

d. This graph passes the vertical line test, so it does represent a function.

e. This graph does not pass the vertical line test, so it does not represent a function.

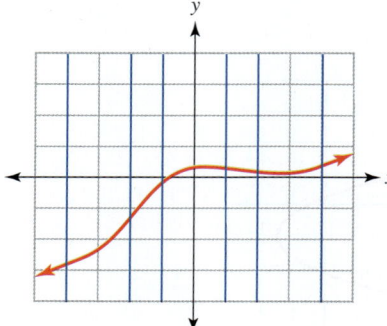

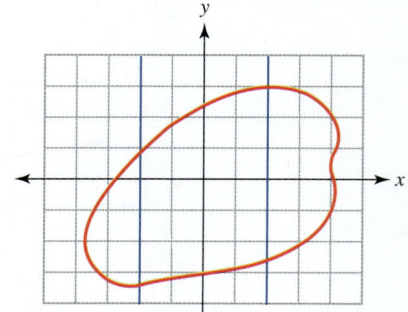

PRACTICE PROBLEM FOR EXAMPLE 2

Determine whether or not the following tables, graphs, and equations represent functions. Explain your reasoning.

a.

Name of Student	Mary	Mark	Karla	Fred	Mark
Gender	Female	Male	Female	Male	Male

b.
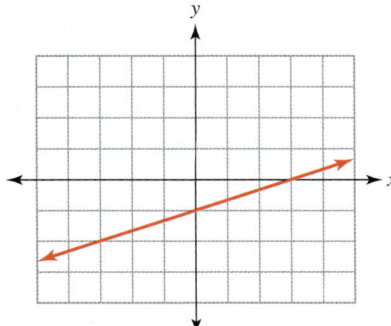

c.
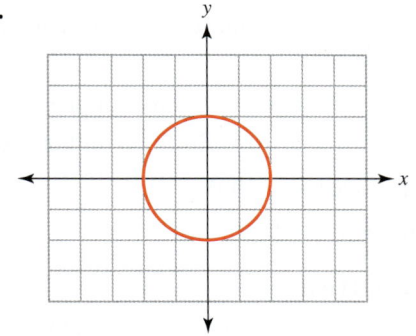

d. $C = 3.59u + 1359.56$

e. $H = 17.125 \pm \sqrt{3.5m}$ This can be entered in the calculator as two equations: one using the plus and the other using the minus.

Function Notation

Function notation was developed as a shorthand way of providing a great deal of information in a very compact form. If variables are defined properly with units and clear definitions, then function notation can be used to communicate what you want to do with the function and what input and/or output values you are considering.

Let's define the following variables:

$$P(t) = \text{Population of Hawaii (in millions)}$$
$$t = \text{Years since 2000}$$

Then the population of Hawaii at time t can be represented by the following function.

$$P(t) = 0.013t + 1.210$$

$P(t)$ is read "P of t" and represents a function named P that depends on the variable t. In real-world applications, the variable in the same position as P represents the output variable, and the variable in the same position as t represents the input variable. You can see this if you consider the P, outside the parentheses, as the output, and the t, inside the parentheses, as the input. There is not much of a difference between this function notation and the equation $P = 0.013t + 1.210$. When you want to use this equation to represent the population as a function of time, it becomes much clearer to communicate using the function notation. If you are given the variable definitions above, you can make the following statements or ask the following questions in a couple of ways.

a. In words:
   i. Use the given equation to determine what the population of Hawaii was in 2005.
   ii. Use the given equation to determine when the population of Hawaii will be 4 million.

b. Using function notation, these same statements can be written as follows:
   i. Find $P(5)$.   ii. Find t such that $P(t) = 4$.

Using the function notation allows you to communicate what the input variable or output variable is equal to without words. $P(5)$ is asking you to substitute 5 for the input variable t and determine the value of the function $P(t)$. $P(t) = 4$ tells you that the output variable P is equal to 4 and directs you to determine the value of the input variable t that gives you a population of 4 million people.

Function notation can be a simple way to communicate information in a short way, but you must be careful when interpreting the information. Be sure to know how the variables are defined and use these definitions as a basis for interpreting any results.

Example 3 Interpreting function notation in an application

Given the following definitions, write sentences interpreting the following mathematical statements.

$$G(t) = \text{Number of guests at a local beach resort during year } t$$
$$P(b) = \text{Profit, in millions of dollars, from the sale of } b \text{ bombers}$$

a. $G(2010) = 2005$

b. $P(10) = 7$

SOLUTION

a. In this notation, it is important to consider the location of each number. The 2010 is inside the parentheses, so it must be the value of t. Therefore, 2010 must be the year. Because 2005 is what $G(t)$ is equal to, this must represent the number of guests at the beach resort. The final interpretation might say, "In 2010, there were 2005 guests at this local beach resort."

b. The number of bombers must be 10, since it is in the parentheses, and 7 must be the profit in millions. The profit from the sale of 10 bombers is 7 million dollars.

PRACTICE PROBLEM FOR EXAMPLE 3

Given the following definitions, write sentences interpreting the following mathematical statements.

$$C(m) = \text{Cost, in hundreds of dollars, for producing } m \text{ Miracle Mops}$$
$$P(t) = \text{Population of Michigan, in millions, } t \text{, years since 2000.}$$

a. $C(2500) = 189$

b. $P(10) = 10.4$

If you are careful in determining which number represents which variable, you can interpret the meaning of the results by referring back to the definitions of each variable. Be sure to pay close attention to the units involved in each problem. If the profit from making bombers were measured in dollars and not millions of dollars, the bombers probably would not be made.

Writing Models in Function Notation

Function notation will change the modeling process given in Section 1.6 only in that you should now write your model in function notation and use that notation in showing your work. This will change step 5 of the modeling process to say, "Write the equation of the model using function notation."

> **Modeling Steps**
> 1. Define the variables and adjust the data (if needed).
> 2. Create a scatterplot.
> 3. Select a model type.
> 4. **Linear Model:** Pick two points and find the equation of the line.
> 5. Write the equation of the model using function notation.
> 6. Check the model by graphing it with the scatterplot.

SECTION 1.7 Functions and Function Notation

Example 4 Models using function notation

The following chart gives the population of Colorado, in millions, for various years.

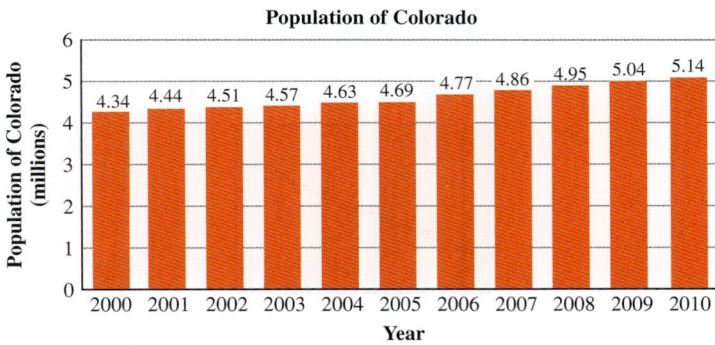

Source: Colorado Department of Local Affairs.

Let $P(t)$ be the population of Colorado, in millions, t years since 2000.

a. Find an equation for a model of these data. Write your model in function notation.

b. Determine a reasonable domain and range for your model.

c. Find $P(15)$ and interpret its meaning in regard to the population of Colorado.

d. Find when $P(t) = 5.35$ and interpret its meaning in regard to the population of Colorado.

SOLUTION

a. Step 1 Define the variables and adjust the data.

$$t = \text{Years since 2000}$$
$$P(t) = \text{Population of Colorado in millions}$$

Step 2 Create a scatterplot.

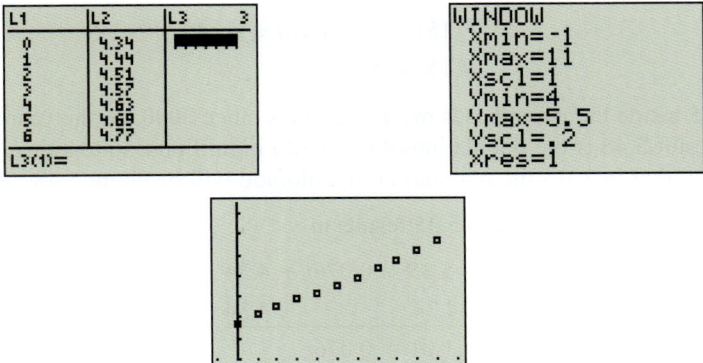

Steps 3 and 4 These data appear to be linear, so pick two points and find the equation of the line. Try the first and ninth points, since a line through these points would appear to go through the data well.

$$(0, 4.34) \quad \text{and} \quad (8, 4.95)$$
$$m = \frac{4.95 - 4.34}{8 - 0}$$
$$m \approx 0.076$$

We have calculated the slope, and the first point, (0, 4.34), is a vertical intercept. Write the equation using these values.

$$P = 0.076t + 4.34$$

Step 5 Write the equation of the model using function notation.

$$P(t) = 0.076t + 4.34$$

Step 6 Check the model by graphing it with the scatterplot.

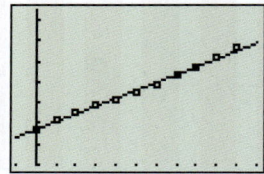

This model fits the data pretty well, with all the points that it misses well balanced above and below the line. If we choose different points, we will usually get a slightly different model. Remember that we are looking for a good eyeball best-fit line.

b. From the data given and the reasonable fit that we obtained with our model, we should be able to expand our domain beyond the data, giving us a possible domain of $[-2, 12]$ or $-2 \leq t \leq 12$. As we look at the graph, the lowest output value is going to be on the left side of the domain, and the highest output value is going to be on the right side of the domain. This tells us that we want to evaluate the function at $t = -2$ and $t = 12$ to get the lowest and highest range values, respectively.

$$P(-2) = 0.076(-2) + 4.34$$
$$P(-2) = 4.188$$

$$P(12) = 0.076(12) + 4.34$$
$$P(12) = 5.252$$

Because $P(-2) = 4.188$, the lowest point, and $P(12) = 5.252$, the highest point, the range corresponding to the domain is $[4.188, 5.252]$, or $4.188 \leq P \leq 5.252$.

c. Substituting 15 for the input variable results in

$$P(15) = 0.076(15) + 4.34$$
$$P(15) = 5.48$$

The 15 inside the parentheses means 15 years since 2000, so the year is 2015. The result 5.48 is the population of Colorado in millions, so together, they indicate that in 2015 the population of Colorado will be about 5.48 million.

d. Setting the function equal to 5.35 results in

$$\begin{aligned} 5.35 &= 0.076t + 4.34 \\ -4.34 & -4.34 \\ \hline 1.01 &= 0.076t \\ \frac{1.01}{0.076} &= \frac{0.076t}{0.076} \\ 13.289 &\approx t \end{aligned}$$

Subtract 4.34 from both sides of the equation.

Divide both sides of the equation by 0.076.

$$P(13.289) = 0.076(13.289) + 4.34$$
$$P(13.289) \approx 5.35$$

Check the answer.

The answer works.

This means that the population of Colorado will reach approximately 5.35 million in about 2013.

The most common letters used to name functions in mathematics are f, g, and h. The functions $f(x)$, $g(x)$, and $h(x)$ will often replace the output variable y. Using different letters helps to distinguish between different equations. When the following equations are written in slope-intercept form, we cannot easily distinguish between them.

$$y = 2x + 7 \qquad y = -4.7x - 8.6 \qquad y = 3x - 6$$

If these equations are written using function notation, we can more easily distinguish between them.

$$f(x) = 2x + 7 \qquad g(x) = -4.7x - 8.6 \qquad h(x) = 3x - 6$$

Now, if we want to evaluate $f(x)$ at $x = 4$, we can write $f(4)$ and the reader would know which function to use and what to substitute in for x.

$$f(4) = 2(4) + 7$$
$$f(4) = 15$$

Example 5 Using function notation

Let

$$f(x) = -3x + 5 \qquad g(x) = 2.5x - 9.7 \qquad h(x) = 4x^2 - 19$$

Find the following.

a. $f(7)$

b. $h(3)$

c. x such that $g(x) = 12.3$

d. x such that $f(x) = 11$

SOLUTION

a. The input value is 7. Substitute 7 for x and solve.

$$f(7) = -3(7) + 5$$
$$f(7) = -21 + 5$$
$$f(7) = -16$$

b. Using the function h, substitute 3 for the input variable x.

$$h(3) = 4(3)^2 - 19$$
$$h(3) = 4(9) - 19$$
$$h(3) = 36 - 19$$
$$h(3) = 17$$

c. Given the value of the function 12.3, set $g(x)$ equal to 12.3 and solve for the input variable.

$$12.3 = 2.5x - 9.7$$
$$\underline{+9.7 \qquad\quad +9.7}$$
$$22 = 2.5x$$
$$\frac{22}{2.5} = \frac{2.5x}{2.5}$$
$$8.8 = x$$
$$g(8.8) = 2.5(8.8) - 9.7 \qquad \text{Check the answer.}$$
$$g(8.8) = 12.3 \qquad\qquad\qquad \text{The answer works.}$$

d. Given the value of the function 11, set $f(x)$ equal to 11 and solve for the input variable.

$$11 = -3x + 5$$
$$\underline{-5 \qquad\qquad -5}$$
$$6 = -3x$$
$$\frac{6}{-3} = \frac{-3x}{-3}$$
$$-2 = x$$
$$f(-2) = -3(-2) + 5 \qquad \text{Check the answer.}$$
$$f(-2) = 11 \qquad \text{The answer works.}$$

PRACTICE PROBLEM FOR EXAMPLE 5

Let

$$f(x) = -4.25x + 5.75 \qquad g(x) = 4x + 8$$

Find the following.

a. $f(3)$

b. x such that $g(x) = 20$

Example 6 **Using function notation**

Use the graph to estimate the following.

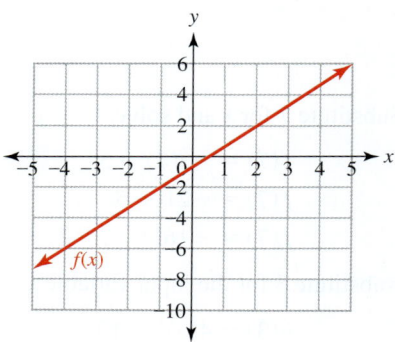

a. $f(-4)$

b. x such that $f(x) = -2$

SOLUTION

a. -4 is inside the parentheses, so it is an input or x-value, and we want to find the output, or y-value. According to the graph, when $x = -4$, the line has a y-value of -6. Therefore, $f(-4) = -6$.

b. The line has an output of -2 when x is -1.

Domain and Range of Functions

When choosing a domain and range for a model in an application, you must consider model breakdown something to be avoided. When you are considering the domain and range of a function that is not in an application, model breakdown will not be considered. This makes the domain and range much less restricted and allows for as broad a domain and range as possible. The only restrictions to the domain of a function will be any real number that results in the function being undefined. Because all nonvertical linear equations are defined for all real numbers, their domain will be all real numbers. Therefore, because a linear graph will continue to go up and down forever, the range of any nonhorizontal linear function will also be all real numbers. All real numbers can also be expressed by using interval notation: $(-\infty, \infty)$.

> **What's That Mean?**
>
> **Infinity**
>
> To infinity and beyond! This well-known saying has helped many people to know the word *infinity*. In mathematics, infinity is a quantity that is unlimited.
>
> We use the infinity symbol, ∞, or the negative infinity symbol, $-\infty$, in interval notation to indicate that an interval is unlimited.

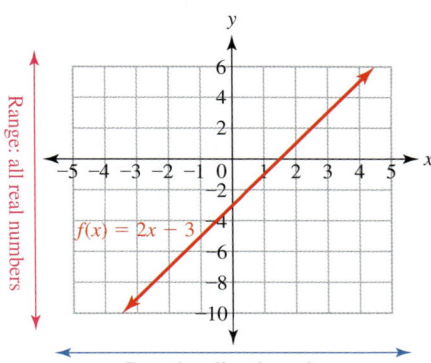

> **DEFINITIONS**
>
> **Domain of a function** The set of all real numbers that make the function defined. Avoid division by zero and negatives under a square root.
>
> **Range of a function** The set of all possible output values resulting from values of the domain.
>
> Domain and range of linear functions that are not vertical or horizontal:
>
> Domain: all real numbers or $(-\infty, \infty)$
>
> Range: all real numbers or $(-\infty, \infty)$

The horizontal line $y = k$ will have a domain of all real numbers but will have a range of only $\{k\}$. This is because no value of x will affect the output of the function, which is always k. A vertical line $x = k$ has a domain and range that are just the reverse. The domain will be $\{k\}$, since that is the only value x can have, but the range will be all real numbers.

Example 7 Domain and range of functions

Determine the domain and range of the following functions.

a. $f(x) = 5x + 2$ **b.** $g(x) = -0.24x + 9$ **c.** $h(x) = 10$

SOLUTION

a. Because this is a linear function, all real number inputs will result in real number outputs. Therefore, the domain is $(-\infty, \infty)$, and its range is also $(-\infty, \infty)$.

b. Because this is also a linear function, its domain is $(-\infty, \infty)$, and its range is $(-\infty, \infty)$.

c. This function represents a horizontal line, so its domain is still $(-\infty, \infty)$, but the range is $\{10\}$, since the only output value for this function is 10.

PRACTICE PROBLEM FOR EXAMPLE 7

Determine the domain and range of the following functions.

a. $f(x) = -2x + 7$ **b.** $g(x) = -8$ **c.** $h(x) = 6x - 2$

1.7 Exercises

For each of the relations in Exercises 1 through 10, specify the input and output variables and their definition and units. Determine whether or not each relation is a function. Explain your answer.

1. $G(a)$ = Grade level of students when they are a years old
2. $S(a)$ = Salary, in dollars, of a person who is a years old
3. $H(a)$ = Heights, in inches, of children attending Mission Meadows Elementary School who are a years old
4. $P(w)$ = Postage, in dollars, it takes to mail a first-class package weighing w ounces
5. $I(t)$ = Interest earned, in dollars, on an investment after t years
6. $P(t)$ = Price, in dollars, of Nike shoes during the year t
7. $S(y)$ = Song at the top of the pop charts during the year y
8. $B(m)$ = Number of students in this class who have a birthday during the mth month of the year
9. $T(t)$ = Amount of taxes, in dollars, you paid in year t
10. $A(m)$ = Number of tourists visiting Arizona, in thousands, during the mth month of 2012

Determine whether the tables in Exercises 11 through 18 represent functions or not. Assume that the input is in the left column or top row. Explain your answer.

11.

Month	Cost (in dollars)
Jan.	5689.35
Feb.	7856.12
May	2689.15
June	1005.36

12.

Year	Number of SBA Loans to Minority-Owned Small Businesses
2000	11,999
2002	14,304
2003	20,183
2004	25,413
2005	29,717

Source: U.S. Small Business Administration.

13.

Hours Playing Poker	Winnings ($)
1	−100
2	−150
4	200
2	125
3	300
5	−650
1	60

14.

Years in College	Credits Earned
1	30
2	75
1	36
3	60
4	120
5	100
1.5	50

15.

Age	Death Rate for HIV (per 100,000)
1–4 years	0.2
5–14 years	0.2
15–24 years	0.5
25–34 years	7.2
35–44 years	13.9
45–54 years	10.9
55–64 years	4.9
65–74 years	2.2
75–84 years	0.6

Source: Centers for Disease Control and Prevention.

16.

Time (in hours)	Cost (in dollars per hour)
1–4	12.00
5–9	15.00
10–14	18.00
15–19	21.00

17.

Day of the Week	Amount Spent on Lunch (in $)
Monday	4.78
Tuesday	5.95
Wednesday	0
Thursday	4.99
Friday	15.26
Monday	5.75
Tuesday	6.33
Wednesday	0
Thursday	4.25
Friday	20.36

18.

Person's Height	5'10"	6'2"	5'5"	5'7"	5'10"	6'1"
Person's Weight (in kg)	86.4	92	70	82	91	90

19. List the domain and range of the relation given in Exercise 11.

20. List the domain and range of the relation given in Exercise 12.

21. List the domain and range of the relation given in Exercise 13.

22. List the domain and range of the relation given in Exercise 16.

23. Use the relation given in Exercise 15 to determine the death rate from HIV for 20-year-olds.

24. Use the relation given in Exercise 15 to determine the death rate from HIV for 30-year-olds.

25. Use the relation given in Exercise 12 to find the number of small business loans made to minority-owned small businesses in 2002.

26. Use the relation given in Exercise 12 to find the number of small business loans made to minority-owned small businesses in 2005.

27. Use the relation in Exercise 17 to determine the amount spent on lunch on Tuesday.

28. Use the relation in Exercise 17 to determine the amount spent on lunch on Friday.

Determine whether the equations in Exercises 29 through 34 represent functions. Explain your answer.

29. $W = -9.2g + 7.5$ **30.** $Q = 2v^2 - 6.5$

31. $8 = x^2 + y^2$ **32.** $y = \frac{23}{5}x - \frac{2}{5}$

33. $K = 5c^3 + 6c^2 - 9$ **34.** $Z = \pm(6d + 9)$

Determine whether the graphs in Exercises 35 through 40 represent functions. Explain your answer.

35.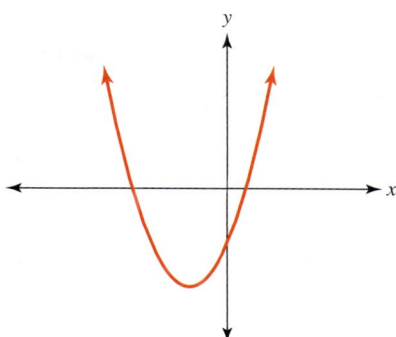

36.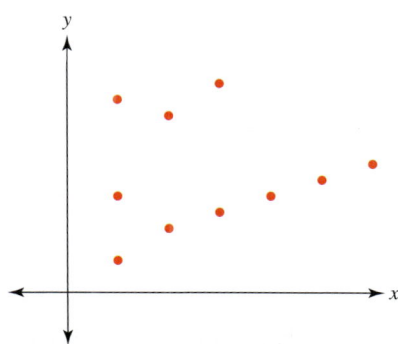

37.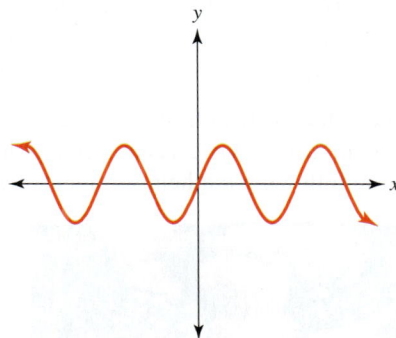

38.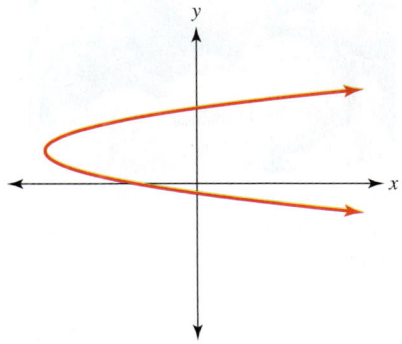

39.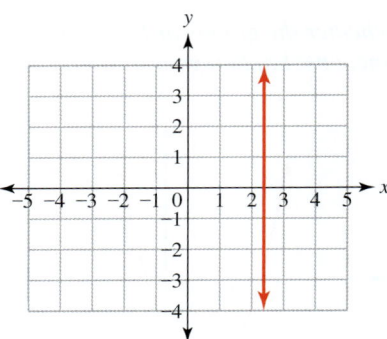

40.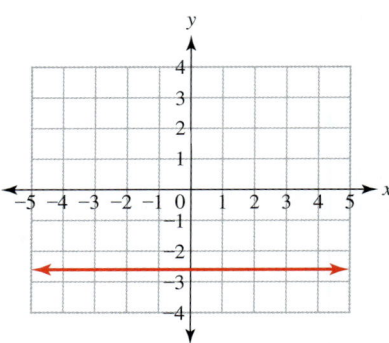

41. $W(d)$ is the weight in kilograms of a person d days after starting a diet. Write a sentence that interprets each of the following mathematical statements.
   a. $W(0) = 86.5$
   b. $W(10) = 82$
   c. $W = 75$ when $d = 30$
   d. $W(100) = 88$

42. $F(t)$ is the value of fresh fruits exported (in millions of dollars) in year t. Write a sentence that interprets each of the following mathematical statements.

Source: Statistical Abstract of the United States, 2001.

   a. $F(1995) = 1973$
   b. $F = 1971$ when $t = 1996$
   c. $F(2000) = 2077$

43. $C(d)$ is the ounces of chocolate consumed in the Clark household on day d of the month. Write a sentence that interprets each of the following mathematical statements.
   a. $C(5) = 8$
   b. $C = 20$ when $d = 15$
   c. $C(30) = 28$

44. $C(h)$ is the number of cyclists on the bike trail h hours after 5:00 A.M. Write a sentence that interprets each of the following mathematical statements.
   a. $C(1) = 12$
   b. $C = 40$ when $h = 2$
   c. $C(9) = 8$

45. $P(s)$ is the population, in millions, of state s in 2006. Write the following statements in function notation.
Source: U.S. Census Bureau.
   a. The population of Ohio was 11.48 million in 2006.
   b. The population of Texas was 23,507,783 in 2006.
   c. The population of Wyoming was 515,004 in 2006.

46. $F(t)$ is the amount of citrus fruits exported from the United States in millions of pounds in year t. Write the following statements in function notation.
Source: U.S. Census Bureau.
   a. In 2001 2392 million pounds of citrus fruits were exported from the United States.
   b. In 2003 2,245,000,000 pounds of citrus fruits were exported from the United States.
   c. In 2005 1.937 billion pounds of citrus fruits were exported from the United States.

For Exercises 47 through 50, the answers may vary.

47. The following bar graph gives the average monthly Social Security benefit for retired workers in dollars for various years.

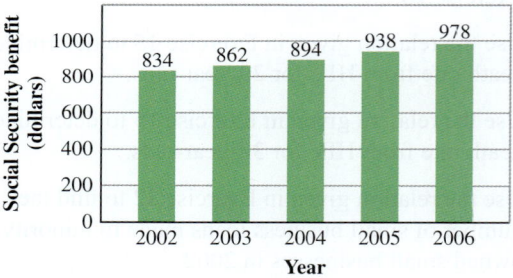

Source: Social Security Administration.

Let $B(t)$ be the average monthly Social Security benefit in dollars t years since 2000.
   a. Find an equation for a model of these data. Write your model in function notation.
   b. Find $B(9)$. Explain its meaning in this situation.
   c. Find t such that $B(t) = 2000$. Interpret your result.
   d. Give a reasonable domain and range for this model.
   e. What is the B-intercept for this model? What does it mean in this situation?

48. The amount that Americans spend on athletic and sports footwear per year is given in the table.

Year	Amount Spent on Athletic & Sports Footwear (in billions of dollars)
2001	13.8
2002	14.1
2003	14.4
2004	14.8
2005	15.0

Source: Statistical Abstract of the United States, 2007.

Let $F(t)$ be the amount spent by Americans on athletic and sports footwear in billions of dollars t years since 2000.

a. Find an equation for a model of these data. Write your model in function notation.
b. Give a reasonable domain and range for this model.
c. Find the F-intercept for this model. Explain its meaning in the terms of the problem.
d. In 2000, Americans spent $13.5 billion on athletic and sports footwear. How does this compare to your result from part c?
e. Find $F(12)$. Interpret your result.
f. Find t such that $F(t) = 20$. Interpret your result.

49. The numbers of people enrolled in Medicare in certain years are given in the table.

Year	2002	2003	2004	2005
Medicare Enrollees (in millions)	40.5	41.2	41.9	42.5

Source: Statistical Abstract of the United States, 2007.

Let $M(t)$ be the number of Medicare enrollees in millions t years since 2000.

a. Find an equation for a model of these data. Write your model in function notation.
b. Give a reasonable domain and range for this model.
c. Find the M-intercept for this model. Explain its meaning in regard to the number of Medicare enrollees.
d. Find $M(10)$. Interpret your result.
e. Find t such that $M(t) = 50$. Interpret your result.

50. The following bar graph gives the number of farms in the United States for various years.

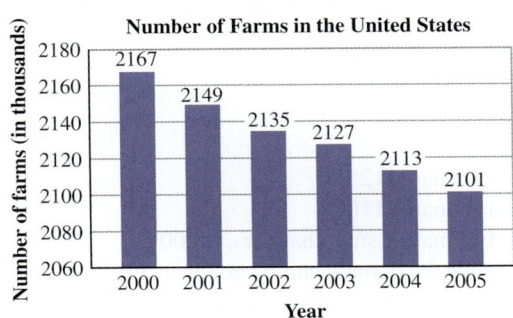

Source: Statistical Abstract of the United States, 2007.

Let $F(t)$ be the number of farms in the United States in thousands t years since 2000.

a. Find a model for the data given. Write your model in function notation.
b. Find $F(11)$. (Remember to explain its meaning in this situation.)
c. Find t such that $F(t) = 1700$. Interpret your result.
d. Give a reasonable domain and range for this model.
e. What is the F-intercept for this model? What does it mean in this situation?

51. $f(x) = 2x - 7$
a. Find $f(5)$.
b. Find $f(-10)$.
c. Find x such that $f(x) = -1$.
d. Give the domain and range for $f(x)$.

52. $g(x) = 5x + 12$
a. Find $g(3)$.
b. Find $g(-7)$.
c. Find x such that $g(x) = 47$.
d. Give the domain and range for $g(x)$.

53. $h(x) = \frac{2}{3}x + \frac{1}{3}$
a. Find $h(15)$.
b. Find $h(-9)$.
c. Find x such that $h(x) = 4$.
d. Give the domain and range for $h(x)$.

54. $f(x) = \frac{1}{5}x + \frac{3}{5}$
a. Find $f(20)$.
b. Find $f(12)$.
c. Find x such that $f(x) = 10$.
d. Give the domain and range for $f(x)$.

55. $g(x) = -18$
a. Find $g(2)$.
b. Find $g(-11)$.
c. Give the domain and range for $g(x)$.

56. $h(x) = 12.4$
a. Find $h(5)$.
b. Find $h(-123)$.
c. Give the domain and range for $h(x)$.

For Exercises 57 through 60, explain what the student did wrong. Give the correct answer.

57. $f(x) = 4x + 8$. Find x such that $f(x) = 20$.

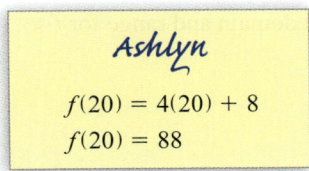

Ashlyn
$f(20) = 4(20) + 8$
$f(20) = 88$

58. $f(x) = 4x + 8$. Find $f(-24)$.

> **JT**
> $-24 = 4x + 8$
> $ -8 -8$
> $\overline{-32 = 4x}$
> $\dfrac{-32}{4} = \dfrac{4x}{4}$
> $-8 = x$

59. $f(x) = 4x + 8$. Give the domain and range for $f(x)$.

> **Alicyn**
> Domain: $-10 \le x \le 10$
> Range: $-10 \le x \le 10$

60. $f(x) = 8$. Give the domain and range for $f(x)$.

> **Wyatt**
> Domain: $x = 8$
> Range: All real numbers

61. $f(x) = 3.2x - 4.8$
  a. Find $f(2)$.
  b. Find $f(-14)$.
  c. Find x such that $f(x) = -10$.
  d. Give the domain and range for $f(x)$.

62. $g(x) = -4.3x - 5$
  a. Find $g(15)$.
  b. Find $g(-20)$.
  c. Find x such that $g(x) = -45.6$.
  d. Give the domain and range for $g(x)$.

63. $h(x) = 14x + 500$
  a. Find $h(105)$.
  b. Find x such that $h(x) = -140$.
  c. Give the domain and range for $h(x)$.

64. $f(x) = 25x - 740$
  a. Find $f(30)$.
  b. Find $f(-19)$.
  c. Find x such that $f(x) = -240$.
  d. Give the domain and range for $f(x)$.

For Exercises 65 through 70, use the graph of the function to answer the questions. Answers may vary.

65.

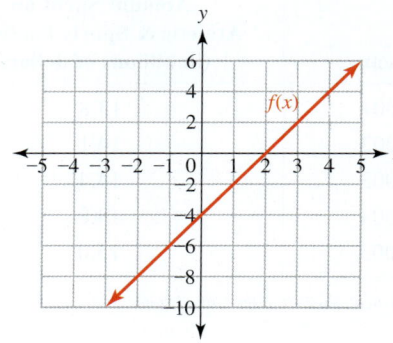

  a. Estimate $f(3)$.
  b. Estimate $f(-2)$.
  c. Estimate x such that $f(x) = -6$.
  d. Give the domain and range for $f(x)$.
  e. Estimate the vertical and horizontal intercepts for $f(x)$.

66.

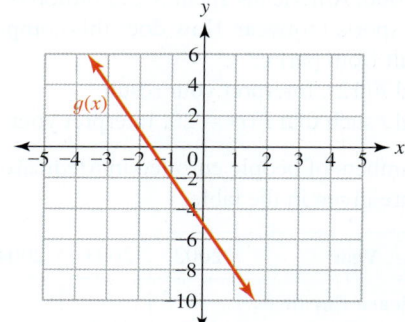

  a. Estimate $g(1)$.
  b. Estimate $g(-3)$.
  c. Estimate x such that $g(x) = 1$.
  d. Give the domain and range for $g(x)$.
  e. Estimate the vertical and horizontal intercepts for $g(x)$.

67.

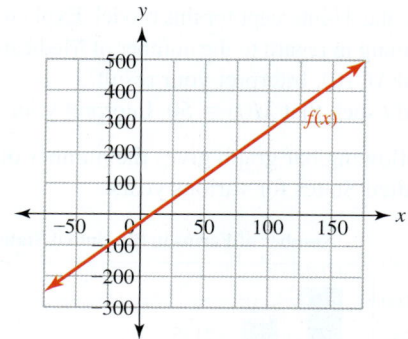

  a. Estimate $f(25)$.
  b. Estimate $f(100)$.
  c. Estimate x such that $f(x) = 200$.
  d. Give the domain and range for $f(x)$.
  e. Estimate the vertical and horizontal intercepts for $f(x)$.

68.

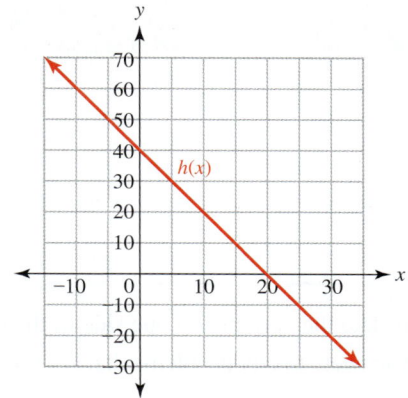

a. Estimate $h(5)$.
b. Estimate $h(20)$.
c. Estimate x such that $h(x) = 20$.
d. Give the domain and range for $h(x)$.
e. Estimate the vertical and horizontal intercept for $h(x)$.

69.

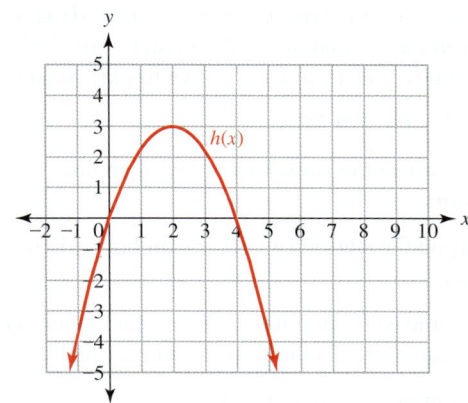

a. Estimate $h(2)$.
b. Estimate $h(5)$.
c. Estimate x such that $h(x) = -1$.
d. Give the domain and range for $h(x)$.
e. Estimate the vertical and horizontal intercepts for $h(x)$.

70.

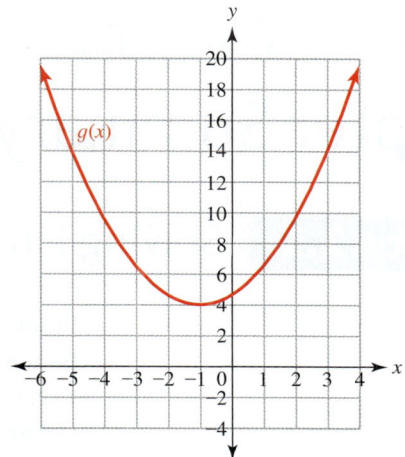

a. Estimate $g(3)$.
b. Estimate $g(-4)$.
c. Estimate x such that $g(x) = 6$.
d. Give the domain and range for $g(x)$.
e. Estimate the vertical and horizontal intercepts for $g(x)$.

Chapter Summary

Section 1.1 Solving Linear Equations

- In **solving** equations, the goal is to isolate the variable on one side of the equation.
- Be sure to write **complete sentence** answers to any problems that are presented in an application.
- **Define all variables** used to create an equation.
- Include **units** for all variable definitions and answers.

Example 1 The number of minutes an hour a local radio station can play music depends on the number of commercials that are played during that hour. Math Rocks 101.7 FM follows the equation below when considering how much time they have to play music.

$$M = 60 - 0.5C$$

where M is the number of minutes of music the station can play in an hour if they play C thirty-second commercials that hour.

a. Find the number of minutes of music the station can play in an hour when they play 12 thirty-second commercials.

b. If the radio station claims to have at least 53 minutes of music each hour, what is the most number of thirty-second commercials the station can play an hour?

SOLUTION a. Since there are 12 thirty-second commercials being played, $C = 12$,

$$M = 60 - 0.5(12)$$
$$M = 54$$

With 12 thirty-second commercials an hour, the station can play 54 minutes of music.

b. If the station needs to play 53 minutes of music, $M = 53$,

$$53 = 60 - 0.5C \quad \text{Subtract 60 from both sides.}$$
$$-7 = -0.5C \quad \text{Divide both sides by } -0.5.$$
$$14 = C$$

The station can play 14 thirty-second commercials an hour if they want to play 53 minutes of music.

Example 2 Hip Hop Math Records is a new record label that is paying Math Dude, a new music artist, $10,000 to record a CD and $1.50 for every CD sold.

a. Write an equation for the amount of money Math Dude will make from this deal depending on how many CDs are sold.

b. Use your equation to find how much money Math Dude will make if he sells 100,000 CDs.

c. Use your equation to find out how many CDs Math Dude must sell to make himself a million dollars.

CHAPTER 1 Summary 113

SOLUTION   a. Let M be the money Hip Hop Math Records will pay Math Dude in dollars. Let C be the number of CDs sold by Math Dude.

$$M = 10000 + 1.5C$$

b. If he sells 100,000 CDs, $C = 100{,}000$,

$$M = 10000 + 1.5(100000)$$
$$M = 160000$$

If Math Dude sells 100,000 CDs, Hip Hop Math Records will pay him $160,000.

c. If Math Dude is going to make a million dollars, we must have $M = 1{,}000{,}000$.

$$1000000 = 10000 + 1.5C$$
$$990000 = 1.5C$$
$$660000 = C$$

For Math Dude to make a million dollars, he must sell 660,000 CDs.

Section 1.2 Using Data to Create Scatterplots

- To create a **scatterplot** from data, determine the independent and dependent variables.
- The **independent variable** will be placed on the horizontal axis. This variable is often called the *input variable*.
- The **dependent variable** will be placed on the vertical axis. This variable is often called the *output variable*.
- An eyeball best-fit line can be used as a **graphical model.**
- The **vertical intercept** is where the graph crosses the vertical axis. It will be a point of the form $(0, b)$.
- The **horizontal intercept** is where the graph crosses the horizontal axis. It will be a point of the form $(a, 0)$.
- The **domain** will include all values of the input variable that will result in reasonable output values.
- The **range** will include all values of the model that come from the inputs in the domain.

Example 3   The gross profit for PETsMART, Inc. is reported in the following table.

Year	2000	2001	2002	2003
Gross Profit (in millions of $)	529.0	673.5	788.0	909.9

Source: CBSMarketwatch.com

a. Create a scatterplot of the data and draw an eyeball best-fit line through the data.

b. Use your graphical model to estimate the gross profit for PETsMART, Inc. in 2004.

c. What is the vertical intercept of your model? What does it mean in this situation?

d. What are a reasonable domain and range for your model?

SOLUTION a. Let t be time in years since 2000.
Let P be the gross profit of PETsMART, Inc. in millions of dollars.

b. According to the graph, in 2004, PETsMART, Inc. had a gross profit of about $1050 million dollars, or $1.05 billion.

c. The vertical intercept for this model is about (0, 550). Therefore, in 2000, PETsMART, Inc. had a gross profit of about $550 million.

d. Domain: $[-1, 4]$; Range: $[400, 1050]$.

Section 1.3 Fundamentals of Graphing and Slope

- The **slope** of a line is the steepness of the line. Slope can be remembered as

$$m = \frac{\text{rise}}{\text{run}} = \frac{\text{change in } y}{\text{change in } x} = \frac{y_2 - y_1}{x_2 - x_1}$$

- The slope of a line is a constant and represents the amount that the output variable changes per a unit change in the input variable.
- The **slope-intercept form** of a line is $y = mx + b$, where m represents the slope and $(0, b)$ is the vertical intercept.
- An equation can be graphed by plotting points.
- A linear equation can be graphed using the slope and y-intercept.

Example 4 Use the graph to find the following.

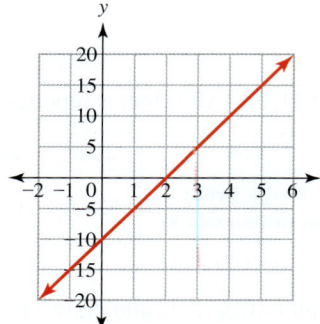

a. Find the slope of the line.
b. Estimate the vertical intercept.
c. Estimate the horizontal intercept.

SOLUTION a. The points $(1, -5)$ and $(4, 10)$ are on the line

$$\frac{10 - (-5)}{4 - 1} = \frac{15}{3} = 5$$

The slope is 5.

b. The vertical intercept is about $(0, -10)$.

c. The horizontal intercept is about $(2, 0)$.

Example 5 Graph the line $y = \frac{2}{3}x - 6$ using the slope and y-intercept.

SOLUTION

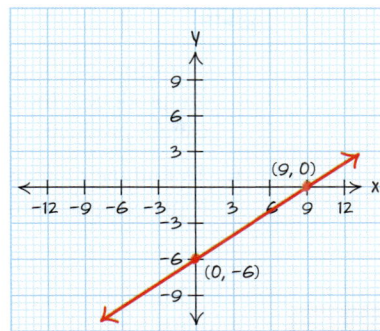

Example 6 Using the equation $M = 10,000 + 1.5C$ found in part a of Example 2, explain the meaning of the slope in the given situation.

SOLUTION The slope is $m = 1.5$. This means that the amount that Math Dude will make from recording and selling the CDs will increase by $1.50 per CD sold.

Section 1.4 Intercepts and Graphing

- The **general form of a line** is $Ax + By = C$ where A, B, and C are integers and A is nonnegative.
- To find an intercept, make the "other" variable zero and solve.
- When interpreting an intercept, be sure to interpret both parts of the point.
- A **horizontal line** has an equation of the form $y = k$.
- A **vertical line** has an equation of the form $x = k$.
- The slope of a horizontal line is zero. The slope of a vertical line is undefined.

Example 7 The cost C in dollars to produce L strands of Christmas lights can be modeled by $C = 3.5L + 10,000$.

a. Find the L-intercept and explain its meaning in regard to the cost.

b. Find the C-intercept and explain its meaning in regard to the cost.

SOLUTION a. To find the L-intercept, let $C = 0$ and solve.

$$0 = 3.5L + 10000$$
$$-10000 = 3.5L$$
$$-2857.14 \approx L$$

Therefore, the L-intercept is $(-2857.14, 0)$. If negative 2857.14 strands of lights are produced, the cost will be zero. This is model breakdown, since a company cannot produce a negative number of strands of lights.

b. To find the C-intercept, let $L = 0$, and we get $(0, 10{,}000)$. This intercept means that if the company produces no strands of lights, they still have costs of $10,000.

Example 8 Find the intercepts and graph the line $2x + y = 4$.

SOLUTION To find the x-intercept, let $y = 0$ and solve.

$$2x + 0 = 4$$
$$2x = 4$$
$$x = 2$$

Therefore, the x-intercept is $(2, 0)$.
To find the y-intercept, let $x = 0$ and solve. This leaves us with $y = 4$, so the y-intercept is $(0, 4)$.
Using these points, we get the graph.

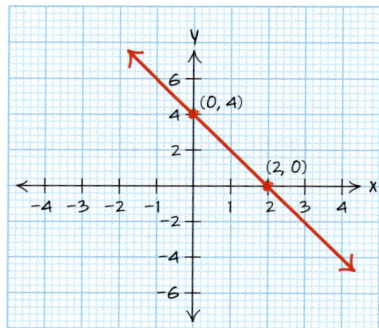

Section 1.5 Finding Equation of Lines

- The slope-intercept form $y = mx + b$ or the point-slope formula $y - y_1 = m(x - x_1)$ can be used to find the equation of a line.
- Use the following steps to find the equation of a line using the point-slope formula.
  1. Use any two points to calculate the slope.
  2. Substitute the slope and a point into the point-slope formula.
  3. Write the equation in slope-intercept form.
  4. Check the equation by plugging in the points to be sure they are solutions.
- Use the following steps to find the equation of a line using the slope-intercept form.
  1. Use any two points to calculate the slope.
  2. Use the slope and a point to find the value of b.
  3. Write the equation in slope-intercept form.
  4. Check the equation by plugging in the points to be sure they are solutions.

- Parallel lines have the same slopes and never intersect.
- Perpendicular lines have opposite reciprocal slopes and intersect at a right angle.

Example 9 A car that is being depreciated is worth $35,000 new and worth $20,000 after 3 years. If linear depreciation is being used, write an equation for the value of the car depending on its age.

SOLUTION First, define the variables.
a = Age of the car in years
v = Value of the car in dollars
The problem gives two values for the car, which result in the points (0, 35,000) and (3, 20,000). Using these two points, calculate the slope and get

$$m = \frac{35000 - 20000}{0 - 3} = -5000$$

Using the point (0, 35,000) as the vertical intercept, we get the equation
$v = -5000a + 35,000$.

Example 10 Determine whether the following set of lines is parallel, perpendicular, or neither.

a. $y = 2x + 7$
   $x + 2y = 20$

b. $y = 0.25x + 5$
   $2x - 8y = 24$

SOLUTION a. First, put the second equation into slope-intercept form to compare the slopes of the two equations.

$$x + 2y = 20$$
$$2y = -x + 20$$
$$y = -\frac{1}{2}x + 10$$

From the slope-intercept form, the slope of the first equation is 2, and the slope of the second equation is $-\frac{1}{2}$. These two slopes are opposite reciprocals, so these lines are perpendicular.

b. First, put the second equation into slope-intercept form.

$$2x - 8y = 24$$
$$-8y = -2x + 24$$
$$y = 0.25x - 3$$

Both lines have a slope of 0.25, so the lines are parallel.

Section 1.6 Finding Linear Models

- To find a **linear model,** follow these steps.
  1. Define the variables and adjust the data.
  2. Create a scatterplot.
  3. Select a model type.
  4. **Linear Model:** Pick two points and find the equation of the line.
  5. Write the equation of the model.
  6. Check the model by graphing it with the scatterplot.

Example 11 The gross profit for PETsMART, Inc. is reported in the following table.

Year	2000	2001	2002	2003
Gross Profit (in millions of $)	529.0	673.5	788.0	909.9

Source: CBSMarketwatch.com

a. Find an equation for a model of these data.

b. Use your model to estimate the gross profit for PETsMART, Inc. in 2005.

c. What is the horizontal intercept of your model? What does it mean in this situation?

d. What are a reasonable domain and range for your model?

SOLUTION a. 1. Define the variables and adjust the data.

t = Time in years since 2000

P = Gross profit of PETsMART, Inc. in millions of dollars

2 and 3. Create a scatterplot and select a model type.

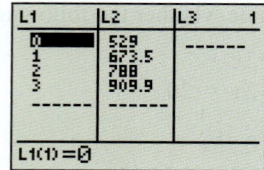 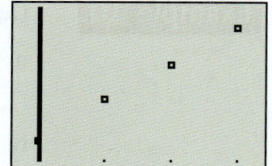

4 and 5. Pick two points and find the equation of the line, and write the equation of the model.

The first and third points seem to fit well, so use them to find the slope.

$$m = \frac{788 - 529}{2 - 0} = 129.5$$

The point (0, 529) is the vertical intercept. Therefore, the equation is $P = 129.5t + 529$.

6. Check the model by graphing it with the scatter-plot.

This model seems to fit well.

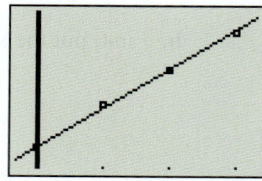

b. 2005 is represented by $t = 5$, so we get

$$P = 129.5(5) + 529 = 1176.5$$

Therefore, in 2005, PETsMART, Inc. should have had a gross profit of about $1176.5 million dollars.

c. To find the horizontal intercept, let $P = 0$ and get

$$0 = 129.5t + 529$$
$$-529 = 129.5t$$
$$-4.085 \approx t$$

Therefore, in about 1996, PETsMART, Inc. made no gross profit. This might be model breakdown.

d. Domain: [−1, 4]; Range: [399.5, 1047].

Section 1.7 Functions and Function Notation

- A **relation** is any relationship between a set of inputs and outputs.
- A **function** is a relation in which each input has exactly one output associated with it.
- The **vertical line test** can be used to determine if a graph represents a function.
- Models should be written using function notation.
- The **domain** and **range** of a linear function that is not in an application and that is not vertical or horizontal are both **all real numbers.**

Example 12 Determine whether the following represent functions. Explain why or why not.

a. $y = 3x + 9$

b.
Inputs	5	8	12	3	5	17	16
Outputs	10	5	4	6	8	2	3

c.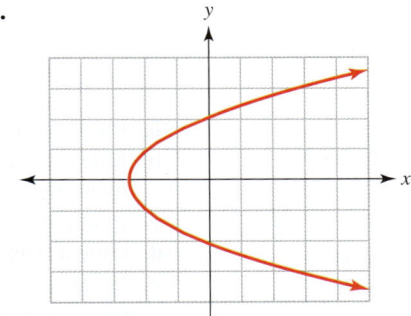

d.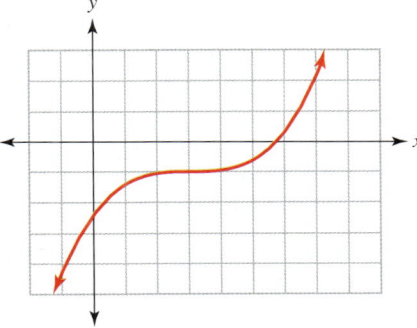

SOLUTION

a. This is a function, since each input will result in only one output.

b. This table does not represent a function because the input value of 5 is associated with the outputs 10 and 8.

c. This graph is not a function because it fails the vertical line test. A vertical line will pass through the graph more than once.

d. This graph does represent a function, since it passes the vertical line test.

Example 13 Let $D(w)$ be the number of deaths in Arkansas due to the West Nile virus during week w of the year.

a. Interpret the notation $D(7) = 6$.

b. Given that there were three deaths during the forty-fifth week of the year, write this in function notation.

SOLUTION

a. During the seventh week of the year, there were six deaths due to the West Nile virus in Arkansas.

b. $D(45) = 3$.

Chapter Review Exercises

At the end of each exercise, the section number in brackets is where the material is covered.
For Exercises 1 through 4, solve the equation.

1. $2x + 5 = 7(x - 8)$ [1.1]
2. $\frac{1}{3}x - \frac{2}{3} = 5$ [1.1]
3. $0.4t + 2.6 = 0.8(t - 8.2)$ [1.1]
4. $-2(x + 3.5) + 4(2x - 1) = 7x - 5(2x + 3)$ [1.1]

5. The height of the grass used in the rough at a local golf course is given by

   $$h = 0.75d + 4.5$$

   where h is the height of the grass in inches d days after the grass was cut and fertilized.

   a. What is the height of the grass 1 week after it was cut and fertilized?
   b. For a coming tournament, the officials want the grass in the rough to be 12 inches high. How many days before the tournament should the golf course cut and fertilize this grass? [1.1]

6. Satellite phone service is available in some airplanes for a $10 connection fee and a $10 per minute charge.
   a. Write an equation for the cost of a satellite phone call.
   b. Use the equation to find the cost of a 3-minute satellite phone call.
   c. If a credit card has only $300 left of available credit, how long a satellite phone call can be made? [1.1]

7.

   a. Use the ad to write an equation for the cost to rent a Bobcat tractor from Pauley's Rental Company.
   b. How much will it cost to rent the Bobcat for 2 hours?
   c. How much will it cost to rent the Bobcat for 3 days if a day is measured as 8 hours each? [1.1]

8. A small manufacturing company makes custom putters for a chain of sporting goods stores. The total cost to produce different numbers of putters is given in the table.

Number of Putters	Total Cost (in dollars)
5	600.00
10	725.00
20	950.00
25	1100.00
50	1700.00

   a. Find an equation for a model of these data.
   b. Find the vertical intercept of your model. Explain its meaning in regard to the cost.
   c. Find the total cost for 100 putters.
   d. Give a reasonable domain and range for this model.
   e. What is the slope of this model? Explain its meaning in regard to the cost. [1.6]

9. The gross profit for Costco Wholesale Corporation is given in the following chart.

   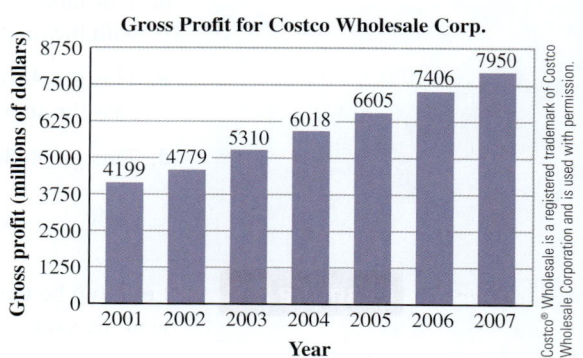

   Source: http://finance.google.com

   a. Find an equation for a model of these data.
   b. Estimate when the gross profit for Costco will be 9 billion dollars.
   c. What was the gross profit for Costco in 2009?
   d. Give a reasonable domain and range for this model.
   e. What is the slope of this model? Explain its meaning in the terms of the problem. [1.6]

10. According to the U.S. Department of Labor Bureau of Labor Statistics, the percentage of full-time workers in private industry who filed an injury case is given in the table.

Years Since 2000	Percent
3	4.7
4	4.5
5	4.4
6	4.2

Source: U.S. Bureau of Labor Statistics.

a. Find an equation for a model of these data. Write your model using function notation.
b. What is the slope for this model? Explain its meaning in this situation.
c. What percentage of full-time workers in private industry will have filed an injury case in 2003?
d. What is the vertical intercept for this model? What does it mean in this situation?
e. What is the horizontal intercept for this model? What does it mean in this situation? [1.7]

11. The numbers of students in kindergarten through twelfth grade in Texas public schools are given in the table.

Year	Students
2000	4,002,227
2001	4,071,433
2002	4,160,968
2003	4,255,821
2004	4,328,028
2005	4,400,644
2006	4,521,043

Source: Texas Education Agency.

a. Find an equation for a model of these data. Write your model using function notation.
b. What is the slope for this model? Explain its meaning in regard to the number of students.
c. How many kindergarten through twelfth grade students can Texas expect to have in its public schools in 2011? [1.7]

12. The number of candles produced by the Holy Light Candle Company can be modeled by

$$C(t) = 0.56t + 4.3$$

where $C(t)$ is the number of candles, in millions, produced t years since 2010.

a. Find the number of candles produced by Holy Light Candle Company in 2015.
b. Find $C(3)$ and interpret its meaning in this situation.
c. Find when $C(t) = 10$ and interpret its meaning in this situation. [1.7]

For Exercises 13 through 16, solve each equation for the indicated variable.

13. Charles's Law: $v = bT$ for T. [1.1]
14. Force: $F = ma$ for m. [1.1]
15. $ax + by = c$ for x. [1.1]
16. $2x - ay = b$ for y. [1.1]

For Exercises 17 through 24, sketch the graphs of the equations by hand using any method you wish. Label the horizontal and vertical intercepts.

17. $y = 3x - 4$ [1.3]
18. $y = 2x + 3$ [1.3]
19. $2x + 3y = 24$ [1.4]
20. $5x - 6y = 42$ [1.4]
21. $y = x^2 + 5$ [1.3]
22. $y = x^2 - 6$ [1.3]
23. $y = \frac{1}{2}x + 7$ [1.3]
24. $y = -\frac{2}{7}x + 4$ [1.3]

For Exercises 25 and 26, write the equation of the line passing through the given points. Write your answer in slope-intercept form, $y = mx + b$.

25. $(2, 7)$ and $(7, 27)$ [1.5]
26. $(4, 9)$ and $(-3, 23)$ [1.5]

For Exercises 27 through 30, write the equation of the line that satisfies the given description. Write your answer in slope-intercept form, $y = mx + b$.

27. The line that passes through the point $(4, 10)$ and is parallel to the line $-3x + y = 12$ [1.5]
28. The line that passes through the point $(2, 16)$ and is parallel to the line $y = -0.5x + 4$ [1.5]
29. The line that passes through the point $(3, 7)$ and is perpendicular to the line $y = -4x + 7$ [1.5]
30. The line that passes through the point $(-2, 9)$ and is perpendicular to the line $-2x + 5y = 35$ [1.5]

31. Use the graph to find the following.

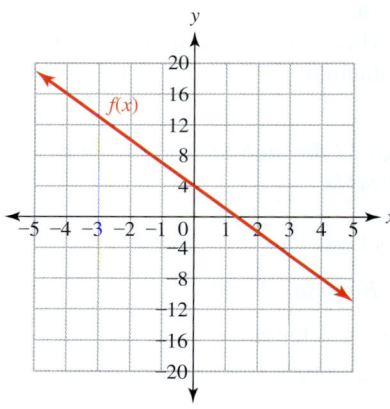

   a. Estimate the y-intercept.
   b. Estimate the x-intercept.
   c. Find the slope of the line. [1.3]
   d. Find $f(4)$.
   e. Find x such that $f(x) = 16$.
   f. Write an equation for $f(x)$. [1.7]

32. Use the graph to find the following.

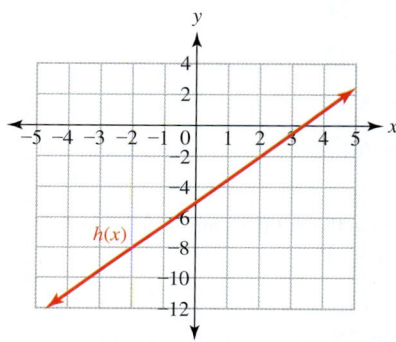

   a. Estimate the vertical intercept.
   b. Estimate the horizontal intercept.
   c. Find the slope of the line. [1.3]
   d. Find $h(-2)$.
   e. Find x such that $h(x) = -2$.
   f. Write an equation for $h(x)$. [1.7]

33. Use the table to find the following.

x	y
-3	4
0	6
3	8
6	10
9	12

   a. The slope of the line passing through the points
   b. The y-intercept
   c. The equation for the line passing through the points [1.5]

34. Use the table to find the following.

x	y
-8	6.6
-2	3.9
0	3
3	1.65
7	-0.15

   a. The slope of the line passing through the points
   b. The y-intercept
   c. The equation for the line passing through the points [1.5]

35. Let $E(h)$ be the cost in dollars for electricity to use holiday lights h hours a day. Explain the meaning of $E(6) = 13$ in this situation. [1.7]

36. The population of Michigan in thousands can be modeled by $P(t) = 30t + 9979$, where t is years since 2000.
   a. Find $P(10)$ and explain its meaning in this situation.
   b. Find when the population of Michigan will reach 11,000 thousand. [1.7]

37. $f(x) = 2x - 8$ [1.7]
   a. Find $f(10)$.
   b. Find x such that $f(x) = 0$.
   c. Give the domain and range for $f(x)$.

38. $h(x) = -\dfrac{2}{3}x + 14$ [1.7]
   a. Find $h(12)$.
   b. Find x such that $h(x) = -8$.
   c. Give the domain and range for $h(x)$.

39. $g(x) = 12$ [1.7]
   a. Find $g(3)$.
   b. Find $g(-20)$.
   c. Give the domain and range for $g(x)$.

40. $f(x) = 1.25x + 4.5$ [1.7]
   a. Find $f(5)$.
   b. Find x such that $f(x) = -8.5$.
   c. Give the domain and range for $f(x)$.

Chapter Test

1. The number of work-related injury cases in U.S. private industry is given in the table.

Year	Number of Injury Cases (in thousands)
2003	4095
2004	4008
2005	3972
2006	3857

   Source: U.S. Bureau of Labor Statistics.

   a. Find an equation for a model of these data.
   b. How many work-related injury cases were there in private industry in 2009?
   c. What is the slope of your model? Explain its meaning in this problem.
   d. Give a reasonable domain and range for this model.
   e. When was the number of work-related injury cases in private industry 5 million?

2. Sketch by hand the graph of the line
   $$y = 4x - 2$$
   Label the vertical and horizontal intercepts.

3. Sketch by hand the graph of the line
   $$y = -\frac{2}{3}x + 5$$
   Label the vertical and horizontal intercepts.

4. Sketch by hand the graph of the line
   $$2x - 4y = 10$$
   Label the vertical and horizontal intercepts.

5. Solve $ax - by = c$ for x.

6. Write the equation for the line passing through the points $(-4, 8)$ and $(6, 10)$. Write the equation in slope-intercept form.

7. Write the equation for the line passing through the point $(5, 8)$ and parallel to the line $y = 2x - 7$. Write the equation in slope-intercept form.

8. The number of paintings that John Clark sells each month can be modeled by
   $$P = 2m + 30$$
   where m is the number of months since he started selling paintings.

   a. In month 6, how many paintings will John sell?
   b. In what month will he sell 50 paintings?

9. The total revenue for Apple, Inc. for the past several years is given in the table.

Years Since 2000	Total Revenue (in millions of $)
4	8279
5	13,931
6	19,315
7	24,006
8	32,479

   Source: http://finance.google.com

   a. Find an equation for a model of these data. Write the model using function notation.
   b. What is the vertical intercept for your model? Explain its meaning in regard to revenue.
   c. Give a reasonable domain and range for this model.
   d. Estimate the total revenue for Apple, Inc. in 2010.
   e. When might the total revenue be 30 billion dollars?
   f. What is the slope of this model? Explain its meaning in regard to revenue.

10. Use the table to find the following.

x	y
-5	16
0	12
5	8
10	4
15	0

    a. The slope of the line passing through these points
    b. The y-intercept
    c. The x-intercept
    d. The equation for the line passing through these points

11. Use the graph to find the following.

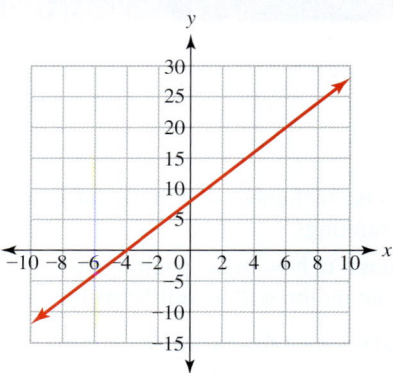

   a. The slope of the line
   b. The vertical intercept
   c. The horizontal intercept
   d. The equation for the line

12. Give the domain and range for the function $f(x) = -4x + 5$.

13. Solve the equation $W = ht^2$ for h.

14. Solve $1.5(x + 3) = 4x + 2.5(4x - 7)$.

15. If $P(t)$ is the population of New York in millions t years since 2000. Explain the meaning of $P(10) = 19.4$ in this situation.

16. Let $H(p) = 0.20p + 5$ be the hours you should study for a test on Chapter 1 if you want a grade of $p\%$ on this test.
   a. Find $H(70)$ and explain its meaning in terms of your test score.
   b. Find how many hours you should study get a 100% on the exam.

17. $f(x) = 7x - 3$
   a. Find $f(4)$.
   b. Find x such that $f(x) = -31$.
   c. Give the domain and range for $f(x)$.

18. $h(x) = \dfrac{4}{7}x + 6$
   a. Find $h(35)$.
   b. Find x such that $h(x) = 4$.
   c. Give the domain and range for $h(x)$.

19. $g(x) = -9$
   a. Find $g(6)$.
   b. Give the domain and range for $g(x)$.

20. Explain how a relation may not be a function.

Chapter 1 Projects

Walking a "Straight" Line

Group Experiment
Three or more people

What you will need
- A tape measure
- A timer (stopwatch)

or, instead of a tape measure and timer,

- A Texas Instruments CBR unit or a Texas Instruments CBL unit with a Vernier motion detector probe

It will be your job to create data that will best be modeled by a linear function. Using the tape measure and timer or the CBR/CBL unit, have someone walk while the others record the time and the distance from a fixed object (e.g., a wall or desk). Collect at least five points. Be sure to check your data so that you are satisfied before you move on so that the data will be a linear function.

Write up

a. Describe how you collected the data. What was the most important thing to do to get a linear model?

b. Create a scatterplot of the data on the calculator or computer and print it out or draw it neatly by hand on graph paper.

c. Find a model to fit the data.

d. What is the slope of your model and what does it represent in this experiment?

e. What is the vertical intercept of your model and what does it represent in this experiment?

f. What are a reasonable domain and range for your model? (Remember to consider any restrictions on the experiment.)

New Flooring Costs

Research Project
One or more people

What you will need
- Prices from a flooring store for either wood flooring or the tile of your choice
- Installation costs per square foot for the flooring of your choice

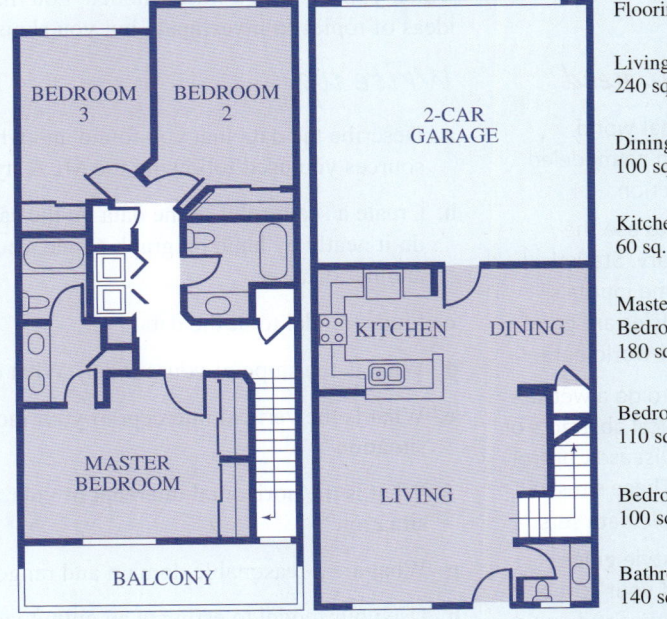

Flooring Areas

Living Room: 240 sq. ft.

Dining Room: 100 sq. ft.

Kitchen: 60 sq. ft.

Master Bedroom: 180 sq. ft.

Bedroom 2: 110 sq. ft.

Bedroom 3: 100 sq. ft.

Bathrooms: 140 sq. ft. total

You are given the task of estimating the cost of flooring for a new house that your family is buying. The house comes with very basic flooring options installed, and you want to decide in how many rooms you want to replace this basic flooring with upgraded wood floors or tile. You should consider both the cost for materials and cost for installation.

Write up

a. Describe the flooring you chose and its cost per square foot for materials and installation. Are there any fixed costs for delivery or other items that you should consider in your calculations?

b. Let f be the number of square feet of flooring to be purchased and installed. Find a function for the cost of materials.

c. Find a function for the cost of installing the flooring.

d. Estimate the cost for the materials if you install them only in the living room, dining room, and kitchen.

e. Estimate the cost for installation only for the living room, dining room, and kitchen.

f. Use the two functions from parts b and c to create a new function that gives you the total cost for materials and installation of the flooring.

g. Estimate the total cost to put your flooring choice in all of the bedrooms.

h. What are a reasonable domain and range for your total cost function?

i. If your budget for flooring is $6000, choose what rooms to install the flooring in and give reasons for your choices. Include how much of your budget you used.

Find Your Own Line

Research Project

One or more people

What you will need

- Find data for a real-world situation that can be modeled with a linear function.
- You might want to use the Internet or a library. Statistical abstracts and some journals and scientific articles are good resources for real-world data.
- You might want to do a website search for statistical abstracts or the Centers for Disease Control and Prevention. These sites will have many possible data sets.
- Follow the MLA style guide for all citations. If your school recommends another style guide, use that.

In this project, you are given the task of finding data for a real-world situation for which you can find a linear model. You may use the problems in this chapter to get ideas of topics to investigate, but you should not use data discussed in this textbook.

Write up

a. Describe the data that you found and where you found the data set. Cite any sources you used following the MLA style guide.

b. Create a scatterplot of the data on the calculator or computer and print it out, or do it neatly by hand on graph paper. Your scatterplot should be linear. If not, find another data set.

c. Find a model to fit the data.

d. What is the slope of your model? What does it represent in this situation?

e. What is the vertical intercept of your model? What does it represent in this situation?

f. What is the horizontal intercept of your model? What does it represent in this situation?

g. What are a reasonable domain and range for your model?

h. Use your model to estimate an output value for an input value that will fit your model but that you did not collect in your original data.

Review Presentation

Research Project

One or more people

What you will need

- This presentation may be done in class or as a video.
- If it is done as a video, you may want to post the video on YouTube or another website.
- You might want to use problems from the homework set or review material in this book.
- Be creative and make the presentation fun for students to watch and learn from.

Create a 5- to 10-minute presentation or video that reviews a section of Chapter 1. The presentation should include the following:

- Examples of the important skills presented in the section
- Explanation of any important terminology and formulas
- Common mistakes made and how to recognize them

Systems of Linear Equations and Inequalities

2

- **2.1** Systems of Linear Equations
- **2.2** Solving Systems of Equations Using the Substitution Method
- **2.3** Solving Systems of Equations Using the Elimination Method
- **2.4** Solving Linear Inequalities
- **2.5** Absolute Value Equations and Inequalities
- **2.6** Solving Systems of Linear Inequalities

The number of cellular telephone subscribers has increased dramatically over the past 20 years. CTIA—The Wireless Association, an industry group, found that in 1985 there were an estimated 203,600 cell phone subscribers in the United States. That number had grown to 233,041,000 in 2006. Comparing different cell phone plans is a valuable skill because choosing the best plan can save money. In this chapter, we will discuss how to analyze comparative data using systems of equations and inequalities. One of the chapter projects on page 214 will use graphing and problem solving to determine which cell phone service plan is the better value.

2.1 Systems of Linear Equations

LEARNING OBJECTIVES
- Identify the solutions of a system of equations using the graph.
- Identify the solutions of a system of equations using numerical methods.
- Explain the meaning of a solution of a system of equations in an application.
- Identify consistent and inconsistent systems of equations.
- Identify dependent lines.

Definition of Systems

In many areas of life, we compare two or more quantities in an attempt to make a good decision about which is best for our situation. This type of decision-making process can often be simplified by using what in mathematics is called a **system of equations.** A system of equations is a set of equations that requires a solution that will work for all of the equations in the set. This is often the case when we are trying to determine when two options are going to be equal.

> **DEFINITIONS**
>
> **System of Linear Equations** A set of two or more linear equations. Each equation in the system can contain one or more variables.
>
> **Solution of a System of Two Linear Equations** An ordered pair (or a set of ordered pairs) that is a solution to every equation in the system. On the graph, this is seen as the intersection points of the lines.

Graphical and Numerical Solutions

If we consider the system of equations

$$y = 2x + 5$$
$$y = 4x - 25$$

we can build tables of ordered pairs that satisfy each equation.

x	$y = 2x + 5$
0	5
5	15
10	25
15	35
20	45

x	$y = 4x - 25$
0	-25
5	-5
10	15
15	35
20	55

By looking at the two tables, we see that the point (15, 35) is a solution to both equations and the solution to the system of equations. By graphing these two equations, we can see that the solution to the system is the point where the two lines intersect.

SECTION 2.1 Systems of Linear Equations

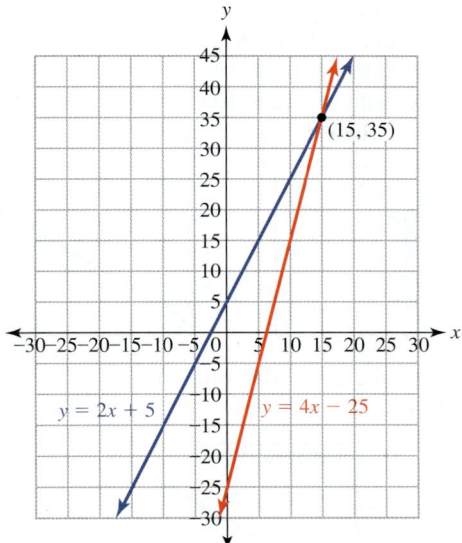

Example 1 Comparing two options by solving a system

Emily is moving to a new apartment and wants to rent a 10-foot truck. She called both U-Haul and Bob's Truck Rental to compare prices. Use Emily's notes to answer the following.

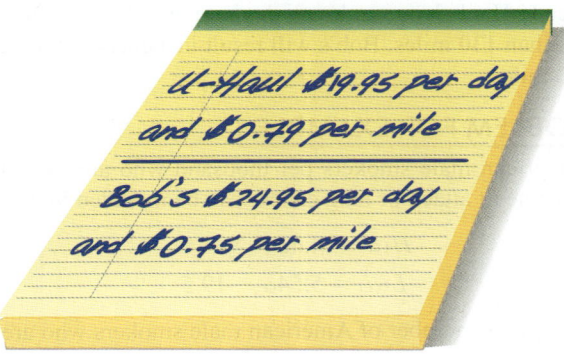

a. Write the equations for these two companies' total rental quotes for one day.

b. Graph the two equations on the same calculator window.

c. Find the distance traveled that will result in the two companies having the same cost.

SOLUTION

a. First, define the variables:

$U(m)$ = Total cost in dollars to rent a 10-foot truck for a day from U-Haul
$B(m)$ = Total cost in dollars to rent a 10-foot truck for a day from Bob's
m = Number of miles traveled in the rented truck

The two equations are

$$U(m) = 19.95 + 0.79m$$
$$B(m) = 24.95 + 0.75m$$

b. Put the function for U-Haul into Y1 and the function for Bob's into Y2 in the Y= screen. Remember to turn the stat plots off. To set an appropriate viewing window, consider the situation. Emily is renting the truck and cannot drive less

Using Your TI Graphing Calculator

Finding an appropriate window:

When you are given two equations for which you do not have data, finding an appropriate window is an important skill.

If the equations represent an application, consider the situation and what input values would make sense. Use these values to determine a reasonable X-min and X-max. Put some input values into the table feature of the calculator and use the resulting outputs to set your Y-min and Y-max.

Using the equations from Example 1, you might have the following:

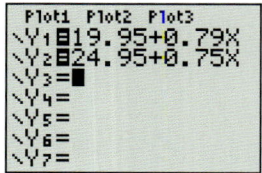

If the equations do not represent an application, you can use the calculator's standard window or use the table to help you find a good window. Plug in some basic values for *x* and let it calculate the *y*-values. Use the values in the table to set an initial window. Once you can see the lines on the screen, you can adjust the window as needed.

than zero miles and will probably not drive more than 200 miles in one day, so set X-min and X-max to 0 and 200, respectively. She knows she will be charged at least $19.95 for the rental, and if she drives the full 200 miles, the cost from each company reaches a maximum of a little less than $180, so set the Y-min and Y-max to 15 and 180, respectively.

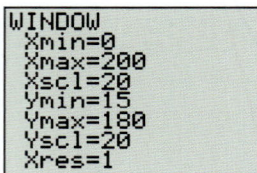 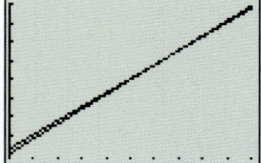

c. Using the trace button on the calculator, the estimate of the solution is about 130 miles at a cost of $122.29.

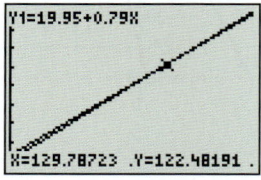

Therefore, U-Haul and Bob's will both rent a 10-foot truck for a day at a cost of about $122.29 if Emily drives the truck 130 miles. To get a more precise solution to this system, zoom in on the graph so that the intersection becomes clearer or solve the system algebraically. The solution to this system shows that U-Haul and Bob's charge the same amount to rent a 10-foot truck for a day and drive it about 130 miles. Since U-Haul charges a lower day rate, they will be cheaper than Bob's for trips shorter than 130 miles. Bob's will result in a cheaper total cost for trips longer than 130 miles.

PRACTICE PROBLEM FOR EXAMPLE 1

The number of American male smokers per thousand who are expected to die in the next 10 years from a heart attack or lung cancer can be modeled by the two functions.

$$H(a) = 4.4a - 9$$
$$L(a) = 5.54a - 19.5$$

where $H(a)$ represents the number of American male smokers who are expected to die in the next 10 years from a heart attack per every 1000 male smokers aged *a* years over 40. $L(a)$ represents the number of American male smokers who are expected to die in the next 10 years from lung cancer per every 1000 male smokers aged *a* years over 40.
Source: Journal of the National Cancer Institute.

a. Graph both equations on the same window.

b. Find the age at which American male smokers have the same risk of dying from a heart attack as they do from lung cancer.

Solving systems of equations by graphing can be done by hand or by using the graphing calculator. In the previous example, we solved the system by graphing the equations on the calculator. In the next example, we will graph the system by hand and find the solution.

Example 2 Solving systems of equations by graphing

Solve the following system by graphing the equations by hand.

$$y = \frac{2}{3}x - 7$$
$$y = -x + 3$$

SOLUTION

First, graph both equations on the same set of axes. The first line has a slope $m = \frac{2}{3}$ and a y-intercept of $(0, -7)$. The second graph has a slope $m = -1$ and a y-intercept of $(0, 3)$. Using these slopes and intercepts, we get the following graph.

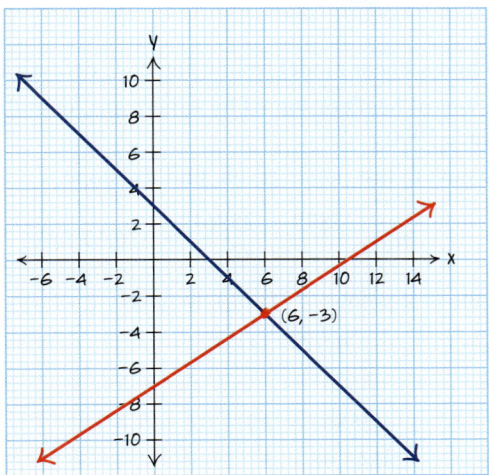

From the graph, the lines intersect at the point $(6, -3)$. Therefore, the solution to the system is $(6, -3)$. This solution should be checked by substituting the x- and y-values into both equations to make sure they are solutions to both equations.

First equation

$$y = \frac{2}{3}x - 7$$

$$-3 \stackrel{?}{=} \frac{2}{3}(6) - 7$$

$$-3 \stackrel{?}{=} 4 - 7$$

$$-3 = -3$$

Second equation

$$y = -x + 3$$

$$-3 \stackrel{?}{=} -(6) + 3$$

$$-3 \stackrel{?}{=} -6 + 3$$

$$-3 = -3$$

PRACTICE PROBLEM FOR EXAMPLE 2

Solve the following system by graphing the equations by hand.

$$y = \frac{1}{2}x - 2$$

$$y = -\frac{1}{4}x + 7$$

Example 3 Using models in a system of equations

The amounts of beef and chicken consumed by Americans are given in the table.

Year	Beef Consumption (millions of pounds)	Chicken Consumption (millions of pounds)
1996	25,861	21,845
1998	26,305	23,254
2000	27,338	25,606
2002	27,878	27,467
2004	28,534	28,678

Source: U.S. Department of Agriculture, National Agricultural Statistics Service.

a. Find a model for the amount of beef consumed by Americans.

b. Find a model for the amount of chicken consumed by Americans.

c. Estimate the year in which the amount of beef consumed will be the same as the amount of chicken consumed.

SOLUTION

a. Using the steps to find a linear model from Chapter 1, define the variables.

$B(t)$ = The amount of beef consumed by Americans, in millions of pounds
t = Years since 1990

Using the first and next to last points given, find the slope.

$$m = \frac{27878 - 25861}{12 - 6} \approx 336.17$$

Use the slope and the first point, (6, 25861), to find b.

$$B(t) = 336.17t + b$$
$$25861 = 336.17(6) + b$$
$$25861 = 2017.02 + b$$
$$23843.98 = b$$

Write the equation for the amount of beef consumed, using function notation.

$$B(t) = 336.17t + 23843.98$$

b. Now, for a model to find the amount of chicken consumed, define the variables.

$C(t)$ = Amount of chicken consumed by Americans, in millions of pounds
t = Years since 1990

Using the first and last points, find the slope.

$$m = \frac{28678 - 21845}{14 - 6} = 854.125$$

Use the slope and the first point, (6, 21845) to find b.

$$C(t) = 854.125t + b$$
$$21845 = 854.125(6) + b$$
$$21845 = 5124.75 + b$$
$$16720.25 = b$$

Write the equation for the amount of chicken consumed using function notation.

$$C(t) = 854.125t + 16720.25$$

c. Graph the two models on one window to see the intersection and trace to estimate the answer.

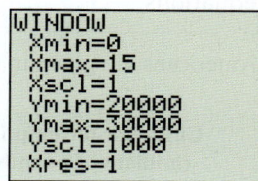
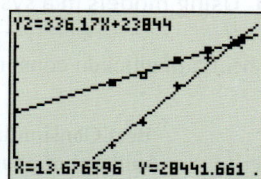

From this graph, the estimate of the solution is (13.68, 28,441.7). In 2004, the amounts of chicken and beef consumed by Americans were approximately 28,442 million pounds.

Using the graph and trace on your calculator is one way to check the solutions to many of the equations we will solve in this book. Another way for you to find solutions

Using Your TI Graphing Calculator

Using two stat plots:

To find a model for a second set of data, you can either erase the first set of data or put the second set of data into the stat lists with the first set and turn on a second stat plot.

In Example 3, the input data for both sets of data are the same years so you only need to add the third list with the chicken consumption data.

to systems of equations is to use the table feature of your calculator. You can also use the table feature to check a solution to a system of equations. First, enter both equations in the Y= screen and then go to the table. Enter the input values you want to check, and as long as the *y* values that are returned are the same, you have found a solution. In Example 2, we graphed and traced to find the approximate solution of (13.68, 28,441.7). Using these two functions and values close to 13.68 in the table, we get

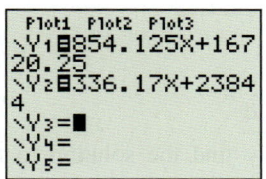

After entering the data, you can now turn on Stat Plot 2 and make sure that Xlist is set to L1 and Ylist is set to L3, where we put the chicken consumption data.

Now set your window to see both sets of data and create your model.

Sometimes it may be best to plot one set of data at a time when deciding on what points to pick for the modeling process.

From this table, a more accurate solution would be about (13.75, 28,465). Because we did not find an exact solution, we averaged the *y*-values to get an approximate solution. In this particular application, we round the answers to a whole number for the year, so the table does confirm that our solution is close.

Example 4 Solving systems of equations numerically

Use the table on the calculator to numerically find the solution to the following systems of equations.

a. $y = 2x + 7$
$y = 5x - 3.5$

b. $y = -9x + 177$
$y = 13x - 65$

SOLUTION

a. Using the table, guess a value for *x* to start and then guess again and again until the *y*-values are closer and eventually the same.

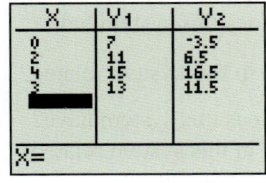

The first guesses were $x = 0$, $x = 2$, and $x = 4$. With the first two guesses, Y1 was greater than Y2. When $x = 4$, the relationship reversed. Now Y1 is less than Y2. This means that the place where they are equal must be between $x = 2$ and $x = 4$.

A guess of $x = 3$ resulted in Y1 being greater than Y2 again, so we will guess again. The answer seems to fall somewhere between 3 and 4.

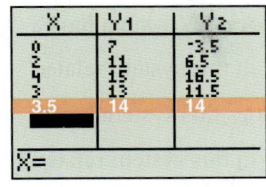

A guess of $x = 3.5$ results in Y1 equal to Y2, so we have found the solution to the system of equations.

From the last table, the solution is (3.5, 14).

b. Using the same guessing technique, we get the table

From this table, the solution is (11, 78).

PRACTICE PROBLEM FOR EXAMPLE 4

Use the table on the calculator to numerically find the solution to the systems of equations.

a. $y = 2x - 8$
   $y = -4x + 28$

b. $y = -\frac{3}{5}x + 9$
   $y = 0.8(7 - x) + 2.8$

Remember that the table feature on your calculator is a good way to get a solution numerically. However, this method can be difficult if you have no idea where the solution is. The table is a great way to check solutions to equations, and we will use it often in the remaining chapters of this book.

Types of Systems

■ CONCEPT INVESTIGATION

What kinds of systems are there?

To start, turn off all your stat plots, set your window to a standard window using ZOOM 6 (ZStandard), and clear all equations from the Y= screen.

a. Graph the following systems one at a time, and answer the questions.

1. $y = 2x - 5$ How many solutions does this system have?
   $y = -5x + 9$

2. $w = \frac{2}{3}p - 2$ How many solutions does this system have?
   $w = \frac{1}{9}p + 3$

b. Graph the following systems one at a time, and answer the questions.

1. $-x + 5y = -15$ How many solutions does this system have?
   $y = \frac{3}{15}x + 2$ How are the equations in this system related?

2. $P = 2.75t - 4.25$ How many solutions does this system have?
   $P = 2.75t + 3$ How are the equations in this system related?

c. Graph the following systems one at a time, and answer the questions.

1. $y = -\frac{1}{3}x + 7$ How many solutions does this system have?
   $21x + 7y = 49$ How are the equations in this system related?

2. $y = -\frac{14}{11}x - \frac{18}{11}$ How many solutions does this system have?
   $9.8x + 7.7y = -12.6$ How are the equations in this system related?

d. Describe in your own words why there are different types of answers to the systems of equations.

This concept investigation shows that systems of two linear equations can have three types of answers.

In mathematics we give the following names to these three types of systems.

- **Consistent system:** A system with at least one solution.
    1. **Independent lines:** The lines are different and intersect in one place.

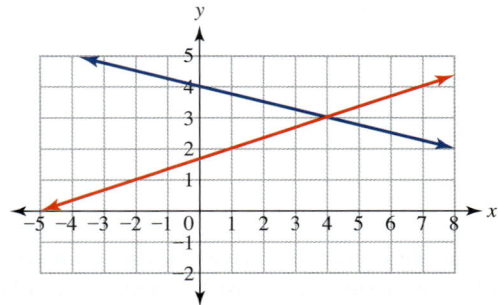

A single solution

   2. **Dependent lines:** The lines are the same and, therefore, intersect in infinitely many places. Each point on the line is a solution for the system. The system is said to have infinitely many solutions. Use caution when looking at the calculator because some lines might seem the same, but they really are not. For the lines to be the same, they must have the same slope and y-intercept.

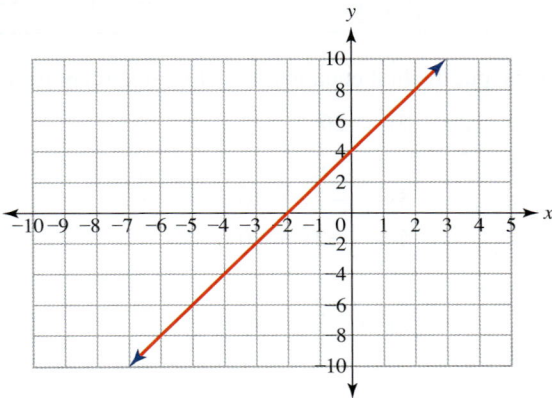

An infinite number of solutions.

Every point on the line is a solution to the system.

- **Inconsistent system:** A system with no solutions. The lines are parallel and, therefore, do not intersect. Use caution when looking at the calculator because some lines might seem parallel, but they really are not. Recall that parallel lines have the same slope but different y-intercepts. If you compare the slopes of the equations, you can determine whether the system is inconsistent.

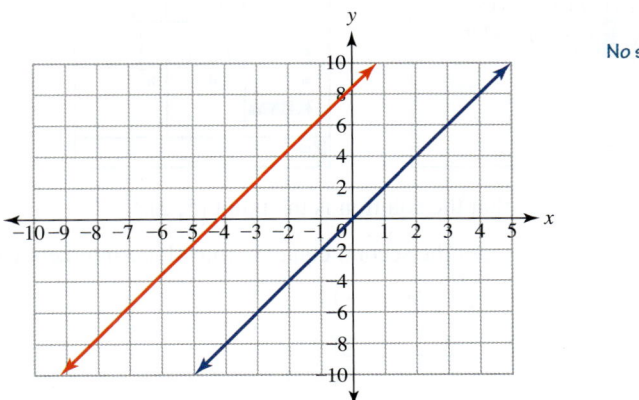

No solution

Getting an idea for the visual representation of these solutions will help in the next section when we solve systems algebraically.

Example 5 Describing types of systems of equations

For each of the following systems of equations, determine whether the system is consistent or inconsistent. If the system is consistent, determine whether the lines are independent or dependent. Give the solution to the system.

a. $y = 3x + 5$
$y = 3x - 2$

b. $2x + 5y = 20$
$y = 4x - 18$

c. $y = \dfrac{1}{3}x + 4$
$x - 3y = -12$

SOLUTION

a. The slopes are both $m = 3$, but the equations do not have the same y-intercept. Therefore, these lines are parallel. The system is inconsistent and has no solutions. If we look at a table of points for these lines, the outputs will always be 7 units apart.

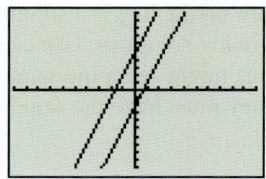

b. To compare these equations, put the first equation into slope-intercept form.

$$2x + 5y = 20$$
$$\underline{-2x \qquad\qquad -2x}$$
$$5y = -2x + 20$$
$$\dfrac{5y}{5} = \dfrac{-2x + 20}{5}$$
$$y = -\dfrac{2}{5}x + 4$$

Compare the equations

$$y = -\dfrac{2}{5}x + 4$$
$$y = 4x - 18$$

The slopes are different, so the system is consistent with independent lines. Using the table, we can find the solutions numerically.

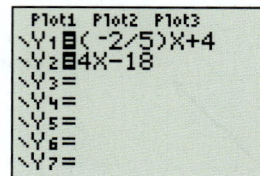

From the table, we can see that the solution is the point (5, 2).

c. To compare these equations, put the second equation into slope-intercept form.

$$x - 3y = -12$$
$$\underline{-x \qquad\qquad -x}$$
$$-3y = -x - 12$$
$$\dfrac{-3y}{-3} = \dfrac{-x - 12}{-3}$$
$$y = \dfrac{1}{3}x + 4$$

Now, if we compare the equations

$$y = \frac{1}{3}x + 4$$

$$y = \frac{1}{3}x + 4$$

we can see that the equations have the same slope and *y*-intercept, so they are the same line. This is a consistent system with dependent lines and has an infinite number of solutions.

Any solution of the equation $y = \frac{1}{3}x + 4$ is a solution to the system.

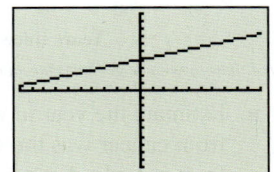

PRACTICE PROBLEM FOR EXAMPLE 5

For each of the following systems of equations, determine whether the system is consistent or inconsistent. If the system is consistent, determine whether the lines are independent or dependent. Give the solution to the system.

a. $5x + y = 3$
  $y = -5x - 7$

b. $y = \frac{1}{4}x + 7$
  $y = 0.125(2x + 40) + 2$

2.1 Exercises

For Exercises 1 through 6, use the graphs to find the solution to the system of equations. Write the solution as a coordinate pair.

1.

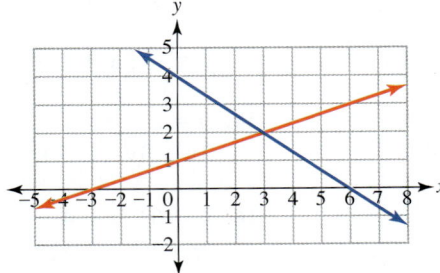

2.

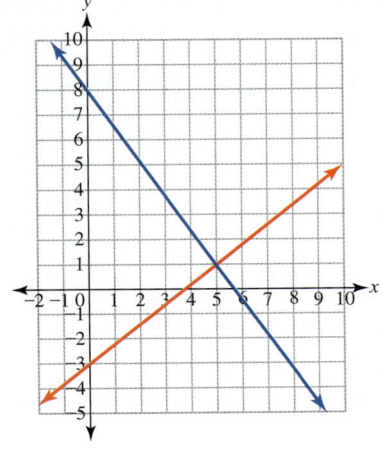

3.

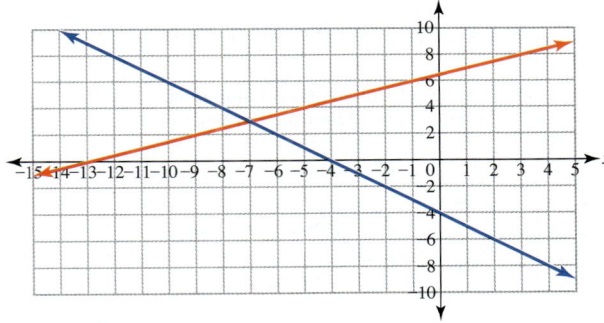

4.

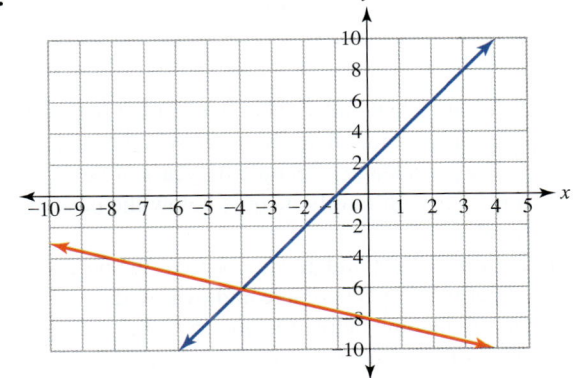

5.

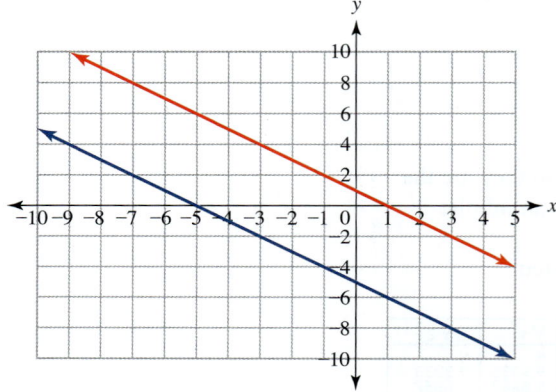

6.

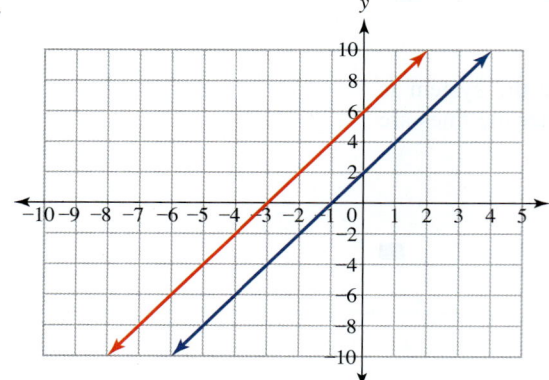

7. Use the graph to answer the questions.

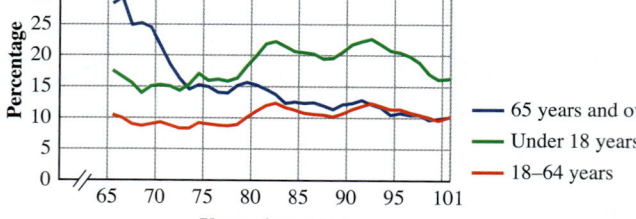

Source: http://www.cdc.gov/nchs/hus.htm

a. Estimate when and at what percent the percentage of people under 18 years of age was the same as the percentage of people 65 years of age and over with family income below the poverty level.

b. Estimate when and at what percent the percentage of people 65 years of age and over with family income below the poverty level was first equal to the percentage of people 18–64 years of age.

8. Use the graph to answer the questions.

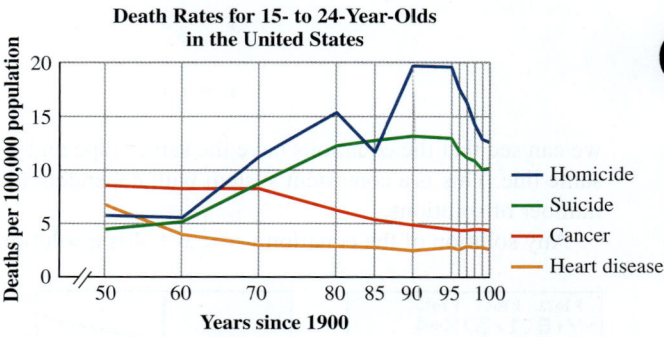

Source: http://www.cdc.gov/nchs/hus.htm

a. Estimate the year in which the number of deaths from cancer was the same as the number of deaths from suicide. Approximate the number of deaths per 100,000 population from each of these causes in that year.

b. Estimate the year in which the number of deaths from heart disease was the same as the number of deaths from homicide. Approximate the number of deaths per 100,000 population from each of these causes in that year.

For Exercises 9 through 16, graph each system by hand and find the solution(s) to the system. Label each system as consistent or inconsistent. If the system is consistent determine whether the lines are independent or dependent.

9. $y = 2x + 4$
$y = -\frac{1}{3}x + 11$

10. $y = x - 2$
$y = \frac{1}{2}x + 2$

11. $y = \frac{2}{3}x - 4$
$y = -\frac{1}{3}x - 10$

12. $y = \frac{1}{5}x + 4$
$y = -\frac{2}{5}x + 1$

13. $y = 0.5x + 2$
$x - 2y = 8$

14. $y = x - 5$
$3x - 3y = -24$

15. $n = \frac{2}{3}m + 7$
$6m - 9n = -63$

16. $p = 0.25t - 4$
$2t + 8p = -32$

17. The amount of manufacturing output per hour in dollars in the United States and Norway is given in the table.

Year	United States ($)	Norway ($)
2000	147.7	105.9
2003	175.5	121.6
2004	187.8	128.8
2005	194.0	132.4

Source: Statistical Abstract of the United States, 2008.

a. Find equations for models of these data.

b. Graph both models on the same calculator screen.

c. Estimate when the amount of manufacturing output per hour in the United States will be equal to the amount of manufacturing output per hour in Norway.

18. The number of women enrolled in U.S. colleges has been increasing steadily since the 1970s. The numbers of women and men enrolled in U.S. colleges are given in the following bar charts.

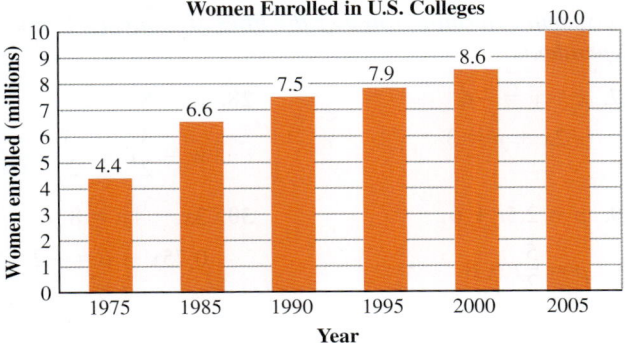

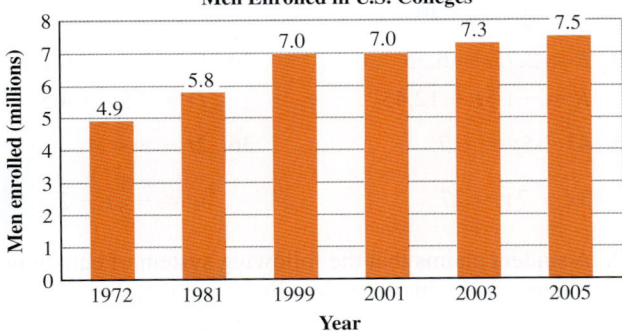

Source: *Statistical Abstract of the United States, 2001 and 2008.*

a. Find equations for models for the numbers of women and men enrolled in U.S. colleges.
b. Graph both models on the same calculator screen.
c. Estimate when the number of women enrolled was the same as the number of men enrolled.

19. The percentages of white male and female Americans 25 years old or older who are college graduates are given in the table.

Year	Males (%)	Females (%)
1980	21.3	12.3
1985	24	16.3
1990	25.3	19
1995	27.2	21
2000	28.5	23.9
2005	29.4	26.8

Source: *Statistical Abstract of the United States, 2001 and 2008.*

a. Find equations for models for the data given.
b. Graph both models on the same calculator screen.
c. Estimate when the percentage of white American men with college degrees will be the same as the percentage of white American women with college degrees.

20. The number of bachelor's degrees earned by blacks and Hispanics in the United States is given in the bar chart.

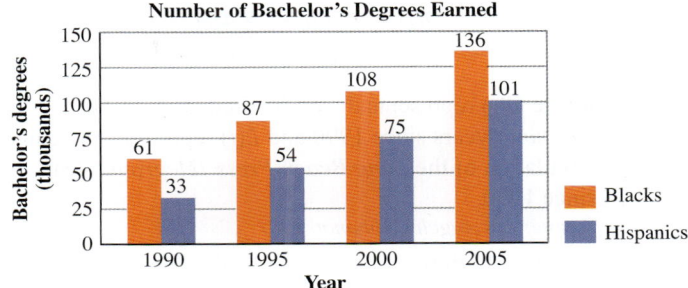

Source: *Statistical Abstract of the United States, 2008.*

a. Find equations for models for the bachelor's degrees earned data given.
b. Graph both models on the same calculator screen.
c. Determine in what year blacks and Hispanics will earn the same number of bachelor's degrees.

21. Hope's Pottery makes clay vases to sell at a local gallery. Hope has determined the following models for her monthly revenue and costs associated with making and selling these vases.

$R(v) = 155v$

$C(v) = 5000 + 65v$

where $R(v)$ represents the monthly revenue in dollars for selling v vases and $C(v)$ represents the monthly costs in dollars for producing and selling v vases.

a. Graph both functions on the same calculator screen.
b. Determine the break-even point for Hope's Pottery. (The break-even point is when the revenue equals the costs.)

22. Jim's Carburetors rebuilds carburetors for local auto repair shops. Jim has determined the following models for his monthly revenue and costs associated with rebuilding these carburetors.

$$R(c) = 65c$$
$$C(c) = 2300 + 17c$$

where $R(c)$ represents the monthly revenue in dollars for rebuilding c carburetors and $C(c)$ represents the monthly costs in dollars for rebuilding c carburetors.

a. Graph both functions on the same calculator screen.
b. Determine the break-even point for Jim's Carburetors.

23. *La Opinion*, a Spanish-language newspaper in Los Angeles County, California, is one of the area's fastest-growing daily newspapers. The *Long Beach Press Telegram* is another large daily newspaper in Los Angeles County. The daily circulation for both of these newspapers is given by the following functions:

$$O(t) = 7982t + 28{,}489$$
$$L(t) = 726t + 97{,}395$$

where $O(t)$ represents the daily circulation of *La Opinion* t years since 1990 and $L(t)$ represents the daily circulation of the *Long Beach Press Telegram* t years since 1990.
Source: Los Angeles Almanac 2001.

   a. Graph both functions on your calculator to estimate when the two newspapers had the same daily circulation.
   b. Compare the slopes of these two functions and describe what they represent.

24. The number of doctoral degrees given to blacks and Asian or Pacific Islanders in the United States can be modeled by the following functions:

$$B(t) = 126t + 1084.5$$
$$A(t) = 95.78t + 1592.9$$

where $B(t)$ represents the number of doctoral degrees earned by blacks t years since 1990 and $A(t)$ represents the number of doctoral degrees earned by Asian and Pacific Islanders t years since 1990.
Source: Statistical Abstract 2008.

   a. Graph both functions on your calculator to estimate when blacks and Asian and Pacific Islanders will earn the same number of doctoral degrees.
   b. Compare the slopes of these two functions and describe what they represent.

25. A local BMW dealer has two salary options for its sales force. The sales force can choose either of the following two salary options:

$$O_1(s) = 250 + 0.03s$$
$$O_2(s) = 0.03s + 200$$

where $O_1(s)$ represents weekly salary option 1 in dollars when s dollars in sales are made per week and $O_2(s)$ represents weekly salary option 2 in dollars when s dollars in sales are made per week. Find what sales level will give a salesperson the same salary with either option.

26. Harry's Flooring has two salary options for its salespeople. Each salesperson can choose which option to base his or her salary on:

$$O_1(s) = 500 + 0.07s$$
$$O_2(s) = 0.07s + 700$$

where $O_1(s)$ represents monthly salary option 1 in dollars when s dollars in sales are made per month and $O_2(s)$ represents monthly salary option 2 in dollars when s dollars in sales are made per month. Find what sales level would result in the two salary options having the same monthly salary.

For Exercises 27 through 36, graph each system by hand and find the solution(s) to the system. Label each system as consistent or inconsistent. If the system is consistent determine whether the lines are independent or dependent.

27. $x + y = -6$
    $-2x + y = 3$

28. $w = 8d + 54$
    $w = \dfrac{3}{2}d + \dfrac{17}{2}$

29. $p = 2.5t + 6$
    $p = \dfrac{5}{2}t - 6$

30. $3x - 4y = 8$
    $0.75x - y = -2$

31. $y = \dfrac{1}{3}x + 5$
    $2x - 6y = -30$

32. $H = 3c + 6.5$
    $-6c + 2H = 13$

33. $R = 2.75t + 6.35$
    $R = -1.5t + 12.45$

34. $D = 3.5c + 6.5$
    $D = 2.5c - 4.5$

35. $4x + 5y = -7$
    $3x - 7y = 27$

36. $2t - w = 4$
    $w = -\dfrac{1}{2}t - 2$

37. A student claims that the following system of equations is consistent with dependent lines, and that there are an infinite number of solutions. Explain why this answer is wrong. Give the correct description of the system.

$$y = 2x + 5$$
$$y = 2x - 3$$

38. A student claims that the following system of equations is inconsistent, and that there are an infinite number of solutions. Explain why this answer is wrong. Give the correct description of the system.

$$y = \dfrac{1}{4}x + 4$$
$$y = 0.25x + 4$$

39. A student says that a consistent system always has only one answer. Is this statement correct? If not explain why.

40. A student says that an inconsistent system has an infinite number of solutions. Explain what is wrong with this statement.

41. Sketch the graph of a consistent system with a solution $(2, 4)$.

42. Sketch the graph of a consistent system with a solution $(-5, -3)$.

43. Sketch the graph of an inconsistent system.

44. Sketch the graph of a consistent system with dependent lines.

For Exercises 45 through 52, solve the systems numerically using the given tables. Label each system as consistent or inconsistent. If the system is consistent determine whether the lines are independent or dependent.

45.

X	Y₁	Y₂
0	5	19
1	8	15
2	11	11
3	14	7
4	17	3
5	20	-1

X=

46.

X	Y₁	Y₂
0	5	-23
-1	2	-19
-2	-1	-15
-3	-4	-11
-4	-7	-7
-5	-10	-3
-6	-13	1

X= -6

47.

X	Y₁	Y₂
0	3	-7
5	8	-2
10	13	3
15	18	8
20	23	13
25	28	18

X=

48.

X	Y₁	Y₂
0	15	4
1	13	2
2	11	0
3	9	-2
4	7	-4
5	5	-6

X=

49.

X	Y₁	Y₂
0	1	1
1	3	3
2	5	5
3	7	7
4	9	9
5	11	11

X=0

50.

X	Y₁	Y₂
0	-4	-4
2	-5.333	-5.333
4	-6.667	-6.667
6	-8	-8
8	-9.333	-9.333
10	-10.67	-10.67

X=

51.

X	Y₁	Y₂
0	5	-5
1	8	-2
2	11	1
3	14	4
4	17	7
5	20	10
6	23	13

X=6

52.

X	Y₁	Y₂
0	9	8
1	11	10
2	13	12
3	15	14
4	17	16
5	19	18
6	21	20

X=0

For Exercises 53 through 60, use the table on the calculator to solve the systems numerically. Label each system as consistent or inconsistent. If the system is consistent determine whether the lines are independent or dependent.

53. $p = 3t + 8$
    $p = 2t + 17$

54. $y = 5x - 20$
    $y = 2x + 1$

55. $2.5x - 0.5y = -23.5$
    $7.5x - 1.5y = -23.5$

56. $R = 0.8t + 6$
    $R = \frac{4}{5}t + 40.5$

57. $C = 1.4n + 74$
    $7n - 5C = -370$

58. $8t + 4p = 23$
    $2p = -4t + 11.5$

59. $y = 2x - 20$
    $y = -4x + 70$

60. $y = 5x - 20$
    $y = -2x + 8$

2.2 Solving Systems of Equations Using the Substitution Method

LEARNING OBJECTIVES

- Solve a system of equations using the substitution method.
- Solve applications using systems of equations and the substitution method.

In Section 2.1, we learned how to solve systems of two linear equations by graphing and numerically using tables. These methods can be done on a graphing calculator. One drawback to solving some systems by graphing or numerically is that we have to have an idea of where the intersection is going to be in order to find it. Another concern when solving systems graphically or numerically is that we may not be able to find the exact answers. To avoid these problems, we can use algebraic methods that will find the exact solution whenever possible.

Substitution Method

There are two basic algebraic ways to solve systems of equations: the **substitution method** and the **elimination method.** Because most of the models that we find in this text are in $y = mx + b$ form, we will learn the substitution method first. The substitution method means to substitute the expression of a variable from one equation into that same variable in the other equation.

> **Substitution Method for Solving Systems of Equations**
> - Used best when a variable is already isolated in at least one equation.
> - Substitute the expression representing the isolated variable from one equation in place of that variable in the other equation.
> - Remember to find the values for both variables.
> - Check the solution in both equations.

Example 1 Solving systems using the substitution method

Solve the system using the substitution method.

$$w = 5k + 7$$
$$w = -3k + 23$$

SOLUTION

Because w is already isolated on the left side of at least one of the equations, substitute the expression from the first equation into the w for the second equation.

$$5k + 7 = -3k + 23$$

Now that we have substituted, we can solve for k.

$$
\begin{array}{r}
5k + 7 = -3k + 23 \\
+3k +3k \\
\hline
8k + 7 = 23 \\
-7 -7 \\
\hline
8k = 16 \\
\dfrac{8k}{8} = \dfrac{16}{8} \\
k = 2
\end{array}
$$

We know that $k = 2$, so we can substitute 2 for k in either equation and find w.

$$w = 5k + 7$$
$$w = 5(2) + 7$$
$$w = 17$$

The solution to the system is the point (2, 17). We should always check the solution in both equations to be sure that it is valid.

First equation	Second equation
$w = 5k + 7$	$w = -3k + 23$
$17 \stackrel{?}{=} 5(2) + 7$	$17 \stackrel{?}{=} -3(2) + 23$
$17 = 17$	$17 = 17$

The point satisfies both equations, so it is the correct answer.

Example 2 Using the substitution method to solve systems with models

Using the models from Section 2.1 for beef and chicken consumption by Americans, find the year when beef and chicken consumption was or will be the same.

$$B(t) = 336.17t + 23844$$
$$C(t) = 854.125t + 16720.25$$

where $B(t)$ is the amount of beef consumed by Americans in millions of pounds t years since 1990 and $C(t)$ is the amount of chicken consumed by Americans in millions of pounds t years since 1990.

SOLUTION
Because both $B(t)$ and $C(t)$ represent amounts of meat consumed by Americans and both are measured in millions of pounds, we can substitute the expression that $B(t)$ equals $C(t)$ in the second equation.

$$B(t) = 336.17t + 23844$$
$$C(t) = 854.125t + 16720.25$$
$$B(t) = C(t)$$

$336.17t + 23844 = 854.125t + 16720.25$ Set the two expressions equal to each other. Solve for t.

$$\frac{-854.125t \qquad\qquad -854.125t}{-517.955t + 23844 = 16720.25}$$

$$\frac{\qquad -23844 \quad -23844}{-517.955t = -7123.75}$$

$$\frac{-517.955t}{-517.955} = \frac{-7123.75}{-517.955}$$

$$t \approx 13.7536$$

Now that we have the value of t, we want to find the amounts of beef and chicken consumed that year. We can do that by substituting this value for t in either equation.

$$B(13.7536) = 336.17(13.7536) + 23844$$
$$B(13.7536) \approx 28467.548$$

or

$$C(13.7536) = 854.125(13.7536) + 16720.25$$
$$C(13.7536) \approx 28467.544$$

Since both equations came to the same approximate answer, we have found the solution to this system. But because t represents a year, we will round it to the nearest whole number—in this case, 14. If we substitute 14 into the variable t for these equations, we will find that they are not exactly equal.

First equation	Second equation
$B(14) = 336.17(14) + 23844$	$C(14) = 854.125(14) + 16720.25$
$B(14) = 28550.38$	$C(14) = 28678$

This gives us the following approximate answer. Americans ate approximately 28,600 million pounds of both beef and chicken in 2004.

PRACTICE PROBLEM FOR EXAMPLE 2

The Pharmaceutical Management Agency of New Zealand (PHARMAC) manages the expenditures on various medical conditions and drugs throughout New Zealand. The amount spent on treating diabetes in New Zealand can be modeled by

$$T(t) = 1.1t + 6.84$$

where $T(t)$ represents the amount in millions of New Zealand dollars spent on treating diabetes t years since 1990. Also, the amount spent on diabetes research in New Zealand can be modeled by

$$R(t) = 1.3t + 0.52$$

where $R(t)$ represents the amount in millions of New Zealand dollars spent on diabetes research in New Zealand t years since 1990.
Source: PHARMAC.

Find the year in which the amount spent in New Zealand on treating diabetes will be the same as the amount spent in New Zealand on diabetes research.

Example 3 Solving systems using the substitution method

Solve the following systems using the substitution method.

a. $p = 2m - 12$
$p - 3m = -17$

b. $4c = 2d + 10$
$6c + 2d = -5$

SOLUTION

a. p is isolated in the first equation, so the expression that p is equal to can be substituted into p for the second equation.

$$p = 2m - 12$$
$$p - 3m = -17$$

Substitute $2m - 12$ for p in the second equation and solve for m.

$$(2m - 12) - 3m = -17$$
$$-1m - 12 = -17$$
$$-1m = -5$$
$$m = 5$$

We know that $m = 5$, so we can substitute 5 for m in either of the equations to find p.

$$p = 2(5) - 12$$
$$p = -2$$

Therefore, we believe that the solution to this system is $(5, -2)$. Check this solution in both equations to be sure that it is valid.

First equation
$-2 \stackrel{?}{=} 2(5) - 12$
$-2 = -2$

Second equation
$-2 - 3(5) \stackrel{?}{=} -17$
$-17 = -17$

b. Neither variable is already isolated, so we can solve the first equation for c, and then that expression can be substituted into c for the second equation.

$$4c = 2d + 10$$
$$\frac{4c}{4} = \frac{2d + 10}{4}$$
$$c = 0.5d + 2.5$$

Substitute this expression into c and solve for d.

$$6(0.5d + 2.5) + 2d = -5$$
$$3d + 15 + 2d = -5$$
$$5d + 15 = -5$$
$$5d = -20$$
$$d = -4$$

It is important to use parentheses around the expression you are substituting into the variable.

We know that $d = -4$, so we can substitute -4 for d in either of the equations to find c.

$$4c = 2(-4) + 10$$
$$4c = -8 + 10$$
$$4c = 2$$
$$c = 0.5$$

Therefore, the solution to this system is $c = 0.5$ and $d = -4$. Check this solution in both equations to be sure that it is valid.

First equation   **Second equation**

$(0.5) \stackrel{?}{=} 2(-4) + 10$ $6(0.5) + 2(-4) \stackrel{?}{=} -5$

$2 = 2$ $-5 = -5$

PRACTICE PROBLEM FOR EXAMPLE 3

Solve the following systems using the substitution method.

a. $y = 3x + 12$
$y = 20x - 158$

b. $n = 4m + 3$
$n - 3m = 8.5$

Example 4   Solving applications using systems of equations

Clear Sign Designs, a local sign manufacturer, wants to give its sales force incentives to make bigger sales. To do this, the company executives are changing the salary structure. The old system was a base salary of $1000 per month and 8% commission on sales made. The new salary structure will consist of a base salary of $500 per month with a 10% commission on sales made.

a. Find equations to represent the new and old salary structures.

b. Find what sales amount will result in the same monthly salary for the sales force.

SOLUTION

a. Define the variables.

$N(s)$ = Salary in dollars for the new salary structure
$O(s)$ = Salary in dollars for the old salary structure
s = Sales made during the month in dollars

Changing the commission percentages into decimals, we get the following two equations.

$$N(s) = 500 + 0.10s$$
$$O(s) = 1000 + 0.08s$$

b. To find the sales amount that will result in the same salary, we want the new salary $N(s)$ to be equal to the old salary $O(s)$. We can substitute the expression $500 + 0.10s$ into the second equation for $O(s)$. Solving this will give us the sales amount that will result in the same monthly salary. Using that sales amount, we can find the actual monthly salary earned.

$$N(s) = 500 + 0.10s$$
$$O(s) = 1000 + 0.08s$$
$$N(s) = O(s)$$

$500 + 0.10s = 1000 + 0.08s$	Set the two expressions equal to each other.
$\underline{-500 - 0.08s \quad -500 - 0.08s}$	
$0.02s = 500$	Solve for s.
$\dfrac{0.02s}{0.02} = \dfrac{500}{0.02}$	
$s = 25000$	

so

$N(25000) = 500 + 0.10(25000)$	Find N when $s = 25000$.
$N(25000) = 3000$	
	Check the solution in the table.

Therefore, if a salesperson has monthly sales of $25,000, the salary will be the same, $3000, with the new and old salary structures.

> **What's That Mean?**
>
> **Simple Interest**
>
> Simple interest for one year is calculated as the amount invested multiplied by the interest rate (as a decimal).
>
> If we have $5000 invested at 4.5% simple interest for one year, the interest will be calculated as
>
> $$5000(0.045) = 225$$
>
> The simple interest is $225. Remember to change the interest rate into a decimal by multiplying by $\dfrac{1}{100}$ or 0.01.

Example 5 — Solving investment problems using the substitution method

Fay Clark is retired and has $500,000 to invest. She wants her investments to earn $24,000 per year in interest so that she can live on the interest. She is considering depositing most of the funds in a very safe bank account that pays 3.5% simple interest per year and the rest in a more risky account that pays 10% simple interest per year.

a. Write a system of equations that will help Fay find the amount she should invest in each investment account.

b. How much should she invest in each account to earn the $24,000 she needs each year?

(*Note*: Some financial accounts, such as bank deposits, are federally insured, while others, such as mutual funds and brokerage accounts, are not federally insured and are, therefore, considered more risky.)

SOLUTION

a. There are two main factors controlling this situation: the amount of money Fay can invest and the total amount of interest income she needs to earn each year. If we set up two equations using these facts, we can solve the system for the amounts in each account. We start by defining the variables.

A = The amount invested in dollars in the account paying 3.5% simple interest
B = The amount invested in dollars in the account paying 10% simple interest

With these definitions and the conditions we were given, we can write the following two equations.

$$A + B = 500000 \quad \text{The two accounts total } \$500{,}000.$$
$$0.035A + 0.10B = 24000 \quad \text{The total interest needs to be } \$24{,}000.$$

b. To find the amount that Fay should deposit in each account, we solve the system. To use the substitution method, we will isolate either variable. For this system, we will isolate the variable A in the first equation and substitute the expression it equals into the second equation.

Subtract B from both sides of the first equation to isolate A.

$$A + B = 500000 \quad \longrightarrow \quad A = 500000 - B$$
$$0.035A + 0.10B = 24000 \quad \longrightarrow \quad 0.035A + 0.10B = 24000$$

Substitute $500000 - B$ for A in the second equation. Solve for B.

$$0.035(500000 - B) + 0.10B = 24000 \quad \text{Substitute } 500000 - B \text{ for } A \text{ in the second equation.}$$
$$17500 - 0.035B + 0.10B = 24000 \quad \text{Distribute the 0.035 through the parentheses.}$$
$$17500 + 0.065B = 24000 \quad \text{Solve for } B.$$
$$\underline{-17500 \quad\quad\quad -17500}$$
$$0.065B = 6500$$
$$\frac{0.065B}{0.065} = \frac{6500}{0.065}$$
$$B = 100000$$

Now that we have found B, we can substitute $\$100{,}000$ for B in the first equation and solve for A.

$$A + 100000 = 500000 \quad \text{Solve for } A.$$
$$A = 400000$$

Check the solution in both equations.

First equation

$400000 + 100000 \stackrel{?}{=} 500000$

$500000 = 500000$

Second equation

$0.035(400000) + 0.10(100000) \stackrel{?}{=} 24000$

$24000 = 24000$

This solution tells us that Fay should deposit $\$400{,}000$ into the account paying 3.5% interest and $\$100{,}000$ into the more risky account paying 10% interest. This will give her a total of $\$24{,}000$ in interest each year.

PRACTICE PROBLEM FOR EXAMPLE 5

George is retiring and has $\$350{,}000$ to invest. George needs $\$18{,}000$ per year more than his Social Security to live on. George has two investments that he plans to use to earn the interest he needs to live on. One investment is conservative and pays 3.6% simple interest annually; the other is less conservative and pays 6% simple interest annually. Determine how much George should deposit into each account to make the interest he needs to live on each year. Note that George must invest the entire $\$350{,}000$.

Consistent and Inconsistent Systems

Recall from Section 2.1 that there are three types of solutions to linear systems. Although most systems will have a single solution, there are times when a system may be consistent with dependent lines or inconsistent and have an infinite number of solutions or no solutions, respectively. In solving a system of equations algebraically, finding out when you have a consistent system with dependent lines or an inconsistent system happens rather suddenly. Keeping an eye open for systems of parallel lines (inconsistent systems) or systems with multiples of the same equation (consistent system with dependent

> **What's That Mean?**
>
> **Consistent System**
>
> A system with at least one solution is called a consistent system. Consistent systems have one of the following:
> **Independent lines** and one solution. This occurs when the equations have different slopes.
> or
> **Dependent lines** and infinitely many solutions. This occurs when one equation is a multiple of the other equation. Once simplified, the two equations are the same.
>
> **Inconsistent System**
>
> A system with no solution is called an inconsistent system. These systems have two parallel lines. The equations have the same slopes but different y-intercepts.

lines) is one way to notice these situations. But these patterns are not always clear, since the equations can look very different yet be remarkably similar or the same.

While solving consistent systems with dependent lines or inconsistent systems, you will notice that all the variables will be eliminated, and you will be left with two numbers set equal to each other. If these remaining numbers are actually equal, then the equations must have been the same. You have a consistent system with dependent lines and an infinite number of solutions. If the remaining numbers are not equal, then the equations were not the same, so the system is inconsistent and there is no solution.

Example 6 — Solving systems and determining the system type

Solve the following systems. Label each system as consistent or inconsistent. If the system is consistent, determine whether the lines are independent or dependent.

a. $r = 4.1t + 6.3$
   $3r - 12.3t = 18.9$

b. $y = 4x + 9$
   $y = 4x - 9$

c. $S = 1500 + 0.05c$
   $S = 0.05c + 750$

SOLUTION

a. This system is set up for substitution because one variable is already isolated.

$$r = 4.1t + 6.3$$
$$3r - 12.3t = 18.9$$

Substitute $4.1t + 6.3$ for r in the second equation. Solve for t.

$3(4.1t + 6.3) - 12.3t = 18.9$ Substitute for r.
$12.3t + 18.9 - 12.3t = 18.9$
$18.9 = 18.9$ These numbers are equal.

Because both variables were eliminated and the remaining numbers are equal, this system is consistent and has dependent lines. All inputs and outputs that solve one equation will solve the other. There are an infinite number of solutions, and any solution to the equation $r = 4.1t + 6.3$ is a solution to the system.

This system is a set of two equations that are in fact the same linear function. If both of these equations are put into slope-intercept form, they are exactly the same.

First equation	Second equation
$r = 4.1t + 6.3$	$3r - 12.3t = 18.9$
	$3r = 12.3t + 18.9$
	$\dfrac{3r}{3} = \dfrac{12.3t + 18.9}{3}$
	$r = 4.1t + 6.3$

b. This system has two equations with the same slope but different vertical intercepts, so they are parallel lines. Therefore, this system is inconsistent and has no solution. If you try the substitution method, both variables will be eliminated, but the remaining numbers will not be equal.

$4x + 9 = 4x - 9$ Substitute for y.
$\underline{-4x -4x }$
$9 \neq -9$ These numbers are not equal.

c. This system of equations is set up for substitution because the variable S is isolated in both equations.

$$S = 1500 + 0.05c$$
$$S = 0.05c + 750$$

SECTION 2.2 Solving Systems of Equations Using the Substitution Method

Substitute for S. Solve for c.

$$1500 + 0.05c = 0.05c + 750 \quad \text{Substitute for S.}$$
$$\underline{-0.05c \quad -0.05c}$$
$$1500 \neq 750 \quad \text{These numbers are not equal.}$$

Because the variables were both eliminated but the remaining numbers are not equal, we have an inconsistent system with no solution.

This system is a set of two parallel lines. If you put the first equation into slope-intercept form, you can see that both equations have the same slope.

PRACTICE PROBLEM FOR EXAMPLE 6

Solve the following systems. Label each system as consistent or inconsistent. If the system is consistent, determine whether the lines are independent or dependent.

a. $a = 1.25b + 4$
$8.75b - 7a = 14$

b. $x = -\dfrac{7}{6}y + \dfrac{2}{3}$
$-3x - 3.5y = -2$

2.2 Exercises

For Exercises 1 through 10, solve the systems using the substitution method. Label each system as consistent or inconsistent. If the system is consistent, determine whether the lines are independent or dependent. Check your answer.

1. $y = x + 7$
$y = 2x + 3$

2. $y = 6x - 20$
$y = 5x - 15$

3. $P = -5t + 20$
$P = 2t - 8$

4. $B = 4c + 7$
$B = -3c - 7$

5. $y = 2x - 5$
$y = \dfrac{1}{3}x + 5$

6. $W = 4p - 20$
$W = \dfrac{1}{2}p + 15$

7. $H = k + 8$
$H = 0.5k + 14$

8. $C = 0.4b + 8$
$C = 0.2b + 10$

9. $y = \dfrac{1}{2}x + \dfrac{3}{2}$
$y = \dfrac{3}{2}x - \dfrac{7}{2}$

10. $y = \dfrac{1}{5}x + \dfrac{11}{5}$
$y = \dfrac{2}{5}x + \dfrac{19}{5}$

11. Hope's Pottery from Exercise 21 in Section 2.1 on page 141 has the revenue and cost functions

$$R(v) = 155v$$
$$C(v) = 5000 + 65v$$

where $R(v)$ represents the monthly revenue in dollars for selling v vases and $C(v)$ represents the monthly costs in dollars for producing and selling v vases. Use the substitution method to algebraically find the break-even point for Hope's Pottery.

12. Jim's Carburetors from Exercise 22 in Section 2.1 on page 141 has the revenue and cost functions

$$R(c) = 65c$$
$$C(c) = 2300 + 17c$$

where $R(c)$ represents the monthly revenue in dollars for rebuilding c carburetors and $C(c)$ represents the monthly costs in dollars for rebuilding c carburetors. Use the substitution method to algebraically find the break-even point for Jim's Carburetors.

13. Optimum Traveling Detail comes to you and details your car at work or home. The owner has determined the following functions for the monthly revenue and cost.

$$R(a) = 175a$$
$$C(a) = 25a + 7800$$

where $R(a)$ represents the monthly revenue in dollars for detailing a automobiles and $C(a)$ represents the monthly costs in dollars for detailing a automobiles. Use the substitution method to algebraically find the break-even point for Optimum Traveling Detail.

14. A Table Affair sells European tablecloths at local farmers markets and can model their weekly revenue and cost using the following functions.

$$R(t) = 120t$$
$$C(t) = 45t + 800$$

where $R(t)$ represents the weekly revenue in dollars for selling t tablecloths and $C(t)$ represents the weekly costs in dollars for selling t tablecloths. Use the substitution method to algebraically find the break-even point for A Table Affair.

15. The number of associate's and master's degrees conferred at U.S. colleges can be modeled by the functions

$$A(t) = 8.61t + 641.25$$
$$M(t) = 12.50t + 543.07$$

where $A(t)$ represents the number of associate's degrees in thousands conferred during the school year ending t years after 2000 and $M(t)$ represents the number of master's degrees in thousands conferred during the school year ending t years after 2000. Find in what year the number of associate's degrees conferred will be the same as the number of master's degrees.
Source: Models derived from data in the Statistical Abstract of the United States, 2008.

16. The number of black and Hispanic students enrolled in U.S. schools can be modeled by the functions

$$B(t) = 107.7t + 10227.8$$
$$H(t) = 337.4t + 10424.2$$

where $B(t)$ represents the number of blacks in thousands enrolled in U.S. schools t years since 2000 and $H(t)$ represents the number of Hispanics in thousands enrolled in U.S. schools t years since 2000. Find in what year the number of blacks equaled the number of Hispanics enrolled in U.S. schools.
Source: Models derived from data in the Statistical Abstract of the United States, 2008.

17. The student-teacher ratio for public and private schools in the United States is given in the chart.

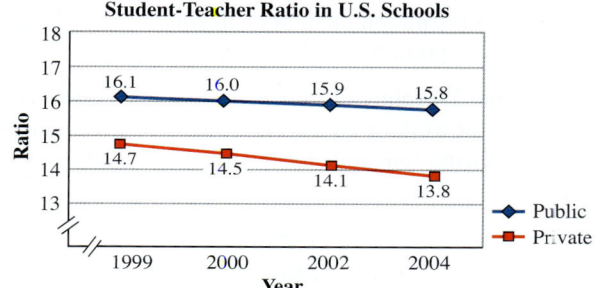

Source: Statistical Abstract of the United States, 2008.

   a. Find equations for models for the student-teacher ratios at public and private schools in the United States.
   b. Determine when the student-teacher ratio will be the same in public and private schools.

18. The number of males and females who participated in high school athletic programs for various school years is given in the table.

School Year	Males	Females
03–04	4,039,253	2,865,299
04–05	4,110,319	2,908,390
05–06	4,206,549	2,953,355
06–07	4,321,103	3,021,807

Source: www.infoplease.com

   a. Find equations for models of these data. (*Hint:* Define the school year in years since school year 1999–2000; that is, 5 represents the school year ending in 2005 (which would mean 2004–2005).)
   b. Determine the school year in which the same number of males and females will participate in high school athletic programs.

19. The percentages of Hispanics and African Americans 25 years old or older who have a college degree are given in the bar chart.

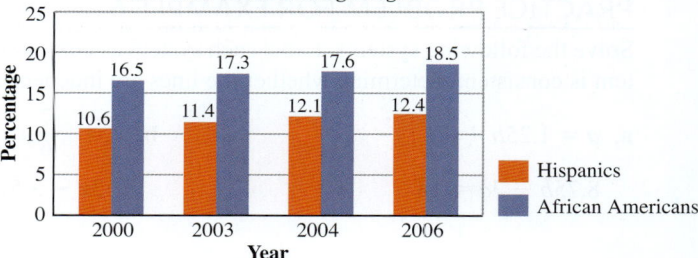

Source: Statistical Abstract of the United States, 2008.

   a. Find an equation for a model for the percentage of Hispanics who have a college degree.
   b. How fast is this percentage growing each year?
   c. Find an equation for a model for the percentage of African Americans who have a college degree.
   d. How fast is this percentage growing each year?
   e. Determine when the percentages of Hispanics and African Americans who have a college degree will be the same.

20. The number of American women of each age per thousand who are expected to die in the next 10 years from colon cancer or breast cancer is given in the table.

Age	Colon Cancer	Breast Cancer
60	4	7
65	6	9
70	8	10
75	11	11
80	14	12

Source: Journal of the National Cancer Institute.

   a. Find equations for models of these data.
   b. Determine the age at which the risk of dying from colon cancer will be the same as the risk of dying from breast cancer.

For Exercises 21 through 30, solve the systems using the substitution method. Label each system as consistent or inconsistent. If the system is consistent determine if the lines are independent or dependent. Check your answer.

21. $y = 2x - 5$
    $3x + y = 10$

22. $W = 7p - 2$
    $-3p + W = -14$

23. $H = 3k + 6$
    $2k + 4H = -4$

24. $C = 2r + 9$
    $5r + 2C = -9$

25. $G = 4a - 5$
    $5a - 2G = 1$

26. $y = x + 3$
    $6x - 4y = 8$

27. $P = \frac{2}{3}t + 5$
    $t + 3P = 21$

28. $H = \frac{3}{5}n + 8$
    $2n + 10H = 56$

29. $3x + 4y = 21$
    $x = 6y - 4$

30. $5x - 8y = -48$
    $x = 2y - 11$

31. As a salesperson for a high-end fashion store, you are given two options for your salary structure. The first option has a base salary of $700 a month plus 5% of all sales made. The second option has a base salary of $1300 plus 3.5% commission on all sales. Find the sales amount that will result in the two options having the same monthly salary.

32. As a salesperson for an electronic parts distributor, you are given two options for your salary structure. The first option has a base salary of $200 a week plus 4% of all sales made. The second option has a base salary of $100 plus 5.5% commission on all sales. Find the sales amount that will result in the two options having the same weekly salary.

33. Damian is investing $150,000 in two accounts to help support his daughter at college. Damian's daughter needs about $9600 each year to supplement her income from her part-time job. Damian decides to invest part of the money in an account paying 5% simple interest and the rest in another account paying 7.2% simple interest.
    a. Write a system of equations that will help Damian find the amount that should be invested in each account.
    b. How much does Damian need to invest in each account to earn enough interest to help support his daughter at college?

34. Juan invests $225,000 in two accounts. He would like to earn $7900 a year to pay some expenses. Juan invests part of the money in an account paying 3% simple interest and the rest in an account paying 4% simple interest.
    a. Write a system of equations that will help Juan find the amount that should be invested in each account.
    b. How much does Juan need to invest in each account to earn $7900 per year?

35. Henry was injured on the job and can no longer work. He received a settlement check from his company for $1.5 million that he will need to live on for the rest of his life. Henry is going to pay off his debts of $125,000 and invest the rest in two accounts. One account pays 5% simple interest, and the other pays 8% simple interest. Henry needs $87,500 a year in interest to continue to live at his current level.
    a. Write a system of equations that will help Henry find the amount that should be invested in each account.
    b. How much should Henry invest in each account to earn the $87,500 he wants?

36. Mona settled a lawsuit for $2.5 million and wants to use the money to pay for a new home and invest the rest to earn enough interest to live on. The house that Mona bought cost $450,000, and she wants $135,000 per year to live on. Mona invests the money in two accounts, one paying 6% simple interest and another paying 7.5% simple interest.
    a. Write a system of equations that will help Mona find the amount that should be invested in each account.
    b. How much should Mona invest in each account to earn the $135,000 she wants?

37. Joan is retiring and has $175,000 to invest. Joan needs to earn $12,000 in interest each year to supplement her Social Security and pension income. Joan plans to invest part of her money in an account that pays 9% simple interest and the rest in a safer account that pays only 5% simple interest. Determine how much Joan should invest in each account to earn the money she needs.

38. Nikki has $600,000 to invest and decides to invest it in two accounts, one paying 5% simple interest and the other paying 6.5% simple interest. Determine how much Nikki should invest in each account if she needs $33,600 in interest per year.

39. Truong invested $40,000 in stocks and bonds and had a total return of $4180 in one year. If his stock investments returned 11% and his bond investments returned 9%, how much did he invest in each?

40. Mike invested $70,000 in stocks and bonds and had a total return of $6305 in one year. If his stock investments returned 9.5% and his bond investments returned 8%, how much did he invest in each?

41. When comparing two tour companies, you find that one charges a $500 base fee plus $25 per person who goes on the tour. The other tour company charges a $700 base fee plus $20 per person who goes on the tour. Find what size group would be charged the same amount from either tour company.

42. A T-shirt printer charges a $150 setup fee and $4.50 per shirt for a T-shirt with a company logo printed on it. Another company charges a $230 setup fee and $4 per shirt for the same T-shirt and logo. Find what size order would result in the same cost from both companies.

For Exercises 43 and 44, compare the two students' work to determine which student did the work correctly. Explain what mistake the other student made.

43. Solve the system of equations. Label each system as consistent or inconsistent. If the system is consistent,

determine whether the lines are independent or dependent.

$$y = -2x + 8$$
$$x + \frac{1}{2}y = 3$$

Tom

$y = -2x + 8$
$x + \frac{1}{2}y = 3$
Substitute
$x + \frac{1}{2}(-2x + 8) = 3$
$x - x + 4 = 3$
$4 \neq 3$
This system is inconsistent and has no solutions.

Matt

$y = -2x + 8$
$x + \frac{1}{2}y = 3$
Substitute
$x + \frac{1}{2}(-2x + 8) = 3$
$x - x + 4 = 3$
$4 = 3$
This system is consistent with dependent lines. There are infinitely many solutions.

44. Solve the system of equations. Label each system as consistent or inconsistent. If the system is consistent, determine whether the lines are independent or dependent.

$$y = 4x + 7$$
$$x + 3y = 47$$

Shania

$y = 4x + 7$
$x + 3y = 47$
Substitute
$x + 3(4x + 7) = 47$
$x + 12x + 21 = 47$
$13x + 21 = 47$
$13x = 26$
$x = 2$

Warrick

$y = 4x + 7$
$x + 3y = 47$
Substitute
$x + 3(4x + 7) = 47$
$x + 12x + 21 = 47$
$13x + 21 = 47$
$13x = 26$
$x = 2$
$y = 4(2) + 7$
$y = 15$
$x = 2 \quad y = 15$
This system is consistent with independent lines. The solution is the point (2, 15).

For Exercises 45 through 62, solve the systems using the substitution method. Label each system as consistent or inconsistent. If the system is consistent, determine whether the lines are independent or dependent. Check your answer.

45. $7d + 4r = 8.17$
    $2r = 8d + 0.98$

46. $1.4h - 2.6t = -10.7$
    $3h = 6.9t - 25.32$

47. $p = \frac{1}{4}r + \frac{7}{4}$
    $p = 0.25r + 1.75$

48. $n = \frac{3}{8}m - \frac{7}{8}$
    $n = 0.375m - 0.875$

49. $b = \frac{3}{10}a + \frac{4}{5}$
    $b = 0.3a + 2.8$

50. $g = -\frac{3}{4}h - \frac{1}{2}$
    $g = -0.75h - 2$

51. $y = 4(x - 7) + 2$
    $y = -3(x + 2) + 5$

52. $f = 7(g + 2) - 8$
    $f = 5(g - 6) + 13$

53. $w = 2(t + 5) - 12$
    $w = 2(t - 3) + 4$

54. $d = -6(c - 4) - 10$
    $d = -6(c + 2) + 26$

55. $6x - 15y = 24$
    $y = \frac{2}{5}x - \frac{8}{5}$

56. $8w + 24y = 14$
    $y = -\frac{1}{3}w + \frac{7}{12}$

57. $10x - 4y = 30$
    $y = \frac{5}{2}x - 14$

58. $7x + 17.5y = 10.5$
    $y = -0.4x + 1.8$

59. $g + 5h = 37$
    $3g + 4h = 34$

60. $q - 2.5z = 23.25$
    $6q + 5z = 49.5$

61. $\frac{2}{3}x + \frac{1}{2}y = -\frac{15}{2}$
    $\frac{3}{4}x - y = \frac{5}{2}$

62. $\frac{3}{7}k - \frac{4}{5}m = \frac{9}{70}$
    $k + \frac{2}{7}m = -\frac{71}{140}$

2.3 Solving Systems of Equations Using the Elimination Method

LEARNING OBJECTIVES

- Solve a system of equations using the elimination method.
- Solve applications using systems of equations and the elimination method.

Although most of the models that we find in this book will be in $y = mx + b$ form, we will sometimes be given a linear model or equation that is not in this form. There are also situations that may be best modeled by using another form. When this is the case, we can isolate a variable in one equation and then use the substitution method. A system that does not have a variable isolated in at least one of the equations might be better suited for the elimination method.

In the elimination method, multiply one or both equations by constant(s) to make one of the variables add to zero (eliminate) when the two equations are added together. Recall that the multiplication property of equality states that we can multiply both sides of an equation by any nonzero constant. The addition property of equality states that we can add any expression to both sides of an equation. The goal will be to get the coefficients of one variable to be the same, but with opposite signs in the two equations. This will allow the variable to be eliminated when the two equations are added together.

> **Elimination Method for Solving Systems of Equations**
> - Used best when both equations are in general form or when no variable is already isolated or easily isolated.
> - Multiply one or more of the equations by a number to make the coefficients of one variable opposite in sign but of the same value.
> - Add the two equations to eliminate the variable, then solve.
> - Remember to find the values for both variables.
> - Check the solution in both equations.

Example 1 Solve a system of equations using the elimination method

Solve the following system using the elimination method.
$$3x + 2y = 10$$
$$5x - 8y = 28$$

SOLUTION

Because the signs of the coefficients of y are already opposites, we can eliminate the variable y by multiplying the first equation by 4.

$$4(3x + 2y) = 4(10) \longrightarrow 12x + 8y = 40$$
$$5x - 8y = 28 \longrightarrow 5x - 8y = 28$$

Add the two equations together to eliminate y.

$$12x + 8y = 40$$
$$\underline{5x - 8y = 28}$$
$$17x = 68$$
$$\frac{17x}{17} = \frac{68}{17}$$
$$x = 4$$

$$3(4) + 2y = 10$$
$$12 + 2y = 10$$
$$2y = -2$$
$$y = -1$$

Substitute 4 into x in either equation to find y.

We found that the solution to this system is $(4, -1)$. Check this solution in both equations to be sure that it is correct.

First equation	Second equation
$3(4) + 2(-1) \stackrel{?}{=} 10$	$5(4) - 8(-1) \stackrel{?}{=} 28$
$10 = 10$	$28 = 28$

In Section 2.2, we worked on applications that involved investments in two accounts with a goal of earning a certain amount of interest per year. Another set of applications that is set up by using systems of equations involves mixtures. When mixing two chemicals or two other substances, we are trying to create a desired amount of the mixture that has a desired quality. For example, we may want 20 milliliters of a 30% saline solution or 80 pounds of granola that cost $2 per pound. In these cases, we have an amount we want, which is similar to the total amount to be invested, and a desired percentage or value for the mixture, which is similar to the interest needed each year.

Example 2 Solving a mixture application using the elimination method

Sally, a college chemistry student, needs 90 ml of a 50% saline solution to do her experiment for lab today. Sally can find only the two saline solutions shown below. How much of each of these solutions does Sally have to combine to get 90 ml of 50% saline solution for her lab experiment?

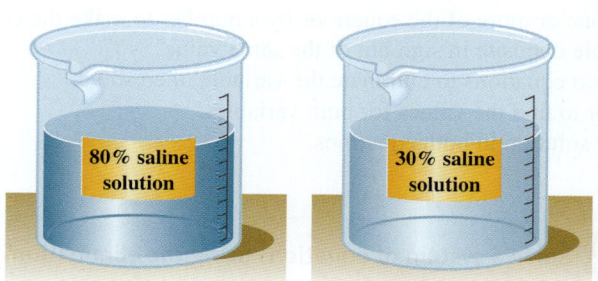

SOLUTION

Sally has two constraints. She needs 90 ml of the solution, and she needs 50% of that to be saline. We define the following variables:

A = The amount of 30% saline solution used (in milliliters)

B = The amount of 80% saline solution used (in milliliters)

Using these variables and the two constraints that are given results in the following two equations.

$$A + B = 90$$ The total mixture should be 90 ml.
$$0.30A + 0.80B = 0.50(90)$$ The saline should be 50% of the total.

Solving this system of equations will give us the amounts of each solution that Sally should use. First, simplify the right side of the second equation.

$$A + B = 90$$
$$0.30A + 0.80B = 45$$

Multiply the first equation by -0.80 to get the coefficients of B to be opposites.

$$-0.80(A + B) = -0.80(90) \longrightarrow -0.80A - 0.80B = -72$$
$$0.30A + 0.80B = 45 \longrightarrow 0.30A + 0.80B = 45$$

Add the two equations together to eliminate the B's.

$$-0.80A - 0.80B = -72$$
$$\underline{0.30A + 0.80B = 45}$$
$$-0.50A = -27 \quad \text{Solve for A.}$$
$$\frac{-0.50A}{-0.50} = \frac{-27}{-0.50}$$
$$A = 54$$

so
$$54 + B = 90 \quad \text{Solve for B.}$$
$$B = 36$$

Check the solution in both equations.

First equation	Second equation
$54 + 36 \stackrel{?}{=} 90$	$0.30(54) + 0.80(36) \stackrel{?}{=} 0.50(90)$
$90 = 90$	$45 = 45$

Sally should use 54 ml of the 30% saline solution and 36 ml of the 80% saline solution to get 90 ml of 50% saline solution.

PRACTICE PROBLEM FOR EXAMPLE 2

Jim Johnson is a local veterinarian who has prescribed a diet of 24% protein for a client's Great Dane. Jim has the two types of dog food shown below, but neither is 24% protein. The client wants 225 pounds of food to feed the Great Dane until the next vet appointment. How much of each type of food should Jim sell his client?

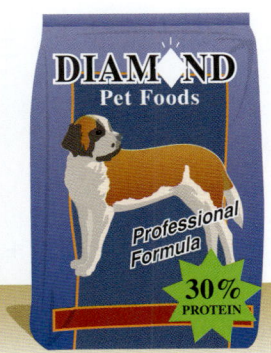

Example 3 Solving systems of equations using the elimination method

Solve the following systems of equations using the elimination method.

a. $2x + 5y = 24$
$3x - 5y = 11$

b. $3g + 5h = 27$
$6g + 7h = 39$

SOLUTION

a. The coefficients of y are already opposite numbers, so we can eliminate the y's by adding the two equations together and then solve for x.

$$2x + 5y = 24$$
$$3x - 5y = 11$$
$$\overline{5x = 35}$$
$$\frac{5x}{5} = \frac{35}{5}$$
$$x = 7$$

Solve for y.

$$2(7) + 5y = 24$$
$$14 + 5y = 24$$
$$5y = 10$$
$$y = 2$$

The solution to this system is (7, 2). Check this solution in both equations to be sure that it is valid.

First equation	Second equation
$2(7) + 5(2) \stackrel{?}{=} 24$	$3(7) - 5(2) \stackrel{?}{=} 11$
$24 = 24$	$11 = 11$

b. The coefficients are not already opposites, so eliminate the h's by multiplying the first equation by 7 and the second by -5 to make the coefficients 35 and -35, respectively.

$$7(3g + 5h) = 7(27) \longrightarrow 21g + 35h = 189$$
$$-5(6g + 7h) = -5(39) \longrightarrow -30g - 35h = -195$$

Add the two equations together to eliminate the h terms.

$$21g + 35h = 189$$
$$\underline{-30g - 35h = -195}$$
$$-9g = -6 \quad \text{Solve for } g.$$
$$\frac{-9g}{-9} = \frac{-6}{-9}$$
$$g = \frac{2}{3}$$

Use this value of g to find h.

$$3\left(\frac{2}{3}\right) + 5h = 27$$
$$2 + 5h = 27$$
$$5h = 25$$
$$h = 5$$

The solution to this system is $g = \frac{2}{3}$ and $h = 5$, that is, $\left(\frac{2}{3}, 5\right)$. Check this solution in both equations to be sure that it is valid.

First equation	Second equation
$3\left(\frac{2}{3}\right) + 5(5) \stackrel{?}{=} 27$	$6\left(\frac{2}{3}\right) + 7(5) \stackrel{?}{=} 39$
$27 = 27$	$39 = 39$

PRACTICE PROBLEM FOR EXAMPLE 3

Solve the following systems of equations using the elimination method.

a. $x + 8y = -35$
$-3x + 5y = -11$

b. $24d + 36g = 180$
$14d + 20g = 96$

Example 4 Solving systems of equations using the elimination method

Solve the following systems of equations using the elimination method.

a. $5p + 2w = 4$
$-4p - w = -3$

b. $-3x + 2y = 7$
$12x - 8y = -28$

SOLUTION

a. This system is set up for elimination because neither equation has a variable isolated. Multiply the second equation by 2 to get the coefficients of w to be opposites.

$$5p + 2w = 4 \longrightarrow 5p + 2w = 4$$
$$2(-4p - w) = 2(-3) \longrightarrow -8p - 2w = -6$$

Add the two equations together to eliminate the w terms.

$$5p + 2w = 4$$
$$\underline{-8p - 2w = -6}$$
$$-3p = -2$$
$$\frac{-3p}{-3} = \frac{-2}{-3}$$
$$p = \frac{2}{3}$$

Use this value of p to find w, or we can redo the elimination process and eliminate the p's this time and solve for w.

$$4(5p + 2w) = 4(4) \longrightarrow 20p + 8w = 16$$
$$5(-4p - w) = 5(-3) \longrightarrow -20p - 5w = -15$$

Add the two equations together to eliminate the p terms.

$$20p + 8w = 16$$
$$\underline{-20p - 5w = -15}$$
$$3w = 1$$
$$w = \frac{1}{3}$$

This system has one solution at the point $p = \frac{2}{3}$ and $w = \frac{1}{3}$, that is, $\left(\frac{2}{3}, \frac{1}{3}\right)$, so it is a consistent system. Check this solution in both equations.

First equation

$$5\left(\frac{2}{3}\right) + 2\left(\frac{1}{3}\right) \stackrel{?}{=} 4$$
$$\frac{10}{3} + \frac{2}{3} \stackrel{?}{=} 4$$
$$\frac{12}{3} \stackrel{?}{=} 4$$
$$4 = 4$$

Second equation

$$-4\left(\frac{2}{3}\right) - \left(\frac{1}{3}\right) \stackrel{?}{=} -3$$
$$-\frac{8}{3} - \frac{1}{3} \stackrel{?}{=} -3$$
$$-\frac{9}{3} \stackrel{?}{=} -3$$
$$-3 = -3$$

b. This system is set up for elimination because no variable is already isolated and the variables are already lined up to be eliminated. Multiply the first equation by 4 to get the coefficients of y (and x in this case) to be opposites.

$$4(-3x + 2y) = 4(7) \longrightarrow -12x + 8y = 28$$
$$12x - 8y = -28 \longrightarrow 12x - 8y = -28$$

Add the two equations together.

$$-12x + 8y = 28$$
$$12x - 8y = -28$$
$$0 = 0$$

All the variable terms have been eliminated, and these numbers are equal.

Because the variables were both eliminated and the remaining numbers are equal, this system is dependent. The two equations are multiples of each other. If the second equation is divided by -4, it will reduce to be the same as the first equation. Thus, this is a consistent system with dependent lines. There are an infinite number of solutions; any solution to the equation $-3x + 2y = 7$ will be a solution to the system.

PRACTICE PROBLEM FOR EXAMPLE 4

Solve the following systems of equations using the elimination method.

a. $15x + 14y = 16$
$-3x + 21y = 7$

b. $\frac{5}{2}x - \frac{3}{2}y = 6$
$10x - 6y = 8$

2.3 Exercises

For Exercises 1 through 10, solve the systems using the elimination method. Label each system as consistent or inconsistent. If the system is consistent, determine whether the lines are independent or dependent. Check your answer.

1. $x + 4y = 11$
$5x - 4y = 7$

2. $6x + 2y = 6$
$-6x - 5y = 21$

3. $2g - 3h = 9$
$-2g - 4h = 26$

4. $7f + g = 20$
$3f - g = 20$

5. $8x - 2y = 36$
$3x + 4y = 23$

6. $p + 4W = -82$
$-3p + W = -14$

7. $7k - 4H = 16$
$2k + 8H = 32$

8. $2G - 3F = -20$
$5G + 9F = -17$

9. $3t + 8k = 20$
$5t + 6k = 26$

10. $4x - 12y = -8$
$10x - 8y = 46$

11. A chemistry student needs to make 20 ml of 15% HCl solution, but he has only the solutions shown below to work with.

How much of each solution should this student use to get the 20 ml of 15% HCl solution?

12. Fred needs 45 ml of 12% sucrose solution to do his science experiment. He needs to mix a 5% sucrose solution with some 30% sucrose solution to get the 12% solution he wants. How much of each solution should Fred use?

13. Kristy is doing her chemistry lab, but the student in front of her used the last of the 5% NaCl solution she needed. Kristy will have to make more 5% solution with the 2% and 10% NaCl solutions left in the lab. If Kristy needs 25 ml for her experiment, how much of each solution should she use?

14. Camren needs 30 ml of 10% HCl solution, but she has only 5% HCl and 30% HCl solutions. How much of each solution should Camren use to make the 30 ml of 10% HCl solution?

15. A store manager wants to create a custom nut mix that contains only 10% peanuts. The manager has two premade nut mixes, one that contains 5% peanuts and another that contains 20% peanuts. How many pounds of each premade mix should the manager use to get 100 lb of a mix that contains only 10% peanuts?

16. A candy store owner has chocolate candies worth $2 per pound and sour candies worth $0.75 per pound. How much of each kind of candies should she combine to get 60 lb of mixed candies worth $1.50 per pound?

For Exercises 17 through 26, solve the systems using the elimination method. Label each system as consistent or inconsistent. If the system is consistent, determine whether the lines are independent or dependent. Check your answer.

17. $w + 8s = -42$
    $-9w - s = -48$

18. $-3x + 7y = -1$
    $3x + 9y = -2$

19. $y = \frac{2}{3}x - 9$
    $-2x + 3y = -27$

20. $g = -0.1f + 5.2$
    $f + 10g = 52$

21. $2.5x + y = 4$
    $x + 0.4y = -5$

22. $7.5p + k = 3$
    $15p + 2k = 7$

23. $W = -3.2c + 4.1$
    $W = 2.4c - 8.8$

24. $H = 2.6t + 4.8$
    $H = -4.6t + 3.5$

25. $p = \frac{7}{4}t + 9$
    $4p - 7t = 36$

26. $y = \frac{7}{2}x + 2$
    $7x - 2y = -4$

27. The local fair offers two options for admission and ride tickets. For option 1, you pay $22 for admission and 50 cents per ride ticket. For option 2, you pay $15 for admission and 75 cents per ride ticket.

    a. Use a system of equations to find the number of rides that will result in the same cost for both options.
    b. If you want to ride a large number of rides, which option should you take?

28. Two car rental agencies advertised a weekend rate in the newspaper. U-rent advertised a weekend rate of $29 plus 15 cents per mile driven. Car Galaxy advertised a weekend rate of $40 plus 10 cents per mile driven.

    a. Use a system of equations to find the number of miles that will result in the same charges from both companies.
    b. If you plan to drive about 200 miles, which company should you rent from?

29. While working on a home project, a man purchased two orders of wood. The first order was for 150 planks and eight 4 × 4's for a total cost of $222.92. The second order was for 45 planks and two 4 × 4's for a total cost of $65.18. Use a system of equations to find the cost for a plank and a 4 × 4.

30. A golf resort offers two getaway options. The Mini Escape includes two nights' stay and two rounds of golf, and the Grand Escape includes three nights' stay and four rounds of golf. The Mini Escape costs $398, and the Grand Escape costs $647. Use a system of equations to find the cost for one night's stay and the cost for one round of golf for these options.

31. In one month, Ana earned $1420 for 165 hours of work. Ana gets paid $8 per hour for her regular time and $12 per hour for any overtime hours. Find the number of regular hours and overtime hours Ana worked that month.

32. In a month, Jerrell earned $3430 for 230 hours worked. Jerrell earns $14 per hour for regular hours and $21 per hour for overtime. Find the number of regular hours and overtime hours Jerrell worked that month.

For Exercises 33 through 36, determine whether the system of equations is better set up to solve using the substitution method or elimination method. Explain your answer.

33. $H = \frac{2}{7}n + 8$
    $n + 7H = 20$

34. $4p + 8r = 11$
    $2p - 4r = 5$

35. $T = 0.5g + 8.5$
    $g - 2T = -17$

36. $B = \frac{2}{8}k - 4$
    $k - 4B = 16$

37. Explain what happens when solving an inconsistent system using the substitution or elimination methods.

38. Explain what happens when solving a consistent system with dependent lines using the substitution or elimination methods.

39. Explain how you can use your calculator to check your answer if you find one solution to the system.

40. Explain how you can use your calculator to check your answer if you determine that the system is inconsistent or consistent with dependent lines.

CHAPTER 2 Systems of Linear Equations and Inequalities

For Exercises 41 through 50, solve each system by elimination or substitution. Label each system as consistent or inconsistent. If the system is consistent, determine whether the lines are independent or dependent. Check your answer.

41. $\frac{5}{3}d - \frac{3}{5}g = 52$
    $\frac{3}{5}d + \frac{5}{3}g = -66$

42. $\frac{2}{5}P + \frac{5}{2}T = 26$
    $\frac{5}{2}P - \frac{2}{5}T = 40.71$

43. $-2x + 11y = 2$
    $11(x - y) = -2$

44. $-8x + 7y = -3$
    $7(x - y) = 3$

45. $-3.5x + y = -16.95$
    $y = -2.4x + 4.88$

46. $4.5c + 2.1b = 4.7$
    $b = 3.51c + 3.43$

47. $W = \frac{3}{7}d + 6$
    $W = \frac{2}{7}d + 8.8$

48. $M = \frac{5}{9}r + 8$
    $M = \frac{7}{9}r - 2.5$

49. $T = 0.5g + 8.5$
    $g - 2T = -17$

50. $B = \frac{2}{8}k - 4$
    $k - 4B = 16$

51. Explain where Frank went wrong while solving the given system of equations.

> **Frank**
> $2x - 6y = 30$
> $-4x + 12y = -60$
>
> **Multiply by 2.**
> $2(2x - 6y) = 30$
> $-4x + 12y = -60$
>
> **Add the equations.**
> $4x - 12y = 30$
> $-4x + 12y = -60$
> $\overline{0 = -30}$
>
> This system is inconsistent and has no solutions.

52. Explain where Hung went wrong while solving the given system of equations.

> **Hung**
> $x - 6y = 14$
> $0.5x + 5y = 15$
>
> **Multiply by -0.5.**
> $-0.5(x - 6y) = -0.5(14)$
> $0.5x + 5y = 15$
>
> **Add the equations.**
> $-0.5x - 6y = -7$
> $0.5x + 5y = 15$
> $\overline{-y = 8}$
> $y = -8$
> $x - 6(-8) = 14$
> $x + 48 = 14$
> $x = -34$
>
> This system is consistent and has the point $(-34, -8)$ as a solution.

For Exercises 53 through 66, solve each system by elimination or substitution. Label each system as consistent or inconsistent. If the system is consistent, determine whether the lines are independent or dependent. Check your answer.

53. $3.7x + 3.5y = 5.3$
    $y = 3.57x + 4.21$

54. $4.1w + 3.7t = 5.1$
    $t = 4.43w + 4.63$

55. $-9c - 7d = 8$
    $c - 7d = 8$

56. $-4r + t = -1$
    $-3r + 8t = 5$

57. $3x + 4y = -15$
    $y = x - 9$

58. $3s - 2t = -2$
    $t = -s - 9$

59. $\frac{4}{3}x + y = 13$
    $x - \frac{7}{6}y = 4$

60. $\frac{7}{4}x + y = -1$
    $x - \frac{5}{3}y = 3$

61. $-8.4x - 2.8y = -58.8$
    $3x + y = 21$

62. $2.3p + 3.5t = 7.8$
    $11.27p + 17.15t = 38.22$

63. $3c - b = -17$
    $b = 3c + 18$

64. $8x - y = 20$
    $y = 8x + 4$

65. $-2.4x + y = 2.55$
    $4.1x + y = -2$

66. $7r - 2c = -8$
    $-2r - 6c = 7$

2.4 Solving Linear Inequalities

LEARNING OBJECTIVES

- Solve linear inequalities algebraically.
- Solve linear inequalities graphically or numerically.
- Solve applications using linear inequalities.

In life, we often want to know when one quantity is less than or greater than another quantity. We may want to know how many questions we can miss and still earn the grade we want on a test. For example, if we get more than 20 questions of 25 on a test, we will pass. We would like to know how many things we can buy and stay within a budget. For example, if we spend less than $300 on groceries each month, we will stay below our budget. The desire to compare quantities leads us to mathematical inequalities. We will work with the four basic inequalities: less than, greater than, less than or equal to, and greater than or equal to.

Introduction to Inequalities

When working with inequalities in algebra, be cautious about which operations are performed. In most cases, using the operations to solve an inequality is exactly the same as using them to solve an equation except that the answer will be expressed differently and will have a different interpretation.

CONCEPT INVESTIGATION

What changes an inequality?

For each inequality, perform the given operations and determine whether the inequality remains true.

a. $20 > 15$

    i. Add 12 to both sides. $20 + 12 > 15 + 12$ $32 > 27$ True
    ii. Subtract 4 from both sides. $20 - 4 > 15 - 4$ $16 > 11$ True
    iii. Multiply both sides by 3. $3(20) > 3(15)$ $60 > 45$ True
    iv. Divide both sides by 5. $\dfrac{20}{5} > \dfrac{15}{5}$ $4 > 3$?

b. $45 \geq 39$

    i. Add -7 to both sides.
    ii. Subtract -9 from both sides.
    iii. Multiply both sides by -2.
    iv. Divide both sides by -3.

c. $36 > -12$

    i. Add 10 to both sides.
    ii. Subtract 15 from both sides.
    iii. Multiply both sides by 9.
    iv. Divide both sides by 4.

What's That Mean?

Inequality Symbols
- Less than: $<$
  $5 < 20$
- Greater than: $>$
  $15 > 8$
- Less than or equal to: $\leq$
  $11 \leq 73$
- Greater than or equal to: $\geq$
  $45 \geq 3$

d. $-15 \leq 7$

  i. Add -5 to both sides.
  ii. Subtract -6 from both sides.
  iii. Multiply both sides by -6.5.
  iv. Divide both sides by -4.

1. Which operations cause the inequality to become false?

2. Write in your own words what operations to be cautious with in working with inequalities.

3. Check your statement with other students in your class or with your instructor.

Solving Inequalities

From the concept investigation above, we can see that when working with inequalities we must watch for operations that will reverse the inequality relationship. When solving inequalities, we will reverse the inequality symbol whenever we multiply or divide both sides of the inequality by a negative number.

In general, it is best to keep the variable on the left side of an inequality symbol, so that the interpretation of the solution will be easier. To check the solution to an inequality, we must check that the number we find will make the two sides equal to one another and then also check that the inequality is facing in the correct direction. We can check the direction of the inequality symbol by picking a value in the solution set and plugging it into the original inequality to check whether it satisfies the inequality.

Example 1 Solving inequalities algebraically

Solve the following inequalities.

a. $-5x + 7 \geq 22$ **b.** $12w + 15 > 3w - 6$ **c.** $\dfrac{b}{-5} \leq -20$

SOLUTION

a. We will solve the inequality the same way we would solve an equation. Note, though, that we have to reverse the inequality symbol any time that we multiply or divide both sides by a negative number.

$$-5x + 7 \geq 22$$
$$\ -7\ \ -7$$
$$-5x \geq 15$$
$$\dfrac{-5x}{-5} \leq \dfrac{15}{-5} \quad \text{Divide both sides by } -5 \text{ and reverse the inequality symbol.}$$
$$x \leq -3$$

Therefore, any number less than or equal to -3 will solve this inequality. To check this solution, check the -3 to be sure that it makes the sides equal to one another and then pick a value less than -3. We will test -5.

Check the number -3 for equality.

$$-5(-3) + 7 \stackrel{?}{=} 22$$
$$22 = 22$$

Check -5 to confirm the direction of the inequality.

$$-5(-5) + 7 \stackrel{?}{\geq} 22$$
$$32 \geq 22$$

The equality and resulting inequality are true, so the solution is correct.

b.

$$12w + 15 > 3w - 6$$
$$\underline{-3w \qquad\quad -3w}$$
$$9w + 15 > -6$$
$$\underline{\quad -15 \quad -15}$$
$$9w > -21$$
$$\frac{9w}{9} > \frac{-21}{9} \qquad \text{Divide both sides by 9. Because we are dividing by a positive number, we will not reverse the inequality symbol.}$$
$$w > \frac{-21}{9}$$

The fraction $\frac{-21}{9}$ is approximately -2.33. To check this solution, test -2.33 for equality and a number greater than -2.33 to test the direction of the inequality. We will test $w = 0$.

Check the number -2.33 for equality.

$$12(-2.33) + 15 \stackrel{?}{=} 3(-2.33) - 6$$
$$-12.96 \approx -12.99$$

Check 0 to confirm the direction of the inequality.

$$12(0) + 15 \stackrel{?}{>} 3(0) - 6$$
$$15 > -6$$

The equality is approximate because we rounded our solution. The equality and resulting inequality are true, so our solution is correct.

c.

$$\frac{b}{-5} \leq -20$$
$$-5\left(\frac{b}{-5}\right) \geq -5(-20) \qquad \text{Multiply both sides by } -5 \text{ and reverse the inequality symbol.}$$
$$b \geq 100$$

Note that we had to reverse the inequality symbol because we multiplied both sides by a negative number. To test this solution, we will check $b = 100$ and pick a number greater than 100. We will test $b = 110$.

Check the number 100 for equality.

$$\frac{100}{-5} \stackrel{?}{=} -20$$
$$-20 = -20$$

Check 110 to confirm the direction of the inequality.

$$\frac{110}{-5} \stackrel{?}{\leq} -20$$
$$-22 \leq -20$$

The equality and the resulting inequality are true, so the solution is correct.

PRACTICE PROBLEM FOR EXAMPLE 1

Solve the following inequalities.

a. $4x + 12 > 10$ **b.** $8h + 9 \leq 10h + 1$ **c.** $\frac{m}{-3} - 8 < 10$

Example 2 Finding errors in solutions of inequalities

George and Martha found the same solution to the inequality

$$7v + 10 < 5(2v - 7)$$

Compare their solutions and determine whether they made any mistakes. If they have, explain what they did wrong.

CHAPTER 2 Systems of Linear Equations and Inequalities

George	Step #
$7v + 10 < 5(2v - 7)$	1
$7v + 10 < 10v - 35$	2
$\quad -10 \quad\quad -10$	
$7v > 10v - 45$	3
$-10v \quad -10v$	
$-3v < -45$	4
$\dfrac{-3v}{-3} > \dfrac{-45}{-3}$	5
$v > 15$	6

Martha	Step #
$7v + 10 < 5(2v - 7)$	1
$7v + 10 < 10v - 35$	2
$\quad -10 \quad\quad -10$	
$7v < 10v - 45$	3
$-10v \quad -10v$	
$-3v < -45$	4
$\dfrac{-3v}{-3} > \dfrac{-45}{-3}$	5
$v > 15$	6

SOLUTION

George made mistakes on both lines 3 and 4. He should not have reversed the inequality symbol when subtracting from both sides. By making both of these errors, he actually ended with the correct solution, but he did not use the correct method. Martha did the problem correctly.

PRACTICE PROBLEM FOR EXAMPLE 2

Tom and Shannon found different answers to the inequality

$$2.5n + 6.5 > -3.2n - 4.3$$

Can you find which one made a mistake and explain what he or she did wrong?

Tom	Step #
$2.5n + 6.5 > -3.2n - 4.3$	1
$2.5n + 6.5 > -3.2n - 4.3$	2
$3.2n \quad\quad\quad 3.2n$	
$5.7n + 6.5 > -4.3$	3
$\quad -6.5 \quad\quad -6.5$	
$5.7n > -10.8$	4
$\dfrac{5.7n}{5.7} > \dfrac{-10.8}{5.7}$	5
$n > -1.895$	6

Shannon	Step #
$2.5n + 6.5 > -3.2n - 4.3$	1
$2.5n + 6.5 > -3.2n - 4.3$	2
$3.2n \quad\quad\quad 3.2n$	
$5.7n + 6.5 > -4.3$	3
$\quad -6.5 \quad\quad -6.5$	
$5.7n > -10.8$	4
$\dfrac{5.7n}{5.7} < \dfrac{-10.8}{5.7}$	5
$n < -1.895$	6

Systems as Inequalities

Many times, when we work with systems of equations, we are not interested in finding when the two equations are equal. Instead, we are interested in finding when one equation is less than or greater than the other. Going back to the example of U-Haul and Bob's Truck Rental from Section 2.1, we are concerned not with when charges from these companies are equal but with when one company's plan is less expensive than the other.

Example 3 Finding the least expensive rental company

If we reconsider the U-Haul and Bob's Truck Rental example from Section 2.1, we get the following.

SECTION 2.4 Solving Linear Inequalities

$U(m)$ = The total cost in dollars to rent a 10-foot truck from U-Haul
$B(m)$ = The total cost in dollars to rent a 10-foot truck from Bob's Truck Rental
m = The number of miles traveled in rented truck

$$U(m) = 19.95 + 0.79m$$
$$B(m) = 24.95 + 0.75m$$

Find the distances that you can drive that will result in Bob's being less expensive than U-Haul.

SOLUTION

Because we want Bob's to be less expensive than U-Haul, we can set the expression for the cost at Bob's less than the expression for the cost at U-Haul.

$$B(m) < U(m)$$
$$24.95 + 0.75m < 19.95 + 0.79m$$
$$\underline{-0.79m \qquad\quad -0.79m}$$
$$24.95 - 0.04m < 19.95$$
$$\underline{-24.95 \qquad\qquad -24.95}$$
$$-0.04m < -5$$
$$\frac{-0.04m}{-0.04} > \frac{-5}{-0.04}$$
$$m > 125$$

Reverse the inequality symbol when dividing by a negative number.

Thus, Bob's will be less expensive than U-Haul for any distances more than 125 miles. We can easily check this solution and the direction of the symbol by using the following table.

X	Y1	Y2
110	107.45	106.85
120	114.95	114.75
125	118.7	118.7
130	122.45	122.65
140	129.95	130.55
150	137.45	138.45

X=110

We can see that for distances greater than 125, Bob's (Y1) costs less than U-Haul (Y2). Therefore, our solution is correct.

Notice in this example that the solution to the system is not a particular set mileage, but a range of miles that makes Bob's less expensive than U-Haul. In solving a system as an inequality, the answer will be an interval of values that make the situation true.

Example 4 Comparing the numbers of male and female athletes

In Exercise 18 of Section 2.2 on page 152, we looked at the number of males and females who participated in high school athletic programs for various school years. The data are given again in the table.

School Year	Males	Females
03–04	4,039,253	2,865,299
04–05	4,110,319	2,908,390
05–06	4,206,549	2,953,355
06–07	4,321,103	3,021,807

Source: www.infoplease.com

a. Find equations for models of these data.

b. Use the models to approximate when there will be more females than males participating in high school athletic programs.

In the first male athlete model, both points missed are below the line.

$M(S) = 0.093S + 3.668$

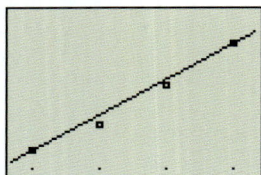

By adjusting the value of b, we can shift the graph down a little to make the line fit the data better.

$M(S) = 0.093S + 3.66$

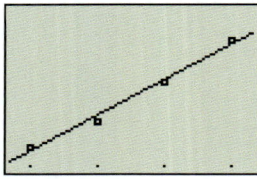

In the first female athlete model, both points missed are below the line.

$F(S) = 0.05S + 2.67$

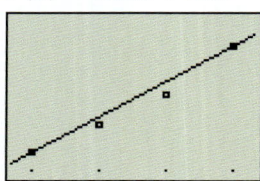

By adjusting the value of b, we can shift the graph down a little to make the line fit the data better.

$F(S) = 0.05S + 2.66$

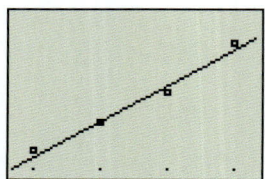

SOLUTION

a. Define the variables.

S = School year in years since school year 1999–2000; that is, 5 represents school year ending in 2005 (which would mean 2004–2005)
$M(S)$ = Number of males in millions participating in high school athletic programs
$F(S)$ = Number of females in millions participating in high school athletic programs

For the model of the number of males, we will use the first and last points.

$$(4, 4.04) \text{ and } (7, 4.32)$$
$$m = \frac{4.32 - 4.04}{7 - 4} = 0.093$$
$$M = 0.093S + b$$
$$4.04 = 0.093(4) + b$$
$$4.04 = 0.372 + b$$
$$3.668 = b$$
$$M = 0.093S + 3.668 \quad \text{This line should be adjusted down a little to make it fit the data better.}$$
$$M(S) = 0.093S + 3.66 \quad \text{The adjusted model. (See left column.)}$$

Now for the model of the female data, we will use the first and last points.

$$(4, 2.87) \text{ and } (7, 3.02)$$
$$m = \frac{3.02 - 2.87}{7 - 4} = 0.05$$
$$F = 0.05S + b$$
$$2.87 = 0.05(4) + b$$
$$2.87 = 0.2 + b$$
$$2.67 = b$$
$$F = 0.05S + 2.67 \quad \text{This line should be adjusted down a little to make it fit the data better.}$$
$$F(S) = 0.05S + 2.66 \quad \text{The adjusted model. (See left column.)}$$

b. Because we want to know when the number of females will be larger than the number of males, we can set up the following inequality.

Females > Males
$$0.05S + 2.66 > 0.093S + 3.66$$
$$\underline{-0.093S \qquad\qquad -0.093S}$$
$$-0.043S + 2.66 > 3.66$$
$$\underline{\qquad -2.66 \quad -2.66}$$
$$-0.043S > 1$$
$$\frac{-0.043S}{-0.043} < \frac{1}{-0.043}$$
$$S < -23.26$$

Check the answer using the calculator table.

X	Y1	Y2
-25	1.41	1.335
-23.26	1.497	1.4968
-20	1.66	1.8

X=

Since S is a school year since 1999–2000, $S = -23.26$ needs to be subtracted from 2000.

$$2000 - 23.26 = 1976.64 \approx 1977$$

Therefore, this inequality is true for school years prior to the school year ending in 1977. This solution can be stated in different ways.

The number of females participating in high school athletic programs was greater than the number of males before the school year 1976–1977.

or

The number of females participating in high school athletic programs was greater than the number of males during the school year 1975–1976 and earlier.

Both statements interpret the inequality correctly. The first does not include the year 1976–1977. The second ends with the previous year, 1975–1976. The solution says that more females participated in high school athletic programs in the past, but this result is model breakdown. In the 1970s, not nearly as many girls participated in high school athletics.

PRACTICE PROBLEM FOR EXAMPLE 4

The percentage of college freshmen who were male and the percentage of college freshmen who were female for various years are given in the chart.

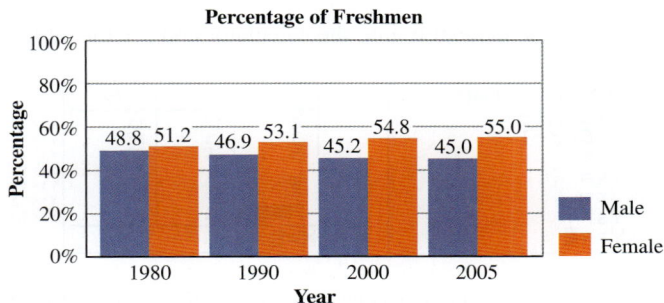

Source: Statistical Abstract of the United States, 2008.

a. Find equations for models of these data.

b. Determine the years for which the number of male freshmen was greater than or equal to the number of female freshman.

When solving an inequality, you might find it easiest to keep the variable on the left side of the inequality symbol. Doing so will make the solution easier to interpret. If you do keep the variable on the left side, you may run into places where you must multiply or divide by a negative number, requiring you to reverse the direction of the inequality symbol. Follow the simple rule of reversing the inequality symbol when multiplying or dividing by a negative number. Solve inequalities using the same skills that you use to solve equations.

Solving Inequalities Numerically and Graphically

We can use a table to estimate the solutions to inequalities by looking for the input value(s) that make the left side of the inequality equal to the right side of the inequality. Once we have the input value(s) that make the sides equal, we will compare the output value(s) on either side of those inputs to see when one is less than or greater than the other.

Start by putting the left side into Y1 and the right side into Y2 of the Y= screen of your calculator. When looking for the input value(s) that make the two sides equal, notice when one side changes from being smaller than the other side to being larger. The value where these two sides are equal must be between these input values. Once you have found the input value that makes the two sides of the inequality equal, look for the inequality relationship that you want to be true. This inequality relationship will be when the left side is less than or greater than the right side.

Example 5 Solving inequalities numerically

Solve the following inequalities numerically using the calculator table.

a. $5x + 7 > 2x + 31$ **b.** $4x + 9 \leq 6x - 1$

SOLUTION

a. First put the two sides of the inequality into the calculator and then use the table to estimate a solution.

From the table, we can see that the two sides of the inequality are equal at $x = 8$ and that the left side (Y1) is greater than the right side (Y2) when $x > 8$. The solution to this inequality is $x > 8$.

b. Put the two sides of the inequality into the calculator and use the table.

We can see on the table that the two sides are equal when $x = 5$ and the left side (Y1) is less than the right side (Y2) when $x > 5$. The solution to this inequality is $x \geq 5$.

PRACTICE PROBLEM FOR EXAMPLE 5

Solve the following inequality numerically using the calculator table.

$$-3x + 8 > x - 4$$

On a graph, a curve that is vertically below another curve is "less than" the other curve. A curve remains less than another curve for any x (input) values that result in the curve staying vertically below the other curve. The same is true for a curve vertically above another curve being "greater than" the other curve. If the two curves cross, the x-value of the intersection will be the endpoint of the solution interval.

We can use the graph below to solve the inequality

$$-\frac{2}{3}x + 4 < \frac{1}{3}x + 1$$

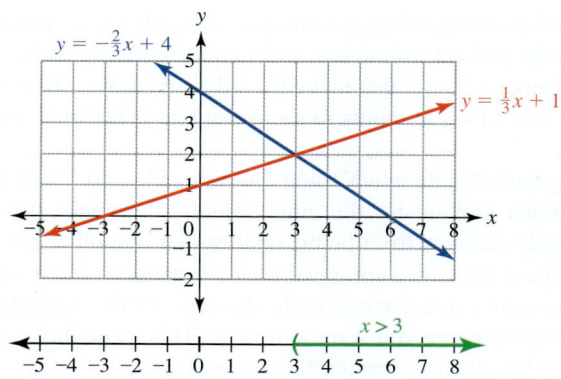

In this graph we can see that the blue line is below the red line when x is greater than 3. The solution to the inequality is $x > 3$.

We can use the calculator to solve graphically by putting one side of the inequality into Y1 and the other side into Y2 and then graphing. Once we find the intersection, we will determine the values of x that make the graph we want less than or greater than the other. The end of the solution interval will always be determined by the places where the graphs intersect. The typical solutions are either less than the x-value or greater than the x-value of the intersection.

SECTION 2.4 Solving Linear Inequalities

Example 6   Solving inequalities graphically

Solve the following inequalities graphically.

a. $-2x + 9 \leq 4x - 33$

b. $2.5x + 4 < 3.5x + 6$

SOLUTION

a. Put the two sides of the inequality $-2x + 9 \leq 4x - 33$ into the calculator and graph them both on the same window. Watch the calculator graph the lines, to know which line is the left side of the inequality and which is the right side of the inequality.

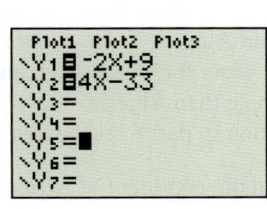 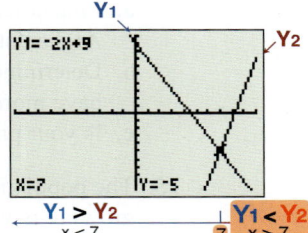

From the graph, we can see that the two sides of the inequality are equal at $x = 7$. The graph of the $-2x + 9$ (Y1) is lower (less than) than the graph of the $4x - 33$ for values of x greater than 7. Therefore, the solution to this inequality is $x \geq 7$. The "greater than or equal to" symbol is used because the original inequality had the "equal to" part included in the inequality symbol.

b. Put each side of the inequality $2.5x + 4 < 3.5x + 6$ into the graphing calculator and graph them both on the same window.

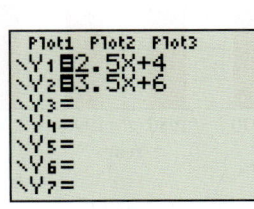 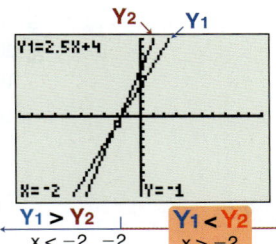

From the graph, we can see that the intersection is at $(-2, -1)$, so the two sides are equal at $x = -2$. The left side (Y1) is less than the right side (Y2) when $x > -2$. The solution to this inequality is $x > -2$.

PRACTICE PROBLEM FOR EXAMPLE 6

Solve the following inequality graphically.

$$1.25x - 2 \geq 2x - 5$$

2.4 Exercises

For Exercises 1 through 10, solve the given inequalities.

1. $5x + 7 > 37$
2. $4x - 10 < 2$
3. $\dfrac{P}{4} \geq 2$
4. $\dfrac{M}{6} < 4.5$
5. $8V + 4 < 5V - 20$
6. $3x + 7 \leq x - 5$
7. $2t + 12 < 5t + 39$
8. $4p - 8 > 15p - 41$
9. $\dfrac{K}{-4} + 7 \leq 21$
10. $\dfrac{V}{-5} - 4 \geq 3$

For Exercises 11 through 12, compare the two students' work to determine which student did the work correctly. Explain what mistake the other student made.

11. Solve $8 - 3x > 20$.

Amy	Michelle
$8 - 3x > 20$	$8 - 3x > 20$
$\quad -8 \quad\quad -8$	$\quad -8 \quad\quad -8$
$-3x < 12$	$-3x > 12$
$\dfrac{-3x}{-3} > \dfrac{12}{-3}$	$\dfrac{-3x}{-3} < \dfrac{12}{-3}$
$x > -4$	$x < -4$

12. Solve $5x + 6 \leq 2x$.

Mark	Marie
$5x + 6 \leq 2x$	$5x + 6 \leq 2x$
$-2x \quad\quad -2x$	$-2x \quad\quad -2x$
$3x + 6 \geq 0$	$3x + 6 \leq 0$
$\quad -6 \quad -6$	$\quad -6 \quad -6$
$3x \leq -6$	$3x \leq -6$
$\dfrac{3x}{3} \geq \dfrac{-6}{3}$	$\dfrac{3x}{3} \leq \dfrac{-6}{3}$
$x \geq -2$	$x \leq -2$

13. The numbers of full-time and part-time faculty in U.S. higher education institutions are given in the graph.

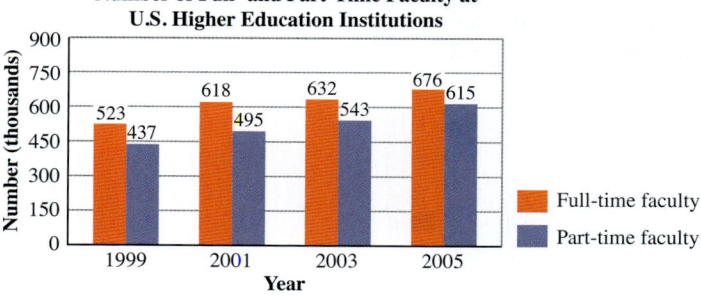

Source: Statistical Abstract of the United States, 2008.

a. Find equations for models of these data.
b. Find when the number of part-time faculty will be greater than the number of full-time faculty.

14. The populations of Afghanistan and Algeria are given in the table.

Year	Afghanistan Population (in thousands)	Algeria Population (in thousands)
1990	14,669	25,093
1995	21,489	28,364
2000	23,898	30,409
2006	31,057	32,930

Source: Statistical Abstract of the United States, 2008.

a. Find equations for models for the populations of Afghanistan and Algeria.
b. Determine the years for which Afghanistan will have more people than Algeria.
c. Is your prediction in part b reasonable?

15. The population on the continent of Africa has been growing rapidly over the past several decades. The populations of Africa and Europe are given in the charts.

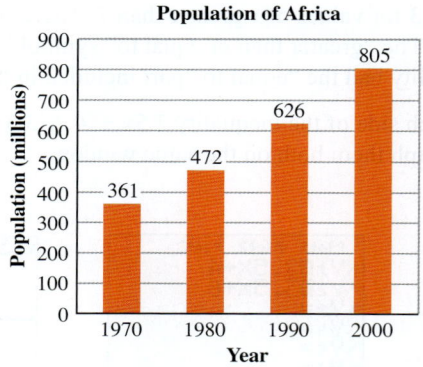

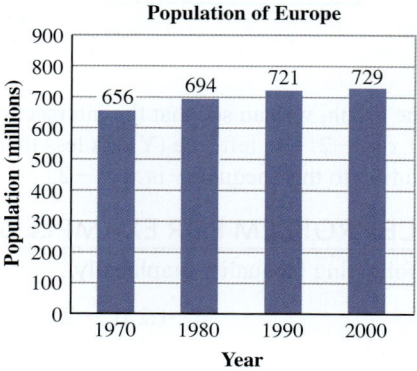

Source: Statistical Abstract of the United States, 2008.

a. Find equations for models for the populations of Africa and Europe.
b. Estimate when Africa will have a greater population than Europe.

16. The percentages of births to teenage mothers in the United States for two states are given in the table.

Year	Idaho	Utah
2002	10.0	7.3
2003	9.6	6.7
2004	9.2	6.4
2005	8.8	6.0

Source: Statistical Abstract of the United States, 2008.

a. Find a model for the percentage of births to teenage mothers in Idaho.
b. What does the slope for the model in part a mean in this context?
c. Find a model for the percentage of births to teenage mothers in Utah.
d. Estimate when the percentage of births to teenage mothers in Utah will be greater than the percentage in Idaho.

For Exercises 17 through 24, indicate what inequality symbol ($<, >, \leq, \geq$) would be used to represent the given phrase.

17. Cheaper than
18. Higher than
19. At least
20. Not more than
21. More expensive than
22. At most
23. Lower than
24. Not less than

25. The revenue and cost functions for a local cabinet manufacturer are

$$R(c) = 450c$$
$$C(c) = 280c + 20000$$

where $R(c)$ represents the monthly revenue in dollars from selling c cabinets and $C(c)$ represents the monthly cost in dollars to manufacture and sell c cabinets. Find the number of cabinets this company must sell each month to break even or make a profit.

26. Dan's Mobile Window Tinting tints car windows. Dan has determined his company's weekly revenue and cost functions as follows:

$$R(w) = 35w + 10$$
$$C(w) = 30w + 350$$

where $R(w)$ represents the weekly revenue in dollars from tinting w windows and $C(w)$ represents the weekly cost in dollars to tint w windows. Find the number of windows Dan must tint each week to break even or make a profit.

27. Optimum Traveling Detail from Exercise 13 in Section 2.2 on page 151 has the revenue and cost functions:

$$R(a) = 175a$$
$$C(a) = 25a + 7800$$

where $R(a)$ represents the monthly revenue in dollars for detailing a automobiles and $C(a)$ represents the monthly costs in dollars for detailing a automobiles. Find the number of automobiles Optimum Traveling Detail must detail to break even or make a profit.

28. A Table Affair from Exercise 14 in Section 2.2 on page 151 has the following weekly revenue and cost functions:

$$R(t) = 120t$$
$$C(t) = 45t + 800$$

where $R(t)$ represents the weekly revenue in dollars for selling t tablecloths and $C(t)$ represents the weekly costs in dollars for selling t table cloths. Find the number of tablecloths A Table Affair must sell to break even or make a profit.

29. Several prepaid cell phone services are advertised in the flyer.

a. Write equations for the cost of these three plans if a person uses m minutes.
b. For what number of minutes will U-R Mobile have the cheapest plan?

30. Several equipment rental services have the following charges for a dump truck rental.

Company	Base Charges ($)	Hourly Rate ($)
Pauley's Equipment	100	65
You Haul It	80	70
Big Red's Equipment	80	75

a. Write equations for the cost of these three companies rentals if a person rents the dump truck for h hours.

b. For what number of hours will You Haul It be the cheapest?

31. The percentage of births to unmarried women in the United States can be modeled by

$$U(t) = 0.71t + 37.1$$

where $U(t)$ represents the percentage of births to unmarried women in the United States t years since 2005. The percentage of births to unmarried women in the United Kingdom t years since 2005 can be modeled by

$$K(t) = 1.3t + 44.7$$

Source: Models derived from data in the Statistical Abstract of the United States, 2008.

Determine when the percentage of births to unmarried women in the United States was greater than the percentage in the United Kingdom.

32. The marriage rate per 1000 population aged 15–64 years in the United States can be modeled by

$$U(t) = -0.19t + 11.48$$

where $U(t)$ represents the marriage rate in the United States t years since 2005. The marriage rate in the United Kingdom t years since 2005 can be modeled by

$$K(t) = -0.15t + 7.66$$

Source: Models derived from data in the Statistical Abstract of the United States, 2008.

Determine when the marriage rate in the United States will be lower than the marriage rate in the United Kingdom.

33. A salesperson earns a base salary of $600 per month plus 21% commission on all sales. Find the minimum amount a salesperson would need to sell to earn at least $1225 for one month.

34. A salesperson earns a base salary of $725 per month plus 17% commission on all sales. Find the minimum amount a salesperson would need to sell to earn at least $1225 for one month.

35. Use the table to find when Y1 < Y2.

36. Use the table to find when Y1 > Y2.

37. Use the table to find when Y1 > Y2.

38. Use the table to find when Y1 > Y2.

For Exercises 39 through 46, solve each inequality numerically, using your calculator table.

39. $8x + 1 > -2x + 51$ **40.** $4x - 8 < -x + 22$

41. $4.5x - 8 \leq 7.5x - 20$ **42.** $0.25x + 12 \geq 0.75x + 2$

43. $-1.5x + 18 \geq 2.5x + 4$

44. $-4.25x + 5.125 \leq 3.6x + 45.25$

45. $\dfrac{1}{4}x - 2 < -\dfrac{1}{2}x - 4.5$

46. $\dfrac{2}{3}x + 4 > \dfrac{1}{3}x + 6$

For Exercises 47 through 50, use the graph to solve the given inequality.

47. Use the graph to find when $f(x) < g(x)$.

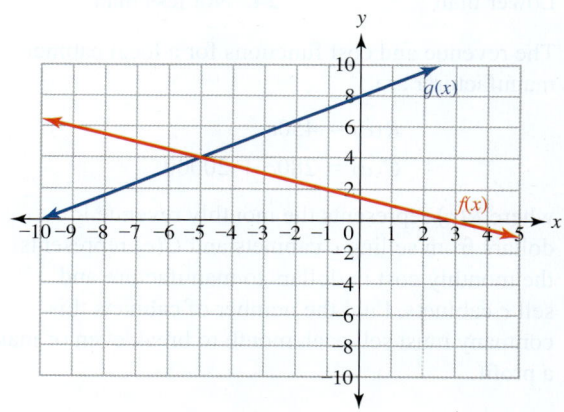

48. Use the graph to find when $f(x) < g(x)$.

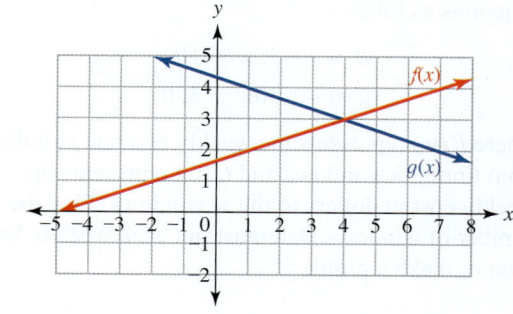

49. Use the graph to find when $f(x) \geq g(x)$.

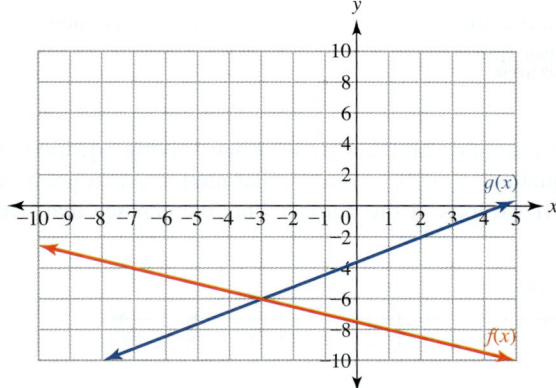

50. Use the graph to find when $f(x) \geq g(x)$.

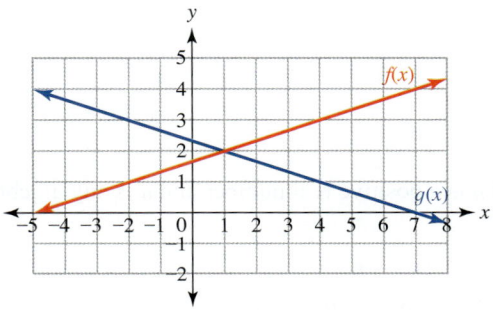

For Exercises 51 through 54, solve the inequality graphically.

51. $5x + 3 > -2x + 17$

52. $4.5x - 8 < 7.5x - 20$

53. $\frac{1}{2}x - 6 < -\frac{1}{3}x + 4$

54. $\frac{2}{5}x + 11 > \frac{4}{5}x + 3$

For Exercises 55 through 68, solve the inequality algebraically. Check your answer.

55. $5 + \frac{3x}{2} \leq -3$

56. $7 + \frac{4x}{3} \geq -15$

57. $5x + 3 \geq 3(x - 2)$

58. $4t - 3 > 2(t + 1)$

59. $\frac{-7d}{3} + 5 < 3$

60. $\frac{-2g}{9} + 12 > 4$

61. $2.7v + 3.69 > 1.5v - 6.5$

62. $3.4b + 2.45 < 0.3b - 8.5$

63. $3.2 + 2.7(1.5k - 3.1) \geq 9.43k - 17.5$

64. $8.7 - 1.4(8.2m + 6) \geq -2.3(7.1m - 4.3)$

65. $\frac{2}{5}(w - 20) \leq -\frac{3}{7}(4w - 9)$

66. $\frac{2}{3}(P + 4) < -\frac{5}{7}(2P - 12)$

67. $2.35x + 7.42 < 1.3x - 4.75$

68. $3.74x - 5.87 > 7.28x + 3.25$

2.5 Absolute Value Equations and Inequalities

LEARNING OBJECTIVES
- Solve absolute value equations.
- Solve absolute value inequalities.
- Solve applications using absolute values.

Absolute Value Equations

The absolute value can be defined in several ways. One way is to use the concept of distance on a number line. The absolute value of a number is defined as the distance from 0 to the number on a real number line. Using distance in the definition may help to see why $|6| = 6$ and $|-6| = 6$. Both 6 and -6 are 6 units from 0 on the real number line.

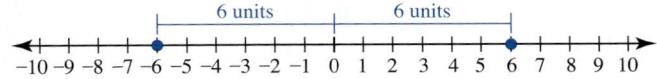

Since distance is always positive, the absolute value will always have a positive result.

Using Your TI Graphing Calculator

To take the absolute value of a number or expression on your graphing calculator, you will use the abs function found in the math menu. To get to the abs function, press MATH and ▶ to get to the NUM submenu.

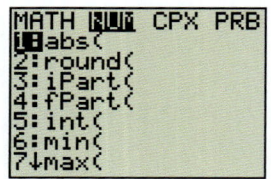

The first option on this submenu is abs(. Pressing 1 or ENTER will bring abs(into your equation or calculation. Remember to close the parentheses at the end of the expression you are taking the absolute value of.

DEFINITION

Absolute Value   The absolute value of a real number n, $|n|$, is the distance from zero to n on a real number line.

Using the distance definition of absolute value guides us when solving equations that contain absolute values. The equation $|x| = 8$ is asking what number(s), x, are 8 units from zero on the number line. If we look at the number line, we can see that both 8 and -8 are 8 units from zero.

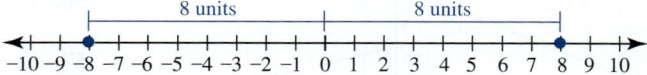

The absolute value equation becomes two equations, one for the positive side of zero and the other for the negative side of zero.

$$|x| = 8$$
$$x = 8 \quad \text{or} \quad x = -8$$

DEFINITION

Absolute Value Equation   If n is a positive real number and u is any algebraic expression, then

$$|u| = n$$
$$u = n \quad \text{or} \quad u = -n$$

If n is negative, then the equation $|u| = n$ has no real solution.
If n is zero, $u = 0$.

When solving an equation that contains an absolute value, isolate the expression containing the absolute value. Then write two equations, one with a positive value and one with a negative value.

Solving Absolute Value Equations
- Isolate the expression containing the absolute value.
- Rewrite the equation as two equations.

$$|u| = n$$
$$u = n \quad \text{or} \quad u = -n$$

- Solve each equation.
- Check the solutions in the original absolute value equation.

Example 1   Solving absolute value equations

Solve the following equations.

a. $|x + 5| = 12$

b. $2|x - 4| + 3 = 9$

c. $|h + 3| + 10 = 4$

SOLUTION

a. The absolute value is already isolated. Rewrite it as two equations and solve.

$$|x + 5| = 12 \quad \text{Rewrite into two equations.}$$

$x + 5 = 12$ or $x + 5 = -12$

$x = 7$ or $x = -17$ Solve by subtracting 5 from both sides.

There are two solutions for this equation. Check them both in the original absolute value equation.

$$\begin{array}{ll} x = 7 & x = -17 \\ |7 + 5| \stackrel{?}{=} 12 & |-17 + 5| \stackrel{?}{=} 12 \\ |12| \stackrel{?}{=} 12 & |-12| \stackrel{?}{=} 12 \\ 12 = 12 & 12 = 12 \end{array}$$

Both solutions work. We can see the two solutions by looking at the graph of both sides of the equation and looking for where the two graphs intersect.

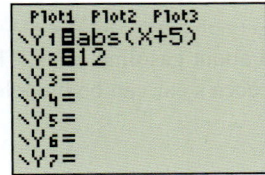 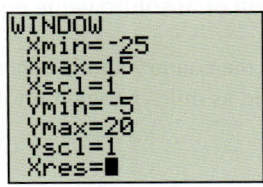 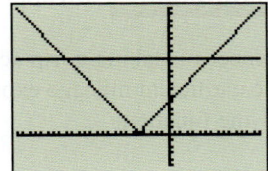

b. First, isolate the absolute value.

$$\begin{aligned} 2|x - 4| + 3 &= 9 \quad \text{Subtract 3 from both sides of} \\ \underline{-3 \quad -3} & \qquad \text{the equation.} \\ 2|x - 4| &= 6 \\ \frac{2|x - 4|}{2} &= \frac{6}{2} \quad \text{Divide both sides of the equation by 2.} \\ |x - 4| &= 3 \end{aligned}$$

Rewrite into two equations and solve.

$$|x - 4| = 3$$

$x - 4 = 3$ or $x - 4 = -3$ Solve by adding 4 to both
$x = 7$ or $x = 1$ sides of the equation.

Check both equations in the original absolute value equation.

$$\begin{array}{ll} x = 7 & x = 1 \\ 2|7 - 4| + 3 \stackrel{?}{=} 9 & 2|1 - 4| + 3 \stackrel{?}{=} 9 \\ 2|3| + 3 \stackrel{?}{=} 9 & 2|-3| + 3 \stackrel{?}{=} 9 \\ 9 = 9 & 9 = 9 \end{array}$$

Both solutions work.

c. First, isolate the absolute value.

$$\begin{aligned} |h + 3| + 10 &= 4 \quad \text{Subtract 10 from both} \\ \underline{-10 \quad -10} & \qquad \text{sides of the equation.} \\ |h + 3| &= -6 \end{aligned}$$

The absolute value equals a negative number, which is not possible. There is no real solution to this equation. A good way to check this result is to graph both sides of the equation and see whether they intersect.

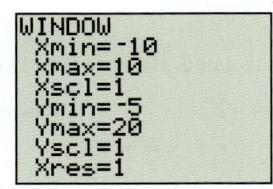
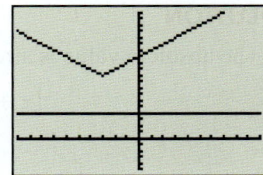

From the graph, we can see the absolute value curve does not intersect the line. This confirms that the result of no real solution is correct.

PRACTICE PROBLEM FOR EXAMPLE 1
Solve the following equations.

a. $|x - 2| = 7$ b. $|d - 12| - 9 = -15$ c. $-3|w + 1| + 12 = -3$

Example 2 An application of absolute value equations

Rebecca is taking a trip across the country and is concerned about gasoline costs. The most efficient mileage expressed as miles per gallon for Rebecca's car can be modeled by the function

$$m(s) = -\frac{1}{3}|s - 60| + 35$$

where m is the gas mileage for Rebecca's car when she is driving at an average speed of s miles per hour, for speeds between 40 and 80 miles per hour. At what speed should Rebecca drive to get gas mileage of 30 miles per gallon during her trip?

SOLUTION

Rebecca wants to get 30 miles per gallon, so $m = 30$. Substitute 30 into m and solve.

$$m(s) = -\frac{1}{3}|s - 60| + 35$$

$$30 = -\frac{1}{3}|s - 60| + 35$$
$$\underline{-35 \qquad\qquad\qquad -35}$$ *Isolate the absolute value by subtracting 35 from both sides.*

$$-5 = -\frac{1}{3}|s - 60|$$

$$-3(-5) = -3\left(-\frac{1}{3}|s - 60|\right)$$ *Multiply both sides by -3 to eliminate the fraction.*

$$15 = |s - 60|$$

$s - 60 = 15$ or $s - 60 = -15$ *Rewrite into two equations and solve.*
$s = 75$ or $s = 45$

Check these solutions in the original equation, using the table on the calculator.

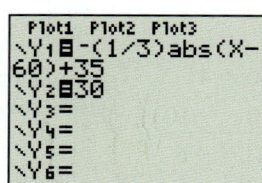

Rebecca can drive either 75 miles per hour or 45 miles per hour on her trip to get gas mileage of 30 miles per gallon.

PRACTICE PROBLEM FOR EXAMPLE 2

Svetlana is driving from San Antonio to Houston, Texas, on Interstate 10 and will pass through Columbus, Texas, on the way. Svetlana's distance from Columbus can be modeled by the function

$$D(t) = |125 - 65t|$$

where $D(t)$ is Svetlana's distance in miles from Columbus, Texas, after driving for t hours. Find the time when Svetlana will be 25 miles away from Columbus.

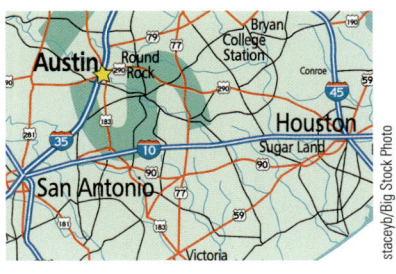

Absolute Value Inequalities Involving "Less Than" or "Less Than or Equal To"

Inequalities that contain absolute values can also be interpreted by using the distance definition of absolute value. The inequality $|x| < 3$ is asking for all values of x that are less than 3 units from zero on the number line.

If we look at the graph of each side of the inequality, we can see the solution set.

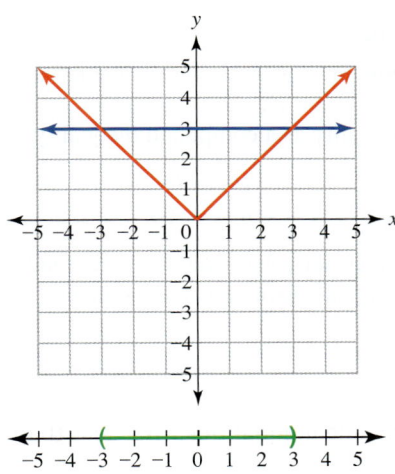

We can see that the red graph of the absolute value is less than 3 when it is below the blue graph of $y = 3$. This happens when x is between -3 and 3.

> **What's That Mean?**
>
> **Compound Inequality**
>
> When an inequality has two inequality symbols, it is called a compound inequality. Compound inequalities can be written in a few different forms:
>
> $$-5 < x < 8$$
>
> or
>
> $$x < 4 \quad \text{or} \quad x > 7$$
>
> We will use compound inequalities when working with absolute value inequalities.

In solving inequalities with absolute values, it is important to keep in mind what the inequality is asking. The absolute value inequality $|x| < 3$ is asking for x-values that are closer than 3 units away from zero. From the above graph, we can see that this interpretation implies that x must be between -3 and 3. Writing this interval with inequalities is similar to the way we rewrote absolute value equations into two equations, but the inequality symbols must be in the correct direction and order.

$$|x| < 3$$
$$-3 < x < 3$$

The last inequality says that x must be between -3 and 3, as seen as the interval on the above graph.

Example 3 Writing an interval using a compound inequality

Write the given interval using a compound inequality.

a.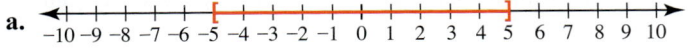

b.

SOLUTION

a. This interval has brackets, so the inequalities will include the "equal to" part. The values in the interval are between -5 and 5. The interval written as an inequality is $-5 \leq x \leq 5$.

b. This interval has parentheses, so the inequalities will not include the "equal to" part. The values in the interval are between -3 and 7. The interval written as an inequality is $-3 < x < 7$.

PRACTICE PROBLEM FOR EXAMPLE 3
Write the given interval using a compound inequality.

When solving absolute value inequalities with a "less than" symbol, write a compound inequality to indicate that the expression must be between the positive number and the opposite of the number. Remember that an absolute value cannot be less than a negative number in the real number system.

> **Solving Absolute Value Inequalities Involving the "Less Than" or "Less Than or Equal To" Symbols**
>
> - Isolate the expression containing the absolute value to the left side of the inequality.
> - Rewrite the absolute value inequality as a compound inequality.
>
> $$|u| < n \qquad\qquad |u| \leq n$$
> $$-n < u < n \qquad -n \leq u \leq n$$
>
> - Solve the compound inequality. (Isolate the variable in the middle.)
> - Check the solution set in the original absolute value inequality.

Example 4 — Solving absolute value inequalities with "less than"

Solve the following inequalities. Give the solution as an inequality and graph the solution set on a number line.

a. $|x| < 4$
b. $|x - 2| \leq 5$
c. $|b + 5| + 15 < 20$

SOLUTION
a. The absolute value is isolated on the left side. Since there is a "less than" symbol, rewrite the absolute value inequality as a compound inequality.

$$|x| < 4$$
$$-4 < x < 4$$

Therefore, x must be between -4 and 4.

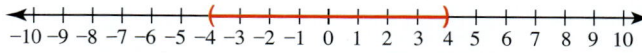

The solution set is the interval from -4 to 4. To confirm this solution set, check the endpoints of the interval and pick a number inside the interval to check the direction of the inequalities. Check each of these values in the original absolute value inequality. For the point inside the interval, we can pick any value we want. We will pick $x = 3$.

SECTION 2.5 Absolute Value Equations and Inequalities 181

Left endpoint	Right endpoint	Point inside the interval
$\lvert -4 \rvert \stackrel{?}{=} 4$	$\lvert 4 \rvert \stackrel{?}{=} 4$	$\lvert 3 \rvert \stackrel{?}{<} 4$
$4 = 4$	$4 = 4$	$3 < 4$

The equations and the inequality are true, so the solution set is correct.

b. The absolute value is isolated on the left side. It uses a "less than or equal to" symbol. Rewrite the absolute value inequality as a compound inequality and solve.

$$\lvert x - 2 \rvert \leq 5$$
$$-5 \leq x - 2 \leq 5 \quad \text{Rewrite as a compound inequality.}$$
$$\underline{+2 \quad\quad +2 \quad +2} \quad \text{Add 2 to all sides of the inequality.}$$
$$-3 \leq x \quad\quad \leq 7$$

Therefore, x must be between or equal to -3 and 7.

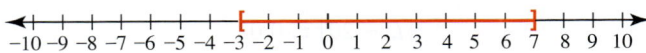

To confirm the solution set, we will check the endpoints of the interval and pick a number inside the interval to check the direction of the inequality symbols. If we pick $x = 4$ and check all of these values in the original absolute value inequality, we get the following:

Left endpoint	Right endpoint	Point inside the interval
$\lvert -3 - 2 \rvert \stackrel{?}{=} 5$	$\lvert 7 - 2 \rvert \stackrel{?}{=} 5$	$\lvert 4 - 2 \rvert \stackrel{?}{\leq} 5$
$\lvert -5 \rvert \stackrel{?}{=} 5$	$\lvert 5 \rvert \stackrel{?}{=} 5$	$\lvert 2 \rvert \stackrel{?}{\leq} 5$
$5 = 5$	$5 = 5$	$2 \leq 5$

The equations for the endpoints and the inequality are true, so the solution set is correct.

c. First, isolate the absolute value. Rewrite it as a compound inequality.

$$\lvert b + 5 \rvert + 15 < 20 \quad \text{Isolate the absolute value.}$$
$$\underline{\quad\quad\quad -15 \quad -15}$$
$$\lvert b + 5 \rvert < 5 \quad \text{Rewrite into a compound inequality and solve.}$$
$$-5 < b + 5 < 5$$
$$\underline{-5 \quad\quad -5 \quad -5} \quad \text{Subtract 5 from all sides.}$$
$$-10 < b \quad\quad < 0$$

Therefore, b must be between -10 and 0.

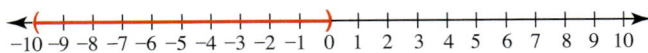

We will confirm the solution set by graphing the left and right sides of the inequality on the calculator and comparing the graphs.

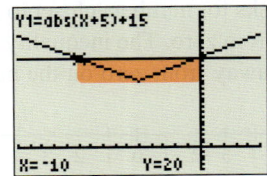

We can see that the absolute value graph is less than 20 when x is between -10 and 0.

PRACTICE PROBLEM FOR EXAMPLE 4

Solve the following inequalities. Give the solution as an inequality, and graph the solution set on a number line.

a. $\lvert x - 8 \rvert < 2$ **b.** $\lvert d + 1 \rvert - 3 \leq 5$

Example 5 — An application of absolute value inequalities with the symbol for "less than"

When a part is made for a machine or other precise application, the accuracy of the size of each part is very important for that part to function properly. The amount the size of a particular part can be larger or smaller than the ideal size is called the tolerance. When a 20-millimeter (mm) bolt is made for an engine, the actual size must be within 0.02 mm of 20 mm. If L is the actual length of the bolt in millimeters, the acceptable lengths of the bolt can be modeled by

$$|L - 20| \leq 0.02$$

Find the acceptable lengths of these bolts.

SOLUTION

Solve the absolute value inequality.

$$|L - 20| \leq 0.02$$
$$-0.02 \leq L - 20 \leq 0.02$$
$$19.98 \leq L \leq 20.02$$

We will confirm this solution set by checking the endpoints of the interval in the original absolute value inequality. We know that the ideal length of 20 mm is inside the interval, so the inequality symbols are correct.

Left endpoint	Right endpoint
$\|19.98 - 20\| \stackrel{?}{=} 0.02$	$\|20.02 - 20\| \stackrel{?}{=} 0.02$
$\|-0.02\| \stackrel{?}{=} 0.02$	$\|0.02\| \stackrel{?}{=} 0.02$
$0.02 = 0.02$	$0.02 = 0.02$

The equations are true, so the bolts must be between or equal to 19.98 mm and 20.02 mm long.

PRACTICE PROBLEM FOR EXAMPLE 5

A rod in an engine is supposed to be 4.125 inches long. The tolerance for the rod is 0.0005 inch. The acceptable lengths of the rod can be modeled by

$$|L - 4.125| \leq 0.0005$$

where L is the actual length of the rod in inches. Find the acceptable lengths for this rod.

Absolute Value Inequalities Involving "Greater Than" or "Greater Than or Equal To"

Where the "less than" symbol asks for numbers closer to zero, the "greater than" symbol asks for numbers farther away from zero. The inequality $|x| > 5$ is asking for the values of x that are more than 5 units away from zero on the number line.

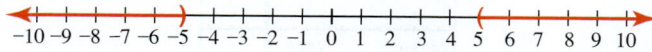

If we look at the graph of each side of the inequality, we can see the solution set.

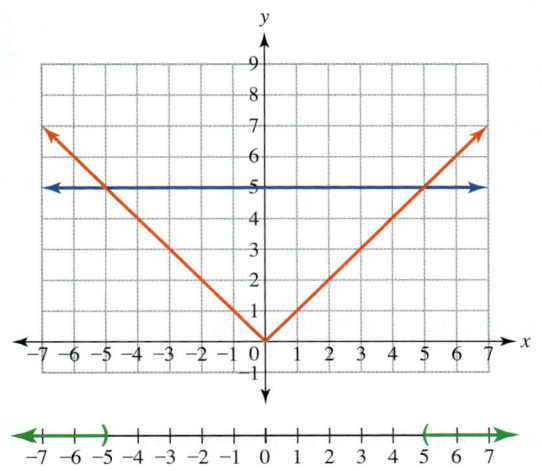

We can see that the red graph of the absolute value is greater than 5 when it is above the blue graph of the line $y = 5$. This occurs when x is less than -5 and when x is greater than 5.

To represent values that are farther away from zero requires an interval on the right and another interval on the left to account for distances in both directions. Therefore, write two inequalities to accommodate the absolute value, with one inequality going left (<) and another going right (>).

$$|x| > 5$$
$$x < -5 \quad \text{or} \quad x > 5$$

Example 6 Writing an interval using a compound inequality

Write the given interval using a compound inequality.

a.

b.

SOLUTION

a. This interval has parentheses, so the inequalities will not include the "equal to" part. The values in the interval are less than -2 or greater than 2. Since a value cannot be in both sides, we use the word "or" between the two inequalities. The interval written as an inequality is $x < -2$ or $x > 2$.

b. This interval has brackets, so the inequalities will include the "equal to" part. The values in the interval are less than -4 or greater than 3. The interval written as an inequality is $x \leq -4$ or $x \geq 3$.

PRACTICE PROBLEM FOR EXAMPLE 6
Write the given interval using a compound inequality.

When solving absolute value inequalities with a "greater than" symbol, write a compound inequality to indicate that the expression must be greater than the positive number and less than the opposite of the number.

Solving Absolute Value Inequalities Involving "Greater Than" or "Greater Than or Equal To"

- Isolate the expression containing the absolute value on the left side of the inequality.
- Rewrite the absolute value inequality into two inequalities, one less than and one greater than.

$$|u| > n \qquad\qquad |u| \geq n$$
$$u < -n \;\text{ or }\; u > n \qquad\qquad u \leq -n \;\text{ or }\; u \geq n$$

- Solve the two inequalities.
- Check the solution set in the original absolute value inequality.

Example 7 — Solving absolute value inequalities with "greater than"

Solve the following inequalities. Give the solution as an inequality. Graph the solution set on a number line.

a. $|x| > 2$ **b.** $|p + 4| > 3$ **c.** $|x - 2| + 7 \geq 10$

SOLUTION

a. The absolute value is already isolated. Rewrite the absolute value inequality as two inequalities.

$$|x| > 2$$
$$x < -2 \quad\text{or}\quad x > 2$$

To check this solution, check the endpoints and a point from each of the intervals. We will check -6 for the left interval and 6 for the right interval.

Left interval endpoint	Right interval endpoint	Left interval test point	Right interval test point
$\|-2\| \stackrel{?}{=} 2$	$\|2\| \stackrel{?}{=} 2$	$\|-6\| \stackrel{?}{>} 2$	$\|6\| \stackrel{?}{>} 2$
$2 = 2$	$2 = 2$	$6 > 2$	$6 > 2$

Each of these points satisfies the original absolute value inequality, so the solution set is correct.

b. The absolute value is already isolated. Rewrite the absolute value inequality as two inequalities and solve.

$$|p + 4| > 3$$
$$p + 4 < -3 \quad\text{or}\quad p + 4 > 3$$
$$p < -7 \quad\text{or}\quad p > -1$$

Rewrite as two inequalities and solve. Subtract 4 from both sides.

Check this solution set using the graph of each side on the calculator.

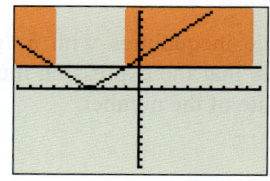

We can see that the absolute value graph is greater than 3 when x is less than -7 or greater than -1.

b. First, isolate the absolute value on the left side of the inequality. Rewrite into two inequalities and solve.

$$|x - 2| + 7 \geq 10 \quad \text{Isolate the absolute value.}$$
$$\underline{\quad -7 \quad -7 \quad}$$
$$|x - 2| \geq 3$$

$x - 2 \leq -3$ or $x - 2 \geq 3$ Rewrite into two inequalities and solve.
$x \leq -1$ or $x \geq 5$ Add 2 to both sides.

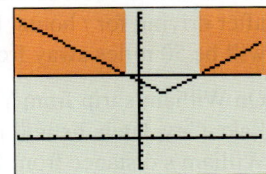

Check this solution set using the graph of each side of the inequality on the calculator.

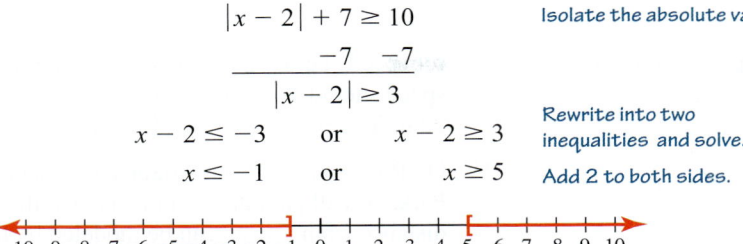

We can see that the absolute value graph is greater than 10 when x is less than -1 or greater than 5.

PRACTICE PROBLEM FOR EXAMPLE 7

Solve the following inequalities. Give the solution as an inequality. Graph the solution set on a number line.

a. $|x + 2| > 7$ **b.** $|g - 3| - 4 \geq 1$

Example 8   An application of absolute value inequalities with "greater than"

A baby's birth weight is considered unusually high or low if it satisfies the inequality

$$|b - 118| > 30$$

where b is a baby's birth weight in ounces. Find the birth weights that would be considered unusually low or high.

SOLUTION

Solve the inequality by rewriting it into two inequalities.

$$|b - 118| > 30$$
$$b - 118 < -30 \quad \text{or} \quad b - 118 > 30$$
$$b < 88 \quad \text{or} \quad b > 148$$

Check this solution set using the endpoints of the intervals and points within each interval.

Left interval endpoint	Right interval endpoint	Left interval test point	Right interval test point
$\|88 - 118\| \stackrel{?}{=} 30$	$\|148 - 118\| \stackrel{?}{=} 30$	$\|70 - 118\| \stackrel{?}{>} 30$	$\|160 - 118\| \stackrel{?}{>} 30$
$30 = 30$	$30 = 30$	$48 > 30$	$42 > 30$

These checks are true, so the solution set is correct. A baby is considered to have an unusually low birth weight when it weighs less than 88 ounces, and it is considered to have an unusually high birth weight when it weighs more than 148 ounces.

PRACTICE PROBLEM FOR EXAMPLE 8

A recent poll asked people how much TV they watched in an average week. If the number of hours h a person watches TV satisfies the inequality

$$\left|\frac{h - 32}{4.5}\right| > 1.96$$

it would be considered unusual. Find the number of hours that would be considered unusual.

2.5 Exercises

For Exercises 1 through 16, solve the following equations. Check your answers.

1. $|x| = 12$
2. $|x| = 20$
3. $|h + 7| = 15$
4. $|m + 9| = 11$
5. $|b - 12| = 8$
6. $|w - 8| = 30$
7. $|k + 10| = -3$
8. $|s - 4| = -15$
9. $2|r + 11| = 36$
10. $8|h - 3| = 4$
11. $|g - 9| + 12 = 8$
12. $|x + 4| + 6 = 1$
13. $-2|d - 8| + 1 = 11$
14. $-4|m + 3| + 7 = 3$
15. $-3|x - 12| = -5$
16. $-5|y + 14| = -36$

17. Ricardo is taking a trip from Arizona to Yosemite National Park in California. The gas mileage in miles per gallon for Ricardo's truck can be modeled by the function

$$m(s) = -\frac{1}{3}|s - 60| + 25$$

where $m(s)$ is the gas mileage for Ricardo's truck when he is driving at an average speed of s miles per hour, for speeds between 40 and 80 miles per hour. At what speed should Ricardo drive to get gas mileage of 20 miles per gallon during his trip?

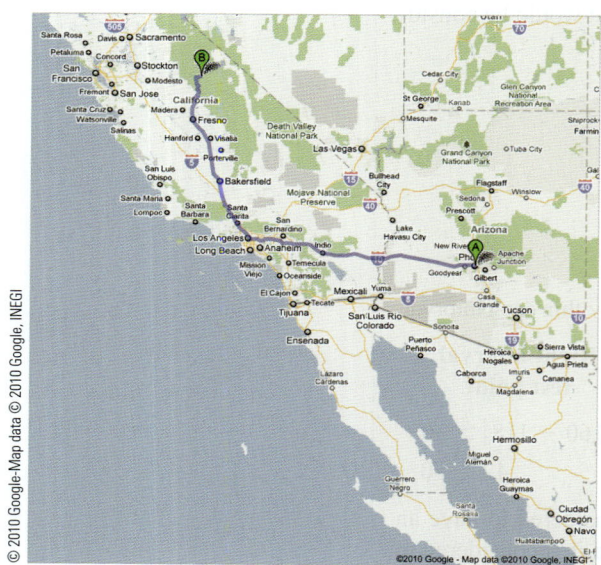

18. William is driving from Yakima, Washington, to Salt Lake City, Utah. The gas mileage in miles per gallon for William's car can be modeled by the function

$$m(s) = -\frac{1}{2}|s - 55| + 24$$

where $m(s)$ is the gas mileage for William's car when he is driving at an average speed of s miles per hour, for speeds between 40 and 80 miles per hour. At what speed should William drive to get gas mileage of 22 miles per gallon during his trip?

19. On Ricardo's trip from Arizona to Yosemite National Park, he will pass through Fresno, California. Ricardo's distance from Fresno can be modeled by the function

$$D(t) = |700 - 60t|$$

where $D(t)$ is Ricardo's distance in miles from Fresno after driving for t hours. Find the time when Ricardo will be 50 miles away from Fresno.

20. On William's trip from Yakima, Washington, to Salt Lake City, Utah, he will pass through Boise, Idaho. William's distance from Boise can be modeled by the function

$$D(t) = |365 - 55t|$$

where $D(t)$ is William's distance in miles from Boise after driving for t hours. Find the time when William will be 30 miles away from Boise.

For Exercises 21 through 30, write the interval shown on the number line using a compound inequality.

21.
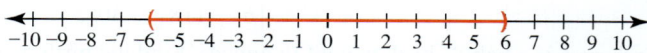

22.
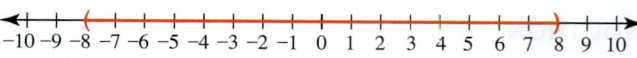

23.
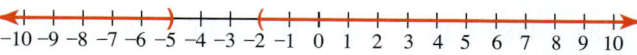

24.
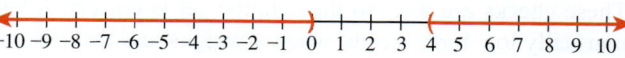

25.

26.
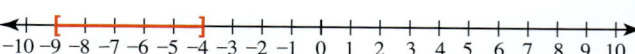

27.

28.

29.
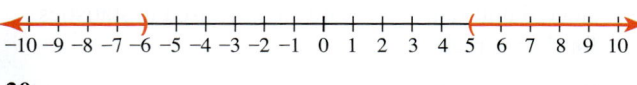

30.
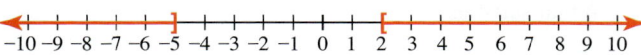

For Exercises 31 through 36, write the solution set using inequality notation for the given graph and inequality.

31.
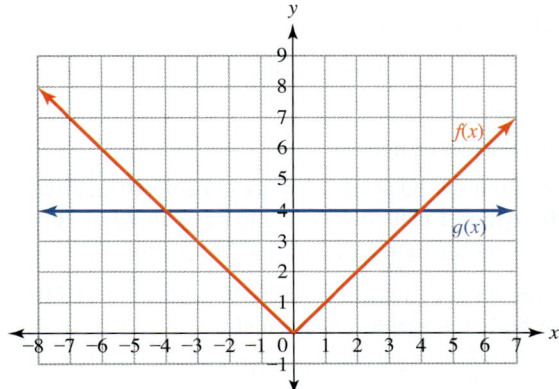

 a. $f(x) < g(x)$ b. $f(x) > g(x)$

32.
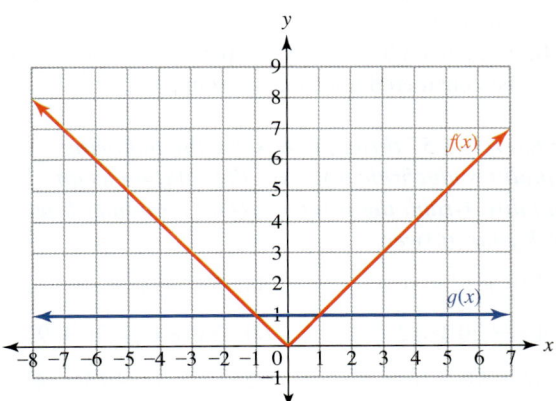

 a. $f(x) < g(x)$ b. $f(x) > g(x)$

33.
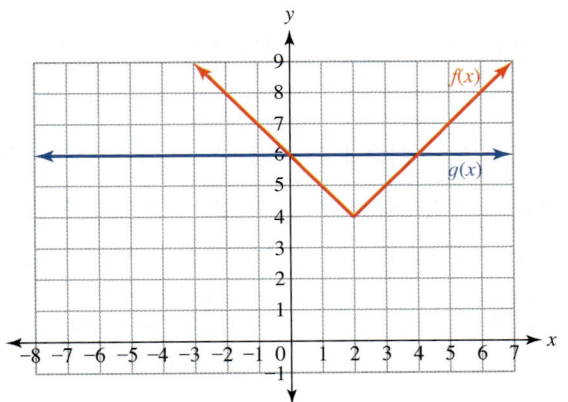

 a. $f(x) \leq g(x)$ b. $f(x) \geq g(x)$

34.
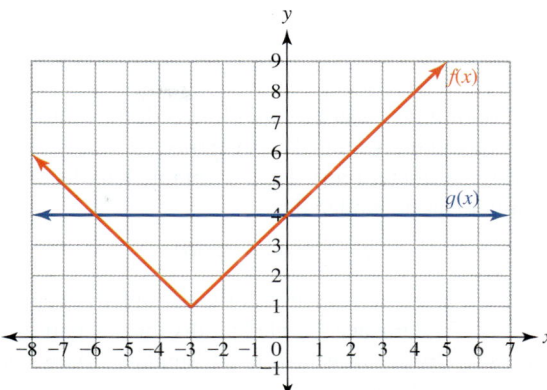

 a. $f(x) \leq g(x)$ b. $f(x) \geq g(x)$

35.
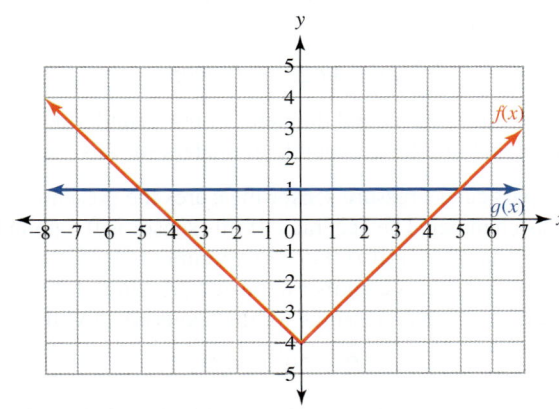

 a. $f(x) < g(x)$ b. $f(x) > g(x)$

36.
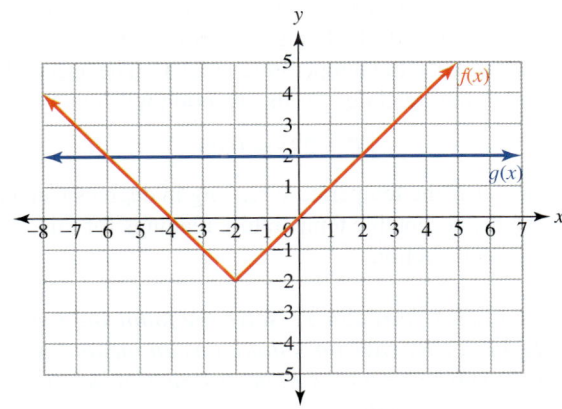

 a. $f(x) \leq g(x)$ b. $f(x) \geq g(x)$

For Exercises 37 through 42, solve the inequality algebraically. Give the solution as an inequality. Graph the solution set on a number line. Check your answer, using a graph.

37. $|x| < 5$

38. $|x| < 7$

39. $|h - 3| \leq 4$

40. $|k+5| \leq 2$

41. $|b-1| < 6$

42. $|p+3| < 3$

43. A coupling for a satellite fuel line is designed to be 3.00 centimeters long. The tolerance for this part is 0.001 centimeter (cm). The acceptable lengths of the coupler can be modeled by
$$|L - 3.00| \leq 0.001$$
where L is the actual length of the coupler in centimeters. Find the acceptable lengths for this part. (The tolerance for a part is how much smaller or larger it can be than the desired size and still be acceptable.)

44. A cylinder for a small engine is designed to be 3.625 inches in diameter. The tolerance for this part is 0.0005 inch. The acceptable diameters of the cylinder can be modeled by
$$|D - 3.625| \leq 0.0005$$
where D is the actual diameter of the cylinder in inches. Find the acceptable diameters for this part.

45. A particular pressure sensor measures the pressure in a gas chamber. The accuracy of this sensor is 0.5 pounds per square inch (psi). For an experiment, the technician requires the gas chamber to be at a pressure of 20 psi. The acceptable pressures for this experiment will satisfy
$$|P - 20| \leq 0.5$$
where P is the measured pressure in pounds per square inch. Find the acceptable pressures for this experiment.

46. A particular thermocouple measures the temperature during a chemical reaction. The accuracy of this thermocouple is 0.05°C. For an experiment, the technician wants to keep the temperature at 22°C. The acceptable temperature readings for this reaction will satisfy
$$|T - 22| \leq 0.05$$
where T is the temperature reading in degrees Celcius. Find the acceptable temperature readings for this chemical reaction.

For Exercises 47 through 52, solve the following inequalities algebraically. Give the solution as an inequality. Graph the solution set on a number line. Check your answer, using a graph.

47. $|x| > 3$

48. $|x| > 7$

49. $|p-4| \geq 2$

50. $|m-3| \geq 1$

51. $|y+6| > 3$

52. $|w+2| > 5$

53. Men's heights have a mean of 69 inches. A height h that satisfies
$$\left|\frac{h-69}{2.5}\right| > 2$$
is considered unusual. Find the heights that are considered unusual. (The word *mean* is used in statistics for the average value.)

54. Women's heights have a mean of 63.6 inches. A height h that satisfies
$$\left|\frac{h-63.6}{2.5}\right| > 2$$
is considered unusual. Find the heights that are considered unusual.

55. The normal mean score of the Stanford Binet IQ test is 100. An IQ score S that satisfies
$$\left|\frac{S-100}{16}\right| > 2$$
is considered unusual.
  a. Find the IQ scores that are considered unusual.
  b. Albert Einstein was said to have an IQ of 160. Was Einstein's IQ unusual? Explain.

56. Human body temperatures have a mean of 98.20°F. A body temperature that satisfies
$$\left|\frac{T-98.20}{0.62}\right| > 2$$
is considered unusual.
  a. Find the body temperatures that are considered unusual.
  b. If a person has a body temperature of 99°F, would it be considered unusual? Explain.

For Exercises 57 through 78, solve the following inequalities algebraically. Give the solution as an inequality. Graph the solution set on a number line. Check your answer.

57. $2|r+6| < 8$

58. $5|t+3| < 20$

59. $4x - 20 > 12$

60. $2x + 6 \leq 5x + 9$

61. $|d-5| + 10 \leq 7$

62. $|x+3| + 15 < 4$

63. $|m+1| - 2 > 6$

64. $-3|s-2| + 2 > -4$

65. $\frac{2}{3}x - 8 > \frac{5}{6}x - 4$

66. $\frac{3}{4}m + \frac{2}{3} \geq 5m + \frac{4}{3}$

67. $-6|p-4| + 3 \leq 16$

68. $-4|x+3| + 9 < 20$

69. $-|x-4| \geq -2$

70. $-|k+11| \geq -36$

71. $5 + 2|2x-4| > 11$

72. $4 - 5|4x+8| > -6$

73. $12 + 8|3d-8| \leq 10$

74. $20 + 3|5x-2| \leq 11$

75. $1.4x + 3 \geq 4.4x - 1$

76. $|2.5b - 3| - 1.4 \geq -0.6$

77. $\left|\frac{1}{2}x + 3\right| \leq 5$

78. $\left|\frac{2}{3}d - 4\right| > 10$

2.6 Solving Systems of Linear Inequalities

LEARNING OBJECTIVES
- Graph a linear inequality with two variables.
- Graph a system of linear inequalities.
- Use a system of linear inequalities to represent an application.

Graphing Linear Inequalities with Two Variables

Many things in life are limited by our circumstances. When we go shopping, we are limited by the amount of money we have as well as by how much of something we need or want. Businesses are often limited by the production in a plant or by the costs associated with production. Dieticians must keep their clients within a certain range of nutrients and, therefore, must balance the types of foods and supplements that they prescribe for a client.

All of these situations can best be looked at as inequalities because a company or person might not need to use the maximum amount of something but may be able to use less. The use of inequalities allows for a set of solutions that will work within the given limitations or constraints. When two or more constraints are considered, a system of inequalities emerges.

Before we consider a system of inequalities, we will investigate an inequality that has two variables and look at its set of solutions. A graph is used to visualize the set of solutions to an inequality with two variables.

■ CONCEPT INVESTIGATION
What is in the solution set?

Use the inequality and the graph to fill in the table.

$$y \geq 2x - 3$$

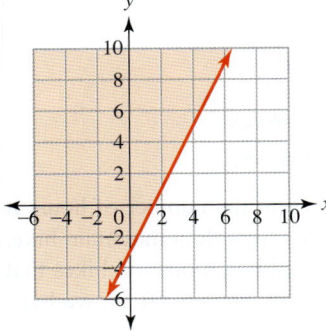

Point	Inequality	Does the point satisfy the inequality?	Is the point in the shaded region?
$(0, 0)$	$0 \stackrel{?}{\geq} (2(0) - 3)$ $0 \geq -3$	Yes	Yes
$(2, 8)$			
$(-4, -2)$			
$(6, 2)$			
$(0, 6)$			

Do all the points in the shaded region satisfy the inequality?

Which of the points in the table above were in the unshaded region? Did that point satisfy the inequality?

Do you think any of the points in the unshaded region satisfy the inequality?

In your own words, describe the solution set to this inequality.

From this concept investigation, we can see that the shaded region represents the solution set to the inequality. The line itself is also a part of this solution set, since the inequality included an "equal to" part. If the inequality does not include an equal to part the line will only border the solution set but not be part of it. When this occurs, we use a dashed line to indicate that the "equal to" part (line) is not included in the solution set.

> **Graphing a Linear Inequality with Two Variables**
> 1. Graph the line as if it were an equation. Use a dashed line if the inequality does not include an "equal to" symbol.
> 2. Pick a point on one side of the line and test the inequality.
>   (If not on the line, the point (0, 0) is a very easy point to use as a test point.)
> 3. If the point satisfies the inequality, shade that side of the line.
> 4. If it does not satisfy the inequality, shade the other side of the line.

Example 1 An application of linear inequalities with two variables

A bicycle plant can produce both beach cruisers and mountain bikes. The plant needs 2 hours to build one beach cruiser and 3.5 hours to produce a mountain bike. If the plant operates up to 60 hours per week, what combinations of bikes can they produce in a week?

SOLUTION

First, define the variables.

$$C = \text{number of beach cruisers built in a week}$$
$$M = \text{number of mountain bikes built in a week}$$

Because each cruiser takes 2 hours to build, $2C$ will represent the number of hours it takes the plant to build cruisers that week. It takes 3.5 hours to build a mountain bike, so $3.5M$ represents the number of hours that plant spends building mountain bikes that week. The maximum number of hours the plant operates in a week is 60, so we have the following inequality.

$$2C + 3.5M \leq 60$$

This inequality can be graphed. Solve for C to get

$$2C + 3.5M \leq 60$$
$$2C \leq -3.5M + 60$$
$$C \leq -1.75M + 30$$

To graph this inequality, we first graph it as if it were an equation.

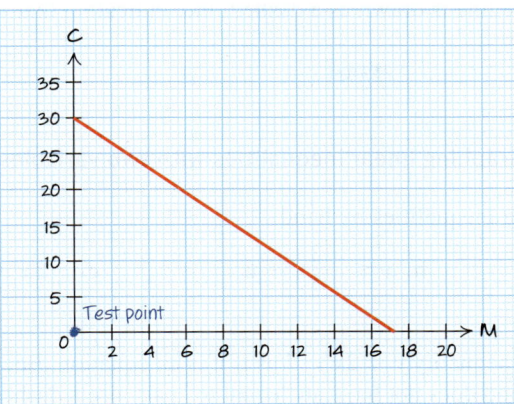

Because this situation requires that the factory work 60 hours or less a week, there will be many possible solutions. To represent all possible solutions, we will shade the side of the line that gives us 60 hours or less each week. To determine which side of the line to shade, test a point on either side of the line to see whether it is a solution to the inequality. If it is a solution, shade the side of the line that the test point is on. If we test the point (0, 0),

$$2(0) + 3.5(0) \stackrel{?}{\leq} 60$$
$$0 \leq 60$$

it satisfies the inequality indicating that the lower portion of the graph should be shaded.

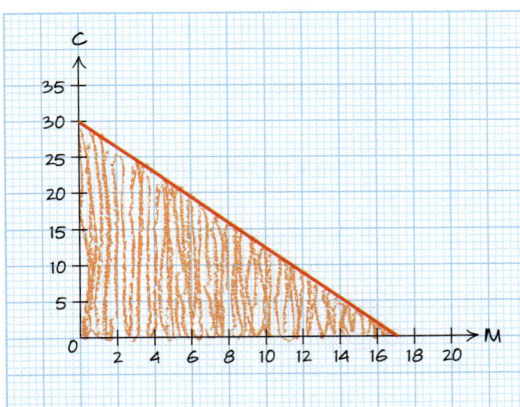

Any point within the shaded region is a solution to this inequality. This means that the plant can build anywhere from no bikes at all to a combination of bikes ranging from no cruisers and 17 mountain bikes to no mountain bikes but 30 cruisers. Any point on the line itself would mean that exactly 60 hours are being used a week. Any point in the shaded region means that fewer than 60 hours are being used. Any point outside the shaded region implies that more than 60 hours would be needed to do the work.

From the example, there is a large set of solutions for this situation. Working with inequalities most often means finding not a single solution, but a set of possible solutions. When graphing an inequality that has two variables, graph the inequality as if it were an equation and then decide which side of the line to shade. Deciding which side of a line to shade can be done by using any point that is not on the line as a test point. If that point is a solution to the inequality, then shade the side that the point is on. If the point is not a solution to the inequality, then test a point on the other side of the line.

In this example, the inequality included things equal to and less than, so the line was drawn as a solid line. If the inequality which is being graphed does not have an "equal to" symbol, then the line is drawn as a dashed line, indicating that it is not a part of the solution set.

Example 2 — Graphing linear inequalities with two variables

Graph the following inequalities by hand.

a. $y < 2x - 7$ **b.** $y \leq -\dfrac{3}{5}x + 6$ **c.** $-3x - y < -7$

SOLUTION

a. First, graph the inequality $y < 2x - 7$ as if it is an equation. Because the inequality is a "less than" and does not have an "equal to" part, use a dashed line.

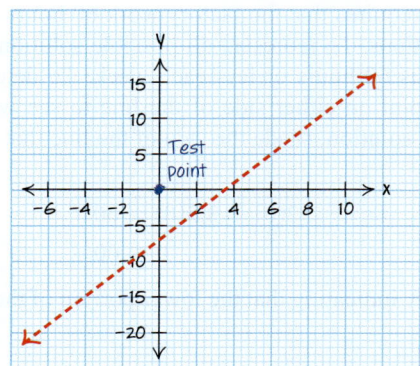

Now, complete the shading to show the solution set. Choose the point (0, 0) as the test point.

$y < 2x - 7$ Plug in the test point (0, 0).
$0 \stackrel{?}{<} 2(0) - 7$
$0 \not< -7$ This inequality is not true, so the test point is not a solution.

The test point (0, 0) does not satisfy the inequality because $0 < -7$ is not true. Shade the opposite side of the graph, the lower portion of the graph. When y is isolated on the left side of the inequality, the "less than" symbol implies shading the lower portion of the graph. Shading gives the following graph.

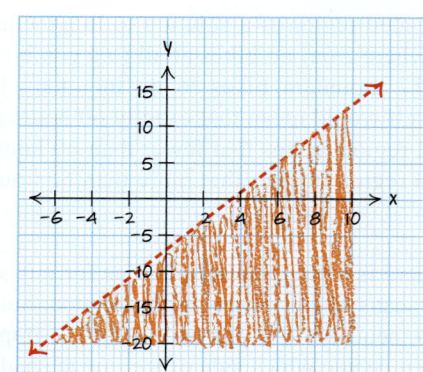

b. Graph the inequality $y \leq -\frac{3}{5}x + 6$ as if it were an equation. This time, the line will be solid, since the symbol does include an "equal to" part.

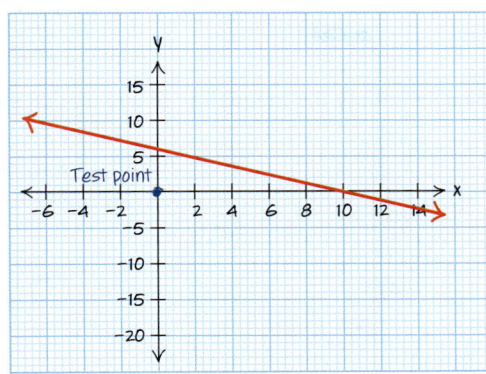

Test the point (0, 0). It satisfies the inequality, so shade the lower portion of the graph.

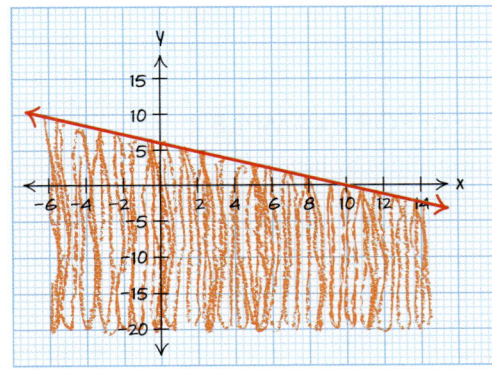

c. Start by putting the inequality $-3x - y < -7$ into slope-intercept form.

$$-3x - y < -7$$
$$-y < 3x - 7$$
$$\frac{-y}{-1} > \frac{3x - 7}{-1}$$
$$y > -3x + 7$$

Remember to reverse the inequality symbol when dividing by a negative number.

This inequality does not have an "equal to" part, so graph the line using dashes to indicate that it is not part of the solution set but only a boundary for it.

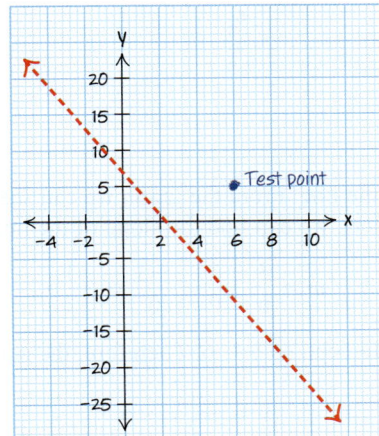

$y > -3x + 7$

Test the point (6, 5), which is clearly above the line.

$$-3x - y < -7$$
$$-3(6) - (5) \stackrel{?}{<} -7$$
Plug the test point (6, 5) into the original inequality.
$$-23 < -7$$
This inequality is true, so the test point is a solution.

The test point (6, 5) does satisfy the inequality. Shade on that side of the line.

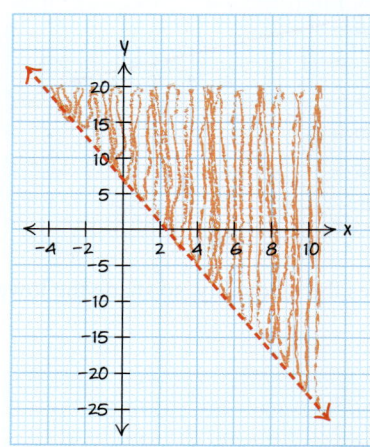

Note that if we pick a point not in the solutions set, it will not satisfy the inequality.

If we test the point (0, 0) that is clearly below the line, we get

$$-3x - y < -7$$
$$-3(0) - (0) \stackrel{?}{<} -7$$
$$0 \not< -7$$

This inequality is not true, so the point (0, 0) is not a solution and therefore is not in the shaded region.

PRACTICE PROBLEM FOR EXAMPLE 2

Graph the following inequalities by hand.

a. $y > 2x - 8$ **b.** $y \leq -\frac{3}{4}x + 5$

Example 3 — Finding an inequality from a graph

Find the inequality for the given graph.

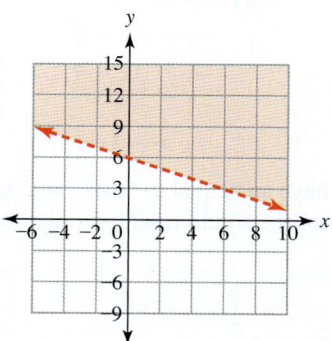

SOLUTION

First, find the equation of the line. From the graph we can see the two points (0, 6) and (6, 3). Using these two points, we can find the slope of the line.

$$m = \frac{3 - 6}{6 - 0} = \frac{-3}{6} = -\frac{1}{2}$$

The point (0, 6) is the y-intercept, so $b = 6$. Therefore, the equation of the line is

$$y = -\frac{1}{2}x + 6$$

Since the line is dashed, the inequality will not have an "equal to" part. The graph is shaded above the line, so we will use the "greater than" symbol.

$$y > -\frac{1}{2}x + 6$$

Check the inequality symbol by checking a point in the shaded region. We will use the point (4, 9).

$$9 \stackrel{?}{>} -\frac{1}{2}(4) + 6$$
$$9 \stackrel{?}{>} -2 + 6$$
$$9 > 4$$

The point from the shaded region checks, so the inequality symbol is facing the correct direction.

PRACTICE PROBLEM FOR EXAMPLE 3
Find the inequality for the following graph.

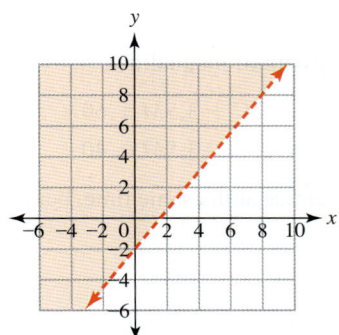

Solving Systems of Linear Inequalities

Now that we can graph an inequality with two variables, we can consider a system of inequalities. Most situations have more than one constraint. The bicycle plant from Example 1 might also have a constraint that they must keep the plant in production at least 40 hours per week. With more than one constraint, the situation calls for a system of inequalities, not just one inequality. When solving a system of inequalities, we will graph each inequality, shading according to the inequality symbol of each line. The area where all the shaded regions overlap will be the solution set to the system of inequalities. If the shaded regions do not overlap, there is no solution to the system.

> **Graphing a System of Linear Inequalities**
> 1. Graph both inequalities on the same set of axes, shading according to the inequality symbol of each line.
> 2. The solution set is the intersection of the shaded regions of all the inequalities in the system.
> 3. If the shaded regions do not intersect, there is no solution.

Example 4 An application of systems of linear inequalities

Example 1 on page 190 discussed a bicycle plant that had the time constraint that they could not exceed 60 hours of production per week. The same plant must stay in production at least 40 hours per week.

a. Create a system of inequalities to model this situation.

b. Graph the solution set for this system.

c. Can the plant produce ten mountain bikes and five cruisers in a week? Explain.

d. Can the plant produce five mountain bikes and eight cruisers a week? Explain.

SOLUTION

a. First, define the variables as in Example 1.

$$C = \text{number of cruisers built in a week}$$
$$M = \text{number of mountain bikes built in a week}$$

The constraint of not exceeding 60 hours per week results in the inequality from Example 1.

$$2C + 3.5M \leq 60$$

The new constraint of not working less than 40 hours a week results in the inequality

$$2C + 3.5M \geq 40$$

This gives the following system of inequalities.

$$2C + 3.5M \leq 60$$
$$2C + 3.5M \geq 40$$

b. Graph both of these inequalities at the same time, being careful to shade the correct sides of each line.

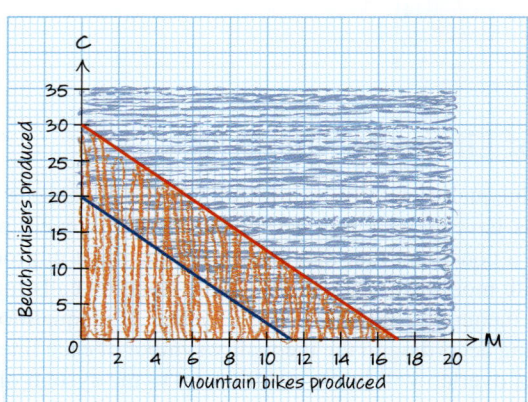

The *red* area shows the solution set for the first equation.

The *blue* area shows the solution set for the second equation.

The overlapping area is the solution set for the system as a whole.

c. Ten mountain bikes and five cruisers would be the point (10, 5) on the graph. This point is inside the area of the graph where the solution sets of the two inequalities overlap. Therefore, this combination would be a reasonable amount of each type of bike to produce a week.

d. Five mountain bikes and eight cruisers would be the point (5, 8) and would lie outside the overlapping section of this system. Therefore, producing this combination of bikes per week would not meet the constraints.

PRACTICE PROBLEM FOR EXAMPLE 4

A small appliance manufacturer has a plant that produces toaster ovens and electric griddles. The plant must keep its weekly costs below $40,000 and cannot be in operation more than 80 hours a week. It takes 1.25 hours and costs the plant $850 to produce 100 toaster ovens. It takes 2 hours and costs the plant $920 to produce 100 electric griddles.

a. Create a system of inequalities to model this situation. (*Hint:* define the variable in hundreds.)

b. Graph the solution set for this system.

c. Can the plant produce 1600 toaster ovens and 2000 electric griddles a week? Explain.

d. Can the plant produce 2000 toaster ovens and 3000 electric griddles a week? Explain.

In Example 4, we needed to graph only the area covered by positive numbers of bikes being produced. Using only positive values is typical of many real-life situations and actually adds two additional inequalities to the system: $M \geq 0$ and $C \geq 0$. We did not write these inequalities this time, but we could have. Common sense told us that a negative number of bikes could not be produced.

Another type of application that we can consider as a system of inequalities is investment questions like those we did in Section 2.2.

Example 5 An investment problem using inequalities

Mary is retiring and has $500,000 to invest in two accounts. One account pays 5% simple interest, and the other pays 3.5% simple interest. How much should she invest in each account to earn at least $21,000 in interest each year?

SOLUTION

Define variables as follows:

$A =$ the amount in dollars invested in the account paying 5% annual interest

$B =$ the amount in dollars invested in the account paying 3.5% annual interest

Because Mary can invest up to $500,000 and must earn at least $21,000 per year in interest, write the following system of inequalities:

$$A + B \leq 500000$$
$$0.05A + 0.035B \geq 21000$$

If we graph this system, we can find the possible solutions to this situation.

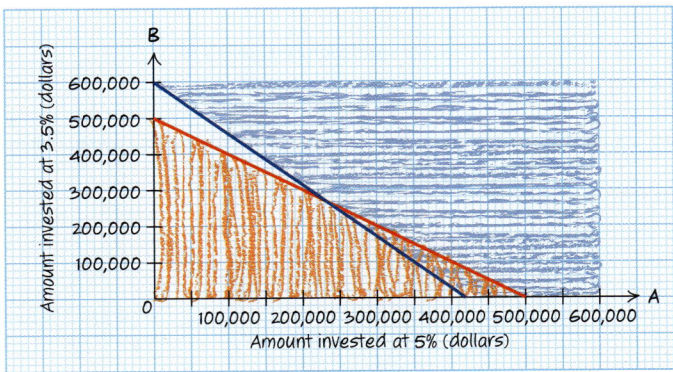

From this graph, we can see that Mary can have many different investment amounts that would earn her the minimum amount of interest she needs and still not use all of her $500,000. Any combination of investments from the overlapping section of the graph can be used to meet Mary's investment goal. Two of the many choices Mary has are as follows:

An investment of $450,000 in the account paying 5% and $50,000 in the account paying 3.5% will satisfy the needs Mary has.
or
An investment of $100,000 in the account paying 5% and $400,000 in the account paying 3.5% will not earn enough interest to meet Mary's needs.

In many systems of equations or inequalities, the points of intersection for the lines of the system are critical values of interest. In the case of Mary's investments, that point of intersection is the investment that meets her needs and invests the total $500,000. This is also the investment with the least amount invested in the more risky account paying 5% and the most money she can invest in the safer account paying 3.5% and still meet her minimum interest needs.

PRACTICE PROBLEM FOR EXAMPLE 5

Don is retiring and has $800,000 to invest in two accounts. One account pays 6.5% annual interest, and the other pays 4.5% annual interest. How much should he invest in each account to earn at least $40,500 in interest each year?

Example 6 Graphing systems of equations

Graph each system of inequalities by hand.

a. $2x + 5y < 15$
$3x - 2y > 6$

b. $y \geq -1.5x + 5$
$y \leq -1.5x + 2$

SOLUTION

a. Solve these inequalities for y, and then graph them.

$2x + 5y < 15$
$5y < -2x + 15$ *Solve the first inequality for y.*
$y < -\dfrac{2}{5}x + 3$

$3x - 2y > 6$ *Solve the second inequality for y.*
$-2y > -3x + 6$
$y < \dfrac{3}{2}x - 3$ *Remember to reverse the inequality symbol when dividing by a negative number.*

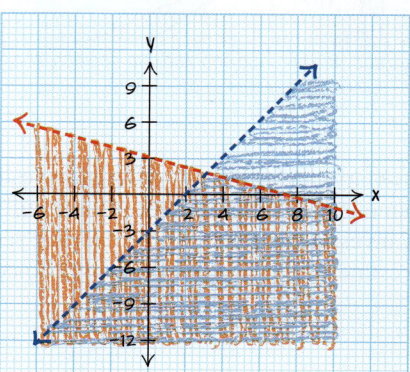

Now graph the system of inequalities.

$y < -\dfrac{2}{5}x + 3$

$y < \dfrac{3}{2}x - 3$

The section of the graph where the two shadings overlap is the solution set for this system of inequalities.

b. Start by graphing the two inequalities.

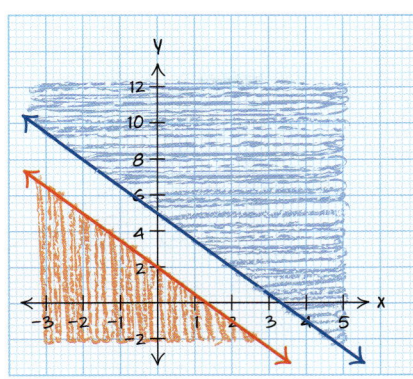

$$y \geq -1.5x + 5$$
$$y \leq -1.5x + 2$$

We can see from the graph that the solution sets for these two inequalities give us no intersection, and therefore, there is no solution to the system.

PRACTICE PROBLEM FOR EXAMPLE 6

Graph the following system of inequalities by hand.

$$y \geq x + 2$$
$$y \leq 2x - 3$$

Graphing these systems can also be done using the graphing calculator. See the instructions in the Using Your Graphing Calculator box on this page or in Appendix C.

Example 7 — Graphing systems of inequalities on a graphing calculator

Graph the following system of inequalities using a graphing calculator.

$$y < -2x + 5$$
$$y > 3x - 5$$

SOLUTION
Both inequalities have y isolated on the left side. The first inequality has a "less than" symbol, so shade below the line. For the second inequality, shade above the line because it has a "greater than" symbol. Both lines should be dashed because these inequalities do not include any "equal to" parts. The calculator does not show dashed lines when shading. This is something to remember when considering possible solutions.

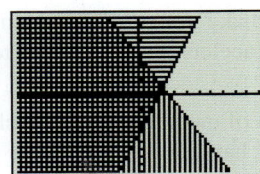 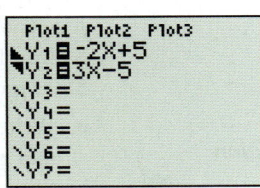

The two shaded regions intersect on the left side, giving the solution set.

PRACTICE PROBLEM FOR EXAMPLE 7

Graph the following system of inequalities using a graphing calculator.

$$2x - 5y > 12$$
$$4y > 7x - 5$$

Using Your TI Graphing Calculator

To graph inequalities on your calculator, you can have the calculator shade above or below the line. Use the cursor buttons to go to the left of the Y1 and press Enter until you get the shade above or below symbol you are looking for.

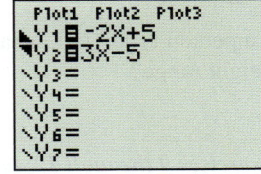

Note that the calculator cannot use shading with dashed lines, so any inequalities without "equal to" in the inequality symbol will graph with a solid line but should be interpreted as a dashed line. Any points on that line should not be considered part of the solution set.

2.6 Exercises

For Exercises 1 through 8, use the following information and graph to answer the questions.

According to the National Institutes of Health, adults are considered underweight or overweight on the basis of a combination of their height and weight. The graph below shows this relationship as a system of inequalities.

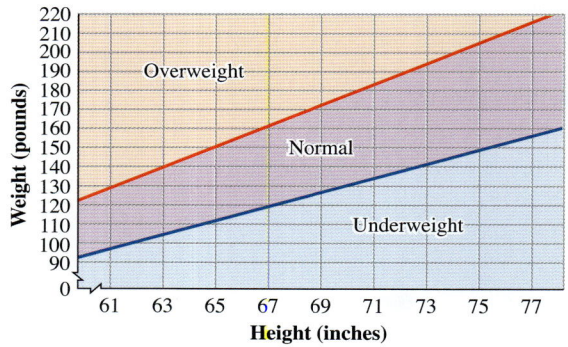

1. What category does a 61" person weighing 105 pounds fall into?

2. What category does a 75" person weighing 210 pounds fall into?

3. Above what weight is a 5' 9" person considered overweight?

4. Under what weight is a 6' person considered underweight?

5. If a 150-pound person is to be considered in the normal weight range, how tall must the person be?

6. If a 130-pound person is to be considered in the normal weight range, how tall must the person be?

7. If a person is 67" tall, what is the person's normal weight range?

8. If a person is 73" tall, what is the person's normal weight range?

For Exercises 9 through 12, use the following information and graph to answer the questions.

Bicycles Galore can manufacture up to 500 bikes a month in its Pittsburgh plant. Cruisers cost $75 to manufacture, and mountain bikes cost $135 to manufacture. Bicycles Galore must keep its total monthly costs below $50,000.

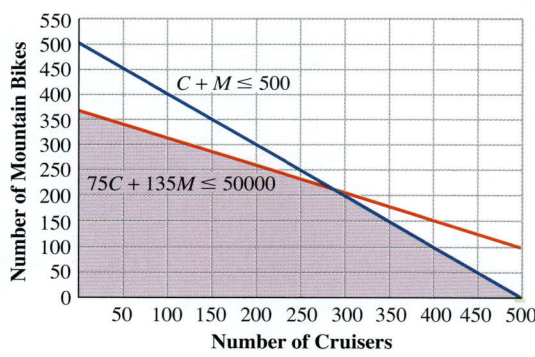

9. What is the greatest number of mountain bikes the company can build per month?

10. What is the maximum number of cruisers the company can build per month?

11. Can Bicycles Galore build 400 cruisers and 100 mountain bikes per month? Explain.

12. Can the company build 150 cruisers and 350 mountain bikes per month? Explain.

13. Bicycles Galore can produce up to 500 bikes per month in their Pittsburgh plant. They can make a profit of $65 per cruiser they build and $120 per mountain bike they build. The company's board of directors wants to make a profit of at least $40,000 per month.

    a. Write a system of inequalities to model this problem and graph it by hand.
    b. Can Bicycles Galore meet a demand by the board of directors for $40,000 in profit per month? Explain.
    c. If the company wants to maximize production at 500 bikes, how many of each bike should the plant produce to make the $40,000 profit? Explain.

14. Theresa has started a small business creating jewelry to sell at boutiques. She can make a necklace in about 45 minutes and a bracelet in about 30 minutes. Theresa can work only up to 25 hours a month making the jewelry. She can make $20 profit from a necklace and $18 profit from bracelets. Theresa wants to earn at least $800 profit.

    a. Write a system of inequalities to model this problem and graph it by hand.
    b. Should Theresa make more necklaces or bracelets? Explain.
    c. If Theresa knows that most of her customers want a bracelet and necklace combination, how many of each should she make to meet her profit goal? Explain.

SECTION 2.6 Solving Systems of Linear Inequalities 201

For Exercises 15 through 26, graph the inequalities by hand.

15. $y > 3x + 2$
16. $y < -4x + 7$
17. $y > \frac{5}{4}x - 2$
18. $y < \frac{3}{7}x + 1$
19. $y \geq \frac{2}{3}x + 6$
20. $y \geq \frac{4}{5}x - 5$
21. $y \leq -2x + 6$
22. $y \leq x + 5$
23. $2x + 5y > 10$
24. $3x + 4y < 8$
25. $12x - 4y > 8$
26. $10x - 5y < 20$

For Exercises 27 through 34, write the inequality for the given graph.

27.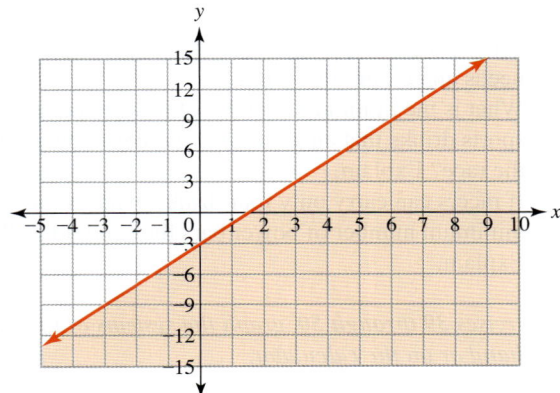

28.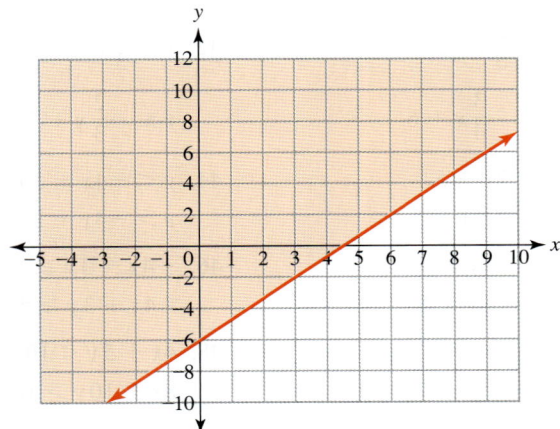

29.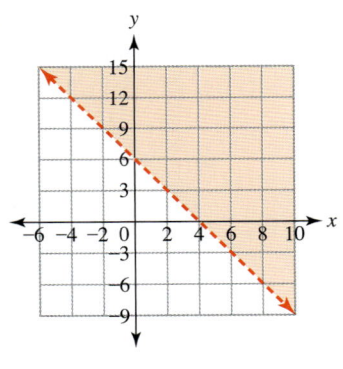

30.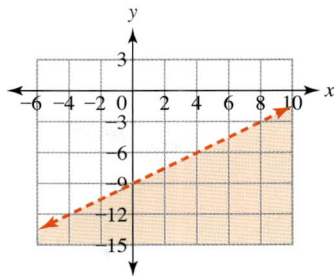

31.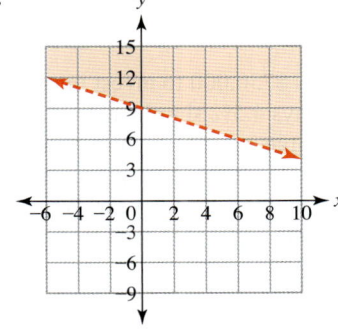

32.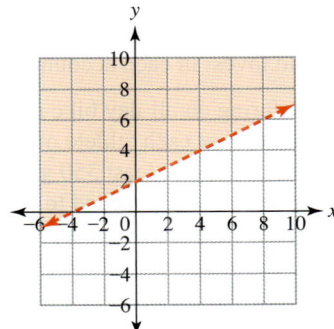

33.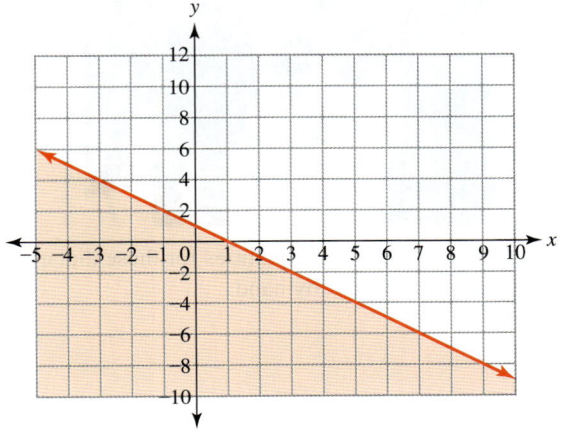

34.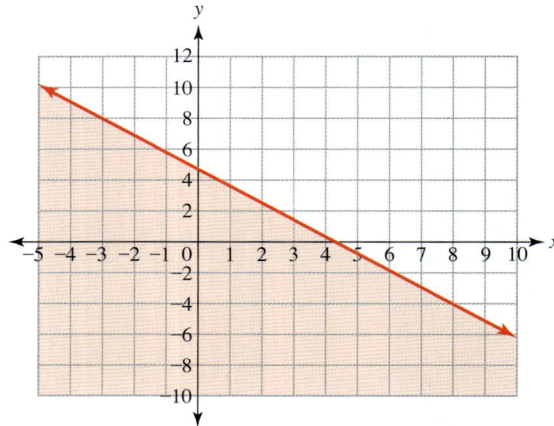

35. Juanita is retiring and has $750,000 to invest in two accounts. One account pays 4.5% simple interest, and the other pays 3.75% simple interest. Juanita wants to know how much she should invest in each account to earn at least $30,000 in interest each year. Write a system of inequalities and graph it by hand to show the possibilities.

36. Casey is retiring and has $2.5 million to invest in stocks and bonds. Casey estimates that the stocks will have an average return of 9.5% and the bonds will pay 7% simple interest. Casey wants $210,000 per year to cover his living and travel expenses each year. Write a system of inequalities and graph it by hand to show the possible amounts he could invest in each account.

37. In preparing meals to eat during an all-day bike race, the team dietician determines that each athlete must consume at least 2000 calories and at least 350 grams of carbohydrates to sustain the energy needed to compete at peak performance. The dietician is providing each athlete with a combination of power bars and sports drinks for the race. Each power bar has 240 calories and 30 grams of carbs. The sports drinks contain 300 calories and 70 grams of carbs.

  a. Write a system of inequalities for these dietary needs and graph it by hand.

  b. What is the minimum number of power bars and sports drinks that an athlete should consume during the race? Explain.

  c. If racers can carry only four drinks, what is the minimum number of power bars they should eat? Explain.

38. The same dietician from Exercise 37 had found that he should increase the calories and carbohydrates for the team to improve their performance. The dietician decides that each athlete must consume at least 2500 calories and at least 420 grams of carbohydrates to sustain the energy needed to compete at peak performance. The dietician is providing each athlete with a combination of power bars and sports drinks for the race. Each power bar has 240 calories and 30 grams of carbs. The sports drinks contain 300 calories and 70 grams of carbs.

  a. Write a system of inequalities for these dietary needs and graph it by hand.

  b. What is the minimum number of power bars and sports drinks that an athlete should consume during the race? Explain.

  c. If racers can carry only four drinks, what is the minimum number of power bars they should eat? Explain.

For Exercises 39 through 50, graph the systems of inequalities by hand or on the calculator.

39. $y > 2x + 4$
$y < -3x + 7$

40. $y < 4x + 5$
$y > -x + 1$

41. $y < \frac{2}{5}x - 3$
$y < -\frac{3}{4}x + 6$

42. $y \geq \frac{1}{3}x + 2$
$y \leq -\frac{2}{3}x + 7$

43. $y \leq 2x + 5$
$y \geq 2x + 8$

44. $y < -4x + 10$
$y > -4x + 2$

45. $2x + 4y \leq 5$
$2x - 4y \leq 5$

46. $-4x - 3y \geq 11$
$4x + 3y \geq 5$

47. $5x + 3y > 7$
$4x + 2y < 10$

48. $4x + 5y < 10$
$8x + 10y > 10$

49. $y > \frac{2}{3}x - 12$
$6x - 4y > 12$

50. $1.5x + 4.6y \leq 1.8$
$y \geq \frac{2}{17}x + 1.2$

Chapter Summary

Section 2.1 Systems of Linear Equations

- A **system of equations** is a set of two or more equations. The solution to a system is a point or points that satisfy all the equations in the system.
- Two ways in which systems can be solved are **graphically** and **numerically.**
- The solution to a system of equations can be seen on a graph where all the graphed lines **intersect** each other.
- Numerically, a system can be solved by looking for input values that give the same output value for every equation in the system.
- There are three possible types of systems of equations:
  1. **Consistent systems** have at least one solution. Consistent systems can have one of the following.
     a. **Independent lines:** The system will have one solution and occur when the equations have different slopes.
     b. **Dependent lines:** The system will have infinitely many solutions and occur when the equations have the same graph.
  2. **Inconsistent systems** have no solution and occur when the graphs do not intersect, such as with parallel lines.

Example 1

Solve the system by graphing.

$$y = 2x - 5$$
$$y = -\frac{1}{2}x + \frac{5}{2}$$

SOLUTION

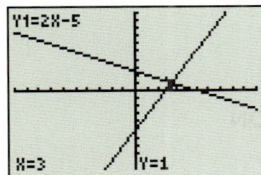

Therefore, this system has the solution (3, 1).

Example 2

Solve the system numerically.

$$y = -3x + 8$$
$$y = 4.5x - 10.75$$

SOLUTION

X	Y1	Y2
0	8	-10.75
1	5	-6.25
2	2	-1.75
3	-1	2.75
2.5	.5	.5

X=

Therefore, this system has the solution (2.5, 0.5).

Example 3 Determine whether the following system is consistent, dependent, or inconsistent.

$$y = 2.75x - 3.5$$
$$11x - 4y = 20$$

SOLUTION

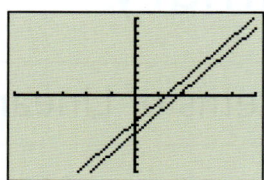

This system is inconsistent and has no solutions because the lines are parallel.

Section 2.2 Solving Systems of Equations Using the Substitution Method

- **Substitution Method**

  Best used when a variable is already isolated in at least one equation.
  1. Substitute the expression representing the isolated variable from one equation in place of that variable in the other equation.
  2. Remember to find the values for both variables.
  3. Check the solution in both equations.

- In solving an **inconsistent system** algebraically, all the variables will be eliminated, and the remaining sides of the equation will not be equal. This means that there will be no solutions to the system.

- In solving a consistent system with **dependent lines** algebraically, all the variables will be eliminated, but the remaining sides of the equation will be equal. This means that any solution to one equation will be a solution to the other, giving infinitely many solutions.

Example 4 Solve the following system by substitution.

$$y = 7x - 15$$
$$3x + 2y = 45$$

SOLUTION

$$y = 7x - 15$$
$$3x + 2y = 45$$

Substitute $7x - 15$ for y in the second equation.

$$3x + 2(7x - 15) = 45$$
$$3x + 14x - 30 = 45$$
$$17x - 30 = 45$$
$$17x = 75$$
$$x \approx 4.412$$

$$y = 7(4.412) - 15$$
$$y = 15.88$$

Remember to find the value of the other variable.

This system has a solution of (4.412, 15.88). Check this solution in both equations.

First equation	Second equation
$15.88 = 7(4.412) - 15$	$3(4.412) + 2(15.88) = 45$
$15.88 \approx 15.884$	$44.996 \approx 45$

Example 5

Solve the following system.

$$y = 3x + 8$$
$$-6x + 2y = 20$$

SOLUTION

$$y = 3x + 8$$
$$-6x + 2y = 20$$

Substitute $3x + 8$ for y in the second equation.

$$-6x + 2(3x + 8) = 20$$
$$-6x + 6x + 16 = 20$$
$$16 \neq 20$$

All the variables were eliminated and the remaining numbers are not equal, so this is an inconsistent system and has no solutions.

Section 2.3 Solving Systems of Equations Using the Elimination Method

- **Elimination Method**

  Best used when neither variable is already isolated.

  1. Multiply one or more of the equations by a number to make the coefficients of one variable opposite in signs, but of the same value.
  2. Add the two equations to eliminate the variable, then solve.
  3. Remember to find the values for both variables.
  4. Check the solution in both equations.

Example 6

Solve the following system by elimination:

$$3x + 5y = 41$$
$$10x - 4y = -8$$

SOLUTION

$$4(3x + 5y) = 4(41) \longrightarrow 12x + 20y = 164$$
$$5(10x - 4y) = 5(-8) \longrightarrow 50x - 20y = -40$$

$$12x + 20y = 164$$
$$\underline{50x - 20y = -40}$$
$$62x = 124$$
$$x = 2$$

$$3(2) + 5y = 41$$
$$y = 7$$

Remember to find the value of the other variable.

Therefore, this system has a solution of $(2, 7)$. Check this equation using the graph.

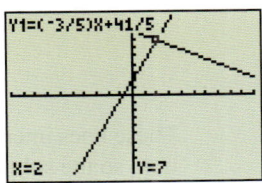

Section 2.4 Solving Linear Inequalities

- When solving **inequalities,** use the same methods as for solving equations, but remember to reverse the inequality symbol whenever multiplying or dividing both sides of the inequality by a negative number.
- Many systems of equations can be solved as inequalities when the situation in the problem implies a "less than or greater than" relationship.
- Use caution when **interpreting the results** from an inequality problem. The solution will include more than one input, so the wording you use should indicate that.
- When solving an inequality **numerically,** look for the input value that makes both sides equal and then whether a number greater than or less than that value will make the inequality relationship true.
- When solving an inequality **graphically,** look for the intersection of the graphs and determine on which side of the intersection the inequality relationship is true.

Example 7 The percentage of people in the United States who believe that the amount of violence on TV does damage to children can be modeled by $V(t) = -2.4t + 73$, where t is time in years since 2000. The percentage of people in the United States who believe that the amount of sex on TV does damage to children can be modeled by $S(t) = 1.7t + 19.8$, where t is time in years since 2000. Find when the percentage of people who believe that violence on TV is damaging will be less than those who believe that the amount of sex on TV is damaging.

SOLUTION
$$V(t) < S(t)$$
$$-2.4t + 73 < 1.7t + 19.8$$
$$-4.1t < -53.2$$
$$t > 12.98$$

In about 2013 and beyond, more people will believe that the amount of violence on TV does damage to children than will believe that the amount of sex on TV does damage to children.

Example 8 Use the table to find when Y1 > Y2.

SOLUTION Y1 = Y2 at $x = -1.5$ and is greater than Y2 when $x > -1.5$, so the solution is $x > -1.5$.

Example 9 Solve the inequality $-2x + 10 < 5x - 8$ graphically.

SOLUTION

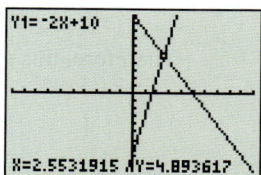

These lines intersect when $x = 2.55$ and $-2x + 10 < 5x - 8$ to the right of the intersection, so the solution is $x > 2.55$.

Section 2.5 Absolute Value Equations and Inequalities

- **Absolute Value:** The absolute value of a real number n, $|n|$, is the distance from zero to n on a real number line.
- **Absolute Value Equation:** If n is a positive real number and u is any algebraic expression,

$$|u| = n$$
$$u = n \quad \text{or} \quad u = -n$$

- If n is negative, then the equation $|n| = n$ has no real solution.
- **Solving absolute value equations:**
  1. Isolate the expression containing the absolute value.
  2. Rewrite the equation into two equations.

  $$|u| = n$$
  $$u = n \quad \text{or} \quad u = -n$$

  3. Solve each equation.
  4. Check the solutions in the original absolute value equation.
- **Solving absolute value inequalities involving less than or less than or equal to:**
  1. Isolate the expression containing the absolute value to the left side of the inequality.
  2. Rewrite the absolute value inequality into a compound inequality.

  $$|u| < n \qquad\qquad |u| \leq n$$
  $$-n < u < n \qquad -n \leq u \leq n$$

  3. Solve the compound inequality.
  4. Check the solution set in the original absolute value inequality.
- **Solving absolute value inequalities involving greater than or greater than or equal to:**
  1. Isolate the expression containing the absolute value to the left side of the inequality.
  2. Rewrite the absolute value inequality into two inequalities, one less than and one greater than.

  $$|u| > n \qquad\qquad |u| \geq n$$
  $$u < -n \quad \text{or} \quad u > n \qquad u \leq -n \quad \text{or} \quad u \geq n$$

  3. Solve the two inequalities.
  4. Check the solution set in the original absolute value inequality.

Example 10 Solve the equation $|x + 5| = 11$.

SOLUTION

$$|x + 5| = 11$$
$$x + 5 = -11 \quad \text{or} \quad x + 5 = 11$$
$$x = -16 \quad \text{or} \quad x = 6$$

Check these solutions in the original equation.

$$x = -16 \qquad\qquad x = 6$$
$$|-16 + 5| = 11 \qquad |6 + 5| = 11$$
$$11 = 11 \qquad\qquad 11 = 11$$

Example 11 Solve the given inequality. Give the solution as an inequality. Graph the solution set on a number line.

$$|x + 3| \leq 4$$

SOLUTION

$$|x + 3| \leq 4$$
$$-4 \leq x + 3 \leq 4$$
$$-7 \leq x \leq 1$$

Check this solution using a graph.

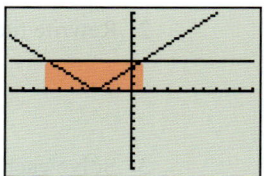

We can see that the absolute value graph is less than 4 when x is between −7 and 1.

Example 12 Solve the given inequality. Give the solution as an inequality. Graph the solution set on a number line.

$$|x - 2| - 1 > 4$$

SOLUTION

$$|x - 2| - 1 > 4$$
$$|x - 2| > 5$$
$$x - 2 < -5 \quad \text{or} \quad x - 2 > 5$$
$$x < -3 \quad \text{or} \quad x > 7$$

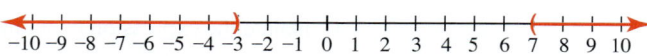

Check this solution using a graph.

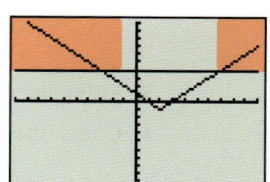

We can see that the absolute value graph is greater than 4 when x is less than −3 or greater than 7.

Section 2.6 Solving Systems of Linear Inequalities

- To **graph a linear inequality with two variables,** graph the line as if it were an equation and shade the side of the line that makes the inequality true. Use a dashed line if the inequality does not include the symbol for equal to.
- To **solve a system of inequalities,** graph all the inequalities on the same set of axes. The region where all the shaded areas overlap is the solution set of the system.

Example 13 Graph the inequality $y < 3x - 12$.

SOLUTION

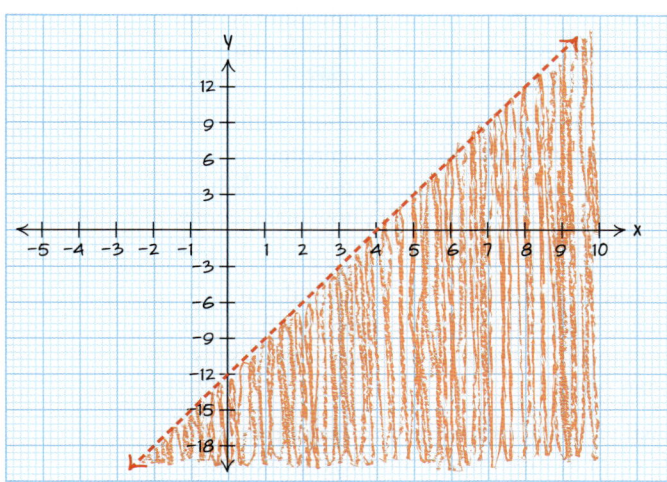

Example 14 Graph the following system of inequalities by hand.

$$y \geq 2x - 6$$
$$y \leq -3x + 7$$

SOLUTION

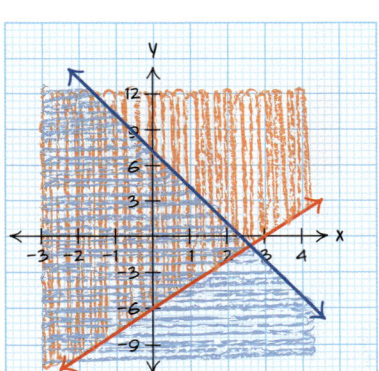

Chapter Review Exercises

At the end of each exercise, the section number [in brackets] indicates the section where the material is covered.

1. Frank's Shoe Repair works out of a kiosk in the local mall and has determined the following models for the monthly revenue and costs associated with repairing shoes.

    $$R(s) = 7.5s$$
    $$C(s) = 235 + 2.25s$$

    where $R(s)$ represents the monthly revenue in dollars for repairing s shoes and $C(s)$ is the monthly costs in dollars for repairing s shoes.

    a. Graph both functions on the same calculator screen.
    b. Determine the break-even point for Frank's Shoe Repair. [2.1]

2. Joanne's Cosmetics Outlet has two salary options for their salespeople. Each salesperson can choose which option to base his or her salary on.

    $$Q_1(s) = 350 + 0.07s$$
    $$Q_2(s) = 0.065s + 450$$

    where $Q_1(s)$ represents monthly salary option 1 in dollars when s dollars in sales are made per month and

$Q_2(s)$ represents monthly salary option 2 in dollars when s dollars in sales are made per month.

a. Graph both functions on the same calculator screen.
b. Find what sales level would result in both salary options having the same monthly salary. [2.1]

3. You decide to market your own custom computer software. You must invest $3232 on computer hardware and spend $3.95 to buy and package each disk. If each program sells for $9.00, how many copies must you sell to break even? [2.2]

4. John has cashews that sell for $4.50 a pound and peanuts that sell for $2.00 a pound. How much of each must he mix to get 100 pounds of a mixture that sells for $3.00 per pound? [2.3]

5. The residual values of cars make up a major part of the long-term value of a car. When considering two sports cars, Stephen found the following data in the 2004 *Kelley Blue Book*. (The residual value is what the car is worth after a certain number of years.)

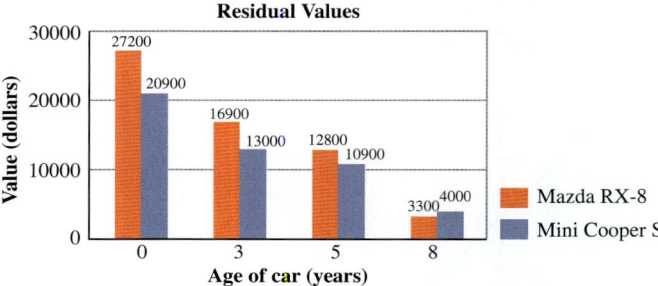

Source: November/December 2004 Kelley Blue Book Residual Values Guide.

a. Find equations for models for the residual values for each car given.
b. Determine when these two cars will have the same value. [2.2]

6. The percentage of births to unmarried women in Germany and the United Kingdom is given in the table.

Year	Germany	United Kingdom
2001	25.0	40.1
2002	26.0	40.6
2003	27.0	41.5
2004	27.9	42.3
2005	29.2	42.9

Source: Statistical Abstract of the United States, 2008.

a. Find a model for the percentage of births to unmarried women in Germany.
b. What does the slope for the model in part a mean in this context?
c. Find a model for the percentage of births to unmarried women in the United Kingdom.

d. Estimate when the percentage of births to unmarried women in Germany will be greater than the percentage in the United Kingdom. [2.4]

For Exercises 7 through 11, solve the system by graphing. Label each system as consistent with independent lines, consistent with dependent lines, or inconsistent.

7.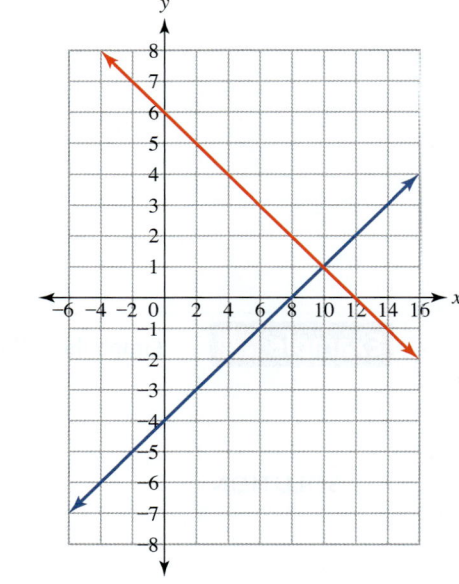

[2.1]

8. $x + 3y = -12$
   $-5x + y = 15$ [2.1]

9. $w = 2d + 7$
   $w = \dfrac{3}{7}d + \dfrac{1}{2}$ [2.1]

10. $p = 1.625t + 9$
    $p = \dfrac{13}{8}t - 5$ [2.1]

11. $5x - 4y = 28$
    $1.25x - y = 7$ [2.1]

12. The Palomar College Foundation has been given an endowment of $3 million to fund scholarships for outstanding math and science students. The foundation wants the endowment to earn enough interest each year to fund 130 $2000 scholarships. They plan to invest part of the money in a safe investment earning 7% simple interest and the rest in another account earning 11% simple interest. How much should they invest in each account to have enough interest to fund the 130 scholarships? [2.2]

13. Brian is an agriculture scientist who is testing different applications of insecticides on tomato plants. For part of the trial he is running now, he needs 150 gallons of a 12% solution of test chemical AX-14. Brian

received only containers with 5% solution and 15% concentrations of test chemical AX-14. How much of each chemical should Brian mix to run his trial? [2.2]

For Exercises 14 through 24, solve the system by elimination or substitution. Label any systems that are consistent with dependent lines or inconsistent.

14. $3x + 4y = -26$
    $y = x - 3$ [2.2]

15. $2w - 5t = -1$
    $3w - 4t = 2$ [2.3]

16. $2.35d + 4.7c = 4.7$
    $c = -7.05d - 21.15$ [2.2]

17. $\frac{5}{6}m + n = 25$
    $m - \frac{4}{5}n = 5$ [2.3]

18. $y = 4.1x - 2.2$
    $y = -2.9x - 7.1$ [2.2]

19. $-7x + 7y = -3$
    $7(x - y) = 3$ [2.3]

20. $4.1w + 3.7t = 5.1$
    $t = 4.43w + 4.63$ [2.2]

21. $2f + 2g = -22$
    $g = -f - 9$ [2.2]

22. $-6x + 15y = 5$
    $15(x - y) = -5$ [2.3]

23. $\frac{w}{2} + \frac{2z}{3} = 32$
    $\frac{w}{4} - \frac{5z}{9} = 40$ [2.3]

24. $-2.7x + y = -13.61$
    $y = -4.28x + 5.69$ [2.2]

25. The percentage of total births that are to teenage mothers in Delaware can be modeled by

    $$D(t) = -0.46t + 12.35$$

    where $D(t)$ represents the percent of total births that are to teenage mothers in Delaware t years since 2000. The percentage of total births that are to teenage mothers in Louisiana t years since 2000 can be modeled by

    $$L(t) = -0.67t + 16.94$$

    Source: Models derived from data in the Statistical Abstract of the United States, 2008.

    Determine when the percentage of total births that are to teenage mothers in Louisiana will be less than the percentage in Delaware. [2.4]

For Exercises 26 through 33, solve the inequality.

26. Use the graph to determine when $f(x) < g(x)$. [2.4]

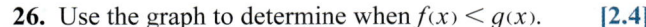

27. Use the table to find when Y1 > Y2. [2.4]

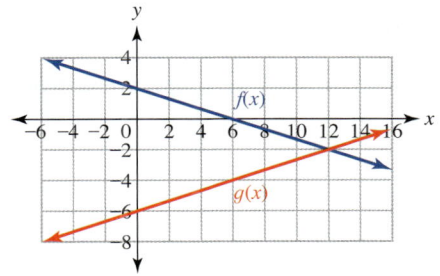

28. $7 - \frac{9x}{11} \leq -8$ [2.4]

29. $-5x + 4 \geq 7(x - 1)$ [2.4]

30. $3t + 4 > -6(4t + 2)$ [2.4]

31. $\frac{3d}{5} + 7 < 4$ [2.4]

32. $-1.5v + 2.84 > -3.2v - 1.48$ [2.4]

33. $1.85 + 1.34(2.4k - 5.7) \geq 3.25k - 14.62$ [2.4]

For Exercises 34 through 37, solve the equations.

34. $|x - 7| = 20$ [2.5]
35. $|2x + 3| - 10 = 40$ [2.5]
36. $|x + 19| = 30$ [2.5]
37. $5|x - 7| + 10 = 95$ [2.5]

For Exercises 38 through 43, solve the inequality. Give the solution as an inequality. Graph the solution set on a number line.

38. $|x| < 8.5$ [2.5]
39. $|x| > 6.5$ [2.5]
40. $|x + 3| \geq 12$ [2.5]
41. $|x - 7| \leq 3$ [2.5]
42. $2|x + 5| - 4 > 16$ [2.5]
43. $-3|x + 5| + 7 \leq 4$ [2.5]

For Exercises 44 through 47, graph the inequality by hand.

44. $y \leq 4x - 10$ [2.6]
45. $y \geq 1.5x - 5$ [2.6]
46. $2x + 5y > 15$ [2.6]
47. $-3x - 4y > 12$ [2.6]

For Exercises 48 through 51, graph the system of inequalities.

48. $y > 2x + 5$
    $y < -x + 7$ [2.6]
49. $y > 1.5x + 2$
    $4.5x - 3y > 6$ [2.6]
50. $4x + 5y \leq 12$
    $2x + 8y \geq 5$ [2.6]
51. $y > \frac{2}{3}x - 7$
    $y < \frac{2}{3}x + 5$ [2.6]

52. A college is selling tickets to a championship basketball game and has up to 12,000 tickets to sell. They can sell both regular admission  tickets and student discount tickets. Student tickets cost $9, and regular admission tickets cost $15. The school wants to make at least $140,000 from selling tickets. Write a system of inequalities to determine the combinations of tickets they can sell to make the money they want. [2.6]

53. Find the inequality for the graph. [2.6]

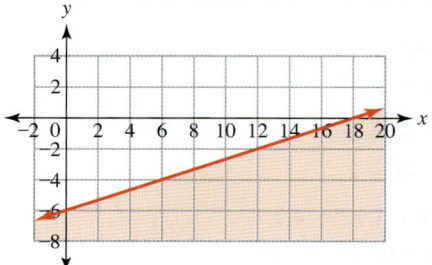

54. Find the inequality for the graph. [2.6]

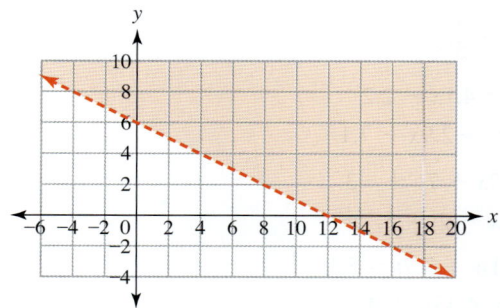

Chapter Test 2

1. The average hourly earnings of production workers in manufacturing industries for California and Massachusetts are given in the table.

Year	California (in dollars per hour)	Massachusetts (in dollars per hour)
2003	15.04	16.53
2004	15.36	16.89
2005	15.70	17.66
2006	15.95	18.26

   Source: Statistical Abstract of the United States, 2008.

   a. Find equations for models of these data.
   b. Find when the average hourly earnings were the same in both states.

2. Christine is ready to retire and has $500,000 to invest. Christine needs to earn $44,500 in interest each year to continue her current lifestyle. She plans to invest the money in two accounts: one paying 12% and a safer account paying 7%.

   a. Write a system of equations that will help you to find the amount Christine needs to invest in each account.
   b. How much should Christine invest in each account to earn the $44,500 she wants?

3. Georgia has up to 20 hours a week to study for her math and history classes. She needs to study at least 12 hours a week to pass her math class. Write a system of inequalities to describe the possible amounts of time she can study for each subject. Graph the system by hand.

4. Wendy is working in a forensics lab and needs 2 liters of an 8% HCl solution to test some evidence. Wendy has only a 5% HCl solution and a 20% HCl solution to work with. Because she knows that the solution must be 8% for the test to be valid, how much of the 5% solution and the 20% solution should she use?

For Exercises 5 through 7, solve the inequality.

5. $5x + 7 < 12x - 8$

6. $3.2m + 4.5 \geq 5.7(2m + 3.4)$

7. $-4.7 + 6.5(a + 2.5) \leq 2.43a - 5$

For Exercises 8 through 11, solve the system.

8. $x + 7y = -2$
   $3x + y = 34$

9. $0.4375w + 4t = 22$
   $-2.4t = 0.2625w - 13.2$

10. $5c + 3d = -15$
    $d = -\dfrac{5}{3}c - 12$

11. $2.68g - 3.45f = 23.87$
    $4.75g + 6.9f = -12.47$

12. Use the graph to find when $f(x) \geq g(x)$.

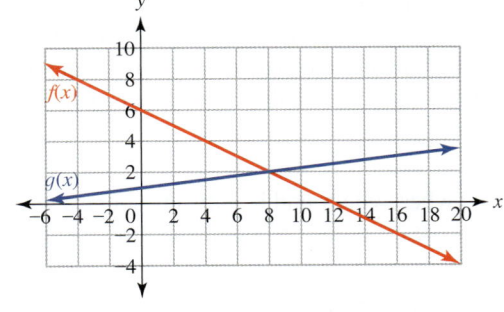

13. Scott has decided to market his custom-built hammock stands. He needs to invest $7500 on tools and spend $395 to buy and build each stand. If each hammock stand sells for $550, how many must Scott sell to break even?

14. Graph by hand: $y > -\dfrac{4}{5}x + 6$

15. The revenue for local telephone service providers can be modeled by

$$L(t) = 3804.32t + 84252.07$$

where $L(t)$ represents the revenue in millions of dollars for local telephone service providers t years since 1990. The revenue in millions of dollars for cellular telephone service providers t years since 1990 can be modeled by

$$C(t) = 6428.29t - 11553$$

Determine when the revenue for cellular telephone providers will be greater than that for local telephone providers.

16. Use the table to find when Y1 < Y2.

X	Y1	Y2
-6	-4	-2.5
-5	-3.5	-2.75
-4	-3	-3
-3	-2.5	-3.25
-2	-2	-3.5
-1	-1.5	-3.75
0	-1	-4

X=0

17. Solve: $|x + 5| + 3 = 17$

18. Solve $|x - 4| - 8 < 5$. Give the solution as an inequality. Graph the solution set on a number line.

For Exercises 19 and 20, graph the system of inequalities.

19. $y \geq 2x - 5$
    $y \leq \dfrac{1}{3}x + 4$

20. $4x - 3y < 6$
    $3x + y > 12$

Chapter 2 Projects

Which Cell Phone Plan Is Cheaper?

Research Project
One or more people

What you will need
- Prices for basic cell phone plans from two or more cell phone service providers

It will be your job to compare two or more cell phone plans from various cell phone service providers. You should visit, phone, or investigate on the Internet two or more cell phone service providers and find their prices for the most basic service plan they offer. Then you will explore which of these companies is cheaper for a variety of users.

Write up

a. Describe how you collected the information. What companies are you comparing? Give a basic description of the plans you are considering.
b. Create a table of data for each plan that includes several different amounts of minutes used for a month and the cost for those minutes.
c. Draw a graph with all of the plans you are comparing that depicts the monthly cost for different amounts of minutes used. (This might not be a straight line.)
d. Determine when each plan will be the cheapest.
e. Is there a number of minutes when two or more plans cost the same amount? If so, find that amount and describe its significance.
f. Describe what else besides cost you might consider in choosing a cell phone service provider.
g. Which of the cell phone service providers would you choose and why?

Find Your Own Comparison

Research Project
One or more people

What you will need
- Find data for two things to compare and analyze using a system of equations.
- You might want to use the Internet or library. Statistical abstracts and some journals and scientific articles are good resources for data.
- Follow the MLA style guide for all citations. If your school recommends another style guide, use that.

You are given the task of finding two real-world situations to compare and analyze using a system of equations. You will need to find two related sets of data that can be modeled by using linear functions. Using these two functions, you should solve the system of equations and explain the results. You may use the problems in this chapter to get ideas of things to investigate, but your data should not be discussed in this textbook.

Write up

a. Describe the data you found and where you found it. Cite any sources you used, following the MLA style guide or your school's recommended style guide.
b. Create a scatter plot of your data.
c. Find equations for models for your sets of data and write a system of equations. Graph your functions on the same graph. Be sure to show the intersection of the two functions.
d. Solve the system of equations.
e. Explain the meaning of the solution to the system of equations you found in part d.
f. Explain any restrictions you would place on this system of equations to avoid model breakdown.

Review Presentation

Research Project
One or more people

What you will need
- This presentation may be done in class or as a video.
- If it is done as a video, you might want to post the video on YouTube or another site.
- You might want to use problems from the homework set or review material in this book.
- Be creative, and make the presentation fun for students to watch and learn from.

Create a five-minute presentation or video that reviews a section of Chapter 2. The presentation should include the following.

- Examples of the important skills presented in the section
- Explanation of any important terminology and formulas
- Common mistakes made and how to recognize them

Equation-Solving Toolbox — Chapters 1-2

Linear Equations

Distributive Property	Use when an equation contains grouping symbols.	$2(x + 4) = 3x - 5$	Section 1.1 and Appendix A
Addition Property of Equality	Use to isolate a term.	$x + 7 = 20$	Section 1.1 and Appendix A
Multiplication Property of Equality	Use to isolate a variable.	$5x = 13$	Section 1.1 and Appendix A

Systems of Equations

Substitution Method	Use when a variable is isolated or can be easily isolated.	$y = 6x + 8$ $2x + 4y = 30$	Section 2.2
Elimination Method	Use when the equations are in general form.	$5x - 7y = 42$ $3x + 20y = 103$	Section 2.3

General Solving Techniques

Graphing	Use when you cannot solve the equation algebraically. Graph each side of the equations and find the intersection(s).		All chapters
Numerically	Use when you cannot solve the equation algbraically. Guess a solution and check it. Best used to check solutions.		All chapters

Modeling Processes — Chapters 1-2

Linear

1. Define the variables and adjust the data.
2. Create a scatterplot.
3. Select a model type.
4. **Linear model:** Pick two points and find the equation of the line. $y = mx + b$
5. Write the equation of the model.
6. Check the model by graphing it with the scatterplot.

Cumulative Review — Chapters 1-2

1. The population of the United States is given in the table.

Year	Population of the United States (in millions)
2003	290
2004	293
2005	296
2006	298
2007	301
2008	304

   Source: U.S. Census Bureau.

   a. Find an equation for a model of these data.
   b. Using your model estimate the population of the United States in 2012.
   c. Give a reasonable domain and range for this model.
   d. What is the vertical intercept for your model? What does it mean in this situation?
   e. Estimate when the U.S. population will reach 325 million.

For Exercises 2 through 4, sketch the graph of the line by hand. Label the vertical and horizontal intercepts.

2. $y = 3x - 4$
3. $y = -\frac{1}{3}x + 2$
4. $5x - 6y = 9$

For Exercises 5 and 6, solve the given equation for the indicated variable.

5. $abc + b = a$ for c.
6. $P = \frac{1}{2}mn^2 + 3$ for m.

7. Write the equation for the line passing through the points $(-2, 6)$ and $(5, 20)$. Write the equation in slope-intercept form.

8. Write the equation for the line passing through the point $(6, 2)$ and parallel to the line $3x + 4y = 7$. Write the equation in slope-intercept form.

9. The number of miles Steve plans to run each week can be modeled by

   $$R = 1.5w + 5$$

   where R is the number of miles run w weeks since he started training.

   a. How many miles does Steve plan to run in week 4?
   b. In what week will Steve run 20 miles?

10. The total revenue for Cisco Systems, Inc. for the past several years is given in the table.

Years Since 2000	Total Revenue (in billions of $)
5	24.8
6	28.5
7	34.9
8	39.5

    Source: http://finance.google.com

    a. Find an equation for a model of these data. Write the model using function notation.
    b. What is the vertical intercept for your model? Explain its meaning in regard to revenue.
    c. Give a reasonable domain and range for this model.
    d. Estimate the total revenue for Cisco Systems, Inc. in 2010.
    e. When might the total revenue be 64 billion dollars?
    f. What is the slope of this model? Explain its meaning in regard to revenue.

11. Use the table to find the following.

x	y
-6	0
0	4
3	6
15	14
24	20

    a. The slope of the line passing through these points
    b. The y-intercept
    c. The x-intercept
    d. The equation for the line passing through these points. Write the equation in slope-intercept form.

12. Use the graph to find the following.

    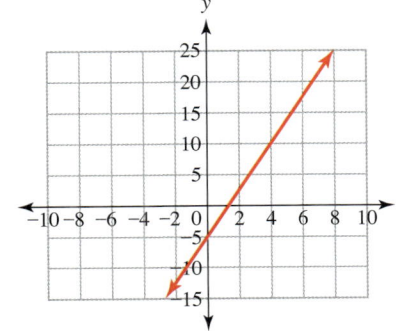

    a. The slope of the line
    b. The vertical intercept
    c. The horizontal intercept
    d. The equation for the line

For Exercises 13 and 14, give the domain and range for the function.

13. $f(x) = 1.5x + 6$
14. $g(x) = 20$

For Exercises 15 through 18, solve the given equation. Check your answer.

15. $4a + 6 = 2a - 18$
16. $1.5(x + 3) = 4x + 2.5(4x - 7)$
17. $\frac{1}{2}x + 10 = \frac{3}{4}x - 6$
18. $p + 3\left(\frac{1}{4}p - 4\right) = \frac{2}{5}$

19. The function $H(d)$ is the number of hours a person sleeps on day d of the month. Explain the meaning of $H(6) = 7$ in this situation.

20. Let $P(d) = -3d + 100$ be the number of people out of 100 remaining on a diet plan d days after starting the diet.
   a. Find $P(14)$ and explain its meaning in terms of the diet.
   b. What is the slope of this function? Explain its meaning in terms of the diet.

21. $f(x) = 9x - 45$
   a. Find $f(3)$.
   b. Find x such that $f(x) = 54$.
   c. Give the domain and range for $f(x)$.

22. $h(x) = -\frac{5}{8}x + 11$
   a. Find $h(32)$.
   b. Find x such that $h(x) = \frac{13}{8}$.
   c. Give the domain and range for $h(x)$.

23. $g(x) = 7$
   a. Find $g(10)$.
   b. Give the domain and range for $g(x)$.

24. Use the graph to find the following.

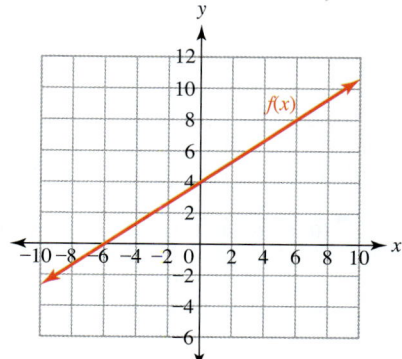

   a. $f(6)$.
   b. Find x such that $f(x) = 0$.
   c. Give the domain and range for $f(x)$.

For Exercises 25 through 30, solve the given system. Check your answer.

25. $2x + 5y = 26$
    $10x - 3y = 18$

26. $c = 3d + 10$
    $c = 2d + 14$

27. $\frac{2}{5}g + \frac{3}{8}h = 2$
    $16g + 15h = 80$

28. $4a + 18b = 8$
    $6a = 7b + \frac{2}{3}$

29. $20x + 25y = -1$
    $35x + 40y = 1$

30. $m = 3n - 12$
    $4m - 12n = 10$

31. Don wants to invest $1,200,000. Don wants to earn $53,250 in interest each year to continue his current lifestyle. He plans to invest the money in two accounts: one paying 5% and a safer account paying 3.5%.
   a. Write a system of equations that will help find the amount Don needs to invest in each account.
   b. How much should Don invest in each account to earn the $53,250 he wants?

32. Hamid needs 4 liters of a 12% HCl solution to run an experiment. Hamid has only a 6% HCl solution and a 24% HCl solution to work with. How much of the 6% solution and the 24% solution should he use?

For Exercises 33 through 36, solve the inequality. Check your answer.

33. $6x + 20 > 15x - 16$
34. $\frac{1}{4}m + \frac{2}{3}(m - 5) \leq \frac{1}{3}(4m + 7)$
35. $-1.4a + 2.34a + 6.1 \geq -0.2a + 8.4$
36. Use the graph to find when $f(x) \geq g(x)$.

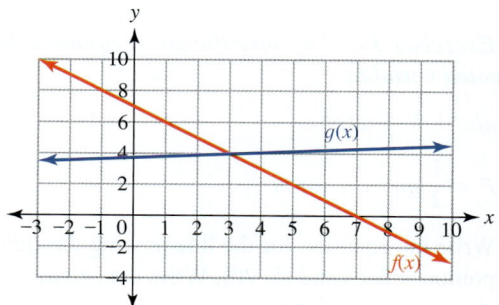

37. Sandra earns $400 per week plus 6% commission on all sales she makes. Sandra needs $565 each week to pay bills. How much in sales does she need to earn at least $565 per week?

For Exercises 38 through 40, sketch a graph by hand of the given inequality.

38. $y > -\frac{2}{3}x + 9$
39. $2x - 4y \geq 11$
40. $6x + y < 12$

41. The percentage of Canadian TV subscribers who use cable services can be modeled by

$$C(t) = -2t + 82$$

where $C(t)$ represents the percentage, as a whole number, of Canadian TV subscribers who use cable services t years since 2000. The percentage of Canadian TV subscribers who use satellite services t years since 2000 can be modeled by

$$S(t) = 2t + 17$$

Determine when the percentage of Canadian TV subscribers that use satellite services will be greater than those that use cable services.

Source: Models derived using data from the Canadian Radio-television and Telecommunications Commission.www.crtc.gc.ca.

42. Use the table to find when Y1 < Y2.

For Exercises 43 through 46, solve the equation. Check your answer.

43. $|x + 2| + 6 = 18$

44. $3|a - 8| - 4 = 44$

45. $|4n + 9| + 20 = 7$

46. $-2|3r + 5| + 15 = 7$

For Exercises 47 through 50, solve the inequality. Give the solution as an inequality. Graph the solution set on a number line.

47. $|x - 2| < 7$

48. $|2x + 3| \leq 15$

49. $|x + 4| \geq 6$

50. $|3x - 7| + 10 > 15$

For Exercises 51 and 52, sketch a graph by hand of the system of inequalities.

51. $y \geq x - 5$
$y \leq \dfrac{1}{5}x + 2$

52. $3x - 5y < 10$
$x + 2y < 12$

Exponents, Polynomials and Functions

3

- **3.1** Rules for Exponents
- **3.2** Combining Functions
- **3.3** Composing Functions
- **3.4** Factoring Polynomials
- **3.5** Special Factoring Techniques

Encryption is the process of transforming information using a system that makes the information unreadable by anyone who does not have the key. Commerce over the Internet would not be nearly as safe without the powerful encryption techniques that are used today. In this chapter, we will discuss how to factor polynomials. One of the chapter projects on page 290 will explore the world of encryption and how factoring is used in this field.

3.1 Rules for Exponents

LEARNING OBJECTIVES

- Use the rules for exponents to simplify expressions.
- Understand and use negative exponents.
- Understand the relationship between rational exponents and radicals.

Rules for Exponents

To work with functions other than linear functions, we should have a good understanding of the basic rules for exponents and how to use them to simplify problems containing exponents. Many of these rules you already know and have used in previous classes and sections of this textbook. This section is a recap to prepare us for more applications in the rest of the textbook. In this section, assume that all variables are not equal to zero. This allows us to ignore the possibility of division by zero, which is not defined.

Recall that the basic concept of an exponent is repeated multiplication.

$$2 \cdot 2 \cdot 2 \cdot 2 \cdot 2 \cdot 2 = 2^6 = 64$$
$$xxxxx = x^5$$
$$xxxyy = x^3 y^2$$
$$3 \cdot 3 \cdot 3 \cdot 3 \cdot 7 \cdot 7 \cdot 7 \cdot 7 \cdot 7 \cdot 7 = 3^4 \cdot 7^6 = 81 \cdot 117649 = 9529569$$

Exponents allow us to write a long expression in a very compact way. When we work with exponents, there are two parts to an exponential expression: the **base** and the **exponent**. The base is the number or variable being raised to a power. The exponent is the power to which the base is being raised.

$$5^3 \quad \text{exponent} \atop \text{base}$$

One of the most common operations we do with exponential expressions is to multiply them together. When we multiply any exponential expressions with the same base, the expressions can be combined into one exponential.

$$x^7 x^2 = xxxxxxx \cdot xx = x^9$$

In this example, we see that we had seven x's multiplied by two more x's, which gives us a total of nine x's multiplied together. Therefore, we can write a final simpler expression, x^9. Combining expressions with the same base leads us to the **product rule for exponents.**

> **The Product Rule for Exponents**
>
> $$x^m x^n = x^{m+n}$$
>
> When multiplying exponential expressions that have the same base, add exponents.
>
> $$x^5 x^3 = x^8$$

When more than one base is included in an expression or multiplication problem, the associative and commutative properties can be used along with the product rule for exponents to simplify the expression or multiplication.

SECTION 3.1 Rules for Exponents

Example 1 — Using the product rule for exponents

Simplify the following expressions.

a. $x^5 x^2 x^3$ **b.** $(3f^4 g^5) 7 f^2 g^7$ **c.** $(a^2 b^5 c)(a^3 b^4 c^3)$

SOLUTION

a. $x^5 x^2 x^3 = x^{10}$ Add the exponents.

b. $(3f^4 g^5)(7f^2 g^7) = 3 \cdot 7 f^4 f^2 g^5 g^7$ Use the commutative property to rearrange the coefficients and bases. Add the exponents of the like bases and multiply the coefficients.
$= 21 f^6 g^{12}$

c. $(a^2 b^5 c)(a^3 b^4 c^3) = a^2 a^3 b^5 b^4 c c^3$ Use the commutative property to rearrange the bases.
$= a^5 b^9 c^4$ Add the exponents of the like bases.

PRACTICE PROBLEM FOR EXAMPLE 1

Simplify the following expressions.

a. $m^3 m^4 m^2$ **b.** $(4w^6 x^2)(8wx^9)$

> **What's That Mean?**
>
> **Associative Property:**
> When multiplying more than two variables or constants together, we can group them in any way we wish.
> $$abc = (ab)c = a(bc)$$
>
> **Commutative Property:**
> When multiplying variables or constants together, we can multiply them in any order we wish.
> $$ab = ba$$
> $$abc = bac = cab = cba$$

It is very important, with the exponent rules, that you notice what operation you are working with in the expression and what operation this results in for the exponents. Since the coefficients are not exponents, the exponent rules will not apply. Be careful that you do not change multiplication of coefficients into addition. The coefficients are not exponents, so they do not follow the rules for exponents.

$2x^5 \cdot 3x^4 = 6x^5 x^4$ Multiply the coefficients.
$= 6x^9$ Add the exponents.
$2x^5 \cdot 3x^4 \neq 5x^9$ Do not add the coefficients. They are not exponents.

The next rule is closely related to the product rule for exponents. When dividing two exponential expressions that have the same base, we can still simplify.

$$\frac{x^5}{x^3} = \frac{xxxxx}{xxx} = \frac{xx}{1} = x^2$$

Because multiplication and division are inverse operations, we can eliminate any variables that are in both the numerator and the denominator of this fraction. That means that we can divide out three x's from the top and bottom, leaving us with only two x's remaining on the top. When we multiplied exponential expressions with the same base, we added exponents. When we divide exponential expressions with the same base, we will subtract exponents. Working with exponents this way leads us to the **quotient rule for exponents**.

> **The Quotient Rule for Exponents**
>
> $$\frac{x^m}{x^n} = x^{m-n}$$
>
> When dividing exponential expressions that have the same base, subtract exponents.
>
> $$\frac{x^7}{x^3} = x^4$$

Remember that coefficients are not exponents, so they will simply be reduced when part of a division problem. Do not divide the coefficients into a decimal form. Leaving the coefficients as a reduced fraction is the standard form these fractions should be left in.

> **Example 2** Using the quotient rule for exponents

Simplify the following expressions.

a. $\dfrac{x^{12}}{x^5}$

b. $\dfrac{a^5 b^3 c^4}{a^3 b^2 c}$

c. $\dfrac{35 m^2 n^4}{7mn}$

d. $\dfrac{10 k^7 p^3}{8 k^4 p}$

SOLUTION

a. $\dfrac{x^{12}}{x^5} = x^7$ *Subtract the exponents.*

b. $\dfrac{a^5 b^3 c^4}{a^3 b^2 c} = a^2 b c^3$ *Subtract the exponents of like bases.*

c. $\dfrac{35 m^2 n^4}{7mn} = 5 m n^3$ *Subtract the exponents of like bases and reduce the coefficients.*

d. $\dfrac{10 k^7 p^3}{8 k^4 p} = \dfrac{5 k^3 p^2}{4} = \dfrac{5}{4} k^3 p^2$ *Subtract the exponents of like bases and reduce the coefficients.*

PRACTICE PROBLEM FOR EXAMPLE 2

Simplify the following expressions.

a. $\dfrac{a^3 b^8 c}{a^2 b^2}$

b. $\dfrac{40 t^{11} w^{14}}{5 t^3 w^9}$

c. $\dfrac{24 b^{18} c^4}{14 b^{10} c^3}$

The next rule for exponents deals with what to do when an exponential expression is raised to another exponent. Changing the exponent to repeated multiplication gives us

$$(x^3)^2 = (x^3)(x^3) = x^6$$
$$(x^2)^5 = x^2 x^2 x^2 x^2 x^2 = x^{10}$$

Because raising to a power is simply repeated multiplication, the **power rule for exponents** follows from the product rule for exponents.

> **The Power Rule for Exponents**
>
> $$(x^m)^n = x^{mn}$$
>
> When raising an exponential expression to another power, multiply the exponents.
>
> $$(x^4)^7 = x^{28}$$

When more than one variable or constant is being raised to a power, we can use the rules for **powers of products and quotients.** When an exponential expression contains more than one variable or a numeric constant that is multiplied or divided, the outside power must be applied to each constant or variable by using the power rule for exponents. It is very important to remember that exponents can be applied only over multiplication and division. When a sum or difference is raised to a power, simply applying the exponent will not work.

Powers of Products and Quotients

$$(xy)^m = x^m y^m \qquad \left(\frac{x}{y}\right)^m = \frac{x^m}{y^m}$$

In raising an expression to a power, that power can be applied over multiplication or division.

$$(xy)^5 = x^5 y^5 \qquad \left(\frac{x}{y}\right)^4 = \frac{x^4}{y^4}$$

Applying exponents does **NOT** work over addition or subtraction.

Incorrect: $(x+y)^2 \neq x^2 + y^2$

Correct: $(x+y)^2 = (x+y)(x+y) = x^2 + 2xy + y^2$

Example 3 Using the power rule for exponents and applying exponents

Simplify the following expressions.

a. $(w^3)^5$
b. $(a^3 b^2 c)^3$
c. $(5m^5 n^3)^2$
d. $\left(\frac{m^2}{n^4}\right)^3$
e. $\left(\frac{2xy^2}{3z^5}\right)^4$

SOLUTION

a. $(w^3)^5 = w^{15}$ Using the power rule, multiply the exponents.

b. $(a^3 b^2 c)^3 = (a^3)^3 (b^2)^3 (c)^3$ Apply the exponent to each base.

$= a^9 b^6 c^3$ Using the power rule, multiply the exponents.

c. $(5m^5 n^3)^2 = (5)^2 (m^5)^2 (n^3)^2$ Apply the exponent to each base and the coefficient.

$= 25 m^{10} n^6$ Using the power rule, multiply the exponents and raise the coefficient 5 to the second power.

d. $\left(\frac{m^2}{n^4}\right)^3 = \frac{(m^2)^3}{(n^4)^3}$ Apply the exponent to each base.

$= \frac{m^6}{n^{12}}$ Using the power rule, multiply the exponents.

e. $\left(\frac{2xy^2}{3z^5}\right)^4 = \frac{2^4 x^4 (y^2)^4}{3^4 (z^5)^4}$ Apply the exponent to each base and the coefficients.

$= \frac{16 x^4 y^8}{81 z^{20}}$ Using the power rule, multiply the exponents and raise the coefficients to the fourth power.

PRACTICE PROBLEM FOR EXAMPLE 3

Simplify the following expressions.

a. $(3x^5 y^2 z)^3$
b. $\left(\frac{4m^4 p^8}{5m^3 p^5}\right)^2$

Using Your TI Graphing Calculator

To enter exponents on the graphing calculator, you can use either of two methods.

If you are squaring a term, you can use the

button on the left-hand side of the calculator. Enter the number or variable that you wish to square and press the button.

If you want to raise a variable or number to any other power, you should use the caret button

^

in the far right column of the calculator. In this case, you will enter the number or variable you want to raise to a power, then press the ^ button, followed by the power you wish to use.

For example, x^5 would be entered as

  5

When expressions get more complicated, it is important to follow the order of operations. Therefore, start by simplifying the inside of any parentheses before distributing any exponents or using the power rule for exponents. Finally, do any remaining multiplication or division using the product and quotient rules for exponents.

Example 4 Combining the rules for exponents

Simplify the following expressions.

a. $(7d^3g^2)^2(5dg^3)^3$

b. $\left(\dfrac{16x^5y^{12}z^9}{2xy^8z^6}\right)^3$

c. $\left(\dfrac{a^4b^3c^5}{a^3bc}\right)^4\left(\dfrac{5ab^3}{bc}\right)^2$

SOLUTION

a. $(7d^3g^2)^2(5dg^3)^3 = [7^2(d^3)^2(g^2)^2][5^3d^3(g^3)^3]$ — Apply the exponents to each base and the coefficients.

$= (49d^6g^4)(125d^3g^9)$ — Using the power rule, multiply the exponents and raise the coefficients to their powers.

$= 6125d^9g^{13}$ — Multiply the coefficients and use the product rule to add exponents.

b. $\left(\dfrac{16x^5y^{12}z^9}{2xy^8z^6}\right)^3 = (8x^4y^4z^3)^3$ — Simplify inside the parentheses by reducing the coefficients and using the quotient rule to subtract exponents.

$= 8^3(x^4)^3(y^4)^3(z^3)^3$ — Apply the exponent to each base and coefficient.

$= 512x^{12}y^{12}z^9$ — Raise the coefficient to the third power and multiply the exponents.

c. $\left(\dfrac{a^4b^3c^5}{a^3bc}\right)^4\left(\dfrac{5ab^3}{bc}\right)^2 = (ab^2c^4)^4\left(\dfrac{5ab^2}{c}\right)^2$ — Simplify inside each parentheses using the quotient rule for exponents. Apply the exponents to each base and coefficient.

$= [a^4(b^2)^4(c^4)^4]\left[\dfrac{5^2a^2(b^2)^2}{c^2}\right]$ — Using the power rule, multiply the exponents and raise the coefficient to the second power.

$= (a^4b^8c^{16})\left(\dfrac{25a^2b^4}{c^2}\right)$ — Multiply the two expressions adding the exponents of like bases.

$= \dfrac{25a^6b^{12}c^{16}}{c^2}$ — Use the quotient rule to simplify.

$= 25a^6b^{12}c^{14}$

PRACTICE PROBLEM FOR EXAMPLE 4

Simplify the following expressions. Be sure to follow the order of operations.

a. $(3x^2y^5)(2x^3y)^3$

b. $\dfrac{(2m^2n^3)^4}{10m^5n^{10}}$

c. $\left(\dfrac{m^2n^3}{mn^2}\right)^4\left(\dfrac{m^5}{n^2}\right)$

Connecting the Concepts

How do we order the operations?

The order of operations are an agreement by mathematicians about what order operations are to be done in.

1. Perform all operations within parentheses or other grouping symbols.
2. Evaluate all exponents.
3. Multiply or divide in order from left to right.
4. Add or subtract in order from left to right.

The order of operations is sometimes remembered by using the acronym

PEMDAS

Negative Exponents and Zero as an Exponent

When using the quotient rule for exponents in some situations, we will get negative numbers when we subtract the denominator's exponent from the numerator's exponent. A result like this makes us think about ways to define how **negative exponents** work. Let's look at a division problem.

$$\frac{x^3}{x^5} = \frac{xxx}{xxxxx} = \frac{1}{xx} = \frac{1}{x^2} \qquad \text{Using the basic definition of exponents.}$$

$$\frac{x^3}{x^5} = x^{3-5} = x^{-2} \qquad \text{Using the quotient rule for exponents.}$$

Because using the basic definition of exponents gives us $\frac{1}{x^2}$ and the quotient rule for exponents gives us x^{-2}, these two expressions must be the same for both of these methods to agree and be reliable. Therefore, we have that

$$x^{-2} = \frac{1}{x^2}$$

The negative part of the exponent represents a reciprocal of the base. Notice that once we take the reciprocal of (flip) the base, the exponent becomes positive. The negative exponent only moves the base; it does not make that base negative. If a base with a negative exponent is in the denominator of a fraction, it will also be a reciprocal and will end up in the numerator of the fraction. Answers without negative exponents are easier to understand and work with, so we will write all our answers with only positive exponents.

Negative Exponents

$$x^{-n} = \frac{1}{x^n}$$

When raising a base to a negative exponent, the reciprocal of that base is raised to the absolute value of the exponent.

$$x^{-3} = \frac{1}{x^3} \qquad \left(\frac{1}{2}\right)^{-1} = 2 \qquad \frac{2}{x^{-4}} = 2x^4$$

Example 5 — Working with negative exponents

Simplify the following expressions. Write all answers without negative exponents.

a. $x^{-5}y^2$

b. $\dfrac{a^5 b}{a^3 b^4}$

c. $-2x^{-4}$

d. $\dfrac{21m^3 n^{-2}}{7m^{-5} n^3}$

e. $\dfrac{25a^3 b^{-7} c^{-2}}{15a^8 b^3 c^5}$

SOLUTION

a. $x^{-5}y^2 = \dfrac{y^2}{x^5}$ 　 The x has a negative exponent, so take the reciprocal of the base x.

b. $\dfrac{a^5 b}{a^3 b^4} = a^2 b^{-3}$ 　 First, subtract the exponents.

$= \dfrac{a^2}{b^3}$ 　 Notice that the result from subtracting exponents is always placed on top of the fraction and then moved if negative exponents remain. Move any bases with negative exponents to the bottom.

c. $-2x^{-4} = \dfrac{-2}{x^4}$ 　 The negative exponent moves the x to the bottom. The negative 2 does not move because it is not an exponent.

d. $\dfrac{21m^3 n^{-2}}{7m^{-5}n^3} = 3m^8 n^{-5}$ First, subtract the exponents. Use caution with the negatives. Again, the results are in the numerator and move to the denominator if they have a negative exponent.

$= \dfrac{3m^8}{n^5}$ Move any bases with negative exponents to the bottom.

e. $\dfrac{25a^3 b^{-7} c^{-2}}{15a^8 b^3 c^5} = \dfrac{5a^{-5} b^{-10} c^{-7}}{3}$ Subtract exponents and reduce the constants.

$= \dfrac{5}{3a^5 b^{10} c^7}$ Move any bases with negative exponents to the bottom.

PRACTICE PROBLEM FOR EXAMPLE 5

Simplify the following expressions. Write all answers without negative exponents.

a. $5c^{-3} d^4$ **b.** $\dfrac{12x^4 y^3 z^8}{10xy^5 z^2}$ **c.** $\dfrac{-2g^{-2} h^3}{g^5 h^{-4}}$

CONCEPT INVESTIGATION

What does an exponent of zero do?

a. Fill in the missing values in the following table.

x	2^x
5	32
4	16
3	
2	
1	
0	
-1	$\dfrac{1}{2}$
-2	
-3	

b. When the exponent is reduced by 1, how does the value of the exponential expression change?

c. What is the value of the exponential when the exponent is zero?

d. Create a table for the exponential expression 5^x. You should include positive, zero, and negative values for x.

e. What was the value for 5^0?

f. Pick any base b ($b \neq 0$) for an exponential and determine the value of b^0.

g. Make a statement about what you think any base to the power of zero should be.

h. Now make one more table for an exponential with a base zero, that is, 0^x.

i. Finally, write a statement about the value of any exponential with exponent zero. Check this statement with other students in your class or your instructor.

This last statement about exponents equal to zero should lead you to the following property of exponents.

Zero as an Exponent

$$x^0 = 1 \qquad x \neq 0$$

When any exponential expression with a base other than zero is raised to the power of zero, the expression will equal 1.

$$25^0 = 1 \qquad (-17.4)^0 = 1$$

$$0^0 \text{ does not exist}$$

The rule for zero as an exponent states that the base must not equal zero. To see why this is necessary, look at the following two patterns.

$$4^0 = 1 \qquad 0^4 = 0$$
$$3^0 = 1 \qquad 0^3 = 0$$
$$2^0 = 1 \qquad 0^2 = 0$$
$$1^0 = 1 \qquad 0^1 = 0$$
$$0^0 = 1? \qquad 0^0 = 0?$$

The last line of these two patterns says that 0^0 equals both 1 and 0, which, of course, it cannot. This is a conflict and is one demonstration of why 0^0 is undefined. Any nonzero expression that is raised to the zero power is 1.

$$\left(\frac{25x^5 y^{16} z^{-8}}{157 a^4 b^{-2}}\right)^0 = 1$$

Example 6 Combining the rules for exponents

Simplify the following expressions. Write all answers without negative exponents.

a. $(2x^3 y^2)^{-2}(3x^{-3} y^6)$

b. $\left(\dfrac{5a^5}{a^3 b^4}\right)^{-1}$

c. $\left(\dfrac{18 m^2 n^{-2}}{9 m^{-4} n^5}\right)^{-3}$

d. $(123 a^{15} b^{-20} c^{30})^0$

SOLUTION

a. $(2x^3 y^2)^{-2}(3x^{-3} y^6) = (2^{-2} x^{-6} y^{-4})(3x^{-3} y^6)$ Apply the outside exponents.

$$= \left(\frac{1}{4} x^{-6} y^{-4}\right)(3 x^{-3} y^6)$$ Take the reciprocal of any numbers with negative exponents.

$$= \frac{3}{4} x^{-9} y^2$$ Multiply the coefficients and use the product rule to add exponents.

$$= \frac{3 y^2}{4 x^9}$$ Take the reciprocal of any variables with negative exponents.

b. $\left(\dfrac{5a^5}{a^3 b^4}\right)^{-1} = \left(\dfrac{5a^2}{b^4}\right)^{-1}$ Simplify the inside of the parentheses. Subtract exponents when you divide.

$$= \frac{5^{-1} a^{-2}}{b^{-4}}$$ Apply the outside exponent.

$$= \frac{b^4}{5 a^2}$$ Take the reciprocal of any bases with negative exponents.

c. $\left(\dfrac{18m^2n^{-2}}{9m^{-4}n^5}\right)^{-3} = (2m^6n^{-7})^{-3}$ Simplify the inside of the parentheses.

$= 2^{-3}m^{-18}n^{21}$ Apply the outside exponent.

$= \dfrac{n^{21}}{2^3 m^{18}}$ Take the reciprocal of any bases with negative exponents.

$= \dfrac{n^{21}}{8m^{18}}$ Multiply out the constant.

d. $(123a^{15}b^{-20}c^{30})^0 = 1$ This expression to the zero power is 1.

PRACTICE PROBLEM FOR EXAMPLE 6

Simplify the following expressions. Write all answers without negative exponents.

a. $(5x^2y^{-3})^{-3}(7x^5y^{-4})^2$ b. $\left(\dfrac{3g^4h^{-5}}{6g^{-2}h^5}\right)^3$ c. $\left(\dfrac{4}{3}a^3b^7c\right)^0$

Rational Exponents

All of the problems that we have worked on so far have had integer exponents. Fractions are often used as exponents in exponential problems, so we will want to know how to work with them. In this section, we will learn the basic meaning of a rational exponent and use them to simplify expressions. In Chapter 5, we will revisit rational exponents and use them to solve some problems.

A rational exponent is another way of writing a radical such as a square root or cube root.

$$\sqrt{25} = 25^{\frac{1}{2}} = 5 \qquad \sqrt[3]{27} = 27^{\frac{1}{3}} = 3$$

The rational exponent $\dfrac{1}{2}$ represents a square root, the rational exponent $\dfrac{1}{3}$ represents a cube root, and so on.

Rational Exponents

$$x^{\frac{1}{n}} = \sqrt[n]{x}$$

Raising a base to a rational exponent with a denominator of n is the same as taking the nth root of the base.

$$8^{\frac{1}{3}} = \sqrt[3]{8} = 2$$

If x is negative, n must be odd. If x is positive, n can be any whole number greater than or equal to 2.

We will use this definition to rewrite some rational exponents into radical form and some radical expressions into exponent form. The power rule for exponents helps us deal with rational exponents that have numerators other than 1.

$$x^{\frac{5}{3}} = (x^5)^{\frac{1}{3}} = \sqrt[3]{x^5}$$

Example 7   Rewriting rational exponents in radical form

Rewrite the following exponents in radical form.

a. $x^{\frac{1}{5}}$ b. $w^{\frac{2}{3}}$ c. $t^{\frac{3}{7}}$

SOLUTION

a. The denominator of the exponent becomes the radical's index. Therefore,

$$\sqrt[5]{x}$$

b. The denominator of the exponent becomes the radical's index. The numerator stays as the exponent of the variable. Therefore,
$$w^{\frac{2}{3}} = (w^2)^{\frac{1}{3}} = \sqrt[3]{w^2}$$

c. The denominator of the exponent is 7, so the index of the radical will be 7. The 3 will stay as the exponent of the variable. Therefore,
$$t^{\frac{3}{7}} = (t^3)^{\frac{1}{7}} = \sqrt[7]{t^3}$$

PRACTICE PROBLEM FOR EXAMPLE 7
Rewrite the following exponents in radical form.

a. $g^{\frac{4}{9}}$ b. $m^{\frac{7}{10}}$

Example 8 Rewriting radicals using rational exponents

Rewrite the following radicals using rational exponents.

a. $\sqrt{5x}$ b. $\sqrt[3]{w^2}$ c. $\left(\sqrt[5]{t}\right)^3$

SOLUTION
a. This is a square root so the index is 2. Therefore, the exponent is $\frac{1}{2}$. We get $(5x)^{\frac{1}{2}}$.

b. The index of the radical is 3, so the denominator of the fraction exponent is 3. We get $w^{\frac{2}{3}}$.

c. The index of the radical is 5, so the denominator of the fraction exponent is 5. We get $t^{\frac{3}{5}}$.

PRACTICE PROBLEM FOR EXAMPLE 8
Rewrite the following radicals using rational exponents.

a. $\sqrt{7a}$
b. $\sqrt[4]{t^3}$

All of the rules that we have learned in this section apply to positive and negative numbers as well as to whole numbers or fractions. When simplifying expressions that contain fractional exponents, simply follow the rules for exponents as if the exponents were whole numbers. Then reduce the fractional exponents if possible.

Example 9 Simplifying expressions with rational exponents

Simplify the following expressions. Write all answers without negative exponents.

a. $\left(25x^4 y^{10}\right)^{\frac{1}{2}}$

b. $\left(\dfrac{24a^5 b^6 c^{-2}}{3a^2 b^{-3} c}\right)^{\frac{1}{3}}$

c. $\left(5g^{\frac{1}{4}} h^{\frac{3}{4}}\right)^2 \left(3g^{\frac{1}{2}} h^{\frac{1}{2}}\right)$

SOLUTION
a. $\left(25x^4 y^{10}\right)^{\frac{1}{2}} = 25^{\frac{1}{2}} (x^4)^{\frac{1}{2}} (y^{10})^{\frac{1}{2}}$ *There is no simplifying to do inside the parentheses, so use the power rule for exponents and apply the fraction exponent.*
$= 5x^2 y^5$

Using Your TI Graphing Calculator

To calculate square roots and other higher roots on the graphing calculator, you can use either of two methods.

If you are taking the square root of a number, the square root function is above the

[x^2]

button on the left-hand side of the calculator. First you must press the [2nd] button, then the [x^2] button.

Then the number you want to take the square root of.

If you want to take a higher root, you can use the caret button

[^]

and a fraction exponent in parentheses. In this case, you will enter the number or variable you want to raise to a power, then press the [^] button, followed by the power you wish to use. For example, $\sqrt[3]{8} = 3^{\frac{1}{3}}$ would be entered as

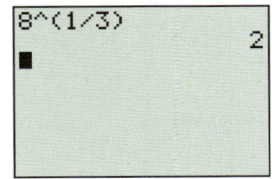

Remember that all fraction exponents must be in parentheses.

b. $\left(\dfrac{24a^5b^6c^{-2}}{3a^2b^{-3}c}\right)^{\frac{1}{3}} = (8a^3b^9c^{-3})^{\frac{1}{3}}$ — Simplify the inside of the parentheses using the quotient rule.

$= 8^{\frac{1}{3}}(a^3)^{\frac{1}{3}}(b^9)^{\frac{1}{3}}(c^{-3})^{\frac{1}{3}}$ — Apply the fraction exponent using the power rule for exponents.

$= 2ab^3c^{-1}$

$= \dfrac{2ab^3}{c}$ — Use the reciprocal of bases with a negative exponent.

c. $\left(5g^{\frac{1}{4}}h^{\frac{3}{4}}\right)^2\left(3g^{\frac{1}{2}}h^{\frac{1}{2}}\right) = \left[5^2\left(g^{\frac{1}{4}}\right)^2\left(h^{\frac{3}{4}}\right)^2\right]\left(3g^{\frac{1}{2}}h^{\frac{1}{2}}\right)$

$= \left(25g^{\frac{1}{2}}h^{\frac{3}{2}}\right)\left(3g^{\frac{1}{2}}h^{\frac{1}{2}}\right)$ — Apply the exponent using the power rule for exponents.

$= 75gh^2$ — Multiply using the product rule for exponents.

PRACTICE PROBLEM FOR EXAMPLE 9

Simplify the following expressions. Write all answers without negative exponents.

a. $(32a^5b^{10})^{\frac{1}{5}}$ **b.** $\left(\dfrac{32n^3m^{-3}}{2nm^{-1}}\right)^{\frac{1}{2}}$

3.1 Exercises

For Exercises 1 through 24, simplify the given expression using the order of operations and exponent rules. Write each answer without negative exponents.

1. $2^5 + 3^4$
2. $4^3 + 7^4$
3. $2^4(2^5)$
4. $5^3(5^8)$
5. $w^2w^5w^3$
6. $15g^3g^7$
7. $\dfrac{7^{23}}{7^{20}}$
8. $\dfrac{6^{30}}{6^{27}}$
9. $\dfrac{z^{12}}{z^8}$
10. $\dfrac{g^{17}}{g^9}$
11. $3^{245}(3^{-242})$
12. $4^{547}(4^{-553})$
13. s^5s^{-3}
14. k^7k^{-2}
15. $4r^5t^{-4}$
16. $12p^{-2}q^6$
17. $-9a^3b^{-7}$
18. $-15g^{-3}h^4$
19. $\dfrac{3}{b^{-2}}$
20. $\dfrac{5c^{-3}}{d^{-2}}$
21. $\dfrac{x^{-3}y^2}{3x^5y^{-7}}$
22. $\dfrac{7c^5d^{-8}}{c^{-9}d^3}$
23. $\dfrac{-4x^{-4}y^2}{x^2y^{-3}}$
24. $\dfrac{-14m^6n^{-4}}{m^{-2}n^{-2}}$

For Exercises 25 through 30, explain what the student did wrong when simplifying the given expression. Give the correct simplified expression.

25. $5^3 + 2^6$

Fred
$5^3 + 2^6 = 7^9 = 40353607$

26. $9^{-58}(9^{60})$

John
$9^{-58}(9^{60}) = 81^2 = 6561$

27. $-7x^2y^{-3}$

Bob
$-7x^2y^{-3} = \dfrac{x^2}{7y^3}$

28. $\dfrac{a^5}{a^{-3}}$

Alice
$\dfrac{a^5}{a^{-3}} = a^2$

29. $m^3 m^4$

Karen
$m^3 m^4 = m^{12}$

30. $5^{-2} a^3 b^{-6}$

Jim
$5^{-2} a^3 b^{-6} = \dfrac{a^3}{-25 b^6}$

For Exercises 31 through 46, simplify the given expression using the order of operations and exponent rules. Write each answer without negative exponents.

31. $\dfrac{200 a^5 b c^3}{25 a^2 b c}$

32. $\dfrac{45 x^7 y}{3 x^5 y}$

33. $3x^2 y (5x^4 y^3)$

34. $5 w^3 z (2 w^2 z^3)$

35. $\left(\dfrac{2}{3} a^3 b^7 c \right)^4$

36. $\left(\dfrac{4}{5} x^5 y^3 z \right)^2$

37. $(2x^2 y^3)^{-2}$

38. $(5 a^3 b c^7)^{-3}$

39. $(a + 5b)^2$

40. $(3m - n)^2$

41. $\left(\dfrac{5 w^3 v^7 x^{-4}}{17 w x^3} \right)^0$

42. $\left(\dfrac{28 a^6 b^{12} c}{148 a b^9 c^4} \right)^0$

43. $\left(\dfrac{3}{5} \right)^{-2}$

44. $\left(\dfrac{2}{3} \right)^{-2}$

45. $\left(\dfrac{1}{3} x^3 y^{-2} \right)^{-4}$

46. $\left(\dfrac{2}{5} a^{-5} b^2 \right)^{-3}$

For Exercises 47 through 50, use the given expression to answer the questions.

47. $\dfrac{(7x^2 y)^{10}}{(7x^2 y)^8}$

   a. What is the base for the exponent 10?
   b. What is the base for the exponent 8?
   c. Can you use the quotient rule for exponents before applying the outside exponents? Explain.
   d. Simplify the expression.

48. $\dfrac{(5ab^3)^{14}}{(5ab^3)^{11}}$

   a. What is the base for the exponent 14?
   b. What is the base for the exponent 11?
   c. Can you use the quotient rule for exponents before applying the outside exponents? Explain.
   d. Simplify the expression.

49. $(2x + 6)^2$

   a. Can you use the power rule for exponents with this base? Explain.
   b. Simplify the expression.

50. $(3t - 7)^2$

   a. Can you use the power rule for exponents with this base? Explain.
   b. Simplify the expression.

For Exercises 51 through 60, simplify the given expression using the order of operations and exponent rules. Write each answer without negative exponents.

51. $\dfrac{(x + 9y)^7}{(x + 9y)^5}$

52. $\dfrac{(2a - c)^{10}}{(2a - c)^8}$

53. $\left(\dfrac{2x^3 y^{-4}}{5xy^5} \right) \left(\dfrac{15 x y^2}{7 x^5 y^{-3}} \right)$

54. $\left(\dfrac{3 g^5 h^{-7}}{4 g^2 h^3} \right) \left(\dfrac{10 g^{-2} h^6}{11 g h^{-2}} \right)$

55. $\left(\dfrac{2 g^{-2} h^{-3}}{5 g h^{-6}} \right)^2$

56. $\left(\dfrac{3 c^5 d^2}{5 c^3 d^2} \right)^{-2}$

57. $(2x^3 y^{-4})^{-3} (3 x^2 y^{-6})^2$

58. $(4a^2 b^{-4})^3 (10 a^5 b^7)^{-2}$

59. $\left(\dfrac{1}{5} a^{-2} b^3 c \right)^{-2} \left(\dfrac{2}{3} a^4 b^{-6} c \right)^{-1}$

60. $\left(\dfrac{2}{3} x y^5 z^{-4} \right)^{-1} \left(\dfrac{5}{7} x^{-3} y^2 z^{-2} \right)^{-3}$

For Exercises 61 and 62, explain what exponent rule the student used for each step of the problem.

61.

Christine
$(2a^2 b)^{-3} (3 a b^3)^4 =$
Step 1. $= (2^{-3} a^{-6} b^{-3})(3^4 a^4 b^{12})$
Step 2. $= 2^{-3} \cdot 3^4 a^{-2} b^9$
Step 3. $= \dfrac{3^4 b^9}{2^3 a^2}$
Step 4. $= \dfrac{81 b^9}{8 a^2}$

   a. What exponent rule did the student use to get from the original expression to step 1?
   b. What exponent rule did the student use to get from step 1 to step 2?
   c. What exponent rule did the student use to get from step 2 to step 3?

62.

Cindy

$$\left(\frac{20x^3y^{-4}}{12xy^3}\right)^{-3} =$$

Step 1. $= \left(\frac{5x^2y^{-7}}{3}\right)^{-3}$

Step 2. $= \frac{5^{-3}x^{-6}y^{21}}{3^{-3}}$

Step 3. $= \frac{3^3y^{21}}{5^3x^6}$

Step 4. $= \frac{27y^{21}}{125x^6}$

a. What exponent rule did the student use to get from the original expression to step 1?
b. What exponent rule did the student use to get from step 1 to step 2?
c. What exponent rule did the student use to get from step 2 to step 3?

For Exercises 63 through 76, rewrite the given radical expressions using exponents.

63. $\sqrt{x}$
64. $\sqrt[3]{k}$
65. $\sqrt[5]{m}$
66. $\sqrt[6]{a}$
67. $\sqrt{c^3}$
68. $\sqrt[4]{m^3}$
69. $\left(\sqrt[3]{t}\right)^2$
70. $\left(\sqrt[6]{p}\right)^5$
71. $\sqrt{5xy}$
72. $\sqrt[3]{2x^2y}$
73. $\sqrt[7]{4m^3n^6p^2}$
74. $\sqrt[4]{10a^3bc^9}$

For Exercises 75 through 84, rewrite the given exponents in radical form.

75. $r^{\frac{1}{3}}$
76. $x^{\frac{1}{2}}$
77. $n^{\frac{2}{3}}$
78. $y^{\frac{3}{5}}$
79. $x^{\frac{1}{3}}y^{\frac{1}{3}}$
80. $(mn)^{\frac{1}{5}}$
81. $r^{\frac{1}{5}}s^{\frac{2}{5}}$
82. $n^{\frac{2}{3}}m^{\frac{1}{3}}p^{\frac{2}{3}}$
83. $(xy^3z)^{\frac{1}{2}}$
84. $(a^2b^4c)^{\frac{1}{5}}$

For Exercises 85 through 92, simplify the given expression using your calculator. Write your answer as fractions if necessary.

85. $27^{\frac{1}{3}}$
86. $16^{\frac{1}{4}}$
87. $1{,}000{,}000^{\frac{1}{6}}$
88. $7776^{\frac{1}{5}}$
89. $(-64)^{\frac{1}{3}}$
90. $\left(\frac{1}{81}\right)^{\frac{1}{4}}$
91. $25^{-\frac{1}{2}}$
92. $2401^{-\frac{1}{4}}$

For Exercises 93 through 104, simplify the given expression using the order of operations and exponent rules. Write each answer without negative exponents.

93. $(8x^3y^9)^{\frac{1}{3}}$
94. $(243t^{20}u^{15})^{\frac{1}{5}}$
95. $(64m^6n^{12})^{\frac{1}{2}}$
96. $(1296a^8b^{20})^{\frac{1}{4}}$
97. $(121a^{-6}b^8)^{\frac{1}{2}}$
98. $(3375f^{12}g^{-21})^{-\frac{1}{3}}$
99. $\left(\frac{49x^5y^3}{25xy^9}\right)^{\frac{1}{2}}$
100. $\left(\frac{-27k^8v^7}{8k^2v}\right)^{\frac{1}{3}}$
101. $\left(\frac{16m^3n^6p}{mn^{-2}p^3}\right)^{\frac{1}{2}}$
102. $\left(\frac{9x^5y^4z^{-7}}{x^3y^{-2}z^{-1}}\right)^{\frac{1}{2}}$
103. $(100ab^3c^2)^{-\frac{1}{2}}(a^6b^{-2})^{\frac{1}{4}}$
104. $(5x^2y^7z)^{\frac{3}{2}}(5x^2y^7z)^{\frac{7}{2}}$

3.2 Combining Functions

LEARNING OBJECTIVES

- Perform arithmetic operations with polynomials.
- Combine functions using addition, subtraction, multiplication and division.
- Combine functions within an application.

The Terminology of Polynomials

In this section we will look at a family of functions called *polynomials*. To start, we will review some basics about polynomials.

The most fundamental component of a polynomial is a **term**. A term can be either a constant, a variable, or a combination of constants and variables multiplied together. The constant part of a term is called the *coefficient*. A **polynomial** is any combination of terms that are added together. The powers of all variables in a polynomial must be positive integers.

SECTION 3.2 Combining Functions   **235**

DEFINITIONS

Term A constant, a variable or the product of any number of constants and variables. Terms can include constants and/or variables raised to exponents.

Examples: $12 \quad -3x \quad 5xy^2$

Coefficient The constant part of any term. The coefficient is usually at the front of any term and includes the sign of the term.

Polynomial Any combination of terms that are added together. The powers of all variables in a polynomial must be positive integers.

Polynomial	Not a Polynomial
$5x^2y$	$3x^2 + \dfrac{5}{x} - 4$
$3m^2 + 2m - 7$	$3\sqrt{x} + 5$

> **What's That Mean?**
>
> **Polynomials**
>
> Many specific names are given to different types of polynomials.
>
> All of the following are different names for specific types of polynomials:
>
> - Monomial: polynomial with one term
> - Binomial: polynomial with two terms
> - Trinomial: polynomial with three terms
>
> The prefixes used in these names are Greco-Roman and mean numbers.
>
> mono = one.
> bi = two.
> tri = three.

Linear functions are an example of polynomials because they are a combination of constants and variables multiplied and added together. Basic linear functions usually have two terms, mx and b.

Example 1   Recognizing terms in a polynomial

Determine the number of terms in the following polynomial expressions. List the terms as either constant terms or variable terms. Give the coefficient of each variable term.

a. $3x^2 - 5x - 9$ **b.** $8t + 6$ **c.** $5m^2 - 3$

d. 104 **e.** $-98xy^2$

SOLUTION

a. The polynomial expression $3x^2 - 5x - 9$ has three terms; -9 is a constant term, and both $3x^2$ and $-5x$ are variable terms because they contain at least one variable. The term $3x^2$ has a coefficient of 3. The term $-5x$ has a coefficient of -5.

b. The linear expression $8t + 6$ has two terms; 6 is a constant term, and $8t$ is a variable term. The term $8t$ has a coefficient of 8.

c. The polynomial expression $5m^2 - 3$ has two terms; -3 is a constant term, and $5m^2$ is a variable term. The term $5m^2$ has a coefficient of 5.

d. The number 104 is a single constant term.

e. The expression $-98xy^2$ is a single variable term with a coefficient of -98. Although two variables are involved, they are multiplied together and are part of one term. Terms are always separated by addition.

Example 2   Recognizing polynomials

Determine whether the given expression is a polynomial. If the expression is not a polynomial, explain why not.

a. $10\sqrt[3]{2m} + 8m + 5$ **b.** $24a^4b - 15a^3b^2$ **c.** $2x + \dfrac{3}{x} - 85$

SOLUTION

a. The expression $10\sqrt[3]{2m} + 8m + 5$ is not a polynomial. If the cube root is rewritten as a fraction exponent, the variable under the cube root will not have an integer exponent.

b. The expression $24a^4b - 15a^3b^2$ is a polynomial. All the terms have variables raised to positive integer exponents.

c. The expression $2x + \dfrac{3}{x} - 85$ is not a polynomial. The division by a variable in the second term could be rewritten as a variable to a negative exponent.

PRACTICE PROBLEM FOR EXAMPLE 2

Determine if the given expression is a polynomial. If the expression is not a polynomial explain why not.

a. $4^w + 5$ b. $4\sqrt{t} + 20t$ c. $8x - 9$

Degree

One important feature of a polynomial is its **degree**. The degree of an individual term is the sum of all the exponents for the variables. The degree of a polynomial is the same as the highest-degree term.

> **DEFINITIONS**
>
> **Degree of a term** The sum of all the exponents of the variables in the term.
> $$\text{Term} = 16x^3y^2 \quad \text{Degree} = 5$$
>
> **Degree of a polynomial** The degree of the highest-degree term.
> $$\text{Polynomial} = 7x^4y + 2x^3y^8 + 5 \quad \text{Degree} = 11$$

Example 3 Identifying the degree of terms and polynomials

For the given expressions, list the degree of each term and the entire polynomial.

a. $3x^2 + 5x + 6$ b. $7t - 4$ c. $2x^2y + 5xy - 7y + 2$
d. $4a^3b^4c - 6ab^2c^2$ e. 12

SOLUTION

a. The term $3x^2$ has degree 2. The term $5x$ has degree 1. The term 6 has degree 0. The polynomial has degree 2.

b. The term $7t$ has degree 1. The term -4 has degree 0. The polynomial has degree 1.

c. The term $2x^2y$ had degree 3 because x has an exponent of 2, and y has an exponent of 1. The term $5xy$ has degree 2. The term $-7y$ has degree 1. The term 2 has degree 0. The polynomial has degree 3.

d. The term $4a^3b^4c$ has degree 8. The term $-6ab^2c^2$ has degree 5. The polynomial has degree 8.

e. The term 12 has degree 0. This is because there are no variables in this term, so there are no variable exponents to sum. The polynomial has degree 0.

PRACTICE PROBLEM FOR EXAMPLE 3

For the given expressions, list the degrees of each term and the entire polynomial.

a. $-0.23x^2 + 4x - 3$ b. $8m^3n + 6mn^2 - 2n^3$ c. $-9s^5t^4u^2 - 7s^3t^2u^2$

Adding and Subtracting Functions

Because functions can represent many things in our lives, it is often helpful to combine them using different operations such as adding, subtracting, multiplying, or dividing. Many functions can be combined to make more complicated functions or to create a new function that gives you the information that you want. By combining functions into one, we do not have to continue working with several functions at once.

To add or subtract two terms, they must be **like terms**. Like terms have the same variables with the same exponents. The next few examples and problems will show some applications of addition and subtraction of polynomials.

> **DEFINITION**
>
> **Like Terms** Terms that have the same variables with the same exponents.

Example 4 Combining revenue and costs to find profit

The revenue and costs for Southwest Airlines can be modeled by the following functions.

$$R(t) = 183t^2 - 581t + 5964$$
$$C(t) = 102t^2 - 100t + 4897$$

where $R(t)$ represents the revenue for Southwest Airlines in millions of dollars and $C(t)$ represents the costs for Southwest Airlines in millions of dollars t years since 2000.
Source: Models derived from data found at finance.google.com.

a. Find the revenue and costs for Southwest Airlines in 2006.

b. Find a new function that will give the profit of Southwest Airlines. (*Hint*: Profit equals revenue minus costs.)

c. Use the new profit function to find the profit in 2006.

SOLUTION

a. 2006 is represented by $t = 6$, so we get

$$R(6) = 9066$$
$$C(6) = 7969$$

Therefore, Southwest Airlines had a revenue of about $9066 million and costs of about $7969 million in 2006.

b. If we define $P(t)$ to be the profit in millions of dollars for Southwest Airlines, we know that profit is revenue minus the costs, so we get

$P(t) = R(t) - C(t)$ *Substitute the revenue and cost functions.*
$P(t) = (183t^2 - 581t + 5964) - (102t^2 - 100t + 4897)$
$P(t) = 183t^2 - 581t + 5964 - 102t^2 + 100t - 4897$ *Distribute the negative and combine like terms.*
$P(t) = 81t^2 - 481t + 1067$

c. Again, 2006 is represented by $t = 6$, so we get $P(6) = 1097$. The profit for Southwest Airlines was about $1097 million in 2006.

PRACTICE PROBLEM FOR EXAMPLE 4

The revenue and profit for the California lottery can be modeled by the following functions.

$$R(t) = 59.6t^2 - 299.4t + 3022.8$$
$$P(t) = 23t^2 - 121t + 1157.2$$

where $R(t)$ represents the revenue in millions of dollars and $P(t)$ represents the profit in millions of dollars from the California lottery t years since 2000.
Source: Models derived from data in Survey of Government Finances 2000–2006.

> **What's That Mean?**
>
> **Function Notation**
>
> When adding, subtracting, multiplying, or dividing functions, you can write these operations in function notation in a few ways.
>
> **Addition**
>
> $$f(x) + g(x) = (f + g)(x)$$
>
> **Subtraction**
>
> $$f(x) - g(x) = (f - g)(x)$$
>
> **Multiplication**
>
> $$f(x)g(x) = (fg)(x)$$
>
> **Division**
>
> $$\frac{f(x)}{g(x)} = f(x) \div g(x) = \left(\frac{f}{g}\right)(x)$$
>
> When combining functions without a meaning for the variables, you are free to combine any functions you are given as long as the input variables are the same. The only caution would be that you can divide as long as you do not divide by zero. This is a typical restriction that is assumed in all cases.

a. Calculate the revenue and profit for the California lottery in 2010.

b. Find a new function for the cost associated with the California lottery. (*Hint*: Remember that profit = revenue − cost.)

c. Using the new cost function, determine the cost for the California lottery in 2015.

It is important to note that the revenue, cost, and profit functions had the same input variable. The output variables were also measured with the same units. You can add or subtract functions for revenue, cost, and profit only when the units are the same. If revenue was measured in millions of dollars and cost was measured in thousands of dollars, you could not simply subtract the two functions to find profit.

Example 5 A geometry example of combining functions

The area of a Norman window (a rectangle with a semicircle on top) can be determined by adding the area functions of the rectangle and semicircle pieces that it is made of. If the rectangular part has a height of 99 inches, find the following:

a. A function for the total area of the Norman window as a function of the width of the window.

b. The area of the Norman window if its width is 48 inches.

SOLUTION

a. The area function will be the sum of the rectangle's area function and the semicircle's area function. We are missing the width of the window, so let's use the following definition:

$$w = \text{Width of the window in inches}$$

The area of the rectangular piece of the window can be calculated as length times width. Since the height of the rectangular piece is 99 inches, we can represent the area with the function.

$$R(w) = 99w \quad \text{Area of a rectangle is length times width.}$$

where $R(w)$ represents the area of the rectangle part in square inches.

The area of the circular piece can be calculated using the formula for the area of a circle, $A = \pi r^2$. In this window, the radius is half of the width of the window. Only half of a circle is part of the window, so we multiply the area by $\frac{1}{2}$. We can represent the area of the half-circle piece with the function

$$C(w) = \frac{1}{2}\pi\left(\frac{w}{2}\right)^2 \quad \text{The area of a half circle is half of pi times radius squared.}$$

where $C(w)$ represents the area of the semicircle part in square inches. To get the total area of the Norman window, we will add these two functions together. This results in a total area function of

$$N(w) = 99w + \frac{1}{2}\pi\left(\frac{w}{2}\right)^2 \quad \text{The total area is the sum of the two functions.}$$

$$N(w) = 99w + \frac{1}{2}\pi\frac{w^2}{4} \quad \text{Simplify.}$$

$$N(w) = \frac{\pi}{8}w^2 + 99w$$

where $N(w)$ represents the total area in square inches of a Norman window with a height of 99 inches and a width of w inches for the rectangle part.

b. Because the width is given as 48 inches, we can substitute 48 for w and find the total area $N(48) = 5656.8$. Therefore, a Norman window with these dimensions will have a total area of 5656.8 square inches.

SECTION 3.2 Combining Functions

Units are the key to combining functions in any application. To determine whether you can combine two functions by addition or subtraction, be sure that the input variables are the same and that the output variables have the same units.

> **Combining Functions in Applications Using Addition or Subtraction**
> - The inputs for both functions must be measured in the same units.
> - The outputs must be measured in the same units.

Example 6 Combining functions using function notation

Combine the following functions using addition or subtraction to write a new function that will give you the result requested. Represent the new function using function notation.

$S(t)$ = Total number of students at Clark State University t years since 2000

$F(t)$ = Number of full-time students at Clark State University t years since 2000

$B(c)$ = Average amount in dollars a student pays for textbooks when taking c credits a semester

$T(c)$ = Average amount in dollars a student pays for tuition and fees when taking c credits a semester

a. The average total amount a student pays for textbooks, tuition, and fees when taking c credits a semester.

b. The number of part-time students at Clark State University t years since 2000.

SOLUTION

a. $B(c) + T(c)$ = Average total amount a student pays for textbooks, tuition, and fees when taking c credits a semester

Since we want the average total amount spent on textbooks as well as tuition and fees, we will add the functions $B(c)$ and $T(c)$ together. The input variables are the same, and the output units are both dollars and it makes sense to add them together.

b. $S(t) - F(t)$ = Number of part-time students at Clark State University t years since 2000

We want the number of part-time students at Clark State University, but we have only a function for the total number of students and another function for the number of full-time students. To find the number of part-time students, we can subtract the number of full-time students from the total number of students. Again, the input variables are the same. The output units are both types of students, so subtracting them makes sense in this situation.

Example 7 Add and subtract functions algebraically

Combine the following functions.

$$f(x) = 5x + 6 \quad g(x) = 2x - 9 \quad h(x) = 3x + 4$$

a. $f(x) + g(x)$ **b.** $g(x) - f(x)$ **c.** $(h - g)(x)$

SOLUTION

a. To add these two functions, combine like terms and simplify.

$$f(x) + g(x) = (5x + 6) + (2x - 9)$$
$$f(x) + g(x) = 7x - 3$$

b. When subtracting functions, be careful to distribute the negative throughout the second function.

$$g(x) - f(x) = (2x - 9) - (5x + 6)$$
$$g(x) - f(x) = 2x - 9 - 5x - 6 \quad \text{Distribute the negative through the parentheses.}$$
$$g(x) - f(x) = -3x - 15 \quad \text{Combine like terms.}$$

c. This function notation is another way to write subtraction of two functions.

$$(h - g)(x) = (3x + 4) - (2x - 9)$$
$$(h - g)(x) = 3x + 4 - 2x + 9 \quad \text{Distribute the negative through the parentheses.}$$
$$(h - g)(x) = x + 13 \quad \text{Combine like terms.}$$

PRACTICE PROBLEM FOR EXAMPLE 7

Combine the following functions.

$$f(x) = 7x + 5 \qquad g(x) = 4x - 8$$

a. $f(x) + g(x)$ **b.** $g(x) - f(x)$

Multiplying and Dividing Functions

Multiplication and division are two more ways in which we can combine functions and polynomial expressions. When using multiplication or division, be sure that the input variables are the same and that when you are combining the output variables, the units make sense. The output units can be the same in computing something like area. On the other hand, the output units can be different, but when they are combined, they have to simplify to an appropriate unit for the problem.

Combining Functions in Applications Using Multiplication or Division
- The inputs for both functions must be the same.
- The outputs must make sense together when combined.

    a. feet · feet = square feet.

    b. pounds · $\dfrac{\text{dollars}}{\text{pound}}$ = ~~pounds~~ · $\dfrac{\text{dollars}}{\text{\sout{pound}}}$ = dollars

    c. hours · $\dfrac{\text{miles}}{\text{hour}}$ = ~~hours~~ · $\dfrac{\text{miles}}{\text{\sout{hour}}}$ = miles

    d. dollars ÷ people = $\dfrac{\text{dollars}}{\text{person}}$ = dollars per person

---

Example 8   Combining functions using function notation

Combine the following functions using addition, subtraction, multiplication, or division to write a new function that will give you the result requested. Represent the new function using function notation.

$U(t) = $ Population of the United States t years since 1900

$F(t) = $ Number of females in the United States t years since 1900

$D(t) = $ U.S. National Debt in dollars t years since 1900

$A(t) = $ Average dollars spent per person on dining out t years since 1900

a. The total amount spent by Americans on dining out t years since 1900.

b. The number of males in the United States t years since 1900.

c. The average amount that each American would have to contribute to pay off the national debt t years since 1900.

SOLUTION

a. $U(t) \cdot A(t)$ = Total amount spent by Americans on dining out, in dollars, t years since 1900

To get the total, we took the average amount spent and multiplied by the total number of people. The units of dollars per person times people simplifies to dollars, which makes sense in this situation.

b. $U(t) - F(t)$ = Number of males in the United States t years since 1900

Subtracting the number of females in the United States from the total population gives us the number of males.

c. $\dfrac{D(t)}{U(t)}$ = Average amount that each American would have to contribute to pay off the national debt t years since 1900

Dividing the national debt by the number of people in the United States gives an average amount for each person.

PRACTICE PROBLEM FOR EXAMPLE 8

Use the following functions to write a new function that will give you the result requested.

$F(t)$ = Number of people employed by Ford Motor Company in year t

$I(t)$ = Average cost, in dollars per employee, for health insurance at Ford Motor Company in year t

$V(t)$ = Total cost, in dollars, for vacations taken by Ford Motor Company nonmanagement employees in year t

$M(t)$ = Number of employees of Ford Motor Company who are in management in year t

a. The total amount spent on health insurance for Ford Motor Company employees in year t.

b. The number of nonmanagement employees at Ford Motor Company in year t.

c. The average cost per nonmanagement employee for vacations at Ford Motor Company in year t.

Unlike addition and subtraction, in which you must have like terms, when you multiply, the terms do not have to be like terms. You will, however, have to use the distributive property and add exponents for any variables that are the same.

Example 9 — An application of multiplying functions

The willingness of digital camera producers to produce cameras depends on the price for which they can sell the cameras. This situation can be modeled by

$$S(p) = 0.015p^2 - 1.5p + 80$$

where $S(p)$ represents the supply of digital cameras in thousands when p is the price in dollars and the price is greater than $50.

a. Find the number of digital cameras that producers are willing to supply when the price is $100.

b. Find a function for the projected revenue if all the cameras that are supplied sell at price p.

c. Estimate the projected revenue if digital cameras are sold for $150 each.

SOLUTION

a. The given price of $100 can be substituted into p to get $S(100) = 80$. Therefore, the digital camera producers are willing to supply 80 thousand cameras if the cameras will sell for $100 each.

Skill Connection

Distributive Property

The distributive property is the basis for multiplication of polynomials. To use the distributive property, multiply each term of the first expression by each term of the second expression in the product.

Example

Perform the indicated operation and simplify.

a. $5x(3x + 8)$
b. $(x + y)(2x + 3y)$

Solution:

a. Use the distributive property to multiply the $5x$ through the parentheses.

$5x(3x + 8)$
$15x^2 + 40x$

b. Use the distributive property to multiply the x and then the y through the second parentheses.

$(x + y)(2x + 3y)$
$2x^2 + 3xy + 2xy + 3y^2$
$2x^2 + 5xy + 3y^2$

In multiplying two binomials, the distributive property can be remembered by using the acronym **FOIL**: Multiply the **F**irst terms, the **O**uter terms, the **I**nner terms, and finally the **L**ast terms.

b. Revenue is calculated by taking the number of items sold and multiplying by the price. In this case, the function $S(p)$ represents the number of digital cameras sold and p represents the price. Using these variables, we get

$$R(p) = \text{price (quantity sold)}$$
$$R(p) = p \cdot S(p)$$
$$R(p) = p(0.015p^2 - 1.5p + 80)$$
$$R(p) = 0.015p^3 - 1.5p^2 + 80p$$

where $R(p)$ represents the revenue in thousands of dollars for selling digital cameras at a price of p dollars.

c. The given price of $150 can be substituted into p to get $R(150) = 28875$. Therefore, the digital camera producers will earn $28,875 thousand if cameras sell for $150 each.

PRACTICE PROBLEM FOR EXAMPLE 9

The willingness of a car manufacturer to produce SUVs depends on the price for which they can sell the vehicles. This can be modeled by

$$S(p) = 0.008p^2 - 0.5p + 30$$

where $S(p)$ represents the supply of SUVs in thousands when p is the price in thousands of dollars and the price is greater than $25,000.

a. Find the number of SUVs this car manufacturer is willing to supply when the price is $45,000.

b. Find a function for the projected revenue if all the SUVs that are supplied sell at price p.

c. Estimate the projected revenue if SUVs are sold for $40,000 each.

Example 10 Combining functions algebraically

Combine the following functions.

$$f(x) = 5x + 6 \qquad g(x) = 2x - 9 \qquad h(x) = 3x + 4$$

a. $f(x)g(x)$
b. $\dfrac{g(x)}{h(x)}$

SOLUTION

a. When multiplying two functions, be sure to use the distributive property.

$$f(x)g(x) = (5x + 6)(2x - 9)$$
$$f(x)g(x) = 10x^2 - 45x + 12x - 54$$
$$f(x)g(x) = 10x^2 - 33x - 54$$

b. For now, we will write the division of two functions as a fraction and we will learn how to simplify these in Chapter 7.

$$\frac{g(x)}{h(x)} = \frac{2x-9}{3x+4}$$

PRACTICE PROBLEM FOR EXAMPLE 10
Combine the following functions.

$$f(x) = 7x + 3 \qquad g(x) = 4x - 5 \qquad h(x) = 9x + 2$$

a. $f(x)g(x)$ 　　　　**b.** $\dfrac{g(x)}{h(x)}$

At times, we will combine polynomial expressions the same way we combine functions. In these problems, the function notation is left out and the operations are done to simplify the expression.

Example 11 Operations with polynomial expressions

Perform the indicated operation and simplify.

a. $(2ab^2 + 5b) + (4a^2b + 3ab^2)$

b. $(x^4 - 3x^2 + 4x) - (3x^2 - 5x + 3)$

c. $(2x + 5)(4x - 8)$

d. $(x + 2)(x^2 + 3x - 7)$

SOLUTION
a. We are adding so combine like terms.

$$(2ab^2 + 5b) + (4a^2b + 3ab^2) = 4a^2b + 5ab^2 + 5b$$

b. First, distribute the negative sign through the second parentheses and then combine like terms.

$$(x^4 - 3x^2 + 4x) - (3x^2 - 5x + 3) = x^4 - 3x^2 + 4x - 3x^2 + 5x - 3$$
$$= x^4 - 6x^2 + 9x - 3$$

c. Using FOIL (distributive property), multiply the $2x$ through and then the 5.

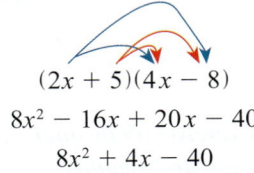

$$(2x + 5)(4x - 8)$$
$$8x^2 - 16x + 20x - 40$$
$$8x^2 + 4x - 40$$

d. Use the distributive property to multiply the x and then to multiply the 2 through and combine like terms.

$$(x + 2)(x^2 + 3x - 7) = x^3 + 3x^2 - 7x + 2x^2 + 6x - 14$$
$$= x^3 + 5x^2 - x - 14$$

PRACTICE PROBLEM FOR EXAMPLE 11
Perform the indicated operation and simplify.

a. $(4x^2y + 3xy - 7y) + (3x^2y - 2y + 10)$

b. $(3m^4n^2 - 5m^2n + 2mn) - (8m^2n - 3mn + 4n^2)$

c. $3(7x + 3)(3x - 5)$

d. $(4x^2 + 5x)(2x^2 - 3x + 7)$

3.2 Exercises

For Exercises 1 through 6, determine the number of terms in the given polynomial expression. List the terms as either constant terms or variable terms. Give the coefficient of each variable term.

1. $5x + 9$
2. $-2x + 4$
3. $-3x^2 + 2x - 8$
4. $0.5t^2 + 5t - 7$
5. $12a^3b^2c + 3abc^2 - 8bc$
6. $m^3n^2 - 3m^2n + 4mn^3 - 4n$

For Exercises 7 through 16, determine whether the given expression is a polynomial. If the expression is not a polynomial, explain why not.

7. $4x + 9$
8. $2x^2 + 5x - 20$
9. $\frac{4}{x} + 20x - 10$
10. $-8x + 3x^{-2} - 10$
11. $14a^2b + 10ab^3 - 9a$
12. $-20m^3n - 10m^2n^2$
13. $5\sqrt{d} + 2$
14. $3s - \sqrt[4]{t}$
15. 108
16. $24y + \sqrt{30}$

For Exercises 17 through 26, list the degrees of each term and the degree of the entire polynomial.

17. $2x^5 - 7$
18. $3x^2 + 5x - 7$
19. $5p^2 + 4p - 87$
20. $1.023\,t + 6.2$
21. $4m + 8$
22. $-14s^3t + 6t^4 - 9$
23. $2a^2b^3 + 3ab^2 - 8b$
24. $23a^4b^2 - 62a^2b + 9b^3$
25. $\frac{2}{3}gh + \frac{1}{4}g^3h^5 - \frac{2}{9}g + 7$
26. $\frac{5}{7}r^2s^3t^2 - \frac{3}{8}r^4st^2 + \frac{4}{9}rs - \frac{5}{11}$

For Exercises 27 through 34, perform the indicated operation and simplify.

27. $(5x + 6) + (2x + 8)$
28. $(4x + 5) + (8x + 3)$
29. $(7x + 6) - (3x + 2)$
30. $(10x - 9) - (4x + 3)$
31. $(5x^2 - 6x - 12) - (-3x^2 - 10x + 8)$
32. $(14x^2 - 4x - 6) - (-2x^2 - 3x + 12)$
33. $(5x^3z^2 - 4x^2z + 6z) - (3x^3z^2 - 17xz + 3z)$
34. $(a^2b + 5ab^2 - 9b) - (3ab^2 + 8b)$

For Exercises 35 through 38, explain what the student did wrong when simplifying the given expression. Give the correct simplified expression.

35. $(2x + 7) - (6x - 9)$

Chie
$(2x + 7) - (6x - 9) =$
$= (2x + 7)(-6x + 9)$
$= -12x^2 + 18x - 42x + 63$
$= -12x^2 - 24x + 63$

36. $(2x + 7) - (6x - 9)$

Rob
$(2x + 7) - (6x - 9) =$
$= 2x + 7 - 6x - 9$
$= -4x - 2$

37. $(5x - 8)^2$

Gordon
$(5x - 8)^2 = 25x^2 + 64$

38. $(4x - 8)(5x + 2)$

Jenni
$(4x - 8)(5x + 2) = 9x - 6$

For Exercises 39 through 48, perform the indicated operation and simplify.

39. $(5ab - 8b)(3a + 4b)$
40. $(3xy + 2x)(5x + 7y)$
41. $(2x + 5)(3x^2 - 5x + 7)$
42. $(4a + 7)(3a^2 - 4a + 2)$
43. $(3x + 7y)^2$
44. $(x + 4y)^2$
45. $(2x^2 + 3x - 9) + (7x + 2)^2$
46. $(5x^2 + 2x - 8) + (3x - 5)^2$

47. $7x(3y^2 + 5y - 6) + 8xy$

48. $(7x - 3)(2x + 8) - 9x^2 + 8$

49. The average number of pounds of fruit and vegetables per person that each American eats can be modeled by
$$T(y) = -4.5y + 712.3$$
pounds per person y years since 2000.
The average number of pounds of fruit per person that each American eats can be modeled by
$$F(y) = -3.5y + 289.6$$
pounds per person y years since 2000.
Source: Based on data from the Statistical Abstract of the United States, 2008.

   a. How many pounds of fruit and vegetables did the average American eat in 2000?
   b. How many pounds of fruit did the average American eat in 2000?
   c. Using your previous results, how many pounds of vegetables did the average American eat in 2000?
   d. Write a new function that will give the average number of pounds of vegetables each American eats per year.
   e. Using your new function, how many pounds of vegetables did each American eat in 2000?
   f. Estimate the number of pounds of vegetables each American will eat in 2010, 2015, and 2020.

50. In Exercise 39 of Section 1.6 on page 92, you found a model for the amount spent by individuals on health care expenses in the United States to be close to
$$I(t) = 17t + 147$$
billions of dollars t years since 2000.
The amount spent by insurance companies on health care expenses in the United States can be modeled by
$$H(t) = 58t + 370$$
billions of dollars t years since 2000.
Source: Based on data from the Statistical Abstract of the United States, 2008.

   a. How much was spent by individuals on health care in 2000?
   b. How much was spent by insurance companies on health care in 2000?
   c. What was the total amount spent by individuals and insurance companies on health care in 2000?
   d. Find a new function that gives the total amount spent on health care by individuals and insurance companies.
   e. Use your new function to determine the total amount spent on health care by individuals and insurance companies in 2010, 2015, and 2020.

51. The per capita consumption of milk products in the United States can be modeled by
$$M(t) = -0.29t + 22.42$$
gallons per person t years since 2000.
The per capita consumption of whole milk in the United States can be modeled by
$$W(t) = -0.21t + 8.09$$
gallons per person t years since 2000.
Source: Based on data from the Statistical Abstract of the United States, 2008.

   a. Use these models to find a new function that gives the per capita consumption of milk products other than whole milk.
   b. How much whole milk was consumed per person in 2005?
   c. How much milk other than whole milk was consumed per person in 2005 and in 2010?
   d. What is the slope of $W(t)$? Explain its meaning in regard to consumption of whole milk?
   e. What is the M-intercept for $M(t)$? Explain its meaning in regard to consumption of milk products?

52. The number of U.S. residents who are of African-American or of Hispanic origin can be modeled by
$$A(t) = 401t - 5240$$
$$H(t) = 11.5t^2 - 1181t + 35414$$
where $A(t)$ represents the number of African-American residents in thousands and $H(t)$ represents the number of residents of Hispanic origin in thousands t years since 1900.
Source: Models derived from data in Statistical Abstract of the United States, 2001.

   a. Find a function that will represent the total number of residents in the United States who are either of African-American or of Hispanic origin.
   b. Use this function to determine the total number of residents who are either of African-American or of Hispanic origin in 2000.

For Exercises 53 through 60, combine the functions by adding, subtracting, multiplying, or dividing, the two functions. Write the function notation for the combination you use and list the units for the inputs and outputs of the new function.

53. Let $U(t)$ be the population of the United States, in millions, t years since 1900. Let $M(t)$ be the number of men in the United States, in millions, t years since 1900. Find a function for the number of women in the United States.

54. Let $S(t)$ be the number of students at Suncoast College, t years since 2010. Let $A(t)$ be the number of students who are involved in the athletics program at Suncoast College, t years since 2010. Find a function for the

number of students at Suncoast College who are not involved in the school's athletics program.

55. Let $S(d)$ be the average speed, in miles per hour, traveled on a cross-country trip during day d of the trip. Let $T(d)$ be the time, in hours, traveled during a cross-country trip on day d of the trip. Find a function for the number of miles traveled on day d of the cross-country trip.

56. Let $P(w)$ be the price of gas in dollars per gallon during week w of the year. Let $G(w)$ be the number of gallons of gas your car needs during week w of the year. Find a function for the total cost of gas used by your car during week w of the year.

57. Let $U(t)$ be the population of the United States, in millions, t years since 1900. Let $D(t)$ be the national debt, in millions of dollars, t years since 1900. Find a function for the average amount of national debt per person.

58. Let $S(d)$ be the number of students in your math class on day d of the semester. Let $C(d)$ be the total number of credits being taken by all students in your math class on day d of the semester. Find a function for the average number of credits a student in your math class is taking on day d of the semester.

59. Let $U(t)$ be the population of the United States, t years since 1900. Let $D(t)$ be the average amount of personal debt in dollars per person in the United States t years since 1900. Find a function for the total amount of personal debt in the United States.

60. Let $P(b)$ be the profit, in thousands of dollars, that Pride Bike Co. makes if they sell b bikes per year. Let $R(b)$ be the revenue, in thousands of dollars, that Pride Bike Co. makes if they sell b bikes per year. Find a function that gives the costs that Pride Bike Co. has if they sell b bikes per year.

61. The percentage of jail inmates who are male in U.S. federal and state prisons can be modeled by $M(t) = -0.24t + 88.64$ percent t years since 2000.
Source: Based on data from Statistical Abstract of the United States, 2008.

   a. Let $F(t)$ be the percentage of jail inmates who are female in U.S. federal and state prisons t years since 2000. Explain why $M(t) + F(t) = 100$.
   b. Substitute the function $M(t)$ into the equation in part a and solve for $F(t)$.
   c. Find $F(5)$ and interpret it in this situation.
   d. What is the slope of $F(t)$? Explain its meaning in this situation.
   e. What is the M-intercept for $M(t)$? Explain its meaning in this situation.

62. The percent of mortgage loans in United States that are considered subprime loans can be modeled by $S(t) = 3.5t - 1$ percent t years since 2000.
Source: Based on data from 2007 Federal Reserve Bank of Chicago, Chicago Fed Letter.

   a. Let $P(t)$ be the percentage of mortgage loans in United States that are not considered subprime t years since 2000. Explain why $S(t) + P(t) = 100$.
   b. Substitute the function $S(t)$ into the equation in part a and solve for $P(t)$.
   c. Find $P(10)$ and interpret it in this situation.
   d. What is the slope of $P(t)$? Explain its meaning in this situation.
   e. What is the P-intercept for $P(t)$? Explain its meaning in this situation.

63. The percentage of occupied housing units that are owner occupied is given in the following chart.

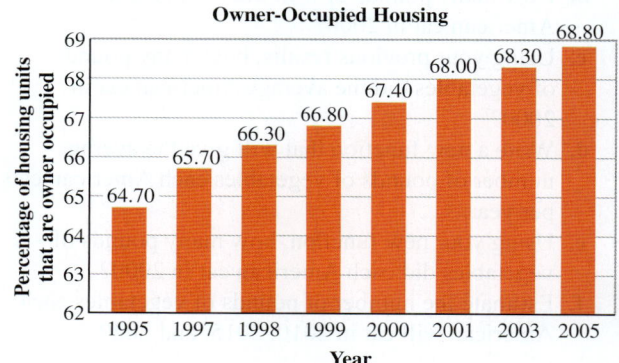

Source: Statistical Abstract of the United States, 2008.

   a. Let $O(t)$ be the percentage of occupied housing units that are owner occupied t years since 1990. Find a model for $O(t)$.
   b. Let $R(t)$ be the percentage of occupied housing units that are renter occupied t years since 1990. Using your model from part a, find a model for $R(t)$.
   c. Find $R(20)$ and explain its meaning in this situation.
   d. What is the slope of $R(t)$? Explain its meaning in this situation.

64. The percentage of all drug prescriptions that are written for brand-name drugs is given in the table.

Year	Percent for Brand-Name Drugs
2002	57.9
2003	55.0
2004	51.9
2005	48.7
2006	45.7

Source: Statistical Abstract of the United States, 2008.

a. Let $B(t)$ be the percentage of all drug prescriptions that are written for brand-name drugs t years since 2000. Find a model for $B(t)$.
b. Let $G(t)$ be the percentage of all drug prescriptions that are written for generic drugs t years since 2000. Using your model from part a, find a model for $G(t)$.
c. Find $G(15)$. Explain its meaning.
d. What is the slope of $G(t)$? Explain its meaning.

65. The area of a Norman window (a rectangle with a semicircle on top) can be determined by adding the area functions of the rectangle and semicircle pieces it is made of. If the rectangle part has a height of 75 inches, find the following.

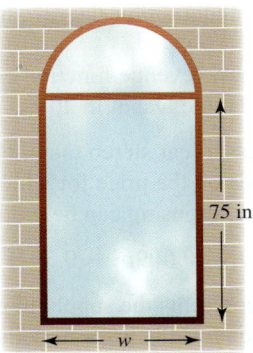

a. A function for the total area of the Norman window.
b. The area of the Norman window if its width is 36 inches.

66. If the rectangle part of a Norman window has a height of 80 inches, find the following.
a. A function for the total area of the Norman window.
b. The area of the Norman window if its width is 45 inches.

67. Let $f(x) = 3x + 8$ and $g(x) = -6x + 9$.
a. Find $f(x) + g(x)$.
b. Find $f(x) - g(x)$.
c. Find $f(x)g(x)$.
d. Find $\dfrac{f(x)}{g(x)}$.

68. Let $f(x) = 19x + 28$ and $g(x) = -17x + 34$.
a. Find $f(x) + g(x)$.
b. Find $g(x) - f(x)$.
c. Find $f(x)g(x)$.
d. Find $\dfrac{f(x)}{g(x)}$.

69. Let $f(x) = \dfrac{1}{2}x + \dfrac{1}{5}$ and $g(x) = 2x - \dfrac{3}{5}$.
a. Find $f(x) + g(x)$.
b. Find $g(x) - f(x)$.
c. Find $f(x)g(x)$.
d. Find $\dfrac{f(x)}{g(x)}$.

70. Let $f(x) = \dfrac{2}{3}x + \dfrac{1}{3}$ and $g(x) = 5x - 3$.
a. Find $f(x) + g(x)$.
b. Find $f(x) - g(x)$.
c. Find $f(x)g(x)$.
d. Find $\dfrac{f(x)}{g(x)}$.

71. Let $f(x) = 3x + 5$ and $g(x) = x^2 + 4x + 10$.
a. Find $f(x) + g(x)$.
b. Find $g(x) - f(x)$.
c. Find $f(x)g(x)$.
d. Find $\dfrac{f(x)}{g(x)}$.

72. Let $f(x) = 4x - 7$ and $g(x) = x^2 - 3x + 12$.
a. Find $f(x) + g(x)$.
b. Find $f(x) - g(x)$.
c. Find $f(x)g(x)$.
d. Find $\dfrac{f(x)}{g(x)}$.

For Exercises 73 and 74, explain what the student did wrong when combining the given functions. Give the correct answer.

$$f(x) = 3x - 10 \quad \text{and} \quad g(x) = 7x + 8$$

73. Find $f(x) + g(x)$.

Jose
$f(x) + g(x) = (3x - 10) + (7x + 8)$
$= 10x - 2$
$0 = 10x - 2$
$2 = 10x$
$\dfrac{2}{10} = x$
$\dfrac{1}{5} = x$

74. Find $f(x)g(x)$.

> **Martha**
> $f(x)g(x) = (3x - 10)(7x + 8)$
> $= 10x - 2$

75. The revenue and costs for goods sold by International Business Machines (IBM) can be modeled by the functions

    $R(t) = 3.1t^2 - 33.7t + 180.9$
    $C(t) = 2.5t^2 - 28.9t + 136.1$

    where $R(t)$ represents the revenue for goods sold in billions of dollars and $C(t)$ represents the costs of selling those goods in billions of dollars t years since 2000.
    Source: Models derived from data found at finance.google.com

    a. Find the revenue and costs for IBM in 2010.
    b. Find a new function that will give the profit of IBM. (*Hint*: Profit equals revenue minus costs.)
    c. Use the new profit function to find the profit in 2010.

76. The revenue and costs for goods sold by Nike, Inc. can be modeled by the functions

    $R(t) = 1.3t + 7.0$
    $C(t) = 0.7t + 4.5$

    where $R(t)$ represents the revenue for goods sold in billions of dollars and $C(t)$ represents the costs of selling those goods in billions of dollars t years since 2000.
    Source: Models derived from data found at finance.google.com

    a. Find the revenue and costs for Nike in 2009.
    b. Find a new function that will give the gross profit of Nike. (*Hint*: Gross profit equals revenue minus costs.)
    c. Use the new profit function to find the profit in 2009.

77. The revenue and profit for Pearson Publishing Company can be modeled by

    $R(t) = 164.1t^2 - 1353.9t + 6380.0$
    $P(t) = 87.9t^2 - 708.9t + 3322.8$

    where $R(t)$ represents the revenue in millions of British pounds and $P(t)$ represents the profit in millions of British pounds for Pearson t years since 2000.
    Source: Models derived from data found at finance.google.com

    a. Calculate the revenue and profit earned by Pearson in 2011.
    b. Find a new function for the costs at Pearson Publishing.
    c. Using the new cost function determine the costs at Pearson Publishing in 2011.

78. The cost and profit for Home Depot, Inc. can be modeled by

    $C(t) = -0.9t^2 + 10.56t + 21.7$
    $P(t) = -0.6t^2 + 7.1t + 5.8$

    where $C(t)$ represents the cost in billions of dollars and $P(t)$ represents the profit in billions of dollars for Home Depot, Inc. t years since 2000.
    Source: Models derived from data found at finance.google.com

    a. Calculate the costs and profit of Home Depot in 2010.
    b. Find a new function for the revenue at Home Depot Inc.
    c. Using the new revenue function estimate the revenue at Home Depot in 2010.

79. The willingness of car stereo manufacturers to produce stereos depends on the price for which the stereos can be sold. This relationship can be modeled by

    $S(p) = 0.009p^2 - 0.5p + 20$

    where $S(p)$ represents the supply of car stereos in thousands when p is the price in dollars and the price is greater than $30.

    a. Find the number of car stereos that manufacturers are willing to supply when the price is $100.
    b. Find a function for the projected revenue if all the car stereos that are supplied sell at price p.
    c. Estimate the projected revenue if car stereos are sold for $90 each.

80. The willingness of sports equipment manufacturers to produce tennis rackets depends on the price for which the tennis rackets can be sold. This relationship can be modeled by

    $S(p) = 0.09p^2 - 4.2p + 51$

    where $S(p)$ represents the supply of tennis rackets in millions when p is the price in dollars and the price is greater than $25.

    a. Find the number of tennis rackets that manufacturers are willing to supply when the price is $40.
    b. Find a function for the projected revenue if all the tennis rackets that are supplied sell at price p.
    c. Estimate the projected revenue if tennis rackets are sold for $75 each.

81. The willingness of car manufacturers to produce minivans depends on the price for which the minivans can be sold. This can be modeled by

$$S(p) = 0.11p^2 - 3.2p + 45$$

where $S(p)$ represents the supply of minivans in thousands when p is the price in thousands of dollars and the price is greater than $14,500.

   a. Find the number of minivans manufacturers are willing to supply when the price is $25,000.
   b. Find a function for the projected revenue if all the minivans that are supplied sell at price p.
   c. Estimate the projected revenue if minivans are sold for $31,000 each.

82. The willingness of a computer manufacturer to produce laptop computers depends on the price for which laptops can be sold. This economic relationship can be modeled by

$$S(p) = 55.7p^2 - 728.6p + 4321.6$$

where $S(p)$ represents the supply of laptop computers when p is the price in hundreds of dollars and the price is greater than $700.

   a. Find the number of laptop computers that manufacturers are willing to supply when the price is $1500.
   b. Find a function for the projected revenue if all the laptop computers that are supplied sell at price p.
   c. Estimate the projected revenue if laptop computers are sold for $3000 each.

83. The total number of immigrants admitted to the United States as permanent residents under refugee acts t years since 1990 can be represented by

$$I(t) = -17854t^2 + 212449t - 501784$$

Of those admitted under refugee acts, the number coming from Europe can be represented by

$$E(t) = -6432t^2 + 74006t - 161816$$

Source: Models derived from data in Statistical Abstract of the United States, 2001.

   a. Estimate the total number of immigrants admitted to the United States as permanent residents under refugee acts in 1995.
   b. Estimate the number of immigrants admitted under refugee acts who were from Europe in 1995.
   c. Find a new function to represent the number of non-European immigrants who were admitted to the United States as permanent residents under refugee acts.
   d. Use your model to determine the number of non-European immigrants who were admitted under refugee acts in 1998.

84. The number of men and women in the United States since 1800 can be modeled by

$$M(t) = 2.86t^2 + 97.22t - 137.68$$
$$F(t) = 3.65t^2 - 48.46t + 4883.84$$

where $M(t)$ represents the number of men in thousands and $F(t)$ represents the number of females in thousands in the United States t years since 1800.
Source: Models derived from data in Statistical Abstract of the United States, 2001.

   a. Estimate the number of males and females in the United States in 2000.
   b. Find a function for the total population in the United States.
   c. Use the new function to estimate the total population in the United States in 2002.

85. The number of U.S. residents who are Caucasian can be modeled by

$$C(t) = 1.693t + 209.107$$

where $C(t)$ represents the number of U.S. residents who are Caucasian in millions t years since 1990.

The number of Caucasian residents who are 18 years old or older can be modeled by

$$O(t) = 0.014t^2 + 1.169t + 157.528$$

where $O(t)$ represents the number of Caucasian residents 18 years old or older in millions t years since 1990.
Source: Models derived from data in Statistical Abstract of the United States, 2001.

   a. Find a new function for the number of Caucasian residents who are under 18 years old.
   b. Estimate the number of Caucasian residents under 18 years old in 2002.

86. The number of U.S. residents who are Caucasian and 65 years old or older can be modeled by

$$R(t) = -0.011t^2 + 0.409t + 28.098$$

where $R(t)$ represents the number of U.S. residents who are Caucasian and 65 years old or older in millions t years since 1990.
Source: Models derived from data in Statistical Abstract of the United States, 2001.

   a. Using the function(s) from Exercise 85 for the number of Caucasian residents, find a new function for the number of Caucasian residents is younger than 65 years old but at least 18 years old.

   b. Estimate the number of Caucasian residents younger than 65 years old but at least 18 years old in 2002.

87. Let $f(x) = 3x + 8$ and $g(x) = -6x + 9$.

   a. Find $f(5) + g(5)$.
   b. Find $f(5) - g(5)$.
   c. Find $f(5)g(5)$.
   d. Find $\dfrac{f(5)}{g(5)}$.

88. Let $f(x) = 19x + 28$ and $g(x) = -17x + 34$.

   a. Find $f(2) + g(2)$.
   b. Find $g(2) - f(2)$.
   c. Find $f(2)g(2)$.
   d. Find $\dfrac{f(2)}{g(2)}$.

89. Let $f(x) = \dfrac{1}{2}x + \dfrac{1}{5}$ and $g(x) = 2x - \dfrac{3}{5}$.

   a. Find $f(2) + g(2)$.
   b. Find $g(2) - f(2)$.
   c. Find $f(2)g(2)$.
   d. Find $\dfrac{f(2)}{g(2)}$.

90. Let $f(x) = \dfrac{2}{3}x + \dfrac{1}{3}$ and $g(x) = 5x - 3$.

   a. Find $f(4) + g(4)$.
   b. Find $f(4) - g(4)$.
   c. Find $f(4)g(4)$.
   d. Find $\dfrac{f(4)}{g(4)}$.

91. Let $f(x) = 3x + 5$ and $g(x) = x^2 + 4x + 10$.

   a. Find $f(5) + g(5)$.
   b. Find $g(5) - f(5)$.
   c. Find $f(5)g(5)$.
   d. Find $\dfrac{f(5)}{g(5)}$.

92. Let $f(x) = 4x - 7$ and $g(x) = x^2 - 3x + 12$.

   a. Find $f(3) + g(3)$.
   b. Find $f(3) - g(3)$.
   c. Find $f(3)g(3)$.
   d. Find $\dfrac{f(3)}{g(3)}$.

For Exercises 93 through 96, determine which of the following the given notation is asking you to find.

   a) *A new function*
   b) *To evaluate a combination of functions to find a single value*
   c) *To solve for x*

93. Find x such that $f(x) = 20$.

94. Find $f(9)g(9)$.

95. Find $f(x) + g(x)$.

96. Find $f(2) + g(2)$.

97. Let $f(x) = 5x + 7$ and $g(x) = 4x + 1$.

   a. Find $g(x) + f(x)$.
   b. Find $g(2) + f(2)$.
   c. Find $f(x)g(x)$.
   d. Find $f(8)g(8)$.

98. Let $f(x) = -3x - 12$ and $g(x) = 6x - 15$.

   a. Find $f(4) - g(4)$.
   b. Find $f(x) - g(x)$.
   c. Find $\dfrac{f(x)}{g(x)}$.
   d. Find $\dfrac{f(5)}{g(5)}$.

99. Let $f(x) = x + 2$ and $g(x) = 4x^2 + x + 1$.

   a. Find $g(x) - f(x)$.
   b. Find $g(2) - f(2)$.
   c. Find $f(x)g(x)$.
   d. Find $f(2)g(2)$.

100. Let $f(x) = 5x - 2$ and $g(x) = 2x^2 - 7x + 6$.

   a. Find $f(4) + g(4)$.
   b. Find $f(x) + g(x)$.
   c. Find $f(4)g(4)$.
   d. Find $f(x)g(x)$.

3.3 Composing Functions

LEARNING OBJECTIVES
- Combine functions using composition.
- Combine functions using composition within an application.

Many things in life require us to do a step-by-step process in which a subsequent step depends on the result of the previous step(s). A simple example of this would be if we are trying to get a drink from a vending machine that does not take dollar bills. If all we have is dollar bills, we will first get change for one or more bills and then use that change to purchase the drink. The process of buying that drink is a combination of getting change and using the vending machine to get the drink. Today most vending machines have combined these two steps into one by combining a change machine and the vending machine into one machine so that dollar bills can be used to make a purchase.

In mathematics, this way of combining functions is called a **composition** of functions. Whenever one function, or the output of the function, is substituted into another function, we are composing the two functions.

Example 1 Composing two functions

A company has analyzed data and come up with the following two functions.

$C(e)$ = Cost of health insurance in dollars for a company with e employees

$E(t)$ = Number of employees at the company in year t

Find a function that will give the company's cost for health insurance in year t.

SOLUTION
If the company wants to know the cost for health insurance for a certain year, they cannot immediately get that information from either of these functions. To use $C(e)$, they need to know the number of employees. They can use $E(t)$ to find the number of employees in year t. Because both e and $E(t)$ represent the number of employees, we have

$$e = E(t)$$

We can substitute $E(t)$ for e and get

$$C(e) = C(E(t))$$

What's That Mean?

Function Notation for Composition

Two notations are used for the composition of functions:

$$(f \circ g)(x)$$
or
$$f(g(x))$$

These are both read as the function f of g of x.

The function $g(x)$ is being substituted into the input of the function $f(x)$.

This substitution is a composition of these two functions. $C(E(t))$ is the cost of health insurance for this company in year t.

This notation can be understood by starting from the inner parentheses and working out. $C(E(t))$ says that t is the input for the function and C is the final output of the function. Therefore, the input is a year t, and the output is the cost C of health insurance for this company.

Example 2 An application of composing functions

NatureJuice, Inc. processes oranges for several Florida citrus growers. NatureJuice, Inc. is concerned with their annual cost projections and is working with the following two functions.

$$C(t) = 135.87t + 25000$$

where $C(t)$ is the annual cost, in dollars, to process t thousand tons of oranges per year, and

$$T(y) = 2.4y - 4500$$

where $T(y)$ is the annual production of oranges, in thousands of tons, in year y.

a. Find a function that will calculate the annual cost to process all oranges in year y.

b. Find the annual cost to process oranges in 2015.

c. When was the annual cost to process oranges $72,000?

SOLUTION

a. NatureJuice, Inc. wants a function that has an output of annual cost but an input of years. The function $T(y)$ depends on the year y and has an output of thousands of tons of oranges. The function $T(y)$ and the variable t both represent thousands of tons of oranges, so

$$T(y) = t$$

Compose $T(y)$ with $C(t)$ to get the annual cost when given the year y. $C(T(y))$ will give the annual cost, in dollars, to process all oranges in year y.

$C(T(y)) = 135.87(2.4y - 4500) + 25000$
$C(T(y)) = 326.088y - 611415 + 25000$ Distribute 135.87.
$C(T(y)) = 326.088y - 586415$ Combine like terms.

b. We can do this problem in two ways. We will do it once using the two functions given separately and then a second time using the newly composed function we found in part a.

  i. Using the two equations separately, first find the number of oranges produced in 2015.

$$T(y) = 2.4y - 4500$$
$$T(2015) = 2.4(2015) - 4500$$
$$T(2015) = 4836 - 4500$$
$$T(2015) = 336$$

Therefore, 336 thousand tons of oranges will be produced in 2015. Now substitute this into the cost equation.

$$C(t) = 135.87t + 25000$$
$$C(336) = 135.87(336) + 25000$$
$$C(336) = 45652.32 + 25000$$
$$C(336) = 70652.32$$

Therefore, the annual cost to process oranges in 2015 will be about $70,652.32.

ii. Using the new composed function from part a, we have

$$C(T(y)) = 326.088y - 586415$$
$$C(T(2015)) = 326.088(2015) - 586415$$
$$C(T(2015)) = 657067.32 - 586415$$
$$C(T(2015)) = 70652.32$$

This is the same result as the first method. It just takes fewer steps once the two functions have been composed. If we had to do this same calculation for several years, perhaps to show a trend in the costs from year to year, the composed function would be much more efficient.

c. Using the new composed function, we have

$$C(T(y)) = 326.088y - 586415$$
$$72000 = 326.088y - 586415$$
$$\underline{586415 \qquad\qquad 586415}$$
$$658415 = 326.088y$$
$$\frac{658415}{326.088} = \frac{326.088y}{326.088}$$
$$2019.13 \approx y$$

In the year 2019, the annual cost to process oranges was about $72,000.

PRACTICE PROBLEM FOR EXAMPLE 2

Consider the following functions for an urban area in the United States.

$$P(t) = 1.2t + 3.4$$

where $P(t)$ is the population of the urban area in millions of people t years since 2000 and

$$A(p) = 15.48p - 104$$

where $A(p)$ is the amount of air pollution in parts per million when p million people live in the urban area.

a. Find the population of this urban area in 2005.

b. Find the amount of air pollution for this urban area in 2005.

c. Find a new function that will give the amount of air pollution for this urban area depending on the number of years since 2000.

d. Use the function that you found in part c to find the air pollution in 2005, 2007, and 2010.

In composing two functions, it is very important to check that the output of one function is the same as the input needed by the other function. This is the only way to compose the two functions together.

> **Combining Functions Using Composition**
> - The input for one function must be the same as the output of the other function.
> - The inputs do **not** have to be the same.

Example 3 Composing functions algebraically

Combine the following functions.

$$f(x) = 3x + 9 \qquad g(x) = 2x - 7 \qquad h(x) = 3x^2 + 5x - 10$$

a. $f(g(x))$ **b.** $(g \circ f)(x)$ **c.** $h(g(x))$

SOLUTION

a. We will take the entire function $g(x)$ and substitute it into the input variable for $f(x)$.

$$f(x) = 3x + 9$$
$$f(g(x)) = 3(2x - 7) + 9$$
$$f(g(x)) = 6x - 21 + 9$$
$$f(g(x)) = 6x - 12$$

b. Substitute $f(x)$ into the input variable for the function $g(x)$.

$$g(x) = 2x - 7$$
$$(g \circ f)(x) = 2(3x + 9) - 7$$
$$(g \circ f)(x) = 6x + 18 - 7$$
$$(g \circ f)(x) = 6x + 11$$

When the two functions in parts a and b are composed in different orders, they do not end with the same function. This implies that you should be cautious when deciding which function goes into which.

c. Substitute $g(x)$ into the input variable for the function $h(x)$. Be sure to substitute $g(x)$ into every x of the $h(x)$ function.

$$h(x) = 3x^2 + 5x - 10$$
$$h(g(x)) = 3(2x - 7)^2 + 5(2x - 7) - 10$$
$$h(g(x)) = 3(2x - 7)(2x - 7) + 5(2x - 7) - 10 \qquad \text{Use FOIL to square the } 2x - 7.$$
$$h(g(x)) = 3(4x^2 - 14x - 14x + 49) + 10x - 35 - 10$$
$$h(g(x)) = 3(4x^2 - 28x + 49) + 10x - 45 \qquad \text{Distribute the 3.}$$
$$h(g(x)) = 12x^2 - 84x + 147 + 10x - 45$$
$$h(g(x)) = 12x^2 - 74x + 102 \qquad \text{Combine like terms.}$$

PRACTICE PROBLEM FOR EXAMPLE 3

Combine the following functions.

$$f(x) = 4x + 5 \qquad g(x) = 5x - 8 \qquad h(x) = 2x^2 + 4x - 1$$

a. $f(g(x))$ **b.** $(h \circ g)(x)$

■ **Connecting the Concepts**

Is composition of functions commutative?

Composing two functions in the opposite order does not always result in the same new function.

$$f(g(x))$$
and
$$g(f(x))$$

Note that the answers to Examples 3a and 3b are not the same. Therefore, composition of functions is not commutative.

SECTION 3.3 Composing Functions 255

Example 4 An application of composing functions

The amount of drywall damaged by a small leak in a pipe can be modeled by the function

$$D(h) = 0.75h + 1$$

where $D(h)$ is the drywall damaged in square feet h hours after the leak started. The cost to repair drywall damage from a water leak can be modeled by

$$C(a) = 20a + 150$$

where $C(a)$ is the cost in dollars to repair a square feet of water damaged drywall.

a. Find the cost to repair 3 square feet of water damage to drywall.

b. Find a new function that will give the cost to repair the drywall damage depending on the number of hours the leak continued.

c. Use the function that you found to determine the cost to repair the drywall damage caused by a leak that has been going on for 2 days.

SOLUTION

a. We are given the amount of damage as 3 square feet, so we can substitute that into the $C(a)$ function.

$$C(3) = 20(3) + 150$$
$$C(3) = 210$$

Therefore, the cost to repair 3 square feet of drywall damage is about $210.

b. We want to find a cost function that depends on the hours a water leak continued. The cost function that we have has an input of square feet of damaged drywall, not hours. The function $D(h)$ has an input of hours and output of drywall damage. Since $D(h)$ represents square feet of drywall damage and the variable a is also square feet of drywall damage, we also have $D(h) = a$, so we can compose the two functions.

$$C(a) = 20a + 150$$
$$C(D(h)) = 20(0.75h + 1) + 150$$
$$C(D(h)) = 15h + 20 + 150$$
$$C(D(h)) = 15h + 170$$

$C(D(h))$ is the cost to repair the drywall damage depending on the number of hours the leak continued.

c. We are given a time of 2 days. Days must be converted to hours, so it can be substituted for h. Because 2 days is 48 hours, we will substitute 48 for h and find the cost.

$$C(D(48)) = 15(48) + 170$$
$$C(D(48)) = 890$$

The drywall damage caused by a leak that continues for 2 days will cost about $890 to repair.

PRACTICE PROBLEM FOR EXAMPLE 4

The number of snowmobiles entering Yellowstone National Park on a winter weekend can be modeled by the function

$$S(h) = -0.04h^2 + 1.11h - 5.76$$

where $S(h)$ is the number of snowmobiles in hundreds h hours after midnight. The carbon monoxide (CO) concentration at Old Faithful can be modeled by

$$C(n) = 0.68n + 0.10$$

where $C(n)$ is the CO concentration at Old Faithful in parts per million (ppm) when n hundred snowmobiles are visiting the park.

a. Find a new function that will give the CO concentration at Old Faithful depending on the number of hours after midnight.

b. Find the CO concentration at Old Faithful at 3:00 P.M.

3.3 Exercises

1. Let $f(x) = 2x + 4$ and $g(x) = 6x + 9$.
   a. Find $f(g(x))$.
   b. Find $g(f(x))$.

2. Let $f(x) = 5x - 6$ and $g(x) = 7x + 3$.
   a. Find $f(g(x))$.
   b. Find $g(f(x))$.

3. Let $f(x) = 3x + 8$ and $g(x) = -6x + 9$.
   a. Find $(f \circ g)(x)$.
   b. Find $(g \circ f)(x)$.

4. Let $f(x) = 19x + 28$ and $g(x) = -17x + 34$.
   a. Find $(f \circ g)(x)$.
   b. Find $(g \circ f)(x)$.

5. Far North Manufacturing is starting to manufacture a new line of toys for this Christmas season. The number of toys they can manufacture each week can be modeled by

   $$T(w) = 500w + 3000$$

   where $T(w)$ represents the number of toys manufactured during week w of production. The total weekly cost for manufacturing these toys can be modeled by

   $$C(t) = 1.75t + 5000$$

   where $C(t)$ represents the total weekly cost in dollars from producing t toys a week.
   a. Use these functions to find a new function that will give the total weekly cost for week w of production.
   b. Find the number of toys produced during the fifth week of production.
   c. Find the total weekly cost to produce 5000 toys.
   d. Find the total weekly cost for the seventh week of production.
   e. In what week will the total weekly cost reach $18,500.00?

6. The number of custom computer systems sold to the biotech industry by a new computer manufacturer m months after starting the business can be modeled by

   $$C(m) = 4m + 2$$

   where $C(m)$ represents the number of computer systems sold during the mth month after starting the business. The monthly profit made by the new computer manufacturer can be modeled by

   $$P(c) = 300c - 6000$$

   where $P(c)$ represents the monthly profit in dollars when c computers are sold during a month.
   a. Use these functions to find a new function that will give the monthly profit for m months after starting the business.
   b. Find the number of computer systems sold during the fifth month after starting the business.
   c. Find the total monthly profit if 20 computer systems are sold in a month.
   d. Find the total monthly profit for the tenth month after the start of business.
   e. In what month will the monthly profit reach $7800.00?

7. West Tech, Inc.'s vacation policy increases each employee's vacation time on the basis of the number of years he or she has been employed at the company. The number of weeks of vacation an employee gets in a year can be modeled by

   $$v(y) = 0.25y + 1.5$$

   weeks of vacation after working for the company for y years. West Tech, Inc.'s cost of an employee's vacation time in a year can be modeled by

   $$C(w) = 1500w + 575$$

   dollars when w weeks of vacation are taken.

a. Find the number of weeks of vacation a West Tech employee will get per year after working with the company for 10 years.
b. What is West Tech's cost for a 10-year employee's vacation?
c. Use these functions to find a new function that determines the cost for vacation taken by an employee who has been with the company for y years. Assume that the employee will take all of his or her vacation time.
d. What is West Tech's cost for a 20-year employee's vacation?
e. What is West Tech's cost for a 30-year employee's vacation?

8. The number of employees working at a small magazine publisher can be modeled by

$$E(m) = 3m + 2$$

where $E(m)$ represents the number of employees m months after starting to publish the magazine. The total cost for health benefits at the publisher can be modeled by

$$C(e) = 1500e + 2000$$

where $C(e)$ represents the total monthly cost for health benefits in dollars when e employees work for the publisher.

a. Find the number of employees at the publisher during the sixth month after the start of publishing the magazine.
b. Find the total cost for health benefits when the publisher has 10 employees.
c. Use these functions to find a new function that determines the total monthly cost for health benefits for the publisher m months after the start of publishing the magazine.
d. What is the total monthly cost of health benefits 8 months after the start of publishing the magazine?
e. Find the total cost for health benefits during the first year of publishing the magazine. (*Hint*: You will need to add the monthly cost for each month of the year.)

9. The number of people who visit a local Renaissance fair can be modeled by

$$A(d) = -400d^2 + 3500d - 1000$$

where $A(d)$ represents the attendance at the Renaissance fair d days after it opens. The profit made by the fair promoters can be modeled by

$$P(a) = 2a - 2500$$

where $P(a)$ represents the profit in dollars for the Renaissance fair on a day when a people attend.

a. Find the number of people in attendance on the fourth day of the Renaissance fair.
b. Find the profit made on the fourth day of the fair.
c. Use these functions to find a new function that will give the profit made at the Renaissance fair d days after the fair opens.
d. How much profit does the fair make on the third day of the fair?
e. If the fair lasts a total of 7 days, find the total amount of profit for the Renaissance fair. (*Hint*: You will need to add up the profit from each day of the fair.)

10. The depth of the water flowing in a local stream w weeks after the start of spring can be modeled by

$$D(w) = -0.75w^2 + 10w + 2$$

where $D(w)$ represents the depths in inches of the stream. The amount of water per minute going over a waterfall at the end of the same stream can be modeled by

$$G(d) = 25d + 1$$

where $G(d)$ represents the amount of water going over the waterfall in gallons per minute when the stream is d inches deep.

a. Find the depth of the stream 2 weeks after the start of spring.
b. Find the amount of water going over the waterfall 2 weeks after the start of spring.
c. Use these functions to find a new function that will give the amount of water going over the waterfall w weeks after the start of spring.
d. Find the amount of water going over the waterfall 8 weeks after the start of spring.

For Exercises 11 through 14, use function notation to show how you would compose the two functions to get the requested new function. List the units for the inputs and outputs of the new function.

11. Let $P(b)$ be the profit, in thousands of dollars, that Pride Bike Co. makes if they sell b bikes per year. Let $B(t)$ be the number of bikes that Pride Bike Co. sells in year t. Find a function for the amount of profit Pride Bike Co. makes in year t.

12. Let $D(v)$ be the amount of environmental damage, in thousands of dollars, done to a national park when v visitors come to the park in a year. Let $V(t)$ be the number of visitors who come to a national park in year t. Find a function that gives the amount of environmental damage at a national park in year t.

13. Let $T(d)$ be the low night-time temperature in degrees Fahrenheit on day d of the week. Let $H(t)$ be the number of homeless who seek space in a local homeless shelter when the low night-time temperature is t degrees Fahrenheit. Find a function for the number of homeless who seek space at the homeless shelter on day d of the week.

14. Let $B(m)$ be the advertising budget in dollars for a new product m months after the product launch. Let $R(a)$ be the monthly revenue in dollars from selling this product when the advertising budget for the month is a dollars. Find a function that gives the monthly revenue from selling this product m months after the product launch.

15. Let $f(x) = 4x + 5$ and $g(x) = 2x + 6$.
    a. Find $f(g(x))$.
    b. Find $g(f(x))$.

16. Let $f(x) = 3x - 8$ and $g(x) = 4x + 7$.
    a. Find $f(g(x))$.
    b. Find $g(f(x))$.

17. Let $f(x) = 7x + 11$ and $g(x) = -8x + 15$.
    a. Find $(f \circ g)(x)$.
    b. Find $(g \circ f)(x)$.

18. Let $f(x) = 11x + 21$ and $g(x) = -15x + 41$.
    a. Find $(f \circ g)(x)$.
    b. Find $(g \circ f)(x)$.

19. Let $f(x) = \frac{1}{2}x + \frac{1}{5}$ and $g(x) = 2x - \frac{3}{5}$.
    a. Find $f(g(x))$.
    b. Find $g(f(x))$.

20. Let $f(x) = \frac{2}{3}x + \frac{1}{3}$ and $g(x) = 5x - 3$.
    a. Find $f(g(x))$.
    b. Find $g(f(x))$.

21. Let $f(x) = 0.68x + 2.36$ and $g(x) = 3.57x + 6.49$.
    a. Find $f(g(x))$.
    b. Find $g(f(x))$.

22. Let $f(x) = -0.6x - 3.2$ and $g(x) = -7x - 4.8$.
    a. Find $f(g(x))$.
    b. Find $g(f(x))$.

23. Let $f(x) = 3x + 5$ and $g(x) = x^2 + 4x + 10$.
    a. Find $f(g(x))$.
    b. Find $g(f(x))$.

24. Let $f(x) = 4x - 7$ and $g(x) = x^2 - 3x + 12$.
    a. Find $f(g(x))$.
    b. Find $g(f(x))$.

25. Let $f(x) = x + 2$ and $g(x) = 4x^2 + x + 1$.
    a. Find $(f \circ g)(x)$.
    b. Find $(g \circ f)(x)$.

26. Let $f(x) = 5x - 2$ and $g(x) = 2x^2 - 7x + 6$.
    a. Find $(f \circ g)(x)$.
    b. Find $(g \circ f)(x)$.

For Exercises 27 through 30, explain what the student did wrong when combining the given functions. Give the correct answer.

$$f(x) = 3x - 10 \quad \text{and} \quad g(x) = 7x + 8$$

27. Find $f(g(x))$.

Lilybell
$f(g(x)) = (3x - 10)(7x + 8)$
$= 21x^2 + 24x - 70x - 80$
$= 21x^2 - 46x - 80$

28. Find $g(f(x))$.

Jeff
$g(f(x)) = 7(3x - 10)$
$= 21x - 70$

29. Find $f(g(x))$.

Colter
$f(g(x)) = 7(3x - 10) + 8$
$= 21x - 70x + 8$
$= 21x - 62$

30. Find $f(g(x))$.

Whitney
$f(g(x)) = 3(g(x)) - 10$
$= 3gx - 10$

31. Let $f(x) = 3x + 8$ and $g(x) = -6x + 9$.
    a. Find $f(g(5))$.
    b. Find $g(f(5))$.

32. Let $f(x) = 19x + 28$ and $g(x) = -17x + 34$.
    a. Find $f(g(2))$.
    b. Find $g(f(2))$.

33. Let $f(x) = \frac{1}{2}x + \frac{1}{5}$ and $g(x) = 2x - \frac{3}{5}$.
    a. Find $f(g(2))$.
    b. Find $g(f(2))$.

34. Let $f(x) = \frac{2}{3}x + \frac{1}{3}$ and $g(x) = 5x - 3$.
    a. Find $f(g(4))$.
    b. Find $g(f(4))$.

35. Let $f(x) = 0.68x + 2.36$ and $g(x) = 3.57x + 6.49$.
    a. Find $(f \circ g)(-3)$.
    b. Find $(g \circ f)(-3)$.

36. Let $f(x) = -0.6x - 3.2$ and $g(x) = -7x - 4.8$.
    a. Find $(f \circ g)(7)$.
    b. Find $(g \circ f)(7)$.

37. Let $f(x) = 3x + 5$ and $g(x) = x^2 + 4x + 10$.
    a. Find $f(g(5))$.
    b. Find $g(f(5))$.

38. Let $f(x) = 4x - 7$ and $g(x) = x^2 - 3x + 12$.
    a. Find $f(g(3))$.
    b. Find $g(f(3))$.

39. Let $f(x) = x + 2$ and $g(x) = 4x^2 + x + 1$.
    a. Find $f(g(2))$.
    b. Find $g(f(2))$.

40. Let $f(x) = 5x - 2$ and $g(x) = 2x^2 - 7x + 6$.
    a. Find $f(g(4))$.
    b. Find $g(f(4))$.

For Exercises 41 through 46, assume that the functions refer to a specific tour company offering European tours and that t represents the year.

- Let $T(t)$ be the total number of people on the tour.
- Let $K(t)$ be the number of children under 12 years old on the tour.
- Let $C(k)$ be the cost for k children under 12 years old to take the tour.
- Let $A(a)$ be the cost for a people over 12 years old to take the tour.
- Let $B(n)$ be the number of buses needed when n people travel on the tour.

Use function notation to show how you would combine the given functions by adding, subtracting, multiplying, dividing, or composing. List the units for the inputs and outputs of the new function.

41. Find a function that gives the number of people over 12 years old who go on the tour in year t.

42. Find a function that gives the cost for all children under 12 years old who traveled on the tour in year t.

43. Find a function that gives the cost for all people over 12 years old who traveled on the tour in year t.

44. Find a function that gives the total cost for all people who traveled on the tour in year t.

45. Find a function that gives the difference between the cost for all children under 12 years old and the cost for all people over 12 years old who went on the tour in year t.

46. Find a function that gives the number of buses needed for the tour in year t.

47. Let $f(x) = 9x + 2$ and $g(x) = 3x + 7$.
    a. Find $f(x)g(x)$.
    b. Find $f(x) + g(x)$.
    c. Find $f(g(x))$.

48. Let $f(x) = -7x + 3$ and $g(x) = 11x + 2$.
    a. Find $g(x) - f(x)$.
    b. Find $f(x)g(x)$.
    c. Find $g(f(x))$.

49. Let $f(x) = -9x + 1$ and $g(x) = x + 8$.
    a. Find $f(3) + g(3)$.
    b. Find $f(4)g(4)$.
    c. Find $f(g(7))$.

50. Let $f(x) = 12x - 5$ and $g(x) = -5x - 7$.
    a. Find $f(g(2))$.
    b. Find $f(6) - g(6)$.
    c. Find $\dfrac{f(6)}{g(6)}$

51. Let $f(x) = \frac{2}{3}x + \frac{1}{3}$ and $g(x) = \frac{1}{4}x + \frac{3}{4}$.
    a. Find $f(g(x))$.
    b. Find $f(42)g(42)$.
    c. Find $f(x) + g(x)$.

52. Let $f(x) = \frac{2}{5}x + \frac{4}{5}$ and $g(x) = \frac{1}{2}x - \frac{1}{2}$.
    a. Find $f(x)g(x)$.
    b. Find $g(f(x))$.
    c. Find $g(100) - f(100)$.

53. Let $f(x) = 1.5x + 4.5$ and $g(x) = 3.5x + 9$.
    a. Find $f(g(x))$.
    b. Find $g(f(6))$.
    c. Find $g(x) - f(x)$.

54. Let $f(x) = -2.4x + 6.2$ and $g(x) = 1.6x + 9$.
    a. Find $f(g(x))$.
    b. Find $f(x)g(x)$.
    c. Find $g(3) - f(3)$.

3.4 Factoring Polynomials

LEARNING OBJECTIVES

- Factor out the GCF.
- Factor by grouping.
- Factor quadratics using the AC method.
- Factor using trial and error.

> **What's That Mean?**
>
> **Factor**
>
> The word *factor* is used in several ways in mathematics.
>
> In multiplying, the expressions being multiplied together are called *factors*.
>
> $$24 = (4)(6)$$
>
> 4 and 6 are called *factors* of 24.
>
> In *factoring*, the act of taking out a common element is sometimes called *to factor out*.
>
> In the phrase
> "*Factor* out the common *factor*,"
> the first word *factor* is an action saying to take out. The second word *factor* is referring to the expression that is going to be taken out.
>
> $$10x^3 + 4x = 2x(5x^2 + 2)$$
>
> $2x$ is the common *factor* that is being *factored* out of the original expression.

Factoring Out the GCF

In Section 3.2, we multiplied two polynomials using the distributive property.

$$a(x + y) = ax + ay$$

When working with polynomials, we sometimes want to "undo" the distributive property by changing a polynomial into a product of two or more factors. The process of changing a polynomial from terms that are added together to factors that are multiplied together is called **factoring**. Factoring is one method that is used to solve some polynomial equations, and it has several other uses in algebra. The basis of factoring is to take a polynomial and rewrite it into simpler parts that are multiplied together.

Standard form	Factored form
$x^2 + 8x + 15$	$(x + 3)(x + 5)$

When the factored form is multiplied out, the expression will return to the standard form.

$$(x + 3)(x + 5) = x^2 + 5x + 3x + 15$$
$$= x^2 + 8x + 15$$

The most basic step in any factoring process is to factor out the common elements from the terms of an expression. When we factor out a common element, it does not go away, but it becomes a factor that is in front of the remaining expression. The common elements of an expression are called a common factor. The terms of the expression $12x + 10$ both are divisible by 2. Therefore, a 2 can be factored out of the expression and be put in front as a factor.

$$12x + 10 = 2(6x + 5)$$

If we multiply the 2 back into the expression $6x + 5$, we would be right back to the original expression, $12x + 10$. The terms in some expressions will have constants, variables, or both variables and constants in common. The **greatest common factor (GCF)** is the largest common factor shared in common with all the terms of the polynomial. When factoring, we will most often factor out the GCF.

Example 1 Factoring out the greatest common factor

Factor out the greatest common factor.

a. $5x^2 + 20x$ **b.** $15w^3z - 9w^2z^2$ **c.** $2x^2 + 6x - 4$
d. $5x(x - 8) + 2(x - 8)$ **e.** $3z(z - 5) - (z - 5)$

SOLUTION

a. Both terms in the expression $5x^2 + 20x$ have at least one x and also are divisible by 5, so we can factor out $5x$ from both terms.

$$5x(x + 4)$$

The first term of this expression had x^2, but one is factored out, leaving a single x.

b. In the expression $15w^3z - 9w^2z^2$, both of the terms have a w^2, have a z, and are divisible by 3. Factoring $3w^2z$ out, we have

$$3w^2z(5w - 3z)$$

c. All of the terms in the polynomial $2x^2 + 6x - 4$ are divisible by 2. Because the last term does not have an x in it, x is not a common factor,

$$2(x^2 + 3x - 2)$$

d. In the expression $5x(x - 8) + 2(x - 8)$, both terms have the expression $(x - 8)$.

$$(x - 8)(5x + 2)$$

e. Both terms in the expression $3z(z - 5) - (z - 5)$ have the expression $(z - 5)$. Remember that the expression $(z - 5)$ has a "hidden" coefficient of -1. When we factor out the $(z - 5)$, a -1 will remain in its place.

$$(z - 5)(3z - 1)$$

PRACTICE PROBLEM FOR EXAMPLE 1

Factor out the greatest common factor.

a. $12t^3 - 20t$ **b.** $6x^2 + 8x - 14$ **c.** $5x(x + 4) - 3(x + 4)$

Factoring by Grouping

Factoring out what is in common will always be the first step in any factoring process. In many cases, there will not be anything in common, and we will move on to other steps.

If a polynomial has four terms, we try a factoring technique called *factoring by grouping*. We will group the first two terms together and factor out the GCF of those two terms. Also we group the last two terms and factor out the GCF. We finish by factoring out the GCF from the remaining expressions.

> **Factor by Grouping**
>
> If a polynomial has four terms, try these steps.
>
> 1. Factor out the greatest common factor.
> 2. Group the first two terms and last two terms.
> 3. Factor out the GCF of the first two terms.
> 4. Factor out the GCF of the last two terms.
> 5. Factor out the GCF of the remaining expressions.

Example 2 Factoring by grouping

Factor by grouping.

a. $2x^2 - 8x + 7x - 28$ **b.** $6a^2 + 8ab + 15a + 20b$

SOLUTION

a. There are four terms in the expression $2x^2 - 8x + 7x - 28$, so we will factor by grouping. There are no common factors for all four terms. Group the first two terms and the last two terms, then factor out the GCFs.

$$2x^2 - 8x + 7x - 28 = (2x^2 - 8x) + (7x - 28) \quad \text{Group the first two and last two terms.}$$
$$= 2x(x - 4) + 7(x - 4) \quad \text{Factor out the GCFs } 2x \text{ and } 7.$$
$$= (x - 4)(2x + 7) \quad \text{Factor out the GCF } (x - 4).$$

To check the factorization, we can multiply out the factored form using the distributive property.

$$(x - 4)(2x + 7) = 2x^2 - 8x + 7x - 28$$

We got the original expression back, so the factored form we found is correct.

b. There are four terms in the expression $6a^2 + 8ab + 15a + 20b$, so we will factor by grouping. There are no common factors for all four terms. Group the first two terms and the last two terms, then factor out the GCFs.

$6a^2 + 8ab + 15a + 20b = (6a^2 + 8ab) + (15a + 20b)$ *Group the first two and last two terms.*

$= 2a(3a + 4b) + 5(3a + 4b)$ *Factor out the GCFs 2a and 5.*

$= (3a + 4b)(2a + 5)$ *Factor out the GCF (3a + 4b).*

To check the factorization, we can multiply out the factored form using the distributive property.

$$(3a + 4b)(2a + 5) = 6a^2 + 8ab + 15a + 20b$$

We got the original expression back, so the factored form we found is correct.

PRACTICE PROBLEM FOR EXAMPLE 2

Factor by grouping.

a. $12x^2 - 9x - 20x + 15$ **b.** $8m^2 + 20mn - 14m - 35n$

Factoring Using the AC Method

One of the most common polynomial functions that we will work with is called a *quadratic*. Quadratic functions are second-degree polynomials of the form

$$f(x) = ax^2 + bx + c$$

where a, b, and c are real numbers and $a \neq 0$. In Chapter 4, we will see how factoring a quadratic in this form will allow us to solve quadratic equations. Many methods are used to factor quadratics, but we will concentrate on one called the AC method. In working with a quadratic in standard form, this method will provide basic steps that will guide us through the factoring process.

> **AC Method of Factoring Quadratics**
> The standard form of a quadratic is $ax^2 + bx + c$.
> 1. Factor out the greatest common factor.
> 2. Multiply a and c together. (Do this step off to the side, in the margin.)
> 3. Find factors of ac that sum to b. (Do this off to the side also.)
> 4. Rewrite the middle (bx) term using the factors from step 3.
> 5. Group and factor out what is in common (the greatest common factor).

Example 3 Factor using the AC method

Factor the following.

a. $x^2 + 8x + 15$ **b.** $6x^2 + x - 35$ **c.** $4x^2 + 7xy + 3y^2$

SOLUTION

a. The quadratic $x^2 + 8x + 15$ is in standard form $ax^2 + bx + c$, so we will use the AC method.

Step 1 Factor out the greatest common factor.
The terms in this quadratic have no common factors.

Step 2 Multiply a and c together. (Do this step off to the side, in the margin.)

$$a = 1 \quad \text{and} \quad c = 15 \quad \text{so} \quad ac = 1(15) = 15$$

Step 3 Find factors of ac that sum to b. (Do this off to the side also.)

List both the positive and negative factors of 15.

$ac = 1(15) = 15$		$b = 8$	
1	15	$1 + 15 = 16$	Not equal to b.
3	5	$3 + 5 = 8$	Equal to b.
-1	-15	$-1 + (-15) = -16$	Not equal to b.
-3	-5	$-3 + (-5) = -8$	Not equal to b.

In this list, the factors 3 and 5 will add up to 8.

Step 4 Rewrite the middle (bx) term, using the factors from step 3.
The factors 3 and 5, found in the previous step, will be used to rewrite the bx term in the expression.

$$x^2 + 3x + 5x + 15$$

Step 5 Group and factor out what is in common.
If we group the first two terms together and the last two terms together, we can factor out some common factors.

$$x^2 + 3x + 5x + 15 = (x^2 + 3x) + (5x + 15) \quad \text{Group first and last two terms together.}$$
$$= x(x + 3) + 5(x + 3) \quad \text{Factor out } x \text{ from the first group and 5 from the second group.}$$
$$= (x + 3)(x + 5) \quad \text{Factor out the } (x + 3).$$

To check the factorization, we can multiply out the factored form using the distributive property.

$$(x + 3)(x + 5) = x^2 + 5x + 3x + 15$$
$$= x^2 + 8x + 15$$

We got the original expression back, so the factored form we found is correct.

b. The quadratic $6x^2 + x - 35$ is in standard form $ax^2 + bx + c$, so we will use the AC method.

Step 1 Factor out the greatest common factor.
The terms in this quadratic have nothing in common.

Step 2 Multiply a and c together. (Do this step off to the side, in the margin.)

$$a = 6 \quad \text{and} \quad c = -35 \quad \text{so} \quad ac = 6(-35) = -210$$

Step 3 Find factors of ac that sum to b. (Do this off to the side also.)
List the factors of -210. Once we have listed the factors of -120, then we will switch the signs of the factors to come up with the entire list of factors for -120.

-210		-210		
1	-210	-1	210	The product must be negative, so one factor must be positive and the other must be negative. Be sure to switch the signs of the factors to find all the factorizations of -120.
2	-105	-2	105	
3	-70	-3	70	
5	-42	-5	42	
6	-35	-6	35	
7	-30	-7	30	
10	-21	-10	21	
14	-15	-14	15	

In this list, the factors -14 and 15 will add up to 1.

Step 4 Rewrite the middle (bx) term, using the factors from step 3.
The factors -14 and 15 will replace the b value in the expression.

$$6x^2 - 14x + 15x - 35$$

Step 5 Group and factor out what is in common.
If we group the first two terms together and the last two terms together, we can factor out some common factors.

$$6x^2 - 14x + 15x - 35 = (6x^2 - 14x) + (15x - 35)$$
$$= 2x(3x - 7) + 5(3x - 7)$$
$$= (3x - 7)(2x + 5)$$

Group the first and last two terms together. Factor out $2x$ from the first group and 5 from the second group. Factor out the $(3x - 7)$.

We will check this factorization by multiplying the factored form out using the distributive property.

$$(3x - 7)(2x + 5) = 6x^2 + 15x - 14x - 35$$
$$= 6x^2 + x - 35$$

This is the same expression that we started with, so the factored form is correct.

c. The polynomial $4x^2 + 7xy + 3y^2$ has two variables, but each variable has a squared term and a first-degree term. We will factor it using the AC method focusing on the x variables, and the y variables will follow along the process.

Step 1 Factor out the greatest common factor.
The terms in this polynomial have nothing in common.

Step 2 Multiply a and c together. (Do this step off to the side, in the margin.)

$$a = 4 \quad \text{and} \quad c = 3 \quad \text{so} \quad ac = 4(3) = 12$$

Step 3 Find factors of ac that sum to b. (Do this off to the side also.)
List the factors of 12.

12		12	
1	12	-1	-12
2	6	-2	-6
3	4	-3	-4

In this list, the factors 3 and 4 will add up to 7.

Step 4 Rewrite the middle (bx) term, using the factors from step 3.
The factors 3 and 4 will replace the b value in the expression.

$$4x^2 + 4xy + 3xy + 3y^2$$

Step 5 Group and factor out what is in common.
If we group the first two terms together and the last two terms together, we can factor out some common factors.

$$4x^2 + 4xy + 3xy + 3y^2 = (4x^2 + 4xy) + (3xy + 3y^2)$$
$$= 4x(x + y) + 3y(x + y)$$
$$= (x + y)(4x + 3y)$$

Group the first two terms and the last two terms. Factor out $4x$ from the first group and $3y$ from the second group. Factor out the $(x + y)$.

We will check this factorization by multiplying the factored form out using the distributive property.

$$(x + y)(4x + 3y) = 4x^2 + 3xy + 4xy + 3y^2$$
$$= 4x^2 + 7xy + 3y^2$$

This is the same expression that we started with, so the factored form is correct.

PRACTICE PROBLEM FOR EXAMPLE 3
Factor

a. $x^2 - 2x - 48$ **b.** $10x^2 + 23x + 12$ **c.** $2x^2 + 11xy + 5y^2$

SECTION 3.4 Factoring Polynomials

Some polynomials that are not quadratics may have something in common in every term of the expression. When this happens, you must remember to do the first step of the AC method and factor out the greatest common factor. After you have taken the common factor out, do not forget that it will be part of the factored form all the way to the end of the process.

Example 4 — Factor using the AC method with a common factor

Factor the following.

a. $40x^2 + 46x + 12$ **b.** $3x^2y + 5xy - 28y$

SOLUTION

a. This quadratic $40x^2 + 46x + 12$ is in standard form $ax^2 + bx + c$, so we will use the AC method.

Step 1 Take out anything in common.
The terms in this quadratic are all divisible by 2, so we will factor out a 2.
$$2(20x^2 + 23x + 6)$$

Step 2 Multiply a and c together. (Do this step off to the side, in the margin.)
$$a = 20 \text{ and } c = 6 \text{ so } ac = 20(6) = 120$$

Step 3 Find factors of ac that sum to b. (Do this off to the side also.)
List the factors of 120.

120			120		
1	120		−1	−120	List both the positive and negative factors.
2	60		−2	−60	
3	40		−3	−40	
4	30		−4	−30	
5	24		−5	−24	
6	20		−6	−20	
8	15		−8	−15	
10	12		−10	−12	

In this list, the factors 8 and 15 will add up to 23.

Step 4 Rewrite the middle (bx) term, using the factors from step 3.
The factors 8 and 15, found in the previous step, will be used to rewrite the bx term in the expression.
$$2(20x^2 + 8x + 15x + 6)$$

Step 5 Group and factor out what's in common.
If we group the first two terms together and the last two terms together, we can factor out some common factors.

$2[(20x^2 + 8x) + (15x + 6)]$ *Group first and last two terms together.*
$= 2[4x(5x + 2) + 3(5x + 2)]$ *Factor out 4x from the first group and 3 from the second group.*
$= 2(5x + 2)(4x + 3)$ *Finally, factor out the (5x + 2).*
Do not forget the 2 that you factored out in step 1.

To check the factorization, we can multiply out the factored form using the distributive property.
$$2(5x + 2)(4x + 3) = 2(20x^2 + 15x + 8x + 6)$$
$$= 2(20x^2 + 23x + 6)$$
$$= 40x^2 + 46x + 12$$

We got the original expression back, so the factored form we found is correct.

■ **Skill Connection**

Using grouping symbols
When factoring out the greatest common factor in step 1 of the AC method, remember to keep the factor in the front of the expression at all times. One way to do this is to write it with brackets for several steps and then complete the factoring process inside those brackets.

$x^2y + 2xy - 3y$
$= y(x^2 + 2x - 3)$
$= y[]$
$= y[]$
$= y[]$
$= y()()$

By writing the common factor in front of each bracket at the beginning, you will not forget it while you concentrate on the other steps of the AC method.

b. The expression $3x^2y + 5xy - 28y$ is not a quadratic in standard form but has an extra variable in every term. Once we factor the common variable out, we will have a quadratic in standard form and will use the AC method to complete the factorization.

$3x^2y + 5xy - 28y$ Step 1 factor out the y.
$= y(3x^2 + 5x - 28)$ Steps 2 and 3 are at the left.
$= y(3x^2 - 7x + 12x - 28)$ Step 4.
$= y[(3x^2 - 7x) + (12x - 28)]$ Step 5.
$= y[x(3x - 7) + 4(3x - 7)]$
$= y(3x - 7)(x + 4)$

Steps 2 and 3
$3(-28) = -84$

-1	84	1	-84
-2	42	2	-42
-3	28	3	-28
-4	21	4	-21
-6	14	6	-14
-7	12	7	-12

We check the factorization using the distributive property.

$$y(3x - 7)(x + 4) = y(3x^2 + 12x - 7x - 28)$$
$$= y(3x^2 + 5x - 28)$$
$$= 3x^2y + 5xy - 28y$$

The multiplication resulted in the original expression, so we have the correct factored form.

PRACTICE PROBLEM FOR EXAMPLE 4
Factor

a. $6x^2 + 15x - 36$ **b.** $16x^3 + 52x^2 + 12x$

When working with the AC method, we may ask whether it matters in which order we rewrite the middle (bx) term during step 4. It does not make a difference mathematically, but many students find it best to put any negative values first so that the negative signs do not cause confusion. This is not always possible, since we may find that two negative numbers are needed to add up to b in step 3.

When this happens, we will have to work with the negative more carefully. One way to deal with this is to always put an addition between the groups and keep any negative signs with the number inside the parentheses.

$$x^2 - 5x + 6 = x^2 - 3x - 2x + 6$$
$$= (x^2 - 3x) + (-2x + 6)$$

Keeping the negative with the number will then cause us to have to factor out a negative number from the second parentheses, so that the two remaining factors match. In this case, we will factor out a -2 from the second group.

$$(x^2 - 3x) + (-2x + 6) = x(x - 3) - 2(x - 3)$$

Now we can see that the two pieces both have $(x - 3)$ in common. It can be factored out to complete the factorization process.

$$x(x - 3) - 2(x - 3) = (x - 3)(x - 2)$$

Remember to use caution when working with negatives, or the factorization might not multiply back together to give the correct original expression.

Example 5   Factor using the AC method with two negatives

Factor the following.

a. $6x^2 - 13x + 5$ **b.** $10x^3 - 28x^2 + 16x$

SOLUTION

a. Use the AC method to factor $6x^2 - 13x + 5$. There is no common factor in every term, so we continue on to the other steps. Steps 2 and 3 will be done in the margin on the next page.

$$6x^2 - 13x + 5 = 6x^2 - 3x - 10x + 5 \quad \text{Step 4.}$$
$$= (6x^2 - 3x) + (-10x + 5) \quad \text{Step 5.}$$
$$= 3x(2x - 1) - 5(2x - 1)$$
$$= (2x - 1)(3x - 5)$$

Steps 2 and 3
$6(5) = 30$

1	30	-1	-30
2	15	-2	-15
3	10	-3	-10
5	6	-5	-6

We check the factorization using the distributive property.

$$(2x - 1)(3x - 5) = 6x^2 - 10x - 3x + 5$$
$$= 6x^2 - 13x + 5$$

When multiplied out, the factored form is the same as the original expression, so we have factored correctly.

b. All the terms in the expression $10x^3 - 28x^2 + 16x$ are divisible by 2 and have an x in common, so we will factor $2x$ out and continue with the AC method. Steps 2 and 3 are in the margin.

$$10x^3 - 28x^2 + 16x = 2x(5x^2 - 14x + 8) \quad \text{Step 1.}$$
$$= 2x(5x^2 - 10x - 4x + 8) \quad \text{Step 4.}$$
$$= 2x[(5x^2 - 10x) + (-4x + 8)] \quad \text{Step 5.}$$
$$= 2x[5x(x - 2) - 4(x - 2)]$$
$$= 2x(x - 2)(5x - 4)$$

Steps 2 and 3
$5(8) = 40$

1	40	-1	-40
2	20	-2	-20
4	10	-4	-10
5	8	-5	-8

We check this factorization using the distributive property.

$$2x(x - 2)(5x - 4) = 2x(5x^2 - 4x - 10x + 8)$$
$$= 2x(5x^2 - 14x + 8)$$
$$= 10x^3 - 28x^2 + 16x$$

The factored form multiplied out results in the original expression, so we have the correct factored form.

PRACTICE PROBLEM FOR EXAMPLE 5

Factor the following.

a. $10x^2 - 43x + 28$ **b.** $90x^3 - 75x^2 + 10x$

Factoring Using Trial and Error

Another method used to factor quadratics and other polynomials is trial and error. This method uses our knowledge of multiplying and adding numbers to help us guess the factors of the quadratic. The first step of this method will also be to take out any common factors. After factoring out anything common, we will use factors of a and c to try to find the factored form of the expression. If $a = 1$ in the standard form, we will have to consider only the factors of c.

$$a = 1$$
$$x^2 + 5x + 6$$
$$(x + \text{factor of } c)(x + \text{factor of } c)$$
$$(x + 2)(x + 3)$$

If a does not equal 1, we will have to consider both the factors of a and c.

$$a \neq 1$$
$$10x^2 + 33x + 20$$
$$(\text{factor of } a \cdot x + \text{factor of } c)(\text{factor of } a \cdot x + \text{factor of } c)$$
$$(2x + 5)(5x + 4)$$

The signs of the factors are another aspect of the factored form that we will have to be cautious about. If both of the terms in the original polynomial expression have plus signs, we will have two plus signs in the factored form. If the polynomial has one or more negative signs, we will have at least one negative sign in the factored form.

$$x^2 + 9x + 20 \qquad x^2 + 2x - 8 \qquad x^2 - 3x + 2 \qquad x^2 - 4x - 5$$
$$(x + 5)(x + 4) \qquad (x + 4)(x - 2) \qquad (x - 2)(x - 1) \qquad (x - 5)(x + 1)$$

It is best not to attempt to memorize every situation. Learn to try something, check it, and try again if it does not work.

Example 6 — Factor using trial and error

Factor the following using trial and error.

a. $x^2 + 7x + 6$ **b.** $x^2 - 4x - 12$

c. $2x^2 + 9x + 4$ **d.** $24x^2 - 2x - 15$

SOLUTION

a. Because a is 1 in the expression $x^2 + 7x + 6$, we have to consider only the factors of c. The factors of 6 are 1 and 6 or 2 and 3. Because 1 and 6 add up to 7, the factorization is

$$x^2 + 7x + 6 = (x + 6)(x + 1)$$

We check this by multiplying the factored form out.

$$(x + 6)(x + 1) = x^2 + x + 6x + 6$$
$$= x^2 + 7x + 6$$

b. Again, a is 1 in the expression $x^2 - 4x - 12$, so we consider only the factors of c. The factors of 12 are 1 and 12, 2 and 6, or 3 and 4. Because we have negatives in the original expression, we will have one or more negatives in the factored form. The factors 2 and 6 can make -4 if the 2 is positive and the 6 is negative. This gives us the factored form.

$$x^2 - 4x - 12 = (x - 6)(x + 2)$$

We check this factored form by multiplying it out.

$$(x - 6)(x + 2) = x^2 + 2x - 6x - 12$$
$$= x^2 - 4x - 12$$

c. Because a is not 1 in the expression $2x^2 + 9x + 4$, we have to consider both the factors of a and c. The factors of 2 are 1 and 2, and the factors of 4 are 1 and 4 or 2 and 2. Using these, we find the factored form.

$$2x^2 + 9x + 4 = (2x + 1)(x + 4)$$

We check this factored form by multiplying it out.

$$(2x + 1)(x + 4) = 2x^2 + 8x + x + 4$$
$$= 2x^2 + 9x + 4$$

d. Again, a is not 1 in the expression $24x^2 - 2x - 15$, so we have to consider the factors of a and c. The factors of 24 are 1 and 24, 2 and 12, 3 and 8, and 4 and 6. The factors of 15 are 1 and 15 or 3 and 5. The negatives in the original expression mean that we will have one or more negatives in the factored form.

SECTION 3.4 Factoring Polynomials

$$24x^2 - 2x - 15$$

Try 1	Try 2
$(2x + 5)(12x - 3)$	$(3x + 5)(8x - 3)$
$24x^2 - 6x + 60x - 15$	$24x^2 - 9x + 40x - 15$
$24x^2 + 54x - 15$	$24x^2 + 31x - 15$
Wrong middle term.	**Wrong middle term.**

Try 3	Try 4
$(4x + 1)(6x - 15)$	$(4x + 3)(6x - 5)$
$24x^2 - 60x + 6x - 15$	$24x^2 - 20x + 18x - 15$
$24x^2 - 54x - 15$	$24x^2 - 2x - 15$
Wrong middle term.	**This one works!**

PRACTICE PROBLEM FOR EXAMPLE 6

Factor

a. $x^2 + 7x + 12$ **b.** $x^2 - 11x + 28$ **c.** $15x^2 - 14x - 8$

Prime Polynomials

Some polynomials will not be able to be factored by using rational numbers. When a polynomial is not factorable, it is called *prime*. When using the AC method, we know that a polynomial is not factorable if in step 3 there is no combination of factors for *ac* that add up to *b*. If this happens, we will stop and say that the polynomial is prime.

When factoring by trial and error, we must try every combination of factors of *a* and *c* to see whether any of the combinations work. If no combination of factors and positive and negative signs work, we will say the polynomial is not factorable over the rational numbers.

$$6x^2 + 10x + 5 \qquad 6(5) = 30$$

1	30	-1	-30
2	15	-2	-15
3	10	-3	-10
5	6	-5	-6

None of the factors of *ac* add up to *b*, so this quadratic is not factorable over the rational numbers. This is a prime polynomial.

> **What's That Mean?**
>
> **Prime numbers and prime polynomials**
>
> Prime numbers and prime polynomials are similar in that neither can be broken up into smaller factors.
>
> The number 13 is prime, since the only factors of 13 are 1 and 13.
>
> The polynomial
>
> $$x^2 + 5x + 2$$
>
> is prime because it is not factorable over the rational numbers.

Example 7 Checking for prime polynomials

Determine whether the given polynomial is prime. If the polynomial is not prime, factor.

a. $x^2 + 25$ **b.** $a^2 + 2a + 20$ **c.** $6m^2 + 9m - 30$

SOLUTION

a. The polynomial $x^2 + 25$ is a quadratic in standard form, so we will try the AC method. In this quadratic, $a = 1$, $b = 0$, and $c = 25$. There is no common factor for every term, so we multiply *a* and *c* to find the factors that will add to *b*.

$$1(25) = 25$$

1	25	-1	-25
5	5	-5	-5

None of the factors of *ac* add up to *b*. This is a prime polynomial.

b. The quadratic $a^2 + 2a + 20$ is in standard form, so we will try the AC method. There is no common factor for every term, so we multiply a and c to find the factors that will add to b.

$$1(20) = 20$$

1	20	−1	−20
2	10	−2	−10
4	5	−4	−5

None of the factors of ac add up to b. This is a prime polynomial.

c. The polynomial $6m^2 + 9m - 30$ is in standard form, so we will try the AC method. All the terms are divisible by 3, so we will factor 3 out and then continue.

$$6m^2 + 9m - 30 = 3(2m^2 + 3m - 10)$$
$$2(-10) = -20$$

1	−20	−1	20
2	−10	−2	10
4	−5	−4	5

None of the factors of ac add up to b. Since we factored out a GCF, the original polynomial was not prime, but the remaining quadratic part $2m^2 + 3m - 10$ is prime.

PRACTICE PROBLEM FOR EXAMPLE 7

Determine whether the given polynomial is prime. If the polynomial is not prime, factor.

a. $4x^2 - 20x + 6$ **b.** $g^2 + 52g - 4$

3.4 Exercises

For Exercises 1 through 16, factor out the greatest common factor.

1. $6x + 8$
2. $12x + 9$
3. $4h^2 + 7h$
4. $5r^2 - 3r$
5. $25x^3 + x^2 - 2x$
6. $10x^3 + 3x^2 - 7x$
7. $4a^2b + 6ab$
8. $18m^3n^2 - 15mn$
9. $4x^2 + 6x - 2$
10. $12x^2 - 9x + 3$
11. $15x^2yz + 6xyz^2 - 3xy^2z$
12. $24a^3b^2c - 18a^2b^2c^2 + 30a^2bc$
13. $3x(x + 5) + 2(x + 5)$
14. $7x(2x - 3) - 5(2x - 3)$
15. $7(5w + 4) - 2w(5w + 4)$
16. $9(6x - 7) - 2x(6x - 7)$

For Exercises 17 through 24, factor by grouping.

17. $x^2 - 8x + 7x - 56$
18. $a^2 - 3a - 6a + 18$
19. $7r^2 + 28r + 2r + 8$
20. $5m^2 + 40m + 3m + 24$
21. $16x^2 + 18x + 24xy + 27y$
22. $30a^2 - 55a - 12ab + 22b$
23. $8m^2 - 28m - 10mn + 35n$
24. $96vw - 56w^2 + 36v - 21w$

For Exercises 25 through 40, factor using the AC method or trial and error. Write "prime" if a polynomial is not factorable using the rational numbers.

25. $x^2 - 4x - 21$
26. $x^2 - 3x - 10$
27. $x^2 + 6x + 5$
28. $x^2 + 9x + 20$
29. $w^2 - 7w - 18$
30. $k^2 - 4k - 32$
31. $t^2 - 11t + 28$
32. $x^2 - 11x + 30$

33. $2x^2 + 13x + 15$

34. $3x^2 + 19x + 28$

35. $7m^2 - 25m + 12$

36. $4t^2 - 43t + 30$

37. $4x^2 - 31x - 45$

38. $8x^2 - 93x - 36$

39. $x^2 + 5x + 3$

40. $p^2 + 2p + 5$

For Exercises 41 and 42, explain what the student did wrong when factoring the given polynomials. Give the correct factorization.

41. Factor $2x^2 - 13x + 21$.

> **Dusty**
> $2x^2 - 13x + 21 = 2x^2 - 6x - 7x + 21$
> $= (2x^2 - 6x) - (7x + 21)$
> $= 2x(x - 3) - 7(x + 3)$
> $= (2x - 7)(x - 3)(x + 3)$

42. Factor $4x^2 + 18x - 70$.

> **Cheryll**
> $4x^2 + 18x - 70 = 4x^2 - 10x + 28x - 70$
> $= (4x^2 - 10x) + (28x - 70)$
> $= 2x(2x - 5) + 14(2x - 5)$
> $= (2x - 5)(2x + 14)$

For Exercises 43 through 72, factor using the AC method or trial and error. Write "prime" if a polynomial is not factorable using the rational numbers.

43. $2x^2 + 16x + 2$

44. $10p^2 - 45p - 40$

45. $6w^2 + 29w + 28$

46. $20k^2 + 43k + 14$

47. $10t^2 - 41t - 18$

48. $77m^2 - 23m - 12$

49. $6x^2 + 2x + 5$

50. $8x^2 + 3x + 4$

51. $12p^2 - 4p - 9$

52. $16w^2 - 40w - 13$

53. $12x^2 - 43x + 56$

54. $20t^2 - 18t + 15$

55. $2m^2 + 14m + 24$

56. $3k^2 + 27k + 42$

57. $6x^2 - 66x + 168$

58. $10w^2 - 80w + 150$

59. $40x^2 + 30x - 45$

60. $24p^2 - 52p - 20$

61. $x^3 - 4x^2 - 21x$

62. $w^3 - 7w^2 + 10w$

63. $10x^3 + 35x^2 + 30x$

64. $6p^3 + 8p^2 + 2p$

65. $x^2y + 13xy + 36y$

66. $3x^2y - 15xy - 42y$

67. $2x^2 + 11xy + 15y^2$

68. $4a^2 + 13ab + 3b^2$

69. $6a^2 - 13ab - 5b^2$

70. $5m^2 - 22mn + 8n^2$

71. $-6mn^2 - 20mn - 16m$

72. $-6ab^2 + 75ab - 225a$

For Exercises 73 through 86, factor using any method. Write "prime" if a polynomial is not factorable using the rational numbers.

73. $24x^2 - 38x + 15$

74. $24p^2 - 41p + 12$

75. $24x^2 - 44x - 40$

76. $-30r^2 - 3r + 63$

77. $20n^2 - 32n - 15mn + 24m$

78. $6a^2 + 10a + 21ab + 35b$

79. $12gf + 28g$

80. $-30x^2y + 35xy^2$

81. $a^2 + 3a - 1$

82. $2w^2 - 7w + 11$

83. $7d^2 + 21d - 70$

84. $5k^2 - 35k + 60$

85. $10x^2 - 20x + 15$

86. $-6ab^2 + 10ab - 6a$

3.5 Special Factoring Techniques

LEARNING OBJECTIVES

- Factor perfect square trinomials.
- Factor difference of squares.
- Factor difference and sum of cubes.
- Factor polynomials that require multiple factoring processes or steps.
- Factor trinomials that are quadratic in form.

In this section, we will look at several other techniques of factoring, along with some special situations in which you can use a pattern to factor. The key to most of these techniques is to recognize the form of the polynomial and follow the pattern to factor it. Many of these problems can be done by using the factoring techniques found in Section 3.4, but they are often solved much more quickly by using these patterns.

Perfect Square Trinomials

A perfect square trinomial comes as a result of squaring a binomial. Using the distributive property and combining like terms, we get the following:

$$(a + b)^2 = (a + b)(a + b) = a^2 + ab + ab + b^2 = a^2 + 2ab + b^2$$
$$(a - b)^2 = (a - b)(a - b) = a^2 - ab - ab + b^2 = a^2 - 2ab + b^2$$

If we look at the resulting trinomials, we can see a pattern for a perfect square trinomial. The first and last terms of the trinomial will be perfect squares, and the middle term will be either plus or minus twice the product of the first and last terms that are being squared. For factoring purposes, the relationship is better seen as follows.

$$a^2 + 2ab + b^2 = (a + b)^2$$
$$a^2 - 2ab + b^2 = (a - b)^2$$

Example 1 Factor perfect square trinomials

Factor the following:

a. $x^2 + 6x + 9$ **b.** $m^2 - 14m + 49$

c. $4a^2 + 20a + 25$ **d.** $9x^2 - 24xy + 16y^2$

SOLUTION

a. First, we must confirm that $x^2 + 6x + 9$ is a perfect square trinomial. The first term is x squared, the last term is 3 squared, and the middle term is twice the product of these two terms ($2 \cdot 3 \cdot x$). This trinomial can be factored into a binomial squared.

$$x^2 + 6x + 9 = (x + 3)^2$$

Check the factorization.

$$(x + 3)^2 = (x + 3)(x + 3)$$
$$= x^2 + 3x + 3x + 9$$
$$= x^2 + 6x + 9$$

b. The first term of $m^2 - 14m + 49$ is m squared, the last term is 7 squared, and the middle term is minus twice the product of these two terms ($2 \cdot 7 \cdot m$).

$$m^2 - 14m + 49 = (m - 7)^2$$

Check the factorization.

$$(m - 7)^2 = (m - 7)(m - 7)$$
$$= m^2 - 7m - 7m + 49$$
$$= m^2 - 14m + 49$$

c. The first term of $4a^2 + 20a + 25$ is $2a$ squared, the last term is 5 squared, and the middle term is twice the product of these two terms ($2 \cdot 2a \cdot 5$).

$$4a^2 + 20a + 25 = (2a + 5)^2$$

Check the factorization.

$$(2a + 5)^2 = (2a + 5)(2a + 5)$$
$$= 4a^2 + 10a + 10a + 25$$
$$= 4a^2 + 20a + 25$$

d. The first term of the $9x^2 - 24xy + 16y^2$ is $3x$ squared, the last term is $4y$ squared, and the middle term is minus twice the product of these two terms ($2 \cdot 3x \cdot 4y$).

$$9x^2 - 24xy + 16y^2 = (3x - 4y)^2$$

Check the factorization.

$$(3x - 4y)^2 = (3x - 4y)(3x - 4y)$$
$$= 9x^2 - 12xy - 12xy + 16y^2$$
$$= 9x^2 - 24xy + 16y^2$$

Connecting the Concepts

What does perfect mean?

Recall from previous chapters and classes that a perfect square is a number or expression that is the result of something being squared.

Perfect square

$25 = 5^2$
$144 = 12^2$
$(x + 5)^2$

A perfect square trinomial is the result of a binomial being squared.
Remember that trinomials have three terms and binomials have two terms.

PRACTICE PROBLEM FOR EXAMPLE 1

Factor the following:

a. $x^2 + 8x + 16$ **b.** $36b^2 - 24b + 4$ **c.** $r^2 - 10rt + 25t^2$

Difference of Squares

A similar pattern that we can use is the difference of two squares. When multiplying two binomials, one that is the sum of two terms and the other that is the difference of those same two terms, we get the difference of two squares.

$$(a + b)(a - b) = a^2 - ab + ab - b^2 = a^2 - b^2$$

From this we can see that a binomial made up of the difference of two perfect squares can be factored by following the pattern

$$a^2 - b^2 = (a + b)(a - b)$$

The factorization is the sum of the two terms multiplied by the difference of the two terms. Although the difference of two squares is factorable, the sum of two squares is not factorable. Many students want to factor the sum of two square as

$$(a + b)(a + b)$$

This is not a correct factorization. We can check this by multiplying the factored form out.

$$(a + b)(a + b) = a^2 + ab + ab + b^2$$
$$= a^2 + 2ab + b^2$$

We can see that the middle term of the final polynomial does not add to zero, so the product does not result in the sum of two squares.

Skill Connection

Using the AC method

Perfect square trinomials and the difference of two squares can be factored nicely with the patterns shown in this section. They can also be factored using the AC method.

When the difference of two squares is factored, the value of b will be zero because the middle term is not there.

$x^2 - 100$
$= x^2 - 10x + 10x - 100$
$= (x^2 - 10x) + (10x - 100)$
$= x(x - 10) + 10(x - 10)$
$= (x - 10)(x + 10)$

It is possible to use the AC method here, but it is much longer than using the pattern.

> **Example 2** Factoring the difference of two squares

Factor the following:

a. $x^2 - 9$

b. $4a^2 - 25$

c. $x^2 + 49$

d. $9m^2 - 16n^2$

SOLUTION

a. The expression $x^2 - 9$ is the difference of x squared and 3 squared, so we can factor it using the difference of two squares pattern.

$$x^2 - 9 = x^2 - 3^2$$
$$= (x + 3)(x - 3)$$

Check the factorization.

$$(x + 3)(x - 3) = x^2 - 3x + 3x - 9$$
$$= x^2 - 9$$

b. The expression $4a^2 - 25$ is the difference of $2a$ squared and 5 squared, so we can factor it using the difference of two squares pattern.

$$4a^2 - 25 = (2a)^2 - 5^2$$
$$= (2a + 5)(2a - 5)$$

Check the factorization.

$$(2a + 5)(2a - 5) = 4a^2 - 10a + 10a - 25$$
$$= 4a^2 - 25$$

c. The expression $x^2 + 49$ is the sum of two squares, not the difference of two squares, and is not factorable using the rational numbers.

d. The expression $9m^2 - 16n^2$ is the difference of $3m$ squared and $4n$ squared, so we can factor it using the difference of two squares pattern.

$$9m^2 - 16n^2 = (3m)^2 - (4n)^2$$
$$= (3m - 4n)(3m + 4n)$$

Check the factorization.

$$(3m - 4n)(3m + 4n) = 9m^2 + 12mn - 12mn - 16n^2$$
$$= 9m^2 - 16n^2$$

PRACTICE PROBLEM FOR EXAMPLE 2

Factor the following:

a. $x^2 - 36$

b. $16a^2 - 81b^2$

c. $m^2 + 25$

Difference and Sum of Cubes

Remember that the *sum* of two squares is not factorable; only the *difference* of two squares is. Two other patterns that we can use are the sum and difference of two cubes. The sign changes in these patterns are very important, and we should be very careful when using these patterns to get the signs correct.

Sum of two cubes

Opposite signs

$$a^3 + b^3 = (a + b)(a^2 - ab + b^2)$$

Same signs

Difference of two cubes

Opposite signs

$$a^3 - b^3 = (a - b)(a^2 + ab + b^2)$$

Same signs

SECTION 3.5 Special Factoring Techniques

Example 3 Factor sum and differences of cubes

Factor the following:

a. $x^3 + 27$ **b.** $m^3 - 8$ **c.** $8a^3 + 125b^3$ **d.** $2p^3 - 54r^3$

SOLUTION

a. The first term of $x^3 + 27$ is x cubed, and the second term is 3 cubed, so we can use the pattern for the sum of two cubes.

$$x^3 + 27 = x^3 + 3^3$$
$$= (x + 3)(x^2 - 3x + 9)$$

Check the factorization.

$$(x + 3)(x^2 - 3x + 9) = x^2 - 3x^2 + 9x + 3x^2 - 9x + 27$$
$$= x^3 + 27$$

b. The first term of $m^3 - 8$ is m cubed, and the second term is 2 cubed, so we can use the pattern for the difference of two cubes.

$$m^3 - 8 = m^3 - 2^3$$
$$= (m - 2)(m^2 + 2m + 4)$$

Check the factorization.

$$(m - 2)(m^2 + 2m + 4) = m^3 + 2m^2 + 4m - 2m^2 - 4m - 8$$
$$= m^3 - 8$$

c. The first term of $8a^3 + 125b^3$ is $2a$ cubed, and the second term is $5b$ cubed, so we can use the pattern for the sum of two cubes.

$$8a^3 + 125b^3 = (2a)^3 + (5b)^3$$
$$= (2a + 5b)(4a^2 - 10ab + 25b^2)$$

Check the factorization.

$$(2a + 5b)(4a^2 - 10ab + 25b^2) = 8a^3 - 20a^2b + 50ab^2 + 20a^2b - 50ab^2 + 125b^3$$
$$= 8a^3 + 125b^3$$

d. For the expression $2p^3 - 54r^3$, we will first factor out the 2 that is in common and then use the pattern for the difference of two cubes.

$$2p^3 - 54r^3 = 2(p^3 - 27r^3)$$
$$= 2[p^3 - (3r)^3]$$
$$= 2(p - 3r)(p^2 + 3pr + 9r^2)$$

Check the factorization.

$$2(p - 3r)(p^2 + 3pr + 9r^2) = 2(p^3 + 3p^2r + 9pr^2 - 3p^2r - 9pr^2 - 27r^3)$$
$$= 2(p^3 - 27r^3)$$
$$= 2p^3 - 54r^3$$

Skill Connection

Perfect Cubes

Knowing some basic numbers that are perfect cubes will help you to recognize the sum and difference of cubes.

$1^3 = 1$
$2^3 = 8$
$3^3 = 27$
$4^3 = 64$
$5^3 = 125$
$6^3 = 216$
$7^3 = 343$
$8^3 = 512$
$9^3 = 729$
$10^3 = 1000$

Remember that any variable that is raised to an exponent that is divisible by 3 can be written as a perfect cube.

$x^{12} = (x^4)^3$
$y^{33} = (y^{11})^3$

PRACTICE PROBLEM FOR EXAMPLE 3

Factor the following:

a. $x^3 + 8$ **b.** $x^3 - 27y^3$ **c.** $250a^3 - 16b^3$

Multistep Factorizations

Some polynomials take several steps to factor completely or require slightly different thinking to find the key to factoring them. Always look for the greatest common factor and factor that out first. Then look for a pattern that you recognize and begin factoring. Several of the patterns that we have discussed can be found in more complicated expressions. The sum or difference of two cubes or the difference of two squares can be found in expressions with exponents that are multiples of 2 or 3.

$$x^6 - y^6 = (x^3)^2 - (y^3)^2$$
$$x^6 - y^6 = (x^2)^3 - (y^2)^3$$

This expression can be looked at as either a difference of two squares or a difference of two cubes. This leads to two different paths to factoring, but they have the same result in the end. Using the difference of two squares first may make it easier to factor completely.

Difference of squares, then sum and difference of cubes.

$$\begin{aligned} x^6 - y^6 &= (x^3)^2 - (y^3)^2 \\ &= (x^3 + y^3)(x^3 - y^3) \\ &= (x + y)(x^2 - xy + y^2)(x - y)(x^2 + xy + y^2) \end{aligned}$$

Difference of cubes, then difference of squares and AC method.

$$\begin{aligned} x^6 - y^6 &= (x^2)^3 - (y^2)^3 \\ &= (x^2 - y^2)(x^4 + x^2y^2 + y^4) \\ &= (x + y)(x - y)(x^2 - xy + y^2)(x^2 + xy + y^2) \end{aligned}$$

Example 4 Multistep factoring

Factor the following:

a. $x^4 - 81$ **b.** $3x^6 - 192y^6$

SOLUTION

a. This expression $x^4 - 81$ is the difference of two squares, since the first term is x^2 squared and the second term is 9 squared.

$$\begin{aligned} x^4 - 81 &= (x^2)^2 - 9^2 \\ &= (x^2 + 9)(x^2 - 9) \\ &= (x^2 + 9)(x + 3)(x - 3) \end{aligned}$$

The second factor is still the difference of two squares, so we can factor again.

Check the factorization.

$$\begin{aligned} (x^2 + 9)(x + 3)(x - 3) &= (x^2 + 9)(x^2 - 3x + 3x - 9) \\ &= (x^2 + 9)(x^2 - 9) \\ &= x^4 - 9x^2 + 9x^2 - 81 \\ &= x^4 - 81 \end{aligned}$$

b. We first factor out the 3 that is in common. The remaining expression $x^6 - 64y^6$ can be viewed as either the difference of two squares or the difference of two cubes. If we consider it to be the difference of two cubes, we end up needing to factor again, using the difference of two squares to completely finish the factoring.

$$\begin{aligned} 3x^6 - 192y^6 &= 3(x^6 - 64y^6) \\ &= 3[(x^2)^3 - (4y^2)^3] \\ &= 3(x^2 - 4y^2)(x^4 + 4x^2y^2 + 16y^4) \\ &= 3(x - 2y)(x + 2y)(x^4 + 4x^2y^2 + 16y^4) \\ &= 3(x - 2y)(x + 2y)(x^2 - 2xy + 4y^2)(x^2 + 2xy + 4y^2) \end{aligned}$$

Factor using the difference of two cubes.

Finish by using the difference of two squares.

Check the factorization.

$$3(x - 2y)(x + 2y)(x^2 - 2xy + 4y^2)(x^2 + 2xy + 4y^2)$$
$$= 3(x^2 - 4y^2)(x^4 + 4x^2y^2 + 16y^4)$$
$$= 3(x^6 + 4x^4y^2 + 16a^2y^4 - 4x^4y^2 - 16a^2y^4 - 64y^6)$$
$$= 3(x^6 - 64y^6)$$
$$= 3x^6 - 192y^6$$

PRACTICE PROBLEM FOR EXAMPLE 4

Factor $t^6 - 64$.

Trinomials in Quadratic Form

Trinomials of one variable are said to be in **quadratic form** when the degree of the highest term is twice that of the next term and the final term is a constant.

$$a(\text{expression})^2 + b(\text{expression}) + c$$

These types of situations take practice to see, so consider the following example and pay close attention to the thinking behind each factorization.

Example 5 Multistep factoring

Factor the following:

a. $3a^6 - a^3 - 10$ **b.** $12w^4 + 52w^2 + 35$

SOLUTION

a. The trinomial $3a^6 - a^3 - 10$ is quadratic in form because the first term has degree twice that of the second term and the third term is a constant. We can use a substitution for a^3 to make the trinomial appear to be quadratic. Then factor and replace the a^3 back into the expression. We can use any variable for this substitution, so for this problem, we will use u. Letting $u = a^3$, we get the following:

$$3a^6 - a^3 - 10 = 3(a^3)^2 - a^3 - 10$$
$$= 3u^2 - u - 10 \quad \text{Substitute in } u \text{ and factor.}$$
$$= (3u + 5)(u - 2) \quad u = a^3$$
$$= (3a^3 + 5)(a^3 - 2) \quad \text{Replace the } u \text{ with } a^3.$$

Check the factorization.

$$(3a^3 + 5)(a^3 - 2) = 3a^6 - 6a^3 + 5a^3 - 10$$
$$= 3a^6 - a^3 - 10$$

b. The trinomial $12w^4 + 52w^2 + 35$ is quadratic in form. Use substitution and then factor the remaining quadratic. Using $u = w^2$, we get the following:

$$12w^4 + 52w^2 + 35 = 12(w^2)^2 + 52w^2 + 35$$
$$= 12u^2 + 52u + 35 \quad \text{Substitute in } u \text{ and factor.}$$
$$= (2u + 7)(6u + 5) \quad u = w^2$$
$$= (2w^2 + 7)(6w^2 + 5) \quad \text{Replace } u \text{ with } w^2.$$

Check the factorization.

$$(2w^2 + 7)(6w^2 + 5) = 12w^4 + 10w^2 + 42w^2 + 35$$
$$= 12w^4 + 52w^2 + 35$$

PRACTICE PROBLEM FOR EXAMPLE 5

Factor $3a^{10} - 5a^5 - 28$.

Choosing a Factoring Method

When we are faced with a polynomial to factor, we should choose a method that is best suited to factor the polynomial. We have looked at four methods of factoring:

- **Factor by grouping**
- **AC method**
- **Trial and error**
- **Pattern recognition**
  - Perfect square trinomials
  - Difference of squares
  - Difference of cubes
  - Sum of cubes
  - Quadratic form

All of these methods start by requiring you to take out the greatest common factor of all the terms. If you do not factor out the GCF, all of these methods become more complicated or impossible.

Determining what method is best depends on the characteristics of the polynomial we are trying to factor.

Factor by Grouping

Factoring by grouping is used when we have four terms in the polynomial.

$$5x^2 + 10x - 3x - 6$$
$$12a^2 - 14a + 30ab - 35b$$

AC method or Trial and Error

Either the AC method or trial and error can be used with quadratics in standard form $ax^2 + bx + c$.

$$x^2 + 5x + 6$$
$$4x^2 - 2x - 15$$

Trial and error is easiest when $a = 1$. After using the AC method, many people start to see more of the common factorization patterns and actually start using more trial and error.

Pattern Recognition

Recognizing patterns can help you to factor some polynomials quickly. The following types of polynomials can all be factored by using a pattern.

Perfect square trinomial	Difference of squares
$x^2 + 6x + 9$	$25x^2 - 16$

Sum of cubes	Difference of cubes	Quadratic form
$x^3 + 8$	$27x^3 - 64$	$x^4 - 5x^2 - 6$

Perfect Square Trinomials

A polynomial in the form of a perfect square trinomial $a^2 + 2ab + b^2$ or $a^2 - 2ab + b^2$ can be factored by using the patterns

$$a^2 + 2ab + b^2 = (a + b)^2$$
$$a^2 - 2ab + b^2 = (a - b)^2$$

Difference of Squares

The difference of two squares $a^2 - b^2$ can be factored by using the pattern

$$a^2 - b^2 = (a + b)(a - b)$$

Sum and Difference of Cubes

The sum or difference of cubes $a^3 + b^3$ or $a^3 - b^3$ can be factored by using the patterns

$$a^3 + b^3 = (a + b)(a^2 - ab + b^2)$$
$$a^3 - b^3 = (a - b)(a^2 + ab + b^2)$$

Quadratic Form

A polynomial that is in quadratic form, that is, $a(\text{expression})^2 + b(\text{expression}) + c$, can be factored by using either the AC method or trial and error.

3.5 Exercises

For Exercises 1 through 8, factor the following perfect square trinomials.

1. $x^2 + 8x + 16$
2. $w^2 + 10w + 25$
3. $9g^2 + 12g + 4$
4. $x^2 - 12x + 36$
5. $4t^2 - 28t + 49$
6. $100d^2 + 20d + 1$
7. $25x^2 + 30xy + 9y^2$
8. $49m^2 - 84mn + 36n^2$

For Exercises 9 through 22, factor the following difference of two squares and difference or sum of two cubes.

9. $x^2 - 36$
10. $m^2 - 49$
11. $9k^2 - 16$
12. $25b^2 - 4$
13. $m^3 - 64$
14. $x^3 - 27$
15. $x^3 + 125$
16. $p^3 + 8$
17. $8x^3 + 27$
18. $27d^3 - 64$
19. $3g^3 - 24$
20. $40x^3 + 625$
21. $50x^2 - 18$
22. $28h^2 - 63$

For Exercises 23 through 26, explain what the student did wrong when factoring the given polynomials. Give the correct factorization.

23. Factor $16a^2 + 81$.

> **Kaitlin**
> $16a^2 + 81 = (4a + 9)(4a - 9)$

24. Factor $8x^3 - 27$.

> **Paul**
> $8x^3 - 27$
> $= (2x)^3 - 3^3$
> $= (2x + 3)(4x^2 - 6x + 9)$

25. Factor $m^2 + 4m + 16$.

> **Sayra**
> $m^2 + 4m + 16 = (m + 4)^2$

26. Factor $8x^3 - 27$.

> **Ted**
> $8x^3 - 27 = (2x)^3 - 3^3$
> $= (2x + 3)(2x - 3)$

For Exercises 27 through 54, completely factor the following polynomials. Write "prime" if a polynomial is not factorable using the rational numbers.

27. $x^2 + 9$
28. $25t^2 + 81$
29. $r^6 - 64$
30. $x^6 - 729$
31. $16b^4 - 625c^4$
32. $81m^4 - 16p^4$
33. $h^4 + 3h^2 - 10$
34. $x^4 - 6x^2 + 5$
35. $g^8 + 6g^4 + 9$
36. $v^{10} - 10v^5 + 25$
37. $2t^{12} + 13t^6 + 15$
38. $3k^6 + 19k^3 + 28$
39. $8x^8 + 12x^4 + 18$
40. $6w^{10} - 27w^5 - 105$
41. $H + 6\sqrt{H} + 9$
42. $x + 2\sqrt{x} - 3$
43. $4t - 20t^{\frac{1}{2}} + 25$
44. $3m - 23m^{\frac{1}{2}} + 14$
45. $7g^4 - 567h^4$
46. $6a^4 - 96b^4$
47. $5w^5x^3z + 25w^3x^2z^2 - 120wxz^3$
48. $9a^5b^3c + 18a^3b^2c^2 - 27abc^3$
49. $24a^3b^2 + 11a^2b^3 - 35ab^4$
50. $21g^7h^2 - 7g^5h^4 + 140g^3h^6$
51. $a^{12} - b^{36}$
52. $x^{12} - y^{36}$
53. $x^{16} - 1$
54. $5w^{16} - 5$

For Exercises 55 through 66, factor using any method. Write "prime" if a polynomial is not factorable using the rational numbers.

55. $20x^2 + 23x - 21$
56. $42x^2 + 24x + 105xy - 60y$
57. $9m^2 - 24m + 16$
58. $18a^2 + 24a - 192$
59. $4t^2 - 3t + 15$
60. $36b^2 - 16$
61. $125x^3 - 64$
62. $40mn^2 + 4mn - 84m$
63. $6r + 7r^{\frac{1}{2}} - 20$
64. $a^6 - 729b^6$
65. $-30x^3 + 58x^2 + 28x$
66. $h^2 + 9$

Chapter Summary

Section 3.1 Rules for Exponents

- Exponents follow these rules and properties
  Product rule for exponents: $x^m x^n = x^{m+n}$
  Quotient rule for exponents: $\dfrac{x^m}{x^n} = x^{m-n}$
  Power rule for exponents: $(x^m)^n = x^{mn}$
  Applying exponents over multiplication and division:
  $(xy)^m = x^m y^m$ and $\left(\dfrac{x}{y}\right)^m = \dfrac{x^m}{y^m}$
  Negative exponents: $x^{-n} = \dfrac{1}{x^n}$ and $\dfrac{1}{x^{-n}} = x^n$
  Zero as an exponent: $x^0 = 1$
  Rational exponents: $x^{\frac{1}{n}} = \sqrt[n]{x}$

Example 1

Simplify the following expressions.

a. $\dfrac{30x^3 y^2 z}{5xy^5 z}$

b. $(4x^3 y^4 z)^2 (3xy^2 z^{-3})^3$

SOLUTION

a. $\dfrac{30x^3 y^2 z}{5xy^5 z}$

$6x^2 y^{-3}$

$\dfrac{6x^2}{y^3}$

b. $(4x^3 y^4 z)^2 (3xy^2 z^{-3})^3$

$(16x^6 y^8 z^2)(27x^3 y^6 z^{-9})$

$432 x^9 y^{14} z^{-7}$

$\dfrac{432 x^9 y^{14}}{z^7}$

Example 2

Rewrite the expression $x^{\frac{1}{3}}$ in radical form.

SOLUTION The denominator of the fraction exponent is 3, so the index of the radical will be 3.

$$\sqrt[3]{x}$$

Example 3

Rewrite the expression $\sqrt[5]{2ab^3}$ using exponents.

SOLUTION The index of the radical is 5, so the fraction exponent will have 5 in the denominator.

$$2^{\frac{1}{5}} a^{\frac{1}{5}} b^{\frac{3}{5}}$$

Section 3.2 Combining Functions

- A **term** is a constant, a variable, or the product of any combination of constants and variables.
- A **polynomial** is any combination of terms added together.
- The **degree** of a term equals the exponents of its variables added together.
- The **degree** of a polynomial is the same as the degree of its highest-degree term.
- **Like terms** are terms that have exactly the same variable part.
- When adding or subtracting polynomials, you can combine only like terms.
- Use the **distributive property** to multiply polynomials.
- When multiplying two terms together, multiply the coefficients and add the exponents of the variables that are the same.
- When working in an application, be sure that your functions make sense when combined.
- To add or subtract two functions in an application, the input units must be the same, and the output units must be the same as well.
- To multiply or divide two functions in an application, the input units must be the same, and the output units must make sense once they are combined.
- When you are not working in an application, you can combine two functions in whatever way you wish (as long as you do not perform an illegal mathematical operation, such as division by zero).

Example 4 Give the degree of each term and the polynomial.

$$5x^3y^2z + 14x^2yz^2 - 3yz + 4z - 9$$

SOLUTION The degrees of the terms are 6, 5, 2, 1, and 0, respectively. The degree of the polynomial is 6.

Example 5 Perform the indicated operation and simplify.

a. $(4x^2 + 8x - 9) + (2x + 6)$

b. $(3x + 7) - (4x - 8)$

c. $3x(4x^2 + 2x + 8)$

d. $(2x + 5)(6x - 4)$

SOLUTION

a. $(4x^2 + 8x - 9) + (2x + 6) = 4x^2 + 10x - 3$

b. $(3x + 7) - (4x - 8) = 3x + 7 - 4x + 8$
$= -1x + 15$

c. $3x(4x^2 + 2x + 8) = 12x^3 + 6x^2 + 24x$

d. $(2x + 5)(6x - 4) = 12x^2 - 8x + 30x - 20$
$= 12x^2 + 22x - 20$

CHAPTER 3 Summary   283

Example 6 Let $f(x) = 4x + 7$ and $g(x) = -5x + 12$.

    a. Find $f(x) + g(x)$.   **b.** Find $(fg)(x)$.

    c. Find $f(4) - g(4)$.

SOLUTION

a. $f(x) + g(x) = (4x + 7) + (-5x + 12)$

$ = -x + 19$

b. $(fg)(x) = (4x + 7)(-5x + 12)$

$ = -20x^2 + 48x - 35x + 84$

$ = -20x^2 + 13x + 84$

c. $f(4) - g(4) = [4(4) + 7] - [-5(4) + 12]$

$ = 23 - (-8)$

$ = 31$

Section 3.3 Composing Functions

- In composing two functions, one function, will become the input for the other function. The notation for composing functions is $f(g(x))$ or $(f \circ g)(x)$.
- To compose two functions, the output unit of one function must be the same as the input unit of the other function.

Example 7 Let $C(m)$ be the cost in thousands of dollars to improve m miles of city streets. Let $M(t)$ be the number of miles of city streets improved in year t. Use function notation to combine these two functions to create a new function that gives the cost to improve city streets in year t.

SOLUTION The number of miles of city streets improved, $M(t)$, can be substituted into m in the function $C(m)$. $C(M(t))$ is the cost in thousands of dollars to improve city streets in year t.

Example 8 Let $f(x) = 4x + 7$ and $g(x) = -5x + 12$.

Find $g(f(x))$.

SOLUTION

$g(f(x)) = -5(4x + 7) + 12$

$g(f(x)) = -20x - 35 + 12$

$g(f(x)) = -20x - 23$

Section 3.4 Factoring Polynomials

- To **factor** a quadratic using the AC method, use these steps.
    1. Factor out the greatest common factor.
    2. Multiply a and c together.

3. Find factors of *ac* that add up to *b*.
4. Rewrite the middle (*bx*) term using the factors from step 3.
5. Group and factor out what is in common (the greatest common factor).

- Factoring by trial and error:

  Consider the factors of *a* and *c* and find a combination of factors that will multiply together to give you the correct factorization.

- A prime polynomial is one that cannot be factored using rational numbers.

Example 9 Factor the expression $6x^2 - 26x - 20$.

SOLUTION Every term in the expression is divisible by 2, so we will factor out a 2 and then continue with steps of the AC method.

$$6x^2 - 26x - 20 = 2(3x^2 - 13x - 10)$$
$$= 2(3x^2 - 15x + 2x - 10)$$
$$= 2[(3x^2 - 15x) + (2x - 10)]$$
$$= 2[3x(x - 5) + 2(x - 5)]$$
$$= 2(x - 5)(3x + 2)$$

Check the factorization.
$$2(x - 5)(3x + 2) = 2(3x^2 + 2x - 15x - 10)$$
$$= 2(3x^2 - 13x - 10)$$
$$= 6x^2 - 26x - 20$$

Example 10 Factor the expression $x^2 - 13x + 40$.

SOLUTION If we use trial and error, the factors of *c* are 1 and 40, 2 and 20, 4 and 10, or 5 and 8. The 5 and 8 sum to 13, and if we use both negatives, we will get the correct signs. Therefore, this expression factors to

$$(x - 5)(x - 8)$$

Check the factorization.
$$(x - 5)(x - 8) = x^2 - 8x - 5x + 40$$
$$= x^2 - 13x + 40$$

Section 3.5 Special Factoring Techniques

- A perfect square trinomial is of the form $a^2 + 2ab + b^2$ or $a^2 - 2ab + b^2$.
- Perfect square trinomials can be factored by using the patterns

$$a^2 + 2ab + b^2 = (a + b)^2$$

or

$$a^2 - 2ab + b^2 = (a - b)^2$$

- The difference of two squares can be factored by using the pattern

$$a^2 - b^2 = (a + b)(a - b)$$

- The sum of two squares cannot be factored using the rational numbers.
- The sum and difference of two cubes can be factored by using the patterns

$$a^3 + b^3 = (a + b)(a^2 - ab + b^2)$$
$$a^3 - b^3 = (a - b)(a^2 + ab + b^2)$$

- An expression that is a quadratic in form may be easier to factor by using substitution.

Example 11

Factor the following expressions.

a. $x^2 + 16x + 64$

b. $36a^2 - 25b^2$

c. $2x^3 - 2$

d. $4x^{12} - 17x^6 - 15$

SOLUTION

a. This expression is a perfect square trinomial and factors to

$$x^2 + 16x + 64 = (x + 8)^2$$

Check the factorization.
$$(x + 8)^2 = (x + 8)(x + 8)$$
$$= x^2 + 8x + 8x + 64$$
$$= x^2 + 16x + 64$$

b. This expression is the difference of two squares, so it factors to

$$36a^2 - 25b^2 = (6a)^2 - (5b)^2$$
$$= (6a + 5b)(6a - 5b)$$

Check the factorization.
$$(6a + 5b)(6a - 5b) = 36a^2 - 30ab + 30ab - 25b^2$$
$$= 36a^2 - 25b^2$$

c. Each term in this expression is divisible by 2, so we will factor out a 2 and then use the difference of two cubes pattern to factor.

$$2x^3 - 2 = 2(x^3 - 1)$$
$$= 2(x - 1)(x^2 + x + 1)$$

Check the factorization.
$$2(x - 1)(x^2 + x + 1) = 2(x^3 + x^2 + x - x^2 - x - 1)$$
$$= 2(x^3 - 1)$$
$$= 2x^3 - 2$$

d. This expression is quadratic in form, since the degree of the first term is twice that of the second term. We will let $w = x^6$ and substitute w into the expression, factor it and then replace w with x^6.

$$4x^{12} - 17x^6 - 15 = 4w^2 - 17w - 15$$
$$= (4w + 3)(w - 5)$$
$$= (4x^6 + 3)(x^6 - 5)$$

Check the factorization.
$$(4x^6 + 3)(x^6 - 5) = 4x^{12} - 20x^6 + 3x^6 - 15$$
$$= 4x^{12} - 17x^6 - 15$$

Chapter Review Exercises

For Exercises 1 through 10, simplify the given expression using the order of operations and exponent rules. Write each answer without negative exponents.

1. $(2x^3y^4z)^3(5x^{-3}y^2z)^{-2}$ [3.1]
2. $(3x^2y^5)(5x^2y^2)$ [3.1]
3. $\left(\dfrac{225a^5b^3c}{15ab^6c^{-4}}\right)$ [3.1]
4. $\left(\dfrac{31m^5n^4}{a^5c^7}\right)^0$ [3.1]
5. $(32x^{10}y^{20})^{\frac{1}{5}}$ [3.1]
6. $(a^3b^{-5}c^{-2})^{-2}$ [3.1]
7. $(3m^{-1}n^3p^{-2})^2(5m^2n^{-6}p^4)$ [3.1]
8. $\left(\dfrac{a^2b^{-3}c^5}{a^3b^5c^7}\right)$ [3.1]
9. $\left(\dfrac{3a^{-1}b^3c^5}{ab^{-2}c^3}\right)^{-2}$ [3.1]
10. $\left(\dfrac{3x^2y^5z}{5xy^3z}\right)\left(\dfrac{15x^3yz^4}{10xz}\right)$ [3.1]

11. Write the following exponents in radical form.
    a. $x^{\frac{1}{4}}y^{\frac{3}{4}}$
    b. $3^{\frac{1}{2}}a^{\frac{1}{2}}$ [3.1]

12. Write the following radicals using rational exponents.
    a. $\sqrt[5]{4x^2}$
    b. $\sqrt{7ab}$ [3.1]

13. Hope's Pottery has the following functions to model her monthly revenue and costs.

    $$R(v) = 155v$$
    $$C(v) = 5000 + 65v$$

    where $R(v)$ represents the monthly revenue in dollars for selling v vases and $C(v)$ represents the monthly costs in dollars for producing and selling v vases. Find a monthly profit function for Hope's Pottery. [3.2]

14. The retail prescription drug sales in the United States is a multi-billion-dollar industry. Data for the total retail prescription drug sales and the mail-order sales is given in the following chart.

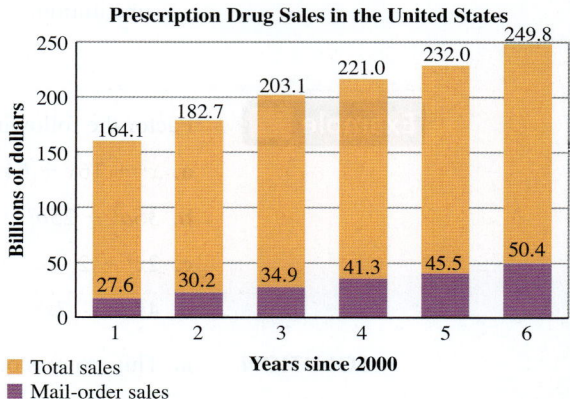

Prescription Drug Sales in the United States
Source: National Association of Chain Drug Stores.

a. Find a model for the total prescription drug sales.
b. Find a model for the mail-order prescription drug sales.
c. What was the total amount of prescription drug sales in 2010?
d. What was the amount of prescription drugs sold through mail order in 2010?
e. Using the two models, find a new function that gives the total amount of prescription drugs sold by non-mail-order retail stores.
f. What was the amount of non-mail-order sales of prescription drugs in 2009 and 2015? [3.2]

15. Johns Hopkins University and the University of Washington are the number one and number two ranked universities in federal obligations for research and development funds. Their funding can be modeled as follows.

    $$J(t) = 16.27t^3 - 321.0t^2 + 2089.32t - 3885.40$$
    $$W(t) = 2.89t^3 - 53.94t^2 + 339.66t - 411.82$$

    where $J(t)$ represents the federal funding for research and development at Johns Hopkins University in millions of dollars and $W(t)$ represents the federal funding for research and development at the University of Washington in millions of dollars t years since 1990.
    Source: Models derived from data in Statistical Abstract of the United States, 2001.

    a. Find the federal funding for research and development at both universities in 1999.

b. Find a new function for the difference in funding at these top-ranked schools.
c. Use the new function to find the difference in federal funding at these universities in 2000. [3.2]

For Exercises 16 through 20, perform the indicated operation and simplify.

16. $(9x - 8) + (-4x + 2)$ [3.2]
17. $(2x^2y + 5xy - 4y^2) - (7xy + 9y^2)$ [3.2]
18. $(3x + 7)(2x - 9)$ [3.2]
19. $(3x + 4y)(5x - 7y)$ [3.2]
20. $(5x - 3)^2$ [3.2]

For Exercises 21 through 23, determine the number of terms in the given polynomial expression. List the terms as either constant terms or variable terms. Give the coefficient of each variable term.

21. $8a^3b^2 - 7a^2b + 19ab$ [3.2]
22. $4m^7np^3 + 24m^5n^4p^3 + 14$ [3.2]
23. $5t + 8$ [3.2]
24. List the degree of each term and the degree of the entire polynomial given in Exercise 21. [3.2]
25. List the degree of each term and the degree of the entire polynomial given in Exercise 22. [3.2]
26. List the degree of each term and the degree of the entire polynomial given in Exercise 23. [3.2]

For Exercises 27 through 30, determine whether the given expression is a polynomial. If the expression is not a polynomial, explain why not.

27. $7x^2y - 4x + 10$ [3.2]
28. $3\sqrt{x} - 5x + 2$ [3.2]
29. 204 [3.2]
30. $2x + 5x^{-1}$ [3.2]

For Exercises 31 through 35, combine the functions by adding, subtracting, multiplying, dividing, or composing the two functions. Write the function notation for the combination you use and list the units for the inputs and outputs of the new function.

31. Let $U(t)$ be the population of the United States, in millions, t years since 1900. Let $C(t)$ be the number of children under 20 years old in the United States, in millions, t years since 1900. Find a function for the number of adults 20 years old or older in the United States. [3.2]

32. Let $C(l)$ be the cost, in thousands of dollars, for Luxury Limousines, Inc. to produce l limousines per year. Let $R(l)$ be the revenue, in thousands of dollars, for Luxury Limousines, Inc. selling l limousines per year. Find a function for the profit of Luxury Limousines, Inc. from producing and selling l limousines per year. [3.2]

33. Let $E(t)$ be the number of employees of Disneyland in year t. Let $S(t)$ be the average number of sick days taken by Disneyland employees in year t. Find a function for the total number of sick days taken by Disneyland employees in year t. [3.2]

34. Let $E(t)$ be the number of employees of Disneyland in year t. Let $W(e)$ be the annual worker's compensation insurance cost in dollars when e employees work at Disneyland during the year. Find a function for the annual worker's compensation insurance cost at Disneyland in year t. [3.3]

35. Let $M(t)$ be the amount spent on treating cancer patients throughout the United States in year t. Let $R(t)$ be the amount spent on cancer research in the United States in year t. Find a function that represents the total amount spent on cancer treatments and research in the United States. [3.2]

36. Let $f(x) = 6x + 3$ and $g(x) = -4x + 8$.
    a. Find $f(x) + g(x)$.   **b.** Find $f(x) - g(x)$.
    c. Find $f(x)g(x)$. [3.2]

37. Let $f(x) = 15x + 34$ and $g(x) = -17x + 34$.
    a. Find $f(x) + g(x)$.   **b.** Find $g(x) - f(x)$.
    c. Find $f(x)g(x)$.   **d.** Find $\dfrac{f(x)}{g(x)}$. [3.2]

38. Let $f(x) = \dfrac{2}{5}x + \dfrac{3}{5}$ and $g(x) = 4x - 7$.
    a. Find $f(x) + g(x)$.   **b.** Find $f(x) - g(x)$.
    c. Find $f(x)g(x)$. [3.2]

39. Let $f(x) = \dfrac{1}{7}x + \dfrac{5}{7}$ and $g(x) = 2x - 7$.
    a. Find $f(4) + g(4)$.   **b.** Find $f(4) - g(4)$.
    c. Find $f(4)g(4)$. [3.2]

40. The number of people in California's labor force and the number of these people who are unemployed are given in the table.

Year	Labor Force	Unemployment
1996	15,398,520	1,167,559
1997	15,747,435	1,061,231
1998	16,224,621	984,728
1999	16,503,004	927,430
2000	16,857,688	838,083
2001	17,246,969	807,290

Source: U.S. Bureau of Labor Statistics.

a. Find a model for the labor force data.
b. Find a model for the number of people who are unemployed.
c. Estimate the number of people in the labor force in 2005.

CHAPTER 3 Exponents, Polynomials and Functions

d. Estimate the number of people who are unemployed in 2005.
e. Using your two models, find a new model for the percentage of the labor force that is unemployed.
f. What percentage of the labor force is unemployed in 1998, 1999, 2000, and 2005? [3.2]

41. Let $f(x) = 6x + 3$ and $g(x) = -4x + 8$.
    a. Find $f(g(x))$. b. Find $g(f(x))$. [3.3]

42. Let $f(x) = 15x + 34$ and $g(x) = -17x + 34$.
    a. Find $f(g(x))$. b. Find $g(f(x))$. [3.3]

43. Let $f(x) = \frac{2}{5}x + \frac{3}{5}$ and $g(x) = 4x - 7$.
    a. Find $(f \circ g)(x)$. b. Find $(g \circ f)(x)$. [3.3]

44. Let $f(x) = \frac{1}{7}x + \frac{5}{7}$ and $g(x) = 2x - 7$.
    a. Find $(f \circ g)(4)$. b. Find $(g \circ f)(4)$. [3.3]

45. Let $f(x) = 0.6x + 2.5$ and $g(x) = 3.5x + 3.7$.
    a. Find $f(-3) + g(-3)$.
    b. Find $g(-3) - f(-3)$.
    c. Find $f(-3)g(-3)$. [3.2]
    d. Find $f(g(-3))$. [3.3]
    e. Find $g(f(-3))$.

46. Let $f(x) = 0.35x - 2.78$ and $g(x) = 2.4x - 6.3$.
    a. Find $f(2) + g(2)$.
    b. Find $g(2) - f(2)$.
    c. Find $f(2)g(2)$. [3.2]
    d. Find $f(g(2))$. [3.3]

For Exercises 47 through 64, factor the given expression using any method you choose.

47. $x^2 + 2x - 35$ 48. $x^2 + 6x + 9$ [3.4]
49. $6x^2 - 16x + 21x - 56$ 50. $5b^2 - 14b - 3$ [3.4]
51. $6p^2 + 23p + 20$ 52. $14k^2 - 21k + 7$ [3.4]
53. $20x^3 - 52x^2 - 24x$ 54. $a^2 - a + 6ab - 6b$ [3.4]
55. $9m^2 - 100$ 56. $25x^2 + 81$ [3.5]
57. $9x^2 + 30x + 25$ 58. $27x^3 - 1$ [3.5]
59. $125w^3 + 8x^3$ 60. $t^6 - 64$ [3.5]
61. $10m^6 - 29m^3 + 21$ 62. $80h^2 - 125$ [3.5]
63. $18a^2b + 84ab + 98b$
64. $3r^4 - 48$ [3.5]

Chapter Test 3

For Exercises 1 through 3, simplify the expressions. Write each answer without negative exponents.

1. $(2b^4c^{-2})^5(3b^{-3}c^{-4})^{-2}$
2. $\left(\dfrac{16b^{12}c^2}{2b^{-3}c^{-4}}\right)^{-\frac{1}{3}}$
3. $\dfrac{25x^{-9}y^{-8}}{35x^{-10}y^{-3}}$

4. The Pharmaceutical Management Agency of New Zealand (PHARMAC) manages the amount spent on various medical conditions and drugs throughout New Zealand. The amount spent on treating diabetes in New Zealand can be modeled by

   $$T(t) = 1.1t + 6.84$$

   where $T(t)$ represents the amount in millions of New Zealand dollars spent treating diabetes t years since 1990. The amount spent on diabetes research in New Zealand can be modeled by

   $$R(t) = 1.3t + 0.52$$

   where $R(t)$ represents the amount in millions of New Zealand dollars spent on diabetes research in New Zealand t years since 1990.
   Source: PHARMAC.

   a. Find $T(9)$ and interpret its meaning.
   b. Find $R(t) = 10$ and interpret its meaning.
   c. Find a new function that represents the amount spent on both research and treatment of diabetes in New Zealand.
   d. Use the function from part c to determine the amount spent on research and treatment of diabetes in 2000.

5. Let $f(x) = 4x + 17$ and $g(x) = 2x - 7$.
   a. Find $f(x) - g(x)$. b. Find $f(x)g(x)$.
   c. Find $f(g(x))$.

6. Let $f(x) = 2.35x + 1.45$ and $g(x) = 2.4x - 6.3$.
   a. Find $(f + g)(4)$. b. Find $fg(-2)$.
   c. Find $(f \circ g)(6)$. d. Find $g(f(0))$.

7. The total number of science/engineering graduate students in doctoral programs can be modeled by

$$S(t) = 1.2t^2 + 4.8t + 376.6$$

where $S(t)$ represents the total number of science/engineering graduate students in thousands t years since 2000.

The number of female science/engineering graduate students in thousands t years since 2000 can be represented by

$$F(t) = 0.4t^2 + 4.1t + 150.9$$

Source: Statistical Abstract of the United States, 2009.

a. Estimate the total number of science/engineering graduate students in doctoral programs in 2010.
b. Find a new function for the number of male science/engineering graduate students in doctoral programs.
c. Using this new model, estimate the number of male science/engineering graduate students in doctoral programs in 2010.

8. The number of murders, $M(t)$ in thousands in the United States t years since 2000 can be represented by

$$M(t) = -0.02t^3 + 0.2t^2 - 0.11t + 13.23$$

The percentage, $P(t)$ of these murders that were committed by using a handgun t years since 2000 can be represented by

$$P(t) = -0.3t^3 + 3.55t^2 - 10.41t + 51.23$$

Source: Models derived from data in Statistical Abstract of the United States, 2009.

a. Estimate the percentage of murders in 2006 that were committed by using a handgun.
b. Find a new function that gives the number of murders committed by using a handgun.
c. Using your new function, estimate the number of murders committed by using a handgun in 2007.

For Exercises 9 through 12, assume that the following functions refer to a specific summer tennis camp and that t represents the year.

- Let $T(t)$ be the total number of children at the camp.
- Let $B(t)$ be the number of boys at the camp.
- Let $C(k)$ be the total cost for k children to go to the camp.
- Let $M(k)$ be the number of matches played at a camp with k children.
- Let $A(t)$ be the average length of time, in minutes, a match takes to play.

Combine the functions by adding, subtracting, multiplying, dividing, or composing the two functions. Write the function notation for the combination you use and list the units for the inputs and outputs of the new function.

9. Find a function that gives the number of girls attending the tennis camp.

10. Find a function for the number of boys' matches at the camp in year t.

11. Find a function for the total cost for the camp in year t.

12. Find the amount of time it takes for all the matches played in a certain year t.

For Exercises 13 through 16, Perform the indicated operation and simplify each of the given expressions.

13. $(4x^2 + 2x - 7) - (3x^2 + 9)$

14. $(4x^2y + 3xy - 2y^2) + (5x^2y - 7xy + 8)$

15. $(2x + 8)(3x - 7)$

16. $(3a - 4b)(2a - 5b)$

For Exercises 17 and 18, determine the number of terms in the given polynomial expression. List the terms as either constant terms or variable terms. Give the coefficient of each variable term.

17. $7x^2 + 8x - 10$

18. $14m^3n^6 - 8m^2n + 205$

19. List the degree of each term and the degree of the entire polynomial given in Exercise 17.

20. List the degree of each term and the degree of the entire polynomial given in Exercise 18.

For Exercises 21 and 22, determine if the given expression is a polynomial. If the expression is not a polynomial, explain why not.

21. $5x^2 + \dfrac{6}{x} - 12$

22. $11a^3b^2 - 17ab + 12b$

For Exercises 23 through 32, factor the given expression using any method.

23. $x^2 - 14x + 45$

24. $2p^3 + 8p^2 - 42p$

25. $12m^2 - 54m + 22mn - 99n$

26. $36b^2 - 49$

27. $t^2 - 20t + 100$

28. $7x^3 - 28x$

29. $8m^3 + 125n^3$

30. $3x^2 + x + 10$

31. $6m^8 - 29m^4 + 28$

32. $a^3 - 125b^3$

Chapter 3 Projects

How Do I Break That Code?

Research Project
One or more people

What you will need
- Follow the MLA style guide for all citations. If your school recommends another style guide, use that.

Factoring plays a critical role in many encryption codes used today and during history. Research how factoring is used in encryption and find an interesting example of the use of factoring in this field.

Write up
Write a one-to two-page paper describing some interesting historical use of factoring in the field of encryption. Use the MLA style guide or your school's recommended style guide.

Review Presentation

Research Project
One or more people

What you will need
- This presentation may be done in class or as a video.
- If it is done as a video, you might want to post the video on YouTube or another site.
- You might want to use problems from the homework set or review material in this book.
- Be creative, and make the presentation fun for students to watch and learn from.

Create a five-minute presentation or video that reviews a section of Chapter 3. The presentation should include the following:

- Examples of the important skills presented in the section
- Explanation of any important terminology and formulas
- Common mistakes made and how to recognize them

Quadratic Functions

- **4.1** Quadratic Functions and Parabolas
- **4.2** Graphing Quadratics in Vertex Form
- **4.3** Finding Quadratic Models
- **4.4** Solving Quadratic Equations by the Square Root Property and Completing the Square
- **4.5** Solving Equations by Factoring
- **4.6** Solving Quadratic Equations by Using the Quadratic Formula
- **4.7** Graphing Quadratics from Standard Form

The U.S. apple industry's profits depend on the number of apples that an orchard can harvest from each tree planted. The competition in the international apple market has increased. This increased competition has made it more important today for the apple industry to maximize its production of apples.

In this chapter, we will discuss quadratic models that will help us to investigate some situations that are not represented by linear data or models. One of the chapter projects, on page 404, will ask you to use the skills from this chapter to investigate how fruit growers can maximize their fruit production.

4.1 Quadratic Functions and Parabolas

LEARNING OBJECTIVES

- Recognize a quadratic from its graph and equation.
- Identify when a quadratic graph is increasing or decreasing.
- Identify the vertex of a parabola and explain its meaning.

Introduction to Quadratics and Identifying the Vertex

In this chapter, we will add another function to our list of possible model types. So far, we have examined only linear models and systems of linear equations. We will now examine data sets that do not follow a linear pattern and then introduce a new function to fit this pattern.

CONCEPT INVESTIGATION
Are the data linear?

1. The average monthly temperature for several months in San Diego, California, is given in the bar chart.

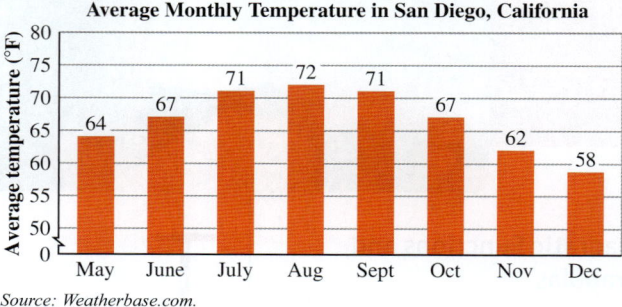

Source: Weatherbase.com.

Create a scatterplot of the data on your calculator and answer the following questions.

Define the variables as follows:

$T(m)$ = Average monthly temperature, in degrees Fahrenheit, in San Diego, California
m = Month of the year; for example, $m = 5$ represents May

a. Do these data follow a linear pattern? If not, describe the shape of the pattern.
b. Graph the function $T(m) = -1(m - 8)^2 + 72$. Describe how well this function fits the data.
c. A meeting planner wants to schedule a conference in San Diego during the hottest month of the year. Using the function as a model for the temperatures in San Diego, find the month for which they should plan the conference.
d. According to the function, what is the average temperature in San Diego during the month of March? Do you think this is reasonable?

2. The number of public branch libraries in the United States for several years is given in the table. Create a scatterplot of the data on your calculator and answer the following questions.

Define the variables as follows:

$B(t)$ = Number of public branch libraries in the United States
t = Time in years since 1990

Year	Public Library Branches
1994	6,223
1995	6,172
1996	6,205
1997	6,332
1998	6,435

Source: Statistical Abstract 2001.

Using Your TI Graphing Calculator

If you are squaring a term, you can use the button on the left side of the calculator. Enter the number or variable that you wish to square and press the button. For example, x^2 would be entered as

When raising a negative number to a power, you will need parentheses around the number for the negative to be raised to the exponent. For example to square -3, you would enter

$(-3)^2$

a. Describe the shape of this data set.
   How is it similar to the pattern of the temperature data in part 1?
b. Graph the function $B(t) = 28.7t^2 - 286.2t + 6899.3$.
   Describe how well this function fits the data.
c. At what point does this graph change directions? What does this point represent in terms of public library branches?
d. According to your model, how many branch libraries were there in 1990?

In the concept investigation, functions
$$T(m) = -1(m - 8)^2 + 72 \quad \text{and} \quad B(t) = 28.7t^2 - 286.2t + 6899.3$$
were given to model the data. These functions are not linear, since they have a squared term and cannot be written in the general form of a line, $Ax + By = C$. These functions are instead called **quadratic functions**. Quadratic functions are most commonly represented by either the **standard form** or the **vertex form.**

What's That Mean?

Quadratic

The Latin prefix *quadri-* is used to indicate the number 4. However, it is also used to indicate the number 2. The word *quadratum* is the Latin word for "square." Since the area of a square with side length x, is x^2, a polynomial with exponent 2 is called a quadratic.

DEFINITIONS

Quadratic function A quadratic function can be written in the standard form or vertex form of a quadratic.

Standard form
$$f(x) = ax^2 + bx + c$$
where a, b, and c are real numbers and $a \neq 0$.

Vertex form
$$f(x) = a(x - h)^2 + k$$
where a, h, and k are real numbers and $a \neq 0$.

What's That Mean?

Vertex

When we discuss a quadratic function, the vertex is one of the most important points on the graph. The following terms can all be used to refer to the vertex of a quadratic function.

- vertex
- maximum point
- minimum point
- highest point
- lowest point
- greatest value
- least value
- largest value
- smallest value

As you can see from the data given in the Concept Investigation, not all data have a linear pattern. Both of the given data sets are curved and form a portion of what is called a **parabola**. Parabolas can be described as a "U shape" or an "upside-down U shape." In Chapter 1, we learned that a line does not change its direction. A parabola will change from increasing to decreasing or from decreasing to increasing.

The point where the parabola turns around and goes in the opposite direction is called the **vertex**. The vertex represents the point where the lowest or minimum value occurs on the graph if the graph opens upward and represents the highest or maximum value on the graph if the graph opens downward. Examples of each of these types of graphs are shown below. Data that show this pattern can be modeled by using a **quadratic function**.

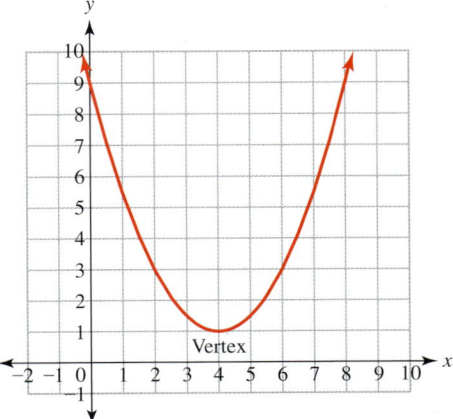

Parabola opens upward.
Vertex: (4, 1)
   Lowest point
   Minimum point
Decreasing when $x < 4$
Increasing when $x > 4$

Connecting the Concepts

Increasing or decreasing?

Remember when reading a graph from left to right that a curve is increasing when it is going up, and it is decreasing when it is going down.

A graph like the one in Example 1 continues on indefinitely from left to right and up or down.

Parabola opens downward.
Vertex: (5, 4)
  Highest point
  Maximum point
Increasing when $x < 5$
Decreasing when $x > 5$

> **DEFINITIONS**
>
> **Parabola** The graph of a quadratic function. A parabola can open either upward or downward.
>
> **Vertex** The point on a parabola where the graph changes direction. The maximum or minimum function value occurs at the vertex of a parabola.

Identifying a Quadratic Function

CONCEPT INVESTIGATION

Is that a quadratic function?

Examine each of the following functions and state if you think it is linear, quadratic, or other. Then graph the function on your calculator to confirm your answer.

1. $f(t) = 3.5t - 7$
2. $R(t) = 5t^2$
3. $P(u) = 4u^2 + 5u - 8$
4. $G(d) = 7(d - 5)^2 + 5$
5. $Q(m) = 3.5(m - 4) - 9.8$
6. $F(x) = (x - 7)(x + 9)$
7. $T(c) = 0.5(c + 4)^3 - 5$
8. $E(w) = 3.5(2)^w$
9. $F(x) = (x + 4)^2$
10. $H(n) = 2n^3 - 5n^2 + n - 15$
11. $K(v) = 5v^2 - 9$

What characteristics do equations of quadratic functions have?

■ **Using Your TI Graphing Calculator**

To enter exponents other than 2 on the graphing calculator, you should use the button in the far right column of the calculator. You will have to enter the number or variable you want to raise to a power, then press the button, followed by the power you wish to use. For example, x^5 would be entered as

You should note that in both forms of a quadratic, the only constant that cannot be zero is a. If a were zero, there would not be a squared term, and the function would no longer be quadratic. All of the other constants, b, c, h, and k, in these forms can be zero.

Any function that can be put into these two forms is a quadratic. A factored quadratic such as $(x + 5)(x - 8)$ does not have an obvious squared term, but if multiplied out, it will have a squared term.

$$(x + 5)(x - 8) = x^2 - 8x + 5x - 40$$
$$= x^2 - 3x - 40$$

Note that the two functions given in the definition box are simply two different forms of the same function. If we multiply out and simplify the vertex form the quadratic will be in standard form. For example if we take the function $f(x) = 2(x - 3)^2 + 5$ that is in vertex form and multiply it out, we can put it into standard form.

$f(x) = 2(x - 3)^2 + 5$	Start in vertex form.
$f(x) = 2(x - 3)(x - 3) + 5$	Multiply out the squared expression.
$f(x) = 2(x^2 - 3x - 3x + 9) + 5$	Simplify.
$f(x) = 2(x^2 - 6x + 9) + 5$	Distribute the 2.
$f(x) = 2x^2 - 12x + 18 + 5$	Simplify.
$f(x) = 2x^2 - 12x + 23$	End in standard form.

The most basic quadratic function is $f(x) = x^2$. The graph of $f(x) = x^2$ has a vertex of $(0, 0)$ and an a value of 1. The vertex of $(0, 0)$ means that the parabola will be centered on the y-axis. Since $a = 1$ and the other constants b and c or h and k are zero, nothing else will affect the output of the function besides the squaring of the input. The graph of $f(x) = x^2$ is the parabola to which we compare all other parabolas.

■ **Skill Connection**

Coefficient of 1

Recall that when a variable term has no visible coefficient, the coefficient is 1.

$$x^2 = 1x^2$$

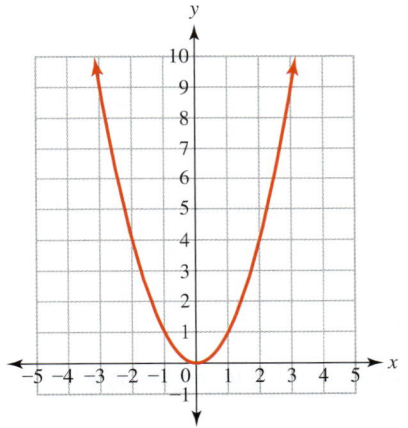

Recognizing Graphs of Quadratic Functions and Identifying the Vertex

Example 1 Reading a quadratic graph

Use the graph of $f(x)$ to estimate the following.

a. For what x-values is this curve increasing? Decreasing? Write your answer as inequalities.

b. Vertex

c. x-intercept(s)

d. y-intercept

e. $f(5) = $?

f. What x-value(s) will make $f(x) = -2$?

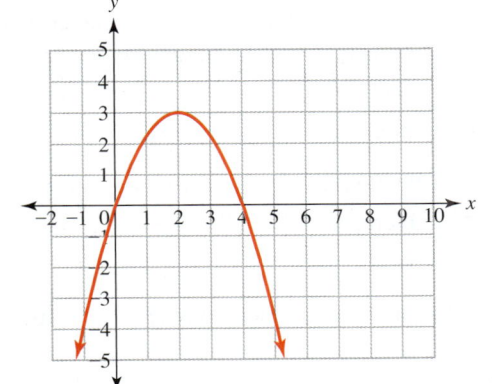

SOLUTION

a. Reading the graph from left to right, we see that the curve is increasing for $x < 2$ and decreasing for $x > 2$.

b. This curve changes from increasing to decreasing when $x = 2$, so the vertex is $(2, 3)$.

c. The curve crosses the x-axis at $x = 0$ and $x = 4$, so $(0, 0)$ and $(4, 0)$ are the x-intercept(s).

d. The curve crosses the y-axis at $y = 0$, so the y-intercept is $(0, 0)$.

e. When $x = 5$, the curve has an output of about $y = -3.5$, so $f(5) = -3.5$.

f. The output of the function is $y = -2$ when the input is about $x = -0.5$ and $x = 4.5$.

Example 2 Reading a quadratic graph

Use the graph of $f(x)$ to estimate the following.

a. For what x-values is this curve increasing? Decreasing? Write your answer as inequalities.

b. Vertex

c. x-intercept(s)

d. y-intercept

e. $f(4) = \ ?$

f. What x-value(s) will make $f(x) = -15$?

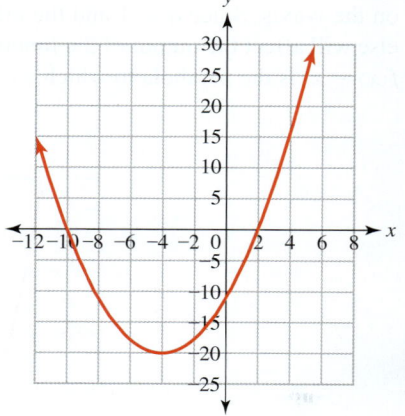

SOLUTION

a. This curve is increasing for $x > -4$ and decreasing for $x < -4$.

b. This curve changes from decreasing to increasing when $x = -4$, so the vertex is $(-4, -20)$.

c. The curve crosses the x-axis at $x = -10$ and $x = 2$, so $(-10, 0)$ and $(2, 0)$ are the x-intercept(s).

d. The curve crosses the y-axis at $y = -11$, so the y-intercept is $(0, -11)$.

e. When $x = 4$, the curve has an output of about $y = 16$, so $f(4) = 16$.

f. The output of the function is $y = -15$ when the input is about $x = -7$ and $x = -1$.

PRACTICE PROBLEM FOR EXAMPLE 2

Use the graph of $f(x)$ to estimate the following.

a. For what x-values is this curve increasing? Decreasing? Write your answer as inequalities.

b. Vertex

c. x-intercept(s)

d. y-intercept

e. $f(2) = \ ?$

f. What x-value(s) will make $f(x) = -10$?

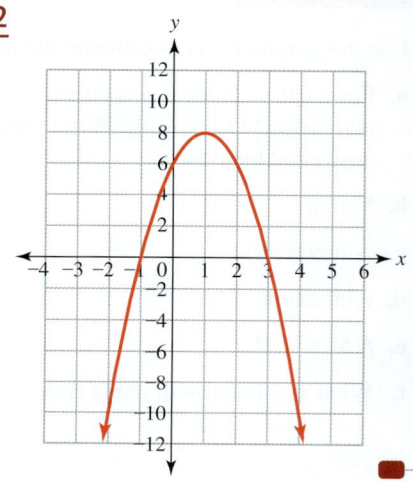

From Examples 1 and 2, a parabola can have two horizontal (x-intercepts) intercepts. The following graphs show how a parabola can actually have two, one, or no horizontal intercepts.

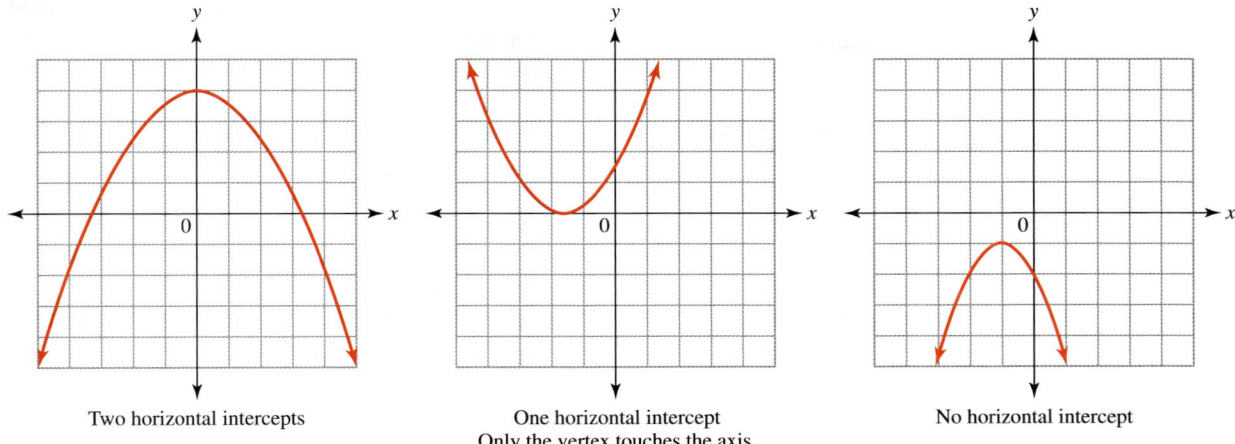

Two horizontal intercepts   One horizontal intercept   No horizontal intercept
                                Only the vertex touches the axis.

In Chapter 1, we also used a table of data to estimate the vertical and horizontal intercepts by looking for when the input and output variables were zero. We can use the same idea with quadratic data if we look carefully for the possibility of more than one horizontal intercept. Remember that the horizontal intercepts may not be whole numbers and, therefore, will be harder to find on a table.

Example 3 Using a table to find intercepts

Use the table to find the vertical and horizontal intercept(s).

a.

Input	Output
−10	114
−7	0
−4	−78
0	−126
3	−120
9	0
12	114

b.

x	y
−8	−7.2
−6	0
−4	4.8
−2	7.2
0	7.2
2	4.8
4	0

SOLUTION

a. The vertical intercept occurs when the input variable is zero. In this table, when the input value is zero, the output is −126. The vertical intercept is (0, −126). The horizontal intercepts will occur when the output variable is zero. In this table the output value is zero twice: when the input value is −7 and again when the input value is 9. The two horizontal intercepts are (−7, 0) and (9, 0).

Input	Output
−10	114
−7	0
−4	−78
0	−126
3	−120
9	0
12	114

b. The vertical intercept is (0, 7.2) and the horizontal intercepts are at (−6, 0) and (4, 0).

x	y
−8	−7.2
−6	0
−4	4.8
−2	7.2
0	7.2
2	4.8
4	0

Example 4 Using scatterplots to investigate data

Create scatterplots for the following sets of data. Determine what type of model would best fit the data. If a quadratic function would be a good fit, give an estimate for the vertex and determine whether it is a maximum or minimum point.

a.

Input	Output
0	19
1	9
2	3
3	1
4	3
5	9
6	19

b.

x	y
-2.3	12.92
-1	12.4
1.2	11.52
2	11.2
3.2	10.72
4.2	10.32
5	10

c.

x	y
-7	-12.5
-6.5	-11.13
-4.3	-8.045
-4	-8
-3	-8.5
-2.6	-8.98

SOLUTION

a.

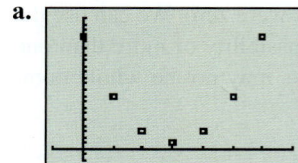 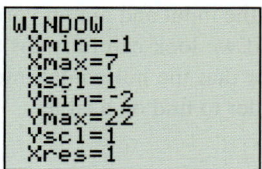

These data decrease and then increase and form the shape of a parabola. Therefore, a quadratic function would be used to model these data. The vertex is the minimum point on the graph and appears to be at about (3, 1).

b.

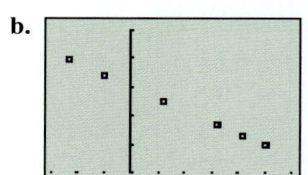 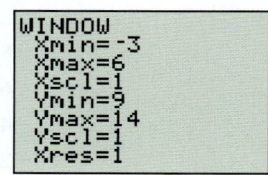

These data are decreasing at a constant rate, so they have a linear pattern. Since these data have a linear pattern, the graph is not a parabola and, therefore, has no vertex.

c.

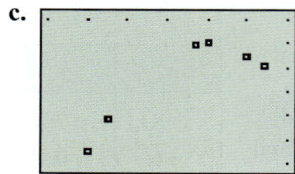 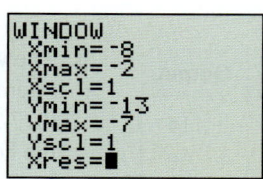

These data increase and then decrease and have the shape of a parabola. Therefore, a quadratic function would be used to model these data. The vertex is the maximum point on the graph and appears to be at about $(-4, -8)$. With this graph you might want to trace the plotted points to determine an estimate for the vertex.

PRACTICE PROBLEM FOR EXAMPLE 4

Create scatterplots for the following sets of data. If a quadratic function would be a good fit, give an estimate for the vertex and determine whether it is a maximum or minimum point.

SECTION 4.1 Quadratic Functions and Parabolas   299

a.
Input	Output
2	10
3	0
4	−6
5	−8
6	−6
7	0
8	10

b.
x	y
−4	−2
−3	2.5
−2	4
−1	2.5
0	−2
1	−9.5

In the exercises for this section, you will be given data that may follow a linear or quadratic pattern. If the data are quadratic, you will be given any needed models to use in your calculator. You can use the trace feature to estimate the vertex of any given model. In the next few sections, you will learn more about quadratic functions and learn a method to model data that follow a quadratic pattern.

4.1 Exercises

For Exercises 1 through 14, determine whether the given function is linear, quadratic, or other. Check your answer by graphing the function on your calculator.

1. $f(x) = 5x^2 + 6x - 9$
2. $h(x) = 2(x + 5)^2 + 3$
3. $g(x) = 4x + 7$
4. $m(t) = 3(t + 4) - 7$
5. $w(b) = b^2 - 7$
6. $h(t) = 12t^2$
7. $f(x) = 4^x + 7$
8. $g(n) = 3n - 12$
9. $c(p) = (p + 3)(p - 4)$
10. $f(x) = (2x + 9)(x - 8)$
11. $h(x) = 5x^2$
12. $g(x) = 2x^3 + 4x^2 - 2x - 8$
13. $d(t) = 3(t + 5)^2$
14. $m(r) = r(r + 3)(r - 7)$

15. Use the graph of $f(x)$ to estimate the following.

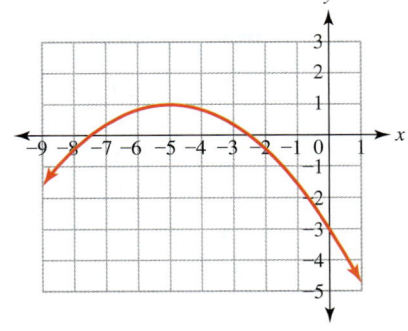

a. Vertex
b. For what x-values is the graph increasing? Write your answer as an inequality.
c. For what x-values is the graph decreasing? Write your answer as an inequality.
d. Horizontal intercept(s)
e. Vertical intercept

16. Use the graph of $f(x)$ to estimate the following.

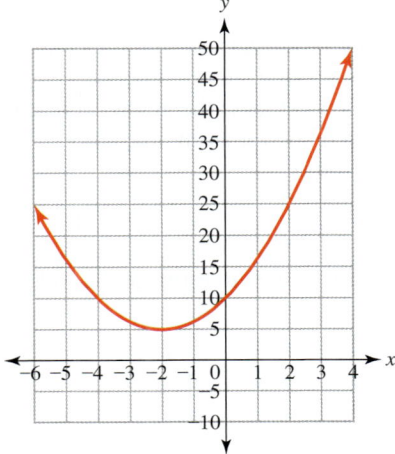

a. Vertex
b. For what x-values is the graph increasing? Write your answer as an inequality.
c. For what x-values is the graph decreasing? Write your answer as an inequality.
d. Horizontal intercept(s)
e. Vertical intercept

17. Use the graph of $f(x)$ to estimate the following.
    a. Vertex
    b. For what x-values is the graph increasing? Write your answer as an inequality.
    c. For what x-values is the graph decreasing? Write your answer as an inequality.
    d. x-intercept(s)
    e. y-intercept
    f. $f(2) = $?

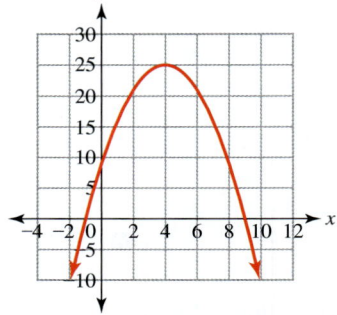

18. Use the graph of $f(x)$ to estimate the following.

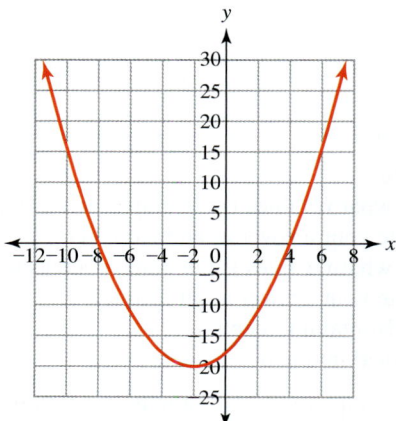

    a. Vertex
    b. For what x-values is the graph increasing? Write your answer as an inequality.
    c. For what x-values is the graph decreasing? Write your answer as an inequality.
    d. x-intercept(s)
    e. y-intercept
    f. $f(6) = $?

19. Use the graph of $f(x)$ to estimate the following.

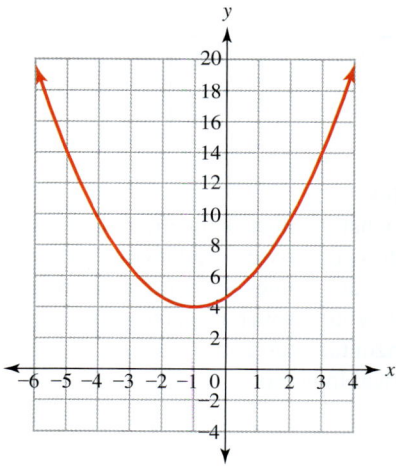

    a. Vertex
    b. For what x-values is the graph increasing? Write your answer as an inequality.
    c. For what x-values is the graph decreasing? Write your answer as an inequality.
    d. Horizontal intercept(s)
    e. Vertical intercept
    f. $f(-4) = $?
    g. What x-value(s) will make $f(x) = 14$?

20. Use the graph of $f(x)$ to estimate the following.

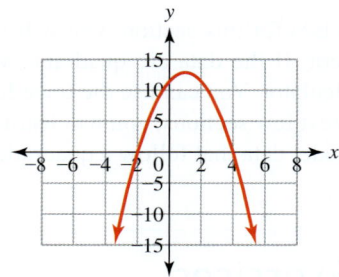

    a. Vertex
    b. For what x-values is the graph increasing? Write your answer as an inequality.
    c. For what x-values is the graph decreasing? Write your answer as an inequality.
    d. x-intercept(s)
    e. y-intercept
    f. $f(-2) = $?
    g. What x-value(s) will make $f(x) = -10$?

For Exercises 21 through 24, use the table to find the horizontal and vertical intercepts.

21.

Input	Output
−6	−9
−4	0
−2	5
0	6
1	5
3	0

22.

Input	Output
−2	20
0	9
2	2
3	0
4	−1
6	0
7	2

23.

Input	Output
−15	15
−10	0
−5	−10
0	−15
5	−15
10	−10
15	0

24.

Input	Output
−2	180
0	80
2	20
4	0
6	20
8	80

Year	Cassette Singles (millions)	Year	Cassette Singles (millions)
1987	5.1	1995	70.7
1988	22.5	1996	59.9
1991	69	1997	42.2
1992	84.6	1998	26.4
1993	85.6	1999	14.2
1994	81.1	2000	1.3

Source: Statistical Abstract of the United States, 2001.

25. The average number of days each month with a high temperature above 70°F in San Diego, California, is given in the table.

Month	Average Number of Days Above 70°F
May	12
June	21
July	30
August	31
September	29
October	26
November	14
December	8

Source: Weatherbase.com.

a. Define variables and create a scatterplot.
b. Would a linear function or a quadratic function better model these data? Explain.
c. If a quadratic model would be better, estimate the vertex and determine whether it is a maximum or minimum point.
d. Using the shape of the distribution, estimate the number of days above 70°F in San Diego, California, during the month of April.

26. The number of cassette singles shipped by major recording media manufacturers is given in the table.

Let $C(t)$ be the number of cassette singles in millions shipped t years since 1980.

a. Create a scatterplot for these data on your calculator.
b. Graph the function $C(t)$ on the same window as your scatterplot.

$$C(t) = -2.336(t - 13)^2 + 85.6$$

c. How well does this model fit the data?
d. Find $C(25)$ and explain its meaning.
e. Estimate a vertex for this model.
f. Explain what the vertex means in terms of cassette singles shipped.
g. Use the graph to estimate the year(s) in which no cassette singles were shipped.

27. The number of cable television systems in the United States during the 1990s is given in the following chart.

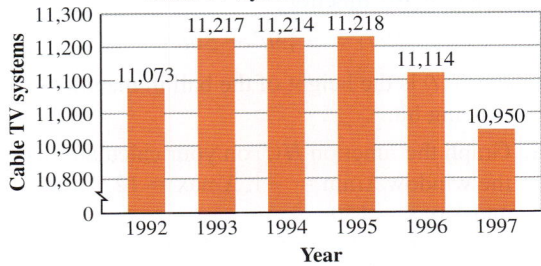

Source: Statistical Abstract of the United States, 2001.

Let $C(t)$ be the number of cable systems in the United States t years since 1990.

a. Create a scatterplot on your calculator for these data.
b. Graph the function $C(t)$ on the same window as your scatterplot.

$$C(t) = -34.625t^2 + 285.482t + 10649.643$$

c. How well does this model fit the given data?
d. Estimate a vertex for this model. Is the vertex a maximum or minimum point for this function?
e. Explain what the vertex means.

28. The United States Golf Association (USGA) requires that golf balls be tested to see whether they fall within the official rules of golf. Under one such test the height of a ball was measured after it was struck by the testing machine. These heights in feet are given in the table.

Time (seconds)	Height of Ball (feet)
0	0.083
1	111.083
2	190.083
3	237.083
4	252.083
5	235.083
6	186.083
7	105.083

Let $H(t)$ be the height of the golf ball in feet t seconds after being hit by the testing machine.

a. Create a scatterplot for these data on your calculator.
b. Graph the function $H(t)$ on the same window as your scatterplot.

$$H(t) = -16t^2 + 127t + 0.083$$

c. Find $H(6.5)$ and explain its meaning.
d. Estimate a vertex for this model. Is the vertex a maximum or minimum point for this function?
e. Explain what the vertex means in regards to the height of the ball.
f. Use the model to estimate when the ball will hit the ground.

29. Out of frustration, a tennis player hits a tennis ball straight up in the air. The height of the ball can be modeled by the function

$$H(t) = -4.9t^2 + 50t + 1$$

where $H(t)$ is the height of the ball in meters t seconds after being hit.

a. Graph the function $H(t)$ on your calculator using the window $X\text{min} = -1$, $X\text{max} = 12$, $Y\text{min} = -40$, and $Y\text{max} = 150$.
b. Use the graph to estimate the vertex of this parabola, and explain its meaning.
c. Find $H(2)$ and explain its meaning.
d. Use the graph to estimate when the tennis ball will hit the ground.

30. Will launched a toy air pressure rocket in his front yard. The height of the rocket can be modeled by the function

$$H(t) = -16(t-2)^2 + 64$$

where $H(t)$ is the height of the rocket in feet t seconds after being launched.

a. Graph the function $H(t)$ on your calculator using the window $X\text{min} = -1$, $X\text{max} = 5$, $Y\text{min} = -5$, and $Y\text{max} = 70$.
b. Use the graph to estimate the vertex of this parabola, and explain its meaning.
c. Find $H(1)$ and explain its meaning.
d. Use the graph to estimate when the rocket will hit the ground.

31. The college bookstore is trying to find the price for school T-shirts that will result in the maximum monthly revenue for the store. By adjusting the price of the T-shirts over several weeks, the bookstore manager has created the following scatterplot for the monthly revenue in dollars from selling T-shirts for d dollars.

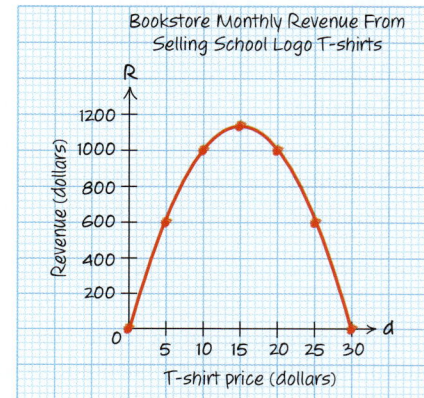

a. Use the graph to estimate the revenue from selling T-shirts for $10 each.
b. Use the graph to estimate the maximum monthly revenue.
c. What price should the bookstore charge to maximize the monthly revenue?
d. Explain why the revenue may go down after the vertex in this situation.

32. A baseball team has 30,000 cheap seats in the stadium and wants to maximize revenue from selling these seats. After some research, marketers have created the following graph for the revenue per game from selling tickets for these cheap seats.

Revenue Per Game From Selling Cheap Seat Tickets

a. Use the graph to estimate the revenue per game from selling the cheap seats for $4 each.
b. Use the graph to estimate the price the baseball team should charge for these cheap seats to maximize revenue per game.
c. Use the graph to estimate what the maximum revenue from selling the cheap seat tickets will be.
d. For what ticket prices will the revenue be increasing?

For Exercises 33 through 40, create a scatterplot for the given set of data and determine what type of model (linear, quadratic, or other) would best fit the data. If quadratic, give an estimate for the vertex and determine whether it is a maximum or minimum point.

33.

Input	Output
−7	−8
−5	8
−3	−8
−1	−56
1	−136
2	−188

34.

Input	Output
1	−57
2	−20
4	6
6	8
8	34
9	71

35.

Input	Output
1	12
2	5
4	−3
6	−3
8	5
9	12

36.

Input	Output
−2	−12
−1	−9.5
0	−7
3	0.5
4	3
6	8

37.

Input	Output
−7	42.8
−2	25.8
1	15.6
5	2
9	−11.6
18	−42.2

38.

Input	Output
−5	−48
−3	8
−2	15
−1	8
1	−48
2	−97

39.

Input	Output
−8	17
−5	19
0	22
7	27
16	32
27	37

40.

Input	Output
−8	−213.6
−5	−48.75
−2	−14.4
0	−20
3	−5.15
5	58.75
8	301.6

For Exercises 41 through 48, sketch a rough graph of a parabola that satisfies the given conditions. Use your graph to answer the question. Answers may vary.

41. Sketch a parabola with a vertex of (2, 4) that faces downward.
  a. How many horizontal intercepts does the graph have?
  b. How many vertical intercepts does the graph have?

42. Sketch a parabola with a vertex of (−3, 7) that faces upward.
  a. How many horizontal intercepts does the graph have?
  b. How many vertical intercepts does the graph have?

43. Sketch a parabola with a vertex of (4, −6) that faces upward.
  a. How many horizontal intercepts does the graph have?
  b. How many vertical intercepts does the graph have?

44. Sketch a parabola with a vertex of (0, 5) that faces up.
  a. How many horizontal intercepts does the graph have?
  b. How many vertical intercepts does the graph have?

45. Sketch a parabola that has only one horizontal intercept and faces upward. Where does the vertex have to be for this graph?

46. Sketch a parabola that has only one horizontal intercept and faces downward. Where does the vertex have to be for this graph?

47. Sketch a parabola that has two horizontal intercepts and faces downward.
  a. Is the vertex above or below the horizontal axis?
  b. Where does the vertex have to be for the parabola to face downward and have no horizontal intercepts?

48. Sketch a parabola that has two horizontal intercepts and faces upward.
  a. Is the vertex above or below the horizontal axis?
  b. Where does the vertex have to be for the parabola to face upward and have no horizontal intercepts?

4.2 Graphing Quadratics in Vertex Form

LEARNING OBJECTIVES

- Identify the axis of symmetry for a parabola.
- Graph a quadratic by hand from the vertex form.
- Determine the domain and range for a quadratic function.

In Section 4.1, we looked at some of the graphical characteristics of quadratic functions such as the general shape and vertex and used several quadratic functions as models for data. In this section, we are going to investigate the vertex form and graphing quadratics.

Vertex Form

Understanding the graph of a function is an important part of the modeling process. We are going to take a close look at the vertex form and see how the constants a, h, and k affect the graph of the function.

■ CONCEPT INVESTIGATION

What do h and k do to the graph?

Consider the vertex form of a quadratic.

$$f(x) = a(x - h)^2 + k$$

We are going to study each component of this function one at a time. For each part of this investigation, consider how the graph is modified as you change one of the constants in the function. First, let's focus on k in $f(x) = x^2 + k$.

1. Graph the following group of functions on the same calculator window. (Use the standard window.)

   a. $f(x) = x^2$ This is the basic quadratic function.
   b. $f(x) = x^2 + 2$
   c. $f(x) = x^2 + 5$
   d. $f(x) = x^2 + 6$
   e. $f(x) = x^2 + 8$

   In these functions, we are considering how a positive k-value changes the graph of a basic quadratic function. Explain what a positive k-value does to the graph.

2. Graph the following group of functions on the standard window.

   a. $f(x) = x^2$
   b. $f(x) = x^2 - 2$
   c. $f(x) = x^2 - 5$
   d. $f(x) = x^2 - 6$
   e. $f(x) = x^2 - 8$

   In these functions, we are considering how a negative k-value changes the graph of a basic quadratic function. Explain what a negative k-value does to the graph.

   Now let's focus on h in $f(x) = (x - h)^2$ and how it affects the graph.

3. Graph the following functions on the same calculator window.

   a. $f(x) = x^2$
   b. $f(x) = (x - 2)^2$
   c. $f(x) = (x - 5)^2$

d. $f(x) = (x - 6)^2$
e. $f(x) = (x - 8)^2$

In these functions, we are considering how a **positive** *h*-value changes the graph of a basic quadratic function. Explain what a **positive** *h*-value does to the graph. (See the Connecting the Concepts.)

4. Graph the following functions on the same calculator window.

   a. $f(x) = x^2$
   b. $f(x) = (x + 2)^2$
   c. $f(x) = (x + 5)^2$
   d. $f(x) = (x + 6)^2$
   e. $f(x) = (x + 8)^2$

   In these functions, we are considering how a **negative** *h*-value changes the graph of a basic quadratic function. Explain what a **negative** *h*-value does to the graph. (Remember that *h* is the only constant in the vertex form that is being subtracted, and therefore, its sign can be confusing.)

5. Graph the following functions and find the vertex of the parabola.

   a. $f(x) = x^2$ $h = 0$ $k = 0$ Vertex (0, 0)
   b. $f(x) = (x - 4)^2 + 3$ $h = $ ___ $k = $ ___   Vertex (___, ___)
   c. $f(x) = (x - 8)^2 - 5$ $h = $ ___ $k = $ ___   Vertex (___, ___)
   d. $f(x) = (x + 2)^2 - 4$ $h = $ ___ $k = $ ___   Vertex (___, ___)
   e. $f(x) = (x + 5)^2 + 2$ $h = $ ___ $k = $ ___   Vertex (___, ___)
   f. $f(x) = (x - 2.5)^2 + 3.5$ $h = $ ___ $k = $ ___ Vertex (___, ___)

   What is the relationship between the vertex form of a quadratic

   $$f(x) = a(x - h)^2 + k$$

   and the vertex of the parabola?

Be cautious when interpreting the value of *h* and *k*. Remember that *h* will appear to have the opposite sign. These two constants control the location of the vertex of the parabola and, therefore, will also give you the values for the maximum or minimum point on the graph.

> **DEFINITION**
>
> **Vertex in the vertex form**   The vertex of a quadratic equation in vertex form can be read directly from the equation.
>
> $$f(x) = a(x - h)^2 + k$$
> $$\text{vertex} = (h, k)$$

To complete our examination of the vertex form, we still have to investigate what the value of *a* does to the graph of a parabola.

■ **CONCEPT INVESTIGATION**

What does *a* do to the graph?

1. Graph the following functions on the same calculator window. Find the vertex.

   a. $f(x) = (x + 2)^2 - 5$ (Note: $a = 1$.)   Vertex (___, ___)
   b. $f(x) = 2(x + 2)^2 - 5$ Vertex (___, ___)
   c. $f(x) = 5(x + 2)^2 - 5$ Vertex (___, ___)
   d. $f(x) = 18.7(x + 2)^2 - 5$ Vertex (___, ___)

   In these functions, we are considering how an *a*-value greater than 1 changes the graph of a quadratic function. Does the value of *a* affect the vertex of the graph? Explain what a positive *a*-value greater than 1 does to the graph.

■ **Connecting the Concepts**

Is *h* tricky?

When you use the vertex form

$$f(x) = a(x - h)^2 + k$$

notice that the constant *h* is being subtracted from the variable *x*. This subtraction can make the sign of *h* confusing.

Consider the expression

$$x - h$$

and how the value of *h* appears once the expression has been simplified.

Substitute the following values for *h* and simplify.

$$h = 1, 3, -5, -3$$

$h = 1$

$$x - (1)$$
$$x - 1$$

The positive *h* value now looks **negative**.

$h = 3$

$$x - (3)$$
$$x - 3$$

Again the positive *h* value of 3 now looks like a **negative** 3.

$h = -5$

$$x - (-5)$$
$$x + 5$$

The negative *h* value looks like a **positive** 5.

$h = -3$

$$x - (-3)$$
$$x + 3$$

Now the negative *h* value of −3 looks like a **positive** 3. Therefore, *h* will appear to have the opposite sign.

2. Graph the following group of functions on the same calculator window.
   a. $f(x) = (x - 4)^2 + 3$
   b. $f(x) = -2(x - 4)^2 + 3$
   c. $f(x) = -5(x - 4)^2 + 3$
   d. $f(x) = -8.7(x - 4)^2 + 3$

   In these functions, we are considering how an *a*-value less than -1 changes the graph of a quadratic function. Explain what a negative *a*-value does to the graph.

   The value of *a* controls more than just whether the parabola faces upward or downward. Let's look at some other values of *a* and see how they can affect the graph.

3. Graph the following group of functions on the same calculator window.
   a. $f(x) = x^2$
   b. $f(x) = 0.5x^2$
   c. $f(x) = -0.3x^2$
   d. $f(x) = -\dfrac{2}{3}x^2$ In your calculator, use parentheses around the fractions.
   e. $f(x) = \dfrac{1}{10}x^2$

   In these functions, we are considering how an *a*-value between -1 and 1 changes the graph of a quadratic function. Explain how these values of *a* change the graph.

Graphing Quadratics in Vertex Form

If we put all of the information we know about *a*, *h*, and *k* together, then we can sketch the graph of a quadratic function by hand. We now know that the vertex of a parabola can be read directly from the vertex form of the quadratic.

$$f(x) = a(x - h)^2 + k$$
$$\text{vertex} = (h, k)$$

Remember to be cautious when reading the *h*-value because its sign will look the opposite of what it is. The value of *a* will affect the width of the graph as well as determine whether the parabola faces upward (positive *a*) or downward (negative *a*).

Effects of the Constants *a*, *h*, and *k* in the Vertex Form

$a > 0$	$a < 0$	$\|a\| < 1$	$\|a\| > 1$
Faces upward	Faces downward	Wider	Narrow
$h > 0$	$h < 0$		
Shifts right	Shifts left		
$k > 0$	$k < 0$		
Shifts up	Shifts down		

What's That Mean?

Symmetry

Symmetry describes the shape of the curve, in particular the fact that one side will be a mirror image, or reflection, of the other side.

Axis of symmetry

The axis of symmetry is an imaginary line that separates the two sides of a symmetric curve.

Unlike the *x*-axis or *y*-axis the axis of symmetry will move to wherever the curve is and divide the two symmetric sides.

One other characteristic of a parabola that is helpful is its **symmetry** about the vertical line going through the vertex. This vertical line is called the **axis of symmetry** and has the equation $x = h$, where the vertex is (h, k). Symmetry basically means that the curve on one side of the axis of symmetry is a reflection of the curve on the other side. If we were to fold the parabola on the axis of symmetry, the sides of the curve would match up.

The symmetry of a parabola makes graphing even easier. If we have points on one side of the axis of symmetry, we can use symmetry to copy them for the other side. In the graph of $f(x) = (x - 4)^2 - 5$ following, we can see that the axis of symmetry is $x = 4$. If we choose the input value 1 unit to the left of the axis of symmetry $x = 3$, the function gives

$$f(3) = (3-4)^2 - 5$$
$$f(3) = -4$$

which represents the point $(3, -4)$. If we choose the input value 1 unit to the right of the axis of symmetry $x = 5$, the function gives

$$f(5) = (5-4)^2 - 5$$
$$f(5) = -4$$

which represents the point $(5, -4)$. These two points are (equidistant) equally distant from the axis of symmetry and are at the same height. Again if we choose the input values 4 units less than and 4 units greater than the axis of symmetry we get two points that are equidistant from the axis of symmetry with the same height.

Every point on the parabola to the left of the axis of symmetry will have a matching point on the right side of the axis of symmetry. These points are called *symmetric points*, since they are equidistant from the axis of symmetry and have the same vertical height.

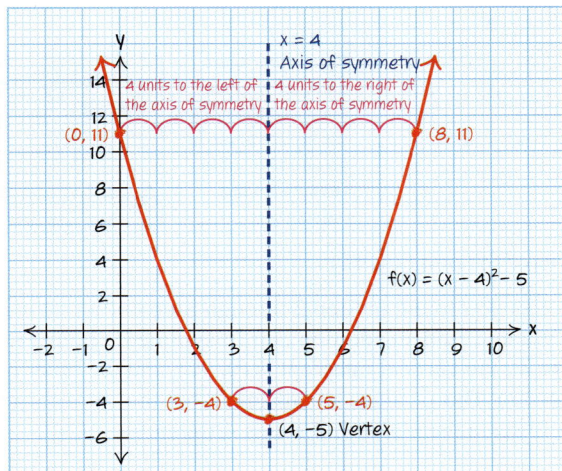

Steps to Graphing a Quadratic Function from the Vertex Form

1. Determine whether the graph opens up or down.
2. Find the vertex and the equation for the axis of symmetry.
3. Find the vertical intercept.
4. Find an extra point by picking an input value on one side of the axis of symmetry and calculating the output value.
5. Plot the points you found in steps 2 through 4. Plot their symmetric points and sketch the graph. (Find an additional pair of symmetric points if needed.)

Example 1 Graphing quadratic functions in vertex form

Sketch the graph of the following quadratic function.

$$f(x) = 2.5(x-6)^2 + 2$$

SOLUTION

Step 1 Determine whether the graph opens up or down.
   The value of a is 2.5. Since a is positive, the graph will open upward, and because it is greater than 1, the graph will be narrow.

Step 2 Find the vertex and the equation for the axis of symmetry.
   Since the quadratic is given in vertex form, the vertex is $(6, 2)$. The axis of symmetry is the vertical line through the vertex $x = 6$.

Step 3 Find the vertical intercept.
To find the vertical intercept, we make the input variable zero.

$$f(x) = 2.5(x - 6)^2 + 2$$
$$f(0) = 2.5(0 - 6)^2 + 2 \quad \text{Be sure to follow the order of operations.}$$
$$f(0) = 2.5(-6)^2 + 2$$
$$f(0) = 2.5(36) + 2$$
$$f(0) = 90 + 2$$
$$f(0) = 92$$

Therefore, the vertical intercept is (0, 92).

Step 4 Find an extra point by picking an input value on one side of the axis of symmetry and calculating the output value.

The axis of symmetry is $x = 6$, so we can choose any x-values less than or greater than 6. We will choose $x = 4$. Substitute 4 into the input of the function to find the output.

$$f(4) = 2.5(4 - 6)^2 + 2$$
$$f(4) = 2.5(-2)^2 + 2$$
$$f(4) = 2.5(4) + 2$$
$$f(4) = 10 + 2$$
$$f(4) = 12$$

Remember that another way to evaluate a function for input values is to use the table feature on the calculator.

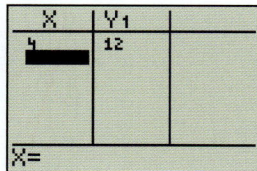

Therefore, we have the point (4, 12).

■ **Connecting the Concepts**

What scale should I use?

When graphing a parabola by hand or using a graphing calculator, the scale that is used can distort the graph. In Example 1, we chose a scale of 20 for the y-axis to accommodate the large y-values. The scale on the x-axis was only 2 because the x-values were much smaller.

If we use the same scale on both axes, we get a more realistic version of the curve. The same scale will allow us to see the narrow width and rapid decrease and increase of the curve.

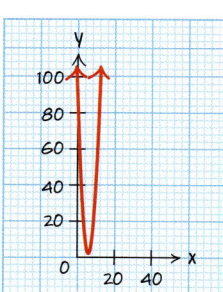

Therefore, pay close attention to the scales on a graph. This way, you know whether the shape has been distorted by using different scales.

Step 5 Plot the points you found in Step 2 through Step 4. Plot their symmetric points and sketch the graph. (Find an addition pair of symmetric points if needed.)

The vertex is (6, 2), the vertical intercept is (0, 92), and the other point we found in step 4 was (4, 12). The x-values are all small, but the y-values get up to 92 and above, so we will scale the x-axis and y-axis differently.

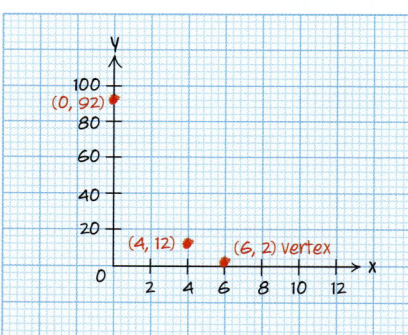

Because (4, 12) is 2 units to the left of the axis of symmetry $x = 6$, the point 2 units to the right of the axis of symmetry, (8, 12), will be its symmetric point. The vertical intercept (0, 92) is 6 units to the left of the axis of symmetry, so its symmetric point will be 6 units to the right of the axis of symmetry at (12, 92).

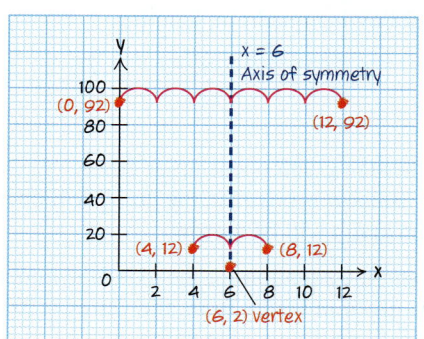

Now that all the points and their symmetric points are plotted, we can complete the graph by connecting the points with a smooth curve.

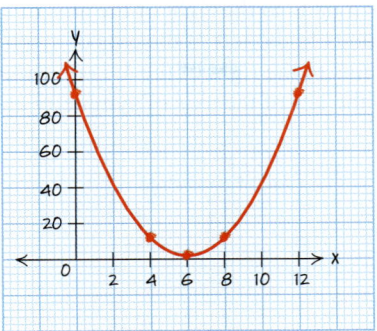

PRACTICE PROBLEM FOR EXAMPLE 1

Sketch the graph of the following quadratic function.

$$f(x) = (x - 4)^2 + 6$$

Example 2 Graphing quadratic functions in vertex form

Sketch the graph of the following quadratic function.

$$f(x) = -0.25(x + 8)^2 + 10$$

SOLUTION

Step 1 Determine whether the graph opens up or down.
   The value of a is -0.25. Because a is negative, the graph will open downward, and because it is less than 1, the graph will be wide.

Step 2 Find the vertex and the equation for the axis of symmetry.
   Since the quadratic is given in vertex form, the vertex is $(-8, 10)$. The axis of symmetry is the vertical line through the vertex $x = -8$.

Step 3 Find the vertical intercept.
   To find the vertical intercept, we make the input variable zero.

$$f(x) = -0.25(x + 8)^2 + 10$$
$$f(0) = -0.25(0 + 8)^2 + 10$$
$$f(0) = -0.25(8)^2 + 10$$
$$f(0) = -0.25(64) + 10$$
$$f(0) = -16 + 10$$
$$f(0) = -6$$

Therefore, the vertical intercept is $(0, -6)$.

Step 4 Find an extra point by picking an input value on one side of the axis of symmetry and calculating the output value.
   The axis of symmetry is $x = -8$, so we can choose any x-value less than or greater than -8. We will choose $x = -3$. Substitute -3 into the input for the function to find the output value.

$$f(x) = -0.25(x + 8)^2 + 10$$
$$f(-3) = -0.25(-3 + 8)^2 + 10$$
$$f(-3) = -0.25(5)^2 + 10$$
$$f(-3) = -0.25(25) + 10$$
$$f(-3) = -6.25 + 10$$
$$f(-3) = 3.75$$

Therefore, we have the point $(-3, 3.75)$.

Step 5 Plot the points you found in steps 2 through 4. Plot their symmetric points and sketch the graph. (Find an additional pair of symmetric points if needed.)
The point $(-3, 3.75)$ is 5 units to the right of the axis of symmetry so its symmetric point $(-13, 3.75)$ is 5 units to the left of the axis of symmetry. The vertical intercept $(0, -6)$ is 8 units to the right of the axis of symmetry so its symmetric point will be 8 units to the left of the axis of symmetry at $(-16, -6)$.

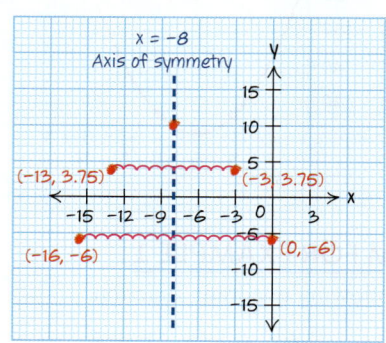

Connect the points with a smooth curve.

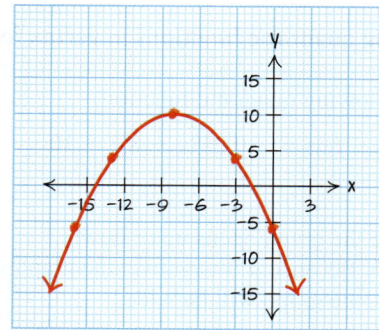

PRACTICE PROBLEM FOR EXAMPLE 2

Sketch the graph of the following quadratic function.

$$f(x) = -0.5(x+2)^2 + 4$$

Example 3 — Graphing quadratic functions in vertex form

Sketch the graph of the following quadratic function.

$$g(x) = 2x^2 - 5$$

SOLUTION

In this case, the quadratic is given in vertex form where $h = 0$, so the function could be written as $g(x) = 2(x - 0)^2 - 5$.

Step 1 Determine whether the graph opens up or down.
The value of a is 2. Because a is positive, the graph will open upward, and because it is more than 1, the graph will be narrow.

Step 2 Find the vertex and the equation for the axis of symmetry.
The vertex for this quadratic is $(0, -5)$. The axis of symmetry is the vertical line through the vertex $x = 0$.

Step 3 Find the vertical intercept.
To find the vertical intercept, we make the input variable zero.

$$f(x) = 2x^2 - 5$$
$$f(0) = 2(0)^2 - 5$$
$$f(0) = -5$$

Therefore, the vertical intercept is $(0, -5)$. Note that the vertex and the vertical intercept are the same point. Therefore, we have only found one point on the parabola.

Step 4 Find an extra point by picking an input value on one side of the axis of symmetry and calculating the output value.

The axis of symmetry is $x = 0$, so we can choose any x-value less than or greater than 0. We will choose $x = 3$. Substitute 3 into the input for the function to find the output value.

$$f(x) = 2x^2 - 5$$
$$f(3) = 2(3)^2 - 5$$
$$f(3) = 13$$

Therefore, we have the point $(3, 13)$.

Step 5 Plot the points you found in steps 2 through 4. Plot their symmetric points and sketch the graph. (Find an additional pair of symmetric points if needed.)

Because we have only two points, we may want to find additional points to make the graph more accurate. To find another point, we pick another input value and calculate the output value. Using $x = 1$, we get

$$f(1) = 2(1)^2 - 5$$
$$f(1) = -3$$
$$(1, -3)$$

The point $(3, 13)$ is 3 units to the right of the axis of symmetry, so its symmetric point $(-3, 13)$ is 3 units to the left of the axis of symmetry. The point $(1, -3)$ is 1 unit to the right of the axis of symmetry, so its symmetric point $(-1, -3)$ is 1 unit to the left of the axis of symmetry.

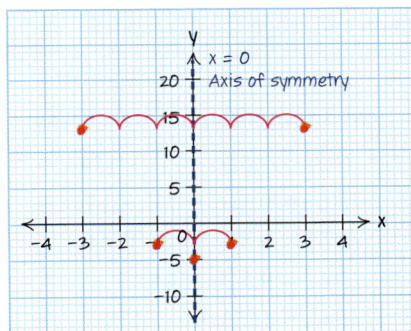

Connect the points with a smooth curve.

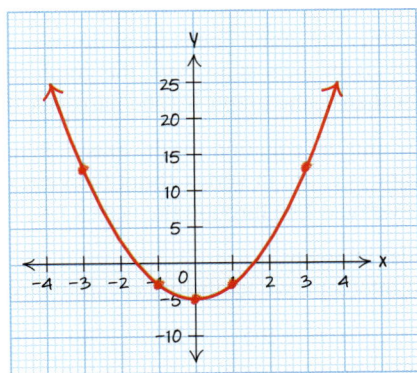

PRACTICE PROBLEM FOR EXAMPLE 3

Sketch the graph of the following quadratic function.

$$f(x) = -x^2 + 8$$

Domain and Range

The domain and range for quadratic models are again restricted to what will make sense in the situation you are modeling. Remember that you might not want to spread out very far in the domain because model breakdown can occur more rapidly with a quadratic model than with a linear one. When you have no data to help guide your domain, you must rely on the situation and any information that will help you to avoid model breakdown. In some cases, only half of the parabola will make sense in the context, so the vertex may be a good starting point or ending point for your domain.

Example 4 Solving applications using the graph

The average number of hours per person per year spent watching television during the late 1990s can be modeled by

$$H(t) = 9.5(t - 7)^2 + 1544$$

where $H(t)$ represents the average number of hours per person per year spent watching television t years since 1990.

a. How many hours did the average person spend watching television in 1996?

b. Sketch a graph of this model.

c. In what year was the average number of hours spent watching television the least?

d. Use your graph to estimate in what year(s) the average number of hours spent watching television was 1800 hours per person per year.

e. Give a reasonable domain and range for this model.

SOLUTION

a. $t = 6$ would represent 1996, so we substitute in 6 and solve for H.

$$H(t) = 9.5(t - 7)^2 + 1544$$
$$H(6) = 9.5(6 - 7)^2 + 1544$$
$$H(6) = 9.5(-1)^2 + 1544$$
$$H(6) = 9.5 + 1544$$
$$H(6) = 1553.5$$

The model predicts that in 1996, the average person spent 1553.5 hours per year watching television.

b. **Step 1** Determine whether the graph opens up or down.
   The value of a is 9.5. Because a is positive, the graph will open upward, and because it is greater than 1, the graph will be narrower than $y = x^2$.

Step 2 Find the vertex and the equation for the axis of symmetry.
   Because the quadratic is given in vertex form, the vertex is (7, 1544). The axis of symmetry is the vertical line through the vertex $t = 7$.

Step 3 Find the vertical intercept.
   To find the vertical intercept, we let the input variable equal zero.

$$H(t) = 9.5(t - 7)^2 + 1544$$
$$H(0) = 9.5(0 - 7)^2 + 1544$$
$$H(0) = 2009.5$$

Therefore, the vertical intercept is (0, 2009.5).

Step 4 Find an extra point by picking an input value on one side of the axis of symmetry and calculating the output value.
The axis of symmetry is $t = 7$, so we will choose $t = 3$. Substitute 3 for the input of the function and calculate the output value.

$$H(t) = 9.5(t - 7)^2 + 1544$$
$$H(3) = 9.5(3 - 7)^2 + 1544$$
$$H(3) = 1696$$

Therefore, we have the point $(3, 1696)$.

Step 5 Plot the points you found in steps 2 through 4. Plot their symmetric points and sketch the graph. (Find an additional pair of symmetric points if needed.) Because $(3, 1696)$ is 4 units to the left of the axis of symmetry, the point $(11, 1696)$ will be its symmetric point on the right side of the axis of symmetry. The vertical intercept $(0, 2009.5)$ is 7 units to the left of the axis of symmetry, so its symmetric point $(14, 2009.5)$ is 7 units to the right.

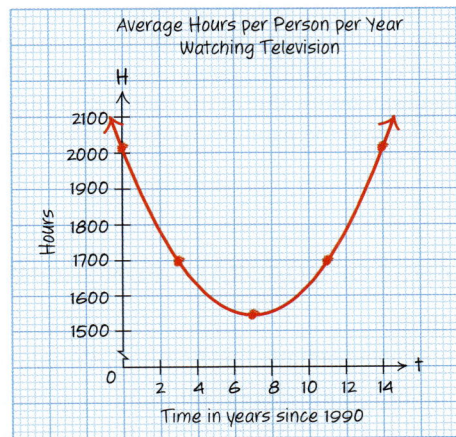

c. The lowest point on the graph is the vertex at $(7, 1544)$. The model says that in 1997, people watched an average of 1544 hours of television per year. This was the minimum amount of television watched during the late 1990s.

d. According to the graph, the average number of hours per person per year spent watching television reached 1800 in year 2 and again in year 12. Since time is measured in years since 1990, 2 and 12 would be the years 1992 and 2002, respectively.

e. The problem states that the function models the late 1990s. Television viewing seems to have continued to grow since the 1990s, so we can extend the domain a couple of years and choose a domain of $5 \leq t \leq 12$. Within this domain, the graph has a lowest output of 1544 at the vertex and grows from there to the highest output when $t = 12$. Substituting $t = 12$ into the function, we will get the highest output.

$$H(t) = 9.5(t - 7)^2 + 1544$$
$$H(12) = 9.5(12 - 7)^2 + 1544$$
$$H(12) = 1781.5$$

Therefore, the range for this model is $1544 \leq H \leq 1781.5$.

PRACTICE PROBLEM FOR EXAMPLE 4

The number of cellular telephone subscribers, in thousands, can be modeled by

$$C(t) = 833.53(t + 1)^2 + 3200$$

where $C(t)$ represents the number of cellular telephone subscribers, in thousands, t years since 1990.

a. How many cellular telephone subscribers were there in 2000?

b. Sketch a graph of this model.

c. According to this model, in what year was the number of cellular telephone subscribers the least?

d. How many cellular telephone subscribers were there in 1985? Does your answer make sense?

e. Give a reasonable domain and range for this model.

When you work with larger numbers, it is usually best to gather as much information about the graph as possible before you start to create the graph. Doing so will allow you to get an idea of how to scale your axes.

The domain of a quadratic function that has no context is all real numbers because no real number will make the function undefined. The range of a quadratic function without a context will start or end with the output value of the vertex because this value is either the lowest or highest output value of the function. If the parabola faces upward, the range will go up to infinity. If the parabola faces downward, the range will go down to negative infinity.

DOMAIN AND RANGE OF QUADRATIC FUNCTIONS

Domain all real numbers, $(-\infty, \infty)$

Range If the vertex of the quadratic function is (h, k), then the range is

$(-\infty, k]$, or $y \leq k$ if the graph is facing downward

$[k, \infty)$, or $y \geq k$ if the graph is facing upward

Example 5 Domain and range of quadratic functions with no context

Give the domain and range for the following functions.

a. $f(x) = 2.5(x - 6)^2 + 2$

b. $f(x) = -0.25(x + 8)^2 + 10$

c. $g(x) = (x - 3)^2 - 5$

SOLUTION

a. We graphed this function in Example 1. Looking at the graph, we can visualize the graph continuing to go upward and getting wider as x increases to infinity and decreases to negative infinity.

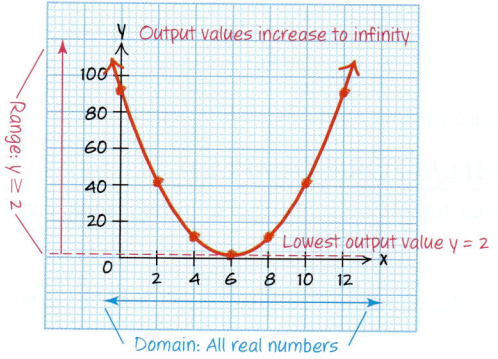

Because the input variable x can be any real number without causing the function to be undefined, the functions domain is all real numbers. The lowest point on the parabola is the vertex, so the lowest output value for the function is $y = 2$, and the outputs continue up to infinity. Thus, we get the following:

Domain: All real numbers or $(-\infty, \infty)$
Range: $y \geq 2$ or $[2, \infty)$

b. We graphed this function in Example 2 on pages 309–310. Looking at the graph, we can visualize the graph continuing to go downward and get wider as x increases to infinity and decreases to negative infinity.

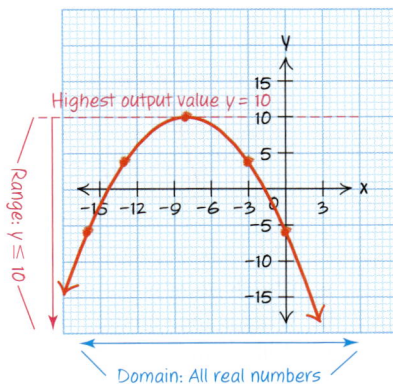

No real number will make this function undefined, so the domain of the function is all real numbers. The highest point on the parabola is the vertex, so the highest output for the function is $y = 10$ and goes down to negative infinity from there, so we get the following:

Domain: All real numbers or $(-\infty, \infty)$
Range: $y \leq 10$ or $(-\infty, 10]$

c. Without graphing the function, we know that no real number will make this function undefined, so the domain is all real numbers. Because the function is in vertex form, we know that the vertex is $(3, -5)$ and that the parabola faces upward, because a is positive. The vertex being the lowest point means that $y = -5$ is the lowest output value on the graph. The outputs will continue to go upward to infinity. Thus, we get the following:

Domain: All real numbers or $(-\infty, \infty)$
Range: $y \geq -5$ or $[-5, \infty)$

PRACTICE PROBLEM FOR EXAMPLE 5

Give the domain and range for the following:

a.

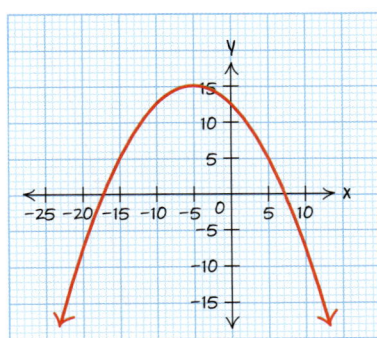

b. $f(x) = 2(x + 7)^2 + 4$

c. $g(x) = -0.3(x - 2.7)^2 - 8.6$

4.2 Exercises

For Exercises 1 through 10, refer to the values of a, h, and k in the vertex form of a quadratic.

$$f(x) = a(x - h)^2 + k$$

1. Does *a*, *h*, or *k* shift the graph to the **right**? Should that value be positive or negative?

2. Does *a*, *h*, or *k* cause the graph to face **downward**? Should that value be positive or negative?

3. Does *a*, *h*, or *k* shift the graph **down**? Should that value be positive or negative?

4. Does *a*, *h*, or *k* affect the **width** of the graph? Considering *a*, *h*, and *k*, what values will result in the graph being wider than the graph of $f(x) = x^2$?

5. Considering *a*, *h*, and *k*, what values will result in the graph being more **narrow** than the graph of $f(x) = x^2$?

6. Considering *a*, *h*, and *k*, what values will result in a parabola that has a **vertex** of $(2, 5)$?

7. Does *a*, *h*, or *k* cause the graph to **face upward**? Should that value be positive or negative?

8. Considering *a*, *h*, and *k*, what values will result in a downward facing parabola with a **vertex** $(0, 3)$?

9. Considering *a*, *h*, and *k*, what values will result in a parabola with a **vertex** at the origin?

10. Considering *a*, *h*, and *k*, what values will result in a parabola with a **vertex** on the horizontal axis?

For Exercises 11 through 18, refer to the given graphs to answer the questions. Assume that all functions are written in the vertex form of a quadratic.

11.
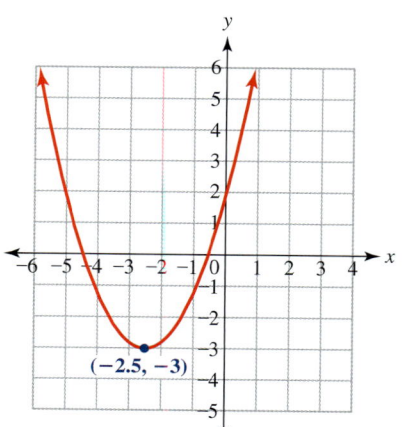

a. Which point is the vertex?
b. What is the equation of the axis of symmetry?
c. What is the symmetric point to the point $(0.5, 4)$?

12.
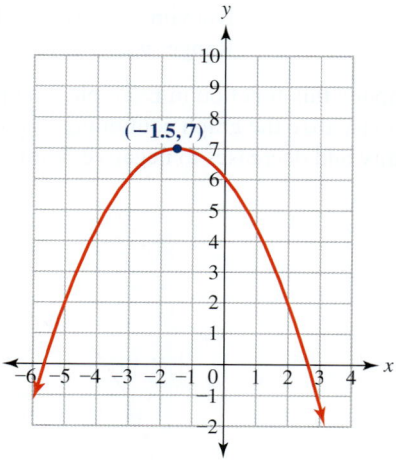

a. Which point is the vertex?
b. What is the equation of the axis of symmetry?
c. What is the symmetric point to the point $(2, 2)$?

13.
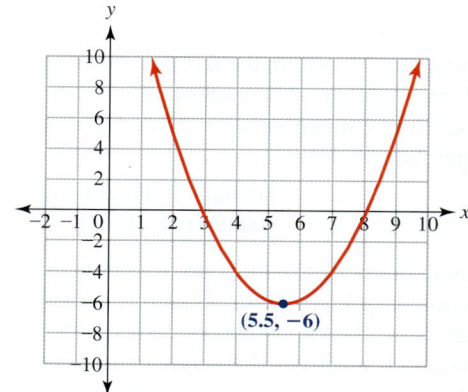

a. Which point is the vertex?
b. What is the equation of the axis of symmetry?
c. What is the symmetric point to the point $(9, 5)$?

14.
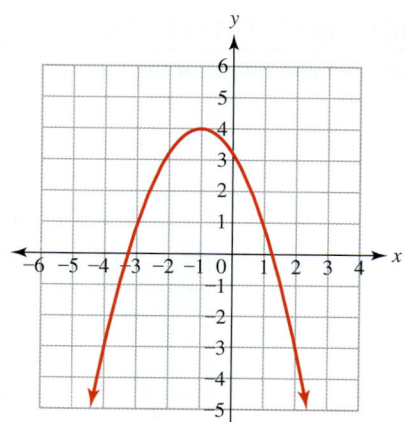

a. Which point is the vertex?
b. What is the equation of the axis of symmetry?
c. What is the symmetric point to the point $(2, -3)$?

15.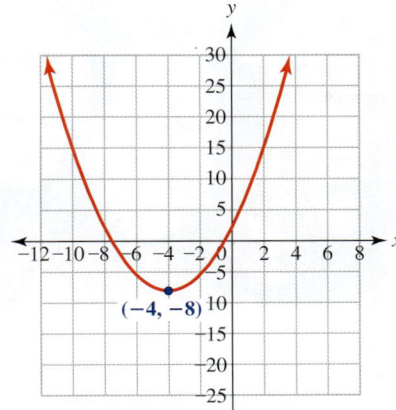

a. Which point is the vertex?
b. What is the equation of the axis of symmetry?
c. Is the value of a positive or negative?
d. In vertex form, what are the values of h and k?
e. What is the symmetric point to the point $(-10, 15)$?

16.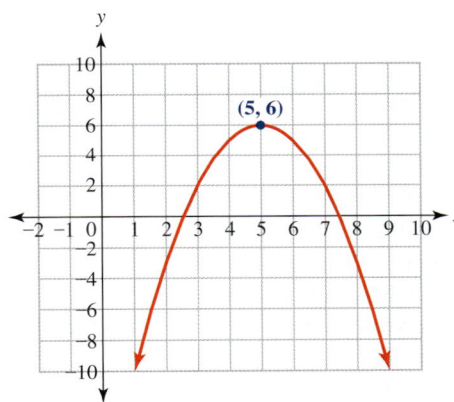

a. Which point is the vertex?
b. What is the equation of the axis of symmetry?
c. Would the value of a be positive or negative?
d. In vertex form, what are the values of h and k?
e. What is the symmetric point to the point $(7, 2)$?

17.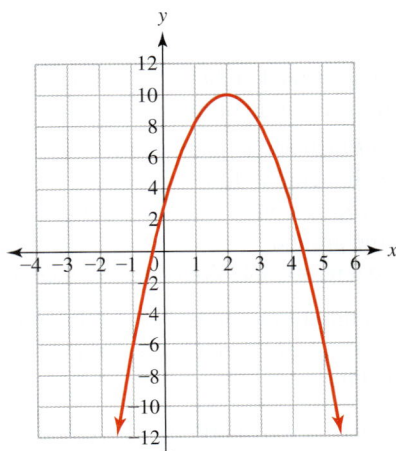

a. Which point is the vertex?
b. What is the equation of the axis of symmetry?

c. Would the value of a be positive or negative?
d. In vertex form, what are the values of h and k?
e. What is the symmetric point to the point $(-1, -6)$?

18.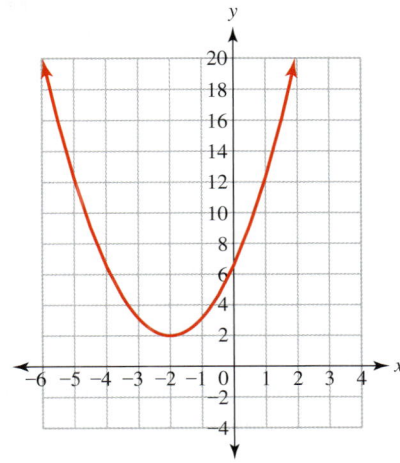

a. Which point is the vertex?
b. What is the equation of the axis of symmetry?
c. Would the value of a be positive or negative?
d. In vertex form, what are the values of h and k?
e. What is the symmetric point to the point $(1.5, 16)$?

For Exercises 19 through 28, sketch the graph of the given functions and label the vertex, vertical intercept, and at least one additional pair of symmetric points.

19. $g(x) = (x - 7)^2 + 9$ **20.** $f(x) = (x - 5)^2 + 3$

21. $f(x) = (x + 4)^2 + 2$ **22.** $w(t) = (t + 3)^2 - 8$

23. $f(x) = -1.5x^2$ **24.** $g(t) = 0.5t^2$

25. $h(x) = x^2 + 4$ **26.** $r(t) = 2t^2 - 12$

27. $g(x) = 4(x - 5)^2$ **28.** $m(c) = 0.3(c + 2)^2$

29. If your income level is less than the poverty threshold, you are considered to be in poverty. The poverty threshold for an individual under 65 years old can be modeled by

$$P(t) = 4.95(t - 57)^2 + 1406$$

where $P(t)$ represents the poverty threshold, in dollars, for an individual in the United States under 65 years old, t years since 1900.
Source: Model based on data from Statistical Abstract of the United States, 2001.

a. What was the poverty threshold in 1990?
b. Find $P(80)$ and explain its meaning.
c. Sketch a graph of this model.
d. According to this model, when did the poverty threshold reach a minimum? Does this minimum make sense?
e. Use your graph to estimate when the poverty threshold was $3000.
f. If the domain for this model is [57, 90], find the range.

30. The total number of people, in thousands, who were below the poverty level can be modeled by

$$N(t) = 155(t - 9)^2 + 33417$$

where $N(t)$ represents the total number of people in the United States who were below the poverty level, in thousands, t years since 1990.

Source: Model based on data from Statistical Abstract of the United States, 2007.

   a. How many people were below the poverty level in 2002?
   b. Find $N(15)$ and explain its meaning.
   c. Sketch a graph of this model.
   d. Estimate the vertex of this model and explain its meaning.
   e. Use your graph to estimate when the number of people in the United States below the poverty level was 37 million.
   f. If the domain for this model is $[3, 18]$, find the range.

31. The amount that personal households and nonprofit organizations have invested in time deposits and savings deposits can be modeled by

$$D(t) = 17(t - 3)^2 + 2300$$

where $D(t)$ represents the amount in billions of dollars that households and nonprofit organizations have invested in time and savings accounts t years since 1990.

Source: Model based on data from Statistical Abstract of the United States, 2010.

   a. Sketch a graph of this model.
   b. How much did households and nonprofit organizations have invested in time and savings accounts in 1996?
   c. In what year were these accounts at their lowest levels?
   d. Use your graph to estimate when the amount that households and nonprofit organizations have invested in time and savings accounts reached $3000 billion.
   e. If the domain for this model is $[0, 18]$, find the range.

32. The percentage of the female population that experiences congestive heart failure based on age in years can be modeled by

$$H(a) = 0.004(a - 26)^2 + 0.007$$

where $H(a)$ represents the percentage, as a whole number, of women a years old who experience congestive heart failure.

Source: Heart Disease and Stroke Statistics, 2004 Update, American Heart Association.

   a. Sketch the graph of this model.
   b. Find $H(30)$ and explain what it means in regard to the percentage of the female population that experiences congestive heart failure.
   c. According to this model, at what age does the lowest percentage of women experience congestive heart failure?
   d. Use your graph to estimate the age at which 5% of women would expect to experience congestive heart failure.
   e. Give a reasonable domain and range for this model.

33. The height of a ball dropped from the roof of a building can be found by using the model

$$h(t) = -16t^2 + 256$$

where $h(t)$ is the height of the ball in feet t seconds after being dropped.

   a. Find $h(2)$ and explain its meaning.
   b. Sketch a graph of this model.
   c. Use your graph to estimate when the ball will hit the ground.
   d. Give a reasonable domain and range for this model. (*Hint*: This model will not work after the ball hits the ground.)

34. A person getting ready to jump out of a plane dropped her altimeter, an instrument that gives the altitude. The height of the altimeter can be modeled by

$$h(t) = -16t^2 + 16000$$

where $h(t)$ is the height of the altimeter in feet t seconds after being dropped.

   a. Find $h(10)$ and explain its meaning.
   b. Sketch a graph of this model.
   c. Use your graph to estimate when the altimeter will hit the ground.
   d. Give a reasonable domain and range for this model.

35. The monthly profit made from selling round-trip airline tickets from New York City to Orlando, Florida, can be modeled by

$$P(t) = -0.000025(t - 120)^2 + 2.75$$

where $P(t)$ is the monthly profit in millions of dollars when the tickets are sold for t dollars each.

a. Find $P(100)$ and explain its meaning.
b. What price will maximize the monthly profit for this route? What is the maximum profit?
c. Sketch a graph of this model.
d. Use your graph to estimate what price the company should set to have a monthly profit of 2 million dollars.
e. Give a reasonable domain and range for this model.

36. An outlet electronics store has calculated from past sales data the revenue from selling refurbished iPod nanos. The weekly revenue from selling these refurbished iPod nanos can be modeled by

$$R(n) = -1.5(n - 60)^2 + 5700$$

where $R(n)$ is the weekly revenue in dollars from selling n refurbished iPod nanos.
a. What is the weekly revenue from selling 30 refurbished iPod nanos?
b. Find the maximum weekly revenue and how many refurbished iPod nanos must be sold per week to make the maximum revenue.
c. Sketch a graph of this model.
d. Use your graph to estimate how many refurbished iPod nanos must be sold in a week to make revenue of $5400.
e. Give a reasonable domain and range of this model.

37. The Centers for Disease Control and Prevention (CDC) collect data on the percentages of deaths caused by influenza (flu) and pneumonia to help alert health officials of flu epidemics that may be occurring in their areas. If the percentage of deaths grows to more than the epidemic threshold, then an epidemic is declared. The epidemic threshold for the first 6 months of 2006 can be modeled by

$$E(w) = -0.006(w - 10)^2 + 8$$

where $E(w)$ is the percentage, as a whole number, of all deaths due to pneumonia and influenza that represents the epidemic threshold during the wth week of the year.
a. Find $E(4)$ and explain its meaning.
b. Sketch a graph of this model.
c. Use your graph to determine if there was a flu epidemic during the 23rd week of 2006 if 6.4% of deaths were caused by pneumonia or influenza.
d. What is the vertex of this model and what does it represent in this context?
e. What is a reasonable domain and range for this model?

38. The total amount of fresh vegetables consumed per person per year can be modeled by

$$V(t) = 0.7(t - 1.5)^2 + 95.4$$

where $V(t)$ is the total amount of fresh vegetables consumed per person in pounds per year t years since 2000.
Source: http://www.ers.usda.gov/data/foodconsumption/, accessed on 12/29/2006.

a. Find $V(5)$ and explain its meaning.
b. Find the vertex of this model and explain its meaning.
c. Sketch a graph of this model for the years 2000 through 2010.
d. Use your graph to estimate the year(s) when the total amount of fresh vegetables consumed per person per year is 125 pounds.
e. If the domain of this model is [0, 10] find the range.

For Exercises 39 through 46, give the domain and range of the quadratic functions.

39. $f(x) = -2(x + 4)^2 + 15$

40. $g(x) = -0.25(x - 5)^2 + 17$

41. $f(x) = 5(x + 4)^2 - 58$

42. $h(x) = 0.3(x - 2.7)^2 + 5$

43.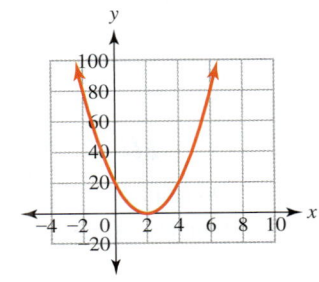

44.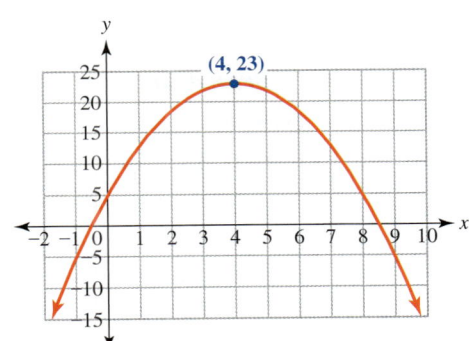

45.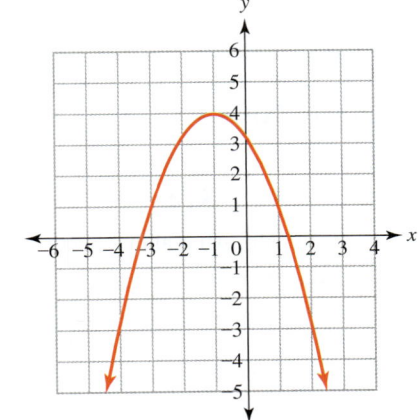

46.

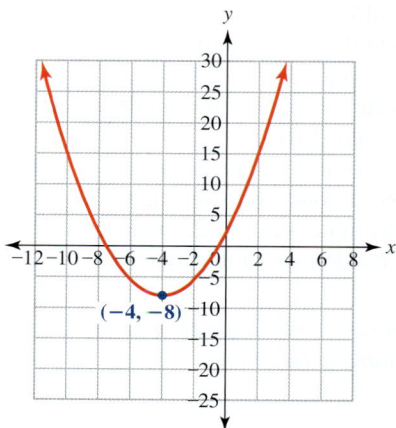

For Exercises 47 through 50, the students were asked to give the domain and range of the given functions. Explain what the student did wrong. Give the correct domain and range of the function.

47. $f(x) = (x + 4)^2 - 9$

> **Eli**
> Domain: $x > -4$
> Range: $y > -9$

48. $f(x) = (x - 6)^2 - 2$

> **Brandan**
> Domain: $[-\infty, \infty]$
> Range: $[-\infty, -2]$

49. $f(x) = -2(x - 3)^2 + 7$

> **Cody**
> Domain: $(-\infty, \infty)$
> Range: $(7, \infty)$

50. $f(x) = -0.5(x + 8)^2 + 3$

> **Rio**
> Domain: All real numbers
> Range: $y \geq 3$

For Exercises 51 through 68, sketch the graph of the given functions and label the vertex, vertical intercept, and at least one additional pair of symmetric points. Give the domain and range of the function.

51. $h(t) = (t - 3)^2 - 4$
52. $f(x) = (x + 4.5)^2 - 3$
53. $k(n) = 2n^2$
54. $b(w) = -3w^2$
55. $f(x) = 0.4x^2 + 2$
56. $g(t) = -0.25t^2 + 6$
57. $f(x) = -2(x + 4)^2$
58. $r(t) = -0.15(t - 7)^2$
59. $q(a) = 3(a + 7)^2 + 2$
60. $B(n) = 4(n - 3)^2 - 4$
61. $T(z) = 0.2(z + 6)^2 - 8$
62. $g(m) = 2.4(m + 4)^2 + 20$
63. $h(x) = -5(x - 17)^2 - 15$
64. $h(d) = -0.4(d + 15)^2 - 8$
65. $s(t) = -0.25(t + 50)^2 + 25$
66. $C(u) = -0.007(u - 400)^2$
67. $p(w) = 123(w - 4)^2 - 2500$
68. $f(x) = 5(x + 13)^2 - 45$

In Exercises 69 through 78, answer the following questions.

a. Find the vertex.
b. Is the parabola wide or narrow as compared to $f(x) = x^2$?
c. Does the parabola face upward or downward?
d. Give a calculator window that will give a good picture of this parabola that fills the calculator screen.

- X Min   X Max
- Y Min   Y Max   *Answers will vary.*

69. $f(x) = 5x^2 + 100$
70. $f(x) = -0.01x^2 + 200$
71. $f(x) = (x + 30)^2 - 50$
72. $f(x) = 20(x - 100)^2 + 250$
73. $f(x) = 0.002(x + 20)^2 + 50$
74. $f(x) = 0.0005(x - 1000)^2$
75. $f(x) = 0.0005(x - 1000)^2 + 1000$
76. $f(x) = -0.0005(x + 10,000)^2 + 5000$
77. $f(x) = -10(x + 25,000)^2 - 10,000$
78. $f(x) = (x - 55.8)^2 - 29.7$

4.3 Finding Quadratic Models

LEARNING OBJECTIVES
- Find a quadratic model for data.
- Use a quadratic model to make predictions.

Finding Quadratic Models

Now we will use all the information we have about graphing quadratics from the vertex form to find models for data that take the shape of a parabola. We will also look more at the domain and range of quadratic functions and what we should be aware of when considering model breakdown. Use the following steps to find a model for quadratic data.

> **Steps to Finding Quadratic Models**
> 1. Define the variables and adjust the data (if needed).
> 2. Create a scatterplot.
> 3. Select a model type (linear or quadratic).
> 4. **Quadratic model:** Pick a vertex and substitute it for h and k in the vertex form $f(x) = a(x - h)^2 + k$.
> 5. Pick another point and use it to find a.
> 6. Write the equation of the model using function notation.
> 7. Check your model by graphing it with the scatterplot.

Example 1 Finding a quadratic model

The average monthly temperature in Anchorage, Alaska, is given in the chart.

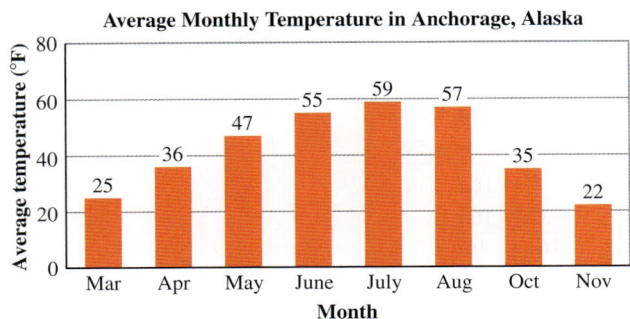

Source: Weatherbase.com.

a. Will a linear or quadratic model fit these data better? Why?

b. Find an equation for a model of these data.

c. Using your model, estimate the average temperatures during September and February. Would either of these months be a good time to travel to Alaska if you wanted to visit when temperatures were mild?

d. The actual average temperature during September is 48°F, and during February is 18°F. How well does your model predict these values?

SOLUTION

a. A quadratic model will fit these data best because the distribution is shaped like a downward-facing parabola.

b. Step 1 Define the variables and adjust the data (if needed).

$T(m)$ = Average temperature in Anchorage, Alaska (in degrees Fahrenheit)

m = Month of the year; e.g., $m = 5$ represents May

The months given in the bar chart must be translated into numerical values as shown in the table.

m	$T(m)$ (degrees Fahrenheit)
3	25
4	36
5	47
6	55
7	59
8	57
10	35
11	22

Step 2 Create a scatterplot.
The inputs go from 3 to 11, and the outputs go from 22 to 59, so spreading out some has the following result.

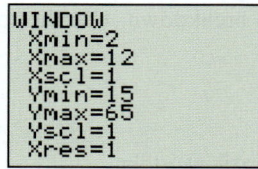

Step 3 Select a model type. We now have linear and quadratic to choose from. Because these data have the shape of a downward-facing parabola, we choose a quadratic model.

Step 4 Quadratic model: Pick a vertex and substitute it for h and k in the vertex form.
The highest point on this scatterplot looks like a reasonable vertex, so we will use it as the vertex.

$$\text{vertex} = (7, 59)$$
$$T(m) = a(m - 7)^2 + 59$$

Step 5 Pick another point and use it to find a.
We can choose the last point in the data set because it is farther away from the vertex and seems to follow a smooth curve.

$$\text{other point} = (11, 22)$$

$$T(m) = a(m - 7)^2 + 59$$
$$22 = a(11 - 7)^2 + 59 \quad \text{Substitute 11 for the month and 22 for the average temperature.}$$
$$22 = a(4)^2 + 59$$
$$22 = 16a + 59 \quad \text{Solve for } a.$$
$$\underline{-59 \qquad\quad -59}$$
$$-37 = 16a$$
$$\frac{-37}{16} = \frac{16a}{16}$$
$$-2.3125 = a$$

Step 6 Write the equation of the model using function notation.
You now have a, h, and k, so you can write the model.

$$T(m) = -2.3(m - 7)^2 + 59$$

Step 7 Check your model by graphing it with the scatterplot.

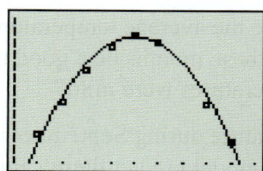

■ Connecting the Concepts

Which point should I pick?

When you are finding a quadratic model, step 5 has you pick another point and use it to find a. When deciding which point to pick, look for a point that is not too close to the vertex you already chose. If the point is too close, it is more likely that the parabola will not come close to the other data points.

If possible, pick a point that appears to have something close to a symmetric point in the data. This way, the parabola will be close to both the point you picked and its symmetric point.

c. September is the ninth month of the year, and February is the second month, so

$$T(m) = -2.3(m - 7)^2 + 59$$
$$T(9) = -2.3(9 - 7)^2 + 59$$
$$T(9) = 49.8$$

and

$$T(2) = -2.3(2 - 7)^2 + 59$$
$$T(2) = 1.5$$

The average temperature in Anchorage, Alaska, during September is 49.8°F, and during February is 1.5°F. February would not be a good month to travel, and September would be questionable for most people.

d. Our estimate for September is fairly accurate, but the estimate for February is not close to the actual value. Since our estimate for February is 16.5°F less than the actual temperature, it probably represents model breakdown in this situation.

PRACTICE PROBLEM FOR EXAMPLE 1

The number of cable television systems in the United States during the 1990s are given in the table.

a. Find an equation for a model of these data.

b. Explain what the vertex means.

c. How many cable television systems does your model predict there will be in 1999?

d. Find the vertical intercept for your model and explain its meaning.

e. Use a table or graph to estimate when there will only be 10,500 cable television systems.

Year	Cable Television Systems
1992	11,075
1993	11,217
1995	11,218
1996	11,119
1997	10,950

Source: Statistical Abstract of the United States, 2001.

In Example 1, the modeling process for quadratics is about the same as that for linear equations. If you can pick a reasonable vertex, the process is not difficult. If your model does not fit well, you might want to check your calculations from step 6. Students most often make mistakes when solving for the value of a.

Recall that when finding a model for data, you can adjust your model to fit the data better by making changes to the constants. With quadratics, that means you can adjust the values of a, h, and k. Changing the value of h will shift the parabola left or right. Changing the value of k will shift the parabola up or down. Finally, changing the value of a will change the width of the parabola. You can change one or more of these values to help you get a good eyeball best fit.

Example 2 Adjusting a quadratic model

Graph the data and the function given on your calculator. Adjust a, h, and/or k to get an eyeball best fit.

a. $f(x) = 2(x - 4)^2 + 3$

x	$f(x)$
-3	77
-1	41
0	27
2	9
4	3
8	27

b. $f(x) = -3(x + 1)^2 - 10$

x	$f(x)$
-5	-34
-3	-10
-2	-7
0	-19
2	-55
5	-154

324 CHAPTER 4 Quadratic Functions

SOLUTION

a. Set a window to fit the data and graph the data and given function.

$f(x) = 2(x - 4)^2 + 3$

x	f(x)
−3	77
−1	41
0	27
2	9
4	3
8	27

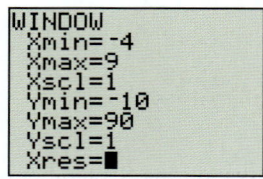 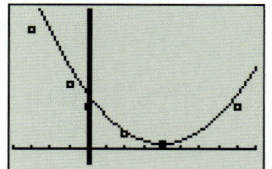

From the graph, it appears that the vertex is correct but the parabola is too narrow. We can adjust the value of a to a number closer to zero to make the parabola wider. We may have to try several values to get a good fit. Since $a = 2$, we will try $a = 1.5$ and graph the adjusted function to test the fit.

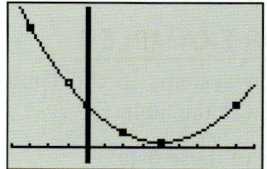

The model now fits well, so the adjusted model is $f(x) = 1.5(x - 4)^2 + 3$.

$f(x) = -3(x + 1)^2 - 10$

x	f(x)
−5	−34
−3	−10
−2	−7
0	−19
2	−55
5	−154

b. Set a window to fit the data and graph the data and given function.

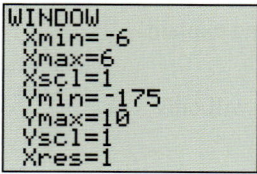 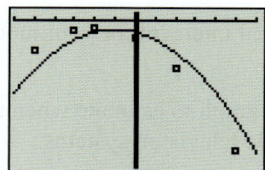

The parabola does not seem to have the correct vertex. The point $(-2, -7)$ looks like a good vertex for these data. We will change h and k to $h = -2$ and $k = -7$ and graph the adjusted function to see whether it fits the data.

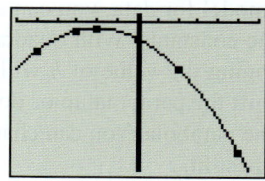

The model now fits well, so the adjusted model is $f(x) = -3(x + 2)^2 - 7$.

PRACTICE PROBLEM FOR EXAMPLE 2

Graph the data and the function given on your calculator. Adjust a, h, and/or k to get an eyeball best fit.

$f(x) = 0.5(x - 3)^2 + 6$

x	f(x)
−2	−6.5
−1	−2
1	4
2	5.5
4	5.5
5	4

Domain and Range

Finding a reasonable domain is exactly the same as for linear models. Try to expand the domain beyond the given data unless model breakdown occurs. You will want to be careful when considering model breakdown, as many quadratic models are not reliable very far beyond the given range of data. Because quadratic models will increase or decrease faster and faster, they usually outpace the actual growth or decay in the situation.

Finding the range still involves the lowest to highest output values of the function within the set domain. Unlike with linear models, be careful not to assume that the lowest and highest outputs of the function will be the outputs from the ends of the domain. Because quadratics have a maximum or minimum point at the vertex, the range will typically include the output value of the vertex as the highest or lowest value. If the vertex is within the domain values, it will be one endpoint of the range, and one of the ends of the domain will provide the other range endpoint. Be sure to give the range stating the lowest output followed by the highest output value.

In Example 1, we saw in part d that the model will work well within the given data but not so well beyond the data. Therefore, we can set the domain to be [3, 11] or $3 \leq m \leq 11$.

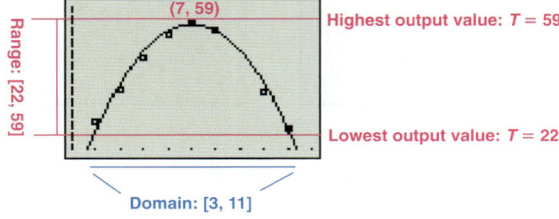

To find the range for this model, we look for the lowest and highest points on this model within the chosen domain. The highest point is clearly the vertex and has a value of 59. The lowest point is on either the right or the left-side of the domain. To determine what the lowest output is, we will substitute the two ends of the domain and find the related output values.

$$T(3) = 22.2$$
$$T(11) = 22.2$$

Because both of these values happen to be the same, the lowest output value is 22.2, so our range will be [22.2, 59] or $22.2 \leq T \leq 59$.

Example 3 Finding a quadratic model

The number of public branch libraries in the United States for several years is given in the table.

Year	Public Library Branches
1994	6223
1995	6172
1996	6205
1997	6332
1998	6435

Source: Statistical Abstract of the United States, 2001.

a. Find an equation for a model of these data.

b. Using your model, estimate the number of branch libraries in 2000.

c. Give a reasonable domain and range for this model.

d. Use the graph or table to estimate in what year(s) there are 6500 branch libraries.
e. Use this model to determine whether it would be reasonable to assume that the number of careers in library science is going to be growing in the future.

SOLUTION

a. **Step 1** Define the variables and adjust the data (if needed).

$B(t)$ = Number of public branch libraries in the United States

t = Time in years since 1990

Adjusting the years to years since 1990 gives us the table.

t	$B(t)$
4	6223
5	6172
6	6205
7	6332
8	6435

Step 2 Create a scatterplot.

The inputs go from 4 to 8, and the outputs go from 6172 to 6435, so spreading out some has the following result.

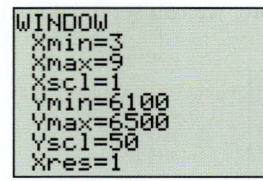
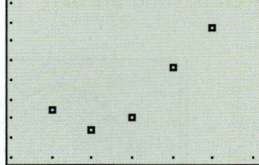

Step 3 Select a model type.

Because the data show the shape of an upward-facing parabola, we choose a quadratic model.

Step 4 Quadratic model: Pick a vertex and substitute it for h and k in the vertex form.

The lowest point on this scatterplot looks like a reasonable vertex, so we will use it for the model.

$$\text{vertex} = (5, 6172)$$
$$B(t) = a(t - 5)^2 + 6172$$

Step 5 Pick another point and use it to find a.

We can choose the last data point because it is farther away from the vertex and seems to follow a smooth curve.

$$\text{other point} = (8, 6435)$$
$$B(t) = a(t - 5)^2 + 6172$$
$$6435 = a(8 - 5)^2 + 6172 \quad \text{Substitute 8 for the year and}$$
$$6435 = a(3)^2 + 6172 \quad \text{6435 for the number of public branch libraries.}$$
$$6435 = 9a + 6172$$
$$\underline{-6172 \quad\quad\; -6172} \quad \text{Solve for } a.$$
$$263 = 9a$$
$$29.22 = a$$

Step 6 Write the equation of the model using function notation.

You now have a, h, and k, so you can write the model.

$$B(t) = 29.22(t - 5)^2 + 6172$$

Step 7 Check your model by graphing it with the scatterplot.

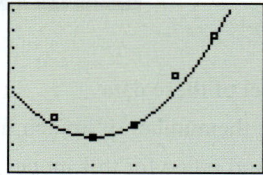

This model fits the data fairly well, but the two points that it misses are both above the model, so moving the model up slightly might make a better fit. We can adjust the model for a better fit by making k bigger. The new model becomes

$$B(t) = 29.22(t - 5)^2 + 6180$$

The graph of this adjusted model is closer overall to all the data points.

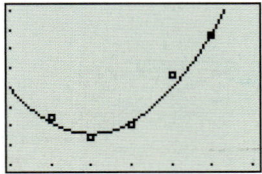

Note that the vertex of our adjusted model differs from the one we picked in step 5. There are several answers for this model that would be considered reasonable in this case. Remember that we are trying to find an "eyeball best fit," so there can be some room for differences in final models.

b. The year 2000 is represented by $t = 10$, so

$$B(10) = 29.22(10 - 5)^2 + 6180$$
$$B(10) = 6910.5$$

In 2000, there were approximately 6911 public branch libraries.

c. It is probably reasonable to assume that there will continue to be more public library branches as populations continue to grow. Therefore, we will spread out the domain to include a few years beyond the given data, giving us a domain of $[3, 10]$ or $3 \leq t \leq 10$.

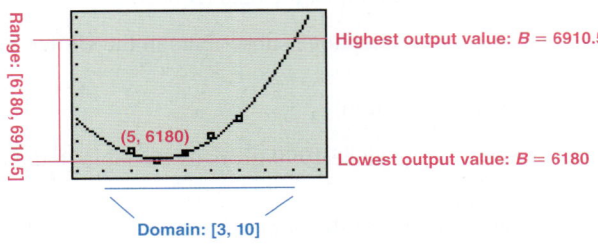

The resulting range will have a minimum value at the vertex of 6180, and the maximum appears to be on one of the ends of the domain. $B(3) = 6296.88$ and $B(10) = 6910.5$, so the maximum value is 6910.5 when $t = 10$, and therefore, the range will be $[6180, 6910.5]$ or $6180 \leq B \leq 6910.5$.

d. Using the table, we get an estimate of $t = 2$ or $t = 8$. There were about 6500 branch libraries in 1992 and again in 1998. Remember that you are solving a quadratic equation and because of its symmetry, you will usually have two answers to this type of question. In some cases, one of those answers may not make sense in the context of the situation and, therefore, would be considered model breakdown. In this case, both answers lie within our domain and seem reasonable.

e. If the number of branch libraries continues to grow, the number of library science careers should grow as well.

Some data have the same curve as a parabola but do not form both sides of the U shape. When data show only half of the parabola shape, you should use caution in determining the domain and range. If the situation does not make sense for the other half of the parabola, model breakdown may occur on the other side of the vertex.

Example 4 Finding a quadratic model

The number of cellular telephone subscribers in the United States is given in the chart.

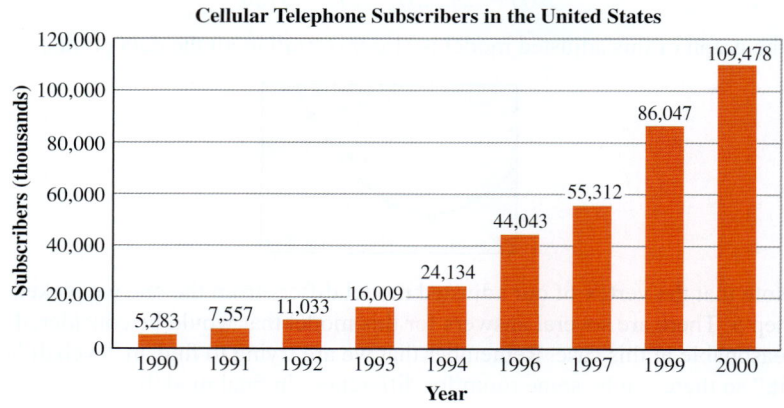

Source: Statistical Abstract of the United States, 2001.

a. Find an equation for a model of these data.

b. Estimate the number of cellular phone subscribers in 1980 and 1995.

c. The actual number of cellular phone subscribers in 1995 was 33,786,000. How does your estimate from part b compare to the actual value?

d. Give a reasonable domain and range for your model.

SOLUTION

a. Step 1 Define the variables and adjust the data (if needed).

$C(t)$ = Number of cellular telephone subscribers in the United States in thousands

t = Time in years since 1990

Adjusting the years to years since 1990 gives us the table in the margin.

Step 2 Create a scatterplot.

The inputs go from 0 to 10, and the outputs go from 5283 to 109,478, so spreading out the X and Ymax and min has the following result.

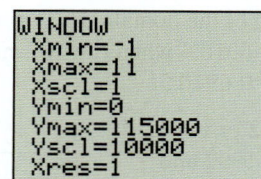 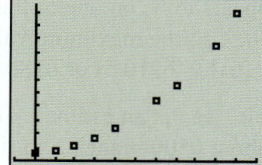

Step 3 Select a model type.

Because the distribution curves upward, the data take on the shape of the right half of an upward-facing parabola. We should choose a quadratic model.

Step 4 Quadratic model: Pick a vertex and substitute it for h and k in the vertex form.

The lowest point on this scatterplot looks like a reasonable vertex, so we will use it for the model.

$$\text{vertex} = (0, 5283)$$
$$C(t) = a(t - 0)^2 + 5283$$
$$C(t) = at^2 + 5283 \qquad \textit{Simplify the function.}$$

Year	Cellular Telephone Subscribers (thousands)
0	5,283
1	7,557
2	11,033
3	16,009
4	24,134
6	44,043
7	55,312
9	86,047
10	109,478

Step 5 Pick another point and use it to find a.
We can choose the seventh point in the data because it is some distance from the vertex and seems to follow a smooth curve that will come close to all the data points.

$$\text{other point} = (6, 44043)$$

$$C(t) = at^2 + 5283$$ *Substitute 6 for the year and 44043 for the number of cellular telephone subscribers.*

$$44043 = a(6)^2 + 5283$$

$$44043 = 36a + 5283$$ *Solve for a.*

$$\underline{-5283 \quad\quad\quad -5283}$$

$$38760 = 36a$$

$$1076.67 = a$$

Step 6 Write the equation for the model using function notation.
You now have a, h, and k, so you can write the model.

$$C(t) = 1076.67t^2 + 5283$$

Step 7 Check your model by graphing it with the scatterplot.
This model is not bad, since it appears to be pretty balanced within the data with several points above and below the graph. Because all the points above the graph are on the left half of the graph and all the points below the graph are on the right side of the graph, we might be able to adjust it by raising the graph and then making the parabola a little wider. These adjustments will mean changing the values of a and k.

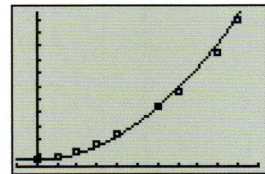

By playing with several possibilities, we might come up with the model

$$C(t) = 1000t^2 + 7000$$

and the following graph.

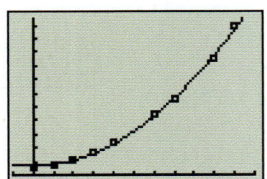

This graph comes closer to more points.

b. 1980 and 1995 are represented by $t = -10$ and $t = 5$, respectively.

$$C(-10) = 107000$$
$$C(5) = 32000$$

In 1980, there were 107,000,000 cellular telephone subscribers in the United States. This does not make sense, since the number would not decrease so drastically during the next 10 years, so we believe that this is model breakdown. The data indicated that only half the parabola would be reasonable in this situation.
In 1995, there were 32,000,000 cellular telephone subscribers.

c. Comparing the actual value in 1995 to our estimate, we are off by a little over 1 million subscribers, but that is not a very large error considering the size of these numbers. We can be satisfied with this result.

d. There were probably fewer cellular telephone subscribers before 1990. This graph trends upward to the left of the vertex. That result leads us to believe that this model is not valid for years to the left of the vertex. It does appear that we could extend the curve farther to the right. So we can use the domain [0, 12], or $0 \leq t \leq 12$.

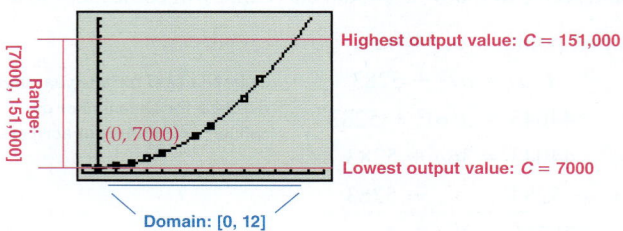

The resulting range will start at the lowest point, the vertex, and go up until it reaches the highest point at the right-hand side of the domain.

$$\text{vertex} = (0, 7000)$$
$$C(12) = 151000$$

This gives a range of [7000, 151000], or $7000 \leq C \leq 151{,}000$.

PRACTICE PROBLEM FOR EXAMPLE 4

Find an equation for a model of the data given in the table.

x	$f(x)$
-6	-33
-4	-45
-2	-49
0	-45
2	-33
4	-13

4.3 Exercises

For Exercises 1 through 6, set a window to fit the given data. Graph the given data and functions and choose which of the functions best fits the given data.

1. $f(x) = 1.5(x + 5)^2 - 15$ $f(x) = 1.8(x + 5)^2 - 15$
   $f(x) = 2.0(x + 5)^2 - 15$ $f(x) = 2.3(x + 5)^2 - 15$

x	$f(x)$
-10	35
-7	-7
-5	-15
-3	-7
1	57
3	113

2. $f(x) = -1.3(x - 3)^2 + 9$ $f(x) = -1.4(x - 3)^2 + 9$
   $f(x) = -1.5(x - 3)^2 + 9$ $f(x) = -1.6(x - 3)^2 + 9$

x	$f(x)$
-5	-80.6
-2	-26.0
0	-3.6
2	7.6
5	3.4
7	-13.4

3. $f(x) = 0.4(x - 4)^2 + 1.2$ $f(x) = 0.4(x - 4)^2 + 2.2$
   $f(x) = 0.4(x - 4)^2 + 3.2$ $f(x) = 0.4(x - 4)^2 + 4.2$

x	$f(x)$
-8	60.8
-3	22.8
0	9.6
4	3.2
7	6.8
10	17.6

SECTION 4.3 Finding Quadratic Models

4. $f(x) = -(x + 5)^2 + 16.2$ $f(x) = -(x + 5)^2 + 15.2$
    $f(x) = -(x + 5)^2 + 14.2$ $f(x) = -(x + 5)^2 + 13.2$

x	f(x)
−10	−11.8
−8	4.2
−6	12.2
−4	12.2
−2	4.2
2	−35.8

5. $f(x) = 2(x + 1.0)^2 - 20$ $f(x) = 2(x + 1.2)^2 - 20$
    $f(x) = 2(x + 1.5)^2 - 20$ $f(x) = 2(x + 1.8)^2 - 20$

x	f(x)
−4	−7.5
−2	−19.5
0	−15.5
2	4.5
4	40.5
6	92.5

6. $f(x) = -3(x - 8.0)^2 + 8$ $f(x) = -3(x - 8.4)^2 + 8$
    $f(x) = -3(x - 9.0)^2 + 8$ $f(x) = -3(x - 9.4)^2 + 8$

x	f(x)
2	−156.3
4	−79.48
6	−26.68
8	2.12
10	6.92
12	−12.28

For Exercises 7 through 12, graph the data and the function on your calculator. Adjust a, h, and/or k to get an eyeball best fit. Answers will vary.

7. $f(x) = 2.7(x - 5)^2 - 15$

x	f(x)
−1	88
0	59
2	15
4	−6
7	2
9	34

8. $f(x) = 0.6(x + 2)^2 + 4$

x	f(x)
−5	7.6
−2	4
0	5.6
2	10.4
5	23.6
7	36.4

9. $f(x) = 2(x - 7)^2 + 15$

x	f(x)
−1	−113
2	−35
4	−3
6	13
8	13
10	−3

10. $f(x) = -1.3(x - 3.1)^2 + 18.6$

x	f(x)
0	−9.2
1	−0.9
2	4.8
3	8.0
4	8.5
5	6.4

11. $f(x) = 4x^2 - 10$

x	f(x)
−6	54
−4	6
−2	−10
0	6
2	54
4	134

12. $f(x) = -(x + 3)^2$

x	f(x)
−9	−11
−6	16
−3	25
0	16
3	−11
6	−56

332 CHAPTER 4 Quadratic Functions

For Exercises 13 through 20, state which of a, h, and/or k in the vertex form you would change to make the parabola fit the data better. Explain your answer.

13.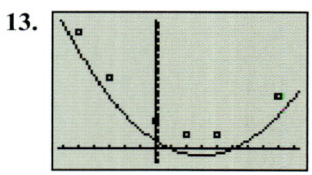
14.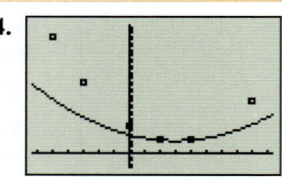
15.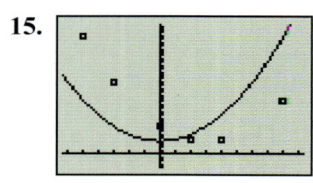
16.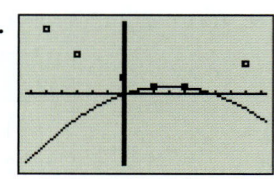
17.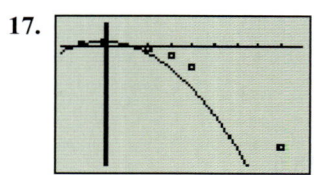
18.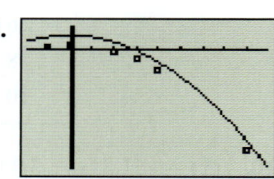
19.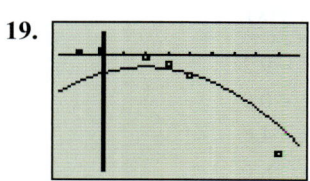
20.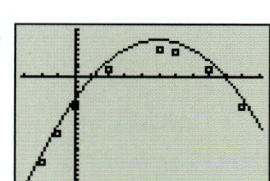

For Exercises 21 through 30, answers will vary.

21. The total revenue from selling cycling gloves is given in the table.

Price (dollars)	Revenue (dollars)
10	400
15	473
20	500
25	476
30	405

a. Find an equation for a model of these data.
b. Explain the meaning of the vertex for your model.
c. What does your model predict the revenue will be if the gloves sell for $22 a pair?
d. Give a reasonable domain and range for your model.

22. The total revenue from selling water bottles is given in the table.

Price (dollars)	Revenue (dollars)
7	1380
8	1440
9	1350
10	1100
11	660

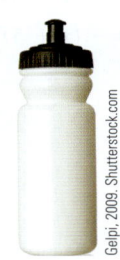

a. Find an equation for a model of these data.
b. Explain the meaning of the vertex for your model.
c. What does your model predict the revenue will be if the water bottles sell for $9.50?
d. Give a reasonable domain and range for your model.

23. The average monthly low temperatures in Anchorage, Alaska, for certain months are given in the table.

Month	Average Low Temperature (in °F)
April	28
May	39
June	47
July	51
August	49
September	41
October	28
November	15

Source: Weatherbase.com.

a. Find an equation for a model of these data.
b. Explain the meaning of the vertex for your model.
c. What does your model predict the average low temperature will be in March?
d. Give a reasonable domain and range for your model.

24. The Reduced-Gravity Program operated by NASA provides astronauts with the unique "weightless" or "zero-g" environment of space flight for tests and training. The program uses a specially modified KC-135A turbojet transport that flies on a parabolic path along the data given in the table.

Time (seconds)	Altitude (thousands of feet)
10	26
20	31
32	32.5
45	31
60	25

Source: NASA Lyndon B. Johnson Space Center Reduced-Gravity Office.

a. Find an equation for a model of these data.
b. Find the altitude of the plane after 25 seconds.
c. Give a reasonable domain and range for your model.

25. The number of Hispanic families living below the federal poverty level in the United States is given for several years in the chart.

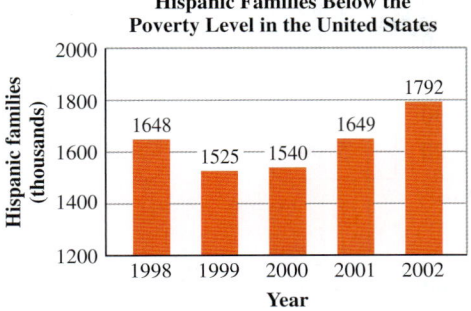

Source: Statistical Abstract of the United States, 2004–2005.

   a. Find an equation for a model of these data.
   b. Estimate the number of Hispanic families below the poverty level in 2005.
   c. Explain the meaning of the vertex.
   d. Give a reasonable domain and range for your model.

26. Of the total number of people who own a home computer, the percentage of those who use their home computers six or seven days a week varies by age. Some of these percentages are given in the table.

Age	Percentage
25	26.7
35	24.2
45	24.3
55	26.6
65	29.7

   Source: Statistical Abstract of the United States, 2001.

   a. Find an equation for a model of these data.
   b. Give a reasonable domain and range for this model.
   c. Estimate the percent of 15-year-olds who use their home computers six or seven days a week.

27. The number of hours per year an average American spent using the Internet is given in the chart.

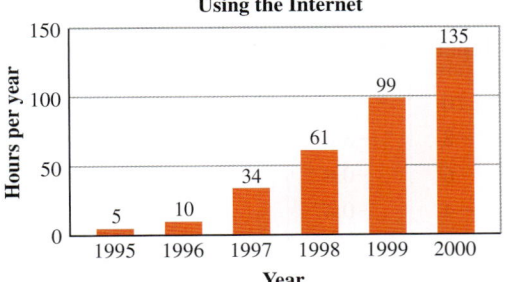

Source: Statistical Abstract of the United States, 2001.

   a. Find an equation for a model of these data.
   b. Determine whether model breakdown occurs near the vertex.
   c. Give a reasonable domain and range for this model.
   d. Estimate the number of hours per year the average person spent on the Internet in 2001.
   e. Use a table or graph to estimate when the average person will spend 365 hours per year on the Internet.

28. In 2007, Mattell and other manufacturers had to recall several products because plants in China had used paint and other materials with high concentrations of lead. In the United States, the use of lead in paint has been decreasing since its peak around 1920. Lead use in paint for several years is given in the table.

Year	Lead Use (thousands of tons)
1940	70
1950	35
1960	10
1970	5
1980	0.01

   Source: Estimated from information at AmericanScientist.org.

   a. Find an equation for a model of these data.
   b. Estimate the amount of lead used in paints in 1955.
   c. Estimate the amount of lead used in paints in 2000.
   d. Give a reasonable domain and range for this model.
   e. Use a table or the graph to estimate when 50 thousand tons of lead were used in paints.

29. A hit at a major league baseball game flew along the following path from home plate.

Horizontal Distance (feet)	Height (feet)
10	15
75	78
110	103
190	136
230	140
310	123
400	65

   a. Find an equation for the model of these data.
   b. Give a reasonable domain and range for this model.
   c. Find the vertex of this model and explain its meaning.
   d. If the center wall of the stadium is 450 feet from home plate and is 10 feet tall, will the ball make it over the wall?

30. A fly ball hit at the same major league baseball game had the following vertical heights.

Time (seconds)	Height (feet)
1	132
3	292
5	324
6	290
7	228
9	4

a. Find an equation for the model of these data.
b. Give a reasonable domain and range for this model.
c. Estimate the height of the ball at 8 seconds.
d. Use your graph to estimate how long it took for the ball to hit the ground.
e. Find the maximum height of the ball and when it reached that height.

For Exercises 31 through 44, find an equation for a model of the given data. Give the domain and range for the model you found. (**Hint:** *These problems do not have a context, so the domain and range will not be restricted.*) Answers will vary

31.

Input	Output
−7	−8
−5	8
−3	−8
−1	−56
1	−136
2	−188

32.

Input	Output
−20	121.2
−10	140.2
4	150.0
15	143.95
30	116.2
50	44.2

33.

Input	Output
1	12
2	5
4	−3
6	−3
8	5
9	12

34.

Input	Output
−2	−12
−1	−9.5
0	−7
3	0.5
4	3
6	8

35.

Input	Output
−100	15,300
−80	−1500
−60	−8700
−40	−6300
−30	−1500
−20	5700

36.

Input	Output
3.5	57.6
7	42.3
10.5	33.1
17.5	33.1
21	42.3
24.5	57.6

37.

Input	Output
−2	−50
0	−22
2	−2
4	10
6	14
8	10

38.

Input	Output
−15	89.9
−12	25.1
−9	3.5
−6	25.1
−3	89.9
0	197.9

39.

Input	Output
−2	−21.4
0	−15.0
2	−8.6
4	−2.2
6	4.2
8	10.6

40.

Input	Output
−3	30
−2	10
−1	−2
1	−2
2	10
3	30

41.

Input	Output
−4	−16
−2	2
0	8
2	2
4	−16
6	−46

42.

Input	Output
−2	192
0	108
2	48
4	12
6	0
8	12

43.

Input	Output
−10	−10
−8	−3.6
−6	−0.4
−4	−0.4
−2	−3.6
0	−10

44.

Input	Output
−3	103.5
−2	46
−1	11.5
0	0
1	11.5
2	46

4.4 Solving Quadratic Equations by the Square Root Property and Completing the Square

LEARNING OBJECTIVES
- Solve a quadratic equation using the square root property.
- Solve a quadratic equation by completing the square.

In this section, we will learn to solve quadratic equations using both the vertex form and the standard form. We will learn several tools that can be used to solve these equations. One thing you should consider is what equation types each tool helps you solve. This way you will know when to use a particular tool.

Solving from Vertex Form

In the previous sections, we were given values for the input variable and were asked to solve the function for the output variable. This has not been too difficult, since it involves simplifying one side of an equation.

When we are given a value for the output variable, solving for the input becomes more difficult. The reason for this difficulty is that the input variable is squared and, therefore, harder to isolate. Isolating the variable will require removing an exponent of 2. This cannot be done by addition, subtraction, multiplication, or division. Therefore, we require an operation that will undo a squared term. That operation is the square root, $\sqrt{\ }$.

One characteristic of numbers to be cautious about is that when we square a real number, squaring will effectively remove any negative sign. We will have to take extra care when we want to undo that operation.

$$x^2 = 25$$

$(5)^2 = 25 \qquad (-5)^2 = 25$
$x = 5 \qquad\qquad x = -5$

We know that 5 squared is 25. Negative 5 squared is also 25. Both must be given as answers to this equation.

$$x^2 = 25$$
$$x = \pm\sqrt{25}$$
$x = 5 \qquad\qquad x = -5$

When using a square root, we must use the plus/minus symbol to represent both answers.

When we write the square root symbol, we are referring to the nonnegative square-root, which is called the *principal square root*. Therefore, $\sqrt{25} = 5$. Because the principal square root only accounts for nonnegative solutions we lose possible negative results, so we must use a plus/minus symbol ($\pm$) to show that there are two possible answers. Using the square root and plus/minus is an example of the **square root property**.

What's That Mean?

Quadratics

- Quadratic **expression** (no equals sign)
$$2x^2 + 3x - 5$$

- Quadratic **equation** (has an equals sign)
$$2x^2 + 3x - 5 = 20$$

- Quadratic **function** (written in function notation)
$$f(x) = 2x^2 + 3x - 5$$

- Parabola: the graph of a quadratic function

DEFINITION

Square root property If $c \geq 0$, the solutions to the equation
$$x^2 = c$$
are
$$x = \pm\sqrt{c}$$
If c is negative, then the equation has no real solutions.

With this property in mind, we are going to look at two basic ways to solve quadratic equations: the square root property and completing the square. First, we will use the square root property to solve quadratics when they are given in vertex form.

Steps to Solving Quadratic Equations Using the Square Root Property

1. Isolate the squared variable expression.
2. Use the square root property to undo the square. Use $\pm$ on the side away from the variable.
3. Rewrite as two equations and solve.
4. Check answer(s) in the original equation. Be sure that answers are within the domain of the model and that they make sense in the context of the problem.

Example 1 Solving quadratic equations using the square root property

Solve $10 = 2(x - 4)^2 - 8$.

SOLUTION

$$10 = 2(x-4)^2 - 8$$
$$\underline{+8 \qquad\qquad\qquad +8}$$
$$18 = 2(x-4)^2$$

Isolate the squared variable expression.

$$\frac{18}{2} = \frac{2(x-4)^2}{2}$$
$$9 = (x-4)^2$$
$$\pm\sqrt{9} = x - 4$$
$$\pm 3 = x - 4$$

Use the square root property. Don't forget the plus/minus symbol.

$$3 = x - 4 \quad \text{or} \quad -3 = x - 4$$
$$7 = x \quad\quad \text{or} \quad\quad 1 = x$$

Rewrite as two equations and solve.

$$10 \stackrel{?}{=} 2(7-4)^2 - 8$$
$$10 = 10$$

Check both answers.

$$10 \stackrel{?}{=} 2(1-4)^2 - 8$$
$$10 = 10$$

Both answers work.

$x = 7$ and $x = 1$ are both valid answers to this equation.

Example 2 Solving applications using the square root property

In the Example 1 Practice Problem of Section 4.3 on page 323, we looked at the number of cable television systems in the United States and found the model

$$C(t) = -33.3(t-4)^2 + 11250$$

where $C(t)$ represents the number of cable systems in the United States t years since 1990.

a. Find the year when there were 10,000 cable systems.

b. Find the horizontal intercepts and explain their meaning.

SOLUTION

a. The quantity 10,000 cable systems is represented by $C(t) = 10000$, so we solve for t.

$$C(t) = -33.3(t-4)^2 + 11250$$
$$10000 = -33.3(t-4)^2 + 11250$$
$$\underline{-11250 \qquad\qquad\qquad\qquad -11250}$$
$$-1250 = -33.3(t-4)^2$$

Isolate the squared variable expression.

$$\frac{-1250}{-33.3} = \frac{-33.3(t-4)^2}{-33.3}$$

SECTION 4.4 Solving Quadratic Equations by the Square Root Property and Completing the Square 337

$$37.538 \approx (t - 4)^2$$
$$\pm\sqrt{37.538} \approx t - 4$$
$$\pm 6.127 \approx t - 4$$

Use the square root property, being sure to use a plus/minus symbol to represent both possible answers.

| $6.127 \approx t - 4$ | or | $-6.127 \approx t - 4$ |
| $10.127 \approx t$ | or | $-2.127 \approx t$ |

Rewrite as two equations and add 4 to both sides.

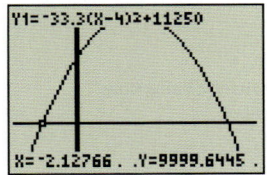 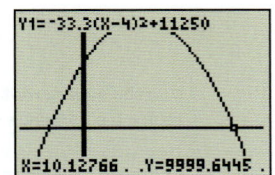

You can check your answers using the graph and trace or the table feature of your calculator.
Both answers work.

Here, $t \approx 10.127$ represents about the year 2000, and $t \approx -2.127$ represents about 1988. The value $t \approx -2.127$ is not within the domain for this model, so we must discard it, leaving us with there being about 10,000 cable systems in the year 2000.

b. To find the horizontal intercepts, we let the output variable equal zero and solve for t.

$$C(t) = -33.3(t - 4)^2 + 11250$$
$$0 = -33.3(t - 4)^2 + 11250$$
$$\underline{-11250 -11250}$$
$$-11250 = -33.3(t - 4)^2$$

Isolate the squared variable expression.

$$\frac{-11250}{-33.3} = \frac{-33.3(t - 4)^2}{-33.3}$$
$$337.84 \approx (t - 4)^2$$
$$\pm\sqrt{337.84} \approx t - 4$$
$$\pm 18.38 \approx t - 4$$

Use the square root property, being sure to use a plus/minus symbol to represent both possible answers.

| $18.38 \approx t - 4$ | or | $-18.38 \approx t - 4$ |
| $22.38 \approx t$ | or | $-14.38 \approx t$ |

Rewrite as two equations and add 4 to both sides.

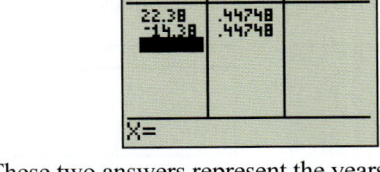

You can check your answers using the table feature of your calculator. The rounding in this solution means that the y-value will not be exactly zero.

These two answers represent the years 2012 and 1975, when, according to this model, there were no cable television systems in the United States. Since these inputs are not in a reasonable domain of the model, the answers are probably model breakdown.

PRACTICE PROBLEM FOR EXAMPLE 2

In Section 4.2, we investigated the average number of hours per person per year spent watching television during the late 1990s. We used the model

$$H(t) = 9.5(t - 7)^2 + 1544$$

where $H(t)$ represents the average number of hours per person per year spent watching television t years since 1990. Determine when the average person will spend 1825 hours per year watching television.

Example 3 Solving quadratic equations using the square root property

Solve the following.

a. $5x^2 + 10 = 255$ **b.** $4(2t + 5)^2 - 82 = -62.64$ **c.** $(x + 5)^2 - 4 = -20$

SOLUTION

a.
$$5x^2 + 10 = 255$$ Isolate the squared variable expression.
$$\underline{ -10 \quad -10}$$
$$5x^2 = 245$$
$$\frac{5x^2}{5} = \frac{245}{5}$$
$$x^2 = 49$$
$$x = \pm\sqrt{49}$$ Use the square root property, being sure to use a plus/minus symbol.
$$x = \pm 7$$
$$5(7)^2 + 10 \stackrel{?}{=} 255$$ Check both answers.
$$255 = 255$$
$$5(-7)^2 + 10 \stackrel{?}{=} 255$$ Both answers work.
$$255 = 255$$

b.
$$4(2t + 5)^2 - 82 = -62.64$$ Isolate the squared variable expression.
$$4(2t + 5)^2 = 19.36$$
$$(2t + 5)^2 = 4.84$$
$$2t + 5 = \pm\sqrt{4.84}$$ Use the square root property, being sure to use a plus/minus symbol.
$$2t + 5 = \pm 2.2$$
$$2t + 5 = 2.2 \quad \text{or} \quad 2t + 5 = -2.2$$ Rewrite as two equations and solve.
$$2t = -2.8 \quad \text{or} \quad 2t = -7.2$$
$$t = -1.4 \quad \text{or} \quad t = -3.6$$
$$4(2(-1.4) + 5)^2 - 82 \stackrel{?}{=} -62.64$$ Check both answers.
$$-62.64 = -62.64$$
$$4(2(-3.6) + 5)^2 - 82 \stackrel{?}{=} -62.64$$
$$-62.64 = -62.64$$ Both answers work.

We can also verify these solutions by graphing both sides of the equation on a calculator and verifying that the two graphs cross at these input values.

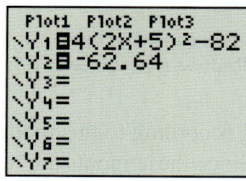 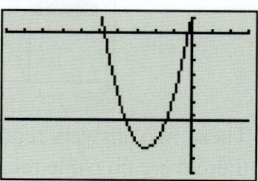

c.
$$(x + 5)^2 - 4 = -20$$ Isolate the squared variable expression.
$$(x + 5)^2 = -16$$
$$x + 5 = \pm\sqrt{-16}$$ Use the square root property.

When we used the square root property, we ended with a negative under the square root, so there are no real solutions to this equation. If we look at the graph of both sides of this equation, we can see that they do not cross and are never equal.

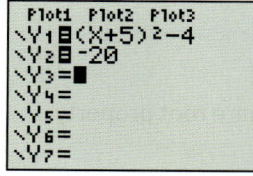 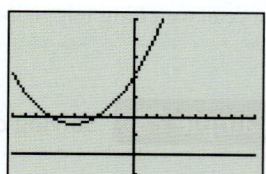

PRACTICE PROBLEM FOR EXAMPLE 3

Solve the following.

a. $7x^2 + 9 = 121$ b. $-3(4x - 7)^2 + 17 = -58$

Completing the Square

If a quadratic equation has both a second-degree term and a first-degree term ($ax^2 + \underline{bx} + c$), the square root property cannot be easily used to solve. If we try to get the squared variable expression by itself, we will have a variable term in the way of the square root. To handle this problem, we use a technique called *completing the square*. Completing the square will transform the equation so that it has a perfect square that can be solved using the square root property.

The idea behind this technique comes from the factoring process that we examined in Chapter 3. Recall from Chapter 3 that a perfect square trinomial will factor to the square of a binomial.

$$a^2 + 2ab + b^2 = (a + b)^2$$
$$a^2 - 2ab + b^2 = (a - b)^2$$

> **What's That Mean?**
>
> **Degree**
>
> Recall from Section 3.2 that the degree of a term is the sum of all the exponents of the variables in the term.
>
> A first-degree term has a single variable with an exponent of 1.
>
> A single-variable second-degree term has a variable that is squared.
>
> Quadratics are all second-degree polynomials, since the highest-degree term has a squared variable.

If we can get one side of a quadratic equation to have the characteristics of a perfect square trinomial, we will be able to factor it into a binomial squared. First, we will consider how we can make an expression into a perfect square trinomial. The expressions

$$x^2 + 10x + 25$$
$$w^2 - 12w + 36$$

are both perfect square trinomials because they factor into a binomial squared.

$$x^2 + 10x + 25 = (x + 5)^2$$
$$w^2 - 12w + 36 = (w - 6)^2$$

If we look at these two examples, we can identify a pattern that can help us complete the square of any quadratic. Consider the following expressions:

$$x^2 + 10x$$
$$w^2 - 12w$$

We know from the examples above that if we add 25 to the first expression and 36 to the second expression, they will become perfect square trinomials. These numbers are related to the coefficient of the first-degree term x or w. In particular, 25 is the square of half of 10, and 36 is the square of half of -12. In general, taking half of the coefficient of the first-degree term and squaring it will give us a constant that will make these expressions a perfect square trinomial.

$$x^2 + bx$$

The constant that will complete the square: $\left(\dfrac{b}{2}\right)^2$

$$x^2 + bx + \left(\dfrac{b}{2}\right)^2$$

$$x^2 + bx + \left(\dfrac{b}{2}\right)^2 = \left(x + \dfrac{b}{2}\right)^2$$

The factored form will include half of b as one of the terms in the binomial.

We will start the process of completing the square by simplifying one side of an equation. This process is easiest when the coefficient of the squared term is 1. If this coefficient is not 1, we will divide both sides of the equation by the coefficient to make it 1. Follow the steps to complete the square in the next example.

> **Steps to Solving Quadratic Equations by Completing the Square**
> 1. Isolate the variable terms on one side of the equation.
> 2. If the coefficient of x^2 is not 1, divide both sides of the equation by the coefficient of x^2.
> 3. Take half of the coefficient of x, then square it. Add this number to both sides of the equation.
> 4. Factor the quadratic into the square of a binomial.
> 5. Solve using the square root property.
> 6. Check your answers in the original equation.

Example 4 Solving quadratic equations by completing the square

Solve by completing the square of $5x^2 + 30x - 35 = 0$.

SOLUTION

Step 1 Isolate the variable terms on one side of the equation.
To isolate the variable terms, we move the constant term to the other side of the equation.

$$5x^2 + 30x - 35 = 0$$
$$\underline{ 35\ \ 35}$$
$$5x^2 + 30x = 35$$

Step 2 If the coefficient of x^2 is not 1, divide both sides of the equation by the coefficient of x^2.
Divide both sides by the coefficient, 5.

$$\frac{5x^2 + 30x}{5} = \frac{35}{5}$$
$$x^2 + 6x = 7$$

Step 3 Take half of the coefficient of x, then square it. Add this number to both sides of the equation.
To complete the square on the left side of the equation, we add a number that will make the resulting trinomial a perfect square trinomial. To find this number, we take half of the coefficient of x and square it. This number is then added to both sides of the equation. In this equation, the coefficient of x is 6, so we get

$$\left(\frac{6}{2}\right)^2 = 9$$

Add this constant to both sides of the equation.

$$x^2 + 6x + 9 = 7 + 9$$
$$x^2 + 6x + 9 = 16$$

Step 4 Factor the quadratic into the square of a binomial.
The left side will factor as a perfect square.

$$x^2 + 6x + 9 = 16$$
$$(x + 3)^2 = 16$$

Step 5 Solve using the square root property.

$$(x + 3)^2 = 16$$
$$x + 3 = \pm\sqrt{16} \qquad \text{Use the square root property.}$$
$$x + 3 = \pm 4$$
$$x + 3 = 4 \qquad x + 3 = -4 \qquad \text{Rewrite as two equations and solve.}$$
$$x = 1 \qquad\ \ \ x = -7$$

SECTION 4.4 Solving Quadratic Equations by the Square Root Property and Completing the Square

Step 6 Check your answers in the original equation.

$$(5(1)^2 + 30(1) - 35) \stackrel{?}{=} 0$$
$$(5 + 30 - 35) \stackrel{?}{=} 0$$
$$0 = 0$$

$$(5(-7)^2 + 30(-7) - 35) \stackrel{?}{=} 0$$
$$(5(49) - 210 - 35) \stackrel{?}{=} 0$$
$$(245 - 210 - 35) \stackrel{?}{=} 0 \qquad \text{Both answers work.}$$
$$0 = 0$$

Completing the square will allow us to solve any quadratic equation using the square root property.

Example 5 Solving quadratic equations by completing the square

Solve the following by completing the square.

a. $x^2 + 10x + 16 = 0$ **b.** $3x^2 - 15x = 12$ **c.** $6x^2 + 8 = 10x$

SOLUTION

a. The coefficient of x^2 is 1, so we do not have to divide. To get the variable terms isolated on one side of the equation, we subtract 16 from both sides.

$$x^2 + 10x + 16 = 0$$
$$\underline{ -16 \quad -16}$$
$$x^2 + 10x = -16$$

Now, we find the constant that will complete the perfect square trinomial. So we will take half of the coefficient of x and square it.

$$\frac{1}{2}(10) = 5$$
$$(5)^2 = 25$$

Add this number to both sides of the equation and factor the quadratic into the square of a binomial.

$$x^2 + 10x = -16$$
$$x^2 + 10x + 25 = -16 + 25$$
$$x^2 + 10x + 25 = 9$$
$$(x + 5)^2 = 9$$

The square root property can be used to solve the equation.

$$(x + 5)^2 = 9$$
$$x + 5 = \pm\sqrt{9} \qquad \text{Use the square root property.}$$
$$x + 5 = \pm 3$$
$$x + 5 = 3 \quad \text{or} \quad x + 5 = -3 \qquad \text{Rewrite as two equations and solve.}$$
$$x = -2 \quad \text{or} \quad x = -8$$
$$((-2)^2 + 10(-2) + 16) \stackrel{?}{=} 0 \qquad \text{Check your answers in the}$$
$$(4 - 20 + 16) \stackrel{?}{=} 0 \qquad \text{original equation.}$$
$$0 = 0$$
$$((-8)^2 + 10(-8) + 16) \stackrel{?}{=} 0$$
$$(64 - 80 + 16) \stackrel{?}{=} 0$$
$$0 = 0 \qquad \text{Both answers work.}$$

b. The variable terms are already isolated, so we start by making the coefficient of x^2 into 1. The coefficient of x^2 is 3, so dividing both sides of the equation by 3 will make this coefficient 1.

$$3x^2 - 15x = 12$$
$$\frac{3x^2 - 15x}{3} = \frac{12}{3}$$
$$x^2 - 5x = 4$$

We want to find the constant that will complete the perfect square trinomial, so we take half of the coefficient of x and square it.

$$\frac{1}{2}(5) = \frac{5}{2}$$
$$\left(\frac{5}{2}\right)^2 = \frac{25}{4}$$

Add this number to both sides of the equation and factor the quadratic into the square of a binomial.

$$x^2 - 5x = 4$$
$$x^2 - 5x + \frac{25}{4} = 4 + \frac{25}{4}$$
$$x^2 - 5x + \frac{25}{4} = \frac{16}{4} + \frac{25}{4}$$
$$\left(x - \frac{5}{2}\right)^2 = \frac{41}{4}$$

Use the square root property to solve.

$$\left(x - \frac{5}{2}\right)^2 = \frac{41}{4}$$

$$x - \frac{5}{2} = \pm\sqrt{\frac{41}{4}} \qquad \text{Use the square root property.}$$

$$x - \frac{5}{2} = \pm\frac{\sqrt{41}}{2}$$

$$x - \frac{5}{2} = \frac{\sqrt{41}}{2} \qquad x - \frac{5}{2} = -\frac{\sqrt{41}}{2} \qquad \text{Rewrite as two equations and solve.}$$

$$x = \frac{5}{2} + \frac{\sqrt{41}}{2} \qquad x = \frac{5}{2} - \frac{\sqrt{41}}{2}$$

$$x \approx 5.702 \qquad x \approx -0.702$$

$$3(5.702)^2 - 15(5.702) \stackrel{?}{=} 12 \qquad \text{Check your answers in the original equation.}$$
$$12.008 \approx 12$$
$$3(-0.702)^2 - 15(-0.702) \stackrel{?}{=} 12$$
$$12.008 \approx 12 \qquad \text{Both answers work.}$$

c. First, we rearrange the terms so that the variable terms are on one side of the equation and the constant term is on the other side.

$$6x^2 + 8 = 10x$$
$$\underline{-10x \quad -10x}$$
$$6x^2 - 10x + 8 = 0$$
$$\underline{-8 \quad -8}$$
$$6x^2 - 10x = -8$$

SECTION 4.4 Solving Quadratic Equations by the Square Root Property and Completing the Square

The coefficient of x^2 is 6, so we will divide both sides of the equation by 6.

$$6x^2 - 10x = -8$$
$$\frac{6x^2 - 10x}{6} = \frac{-8}{6}$$
$$x^2 - \frac{5}{3}x = -\frac{4}{3}$$

We want to find the constant that will complete the perfect square trinomial, so we take half of the coefficient of x and square it.

$$\frac{1}{2}\left(\frac{5}{3}\right) = \frac{5}{6}$$
$$\left(\frac{5}{6}\right)^2 = \frac{25}{36}$$

Add this number to both sides of the equation and factor the quadratic into the square of a binomial

$$x^2 - \frac{5}{3}x + \frac{25}{36} = -\frac{4}{3} + \frac{25}{36}$$
$$x^2 - \frac{5}{3}x + \frac{25}{36} = -\frac{48}{36} + \frac{25}{36} \quad \textit{Get like denominators.}$$
$$\left(x - \frac{5}{6}\right)^2 = -\frac{23}{36}$$

We can use the square root property to solve the equation.

$$\left(x - \frac{5}{6}\right)^2 = -\frac{23}{36}$$
$$x - \frac{5}{6} = \pm\sqrt{-\frac{23}{36}}$$

There is a negative number under the square root, so there is no real solution to this equation. If we graph the two sides of this equation in our calculator, we can see that the two sides never touch and thus are not equal in the real number system.

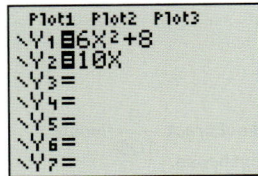
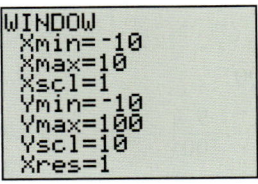
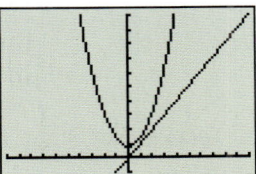

PRACTICE PROBLEM FOR EXAMPLE 5

Solve the following by completing the square.

a. $x^2 + 8x - 20 = 0$

b. $4x^2 - 20x = -8$

c. $2x^2 + 8x = -60$

Converting to Vertex Form

To put a quadratic function into vertex form, use the technique of completing the square. The process is very similar, but instead of adding the constant that will complete the square to both sides, you will add and subtract the constant on one side.

CHAPTER 4 Quadratic Functions

> **Example 6** Converting a quadratic function into vertex form

Convert to vertex form.

a. $f(x) = 2x^2 + 20x - 6$ **b.** $g(x) = 5x^2 + 3x + 20$

SOLUTION

a. Converting a function to vertex form will be similar to completing the square, as we did earlier. Because we are not solving, we will keep everything on one side of the equation, leaving the function notation alone.

$f(x) = 2x^2 + 20x - 6$ *Group the variable terms.*
$f(x) = (2x^2 + 20x) - 6$ *Factor out the coefficient*
$f(x) = 2(x^2 + 10x) - 6$ *of the squared term.*

Now we want to complete the square of the variable terms in parentheses. Find the constant that will complete the square.

$$\frac{1}{2}(10) = 5$$

$$5^2 = 25$$

Add and subtract 25 inside the parentheses. Then take out the subtracted 25, keeping the multiplication by 2. Finally, you can simplify and factor.

$f(x) = 2(x^2 + 10x + 25 - 25) - 6$ *Add and subtract 25 inside the parentheses.*
$f(x) = 2(x^2 + 10x + 25) - 2(25) - 6$
$f(x) = 2(x^2 + 10x + 25) - 50 - 6$ *Bring out the subtracted 25, keeping the multiply by 2 from the parentheses.*
$f(x) = 2(x + 5)^2 - 56$ *Simplify and factor.*

b.
$g(x) = 5x^2 + 3x + 20$
$g(x) = (5x^2 + 3x) + 20$ *Group the variable terms.*
$g(x) = 5\left(x^2 + \frac{3}{5}x\right) + 20$ *Factor out the coefficient of the squared term.*

Find the constant that will complete the square.

$$\frac{1}{2}\left(\frac{3}{5}\right) = \frac{3}{10}$$

$$\left(\frac{3}{10}\right)^2 = \frac{9}{100}$$

$g(x) = 5\left(x^2 + \frac{3}{5}x + \frac{9}{100} - \frac{9}{100}\right) + 20$ *Add and subtract $\frac{9}{100}$ inside the parentheses.*

$g(x) = 5\left(x^2 + \frac{3}{5}x + \frac{9}{100}\right) - 5\left(\frac{9}{100}\right) + 20$ *Bring out the subtracted $\frac{9}{100}$, keeping the multiply by 5 from the parentheses.*

$g(x) = 5\left(x^2 + \frac{3}{5}x + \frac{9}{100}\right)^2 - \frac{9}{20} + 20$ *Simplify.*

$g(x) = 5\left(x^2 + \frac{3}{5}x + \frac{9}{100}\right)^2 - \frac{9}{20} + \frac{400}{20}$

$g(x) = 5\left(x + \frac{3}{10}\right)^2 + \frac{391}{20}$ *Factor.*

PRACTICE PROBLEM FOR EXAMPLE 6

Convert to vertex form.

a. $f(x) = 0.25x^2 + 10x - 3$ **b.** $h(x) = -3x^2 + 4x + 10$

SECTION 4.4 Solving Quadratic Equations by the Square Root Property and Completing the Square

Graphing from Vertex Form with x-Intercepts

Now that we know how to solve a quadratic in vertex form, we can find the horizontal intercepts of a quadratic in vertex form, and these can be part of our graphs.

> **Steps to Graphing a Quadratic Function from the Vertex Form**
> 1. Determine whether the graph opens up or down.
> 2. Find the vertex and the equation for the axis of symmetry.
> 3. Find the vertical intercept.
> 4. Find the horizontal intercepts (if any).
> 5. Plot the points you found in steps 2 through 4. Plot their symmetric points, and sketch the graph. (Find an additional pair of symmetric points if needed.)

Example 7 Graphing a quadratic function in vertex form including horizontal intercepts

Sketch the graph of $f(x) = -1.5(x - 2.5)^2 + 45.375$. Give the domain and range of the function.

SOLUTION

Step 1 Determine whether the graph opens up or down.
The value of a is -1.5. Because a is negative, the graph will open downward.

Step 2 Find the vertex and the equation for the axis of symmetry.
Because the quadratic is given in vertex form, the vertex is $(2.5, 45.375)$.
The axis of symmetry is the vertical line through the vertex $x = 2.5$.

Step 3 Find the vertical intercept.
To find the vertical intercept, we make the input variable zero.

$$f(x) = -1.5(x - 2.5)^2 + 45.375$$
$$f(0) = -1.5(0 - 2.5)^2 + 45.375$$
$$f(0) = -1.5(-2.5)^2 + 45.375$$
$$f(0) = -1.5(6.25) + 45.375$$
$$f(0) = 36$$

Therefore, the vertical intercept is $(0, 36)$.

Step 4 Find the horizontal intercepts (if any).
Horizontal intercepts occur when the output variable is equal to zero, so substitute zero for the output variable and solve.

$$f(x) = -1.5(x - 2.5)^2 + 45.375$$
$$0 = -1.5(x - 2.5)^2 + 45.375 \quad \text{Isolate the squared variable expression.}$$
$$\underline{-45.375 \qquad\qquad\qquad -45.375}$$
$$-45.375 = -1.5(x - 2.5)^2$$
$$\frac{-45.375}{-1.5} = \frac{-1.5(x - 2.5)^2}{-1.5}$$
$$30.25 = (x - 2.5)^2 \quad \text{Use the square root property.}$$
$$\pm\sqrt{30.25} = x - 2.5$$
$$\pm 5.5 = x - 2.5$$
$$5.5 = x - 2.5 \qquad -5.5 = x - 2.5 \quad \text{Write two equations and solve.}$$
$$8 = x \qquad\qquad\qquad -3 = x$$

Therefore, we have the horizontal intercepts $(-3, 0)$ and $(8, 0)$.

Step 5 Plot the points you found in steps 2 through 4. Plot their symmetric points and sketch the graph. (Find an additional pair of symmetric points if needed.)
The vertical intercept (0, 36) is 2.5 units to the left of the axis of symmetry $x = 2.5$, so its symmetric point will be 2.5 units to the right of the axis of symmetry at (5, 36).

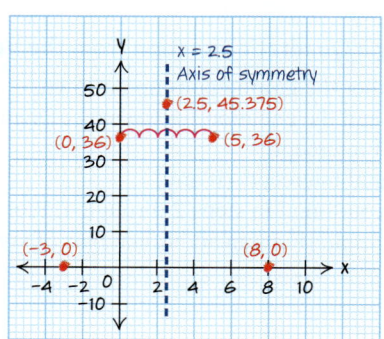

Connect the points with a smooth curve. The input variable x can be any real number without causing the function to be undefined, so the domain of the function is all real numbers or $(-\infty, \infty)$. The graph decreases to negative infinity, and the highest output, $y = 45.375$, is at the vertex so the range is $(-\infty, 45.375]$ or $y \leq 45.375$.

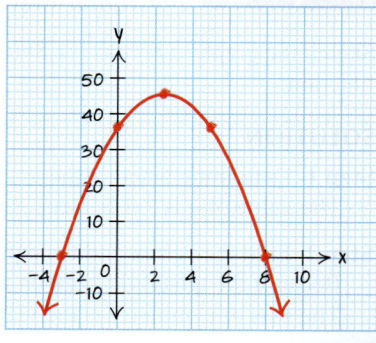

PRACTICE PROBLEM FOR EXAMPLE 7

Sketch the graph of $f(x) = 2(x + 1)^2 - 32$. Give the domain and range for the function.

4.4 Exercises

For Exercises 1 through 16, solve the equation using the square root property. Check your answers.

1. $x^2 = 100$
2. $x^2 = 70$
3. $2x^2 = 162$
4. $5x^2 = 180$
5. $x^2 + 12 = 181$
6. $4x^2 - 43 = 1764$
7. $(x - 5)^2 = 49$
8. $(x + 4.5)^2 = 156.25$
9. $4(x + 7)^2 = 400$
10. $-9(x - 5)^2 = -369$
11. $3(x - 7)^2 + 8 = 647.48$
12. $-2(x + 3)^2 + 11 = -151$
13. $-5(x - 6)^2 - 15 = -30$
14. $2.5(x - 6.5)^2 - 20 = 42.5$
15. $-0.5(x + 4)^2 - 9 = -20$
16. $-0.25(x - 8)^2 + 18 = 6.25$

For Exercises 17 and 18, the students were asked to solve the given equation using the square root property. Explain what the student did wrong. Give the correct answer.

17. $27 = (x + 4)^2 - 9$

James

$27 = (x + 4)^2 - 9$
$ + 9 + 9$
$36 = (x + 4)^2$
$\sqrt{36} = x + 4$
$6 = x + 4$
$-4 -4$
$2 = x$

18. $95 = -4(m-3)^2 - 5$

> **Mary**
>
> $95 = -4(m-3)^2 - 5$
> $\underline{+5 \qquad\qquad\qquad +5}$
> $100 = -4(m-3)^2$
> $\pm\sqrt{100} = -2(m-3)$
> $\pm 10 = -2(m-3)$
>
> $10 = -2(m-3) \qquad -10 = -2(m-3)$
>
> $\dfrac{10}{-2} = m - 3 \qquad\qquad \dfrac{-10}{-2} = m - 3$
>
> $-5 = m - 3 \qquad\qquad 5 = m - 3$
>
> $-2 = m \qquad\qquad\qquad 8 = m$

19. The number of cassette singles shipped by major recording media manufacturers can be modeled by

    $$C(t) = -2.336(t - 13)^2 + 85.6$$

    where $C(t)$ is the number of cassette singles in millions shipped t years since 1980.
    Source: Model based on data from Statistical Abstract of the United States, 2001.

    a. Find how many cassette singles were shipped in 1998.
    b. Find the year when 50 million cassette singles were shipped.
    c. What is the vertex of this model and what does it represent?
    d. When do you think model breakdown occurs for this model?

20. The poverty threshold for an individual under 65 years old can be modeled by

    $$P(t) = 4.95(t - 57)^2 + 1406$$

    where $P(t)$ represents the poverty threshold, in dollars, for an individual in the United States under 65 years old t years since 1900.
    Source: Model based on data from Statistical Abstract of the United States, 2001.

    a. What was the poverty threshold in 1995?
    b. What year was the poverty threshold $11,500?
    c. What year was the poverty threshold $3000?
    d. What is the vertex of this model and what does it represent?
    e. When do you think model breakdown occurs for this model?

21. The amount that personal households and nonprofit organizations have invested in time and savings deposits can be modeled by

    $$D(t) = 17(t - 3)^2 + 2300$$

    where $D(t)$ represents the amount in billions of dollars the households and nonprofit organizations have deposited in time and savings accounts t years since 1990.
    Source: Model based on data from the Statistical Abstract of the United States, 2010.

    a. How much did personal households and nonprofit organizations have invested in time and savings deposits in 2000?
    b. Find in what year personal households and nonprofit organizations had $7500 billion invested in time and savings deposits.
    c. What is the vertex of this model, and what does it represent?

22. The total number of people, in thousands, age 65 years old and over who were below the poverty level can be modeled by

    $$N(t) = -59(t - 12)^2 + 3890$$

    where $N(t)$ represents the total number of people 65 years old and over in the United States who were below the poverty level, in thousands, t years since 1980.
    Source: Model based on data from Statistical Abstract of the United States, 2001.

    a. How many people 65 years old and over were below the poverty level in 1990?
    b. When were there 1 million people 65 years old and over below the poverty level?
    (Hint: 1 million = 1000 thousand)
    c. What is the vertex of this model, and what does it represent?

23. The poverty threshold for a family of four is given in the table.

Year	Poverty Threshold (in $)
1965	3022
1970	3223
1975	3968
1980	5500
1985	8414

    Source: Statistical Abstract of the United States, 2001.

    a. Find an equation for a model of these data.
    b. Estimate the poverty threshold for a family of four in 1990.
    c. Give a reasonable domain and range for your model.
    d. In what year was the poverty threshold $5000?

24. A basketball player shoots a basket from 12 feet away. The ball flies along a parabolic path given by the data in the table on the following page.

    a. Find an equation for a model of these data.
    b. What is the height of the ball when it was 8 feet from the basket?

Distance from Basket (feet)	Height of Ball (feet)
12	6.5
10	10.3
9	11.7
5	14.2
4	14
0	10

  c. Where did the ball reach its maximum height?
  d. When was the ball 11 feet high?

25. If a tank holds 2000 gallons of water, which drains from the bottom of the tank in 20 minutes, then Torricelli's law gives the volume in gallons of water remaining in the tank after t minutes as

$$V(t) = 2000\left(1 - \frac{t}{20}\right)^2$$

Evangelista Torricelli (1608–1647)

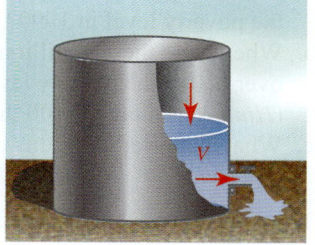

  a. Find the volume of water remaining in the tank after 5 minutes.
  b. When will there be only 500 gallons of water remaining in the tank?
  c. When will there be only 20 gallons of water remaining in the tank?

26. If a tank holds 5000 gallons of water, which drains from the bottom of the tank in 80 minutes, then Torricelli's law gives the volume in gallons of water remaining in the tank after t minutes as

$$V(t) = 5000\left(1 - \frac{t}{80}\right)^2$$

  a. Find the volume of water remaining in the tank after 1 hour.
  b. When will there be only 2450 gallons of water remaining in the tank?
  c. When will there be only 100 gallons of water remaining in the tank?

For Exercises 27 through 44, solve by completing the square. Check your answers.

27. $x^2 + 6x = 7$
28. $t^2 + 10t = 24$
29. $k^2 - 16k = 3$
30. $m^2 - 12m = -27$
31. $x^2 + 4x + 7 = 0$
32. $h^2 + 2h - 5 = 0$
33. $t^2 + 11t = 4$
34. $p^2 + 7p - 12 = 0$
35. $x^2 - 9x + 14 = 0$
36. $x^2 - 20x + 51 = 0$
37. $m^2 + 8m + 20 = 0$
38. $k^2 - 10k + 40 = 5$
39. $3x^2 + 12x - 15 = 0$
40. $4t^2 + 80t + 20 = -56$
41. $5x^2 + 7x = 0$
42. $3h^2 + 9h - 14 = 0$
43. $-7x^2 + 4x + 20 = 0$
44. $3x^2 - 10x - 7 = 0$

For Exercises 45 through 48, use the following information and formula.

Money in a savings account that earns interest at r percent compounded annually will grow according to the formula

$$A = P(1 + r)^t$$

where

  A = *Account balance in dollars*
  t = *Time in years*
  P = *Principal in dollars (amount deposited)*
  r = *Interest rate compounded annually*

45. Find the interest rate that will turn a deposit of $800 into $900 in 2 years.

46. Find the interest rate that will turn a deposit of $4000 into $4400 in 2 years.

47. Find the interest rate that will turn a deposit of $2000 into $2150 in 2 years.

48. Find the interest rate that will turn a deposit of $15,000 into $17,500 in 2 years.

49. The KVLY-TV mast is a television transmitter in North Dakota used by Fargo station KVLY channel 11. At 2063 feet, it is currently the world's tallest supported structure on land. The mast is held in place by several guy wires that are attached on the sides of the mast. One of the guy wires is anchored in the ground 1000 ft from the base of the mast and it attaches to the mast 2000 ft up from the base. How long is the guy wire? (*Hint:* Use the Pythagorean Theorem.)

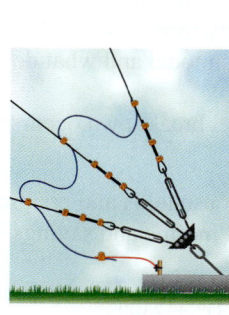
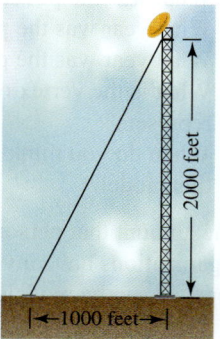

50. Rodalfo has a 30-foot ladder and puts the base of the ladder 12 feet from the house. How far will the ladder reach?

51. Abida is a painter and has a 25-foot ladder that she wants to lean against the house 20 feet up. How far from the house should she put the base of the ladder?

52. Kasha installs awnings above windows on homes and needs her 30-foot ladder to reach 22 feet up the side of a home. How far from the house should she put the base of the ladder?

For Exercises 53 through 70, convert the function into vertex form.

53. $f(x) = x^2 + 6x + 8$
54. $g(x) = x^2 - 10x - 7$
55. $g(t) = t^2 - 8t - 20$
56. $s(p) = p^2 + 14p + 35$
57. $f(x) = x^2 - 7x + 10$
58. $k(m) = m^2 + 5m - 12$
59. $h(x) = 3x^2 + 12x + 24$
60. $m(d) = 4d^2 + 24d - 30$
61. $f(t) = 2t^2 - 16t - 12$
62. $k(p) = 5p^2 - 50p + 20$
63. $f(x) = 4x^2 + 5x - 20$
64. $m(p) = 6p^2 + 10p + 21$
65. $g(x) = 0.5x^2 + 7x - 30$
66. $d(m) = 0.25m^2 - 9m + 13$
67. $f(x) = 0.2x^2 - 7x - 10$
68. $h(t) = \frac{1}{3}t^2 + 6t + 8$
69. $c(p) = \frac{2}{7}p^2 - 5p - \frac{3}{7}$
70. $g(x) = \frac{3}{8}x^2 + \frac{1}{8}x - \frac{7}{8}$

For Exercises 71 through 86, sketch the graph of the given functions and label the vertex, vertical intercept, and horizontal intercepts (if any). Give the domain and range of the function.

71. $f(x) = (x - 5)^2 - 16$
72. $g(x) = (x - 3)^2 - 9$
73. $h(x) = (x + 2)^2 - 5$
74. $f(x) = (x + 6)^2 - 11$
75. $g(x) = 3(x - 7)^2 + 10$
76. $h(x) = 2(x + 2)^2 + 5$
77. $g(x) = -2(x - 4)^2 + 18$
78. $f(x) = -2(x + 6)^2 + 20$
79. $h(x) = -0.5(x + 3)^2 + 32$
80. $h(x) = -0.3(x - 10)^2 - 5$
81. $f(x) = \frac{1}{3}(x - 5)^2 - 12$
82. $g(x) = \frac{1}{5}(x + 7)^2 + 10$
83. $f(x) = \frac{2}{5}(x - 6)^2 + \frac{1}{5}$
84. $g(x) = \frac{3}{4}(x + 8)^2 - 27$
85. $g(x) = 4(x + 3.5)^2 - 16$
86. $f(x) = 5(x + 1.5)^2 - 6$

For Exercises 87 through 100, sketch the graph of the function you converted to vertex form in the given exercise. Label the vertex, vertical intercept, and horizontal intercepts (if any). Give the domain and range of the function.

87. Exercise 53
88. Exercise 54
89. Exercise 55
90. Exercise 56
91. Exercise 57
92. Exercise 58
93. Exercise 59
94. Exercise 60
95. Exercise 61
96. Exercise 62
97. Exercise 63
98. Exercise 64
99. Exercise 65
100. Exercise 66

4.5 Solving Equations by Factoring

LEARNING OBJECTIVES
- Solve a polynomial equation by factoring.
- Find a quadratic equation from a graph.

Solving by Factoring

In many situations, we may be given an equation that is easy to factor. Factoring can be used to solve some polynomial equations. We saw in Chapter 3 that the basis of factoring is to take a polynomial and break it into simpler pieces that are multiplied together.

Standard Form
$f(x) = x^2 + 8x + 15$

Factored Form
$f(x) = (x + 3)(x + 5)$

If we multiply the factored form out and simplify, it will become the standard form.

$$f(x) = (x + 3)(x + 5)$$
$$f(x) = x^2 + 5x + 3x + 15$$
$$f(x) = x^2 + 8x + 15$$

The main reason factoring an expression is useful is the **product property of zero**, which says that when you multiply any number by zero, the answer is zero.

> **DEFINITIONS**
>
> **Product property of zero** $a \cdot 0 = 0$.
>
> **Zero factor property** If $ab = 0$, then $a = 0$, $b = 0$, or both, where a and b are real numbers.

The **zero factor property** follows from the product property of zero and states that if two numbers are multiplied together and the answer is zero, then one or both of those factors must be zero. To understand this simple property, do the following arithmetic in your head.

- First pick any number.
- Now multiply that number by zero.
- You should now have zero as the result.

Okay. Start over. This time:

- Pick any number.
- Now multiply by any number you want, but the result must be zero.
- You should have multiplied by zero.

If you do these two operations with any nonzero number, you will get the same results every time. This statement seems obvious, but it is very helpful in algebra. Let's see why.

CONCEPT INVESTIGATION

How does factoring help us solve?

1. Solve the following equations for x. Note that all of these problems include multiplication.

   a. $3x = 0$ **b.** $-9x = 0$
   c. $5(x + 2) = 0$ **d.** $-13(x - 6) = 0$

2. For each of the following equations, list the two factors being multiplied together, and tell which one must equal zero for the product to equal zero.

   a. $3x = 0$ 3 and x are the factors.
                       3 cannot equal zero, so x must equal zero.
   b. $-9x = 0$

   c. $5(x + 2) = 0$ 5 and $(x + 2)$ are the factors.
                           $5 \neq 0$, so $(x + 2)$ must equal zero.
   d. $-13(x - 6) = 0$

   Note that in all of the equations given, there has been only one factor that could have equaled zero.

3. In the following equations, list the factors being multiplied together and state which can equal zero. Then solve.

   a. $(x + 5)(x - 9) = 0$ $(x + 5)$ and $(x - 9)$; both can equal zero.
                                  $x + 5 = 0$ or $x - 9 = 0$
                                  $x = -5$,   or $x = 9$
   b. $(x - 3)(x - 12) = 0$

c. $(x + 17)(x + 2) = 0$ **d.** $(2x - 5)(x - 4) = 0$
e. $3(x - 7)(x + 15) = 0$ **f.** $(x + 3)(x - 7)(x + 15) = 0$

In this concept investigation, equations that involve a product that is set equal to zero can be solved by setting each factor equal to zero one at a time rather than the entire equation at once. If we factor a polynomial into simpler parts, we will be able to solve the simpler equations.

It is very important to note that we can use the factored form to solve an equation only if it is set equal to zero. If the quadratic equation is not equal to zero, we must first move everything to one side so that it equals zero and then factor.

> **Steps to Solving Polynomial Equations by Factoring**
> 1. Set the polynomial equal to zero.
> 2. Factor the polynomial completely.
> 3. Set each of the factors equal to zero and solve.
> 4. Check your answers in the original equation.

> ■ **Skill Connection**
>
> **AC Method of Factoring**
> $$ax^2 + bx + c$$
> 1. Factor out any common factors.
> 2. Multiply a and c together.
> 3. Find factors of ac that add up to b.
> 4. Rewrite the middle (bx) term using the factors from step 3.
> 5. Group and factor out what is in common.
>
> To review factoring, see Chapter 3.

Example 1 Solving polynomial equations by factoring

Solve the following by factoring.

a. $x^2 + 7x - 40 = 20$ **b.** $2x^2 + 27 = 21x$ **c.** $x^3 + 8x^2 + 15x = 0$

SOLUTION

a.
$$x^2 + 7x - 40 = 20$$ Set the quadratic equal to zero and factor.
$$x^2 + 7x - 60 = 0$$ Steps 2 and 3 are in the margin on the right.
$$x^2 - 5x + 12x - 60 = 0$$ Step 4.
$$(x^2 - 5x) + (12x - 60) = 0$$ Step 5.
$$x(x - 5) + 12(x - 5) = 0$$
$$(x - 5)(x + 12) = 0$$
$$x - 5 = 0 \quad x + 12 = 0$$ Set the factors equal to zero and finish solving.
$$x = 5 \quad x = -12$$ Check both answers.

$$(5)^2 + 7(5) - 40 \stackrel{?}{=} 20$$
$$20 = 20$$

$$(-12)^2 + 7(-12) - 40 \stackrel{?}{=} 20$$ Both answers work.
$$20 = 20$$

Step 2 $1(-60) = -60$

1	−60	−1	60
2	−30	−2	30
3	−20	−3	20
4	−15	−4	15
5	−12	−5	12 Step 3
6	−10	−6	10

b.
$$2x^2 + 27 = 21x$$ Set the quadratic equal to zero and factor.
$$2x^2 - 21x + 27 = 0$$ Steps 2 and 3 are in the margin on the right.
$$2x^2 - 3x - 18x + 27 = 0$$ Step 4.
$$(2x^2 - 3x) + (-18x + 27) = 0$$ Step 5.
$$x(2x - 3) - 9(2x - 3) = 0$$
$$(2x - 3)(x - 9) = 0$$ Set the factors equal to zero and finish solving.
$$2x - 3 = 0 \quad x - 9 = 0$$
$$x = \frac{3}{2} \quad x = 9$$

Step 2 $2(27) = 54$

1	54	−1	−54
2	27	−2	−27
3	18	−3	−18 Step 3
6	9	−6	−9

$$2\left(\frac{3}{2}\right)^2 + 27 \stackrel{?}{=} 21\left(\frac{3}{2}\right)$$ Check both answers.

$$31.5 = 31.5$$

$$2(9)^2 + 27 \stackrel{?}{=} 21(9)$$ Both answers work.

$$189 = 189$$

c.
$$x^3 + 8x^2 + 15x = 0$$
$$x(x^2 + 8x + 15) = 0$$ Take out the x in common and factor the remaining quadratic.
$$x(x^2 + 3x + 5x + 15) = 0$$
$$x[(x^2 + 3x) + (5x + 15)] = 0$$ The factored-out x will remain at the front of each step until the factoring is complete.
$$x[x(x + 3) + 5(x + 3)] = 0$$
$$x(x + 3)(x + 5) = 0$$

$$x = 0 \quad x + 3 = 0 \quad x + 5 = 0$$ Now rewrite into three equations and solve each separately.
$$x = 0 \quad\quad x = -3 \quad\quad x = -5$$

$$(0)^3 + 8(0)^2 + 15(0) \stackrel{?}{=} 0$$ Check all three answers in the original equation.
$$0 = 0$$
$$(-3)^3 + 8(-3)^2 + 15(-3) \stackrel{?}{=} 0$$
$$0 = 0$$
$$(-5)^3 + 8(-5)^2 + 15(-5) \stackrel{?}{=} 0$$ All three answers work.
$$0 = 0$$

PRACTICE PROBLEM FOR EXAMPLE 1

Solve the following by factoring.

a. $x^2 + 13x = -36$ **b.** $2x^3 + 11x^2 + 21x - 5 = 6x - 5$

Example 2   Applications of solving by factoring

The profit a local photographer makes from selling n copies of a photograph can be modeled by

$$P(n) = n(40 - 2n)$$

where $P(n)$ represents the amount of profit in dollars from selling n copies of a photograph. Find the number of copies the photographer must sell of a photo to make a profit of $150.

AP Photo/Matt Sayles

SOLUTION

The $150 is a desired profit, so it must take the place of $P(n)$.

$$150 = n(40 - 2n)$$ Substitute 150 for P.
$$150 = 40n - 2n^2$$ Distribute the n.
$$\underline{+2n^2 \quad\quad\quad +2n^2}$$
$$2n^2 + 150 = 40n$$
$$\underline{-40n \quad\quad -40n}$$ Get everything to one side and make it equal to zero. Factor using AC method.
$$2n^2 - 40n + 150 = 0$$
$$2(n^2 - 20n + 75) = 0$$ Step 1.
$$\frac{2(n^2 - 20n + 75)}{2} = \frac{0}{2}$$ Because this is an equation, we can divide both sides by 2.

SECTION 4.5 Solving Equations by Factoring   353

$$n^2 - 20n + 75 = 0$$ Steps 2 and 3 are in the margin on the right.
$$n^2 - 15n - 5n + 75 = 0$$ Step 4.
$$(n^2 - 15n) + (-5n + 75) = 0$$ Step 5.
$$n(n - 15) - 5(n - 15) = 0$$
$$(n - 15)(n - 5) = 0$$

Step 2	1(75)
	1 75
	3 25
	5 15
	−1 −75
	−3 −25
Step 3	−5 −15

$n - 15 = 0$ $n - 5 = 0$ Set the two factors equal to zero
$n = 15$ $n = 5$ using the zero factor property.

$150 \stackrel{?}{=} 15(40 - 2(15))$ Check both answers.
$150 \stackrel{?}{=} 15(10)$
$150 = 150$
$150 \stackrel{?}{=} 5(40 - 2(5))$
$150 \stackrel{?}{=} 5(30)$
$150 = 150$ Both answers work.

To make a profit of $150, the photographer must sell 5 or 15 photographs. If he sells between 5 and 15, he will make more than $150 in profit.

PRACTICE PROBLEM FOR EXAMPLE 2

The average cost for building a small plane at Private Planes 101 depends on the number of planes built in a year. This average cost can be modeled by

$$A(p) = p(31 - 5p) + 194$$

where $A(p)$ represents the average cost in thousands of dollars per plane to build p planes in a year. How many planes does Private Planes 101 have to build in a year to have an average cost per plane of $200,000?

Often, the factored form can be the easiest way to model a situation. In business, when a company changes the price of an item, this change will affect both the price and the quantity of that item that is sold. For example, if a company can sell 2000 bikes at a price of $1500.00, the company may be able to sell only 1900 bikes if they raise the price to $1700.00.

Through experience and research, companies can get an idea of how price changes will increase or decrease the quantity they sell and, therefore, also increase or decrease their revenue. Remember that revenue can be calculated by multiplying the price by the quantity sold.

Example 3   Applications of the factored form of a quadratic

A movie theater sells an average of 35,000 tickets per week when the price per ticket averages $7. From researching other theaters in the area, the managers believe that for each $0.50 increase in the ticket price, they will lose about 1400 ticket sales per week.

a. What will the weekly revenue for this theater be if management increases the ticket price to $8?

b. Find an equation for the model of the weekly revenue at this theater if management increases the ticket price $0.50 x times.

c. Use your model to determine the weekly revenue if the ticket price is $9.

d. Use your model to determine the weekly revenue if the ticket price is $12.

SOLUTION

a. If the theater increases the ticket price to $8, that change in price will represent two $0.50 increases, so the quantity sold will decrease by $2(1400) = 2800$. Therefore, they will expect to sell $35000 - 2800 = 32200$ tickets at $8 each for a revenue of

$$\text{revenue} = (\text{price})(\text{quantity})$$
$$\text{revenue} = (8)(32200)$$
$$\text{revenue} = 257600$$

Therefore, at $8 per ticket, the theater expects to have a weekly revenue of $257,600.

b. Define variables

$$R(x) = \text{Weekly revenue in dollars for this theater}$$
$$x = \text{Number of \$0.50 increases in price}$$

The new price will be represented by $7 + 0.50x$. We also know that the quantity sold will decrease by 1400 per $0.50 increase. Therefore, the quantity sold can be represented by $35000 - 1400x$. The weekly revenue in dollars can be represented by

$$R(x) = (7 + 0.5x)(35000 - 1400x)$$

c. If the ticket price is $9, there were four $0.50 increases, so $x = 4$, and we get

$$R(4) = (7 + 0.5(4))(35000 - 1400(4))$$
$$R(4) = 264600$$

If the theater increases the price to $9 per ticket, weekly revenue will be $264,600.00.

d. If the ticket price is $12, the increase was $5, or ten $0.50 increases. Therefore, with $x = 10$, we get

$$R(10) = (7 + 0.5(10))(35000 - 1400(10))$$
$$R(10) = 252000$$

If the theater increases the ticket price to $12 per ticket, weekly revenue will be $252,000. At this price, the theater is making less revenue because fewer people are buying tickets.

PRACTICE PROBLEM FOR EXAMPLE 3

A small restaurant is planning to increase the fee it charges for valet parking. Currently, the restaurant charges $3 per car, and the valet parks about 700 cars per week. For every dollar the restaurant raises the price, the restaurant expects about 100 fewer cars to use the valet parking service.

a. What will the weekly revenue from valet parking be if the restaurant increases the fee to $4?

b. Find an equation for the model of the weekly revenue from valet parking at this restaurant if the owners increase the fee x dollars.

c. Use your model to determine the weekly revenue if the fee is $5.

d. Use your model to determine the weekly revenue if the fee is $6.

e. What fee brings the restaurant the highest weekly revenue from valet parking?

Finding an Equation from the Graph

The zero factor property can also be used to help find equations by looking at their graphs. As we solved different equations by factoring, we always set the equation equal to zero and then factor. If we set a function equal to zero and solve, we are looking for the horizontal intercepts of the function. When we solve using factoring, we are actually finding the horizontal intercepts of the graph related to that equation. We can use the connection between the horizontal intercepts and the factors of an expression to find an equation using its graph.

The horizontal intercepts of a graph also represent the zeros of the function. In this graph, the horizontal intercepts are $(-2, 0)$ and $(-8, 0)$, so the zeros of the equation are and $x = -2$ and $x = -8$. Taking the zeros and working backwards through the process of solving by factoring, we will get an equation for the graph.

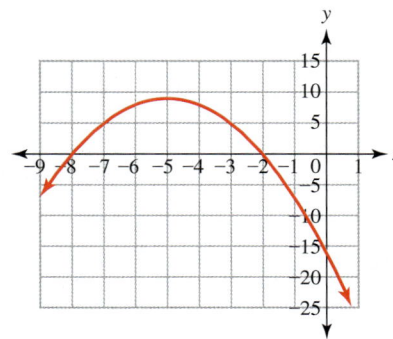

$$x = -8 \qquad x = -2$$
$$x + 8 = 0 \qquad x + 2 = 0$$
$$0 = (x + 8)(x + 2)$$
$$y = (x + 8)(x + 2)$$

This equation in factored form could be missing a constant that was factored out. To find this constant, we can use another point from the graph and solve for the missing constant. If we let a represent the missing constant and use the point $(-3, 5)$, we can solve for a.

$$y = a(x + 8)(x + 2)$$
$$5 = a(-3 + 8)(-3 + 2)$$
$$5 = a(5)(-1)$$
$$5 = -5a$$
$$\frac{5}{-5} = \frac{-5a}{-5}$$
$$-1 = a$$

Substituting in the -1 for a and simplifying, we get

$$y = -1(x + 8)(x + 2)$$
$$y = -x^2 - 10x - 16$$

In the graph we can see that a negative a-value makes sense because the graph is facing downward. After simplifying, the equation also matches the y-intercept of $(0, -16)$ found in the graph. The horizontal intercepts cannot tell us everything about the equation. Once we have the factors, we can use another point on the graph to find any missing constant factors.

> **Steps to Finding a Quadratic Equation from the Graph**
> 1. Find the horizontal intercepts of the graph and write them as zeros of the equation.
> 2. Undo the solving process to find the factors of the equation.
> 3. Use another point from the graph to find any constant factors.
> 4. Multiply out the factors and simplify.

Example 4 Finding a quadratic equation from the graph

Use the graph to find an equation for the quadratic.

a.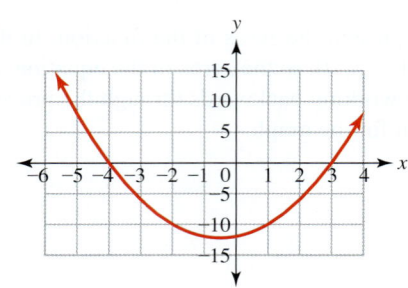

b.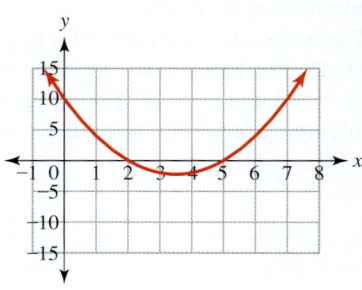

c.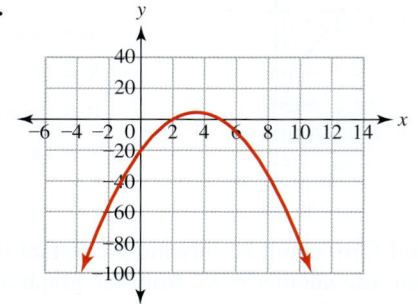

SOLUTION

a. **Step 1** Find the horizontal intercepts of the graph and write them as zeros of the equation. The horizontal intercepts for this graph are located at $(-4, 0)$ and $(3, 0)$, so we know that $x = -4$ and $x = 3$ are the zeros of the quadratic equation.

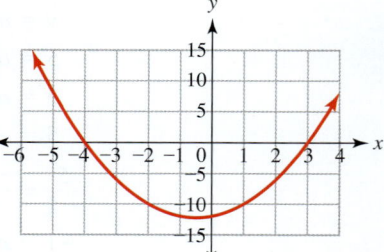

Step 2 Undo the solving process to find the factors of the equation.
Therefore, we have:

$$x = -4 \qquad x = 3$$
$$x + 4 = 0 \qquad x - 3 = 0$$

This gives us the factors $(x + 4)$ and $(x - 3)$ and the equation
$$y = (x + 4)(x - 3)$$

Step 3 Use another point from the graph to find any constant factors.
Using the point $(1, -10)$, we get

$$y = a(x + 4)(x - 3)$$
$$-10 = a(1 + 4)(1 - 3)$$
$$-10 = a(5)(-2)$$
$$-10 = -10a$$
$$1 = a$$

The constant factor is 1.

Step 4 Multiply out the factors and simplify.
Therefore, we have the equation

$$y = 1(x + 4)(x - 3)$$
$$y = x^2 + x - 12$$

The y-intercept for this quadratic would be $(0, -12)$, which agrees with the graph, so we have the correct equation.

b. Step 1 Find the horizontal intercepts of the graph and write them as zeros of the equation.

The horizontal intercepts for this graph are $(2, 0)$ and $(5, 0)$, so we know that $x = 2$ and $x = 5$ are the zeros of the quadratic equation.

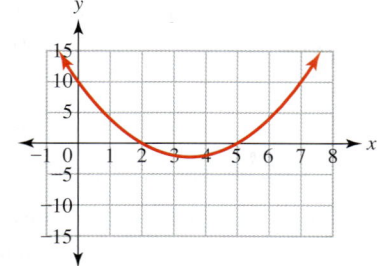

Step 2 Undo the solving process to find the factors of the equation.
Therefore, we have

$$x = 2 \qquad x = 5$$
$$x - 2 = 0 \qquad x - 5 = 0$$

This gives us the factors $(x - 2)$ and $(x - 5)$ and the equation

$$y = (x - 2)(x - 5)$$

Step 3 Use another point from the graph to find any constant factors.
Using the point $(7, 10)$, we get

$$y = a(x - 2)(x - 5)$$
$$10 = a(7 - 2)(7 - 5)$$
$$10 = a(5)(2)$$
$$10 = 10a$$
$$1 = a$$

The constant factor is 1.

Step 4 Multiply out the factors and simplify.
Therefore, we have the equation

$$y = 1(x - 2)(x - 5)$$
$$y = x^2 - 7x + 10$$

The y-intercept for this quadratic would be $(0, 10)$, which agrees with the graph, so we have the correct equation.

c. Step 1 Find the horizontal intercepts of the graph and write them as zeros of the equation.

The horizontal intercepts for this graph are $(2, 0)$ and $(5, 0)$, so we know that $x = 2$ and $x = 5$ are the zeros of the quadratic equation.

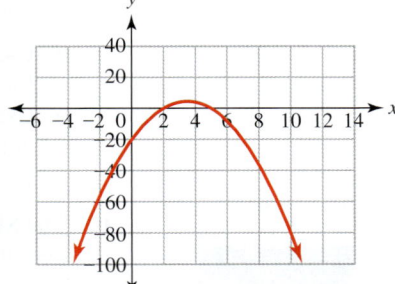

Step 2 Undo the solving process to find the factors of the equation.
Therefore, we have

$$x = 2 \qquad x = 5$$
$$x - 2 = 0 \qquad x - 5 = 0$$

This gives us the factors $(x - 2)$ and $(x - 5)$ and the equation

$$y = (x - 2)(x - 5)$$

Step 3 Use another point from the graph to find any constant factors.
Using the point $(10, -80)$, we get

$$y = a(x - 2)(x - 5)$$
$$-80 = a(10 - 2)(10 - 5)$$
$$-80 = 40a$$
$$-2 = a$$

The constant factor is -2.

Step 4 Multiply out the factors and simplify.
Therefore, we have the equation

$$y = -2(x - 2)(x - 5)$$
$$y = -2x^2 + 14x - 20$$

The y-intercept for this quadratic would be $(0, -20)$, which agrees with the graph, so we have the correct equation.

PRACTICE PROBLEM FOR EXAMPLE 4

Use the graph to find an equation for the quadratic.

4.5 Exercises

For Exercises 1 through 10, solve and check.

1. $(x + 3)(x - 2) = 0$
2. $(x - 8)(x - 4) = 0$
3. $2(w + 7)(3w + 10) = 0$
4. $8(k - 46)(k + 23) = 0$
5. $x(x - 4)(x + 5) = 0$
6. $x(x + 1)(5x + 6) = 0$
7. $4x(x - 9)(x + 7) = 0$
8. $-3x(2x + 5)(4x - 9) = 0$
9. $4.5w(3w - 8)(7w + 5)(w - 6) = 0$
10. $-2m(m + 3.8)(m - 7.5)(2m - 9)(m + 8) = 0$

11. A math professor vacationing in Kauai jumped off a 20-foot waterfall into a pool of water below. The height of the math professor above the pool of water can be modeled by

$$H(t) = -16t^2 + 4t + 20$$

where $H(t)$ represents the height of the professor above the pool in feet t seconds after jumping off the waterfall.

a. What was the height of the professor above the pool $\frac{1}{2}$ second after jumping off the waterfall?
b. Use factoring to determine how many seconds it took for the professor to hit the pool of water below.
c. Use factoring to determine how many seconds after jumping the professor was 8 feet above the pool of water.

12. The Skywalk is a glass-bottomed platform that hangs over the edge of the Grand Canyon and is suspended 4000 feet above the floor of the canyon. If a tourist shot an arrow straight up in the air from the observation platform, the arrow's height could be modeled by

$$H(t) = -16t^2 + 240t + 4000$$

where $H(t)$ is the height of the arrow above the canyon floor in feet t seconds after being shot.

a. Find $H(5)$ and explain its meaning.
b. Use factoring to determine how many seconds before the arrow was at a height of 2400 feet.
c. Use factoring to determine after how many seconds the arrow would hit the canyon floor.

13. The average profit made by a local company that manufactures custom car parts for racing teams can be modeled by

$$A(p) = p(80 - 4p)$$

where $A(p)$ represents the average profit in dollars per part when p parts are produced. Use factoring to find the following. Check your answers using the table.

a. Find the number of parts that have to be produced for the company to earn $144 profit per part.
b. Find the number of parts that have to be produced for the company to earn $300 profit per part.
c. Find the number of parts that have to be produced for the company to earn $400 profit per part.

14. The average profit for mixing specialized metallic paints for a small manufacturing company can be modeled by

$$P(q) = q(105 - 7q)$$

where $P(q)$ represents the average profit in dollars per quart mixed when q quarts are mixed for one order. Use factoring to find the number of quarts that must be ordered and mixed for the average profit to be $350 per quart. Check your answer with the table.

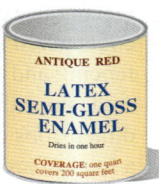

15. The average cost to manufacture custom tooling machines can be modeled by

$$C(m) = m(108 - 6m)$$

where $C(m)$ represents the average cost in dollars per machine when m machines are made.

a. When will the average cost per machine be $390? Check your answer with the table.
b. When will the average cost per machine be $480? Check your answer with a graph.

16. An open box can be made from a rectangular piece of cardboard that is 36 by 30 inches by cutting a square from each corner and folding up the sides. The volume of the open box can be modeled by

$$V(x) = 4x^3 - 132x^2 + 1080x$$

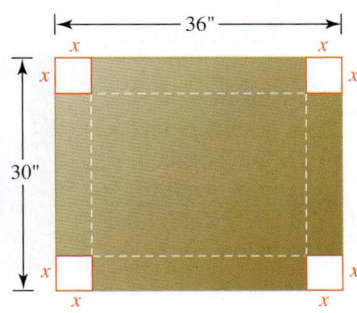

where $V(x)$ is the volume of the open box in cubic inches when squares of length x inches are cut from each corner.

a. Completely factor the volume formula.
b. Find $V(4)$ and explain its meaning.
c. Use the graph to estimate the size of square that should be cut out of each corner to get a box with the greatest volume.

For Exercises 17 through 32, solve by factoring. Check your answers.

17. $x^2 - 4x - 21 = 0$
18. $x^2 + 6x + 5 = 0$
19. $h^2 + 12h + 27 = 0$
20. $p^2 + 3p - 28 = 0$
21. $x^2 - 36 = 0$
22. $t^2 - 64 = 0$
23. $5x^2 - 80 = 0$
24. $-3m^2 + 75 = 0$
25. $x^2 + 50 = 150$
26. $4x^2 - 20 = 5$
27. $5x^2 + 20x = 0$
28. $2x^2 - 9x = 0$
29. $6x^2 = 10x$
30. $8x^2 = 6x$
31. $x^2 + 9x + 20 = 0$
32. $w^2 - 7w - 18 = 0$

For Exercises 33 and 34, the students were asked to solve the given equation by factoring. Explain what the student did wrong. Give the correct answer.

33. $x^2 + 5x + 6 = 2$

> *Elijah*
> $x^2 + 5x + 6 = 2$
> $(x + 2)(x + 3) = 2$
> $x + 2 = 2 \qquad x + 3 = 2$
> $x = 0 \qquad x = -1$

34. $4x^2 = 6x$

> *Amelia*
> $4x^2 = 6x$
> $\dfrac{4x^2}{6x} = \dfrac{6x}{6x}$
> $\dfrac{x}{3} = 1$
> $x = 3$

For Exercises 35 through 54, solve by factoring. Check your answers.

35. $t^2 - 11t + 21 = -7$
36. $2x^2 + 13x + 10 = -5$
37. $7m^2 - 25m + 18 = 6$
38. $3t^2 - 7t - 5 = 5$
39. $4x^2 - 31x - 45 = 0$
40. $6w^2 + 29w + 28 = 0$

41. $-41t + 10t^2 = 18$

42. $11x + x^2 = -28$

43. $x^2 + 7x - 9 = 7x + 16$

44. $x^2 - 3x + 8 = 24 - 3x$

45. $28p^2 + 3p + 60 = 100$

46. $12t^2 - 23t + 40 = 30$

47. $55 + 55h = 10h^2 - 50$

48. $6w^2 - 15w = 4w - 10$

49. $x^3 - 4x^2 - 21x = 0$

50. $x^3 - 4x^2 + 4x = 0$

51. $w^3 - 7w^2 + 15w = 5w$

52. $w^3 - 3w^2 - 3w + 9 = 7w + 9$

53. $10x^3 + 30x^2 + 40x + 15 = -5x^2 + 10x + 15$

54. $14x^3 + 35x^2 - 20x + 10 = 6x^2 - 5x + 10$

55. A school sells candy as a fund-raiser for their sports programs. They sell about 5500 candy bars at a profit of $0.75 each. Each $0.50 increase in the price will increase the profit of each candy bar by $0.50. However, they will sell about 1,000 fewer candy bars.

   a. What is the profit for this school if the profit is $0.75 per candy bar?
   b. Find an equation for the model of the profit this school makes if they increase the price of each candy bar by $0.50 x times.
   c. Use your model to determine the profit if the price per candy bar is increased by $1.
   d. Use your model to determine the profit if the price per candy bar is increased by $2.
   e. What price increase will bring the school the highest profit? Explain.

56. A church is bringing a group of its members to a "clean comedians" show. If the church brings 15 people, it will cost $10 each. For each additional person the church brings, the price per ticket will go down $0.20. Groups are limited to 35 people because of the seating capacity of the auditorium.

   a. What will it cost the church to bring 20 members?
   b. Find an equation for the model of the total cost of bringing a group to the show. (Define x as the number of members beyond 15.)
   c. Use your model to find the total cost of bringing a group of 25 people.
   d. If the church can afford to pay only $210 for this trip, how many people can the church bring?

57. A sporting goods store sells about 100 boxes of golf balls a week when the price is $15 per box. If the store decreases the price by $1, they will sell an additional 20 boxes.

   a. What is the store's weekly revenue from selling boxes of golf balls for $15?
   b. What is the store's weekly revenue from selling boxes of golf balls for $17?
   c. Find an equation for the model of the store's weekly revenue from selling boxes of golf balls if the store lowers the price by $x.
   d. Use your model to find the weekly revenue from selling boxes of golf balls if the price is $8.
   e. What price will maximize the store's weekly revenue from selling boxes of golf balls?

58. A local college has 12,000 students who each pay $800 per semester for tuition. For every $25 increase in tuition, the college will lose 300 students.

   a. How much revenue will the college make per semester if the tuition is raised to $850?
   b. Find an equation for the model of the college's revenue per semester when the tuition is raised $25 x times.
   c. Use your model to find the college's revenue if their tuition is raised to $1000.
   d. What should the tuition be increased to so that the college makes the most revenue?

59. A nature conservancy group wants to protect a rectangular nesting ground next to a river to save an endangered bird species. The group used 140 feet of fencing and fenced only three sides of the rectangular area, since the river made the fourth side. What are the dimensions of the enclosure if they were able to protect a total of area of 2400 square feet?

60. A walkway is being installed around a rectangular swimming pool. The pool is 30 feet by 12 feet, and the total area of the pool and the walkway is 1288 square feet. What is the width of the walkway?

61. A framed painting has overall dimensions of 32 inches by 29 inches. The area of the painting itself is 550 square inches. Find the width of the frame.

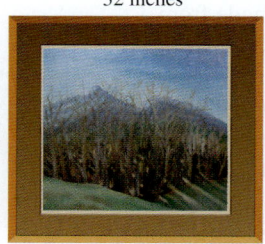

32 inches

29 inches

62. A framed painting has overall dimensions of 36 inches by 30 inches. The area of the painting itself is 832 square inches. Find the width of the frame.

SECTION 4.5 Solving Equations by Factoring **361**

For Exercises 63 through 70, use the graph to find the equation for the quadratic. Write the equation in standard form $y = ax^2 + bx + c$.

63.

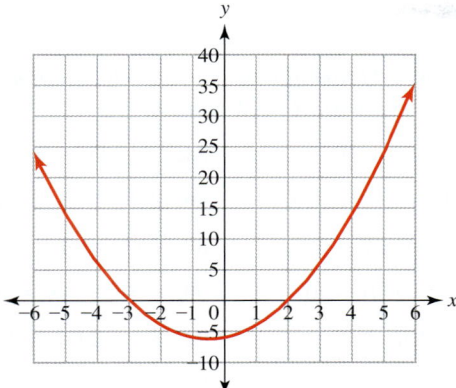

64.

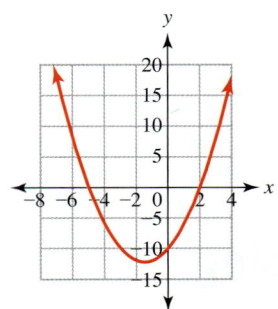

65.

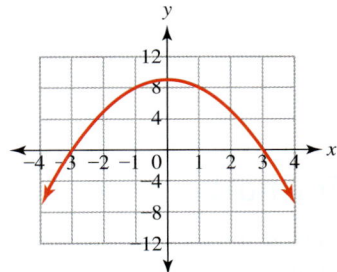

66.

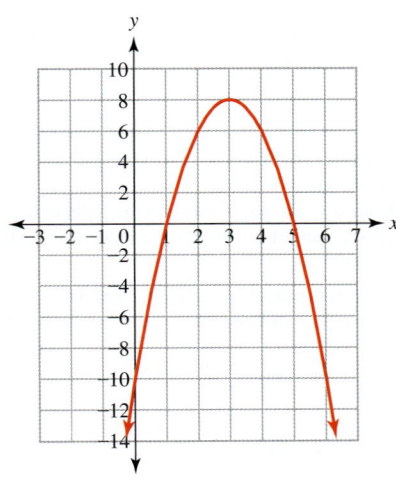

67.

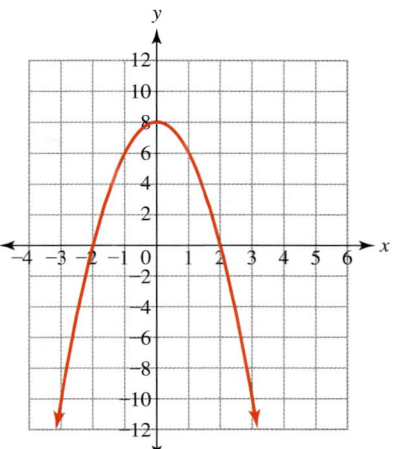

68.

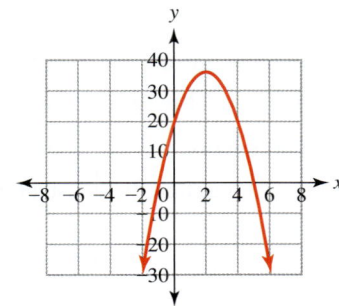

69.

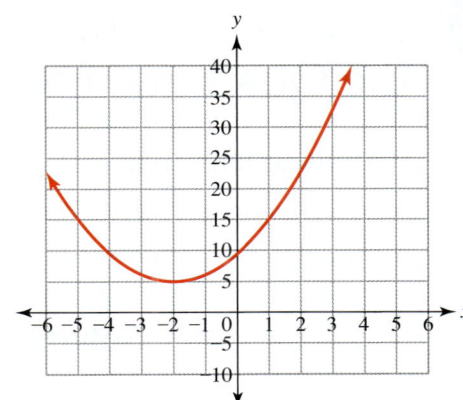

70.

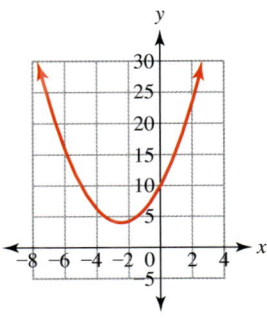

For Exercises 71 through 82, use the given information to find the equation of the quadratic function. Write the equation in standard form $f(x) = ax^2 + bx + c$.

71. The zeros of the function are $x = 3$ and $x = -7$, and the constant multiplier is one.

72. The zeros of the function are $x = -5$ and $x = -9$, and the constant multiplier is one.

73. The zeros of the function are $x = 2$ and $x = 4$, and the constant multiplier is one.

74. The zeros of the function are $x = \frac{1}{2}$ and $x = -8$, and the constant multiplier is one.

75. The zeros of the function are $x = \frac{2}{3}$ and $x = 4$, and the constant multiplier is one.

76. The zeros of the function are $x = 6$ and $x = -1$. Use the fact that $f(2) = -36$ to find the constant multiplier.

77. The zeros of the function are $x = 4$ and $x = 2$. Use the fact that $f(0) = 40$ to find the constant multiplier.

78. The zeros of the function are $x = 15$ and $x = -12$. Use the fact that $f(3) = -45$ to find the constant multiplier.

79. The zeros of the function are $x = \frac{1}{4}$ and $x = -2$. Use the fact that $f(2) = -21$ to find the constant multiplier.

80. The zeros of the function are $x = \frac{2}{5}$ and $x = 9$. Use the fact that $f(4) = 13.5$ to find the constant multiplier.

81. The zeros of the function are $x = -\frac{1}{3}$ and $x = \frac{9}{2}$. Use the fact that $f(3) = 20$ to find the constant multiplier.

82. The zeros of the function are $x = -\frac{2}{5}$ and $x = \frac{3}{4}$. Use the fact that $f(2) = 18$ to find the constant multiplier.

For Exercises 83 through 98, solve each equation using any method you wish. Check your answers.

83. $4x^2 - 100 = 0$

84. $r^2 + 8r - 9 = 0$

85. $2n^3 + 5n^2 - 7n = 0$

86. $2(k - 4)^2 + 10 = 28$

87. $3(h + 5) = 22$

88. $5w^2 = 80$

89. $p^2 - 7p + 10 = 0$

90. $-3r^2 - 48 = 0$

91. $6b^2 + 18b = 0$

92. $5x + 24 = -16$

93. $4h^2 + h + 10 = 0$

94. $5c^3 - 45c = 0$

95. $4m^2 + 5m = 4m^2 + 10$

96. $30x^3 - 82x^2 + 56x = 0$

97. $-3(y + 5)^2 + 21 = -9$

98. $(d + 7)(d - 4) = -24$

4.6 Solving Quadratic Equations by Using the Quadratic Formula

LEARNING OBJECTIVES

- Solve a quadratic equation using the quadratic formula.
- Solve systems of equations with quadratics.

Solving by the Quadratic Formula

In this section, we will discuss a third method for solving quadratics. Many models that we may encounter have numbers that are not going to be easy to factor. We are now going to generate a formula based on completing the square and the square root property. If we start with a standard quadratic form, we can solve for x while leaving the coefficients and constants unknown. The only requirement is that $a \neq 0$, or the division would be undefined and the equation would not be a quadratic.

$ax^2 + bx + c = 0$ We start with the standard qudratic equal to zero.

$ax^2 + bx = -c$ Move the constant to the other side of the equal sign.

$x^2 + \frac{b}{a}x = \frac{-c}{a}$ Divide by a to make the leading coefficient 1. Note: a cannot equal zero, so that division by a is defined.

$x^2 + \frac{b}{a}x + \frac{b^2}{4a^2} = \frac{-c}{a} + \frac{b^2}{4a^2}$ Add a constant that will complete the square (half of the coefficient of x, squared).

$\left(x + \frac{b}{2a}\right)^2 = \frac{b^2 - 4ac}{4a^2}$ Factor and simplify.

$x + \frac{b}{2a} = \pm\sqrt{\frac{b^2 - 4ac}{4a^2}}$ Use the square root property.

SECTION 4.6 Solving Quadratic Equations by Using the Quadratic Formula

$$x + \frac{b}{2a} = \pm \frac{\sqrt{b^2 - 4ac}}{2a}$$ Simplify the denominator of the radical.

$$x = \frac{-b}{2a} \pm \frac{\sqrt{b^2 - 4ac}}{2a}$$ Get x by itself.

$$x = \frac{-b \pm \sqrt{b^2 - 4ac}}{2a}$$

Because we solved the quadratic equation for any values for a, b, and c, the mathematics will be true for any real values of $a \neq 0$, b, and c. Therefore, in general, we have solved any quadratic from standard form. The final result is called the **quadratic formula.**

$$x = \frac{-b \pm \sqrt{b^2 - 4ac}}{2a} \qquad a \neq 0$$

This formula can be used to solve any quadratic equation in standard form as long as the equation equals zero.

> **What's That Mean?**
>
> **Quadratic Formula**
>
> Be careful with the difference between the following:
>
> - Quadratic **formula:** a formula that is used to solve quadratic equations.
>
> $$x = \frac{-b \pm \sqrt{b^2 - 4ac}}{2a}$$
>
> - Quadratic **equation:** what is being solved.
>
> $$2x^2 + 3x - 5 = 20$$

> **Steps to Solving Quadratic Equations Using the Quadratic Formula**
>
> 1. Set the quadratic equal to zero.
> 2. Put the quadratic into standard form.
> 3. Substitute the values of a, b, and c into the quadratic formula.
>
> $$x = \frac{-b \pm \sqrt{b^2 - 4ac}}{2a}$$
>
> 4. Simplify the quadratic formula.
> 5. Check answers in the original equation.

Example 1 Using the quadratic formula

Solve the following quadratic equations. Round your answers to three decimal places. Check your answers.

a. $3x^2 - 6x - 24 = 0$ b. $4t^2 - 8t + 5 = 50$ c. $3.4x^2 + 4.2x - 7.8 = 0$

SOLUTION

a.
$$3x^2 - 6x - 24 = 0$$
$$a = 3 \quad b = -6 \quad c = -24$$

The equation is in standard form and equal to zero, so use the quadratic formula.

$$x = \frac{-(-6) \pm \sqrt{(-6)^2 - 4(3)(-24)}}{2(3)}$$

```
(-6)²-4(3)(-24)
              324
√(Ans)
               18
```

$$x = \frac{6 \pm \sqrt{324}}{6}$$

$$x = \frac{6 \pm 18}{6}$$

$$x = \frac{6 + 18}{6} \qquad x = \frac{6 - 18}{6}$$ Separate into two equations. Simplify.

$$x = \frac{24}{6} \qquad x = \frac{-12}{6}$$

$$x = 4 \qquad x = -2$$

$$3(4)^2 - 6(4) - 24 \stackrel{?}{=} 0$$ Check both answers.
$$48 - 24 - 24 \stackrel{?}{=} 0$$
$$0 = 0$$

$$3(-2)^2 - 6(-2) - 24 \stackrel{?}{=} 0$$
$$12 + 12 - 24 \stackrel{?}{=} 0$$
$$0 = 0$$ Both answers work.

b.
$$4t^2 - 8t + 5 = 50$$ Set the equation equal to zero.
$$4t^2 - 8t - 45 = 0$$
$$a = 4 \quad b = -8 \quad c = -45$$ Use the quadratic formula.
$$t = \frac{-(-8) \pm \sqrt{(-8)^2 - 4(4)(-45)}}{2(4)}$$
$$t = \frac{8 \pm \sqrt{784}}{8}$$
$$t = \frac{8 \pm 28}{8}$$
$$t = \frac{8 + 28}{8} \qquad t = \frac{8 - 28}{8}$$ Separate into two equations.
$$t = \frac{36}{8} \qquad t = \frac{-20}{8}$$ Simplify.
$$t = \frac{9}{2} \qquad t = \frac{-5}{2}$$

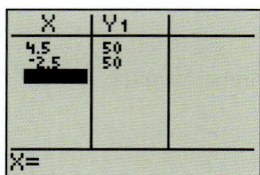

Check both answers using the table.

Both answers work.

c.
$$3.4x^2 + 4.2x - 7.8 = 0$$
$$a = 3.4 \quad b = 4.2 \quad c = -7.8$$
$$x = \frac{-(4.2) \pm \sqrt{(4.2)^2 - 4(3.4)(-7.8)}}{2(3.4)}$$
$$x = \frac{-4.2 \pm \sqrt{123.72}}{6.8}$$ Separate into two equations. Simplify.
$$x \approx \frac{-4.2 \pm 11.1229}{6.8}$$
$$x \approx \frac{-4.2 + 11.1229}{6.8} \qquad x \approx \frac{-4.2 - 11.1229}{6.8}$$
$$x \approx 1.018 \qquad x \approx -2.253$$

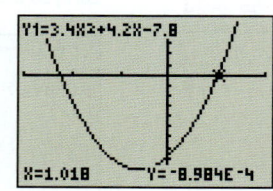

Check both answers using the graph.

Both answers work.

PRACTICE PROBLEM FOR EXAMPLE 1

Solve the following quadratic equations. Round your answers to three decimal places.

a. $2x^2 + 13x + 15 = 0$
b. $1.5x^2 + 2 = -6.5x$

The quadratic formula will find all possible solutions of every quadratic equation, making it a very powerful tool. When using the quadratic formula, be very careful with your calculations. Doing each calculation one step at a time on the calculator will lessen the number of arithmetic errors you make. In the formula itself, you should note that it starts with the opposite of b and has a plus/minus sign in front of the square root. A common mistake is to put the $2a$ in the denominator only under the square root and not under the $-b$ as well.

Example 2 Using the quadratic formula

In Section 4.1, we considered data for the number of public branch libraries in the United States and were given the model

$$B(t) = 28.7t^2 - 286.2t + 6899.3$$

where $B(t)$ represents the number of public branch libraries in the United States t years since 1990. In what year were there 6500 public branch libraries in the United States?

Year	Public Library Branches
1994	6,223
1995	6,172
1996	6,205
1997	6,332
1998	6,435

SOLUTION

Because 6500 represents a number of branch libraries, it will be substituted into $B(t)$. The numbers in this model are large, so we will use the quadratic formula to solve.

$$6500 = 28.7t^2 - 286.2t + 6899.3$$
$$\underline{-6500 \qquad\qquad\qquad -6500}$$
$$0 = 28.7t^2 - 286.2t + 399.3$$ *Set the equation equal to zero.*

$a = 28.7 \quad b = -286.2 \quad c = 399.3$ *Use the quadratic formula.*

$$t = \frac{-b \pm \sqrt{b^2 - 4ac}}{2a}$$ *Remember that $-b$ means the "opposite of b."*

$$t = \frac{-(-286.2) \pm \sqrt{(-286.2)^2 - 4(28.7)(399.3)}}{2(28.7)}$$

$$t = \frac{286.2 \pm \sqrt{36070.8}}{57.4}$$

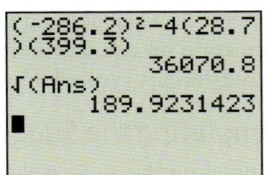

$$t \approx \frac{286.2 \pm 189.92}{57.4}$$

$$t \approx \frac{286.2 - 189.92}{57.4} \qquad t \approx \frac{286.2 + 189.92}{57.4}$$ *Separate into two equations. Simplify.*

$$t \approx \frac{96.28}{57.4} \qquad t \approx \frac{476.12}{57.4}$$

$$t \approx 1.68 \qquad t \approx 8.29$$

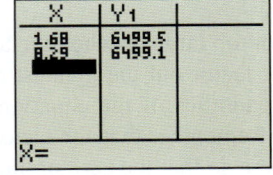

Check both answers using the table.

The answers check.

In 1992 and 1998, there were approximately 6500 public branch libraries in the United States. Both of these years are reasonable because of the fall in branch libraries during the early 1990s and the increase in the mid to late 1990s.

PRACTICE PROBLEM FOR EXAMPLE 2

The number of people who were unemployed in Hawaii during the 1990s can be modeled by

$$U(t) = -726.25t^2 + 9824.85t + 3763.21$$

where $U(t)$ represents the number of people unemployed in Hawaii t years since 1990. According to this model, when were there 30,000 people unemployed in Hawaii?

Determining Which Algebraic Method to Use When Solving a Quadratic Equation

When we are faced with an equation to solve, we should first decide which type of equation we are trying to solve (linear, quadratic, or higher-degree polynomial) and then choose a method that is best suited to solve that equation. We have looked at four algebraic methods to solve a quadratic equation:

- Square root property
- Completing the square
- Factoring
- Quadratic formula

Determining what method is best depends on the characteristics of the equation we are trying to solve.

Square Root Property

The square root property works best when the quadratic is in vertex form or when there is a squared variable term but no other variable terms.

$$(x + 5)^2 - 9 = 16$$
$$x^2 + 13 = 49$$
$$6x^2 - 4 = 18$$

In all of these equations, the square root property would be a good method to use. Remember, when solving using the square root property, to isolate the squared variable expression on one side before using the square root property. Also do not forget to use the plus/minus symbol to indicate all possible answers.

Completing the Square

Completing the square works well for equations that have both a squared term and a first-degree term. This method is usually easiest if the numbers are not too large and the leading coefficient is 1.

$$x^2 + 4x + 9 = 0$$
$$x^2 - 5x = 20$$

After completing the square, we will again use the square root property to solve.

Factoring

Factoring works best when the numbers are not too large or when the terms are higher than second degree. Always remember to first factor out the greatest common factor. Although factoring does not work with all quadratics, if the equation factors easily, it can be one of the fastest solution methods. Remember that for factoring to be used when solving, the equation must be equal to zero so that the zero product property can be used.

$$x^2 + 5x + 6 = 0$$
$$x^3 + x^2 - 6x = 0$$

Equations that cannot be factored may still have solutions, so use one of the other methods, such as completing the square or the quadratic formula. Occasionally, an equation can be factored partially, and then the separate pieces can be solved using another method.

Quadratic Formula

The quadratic formula will work with any quadratic, but it is easiest if the quadratic starts out in standard form. Because the quadratic formula basically requires substituting values for a, b, and c and then simplifying an arithmetic expression, the formula will work equally well for large or small numbers. Whenever a quadratic equation has decimals or fractions, the quadratic formula is probably the best method to choose.

$$5x^2 + 16x - 85 = 0$$
$$0.25x^2 - 3.4x + 9 = 0$$

Remember that some quadratic equations will have no real solutions. In solving with the square root property or the quadratic formula, a negative under the square root will indicate no real solutions.

Example 3 examines the thought process involved in determining which method to use to solve a quadratic equation. Solve each part, keeping the four methods we have learned in mind.

Example 3 Choosing a method to solve equations

Solve the following equations using any method you wish. Check your answers using the calculator table or graph.

a. $5x^3 + 37x^2 - 5x = -19x$ **b.** $x^2 + 12x + 7 = 0$ **c.** $5x^2 - 30 = 40$

d. $2x(x + 5) = 4x$ **e.** $x^3 + 8x^2 + 5x = 0$

SOLUTION

a. This equation has a third-degree term, so we will try to factor it and break it up into smaller pieces that will be easier to solve. The first step will be to set one side of the equation equal to zero.

$$5x^3 + 37x^2 - 5x = -19x$$

$5x^3 + 37x^2 + 14x = 0$	Set the equation equal to zero.
$x(5x^2 + 37x + 14) = 0$	Factor.
$x(5x^2 + 2x + 35x + 14) = 0$	
$x[(5x^2 + 2x) + (35x + 14)] = 0$	
$x[x(5x + 2) + 7(5x + 2)] = 0$	
$x(x + 7)(5x + 2) = 0$	

$x = 0 \qquad x + 7 = 0 \qquad 5x + 2 = 0$ Separate into three
$x = 0 \qquad x = -7 \qquad x = -\dfrac{2}{5}$ equations and solve.

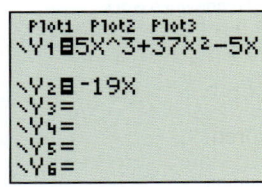

Check all three answers using the table.

All the answers work.

b. The expression on the left side does not factor nicely. (try it!), so we must use another method. Since the equation has a squared term and a first-degree term, we will not be able to use the square root property directly. Therefore, we could complete the square and then use the square root property.

$$x^2 + 12x + 7 = 0 \qquad \text{Complete the square.}$$
$$x^2 + 12x = -7$$
$$x^2 + 12x + 36 = -7 + 36$$
$$(x + 6)^2 = 29$$
$$x + 6 = \pm\sqrt{29} \qquad \text{Use the square root property.}$$

$x + 6 = \sqrt{29} \qquad x + 6 = -\sqrt{29}$ Separate into two
$x = -6 + \sqrt{29} \qquad x = -6 - \sqrt{29}$ equations and solve.
$x \approx -0.615 \qquad x \approx -11.385$

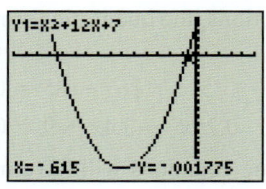

Check both answers using the graph.

Both answers work.

c. This equation has no first-degree term, so the square root property will be a good method to choose. First isolate the squared variable expression.

$$5x^2 - 30 = 40$$ Isolate the squared variable expression.
$$5x^2 = 70$$
$$x^2 = 14$$
$$x = \pm\sqrt{14}$$ Use the square root property.

$x = \sqrt{14}$ $x = -\sqrt{14}$ Separate into two equations and solve.
$x \approx 3.742$ $x \approx -3.742$

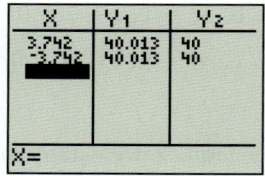

Check both answers using the table.

Both answers work.

d. This equation looks like it is already factored. Since it does not equal zero, we will have to multiply it out and set it equal to zero before we can factor and solve.

$$2x(x + 5) = 4x$$ Multiply out using the distributive property.
$$2x^2 + 10x = 4x$$
$$2x^2 + 6x = 0$$ Set the equation equal to zero.
$$2x(x + 3) = 0$$ Factor.

$2x = 0$ $x + 3 = 0$ Separate into two equations and solve.
$x = 0$ $x = -3$

 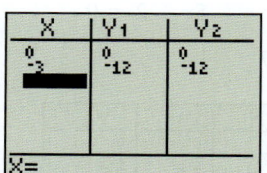

Check both answers using the table.

Both answers work.

e. This equation has a third-degree term, so it should be factored.

$$x^3 + 8x^2 + 5x = 0$$ Factor out the common term.
$$x(x^2 + 8x + 5) = 0$$

The remaining quadratic expression does not factor nicely. We will set each factor equal to zero and then solve the remaining quadratic equation using the quadratic formula.

$x = 0$ or $x^2 + 8x + 5 = 0$ Separate into two equations and solve.
               $a = 1$ $b = 8$ $c = 5$

$x = 0$ $x = \dfrac{-(8) \pm \sqrt{(8)^2 - 4(1)(5)}}{2(1)}$ Use the quadratic formula.

$x = 0$ $x = \dfrac{-8 \pm \sqrt{44}}{2}$

$x = 0$ $x = \dfrac{-8 \pm 2\sqrt{11}}{2}$ Simplify the radical.

$$x = 0 \qquad x = -4 \pm \sqrt{11}$$
$$x = 0 \qquad x = -4 + \sqrt{11} \qquad x = -4 - \sqrt{11}$$
$$x = 0 \quad \text{or} \quad x \approx -0.683 \quad \text{or} \quad x \approx -7.317$$

This equation appears to have three answers. Check them in the table.

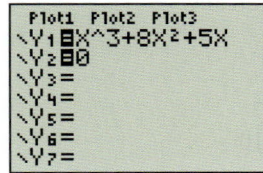

All answers work.

PRACTICE PROBLEM FOR EXAMPLE 3

Solve the following equations.

a. $6x^3 - 7x^2 = 20x$ **b.** $4x^2 + 10x - 40 = 2(5x + 12)$

Solving Systems of Equations with Quadratics

Now that we can solve quadratics in several ways, we will consider systems of equations that contain quadratics. In Chapter 2, we learned that we could solve systems using three different methods: graphing, substitution, and elimination.

Using graphs and tables to estimate solutions to systems of equations that involve functions other than lines is the same process as with linear systems. When using the graph, we look for the place(s) where the two graphs intersect. When using a table, we find input value(s) that make the outputs for both equations equal. When looking for these input value(s), notice when the output from one equation changes from being smaller than the outputs from the other equation to being larger than the other. The intersection of the two equations must be between these input values.

Example 4 Using graphs and tables to solve systems of equations

Use the given graph or table to estimate the solutions to the systems of equations.

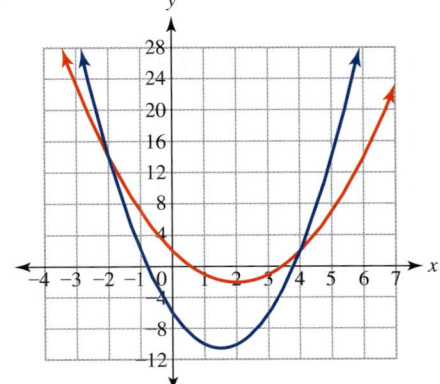

SOLUTION

a. From the graph, we can see that the two graphs intersect at about the points $(4, 2)$ and $(-2, 14)$.

b. The table shows that the two equations are equal at the points $(2, 10)$ and $(5, 64)$.

c. This table does not show any exact places where the equations are equal. The outputs for the equation in Y1 is greater than Y2 at $x = -4.4$, but Y1 is less than Y2 at $x = -4.375$. So there must be an intersection between these values. One estimate for this intersection could be $(-4.38, 0.3)$. The two equations must also intersect between $x = -0.12$ and $x = -0.11$, so we might estimate this intersection at about $(-0.115, 2.43)$. Remember that these are only estimates. Many other answers could be reasonable.

Although graphs and tables can be used to find solutions to these systems, we can also use algebraic methods to solve. In solving systems that contain quadratics algebraically, substitution is often the best choice.

Example 5 — Solving systems of equations algebraically

Solve the following systems of equations.

a. $y = 4x + 9$
$y = x^2 + 5x + 3$

b. $y = 5x^2 + 2x + 7$
$y = 2x^2 - 3x + 10$

SOLUTION

a. $y = 4x + 9$
$y = x^2 + 5x + 3$

Substitute $4x + 9$ for y in the second equation and solve for x.

$4x + 9 = x^2 + 5x + 3$ Substitute for y.
$0 = x^2 + x - 6$ Set the equation equal to zero.
$0 = (x + 3)(x - 2)$ Factor.
$x + 3 = 0 \quad x - 2 = 0$ Separate into two equations and solve.
$x = -3 \quad x = 2$
$y = 4(-3) + 9$ Use the values of x to find the corresponding values of y.
$y = -3$
$y = 4(2) + 9$
$y = 17$
$(-3, -3) \quad (2, 17)$ These two points are the solutions to the system of equations.

These answers can be checked by using either the table feature or the graph feature on the calculator. First, put both equations into the Y= screen, and then graph them using a window that will include both answers. Another option is to go to the table and input both x-values to see whether the y-values are the same. Remember that some rounding error can occur if the solutions were not exact.

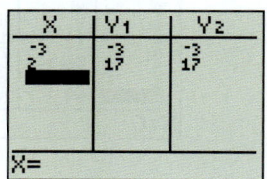 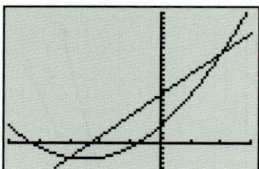

b. $y = 5x^2 + 2x + 7$
$y = 2x^2 - 3x + 10$

Substitute for y and solve for x.

$5x^2 + 2x + 7 = 2x^2 - 3x + 10$ Substitute for y and simplify.
$3x^2 + 5x - 3 = 0$ You are left with a quadratic equation that can be solved by using the quadratic formula.
$a = 3 \quad b = 5 \quad c = -3$

$x = \dfrac{-(5) \pm \sqrt{(5)^2 - 4(3)(-3)}}{2(3)}$

$x = \dfrac{-5 \pm \sqrt{61}}{6}$

$x \approx \dfrac{-5 \pm 7.8102}{6}$

$x \approx 0.468 \quad x \approx -2.135$

Use the *x*-values to find the corresponding *y*-values.

$$y = 5(0.468)^2 + 2(0.468) + 7$$
$$y = 9.031$$
$$y = 5(-2.135)^2 + 2(-2.135) + 7$$
$$y = 25.521$$
$$(0.468, 9.03) \quad (-2.135, 25.52)$$

There are two answers to this system.

Check the answers using the table or graph.

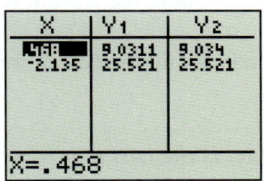 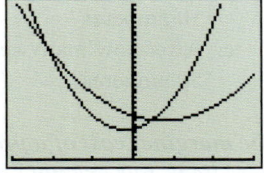

Both answers work.

PRACTICE PROBLEM FOR EXAMPLE 5

Solve the following system of equations.

$$y = 3x^2 + 5x - 74$$
$$y = 6x^2 + 60x + 126$$

We may be asked to find a solution of a quadratic equation that does not exist in the real numbers. A quadratic equation can have no real solutions when we are asked to find a value that the function will never reach. No real solutions will occur in an upward-facing parabola when the output value you are looking for is below the vertex or in a downward-facing parabola when the output value is above the vertex.

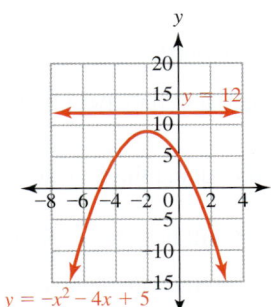

$$12 = -x^2 - 4x + 5$$

Neither of these will have real number solutions.

$$3 = (x + 2)^2 + 6$$

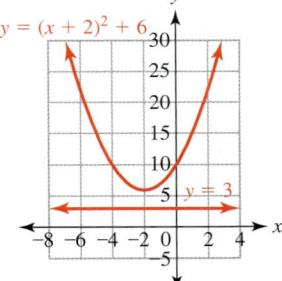

In using the quadratic formula, the value of the **discriminant**, $b^2 - 4ac$, determines whether the equation will have real solutions or not. When the discriminant is negative, the quadratic equation will have no real solutions. In the real number system, we cannot take the square root of negative numbers. Using the quadratic formula to solve the first equation above, we get

$$12 = -x^2 - 4x + 5$$
$$0 = -x^2 - 4x - 7$$
$$a = -1 \quad b = -4 \quad c = -7$$
$$x = \frac{-(-4) \pm \sqrt{(-4)^2 - 4(-1)(-7)}}{2(-1)}$$
$$x = \frac{4 \pm \sqrt{-12}}{-2}$$

For our work in this chapter, when we get a negative under the square root, we will simply state that there are no real solutions.

4.6 Exercises

For Exercises 1 through 10, solve each equation using the quadratic formula.

1. $x^2 + 8x + 15 = 0$
2. $x^2 + 5x + 6 = 0$
3. $x^2 - 7x + 10 = 0$
4. $x^2 - 12x - 45 = 0$
5. $3x^2 + 9x - 12 = 0$
6. $5x^2 + 11x - 42 = 0$
7. $2x^2 - 9x = 5$
8. $7x^2 + 15x - 9 = 11$
9. $-4x^2 + 7x - 8 = -20$
10. $-3x^2 - 8x + 1000 = 360$

11. In the United States, the use of lead in paint has been decreasing since its peak around 1920. The amount of lead used in paint can be modeled by

$$L(t) = 0.0572t^2 - 8.5587t + 320.3243$$

where $L(t)$ represents lead use in thousands of tons t years since 1900.
Source: Model based on data from Statistical Abstract of the United States, 2001.

   a. How much lead was used in 1955?
   b. When were 5500 tons of lead used in paints? (*Hint:* 5500 tons = 5.5 thousand tons.)
   c. In what year were 51 thousand tons of lead used in paint?

12. The number of Americans who participate in aerobic exercise in thousands can be modeled by

$$E(a) = -7.5a^2 + 523.1a - 3312.3$$

where $E(a)$ represents the number of people in thousands who participated in aerobic exercise at age a years.

   a. Find the number of 20-year-old Americans who participate in aerobic exercise.
   b. At what age(s) do 3 million Americans participate in aerobic exercise? (*Hint:* 3 million = 3000 thousand.)

13. The amount of electricity flowing through a surface is called the current I and is measured in amperes (A). The current through a particular wire can be modeled by

$$I(t) = 3t^2 - 4t + 5$$

where $I(t)$ is the current in amperes after t seconds.

   a. Find the current in the wire after 0.5 second.
   b. After how many seconds would the current reach 20 amperes?
   c. After how many seconds would the current reach 60 amperes?

14. The current through a particular wire can be modeled by

$$I(t) = 2t^2 - 6t + 7$$

where $I(t)$ is the current in amperes after t seconds.

   a. Find the current in the wire after 0.25 second.
   b. After how many seconds would the current reach 10 amperes?
   c. After how many seconds would the current reach 50 amperes?

The marginal cost of a product is the cost to produce one more item. For example, the marginal cost of 100 items is the cost to produce the 101st item.

15. The marginal cost of producing a particular kind of dress shoe can be modeled by

$$M(n) = 0.00045n^2 + 0.02n + 5$$

where $M(n)$ is the marginal cost in dollars per pair of dress shoes when n dress shoes are produced.

   a. Find the marginal cost of producing the 201st pair of shoes. (Find $M(200)$.)
   b. Find when the marginal cost will be $20 per pair of shoes.

16. The marginal cost of producing a cell phone can be modeled by

$$M(n) = 0.0000015n^2 + 0.0001n + 8$$

where $M(n)$ is the marginal cost in dollars per phone when n phones are produced.

   a. Find the marginal cost of producing the 5001st phone. (Find $M(5000)$.)
   b. Find when the marginal cost will be $30 per phone.

17. The marginal cost in dollars of producing a bike can be modeled by

$$M(n) = 0.0005n^2 + 0.07n + 50$$

where $M(n)$ is the marginal cost in dollars per bike when n bikes are produced.

   a. Find the marginal cost of producing the 401st bike. (Find $M(400)$.)
   b. Find when the marginal cost will be $700 per bike.

18. The marginal cost in dollars of producing a laptop computer can be modeled by

$$M(n) = 0.0007n^2 + 0.04n + 85$$

where $M(n)$ is the marginal cost in dollars per laptop when n laptops are produced.

   a. Find the marginal cost of producing the 801st laptop. (Find $M(800)$.)
   b. Find when the marginal cost will be $825 per laptop.

SECTION 4.6 Solving Quadratic Equations by Using the Quadratic Formula 373

19. A local rocketry club held a competition for the highest launch of a model rocket. From the power produced by the motor, the height t seconds after launch is modeled by

$$h(t) = -16t^2 + 200t + 2$$

where $h(t)$ represents the height in feet t seconds after the rocket is launched.

a. Find the height of the rocket 1 second after launch.
b. When will the rocket first reach a height of 450 feet?
c. When will the rocket first reach a height of 600 feet?
d. When will the rocket first reach a height of 700 feet?

20. The height of a ball thrown straight up into the air can be modeled by

$$h(t) = -16t^2 + 30t + 3.5$$

where $h(t)$ is the height of the ball in feet t seconds after it is thrown.

a. What is the height of the ball after 0.5 second?
b. When will the ball be at a height of 15 feet?
c. Will the ball ever reach a height of 20 feet?

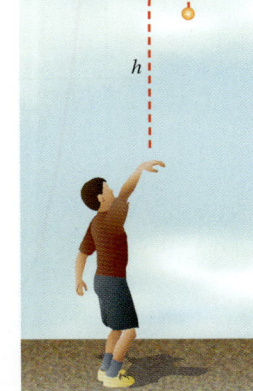

21. The number of hours per year an average person spent using the Internet can be modeled by

$$I(t) = 3.8t^2 - 29.5t + 56.0$$

where $I(t)$ represents the number of hours per year an average person spent using the Internet t years since 1990.
Source: Model based on data from Statistical Abstract of the United States, 2001.

a. How many hours did the average person spend using the Internet in 2000?
b. When will/did the average person spend 100 hours per year on the Internet?
c. When will the average person spend an average of 1 hour per day on the Internet?

22. The number of cable television systems in the United States during the 1990s can be modeled by

$$C(t) = -34.6t^2 + 285.5t + 10649.6$$

where $C(t)$ represents the number of cable television systems in the United States t years since 1990.

a. Find the number of cable television systems in the United States in 1995.
b. Find the year(s) when there were 11,000 cable television systems in the United States.

In Exercises 23 through 68, solve the given equations using any of the methods you have learned. Round all answers to two decimal places when necessary. Check your answers graphically or with a table.

23. $a^2 + 2a = 15$
24. $x^2 + 8x = 30$
25. $5t^2 - 14 = 0$
26. $-6m^2 + 56 = 0$
27. $\frac{1}{7}x^2 - \frac{5}{7}x = 0$
28. $\frac{2}{5}x^2 + \frac{3}{5}x = 0$
29. $4.7x^2 - 2.6x = 0$
30. $3.5w^2 + 8.2w = 0$
31. $5x - 12 = 80$
32. $4x + 45 = 5$
33. $5x^2 + 3x = 0$
34. $4t^2 - 4t = 0$
35. $(x - 9)^2 + 8 = 24$
36. $(p + 5)^2 - 4 = 0$
37. $3(4x - 5)^2 + 20 = 47$
38. $7(2k - 7)^2 - 11 = 17$
39. $33 = -7(4 - w)^2 + 59$
40. $47 = -4(6 - x)^2 + 63$
41. $r^2 + 1.4r - 14.9 = 0$
42. $k^2 + 2.3k - 11.4 = 0$
43. $d^2 + 2d - 35 = 0$
44. $x^2 + 11x + 18 = 0$
45. $b^2 - 3b = 28$
46. $v^2 + 14v = -48$
47. $23 = -2(s + 9)^2 + 5$
48. $105 = 5(17 + d)^2 - 20$
49. $\frac{2}{3}c^2 - \frac{5}{6} = \frac{1}{6}c - 2$
50. $\frac{3}{4}x^2 + \frac{1}{2}x - 7 = \frac{1}{4}x - 3$
51. $120 = -28f + 7f^2 - 939$
52. $9.9x^2 + 57.4x - 134.8 = 0$
53. $(3p - 4)(p + 3) = 15$
54. $(z + 9)(5z - 4) = 18$
55. $3x^2 + 4x + 20 = 0$
56. $-2x^2 + 7x - 18 = 0$
57. $\frac{3}{2}(x - 8)^2 + 10 = 1$
58. $\frac{2}{3}(x - 5)^2 - 7 = 10$
59. $3x^3 - 15x^2 = 252x$
60. $1.75x^3 + 2.5x^2 = 6.25x$
61. $7(x + 3) - 8 = 20$
62. $-3(r - 8) + 7 = -15$
63. $-1.5(d + 5) - 7 = 2d + 8$
64. $-2.3(p + 7) + 2 = 4.3p + 20$
65. $x^2 + 6x + 25 = 0$
66. $w^2 - 12x - 16 = 0$
67. $\frac{1}{4}x^2 - \frac{3}{4}x + 7 = 13$
68. $\frac{11}{21}x^2 - \frac{9}{21}x - \frac{4}{21} = 0$

For Exercises 69 through 76, estimate the solutions to the systems using the graphs. Write your answers as points.

69.

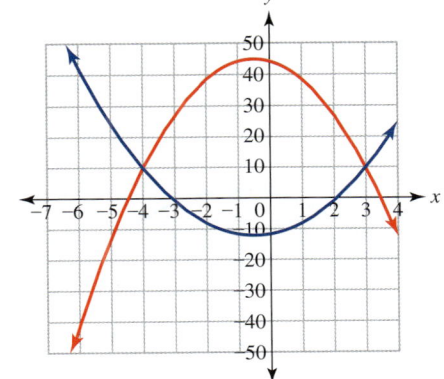

70.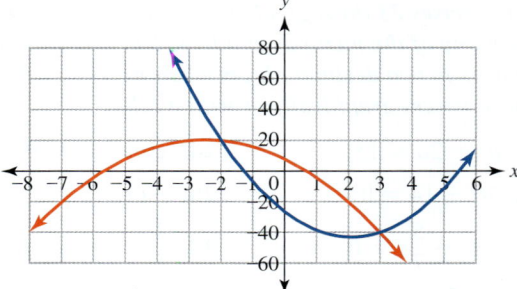

71.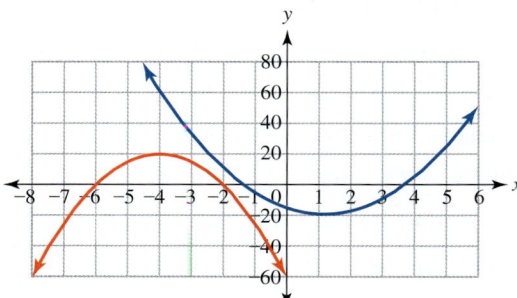

72.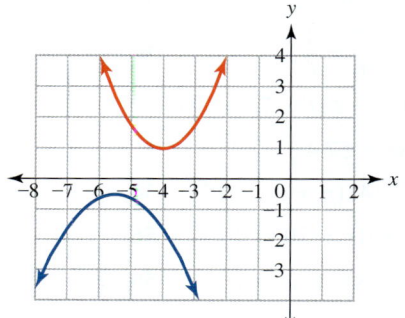

73.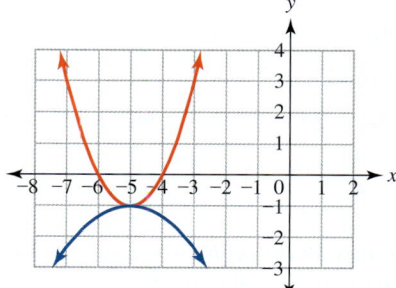

74.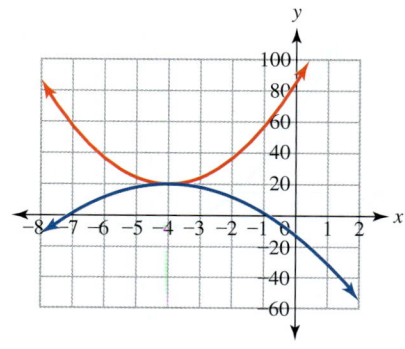

75.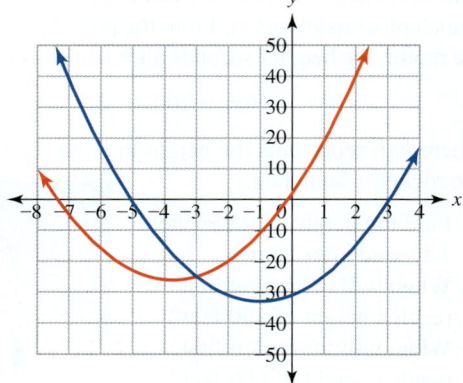

76.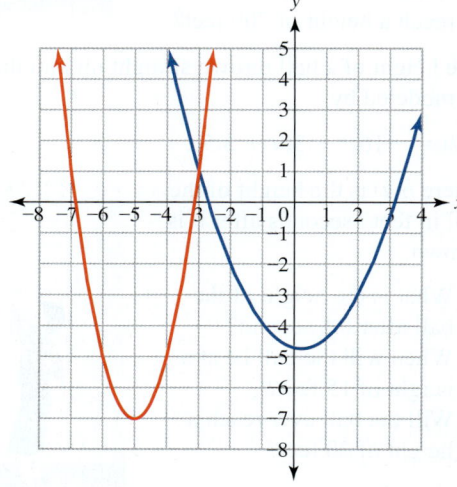

For Exercises 77 through 80, estimate the solution(s) to the system using the given tables. Write your answers as points.

77.

X	Y1	Y2
-6	48	-42
-4	12	12
-2	-8	42
0	-12	48
2	0	30
3	12	12
4	28	-12

Y1 ■ 2X² + 2X − 12

78.

X	Y1	Y2
2.176	17.292	17.797
2.188	17.525	17.499
2.203	17.819	17.125
0	-6	48
-2.735	18.451	18.804
-2.743	18.61	18.609
-2.765	19.051	18.069

Y1 ■ 4X² + 2X − 6

79.

X	Y1	Y2
-3.9	1.41	.1
-3.8	.84	.2
-3.7	.29	.3
-3.6	-.24	.4
2.6	5.96	6.6
2.7	6.69	6.7
2.8	7.44	6.8

Y1 ■ X² + 2X − 6

80.

X	Y1	Y2
2	27	9
3	17	9
4	11	9
5	9	9
6	11	9
7	17	9
8	27	9

Y1 ■ 2(X−5)² + 9

In Exercises 81 through 98, solve the system of equations algebraically. Check your answer(s) graphically or with a table.

81. $y = x^2 + 3x - 9$
$y = 5x - 8$

82. $y = 0.25x^2 + 5x - 3.4$
$y = -4.5x + 7.5$

83. $y = x^2 + 5x - 3$
$y = 2x - 10$

84. $y = 2x^2 - 3x + 7$
$y = x - 12$

85. $y = 3x^2 + 5x - 9$
$y = -0.5x^2 - 3x + 15$

86. $y = 3x^2 + 4x - 20$
$y = 2x^2 + 6x + 15$

87. $y = x^2 - 4x + 11$
    $y = -x^2 + 7x - 4$

88. $y = 4x^2 + 2x - 7$
    $y = -2x^2 + 5x + 12$

89. $y = -1.8x^2 - 2.3x + 4.7$
    $y = 2.5x^2 + 3.4x - 8.5$

90. $y = -0.3x^2 + 5x - 2.6$
    $y = 0.5x^2 - 3x - 7.5$

91. $y = x^2 + 6x - 20$
    $y = -x^2 - 6x - 38$

92. $y = 5x^2 + 20x - 40$
    $y = -3x^2 - 12x - 72$

93. $y = x^2 + 2x - 8$
    $y = -2x^2 + 2x - 18$

94. $y = -x^2 + 4x + 1$
    $y = x^2 + 3x + 14$

95. $y = x^2 - 10x + 30$
    $y = -x^2 + 6x - 15$

96. $y = -0.5x^2 + 4x - 8$
    $y = 0.3x^2 + 5x + 7$

97. $y = 6x^2 + 2x - 9$
    $y = 9x^2 + 2x - 15$

98. $y = 4x^2 + 9x - 12$
    $y = -2x^2 + 20x + 23$

4.7 Graphing Quadratics from Standard Form

LEARNING OBJECTIVES

- Graph a quadratic from standard form.
- Graph quadratic inequalities in two variables.

Graphing from Standard Form

Because quadratic functions are often written in standard form rather than in vertex form, it is important that we learn to graph from the standard form. Since all quadratics have a vertex and a vertical intercept, these two will be the starting points for graphing.

A quadratic function in standard form $f(x) = ax^2 + bx + c$ gives us several key pieces of information about the graph. The value of a in the standard form affects the graph of the quadratic the same way that it affects the vertex form. If a is positive, the parabola will open upward. If a is negative, the parabola will open downward. Also a will affect whether the graph is wide or narrow. The value c will also give us information about the graph. When we substitute zero into the input variable for the function, the output will be c.

Let

$$f(x) = ax^2 + bx + c \quad \text{and} \quad x = 0$$
$$f(0) = a(0)^2 + b(0) + c$$
$$f(0) = c$$

which means that $(0, c)$ will be the vertical intercept for the graph.

$$f(x) = 4x^2 + 3x - 12$$
$$f(0) = 4(0)^2 + 3(0) - 12$$
$$f(0) = -12$$
$$(0, -12) \quad \text{The vertical intercept.}$$

Another important relationship for parabolas is that between the vertex and the standard form of a quadratic function. There are several ways to see this relationship, one being to consider the quadratic formula and symmetry.

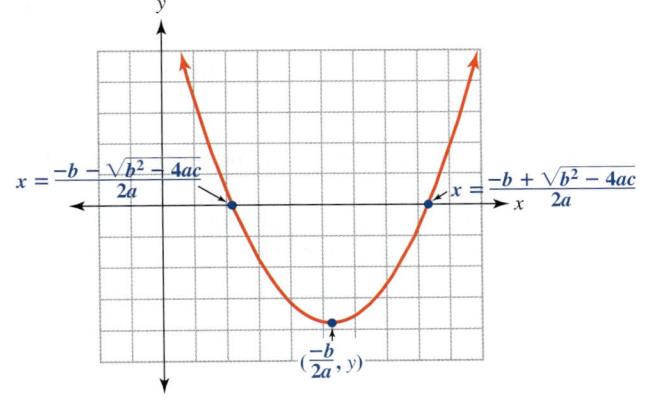

> **Connecting the Concepts**
>
> **Is a the same in both quadratic forms?**
>
> The a-value in the vertex form is the same as the a-value in standard form. If we multiply out the vertex form and simplify into the standard form, the a-value does not change.
>
> Example:
>
> $$f(x) = 5(x - 2)^2 + 9$$
> $$f(x) = 5(x - 2)(x - 2) + 9$$
> $$f(x) = 5(x^2 - 4x + 4) + 9$$
> $$f(x) = 5x^2 - 20x + 29$$
>
> The fact that the a-value is the same in both forms of a quadratic means that it will affect the graph of the parabola in the same ways.
>
> Positive a: upward facing
>
> Negative a: downward facing
>
> $|a| < 1$: wide parabola
>
> $|a| > 1$: narrow parabola

From the graph, the two x-intercepts are located at the x-values given by the quadratic formula. The symmetry of the graph indicates that the x-value of the vertex must be midway between these two points. If we take the two x-values and add them together, the square root portions will cancel out because one is positive and the other is negative. Then we divide by 2 to get the average.

$$x = \frac{-b - \sqrt{b^2 - 4ac}}{2a} \qquad x = \frac{-b + \sqrt{b^2 - 4ac}}{2a}$$ Horizontal intercepts are the x-values.

$$x = \frac{-b}{2a} - \frac{\sqrt{b^2 - 4ac}}{2a} \qquad x = \frac{-b}{2a} + \frac{\sqrt{b^2 - 4ac}}{2a}$$

$$\frac{x + x}{2} = \frac{\frac{-b}{2a} - \frac{\sqrt{b^2 - 4ac}}{2a} + \frac{-b}{2a} + \frac{\sqrt{b^2 - 4ac}}{2a}}{2}$$ Add them and divide by 2.

$$\frac{2x}{2} = \frac{2\left(\frac{-b}{2a}\right)}{2}$$ The radical parts cancel.

$$x = \frac{-b}{2a}$$ Reduce.

This result indicates that the x-value for the vertex will be $x = \frac{-b}{2a}$.

Another way to discover this relationship is to consider that the only point on a parabola without a symmetric point is the vertex. With this information, the only way to come to a single solution from the quadratic formula is to have the square root after the plus/minus symbol be zero. This way, we are not adding or subtracting anything to or from the remaining parts of the formula. If we consider the square root to be zero, we are left with

$$x = \frac{-b \pm \sqrt{0}}{2a}$$

$$x = \frac{-b}{2a}$$

This x-value will then represent the one input value that does not have a symmetric point. Hence, the vertex has the input value $x = \frac{-b}{2a}$. Remember that the vertex is a point on the graph, so find the y-value of the point. To find the y-value of the vertex, simply evaluate the equation at $x = \frac{-b}{2a}$.

$$f(x) = 4x^2 + 24x - 20$$
$$x = \frac{-b}{2a} = \frac{-(24)}{2(4)} = \frac{-24}{8} = -3$$ Find the x-part.
$$f(-3) = 4(-3)^2 + 24(-3) - 20$$
$$f(-3) = -56$$ Substitute the x-part into the equation to find the y-part.
$$(-3, -56)$$ The vertex.

> **Example 1** Finding the vertical intercept and vertex of quadratics

Find the vertical intercept and vertex of the following quadratics. State whether the vertex is a minimum or maximum point on the graph.

a. $f(x) = 3x^2 + 24x + 8$ **b.** $g(t) = t^2 - 5t - 11$

c. $h(m) = -3(m + 4)^2 - 8$

SOLUTION

a. The function $f(x) = 3x^2 + 24x + 8$, is a quadratic in standard form. When $x = 0$, the function equals c, which is 8.

$$f(x) = 3x^2 + 24x + 8$$
$$f(0) = 3(0)^2 + 24(0) + 8$$
$$f(0) = 8$$
$$(0, 8)$$

The vertical intercept is $(0, 8)$.

The input value of the vertex can be found by using the formula $x = \frac{-b}{2a}$.

$$a = 3 \quad b = 24$$
$$x = \frac{-(24)}{2(3)}$$
$$x = -4$$

Because the input value of the vertex is $x = -4$, we can find the output value by substituting -4 for x.

$$f(-4) = 3(-4)^2 + 24(-4) + 8$$
$$f(-4) = -40$$
$$(-4, -40)$$

The vertex is the point $(-4, -40)$. The a-value of this quadratic function is positive, so this parabola faces upward. Therefore, this vertex is a minimum point.

b. The function $g(t) = t^2 - 5t - 11$, is a quadratic in standard form. When $t = 0$, the function equals c, which is -11. The vertical intercept is $(0, -11)$.

The t-value of the vertex can be found by using the formula $t = \frac{-b}{2a}$.

$$a = 1 \quad b = -5$$
$$t = \frac{-(-5)}{2(1)}$$
$$t = 2.5$$

Because the input value of the vertex is $t = 2.5$, we can find the output value by substituting 2.5 for t.

$$f(2.5) = (2.5)^2 - 5(2.5) - 11$$
$$f(2.5) = -17.25$$
$$(2.5, -17.25)$$

The vertex is the point $(2.5, -17.25)$. The a-value of this quadratic function is positive, so this parabola faces upward. Therefore, this vertex is a minimum point.

c. The function $h(m) = -3(m + 4)^2 - 8$, is a quadratic in vertex form. The vertical intercept must be calculated carefully by substituting in zero for m.

$$h(m) = -3(m + 4)^2 - 8$$
$$h(0) = -3(0 + 4)^2 - 8$$
$$h(0) = -3(16) - 8$$
$$h(0) = -56$$
$$(0, -56)$$

The vertical intercept is $(0, -56)$. Note that the value we calculated is not the constant in the function. Instead, we must substitute zero into the equation and calculate to get the correct vertical intercept for this quadratic function.

Since the function is in vertex form, we do not have to calculate the vertex. The vertex will be (h, k), so the vertex is $(-4, -8)$. The a-value is negative, so this parabola faces downward. Therefore, the vertex is the maximum point on the graph.

PRACTICE PROBLEM FOR EXAMPLE 1

Find the vertical intercept and vertex of the following quadratics. State whether the vertex is a minimum or maximum point on the graph.

a. $f(x) = 4x^2 + 40x + 7$ **b.** $k(x) = -2(x - 3)^2 + 12$

c. $h(t) = -7t^2 - 15t - 6$

The vertical intercept, vertex, and symmetry will all help us to graph quadratic functions. The process will be basically the same as graphing from vertex form. We will find the vertex and vertical intercept, and we will use the axis of symmetry to find symmetric points. Now that we know how to solve quadratic equations, we will also find the horizontal intercepts, if there are any, and use them as an additional pair of symmetric points. In general, at least five points should be plotted for a graph of a parabola. If we want more details, we can always find more symmetric points by plugging in values for x and finding y.

> **Steps to Graphing a Quadratic Function from the Standard Form**
> 1. Determine whether the graph opens up or down.
> 2. Find the vertex and the equation of the axis of symmetry.
>
> $$\left(\text{Use } x = \frac{-b}{2a} \text{ to find the } x\text{-part of the vertex.}\right)$$
>
> 3. Find the vertical intercept.
> 4. Find the horizontal intercepts (if any).
> 5. Plot the points you found in steps 2 through 4. Plot their symmetric points and sketch the graph. (Find an additional pair of symmetric points if needed.)

Example 2 Graphing quadratic functions

Sketch a graph of the following:

a. $f(x) = 2x^2 - 12x + 5$ **b.** $f(x) = -0.5x^2 + 4x - 10$

SOLUTION

a. $f(x) = 2x^2 - 12x + 5$

Step 1 Determine whether the graph opens up or down.
Because a is positive in this quadratic, we know that the parabola will open upward.

Step 2 Find the vertex and the equation for the axis of symmetry.
This function is in standard form for a quadratic, so we can use the formula $x = \frac{-b}{2a}$ to find the x-value of the vertex.

$$x = \frac{-(-12)}{2(2)}$$

$$x = 3$$

Now that we have the input value for the vertex, we can substitute it into the function and find the output value for the vertex.

$$f(x) = 2x^2 - 12x + 5$$
$$f(3) = 2(3)^2 - 12(3) + 5$$
$$f(3) = 18 - 36 + 5$$
$$f(3) = -13$$

The vertex is $(3, -13)$. The axis of symmetry is the vertical line through the vertex. The axis of symmetry is $x = 3$.

Step 3 Find the vertical intercept.

Because this quadratic is in standard form, we know that the vertical intercept has an output value equal to the constant c. In this case, the vertical intercept is $(0, 5)$.

Step 4 Find the horizontal intercepts (if any).

Horizontal intercepts will happen when the output variable is equal to zero. Substitute zero for the output variable and solve.

$$f(x) = 2x^2 - 12x + 5$$
$$0 = 2x^2 - 12x + 5$$
$$a = 2 \quad b = -12 \quad c = 5 \quad \text{Use the quadratic formula.}$$
$$x = \frac{-b \pm \sqrt{b^2 - 4ac}}{2a}$$
$$x = \frac{-(-12) \pm \sqrt{(-12)^2 - 4(2)(5)}}{2(2)}$$
$$x = \frac{12 \pm \sqrt{144 - 40}}{4}$$
$$x = \frac{12 \pm \sqrt{104}}{4}$$
$$x = \frac{12 \pm 10.2}{4}$$
$$x = \frac{12 + 10.2}{4} \quad x = \frac{12 - 10.2}{4}$$
$$x = 5.55 \quad x = 0.45$$

The horizontal intercepts are $(5.55, 0)$ and $(0.45, 0)$.

Step 5 Plot the points you found in steps 2 through 4. Plot their symmetric points and sketch the graph. (Find an additional pair of symmetric points if needed.)

We now have the vertex $(3, -13)$, the vertical intercept $(0, 5)$, and the horizontal intercepts $(5.55, 0)$ and $(0.45, 0)$. Plotting these points and using the axis of symmetry to plot their symmetric points, we get the following graph.

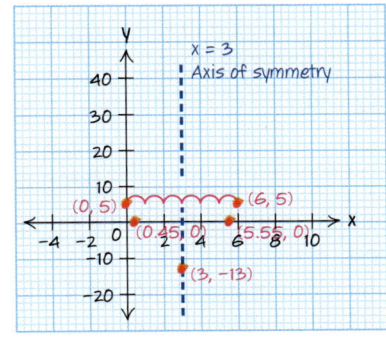

Sketching a smooth curve through the points gives us the following graph.

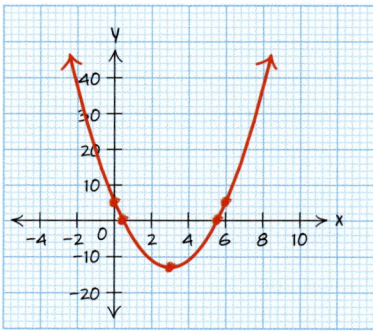

b. $f(x) = -0.5x^2 + 4x - 10$

Step 1 Determine whether the graph opens up or down.
Because a is negative in this quadratic, we know that the parabola will open downward.

Step 2 Find the vertex and the equation for the axis of symmetry.
This function is in standard form for a quadratic, so we can use the formula $x = \dfrac{-b}{2a}$ to find the x-value of the vertex.

$$x = \dfrac{-(4)}{2(-0.5)}$$
$$x = 4$$

Now that we have the input value for the vertex, we can substitute it into the function and find the output value for the vertex.

$$f(x) = -0.5x^2 + 4x - 10$$
$$f(4) = -0.5(4)^2 + 4(4) - 10$$
$$f(4) = -8 + 16 - 10$$
$$f(4) = -2$$

The vertex is $(4, -2)$. The axis of symmetry is the vertical line through the vertex. The axis of symmetry is $x = 4$.

Step 3 Find the vertical intercept.
Because this quadratic is in standard form, we know that the vertical intercept has an output value equal to the constant c. In this case, the vertical intercept is $(0, -10)$.

Step 4 Find the horizontal intercepts (if any).
Horizontal intercepts will happen when the output variable is equal to zero, so substitute zero for the output variable and solve.

$$f(x) = -0.5x^2 + 4x - 10$$
$$0 = -0.5x^2 + 4x - 10 \quad \text{Set the function equal to zero.}$$
$$a = -0.5 \quad b = 4 \quad c = -10$$
$$\text{Use the quadratic formula.}$$
$$x = \dfrac{-(4) \pm \sqrt{(4)^2 - 4(-0.5)(-10)}}{2(-0.5)}$$
$$x = \dfrac{4 \pm \sqrt{-4}}{-1} \quad \text{A negative discriminant indicates no real solutions.}$$

There are no real solutions when this function is set equal to zero, so there are no horizontal intercepts.

Step 5 Plot the points you found in steps 2 through 4. Plot their symmetric points and sketch the graph. (Find an additional pair of symmetric points if needed.)

We now have the vertex $(4, -2)$ and the vertical intercept. There are no horizontal intercepts, so we will want an additional pair of symmetric points to give us more points to sketch our curve through. To find another point, we can pick an additional input value on one side of the axis of symmetry and find the output value.

$$x = 2$$
$$f(2) = -0.5(2)^2 + 4(2) - 10$$
$$f(2) = -4$$
$$(2, -4)$$

Pick any input value we don't already have and substitute to find the output value.

Plotting all the points that we have found and using the axis of symmetry to plot their symmetric points, we get the following graph.

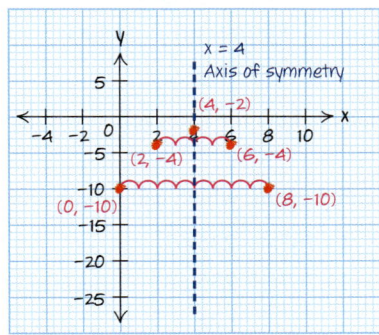

We now have the vertex $(4, -2)$ and two sets of symmetric points: $(2, -4)$, $(6, -4)$ and $(0, -10)$, $(8, -10)$. Sketching a smooth curve through these points gives us the final graph.

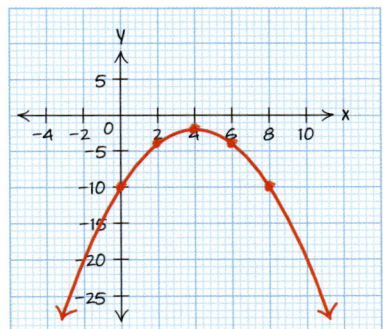

PRACTICE PROBLEM FOR EXAMPLE 2

Sketch a graph of the following:

a. $f(x) = 0.25x^2 + 6x - 15$ **b.** $f(x) = 3x^2 + 6x + 7$

In part b of Example 2, the parabola had no horizontal intercepts. If we thought ahead about the fact that the vertex is below the x-axis and the parabola is facing downward, we could have known that there would be no horizontal intercepts. After finding the vertex and deciding whether the graph faces upward or downward, we can determine whether the parabola has any horizontal intercepts and, therefore, whether we should solve for them.

Example 3 Applications of the vertex

The monthly profit for an amusement park can be modeled by

$$P(t) = -0.181t^2 + 13.767t - 235.63$$

where $P(t)$ represents the monthly profit in millions of dollars when tickets are sold for t dollars each.

a. What is the amusement park's monthly profit if it sells tickets for $30 each?

b. How much should the amusement park charge for tickets if it wants a monthly profit of $25 million?

c. Find the vertex for this model and explain its meaning for the amusement park.

SOLUTION

a. $30 is a ticket price, so it can be substituted into t.

$$P(t) = -0.181t^2 + 13.767t - 235.63$$
$$P(30) = -0.181(30)^2 + 13.767(30) - 235.63$$
$$P(30) = 14.48$$

If the amusement park sells tickets for $30 each, it will have a monthly profit of about $14.48 million.

b. The $25 million is a monthly profit, which can be substituted into P.

$$P(t) = -0.181t^2 + 13.767t - 235.63$$
$$25 = -0.181t^2 + 13.767t - 235.63 \quad \text{Substitute 25 for } P.$$
$$\underline{-25 \qquad\qquad\qquad\qquad -25}$$
$$0 = -0.181t^2 + 13.767t - 260.63 \quad \text{Set the equation equal to zero and use the quadratic formula.}$$
$$t \approx 35.51 \qquad t \approx 40.55$$

If the amusement park charges $35.51 or $40.55 per ticket, it will have a monthly profit of about $25 million. We hope that the park would choose the $35.51 to make more people happy.

c. The vertex will have an input of

$$t = \frac{-b}{2a} = \frac{-13.767}{2(-0.181)}$$
$$t \approx 38.03$$

Using a ticket price of $38.03, we get

$$P(t) = -0.181t^2 + 13.767t - 235.63$$
$$P(38.03) = -0.181(38.03)^2 + 13.767(38.03) - 235.63$$
$$P(38.03) = 26.15$$
$$(38.03, 26.15)$$

If the amusement park charges $38.03 per ticket, it will make its maximum profit of $26.15 million per month.

PRACTICE PROBLEM FOR EXAMPLE 3

The net sales for computer manufacturer Gateway, Inc. can be modeled by

$$S(t) = 75.85t^2 - 1204.22t + 4725.95$$

where $S(t)$ represents the annual net sales in millions of dollars for Gateway, Inc. t years since 1980.

Source: Model derived from data found in Gateway, Inc. annual reports at Gateway.com.

a. Using this model, estimate the annual net sales for Gateway, Inc. in 1995.

b. When were Gateway's net sales $9000 million?

c. Find the vertex for this model and explain its meaning.

SECTION 4.7 Graphing Quadratics from Standard Form

Graphing Quadratic Inequalities in Two Variables

Graphing quadratic inequalities can be done by using the same techniques as graphing linear inequalities. First, graph the inequality as if it were an equality using a solid curve if there is an "equal to" part and a dashed curve if it is only "less than" or "greater than" but not "equal to." This separates the graph into two sections: points that are above the parabola and points that are below the parabola. Then decide which section satisfies the inequality and, therefore, should be shaded. So pick a point that is not on the curve and check to see whether it satisfies the inequality. Finally, shade the side of the curve that does satisfy the inequality.

Example 4 Graphing quadratic inequalities

Graph the following inequalities.

a. $y < x^2 + 3x - 10$ **b.** $y \geq -2x^2 + 11x - 12$

SOLUTION

a. First, we will sketch the graph of the quadratic using a dashed curve because there is no "equal to" part.

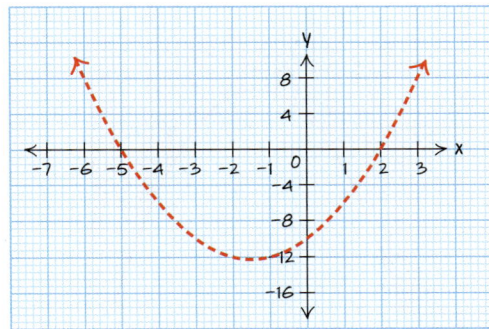

Doing so separates the grid into two sections: the points above the parabola and the points below the parabola.

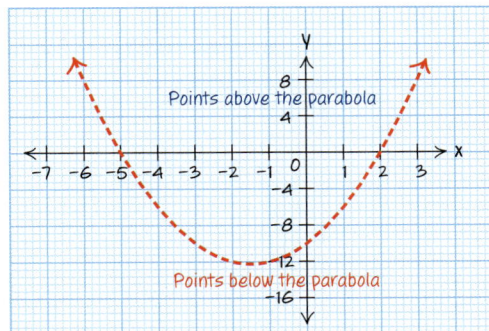

Now if we pick (0, 0) as a test point, we find that it does not satisfy the inequality, so we will shade the points below the parabola.

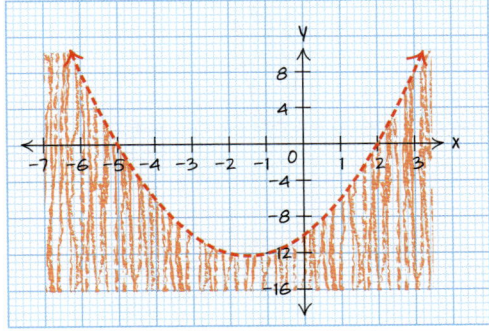

b. First, we will sketch the graph using a solid line because it is a "greater than or equal to" inequality.

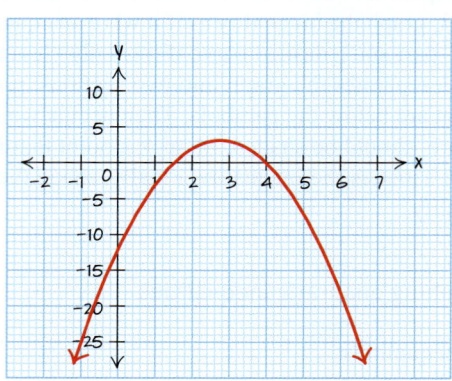

If we pick $(0, 0)$ for our test point, we find that it does satisfy the inequality, so we should shade above the curve.

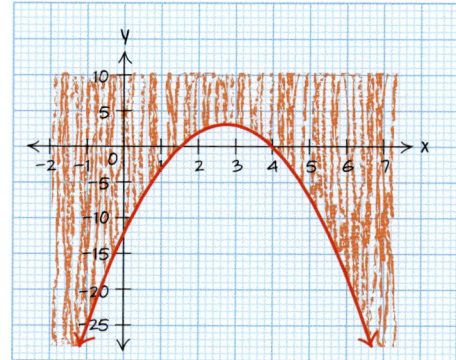

PRACTICE PROBLEM FOR EXAMPLE 4
Graph the following inequalities.

a. $y > 2x^2 - 12x - 8$ **b.** $y \leq -0.25x^2 - 3x + 5$

4.7 Exercises

For Exercises 1 through 4, without using your calculator, match the given equations with the appropriate graph. State which equation does not match any of the graphs. Explain.

1. $y = -2x^2 - 6x - 7$

$y = -0.5x^2 - x + 7$

$y = x^2 + 4x + 4$

$y = x^2 - 5x - 6$

Graph A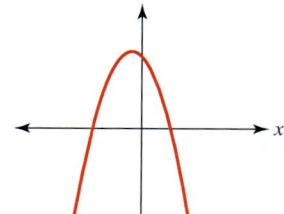

Graph B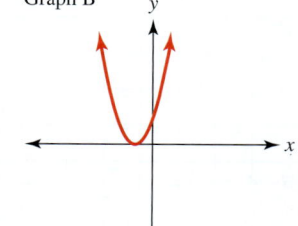

Graph C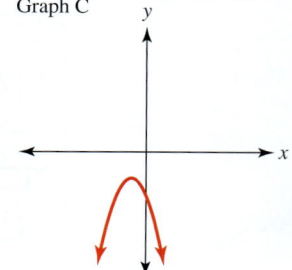

2. $y = x^2 - 6x + 10$
   $y = x^2 - 3x - 4$
   $y = -0.3x^2 + 2x + 3$
   $y = -3x^2 + 12x - 8$

Graph A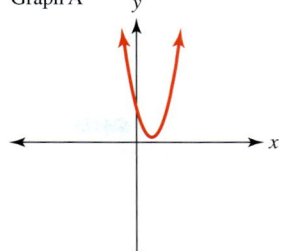
Graph B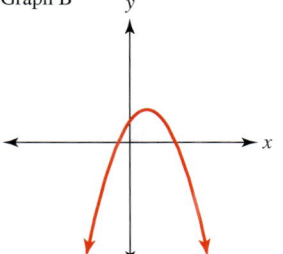
Graph C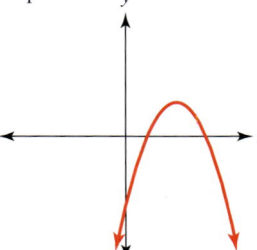

3. $y = -2(x - 3)^2 - 9$
   $y = 2(x - 3)^2 - 9$
   $y = 2(x + 4)^2 + 1$
   $y = 2(x + 4)^2 - 9$

Graph A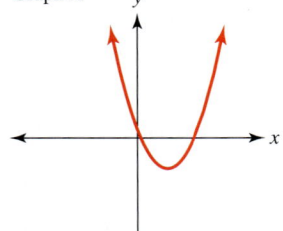
Graph B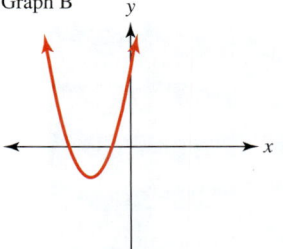
Graph C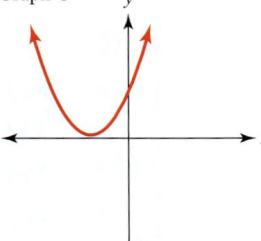

4. $y = x^2$
   $y = x^2 - 7$
   $y = -x^2 + 20$
   $y = -x^2$

Graph A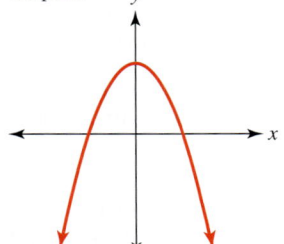
Graph B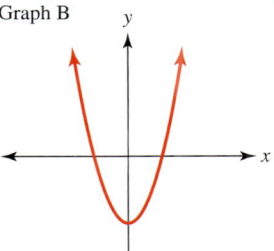
Graph C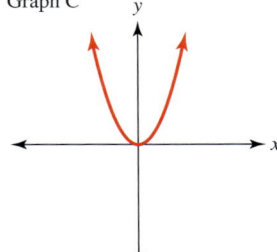

For Exercises 5 through 14, find the vertex, vertical intercept, and horizontal intercept(s) of each quadratic function.

5. $f(x) = x^2 + 6x + 8$
6. $g(x) = x^2 + 12x - 28$
7. $h(x) = 3x^2 - 18x + 15$
8. $f(x) = -4x^2 + 10x + 20$
9. $h(x) = 4(x - 8)^2 - 20$
10. $f(x) = -2(x + 10)^2 + 15$
11. $g(x) = 5x^2 + 12x + 10$
12. $h(x) = -3x^2 + 11x - 20$
13. $f(x) = 1.5x^2 - 6x + 4$
14. $g(x) = 0.75x^2 + 10x - 15$

15. The cost to produce football uniforms for a school can be modeled by

$$C(u) = \frac{1}{4}u^2 - 25u + 3500$$

where $C(u)$ represents the cost in dollars to produce u football uniforms.

a. Find the cost to produce 30 uniforms.
b. Find the vertex and describe its meaning.
c. How many uniforms can a school get with a budget of $1600?

16. The cost to produce backpacks with a school's logo can be modeled by

$$C(b) = \frac{2}{5}b^2 - 12b + 500$$

where $C(b)$ represents the cost in dollars to produce b backpacks.

a. Find the cost to produce 20 backpacks.
b. Find the vertex and describe its meaning.
c. How many backpacks can a school get with a budget of $3000?

17. The revenue from selling digital cameras can be modeled by

$$R(c) = -3c^2 + 90c$$

where $R(c)$ represents the revenue in thousands of dollars from selling c thousand digital cameras.

a. Find the revenue from selling 5 thousand digital cameras.
b. How many cameras must the company sell to have a revenue of $600,000?
c. How many digital cameras must it sell to maximize its revenue?

18. The revenue from selling sunglasses can be modeled by

$$R(s) = -1.5s^2 + 30s$$

where $R(s)$ represents the revenue in hundreds of dollars from selling s hundred pairs of sunglasses.

a. Find the revenue from selling 9 hundred pairs of sunglasses.
b. How many pairs of sunglasses must the company sell to have a revenue of $10,000?
   (Hint: $10,000 = $100 hundred)
c. How many pairs of sunglasses must the company sell to maximize its revenue?

19. The net sales for Home Depot can be modeled by

$$N(t) = 376.5t^2 + 548.1t + 2318.4$$

where $N(t)$ represents the net sales in millions of dollars for Home Depot t years since 1990.
Source: Model derived from data found in Home Depot annual reports.

a. Using this model estimate the annual net sales for Home Depot in 1995.
b. When were Home Depot's net sales 30,000 million dollars?
c. Find the vertex for this model and explain its meaning.

20. The net income for Dell Corporation can be modeled by

$$I(t) = -145.86t^2 + 3169.342t - 15145.2$$

where $I(t)$ represents the net income in millions of dollars t years since 1990.
Source: Model derived from data found in Dell's 2002 annual report.

a. Using this model, estimate the net income for Dell in 2000.
b. When did Dell's net income reach 1500 million dollars?
c. Find the vertex for this model and explain its meaning.

21. The net income for clothing and gear retailer Quiksilver can be modeled by

$$I(t) = -1.5t^2 + 32.3t - 138.8$$

where $I(t)$ represents the net income in millions of dollars t years since 1990.
Source: Model derived from data found in Quiksilver annual report 2001.

a. Using this model, estimate the net income for Quiksilver in 1999.
b. When did Quiksilver's net income reach 18 million dollars?
c. Find the vertex for this model and explain its meaning.

22. The average fuel consumption of vehicles driving the roads can be modeled by

$$F(t) = 0.425t^2 - 16.431t + 840.321$$

where $F(t)$ represents the gallons per vehicle per year t years since 1970.

a. What was the average fuel consumption in 1975?
b. What was the average fuel consumption in 1985?
c. Find the vertex for this model and explain its meaning.

23. A baseball is hit so that its height in feet t seconds after it is hit can be represented by

$$h(t) = -16t^2 + 40t + 4$$

a. What is the height of the ball when it is hit?
b. When does the ball reach a height of 20 feet?
c. When did the ball reach its maximum height?
d. What is the ball's maximum height?
e. If the ball does not get caught, when does it hit the ground?

24. A baseball is hit so that its height in feet t seconds after it is hit can be represented by

$$h(t) = -16t^2 + 56t + 4$$

a. What is the height of the ball when it is hit?
b. When does the ball reach a height of 25 feet?
c. When does the ball reach its maximum height?
d. What is the ball's maximum height?

25. A baseball is hit so that its height in feet t seconds after it is hit can be represented by

$$h(t) = -16t^2 + 60t + 4.2$$

   a. What is the height of the ball when it is hit?
   b. When does the ball reach a height of 40 feet?
   c. What is the ball's maximum height?
   d. If it is not caught, when does the ball hit the ground?

26. A baseball is hit so that its height in feet t seconds after it is hit can be represented by

$$h(t) = -16t^2 + 65t + 3.5$$

   a. What is the height of the ball when it is hit?
   b. When does the ball reach a height of 50 feet?
   c. What is the ball's maximum height?
   d. If it is not caught, when does the ball hit the ground?

27. The average high temperature in Melbourne, Australia, can be modeled by

$$H(m) = 0.9m^2 - 13m + 104$$

   where $H(m)$ represents the average high temperature (in degrees Fahrenheit) in Melbourne during month m of the year.

   a. Find $H(6)$ and explain its meaning.
   b. Find the vertex for this model and explain its meaning.
   c. During what month(s) is the average temperature in Melbourne 60°F?

28. The average high temperature in Paris, France, can be modeled by

$$H(m) = -1.5m^2 + 21.6m - 3.2$$

   where $H(m)$ represents the average high temperature (in degrees Fahrenheit) in Paris during month m of the year.

   a. Find $H(4)$ and explain its meaning.
   b. Find the vertex for this model and explain its meaning.
   c. During what month(s) is the average temperature in Paris 70°F?

In Exercises 29 through 50:

   a. *Find the vertex.*
   b. *Find the vertical and horizontal intercept(s).*
   c. *Sketch a graph of the function.*
   d. *Give the domain and range of the function.*

29. $f(x) = x^2 + 2x - 15$

30. $g(x) = x^2 + 6x + 3$

31. $m(b) = -b^2 + 11b - 24$

32. $D(z) = -z^2 - 12z + 43$

33. $g(s) = 2s^2 - 62s + 216$

34. $h(t) = 4t^2 + 30t - 17$

35. $f(x) = 2x^2 + 5$

36. $b(w) = -3w^2 + 8$

37. $d(p) = -1.5p^2 - 3$

38. $g(n) = \dfrac{2}{3}n^2 + 1$

39. $h(x) = -\dfrac{1}{4}x^2 - 5$

40. $s(m) = -\dfrac{1}{2}x^2 + 12$

41. $p(k) = -5k^2 - 17.5k - 12.5$

42. $f(x) = -3x^2 - 14.6x - 11.2$

43. $h(w) = 0.4w^2 - 3.6w - 44.8$

44. $f(x) = 0.25x^2 + 4x + 6.5$

45. $p(x) = \dfrac{2}{5}x^2 - 2x - \dfrac{3}{5}$

46. $h(t) = \dfrac{1}{3}t^2 - 2t + 4$

47. $Q(p) = -0.3p^2 - 2.4p + 82$

48. $M(a) = -0.25a^2 - 3.5a - 27.5$

49. $W(g) = -0.3(g + 2)^2 + 17$

50. $f(x) = 3(x - 4)^2 - 18$

For Exercises 51 through 64, graph the inequalities.

51. $y < x^2 + 8x + 15$

52. $y > x^2 - 4x - 21$

53. $y \leq -2(x + 3)^2 + 10$

54. $y \leq -3(x - 4)^2 + 7$

55. $y \geq 1.5x^2 + 6x + 10$

56. $y < 0.5x^2 - 8x - 11$

57. $y \geq -0.25x^2 + 3x - 2$

58. $y \leq 4x^2 - 20x - 16$

59. $y > 0.3(x - 4)^2 - 6$

60. $y < (x + 8)^2 - 9$

61. $y \leq 2.5x^2$

62. $y \geq -1.75x^2 + 5$

63. $y > 2(x + 7)(x - 4)$

64. $y < 0.5(x + 2)(x - 6)$

For Exercises 65 through 68, find the inequality represented by the given graph.

65.

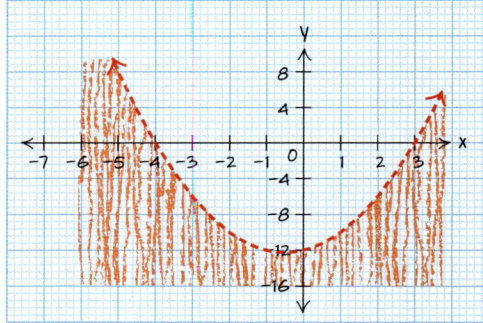

66.

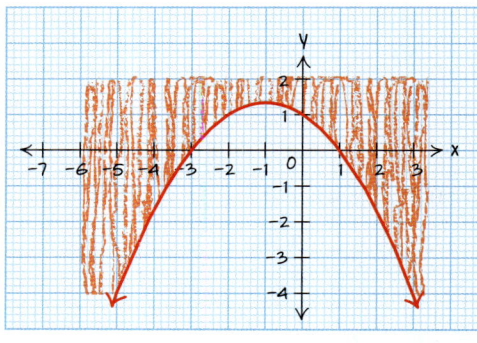

67.

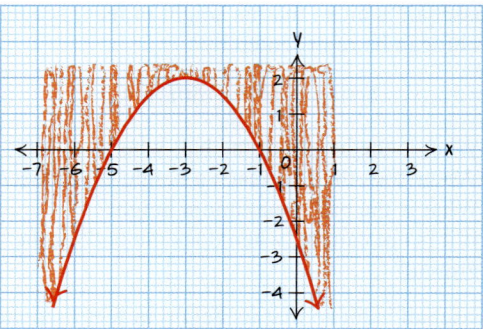

68.

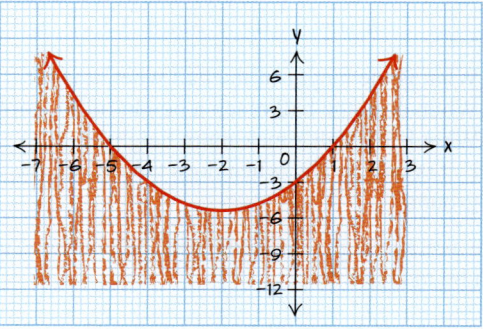

Chapter Summary

Section 4.1 Quadratic Functions and Parabolas

- The graph of a quadratic function is called a **parabola** and is either upward- or downward-facing.
- The **vertex** of a parabola is the maximum or minimum point on the graph and is the point where the graph changes directions.
- A **quadratic function** can be written in two forms:

$$\text{Standard form: } f(x) = ax^2 + bx + c$$

or

$$\text{Vertex form: } f(x) = a(x - h)^2 + k$$

where a, b, c, h, and k are real numbers and $a \neq 0$.

Example 1

Create a scatterplot for the given data and estimate the vertex.

x	0	1	2	3	4	5	6	7
y	14	10.5	8	6.5	6	6.5	8	10.5

SOLUTION

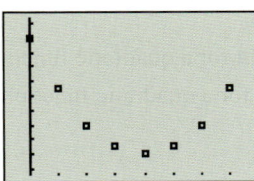

This distribution takes the shape of a parabola, and the vertex is a minimum point at about (4, 6).

Example 2

Use the graph to estimate the following.

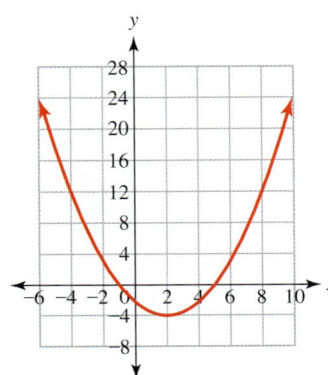

a. The vertex.
b. The vertical intercept.
c. The horizontal intercept(s).

SOLUTION

a. Vertex: $(2, -4)$
b. Vertical intercept: $(0, -2)$
c. Horizontal intercepts: $(-1, 0)$ and $(5, 0)$

Section 4.2 Graphing Quadratics in Vertex Form

- The **vertex form of a quadratic** is
$$f(x) = a(x - h)^2 + k$$
where a, h, and k are real numbers and $a \neq 0$.
- The **value of a** will determine whether the parabola faces upward or downward as well as how wide or narrow the graph is.
- The **value of h** will determine how far the vertex is moved to the left or right from the origin.
- The **value of k** will determine how far the vertex is moved up or down from the origin.
- The **vertex** of the parabola will be (h, k).
- The **axis of symmetry** is the vertical line through the vertex and has the equation $x = h$.
- The **symmetry** of the parabola helps to sketch the graph more accurately and more quickly.
- To **graph a quadratic function from the vertex form,** follow these five steps.
  1. Determine whether the graph opens up or down.
  2. Find the vertex and the equation of the axis of symmetry.
  3. Find the vertical intercept.
  4. Find an extra point by picking an input value on one side of the axis of symmetry and calculating the output value.
  5. Plot the points you found in steps 2 through 4. Plot their symmetric points and sketch the graph. (Find an additional pair of symmetric points if needed.)
- The **domain of a quadratic model** will be restricted only by the context. Avoid model breakdown.
- The **range of a quadratic model** is the output values of the model that come from the domain.
- The **domain** for a quadratic function with no context will be all real numbers.
- The **range** for a quadratic function with no context will be either $(-\infty, k]$ if $a < 0$ or $[k, \infty)$ if $a > 0$.

Example 3

Sketch the graph of $f(x) = -2(x + 5)^2 + 10$. Label the vertex, the vertical intercept, and at least one other pair of symmetric points. Give the domain and range of the function.

SOLUTION The vertex is $(-5, 10)$. $f(0) = -40$, so the vertical intercept is $(0, -40)$. The graph faces downward because a is negative. One pair of symmetric points is $(-2, -8)$ and $(-8, -8)$.

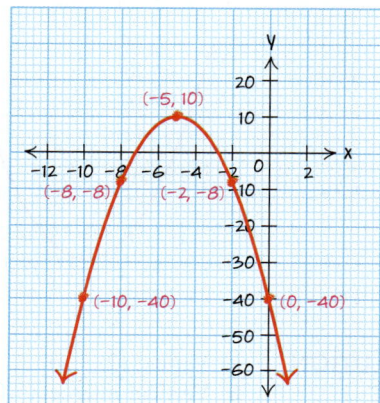

Domain: All real numbers
Range: $y \leq 10$

Example 4 Give the domain and range of $f(x) = 4(x + 7)^2 - 45$.

SOLUTION Domain: All real numbers
Range: $[-45, \infty)$

Section 4.3 Finding Quadratic Models

- To find a **quadratic model,** follow these seven steps.
    1. Define the variables and adjust the data (if needed).
    2. Create a scatterplot.
    3. Select a model type.
    4. **Quadratic model:** Pick a vertex and substitute it for h and k in the vertex form $f(x) = a(x - h)^2 + k$.
    5. Pick another point and use it to find a.
    6. Write the equation of the model using function notation.
    7. Check the model by graphing it with the scatterplot.
- In the **context of an application,** the **domain** of a quadratic model expands beyond the data. Be sure to avoid inputs that will cause model breakdown to occur.
- In the **context of an application,** the **range** of a quadratic model is the lowest to highest points on the graph within the domain.

Example 5 The number of solar flares each year appears to be cyclical. The data from the latest cycle are given in the table.

a. Find an equation for a model of these data.

b. What is the vertex? What does it represent?

c. According to your model, how many solar flares were there in 2001?

d. Give a reasonable domain and range for your model.

Year	Number of Solar Flares
1997	790
1998	2423
1999	3963
2000	4474
2002	3223
2003	1552

SOLUTION a. $S(t)$ = Number of solar flares during the year
t = Years since 1995

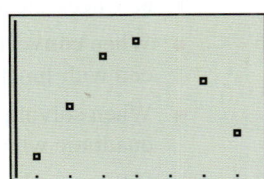

The vertex could be (5, 4474), so we have

$$S(t) = a(t - 5)^2 + 4474$$

Using (2, 790), we get

$$a = -409 \quad \text{and} \quad S(t) = -409(t-5)^2 + 4474$$

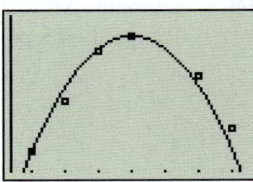

This model could be moved over to the right to make a better fit, so we can adjust the model to

$$S(t) = -409(t-5.2)^2 + 4474$$

This gives the following graph.

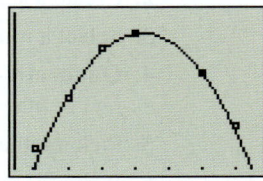

b. The vertex is (5.2, 4474). The maximum number of solar flares occurred in about 2000 with 4474 solar flares.

c. $S(6) = 4212.24$, so in 2001, there were about 4212 solar flares.

d. Domain: [2, 8], range: [286, 4474]. Note that if we go beyond this domain, model breakdown occurs because the model predicts a negative number of solar flares.

Section 4.4 Solving Quadratic Equations by the Square Root Property and Completing the Square

- When **solving** quadratic equations in vertex form, isolate the squared variable expression and use the **square root property** to remove the exponent. Don't forget the plus/minus symbol.
- When **solving** quadratic equations using **completing the square**, follow these six steps.
  1. Get all the variable terms alone on one side of the equation.
  2. If the coefficient of x^2 is not 1, divide both sides by the coefficient of x^2.
  3. Take half of the coefficient of x and then square it. Add this number to both sides of the equation.
  4. Factor the quadratic into the square of a binomial.
  5. Solve using the square root property.
  6. Check the answers in the original equation.
- When **converting to vertex form**, use the same steps as completing the square, but end with function notation.
- When solving a quadratic, if a negative number is under the square root, the quadratic will have **no real solutions.**
- When **graphing a quadratic in vertex form**, include the horizontal intercepts. Use these five steps.
  1. Determine whether the graph opens up or down.
  2. Find the vertex and the equation for the axis of symmetry.

CHAPTER 4 Summary

3. Find the vertical intercept.
4. Find the horizontal intercepts (if any).
5. Plot the points you found in steps 2 through 4. Plot their symmetric points and sketch the graph. (Find an additional pair of symmetric points if needed.)

Example 6 Solve the following equations.

a. $15 = 2(x + 5)^2 - 17$

b. $x^2 - 6x + 20 = 47$

SOLUTION a. Because this equation is in vertex form, we would use the square root property to solve.

$$15 = 2(x + 5)^2 - 17$$
$$32 = 2(x + 5)^2$$
$$16 = (x + 5)^2$$
$$\pm 4 = x + 5$$
$$4 = x + 5 \qquad -4 = x + 5$$
$$x = -1 \qquad x = -9$$

Check your answers.

$$15 \stackrel{?}{=} 2((-1) + 5)^2 - 17 \qquad 15 \stackrel{?}{=} 2((-9) + 5)^2 - 17$$
$$15 \stackrel{?}{=} 2(4)^2 - 17 \qquad 15 \stackrel{?}{=} 2(-4)^2 - 17$$
$$15 \stackrel{?}{=} 32 - 17 \qquad 15 \stackrel{?}{=} 32 - 17$$
$$15 = 15 \qquad 15 = 15$$

b. This equation is not in vertex form, so we will use completing the square.

$$x^2 - 6x + 20 = 47$$
$$x^2 - 6x = 27$$
$$x^2 - 6x + 9 = 36$$
$$(x - 3)^2 = 36$$
$$x - 3 = \pm 6$$
$$x - 3 = 6 \qquad x - 3 = -6$$
$$x = 9 \qquad x = -3$$

Check your answers.

$$(9)^2 - 6(9) + 20 \stackrel{?}{=} 47 \qquad (-3)^2 - 6(-3) + 20 \stackrel{?}{=} 47$$
$$47 = 47 \qquad 47 = 47$$

Example 7 Convert $f(x) = 3x^2 + 12x - 21$ into vertex form.

SOLUTION

$$f(x) = 3x^2 + 12x - 21$$
$$f(x) = (3x^2 + 12x) - 21$$
$$f(x) = 3(x^2 + 4x) - 21$$
$$f(x) = 3(x^2 + 4x + 4 - 4) - 21$$
$$f(x) = 3(x^2 + 4x + 4) - 3(4) - 21$$
$$f(x) = 3(x + 2)^2 - 12 - 21$$
$$f(x) = 3(x + 2)^2 - 33$$

Section 4.5 Solving Quadratic Equations by Factoring

- To solve a quadratic using **factoring,** the equation must be set equal to zero and put into standard form. Then factor, set each of the factors equal to zero, and solve.
- When finding an equation from the graph, set x equal to the zeros and undo the solving process to find the factors. Use another point to find any constant multiplier.

Example 8 Solve the following equations.

a. $2x^2 - x - 4 = 11$

b. $2x^3 - 26x^2 + 80x = 0$

SOLUTION a.
$$2x^2 - x - 4 = 11$$
$$2x^2 - x - 15 = 0$$
$$2x^2 - 6x + 5x - 15 = 0$$
$$(2x^2 - 6x) + (5x - 15) = 0$$
$$2x(x - 3) + 5(x - 3) = 0$$
$$(2x + 5)(x - 3) = 0$$
$$2x + 5 = 0 \qquad x - 3 = 0$$
$$x = -\frac{5}{2} \qquad x = 3$$

Check your answers.

$$2x^2 - x - 4 = 11 \qquad\qquad 2x^2 - x - 4 = 11$$
$$2\left(-\frac{5}{2}\right)^2 - \left(-\frac{5}{2}\right) - 4 = 11 \qquad 2(3)^2 - (3) - 4 = 11$$
$$2\left(\frac{25}{4}\right) + \frac{5}{2} - 4 = 11 \qquad\qquad 2(9) - 3 - 4 = 11$$
$$15 - 4 = 11 \qquad\qquad 18 - 3 - 4 = 11$$
$$11 = 11 \qquad\qquad 11 = 11$$

b.
$$2x^3 - 26x^2 + 80x = 0$$
$$2x(x^2 - 13x + 40) = 0$$
$$2x[x^2 - 5x - 8x + 40] = 0$$
$$2x[(x^2 - 5x) + (-8x + 40)] = 0$$
$$2x[x(x - 5) - 8(x - 5)] = 0$$
$$2x(x - 5)(x - 8) = 0$$
$$2x = 0 \qquad x - 5 = 0 \qquad x - 8 = 0$$
$$x = 0 \qquad x = 5 \qquad x = 8$$

Check your answers.

$$2(0)^3 - 26(0)^2 + 80(0) \stackrel{?}{=} 0$$
$$0 = 0$$

$$2(5)^3 - 26(5)^2 + 80(5) \stackrel{?}{=} 0$$
$$250 - 650 + 400 \stackrel{?}{=} 0$$
$$0 = 0$$

$$2(8)^3 - 26(8)^2 + 80(8) \stackrel{?}{=} 0$$
$$1024 - 1664 - 640 \stackrel{?}{=} 0$$
$$0 = 0$$

Example 9 Use the graph to find an equation for the quadratic.

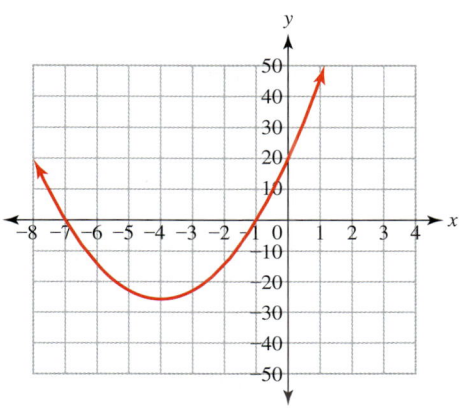

SOLUTION

$$x = -1 \qquad x = -7$$
$$x + 1 = 0 \qquad x + 7 = 0$$
$$y = a(x + 1)(x + 7)$$
$$20 = a(0 + 1)(0 + 7)$$
$$20 = 7a$$
$$\frac{20}{7} = a$$
$$y = \frac{20}{7}x^2 + \frac{160}{7}x + 20$$

Section 4.6 Solving Quadratic Equations by Using the Quadratic Formula

- The **quadratic formula** can be used to solve any quadratic equation that is in standard form.

$$ax^2 + bx + c = 0$$
$$x = \frac{-b \pm \sqrt{b^2 - 4ac}}{2a}$$

- Be sure that your equation is set equal to zero before using the quadratic formula.
- A **system of equations with quadratics** can be solved **graphically** by looking for the intersections of the graphs.
- A **system of equations with quadratics** can be solved **numerically** by looking for the input values that give the same output values in both equations.
- A **system of equations with quadratics** can be solved **algebraically** by using the substitution method.
- Whenever the quadratic formula results in a negative under the square root, there will be **no real solutions** to the equation.

Example 10 Solve the following quadratics.

a. $5x^2 + 3x - 20 = 10$

b. $4x^2 - 2x + 15 = -30$

SOLUTION

a. $5x^2 + 3x - 20 = 10$

$$5x^2 + 3x - 30 = 0$$

$$a = 5 \quad b = 3 \quad c = -30$$

$$x = \frac{-(3) \pm \sqrt{(3)^2 - 4(5)(-30)}}{2(5)}$$

$$x = \frac{-3 \pm \sqrt{609}}{10}$$

$$x \approx \frac{-3 \pm 24.678}{10}$$

$$x \approx \frac{-3 + 24.678}{10} \qquad x \approx \frac{-3 - 24.678}{10}$$

$$x \approx 2.168 \qquad x \approx -2.768$$

Check the answers with the table.

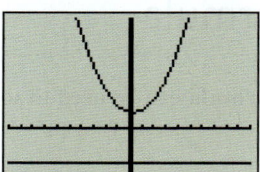

b. $4x^2 - 2x + 15 = -30$

$$4x^2 - 2x + 45 = 0$$

$$a = 4 \quad b = -2 \quad c = 45$$

$$x = \frac{-(-2) \pm \sqrt{(-2)^2 - 4(4)(45)}}{2(4)}$$

$$x = \frac{2 \pm \sqrt{-716}}{8}$$

Verify with the graph.

Because the number under the square root is negative, there are no real solutions to this equation. This result is verified by the graph, which shows that the parabola never gets down to -30.

Example 11 Estimate the solutions to the system of equations given in the graph.

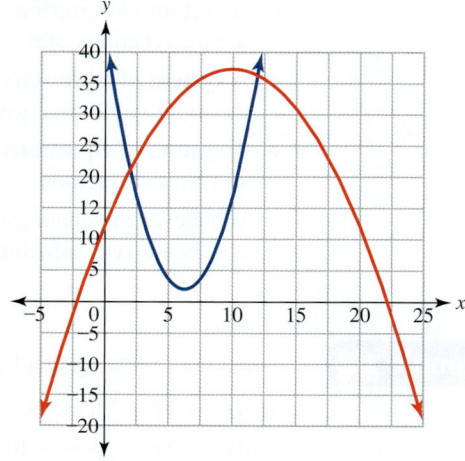

SOLUTION These graphs appear to cross at about $(2, 21)$ and $(12, 36)$.

Example 12 Solve the following system of equations.

$$y = 3x + 5$$
$$y = x^2 + 2x - 1$$

SOLUTION Using the substitution method,

$$3x + 5 = x^2 + 2x - 1$$
$$0 = x^2 - x - 6$$
$$0 = (x + 2)(x - 3)$$
$$x = -2 \qquad x = 3$$
$$y = 3(-2) + 5 \qquad y = 3(3) + 5 \qquad \text{Find } y.$$
$$y = -1 \qquad y = 14$$

The answers are $(-2, -1)$ and $(3, 14)$ and are verified in the table.

Section 4.7 Graphing Quadratics from Standard Form

- When **graphing a quadratic in standard form**, follow these five steps.
  1. Determine whether the graph opens up or down.
  2. Find the vertex and the equation for the axis of symmetry.
  3. Find the vertical intercept.
  4. Find the horizontal intercepts (if any).
  5. Plot the points you found in steps 2 through 4. Plot their symmetric points and sketch the graph. (Find an additional pair of symmetric points if needed.)
- The x value of the **vertex** can be found in the middle of the x values of any pair of symmetric points or by using the input value.

$$x = \frac{-b}{2a}$$

- To graph a **quadratic inequality**, sketch the graph as if it were an equation, using a dashed line if there is no "equal to" part, a solid line if there is an "equal to" part, and then shade the side of the curve that makes the inequality true. Using a test point that is not on the parabola may help to determine which side to shade.

Example 13 Graph the function $f(x) = -0.25x^2 + 5x + 12$.

SOLUTION This graph faces downward because a is negative and is somewhat wide because $|a| < 1$.

The vertical intercept is $(0, 12)$.

The vertex has an input of $x = \dfrac{-5}{2(-0.25)} = 10$ and $f(10) = 37$, so the vertex is $(10, 37)$.

Because the graph faces downward and has a vertex above the x-axis, the graph will have x-intercepts. We can use the quadratic formula to find them. The answers are $(-2.17, 0)$ and $(22.17, 0)$.

Another pair of symmetric points is $(5, 30.75)$ and $(15, 30.75)$. Notice that the x-value of the vertex is 10 and is in the middle of the x-values of both of these pairs of symmetric points.

Using all these points, we get the following sketch.

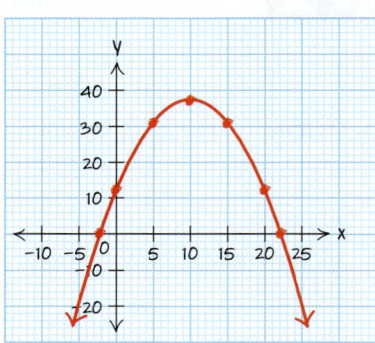

Example 14 Graph the inequality $y < x^2 + 4x - 12$.

SOLUTION We will sketch the graph with a dashed line, since there is no "equal to" part in the inequality symbol. Then we will shade below the curve.

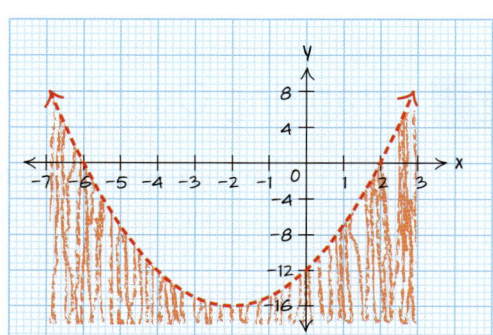

Chapter Review Exercises 4

1. Use the graph to estimate the following.

   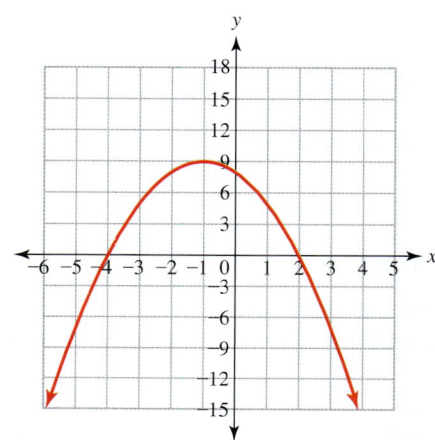

   a. Vertex.
   b. For what x-values is the graph increasing? Write your answer as an inequality.
   c. For what x-values is the graph decreasing? Write your answer as an inequality.
   d. Horizontal intercept(s).
   e. Vertical intercept. [4.1]

2. Use the graph from Exercise 1 to answer the following.
   a. Write the equation for the axis of symmetry.
   b. If $f(x)$ is written in vertex form, would a be positive or negative?
   c. If $f(x)$ is written in vertex form, what would the values of h and k be?
   d. $f(1) =$?
   e. What x-value(s) will make $f(x) = -12$? [4.2]

3. Use the graph to estimate the following.

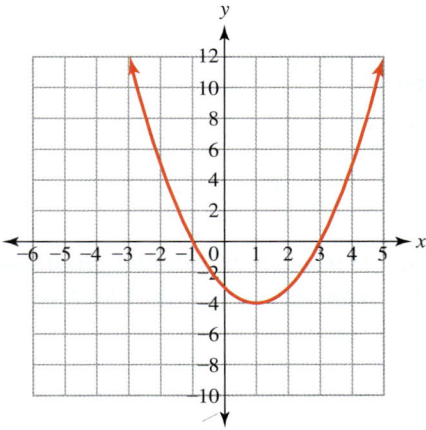

a. Write the equation for the axis of symmetry.
b. If $f(x)$ is written in vertex form, would a be positive or negative?
c. If $f(x)$ is written in vertex form, what would the values of h and k be?
d. $f(-1.5) = $?
e. What x-value(s) will make $f(x) = 3$? [4.2]

4. Use the graph from Exercise 3 to estimate the following.
a. Vertex.
b. For what x-values is the graph increasing? Write your answer as an inequality.
c. For what x-values is the graph decreasing? Write your answer as an inequality.
d. Horizontal intercept(s).
e. Vertical intercept. [4.1]

5. The population of North Dakota can be modeled by

$$P(t) = -0.89(t - 5.6)^2 + 643$$

where $P(t)$ represents the population in thousands t years since 1990.

Source: Model derived from data in Statistical Abstract of the United States, 2001.

a. Find the population of North Dakota in 1998.
b. Sketch a graph of this model. [4.2]
c. Find the vertex of this model and explain its meaning.

d. When was the population of North Dakota 640 thousand?
e. If the domain of this function is $0 \leq t \leq 12$, find the range. [4.4]

6. The number of murders, $M(t)$, in thousands in the United States t years since 1990 can be represented by

$$M(t) = -0.42(t - 2.5)^2 + 23$$

Source: Model derived from data in Statistical Abstract of the United States, 2001.

a. Estimate the number of murders in the United States in 1996.
b. Find the vertex of this model and explain its meaning.
c. In what year were there 14.5 thousand murders in the United States?
d. If the domain of this function is $[0, 8]$, find the range. [4.4]

For Exercises 7 through 12, sketch the graph and label the vertex, vertical intercept, and the horizontal intercept(s) for the given functions. Give the domain and range.

7. $h(x) = (x + 3)^2 - 16$ [4.2]
8. $f(x) = 1.25(x - 4)^2 - 20$ [4.2]
9. $g(x) = -\dfrac{1}{4}(x - 8)^2 + 9$ [4.2]
10. $h(x) = -2(x + 4.5)^2 + 12.5$ [4.2]
11. $g(x) = -0.5(x + 7)^2 - 3$ [4.2]
12. $f(x) = 3(x - 8)^2 + 4$ [4.2]

13. The number of juveniles arrested for possession of drugs through the 1990s is given in the table. [4.3]

Year	Number of Arrests
1992	47,901
1994	92,185
1995	115,159
1996	116,225
1997	124,683
1998	118,754
1999	112,640

Source: Statistical Abstract of the United States, 2001.

a. Find an equation for a model of these data.
b. Give a reasonable domain and range for this model.
c. Estimate the number of juveniles arrested for possession of drugs in 2000.
d. If this trend continues, in what year will the number of juvenile arrests be at 50 thousand again?

14. The U.S. Department of Commerce obligations for research and development for several years is given in the table. [4.3]

Year	R & D Obligations (millions of dollars)
1995	1136
1996	1068
1997	1003
1999	990
2000	1041
2001	1127

Source: Statistical Abstract of the United States, 2001.

a. Find an equation for a model of these data.
b. Give a reasonable domain and range for this model.
c. Estimate the research and development obligations for the Department of Commerce in 1998.
d. Estimate when the Department of Commerce research and development obligations will be $1.5 billion ($1,500 million).

15. The median sales price in thousands of dollars of new houses in the western United States t years since 1990 can be modeled by

$$P(t) = 1.073t^2 - 5.84t + 144.68$$

Source: Model derived from data in Statistical Abstract of the United States, 2001.

a. Estimate the median price in 1995.
b. When will the median price of a new home in the West reach $250,000?
c. Find the vertex of this model and explain its meaning. [4.7]

16. The median asking price in thousands of dollars of houses in Memphis, Tennessee, t years since 2000 can be modeled by

$$P(t) = -1.43t^2 + 19.89t + 101.68$$

Source: Model derived from data found at homepricetrend.com.

a. Estimate the median price in 2004.
b. When was the median price of a house in Memphis $165,000?
c. Find the vertex of this model and explain its meaning. [4.7]

For Exercises 17 through 24, solve the following quadratic equations using the square root property. Check your answers.

17. $t^2 = 169$ [4.4]
18. $-6m^2 + 294 = 0$ [4.4]
19. $4(x + 7)^2 - 36 = 0$ [4.4]
20. $-0.25(x - 6)^2 + 8 = 0$ [4.4]
21. $3x^2 + 75 = 0$ [4.4]
22. $-4(t + 3)^2 - 100 = 0$ [4.4]
23. $\frac{1}{2}(c + 6)^2 - \frac{5}{2} = 0$ [4.4]
24. $\frac{2}{7}(p - 4)^2 - \frac{3}{14} = 0$ [4.4]

For Exercises 25 through 28, solve the given quadratic equations by completing the square. Check your answers.

25. $x^2 + 26x = 30$ [4.4]
26. $x^2 - 12x - 13 = 0$ [4.4]
27. $3x^2 + 15x = 198$ [4.4]
28. $4x^2 - 6x + 20 = 0$ [4.4]

For Exercises 29 and 30, use completing the square to convert the given functions to vertex form.

29. $f(x) = x^2 + 8x + 11$ [4.4]
30. $g(x) = -6x^2 + 20x - 18$ [4.4]

For Exercises 31 through 40, solve the following quadratic equations by factoring. Check your answers.

31. $t^2 - 12t + 20 = 0$ [4.5]
32. $p^2 + 6p = 27$ [4.5]
33. $6x^2 - 8x = 0$ [4.5]
34. $3x^2 - x = 2$ [4.5]
35. $m^2 - 64 = 0$ [4.5]
36. $8m^2 - 50 = 0$ [4.5]
37. $9x^2 - 24x + 5 = -11$ [4.5]
38. $2x^2 - 8x - 120 = 0$ [4.5]
39. $x^3 + 7x^2 + 10x = 0$ [4.5]
40. $12h^3 - 60h^2 = 168h$ [4.5]

41. Use the graph to find an equation for the quadratic.

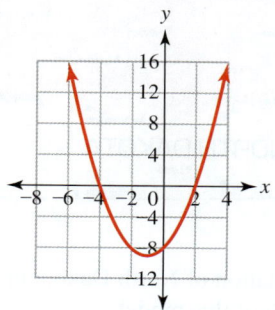

[4.5]

42. Use the graph to find an equation for the quadratic.

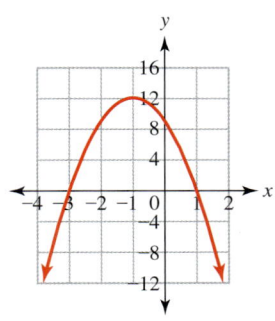

[4.5]

For Exercises 43 through 50, solve the following quadratic equations using the quadratic formula.

43. $x^2 - 4x - 12 = 0$ [4.6]

44. $2t^2 + 11t - 63 = 0$ [4.6]

45. $x^2 + 6x + 18 = 0$ [4.6]

46. $3x^2 + 9x - 20 = 14$ [4.6]

47. $-4.5x^2 + 3.5x + 12.5 = 0$ [4.6]

48. $3.25n^2 - 4.5n - 42.75 = 0$ [4.6]

49. $3x^2 + 8x = -2$ [4.6]

50. $2a^2 + 6a - 10 = 6a + 20$ [4.6]

For Exercises 51 through 54, sketch the graph. Label the vertex and vertical and horizontal intercept(s). Give the domain and range.

51. $f(x) = -2.7x^2 + 16.2x - 21.8$ [4.7]

52. $h(n) = 0.3n^2 + 3n + 14.5$ [4.7]

53. $f(x) = 2(x + 9)^2 - 15$ [4.7]

54. $f(x) = -3x^2 + 12x + 16$ [4.3]

For Exercises 55 through 58, solve the following systems of equations. Write your answers as points. Check your answers.

55. $y = 5x - 9$
    $y = -2x^2 + 7x + 15$ [4.6]

56. $y = -4x + 3$
    $y = x^2 - 8x - 9$ [4.6]

57. $y = 6x^2 - 3x + 10$
    $y = 4x^2 + 8x + 1$ [4.6]

58. $y = 0.25x^2 + 4x - 7$
    $y = 0.5x^2 + 3x - 9$ [4.6]

59. Use the table to estimate the solution(s) to the following system. [4.7]

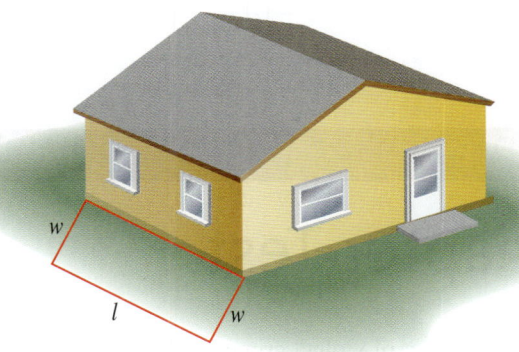

60. Hitomi has 100 feet of lights he wants to put around a rectangular area next to his house. The house will form one side of the area, as shown in the figure.

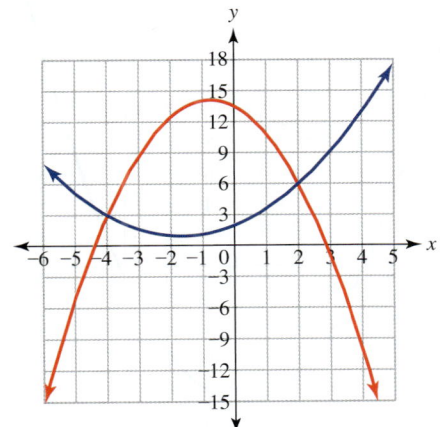

a. What is the largest possible rectangular area?
b. What length and width give the largest possible area? [4.6]

For Exercises 61 through 64, graph the inequalities.

61. $y < 2x^2 + 24x - 10$ [4.7]

62. $y \geq -0.5x^2 - 4x - 6$ [4.7]

63. $y \leq 2.5(x - 4)^2 - 18$ [4.7]

64. $y > -0.8(x + 1)^2 - 3$ [4.7]

65. Use the graph to solve the system of equations. Write your answers as points. [4.6]

66. Use the graph to solve the system of equations. Write your answers as points. [4.6]

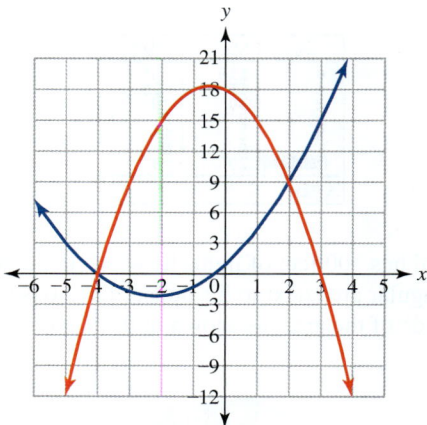

Chapter Test   4

1. The number of violent crimes in the United States can be modeled by
$$V(t) = 13(t - 4)^2 + 1370$$
where $V(t)$ represents the number of violent crimes in thousands t years since 2000.
Source: Model derived from data in Statistical Abstract of the United States, 2009.
   a. Estimate the number of violent crimes in 2008.
   b. Sketch a graph of this function.
   c. Determine the vertex and explain its meaning.
   d. In what year will there be 1.5 million violent crimes in the United States?

2. Use the graph to solve the system of equations. Write your answers as points.

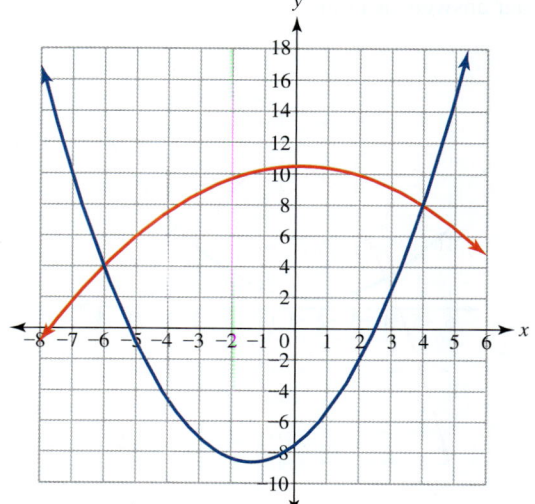

For Exercises 3 and 4, solve the following quadratic equations by factoring. Check your answers.

3. $3x^2 - 5x + 5 = 33$

4. $8x^2 - 34x = -35$

5. The total outlays for national defense and veterans benefits by the United States for several years is given in the table.

Year	Total Outlays (billions of dollars)
1993	326.8
1995	310
1997	309.8
1999	320.2
2000	337.4

Source: Statistical Abstract of the United States, 2001.

   a. Find an equation for a model of these data.
   b. Give a reasonable domain and range for this model.
   c. Estimate the total outlays for national defense and veterans benefits in 1998.
   d. When will the total outlays for national defense and veterans benefits reach half a trillion dollars? (*Hint*: Half a trillion = 500 billion.)

6. The U.S. commercial space industry revenue for satellite manufacturing for several years is given in the chart.

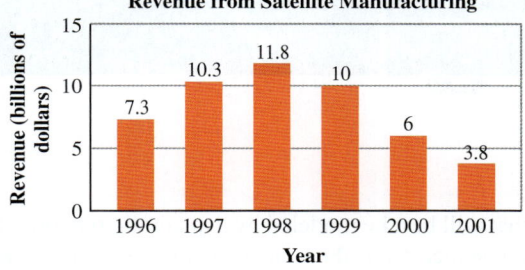

U.S. Commercial Space Industry Revenue from Satellite Manufacturing

Source: Statistical Abstract of the United States, 2004–2005.

a. Find an equation for a model of these data.
b. Give a reasonable domain and range for this model.
c. Estimate the commercial space industries revenues for satellite manufacturing in 2000.
d. Give the vertex for your model and explain its meaning.

For Exercises 7 through 9, sketch the graph. Label the vertex and the vertical and horizontal intercepts. Give the domain and range.

7. $f(x) = -1.5x^2 + 12x - 20.5$

8. $g(m) = 0.4m^2 + 1.6m + 11.6$

9. $f(x) = -0.4(x + 20)^2 - 17$

10. Solve the system of equations. Write the answers as points.
$$y = 2x^2 + 5x - 9$$
$$y = -7x^2 + 3x + 2$$

11. The number of privately owned single-unit houses built in the late 1990s can be modeled by
$$H(t) = -52t^2 + 916.2t - 2731.2$$
where $H(t)$ represents the number of privately owned single-unit houses built in thousands t years since 1990.
Source: Model derived from data in the Statistical Abstract of the United States, 2001.

a. Estimate the number of single-unit houses built in 2000.
b. In what year were 1000 thousand single-unit homes built?

12. Jay has a 20-foot ladder that he leaned against the wall of his house. Jay put the base of the ladder 4 feet away from the house. How far up the house is the top of the ladder?

13. Use the table to estimate the solution(s) to the system of equations.

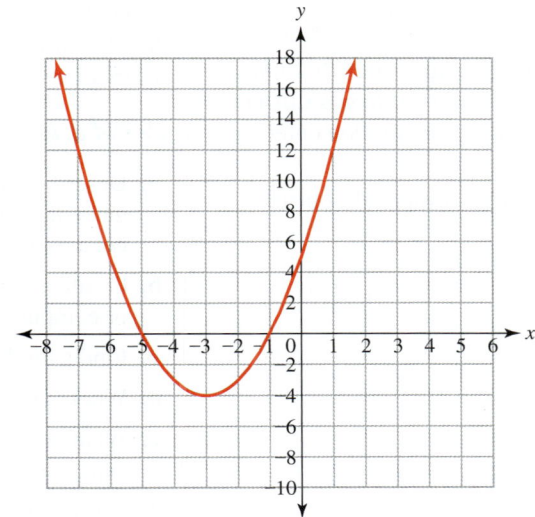

14. Use the graph to estimate the following:

a. Vertex
b. For what x-values is the graph increasing? Write your answer as an inequality.
c. For what x-values is the graph decreasing? Write your answer as an inequality.
d. An equation for the axis of symmetry
e. Vertical intercept
f. Horizontal intercept(s)
g. Write an equation for the function.

15. Solve $x^2 + 12x + 30 = 0$ by completing the square.

For Exercises 16 and 17, solve the given equations using any method you wish. Check your answers.

16. $4(x - 9)^2 - 20 = 124$

17. $8x^2 + 15 = 65$

18. Complete the square to convert the function $f(x) = 3x^2 + 24x - 30$ to vertex form.

19. Sketch the graph of $y < 0.5x^2 - 3x + 9$.

Chapter 4 Projects

Can You Move Like a Parabola?

Group Experiment
Two or more people

What you will need
- A Texas Instruments CBR unit or a Texas Instruments CBL unit with a Vernier motion detector probe.
- A book or other light object.

It will be your job to create data that will best be modeled by a quadratic function. Using the CBR/CBL unit, measure the distance from the unit over time and have someone move a book or other object in front of the motion detector to create data that have the shape of a parabola. It might take several tries before you collect data that are reasonable. Be sure to check your data so that before you move on, you are satisfied that the observations will be modeled by a quadratic function.

Write up

a. Describe how you collected the data. What was the most important thing to do to get a quadratic model? Did you select only part of the data that were collected to get a good quadratic model? Explain.

b. Create a scatterplot using the values that you found in the table on the calculator or computer, and print it out or complete it neatly by hand.

c. Find a model to fit the data.

d. What is the vertex of your model, and what does it represent in this experiment?

e. What is the vertical intercept of your model, and what does it represent in this experiment?

f. Does your parabola face upward or downward? How would you have to change your experiment to get the graph to face the other direction?

g. What are a reasonable domain and range for your model? (Remember to consider any restrictions on the experiment.)

Maximize Your Fruit

Written Project
One or more people

You are given the task of estimating the number of trees that will maximize the fruit production per acre of a local orchard. By researching other local orchards, you discover that at 400 trees per acre, each tree will produce an average of 30 pounds of fruit. For each tree that is added to the acre, the average production *per tree* goes down 0.05 pound.

Write up

a. Use the given information to build a table of values for this situation. Let t be the number of trees above 400 planted on an acre and find the total amount of fruit each *acre* will produce. Find at least eight sets of values.

b. Create a scatterplot using the values that you found in the table on the calculator or computer, and print it out or complete it neatly by hand.

c. Find a function for the total production per acre when t trees over 400 are planted on one acre.

d. Find the total production per acre when 400 trees are planted.

e. Estimate how many trees should be planted per acre to produce 12,400 pounds of fruit.

f. Find the number of trees per acre that will maximize the fruit production.

g. Find a reasonable domain and range for your model.

What Goes Up Must Come Down?

Group Experiment
Two or more people

What you will need
- A Texas Instruments CBR unit or a Texas Instruments CBL unit with a Vernier motion detector probe.
- A volleyball or other type of ball.

In this project, you are going to explore the path of an object that is tossed into the air and its speed and acceleration during that time. Place the CBL/CBR unit on the ground and have someone toss the ball into the air while you record the height of the ball as time goes by. (Be careful that the ball does not land on the CBL/CBR unit.) Do the experiment again, this time recording the velocity (speed) of the ball over time. Do it a third time, recording the ball's acceleration over time.

Write up
a. Create tables of the height-versus-time data, the velocity-versus-time data, and the acceleration-versus-time data. (You will have to collect a large number of points for this project.)

b. Create scatterplots of these three sets of data on your calculator or computer, and print them out or complete them neatly by hand on graph paper.

c. What kind of model will best fit the height-versus-time data? Create a model for these data.

d. What kind of model will best fit the velocity-versus-time data? Create a model for these data.

e. What kind of model will best fit the acceleration-versus-time data? Create a model for these data.

f. What is the maximum height the ball reached?

g. What is the maximum velocity of the ball? Was the velocity ever zero? When did these velocities occur?

h. What was the maximum acceleration of the ball? When did this occur?

i. Use your models to estimate the height, velocity, and acceleration of the ball after 1.5 seconds.

j. Explain why you think the graphs of these different values are the shapes they are.

Find Your Own Parabola

Research Project
One or more people

What you will need
- Find data for a real-world situation that can be modeled with a quadratic function.
- You might want to use the Internet or library. Statistical abstracts and some journals and scientific articles are good resources for real-world data.
- Follow the MLA style guide for all citations. If your school recommends another style guide, use that.

In this project, you are given the task of finding data for a real-world situation to which you can apply a quadratic model. You may use the problems in this chapter to get ideas of things to investigate, but your data should not be taken from this textbook.

Write up
a. Describe the data you found and where you found them. Cite any sources you used following the MLA style guide or your school's recommended style guide.

b. Create a scatterplot of the data on the calculator or computer, and print it out or complete it neatly by hand on graph paper.

c. Find a model to fit the data.

d. What is the vertex of your model, and what does it represent?

e. What is the vertical intercept of your model, and what does it represent?

f. What are a reasonable domain and range for your model?

g. Use your model to estimate an output value of your model for an input value that you did not collect in your original data.

Cumulative Review — Chapters 1-4

For Exercises 1 through 10, solve the given equation using any method you have learned. Check your answers.

1. $10m + 4 = 3m + 32$
2. $t^2 - 11t + 10 = 0$
3. $4(x - 7)^2 + 30 = 46$
4. $\frac{7}{15}a + \frac{3}{5} = \frac{2}{3}(a - 4) - \frac{5}{3}$
5. $3x^2 - 8x = 0$
6. $3.4m^2 - 4.6m = 108$
7. $\frac{1}{3}(c - 9) - 6 = \frac{5}{6}c - 15$
8. $2n^3 + 12n^2 - 432n = 0$
9. $(x + 3)(x - 7) = -16$
10. $3.8h - 4.2 = 7.5h + 27.9$

For Exercises 11 and 12, solve the given equation for the indicated variable.

11. $4mn - 5 = n$ for m
12. $\frac{2}{5}a^2 b + 7a = c$ for b

For Exercises 13 through 18, solve the given system. Check your answer.

13. $4x - 7y = -8$
    $3x + 10y = 55$
14. $y = 3x - 7$
    $y = x^2 + 6x - 17$
15. $y = 2.5x - 9$
    $-10x + 4y = 20$
16. $m = 7n + 8$
    $3m + 2n = 1$
17. $b = 3a^2 + 3a - 18$
    $b = -8a^2 + 4a + 30$
18. $h = 1.4g + 2.6$
    $7g - 5h = -13$

19. Write the equation of the line passing through the points (4, 8) and (10, 35). Write the equation in slope-intercept form.

20. Write the equation of the line passing through the point (12, 4.5) and perpendicular to the line $y = 8x - 30$. Write the equation in slope-intercept form.

21. Write the equation of the line passing through the point (−3, 10) and parallel to the line $2x + 5y = 11$. Write the equation in slope-intercept form.

22. It costs $138.75 to have 25 custom-printed tote bags created. One hundred bags cost $405.
    a. Assuming that the cost is linear write an equation for the cost of custom-printed tote bags depending on the number of bags ordered.
    b. How much will 200 bags cost?
    c. What is the slope of your equation? Explain its meaning.

23. The population of Pittsburgh, Pennsylvania, is given in the table.

Year	Population (millions)
2002	2.41
2003	2.40
2004	2.39
2005	2.37
2006	2.36
2007	2.35

Source: Statistical Abstract of the United States, 2009.

a. Find an equation for a model of these data.
b. Give a reasonable domain and range for this model.
c. Estimate the population of Pittsburgh in 2012.
d. What is the slope of this model? Explain its meaning in regard to the population.
e. Estimate the year the population of Pittsburgh will be only 2.25 million.

24. The total revenue from four-person tents is given in the table.

Price (dollars)	Revenue (dollars)
30	1640
40	2360
55	3010
70	3250
90	2870

a. Find an equation for a model of these data. Write the model using function notation.
b. Explain the meaning of the vertex for your model.
c. What does your model predict the revenue will be if they sell the tents for $60?
d. Give a reasonable domain and range for your model.

25. The net summer electricity capacity in the United States is given in the table.

Year	Net Summer Capacity (million kilowatts)
2003	948
2004	963
2005	978
2006	986
2007	999

Source: Statistical Abstract of the United States, 2009.

a. Find an equation for a model of these data. Write the model using function notation.
b. Give a reasonable domain and range for this model.
c. Estimate the net summer electricity capacity in 2015.
d. What is the slope of this model? Explain its meaning.
e. What is the vertical intercept for your model? Explain its meaning.

For Exercises 26 through 33, sketch a graph of the given equation or inequality. Label the vertical and horizontal intercepts. If quadratic label the vertex.

26. $y = \frac{2}{3}x - 6$
27. $y = -2x + 5$
28. $y = 0.5(x - 6)^2 - 8$
29. $y = -2x^2 + 10x + 12$
30. $4x + 7y = 12$
31. $x - 5y < 20$
32. $y \geq \frac{5}{8}x + 7$
33. $y < x^2 - 5x - 6$

34. Let $f(x) = 2x + 8$ and $g(x) = 7x - 18$.
   a. Find $f(x) - g(x)$.
   b. Find $f(x)g(x)$.
   c. Find $f(g(x))$.

35. Let $f(x) = 3x - 4$ and $g(x) = 2x^2 + 5x - 9$.
   a. Find $f(x) - g(x)$.
   b. Find $f(x)g(x)$.
   c. Find $f(g(x))$.
   d. Find $g(f(x))$.

36. Let $f(x) = 4.8x + 1.4$ and $g(x) = -1.6x + 3.2$.
   a. Find $(f + g)(5)$.
   b. Find $fg(-3)$.
   c. Find $f(g(2))$.
   d. Find $g(f(1))$.

37. Let $f(x) = x - 8$ and $g(x) = x^2 - 7x - 20$.
   a. Find $(f + g)(8)$.
   b. Find $fg(5)$.
   c. Find $f(g(12))$.
   d. Find $g(f(6))$.

38. $f(x) = -\frac{4}{3}x + 5$
   a. Find $f(21)$.
   b. Find x such that $f(x) = \frac{41}{3}$.
   c. Give the domain and range for $f(x)$.

39. $g(x) = 20$
   a. Find $g(9)$.
   b. Give the domain and range for $g(x)$.

40. $h(x) = 4x + 20$
   a. Find $h(-8)$.
   b. Find x such that $h(x) = 75$.
   c. Give the domain and range for $h(x)$.

41. The function $L(m)$ is the number of laps an eighth grader has run during the first m minutes of a jogathon. Explain the meaning of $L(8) = 3.5$ in this situation.

42. Use the table to find the following.
   a. The y-intercept.
   b. The x-intercept.
   c. The slope of the line passing through these points.
   d. The equation for the line passing through these points. Write the equation in slope-intercept form.

x	y
−8	13.8
−2	10.2
0	9.0
9	3.6
15	0

43. Use the graph to find the following.

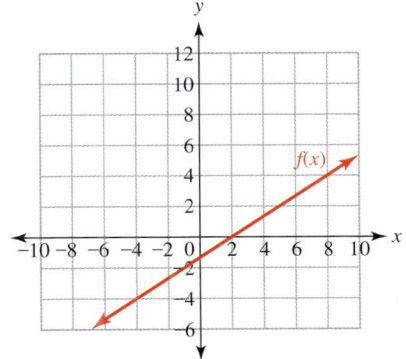

   a. $f(8)$.
   b. x such that $f(x) = -4$.
   c. Give the domain and range of $f(x)$.

44. Use the graph to find the following.

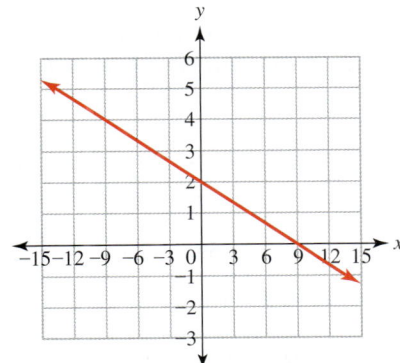

   a. The slope of the line.
   b. The vertical intercept.
   c. The horizontal intercept.
   d. The equation for the line.

45. Greg is investing $900,000 to earn $25,800 in interest each year to fund his son's college education. He plans to invest the money in two accounts: one paying 3% and a safer account paying 2.5%.
   a. Write a system of equations that will help to find the amount Greg needs to invest in each account.
   b. How much should Greg invest in each account to earn the $25,800 he wants?

46. The Vermont Coffee Company is making 200 pounds of their uptown blend by mixing two types of coffee. The first type is a bold roast that cost $8.99 per pound. The second is a mild roast that costs $11.99 per pound. The uptown blend sells for $10.19 per pound. Find how many pounds of each type of coffee should be used to make the mix.

47. Deeva earns $300 per week plus 4% commission on all sales she makes. Deeva needs $410 each week to pay bills. How much in sales does she need to earn at least $410 per week?

For Exercises 48 and 49, determine the number of terms in the given polynomial expression. List the terms as either constant terms or variable terms. Give the coefficient of each variable term.

48. $24x^3y - 7x^2y^2 + 12y^3 - 10$

49. $-6x^5 + 4x^2 - 2x + 4$

50. List the degree of each term and the degree of the entire polynomial given in Exercise 48.

51. List the degree of each term and the degree of the entire polynomial given in Exercise 49.

For Exercises 52 and 53, determine whether the given expression is a polynomial. If the expression is not a polynomial, explain why not.

52. $\frac{2}{9}x^4 + 3x + \sqrt{10}$ **53.** $5xy^{-2} + 7xy + 8$

For Exercises 54 through 59, solve the inequality. Give the solution as an inequality. Graph the solution set on a number line.

54. $4x - 11 > 10x + 1$ **55.** $\frac{1}{4}(x+3) - \frac{5}{12} \le \frac{1}{2}x + \frac{5}{6}$

56. $|x - 12| < 9$ **57.** $|-3x + 5| \le 11$

58. $|x - 9| \ge 3$ **59.** $-5|x - 4| + 20 < -15$

60. Use the table to find when Y1 < Y2.

X	Y1	Y2
2	10	46
4	14	38
7	20	26
8	22	22
9	24	18
12	30	6
15	36	-6

X=15

For Exercises 61 and 62, solve the equation. Check your answer.

61. $|2m - 5| + 7 = 22$ **62.** $|a - 9| - 4 = -2$

For Exercises 63 and 64, sketch a graph by hand of the system of inequalities.

63. $y \ge 2x - 8$
$y \le -\frac{1}{3}x + 2$

64. $2x - 4y \le 16$
$-2x + y \ge 6$

For Exercises 65 through 70, simplify the expressions. Write each answer without negative exponents.

65. $\left(\frac{3}{5}m^4n^5\right)^4$ **66.** $(10x^3y^7)(-4xy^2)$

67. $\frac{24g^3h^6}{16gh^2}$ **68.** $\frac{-14x^4yz^{-3}}{20xy^{-5}z^{-3}}$

69. $(-27a^6b^{-3})^{\frac{1}{3}}(16a^{12}b^{-20})^{-\frac{1}{4}}$ **70.** $\left(\frac{72m^3n^{12}}{50m^{-1}n^8}\right)^{\frac{1}{2}}$

71. The Gross Domestic Product (GDP) per employed person in Sweden can be modeled by

$$S(t) = 1.88t + 54.28$$

where $S(t)$ is Sweden's GDP per employed person in thousands of dollars t years since 2000. The Netherlands' GDP per employed person in thousands of dollars t years since 2000 can be represented by

$$N(t) = 1.01t + 61.3$$

Source: Statistical Abstract of the United States, 2009.

  a. Find $S(6)$ and interpret its meaning.
  b. Find when Sweden's GDP per employed person will be greater than that of the Netherlands.

72. The number of loans to African American–owned small businesses, $A(t)$, in thousands by the Small Business Administration (SBA) t years since 2000 can be represented by

$$A(t) = 1.26t - 0.04$$

The number of loans to Hispanic American–owned small businesses, $H(t)$, in thousands by the SBA t years since 2000 can be represented by

$$H(t) = 1.33t + 2.31$$

Source: Models derived from data in Statistical Abstract of the United States, 2009.

  a. Estimate the number of loans to African American–owned small businesses in 2007.
  b. Find a new function that gives the total number of loans given to African American–owned and Hispanic American–owned small businesses.
  c. Using your new function, estimate the number of loans given to African American–owned and Hispanic American–owned small businesses in 2010.

For Exercises 73 through 78, factor the given expression using any method.

73. $a^2 + 11a + 24$ **74.** $3n^3 - 27n^2 + 42n$

75. $16x^2 - 40xy - 14x + 35y$ **76.** $25k^2 - 49$

77. $12t^2 - 19t - 21$ **78.** $5z^3 - 40$

79. The number of burglaries is related to population of a city by

$$B(p) = 0.6p + 0.2$$

where $B(p)$ represents the number of burglaries in thousands when p thousand people live in a certain city. The population in thousands of this city t years since 2000 can be modeled by

$$P(t) = 2t + 48$$

a. Estimate the population of this city in 2015.
b. Estimate the number of burglaries committed in this city in 2015.
c. Using the two functions, find a new function that gives the number of burglaries in this city t years since 2000.
d. Use the new model to estimate the number of burglaries in this city in 2020.

80. Complete the square to convert the function $f(x) = x^2 + 10x - 14$ to vertex form.

Exponential Functions 5

- **5.1** Exponential Functions: Patterns of Growth and Decay
- **5.2** Solving Equations Using Exponent Rules
- **5.3** Graphing Exponential Functions
- **5.4** Finding Exponential Models
- **5.5** Exponential Growth and Decay Rates and Compounding Interest

How could such a small animal cause so much damage? Rabbits like the ones pictured here cause an estimated $600 million in losses to Australian farmers every year. Having few natural predators, rabbits eat vegetation that other animals could feed on, endangering sheep and cattle and contributing to the extinction of many native plants and animal species. In this chapter, we will study exponential growth and decay patterns and how to model these situations mathematically. One of the chapter projects will ask you to use these functions to investigate the exponential growth of the Australian rabbit population.

5.1 Exponential Functions: Patterns of Growth and Decay

LEARNING OBJECTIVES

- Determine a pattern of exponential growth or decay from data or an application.
- Find an exponential model using a pattern.

Exploring Exponential Growth and Decay

In this chapter, we will add yet another type of function to our list of possible models. So far, we have found linear and quadratic models and have worked with systems of equations. Now we will discuss another function that is found in many areas of science and life called an *exponential function*.

> *The mathematics of uncontrolled growth are frightening. A single cell of the bacterium E. coli would, under ideal circumstances, divide every twenty minutes. That is not particularly disturbing until you think about it, but the fact is that bacteria multiply geometrically: one becomes two, two become four, four become eight, and so on. In this way it can be shown that in a single day, one cell of E. coli could produce a super-colony equal in size and weight to the entire planet Earth.*
>
> Michael Crichton, *The Andromeda Strain* (New York: 1969), p. 247.

This quote might sound unrealistic. However, with the assumption of "ideal circumstances," the mathematics of this type of situation is amazing. Although Crichton calls this type of growth *geometric*, today it is more often called *exponential growth*. Later in this section, we will investigate this situation and see just how fast E. coli bacteria grow in "ideal circumstances." First let's consider a simpler version of this situation.

■ CONCEPT INVESTIGATION

How fast are those bacteria growing?

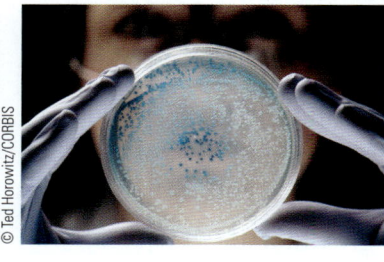

Bacterial growth involves the process of doubling and then doubling again and again. Let's assume that we are doing an experiment in a biology lab starting with a single bacteria cell that in the current environment will split into two new cells every hour.

1. Complete the table with the number of bacteria, B, after h hours. Remember that the number of bacteria is doubling every hour.

h (Hours)	B (Bacteria)
0	1
1	2
2	4
3	
4	
5	
6	

2. Create a scatterplot for these data on your calculator. Describe the shape of the graph.

3. Now rewrite the table from part a, writing out the calculations, not the final results.

Hours	Bacteria
0	1
1	1 · 2
2	1 · 2 · 2
3	
4	
5	
6	

4. From this table, this situation is calculated by using repeated multiplication. Recall that exponents are a shorter way of representing repeated multiplication. Use exponents to rewrite this table and find a pattern for a model that will give the number of bacteria after h hours.

Hours	Bacteria
0	1
1	$1(2)$
2	$1(2)^2$
3	$1(2)^3$
4	
5	
6	
h	

5. Graph your model with the data. How well does your model fit the data?

6. Use your model to find the number of bacteria after 12 hours and after 24 hours.

Using Your TI Graphing Calculator

To enter an exponential function, you will need to have a variable in the exponent. This means that you will need to use the caret button, 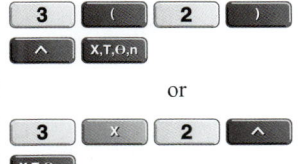, that is just above the division button on the right side of your calculator.

Example:
To enter the exponential function

$$f(x) = 3(2)^x$$

into the Y= screen of your calculator, you would use the following key strokes.

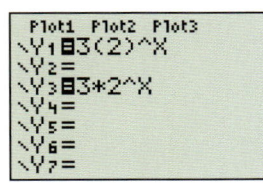

or

[3] [×] [2] [^]
[X,T,Θ,n]

```
Plot1 Plot2 Plot3
\Y1■3(2)^X
\Y2=
\Y3■3*2^X
\Y4=
\Y5=
\Y6=
\Y7=
```

Because of the order of operations, the exponent x will be performed first, so the parentheses will not be needed. Many students find it safer to put in the parentheses to avoid any confusion.

An **exponential function** is based on a pattern of repeated multiplication that leads to using exponents to simplify the expression. Because the exponent in this type of model is what changes from one input value to the other, there is a variable in the exponent rather than a constant. It is this variable in the exponent that makes the function exponential rather than a parabola or other type of function.

DEFINITIONS

Exponential function A basic exponential function can be written in the form
$$f(x) = a \cdot b^x$$
where a and b are real numbers, $a \neq 0$, $b > 0$, and $b \neq 1$.

Base The constant b is called the base of an exponential function.

Finding a model for exponential data can be done in several ways. However, a common approach involves using a real life situation and looking for a pattern. This will take practice and, in some cases, a new way of thinking. Read through the following examples and problems. Pay close attention to how the final pattern is found that leads to a model.

416 CHAPTER 5 Exponential Functions

> **Example 1** **Exponential growth pattern**

A rumor is spreading across a local college campus that there will be no finals for any classes this summer. At 8:00 A.M. today, 7 people have heard the rumor. Assume that after each hour, 3 times as many students have heard the rumor. For example, 21 people have heard the rumor by 9:00 A.M. Let $R(t)$ represent the number of people who have heard the rumor t hours after 8:00 A.M.

a. Write an equation for a model of $R(t)$.

b. If the college has about 15,000 students, estimate numerically the time at which all of them would have heard the rumor.

c. How many people have heard the rumor by 8:00 P.M.?

SOLUTION

a. First, we need to define the variables and to create a table.

$$R(t) = \text{Number of people who have heard the rumor}$$
$$t = \text{Time in hours after 8:00 A.M.}$$

We start with 7 people at 8:00 A.M. Each hour, 3 times as many people have heard the rumor. We can start building the table by writing out the calculations and simplify later.

t	$R(t)$	
0	7	*Start with 7 people.*
1	7 · 3	*One hour later, 3 times as many people.*
2	7 · 3 · 3	*Another hour later, 3 times as many people.*
3	7 · 3 · 3 · 3	*Every hour after, 3 times as many people*
4	7 · 3 · 3 · 3 · 3	*have heard the rumor.*
5	7 · 3 · 3 · 3 · 3 · 3	
6	7 · 3 · 3 · 3 · 3 · 3 · 3	

Because each hour, 3 times as many people have heard the rumor, we have repeated multiplication by 3. Each row in the table can be simplified by using exponents.

t	$R(t)$
0	7
1	7(3)
2	7(3)^2
3	7(3)^3
4	7(3)^4
5	7(3)^5
6	7(3)^6

In looking at this pattern, notice that the 7 and 3 are the same in each expression after the first. The base for each exponent is 3, so the base b for the exponential function equals 3. The only part of each expression that is changing is the exponent itself. Note that the input value that t has for each row is the same as the exponent in the resulting expression.

t	$R(t)$
0	7
1	$7(3)^1$
2	$7(3)^2$
3	$7(3)^3$
4	$7(3)^4$
5	$7(3)^5$
6	$7(3)^6$

Because the input value is the same as the exponent in each row of the table, if we consider t to be the input, the exponent will also be t. With this in mind, we can express our model as $R(t) = 7(3)^t$.

If we graph these data and our model, we get the following.

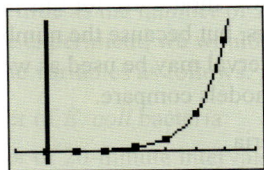 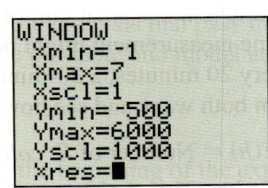

Skill Connection

Checking a model

Remember that whenever we find a model for data, we check that model by graphing the data and model on the same graph to verify a good fit.

b. Using the table we get the following.

The model reaches 15,309 people having heard the rumor when $t = 7$, so by 3:00 P.M. all of the students at the college should have heard the rumor.

c. 8:00 P.M. that night is 12 hours after 8:00 A.M., so we can substitute $t = 12$ into our model.

$$R(12) = 7(3)^{12}$$
$$R(12) = 3720087$$

This means that 3,720,087 people will have heard this rumor by 8:00 P.M. that night. This is clearly model breakdown, since 3.7 million people will not have heard this rumor in 12 hours.

This exponential model is growing so quickly that its domain must be carefully determined to avoid obvious model breakdown. We will discuss the domain for an exponential function in Section 5.3.

PRACTICE PROBLEM FOR EXAMPLE 1

A loan shark lends desperate people money at a very high rate of interest. You have found yourself in trouble and need to borrow $5000 to pay your legal bills. One loan shark will lend you the money under the rules stated in the IOU shown at the right.

a. Find an equation for a model that will give the balance on this loan after w weeks have gone by.

b. Find the balance on this loan if you keep the money for 4 weeks.

c. Estimate numerically after how many weeks the balance of the loan will exceed 1 million dollars.

IOU

Original loan amount $5000.00
Each week the balance will double until paid in full.

There were two ways to look at this pattern. Most people are comfortable with measuring time in hours, while some are comfortable with measuring time in 20-minute intervals. Neither of these models is better than the other, but we must use caution when using and interpreting our models. Please note that these are not the only two possible models, but they are two common ones.

PRACTICE PROBLEM FOR EXAMPLE 2

Suppose that under "ideal conditions," human beings could double their population size every 50 years and that there were 6 billion humans on the earth in the year 2000.

Population density map of the world. The brightest areas are the most densely populated.

a. Find an equation for a model for the world population.

b. Estimate the world population in 2500.

c. Estimate graphically when the world population would reach 10 billion.

So far, we have seen several examples of exponential growth. The opposite of this type of growth is exponential decay. Exponential decay plays a large role in sciences such as archeology, in which scientists use exponential decay of carbon-14 to date historical objects. The exponential decay of radioactive elements is also a large part of the concerns people have about nuclear power plants. The basic concept driving many exponential decay problems is the half-life of the element. The half-life is simply a measurement of how long it takes before only half of an initial quantity remains. The half-lives of different elements have large ranges of values, such as radon-222, which has a half-life of only 3.825 days, and rubidium-87, with a half-life of 49 billion years. To get a better idea of how this works, let's look at a few examples and problems dealing with exponential decay.

■ **Skill Connection**

Increasing or Decreasing

In considering a table of data, if the outputs are getting larger as the inputs get larger, it is an increasing pattern. If the outputs are getting smaller as the inputs get larger, it is a decreasing pattern.

Example 3 Exponential decay: half-life

Carbon-14 is an isotope that is found in all living creatures. Once a creature dies and stops taking in new carbon, the existing carbon-14 decays and is no longer replaced. Because carbon-14 has a half-life of 5700 years, we can use the amount of carbon-14 that is left in an artifact to determine the age of the formerly living thing. Let's assume that an artifact, such as a bark sandal found in Oregon's Fort Rock Cave, started with 300 atoms of carbon-14 at the time the wood was harvested.

a. Find an equation for a model of the amount of carbon-14 remaining in the artifact.

b. Estimate the amount of carbon-14 that should be in the artifact if it is 34,200 years old.

SOLUTION

a. We know that the original amount of carbon-14 was 300 atoms and that half of that amount will be left after 5700 years. With this information we define variables and create a table to help us find a pattern.

$$C(t) = \text{Amount of carbon-14 atoms in the artifact}$$
$$t = \text{Time in years since the wood was harvested}$$

We do not know what happens after 1 year. We do know that after 5700 years half of the carbon-14 remains. Because we know only what happens every 5700 years, we will count the years by 5700s.

t	$C(t)$
0	300
5700	$300\left(\frac{1}{2}\right)$
11,400	$300\left(\frac{1}{2}\right)\left(\frac{1}{2}\right)$
17,100	$300\left(\frac{1}{2}\right)\left(\frac{1}{2}\right)\left(\frac{1}{2}\right)$

t	$C(t)$
0	300
5700	$300\left(\frac{1}{2}\right)$
11,400	$300\left(\frac{1}{2}\right)^2$
17,100	$300\left(\frac{1}{2}\right)^3$

> **What's That Mean?**
>
> **Exponential Words**
>
> There are many words or phrases that imply that a situation will be modeled by an exponential function. Here are just a few of them:
> - Exponential growth
> - Exponential decay
> - Half-life
> - Growth or decay by a factor of:
>   fraction
>   percentage
> - Double every ...
> - Triple every ...
> - Repeated multiplication

Just as in the previous problems, the constant 300 and the base $\frac{1}{2}$ are the same in each row of the table. The only change is in the exponent. This means that we need to find the connection between the input values and the exponents. If we look carefully, we can note that the inputs are multiples of 5700. If we divide each input by 5700, we will get the exponent on that row.

t	$C(t)$
0	300
5700	$300\left(\frac{1}{2}\right)^1$ (5700 ÷ 5700 = 1)
11,400	$300\left(\frac{1}{2}\right)^2$ (11400 ÷ 5700 = 2)
17,100	$300\left(\frac{1}{2}\right)^3$ (17100 ÷ 5700 = 3)

If we let t represent the input value and divide it by 5700, we get the model

$$C(t) = 300\left(\frac{1}{2}\right)^{\frac{t}{5700}}$$

Check the model.

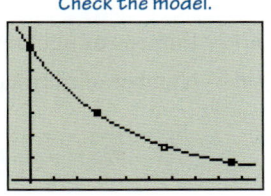

b. 34,200 years old can be represented by $t = 34200$, so we get

$$C(34200) = 300\left(\frac{1}{2}\right)^{\frac{34200}{5700}}$$

$$C(34200) = 300\left(\frac{1}{2}\right)^6$$

$$C(34200) = 4.6875$$

This means that after 34,200 years, the artifact should have only about 4 or 5 atoms of carbon-14 remaining.

PRACTICE PROBLEM FOR EXAMPLE 3

Uranium-238 is a radioactive element that occurs naturally in most types of granite and sometimes in soil. As it undergoes radioactive decay, a chain of elements is formed. As each element decays, it gives off another element until it reaches a stable element. Because of this radioactive decay, uranium-238 decays through several elements until it changes into stable, nonradioactive lead-206. Thorium-234 is the first by-product of the decay of uranium-238 and has a half-life of 24.5 days.

a. Find an equation for a model for the percent of a thorium-234 sample left after d days.

b. Estimate the percentage of thorium-234 left after 180 days.

c. Estimate numerically how many days there will be before only 40% of the sample is left.

Example 4   Exponential decay pattern

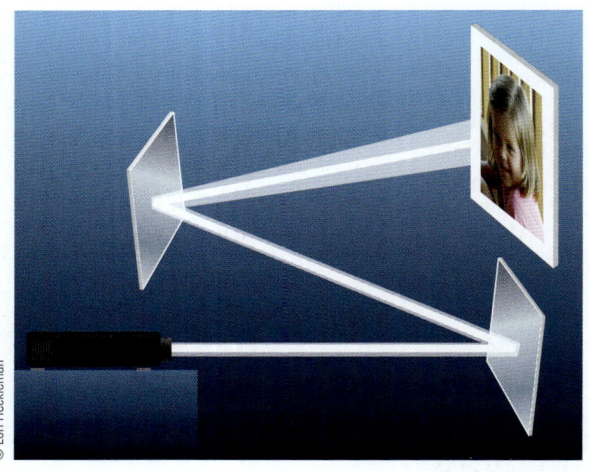

The brightness of light is measured with a unit called a *lumen*. Sharp makes a high-end Conference Series XG-P25X LCD Projector that has a brightness of 4000 lumens. Using a series of several mirrors that reflect only $\frac{3}{5}$ of the light that hits it, a stage technician is trying to project a series of photos onto several walls of a concert stage. The technician is concerned with the brightness of the light that will remain after using several mirrors to place the projected photos in the right places.

a. Find an equation for a model that will tell the technician the remaining lumens that will be projected after m mirrors have reflected the image.

b. Use the model to determine the lumens remaining after five mirrors.

c. If the technician knows that he needs a minimum of 500 lumens to have a good image, use the table to find the maximum number of mirrors he can use with this projector.

SOLUTION

a. The projector will initially project a light with 4000 lumens. After each mirror reflects the light, only $\frac{3}{5}$ of that light will remain. Let's define the following variables:

$$L(m) = \text{Lumens of light remaining}$$
$$m = \text{Number of mirrors used to reflect the light}$$

m	$L(m)$
0	4000
1	$4000\left(\frac{3}{5}\right)$
2	$4000\left(\frac{3}{5}\right)^2$
3	$4000\left(\frac{3}{5}\right)^3$
m	$4000\left(\frac{3}{5}\right)^m$

This pattern results in the model

$$L(m) = 4000\left(\frac{3}{5}\right)^m$$

Check the model.

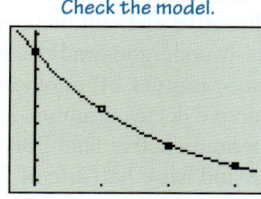

b. Five mirrors are represented by $m = 5$, so we get

$$L(5) = 4000\left(\frac{3}{5}\right)^5$$

$$L(5) = 311.04$$

If the technician uses five mirrors to place the photo projection, it will have a brightness of only 311.04 lumens.

c. Using the table feature on the calculator, we get

X	Y1
1	2400
2	1440
3	864
4	518.4
5	311.04
6	186.62
7	111.97

Y1=4000(3/5)^X

From this table, we can conclude that the technician can use only up to four mirrors to have the 500 lumens necessary to project a good image.

Recognizing Exponential Patterns

Whenever data are given without a situation to consider, remember that the basic concept of an exponential model is repeated multiplication by the same number, the base. When investigating data, look for the initial amount and the base.

Example 5 Finding an exponential model given data

Use the following tables to find exponential models of the given data.

a.

x	$f(x)$
0	15
1	60
2	240
3	960
4	3840

b.

x	$f(x)$
0	16
1	36
2	81
3	182.25
4	410.06

c.

t	$h(t)$
0	1500
4	300
8	60
12	12
16	2.4

SOLUTION

a. For each of these tables, we are given the initial value when the input variable is zero. This initial value is the value of a in the exponential model. We can find the base by dividing each consecutive output value by the previous output value.

x	$f(x)$	base
0	15	
1	60	$\frac{60}{15} = 4$
2	240	$\frac{240}{60} = 4$
3	960	$\frac{960}{240} = 4$
4	3840	$\frac{3840}{960} = 4$

By performing these divisions, the common multiplier is 4 for each row. If we rewrite each output showing how to calculate it using the base 4, we should see the pattern and therefore the appropriate model.

x	$f(x)$	base
0	15	15
1	60	$15(4)$
2	240	$15(4)^2$
3	960	$15(4)^3$
4	3840	$15(4)^4$

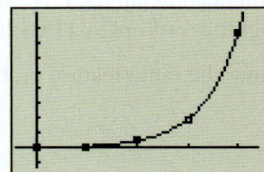

This leads us to the model $f(x) = 15(4)^x$.

b.

x	$f(x)$	base
0	16	$\frac{36}{16} = 2.25$
1	36	$\frac{81}{36} = 2.25$
2	81	$\frac{182.25}{81} = 2.25$
3	182.25	$\frac{410.0625}{182.25} = 2.25$
4	410.0625	

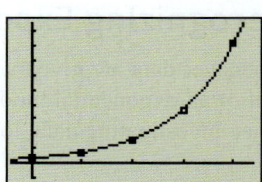

Again, by looking for the common multiplier, we have found a base. Because the initial value when $x = 0$ is 16, we get the model $f(x) = 16(2.25)^x$.

c. Be careful when looking at this example since the inputs are not 1 unit apart. Notice that each input is increasing by 4 units. This will change the pattern we find for this model.

t	$h(t)$	base
0	1500	$\frac{300}{1500} = 0.2$
4	300	$\frac{60}{300} = 0.2$
8	60	$\frac{12}{60} = 0.2$
12	12	$\frac{2.4}{12} = 0.2$
16	2.4	

Again, by looking for the common multiplier, we have found a base. Because the initial value when $t = 0$ is 1500, we know the constant. To find a pattern, it would be helpful to build another table showing the calculations.

t	$h(t)$
0	1500
4	$1500(0.2)$
8	$1500(0.2)^2$
12	$1500(0.2)^3$
16	$1500(0.2)^4$

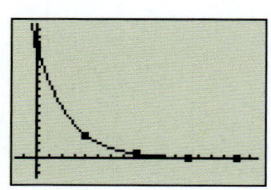

The inputs are not the same as the exponents. We need to find a pattern to show how to relate the input values to the changing exponents. We should be able to see that the exponents are the input values divided by 4. With this fact, we get the model $h(t) = 1500(0.2)^{\frac{t}{4}}$.

PRACTICE PROBLEM FOR EXAMPLE 5

Use the following tables to find exponential models of the given data.

a.

x	$f(x)$
0	10
1	31
2	96.1
3	297.91
4	923.521

b.

x	$f(x)$
0	16
3	14.4
6	12.96
9	11.664
12	10.4976

5.1 Exercises

For Exercises 1 through 10, determine whether the given function is linear, quadratic, exponential, or other. Check your answer by graphing the function on your calculator.

1. $f(x) = 6^x$
2. $g(x) = 5x + 7$
3. $h(x) = 4(0.2^x)$
4. $p(t) = 3(t-4)^2 - 8$
5. $f(x) = 7x^2$
6. $M(d) = 4\left(\frac{1}{3}\right)^d + 5$
7. $R(n) = 2(n-8)(n+4)$
8. $g(x) = -40\left(\frac{2}{3}\right)^x$
9. $h(t) = \dfrac{3t+5}{2t-8}$
10. $P(b) = -b + 6(b+8)$

11. *Lactobacillus acidophilus* is a bacteria that doubles every hour when grown in milk under ideal circumstances. If a sample of 30 bacteria is allowed to grow in these ideal circumstances, answer the following.
    a. Write an equation for a model for the number of *L. acidophilus* bacteria after *h* hours have passed.
    b. Estimate the number of bacteria present after 12 hours.
    c. Estimate the number of bacteria present after 24 hours.
    d. Estimate numerically when the number of bacteria will reach 1 million.

12. Under ideal circumstances, a certain type of bacteria can triple every hour. If a sample of 20 bacteria is allowed to grow in these ideal circumstances, answer the following.
    a. Find an equation for a model for the number of bacteria after *h* hours have passed.
    b. Estimate the number of bacteria present after 5 hours.
    c. Estimate the number of bacteria present after 10 hours.
    d. Estimate numerically when the number of bacteria will reach 1 million.

13. The number of *E. coli* bacteria doubles every 30 minutes when they are growing in an environment that is about 59°F. If a sample of 3 bacteria is allowed to grow in these circumstances, answer the following.
    a. Find an equation for a model for the number of bacteria after *h* hours have passed.
    b. Estimate the number of bacteria present after 5 hours.
    c. Estimate graphically when the number of bacteria will reach 1 million.

14. Under ideal circumstances, a certain type of bacteria can triple every 20 minutes. If a sample of 5 bacteria is allowed to grow in ideal circumstances, answer the following.
    a. Find an equation for a model for the number of bacteria after *h* hours have passed.
    b. Estimate the number of bacteria present after 7 hours.
    c. Estimate graphically when the number of bacteria will reach 1 million.

15. *Streptococcus lactis* is a bacteria that can double every 26 minutes when grown in milk under ideal circumstances. If a sample of 8 of these bacteria is allowed to grow in these ideal circumstances, answer the following.
    a. Find an equation for a model for the number of bacteria after *n* 26-minute intervals have passed.
    b. Estimate the number of bacteria present after 6.5 hours.

16. *S. lactis* is a bacteria that can double every 48 minutes when grown in a lactose broth under ideal circumstances. If a sample of 20 of these bacteria is allowed to grow in these ideal circumstances, answer the following.
    a. Find an equation for a model for the number of bacteria after n 48-minute intervals have passed.
    b. Estimate the number of bacteria present after 4 hours.

17. Under ideal circumstances, a certain type of bacteria can triple every 30 minutes. If a sample of 8 of these bacteria is allowed to grow in these ideal circumstances, answer the following.
    a. Find an equation for a model for the number of bacteria after h hours have passed.
    b. Estimate the number of bacteria present after 10 hours.

18. Under ideal circumstances, a certain type of bacteria can double every 10 minutes. If a sample of 15 of these bacteria is allowed to grow in these ideal circumstances, answer the following.
    a. Find an equation for a model for the number of bacteria after h hours have passed.
    b. Estimate the number of bacteria present after 4 hours.

19. *Treponema pallidum* is a bacteria that doubles every 33 hours when grown in rabbit testes under ideal circumstances. If a sample of 12 of these bacteria is allowed to grow in these ideal circumstances, answer the following.
    a. Find an equation for a model for the number of bacteria after h hours have passed.
    b. Estimate the number of bacteria present after 1 week.

20. *Mycobacterium tuberculosis* is a bacteria that doubles every 15 hours when grown in a synthetic medium under ideal circumstances. If a sample of 9 of these bacteria is allowed to grow in these ideal circumstances, answer the following.
    a. Find an equation for a model for the number of bacteria after h hours have passed.
    b. Estimate the number of bacteria present after 2 weeks.

21. A building fire in a room with fire sprinklers doubles in area about every 6 minutes. If a fire covering 2 m² continues to burn, answer the following.
    a. Find an equation for a model for the number of square meters burnt after n 6-minute intervals have passed.
    b. Estimate the number of square meters burnt after half an hour.

22. A building fire in a room without fire sprinklers doubles in area about every 3.5 minutes. If a fire covering 2 m² continues to burn, answer the following.
    a. Find an equation for a model for the number of square meters burnt after n 3.5-minute intervals have passed.
    b. Estimate the number of square meters burnt after 35 minutes.

23. The U.S. Census Bureau reported that the number of centenarians (people older than 100) is increasing rapidly. According to the report, the number of Hispanic centenarians is doubling about every 7.5 years. In 1990, there were approximately 2072 Hispanic centenarians.
    Source: National Institute on Aging Journal, June 1999.
    a. If this trend continues, find an equation for a model that would predict the number of Hispanic centenarians.
    b. Use your model to estimate the number of Hispanic centenarians in 2050.

24. According to the same U.S. Census report discussed in Exercise 23, the number of centenarians, in general, will double every 10 years. In 1990, there were about 37,000 centenarians in the United States.
    Source: National Institute on Aging Journal, June 1999.
    a. If this trend continues, find an equation for a model that would predict the number of centenarians.
    b. Use your model to estimate the number of centenarians in 2050.
    c. According to census estimates, the number of centenarians in 2050 could be as high as 4.2 million people. How does your estimate compare?

Use this article to answer Exercises 25 and 26.

POPULATION NEWS

India: Population is expected to double every 33 years. In 1971, there were about 560 million people in India. In 2000, India's population broke 1 billion.

Delhi: Is a territory in India and is one of the fastest-growing areas in India. The Delhi territory's population has doubled about every 20 years since 1950.

Source: Census India 2001, CensusIndia.net.

25. a. Using this information, find an equation for a model for the population of India.
    b. Using your model, estimate the population of India in 2015.
    c. Estimate graphically when India's population will reach 2 billion.

26. a. Using this information, find an equation for a model for the population of Delhi, India, if in 1950 there were approximately 2.2 million people.
    b. Using your model, estimate the population of Delhi, India, in 2011.
    c. In 2011, the population of Delhi was about 19 million people. How does this compare with your projections for 2011?

27. **Internet Search:** Use the Internet to find how long researchers project it will take the U.S. population to double.

28. **Internet Search:** Use the Internet to find how long researchers project it will take the population of France to double.

29. Suppose you are given the two different options presented in the offer letter.

    Joe Smith Associates

    123 Bank Ave.   Dallas, TX 50344

    Dear Sir,

    We are offering two salary options at this time.

    Option 1: $2000 each week.

    Option 2: You earn 1 penny the first week, 2 pennies the next week, and so on, doubling your salary each week.

    The job is going to last 25 weeks.

    Thank you,
    Joe L. Smith

    ---

    a. Find the total salary for the 25 weeks if you take option 1.
    b. Before you calculate the total for option 2, decide which of the two options you would want at this point.
    c. Find an equation for a model for the salary you will earn in week w of this temporary job under option 2.
    d. Find the total amount of salary you will earn using option 2. (*Note:* You will need to add up all 25 weeks' salary.)
    e. Now which salary option would you choose for this job?
    f. What option would be best for a 20-week job?

30. You are given the following two salary options for a 20-week temporary job.

    Option 1: You earn $5000 the first week, $6000 the second week, and so on, adding $1000 to your salary each week.

    Option 2: You earn 1 penny the first week, 3 pennies the next week, and so on, tripling your salary each week.

    a. Find an equation for a model for the salary you will earn in week w of this temporary job under option 1. (*Note:* You are adding an amount not multiplying so your model will be linear.)
    b. Find the total salary for the 20 weeks if you take option 1.
    c. Find an equation for a model for the salary you will earn in week w of this temporary job under option 2.
    d. Find the total amount of salary you will earn using option 2. (*Note:* You will need to add up all 20 weeks' salary.)
    e. Which salary option would you choose for this job?
    f. What option would be best for a 15-week job?

31. Recent advances in the forensic dating of skeletal remains has led forensic pathologists to monitor the decay of certain isotopes found in humans. The isotope lead-210 has a half-life of 22 years and starts to decay after someone has died. Stuart Black of Reading University in the United Kingdom used this isotope to date a body that was found floating down London's River Thames in September 2001.

    a. Find an equation for a model for the percent of a lead-210 sample left after t years.
    b. Estimate the percent of lead-210 left after 60 years.
    c. Estimate graphically the number of years after which only 5% of the sample will be left.

32. Another isotope that forensic pathologists use to date skeletal remains is polonium-210. This element has a half-life of only 138 days and can be used to fine-tune a date of death to within a two-week period.

    a. Find an equation for a model for the percent of polonium-210 atoms left d days after the death of a person.
    b. Estimate the percent of polonium-210 left 1 year after the death.
    c. Estimate numerically the time of death for a skeleton that has 10% of the polonium-210 expected in a body at the time of death.

33. The U.S. Surgeon General estimates that radon is responsible for nearly 30,000 lung cancer deaths in the United States each year. Radon-222 is a gas that occurs naturally as uranium-238 decays in the soil. Because radon-222 is a gas, it gets released into the air and can enter homes through cracks in foundations or basements. This gas is colorless and odorless, so it cannot be detected easily by humans. Radon detectors (shown below) have now made it easier for homeowners to keep their homes safe from extreme radon levels. Radon-222 has a half-life of 3.825 days.

a. Find an equation for a model for the percent of a radon-222 sample left after d days.
b. Estimate the percent of radon-222 left after 30 days.
c. Estimate graphically the number of days after which only 5% of the radon-222 sample will be left.

34. A sample of 300 polonium-218 atoms is being stored for an experiment that will take place in 2 hours. Polonium-218 has a half-life of 3.05 minutes.

a. Find an equation for a model for the number of polonium-218 atoms left after m minutes.
b. Find the number of polonium-218 atoms remaining at the beginning of the experiment.
c. Estimate numerically when there were 100 polonium-218 atoms left.

For Exercises 35 through 40, use the graphs of the exponential functions to answer the questions.

35. Given the graph of $f(z)$, estimate the following.

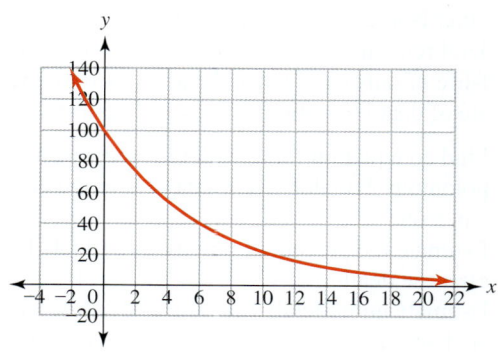

a. Is the graph increasing or decreasing?
b. Is this an example of exponential growth or exponential decay?
c. $f(10) = $?
d. What x-value(s) will make $f(x) = 40$?
e. Estimate the y-intercept.

36. Given the graph of $f(x)$, estimate the following.

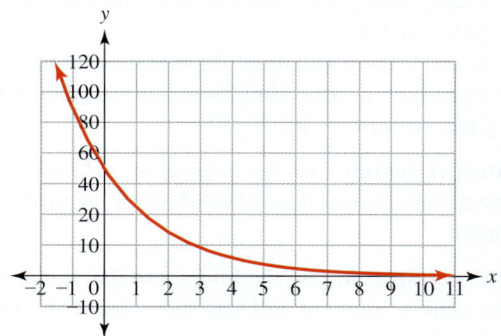

a. Is the graph increasing or decreasing?
b. Is this an example of exponential growth or exponential decay?
c. $f(2) = $?
d. What x-value(s) will make $f(x) = 10$?
e. Estimate the y-intercept.

37. Given the graph of $h(x)$, estimate the following.

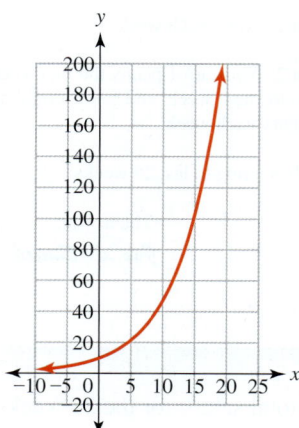

a. Is the graph increasing or decreasing?
b. Is this an example of exponential growth or exponential decay?
c. $h(5) = $?
d. What x-value(s) will make $h(x) = 100$?
e. Estimate the y-intercept.

38. Given the graph of $f(x)$, answer the following.

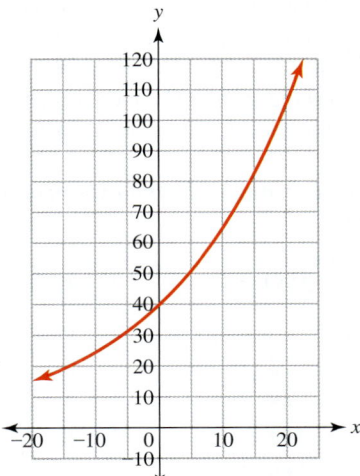

a. Is the graph increasing or decreasing?
b. Is this an example of exponential growth or exponential decay?
c. $f(-15) = $?
d. What x-value(s) will make $f(x) = 80$?
e. Estimate the vertical intercept.

39. Given the graph of $g(x)$, find the following.

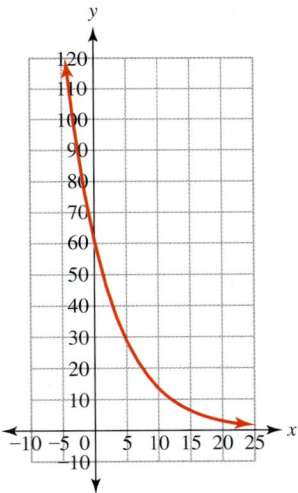

a. Is the graph increasing or decreasing?
b. Is this an example of exponential growth or exponential decay?
c. $g(7.5) = $?
d. What x-value(s) will make $g(x) = 30$?
e. Estimate the vertical intercept.

40. Given the graph of $g(x)$, find the following.

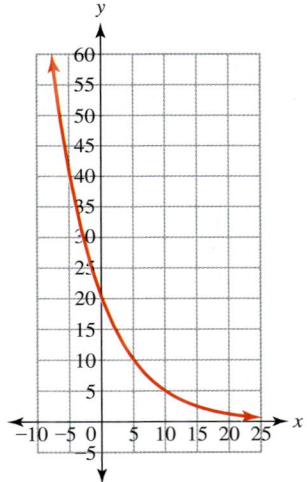

a. Is the graph increasing or decreasing?
b. Is this an example of exponential growth or exponential decay?
c. $g(-5) = $?
d. What x-value(s) will make $g(x) = 5$?
e. Estimate the vertical intercept.

For Exercises 41 through 50, find exponential models for the given data.

41.

x	$f(x)$
0	25
1	100
2	400
3	1600
4	6400

42.

x	$f(x)$
0	12
1	36
2	108
3	324
4	972

43.

x	$f(x)$
0	-35
1	-245
2	-1715
3	-12,005
4	-84,035

44.

x	$f(x)$
0	-20
1	-100
2	-500
3	-2500
4	-12,500

45.

x	$f(x)$
0	2000
1	400
2	80
3	16
4	3.2

46.

x	$f(x)$
0	360
1	180
2	90
3	45
4	22.5

47.

x	f(x)
0	32
1	48
2	72
3	108
4	162

48.

x	f(x)
0	81
1	135
2	225
3	375
4	625

49.

x	f(x)
0	6400
1	800
2	100
3	12.5
4	1.5625

50.

x	f(x)
0	2400
1	1800
2	1350
3	1012.5
4	759.375

For Exercises 51 through 56, use the given exponential functions, $f(x) = a \cdot b^x$, to determine the values of the base b and the coefficient a.

51. $f(x) = 5(3^x)$

52. $g(x) = 2(6^x)$

53. $f(x) = 200\left(\dfrac{1}{5}\right)^x$

54. $h(x) = 1400(0.75)^x$

55. $f(x) = -4(2)^x$

56. $g(x) = -300(0.5)^x$

57. Using the model for *E coli* bacteria you found in Exercise 13, answer the following questions.
   a. What is the base for the exponential model you found?
   b. What is the value for the coefficient *a* in the model you found?
   c. What does the value for the coefficient *a* represent in regard to the *E coli* bacteria?

58. Using the model for bacteria you found in Exercise 14, answer the following questions.
   a. What is the base for the exponential model you found?
   b. What is the value for the coefficient *a* in the model you found?
   c. What does the value for the coefficient *a* represent in regard to the bacteria?

59. Using the model for *T. pallidum* bacteria you found in Exercise 19, answer the following questions.
   a. What is the base for the exponential model you found?
   b. What is the value for the coefficient *a* in the model you found?
   c. What does the value for the coefficient *a* represent in regard to the *T. pallidum* bacteria?

60. Using the model for *M. tuberculosis* bacteria you found in Exercise 20, answer the following questions.
   a. What is the base for the exponential model you found?
   b. What is the value for the coefficient *a* in the model you found?
   c. What does the value for the coefficient *a* represent in regard to the *M. tuberculosis* bacteria?

For Exercises 61 through 72, find exponential models for the given data.

61.

x	f(x)
0	3
5	12
10	48
15	192
20	768

62.

x	f(x)
0	7
2	42
4	252
6	1512
8	9072

63.

x	f(x)
0	−7
10	−21
20	−63
30	−189
40	−567

64.

x	f(x)
0	−8
15	−32
30	−128
45	−512
60	−2048

65.

x	f(x)
0	1701
6	567
12	189
18	63
24	21

66.

x	f(x)
0	4352
3	3264
6	2448
9	1836
12	1377

67.

x	f(x)
2	80
3	320
4	1280
5	5120
6	20,480

68.

x	f(x)
2	11
3	77
4	539
5	3773
6	26,411

69.

x	f(x)
5	2
6	12
7	72
8	432
9	2592

70.

x	f(x)
7	−5
8	−30
9	−180
10	−1080
11	−6480

71.

x	$f(x)$
4	3584
5	896
6	224
7	56
8	14

72.

x	$f(x)$
3	-7776
4	-1296
5	-216
6	-36
7	-6

For Exercises 73 through 80, determine whether the given exponential function is going to be increasing or decreasing. Check your answer by graphing the function on your calculator.

73. $f(x) = 11(3^x)$

74. $g(x) = 20(5^x)$

75. $f(x) = 10(0.3)^x$

76. $g(x) = 20(0.7)^x$

77. $f(x) = 1.5(1.25)^x$

78. $h(x) = 0.8(1.5)^x$

79. $f(x) = 5000\left(\frac{1}{4}\right)^x$

80. $g(x) = 24000\left(\frac{4}{7}\right)^x$

5.2 Solving Equations Using Exponent Rules

LEARNING OBJECTIVES

- Solve exponential equations using inspection.
- Solve power equations using exponent rules.

Recap of the Rules for Exponents

In Section 3.1, we went over the rules for exponents. In this section, we will use some of these rules to solve different types of equations.

Rules for Exponents

1. $x^m \cdot x^n = x^{m+n}$
2. $(xy)^m = x^m y^m$
3. $\dfrac{x^m}{x^n} = x^{m-n} \quad x \neq 0$
4. $\left(\dfrac{x}{y}\right)^m = \dfrac{x^m}{y^m} \quad y \neq 0$
5. $(x^m)^n = x^{mn}$
6. $x^0 = 1 \quad 0^0 =$ undefined
7. $x^{-n} = \dfrac{1}{x^n} \quad x \neq 0$
8. $x^{\frac{1}{n}} = \sqrt[n]{x} \quad$ if n is even $x \geq 0$

Solving Exponential Equations by Inspection

Some equations can be solved by inspection or trial and error. In the concept investigation following, we will investigate a property of equality that will help us to solve some exponential equations using inspection.

CONCEPT INVESTIGATION

Are those exponents the same?

1. Complete the table.

$2^2 = 4$	$3^2 = 9$	$4^2 = 16$	$5^2 = 25$	$10^2 = 100$
$2^3 = 8$	$3^3 = 27$	$4^3 = 64$	$5^3 = 125$	$10^3 = 1000$
$2^4 = 16$	$3^4 =$ ___	$4^4 =$ ___	$5^4 =$ ___	$10^4 =$ ___
$2^5 =$ ___	$3^5 =$ ___	$4^5 =$ ___	$5^5 =$ ___	$10^5 =$ ___

2. Use the table from part 1 to rewrite the following numbers as powers of the given base.
   a. 32 as a power of 2.
   b. 625 as a power of 5.
   c. 1024 as a power of 4.
   d. 1000 as a power of 10.

3. Use the table from part 1 and your answers to part 2 to find the value of x that is a solution to the given exponential equations.
   a. $2^x = 32$
   b. $5^x = 625$
   c. $4^x = 1024$
   d. $10^x = 1000$

4. Describe the relationships between the exponents m and n in the equation
$$b^m = b^n$$

What's That Mean?

Inspection

To inspect something means to examine it carefully for possible defects or a solution to a defect. In math, to solve a problem by inspection means to examine a problem carefully, looking for the answer.

Often, rewriting the equation in a different form can help us to see an answer more easily.

In the concept investigation, we looked at a property of equality that is very helpful. If two exponential expressions have the same base and are equal to one another, then the exponents must also be equal.

$3^x = 9$	If possible, make the bases the same.
$3^x = 3^2$	Because the bases are the same, the exponents
$x = 2$	themselves must be equal.

We can use this basic property of equality to solve some exponential equations.

Example 1 — Solving exponential equations by inspection

Solve the following exponential equations by inspection or trial and error.

a. $2^x = 16$
b. $3^x = 243$
c. $\dfrac{1}{1000} = 10^t$
d. $2^x = \dfrac{1}{8}$

SOLUTION

a.
$2^x = 16$	Write 16 as a power of 2.
$2^x = 2^4$	Because the bases are the same, the exponents must be equal.
$x = 4$	
$2^4 \stackrel{?}{=} 16$	Check the answer.
$16 = 16$	The answer works.

b. Many people do not know what power of 3 will give us 243. In this case, we will use trial and error and calculate different powers of 3 until we find the power that gives us 243.

$3^x = 243$	Write 243 as a power of 3.
$3^x = 3^5$	Because the bases are the same, the
$x = 5$	exponents must be equal.
$3^5 \stackrel{?}{=} 243$	Check the answer.
$243 = 243$	The answer works.

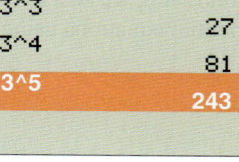

What's That Mean?

Trial and error

When trying to find a solution to some equations, we may try several possible values to find the correct answer. This process is sometimes called *trial and error*.

SECTION 5.2 Solving Equations Using Exponent Rules

c. This exponential equation has a base of 10, but the other side of the equation is a fraction. This means that a negative exponent will be needed to get the reciprocal of the fraction.

$\dfrac{1}{1000} = 10^t$ *Write both sides using the same base.*

$\dfrac{1}{10^3} = 10^t$ *The bases on both sides of the equal sign must be the same and be in the same location. In this problem, we will put both bases in the numerator.*

$10^{-3} = 10^t$ *We will need negative exponents to get the reciprocal of the fraction so that the 10 cubed will come up to the numerator.*

$-3 = t$

$\dfrac{1}{1000} \stackrel{?}{=} 10^{-3}$ *Check the answer.*

$\dfrac{1}{1000} = \dfrac{1}{1000}$ *The answer works.*

d. Again this exponential equation has a fraction, so a negative exponent will be needed to get a reciprocal of the fraction and make the bases the same.

$2^x = \dfrac{1}{8}$ *Write both sides using the same base.*

$2^x = \dfrac{1}{2^3}$

$2^x = 2^{-3}$ *We will need negative exponents to get the reciprocal.*

$x = -3$

$2^{-3} \stackrel{?}{=} \dfrac{1}{8}$ *Check the answer.*

$\dfrac{1}{8} = \dfrac{1}{8}$ *The answer works.*

PRACTICE PROBLEM FOR EXAMPLE 1

Solve the following exponential problems by inspection or trial and error.

a. $10^x = 100000$ **b.** $4^t = 4096$ **c.** $2^x = \dfrac{1}{4}$

Some exponential equations have more than the base and exponent on one side of the equation. When this happens, we must first isolate the exponential part, and then we can try to solve using inspection or trial and error.

Example 2 Solving exponential equations

Solve the following exponential equations by inspection or trial and error.

a. $5(3^t) = 45$ **b.** $6(4^x) - 34 = 350$ **c.** $640(2^t) + 9 = 29$

SOLUTION

a. $5(3^t) = 45$

$\dfrac{5(3^t)}{5} = \dfrac{45}{5}$ *First isolate the exponential part by dividing by 5.*

$3^t = 9$

$3^t = 3^2$ *Write both sides using the same base.*

$t = 2$ *Because the bases are the same, the exponents must be equal.*

$5(3^2) \stackrel{?}{=} 45$ *Check the answer.*

$5(9) \stackrel{?}{=} 45$

$45 = 45$ *The answer works.*

b.
$$6(4^x) - 34 = 350$$
$$+34 +34$$
$$6(4^x) = 384$$
$$\frac{6(4^x)}{6} = \frac{384}{6}$$
$$4^x = 64$$
$$4^x = 4^4$$
$$x = 4$$

First isolate the exponential part by adding 34 to both sides and dividing both sides by 6.

Check the answer using the calculator.

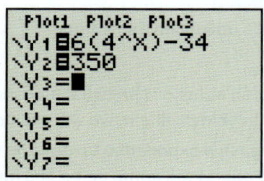 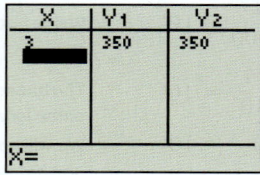

The answer works.

c.
$$640(2^t) + 9 = 29$$
$$-9 -9$$
$$640(2^t) = 20$$
$$\frac{640(2^t)}{640} = \frac{20}{640}$$
$$2^t = \frac{1}{32}$$
$$2^t = \frac{1}{2^5}$$
$$2^t = 2^{-5}$$
$$t = -5$$

First isolate the exponential part by subtracting 9 from both sides. Divide both sides by 640.

Reduce the fraction, since the decimal form may not be easy to recognize as a power of 2.

Check the answer using the calculator.

 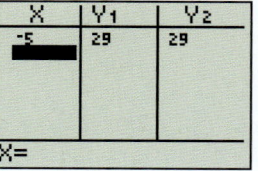

The answer works.

PRACTICE PROBLEM FOR EXAMPLE 2

Solve the following exponential problems by inspection or trial and error.

a. $-2(5^x) - 74 = -324$ **b.** $162(3^x) + 30 = 36$

In some situations, we may be able to find a solution to an application problem using inspection or trial and error. In this section, we will see a few examples of solving applications. In Chapter 6, we will learn to solve more exponential equations and applications.

Example 3 Applications of exponentials

In Example 2 of Section 5.1 on page 418, we found the model

$$E(n) = 1(2)^n$$

where $E(n)$ is the number of *E. coli* bacteria in a sample after n 20-minute intervals. Use this model to find the number of 20-minute intervals before there are 2048 *E. coli* bacteria in the sample.

SOLUTION

Since we want to find the time it takes before there are 2048 *E. coli* bacteria, we want $E(n) = 2048$, so we substitute and solve.

Using trial and error:

$$2048 = 1(2)^n$$
$$2048 = 2^n$$
$$2^{11} = 2^n$$
$$11 = n$$

2^9	256
2^10	512
	1024
2^11	2048

$$2048 \stackrel{?}{=} 2^{11}$$
$$2048 = 2048$$

Check the answer.

The answer woks.

It will take eleven 20-minute intervals for there to be 2048 *E. coli* bacteria in the sample.

PRACTICE PROBLEM FOR EXAMPLE 3

In Example 1 Practice Problem in Section 5.1 on page 417, we found the model

$$B(w) = 5000(2^w)$$

where $B(w)$ is the balance due in dollars after w weeks of not making payments. Find when you will owe $160,000 if you don't make any payments to the loan shark.

Solving Power Equations

We will use rational exponents to solve equations that have variables raised to different powers. Raising both sides of an equation to the reciprocal exponent will help us to eliminate exponents of variables we are trying to solve for. Remember that a fraction exponent is another way of writing a radical, so we are actually taking the appropriate root of both sides. The square root property that we learned in Chapter 4 was an example of this process. Recall that the square root property required us to use a plus/minus symbol to account for both possible solutions. The plus/minus symbol is necessary whenever we take an even root or raise both sides to a fractional exponent where the denominator is an even number.

Odd power.
$$x^5 = 32$$
$$(x^5)^{\frac{1}{5}} = (32)^{\frac{1}{5}}$$
$$x = 2$$

Even power requires plus/minus symbol.
$$x^4 = 81$$
$$(x^4)^{\frac{1}{4}} = \pm(81)^{\frac{1}{4}}$$
$$x = \pm 3$$

Raising both sides of an equation to the reciprocal power will find only the real number solutions to the equation. In Chapter 8, we will learn about other types of solutions to equations.

Example 4 Solving power equations

Solve the following equations.

a. $78125 = x^7$
b. $150m^4 = 22509.375$
c. $\dfrac{t^6}{500} = 20$
d. $3t^5 - 17 = 3055$

SOLUTION

a.
$$78125 = x^7 \quad \text{We need to undo an exponent of 7.}$$
$$(78125)^{\frac{1}{7}} = (x^7)^{\frac{1}{7}} \quad \text{Raise both sides of the equation to the reciprocal of 7.}$$
$$5 = x$$
$$78125 = 5^7 \quad \text{Check the answer.}$$
$$78125 = 78125 \quad \text{The answer woks.}$$

b.
$$150m^4 = 22509.375$$
$$\dfrac{150m^4}{150} = \dfrac{22509.375}{150} \quad \text{Isolate the variable factor by dividing by 150.}$$
$$m^4 = 150.0625 \quad \text{Raise both sides to the reciprocal power.}$$
$$(m^4)^{\frac{1}{4}} = \pm(150.0625)^{\frac{1}{4}} \quad \text{Because the exponent is an even power, we need to use a plus/minus symbol.}$$
$$m = \pm 3.5$$
$$150(3.5)^4 \stackrel{?}{=} 22509.375 \quad \text{Check the answers.}$$
$$22509.375 = 22509.375$$
$$150(-3.5)^4 \stackrel{?}{=} 22509.375 \quad \text{The answers work.}$$
$$22509.375 = 22509.375$$

Using Your TI Graphing Calculator

To calculate square roots and other higher roots on the graphing calculator, you can use either of two methods.

If you are taking the square root of a number, the square root function is above

button on the left side of the calculator. First you must press the 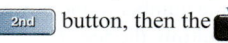 button, then the x^2 button, and then the number you want to take the square root of.

If you want to take a higher root, you can use the caret button

and a fraction exponent in parentheses. In this case, you will need to enter the number or variable you want to raise to a power, then press the ^ button, followed by the power you wish to use.

For example, $\sqrt[3]{8} = 8^{\frac{1}{3}}$ would be entered as

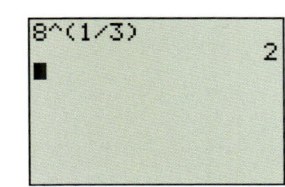

Using Your TI Graphing Calculator

Be careful when entering negative numbers into your calculator. Always use the negative button, (−), to the left of the ENTER button, not the subtraction button.

Also be careful with negative numbers that are being raised to a power. You must use parentheses in your calculator for it to raise the negative to the power. If you do not use parentheses, the calculator will raise the number to the power and then multiply the result by negative 1. The calculator follows the order of operations in a very specific way and might cause errors if you do not use parentheses.

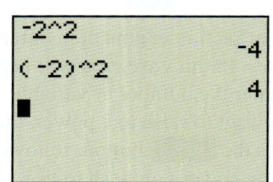

c.
$$\frac{t^6}{500} = 20 \quad \text{Isolate the variable factor by multiplying by 500.}$$
$$500\left(\frac{t^6}{500}\right) = 500(20)$$
$$t^6 = 10000$$
$$(t^6)^{\frac{1}{6}} = \pm(10000)^{\frac{1}{6}} \quad \text{Raise both sides to the reciprocal power.}$$
$$t \approx \pm 4.642 \quad \text{Because the exponent is an even power, we need to use a plus/minus symbol.}$$
Check the answers using the calculator.

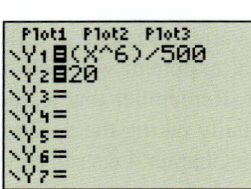 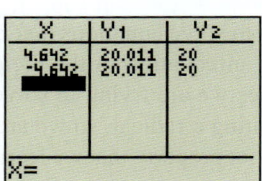

The Y1 and Y2 values are not exactly the same because of rounding.

The answers work.

d.
$$-3t^5 - 17 = 3055 \quad \text{Isolate the variable term.}$$
$$-3t^5 = 3072 \quad \text{Divide by } -3.$$
$$t^5 = -1024$$
$$(t^5)^{\frac{1}{5}} = (-1024)^{\frac{1}{5}} \quad \text{Raise both sides to the reciprocal power.}$$
$$t = -4 \quad \text{Check the answer on the calculator.}$$

The answer works.

PRACTICE PROBLEM FOR EXAMPLE 4

Solve the following equations.

a. $-6x^9 + 148500 = 30402$ **b.** $\dfrac{p^4}{30} = 7$ **c.** $2m^5 - 61 = 847.70848$

Remember when you are solving equations to consider the type of equation you are working with. We use reciprocal powers to solve equations that have numbers as the exponents. If the variable is in the exponent, the equation is an exponential and will not be solved by using reciprocal exponents. In the next chapter, we will learn a way to solve more exponential equations that are not as easy to solve by using inspection or trial and error.

Example 5 Identifying exponential and power equations

Determine whether the following equations are exponential or power equations.

a. $3^x + 5 = 32$ **b.** $x^7 + 616 = 17000$ **c.** $4(5^x) - 23 = 1435$

SOLUTION

a. Since the variable x is in the exponent, this is an exponential equation.

b. Since the variable x is being raised to a power, this is a power equation.

c. Again the variable x is in the exponent, so this is an exponential equation.

PRACTICE PROBLEM FOR EXAMPLE 5

Determine whether the following equations are exponential or power equations.

a. $5x^8 - 16 = 28$ **b.** $3(8^x) - 20 = 40$

5.2 Exercises

For Exercises 1 through 10, determine whether the given equation is a power equation or an exponential equation. Do not solve these equations.

1. $12^x = 45$
2. $7^x = 265$
3. $x^5 = 436$
4. $m^8 = 0.2586$
5. $3x^7 + 25 = 6986$
6. $154 + 36p^5 = 0.568$
7. $4(6^x) + 8(6^x) = 4586$
8. $2.3\left(\dfrac{1}{3}\right)^x - 4.8\left(\dfrac{1}{3}\right)^x - 96.8 = 42.3658$
9. $4.5m^{\frac{2}{3}} + 86 = 3316$
10. $2.8b^{\frac{3}{4}} = 2535 - 1.7b^{\frac{3}{4}}$

For Exercises 11 through 36, solve the following exponential equations using inspection. Check your answers.

11. $2^x = 8$
12. $4^x = 16$
13. $5^c = 3125$
14. $10^t = 10{,}000$
15. $\dfrac{1}{9} = 3^t$
16. $5^x = \dfrac{1}{25}$
17. $3^x = \dfrac{1}{81}$
18. $2^r = \dfrac{1}{8}$
19. $(-2)^d = -32$
20. $(-5)^x = -125$
21. $(-6)^w = 36$
22. $(-7)^x = 49$
23. $\left(\dfrac{1}{2}\right)^x = \dfrac{1}{16}$
24. $\left(\dfrac{1}{3}\right)^t = \dfrac{1}{9}$
25. $\left(\dfrac{1}{2}\right)^x = 16$
26. $\left(\dfrac{1}{3}\right)^t = 9$
27. $\left(\dfrac{1}{2}\right)^x = 32$
28. $\left(\dfrac{1}{5}\right)^t = 25$
29. $10^x = 1000$
30. $10^c = 100{,}000{,}000$
31. $10^x = 1$
32. $3^x = 1$
33. $(-5)^m = 1$
34. $(-7)^x = 1$
35. $\left(\dfrac{2}{3}\right)^t = 1$
36. $\left(\dfrac{4}{7}\right)^x = 1$

For Exercises 37 and 38, the students were asked to solve the given equation. Explain what the student did wrong. Give the correct answer.

37. $4^x = \dfrac{1}{16}$

> **Tom**
> $4^x = \dfrac{1}{16}$
> $4^x = 4^2$
> $x = 2$

38. $3^x = 27$

> **Shannon**
> $3^x = 27$
> $\dfrac{1}{3}(3^x) = \dfrac{1}{3}(27)$
> $x = 9$

39. The number of *L. acidophilus* bacteria in a sample can be estimated with the model
$$B(h) = 100(2^h)$$
where $B(h)$ is the number of bacteria after h hours.
  a. Use the model to estimate the number of bacteria after 8 hours.
  b. Use the model to estimate how long until there will be 800 bacteria.

40. The number of *E. coli* bacteria on a kitchen counter can be estimated with the model
$$B(n) = 45(2^n)$$
where $B(n)$ is the number of bacteria after n 20-minute intervals.
  a. Use the model to estimate the number of bacteria after 4 hours.
  b. Use the model to estimate how long until there will be 1440 bacteria.

41. The number of people who have heard the rumor that there are no finals this semester (not true) can be modeled by
$$R(h) = 5(3^h)$$
where $R(h)$ is the number of people who have heard the rumor after h hours.
  a. Use the model to estimate the number of people who have heard the rumor after 8 hours.
  b. Use the model to estimate how long until 405 people have heard the rumor.

42. The number of people in a school that have caught a cold can be modeled by
$$C(d) = 2(5^d)$$
where $C(d)$ is the number of people who have caught a cold after d days.
  a. Use the model to estimate the number of people who have caught this cold after 5 days.
  b. Use the model to estimate how long it takes for 250 people to catch this cold.

For Exercises 43 through 60, solve the exponential equations using inspection or trial and error. Check your answers.

43. $5(2^x) = 40$
44. $7(3^x) = 189$
45. $-2(5^c) = -250$
46. $-4(3^t) = -36$
47. $3^c + 5 = 32$
48. $2^x + 30 = 62$
49. $7^t + 8 = 57$
50. $5^m - 200 = 425$
51. $-4(5^m) - 9 = -109$
52. $-6(2^t) - 25 = -409$
53. $3(6^x) + 2(6^x) = 180$
54. $4(5^k) + 8(5^k) = 1500$
55. $10(2^x) - 7(2^x) = 48$
56. $5(3^x) - 4(3^x) = 81$
57. $7(2^x) = 48 + 4(2^x)$
58. $12(6^t) = -252 + 19(6^t)$
59. $13(7^w) + 20 = 5(7^w) + 2764$
60. $4(3^k) - 40 = -6(3^k) + 230$

For Exercises 61 through 68, solve the power equations. Check your answers.

61. $x^5 = 32$
62. $t^3 = 64$
63. $x^6 = 15625$
64. $p^4 = 2401$
65. $12w^4 = 7500$
66. $7t^8 = 45927$
67. $-7x^3 = 1512$
68. $-2.5m^7 = 320$

For Exercises 69 through 72, the students were asked to solve the given equation. Explain what the student did wrong. Give the correct answer.

69. $x^4 = 1296$
70. $n^8 = 0.05764801$

Warrick
$x^4 = 1296$
$(x^4)^{\frac{1}{4}} = 1296^{\frac{1}{4}}$
$x = 6$

Marie
$n^8 = 0.05764801$
$(n^8)^{\frac{1}{8}} = 0.05764801^{\frac{1}{8}}$
$n = 0.7$

71. $r^3 = 216$
72. $4p^5 = 972$

Frank
$r^3 = 216$
$(r^3)^{\frac{1}{3}} = 216^{\frac{1}{3}}$
$r = 72$

Amy
$4p^5 = 972$
$(4p^5)^{\frac{1}{5}} = 972^{\frac{1}{5}}$
$4p \approx 3.9585$
$p \approx 0.9896$

For Exercises 73 through 76, use the following formulas.
Volume of the sphere: $V = \frac{4}{3}\pi r^3$
Volume of a cube: $V = s^3$

73. **a.** Find the volume of an NBA basketball that has a radius of 12 cm.
    b. Find the radius of an International Basketball Federation basketball that has a volume of 8011 cm³.

74. **a.** Find the volume of a volleyball that has a radius of 10.5 cm.
    b. Find the radius of a mini basketball that has a volume of 2943 cm³.

75. Find the length of a side of a cube that has a volume of 91.125 in³.
76. Find the length of a side of a cube that has a volume of 13,284 cm³.

For Exercises 77 through 86, solve the power equations. Check your answers.

77. $3x^5 + 94 = 190$
78. $4m^7 + 20 = 380$
79. $5x^6 - 30 = 1475$
80. $6t^8 + 18 = 393234$
81. $96 - 24c^4 = 40$
82. $74 - 3x^{12} = -2457$
83. $\dfrac{4x^3 + 5}{8} = 63.125$
84. $\dfrac{2m^5 + 38}{16} = 32.75$
85. $\dfrac{7b^4 + 58}{8} = 1141.25$
86. $\dfrac{2.4x^6 - 19}{1000} = 282.3386$

For Exercises 87 through 102, determine whether the equation is a power equation or an exponential equation and solve. Check your answers.

87. $3^k = 81$
88. $6^b = 216$
89. $150r^5 = 1725$
90. $13p^7 = 101.4$
91. $3(4^x) + 20 = 212$
92. $8(3^x) + 74 = 722$
93. $45 - 3.5g^5 = -67$
94. $14 - 7.8x^5 = 1836.5$
95. $\dfrac{3.6h^8 - 56}{33} = 58.168$
96. $\dfrac{23.5h^4 - 75}{56} = 61.63$
97. $\left(\dfrac{1}{5}\right)^x = \dfrac{1}{125}$
98. $\left(\dfrac{1}{4}\right)^t = \dfrac{1}{16}$
99. $5x^6 + 20 = 3x^6 + 1478$
100. $\dfrac{1}{3}x^8 - 24 = \dfrac{5}{3}x^8 - 2239464$
101. $5\left(\dfrac{1}{2}\right)^x - \dfrac{2}{32} = 3\left(\dfrac{1}{2}\right)^x - \dfrac{3}{64}$
102. $7\left(\dfrac{1}{3}\right)^x + 142 = 2\left(\dfrac{1}{3}\right)^x + 547$

5.3 Graphing Exponential Functions

LEARNING OBJECTIVES
- Relate the parameters of an exponential equation to its graph.
- Sketch the graph of an exponential function by hand.
- Find the horizontal asymptote of an exponential function of the form $f(x) = a \cdot b^x + c$

In the next section, we will use the concept of an exponential function, the rules for exponents, and the solving techniques that we reviewed in Section 5.2 to develop a method for finding an exponential model. It is crucial that we recognize the graph of an exponential function because we always start with a scatterplot of data and choose a model from the shape the data takes. With this in mind, let's look at a few exponential graphs and describe their basic characteristics.

Recall from Section 5.1 that a basic exponential function has the form

$$f(x) = a \cdot b^x$$

where a, and b are real numbers, $a \neq 0$, $b > 0$, and $b \neq 1$.

Exploring Graphs of Exponentials

CONCEPT INVESTIGATION
What do a and b do to the graph?

Consider the standard form of an exponential function $f(x) = a \cdot b^x$. We are going to investigate one part of this function at a time. For each section of this investigation, consider how the graph changes as you change one of the constants in the function.

1. Graph the following exponential functions on the same calculator window. (Use the window Xmin = −7, Xmax = 7, Ymin = −10, and Ymax = 100. Remember to turn STATPLOT off.)

    a. $f(x) = 2^x$
    b. $f(x) = 5^x$
    c. $f(x) = 10^x$
    d. $f(x) = 22.5^x$

    Describe the basic shape that all of these graphs have.
    In these examples, we are considering the shape of the graph of a basic exponential function when the value of the base b is greater than 1. Describe how increasing the value of b affects the graph.

2. Graph the following exponential functions on the same calculator window. (Use the window Xmin = −7, Xmax = 7, Ymin = −10, and Ymax = 100.)

    a. $f(x) = \left(\dfrac{1}{2}\right)^x$
    b. $f(x) = \left(\dfrac{1}{5}\right)^x$
    c. $f(x) = (0.1)^x$
    d. $f(x) = (0.025)^x$

    In these functions, we are considering how having $0 < b < 1$ changes the graph of a basic exponential function. Recall that the base of an exponential function must be positive and cannot equal 1. Describe what a value of b less than 1 does to the graph.

3. Graph the following exponential functions on the same calculator window. (Use the window Xmin = -7, Xmax = 7, Ymin = -10, and Ymax = 100.)

   a. $f(x) = (2)^x$
   b. $f(x) = 5(2)^x$
   c. $f(x) = 10(2)^x$
   d. $f(x) = 8\left(\frac{1}{2}\right)^x$

   In these functions, we are considering how a positive a value changes the graph of a basic exponential function. Describe what a positive a-value does to the graph.

4. Graph the following exponential functions on the same calculator window. (Use the window Xmin = -7, Xmax = 7, Ymin = -100, and Ymax = 10.)

   a. $f(x) = -1(2)^x$
   b. $f(x) = -5(2)$
   c. $f(x) = -10(2)^x$
   d. $f(x) = -8\left(\frac{1}{2}\right)^x$

   In these functions, we are considering how a negative a-value changes the graph of a basic exponential function. Describe what a negative a-value does to the graph.

What's That Mean?

Asymptote

A straight line that is approached by a curve as a variable in the equation approaches infinity or negative infinity.

In this chapter, we see horizontal asymptotes for exponential functions. In later chapters, we will see vertical asymptotes for other functions.

Using all of the information from this concept investigation should allow us to sketch a graph of a basic exponential function and adjust any models that we are going to make. Note that the graph of an exponential function of the form $f(x) = ab^x$ never crosses the horizontal axis. The graph does get increasingly close to the horizontal axis but does not touch it. This is because an exponential function of this form can never equal zero. This makes the horizontal axis a **horizontal asymptote** for this graph. A horizontal asymptote is any horizontal line that a graph gets increasingly close to as the values of x get increasingly large positive or negative. Other functions that we will study may have horizontal asymptotes along with vertical asymptotes.

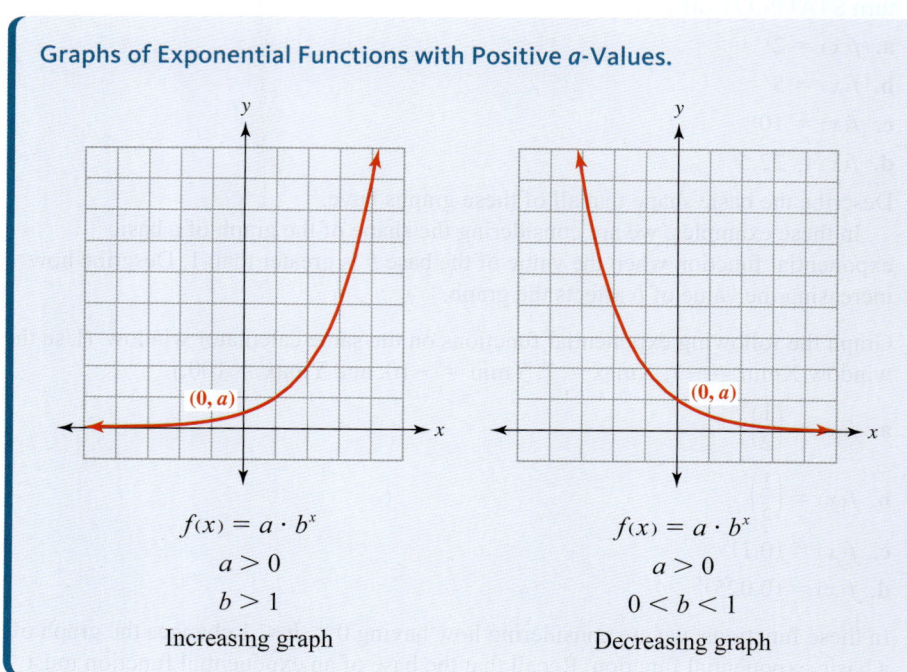

Graphs of Exponential Functions with Positive a-Values.

$f(x) = a \cdot b^x$
$a > 0$
$b > 1$
Increasing graph

$f(x) = a \cdot b^x$
$a > 0$
$0 < b < 1$
Decreasing graph

Graphs of Exponential Functions with Negative a-Values.

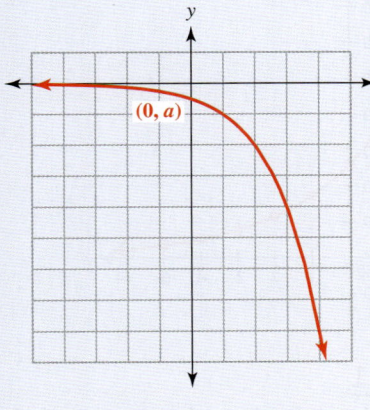

$f(x) = a \cdot b^x$
$a < 0$
$b > 1$
Decreasing graph

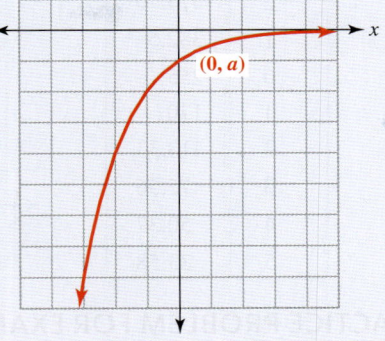

$f(x) = a \cdot b^x$
$a < 0$
$0 < b < 1$
Increasing graph

Example 1 Graphing exponential functions with positive *a*-values

Sketch the graph of the following functions by hand. Explain what the values of a and b tell you about the graph.

a. $f(x) = 5(3)^x$

b. $g(x) = 255\left(\dfrac{2}{3}\right)^x$

SOLUTION

a. In the function $f(x) = 5(3)^x$, $a = 5$, so the vertical intercept is $(0, 5)$, and the graph will be above the horizontal axis. The base $b = 3$ is greater than 1, so this graph is increasing, showing exponential growth. By trying a few input values, we get a table of points. Plotting these points, we get the following graph

x	f(x)
−1	1.67
0	5
1	15
2	45
3	135

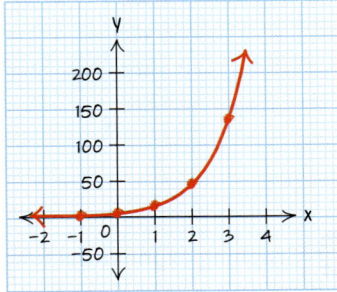

b. In the function $g(x) = 255\left(\dfrac{2}{3}\right)^x$, $a = 255$, so the vertical intercept is $(0, 255)$, and the graph will be above the horizontal axis. The base $b = \dfrac{2}{3}$ is less than 1, so this graph is decreasing, showing exponential decay. By trying a few input values, we get a table of points. Using these points, we get the following graph.

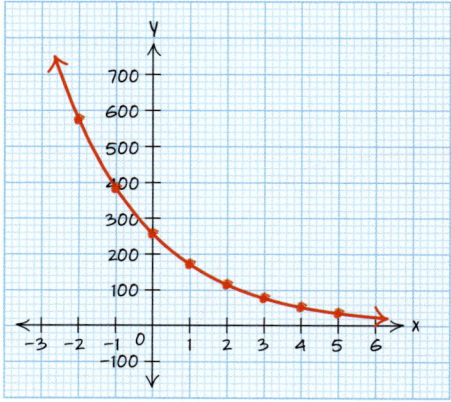

x	$g(x)$
−2	573.75
−1	382.5
0	255
1	170
2	113.33
3	75.56
4	50.37
5	33.58

PRACTICE PROBLEM FOR EXAMPLE 1

Sketch the graph of the following functions by hand. Explain what the values of a and b tell you about the graph.

a. $f(x) = 8(1.25)^x$ **b.** $g(x) = 300(0.5)^x$

From the concept investigation above, we saw that when a is a negative number, the graph is below the horizontal axis. The graph has actually been reflected (flipped) over the horizontal axis. The graph is a mirror image of the same graph with the opposite a-value. This reflection makes graphing an exponential function with a negative a-value easier. If a is negative, we can imagine the graph as if a were actually positive and then flip the graph over the horizontal axis. If we consider the function $f(x) = -10(2)^x$, we see that $a = -10$, so the graph will be below the horizontal axis. To start graphing, we could pretend that a is positive and sketch the graph. Since $b = 2$ is greater than 1, the graph would be exponential growth. Using a positive a-value and the fact that this is exponential growth gives us the blue dashed curve. Flipping the blue dashed curve over the horizontal axis gives us the final graph of $f(x) = -10(2)^x$.

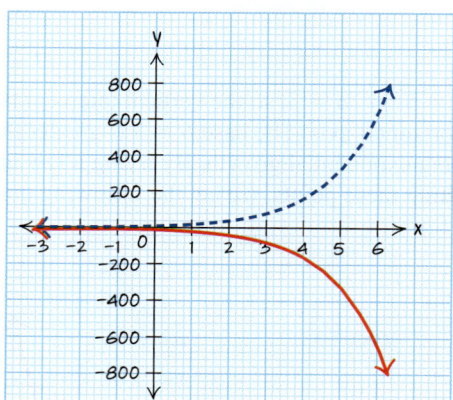

Example 2 Graphing exponential functions with negative a-values

Sketch the graph of the following functions by hand. Explain what the values of a and b tell you about the graph.

a. $h(x) = -3(5)^x$ **b.** $W(x) = -50(0.2)^x$

SOLUTION

a. In the function $h(x) = -3(5)^x$, $a = -3$, so the vertical intercept is $(0, -3)$, and the graph will be below the horizontal axis. The base $b = 5$ is greater than 1, so this graph should show exponential growth, but it has been reflected over the horizontal axis by the negative *a*-value. By trying a few input values, we get a table of points. Using these points, we get the following graph.

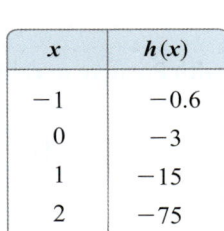

x	h(x)
−1	−0.6
0	−3
1	−15
2	−75

b. In the function $W(x) = -50(0.2)^x$, $a = -50$, so the vertical intercept is $(0, -50)$, and the graph will be below the horizontal axis. The base $b = 0.2$ is less than 1, so this graph should show exponential decay, but it has also been reflected over the horizontal axis by the negative a value. By trying a few input values, we get a table of points. Using these points, we get the following graph.

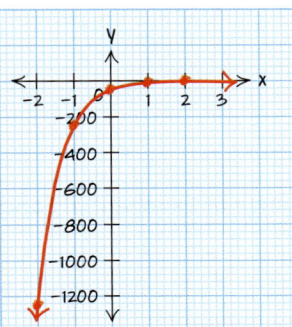

x	W(x)
−2	−1250
−1	−250
0	−50
1	−10
2	−2

PRACTICE PROBLEM FOR EXAMPLE 2

Sketch the graph of the following functions by hand. Explain what the values of *a* and *b* tell you about the graph.

a. $f(x) = -2(2.5)^x$ **b.** $h(x) = -145(0.5)^x$

Domain and Range of Exponential Functions

Exponential functions, like the linear and quadratic functions that we have already studied, have a pretty standard domain and range when the domain is not restricted by a real-world context. Consider a basic graph of an exponential function.

In this graph, notice that the function is valid for all input values but does not hit all outputs. This exponential growth function does not have any negative output values. A basic exponential function cannot equal zero, so that cannot be part of the range for the function. The curve will get as close to the horizontal axis as possible but will never touch it. Recall that for this graph, the horizontal axis is called a *horizontal asymptote*. Because this exponential function has only positive output values, the range will be only the positive real numbers.

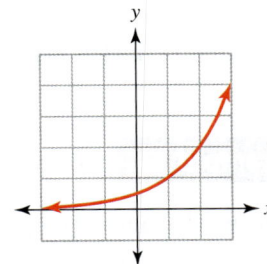

In general, the domain of an exponential function that has no real-world context will be all real numbers, or $(-\infty, \infty)$. The range for an exponential function of the form $f(x) = a \cdot b^x$ will either be all positive or all negative real numbers, $(-\infty, 0)$ or $(0, \infty)$. The range will be positive whenever the value of a is positive and will be negative whenever the value of a is negative. Making a quick sketch of the graph of the function can help to determine the correct domain and range.

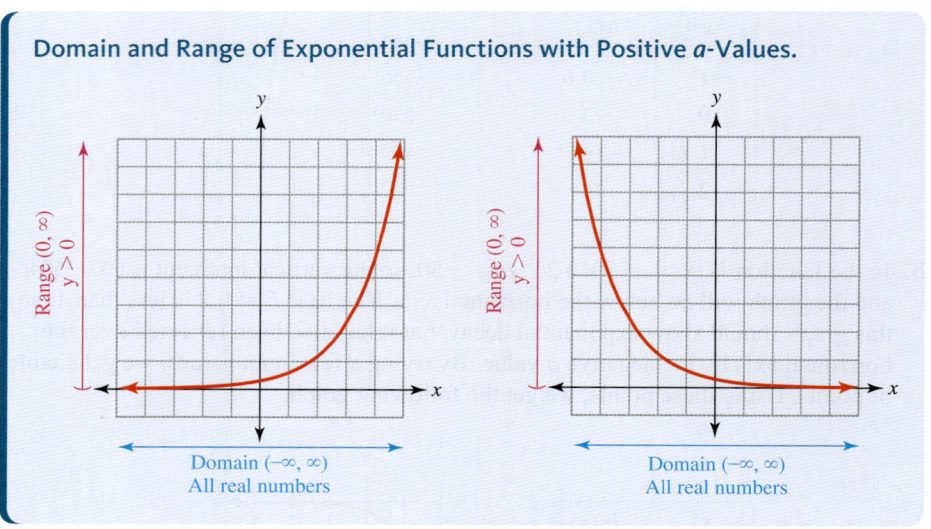

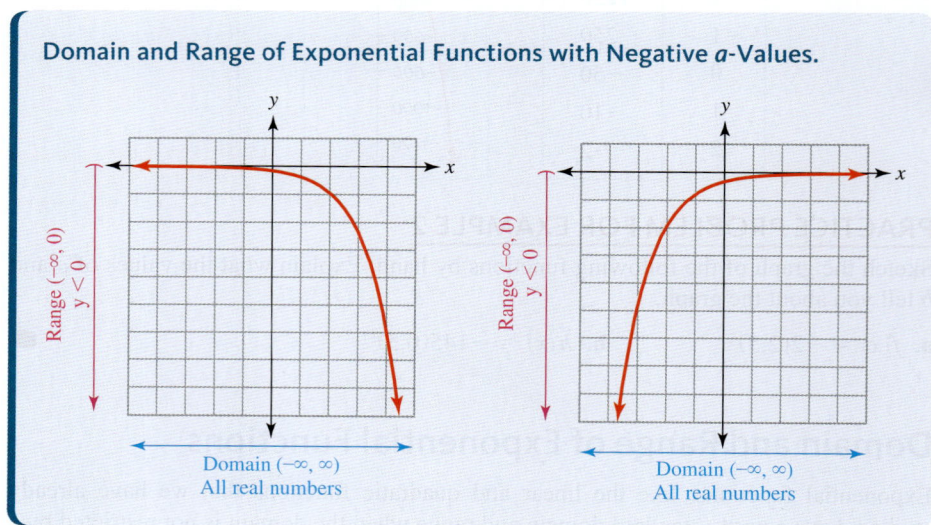

Remember that the domain and range in an application should take into consideration possible model breakdown in the situation. Model breakdown will limit the domain and range for the models that we find in later sections of this chapter.

Example 3 Domain and range of exponential functions

Find the domain and range of the following exponential functions.

a. $f(x) = 5(1.25)^x$

b. $g(x) = -30(0.7)^x$

c. $h(x) = 400(0.8)^x$

SOLUTION

a. The function $f(x) = 5(1.25)^x$ will be exponential growth because the base is larger than 1. It will also be positive, since a is positive. A quick sketch of the graph would give us the following:

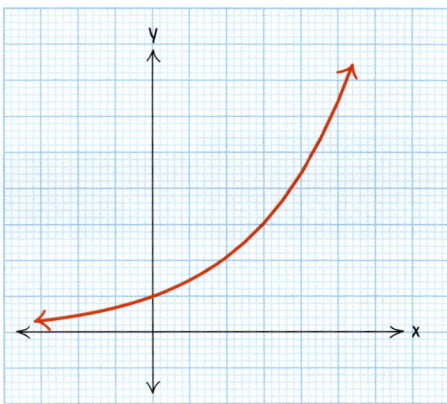

The sketch of the graph confirms the following domain and range.

Domain: All real numbers or $(-\infty, \infty)$

Range: $y > 0$ or $(0, \infty)$

b. The function $g(x) = -30(0.7)^x$ would be exponential decay because the base is less than 1, but a is negative, so the graph will be reflected over the horizontal axis. A quick sketch of the graph would give us the following:

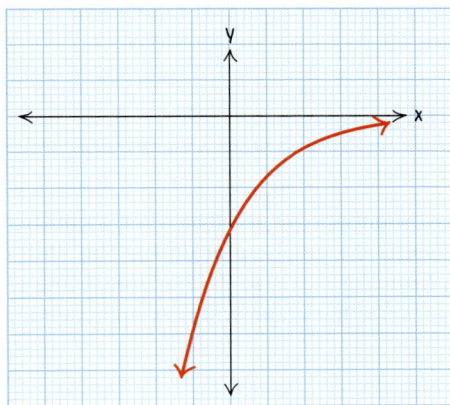

The sketch of the graph confirms the following domain and range.

Domain: All real numbers or $(-\infty, \infty)$

Range: $y < 0$ or $(-\infty, 0)$

c. The function $h(x) = 400(0.8)^x$ will be exponential decay, since the base is less than 1, and positive, since a is positive. With this information, we know that the function has the following domain and range.

Domain: All real numbers or $(-\infty, \infty)$

Range: $y > 0$ or $(0, \infty)$

PRACTICE PROBLEM FOR EXAMPLE 3

Find the domain and range of the following exponential functions.

a. $f(x) = -3(2.1)^x$

b. $h(x) = 250(0.5)^x$

Exponentials of the Form $f(x) = a \cdot b^x + c$

The exponential functions we have worked with so far have all been of the form $f(x) = a \cdot b^x$. Now we will consider exponentials of the form $f(x) = a \cdot b^x + c$. These exponentials have a constant that has been added to the basic exponential form that we have already graphed. Use the concept investigation to see what this constant does to the graph of an exponential.

CONCEPT INVESTIGATION

What does c do to the graph?

Consider the form of an exponential function $f(x) = a \cdot b^x + c$. We already know how a and b affect the graph. Now we will investigate how c affects the graph.

1. Graph the following exponential functions on the same calculator window. (Use the window Xmin = -7, Xmax = 4, Ymin = -2, and Ymax = 20. Remember to turn STATPLOT off.)

   a. $f(x) = 2^x$

   b. $f(x) = 2^x + 3$

   c. $f(x) = 2^x + 8$

   Describe how the value of c affects the graph.

2. Graph the following exponentials on your calculator and write the equation for their horizontal asymptote. Remember that a horizontal line has an equation of the form $y = k$.

   a. $f(x) = 1.5^x + 7$

   b. $g(x) = 200(0.5)^x + 10$

   c. $h(x) = -40(0.8)^x - 5$

   d. $f(x) = -3(4^x) - 7.5$

From this concept investigation, we see that when a constant is added to the basic exponential function, it shifts the graph up or down. This shift changes the horizontal asymptote of the graph from the x-axis to the horizontal line $y = c$. The range of the function will also change, since the graph will now approach the value of c rather than zero.

Example 4 — Horizontal aysmptote, domain, and range of exponential functions

Sketch a graph of the given exponential function. Write the equation for the graph's horizontal asymptote. Give the domain and range.

a. $f(x) = 3(2)^x + 4$

b. $g(x) = -400(0.5)^x - 80$

SOLUTION

a. In the function $f(x) = 3(2)^x + 4$, a is positive, so the graph will be above the horizontal axis. Substituting $x = 0$, we get $f(0) = 7$, so the vertical intercept is $(0, 7)$. The base $b = 2$ is greater than 1, so this graph is increasing, showing exponential growth. The constant $c = 4$ is added to the function, so the horizontal asymptote is $y = 4$. By trying a few input values, we get a table of points. Plotting these points, we get the following graph

x	$f(x)$
-1	5.5
0	7
1	10
2	16
3	28

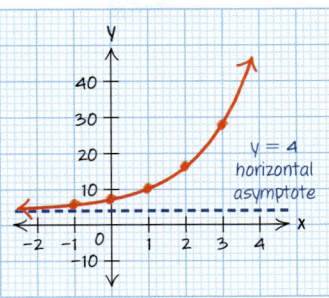

According to the graph and the information we collected the function has the following domain and range.

$$\text{Domain:} \quad \text{All real numbers} \quad \text{or} \quad (-\infty, \infty)$$
$$\text{Range:} \quad y > 4 \quad \text{or} \quad (4, \infty)$$

b. In the function $g(x) = -400(0.5)^x - 80$, a is negative, so the graph will be reflected over the horizontal axis. Substituting $x = 0$, we get $g(0) = -480$, so the vertical intercept is $(0, -480)$. The base $b = 0.5$ is less than 1, so this graph should show exponential decay, but it has been reflected over the horizontal axis because of the negative a-value. The graph has been shifted down 80 units because of the constant $c = -80$. The horizontal asymptote is $y = -80$. By trying a few input values, we get a table of points. Using these points, we get the following graph.

x	$f(x)$
-1	-880
0	-480
1	-280
2	-180
3	-130

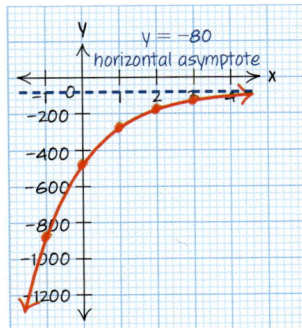

According to the graph and the information we collected, the function has the following domain and range.

$$\text{Domain:} \quad \text{All real numbers} \quad \text{or} \quad (-\infty, \infty)$$
$$\text{Range:} \quad y < -80 \quad \text{or} \quad (-\infty, -80)$$

PRACTICE PROBLEM FOR EXAMPLE 4

Sketch a graph of the given exponential function. Write the equation for the graph's horizontal asymptote. Give the domain and range.

a. $f(x) = 100(0.2)^x + 20$

b. $g(x) = -4(3)^x - 2$

5.3 Exercises

For Exercises 1 through 10, use the graphs of the exponential functions $f(x) = ab^x$ to give the following information.

  a. Is a positive or negative? Explain.
  b. Is b greater than or less than 1? Explain.
  c. Is the graph increasing or decreasing?
  d. Domain of the function
  e. Range of the function

1.
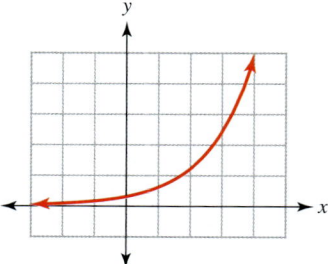

2.
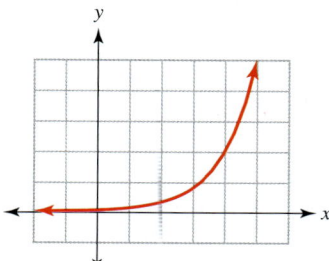

3.
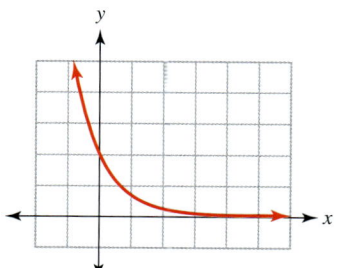

4.
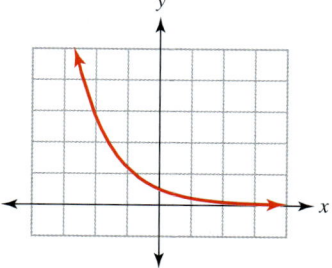

5.
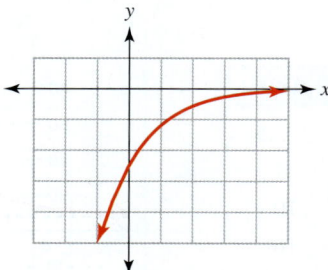

6.
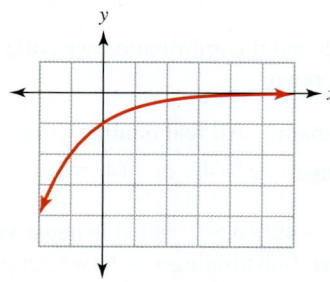

7.
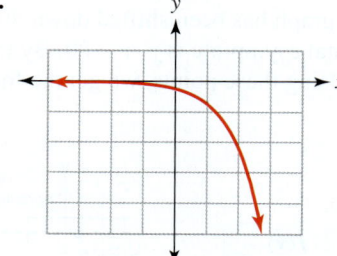

8.
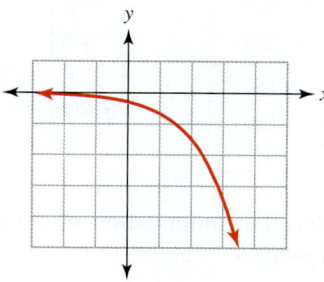

9.
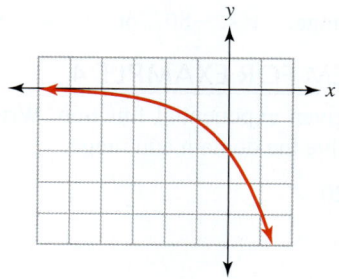

10.
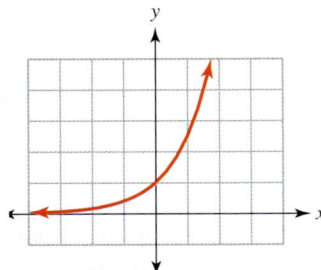

For Exercises 11 through 20, sketch the graph of the functions by hand. Explain what the values of a and b tell you about this graph. Label all intercepts. *Give the domain and range of the function.*

11. $f(x) = 7(2)^x$
12. $h(x) = 5(3)^x$
13. $g(x) = 3(1.2)^x$
14. $r(x) = 4(1.5)^x$
15. $h(x) = 12(1.4)^x$
16. $f(x) = 20(1.15)^x$
17. $p(t) = 140\left(\frac{1}{2}\right)^t$
18. $r(t) = 100\left(\frac{2}{5}\right)^t$
19. $f(x) = 250\left(\frac{1}{4}\right)^x$
20. $g(x) = 400\left(\frac{3}{4}\right)^x$

21. A sample of *E. coli* bacteria doubles every 20 minutes. The number of bacteria, $B(n)$, in the sample after n 20-minute intervals can be modeled by the function
$$B(n) = 4(2)^n$$
 a. Sketch a graph by hand of this model.
 b. What is the vertical intercept? What does it represent in regard to *E. coli* bacteria.

22. The number of *L. acidophilus* bacteria, $B(h)$, in a sample after h hours can be estimated with the model
$$B(h) = 100(2^h)$$
 a. Sketch a graph by hand of this model.
 b. What is the vertical intercept? What does it represent in regard to *L. acidophilus* bacteria.

23. The isotope radium-203 (^{203}Ra) has a half-life of 1 ms. The number of atoms, $N(t)$, of radium-203 remaining in a sample after t milliseconds can be modeled by the function
$$N(t) = 5000\left(\frac{1}{2}\right)^t$$
 a. Sketch a graph by hand of this model.
 b. What is the vertical intercept? What does it represent in regard to radium-203?

24. The isotope strontium-96 (^{96}Sr) has a half-life of 1 s. The number of atoms, $N(t)$, of strontium-96 remaining in a sample after t seconds can be modeled by the function
$$N(t) = 2000\left(\frac{1}{2}\right)^t$$
 a. Sketch a graph by hand of this model.
 b. What is the vertical intercept? What does it represent in regard to strontium-96?

For Exercises 25 through 40, sketch the graph of the functions by hand. Explain what the values of a and b tell you about this graph. Label all intercepts. *Give the domain and range of the function.*

25. $f(x) = -2(1.4)^x$
26. $h(x) = -6(1.25)^x$
27. $g(t) = -3(0.7)^t$
28. $f(x) = -8(0.6)^x$
29. $h(m) = 0.5(2.5)^m$
30. $m(t) = 0.75(4)^t$
31. $j(w) = -0.5(4)^w$
32. $g(x) = -10(2)^x$
33. $h(t) = -0.4(1.5)^t$
34. $m(t) = -0.5(1.75)^t$
35. $c(n) = 550\left(\frac{3}{4}\right)^n$
36. $f(t) = 200\left(\frac{3}{5}\right)^t$
37. $f(x) = -500\left(\frac{1}{5}\right)^x$
38. $g(x) = -1000\left(\frac{1}{10}\right)^x$
39. $f(x) = -700(0.95)^x$
40. $h(t) = -250(0.85)^t$

For Exercises 41 through 44, the students were asked to give the domain and range for the given function. Explain what the student did wrong. Give the correct answer.

41. $f(x) = 7(2.5)^x$
 Student answer:
 Domain: $(0, \infty)$, range: $(7, \infty)$

42. $f(x) = 10(1.25)^x$
 Student answer:
 Domain: $(-\infty, \infty)$, range: $[0, \infty]$

43. $f(x) = 200(0.3)^x$
 Student answer:
 Domain: $(-\infty, \infty)$, range: $(-\infty\ 0,)$

44. $f(x) = -40(1.5)^x$
 Student answer:
 Domain: $(-\infty, \infty)$, range: $(0, -\infty)$

For Exercises 45 through 48, use the graphs of the exponential functions $f(x) = ab^x + c$ to give the following information.
 a. *The equation for the horizontal asymptote*
 b. *Domain of the function*
 c. *Range of the function*

45.
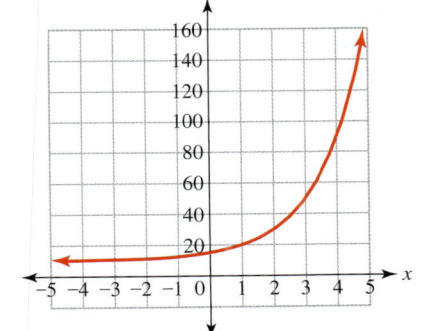

46.
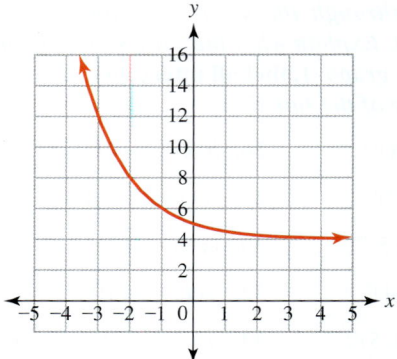

47.
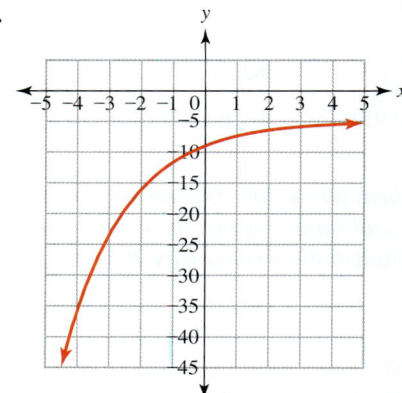

48.
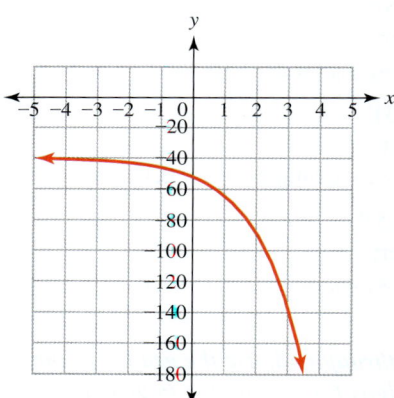

For Exercises 49 through 54, use the given exponential functions $f(x) = ab^x + c$ to give the following information.
  a. The equation for the horizontal asymptote
  b. Domain of the function
  c. Range of the function

49. $f(x) = 4(5)^x + 3$

50. $f(x) = 100(0.9)^x + 12$

51. $f(x) = 7(4)^x$

52. $f(x) = -20(3)^x$

53. $f(x) = -40(1.25)^x - 6$

54. $f(x) = -4500(0.3)^x - 100$

For Exercises 55 through 64, sketch the graph of the functions by hand. Give the equation for the horizontal asymptote. Label all intercepts. Give the domain and range of the function.

55. $f(x) = 200(0.4)^x + 30$

56. $g(x) = 800(0.1)^x + 15$

57. $h(x) = 20(1.5)^x + 3$

58. $f(t) = 4(5)^t + 10$

59. $h(t) = -3(2)^x - 12$

60. $f(x) = -9(1.75)^x - 20$

61. $g(x) = -300(0.9)^x - 50$

62. $h(x) = -450(0.3)^x - 27$

63. $f(x) = 750\left(\dfrac{2}{5}\right)^x + 25$

64. $g(x) = 1200\left(\dfrac{3}{7}\right)^x + 100$

For Exercises 65 through 74, sketch the graph of the functions by hand. Label all intercepts. Give the domain and range of the function.

65. $f(x) = 3x + 5$

66. $g(x) = \dfrac{2}{3}x - 6$

67. $f(x) = (x - 4)^2 + 3$

68. $h(x) = 2x^2 + 6x - 8$

69. $f(x) = 6(3)^x$

70. $g(x) = -20(2)^x$

71. $f(x) = 5 + 2(x + 7)$

72. $g(x) = 300\left(\dfrac{1}{2}\right)^x + 30$

73. $h(x) = \dfrac{1}{3}(x + 6)^2 - 75$

74. $f(x) = -4(1.8)^x - 9$

5.4 Finding Exponential Models

LEARNING OBJECTIVES
- Find an exponential equation from two points.
- Create an exponential model as appropriate.
- Determine the domain and range of an exponential function.

Finding Exponential Equations

The exponential form $y = a \cdot b^x$ requires values for a and b. This requirement is similar to the linear equation when we found m and b or the quadratic equation when we found a, h and k. For exponential equations, we will use a system of equations and a version of the elimination method to solve for b. This process will require us to use the solving techniques for power equations we learned in Section 5.2.

> **Steps to Finding Exponential Equations**
> 1. Use two points to write two equations using the form $f(x) = a \cdot b^x$.
> 2. Divide the two equations to eliminate the a's and solve for b.
> 3. Use b and one of the equations to find a.
> 4. Write the equation using the values of a and b you found.

Example 1 Finding an exponential equation

Write the exponential equation that passes through the given points.

a. $(2, 24)$ and $(5, 192)$ **b.** $(1, 900)$ and $(3, 100)$

SOLUTION

a. First, use the two given points to write two equations.

$$24 = ab^2 \qquad 192 = ab^5$$

Divide the two equations to eliminate the a's and solve for b.

$$\frac{192}{24} = \frac{ab^5}{ab^2} \quad \text{Divide the two equations. Use the quotient rule of exponents to reduce.}$$

$$8 = b^3 \quad \text{We are left with a power equation.}$$

$$8^{\frac{1}{3}} = (b^3)^{\frac{1}{3}} \quad \text{Raise both sides of the equation to the reciprocal exponent and simplify.}$$

$$2 = b$$

Use b and one equation to find a.

$$24 = ab^2 \quad \text{Substitute } b = 2 \text{ into one of the equations. Simplify and solve.}$$

$$24 = a(2)^2$$

$$24 = 4a$$

$$6 = a$$

Write the equation.

$$y = 6(2)^x$$

> **■ Connecting the Concepts**
>
> **Is that a system of equations?**
>
> When we write two exponential equations and use them to solve for a and b, we are solving a system of equations.
>
> Because the variables are being multiplied together division is needed to eliminate a variable. This is similar to the elimination method used in Chapter 2. The difference is that the variable terms were being added together so that we could eliminate a variable by adding the opposite.

b. Write two equations using the points given.

$$900 = ab^1 \qquad 100 = ab^3$$

Divide the two equations to eliminate a and solve for b.

$$\frac{100}{900} = \frac{ab^3}{ab}$$

$$\frac{1}{9} = b^2 \qquad \text{Divide the two equations and reduce, using the quotient rule of exponents.}$$

$$\pm\left(\frac{1}{9}\right)^{\frac{1}{2}} = (b^2)^{\frac{1}{2}} \qquad \text{Use reciprocal exponents to solve for } b. \text{ The even exponent means that we need the plus/minus symbol.}$$

$$\pm\frac{1}{3} = b$$

$$\frac{1}{3} = b \qquad \text{The base of an exponential must be positive.}$$

Use b and one of the equations to find a.

$$900 = a\left(\frac{1}{3}\right)^1$$

$$3(900) = 3\left(\frac{1}{3}a\right)$$

$$2700 = a$$

Write the equation.

$$y = 2700\left(\frac{1}{3}\right)^x$$

PRACTICE PROBLEM FOR EXAMPLE 1

Write the exponential equation that passes through the given points.

a. (3, 3375) and (5, 1898.4375)   **b.** (3, 576) and (6, 36864)

Finding an exponential equation is part of finding exponential models. We will pick two points to use from the scatterplot of the data.

Finding Exponential Models

The steps to model exponential data are similar to the steps in the other modeling processes we have done. We will use the following steps to model data that follow an exponential pattern. Notice that all the steps, except steps 4 and 5, are the same as those in linear and quadratic models.

■ **Connecting the Concepts**

Which point should I pick?

In finding an exponential model, step 4 has you pick two points. When deciding which points to pick, look for a point that is in the flat portion of the graph and another point in the steeper portion.

Modeling Steps for Exponentials (with no given vertical intercept)

1. Define the variables and adjust the data.
2. Create a scatterplot.
3. Select a model type.
4. **Exponential model:** Pick two points, write two equations using the form $f(x) = a \cdot b^x$, and solve for b.
5. Use b and one equation to find a.
6. Write the equation of the model using function notation.
7. Check the model by graphing it with the scatterplot.

Example 2   Finding an exponential model

Software development is undergoing a major change from a closed software development process, such as that at Microsoft, to a process that uses open source software, such as OpenOffice, Linux, and Mozilla Firefox. The total numbers of open source lines of code in millions are given in the following bar chart.

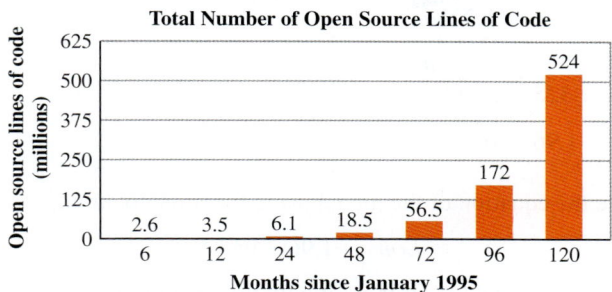

Source: Data estimated from information found at www.riehle.org.

a. Find an equation for a model for these data.

b. Estimate the total number of source lines of code 12 years after January 1995.

SOLUTION

a. **Step 1** Define the variables and adjust the data.

$L(m)$ = Total number of source lines of code in millions
m = Months since January 1995

These data do not have to be adjusted.

Step 2 Create a scatterplot.

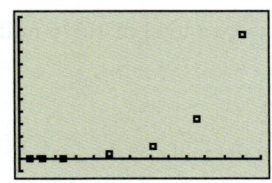

Step 3 Select a model type.
The graph is curved up much like a quadratic, but it seems to be flatter on the left side of the data than a quadratic should be. A quadratic model would go up to the left of a vertex and, therefore, would not be a good choice here. Because of this shape and type of growth, we should choose an exponential model.

Step 4 Exponential model: Pick two points, write two equations using the form $f(x) = a \cdot b^x$, and solve for b.
We can pick the first and last points of these data because they appear to fall along the exponential path. One is in the flat section of the curve, while the other is in the steeper part.

$(6, 2.6)$ and $(120, 524)$ — Use the two points to write two equations in standard form.

$2.6 = a \cdot b^6 \qquad 524 = a \cdot b^{120}$ — We now have a system of two equations with the variables multiplied together.

$\dfrac{524}{2.6} = \dfrac{a \cdot b^{120}}{a \cdot b^6}$ — Therefore, we divide the two equations and solve for b.

$201.54 \approx b^{114}$

$\pm (201.54)^{\frac{1}{114}} \approx (b^{114})^{\frac{1}{114}}$ — Use the reciprocal exponent to isolate b.

$\pm 1.048 \approx b$

$1.048 \approx b$ — The base of an exponential must be positive.

Step 5 Use b and one equation to find a.

$$2.6 = a(1.048)^6 \quad \text{Substitute } b \text{ into one equation and solve for } a.$$
$$2.6 \approx 1.325a$$
$$\frac{2.6}{1.325} \approx \frac{1.325a}{1.325}$$
$$1.96 \approx a$$

Step 6 Write the equation of the model using function notation.

$$L(m) = 1.96(1.048)^m$$

Step 7 Check the model by graphing it with the scatterplot.

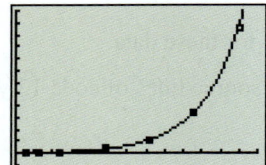

This model fits the data very well. It follows the flat pattern on the left side and then grows along with the data as we look to the right.

b. Twelve years after January 1995 is represented by $m = 144$, so we can substitute 144 into the model and get

$$L(144) = 1.96(1.048)^{144}$$
$$L(144) = 1676$$

Therefore, 12 years after January 1995, there were a total of 1,676 million source lines of code.

PRACTICE PROBLEM FOR EXAMPLE 2
Find an equation for a model for the data given in the table.

x	$f(x)$
2	4.8
3	7.0
5	14.7
6	21.4
8	44.9
10	94.5
12	198.7

Domain and Range for Exponential Models

In working with exponential models in real-world applications, the domain and range will again depend on the situation and require us to avoid model breakdown. Because exponential functions grow or decay very quickly, the domain will usually be very restricted to values close to the original data. In an application, always try to expand the domain beyond the data, but be cautious of output values that get too extreme for the situation given in the problem.

The range for an exponential model must come from the domain. Because the endpoints of the graph will always be its lowest and highest values, the range should always come from the endpoints of the domain just as they did for lines. Remember to always write the domain and range as lowest value to highest value.

Example 3 Domain and range of an exponential model

What would a reasonable domain and range be for the model of source lines of code found in Example 2 on page 452?

SOLUTION

In this case, the total number of source lines of code would be low before January 1995 and continue to grow for some time after 2005, so a reasonable domain would probably be [3, 144]. In this case, the lowest point is on the left endpoint of $m = 3$, and the highest point is on the right endpoint at $m = 144$. Substituting in these values for m gives us a range of [2.26, 1676]. Because the curve shows fairly fast growth, it will most likely not be able to sustain that kind of growth for long. Be careful when picking a domain. Do not extend too far in either direction, as this may result in model breakdown.

When finding an exponential model, having a vertical intercept to take from the data can simplify the modeling steps. When you have a vertical intercept to take from the data, you will not need two equations to solve for b. The output value of the vertical intercept will be the a-value for the model, and you can use another point and an equation to find b.

Modeling Steps for Exponentials (with a given vertical intercept)

1. Define the variables and adjust the data (if needed).
2. Create a scatterplot.
3. Select a model type.
4. **Exponential model:** Pick the vertical intercept and one other point.
5. Substitute the output value of the vertical intercept into a and use the other point to find b.
6. Write the equation of the model using function notation.
7. Check the model by graphing it with the scatterplot.

Connecting the Concepts

How do I recognize an intercept?

Remember when reading data that the intercepts will occur when one of the variables equals zero.

The vertical intercept will occur whenever the input variable is zero.

The horizontal intercept(s), if it exists, will occur whenever the output variable is zero.

In Example 4, we can see that the year zero is given, so the vertical intercept for this model is (0, 2).

Example 4 Finding an exponential model given a vertical intercept

The invasion of certain weeds can be a devastating problem for an agricultural area. The numbers of buckthorn plants found on a certain acre of land during a 4-year period are given in the table.

Years	Number of plants
0	2
0.5	6
1	16
1.5	45
2	126
2.5	355
3	1001

a. Find an equation for a model for these data.

b. Estimate the number of buckthorn plants after 4 years.

c. Give a reasonable domain and range for this model.

SOLUTION

a. Step 1 Define the variables and adjust the data.

$$P(t) = \text{Number of buckthorn plants on this acre}$$
$$t = \text{Years since the start of the invasion}$$

We do not need to adjust the data, since the data are already reasonable.

Step 2 Create a scatterplot.

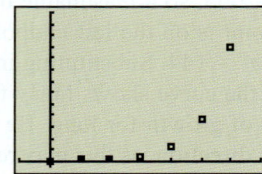 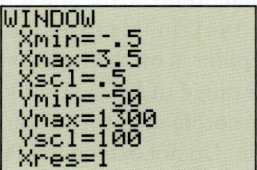

Step 3 Select a model type.

This graph is again flat on the left side and rises quickly to the right, so it seems to be exponential.

Step 4 **Exponential model:** Pick the vertical intercept and one other point.

The vertical intercept is given as (0, 2), and another point could be the last one at (3, 1001). If we choose two points that are too close together, we might not get a reasonable value for b. In this case, one of the last few points in the data should work well.

Step 5 Substitute the output value of the vertical intercept into a and use the other point to find b.

$$P(t) = 2b^t \quad \text{Substitute the vertical intercept into a.}$$
$$1001 = 2b^3 \quad \text{Substitute the other point in for } P(t) \text{ and } t.$$
$$500.5 = b^3 \quad \text{Isolate the variable factor.}$$
$$(500.5)^{\frac{1}{3}} = (b^3)^{\frac{1}{3}} \quad \text{Solve for } b \text{ using the reciprocal exponent.}$$
$$7.94 \approx b$$

Step 6 Write the equation of the model using function notation.

$$P(t) = 2(7.94)^t$$

Step 7 Check the model by graphing it with the scatterplot.

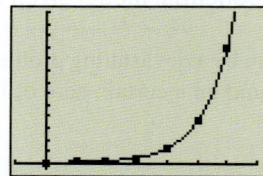

This seems to be a great fit for these data.

b. The fourth year can be represented as $t = 4$, so we get

$$P(4) = 2(7.94)^4$$
$$P(4) = 7949$$

This means that in year 4, there were approximately 7949 buckthorn plants on the acre of land.

c. It makes sense to start our domain at $t = 0$, since that is the start of the invasion. Because the amounts of land and water available to the weeds are limited, we should not extend the domain much beyond the data. The answer for part *b* seems like a possible limit to this situation, so we will set the end of the domain to be $t = 4$. Therefore, for the domain of [0, 4], we get a range of [2, 7949].

PRACTICE PROBLEM FOR EXAMPLE 4

Find an equation for a model for the data given in the table.

x	f(x)
0	145
3	74.2
5	47.5
8	24.3
10	14.6
14	6.4
15	5.1

Example 4 shows us the second option for finding an exponential model. This method is easier because we already have a value for a in the given vertical intercept, so we need to use only one equation to help us find b.

5.4 Exercises

For Exercises 1 through 4, set a window to fit the given data. Graph the given data and functions and choose which of the functions best fits the given data.

1. $f(x) = 10(1.4)^x$ $f(x) = 10(1.5)^x$
  $f(x) = 25(1.4)^x$ $f(x) = 25(1.5)^x$

x	f(x)
0	25
1	38
3	84
5	190
6	285

2. $f(x) = 6.8(2.1)^x$ $f(x) = 6.8(2.3)^x$
  $f(x) = 7.8(2.1)^x$ $f(x) = 7.8(2.3)^x$

x	f(x)
0	6.8
2	30
5	277.7
9	5401
11	23819

3. $f(x) = 850(0.7)^x$ $f(x) = 850(0.9)^x$
  $f(x) = 900(0.7)^x$ $f(x) = 900(0.9)^x$

x	f(x)
2	729
8	387
15	185
23	80
30	38

4. $f(x) = 200(0.6)^x$ $f(x) = 200(0.8)^x$
  $f(x) = 240(0.6)^x$ $f(x) = 240(0.8)^x$

x	f(x)
1	120
3	43
5	16
6	9
8	3

For Exercises 5 through 8, state which of a and/or b in the exponential form you would change to make the function fit the data better. Explain your answer.

5.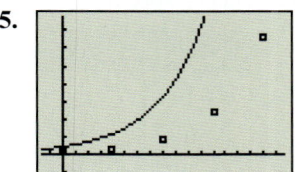

6.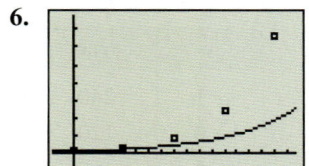

7.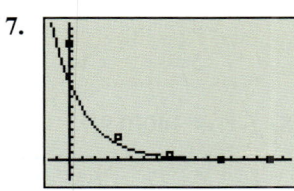

8.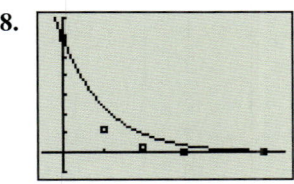

For Exercises 9 through 18, graph the data and the function on your calculator and adjust a and/or b to get an eye-ball best fit. Answers will vary.

9. $f(x) = 5.5(1.2)^x$

x	f(x)
0	3.5
2	5
5	9
8	15
10	22
14	45

10. $f(x) = 6(1.6)^x$

x	f(x)
−2	3
0	7
3	29
5	73
7	188
11	232

11. $f(x) = 100(0.45)^x$

x	f(x)
−1	173
0	78
1	35
3	7
4	3.2
6	0.65

12. $f(x) = 450(0.75)^x$

x	f(x)
−1	567
0	500
2	281
3	211
5	119
7	67

13. $f(x) = 8(1.5)^x$

x	f(x)
−2	5
0	8
1	10
3	18
5	30
8	65

14. $f(x) = 12(1.3)^x$

x	f(x)
−1	11
0	12
2	14.5
5	19
9	23
14	45

15. $f(x) = 120(0.4)^x$

x	f(x)
−1	240
0	120
1	60
3	15
4	7.5
6	2

16. $f(x) = 90(0.8)^x$

x	f(x)
−1	129
0	90
1	63
2	44
4	22
5	15

17. $f(x) = 6(1.4)^x$

x	f(x)
−1	4
0	5
2	7
3	8.5
5	12.5
10	31

18. $f(x) = 140(0.9)^x$

x	f(x)
−1	183
1	120
3	77
5	49
7	31
10	16

19. The growth of the Internet has outpaced anything most people had imagined. Data for the number of Internet hosts in the early 1990s is given in the chart.

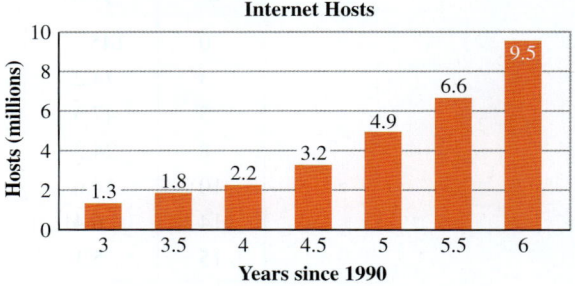

Source: Matthew Gray, Massachusetts Institute of Technology.

a. Find an equation for a model of these data.
b. Estimate the number of Internet hosts in 1997.
c. Give a reasonable domain and range for this model.
d. Estimate graphically when the number of Internet hosts reached 35 million.

20. In 1959, the United States started building intercontinental ballistic missiles (ICBMs) to act as a deterrent during the Cold War. The number of ICBMs in the U.S. arsenal is given in the chart.

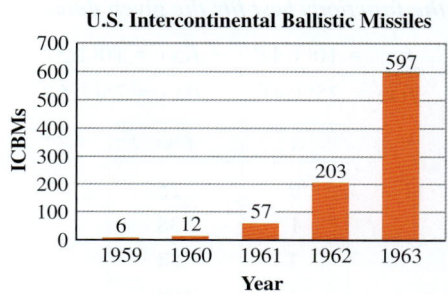

Source: Natural Resources Defense Council.

a. Find an equation for a model of these data.
b. Estimate the number of ICBMs in the U.S. arsenal in 1964.
c. Give a reasonable domain and range for this model.
d. Estimate numerically the year in which there would be 2500 ICBMs in the U.S. arsenal.

21. The number of nuclear warheads in the U.S. arsenal from 1945 to 1960 was growing at an exponential rate. The number of stockpiled warheads in the United States is given in the table.

Year	Stockpiled Warheads
1945	6
1947	32
1949	235
1951	640
1953	1436
1955	3057
1957	6444
1959	15,468

Source: Natural Resources Defense Council.

a. Find an equation for a model of these data.
b. Estimate the number of stockpiled warheads in 1950.
c. Give a reasonable domain and range for this model.
d. Estimate graphically the year in which the number of stockpiled warheads surpassed 50,000.

22. The number of weblogs has been growing very rapidly since early 2003. The estimated numbers of weblogs for years since January 2003 are given in the table.

Time (years)	Weblogs (millions)	Time (years)	Weblogs (millions)
0.5	0.75	2.5	10.4
1	1.7	3	24
1.5	2.8	3.5	44
2	6		

Source: Data estimated from information at www.sifry.com and Technorati.

a. Find an equation for a model of these data.
b. Estimate the number of weblogs in January 2008.
c. Give a reasonable domain and range for this model.
d. Estimate graphically the year in which the number of weblogs surpassed 100 million.

23. While running an experiment in physics class, a student tested the volume of water remaining in a cylinder after a small hole was made in the bottom. The data collected by the student are given in the table.

Time (seconds)	Volume (liters)	Time (seconds)	Volume (liters)
30	13.24	180	5.32
60	11.45	210	4.34
90	9.96	240	3.59
120	7.79	270	3.20
150	6.10	300	2.66

a. Find an equation for a model of these data.
b. Estimate the water remaining after 6 minutes.
c. Give a reasonable domain and range for this model.
d. Estimate numerically how long it took for there to be only 1 liter of water remaining in the cylinder.

24. One concern all people should share is the overuse of our natural resources due to larger populations. The amount of organic material in our soils is crucial to the growth of food and other plants. With overuse, the percentage of soil that is organic material decreases to dangerous levels. Data demonstrating this loss after years of overuse are given in the table.

Year	Percent of Organic Material in Soil
0	3
25	2.1
75	0.9
125	0.4
175	0.2

Source: Data estimated from information from the University of Wisconsin.

a. Find an equation for a model of these data.
b. Estimate the percent of soil that is organic material after 100 years of overuse.
c. Give a reasonable domain and range for this model.

25. After a significant rainfall event, the river gage level will typically decrease exponentially until it reaches its normal level. The following river gage levels were taken after such a rainfall event.

Days After Rainfall Event	River Gage Height (feet above normal)
0	14
1	8
2	4
3	2.4
4	1.25
5	0.7
6	0.4
7	0.25

a. Find an equation for a model of these data.
b. How far above normal will the gage height be 10 days after the rainfall event?
c. Give a reasonable domain and range for this model.

26. Students in a chemistry class were asked to run an experiment that would simulate the decay of an imaginary element. The students were given 100 dice to roll all at once and after each roll they would remove any die that showed a 5 on top. The students in one group collected the following data.

a. Find an equation for a model of these data.
b. How many dice will remain after 10 rolls?

Number of Rolls	Remaining Dice
0	100
1	84
2	69
3	55
4	48
5	42
6	33
7	28

460 CHAPTER 5 Exponential Functions

For Exercises 27 through 34, find an equation for a model for the given data. Give the domain and range for the model. Answers may vary.

27.
x	$f(x)$
0	4
2	7
5	16
7	28
11	85
13	148

28.
x	$f(x)$
0	7
3	38
4	67
9	1134
11	3514
13	10883

29.
x	$f(x)$
0	94
2	39
4	16
7	4
10	1
13	0.25

30.
x	$f(x)$
0	127
3	73
4	60
6	42
9	24
11	16

31.
x	$f(x)$
5	14
7	23
12	76
15	155
18	318
20	512

32.
x	$f(x)$
2	7
5	90
7	477
9	2522
11	13339
12	30680

33.
x	$f(x)$
−1	855.56
3	73.75
7	6.19
10	0.97
11	0.53
13	0.15

34.
x	$f(x)$
−3	1112.20
−1	256.25
2	28.34
5	3.13
8	0.35
11	0.04

For Exercises 35 and 36, decide whether the student found a reasonable model for the data provided. If not, explain what the student did wrong. Answers may vary.

The revenue for a company is increasing exponentially. Revenues for this company are given in the table.

Years Since 2010	1	3	5	6	8
Revenue (million $)	1.44	2.07	2.99	3.58	5.16

35.

Harold

Let

t = Years since 2010

$R(t)$ = Revenue for this company in millions of dollars

(1, 1.44)   (8, 5.16)

$1.44 = ab^1$ $5.16 = ab^8$

$$\frac{5.16}{1.44} = \frac{ab^8}{ab}$$

$3.58 \approx b^7$

$\frac{1}{7}(3.58) \approx \frac{1}{7}(b^7)$

$0.512 \approx b$

$1.44 = ab^1$

$1.44 = a(0.512)^1$

$\frac{1.44}{0.512} \approx a$

$2.81 \approx a$

$R(t) = 2.81(0.512)^t$

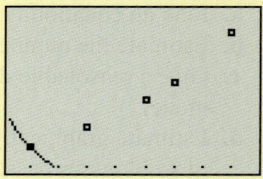

36.

Warren

Let

t = Years since 2010

$R(t)$ = Revenue for this company in millions of dollars

(1, 1.44)   (8, 5.16)

$1.44 = ab^1$ $5.16 = ab^8$

$$\frac{5.16}{1.44} = \frac{ab^8}{ab}$$

$3.58 \approx b^7$

$(3.58)^{\frac{1}{7}} \approx (b^7)^{\frac{1}{7}}$

$1.2 \approx b$

$1.44 = ab^1$

$1.44 = a(1.2)^1$

$1.44 - 1.2 \approx a$

$0.24 \approx a$

$R(t) = 0.24(1.2)^t$

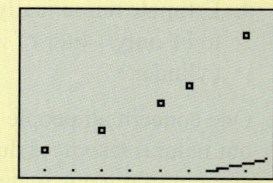

For Exercises 37 through 44, find an equation for a model for the given data. Give the domain and range for the model. Answers may vary.

37.

x	$f(x)$
0	-17
1	-21
4	-38
6	-56
10	-124
13	-226

38.

x	$f(x)$
0	-23
5	-57
9	-120
15	-354
25	-2190
40	-33805

39.

x	$f(x)$
0	-56
1	-25.2
3	-5.103
5	-1.033
7	-0.2093
9	-0.0424

40.

x	$f(x)$
0	-89
2	-46
5	-17.2
6	-12.4
8	-6.4
11	-2.4

41.

x	$f(x)$
-5	-84.25
-3	-129.5
1	-306.3
4	-583.96
8	-1380.61
10	-2122.82

42.

x	$f(x)$
-6	-3.66
-2	-16.17
2	-71.49
5	-217.93
7	-458.20
11	-2025.48

43.

x	$f(x)$
-4	-1129.95
-1	-282.54
3	-44.51
5	-17.67
7	-7.01
9	-2.78

44.

x	$f(x)$
-2	-28291.02
-1	-9053.13
2	-296.65
4	-30.38
6	-3.11
9	-0.10

For Exercises 45 through 50, decide whether the following situations are most likely linear, quadratic, or exponential. Explain your reasoning. Answers may vary.

45. The number of text messages since 2000.

46. The number of CDs sold since 1991.

47. The number of music downloads since 2000.

48. The growth of manga comics since 2005.

49. The population of New Hampshire since 1990.

50. The average temperature in Minnesota from September to May.

51. The approximate number of visitors to MySpace.com grew rapidly during its first months as a Website. Some growth milestones are listed in the table.

Months After Site Started	10	13	18	22	24	25
Visitors (millions)	1	2	5	10	15	20

Source: http://blogs.zdnet.com.

a. Find an equation for a model of these data.
b. Give a reasonable domain and range for this model.
c. Estimate graphically when MySpace.com reached 50 million visitors.

52. The approximate number of visitors to YouTube.com grew rapidly during its first months as a Website. Some growth milestones are listed in the table.

Months After Site Started	8	9	10	13	14	16
Visitors (millions)	1	2	5	10	15	20

Source: http://blogs.zdnet.com.

a. Find an equation for a model of these data.
b. Give a reasonable domain and range for this model.
c. Estimate graphically when YouTube.com reached 50 million visitors.

53. The population of Georgia is given in the table.

Years	2004	2005	2006	2007	2008
Population (millions)	8.91	9.09	9.32	9.52	9.69

a. Find an equation for a model of these data.
b. Give a reasonable domain and range for this model.
c. Estimate when the population of Georgia will reach 11 million.

54. The cost to produce netbooks is given in the table.

Netbooks Produced (hundreds)	10	15	20	25	30
Cost (thousand $)	563	750	863	900	860

a. Find an equation for a model of these data.
b. Give a reasonable domain and range for this model.
c. Estimate the number of netbooks that can be created with a budget of $800,000.

CHAPTER 5 Exponential Functions

For Exercises 55 through 60, find a linear, quadratic, or exponential model for the given data. Answers may vary.

55.

x	−7	−4	−2	0	3	5
f(x)	17	−10	2	38	137	233

56.

x	−3	−1	2	4	7	11
f(x)	−7	−9.8	−14	−16.8	−21	−26.6

57.

x	−2	−1	1	3	4	6
f(x)	163.3	114.3	56	27.4	19.2	9.4

58.

x	−2	0	1	3	5	6
f(x)	25.5	13.5	9	3	1	1.5

59.

x	−1	1	3	6	7	9
f(x)	0.1	0.5	0.9	1.5	1.7	2.1

60.

x	−1	1	3	4	6	7
f(x)	0.75	12	192	768	12288	49152

5.5 Exponential Growth and Decay Rates and Compounding Interest

LEARNING OBJECTIVES
- Define a growth rate.
- Show the relationship between growth rates and exponential functions.
- Use the formulas for compounding interest.

Exponential Growth and Decay Rates

Exponential growth and decay are often measured as a percentage change per year or other time period. Since a percentage is calculated by using multiplication, if the same percentage growth or decay occurs over several years, we will have repeated multiplication and, thus, an exponential pattern. The percentage change is often called a **growth rate**. Use the following concept investigation to compare two different growth patterns.

CONCEPT INVESTIGATION
What pattern grows faster?

In this investigation, we will consider two salary options for a company's employees.

Option 1: Starting salary of $50,000 with a $2000 raise each year
Option 2: Starting salary of $50,000 with a 4% raise each year

1. Complete the following table of calculations.

Years on the Job	Option 1 Salary (dollars)	Option 2 Salary (dollars)
0	50,000	50,000
1	50,000 + 2,000 = 52,000	50,000 + 50,000(0.04) = 52,000
2	52,000 + 2,000 = 54,000	52,000 + 52,000(0.04) = 54,080
3		
4		
5		
6		

■ Connecting the Concepts

How does our salary grow?

When calculating both of these salary options, we need to remember to use the previous year's salary each time we do the calculation.

With option 2, we take the previous year's salary and add the 4% growth to it. We are calculating the 4% growth by multiplying the previous year's salary by 0.04. Because the 4% is based on a new amount each year, the growth is larger each year.

2. According to these calculations, which of the salary options is best for the employees?

   Option 1 has a linear growth pattern because we are adding the same amount with each change in the number of years. Option 2 is not linear, since we are not adding the same amount with each change in the years. This second option is actually exponential because we are multiplying by the same percentage each year.

3. Find an equation for a linear model for the option 1 salary.

4. Find an equation for an exponential model for the option 2 salary. (You can use the modeling process from Section 5.4.)

5. Graph both salary option models on the same calculator screen and describe what happens over a 40-year career.

When you work with a percentage growth, the growth rate can be seen in the base of the exponential function. Consider the next example and the exponential model you found in the concept investigation to see how the growth rate becomes a part of the base.

Example 1 Exponential growth rate in nature

An ant colony can grow at a very rapid rate. One such colony studied by Deborah Gordon started with about 450 worker ants and grew 79% per year.
Source: Information taken from "The Development of Organization in an Ant Colony" by Deborah M. Gordon, American Scientist, Jan–Feb 1995.

a. Find an equation for a model for the number of worker ants in the colony.

b. Estimate the number of worker ants in the colony after 5 years.

c. In the sixth year of the study, the colony had about 10,000 worker ants. Does your model predict this many worker ants? Explain possible reasons why or why not.

SOLUTION

a. To find an equation for a model, either we need two points to work with or we need to see a pattern to follow to build the model. We can start by building a small table of values from the information given. Because the colony grew by 79% each year, we can find the size of the colony by calculating the growth and adding that to the original population.

Let's start by defining variables.

$A(t)$ = The number of worker ants in the colony
t = Years after the study began

t	Calculation	$A(t)$
0		450
1	$450 + 0.79(450)$	805.5
2	$805.5 + 0.79(805.5)$	1441.8
3	$1441.8 + 0.79(1441.8)$	2580.9

The pattern here might be hard to see, so we will use two points and find the model. Since the first point $(0, 450)$ is a vertical intercept, we will use it to simplify the process.

$$(0, 450) \quad \text{and} \quad (2, 1441.8)$$
$$450 = ab^0 \qquad 1441.8 = ab^2$$
$$a = 450 \qquad 1441.8 = ab^2$$
$$1441.8 = 450\, b^2$$
$$\frac{1441.8}{450} = \frac{450\, b^2}{450}$$
$$3.204 = b^2$$
$$\pm\sqrt{3.024} = b$$
$$\pm 1.79 \approx b$$

Therefore, our model will be $A(t) = 450(1.79)^t$.

b. After 5 years, $t = 5$, so we get

$$A(5) = 450(1.79)^5$$
$$A(5) \approx 8269.5$$

Therefore, after 5 years, there will be about 8270 worker ants in the colony.

c. After 6 years, $t = 6$, so we get

$$A(6) = 450(1.79)^6$$
$$A(6) \approx 14802.4$$

Our model gives a much higher number of ants than the actual study, which seems to indicate that the colony's growth slowed. This might have occurred because of a limit in food or space available.

From this example and the model found in the concept investigation, we can see that the growth rate becomes part of the base of the exponential. In the concept investigation, the growth rate of 4% = 0.04 became part of the 1.04 base of the exponential model. In Example 1, we see the 79% = 0.79 also become part of the 1.79 base for the exponential model. In both cases, we see that there is a 1 included in the base along with the growth rate. This 1 is there to account for the original population in the calculations, while the growth rate accounts for the amount of growth each year. With these examples, we can see that the base of the exponential function can be represented by $b = 1 + r$, where r is the growth rate as a decimal. This same base works for exponential decay, except that the rate is considered negative, so the base will end up being smaller than 1.

> **DEFINITION**
>
> **Growth or decay rate**   The percentage change in a quantity per 1 unit of time is called the growth or decay rate r.
>
> $$b = 1 + r$$
> $$r = b - 1$$
>
> where r is a percentage written as a decimal and b is the base of an exponential function of the form $f(x) = a \cdot b^x$.

Using the relationship between a growth rate and the base of an exponential function, we can find a growth rate from a given function. If we are given a starting value for the output and a growth rate, we can also write an exponential model. Since a is the starting value, or vertical intercept, we can easily find a and b to write the model. In Example 1, we knew that we had a starting population of 450 ants and a growth rate of 79%, so we could have written the model $A(t) = 450(1.79)^t$ without doing all the work, using the two points.

Example 2   Exponential growth rate in a recovering population

The California sea otter population from 1914 to 1975 was growing at an exponential rate. The population during these years could be modeled by

$$O(t) = 31.95(1.057)^t$$

where $O(t)$ represents the number of California sea otters t years since 1900.
Source: Model derived from data from the National Biological Service.

a. Use this model to predict the number of California sea otters in 1960.

b. According to this model, what was the growth rate during this time period?

SOLUTION

a. 1960 is represented by $t = 60$, so we get $O(60) = 889$. Thus, in 1960, there were approximately 889 California sea otters.

b. Since the base is 1.057 the growth rate can be found by using the formula $r = b - 1$.

$$r = 1.057 - 1$$
$$r = 0.057$$

Therefore, the California sea otter population was growing at a rate of 5.7% per year.

PRACTICE PROBLEM FOR EXAMPLE 2

According to the CIA World Factbook 2008, the population of Australia can be modeled by

$$P(t) = 19(1.008)^t$$

where $P(t)$ is the population of Australia in millions, t years since 2000.
Source: Model based on information found in the CIA World Factbook 2008.

a. Use this model to estimate the population of Australia in 2010.

b. According to this model, what is the growth rate of Australia's population?

Example 3 Finding an exponential model given a growth rate

The Gross Domestic Product (GDP) of Germany in 2005 was approximately 2.9 trillion U.S. dollars (US$) and has been growing by a rate of about 2.5% per year.
Source: Estimated from the CIA World Factbook and www.data.un.org.

a. Find an equation for a model for the GDP of Germany.

b. Use your model to estimate the GDP of Germany in 2010.

SOLUTION

a. First, we will define our variables.

$G(t)$ = Gross Domestic Product of Germany in trillion US$.
t = Time in years since 2005

Since we know that the GDP in 2005 is 2.9 trillion US$, the value of a in our model will be 2.9. The growth rate is 2.5%, so we can find the base b using the formula.

$$b = 1 + 0.025$$
$$b = 1.025$$

Therefore, our model is $G(t) = 2.9(1.025)^t$.

b. 2010 would be $t = 5$, so we get

$$G(5) = 2.9(1.025)^5$$
$$G(5) \approx 3.281$$

In 2010, the German GDP will be about 3.28 trillion US$.

PRACTICE PROBLEM FOR EXAMPLE 3

A survey of Boulder, Colorado, residents asked about the optimal size for growth. The results of this survey stated that most residents thought that a growth in population at a rate of 10% per year was desirable. In the year 2000, there were approximately 96,000 people in Boulder.
Source: Census 2000.

a. Find an equation for a model for the population of Boulder, Colorado, if the growth rate is 10% per year.

b. Use your model to estimate the population of Boulder, Colorado, in 2002 to see what the population of Boulder would be if the 10% growth rate had been achieved.

Example 4 — Finding an exponential decay model given a growth rate

In 2008, South Africa had a population of about 43.8 million, but that population was estimated to be decreasing by approximately 0.5% per year.
Source: CIA World Factbook 2008.

a. Find an equation for a model for the population of South Africa.

b. Use your model to estimate the population of South Africa in 2020.

SOLUTION

a. Since the population for 2008 is given, we will use that as our starting year. We will start by defining the variables.

$$P(t) = \text{Population of South Africa in millions}$$
$$t = \text{Time in years since 2008}$$

Because we have the starting population, we know that $a = 43.8$. The population is decaying, so the growth rate will be represented by a negative number. Therefore, $r = -0.005$, and we have a base of

$$b = 1 - 0.005$$
$$b = 0.995$$

Using these values for a and b, we have the following model.

$$P(t) = 43.8(0.995)^t$$

b. The year 2020 will be represented by $t = 12$, so we get

$$P(12) \approx 41.24$$

Therefore, the population of South Africa in 2020 will be about 41.24 million.

PRACTICE PROBLEM FOR EXAMPLE 4

In 2008, Trinidad and Tobago had a population of about 1 million, but that population was estimated to be decreasing by approximately 0.9% per year.
Source: CIA World Factbook 2008.

a. Find an equation for a model for the population of Trinidad and Tobago.

b. Use your model to estimate the population of Trinidad and Tobago in 2020.

Compounding Interest

Exponential equations are found in all kinds of scientific and natural situations as well as in a very common business situation: compounding interest. The idea of compounding is that the interest paid to you by a bank for a savings account or certificate of deposit is often compounded daily. Interest that you pay on a loan or credit card balance is also often compounded daily. In the past, many banks would compound interest for savings accounts only annually or monthly to reduce the amount of interest paid to the customers. Understanding the function that is used for compounding interest problems is useful when one is faced with making decisions in many business and financing situations.

For compounding interest problems, we use the formula

$$A = P\left(1 + \frac{r}{n}\right)^{nt}$$

where the variables represent the following.

$A = $ Amount in the account
$P = $ Principal (the amount intially deposited)
$r = $ Annual interest rate written as a decimal
$n = $ Number of times the interest is compounded in one year
$t = $ Time the investment is made for, in years

> **What's That Mean?**
>
> **Number of compounds**
>
> The most common interest rates today are compounded daily or annually. Here is a list of some different annual compoundings and how many compounds occur in 1 year.
>
> | | |
> |---|---|
> | Annually | $n = 1$ |
> | Semiannually | $n = 2$ |
> | Quarterly | $n = 4$ |
> | Monthly | $n = 12$ |
> | Weekly | $n = 52$ |
> | Daily | $n = 365$ |

To use this formula, you need to identify each quantity and substitute it into the appropriate variable. Then be sure to use the order of operations when calculating the result.

CONCEPT INVESTIGATION

What is the difference between simple and compound interest?

In this investigation, we will compare two investments: one that earns simple interest and one that earns annually compounded interest. With simple interest, we use the formula $A = P + Prt$ to calculate the account balance A after P dollars have been invested for t years. With annually compounding interest, we will use the formula

$$A = P\left(1 + \frac{r}{n}\right)^{nt}$$

and n will be 1, since the interest will compound only once a year. Since $n = 1$, the formula simplifies to $A = P(1 + r)^t$.

1. If we assume a $100 investment earning 10% interest, complete the following table of calculations.

Years	Simple Interest	Interest
0	100	100
1	$100 + 100(0.10)(1) = 110$	$100(1 + 0.10)^1 = 110$
2		
5		
10		
20		
50		

2. How do simple interest and compounding interest compare?
3. What happens when you increase the number of compounding periods per year?

Example 5 Using the compounding interest formula

If $5000 is invested in a savings account that pays 3% annual interest compounded daily, what will the account balance be after 4 years?

SOLUTION

First, identify each given quantity and decide what we are solving for.

$5000 is the initial deposit, so it is represented by $P = 5000$.
3% is the annual interest rate, so it is represented by $r = 0.03$.
"Compounded daily" tells us that there will be 365 compounds in one year, so we have $n = 365$.
Four years is the time the investment is made for, so it is represented by $t = 4$.

We are asked to find the amount in the account, so we need to solve for A. All of this gives us the following equation.

$$A = 5000\left(1 + \frac{0.03}{365}\right)^{365(4)}$$

$$A \approx 5000(1 + 0.000082192)^{1460}$$

$$A \approx 5000(1.127491292)$$

$$A \approx 5637.46$$

After 4 years the account will have a balance of $5637.46.

Using Your TI Graphing Calculator

When entering the compound interest formula into your graphing calculator, you need to be very careful to use parentheses when neccessary.

Notice the extra parentheses around the exponent. You may want to do this calculation in pieces so that you don't miss any part.

Calculate the exponent first. Then calculate the base for the exponent. Raise the base to the exponent you found. Multiply by the principal.

PRACTICE PROBLEM FOR EXAMPLE 5

An initial deposit of $30,000 is placed in an account that earns 8% interest. Find the amount in the account after 10 years if the interest is compounded:

a. monthly **b.** weekly **c.** daily

One other type of compounding growth is discussed in some areas of business and finance. Continuous compounding is the idea that the interest is always being compounded on itself, resulting in a higher return. In dealing with this type of problem, the compounding interest equation that we have been using will no longer work, since n would have to be infinity. Use the following concept investigation to discover what happens to this formula when n goes to infinity.

CONCEPT INVESTIGATION

What about continuous compounding?

In this investigation, we will consider an investment of $1 at 100% interest for 1 year. Let the number of compounds per year increase toward infinity.

1. Complete the following table using your calculator. If you use the table feature, you will need to move your cursor to the Y1 column to see more decimal places.

n	$A = 1\left(1 + \dfrac{1}{n}\right)^n$
1	2
100	2.7048138
1,000	
10,000	
100,000	
1,000,000	
5,000,000	

2. Above the division button 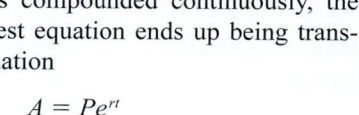 is the number e. Press and write down the number you get.

3. How does the number e compare with the number you found as n went toward infinity?

The number e is a special number that occurs often in nature and science as well as in continuously compounding interest situations.

■ What's That Mean?

Pi and e

Most people know the number π as approximately 3.14 and that it is actually a number whose decimal expansion is infinite and nonrepeating. The number e is also an infinite and nonrepeating decimal that we find in many applications.

DEFINITION

The number e An irrational number represented by the letter e.

$$e \approx 2.718281828$$

If we look at the graph of the following exponential functions, we can see that the number e lies between 2 and 3.

For now, we will use the number e for continuously compounding interest problems, and we will see it again in the next chapter.

When interest is compounded continuously, the compounding interest equation ends up being transformed into the equation

$$A = Pe^{rt}$$

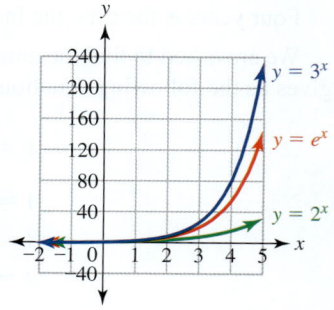

SECTION 5.5 Exponential Growth and Decay Rates and Compounding Interest

where A is the account balance after the principal P has been invested at an annual rate r compounded continuously for t years. In this equation, $e \approx 2.718281828$, and we use the e^x key [LN] on the calculator to perform the calculations.

Example 6 Continuously compounding interest

If $2000 is invested in a savings account that pays 5% annual interest compounded continuously, what will the account balance be after 7 years?

SOLUTION

We know the following values,

$$P = 2000 \qquad r = 0.05 \qquad t = 7$$

A is the missing quantity that we are being asked to solve for.

$$A = 2000 \, e^{0.05(7)}$$

$$A \approx 2838.14$$

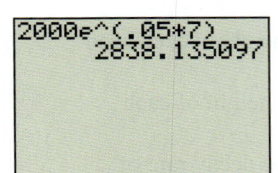

Therefore, after 7 years, this account will have $2,838.14 in it.

PRACTICE PROBLEM FOR EXAMPLE 6

An initial deposit of $80,000 is placed in an account paying 5% annual interest compounded continuously. What will the account balance be after 12 years?

5.5 Exercises

For Exercises 1 through 10, find the growth or decay rate of the given exponential model.

1. $f(x) = 45(1.03)^x$
2. $f(x) = 20(1.07)^x$
3. $g(x) = 130(1.25)^x$
4. $h(x) = 3(1.46)^x$
5. $f(x) = 2(3.5)^x$
6. $g(x) = 7(4.75)^x$
7. $h(x) = 25(0.95)^x$
8. $f(x) = 150(0.92)^x$
9. $f(x) = 750(0.36)^x$
10. $g(x) = 5000(0.001)^x$

11. The white-tailed deer population in the northeastern United States has been growing since the early 1980s. The population of white-tailed deer can be modeled by

$$W(t) = 1.19(1.08)^t$$

where $W(t)$ represents the number of white-tailed deer in millions in the Northeast t years since 1980.
Source: Model derived from data from the National Biological Service.

   a. Estimate the white-tailed deer population in 1990.
   b. According to this model, what is the growth rate of the white-tailed deer population?
   c. What could cause this population to stop growing at this rate?

12. According to United Nations records, the world population in 1975 was approximately 4 billion. Statistics indicate that the world population since World War II has been growing exponentially. If we assume exponential growth, the world population can be modeled by

$$P(t) = 4(1.019)^t$$

where $P(t)$ is the world population in billions and t is the time in years since 1975.

   a. Estimate the world population in 2015.
   b. According to this model, what is the growth rate of the world population?

13. The population of the Virgin Islands can be modeled by

$$V(t) = 109.8(0.99939)^t$$

where $V(t)$ is the population of the Virgin Islands in thousands t years since 2010.
Source: CIA World Factbook 2010.

   a. Estimate the population of the Virgin Islands in 2015.
   b. According to this model, what is the decay rate of the Virgin Islands' population?

14. Dr. Arnd Leike, a professor of physics at the University Muenchen in Germany, won a 2002 Ig Nobel Prize in Physics for his investigation into the "exponential decay" of a beer's head (the foam on top when poured). After testing several beers for the rate of decay of the head, he came to the conclusion that a beer could be

identified by its unique head decay rate. The following functions are for three of the beers that Dr. Leike tested.

$$E(s) = 16.89(0.996)^s$$
$$B(s) = 13.25(0.991)^s$$
$$A(s) = 13.36(0.993)^s$$

where $E(s)$ represents the height of the head of an Erdinger beer, $B(s)$ is the height of the head of a Budweiser Budvar, and $A(s)$ is the head height of an Augustinebrau. All head heights are in centimeters s seconds after being poured.
Source: European Journal of Physics, Volume 23, 2002.

a. Use these models to estimate the height of each head at the end of the pour.
b. Use these models to estimate the height of each head after 200 seconds.
c. Use these models to determine the decay rate of each head of beer.

15. Write an exponential model for a population that started with 40 animals and is growing at an annual rate of 3% per year.

16. Write an exponential model for a population that started with 10 animals and is growing at an annual rate of 1.75% per year.

17. Write an exponential model for a population that started with 200 animals and is shrinking at an annual rate of 2% per year.

18. Write an exponential model for a population that started with 520 animals and is shrinking at an annual rate of 6.2% per year.

19. A 13-year research study found that the humpback whale population off the coast of Australia was increasing at a rate of about 14% per year. In 1981, the study estimated the population to be 350 whales.

a. Find an equation for a model for the humpback whale population.
b. Give a reasonable domain and range for the model.
c. Estimate the humpback whale population in 1990.

20. The population of Egypt has been growing at a rate of 1.68% per year. In 2008, the population of Egypt was about 81.7 million.
Source: CIA World Factbook 2008.

a. Assuming exponential growth, find an equation for a model for the population of Egypt.
b. Estimate Egypt's population in 2020.

21. The population of Denmark has been growing at a natural rate of 0.295% per year. In 2008, the population of Denmark was about 5.5 million.
Source: CIA World Factbook 2008.

a. Assuming exponential growth, find an equation for a model for the population of Denmark.
b. Estimate Denmark's population in 2020.

22. The sandhill crane is a migratory bird in the United States that is being watched for its population trends. The National Biological Service estimates the population's present trend to be a growth of approximately 4.3% per year. In 1996, the estimated population was 500,000.
Source: National Biological Service.

a. Assuming that this trend continues, find an equation for a model for the sandhill crane population.
b. Estimate the population in 2002.

For Exercises 23 through 30, identify whether you should use the compounding interest formula

$$A = P\left(1 + \frac{r}{n}\right)^{nt}$$

or the continuously compounding interest formula

$$A = Pe^{rt}$$

for the given interest rates.

23. 6% compounded daily.
24. 2% compounded continuously.
25. 4.2% compounded weekly.
26. 3.75% compounded quarterly.
27. 11% compounded monthly.
28. 1.25% compounded semiannually.
29. 4.5% compounded continuously.
30. 9% compounded annually.

For Exercises 31 through 36 identify the given values of A, P, r, n, and t. Identify which variable is missing and would need to be solved for. Do not solve.

31. An investment of $20,000 is put in an account paying 3% annual interest compounded daily. Find the balance of the account after 5 years.

32. An investment of $150,000 is put in an account paying 6.5% annual interest compounded weekly. Find the balance of the account after 4 years.

33. How long will it take an investment of $4000 to double if it is invested in an account paying 2.75% annual interest compounded monthly?

34. How long will it take an investment of $20,000 to triple if it is invested in an account paying 4.5% annual interest compounded annually?

35. What interest rate compounded daily must an account earn to make $500 double in 10 years?

36. What interest rate compounded quarterly must an account earn to make $9,000 double in 12 years?

Use the bank brochure to answer Exercises 37 through 48.

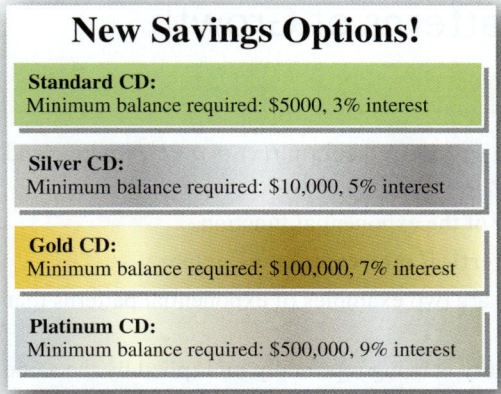

New Savings Options!

Standard CD: Minimum balance required: $5000, 3% interest

Silver CD: Minimum balance required: $10,000, 5% interest

Gold CD: Minimum balance required: $100,000, 7% interest

Platinum CD: Minimum balance required: $500,000, 9% interest

For Exercises 37 through 40, an initial deposit of $10,000 is placed in a Silver CD. Find the amount in the account after 10 years if the interest is compounded in the given way.

37. Compounded monthly.

38. Compounded weekly.

39. Compounded daily.

40. Compounded continuously.

For Exercises 41 through 44, an initial deposit of $500,000 is placed in a Platinum CD. Find the amount in the account after 15 years if the interest is compounded in the given way.

41. Compounded continuously.

42. Compounded daily.

43. Compounded hourly.

44. Compounded monthly.

For Exercises 45 through 48, an initial deposit of $100,000 is placed in a Gold CD. Find the amount in the account after 20 years if the interest is compounded in the given way.

45. Compounded quarterly.

46. Compounded monthly.

47. Compounded continuously.

48. Compounded weekly.

49. Edith has two investment options for $400,000 she inherited. One account pays 4% interest compounded annually. The second account pays 3.95% interest compounded continuously. Which account will have the larger balance after 5 years?

50. Karina has two investment options for $50,000 she is saving for her future college expenses. One account pays 3.2% interest compounded quarterly. The second account pays 3.1% compounded daily. Which account will have the larger balance after 3 years?

Chapter Summary

Section 5.1 — Exponential Functions: Patterns of Growth and Decay

- An **exponential function** can be written in the form $f(x) = a \cdot b^x$, where a and b are real numbers such that $a \neq 0$, $b > 0$, and $b \neq 1$.
- The constant b is called the **base** of the exponential function.
- $f(0) = a$, and thus, $(0, a)$ is the **vertical intercept** for the graph.
- **Exponential growth and decay** are two examples of exponential patterns that are found in the world.
- Exponentials occur when a quantity is being repeatedly multiplied by the same constant.
- When looking for an exponential pattern, try to find the quantity that is used repeatedly as the multiplier. Once found, it will help find the base b.

Example 1

A biologist is studying the growth of 50 salmonella bacteria. When given the right food source and temperature range, these bacteria double in number every 15 minutes.

a. Find an equation for a model for the number of salmonella bacteria after h hours.

b. Use your model to estimate the number of salmonella bacteria after 3 hours.

c. Estimate graphically when the number of salmonella bacteria will reach 500,000.

SOLUTION

a. h = Number of hours since starting with 50 salmonella bacteria
$S(h)$ = Number of salmonella bacteria

Because the bacteria double every 15 minutes, they will double 4 times every hour. By building a table, we can find a pattern.

h	0	1	2	3	h
$S(h)$	50	$50(2)^4$	$50(2)^8$	$50(2)^{12}$	$50(2)^{4h}$

From the pattern, we have the model $S(h) = 50(2)^{4h}$.

b. $S(3) = 50(2)^{4(3)} = 204800$, so after 3 hours, there will be about 204,800 salmonella bacteria.

c. Graphing the function and 500,000, we can trace to find an estimate.

Therefore, there will be about 500,000 salmonella bacteria after about 3.3 hours.

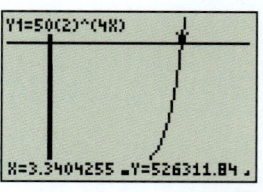

Section 5.2 — Solving Equations Using Exponent Rules

- When **solving exponential equations by inspection**, try to get the bases on each side of the equation the same and set the exponents equal to one another.
- When **solving power equations**, get the variable being raised to a power by itself and use the reciprocal exponent of both sides to undo it. Remember to use the plus/minus symbol whenever the reciprocal has an even denominator.

Example 2 Solve the following equations.

a. $2^x = 16$

b. $3x^7 = 126$

SOLUTION

a. $2^x = 16$

$2^x = 2^4$

$x = 4$

b. $3x^7 = 126$

$x^7 = 42$

$(x^7)^{\frac{1}{7}} = (42)^{\frac{1}{7}}$

$x = 1.706$

Section 5.3 Graphing Exponential Functions

- The graph of an exponential function is affected by the values of a and b: $f(x) = a \cdot b^x$.
- The vertical intercept for the graph of an exponential function is $(0, a)$.
- If a is negative, the graph of the function will be reflected over the horizontal axis.
- If $a > 0$ and $0 < b < 1$, then the graph will be decreasing. The smaller the value of b, the steeper the decrease.
- If $a > 0$ and $b > 1$, then the graph will be increasing. The larger the value of b, the steeper the increase.
- The domain of an exponential function without an application will be all real numbers.
- The range of an exponential function of the form $f(x) = a \cdot b^x$ without an application will be either the positive real numbers or the negative real numbers depending on the sign of a.
- When a constant is added to an exponential function, the graph is shifted up or down.
- An exponential function of the form $f(x) = a \cdot b^x + c$ has a horizontal asymptote at the line $y = c$.

Example 3 Sketch a graph by hand of the function $f(x) = 4(1.2)^x$.

SOLUTION $a = 4$, so the vertical intercept is $(0, 4)$, and the graph will be above the horizontal axis. The base $b = 1.2$, so this graph is increasing, showing exponential growth. By trying a few input values, we get a table of points. Using these points to plot the graph, we get the following graph

x	$f(x)$
3	6.9
6	11.9
9	20.6

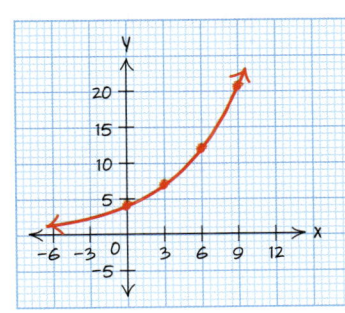

Example 4 Give the domain and range of the given functions.

a. $f(x) = -2(1.7)^x$ b. $f(x) = 45(0.2)^x + 30$

SOLUTION
a. Domain: All real numbers, range: $(-\infty, 0)$.

b. Domain: All real numbers, range: $(30, \infty)$

Section 5.4 Finding Exponential Models

- To find an exponential model, follow these seven steps:
  1. Define the variables and adjust the data.
  2. Create a scatterplot.
  3. Select a model type.
  4. **Exponential model:** Pick two points, write two equations using the form $f(x) = a \cdot b^x$ and solve for b.
  5. Use b and one equation to find a.
  6. Write the equation of the model using function notation.
  7. Check the model by graphing it with the scatterplot.
- If the data you are modeling show the vertical intercept, you can use that intercept for a and one equation to help find b.
- In an application, the domain of an exponential model expands beyond the data. Be careful to avoid inputs that will cause model breakdown to occur. The range will be the lowest to highest points on the graph within the domain.

Example 5 The gross profit for eBay, Inc. is given in the table.

Year	Gross Profit (in millions of $)
1997	33
1998	70
1999	167
2000	336
2002	1000.2
2003	1749

Source: SEC Financial Filings from eBay, Inc., 2004.

a. Find an equation for a model of these data.

b. Estimate the gross profit for eBay, Inc. in 2001.

c. Give a reasonable domain and range for the model.

SOLUTION
a. $P(t)$ = Gross profit for eBay, Inc. in millions of dollars
t = Time in years since 1990

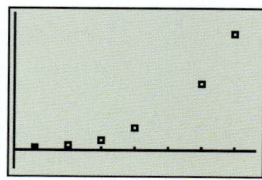

We can use the points (8, 70) and (12, 1000.2).

$$70 = a \cdot b^8 \qquad 1000.2 = a \cdot b^{12}$$

$$\frac{1000.2}{70} = \frac{a \cdot b^{12}}{a \cdot b^8}$$

$$14.289 \approx b^4$$

$$\pm(14.289)^{\frac{1}{4}} \approx (b^4)^{\frac{1}{4}}$$

$$\pm 1.94 \approx b$$

$$1.94 \approx b$$

$$70 = a \cdot (1.94)^8$$

$$0.349 = a$$

We have the model $P(t) = 0.349(1.94)^t$.

b. $P(11) = 511.26$, so eBay, Inc. had a gross profit of about $511.3 million in 2001.

c. Domain: [6, 13], range: [18.6, 1924.2]. Beyond this domain, model breakdown may occur.

Section 5.5 Exponential Growth and Decay Rates and Compounding Interest

- The **growth rate**, r, is the percentage change in a quantity per one unit of time.

$$b = 1 + r$$
$$r = b - 1$$

  where b is the base of the exponential function of the form $f(x) = a \cdot b^x$.

- The **compound interest** formula is

$$A = P\left(1 + \frac{r}{n}\right)^{nt}$$

  where A is the account balance, P is the principal invested, r is the interest rate (as a decimal), n is the number of times the account compounds per year, and t is the time in years of the investment. This formula works if n is any finite number.

- If the interest rate is **compounding continuously,** use the formula $A = Pe^{rt}$, where A is the account balance, P is the principal invested, r is the interest rate (as a decimal), and t is the time in years of the investment.

Example 6
If a population of 200 is decaying at 3% per year, find an exponential model for the size of the population after t years.

SOLUTION Let $P(t)$ be the size of the population after t years. Since the population starts with 200, we know that $a = 200$. We are given a decay rate, so the base is

$$b = 1 + (-0.03)$$
$$b = 0.97$$

Therefore, the model would be $P(t) = 200(0.97)^t$.

Example 7
$5000 is invested in an account for 30 years.

a. Find the balance in the account if the money earns 4% interest compounded monthly.

b. Find the balance in the account if the money earns 4% interest compounded continuously.

SOLUTION a. $A = 5000\left(1 + \dfrac{0.04}{12}\right)^{12(30)} = 16567.49$.

The account will have $16,567.49 after 30 years if compounded monthly.

b. $A = 5000 \cdot e^{0.04(30)} = 16600.58$. The account will have $16,600.58 after 30 years if compounded continuously.

Chapter Review Exercises

1. The population of Africa is estimated to be doubling about every 28 years. The population of Africa in 1996 was about 731.5 million.
   Source: U.S. Department of Commerce, Bureau of the Census.
   a. Using this information, find an equation for a model for the population of Africa.
   b. Using your model, estimate the population of Africa in 2005. [5.1]

2. The flu is traveling through a large city at an alarming rate. Every day, the number of people who have flu symptoms is 3 times what it was the previous day. Assume that one person started this flu epidemic.
   a. Find an equation for a model for the number of people with flu symptoms after d days.
   b. Use your model to estimate the number of people with flu symptoms after two weeks. [5.1]

3. Polonium-210 has a half-life of 138 days.
   a. Find an equation for a model for the percent of a polonium-210 sample left after d days.
   b. Estimate the percent of polonium-210 left after 300 days. [5.1]

4. Thorium-228 has a half-life of 1.9 years.
   a. Find an equation for a model for the percent of a thorium-228 sample left after t years.
   b. Estimate the percent of thorium-228 left after 50 years. [5.1]

5. Find an exponential model for the data. [5.1]

x	$f(x)$
0	5000
1	4000
2	3200
3	2560
4	2048

6. Find an exponential model for the data. [5.1]

x	$f(x)$
0	0.2
1	1.52
2	11.552
3	87.7952
4	667.24352

For Exercises 7 through 18, solve the given equation.

7. $6^x = 216$ [5.2]

8. $4^x = 1024$ [5.2]

9. $6(7^x) = 14406$ [5.2]

10. $5(3^x) = 10935$ [5.2]

11. $2x^5 + 7 = 1057.4375$ [5.2]

12. $3x^4 - 600 = 1275$ [5.2]

13. $-\dfrac{1}{3}x^7 + 345 = -92967$ [5.2]

14. $-4x^8 + 2768 = -6715696$ [5.2]

15. $2^x = \dfrac{1}{32}$ [5.2]

16. $(-3)^x = \dfrac{1}{81}$ [5.2]

17. $\left(\dfrac{2}{3}\right)^x = \dfrac{27}{8}$ [5.2]

18. $\left(\dfrac{1}{4}\right)^x = 1024$ [5.2]

CHAPTER 5 Review Exercises 477

19. Use the graph of the exponential to answer the following. [5.2]

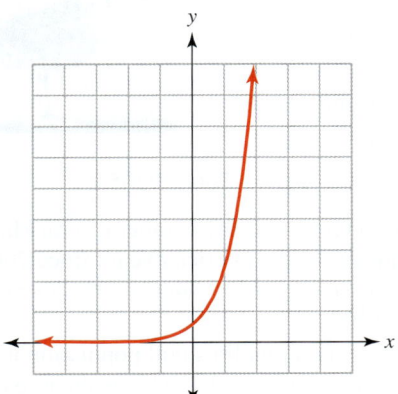

a. Is the value of *a* positive or negative? Explain.
b. Is the value of *b* less than or greater than 1? Explain.
c. Is this graph increasing or decreasing?
d. Give the domain and range for the function.

20. Use the graph of the exponential to answer the following. [5.2]

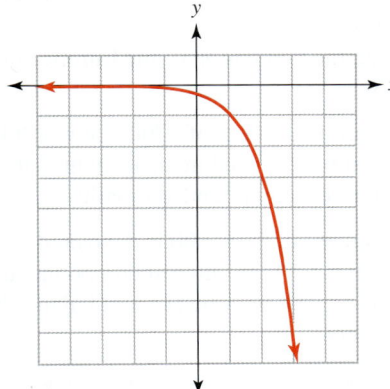

a. Is the value of *a* positive or negative? Explain.
b. Is the value of *b* less than or greater than 1? Explain.
c. Is this graph increasing or decreasing?
d. Give the domain and range for the function.

21. Use the graph of the exponentials $f(x) = ab^x$ and $g(x) = cd^x$ to answer the following. [5.2]

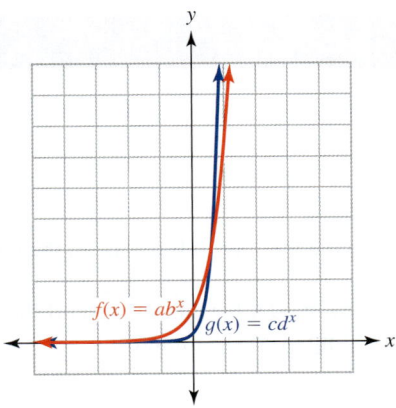

a. Is the value of *a* bigger than or less than the value of *c*? Explain.
b. Is the value of *b* less than or greater than the value of *d*? Explain.

22. Use the graph of the function $f(x)$ to answer the following. [5.1]

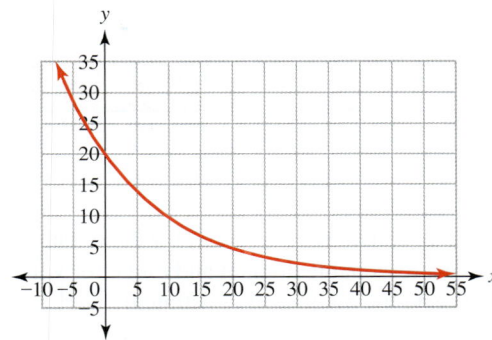

a. Is this graph increasing or decreasing?
b. Is this an example of exponential growth or decay?
c. $f(20) = $?
d. What *x*-value(s) will make $f(x) = 10$?
e. Estimate the vertical intercept.

23. Without graphing the function $f(x) = 2(1.4)^x$, answer the following. [5.2]

a. Is this graph increasing or decreasing? Explain.
b. Is this an example of exponential growth or decay? Explain.
c. What is the *y*-intercept?

For Exercises 24 through 29, sketch the graph of the function. Explain what the values of a and b tell you about this graph.

24. $f(x) = 5(1.3)^x$ [5.3]

25. $g(t) = -0.25(0.5)^t$ [5.3]

26. $f(x) = 200(0.2)^x$ [5.3]

27. $h(x) = -150(0.6)^x$ [5.3]

28. $f(x) = 500(0.4)^x + 20$ [5.3]

29. $g(t) = -3(1.25)^t - 10$ [5.3]

30. Give the domain and range of the function in Exercise 24. [5.3]

31. Give the domain and range of the function in Exercise 25. [5.3]

32. Give the domain and range of the function in Exercise 28. [5.3]

33. Give the domain and range of the function in Exercise 29. [5.3]

34. The number of nonstrategic nuclear warheads stockpiled by the United States during the 1950s is given in the chart. **[5.4]**

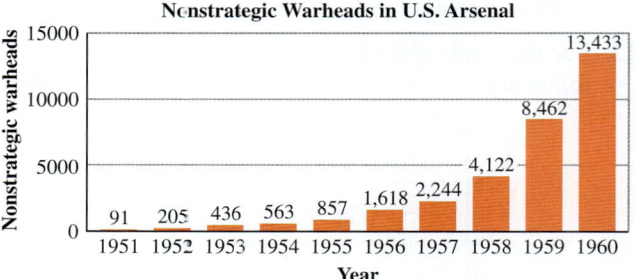

Nonstrategic Warheads in U.S. Arsenal

Source: Natural Resources Defense Council.

  a. Find an equation for a model for these data.
  b. Give a reasonable domain and range for your model.
  c. Estimate the number of nonstrategic warheads stockpiled by the United States in 1965.

35. The following data were collected during an experiment designed to measure the decay of the radioactive material barium-137. Every 60 seconds, the number of counts per minute was taken from a Geiger counter.

Time (seconds)	Counts per Minute
0	3098
60	2480
120	1856
180	1546
240	1220
300	914
360	596

  a. Find an exponential model for these data.
  b. Estimate the number of counts per minute this sample would have after 10 minutes.
  c. Give a reasonable domain and range for this model. **[5.4]**

36. The number of Franklin's gulls in North America has been in significant decline in the past several years. The population of Franklin's gulls can be modeled by

$$F(t) = 750(0.9405^t)$$

where $F(t)$ represents the number of Franklin's gulls in thousands in North America t years since 2000.
Source: Model derived from data from the National Biological Service.

  a. Estimate the Franklin's gull population in 2005.
  b. According to the model, what is the rate of decay for the Franklin's gull population?
  c. What might cause this population to stop declining at this rate? **[5.5]**

37. The population of the European Union was about 491 million in 2008 and was growing at a rate of approximately 0.12% per year.
Source: CIA World Factbook 2008.

  a. Find an equation for a model of the population of the European Union.
  b. Estimate the population of the European Union in 2020. **[5.5]**

38. What will an investment of $7000 be worth in 10 years if it is deposited in an account that pays 6% interest compounded monthly? **[5.5]**

39. If an investment of $100,000 is deposited in an account that pays 8.5% interest compounded continuously, what will the account balance be after 25 years? **[5.5]**

40. If $50,000 is deposited in an account paying 4% interest compounded daily, what will the account balance be after 15 years? **[5.5]**

Chapter Test

1. At 12:00 noon, there are 5 million bacteria on the bathroom door handle. Under these conditions, the number of bacteria on the handle is doubling every hour.

  a. Find an equation for a model for the number of bacteria on the bathroom door handle h hours after 12:00 noon.
  b. Estimate the number of bacteria on the handle at 6:00 P.M.
  c. Find a new model for this situation if the bacteria are doubling every 15 minutes.

CHAPTER 5 Test

2. Thallium-210 has a half-life of 1.32 minutes.
   a. Find an equation for a model for the percent of a thallium-210 sample left after m minutes.
   b. Estimate the percent of thallium-210 left after 15 minutes.

3. How much will an investment of $1000 be worth after 12 years if it is deposited in an account that pays 3.5% interest compounded monthly?

4. How much will an investment of $40,000 be worth after 35 years if it is deposited in an account that pays 6% compounded continuously?

5. Use the graph of the exponential to answer the following.

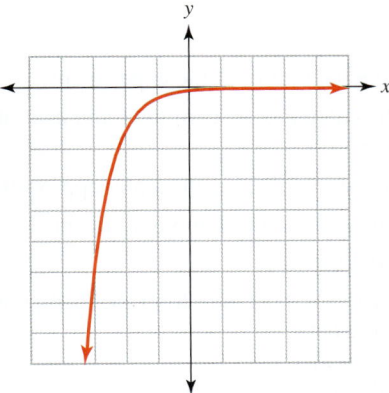

   a. Is the value of a positive or negative? Explain.
   b. Is the value of b less than or greater than 1? Explain.
   c. Give the domain and range for the function.

6. Use the graph of the function $f(x)$ to answer the following.

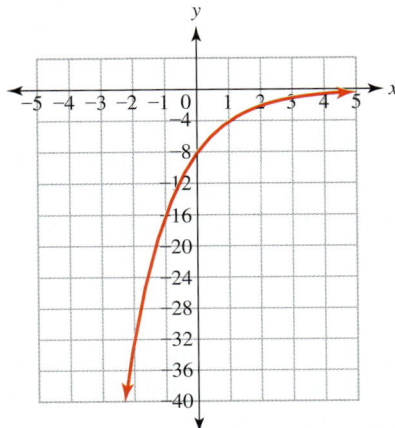

   a. Is this graph increasing or decreasing?
   b. $f(1) = $?
   c. What x-value(s) will make $f(x) = -32$?
   d. Estimate the vertical intercept.

7. The number of CD singles shipped in the mid-1990s is given in the table.

Year	CD Singles (mllions)
1993	7.8
1994	9.3
1995	21.5
1996	43.2
1997	66.7

   Source: Statistical Abstract of the United States, 2001.

   a. Find an equation for a model for these data.
   b. Give a reasonable domain and range for this model.
   c. Estimate the number of CD singles shipped in 2000.

8. The number of professionals, such as lawyers, dentists, doctors, and accountants, is growing rapidly in developing countries. In 2009, there were about 4 million professionals in developing countries. This number is growing at a rate of about 6.5% a year. *Source: BusinessWeek August 24 & 31, 2009.*

   a. Find an equation for a model for the number of professionals in developing countries.
   b. Estimate the number of professionals in developing countries in 2013.

For Exercises 9 through 12, sketch the graph of the function. Explain what the values of a and b tell you about this graph. Give the domain and range of each function.

9. $f(x) = -2(3.1)^x$ 10. $g(t) = 40(0.8)^t$

11. $f(x) = 200(0.75)^x + 50$

12. $h(x) = -300(0.9)^x - 20$

For Exercises 13 through 19, solve the given equation.

13. $12(4^x) = 12288$ 14. $2(3^x) - 557 = -503$

15. $-6(5^x) - 510 = -4260$ 16. $x^3 = 216$

17. $-3x^5 + 650 = 3722$ 18. $4.5x^6 - 2865 = 415.5$

19. $300x^4 - 580 = 218768.48$

Chapter 5 Projects

How Hot Is That Water?

Group Experiment
Three or more people

What you will need
- A Texas Instruments CBL unit with a Vernier temperature probe.
- A cup of *hot* water.
- About 45 minutes to an hour.

In this project, you will test the temperature of a cup of hot water over time. Using a Texas Instruments CBL unit and the temperature probe, measure the room temperature at the beginning of the experiment. Then put the probe into the hot water, allowing the probe to warm up for a couple of seconds before you start to collect temperature data from the water. Try to collect data for at least 45 minutes or longer. Be sure to keep the cup in a location that will not get too much temperature change during the experiment (away from air conditioning vents or open doors).

Write up

a. What was the room temperature at the beginning of the experiment?
b. Create a scatterplot of the data on the calculator or computer and print it out or draw it neatly by hand on graph paper.
c. As more and more time goes by, what temperature is the water approaching? Why?
d. What kind of model would best fit these data? Explain your reasons.
e. Complete the following steps to model the data.
   1. Adjust the data by subtracting the room temperature from each temperature value.
   2. Find an equation for a model for the data.
   3. Add the room temperature to your model.
   4. Graph the model with the original data you collected.
f. What is the vertical intercept of your model and what does it represent in this experiment?
g. What are a reasonable domain and range for your model? (Remember to consider any restrictions on the experiment.)

Now That's a Lot of Rabbits!

Research Project
One or more people

What you will need
- Find information regarding the rabbit problem in Australia.
- Follow the MLA style guide for all citations. If your school recommends another style guide, use that.

In this project, you will investigate the rabbit problem in Australia. In 1859, 24 wild rabbits were brought to Australia and released. This original population of 24 rabbits grew to an estimated 600,000,000 rabbits by 1950. This was, and still is, devastating to the Australian economy and natural resources.

Write up

a. Research the rabbit problem in Australia and describe the reasons for such a large growth in the number of rabbits. Cite any sources you used following the MLA style guide or your school's recommended style guide.
b. Assume that the growth in the rabbit population is exponential over these years. Find an equation for a model to describe the number of rabbits in Australia during this time period.
c. Did the population of rabbits in Australia continue to grow at such a fast pace after 1950? Why or why not?
d. What are a reasonable domain and range for your model?
e. What tactics have been used to try to stop the rabbit population from growing? How successful were these tactics?

Find Your Own Exponential

Research Project
One or more people

What you will need

- Find data for a real-world situation that can be modeled with an exponential function.
- You might want to use the Internet or library. Statistical abstracts and some journals and scientific articles are good resources for real world data.
- Follow the MLA style guide for all citations. If your school recommends another style guide, use that.

In this project, you are given the task of finding data for a real-world situation for which you can find an exponential model. You may use the problems in this chapter to get ideas of things to investigate, but your data should not be taken from this textbook.

Write up

a. Describe the data you found and where you found them. Cite any sources you used following the MLA style guide or your school's recommended style guide.

b. Create a scatterplot of the data on the calculator or computer and print it out or draw it neatly by hand on graph paper.

c. Describe what about the data you collected and/or the situation led you to use an exponential model.

d. Find an equation for a model to fit the data.

e. What is the vertical intercept of your model and what does it represent in this real-world situation?

f. What are a reasonable domain and range for your model?

g. Use your model to estimate an output value of your model for an input value that you did not collect in your original data.

h. Use your model to estimate the input value for which your model will give you a specific output value that you did not collect in your original data.

Review Presentation

Research Project
One or more people

What you will need

- This presentation may be done in class or as a video.
- If it is done as a video, you may want to post the video on YouTube or another site.
- You might want to use problems from the homework set or review material in this book.
- Be creative, and make the presentation fun for students to watch and learn from.

Create a five-minute presentation or video that reviews a section of Chapter 5. The presentation should include the following:

- Examples of the important skills presented in the section
- Explanation of any important terminology and formulas
- Common mistakes made and how to recognize them

Logarithmic Functions

6

6.1 Functions and Their Inverses

6.2 Logarithmic Functions

6.3 Graphing Logarithmic Functions

6.4 Properties of Logarithms

6.5 Solving Exponential Equations

6.6 Solving Logarithmic Equations

Earthquakes can cause catastrophic destruction to a large area. An earthquake can be described by using many different terms: the dollar amount of damage, the number of lives lost, how far away it was felt. The most common way to measure the size of an earthquake is with magnitude. The magnitude of an earthquake is an expression of the intensity of the quake in numbers that most people can comprehend. In this chapter, you will find how logarithms are used to calculate the magnitude of earthquakes.

6.1 Functions and Their Inverses

LEARNING OBJECTIVES

- Describe and find inverse functions.
- Use the horizontal line test to verify one-to-one functions.

Introduction to Inverse Functions

In Chapter 5, we investigated exponential functions, but we have not yet solved for the input variable because it is in the exponent. In this section, we will discuss the concept of an **inverse function** and build some basic rules so that we can define the inverse for exponential functions in the next section. These will be used throughout the rest of this chapter to solve exponential functions.

CONCEPT INVESTIGATION
Which way do you want to go?

Temperature is measured in different ways around the world. In the United States, we commonly measure temperature in degrees Fahrenheit, but in most other countries, temperature is measured in degrees Celsius. This means that people often have to switch a given temperature from one unit to another. This is especially true for people who travel to different countries and need temperatures in different units. The function

$$C(F) = \frac{5}{9}(F - 32)$$

has the temperature in degrees Fahrenheit as the input value and the temperature in degrees Celsius as the output value.

Today's Weather
London
Cloudy with moderate showers at times
High 19°C
Low 12°C

Birmingham
Cloudy with light showers
High 20°C
Low 14°C

1. If you know that the temperature is 68°F outside, calculate the temperature in degrees Celsius.

   If you were given the temperature in degrees Celsius, it would be convenient to have a function that had an input variable that took degrees Celsius and gave you the value in degrees Fahrenheit. To find such a function, we can simply solve the above function for F.

2. Solve the function $C = \frac{5}{9}(F - 32)$ for F.

3. Use the function that you just found to change 20 degrees Celsius into degrees Fahrenheit.

The function that you found in part 2 of the concept investigation undoes the operations that were done in the original given function. These functions are an example of what we call *inverses*. When one function "undoes" the operations of the other function, you are looking at an inverse. Finding inverses for linear equations is not a difficult process.

> **Finding an Inverse Function in a Real-World Problem**
> 1. Write without function notation.
> 2. Solve for the "other" variable, that is, the original input variable. This will make it the new output variable.
> 3. Rewrite in function notation.

Example 1 Finding and using an inverse in an application

In Chapter 1, we looked at a function to calculate the cost to rent a 10-foot truck from U-Haul.

$$U(m) = 0.79m + 19.95$$

where $U(m)$ represented the cost in dollars to rent a 10-foot truck from U-Haul and drive it m miles.

a. What are the input and output variables for this function? How would this function be used?

b. Find the inverse of this function.

c. What are the input and output variables for the inverse? How would this function be used?

d. Find the cost to travel 100 miles in this truck.

e. Find the number of miles you can travel in this truck for a cost of $150.

SOLUTION

a. In this function, m miles is the input variable, and U dollars is the output variable. If you wanted to know the cost to drive m miles, you would substitute the number of miles and get the cost in dollars.

b. To find the inverse of this function, we basically want to make m the output and U the input. To do this, it is easiest not to use function notation until the end of the process.

$$U(m) = 0.79m + 19.95$$
$$U = 0.79m + 19.95 \quad \text{Write without function notation.}$$
$$U - 19.95 = 0.79m \quad \text{Solve for } m.$$
$$\frac{U - 19.95}{0.79} = m \quad \text{Divide by 0.79.}$$
$$\frac{U}{0.79} - \frac{19.95}{0.79} = m$$
$$1.27U - 25.25 = m$$
$$m(U) = 1.27U - 25.25 \quad \text{Rewrite in function notation.}$$

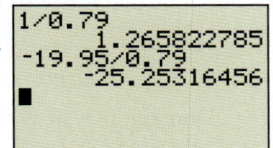

c. For the inverse, the input variable is now U dollars, and the output is now m miles. This function would be used if you had a cost in mind and wanted to know how many miles you could drive the truck.

d. The number of miles is given, so it would be best to use the original function, where the input is m = number of miles.

$$U(m) = 0.79m + 19.95$$
$$U(100) = 0.79(100) + 19.95$$
$$U(100) = 98.95$$

Therefore, it cost $98.95 to rent this truck and drive it 100 miles.

e. The cost is the given quantity, so it is easiest to use the inverse function, where the input is U = the cost of the rental.

$$m(U) = 1.27U - 25.25$$
$$m(150) = 1.27(150) - 25.25$$
$$m(150) = 165.25$$

Therefore, you can drive about 165 miles for a rental cost of $150.

PRACTICE PROBLEM FOR EXAMPLE 1

A team of engineers is trying to pump down the pressure in a vacuum chamber. They know that the following equation represents the pressure in the chamber.

$$P(s) = 35 - 0.07s$$

where $P(s)$ is the pressure in pounds per square inch (psi) of a vacuum chamber after s seconds.

a. Find the inverse for this function.

b. Use the inverse function to estimate the time it will take to pump down this vacuum chamber to 5 psi.

c. If the original function had a domain of $0 \leq s \leq 500$ and a range of $0 \leq P \leq 35$, what are the domain and range of the inverse?

Inverses basically take the inputs and outputs of one function and reverse their roles. The input of one function will become the output of the inverse and vice versa. This also means that the domain and range of an inverse are simply the domain and range of the original function—but reversed. The domain of the function is the range of its inverse, and the range of the function is the domain of the inverse.

In the U-Haul example, the domain for the function may be something like $0 \leq m \leq 400$ miles driven for the day. With this domain, the range would be the lowest to highest cost for those miles, giving us $19.95 \leq U \leq 335.95$. Since the inverse of this function has the cost as the input variable, the domain will be the costs $19.95 \leq U \leq 335.95$, and the range will be the possible miles driven $0 \leq m \leq 400$. The domain and range for the inverse simply switched roles.

> **Domain and Range of Inverse Functions**
> The domain and range of inverse functions will switch roles.
>
> **Inverse Functions**
>
> $U(m)$ $\qquad\qquad$ $m(U)$
>
> Domain: $0 \leq m \leq 400$ $\qquad$ Domain: $19.95 \leq U \leq 335.95$
>
> Range: $19.95 \leq U \leq 335.95$ $\qquad$ Range: $0 \leq m \leq 400$

It is important in a real-world problem that you keep the units and definitions of each variable intact so that their meanings do not get lost. The definition of the variables is what will control the meaning of the input and outputs of the function and its inverse. The U, in the U-Haul function and its inverse, represents the cost in dollars, and the m, in both functions, represents miles driven.

In working with inverses in a real-world situation, the meanings of the variables play an important role in making the equation apply to the situation proposed in the problem. In problems without a context, we do not define the variables, so most often x is the input variable for a function and $f(x) = y$ is the output for the function. When they are finding an inverse function, because mathematicians want the input variable of all functions to be x, they will interchange the variables. To keep the original function and its inverse distinct, mathematicians use different notation when working with inverses without a context. If the original function is $f(x)$, the inverse will have a negative 1 placed as a superscript to the f to indicate that it is the inverse of the function $f(x)$. Therefore, the inverse function will be written as $f^{-1}(x)$, which is read "f inverse of x."

> **What's That Mean?**
>
> **Inverse Notation**
>
> Function notation for inverses can cause some confusion for students. The -1 that is used in the notation to show that the function is an inverse is not an exponent. Recall that a negative exponent would result in a reciprocal of the base. The negative 1 in the function notation symbolizes the inverse relation, not a reciprocal. If
>
> $$f(x) = 2x + 10$$
>
> then the inverse of $f(x)$ is
>
> $$f^{-1}(x) = \frac{x-10}{2} = \frac{1}{2}x - 5$$
>
> Notice that this is not the reciprocal of the original function.
>
> $$\frac{1}{f(x)} = \frac{1}{2x+10}$$
>
> Always remember that
>
> $$f^{-1}(x) \neq \frac{1}{f(x)}$$
>
> If we want $\frac{1}{f(x)}$, we write
>
> $$[f(x)]^{-1} = \frac{1}{f(x)}$$

Finding an Inverse Function in a Problem Without a Context

1. Write without function notation by replacing $f(x)$ with y.
2. Solve for x, the original input variable. This will make it the new output variable.
3. Interchange the variables x and y.
4. Rewrite in function notation. Use $f^{-1}(x)$ to designate it as the inverse. Note that the -1 in the inverse notation, $f^{-1}(x)$, is not an exponent. It simply indicates an inverse in function notation.

When you are finding an inverse, steps 2 and 3 above can be swapped, and it will not affect the inverse function.

Example 2 Finding an inverse function

Find the inverse for the following functions.

a. $f(x) = 2x - 8$ **b.** $g(x) = \dfrac{1}{3}x + 7$

SOLUTION

a. To find the inverse, we still want to reverse the roles of the two variables involved. This is most easily done without using function notation at first and then going back to function notation at the end of the problem.

Step 1 Write without function notation by replacing $f(x)$ with y.

$$f(x) = 2x - 8$$
$$y = 2x - 8$$

Step 2 Solve for x. (This will reverse the role of the two variables.)

$$y = 2x - 8$$
$$y + 8 = 2x$$
$$\dfrac{y+8}{2} = \dfrac{2x}{2} \quad \text{Divide both sides by 2.}$$
$$0.5y + 4 = x$$

Step 3 Interchange the variables x and y. We do this only when there is no context because the input is usually denoted by x.

$$0.5x + 4 = y$$

Step 4 Rewrite in function notation. Use $f^{-1}(x)$ to designate it as the inverse.

$$f^{-1}(x) = 0.5x + 4$$

b. We will follow the same four steps.

$$g(x) = \dfrac{1}{3}x + 7$$
$$y = \dfrac{1}{3}x + 7 \quad \text{Write without function notation.}$$
$$y - 7 = \dfrac{1}{3}x \quad \text{Solve for x.}$$
$$3(y - 7) = 3\left(\dfrac{1}{3}x\right)$$
$$3y - 21 = x$$
$$3x - 21 = y \quad \text{Interchange the variables x and y.}$$
$$g^{-1}(x) = 3x - 21 \quad \text{Put back into function notation using the} -1.$$

PRACTICE PROBLEM FOR EXAMPLE 2

Find the inverse for the following functions.

a. $f(x) = -4x + 9$ **b.** $g(t) = 2.5t - 3.5$

When you are working with linear functions, the "undoing" process of inverses can clearly be seen. If we look at a function and its inverse, the inverse uses the opposite operations, but they are done in reverse order.

$f(x) = 2x + 1$	$f^{-1}(x) = \frac{1}{2}(x - 1)$
Input x	Input x
x	x
Multiply by 2	Subtract 1 (to undo the addition)
$2x$	$x - 1$
Add 1	Divide by 2 (to undo the multiplication)
$2x + 1$	$\frac{(x-1)}{2} = \frac{1}{2}(x - 1)$
$f(x) = 2x + 1$	$f^{-1}(x) = \frac{1}{2}(x - 1)$

Notice that the steps are in reverse order and undo each other.

One-to-One Functions

Recall that a function was defined as a relation in which each input x results in exactly one output y. In an inverse of a function, the variables x and y are swapped, so this relationship of each input to exactly one output must work in reverse for the inverse to be a function. Therefore, inverse functions are possible only if each output of an original function can be brought back to a single input value that it came from. When an output is associated with more than one input, it becomes a problem. An example would be the function $f(x) = x^2$, which has the same output of 4 when $x = 2$ as when $x = -2$, so if we want to go back from an output of 4 to the original input, $x = 2$ and $x = -2$ are both options, and choosing which one it should be is a problem. Because we want an inverse to be a function, this problem of one output coming from two different inputs eliminates the possibility of $f(x) = x^2$ having an inverse function. If an inverse is not a function, we are not interested. Many types of functions do not have inverse functions.

Functions that have only one input for any one output are called **one-to-one functions;** that is, each input goes to exactly one output, and each output comes from exactly one input. A function that is one-to-one will have an inverse function. A one-to-one function is easiest to identify by using its graph. If a function is one-to-one, it will pass **the horizontal line test,** much as a function must pass the vertical line test discussed earlier.

> ### DEFINITIONS
>
> **One-to-one function**   A function in which each input corresponds to only one output and each output corresponds to only one input.
>
> **The horizontal line test for a one-to-one function**   If any horizontal line intersects the graph of a function at most once, then that graph is a one-to-one function.

Example 3 Using the horizontal line test

Graph the following functions. Use the horizontal line test to determine whether or not each function is a one-to-one function.

a. $f(x) = -2.4x + 9$ **b.** $g(x) = 2.5x^2 + 3x - 5$
c. $h(t) = 3.4(1.4^t)$ **d.** $f(x) = 5$

SOLUTION

Each of the given equations can be graphed on a graphing calculator or by hand. You can visually use the vertical and horizontal line tests to determine whether they are one-to-one.

a.
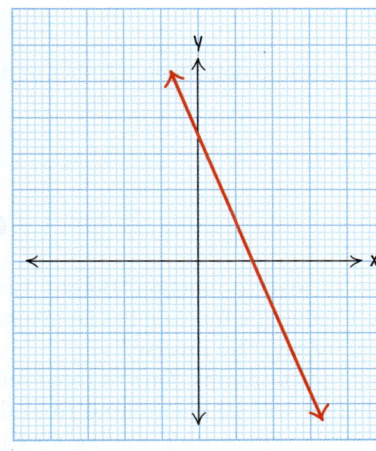
This function passes the horizontal line test because each horizontal line that you could draw across this graph would hit the function only once. Therefore, this is a one-to-one function.

b.
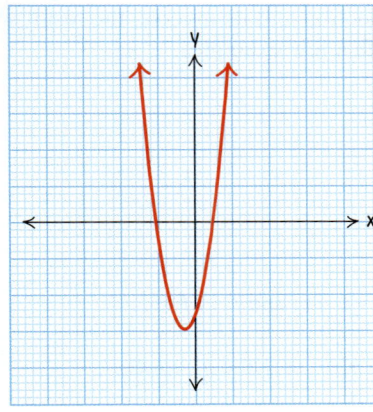
This function fails the horizontal line test because almost any horizontal line that you draw through this graph hits the graph more than once. Therefore, this is not a one-to-one function.

c.
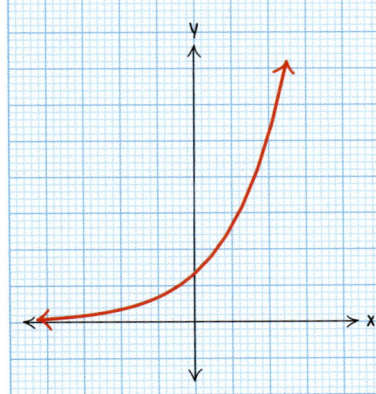
This function passes the horizontal line test because each horizontal line that you could draw across this graph would hit the function only once. Therefore, this is a one-to-one function.

d.

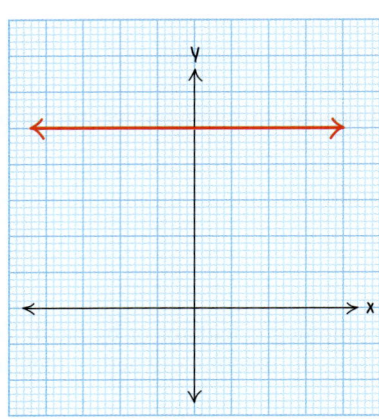

This line is horizontal, so when it is tested by using the horizontal line test, it fails to be one-to-one. Horizontal lines are not one-to-one.

PRACTICE PROBLEM FOR EXAMPLE 3

Graph the following functions. Use the horizontal line test to determine whether or not each function is a one-to-one function.

a. $f(t) = \frac{2}{3}t - 7$ **b.** $g(w) = 0.5w^3 + 3w^2 - 7$

c. $g(x) = -5(0.7)^x$

Functions that are not one-to-one will not have an inverse function. Before using the steps to find an inverse function, we will check whether the function has an inverse. If the function does not have an inverse, we will simply say, "No inverse."

Functions and their inverses have several special relationships. One of these relationships comes from the fact that an inverse function will take an output from the original function and give back the input we originally started with. Ending up with the original input that we started with is a result of the inverse function "undoing" what the original function did. When we compose a function and its inverse, the result will always be the input variable. This is best seen with a few examples.

Example 4 Composing inverses

Perform the following compositions and simplify.

a. Let $f(x) = 2x + 6$ and $f^{-1}(x) = \frac{1}{2}x - 3$. Find $f \circ f^{-1}(x)$ and $f^{-1} \circ f(x)$.

b. Let $g(x) = 5x - 20$ and $g^{-1}(x) = \frac{1}{5}x + 4$. Find $g(g^{-1}(x))$ and $g^{-1}(g(x))$.

SOLUTION

a. $f \circ f^{-1}(x) = 2\left(\frac{1}{2}x - 3\right) + 6$ $\quad\quad f^{-1} \circ f(x) = \frac{1}{2}(2x + 6) - 3$

$f \circ f^{-1}(x) = x - 6 + 6$ $\quad\quad\quad\quad\quad\quad f^{-1} \circ f(x) = x + 3 - 3$

$f \circ f^{-1}(x) = x$ $\quad\quad\quad\quad\quad\quad\quad\quad f^{-1} \circ f(x) = x$

b. $g(g^{-1}(x)) = 5\left(\frac{1}{5}x + 4\right) - 20$ $\quad\quad g^{-1}(g(x)) = \frac{1}{5}(5x - 20) + 4$

$g(g^{-1}(x)) = x + 20 - 20$ $\quad\quad\quad\quad g^{-1}(g(x)) = x - 4 + 4$

$g(g^{-1}(x)) = x$ $\quad\quad\quad\quad\quad\quad\quad g^{-1}(g(x)) = x$

PRACTICE PROBLEM FOR EXAMPLE 4

Perform the following compositions and simplify.

Let $f(x) = \frac{2}{5}x + 4$ and $f^{-1}(x) = \frac{5}{2}x - 10$. Find $f \circ f^{-1}(x)$ and $f^{-1} \circ f(x)$.

■ Skill Connection

Composing Functions

Recall from Chapter 3 that the composition of functions is the process of making one function the input to the other function.

SC Example:

Let

$f(x) = 2x + 5$

$g(x) = 7x - 9$

Find $f(g(x))$.

$f(g(x)) = 2(7x - 9) + 5$

$f(g(x)) = 14x - 18 + 5$

$f(g(x)) = 14x - 13$

SECTION 6.1 Functions and Their Inverses

The graphs of inverses will also have a simple relationship to the graph of the original function. Because an inverse function basically switches the roles of the input and outputs of the original function, the graph of the inverse will do just that. This results in a graph that is a reflection of the original graph over the line $y = x$. To graph an inverse, you can use techniques you already know, or you can use this reflection property.

Example 5 Graphing functions and their inverses

Graph the following functions and their inverses. Include the graph of the line $y = x$.

a. $f(x) = 2x + 6, \quad f^{-1}(x) = 0.5x - 3$

b. $h(x) = -\dfrac{2}{9}x - 6, \quad h^{-1}(x) = -\dfrac{9}{2}x - 27$

c. $g(t) = t^3, \quad g^{-1}(t) = \sqrt[3]{t}$

SOLUTION

a. Because both of these functions are linear, they can be graphed easily. To see how the points on $f(x)$ and $f^{-1}(x)$ are related, we will build a small table of points.

$f(x) = 2x + 6$	$f^{-1}(x) = 0.5x - 3$
$(-10, -14)$	$(-14, -10)$
$(-6, -6)$	$(-6, -6)$
$(-3, 0)$	$(0, -3)$
$(0, 6)$	$(6, 0)$
$(5, 16)$	$(16, 5)$

Notice that each point for the function has the x and y swapped in the corresponding point for the inverse function.

Putting both graphs on the same axis and showing the line $y = x$, we get the following.

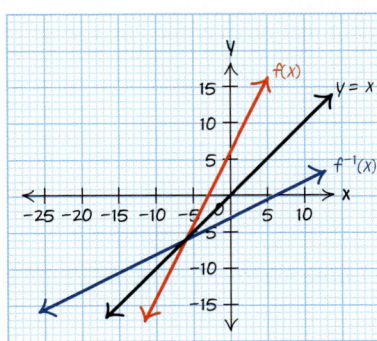

b. These functions are both linear, so we can graph them easily.

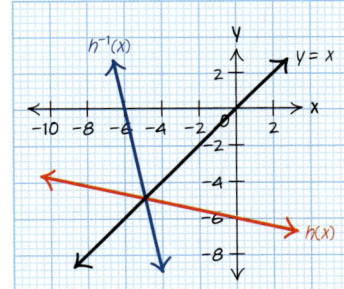

c. These functions are harder to graph by hand. We will plot several points for the original function and use the reflection properties of inverses to get a graph of the inverse. Therefore, for each point that we find for the original function $g(x) = x^3$, we will reverse the x- and y-values to find a point on the inverse function $g^{-1}(x) = \sqrt[3]{x}$.

Plotting the points for the function and the inverse, we get the following.

$g(x) = x^3$	$g^{-1}(x) = \sqrt[3]{x}$
$(-1.4, -2.744)$	$(-2.744, -1.4)$
$(-1, -1)$	$(-1, -1)$
$(-0.4, -0.064)$	$(-0.064, -0.4)$
$(0, 0)$	$(0, 0)$
$(0.2, 0.008)$	$(0.008, 0.2)$
$(0.6, 0.216)$	$(0.216, 0.6)$
$(1, 1)$	$(1, 1)$
$(1.2, 1.728)$	$(1.728, 1.2)$

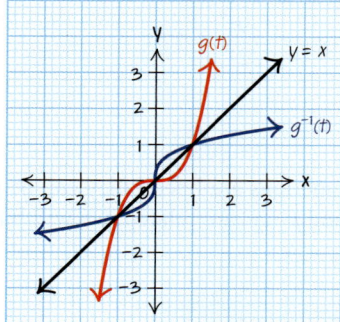

In graphing these on a graphing calculator, it is best to use the ZOOM Square feature so that the graph will not be distorted by the shape of the calculator screen.

PROPERTIES OF INVERSES

Composing inverse functions In composing a function and its inverse, the composition will simplify to the input variable.

$$f(f^{-1}(x)) = x \quad \text{and} \quad f^{-1}(f(x)) = x$$

Graphs of inverse functions The graph of the inverse function will be a reflection of the original function over the line $y = x$.

6.1 Exercises

For Exercises 1 through 10, use the horizontal line test to determine whether or not the graph represents a one-to-one function.

1.
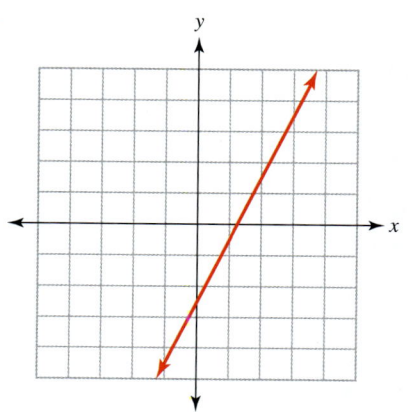

2.
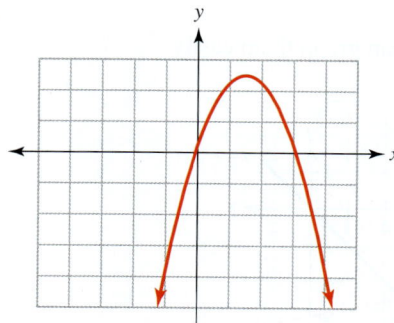

3.
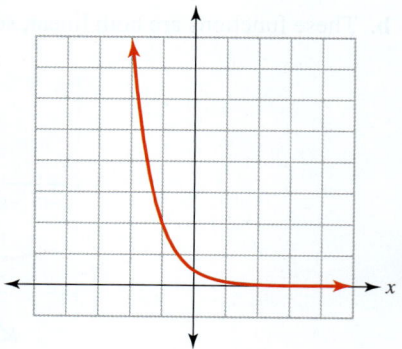

SECTION 6.1 Functions and Their Inverses **493**

4.
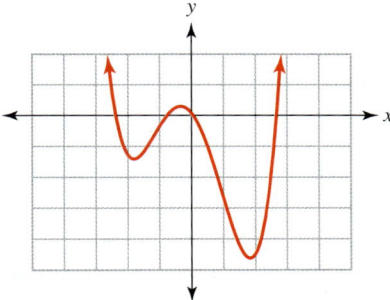

5.
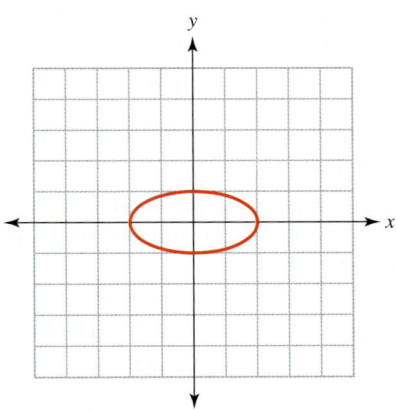

6.
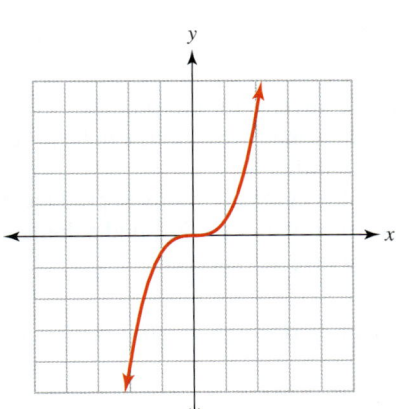

7.
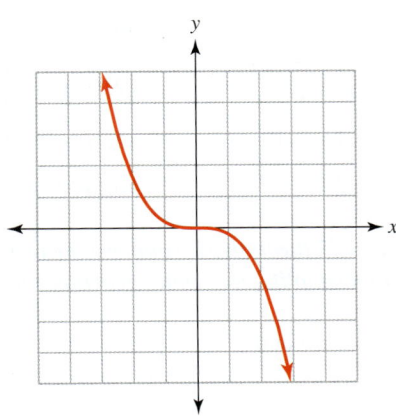

8.
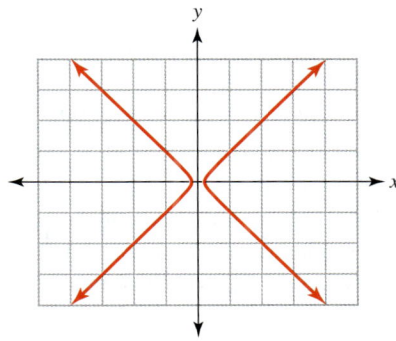

9.
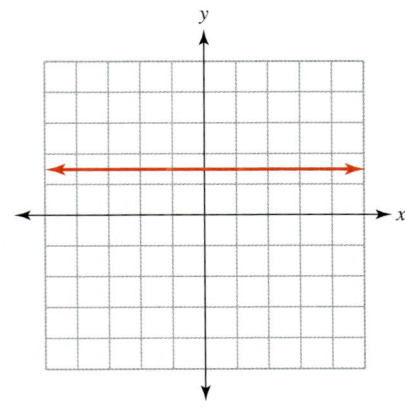

10.
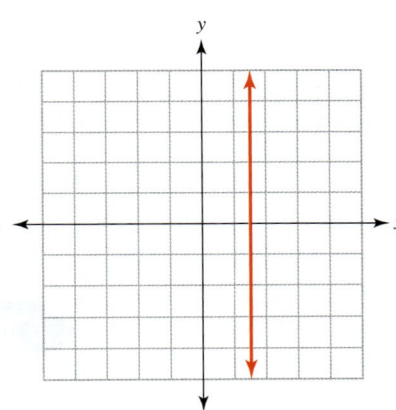

For Exercises 11 through 26, use the horizontal line test to determine whether or not the function is one-to-one.

11. $f(x) = \frac{1}{2}x + 3$ **12.** $g(x) = -\frac{2}{3}x + 7$

13. $h(x) = 2.5x^2 + 3x - 9$

14. $f(x) = -0.4(x - 3.5)^2 + 8$

15. $f(x) = 2x^3 + 4$ **16.** $g(x) = -3x^5 + 1$

17. $f(x) = 4x^3 + 2x^2 - 5x - 4$

18. $h(x) = -0.5x^3 + 0.25x^2 - 4$ **19.** $f(x) = 2x^4$

20. $g(x) = -3x^4 + 3x^3 - 2x^2 - 2$

21. $g(x) = 3(1.2)^x$

22. $h(x) = -25(0.7)^x$

23. $f(x) = 100(0.4)^x + 20$ **24.** $g(x) = -30(1.5)^x - 12$

25. $f(x) = 20$ **26.** $g(x) = -14.8$

27. The number of homicides, $N(t)$, of 15- to 19-year-olds in the United States t years after 1990 can be represented by the function
$$N(t) = -315.9t + 4809.8$$
Source: Based on data from Statistical Abstract of the United States, 2001.
   a. Find the inverse for this model.
   b. If the model has a domain of $[-1, 12]$ and a range of $[1019, 5125.7]$, give the domain and range for the inverse.
   c. Estimate the year in which there were 2000 homicides of 15- to 19-year-olds.

28. The average number of pounds of fruits and vegetables per person that each American eats can be modeled by
$$T(y) = 6.056y + 601.39$$
pounds per person y years since 1980.
   a. Find an inverse for this model.
   b. Estimate when the average number of pounds of fruits and vegetables per person reached 650.
   c. Give a reasonable domain and range for this inverse.

29. The function $P(b) = 5.5b - 2500$ represents the profit in dollars from selling b books.
   a. Given a domain for $P(b)$ of $0 \leq b \leq 5000$, find the range for this model.
   b. Find an inverse for this model.
   c. Estimate the profit from selling 1000 books.
   d. Estimate the number of books you would need to sell to make $5000 profit.
   e. Give the domain and range for the inverse.

30. The function $C(n) = 2.5n + 1300$ represents the cost in dollars from producing n Bluetooth headsets.
   a. Given a domain for $C(n)$ of $0 \leq n \leq 2500$, find the range for this model.
   b. Find an inverse for this model.

   c. Estimate the cost from producing 1500 headsets.
   d. Estimate the number of Bluetooth headsets you could produce with a budget of $9800.
   e. Give the domain and range for the inverse.

31. The population of the United States during the 1990s can be estimated by the equation
$$P(t) = 2.57t + 249.78$$
where $P(t)$ is the population in millions t years since 1990.
Source: Model based on data from Statistical Abstract of the United States, 2001.
   a. Find an inverse for this model.
   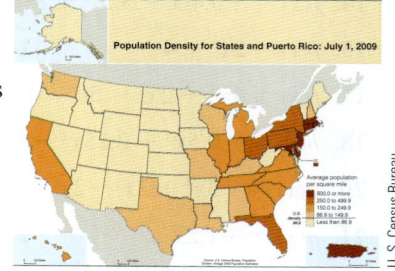
   Population density of the United States
   b. Estimate the year when the U.S. population reached 260 million people.
   c. What are the input and output variables for the inverse?

32. The population of Colorado in millions t years since 1990 can be modeled by
$$P(t) = 0.0875t + 3.285$$
Source: Model based on data from Statistical Abstract of the United States, 2001.
   a. Find an inverse for this model.
   b. Estimate when Colorado's population reached 3.5 million people.
   c. Give a reasonable domain and range for the inverse.

33. The amount of money Hip Hop Math Records will pay Math Dude can be modeled by
$$M(c) = 10000 + 1.5c$$
where $M(c)$ is the money in dollars paid to Math Dude when c CDs are sold.
   a. Find an inverse for this model.
   b. Find how many CDs must be sold for Math Dude to earn $40,000.
   c. If the record company believes they can sell between 10,000 and 50,000 CDs, find a domain and range for the inverse function.

34. The time it takes to produce toy trains can be modeled by
$$H(t) = 0.2t + 0.5$$
where $H(t)$ is the time in hours it takes to produce t toy trains.
   a. Find an inverse for this model.
   b. Find how many toy trains can be produced in an 8-hour shift.
   c. If the toy company wants to produce between 25 and 100 toy trains a day, find a domain and range for the inverse function.

For Exercises 35 through 46, find the inverse of each function

35. $f(x) = 3x + 5$ **36.** $f(x) = 5x + 7$

37. $g(t) = -4t + 8$ **38.** $h(x) = -6x - 10$

39. $h(x) = \frac{2}{3}x - 9$ **40.** $f(x) = -\frac{5}{7}x + 4$

41. $f(x) = \frac{1}{5}x + \frac{3}{5}$ **42.** $g(x) = \frac{1}{4}x - \frac{3}{4}$

43. $h(x) = 0.4x - 1.6$ **44.** $f(x) = 0.6x + 3.6$

45. $P(t) = -2.5t - 7.5$ **46.** $W(x) = 2.4x + 3.7$

For Exercises 47 through 52, compose the two given functions and simplify. Determine whether they are inverses.

47. $f(x) = 3x - 9$ and $g(x) = \frac{1}{3}x + 3$

48. $f(x) = 5x - 30$ and $g(x) = \frac{1}{5}x + 6$

49. $f(x) = 4x + 12$ and $g(x) = 0.25x - 3$

50. $h(x) = 0.6x + 5$ and $g(x) = 0.4x - 2$
51. $f(x) = \frac{1}{7}x + 21$ and $h(x) = 7x + 3$
52. $g(x) = -8x - 12$ and $h(x) = -\frac{1}{8}x - 1\frac{1}{2}$

For Exercises 53 through 60, graph the given function and its inverse on the same set of axes. Use a table of data if necessary. Include the graph of the line $y = x$.

53. $f(x) = 3x - 9$
54. $f(x) = 4x + 12$
55. $g(t) = -4t + 8$
56. $h(x) = -2x + 3$
57. $g(x) = \frac{2}{3}x - 6$
58. $f(x) = \frac{1}{5}x - 4$
59. $h(x) = -\frac{3}{4}x + 2$
60. $f(x) = -\frac{5}{7}x + 4$

For Exercises 61 through 72, graph the given function and its inverse on the separate set of axes. Use a table of data if necessary.

61. $f(x) = 2^x$
62. $f(x) = 3^x$
63. $h(x) = 150(0.5)^x$
64. $g(x) = 100(0.25)^x$
65. $f(x) = -20(1.25)^x$
66. $g(x) = -5(2)^x$
67. $h(x) = 2(3)^x + 5$
68. $f(x) = 4(1.5)^x + 10$
69. $f(x) = -2(2.5)^x - 8$
70. $g(x) = -6(4)^x - 20$
71. $f(x) = 40\left(\frac{1}{10}\right)^x + 15$
72. $g(x) = 900\left(\frac{1}{3}\right)^x + 20$

For Exercises 73 through 78, use the given graph to sketch the graph of the inverse function. Do not find the equation of the given function.

73.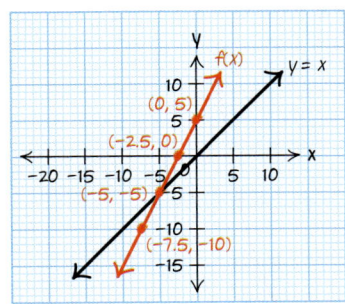

74.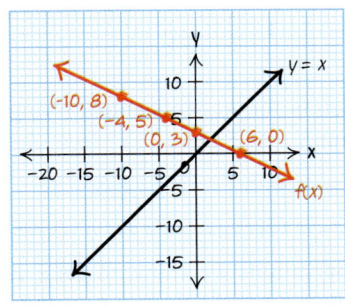

75.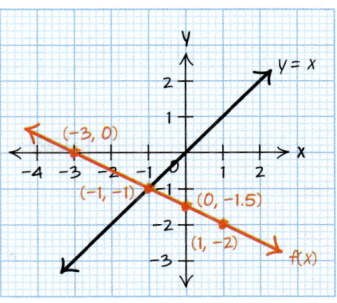

76.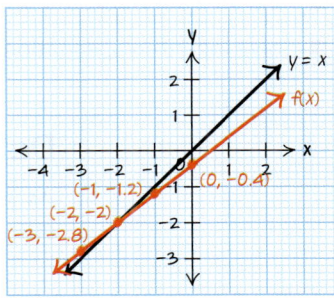

77.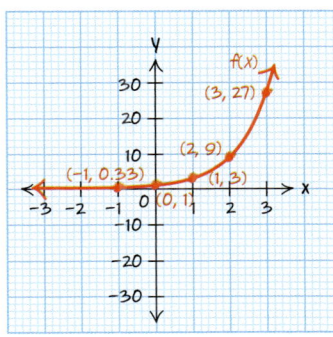

78.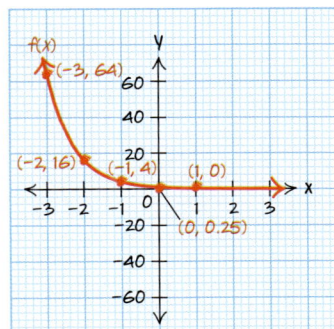

For Exercises 79 through 82, determine whether the given information describes a function. If it describes a function, determine whether it has an inverse function. Explain your reasoning.

79. An equation of the form $y = mx + b$, where m is a nonzero real number.

80. An equation of the form $y = mx + b$, where m is zero.

81. An equation of the form $x = c$, where c is a real number.

82. An equation of the form $y = ax^2 + bx + c$, where a, b, and c are real numbers and $a \neq 0$.

6.2 Logarithmic Functions

LEARNING OBJECTIVES
- Relate logarithmic functions to their inverses.
- Use logarithm properties to evaluate logarithms.
- Use the change of base formula.
- Find inverses of exponential functions.
- Solve logarithmic equations.

Definition of Logarithms

As we can see from the horizontal line test, most linear functions have inverses, but quadratic functions do not.

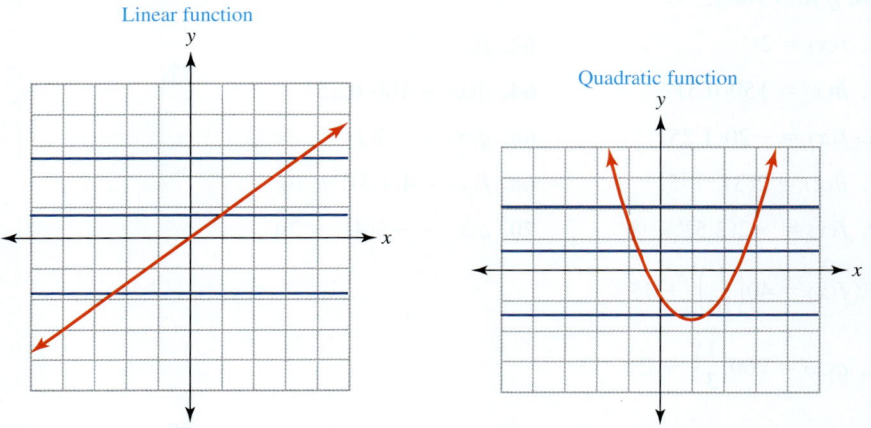

The exponential function also passes the horizontal line test, so it also must have an inverse function.

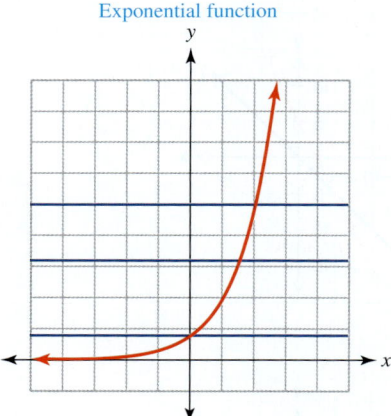

When we try to find the inverse function for an exponential, trying to solve for the input variable leads us to a basic problem.

$$f(x) = 2^x \quad \text{Get out of function notation and solve for x.}$$
$$y = 2^x \quad \text{We have no technique to get a variable out of the exponent.}$$

There is no way, using the arithmetic operations we know, to get a variable out of an exponent. This problem leads us to a new type of function that, at the most basic level, can be defined as the inverse function of an exponential. We use the symbol $\log_b$, read "log base b," and call this new function a **logarithm.** A logarithm at its core asks the question, "What exponent of the base b gives us what's inside the log?"

> **DEFINITIONS**
>
> **Logarithmic function** An inverse for an exponential function. A basic logarithmic function can be written in the form
>
> $$f(x) = \log_b x \quad \text{read} \quad \text{log base } b \text{ of } x.$$
>
> where b is a real number greater than zero and not equal to 1. By definition, logarithms and exponentials have an inverse relationship.
>
> $$f(x) = y = b^x \qquad f^{-1}(x) = y = \log_b x$$
>
> **Base** The constant b is called the base of the logarithm function. If no base is given, then that log is assumed to have base 10.
>
> A logarithm asks the key question "What exponent of the base b gives us what's inside the log?"

Example 1   Evaluating basic logarithms

Evaluate the following logarithms.

a. $\log_5 25$ **b.** $\log_2 16$ **c.** $\log 1000$

SOLUTION

a. With each of these logarithms, we will want to ask ourselves the key question "What exponent of 5 will give us 25?" In this case, we know that $5^2 = 25$, so we know that an exponent of 2 will give us 25. Therefore, we have $\log_5 25 = 2$.

b. Ask, "What exponent of 2 will give me 16?" We should know that $2^4 = 16$, so an exponent of 4 will give you 16. Therefore, we have that $\log_2 16 = 4$.

c. In this case, no base is written in the logarithmic expression. This implies that this is a common logarithm and always has the base 10. Therefore, the question becomes "What exponent of 10 will give me 1000?" Because $10^3 = 1000$, we know that $\log 1000 = 3$.

PRACTICE PROBLEM FOR EXAMPLE 1

Evaluate the following logarithms.

a. $\log 10000$ **b.** $\log_7 49$

Properties of Logarithms

As you can see from Example 1, evaluating logarithms is pretty simple if you can recognize the exponent that you need. Some logarithms have the same value for any base. In particular, in taking the logarithm of 1, no matter what the base is, the logarithm will always equal zero. This follows from the exponent rule that says anything, except zero, to the zero power will be 1.

$$b^0 = 1 \quad \text{so} \quad \log_b 1 = 0$$

Another property says that if you take the logarithm of the base to a power, you will get that power back. Using the basic question a logarithm asks, for $\log_4 4^9$, we get "What exponent of 4 will give me 4^9?" An exponent of 9 is the only exponent that would make sense. These properties are all related directly to the definition of a logarithm.

> **Properties of Logarithms**
> 1. The logarithm of 1 for any base logarithm will always equal zero.
> $$\log_b 1 = 0 \quad \text{since} \quad b^0 = 1$$
> 2. The logarithm of its base is always equal to 1.
> $$\log_b b = 1 \quad \text{since} \quad b^1 = b$$
> 3. The logarithm of its base to a power is just that power.
> $$\log_b (b^m) = m \quad \text{since} \quad b^m = b^m$$

Example 2 Evaluating logarithms using the properties

Evaluate the following logarithms using the properties of logarithms.

a. $\log 10^5$ **b.** $\log_3 1$ **c.** $\log_9 9$

SOLUTION

a. The base of this logarithm is understood to be 10. We have the log of its base raised to a power, so the logarithm will be equal to the power.

$$\log 10^5 = 5$$

b. We are taking the log of 1, so the log equals zero.

c. This is a log of its own base, so the log equals 1.

PRACTICE PROBLEM FOR EXAMPLE 2

Evaluate the following logarithms using the properties of logarithms.

a. $\log_7 7^8$ **b.** $\log_4 1$

Although logarithms can be written with any positive base other than 1, there are two logarithms that are used the most. The **"common" logarithm** is base 10 and is written as "log" without a base present. This logarithm is used often in science when very large or very small numbers are needed. One other base that is often used in science and business is base e. When a logarithm has a base of e, the logarithm is written as "ln," and a base of e is assumed. A logarithm with base e is most often called the **natural logarithm**. One of the nicest features of calculators is that the common and natural logarithms can be found with the touch of a button.

■ **Connecting the Concepts**

What is the base of ln?

Recall from Section 5.5 that we introduced the number

$$e \approx 2.718281828$$

Because the number e is used in many areas of math and science, a logarithm with base e was needed. Instead of writing

$$\log_e$$

we give a logarithm base e a special symbol:

$$\ln$$

A logarithm base e is often called the *natural logarithm*.

Example 3 Evaluating logarithms using a calculator

Use your calculator to evaluate the following logarithms. Round answers to three decimal places.

a. $\log 500$ **b.** $\ln 13.4$

SOLUTION

Be sure to use the correct button on your calculator.

a. $\log 500 \approx 2.699$

b. $\ln 13.4 \approx 2.595$

```
log(500)
          2.698970004
ln(13.4)
          2.595254707
```

Change of Base Formula

If you want to calculate a logarithm with a base other than 10 or e, you can start by trial and error, using different exponents of the base, or you can change the base of the logarithm to use a common or natural logarithm calculation on your calculator.

CONCEPT INVESTIGATION

What do I do with that base?

Since we have both log base 10 and log base e buttons on our calculator but no other bases, we need a way to calculate logarithms of bases other than 10 or e. Answer the following questions to help discover the way to change a base of a logarithm.

1. We will start with $\log_5 25$. We know that this logarithm asks, "What exponent of 5 will give me 25?" We know that 5 squared gives us 25, so $\log_5 25 = 2$. Now we will try a few other calculations with log base 10 to see whether we can find this same answer on our calculator.

$\log 25 + \log 5$	$\log 25 - \log 5$
2.097	
$\log 25 \cdot \log 5$	$\dfrac{\log 25}{\log 5}$

   Which of these calculations were equal to 2?

2. Now let's consider $\log_3 81$. We know that this logarithm asks, "What exponent of 3 will give me 81?" We know that 3 raised to the fourth power will give us 81, so $\log_3 81 = 4$. Now we will try a few other calculations with ln to see whether we can find this same answer on our calculator.

$\ln 81 + \ln 3$	$\ln 81 - \ln 3$
5.493	
$\ln 81 \cdot \ln 3$	$\dfrac{\ln 81}{\ln 3}$

   Which of these calculations were equal to 4?

3. Now let's try one we don't know the answer to: $\log_7 250$. Since we do not know what this logarithm equals, we will need to do each calculation and check it by raising the base seven to that power to see whether or not we get 250. Either log or ln will work here; this time we will use ln.

$\ln 250 + \ln 7$	$\ln 250 - \ln 7$
7.467	
Check this answer.	
$7^{7.467} \approx 2044846.2$	
$\ln 250 \cdot \ln 7$	$\dfrac{\ln 250}{\ln 7}$

   Which calculation gave us the result we were looking for?

Using Your TI Graphing Calculator

To evaluate logarithms on your calculator, you use the **LOG** or **LN** button on the left side of the calculator.

These buttons will calculate only common or natural logarithms. Be sure to use the one with the base that you want. Above each button is the exponential that it is related to. The LOG button has 10^x above it and has a base of 10. The LN button has an e^x above it and has a base of e.

When you use these buttons, they will automatically start the expression with a left parenthesis. It is a good idea to end with a right parenthesis so that you get the values you want. This will be more important when you get to later sections and are doing more complicated calculations.

```
ln(250)+ln(7)
         7.467371067
7^Ans
         2044846.219
```

From the concept investigation, we can see that by using division, we can change from one base to a more convenient base. The order in which you divide is important, so be careful to take the log of the inside and divide by the log of the old base. The change of base formula is a simple way to change a logarithm from one base to another.

> **Change of Base Formula for Logarithms**
>
> $$\log_b x = \frac{\log_c x}{\log_c b}$$
>
> To change the base of a logarithm from base b to another base c, take the $\log_c$ of the inside and divide by the $\log_c$ of the old base.
>
> Most often, we change to base 10 or base e because our calculators can evaluate these bases. Therefore, the change of base formula most often looks like
>
> $$\log_b x = \frac{\log x}{\log b} \quad \text{or} \quad \log_b x = \frac{\ln x}{\ln b}$$

Example 4 Using the change of base formula

Evaluate the following logarithms. Round your answers to three decimal places.

a. $\log_5 114$ **b.** $\log_2 0.365$

SOLUTION

Because these logarithms are not of base 10 or e, we will need to change their base to use our calculator to evaluate them. We can change the base to either 10 or e for either of these examples. We change one to base 10 and the other to base e.

a. Using a base 10 logarithm, we get

$\log_5 114$

$\log_5 114 = \dfrac{\log 114}{\log 5}$ Use the change of base formula.

$\log_5 114 \approx 2.943$

b. Using a natural logarithm, we get

$\log_2 0.365$

$\log_2 0.365 = \dfrac{\ln 0.365}{\ln 2}$ Use the change of base formula.

$\log_2 0.365 \approx -1.454$

PRACTICE PROBLEM FOR EXAMPLE 4

Evaluate the following logarithms on your calculator. Round your answers to three decimal places.

a. $\log_3 278$ **b.** $\log_{17} 11$

In Example 4, we use the common logarithm, log, to perform one calculation and the natural logarithm, ln, to perform the other calculation. When using the change of base formula, we can use any base we wish. In part a of Example 4, we could have used the natural logarithm and ended with the same result.

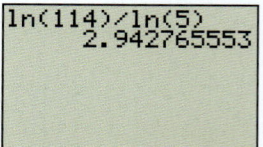

The goal of this section is to help you understand the concept of a logarithm and how it relates to exponential functions. One of the most important skills that you need to learn is to rewrite a problem given in exponential form into logarithm form and vice versa. When working with logarithms and exponentials, remember that the bases are related and that the exponent in an exponential is the same as the result from the logarithm.

Logarithm and Exponential Forms

$$\log_b n = m$$
$$n = b^m$$

The base of the logarithm is the same as the base of the related exponential function. The result from the logarithm is the exponent in the exponential function.

When rewriting a logarithm, think, "The base raised to the outside equals the inside."

$$\log_5 125 = 3$$
$$125 = 5^3$$

The domain and range of exponential and logarithmic functions are related because these functions are inverses of one another.

$$f(x) = b^x \qquad f^{-1}(x) = \log_b x$$

Domain: $(-\infty, \infty)$ Domain: $(0, \infty)$

Range: $(0, \infty)$ Range: $(-\infty, \infty)$

Example 5 Rewriting in logarithm or exponential form

Rewrite each exponential equation into logarithm form.

a. $7^5 = 16807$

b. $4.5^3 = 91.125$

Rewrite each logarithm equation into exponential form.

c. $\log_3 243 = 5$ **d.** $\log_{2.5} 15.625 = 3$

SOLUTION

a. The base is 7, so the base of the logarithm will also be 7. The exponent is 5, so the logarithm will equal 5.

$$\log_7 16807 = 5$$

b. The base is 4.5, so the base of the logarithm will also be 4.5. The exponent is 3, so the logarithm will equal 3.

$$\log_{4.5} 91.125 = 3$$

c. The base is 3, so the base of the exponential will also be 3. The logarithm equals 5, so the exponent will equal 5.

$$243 = 3^5$$

d. The base is 2.5, so the base of the exponential will also be 2.5. The logarithm equals 3, so the exponent will equal 3.

$$15.625 = 2.5^3$$

PRACTICE PROBLEM FOR EXAMPLE 5

Rewrite the exponential equation into logarithm form.

a. $4^8 = 65536$

Rewrite the logarithm equation into exponential form.

b. $\log_6 1296 = 4$

Inverses

Rewriting exponential functions into logarithmic form and logarithm functions into exponential form is actually writing their inverse functions.

Logarithms and Exponentials as Inverses

$$f(x) = b^x \qquad\qquad f(x) = \log_b x$$
$$\text{or}$$
$$f^{-1}(x) = \log_b x \qquad\qquad f^{-1}(x) = b^x$$

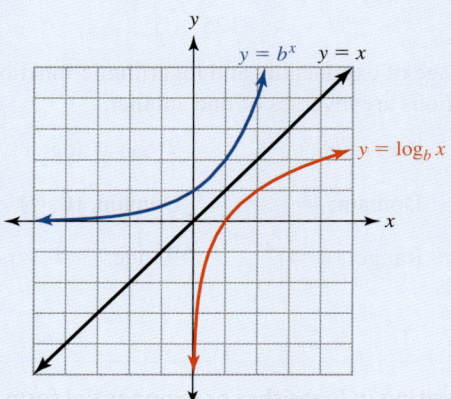

Example 6 — Finding inverses for logarithms and exponentials

Find the inverses for the following functions.

a. $f(x) = 3^x$ **b.** $g(t) = \log t$ **c.** $h(x) = 5(4)^x$ **d.** $f(x) = e^x$

SOLUTION

We need to follow the same steps as finding an inverse in a problem without a context that we learned in Section 6.1. But when we solve, we will rewrite exponentials into logarithmic form and logarithms into exponential form to isolate the variable we want.

a. Because this function is an exponential, the inverse will be a logarithm.

$$f(x) = 3^x$$
$$y = 3^x \qquad \text{Rewrite the function without function notation.}$$
$$\log_3 y = x \qquad \text{Solve for } x \text{ by writing it in logarithm form.}$$
$$\frac{\log y}{\log 3} = x \qquad \text{Rewrite using the change of base formula.}$$
$$\frac{\log x}{\log 3} = y \qquad \text{Interchange the variables.}$$
$$f^{-1}(x) = \frac{\log x}{\log 3} \qquad \text{Write in function notation.}$$

b. Because this function is a logarithm, the inverse will be an exponential.

$$g(t) = \log t$$
$$y = \log t \qquad \text{Rewrite the function without function notation.}$$
$$10^y = t \qquad \text{Solve for } t \text{ by writing it in exponential form.}$$
$$10^t = y \qquad \text{Interchange the variables.}$$
$$g^{-1}(t) = 10^t \qquad \text{Write in function notation.}$$

c. Because this function is an exponential, the inverse will be a logarithm.

$$h(x) = 5(4)^x$$
$$y = 5(4)^x \quad \text{Rewrite the function without function notation.}$$
$$\frac{y}{5} = 4^x \quad \text{Isolate the exponential base.}$$
$$\log_4\left(\frac{y}{5}\right) = x \quad \text{Solve for } x \text{ by writing in logarithmic form.}$$
$$\frac{\log\left(\frac{y}{5}\right)}{\log 4} = x \quad \text{Rewrite using the change of base formula.}$$
$$\frac{\log\left(\frac{x}{5}\right)}{\log 4} = y \quad \text{Interchange the variables.}$$
$$h^{-1}(x) = \frac{\log\left(\frac{x}{5}\right)}{\log 4} \quad \text{Write in function notation.}$$

d. Since the base of the exponential is e, the inverse will be the natural logarithm.

$$f(x) = e^x$$
$$f^{-1}(x) = \ln x$$

PRACTICE PROBLEM FOR EXAMPLE 6

Find the inverses for the following functions.

a. $f(x) = 7^x$ **b.** $g(t) = \log_2 t$ **c.** $h(x) = 3(4)^x$

Solving Logarithmic Equations

Logarithm equations can be solved by rewriting the logarithm into exponential form.

Example 7 Solving logarithm equations

Solve each logarithm equation by rewriting it in exponential form.

a. $\log x = 2$ **b.** $\log_4 x = 5$

SOLUTION

a. $\log x = 2$
$\quad 10^2 = x \quad$ Rewrite as an exponential and calculate.
$\quad x = 100$

b. $\log_4 x = 5$
$\quad 4^5 = x \quad$ Rewrite as an exponential and calculate.
$\quad x = 1024$

PRACTICE PROBLEM FOR EXAMPLE 7

Solve each logarithm equation by rewriting it in exponential form.

a. $\ln x = 4$ **b.** $\log_2(5x) = 6$

504 CHAPTER 6 Logarithmic Functions

6.2 Exercises

For Exercises 1 through 6, determine the base of each logarithm.

1. $\log_5 125$
2. $\log_2 32$
3. $\log 1000$
4. $\log 0.0001$
5. $\ln 45$
6. $\ln 543$

For Exercises 7 through 22, evaluate each logarithm without a calculator.

7. $\log_3 27$
8. $\log_2 32$
9. $\log_5 125$
10. $\log_4 16$
11. $\log_2 256$
12. $\log_3 243$
13. $\log 1000$
14. $\log 100$
15. $\log_5 \left(\dfrac{1}{5}\right)$
16. $\log_2 \left(\dfrac{1}{16}\right)$
17. $\log 0.1$
18. $\log 0.0001$
19. $\log_7 \left(\dfrac{1}{49}\right)$
20. $\log_3 \left(\dfrac{1}{27}\right)$
21. $\log_5 \left(\dfrac{1}{25}\right)$
22. $\log_4 \left(\dfrac{1}{16}\right)$

For Exercises 23 through 32, evaluate each logarithm using the properties of logarithms.

23. $\log_7 1$
24. $\ln 1$
25. $\log_8 (8^4)$
26. $\log_4 (4^6)$
27. $\ln (e^3)$
28. $\log_3 (3^7)$
29. $\ln e$
30. $\log 10$
31. $\log_{19} 19$
32. $\log_6 6$

For Exercises 33 through 40, use a calculator to approximate the value of the logarithms. Round answers to three decimal places.

33. $\log 125$
34. $\log 3000$
35. $\log 275$
36. $\log 24000$
37. $\ln 45$
38. $\ln 543$
39. $\ln 120$
40. $\ln 35000$

41. a. Rewrite $\log_3 40$ using log base 10 and evaluate.
    b. Rewrite $\log_3 40$ using log base e and evaluate.
42. a. Rewrite $\log_8 347$ using log base 10 and evaluate.
    b. Rewrite $\log_8 347$ using log base e and evaluate.
43. a. Rewrite $\log_5 63$ using log base 10 and evaluate.
    b. Rewrite $\log_5 63$ using log base e and evaluate.
44. a. Rewrite $\log_9 875$ using log base 10 and evaluate.
    b. Rewrite $\log_9 875$ using log base e and evaluate.

For Exercises 45 through 54, use a calculator to approximate the value of the logarithms. Round answers to three decimal places.

45. $\log_7 25$
46. $\log_5 43.2$
47. $\log_2 0.473$
48. $\log_9 0.68$
49. $\log_{14} 478$
50. $\log_{25} 5860$
51. $\log_{11} 0.254$
52. $\log_{17} 0.375$
53. $\log_{12} 36478$
54. $\log_{15} 28457$

For Exercises 55 through 68, rewrite the logarithm equations into exponential form.

55. $\log 1000 = 3$
56. $\log 10000 = 4$
57. $\log 0.01 = -2$
58. $\log 0.0001 = -4$
59. $\log_2 (8) = 3$
60. $\log_5 (25) = 2$
61. $\log_3 (81) = 4$
62. $\log_7 (49) = 2$
63. $\ln (e^5) = 5$
64. $\ln (e^7) = 7$
65. $\log_3 \left(\dfrac{1}{9}\right) = -2$
66. $\log_2 \left(\dfrac{1}{32}\right) = -5$
67. $\log_5 \left(\dfrac{1}{25}\right) = -2$
68. $\log_4 \left(\dfrac{1}{64}\right) = -3$

For Exercises 69 through 82, rewrite each exponential equation into logarithm form.

69. $2^{10} = 1024$
70. $3^5 = 243$
71. $5^4 = 625$
72. $4^6 = 4096$
73. $25^{0.5} = 5$
74. $81^{0.25} = 3$
75. $10^5 = 100{,}000$
76. $10^7 = 10{,}000{,}000$
77. $\left(\dfrac{1}{5}\right)^4 = \dfrac{1}{625}$
78. $\left(\dfrac{1}{7}\right)^3 = \dfrac{1}{343}$
79. $\left(\dfrac{1}{2}\right)^3 = \dfrac{1}{8}$
80. $\left(\dfrac{1}{3}\right)^4 = \dfrac{1}{81}$
81. $3^{2x} = 729$
82. $b^y = c$

For Exercises 83 through 96, find the inverse of each function.

83. $f(x) = 7^x$
84. $B(t) = e^t$
85. $h(c) = 10^c$
86. $g(t) = 2^t$
87. $m(r) = 5^r$
88. $f(x) = 12^x$
89. $f(x) = \log_3 x$
90. $f(x) = \log_7 x$
91. $g(x) = \ln x$
92. $h(t) = \log_9 t$
93. $f(x) = 3(4)^x$
94. $g(x) = 6(5)^x$
95. $h(x) = \dfrac{1}{2}(9)^x$
96. $f(t) = \dfrac{1}{3}(2)^t$

For Exercises 97 through 110, solve each logarithm equation by rewriting it in exponential form.

97. $\log x = 3$
98. $\log x = 5$
99. $\ln t = 2$
100. $\ln t = 4$
101. $\log_3 w = 2.4$
102. $\log_9 x = 0.5$
103. $\log_{16} m = \frac{1}{2}$
104. $\log_8 x = \frac{1}{3}$
105. $\log_{12} x = \frac{1}{5}$
106. $\log_{11} x = \frac{2}{3}$
107. $\log(8t) = 2$
108. $\log(2x) = 0.5$
109. $\ln(3x) = 4$
110. $\log_7(4x) = 3$

6.3 Graphing Logarithmic Functions

LEARNING OBJECTIVES
- Graph logarithm functions.
- Find the domain and range of logarithmic functions.

Graphing Logarithmic Functions

In Section 6.2, we learned that logarithmic functions are defined as inverses of exponential functions. This inverse relationship helps us to investigate the graph of logarithm functions using the information we know about the related exponential graphs. Recall from Section 6.1 that the graph of an inverse function is a reflection of the graph of the original function over the line $y = x$. With this relationship in mind, lets look at some exponential graphs and their related logarithm graphs.

CONCEPT INVESTIGATION
How do we build a logarithm graph?

1. Fill in the following table of values for the function $f(x) = 2^x$.

x	$f(x) = 2^x$
-2	$2^{-2} = \frac{1}{4}$
-1	
0	
1	
2	
3	
4	
5	
6	
7	
8	

2. What kind of scale should we use on the x-axis for this graph?
    0.5   1   2   5   10   other: _____

3. What kind of scale should we use on the y-axis for this graph?
    0.5   1   2   5   10   other: _____

4. Sketch the graph of the function in part 1.

Skill Connection

Scale

Remember that the scale is the spacing on the axes. The spacing on each axis must be consistent so that each space between tick marks represents the same number of units.

5. Use the table you completed in part 1 to create a table for the inverse function. (Remember that the x- and y-values will be swapped.)

x	$f^{-1}(x) = \log_2 x = \dfrac{\log x}{\log 2}$
$\dfrac{1}{4}$	-2

6. What kind of scale should we use on the x-axis for this graph?

   0.5   1   2   5   10   other: _____

7. What kind of scale should we use on the y-axis for this graph?

   0.5   1   2   5   10   other: _____

8. Are these scales related to the scales you chose in parts 2 and 3? If so, how?

9. Sketch the graph of the function in part 5.

From the concept investigation, using the related exponential graph to sketch the graph of a logarithm is a reasonable approach. If we look at the characteristics of an exponential graph, we can find the basic characteristics of a logarithmic function's graph.

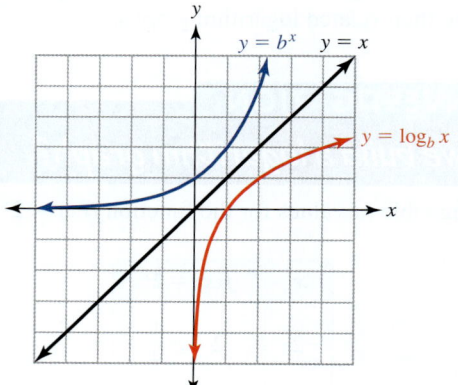

When the base is greater than 1, the graphs have the following characteristics.

Characteristics of Exponential Graph $b > 1$	Characteristics of Logarithmic Graphs $b > 1$
No x-intercept	No y-intercept
y-intercept (0, 1)	x-intercept (1, 0)
The x-axis is a horizontal asymptote.	The y-axis is a vertical asymptote.
The graph increases slowly and then rapidly.	The graph increases rapidly and then slowly.

The growth of a logarithmic function gets slower and slower as the input values increase. If you consider the common logarithm, log x, it takes a large value of x to get a relatively small y-value.

x	1	10	100	1,000	10,000
$f(x) = \log x$	0	1	2	3	4

As you can see from this table, to get an output of 3, the input must be 1000. To get an output of 4, you need the input to be 10,000. The very slow growth of a logarithm function makes graphing these functions harder. Looking at the graph of log x on a standard window and even on an expanded window shows how slowly the graph grows as x increases.

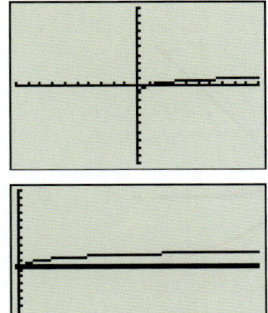
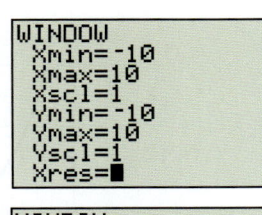
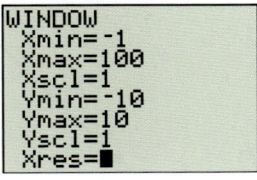

Example 1 Graphing a logarithmic function with base greater than 1

Sketch the graph of $f(x) = \log_3 x$.

SOLUTION

Because this logarithm has a base of 3, the inverse will be the exponential function

$$f^{-1}(x) = 3^x$$

We will use this exponential function to create a table of points to plot for the graph.

Points for a Graph of the Inverse

x	$f^{-1}(x) = 3^x$
-2	$\frac{1}{9}$
-1	$\frac{1}{3}$
0	1
1	3
2	9
3	27
4	81
5	243

Points for the Logarithm Graph

x	$f(x) = \log_3 x$
$\frac{1}{9}$	-2
$\frac{1}{3}$	-1
1	0
3	1
9	2
27	3
81	4
243	5

To plot the logarithm graph, we need a large scale for the x-axis and a small scale for the y-axis.

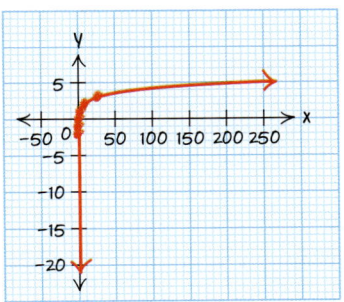

PRACTICE PROBLEM FOR EXAMPLE 1
Sketch the graph of $f(x) = \log_5 x$.

Now we will consider the graphs of logarithms with bases between 0 and 1.

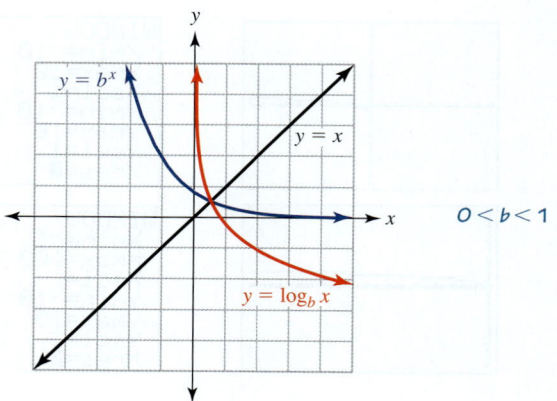

When the base is between 0 and 1, the graphs have the following characteristics.

Characteristics of Exponential Graphs $0 < b < 1$	Characteristics of Logarithmic Graphs $0 < b < 1$
No x-intercept	No y-intercept
y-intercept $(0, 1)$	x-intercept $(1, 0)$
The x-axis is a horizontal asymptote.	The y-axis is a vertical asymptote.
The graph decreases rapidly and then slowly.	The graph decreases rapidly and then slowly.

Example 2   Graphing a logarithmic function with base less than 1

Sketch the graph of $f(x) = \log_{0.5} x$.

SOLUTION

Because this logarithm has a base of 0.5, the inverse will be the exponential function
$$f^{-1}(x) = (0.5)^x$$
We will use this exponential function to create a table of points to plot for the graph.

Points for a Graph of the Inverse

x	$f^{-1}(x) = (0.5)^x$
-10	1024
-8	256
-6	64
-4	16
-2	4
0	1
2	0.25
4	0.0625

Points for the Logarithm Graph

x	$f(x) = \log_{0.5} x$
1024	-10
256	-8
64	-6
16	-4
4	-2
1	0
0.25	2
0.0625	4

To plot the logarithm graph, we need a large scale for the x-axis and a small scale for the y-axis.

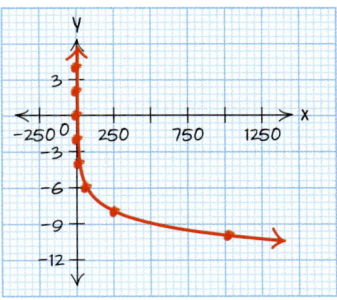

PRACTICE PROBLEM FOR EXAMPLE 2
Sketch the graph of $f(x) = \log_{0.25} x$.

We can see from both of these examples that basic logarithm graphs have a few things in common. They have the y-axis as a vertical asymptote, they have an x-intercept of $(1, 0)$, and they grow or decay rapidly at first and then slowly as the inputs increase.

Domain and Range of Logarithmic Functions

There are several ways to think about the domain and range of a logarithmic function. The inverse relationship between logarithmic and exponential functions can be used to relate the domain and range of exponential functions to the domain and range of logarithmic functions. We can also consider the graphs of logarithmic functions to find their domain and range.

In Chapter 5, we found that the domain and range of an exponential function of the form $f(x) = a \cdot b^x$ were all real numbers, $(-\infty, \infty)$, and all positive real numbers, $(0, \infty)$, respectively. Since logarithms are the inverse of exponentials, the domain and range are simply switched. Therefore, the domain for a logarithm function of the form $f(x) = \log_b x$ will be all positive real numbers or $(0, \infty)$, and the range will be all real numbers or $(-\infty, \infty)$.

We can confirm this domain and range using the graphs of logarithms of the form $f(x) = \log_b x$.

Connecting the Concepts

Why is that domain restricted?

So far, all the functions we have studied have had a domain of all real numbers. The log function is the first function we study that has a restricted domain.

The log function is undefined for negative input values, so those values cannot be part of the domain.

In Chapters 7 and 8, we will study two more functions that also have restricted domains.

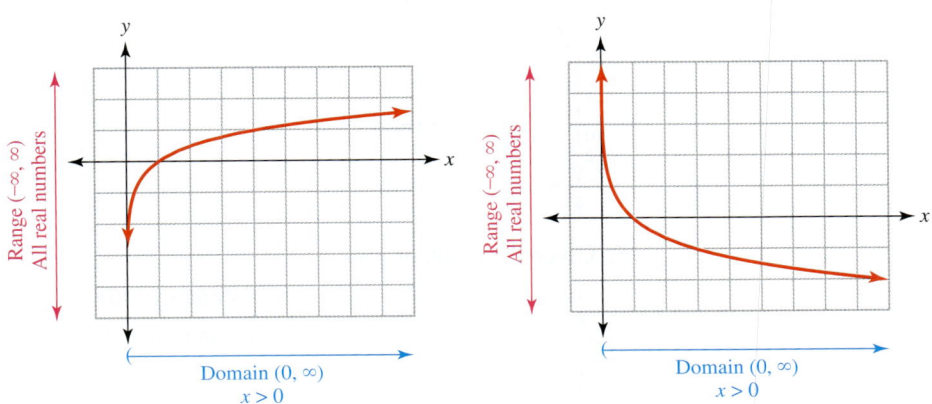

The domain is restricted to only the positive real numbers because we cannot take the logarithm of zero or of a negative number.

Example 3 Finding the domain and range of a logarithmic function

Give the domain and range of $f(x) = \log_7 x$.

SOLUTION
Because this is a logarithm function of the form $f(x) = \log_b x$, we know the following:

Domain: All positive real numbers, or $(0, \infty)$
Range: All real numbers, or $(-\infty, \infty)$

6.3 Exercises

For Exercises 1 through 8, use the given graphs to estimate the answers.

1. Given the graph of $f(x)$,

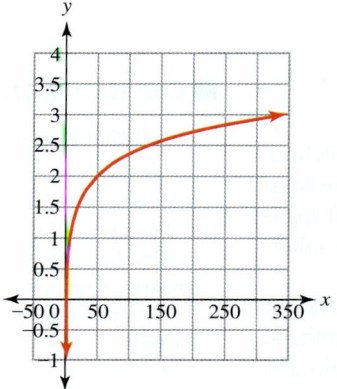

   a. Estimate $f(50)$.
   b. Estimate $f(x) = 2.5$.

2. Given the graph of $g(x)$,

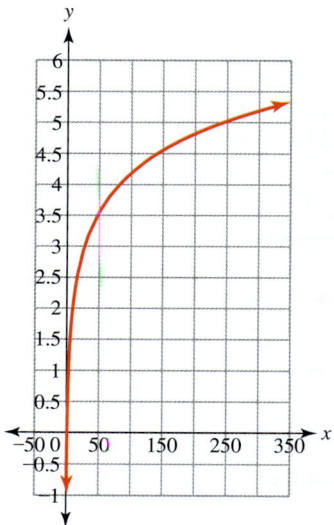

   a. Estimate $g(150)$.
   b. Estimate $g(x) = 5$.

3. Given the graph of $h(x)$,

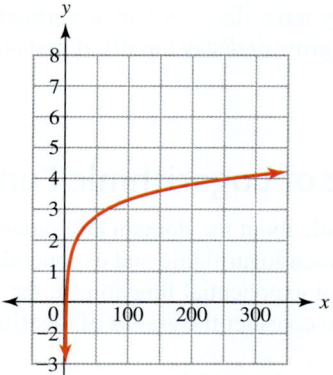

   a. Estimate $h(250)$.
   b. Estimate $h(x) = 2$.

4. Given the graph of $f(x)$,

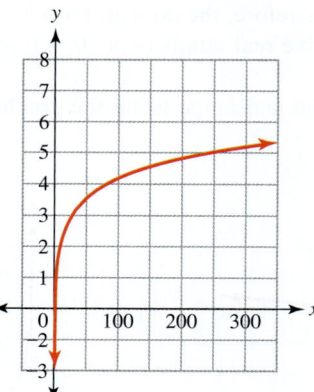

   a. Estimate $f(250)$.
   b. Estimate $f(x) = 4$.

5. Given the graph of $g(x)$,

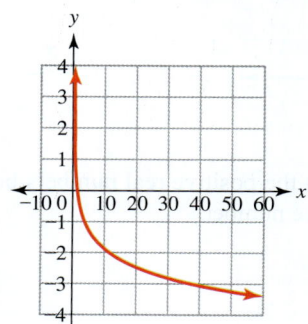

   a. Estimate $g(40)$.
   b. Estimate $g(x) = -2$.

SECTION 6.3 Graphing Logarithmic Functions

6. Given the graph of $f(x)$,

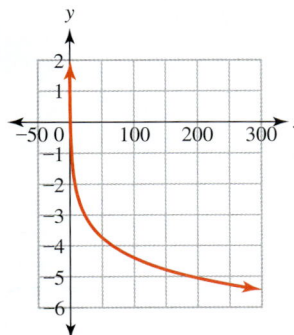

   a. Estimate $f(50)$.
   b. Estimate $f(x) = -5$.

7. Given the graph of $h(x)$,

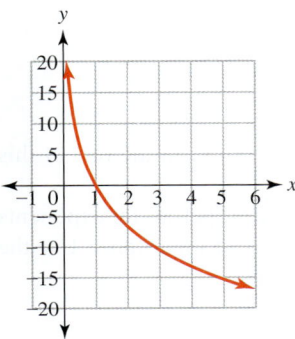

   a. Estimate $h(3)$.
   b. Estimate $h(x) = -15$.

8. Given the graph of $f(x)$,

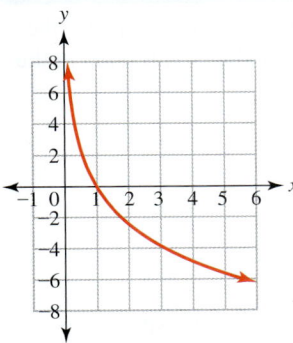

   a. Estimate $f(5)$.
   b. Estimate $f(x) = -4$.

9. Is the base of the logarithm graphed in Exercise 1 greater than or less than 1? Explain.

10. Is the base of the logarithm graphed in Exercise 2 greater than or less than 1? Explain.

11. Is the base of the logarithm graphed in Exercise 5 greater than or less than 1? Explain.

12. Is the base of the logarithm graphed in Exercise 6 greater than or less than 1? Explain.

For Exercises 13 through 26, sketch the graph of the given function. Generate a table of values. Clearly label and scale each axis.

13. $f(x) = \log x$ 14. $f(x) = \ln x$ 15. $g(x) = \log_4 x$

16. $h(x) = \log_6 x$ 17. $f(x) = \log_9 x$ 18. $g(x) = \log_8 x$

19. $f(x) = \log_{20} x$ 20. $h(x) = \log_{25} x$ 21. $g(x) = \log_{0.2} x$

22. $f(x) = \log_{0.4} x$ 23. $h(x) = \log_{0.9} x$ 24. $g(x) = \log_{0.8} x$

25. $h(x) = \log_{0.6} x$ 26. $f(x) = \log_{0.1} x$

27. Give the domain and range of the function graphed in Exercise 1.

28. Give the domain and range of the function graphed in Exercise 2.

29. Give the domain and range of the function graphed in Exercise 5.

30. Give the domain and range of the function graphed in Exercise 6.

31. Give the domain and range of the function given in Exercise 7.

32. Give the domain and range of the function given in Exercise 8.

33. Give the domain and range of the function given in Exercise 13.

34. Give the domain and range of the function given in Exercise 14.

35. Give the domain and range of the function given in Exercise 19.

36. Give the domain and range of the function given in Exercise 20.

37. Give the domain and range of the function given in Exercise 21.

38. Give the domain and range of the function given in Exercise 22.

For Exercises 39 through 44, find the inverse of each function.

39. $f(x) = 2x + 12$ 40. $g(x) = -4x + 20$

41. $f(x) = 4^x$ 42. $h(x) = 6^x$

43. $g(x) = \log_9 x$ 44. $f(x) = \log_{0.4} x$

For Exercises 45 through 52, sketch the graph of the given function.

45. $f(x) = \frac{2}{3}x - 9$ 46. $g(x) = \frac{1}{2}x + 4$

47. $f(x) = 5(1.2)^x$

48. $h(x) = 4(2.5)^x$

49. $g(x) = 2(x + 3)^2 - 8$

50. $h(x) = -0.5(x - 4)^2 + 10$

51. $f(x) = \log_5 x$

52. $g(x) = \log_{0.25} x$

For Exercises 53 through 62, evaluate each logarithm. Round answers to three decimal places.

53. $\log 300$
54. $\log 17.5$
55. $\ln 22$
56. $\ln 480$
57. $\log_5 630$
58. $\log_8 12.5$
59. $\log_{4.5} 1256$
60. $\log_{0.4} 167$
61. $\log_6 0.45$
62. $\log_7 0.28$

For Exercises 63 through 68, solve each logarithm equation by rewriting it in exponential form.

63. $\log_8 x = 3$
64. $\log_6 x = 4$
65. $\log x = 7$
66. $\ln x = 3$
67. $\log_4 x = -2$
68. $\log_2 x = -3$

6.4 Properties of Logarithms

LEARNING OBJECTIVE

■ Use the properties of logarithms to simplify and expand logarithm expressions.

This section will cover the properties of logarithms that we will use in the rest of this chapter to solve exponential and logarithm problems.

Because logarithms are inverses of exponential functions, the properties for exponents that we learned in Section 3.1 will have their related properties for logarithms. Use the concept investigation to find the related properties of logarithms.

CONCEPT INVESTIGATION
Are these logs the same?

1. Evaluate the following logarithm expressions using your calculator. Compare the results for the expression in the left column to the corresponding expression in the right column.

   1a. $\ln(7 \cdot 2) =$ **1b.** $\ln 7 + \ln 2 =$

   2a. $\log(8 \cdot 3) =$ **2b.** $\log 8 + \log 3 =$

   3a. $\ln(5 \cdot 11) =$ **3b.** $\ln 5 + \ln 11 =$

   4a. $\log 100 =$ **4b.** $\log 10 + \log 10 =$

   Describe the relationship between these two columns of logarithm expressions.

2. Evaluate the following logarithm expressions using your calculator. Compare the results for the expression in the left column to the corresponding expression in the right column.

   1a. $\log\left(\dfrac{30}{6}\right) =$ **1b.** $\log 30 - \log 6 =$

   2a. $\ln\left(\dfrac{45}{5}\right) =$ **2b.** $\ln 45 - \ln 5 =$

   3a. $\ln\left(\dfrac{5000}{45}\right) =$ **3b.** $\ln 5000 - \ln 45 =$

   4a. $\log\left(\dfrac{1000}{10}\right) =$ **4b.** $\log 1000 - \log 10 =$

   Describe the relationship between these two columns of logarithm expressions.

These two relationships are stated in the **product property for logarithms** and the **quotient property for logarithms**.

SECTION 6.4 Properties of Logarithms 513

> **The Product Property for Logarithms**
>
> $$\log_b(mn) = \log_b m + \log_b n$$
>
> A logarithm of any base with multiplication inside can be written as two separate logarithms added together.
>
> $$\log_b(5 \cdot 9) = \log_b 5 + \log_b 9$$

> **The Quotient Property for Logarithms**
>
> $$\log_b\left(\frac{m}{n}\right) = \log_b m - \log_b n$$
>
> A logarithm of any base with division inside can be written as two separate logarithms subtracted from one another.
>
> $$\log_b\left(\frac{50}{8}\right) = \log_b 50 - \log_b 8$$

These two properties allow us to either take apart, that is, to expand, a complex logarithm or put two logarithms together into one. In advanced math classes, we sometimes want to expand a complex logarithm into several simpler logarithms. In this class, though, we will be putting logarithms together to become one logarithm so that we can solve a simpler equation.

Example 1 Expanding logarithms

Expand the following logarithms as separate simpler logarithms.

a. $\ln(5 \cdot 17)$ **b.** $\log\left(\dfrac{35}{8}\right)$

c. $\log_4(13xy)$ **d.** $\log_5\left(\dfrac{mn}{9}\right)$

SOLUTION
To separate these logarithms into simpler logarithms, we can use the product and quotient properties for logarithms.

a. $\ln(5 \cdot 17) = \ln 5 + \ln 17$ *Make separate logarithms separated by addition.*

b. $\log\left(\dfrac{35}{8}\right) = \log 35 - \log 8$ *Make separate logarithms separated by subtraction.*

c. $\log_4(13xy) = \log_4 13 + \log_4 x + \log_4 y$

d. $\log_5\left(\dfrac{mn}{9}\right) = \log_5 m + \log_5 n - \log_5 9$

PRACTICE PROBLEM FOR EXAMPLE 1
Expand the following logarithms as separate simpler logarithms.

a. $\ln(5ab)$ **b.** $\log\left(\dfrac{7n}{m}\right)$

Example 2 Combining logarithms

Write the following logarithms as a single logarithm.

a. $\log 8 + \log 7$
b. $\ln(x+2) + \ln(x+5)$
c. $\ln(3x) - \ln(5z)$
d. $\log_7(2m) - \log_7 n$

SOLUTION
Combining these logarithms will use the product and quotient properties for logarithms.

a. $\log 8 + \log 7 = \log(8 \cdot 7) = \log 56$ — Multiply the insides of the logarithms.

b. $\ln(x+2) + \ln(x+5) = \ln[(x+2)(x+5)]$ — Multiply the insides of the logarithms using the distributive property.
$= \ln(x^2 + 7x + 10)$

c. $\ln(3x) - \ln(5z) = \ln\left(\dfrac{3x}{5z}\right)$ — Divide the insides of the logarithms.

d. $\log_7(2m) - \log_7 n = \log_7\left(\dfrac{2m}{n}\right)$ — Divide the insides of the logarithms. The base stays the same.

PRACTICE PROBLEM FOR EXAMPLE 2
Write the following logarithms as a single logarithm.

a. $\ln(18s) - \ln(3t)$
b. $\log_7 5 + \log_7 m + \log_7(3n)$
c. $\log(x+3) + \log(x-7)$

■ CONCEPT INVESTIGATION
What happens to the power?

Evaluate the following logarithm expressions on your calculator. Compare the results for the expression in the left column to the expression in the right column.

1a. $\log(7^3) =$
1b. $3 \log 7 =$

2a. $\ln(8^5) =$
2b. $5 \ln 8 =$

3a. $\log(7^{-2}) =$
3b. $-2 \log 7 =$

Describe the relationship between these two columns of logarithm expressions.

This relationship is one of the most useful properties for logarithms and will be used in a very important role in most exponential and logarithm equations. This relationship is stated in the **power property for logarithms.**

The Power Property for Logarithms

$$\log_b(m^n) = n \log_b m$$

A logarithm of any base with a power inside can be written as that logarithm with that power now being multiplied to the front.

$$\log_b(5^4) = 4 \log_b 5$$

Using the power property for logarithms together with the product and quotient properties allows us to combine or simplify logarithms of all kinds.

SECTION 6.4 Properties of Logarithms 515

Example 3 Expanding logarithms with exponents

Expand the following logarithms as separate simpler logarithms without any exponents.

a. $\ln(5x^2)$ **b.** $\log\left(\dfrac{2x^3}{5}\right)$ **c.** $\log_4(9x^3y^2)$

d. $\log_5\left(\dfrac{m^2n^5}{2z^3}\right)$ **e.** $\log(\sqrt{5xy})$

SOLUTION

To separate these logarithms into simpler logarithms, we can use the product and quotient properties for logarithms.

a. $\ln 5(x^2) = \ln 5 + \ln x^2$ Use the product property for logarithms to write as separate logs with addition.

$= \ln 5 + 2\ln x$ Use the power property for logarithms to bring the exponent down.

b. $\log\left(\dfrac{2x^3}{5}\right) = \log 2 + 3\log x - \log 5$

c. $\log_4(9x^3y^2) = \log_4 9 + 3\log_4 x + 2\log_4 y$

d. $\log_5\left(\dfrac{m^2n^5}{2z^3}\right) = \log_5(m^2) + \log_5(n^5) - \log_5 2 - \log_5(z^3)$ Expand using the product and quotient properties.

$= 2\log_5 m + 5\log_5 n - \log_5 2 - 3\log_5 z$ Bring the exponents down using the power property.

e. $\log(\sqrt{5xy}) = \log\left((5xy)^{\frac{1}{2}}\right)$ Change the square root into a fraction exponent.

$= \dfrac{1}{2}\log(5xy)$ Use the power property to bring the exponent down to the front of the log.

$= \dfrac{1}{2}(\log 5 + \log x + \log y)$ Use the product property to expand the log.

PRACTICE PROBLEM FOR EXAMPLE 3

Expand the following logarithms as separate simpler logarithms without any exponents.

a. $\log_4(5xy^3)$ **b.** $\log\left(\dfrac{3m^2}{2n^3}\right)$ **c.** $\ln(\sqrt{2xy})$

Example 4 Combining logarithms with exponents

Write the following logarithms as a single logarithm with a coefficient of 1.

a. $\log 8 + 3\log x + 4\log y$ **b.** $\ln 3 + 2\ln 5a - \ln b$

c. $\ln x - 5\ln z + 2\ln y$ **d.** $\log_7 2 + \log_7 m - 3\log_7 n + \log_7 5$

SOLUTION

Combining these logarithms will also use the product and quotient properties for logarithms.

a. $\log 8 + 3\log x + 4\log y = \log 8 + \log x^3 + \log y^4$ Use the power property to bring the coefficents into the logs as exponents.

$= \log 8x^3y^4$ Combine the logs using the product property.

b. $\ln 3 + 2 \ln 5a - \ln b = \ln 3 + \ln (5a)^2 - \ln b$ — Use the power property to bring the coefficient into the log as an exponent.

$\qquad = \ln 3 + \ln (25a^2) - \ln b$ — Both the 5 and a are now squared. Simplify the constants.

$\qquad = \ln \left(\dfrac{75a^2}{b}\right)$ — Combine the logs using the product and quotient properties.

c. $\ln x - 5 \ln z + 2 \ln y = \ln \left(\dfrac{xy^2}{z^5}\right)$

d. $\log_7 2 + \log_7 m - 3 \log_7 n + \log_7 5 = \log_7 \left(\dfrac{10m}{n^3}\right)$

PRACTICE PROBLEM FOR EXAMPLE 4

Write the following logarithms as a single logarithm with a coefficient of 1.

a. $2 \log_5 m + 3 \log_5 n + \log_5 7 - 2 \log_5 p$

b. $\log x + \dfrac{1}{2} \log y - 3 \log z$

c. $5 \ln x + 3 \ln 5z + 5 \ln y - 5 \ln 2$

These properties for logarithms together with the properties discussed in Section 6.2 allow us to work with logarithms, expanding or simplifying them as needed.

6.4 Exercises

For Exercises 1 through 28, expand the logarithms as separate simpler logarithms with no exponents.

1. $\log (5x)$
2. $\log (9y)$
3. $\ln (xy)$
4. $\log_3 (2ab)$
5. $\ln (4ab^3)$
6. $\log (5xy^2)$
7. $\ln (2h^2k^3)$
8. $\log_7 (3x^3y^4)$
9. $\log_3 \left(\dfrac{1}{2}ab^2c^3\right)$
10. $\ln \left(\dfrac{2}{3}x^2yz^5\right)$
11. $\log \left(\dfrac{x}{y}\right)$
12. $\log_8 \left(\dfrac{x}{8}\right)$
13. $\log \left(\dfrac{12}{m}\right)$
14. $\ln \left(\dfrac{3x}{y}\right)$
15. $\log \left(\dfrac{2x^2}{y}\right)$
16. $\log \left(\dfrac{5x^3y}{2z^2}\right)$
17. $\ln \left(\dfrac{3x^4y^3}{z}\right)$
18. $\log_2 \left(\dfrac{4a^5b^2}{c^3}\right)$
19. $\log (\sqrt{5x})$
20. $\ln (\sqrt{6a})$
21. $\ln (\sqrt{7ab})$
22. $\log \sqrt{5xy}$
23. $\log_9 (\sqrt{2w^2z^3})$
24. $\log_9 (\sqrt{7a^5b^3})$
25. $\ln (\sqrt[5]{m^2p^3})$
26. $\log_7 (\sqrt[3]{3x^3y^4})$
27. $\log_{15} \left(\dfrac{\sqrt{3x^4y^3}}{z^5}\right)$
28. $\ln \left(\dfrac{\sqrt{5a^3b^5}}{c^4}\right)$

For Exercises 29 through 54, write the logarithms as a single logarithm with a coefficient of 1.

29. $\ln x + \ln y$
30. $\log 4 + \log a$
31. $\log a^2 + \log b^3$
32. $\log 2x^4 + \log 5y$
33. $\ln x - \ln y$
34. $\log a - \log b$
35. $\log 5 + 2 \log x + \log y + 3 \log z$
36. $\log 3 + 3 \log x + 4 \log y - \log 7$
37. $\log_3 a + 2 \log_3 b - 5 \log_3 c$
38. $\ln 5 + 2 \ln x + 3 \ln z - 4 \ln y$
39. $\log_5 7 + \dfrac{1}{2} \log_5 x + \dfrac{1}{2} \log_5 y$
40. $\log 4 + \dfrac{1}{2} \log a - \dfrac{1}{2} \log b$
41. $\dfrac{1}{2} \ln 7 + \dfrac{1}{2} \ln a + \dfrac{3}{2} \ln b - 4 \ln c$
42. $\log_9 5 + \dfrac{2}{3} \log_9 x + \dfrac{2}{3} \log_9 y$
43. $5 \ln 7 + 2 \ln a + 4 \ln b - 3 \ln c - 2 \ln d$
44. $\log 3x + 3 \log x - 2 \log 7y$
45. $\log_3 a^2 + 2 \log_3 bc - 5 \log_3 3c$
46. $\ln 5x^3 + 2 \ln x + 3 \ln z - 4 \ln yz$
47. $\log_5 7 + 2 \log_5 xy + \log_5 xy$
48. $\log x + 3 \log xy + 2 \log y$
49. $\dfrac{1}{2} \log 5 + \dfrac{1}{2} \log 3x + \dfrac{5}{2} \log y - 4 \log z$
50. $\dfrac{1}{2} \ln 7 + \dfrac{1}{2} \ln 5a + \dfrac{3}{2} \ln ab - 3 \ln c$
51. $\log (x + 6) + \log (x - 2)$
52. $\log (x + 1) + \log (x - 5)$
53. $\ln (x - 3) + \ln (x - 7)$
54. $\ln (x - 10) + \ln (x + 4)$

For Exercises 55 and 56, the students were asked to expand the given logarithm. Explain what the student did wrong.

55. $\log(x + 5)$

> **Sharma**
> $\log(x + 5) = (\log x)(\log 5)$

56. $\ln(x - y)$

> **Mark**
> $\ln(x - y) = \dfrac{\ln x}{\ln y}$

For Exercises 57 and 58, the students were asked to write the logarithms as a single logarithm with a coefficient of one. Explain what the student did wrong.

57. $\log x + \log y$

> **Hung**
> $\log x + \log y = \log x \log y$

58. $\log a - \log b$

> **Michelle**
> $\log a - \log b = \dfrac{\log a}{\log b}$

For Exercises 59 through 68, use the given information and the properties of logarithms to find a requested value.

59. If $\log_b 2 = 3$ and $\log_b 16 = 4$, find $\log_b 32$.

60. If $\log_b 5 = 10$ and $\log_b 20 = 4$, find $\log_b 100$.

61. If $\log_b 2 = 3$ and $\log_b 16 = 4$, find $\log_b 8$.

62. If $\log_b 5 = 10$ and $\log_b 20 = 4$, find $\log_b 4$.

63. If $\log_b 30 = 8$ and $\log_b 2 = 14$, find $\log_b 15$.

64. If $\log_b 30 = 8$ and $\log_b 2 = 14$, find $\log_b 60$.

65. If $\log_b 100 = 4$ and $\log_b 500 = 6$, find $\log_b 50000$.

66. If $\log_b 100 = 4$ and $\log_b 500 = 6$, find $\log_b 5$.

67. If $\log_b 5 = 10$ and $\log_b 20 = 4$, find $\log_b 0.25$.

68. If $\log_b 20 = 11$ and $\log_b 40 = 13$, find $\log_b 0.5$.

6.5 Solving Exponential Equations

LEARNING OBJECTIVES
- Solve exponential equations.
- Solve applications of compounding interest formulas.

Solving Exponential Equations

In this section, we will use the properties for exponentials and logarithms to solve problems involving exponentials. There are two basic techniques used to solve exponential equations. We will find that the power property for logarithms and the change of base formula are some of the most important properties that we use in the solving process. Recall the following rules for exponents and properties of logarithms.

Rules for Exponents

1. $x^m \cdot x^n = x^{m+n}$
2. $(xy)^m = x^m y^m$
3. $\dfrac{x^m}{x^n} = x^{m-n}$ $x \neq 0$
4. $\left(\dfrac{x}{y}\right)^m = \dfrac{x^m}{y^m}$ $y \neq 0$
5. $(x^m)^n = x^{mn}$
6. $x^0 = 1$ $0^0 =$ undefined
7. $x^{-n} = \dfrac{1}{x^n}$ $x \neq 0$
8. $x^{\frac{1}{n}} = \sqrt[n]{x}$ if n is even $x \geq 0$

Properties for Logarithms

1. $\log_b(mn) = \log_b m + \log_b n$
2. $\log_b(m^n) = n \log_b m$
3. $\log_b\left(\dfrac{m}{n}\right) = \log_b m - \log_b n$
4. $\log_b x = \dfrac{\log x}{\log b}$
5. $\log_b(b) = 1$
6. $\log_b(1) = 0$

Remember that when working with equations, what we do to one side of an equation we must also do to the other side of the equation. Because a logarithm is a one-to-one function, we can take the logarithm of both sides of an equation much as we would square both sides. When taking the logarithm of both sides, we are not multiplying, but each side is now the input to the logarithm. Taking the logarithm of both sides of an exponential equation and then using the power property for logarithms is one option. Writing an exponential equation in logarithm form and then using the change of base formula is another option when solving. In all of these problems, we use either log or ln, since these are logarithms that we can calculate on our calculator.

Example 1 — Solving an exponential equation: Two options

Solve $4^x = 20$. Round your answer to three decimal places.

SOLUTION

Option 1: Take the log of both sides. Use the power property for logs.

$4^x = 20$ — Take the log of both sides.

$\log(4^x) = \log(20)$ — Use the power property for logs to bring down the variable x.

$x \log 4 = \log 20$

$\dfrac{x \log 4}{\log 4} = \dfrac{\log 20}{\log 4}$ — Divide both sides by log 4.

$x = \dfrac{\log 20}{\log 4}$

$x \approx 2.161$

Option 2: Rewrite in log form. Use the change of base formula.

$4^x = 20$ — Rewrite in log form.

$x = \log_4 20$

$x = \dfrac{\log 20}{\log 4}$ — Use the change of base formula.

$x \approx 2.161$

Either of these methods works well for solving exponential equations. Using rewriting in logarithm form and using the change of base formula can lead to fewer mistakes with more complicated equations, so we will use this option more often.

In Section 6.2, we introduced the change of base formula, but we never demonstrated how it can be derived by using the properties for logarithms. If we take $\log_3 400 = x$, we can use the same solving technique that we used in option 1 of Example 1 to show how the change of base formula comes about.

$\log_3 400 = x$ — We first rewrite the log in exponential form.

$3^x = 400$

$\log(3^x) = \log 400$ — Take the log of both sides of the equation and use the power property for logs to bring the exponent down.

$x \log 3 = \log 400$

$x = \dfrac{\log 400}{\log 3}$ — Divide both sides by log 3 to isolate x. This last result is the same as the change of base formula.

This same solving technique can be done without numbers to give us the change of base formula in abstract form, as in the definition found in Section 6.2.

Steps to Solving Exponential Equations (Option 1)

1. Isolate the base and exponent on one side of the equation.
2. Take the logarithm of both sides of the equation.
3. Use the power property for logarithms to bring the exponent down to the front of the log.
4. Solve the equation by isolating the variable.

Skill Connection

Choosing a base

In working with exponential equations, it is best to use a logarithm base that is convenient to calculate and find approximate values for.

Within any problem you solve, you can use any base logarithm you wish. You should stay consistent throughout the problem.

SC Example: Solve

$9^x = 62$

Solution:
Using log:

$\log(9^x) = \log 62$

$x \log 9 = \log 62$

$x = \dfrac{\log 62}{\log 9}$

$x \approx 1.878$

Using ln:

$\ln(9^x) = \ln 62$

$x \ln 9 = \ln 62$

$x = \dfrac{\ln 62}{\ln 9}$

$x \approx 1.878$

In this example,

$x = \dfrac{\log 62}{\log 9} = \dfrac{\ln 62}{\ln 9}$

are exact answers, since they have not been rounded during calculation.

$x \approx 1.878$

is an approximate solution, since it has been rounded after calculation.

SECTION 6.5 Solving Exponential Equations **519**

> **Steps to Solving Exponential Equations (Option 2)**
> 1. Isolate the base and exponent on one side of the equation.
> 2. Rewrite the exponential in logarithm form.
> 3. Use the change of base formula to write the log in a convenient base.
> 4. Solve the equation by isolating the variable.

Example 2 Solving an exponential model

As the population ages, more and more people will develop the symptoms of dementia resulting from Alzheimer's disease. According to medical studies shown at Brain.com, the number of people worldwide who suffer from Alzheimer's disease could double in the next 25 years. Data from past numbers and projections for the future are given in the bar graph.

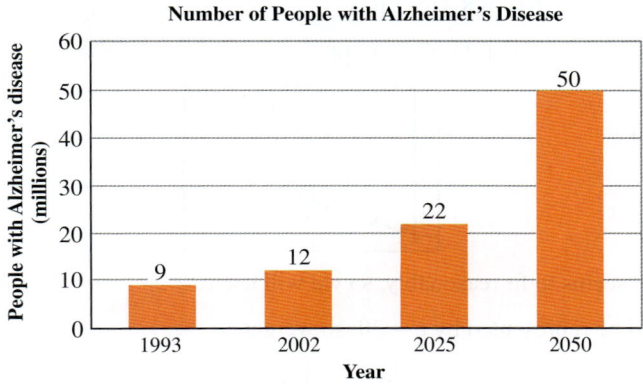

Source: Data estimated from information given at www.brain.com.

a. Write an equation for a model of these data.

b. Estimate the number of people with Alzheimer's disease in 2010.

c. Find when there will be 30 million people with Alzheimer's.

SOLUTION

a. Define the variables.

$A(t)$ = Number of people in the world with Alzheimer's disease in millions

t = Time in years since 1990

With these definitions, we get the following adjusted data.

t	$A(t)$
3	9
12	12
35	22
60	50

These data give us the following scatterplot.

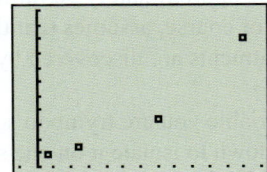 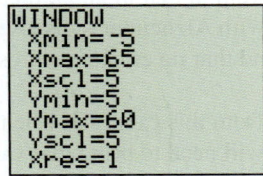

Because of the description given in the problem and the sudden rise in the data, we can choose an exponential model. Using the points (12, 12) and (60, 50), we get

$(12, 12)$ and $(60, 50)$

$12 = a \cdot b^{12}$ $\quad 50 = a \cdot b^{60}$ — Use the two points to write two equations in standard form.

$\dfrac{50}{12} = \dfrac{a \cdot b^{60}}{a \cdot b^{12}}$ — Divide the two equations and solve for b.

$4.167 = b^{48}$

$(4.167)^{1/48} = (b^{48})^{1/48}$ — Use the power rule for exponents to isolate b.

$1.03 = b$

$12 = a(1.03)^{12}$

$12 = 1.426a$

$\dfrac{12}{1.426} = \dfrac{1.426a}{1.426}$

$8.415 = a$

$A(t) = 8.415(1.03)^t$

This model gives us the following graph.

This model seems to fit reasonably, so we will use it.

b. 2010 would be represented by $t = 20$, so we get

$$A(20) = 15.198$$

Thus, in 2010, there will be approximately 15 million people in the world with Alzheimer's disease.

c. 30 million people with Alzheimer's disease would be represented by $A(t) = 30$, so we get

$30 = 8.415(1.03)^t$

$\dfrac{30}{8.415} = \dfrac{8.415(1.03)^t}{8.415}$ — Start by isolating the exponential part.

$3.565 \approx 1.03^t$

$\log_{1.03} 3.565 \approx t$ — Rewrite the exponential in logarithm form.

$\dfrac{\log 3.565}{\log 1.03} \approx t$ — Use the change of base formula.

$43 \approx t$

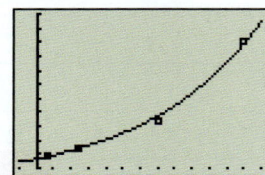

Check the answer, using the calculator.

Therefore, we find that in 2033, there will be approximately 30 million people worldwide with Alzheimer's disease. This, of course, assumes that the current trend continues and that no effective cures or treatments are discovered by then.

As you see from this example, when the variable you are trying to solve for is in the exponent, you will need to bring that variable down to isolate it on one side of the equation. This is why the power property for logarithms is so crucial. By taking the logarithm of both sides of the equation, we can use the power property for logarithms to bring the variable in the exponent down and make it multiplication. It is crucial to first get the exponential part of the equation by itself before you take the logarithm of both sides.

Example 3 Solving exponential equations

Solve the following exponential equations. Round your answers to three decimal places.

a. $5^{3x} = 5.6$ **b.** $3^x - 9 = 21$ **c.** $2(8)^{x-2} = 24$ **d.** $10^{x^2-2} = 100$

SOLUTION

a.
$$5^{3x} = 5.6$$ The base and exponent are already isolated.
$$\log_5 5.6 = 3x$$ Rewrite the exponential in logarithm form.
$$\frac{\log 5.6}{\log 5} = 3x$$ Use the change of base formula.
$$\frac{\log 5.6}{3 \log 5} = x$$ Isolate x by dividing both sides by 3.
$$x \approx 0.357$$

Check the answer, using the calculator.

b.
$$3^x - 9 = 21$$ Isolate the base and exponent by adding 9 to both sides.
$$+9 +9$$
$$3^x = 30$$ Take the logarithm of both sides.
$$\ln(3^x) = \ln 30$$ Use the power property for logarithms to bring down the exponent.
$$x \ln 3 = \ln 30$$
$$\frac{x \ln 3}{\ln 3} = \frac{\ln 30}{\ln 3}$$ Solve for x.
$$x \approx 3.096$$

Check the answer, using the calculator.

c.
$$2(8)^{x-2} = 24$$
$$\frac{2(8)^{x-2}}{2} = \frac{24}{2}$$ Isolate the base and exponent by dividing both sides by 2.
$$8^{x-2} = 12$$
$$x - 2 = \log_8 12$$ Rewrite in logarithm form.
$$x - 2 = \frac{\log 12}{\log 8}$$ Use the change of base formula.
$$ +2 +2$$ Isolate x.
$$x = \frac{\log 12}{\log 8} + 2$$ This is the exact solution.
$$x \approx 3.195$$ This is an approximate solution.

Using Your TI Graphing Calculator

When dividing a number by more than one quantity, you will need to use parentheses for the calculator to perform the operation you want to do.

To correctly calculate the fraction

$$\frac{\log 5.6}{3 \log 5}$$

you will need to put parentheses around the entire denominator.

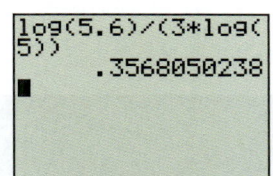

Without the parentheses around the denominator, the calculator will follow the order of operations and first divide by 3 and then multiply by log 5.

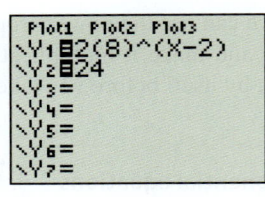

Check the answer, using the calculator. Be sure to use parentheses around the exponent.

d.
$$10^{x^2-2} = 100$$
$$x^2 - 2 = \log 100 \quad \text{Rewrite in logarithm form.}$$
$$x^2 - 2 = 2 \quad \text{Evaluate the logarithm.}$$
$$x^2 = 4 \quad \text{We have a quadratic left to solve.}$$
$$x = \pm\sqrt{4} \quad \text{Remember to use the plus/minus symbol when using the square root property.}$$
$$x = \pm 2$$

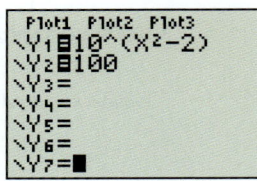

Check the answer, using the calculator. Be sure to use parentheses around the exponent.

PRACTICE PROBLEM FOR EXAMPLE 3

Solve the following exponential equations. Round your answers to three decimal places.

a. $2^{5x} - 12 = 7$ **b.** $7^{2x+3} = 81$

Example 4 — Solving an exponential application

In Section 5.5, we were given the following model for the California sea otter population.
$$O(t) = 31.95(1.057)^t$$
where $O(t)$ represents the number of California sea otters t years since 1900.
Source: Model derived from data from the National Biological Service.

According to this model, when were there 1500 California sea otters?

SOLUTION

1500 California sea otters are represented by $O(t) = 1500$, so we get

$$1500 = 31.95(1.057)^t$$
$$\frac{1500}{31.95} = \frac{31.95(1.057)^t}{31.95} \quad \text{Isolate the base and exponent.}$$
$$46.95 \approx 1.057^t$$
$$\log 46.95 \approx \log 1.057^t \quad \text{Take the logarithm of both sides.}$$
$$\log 46.95 \approx t \log 1.057 \quad \text{Use the power property for logarithms.}$$
$$\frac{\log 46.95}{\log 1.057} \approx t \quad \text{Isolate the variable } t.$$
$$69.4 \approx t$$

Check the answer, using a calculator. Use more decimal places to confirm that the solution is correct.

We know that in 1969, there were approximately 1500 California sea otters.

PRACTICE PROBLEM FOR EXAMPLE 4

In Section 5.5, we found the following model for the population of Boulder, Colorado.

$$P(t) = 96000(1.1)^t$$

where $P(t)$ represents the population of Boulder, Colorado, t years since 2000.
Source: Model derived from data in Census 2000.

According to this model, when will the population of Boulder, Colorado, reach 1 million?

Compounding Interest

Recall, from Section 5.5, the following formulas for compounding interest problems.

Compounding Interest

$$A = P\left(1 + \frac{r}{n}\right)^{nt}$$

A = Amount in the account
P = Principal (the amount initially invested)
r = Annual interest rate written as a decimal
n = Number of times the interest is compounded in one year
t = Time the investment is made for, in years

Compounding Continuously

$$A = Pe^{rt}$$

In many situations, we might want to know how long it will take to double an investment or at what rate we need to invest money in order for it to double in a certain amount of time. The compounding interest formulas can be used to answer these questions. When solving for time, we will be solving an exponential equation. When solving for the rate, we will be solving a power equation. When solving the equation for compounding continuously, we will be solving an exponential equation.

Example 5 Find a rate given doubling time

JP Manufacturing has $100,000 to invest. They need to double their money in 10 years to replace a piece of equipment that they rely on for their business. Find the annual interest rate compounded daily that they need in order to double their money in the 10 years.

SOLUTION
In this case, we have the following given amounts.

$100,000 is to be invested, so $P = 100000$

The investment is for 10 years, so $t = 10$.

They need to double their money, so the account balance will need to be $200,000, so $A = 200000$.

The interest rate is to compound daily, so $n = 365$.

This leaves us to find the interest rate r, so we need to solve a power equation.

$$200000 = 100000\left(1 + \frac{r}{365}\right)^{365(10)}$$ Substitute the values of P, n, A, and t.

$$200000 = 100000\left(1 + \frac{r}{365}\right)^{3650}$$ Simplify the exponent.

$$\frac{200000}{100000} = \frac{100000\left(1 + \frac{r}{365}\right)^{3650}}{100000}$$ Isolate the parentheses.

$$2 = \left(1 + \frac{r}{365}\right)^{3650}$$

$$2^{\frac{1}{3650}} = \left(\left(1 + \frac{r}{365}\right)^{3650}\right)^{\frac{1}{3650}}$$ Use a reciprocal exponent to eliminate the exponent.

$$1.000189921 = 1 + \frac{r}{365}$$ Solve for r.

$$\begin{array}{r} 1.000189921 = 1 + \frac{r}{365} \\ -1 -1 \phantom{+\frac{r}{365}} \\ \hline 0.000189921 = \frac{r}{365} \end{array}$$

When you subtract 1, the calculator will usually give you the result in scientific notation.

$$365(0.000189921) = 365\left(\frac{r}{365}\right)$$

$$0.0693 \approx r$$

Check the answer.

Therefore, JP Manufacturing needs to find an account that will pay at least 6.93% compounded daily to double the initial amount in 10 years.

PRACTICE PROBLEM FOR EXAMPLE 5

What interest rate compounded monthly do you need for $60,000 to double in 15 years?

Example 6 — Finding the doubling time given an interest rate

Many people use U.S. government savings bonds as a long-term investment. When you purchase a series I or EE savings bond, you pay half of the face value of the bond. The bond matures when it has earned enough interest to be worth the entire face value of the bond. If you invested $4000 in series EE bonds, how long would it take the bonds to mature if the bonds pay an average of 4% interest compounded semiannually?

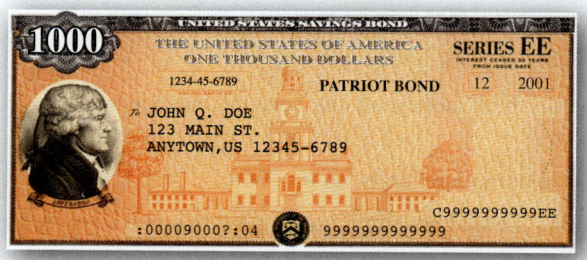

SOLUTION

Because we invested $4000, we will want the bonds to grow until the value doubles that amount, so we have the following values.

$$P = 4000 \quad A = 8000$$
$$r = 0.04 \quad n = 2$$

We need to find the time it takes to double the investment, so we need to solve for t.

$$A = P\left(1 + \frac{r}{n}\right)^{nt}$$

$$8000 = 4000\left(1 + \frac{0.04}{2}\right)^{2t}$$

$2 = \left(1 + \frac{0.04}{2}\right)^{2t}$ Isolate the base and exponent.

$2 = 1.02^{2t}$ Simplify the base.

$\log_{1.02} 2 = 2t$ Rewrite in logarithm form.

$\dfrac{\log 2}{\log 1.02} = 2t$ Use the change of base formula.

$\dfrac{\frac{\log 2}{\log 1.02}}{2} = \dfrac{2t}{2}$ Isolate the variable.
                       Check the answer.

$\dfrac{\frac{\log 2}{\log 1.02}}{2} = t$

$17.5 \approx t$

```
4000(1+0.04/2)^(
2*17.5)
          7999.558211
```

It will take about 17.5 years for the $4000 to double at 4% compounded semiannually.

PRACTICE PROBLEM FOR EXAMPLE 6

How long will it take for $10,000 to triple if it is invested in an account that pays 5% compounded weekly?

Example 7   Finding the doubling time given continuously compounded interest rate

If $2000 is invested in a savings account that pays 5% annual interest compounded continuously, how long will it take for the investment to double?

SOLUTION

The interest is being compounded continuously, so we will use the formula $A = Pe^{rt}$. We know the following values.

$$P = 2000 \quad r = 0.05 \quad A = 4000$$

t is the missing quantity that we are being asked to solve for. We will use ln since e is the base of the exponential.

$4000 = 2000e^{0.05t}$

$2 = e^{0.05t}$

$\ln 2 = 0.05t$

$\dfrac{\ln 2}{0.05} = t$

$13.86 \approx t$

Check the answer.
```
2000e^(0.05*13.8
6)
          3999.411321
```

Therefore, it will take about 13.86 years for this investment to double.

PRACTICE PROBLEM FOR EXAMPLE 7

If $7000 is invested in a savings account that pays 8% annual interest compounded continuously, how long will it take for the investment to double?

6.5 Exercises

For Exercises 1 through 10, solve the exponential equations. Round all answers to three decimal digits.

1. $3^w = 125$
2. $7^x = 240$
3. $5^x = 373$
4. $1.5^x = 48$
5. $6^x = 0.48$
6. $9^x = 0.246$
7. $0.4^x = 0.13$
8. $0.75^x = 0.31$
9. $0.6^x = 24$
10. $0.12^x = 140$

11. The use of supercomputers at universities all over the country has been a vital part of research and development for many areas of education. The Cray C90 housed at Rutgers University in New Jersey has seen dramatic increases in its use per year since 1987. The number of hours the Cray C90 has been used per academic year are given in the table.

Academic Year	Cray C90 Hours
1987	32
1988	100
1989	329
1990	831
1991	1,685
1992	2,233
1993	3,084
1994	8,517
1995	15,584
1996	27,399

Source: Rutgers University High Performance Computing.

a. Find an equation for a model of these data.
b. Give a reasonable domain and range for this model.
c. Estimate the number of hours the Cray C90 was used in 1997.
d. According to your model, when will the Cray C90 be used 500,000 hours per year?

12. The number of weeds in the author's backyard started to decline rapidly after being treated with weed killer. The approximate number of weeds remaining is recorded in the table.

Days	Weeds Remaining
1	900
3	345
5	130
6	80
7	50

a. Find an equation for a model of these data.
b. Give a reasonable domain and range for this model.
c. Estimate the number of weeds remaining after 4 days.
d. According to your model, when will there be only 10 weeds remaining?

For Exercises 13 through 30, solve the following exponential equations. Round all answers to three decimal digits.

13. $7^{x-3} = 16807$
14. $3^{t-7} = 6561$
15. $4^{x+7} = 3$
16. $5^{x+2} = 235$
17. $604(0.4)^x = 0.158$
18. $40(0.3)^x = 609663$
19. $5(3.2)^x = 74.2$
20. $4(1.5)^x = 30.375$
21. $2500\left(\frac{1}{5}\right)^x = 0.032$
22. $81000\left(\frac{1}{9}\right)^x = 0.015$
23. $2^t - 58 = 6$
24. $4^t - 5700 = 59836$
25. $1.5^h + 20 = 106.5$
26. $3.4^t + 8 = 47.304$
27. $-3(2.5)^m - 89 = -3262$
28. $-5(1.4)^x - 548 = -6475$
29. $3(2)^{n+2} = 96$
30. $5(4)^{3x+2} - 830 = 14.5$

31. According to United Nations records, the world population in 1975 was approximately 4 billion. Statistics indicate that the world population since World War II has been growing at a rate of 1.9% per year. Assuming exponential growth, the world population can be modeled by

$$P(t) = 4(1.019)^t$$

where $P(t)$ is the world population in billions and t is the time in years since 1975. When will the world population reach 10 billion?

32. In Exercise 19 in Section 5.5 on page 470, you found a model such as

$$W(t) = 350(1.14)^t$$

where $W(t)$ is the number of humpback whales off the coast of Australia t years since 1981. According to this model, when will the whale population reach 10,000?

33. The white-tailed deer population in the northeastern United States has been growing since the early 1980s. The population of white-tailed deer can be modeled by

$$W(t) = 1.19(1.08)^t$$

where $W(t)$ represents the number of white-tailed deer in millions in the Northeast t years since 1980. When will the white-tailed deer population reach 5 million?
Source: Model derived from data from the National Biological Service.

34. Dr. Arnd Leike, a professor of physics at the University of Munich in Germany, won a 2002 Ig Nobel Prize in Physics for his investigation into the "exponential decay" of a beer's head (the foam on top when poured). After testing several beers for the rate of decay of the head, he came to the conclusion that a beer could be identified by its unique head decay rate. The following functions are for three of the beers that Dr. Leike tested.

$$E(s) = 16.89(0.996)^s$$
$$B(s) = 13.25(0.991)^s$$
$$A(s) = 13.36(0.993)^s$$

where $E(s)$ represents the height of the head of an Erdinger beer, $B(s)$ is the height of the head of a Budweiser Budvar, and $A(s)$ is the head height of an Augustinebrau. All head heights are in centimeters s seconds after being poured.
Source: European Journal of Physics, Volume 23, 2002.

   a. You might not want to drink a beer until the head has decayed to only 1 cm. How long will it take the head on the Erdinger beer to reach that height?
   b. How long will it take the head on the Budweiser Budvar beer to decay to only 1 cm?
   c. How long will it take the head on the Augustinebrau beer to decay to only 1 cm?

35. In Example 4 in Section 5.5 on page 466, we found the following model for the population of South Africa.

$$P(t) = 43.8(0.995)^t$$

where $P(t)$ is the population of South Africa in millions t years since 2008.

   a. According to this model, when will South Africa's population be only 40 million?
   b. If this trend continues, how long will it take South Africa's population to be half of what it was in 2008?

36. In Exercise 21 in Section 5.5 on page 470, we found a model for the population of Denmark

$$P(t) = 5.5(1.00295)^t$$

where $P(t)$ is the population of Denmark in millions t years since 2008.

   a. According to this model, when will Denmark's population reach 6 million?
   b. How long will it take Denmark's population to double if it continues to grow at this rate?

For Exercises 37 through 48, solve the exponential equations. Round all answers to three decimal digits.

37. $3^{5x-4} = 729$

38. $4^{0.5x-7} = 4096$

39. $3(4.6)^{2m+7} = 17.3$

40. $7(1.8)^{3x+5} = 12.8$

41. $2^{x^2+6} = 1024$

42. $10^{x^2-5} = 0.10$

43. $\left(\dfrac{1}{2}\right)^{-x^2+4} = 4$

44. $\left(\dfrac{1}{4}\right)^{x^2-3x} = 16$

45. $10(2)^t - 450 = 6(2)^t + 15934$

46. $-4(3)^m + 4300 = (3)^m - 6635$

47. $5(4.3)^x + 7 = 2(4.3)^x + 89$

48. $7(2.5)^x + 12 = 4(2.5)^x + 49$

49. Cogs R Us has $40,000 to invest. They need to double their money in 8 years to replace a piece of equipment that they rely on for their business. Find the annual interest rate compounded daily that they need in order to double their money in the 8 years.

50. John and Mary hope to pay for their child's college education. They have $20,000 to invest and believe that they will need about 4 times that much money in about 17 years to pay for college. Find the annual interest rate compounded daily that they need in order to have the money they need in the 17 years.

51. What interest rate compounded continuously do you need to earn to make $5000 double in 12 years?

52. What interest rate compounded continuously do you need to earn to make $7000 double in 9 years?

53. What interest rate compounded monthly do you need to earn to make $50,000 double in 7 years?

54. What interest rate compounded monthly do you need to earn to make $46,000 double in 10 years?

55. How long will it take an investment of $5000 to double if it is invested in an account paying 7% interest compounded monthly?

56. How long will it take an investment of $4000 to double if it is invested in an account paying 5% interest compounded daily?

57. How long will it take an investment of $10,000 to double if it is invested in an account paying 9% interest compounded continuously?

58. How long will it take an investment of $8,000 to double if it is invested in an account paying 4.5% interest compounded continuously?

59. How long will it take an investment of $8000 to triple if it is invested in an account paying 2% interest compounded daily?

60. How long will it take an investment of $4000 to triple if it is invested in an account paying 6% interest compounded daily?

61. How long will it take an investment of $7000 to triple if it is invested in an account paying 2.5% interest compounded continuously?

62. How long will it take an investment of $12,000 to triple if it is invested in an account paying 5.75% interest compounded continuously?

528 CHAPTER 6 Logarithmic Functions

In Exercises 63 through 68, find the inverse for the given function.

63. $f(x) = 5(3)^x$
64. $g(x) = 3.4(10)^x$
65. $h(x) = -2.4(4.7)^x$
66. $h(x) = -3.5(1.8)^x$
67. $g(x) = -3.4e^x$
68. $f(x) = 4.2e^x$

For Exercises 69 through 76, solve the equations. Round all answers to three decimal digits.

69. $4n^2 + 3n - 8 = 0$
70. $-2b^2 - 5b + 10 = 3$
71. $7d^5 + 20 = 300$
72. $4w^7 - 8400 = -260$
73. $3(1.6)^x - 20 = 400$
74. $-7(0.8)^x + 10 = 2$
75. $4(n - 3)^2 + 10 = 30$
76. $-2(k + 4)^2 + 8 = -14$

For Exercises 77 through 82, solve the system of equations by graphing the equations on your graphing calculator.

77. $y = 2.5(3)^x$
    $y = 150(0.8)^x$
78. $y = 2(5)^x$
    $y = 300(0.6)^x$
79. $y = -5(0.8)^x$
    $y = -1.8x - 20$
80. $y = 1.5(1.8)^x$
    $y = 3x + 15$
81. $y = 105(0.7)^x$
    $y = -1.5x^2 + 5x + 50$
82. $y = -24(1.2)^x$
    $y = 3x^2 + 2x - 75$

6.6 Solving Logarithmic Equations

LEARNING OBJECTIVES

- Solve logarithmic equations.
- Use logarithms in applications.

Applications of Logarithms

In this section, we will learn to use the properties for both exponentials and logarithms to solve equations involving logarithms. Logarithms are used most often in areas of science that require us to work with very large or extremely small numbers.

Solving a logarithm equation is very similar to solving an exponential equation except that you will rewrite the logarithm into exponential form instead of the other way around.

> **Steps to Solving Logarithmic Equations**
> 1. Isolate the logarithm(s) on one side of the equation.
> 2. Combine logarithms into a single logarithm if necessary.
> 3. Rewrite the logarithm in exponential form.
> 4. Solve the equation by isolating the variable.

Example 1 Solving a logarithmic equation

Solve the logarithmic equation $\log(x + 20) = 2$.

SOLUTION

This equation starts out with a logarithm isolated on the left side of the equation, and it is equal to a constant. In this case, we can rewrite the logarithm in exponential form and then solve.

$$\log(x + 20) = 2$$
$$10^2 = x + 20 \quad \text{Rewrite the logarithm into exponential form.}$$
$$100 = x + 20$$
$$80 = x$$
$$\log(80 + 20) = 2 \quad \text{Check the answer.}$$
$$\log(100) = 2$$
$$2 = 2 \quad \text{The answer works.}$$

PRACTICE PROBLEM FOR EXAMPLE 1
Solve the logarithmic equation $\log(x + 4) = 4$.

One area that uses logarithms is the measurement of earthquakes. The size of an earthquake is most often measured by its magnitude. In 1935, Charles Richter defined the magnitude of an earthquake to be

$$M = \log\left(\frac{I}{S}\right)$$

where I is the intensity of the earthquake (measured by the amplitude in centimeters of a seismograph reading taken 100 km from the epicenter of the earthquake) and S is the intensity of a "standard earthquake" (whose amplitude is 1 micron = 10^{-4} cm). (A standard earthquake is the smallest measurable earthquake known.)

Example 2 Finding an earthquake's magnitude

If the intensity of an earthquake were 40,000 cm, what would its magnitude be?

SOLUTION
The intensity of this earthquake is 40,000 cm, so $I = 40,000$. $S = 10^{-4}$ cm, so we can substitute these values into the magnitude formula and calculate the magnitude.

$$M = \log\left(\frac{40000}{10^{-4}}\right)$$
$$M \approx 8.6$$

Therefore, an earthquake that has an intensity of 40,000 cm has a magnitude of 8.6.

Example 3 Finding an earthquake's intensity

What is the intensity of an earthquake of magnitude 6.5?

SOLUTION
Because we know that the intensity of a standard earthquake is 10^{-4}, we can substitute this for S and 6.5 for M and solve for the intensity of the earthquake.

$$6.5 = \log\left(\frac{I}{10^{-4}}\right)$$
$$10^{6.5} = \frac{I}{10^{-4}} \quad \text{Rewrite the logarithm in exponential form.}$$
$$10^{-4}(10^{6.5}) = \left(\frac{I}{10^{-4}}\right)10^{-4} \quad \text{Add the exponents together.}$$
$$10^{2.5} = I$$
$$316.23 \approx I$$

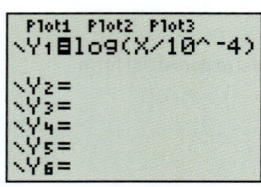
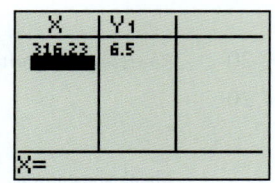

Check the answer, using the calculator table.

Therefore, the intensity of a magnitude 6.5 earthquake is about 316.23 cm. A magnitude 6.5 earthquake hit central California in 2003, killing two people and destroying several buildings. If this earthquake had hit in a more developed area, the destruction could have been much worse.

PRACTICE PROBLEM FOR EXAMPLE 3

What is the intensity of an earthquake with magnitude 8?

Another application is the use of common logarithms in chemistry to calculate the pH of a solution. The pH of a solution is a measurement of how acidic or alkaline a solution is. A neutral solution will have a pH value of 7, and an acidic solution such as vinegar (pH of about 3) or your stomach acid (pH of about 1) will have pH values less than 7. Alkaline solutions such as lye (pH of about 9), used to make soap, have pH values greater than 7. The pH of a solution can be calculated by taking the negative log of the hydrogen ion concentration.

$$pH = -\log(H^+)$$

The hydrogen ion concentration has the unit M, which stands for the molarity of the solution.

Example 4   Finding the pH of acids and bases

What is the pH of an aqueous solution when the concentration of hydrogen ions is 5.0×10^{-4} M?

SOLUTION

Because the hydrogen ion concentration is given as 5.0×10^{-4} M, we can substitute that in for H^+ and get

$$pH = -\log(5.0 \times 10^{-4})$$
$$pH = 3.30$$

Therefore, an aqueous solution with a hydrogen ion concentration of 5.0×10^{-4} M has a pH of 3.30.

Example 5   Finding a hydrogen ion concentration

Find the hydrogen ion concentration of the solution in the beaker shown.

SECTION 6.6 Solving Logarithmic Equations

SOLUTION

Because we know the pH value, we can substitute 4.7 for pH in the formula and solve for H.

$$4.7 = -\log(H^+)$$ Multiply both sides by -1 to isolate the logarithm.
$$-4.7 = \log(H^+)$$ Rewrite the logarithm in exponential form.
$$10^{-4.7} = H^+$$ Write the solution using scientific notation.
$$1.995 \times 10^{-5} = H^+$$ Check the solution, using the calculator table.

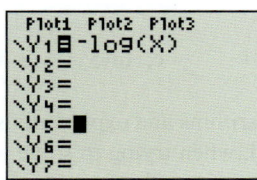 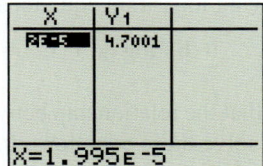

Therefore, a solution with a pH value of 4.7 will have a hydrogen ion concentration of 1.995×10^{-5} M.

PRACTICE PROBLEM FOR EXAMPLE 5

Find the hydrogen ion concentration of a solution with pH $= 2.3$.

Solving Other Logarithmic Equations

Example 6 Solving logarithmic equations

Solve the following logarithm equations. Round your answers to three decimal places.

a. $\log x = \log 5$ **b.** $\ln t = 3$ **c.** $\log(2x) + 7 = 12$

SOLUTION

a. In this case, we have a logarithm that is equal to another logarithm of the same base, so we know that the inside of the logarithms must be equal. Therefore, $x = 5$ is the solution to this equation.

b. We have a single logarithm equal to a constant, so we rewrite the logarithm into exponential form and solve.

$$\ln t = 3$$
$$e^3 = t$$ Rewrite the logarithm into exponential form.
$$20.086 \approx t$$

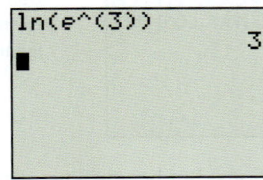

Check the solution.

In this case, e^3 is the exact answer and is sometimes the required or needed solution. 20.086 is an approximate answer and is suitable for many applications.

c. We must first start by isolating the logarithm on one side of the equation, and we then rewrite the logarithm in exponential form and solve.

$$\log(2x) + 7 = 12$$
$$\underline{ -7 \quad -7}$$ Isolate the logarithm.
$$\log(2x) = 5$$
$$10^5 = 2x$$ Rewrite the logarithm in exponential form.
$$100000 = 2x$$ Solve for x.
$$50000 = x$$

Skill Connection

In chemistry, hydrogen ion concentrations are typically written using scientific notation. In Example 4, we find the hydrogen ion concentration

$$10^{-4.7} = H^+$$

To convert this to scientific notation, we will use a calculator.

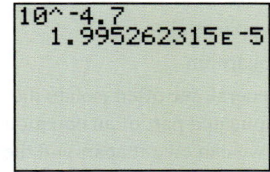

The E in the result indicates scientific notation and is written as

$$1.995 \times 10^{-5}$$

The number after the E becomes the exponent of 10.

Connecting the Concepts

How do I identify the first step?

Now is a good time to recognize a common first step in solving many of the problems that we have studied in this textbook.

Isolate the . . .
- Variable term
- Squared term (when in vertex form)
- Exponential
- Logarithm

In general, we often isolate the complicated part of an equation so that we can take it apart and then continue solving.

If you remember to isolate the complicated part of an equation, all you will need from there is what undoes that particular type of function.

For example:
- Variable term: multiply or divide
- Squared term: square root property
- Exponential: rewrite in logarithm form
- Logarithm: rewrite in exponential form

PRACTICE PROBLEM FOR EXAMPLE 6

Solve the following logarithm equations.

a. $\log x = 3.5$ **b.** $\log(x + 12) = \log(20)$ **c.** $\ln(5x) + 8 = 3$

You can see that the relationship between logarithms and exponentials will play a big role in how we solve these equations. In general, when trying to solve an equation that has a variable inside a logarithm, you will want to isolate the logarithm and then rewrite the logarithm in exponential form. This will get the variable out of the logarithm and allow you to solve the equation. If there is more than one logarithm in the equation, first combine them into a single log using the properties for logarithms.

Example 7 Solving logarithmic equations

Solve the following logarithmic equations. Round your answers to three decimal places.

a. $\log_6(x) + \log_6(2x) = 2$ **b.** $\log(2x) + \log(x - 4) = 1$

c. $\log_2(3x^2) - \log_2 x = 5$ **d.** $\log_2(x + 5) + \log_2(x + 4) = 1$

SOLUTION

a. First you can combine the logarithms together using the product property for logarithms and then rewrite the logarithm in exponential form so that you can solve for x.

$$\log_6(x) + \log_6(2x) = 2$$
$$\log_6(2x^2) = 2 \qquad \text{Combine the logarithms.}$$
$$6^2 = 2x^2 \qquad \text{Rewrite the logarithm in exponential form.}$$
$$36 = 2x^2$$
$$18 = x^2$$
$$\pm\sqrt{18} = x \qquad \text{These two answers will need to be checked.}$$
$$\pm 4.243 \approx x$$

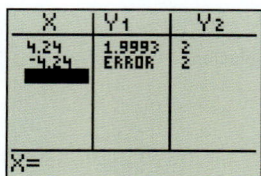 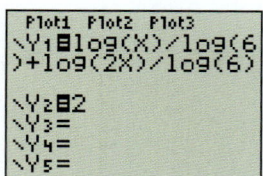

Check the answers. Use the change of base formula to enter the logarithms.

In this case, we have two possible solutions, but because we cannot evaluate the logarithm of a negative number, $x = -4.24$ cannot be a solution. Therefore, $x = 4.24$ is our only valid solution.

b. Combine the logarithms and then rewrite in exponential form.

$$\log(2x) + \log(x - 4) = 1$$
$$\log(2x^2 - 8x) = 1 \qquad \text{Combine the logarithms.}$$
$$10 = 2x^2 - 8x \qquad \text{Rewrite the logarithm in exponential form.}$$
$$0 = 2x^2 - 8x - 10 \qquad \text{We have a quadratic left, so we can factor or use the quadratic formula.}$$
$$0 = 2(x - 5)(x + 1)$$
$$x - 5 = 0 \qquad x + 1 = 0 \qquad \text{These two answers will need to be checked.}$$
$$x = 5 \qquad\quad x = -1$$

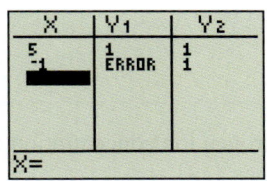

Check the answers.
Only one answer works.

Again we have two possible solutions, but when $x = -1$ is substituted into the logarithm, it is impossible to evaluate. Therefore, $x = 5$ is our only solution.

c. In this situation, we will need to use the quotient property for logarithms to combine the two logarithms that are subtracted.

$$\log_2(3x^2) - \log_2 x = 5 \quad \text{Combine the logarithms.}$$
$$\log_2\left(\frac{3x^2}{x}\right) = 5 \quad \text{Simplify the fraction.}$$
$$\log_2(3x) = 5$$
$$2^5 = 3x \quad \text{Rewrite in exponential form.}$$
$$32 = 3x$$
$$10.67 \approx x$$

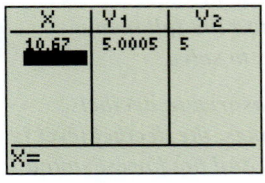

Check the answer.
The answer works.

d. $\log_2(x + 5) + \log_2(x + 4) = 1$

$$\log_2(x^2 + 9x + 20) = 1 \quad \text{Combine the logarithms, using the product property, and then rewrite in exponential form.}$$
$$x^2 + 9x + 20 = 2^1$$
$$x^2 + 9x + 18 = 0$$
$$(x + 3)(x + 6) = 0$$
$$x + 3 = 0 \qquad x + 6 = 0$$
$$x = -3 \qquad x = -6$$

Check the answers.
Only one answer works.

Although we found two answers for this problem, only $x = -3$ works in this case. When you substitute $x = -6$ into the equation, it is undefined. The negative 3 does work, because once it is substituted, we get a positive value inside each logarithm.

■ Connecting the Concepts

Why doesn't that answer work?

Recall from Section 6.3 that a logarithm has a restricted domain because we cannot take the logarithm of a negative number.

In Example 7d, the $x = -6$ makes the inside of the logarithms negative and thus undefined. This means that it is not a valid answer.

PRACTICE PROBLEM FOR EXAMPLE 7

Solve the following logarithm equations.

a. $\log(2x) + \log(3x) = 3$

b. $\ln(3x + 5) = 4$

c. $\log_3(x + 11) + \log_3(x + 5) = 3$

d. $\log_4(3x) + \log_4(2x - 7) = 2$

6.6 Exercises

For Exercises 1 through 10, solve each logarithm equation. Check that all solutions are valid.

1. $\log x = 4$
2. $\log_3 x = 5$
3. $\ln x = 5$
4. $\ln x = 3$
5. $\log_2(4x) = 6$
6. $\log_6(3x) = 2$
7. $\log(5x + 2) = 2$
8. $\log(2x + 7) = 3$
9. $\ln(3x - 5) = 2$
10. $\ln(4x + 5) = 4$

11. If the intensity of an earthquake were 2000 cm, what would its magnitude be?

12. In May 2003, Japan experienced an earthquake that had an intensity of 1000 cm. What was the magnitude of this earthquake?

13. If the intensity of an earthquake were 500,000 cm, what would its magnitude be?

14. The strongest recorded earthquake was in Chile in 1960 and had an intensity of 316227.77 cm. What was its magnitude?

15. The deadliest earthquake ever recorded caused an estimated 830,000 fatalities. This occurred in January 1556 in Shensi, China. What was the intensity of this earthquake if its magnitude was estimated as 8.0?

16. Many of the strongest earthquakes in the world take place off the coast of Alaska in the Andreanof Islands or in Prince William Sound. In 1957, a 9.1 magnitude earthquake hit this area. What was the intensity of that earthquake?

17. The deadliest earthquake in 2003 was in Southeastern Iran and had a magnitude of 6.6. What was its intensity?

18. On May 12, 2008, China had a 7.9 magnitude earthquake. What was the intensity of that earthquake?

19. What is the pH of Scope mouthwash if its hydrogen ion concentration is 1.0×10^{-7} M?

20. What is the pH of the aqueous solution shown in the beaker?

21. Find the hydrogen ion concentration of car battery acid, which has a pH = 1.

22. Find the hydrogen ion concentration of Pepsi, which has a pH = 3.

23. Find the hydrogen ion concentration of an oven cleaner that has a pH = 14.

24. Find the hydrogen ion concentration of a healthy person's stomach acid, which has a pH around 1.36.

25. Blood plasma should have a pH level between 7.35 and 7.45. Find the hydrogen ion concentrations that would fall in this pH range.

26. Swiss cheese that is 1 day into production should have a pH level between 5.2 and 5.4. Find the hydrogen ion concentrations that would fall in this pH range.

For Exercises 27 through 32, use the following information about sound levels to solve.

The loudness of a sound is measured in decibels. Like measurement of earthquakes, the decibel level is calculated by using logarithms and the lowest sound intensity, 10^{-12} watt per square meter, that a human ear can detect. The loudness of a sound in decibels can be found by using the function

$$L(I) = 10 \log\left(\frac{I}{10^{-12}}\right)$$

when the sound has the intensity of I watts per square meter.

27. The sound of a chain saw has an intensity of 10^{-1} watt per square meter. How many decibels does this chain saw make?

28. The sound of fireworks has an intensity of 100 watts per square meter. What is the decibel level of a fireworks show?

29. The sound of raindrops has an intensity of 10^{-8} watt per square meter. What is the decibel level of raindrops?

30. The sound produced by a hair dryer has an intensity of 10^{-3} watt per square meter. What is the decibel level for a hair dryer?

31. The sound of a jackhammer reaches 120 decibels. What is the intensity of this sound?

32. The sound of an iPod at peak volume is 115 decibels. What is the intensity of this sound?

For Exercises 33 through 46, solve each logarithm equation. Round your answers to three decimal places. Check your answers.

33. $\log(2x) + 5 = 7$
34. $\log(3x) - 30 = -28$
35. $\log_4(-2h) + 40 = 47$
36. $\log_3(-5t) + 13 = 17$
37. $\ln(3t) + 5 = 3$
38. $\ln(2t) + 8 = 2$
39. $\ln(r + 4) = 2$
40. $\log_4(m - 2) = 2$
41. $\log(2p + 1) = -0.5$
42. $\log(x - 3) = -2$
43. $\log_5(3x) + \log_5 x = 4$
44. $\log_6(4x) + \log_6 x = 2$
45. $\log_3(x + 2) + \log_3(x - 3) = 2$
46. $\log_2(2x) + \log_2(5x - 4) = 6$

For Exercises 47 through 50, the students were asked to solve the given equation. Explain what the student did wrong.

47. $\ln(x + 2) = 2$

> **Matt**
> $\ln(x + 2) = 2$
> $10^2 = x + 2$
> $100 = x + 2$
> $98 = x$

48. $\log x + \log(3x) = 2$

> **Shannon**
> $\log x + \log(3x) = 2$
> $\log(4x) = 2$
> $10^2 = 4x$
> $100 = 4x$
> $\dfrac{100}{4} = x$
> $25 = x$

49. $\log_2(x + 3) + 5 = 4$

> **Frank**
> $\log_2(x + 3) + 5 = 4$
> $\log_2(x + 8) = 4$
> $2^4 = x + 8$
> $16 = x + 8$
> $8 = x$

50. $\log_6(5x^2) - \log_6 x = 1$

> **Marie**
> $\log_6(5x^2) - \log_6 x = 1$
> $\log_6(5x^2 - x) = 1$
> $6^1 = 5x^2 - x$
> $0 = 5x^2 - x - 6$
> $0 = (5x - 6)(x + 1)$
> $0 = 5x - 6 \qquad 0 = x + 1$
> $6 = 5x \qquad -1 = x$
> $\dfrac{6}{5} = x \qquad -1 = x$

For Exercises 51 through 62, solve each logarithm equation. Round your answers to three decimal places. Check your answers.

51. $\log(5t^3) - \log(2t) = 2$
52. $\log_4(6x^3) - \log_4(3x) = 5$
53. $\log_5(x) + \log_5(2x + 7) - 3 = 0$
54. $\log_5(c + 12) + \log_5(c + 8) - 2 = -1$
55. $\log_4(-2x + 5) + \log_4(x + 21.5) = 4$
56. $\log_2(-x + 4) + \log_2(x + 3.5) = 3$
57. $3 \log_2(2x) + \log_2(16) = 7$
58. $\ln(3x) + \ln(x + 5) = 4$
59. $\ln(3x^2 + 5x) - 3 = 2$
60. $\log(x^2 - 3x) - 4 = 1$
61. $2 \log(3x) + 8 = 2$
62. $5 \log(2x) + 4 = 3$

For Exercises 63 through 70, solve each equation. Round your answers to three decimal places. Check your answers.

63. $3^x = 27$
64. $5x + 12 = 2x - 8$
65. $4x^2 + 6x - 10 = 0$
66. $\ln x + 7 = 5$
67. $t^3 + 6t^2 = -5t$
68. $\log(x + 3) + \log(x - 2) = 1$
69. $3(d + 4) - 14 = 0$
70. $\left(\dfrac{1}{3}\right)^t = 81$

Chapter Summary

Section 6.1 Functions and Their Inverses

- An **inverse function** will "undo" what the original function did to the input values.
- To find an inverse function within an application, solve the function for the original input variable and rewrite using function notation showing the two variables changing from input to output and vice versa.
- To find an inverse function when not in a context, follow these steps.
  1. Take it out of function notation. (Replace $f(x)$ with y in the equation.)
  2. Solve for the original input variable.
  3. Interchange the variables. This will keep x as the input variable.
  4. Rewrite in function notation using the inverse notation $f^{-1}(x)$.
- **One-to-one functions** have exactly one input for each output. All one-to-one functions have inverses.
- Test whether a function is one-to-one by using the horizontal line test.
- The domain and range of inverse functions will be the same, but switched. The domain becomes the range of the inverse, and the range becomes the domain.

Example 1

The number of minutes an hour a local radio station can play music depends on the number of commercials played during that hour. Math Rocks 101.7 FM uses the following equation when considering how much time they have to play music.

$$M(C) = 60 - 0.5C$$

where M is the number of minutes of music the station can play in an hour if it plays C 30-second commercials that hour.

a. Find an inverse for this function.

b. Find the number of commercials the radio station can play if it wants to have 50 minutes of music each hour.

c. If the domain for the original function is [2, 30], find the domain and range for the inverse function.

SOLUTION

a.
$$M(C) = 60 - 0.5C$$
$$M = 60 - 0.5C$$
$$M - 60 = -0.5C$$
$$-2M + 120 = C$$
$$C(M) = -2M + 120$$

b. $C(50) = 20$, so if the radio station wants to play 50 minutes of music each hour, it can play 20 thirty-second commercials each hour.

c. Domain: [45, 59], range: [2, 30].

CHAPTER 6 Summary 537

Example 2 Find the inverse for the function $g(x) = -3x + 5$.

SOLUTION

$$g(x) = -3x + 5$$
$$y = -3x + 5$$
$$y - 5 = -3x$$
$$-\frac{y}{3} + \frac{5}{3} = x$$
$$-\frac{x}{3} + \frac{5}{3} = y$$
$$g^{-1}(x) = -\frac{x}{3} + \frac{5}{3}$$

Section 6.2 Logarithmic Functions

- A **logarithm** is defined as the inverse of an exponential function. A logarithmic function is of the form $f(x) = \log_b x$, where b is called the **base** of the logarithm and b must be positive and not equal to 1.
- A logarithm asks the question, "What exponent of the base b will give me what's inside the log?"
- Some properties for logarithms are as follows.

    $\log_b 1 = 0$. The logarithm of 1 is always zero.

    $\log_b b = 1$. The logarithm of its own base is always 1.

    $\log_b(b^m) = m$. The logarithm of its base to a power is just that power.
- The **common logarithm** is log base 10. This logarithm is written as "log" with no base noted.
- The **natural logarithm** is log base e. This logarithm is written as "ln" with no base noted.
- The **change of base formula** can be used to convert a logarithm from one base to another: $\log_b x = \dfrac{\log_c x}{\log_c b}$.
- Note that in the change of base formula, c is typically 10 or e, since these logs are on a calculator.
- Logarithms and exponentials are inverses of one another and can be used to solve equations. Often, rewriting a logarithm into exponential form or an exponential into logarithm form will help solve an equation.

Example 3 Evaluate the following logarithms.

    a. $\log_3 27$

    b. $\log 275$

    c. $\ln 13$

    d. $\log_6 145$

SOLUTION
    a. $\log_3 27 = 3$

    b. $\log 275 = 2.439$

    c. $\ln 13 = 2.565$

    d. $\log_6 145 = \dfrac{\log 145}{\log 6} = 2.778$

Example 4 Solve the logarithmic equation by rewriting it into exponential form.

a. $\log(3x) = 2$ b. $\ln(x + 5) = 4$ c. $\log_4(x) = 3$

SOLUTION

a. $\log(3x) = 2$
$10^2 = 3x$
$\dfrac{100}{3} = x$

b. $\ln(x + 5) = 4$
$e^4 = x + 5$
$e^4 - 5 = x$
$49.6 \approx x$

c. $\log_4(x) = 3$
$4^3 = x$
$64 = x$

Section 6.3 Graphing Logarithmic Functions

- The basic characteristics of logarithmic graphs are as follows:

  No y-intercept

  x-intercept of $(1, 0)$

  The y-axis is a vertical asymptote.

  When the base is greater than 1, the graph increases quickly for inputs close to zero and then more slowly as the inputs increase.

  When the base is between 0 and 1, the graph decreases rapidly for inputs close to zero and then more slowly as the inputs increase.

- The domain of basic logarithmic functions is all positive real numbers, or $(0, \infty)$.
- The range of basic logarithmic functions is all real numbers, or $(-\infty, \infty)$.

Example 5 Sketch the graph of $f(x) = \log_5 x$.

SOLUTION Because this logarithm has a base of 5, the inverse will be the exponential function $f^{-1}(x) = 5^x$. Using this exponential function to create a table of points, we will plot the graph.

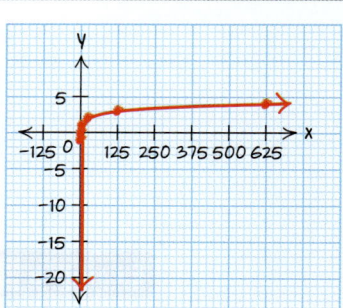

Points for a Graph of the Inverse

x	$f^{-1}(x) = 5^x$
-1	0.2
0	1
1	5
2	25
3	125
4	625

Points for the Logarithm Graph

x	$f(x) = \log_5 x$
0.2	-1
1	0
5	1
25	2
125	3
625	4

CHAPTER 6 Summary 539

Section 6.4 Properties of Logarithms

- The three main properties of logarithms are as follows.

  The product property for logarithms: $\log_b(mn) = \log_b m + \log_b n$.

  The quotient property for logarithms: $\log_b\left(\dfrac{m}{n}\right) = \log_b m - \log_b n$.

  The power property for logarithms: $\log_b(m^n) = n \log_b m$.

- Using these properties for logarithms allows us either to break a complicated logarithm apart into smaller logarithms or to combine several logarithms of the same base together as a single log.

Example 6 Expand the following logarithms into separate simpler logarithms with no exponents.

   a. $\log(5x^2 y)$ b. $\ln\left(\dfrac{4x^3 y^2}{z^5}\right)$ c. $\log_5\left(\sqrt{6x^3 yz}\right)$

SOLUTION

a. $\log(5x^2 y)$

$\log 5 + \log x^2 + \log y$

$\log 5 + 2 \log x + \log y$

b. $\ln\left(\dfrac{4x^3 y^2}{z^5}\right)$

$\ln 4 + \ln x^3 + \ln y^2 - \ln z^5$

$\ln 4 + 3 \ln x + 2 \ln y - 5 \ln z$

c. $\log_5\left(\sqrt{6x^3 yz}\right)$

$\dfrac{1}{2}\log_5(6x^3 yz)$

$\dfrac{1}{2}(\log_5 6 + \log_5 x^3 + \log_5 y + \log_5 z)$

$\dfrac{1}{2}(\log_5 6 + 3\log_5 x + \log_5 y + \log_5 z)$

Example 7 Combine the following logarithms into one logarithm with a coefficient of 1.

   a. $\log 7 + 4 \log x + 2 \log y - \log z$

   b. $\ln 3 + \dfrac{1}{2} \ln x + \ln y$

SOLUTION

a. $\log 7 + 4 \log x + 2 \log y - \log z = \log\left(\dfrac{7x^4 y^2}{z}\right)$

b. $\ln 3 + \dfrac{1}{2} \ln x + \ln y = \ln(3y\sqrt{x})$

Section 6.5 Solving Exponential Equations

- When **solving exponential equations,** use one of the following solving techniques.

  Get the exponential term by itself, take the logarithm of both sides of the equation, and then use the power property for logarithms to make the exponent a coefficient. This will allow you to finish solving the equation.

  Get the exponential term by itself and rewrite it in logarithm form. Use the change of base formula if needed and finish solving for the variable.

Example 8 Solve the following equations.

a. $5 \cdot 3^{4x} = 75$ b. $4^{2x+1} + 7 = 115$

SOLUTION

a. $5 \cdot 3^{4x} = 75$

$3^{4x} = 15$

$\log(3^{4x}) = \log 15$

$4x \log 3 = \log 15$

$x = \dfrac{\log 15}{4 \log 3}$

$x \approx 0.616$

Check the answer.

X	Y1	Y2
.616	74.92	75

X=

b. $4^{2x+1} + 7 = 115$

$4^{2x+1} = \log 108$

$2x + 1 = \log_4(108)$

$2x + 1 = \dfrac{\log 108}{\log 4}$

$2x = \dfrac{\log 108}{\log 4} - 1$

$x = \dfrac{\dfrac{\log 108}{\log 4} - 1}{2}$

$x \approx 1.1887$

Check the answer.

X	Y1	Y2
1.1887	114.99	115

X=

Section 6.6 Solving Logarithm Equations

- When **solving logarithmic equations,** follow these steps.
  Combine all logarithms into a single logarithm, using the log properties.
  Isolate the logarithm.
  Rewrite the logarithm into exponential form.
  Finish solving the equation.
  Check the answer in the original problem.

- The magnitude of an earthquake can be found by using the formula $M = \log\left(\dfrac{I}{S}\right)$, where M is the magnitude of the earthquake on the Richter scale, I is the intensity of the earthquake in centimeters, and S is the intensity of a standard earthquake, 10^{-4} cm.

- The pH of a solution can be found using the formula $pH = -\log(H^+)$, where H^+ is the hydrogen ion concentration of the solution.

Example 9 Solve the following equations.

a. $\ln(3x + 5) = 4$ **b.** $\log(5x) + \log(2x + 7) = 2$

SOLUTION

a. $\ln(3x + 5) = 4$

$e^4 = 3x + 5$

$\dfrac{e^4 - 5}{3} = x$

$16.53 \approx x$

Check the answer.

b. $\log(5x) + \log(2x + 7) = 2$

$\log(10x^2 + 35x) = 2$

$10x^2 + 35x = 100$

$10x^2 + 35x - 100 = 0$

$x \approx -5.36 \qquad x \approx 1.86$

$x \approx 1.86$

Check the answer.

Only one answer works this time. The -5.36 makes the inside of a log negative, and thus, it cannot be evaluated.

Example 10 The pH of vinegar is 3. Find the hydrogen ion concentration.

SOLUTION

$3 = -\log(H^+)$

$-3 = \log(H^+)$

$10^{-3} = H^+$

The hydrogen ion concentration of vinegar is 0.001 M.

Chapter Review Exercises

For Exercises 1 through 7, find the inverse for each function.

1. $f(x) = 1.4x - 7$ [6.1]

2. $g(t) = -2t + 6$ [6.1]

3. $h(x) = 5^x$ [6.2]

4. $f(x) = 3.5(6)^x$ [6.2]

5. $g(x) = 2e^x$ [6.2]

6. $f(x) = \log_5 x$ [6.2]

7. $h(t) = \log_{0.2} t$ [6.2]

8. The AEM Toy Company can manufacture $D(e) = 400e + 50$ dolls per day when e employees are working.

   a. Find an inverse for the given function.

   b. How many dolls can the AEM Toy Company manufacture if 10 employees are working on a particular day?

   c. How many employees must work for the company to manufacture 1000 dolls in a day?

   d. If the company can employ between 5 and 30 employees a day, give a domain and range for the inverse function. [6.1]

9. The AEM Toy Company has a profit of $P(d) = 4.5d - 300$ dollars when d dolls are sold a month.

   a. Find an inverse for the given function.

   b. If AEM Toy Company wants to have $12,000 in profit a month, how many dolls do they need to sell?

   c. If the company usually sells between 500 and 4000 dolls a month, give a domain and range for the inverse function. [6.1]

For Exercises 10 and 11, determine whether or not the graph represents a one-to-one function.

10.

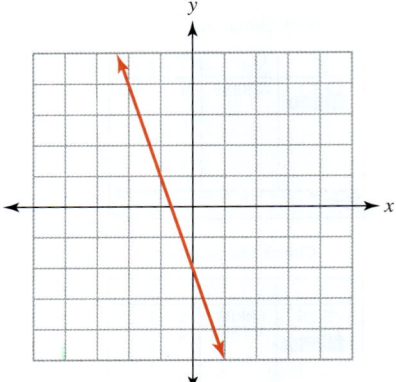

11.

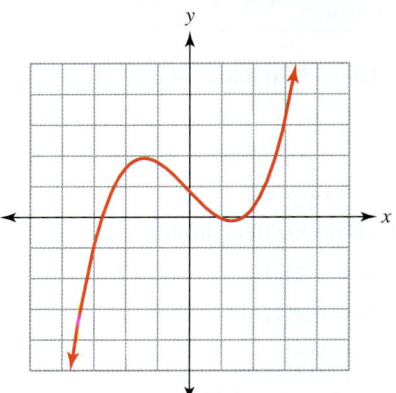

For Exercises 12 through 15, evaluate each logarithm without a calculator.

12. $\log_4 64$ [6.2]
13. $\log_3 1$ [6.2]
14. $\log_8 8^5$ [6.2]
15. $\log_2 \left(\dfrac{1}{8}\right)$ [6.2]

For Exercises 16 and 17, sketch a graph of the function. Clearly label and scale the axes.

16. $f(x) = \log_{2.5}(x)$ [6.3]
17. $f(x) = \log_{0.4}(x)$ [6.3]
18. Give the domain and range of the function in Exercise 16. [6.3]
19. Give the domain and range of the function in Exercise 17. [6.3]

For Exercises 20 through 26, write the logarithms as separate simpler logarithms with no exponents.

20. $\log(7xy)$ [6.4]
21. $\log_5(3a^3b^4)$ [6.4]
22. $\ln(2x^3y^4)$ [6.4]
23. $\log_7\left(\sqrt[5]{3x^3y}\right)$ [6.4]
24. $\log_3\left(\sqrt{5ab^2c^3}\right)$ [6.4]
25. $\log\left(\dfrac{3x^3y^5}{z^4}\right)$ [6.4]
26. $\log\left(\dfrac{4a^5b}{c^3}\right)$ [6.4]

For Exercises 27 through 31, write the logarithms as a single logarithm with a coefficient of 1.

27. $\log 4 + 3 \log x + 2 \log y + \log z$ [6.4]
28. $\log 2 + 4 \log x + 5 \log y - 2 \log z$ [6.4]
29. $\log_3 a + 3 \log_3 b - 2 \log_3 c$ [6.4]
30. $\ln 7 + 2 \ln x + \ln z - 5 \ln y - 3 \ln z$ [6.4]
31. $\log(x + 2) + \log(x - 7)$ [6.4]

32. The number of Franklin's gulls in North America has been in significant decline in the past several years. The population of Franklin's gulls can be modeled by

$$F(t) = 750(0.9405^t)$$

where $F(t)$ represents the number of Franklin's gulls in thousands in North America t years since 2000.
Source: Model derived from data from the National Biological Service.

   a. Estimate the Franklin's gull population in 2005.
   b. When will the Franklin's gull population reach 500,000?
   c. What might cause this population to stop declining at this rate? [6.5]

33. How long will it take an investment of $7000 to double if it is invested in an account that pays 6% interest compounded monthly? [6.5]

34. How long will it take an investment of $100,000 to double if it is invested in an account that pays 8.5% interest compounded continuously? [6.5]

For Exercises 35 through 42, solve the following exponential equations.

35. $4^w = 12.5$ [6.5]
36. $3^t = \dfrac{1}{243}$ [6.5]
37. $6^{3x-5} = 204$ [6.5]
38. $7^{x+2} = 47$ [6.5]
39. $17(1.9)^x = 55.4$ [6.5]
40. $4\left(\dfrac{1}{3}\right)^x - 8 = 28$ [6.5]
41. $3(5)^{3x+2} - 700 = 11$ [6.5]
42. $5^{x^2+4} = 15625$ [6.5]

43. What was the intensity of the magnitude 5.7 earthquake that hit Round Valley, California, in November 1984? [6.6]

44. What is the pH of cranberry juice if its concentration of hydrogen ion is 3.7×10^{-3} M? [6.6]

45. Find the hydrogen ion concentration of a shampoo with a pH $= 5.6$. [6.6]

For Exercises 46 through 53, solve each logarithm equation. Check that all solutions are valid.

46. $\log x = -4$ [6.6]

47. $\log_8 t = 5$ [6.6]

48. $\log(10x + 20) = 5$ [6.6]

49. $\ln(4.3t) + 7.5 = 3$ [6.6]

50. $\log_5(2x) + \log_5 x = 4$ [6.6]

51. $\log_3(x - 4) + \log_3(x + 1) = 3$ [6.6]

52. $\log_4(x + 4) + \log_4(x + 7) = 1$ [6.6]

53. $\log(8x^5) - \log(2x) = 3$ [6.6]

Chapter Test

1. How long will it take an investment of $1000 to double if it is invested in an account that pays 3.5% interest compounded monthly?

2. What was the intensity of the magnitude 7.0 earthquake that hit Papua, Indonesia, in February 2004?

For Exercises 3 through 6, solve the equation.

3. $\log_{16}(2x + 1) = -0.5$

4. $\log_7 x + \log_7(2x + 7) - 3 = 0$

5. $20 = -3 + 4(2^x)$ 6. $8 = 3^{7x-1}$

7. Is this a graph of a one-to-one function? Explain why or why not.

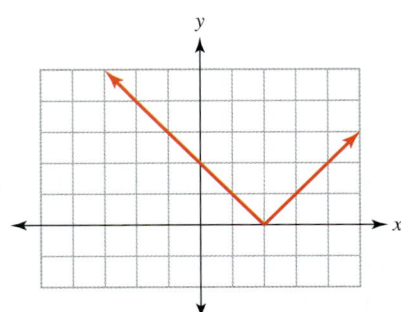

8. Write as a single logarithm with coefficient 1.
$$4 \log(2a^3) + 5 \log(ab^3)$$

9. Write as a single logarithm with coefficient 1.
$$\log 5 + \log x + 3 \log y - 4 \log z$$

10. Write the logarithm as separate simpler logarithms with no exponents.
$$\ln(4a^3bc^4)$$

11. Write the logarithm as separate simpler logarithms with no exponents.
$$\log\left(\frac{3xy^5}{\sqrt{z}}\right)$$

12. Sketch the graph of $f(x) = \log_{12} x$.

13. Give the domain and range of the function given in Exercise 12.

14. The number of CD singles shipped in the mid-1990s is given in the table.

Year	CD singles (millions)
1993	7.8
1994	9.3
1995	21.5
1996	43.2
1997	66.7

Source: Statistical Abstract of the United States, 2001.

a. Write an equation for a model of these data.
b. Give a reasonable domain and range for this model.
c. Estimate the number of CD singles shipped in 2000.
d. When were only 5 million CD singles shipped?

For Exercises 15 through 18, find the inverse of the function.

15. $f(x) = 5x + 2$

16. $g(x) = \log_{15} x$

17. $f(x) = 5^x$

18. $h(x) = 2(7)^x$

19. Find the hydrogen ion concentration of a solution with pH $= 2.58$.

20. Evaluate $\log_3 200$.

Chapter 6 Projects

How Fast Is That Tumor Growing?

Written Project
One or more people

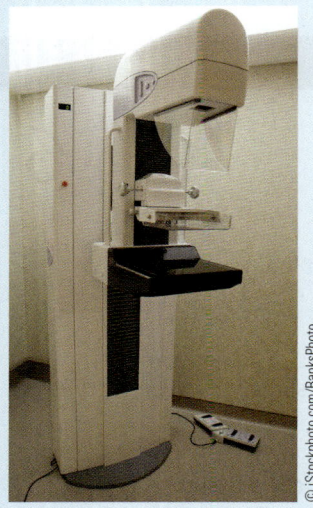

In this project, you will be asked to investigate the growth of an average breast cancer tumor. A simple model of the growth of a cancerous tumor is one in which the number of cancer cells doubles on average every 100 days. Current mammography technology can detect a tumor that is 1 cubic centimeter (cc) in size. A tumor that is 1 cc consists of about 1 billion cells. A tumor that is 1 liter (1000 cc) is often considered lethal. Use this information to answer the following questions.

Write up

a. If a tumor starts from a single cell that has undergone a malignant transformation, find a model for the number of cells that are present as time goes by.

b. Create a graph of your model on the calculator or computer and print it out or draw it very neatly by hand on graph paper. Remember that we are going to look at very large numbers of cells, so scale your graph carefully.

c. Use your model to estimate how long after that initial malignant transformation occurred until the tumor is detectable by using a mammogram.

d. If the tumor is not detected or treated, how long will it take for the tumor to reach a typically lethal size?

e. What is the range in time that mammography can detect a tumor before it reaches a lethal size?

f. With this time range in mind, how often should a woman get a mammogram? Explain your reasoning.

Review Presentation

Research Project
One or more people

What you will need
- This presentation may be done in class or as a video.
- If it is done as a video, you might want to post the video on YouTube or another site.
- You might want to use problems from the homework set or review material in this book.
- Be creative, and make the presentation fun for students to watch and learn from.

Create a 5-minute presentation or video that reviews a section of Chapter 6. The presentation should include the following:

- Examples of the important skills presented in the section
- Explanation of any important terminology and formulas
- Common mistakes made and how to recognize them

Equation-Solving Toolbox — Chapters 1-6

Linear Equations

Name	Uses	Example	Where to Find It
Distributive Property	Use when an equation contains grouping symbols	$2(x + 4) = 3x - 5$	Section 1.1 and Appendix A
Addition Property of Equality	Use to isolate a term.	$x + 7 = 20$	Section 1.1 and Appendix A
Multiplication Property of Equality	Use to isolate a variable.	$5x = 13$	Section 1.1 and Appendix A

Systems of Equations

Name	Uses	Example	Where to Find It
Substitution Method	Use when a variable is isolated or can be easily isolated.	$y = 6x + 8$ $2x + 4y = 30$	Section 2.2
Elimination Method	Use when the equations are in general form.	$5x - 7y = 42$ $3x + 20y = 103$	Section 2.3

Quadratic Equations

Name	Uses	Example	Where to Find It
Square Root Property	Use when there is a squared term but no first-degree term.	$3(x - 4)^2 + 5 = 0$	Section 4.4
Completing the Square	Use if the vertex form is required.	$x^2 + 6x + 4 = 0$	Section 4.4
Factoring	Use when the quadratic has small coefficients that factor easily.	$x^2 + 7x + 10 = 0$	Sections 3.4, 3.5, and 4.5
Quadratic Formula $x = \dfrac{-b \pm \sqrt{b^2 - 4ac}}{2a}$	Use when there are fractions, decimals, or large numbers. The quadratic formula will always give you the answers.	$11x^2 + 42x - 8 = 0$	Section 4.6

Exponential Equations

Name	Uses	Example	Where to Find It
Rewrite in Logarithmic Form	Use when the equation has a variable in the exponent. Isolate the base and exponent part first.	$5^x + 20 = 300$	Sections 6.2 and 6.5

Logarithmic Equations

Name	Uses	Example	Where to Find It
Rewrite in Exponential Form	Use when the equation contains a logarithm. Isolate the logarithm first.	$\log(x + 2) = 4$	Sections 6.2 and 6.6

General Solving Techniques

Name	Uses	Example	Where to Find It
Graphing	Use when you cannot solve the equation algebraically. Graph each side of the equation and find the intersection(s).		All chapters
Numerically	Use when you cannot solve the equation algebraically. Estimate a solution and check it. Best used to check solutions.		All chapters

Modeling Processes — Chapters 1-6

Linear	Quadratic	Exponential
1. Define the variables and adjust the data (if needed).	1. Define the variables and adjust the data (if needed).	1. Define the variables and adjust the data (if needed).
2. Create a scatterplot.	2. Create a scatterplot.	2. Create a scatterplot.
3. Select a model type.	3. Select a model type.	3. Select a model type.
4. **Linear model:** Pick two points and find the equation of the line. $y = mx + b$	4. **Quadratic model:** Pick a vertex and substitute it for h and k in the vertex form $f(x) = a(x - h)^2 + k$	4. **Exponential model:** Pick two points, write two equations using the form $f(x) = a \cdot b^x$, and solve for b.
5. Write the equation of the model using function notation.	5. Pick another point and use it to find a.	5. Use b and one equation to find a.
6. Check the model by graphing it with the scatterplot.	6. Write the equation of the model using function notation.	6. Write the equation of the model using function notation.
	7. Check the model by graphing it with the scatterplot.	7. Check the model by graphing it with the scatterplot.

Cumulative Review — Chapters 1-6

For Exercises 1 through 16, solve the given equation using any method you have learned. Check your answers in the original equation.

1. $3(4.2)^x + 7 = 23$
2. $3x + 18 = 7x - 6$
3. $3(t - 5)^2 - 18 = 30$
4. $\ln(2c - 7) = 5$
5. $\frac{2}{5}b + \frac{4}{5} = \frac{1}{5}(b + 3)$
6. $2.5n^2 + 1.4n - 3.8 = 0$
7. $4h^3 + 16h^2 - 32h = 0$
8. $2^{3x-7} = 45$
9. $8.4g + 7.6 = 3.2(g - 4.2)$
10. $(x + 5)(x - 4) = -6$
11. $\log(x + 2) + \log(x - 7) = 2$
12. $e^{x+5} = 89$
13. $|b + 7| = 20$
14. $|2t - 12| = 6$
15. $4(3^x) + 25 = 8(3^x) - 299$
16. $2.5x^8 - 3000 = -2360$

For Exercises 17 through 22, solve the given systems. Check your answer.

17. $y = 5x - 20$
    $2x - 3y = -31$
18. $5x + 12y = 66.4$
    $-4x + 10y = 29.2$
19. $y = 0.6x + 3$
    $3x - 5y = -15$
20. $h = 3g + 36$
    $h = 2g^2 + 7g - 12$
21. $y = x^2 + 3x - 12$
    $y = 4x^2 - 7x - 9$
22. $y = 3x^2 - 4x - 10$
    $y = -2x^2 + 6x - 18$

23. The gross profit for UTstarcom, Inc., a leading telecommunications company in China and the world, is given in the table.

Year	Gross profit (millions of $)	Year	Gross profit (millions of $)
1996	13.2	2000	128.2
1997	26.8	2001	224.5
1998	41.0	2002	345.5
1999	74.8	2003	636.2

Source: UTstarcom Inc. Annual Reports 2000 and 2003.

a. Find an equation for a model for these data.
b. Give a reasonable domain and range.
c. According to your model when will UTstarcom, Inc. reach 2 billion in gross profit?

24. The number of cocaine-related emergency department episodes for people over 35 years old are given in the table.

Year	Number of Episodes
1990	23,054
1991	30,582
1995	57,341
1996	68,717
1997	74,600
1998	83,730
1999	85,869
2000	93,357
2001	106,810

Source: Substance Abuse and Mental Health Services Administration.

a. Find an equation for a model of these data.
b. Use your model to estimate the number of cocaine-related emergency department episodes in 2003.
c. Find when there will be 140,000 cocaine-related emergency department episodes.
d. What is the slope of your model and what does it mean in this context?
e. Give a reasonable domain and range for your model.

25. $f(x) = 4x - 7$
    a. Find $f(9)$.
    b. Find x such that $f(x) = -2$.
    c. Give the domain and range of $f(x)$.

26. Let $f(x) = 3.5x + 4$ and $g(x) = -2x + 5.5$.
    a. Find $f(x) + g(x)$.   b. Find $f(g(x))$.
    c. Find $f(x)g(x)$.

27. Let $f(x) = 4x - 15$ and $g(x) = -6x + 3$.
    a. Find $f(3) - g(3)$.   b. Find $f(g(5))$.
    c. Find $f(4)g(4)$.

28. Let $f(x) = 3x^2 + 5x - 2$ and $g(x) = 4x - 7$.
    a. Find $f(x) + g(x)$.   b. Find $f(g(x))$.
    c. Find $f(x)g(x)$.

29. Let $f(x) = -4x^2 + 7x - 3$ and $g(x) = 5x + 8$.
    a. Find $f(5) - g(5)$.   b. Find $f(g(2))$.
    c. Find $f(3)g(3)$.

548 CHAPTER 6 Logarithmic Functions

30. Let $V(t)$ be the number of visitors to Yellowstone National Park in year t. Let $L(p)$ be the amount of litter in tons left by p people visiting the park. Write the function notation to combine these two functions to create a new function that gives the total amount of litter left in Yellowstone National Park in year t.

31. The number of women participating in triathlons has been increasing rapidly over the past 15 years. The percentage, $M(t)$, of members of USA Triathlon (the official triathlon association of the United States) who are men t years since 1990 can be modeled by

$$M(t) = -1.31t + 86.06$$

The percentage, $W(t)$, of USA Triathlon members who are women t years since 1990 can be modeled by

$$W(t) = 1.31t + 13.94$$

Source: Models derived from data obtained from USA Triathlon.

a. Estimate the percentage of USA Triathlon members that were women in 1995.
b. Find when the percentage of USA Triathlon members that are women will be greater than that of men.

For Exercises 32 through 35, solve the given inequalities.

32. $2x - 7 < 5x - 15$
33. $3(4x - 8) \geq 2x + 8$
34. $|x - 7| > 5$
35. $|3x + 5| \leq 8$

36. Use the table to find the following.

x	y
-2	10
0	5
2	0
4	-5
6	-10

a. The slope of the line passing through the points.
b. The x-intercept.
c. The y-intercept.
d. The equation of the line passing through the points.

37. Write the equation of the line passing through the point $(24, 4)$ and parallel to the line $2x - 3y = 30$. Write the equation in slope-intercept form.

38. Write the inequality for the following graph.

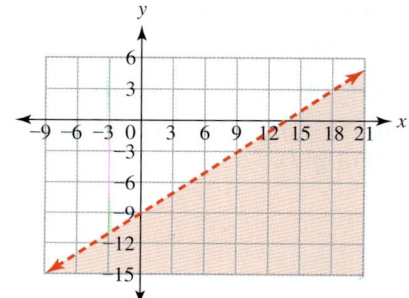

39. The population of Detroit, Michigan, has been declining rapidly since 2005. In 2006, Detroit had approximately 4487 thousand residents. In 2008, Detroit had approximately 4425 thousand residents.

a. Assuming that the population is decreasing at a constant rate, find an equation for a model of the population of Detroit.
b. Estimate the population of Detroit in 2015 if this decline continues.
c. What is the slope of your model? Explain its meaning.

40. Sarah needs 30 ml of 40% HCl solution but has only some 15% HCl solution and some 60% HCl solution. How much of each of these solutions does Sarah need to use to get the 30 ml of 40% HCl solution?

41. The number of women who are attempting triathlons in the United States is increasing, as shown in the chart.

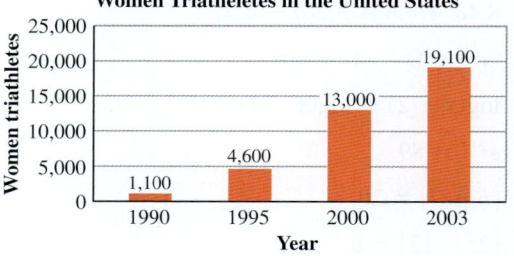

Source: USAToday.com.

a. Find an equation for a quadratic model for these data.
b. Give a reasonable domain and range for the model.
c. Use the model to estimate the number of women triathletes in the United States in 2005.
d. Give the vertex of your model and explain what it means in this context.
e. Find when there will be 40,000 women triathletes in the United States.

For Exercises 42 through 53, sketch the graph of the given equation or inequality. Label the vertical and horizontal intercepts. If the graph is quadratic, label the vertex.

42. $y = \dfrac{2}{5}x - 8$
43. $y = -0.5(x + 4)^2 + 20$
44. $y = -5(2)^x$
45. $5x - 4y = -3$
46. $y = x^2 + 4x + 15$
47. $y = 100(0.4)^x$
48. $y = \log_5 x$
49. $y = 3(1.8)^x$
50. $y = 490\left(\dfrac{1}{7}\right)^x + 12$
51. $y < -2x + 12$
52. $3x - 4y \leq 20$
53. $y > 2(x - 3)^2 + 4$

For Exercises 54 through 58, perform the indicated operations and simplify.

54. $(4d - 8) + (7d + 2)$
55. $(2x + 5)(7x - 8)$
56. $(3x^2 + 4x - 12) - (x^2 + 9x - 20)$
57. $(3t - 4)^2$
58. $4p^2 + (2p + 3)(5p - 8)$

59. Sketch the graph by hand of the system of inequalities.

$$y \le \tfrac{1}{4}x + 2$$
$$y \ge -\tfrac{2}{5}x + 4$$

For Exercises 60 through 64, factor the given expression using any method.

60. $2x^2 - 11x - 21$
61. $25r^2 - 64t^2$
62. $42g^3 - 7g^2 - 280g$
63. $t^3 + 343$
64. $4x^2 - 12x + 9$

65. What is the degree of the polynomial given in Exercise 60?

66. Complete the square to convert the function $f(x) = 3x^2 + 24x - 10$ to vertex form.

67. The average ticket price for a professional baseball game reached $19.82 in 2004. The prices for previous years can be modeled by

$$P(t) = 0.0237t^2 + 0.507t + 8.384$$

where $P(t)$ represents the average price in dollars for a baseball ticket t years since 1990.

a. Use the model to estimate the average price in 2010.
b. Give the vertex of this model, and explain what it means in this context.
c. Find when the average ticket price will reach $40.77, the price the Boston Red Sox charged in 2004.

68. Alice is running an experiment using 5000 atoms of thorium-233, which has a half-life of about 22 minutes.

a. Find a model for the number of thorium-233 atoms remaining after m minutes of the experiment.
b. Use your model to estimate when there will be 1000 thorium-233 atoms remaining.
c. If the experiment is supposed to last 3 hours, how many thorium-233 atoms will remain at the end of the experiment?

For Exercises 69 and 70, find the inverse of the functions.

69. $f(x) = 3x + 12$
70. $g(x) = 4(3)^x$

71. Milk has a pH of 6. Find its hydrogen ion concentration. Use the formula $pH = -\log(H^+)$.

72. The Muslim population in Israel was 1.24 million in 2008. This population is growing at a rate of 2.8% per year.
Source: Online Edition Jerusalem Post. www.jpost.com November 26, 2009.

a. Find a model to represent the Muslim population in Israel.
b. Use your model to estimate the Muslim population in Israel in 2012.
c. When does the model predict the Muslim population in Israel will reach 2 million?

For Exercises 73 through 78, simplify the expressions. Write each without negative exponents.

73. $(4m^5n^3)(15mn^2)$
74. $(-3g^3h^{-4})^4$
75. $3(17x^3y^7)^0 + 5(12x^4y^6)^0$
76. $\dfrac{300x^4y^{-3}z^2}{125x^2y^4z^{-5}}$
77. $(125a^3b^9)^{-\tfrac{1}{3}}(4ab^{-5})^2$
78. $\left(\dfrac{16m^{-5}n^3}{81m^{-1}n^{-5}}\right)^{-\tfrac{1}{4}}$

For Exercises 79 through 81, use the graph of the exponential $f(x) = a \cdot b^x$ to answer the following.

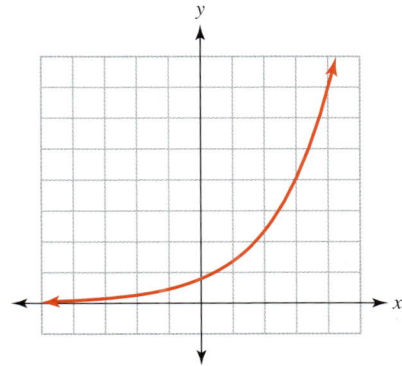

79. Is the value of a positive or negative? Explain your reasoning.

80. Is the value of b less than 1 or greater than 1? Explain your reasoning.

81. Give the domain and range of $f(x)$.

For Exercises 82 through 86, use the graph of the quadratic $f(x) = a(x - h)^2 + k$ to answer the following.

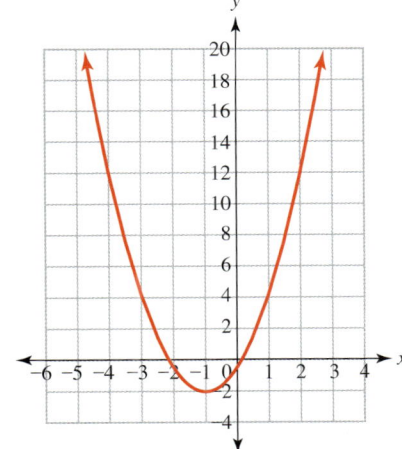

82. Is the value of a positive or negative? Explain your reasoning.

83. Estimate the values of h and k. Explain your reasoning.

84. $f(2.5) = $?

85. Find x such that $f(x) = 12$.

86. Give the domain and range of $f(x)$.

For Exercises 87 through 91, use the graph of the line $f(x) = mx + b$ to answer the following.

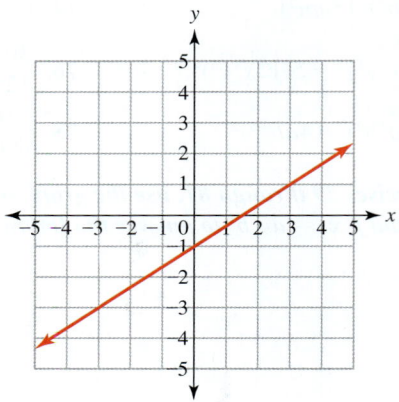

87. Is the value of *m* positive or negative? Explain your reasoning.
88. Estimate the value of *b*. Explain your reasoning.
89. Find the slope of the line.
90. Find *x* such that $f(x) = 1$.
91. Give the domain and range of $f(x)$.

Rational Functions

7

- **7.1** Rational Functions and Variation
- **7.2** Simplifying Rational Expressions
- **7.3** Multiplying and Dividing Rational Expressions
- **7.4** Adding and Subtracting Rational Expressions
- **7.5** Solving Rational Equations

It is extremely important to the survival of ships near the world's coastlines that lighthouses give off intense light that can be seen through the night and through inclement weather. The light from a lighthouse must be seen from far away to be effective, but the intensity of light lessens the greater the distance a ship is from shore. In this chapter, we will investigate inversely proportional relationships such as this one. One of the chapter projects will ask you to model the mechanics of a lighthouse by testing the intensity of light coming from a flashlight at various distances.

7.1 Rational Functions and Variation

LEARNING OBJECTIVES
- Identify a rational function.
- Set up direct and inverse variation problems.
- Find rational functions that model an application.
- Find the domain of a rational function in an application.
- Find the domain of a rational function.

Rational Functions

In this chapter, we will study another type of function that is found in many areas of business, physics, and other sciences. **Rational functions** are functions that contain fractions involving polynomials. These functions can be simple or very complex. Rational functions often result from combining two functions together using division. In this section, we will learn some of the basics about different characteristics of rational functions and real world situations in which they occur.

■ CONCEPT INVESTIGATION

Are we there yet?

Let's start by considering a simple situation involving driving a car. John is planning to take a 200-mile car trip. Let's consider how long this trip will take John if he travels at different average speeds.

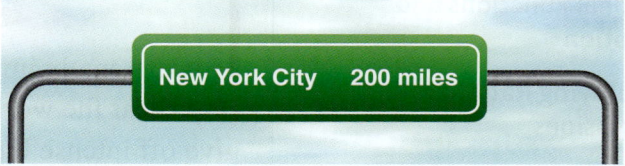

1. Fill in the following table with the times it will take John to travel 200 miles if he drives at an average rate (speed) of r miles per hour.

Rate (r) (in mph)	Time (t) (in hours)
10	20
25	8
40	
50	
80	
100	
200	

2. What happens to the time it takes to travel the 200 miles as the average speed gets slower?

3. What happens to the time it takes to travel the 200 miles as the average speed gets faster?

4. On your graphing calculator, create a scattergram of the data in the table. Does the scattergram agree with your answers from parts 2 and 3?

5. Solve the equation $D = rt$ for time and use that to write a model for the time it takes John to travel 200 miles. (Verify your model by graphing it with your data.)

6. Are the last two rates given in the table reasonable in this situation?

7. What are a reasonable domain and range for the model in this situation?

A function such as $t(r) = \dfrac{200}{r}$ is a simple rational function because it has a variable in the denominator of the fraction. Any expression of the form

$$\frac{P(x)}{Q(x)}$$

where $P(x)$ and $Q(x)$ are polynomials and $Q(x) \neq 0$ is called a **rational expression.** Notice that $Q(x)$ cannot equal zero, or you would have division by zero, and the rational expression would be undefined. Division by zero is always a concern in working with rational expressions.

In the Concept Investigation, you should note that it does not make sense for John to average zero miles per hour, since he would then never travel the 200 miles.

> **What's That Mean?**
>
> **Rational**
>
> Recall that a rational number is defined as any number that can be written as a fraction of two integers. A ratio is a comparison of two things and is often written as a fraction. A rational expression is a fraction that contains polynomials.
>
> As with all fractions, the denominator of a rational expression cannot equal zero.

DEFINITION

Rational expressions An expression of the form

$$\frac{P(x)}{Q(x)}$$

where $P(x)$ and $Q(x)$ are polynomials and $Q(x) \neq 0$ is called a rational expression. The domain of a rational expression consists of those values of x such that $Q(x) \neq 0$.

$$\frac{5x + 2}{x - 6} \quad \text{and} \quad \frac{3}{x}$$

are rational expressions.

The domain of a rational function in the context of an application must avoid any values that make the denominator zero or turn the situation into something that does not make sense, like an average speed of 200 mph in the concept investigation. In an application, the domain will usually be restricted to a small area of the graph. The range in an application will typically be the lowest to the highest points on the graph within the domain. As you work with a situation, set the domain according to what makes sense in the situation. Be cautious with any restrictions stated in the problem that would limit the domain in some way.

Example 1 Cost per student

A group of students in the chess club wants to rent a bus to take them to the national chess competition. The bus is going to cost $1500 to rent and can hold up to 60 people.

a. Write a model for the cost per student to rent the bus if s students take the bus and each student pays an equal share.

b. How much would the cost per student be if 30 students take the bus?

c. How much would the cost per student be if 60 students take the bus?

d. What would a reasonable domain and range be for this model? Explain.

SOLUTION

a. Let $C(s)$ be the cost per student in dollars for s students to take the bus to the national chess competition. Because each student is going to pay an equal amount, we might consider a few simple examples: If only one student takes the bus, that student would have to pay $1500. If two students take the bus, they will have to pay $\frac{1500}{2} = 750$ dollars each. We are taking the total cost of $1500 and dividing it by the number of students taking the bus. This pattern would continue, and we would get the following function.

$$C(s) = \frac{1500}{s}$$

b. If 30 students take the bus, we can substitute 30 for s and calculate $C(30)$.

$$C(30) = \frac{1500}{30}$$
$$C(30) = 50$$

Therefore, if 30 students take the bus, it will cost $50 per person.

c. Substituting in $s = 60$, we get

$$C(60) = \frac{1500}{60}$$
$$C(60) = 25$$

Therefore, if 60 students take the bus, it will cost $25 per person.

d. In an application problem, we will continue to avoid model breakdown when setting a domain. Because the bus can hold only up to 60 people, we must limit the domain to positive numbers up to 60. This means that we could have a possible domain of [1, 60]. With this domain, the range would be [25, 1500]. Of course, there are other possible domains and ranges, but these would be considered reasonable.

Direct and Inverse Variation

The function $C(s) = \frac{1500}{s}$, which we found in part a of Example 1, is an example of **inverse variation,** and it could be stated that the cost, $C(s)$, varies inversely with the number of students, s. That is, when one value increases, the other decreases. In Example 1, the more students who take the bus, the lower the per student cost will be.

The model $t(r) = \frac{200}{r}$, found in the concept investigation, is also an example of inverse variation of t and r.

Variation occurs when two or more variables are related to one another using multiplication or division. When two variables are related and both either increase together or decrease together, we call it **direct variation.** The equation $D = 60t$ is an example of direct variation; when the value of t increases, so does the value of D.

> **What's That Mean?**
>
> **Types of Variation**
>
> When it comes to variation, there are a couple of phrases we use to describe the relationship between the two variables.
>
> **Direct Variation:** $y = kx$
> - varies directly with
> - is directly proportional to
>
> **Inverse Variation:** $y = \frac{k}{x}$
> - varies inversely with
> - is inversely proportional to

> **DEFINITIONS**
>
> **Direct variation** The variable y varies directly with x if
>
> $$y = kx$$
>
> The variable y varies directly with x^n if
>
> $$y = kx^n$$

Inverse variation The variable y varies inversely with x if

$$y = \frac{k}{x}$$

The variable y varies inversely with x^n if

$$y = \frac{k}{x^n}$$

In each case, $k \neq 0$ and $n > 0$. The k is called the variation constant (or the constant of proportionality). The phrases *varies directly* and *varies inversely* can also be stated as *directly proportional to* and *inversely proportional to*, respectively.

A linear equation with a vertical intercept of $(0, 0)$ is a simple representation of direct variation. The variation constant is the slope of the line.

Example 2 Cost for car repair labor

The cost for labor at an auto repair shop is directly proportional to the time the mechanic spends working on the car. If a mechanic works on the car for five hours, the labor cost is $325.

a. Write a model for the labor cost at this auto repair shop.

b. What is the labor cost for two hours of work?

c. If Sam were charged $292.50 for labor on a recent repair, how many hours did the mechanic work on the car?

SOLUTION

a. Let $C(h)$ be the labor cost in dollars for h hours of work done by the mechanic. Because the cost is directly proportional to the hours worked, the cost will be equal to a constant times the hours.

$$C(h) = kh$$

We are told that for five hours of work, the cost was $325, so we substitute these values and solve for k.

$$325 = k(5)$$
$$\frac{325}{5} = \frac{k(5)}{5} \quad \text{Solve for } k.$$
$$65 = k$$

Now that we know that $k = 65$, we can write the model for the cost as

$$C(h) = 65h$$

b. Substitute $h = 2$ into our model and calculate the cost.

$$C(2) = 65(2)$$
$$C(2) = 130$$

If a mechanic works on the car for two hours, the labor cost will be $130.

c. Sam was charged $292.50 for labor, so we substitute this value in for $C(h)$ and solve for h.

$$292.50 = 65h$$
$$\frac{292.50}{65} = \frac{65h}{65}$$
$$4.5 = h$$

The mechanic worked on Sam's car for 4.5 hours, resulting in a labor cost of $252.50.

PRACTICE PROBLEM FOR EXAMPLE 2

The number of calories burned by a 130-pound woman while rowing is directly proportional to the time spent rowing. If Sheila rows for 10 minutes during a competition, she burns about 118 calories.

a. Write a model for the calories c burned from rowing in this competition for m minutes.

b. How many calories are burned during a 14-minute event?

In Example 2 and the Practice Problem, we saw that as the time increases, so does the cost or the number of calories. With direct variation, as the input increases, the output also increases. When variables are inversely proportional, they change in opposite ways. As the input increases, the output will decrease. As the input decreases, the output will increase. Note that an equation that represents variables that vary inversely is also a rational function.

Example 3 | Illumination from a light source

The illumination of a light source is inversely proportional to the square of the distance from the light source. A certain light has an illumination of 50 foot-candles at a distance of 5 feet from the light source.

a. Write a model for the illumination of this light.

b. What is the illumination of this light at a distance of 10 feet from the light source?

c. What is the illumination of this light at a distance of 100 feet?

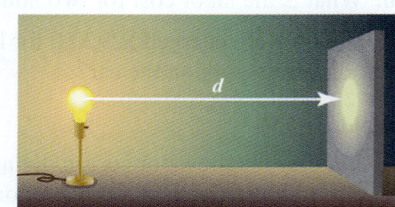

SOLUTION

a. Let $I(d)$ be the illumination of the light in foot-candles and let d be the distance from the light source in feet.

Because we are told that the illumination, $I(d)$, is inversely proportional to the square of the distance, d^2, from the light source, the illumination will be equal to a constant divided by distance squared.

$$I(d) = \frac{k}{d^2}$$

We still need to find the variation constant k. Because we are told that at 5 feet from the light source, the illumination is 60 foot-candles, we can substitute these values in and find k.

$$I(5) = 50$$

$50 = \dfrac{k}{5^2}$ Substitute the given values for I and d.

$50 = \dfrac{k}{25}$ Multiply both sides by 25 to solve for k.

$$1250 = k$$

Now that we know that $k = 1250$, we can write the model for illumination as

$$I(d) = \frac{1250}{d^2}$$

> **What's That Mean?**
>
> **Measurements for Light**
>
> The **illumination** (illuminance) of a light is the amount of light cast onto an object.
>
> A **foot-candle** is a unit of measure that originated with how much light a standard candle would cast on an object 1 foot away.

b. We are given the distance of 10 feet, so we can substitute $d = 10$ and calculate the illumination.

$$I(10) = \frac{1250}{10^2}$$

$$I(10) = \frac{1250}{100}$$

$$I(10) = 12.5$$

Therefore, the illumination at 10 feet from the light source is 12.5 foot-candles.

c. We are given the distance of 100 feet, so we can substitute $d = 100$ and calculate the illumination.

$$I(100) = \frac{1250}{100^2}$$

$$I(100) = \frac{1250}{10000}$$

$$I(100) = 0.125$$

Therefore, the illumination at 100 feet from the light source is 0.125 foot-candle.

PRACTICE PROBLEM FOR EXAMPLE 3

In physics, we learn that if a fixed amount of force is applied to an object, the amount of acceleration a in meters per second squared of that object is inversely proportional to the mass in kilograms of that object. When a fixed amount of force is applied to a 5-kg mass, it is accelerated at 10 m/s² (m/s² is meters per second squared).

a. Write a model for the acceleration a of a mass m when this fixed amount of force is applied.

b. What is the acceleration of a 15-kg mass when this force is applied?

In any direct or inverse variation problem, we need one example of the relationship between the two variables to find the variation constant. Then we will be able to write the formula for that particular situation. In Example 3, the variation constant will change on the basis of the strength of the light source we are working with.

Example 4 Inverse variation

a. If y varies inversely with $\sqrt{x}$ and $y = 58$ when $x = 4$, write an equation to represent this relationship.

b. Find y if $x = 9$.

SOLUTION

a. Because y varies inversely with $\sqrt{x}$, we know that y will equal a constant divided by $\sqrt{x}$.

$$y = \frac{k}{\sqrt{x}}$$

We also know that $y = 58$ when $x = 4$. We can substitute these values into y and x and solve for the constant k.

$$58 = \frac{k}{\sqrt{4}}$$

$$58 = \frac{k}{2}$$

$$116 = k$$

Now that we know $k = 116$, we can write the final equation

$$y = \frac{116}{\sqrt{x}}$$

b. We will substitute $x = 9$ and find y.

$$y = \frac{116}{\sqrt{9}}$$

$$y = \frac{116}{3}$$

$$y = 38\frac{2}{3}$$

PRACTICE PROBLEM FOR EXAMPLE 4

a. If y varies inversely with x^4 and $y = 143.75$ when $x = 2$, find an equation to represent this relationship.

b. Find y if $x = 5$.

Another type of situation that involves rational functions occurs when we are combining two functions using division. In the next example, we will investigate one of these situations.

Example 5 Finding an average

In the state of California, the student-teacher ratio dropped steadily from 1995 to about 2007. The total number of students in the K–12 public schools in California can be modeled by

$$S(t) = 112.25t + 4930.6$$

where $S(t)$ is the number of students in thousands in the California K–12 public schools and t is time in years since 1990. The number of teachers in California's K–12 public schools can be modeled by

$$T(t) = 11.75t + 175.6$$

where $T(t)$ is the number of teachers in thousands in the California K–12 public schools and t is time in years since 1990.
Source: Models derived from data found at the California Department of Education.

a. Estimate the number of students in California's K–12 public schools in 1995.

b. Estimate the number of teachers in California's K–12 public schools in 1995.

c. Find a model for the average number of students per teacher in California's K–12 public schools.

d. Estimate the average number of students per teacher in California's K–12 public schools during the year 2000.

e. Use the graph of the model found in part c to estimate when the number of students per teacher in California's K–12 public schools was 20.

SOLUTION

a. We are given the year 1995, so we can substitute $t = 5$ and get

$$S(5) = 112.25(5) + 4930.6$$

$$S(5) = 5491.85$$

Therefore, in 1995, there were approximately 5491.85 thousand students in California's K–12 public schools.

b. We are given 1995, but this time we substitute $t = 5$ into the $T(t)$ function.

$$T(5) = 11.75(5) + 175.6$$
$$T(5) = 234.35$$

Therefore, in 1995, there were approximately 234.35 thousand teachers in California's K–12 public schools.

c. We are asked for a model that would give the average number of students per teacher in California's K–12 public schools. We take the number of students and divide it by the number of teachers. Let $A(t)$ be the average number of students per teacher in California's K–12 public schools t years since 1990.

$$A(t) = \frac{S(t)}{T(t)} = \frac{112.25t + 4930.6}{11.75t + 175.6}$$

d. We are asked for the average number of students in 2000, so we can substitute $t = 10$ into our new model $A(t)$.

$$A(10) = \frac{112.25(10) + 4930.6}{11.75(10) + 175.6}$$
$$A(10) \approx 20.65$$

Therefore, in 2000, there was an average of 20.65 students per teacher in California's K–12 public schools.

e. To find the graph of this function, we can put the function into Y1 and then use the table to get an idea of the inputs that get us close to the output of 20 that we are interested in.

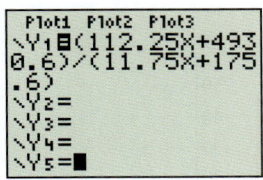 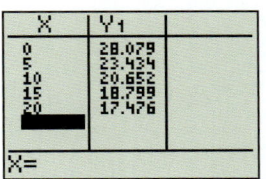 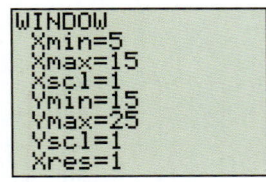

Looking at the graph and using TRACE, we get the following.

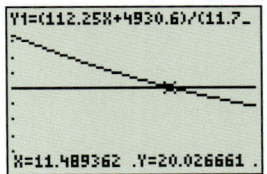

We can estimate that when $t \approx 11.5$, the number of students per teacher is about 20. According to this model, around the year 2002 the average number of students per teacher in California's K–12 public schools was 20.

PRACTICE PROBLEM FOR EXAMPLE 5

The number of people in the United States who are enrolled in Medicare can be modeled by

$$M(t) = 0.39t + 35.65$$

where $M(t)$ represents the number of people enrolled in Medicare in millions t years since 1990. Disbursements by the Medicare program can be modeled by

$$D(t) = -3160.64t^2 + 52536.5t - 4548.31$$

where $D(t)$ represents the amount of money Medicare disbursed to people in millions of dollars t years since 1990.

a. Find a model for the average amount disbursed per person enrolled in Medicare t years since 1990.

b. Estimate the average amount disbursed to a Medicare recipient in 1999.

c. Use the graph of your model to estimate the year when the average amount disbursed to Medicare recipients reached a maximum.

Using Your TI Graphing Calculator

When entering rational functions into your calculator, you must use parentheses around the entire numerator and again around the entire denominator.

$$f(x) = \frac{5 + x}{x + 9}$$

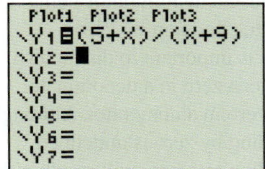

Connecting the Concepts

How do we find a good window?

Using the table in our calculator, we can evaluate the function for different values of the input that make sense in the problem. The table will then give us an idea of what output values we need to look at to see the graph.

In Example 5, the inputs represent years since 1990, so small values of t make sense in the problem. Putting those into the table gives us values for the output. From the table, we can see that the outputs range in value from about 17 to 28. If we set a window around these values, we will be able to get a reasonable graph on the screen.

■ Connecting the Concepts

How do we avoid the undefined?

Remember that when considering the domain of a function, we want to remove any input values that would make the output undefined.

A fraction would be considered undefined whenever the denominator is zero. This is because a denominator being zero implies division by zero.

It is important to distinguish between zero in a denominator and zero in a numerator. Although dividing by zero is undefined, dividing into zero will result in the answer being zero.

$$\frac{5}{0} = \text{undefined}$$

$$\frac{0}{5} = 0$$

When working with problems like those in Example 3, we need to remember the rules for when we can divide two functions. When the functions are in context, be sure that the units of the input variables are the same and that when the units of the outputs are divided, the resulting unit makes sense in the situation of the problem.

Domain of a Rational Function

When considering the domain of a rational function, we will mainly be concerned with excluding values from the domain that would result in the denominator being zero. The easiest way to determine the domain of a rational function is to set the denominator equal to zero and solve. The domain then becomes all real numbers except those values that make the denominator equal zero.

Any place where the denominator is zero would result in a vertical asymptote or a hole with a missing value. Recall that basic logarithmic functions had the y-axis as a vertical asymptote. Vertical asymptotes are also similar to the horizontal asymptotes that we saw in the graphs of exponential functions. The graph of a function will not touch a vertical asymptote but instead will get as close as possible and then jump over it and continue on the other side. Whenever an input value makes the numerator and denominator both equal to zero, a hole in the graph will occur instead of a vertical asymptote. Consider the two graphs below to see when a vertical asymptote occurs and when a hole occurs.

$$f(x) = \frac{3}{x-2} \qquad\qquad g(x) = \frac{x-2}{x^2-4}$$

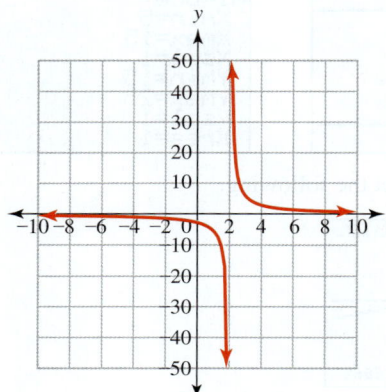
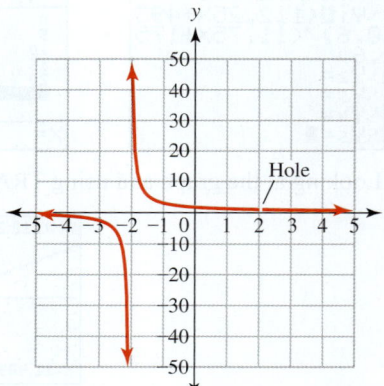

$$f(2) = \frac{3}{0} = \text{undefined} \qquad\qquad g(2) = \frac{0}{0} = \text{undefined}$$

Denominator = zero   Denominator and numerator = zero

Asymptote   Hole

$$g(-2) = -\frac{4}{0} = \text{undefined}$$

Denominator = zero

Asymptote

A vertical asymptote occurs when an input value makes the denominator equal zero but the numerator does not equal zero. A hole occurs in a graph when an input value makes both the numerator and denominator equal zero.

Example 6   Finding the domain of a rational function

Find the domain of the following rational functions. Determine whether the excluded values represent where a vertical asymptote or a hole appear in the graph.

a. $f(x) = \dfrac{5}{x}$

b. $g(x) = \dfrac{5+x}{x+9}$

c. $h(x) = \dfrac{x+4}{(x+4)(x-7)}$

d. $f(x) = \dfrac{3x+2}{x^2+5x+6}$

e.

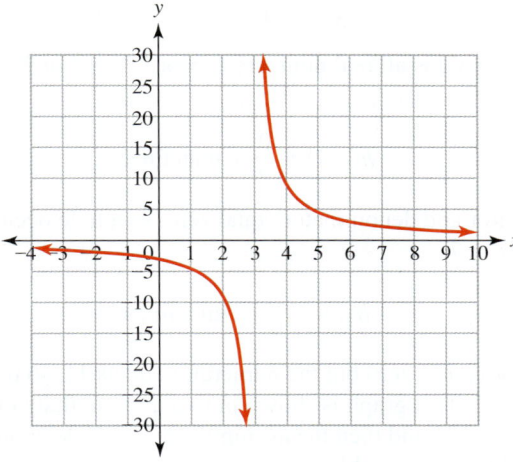

SOLUTION

a. Because the denominator of the function $f(x) = \dfrac{5}{x}$ would be zero when $x = 0$, we have a domain of all real numbers except zero. This can also be written simply as $x \neq 0$. When $x = 0$,

$$f(0) = \dfrac{5}{0} = \text{undefined}$$

The denominator is zero but the numerator is not, so a vertical asymptote occurs when $x = 0$. Looking at the graph of $f(x)$, we see that the function jumps over the input value $x = 0$, and there is a vertical asymptote in its place.

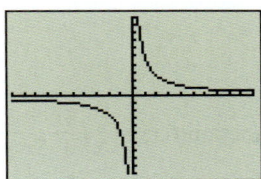

b. The denominator of the function $g(x) = \dfrac{5+x}{x+9}$ would be zero when $x = -9$, so its domain is all real numbers such that $x \neq -9$. When $x = -9$,

$$g(-9) = \dfrac{-4}{0} = \text{undefined}$$

The denominator is zero but the numerator is -4, so there is a vertical asymptote at $x = -9$. Looking at this graph again, we see a vertical asymptote. Pay attention to the way in which this function must be entered into the calculator with parentheses around the numerator and another set of parentheses around the denominator of the fraction.

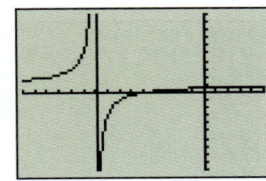 

Using Your TI Graphing Calculator

In graphing rational functions on your calculator, be aware that when most window settings are used, the calculator will try to connect the points on either side of a vertical asymptote. This will result in a vertical line being drawn where there should be no line. The graph of the function

$$f(x) = \frac{5}{x-4}$$

is as follows.

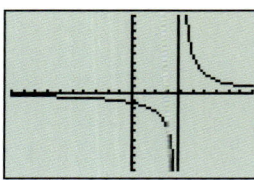

This graph should have a gap or vertical asymptote at $x = 4$. Instead, the calculator is showing a vertical line that is not really part of the graph of this function. If we use a different window, the graph is as follows.

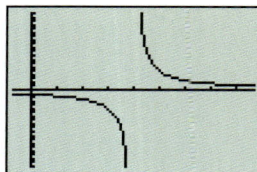

Notice that now the vertical asymptote is shown as a gap in the graph, not as a vertical line.

A hole in the graph is even harder to see on the graphing calculator. This is a good reason to know when a hole or vertical asymptote is going to appear so that you can confirm it on the calculator and not always trust what you see.

c. If you set the denominator of the function $h(x) = \dfrac{x+4}{(x+4)(x-7)}$ equal to zero, you get

$$(x+4)(x-7) = 0$$

$x + 4 = 0 \quad\quad x - 7 = 0$

$x = -4 \quad\quad\quad x = 7$

Therefore, the domain is all real numbers except $x = -4$ or 7. When $x = -4$,

$$h(-4) = \frac{0}{0} = \text{undefined}$$

Both the numerator and denominator equal zero, so a hole occurs in the graph when $x = -4$. When $x = 7$,

$$h(7) = \frac{11}{0} = \text{undefined}$$

The denominator equals zero but the numerator equals 11, so there is a vertical asymptote at $x = 7$. This graph is shown in two parts so that you can see the hole that appears at $x = -4$ and then the asymptote at $x = 7$. Without setting up two windows, it is almost impossible to see the hole.

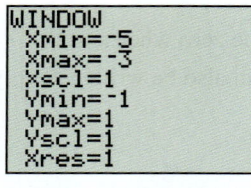  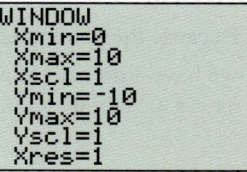

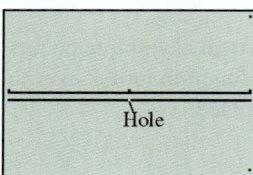 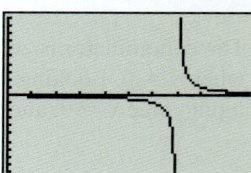

d. Set the denominator of the function $f(x) = \dfrac{3x+2}{x^2+5x+6}$ equal to zero.

$$x^2 + 5x + 6 = 0$$
$$(x+3)(x+2) = 0$$
$x + 3 = 0 \quad\quad x + 2 = 0$
$x = -3 \quad\quad\quad x = -2$

Therefore, the domain is all real numbers except $x = -3$ or -2. For both $x = -3$ and -2, the denominator equals zero but the numerator does not. Therefore, there are vertical asymptotes at $x = -3$ and $x = -2$. This graph has an interesting shape, but it does have two vertical asymptotes. Again the numerator and denominator of the fraction need parentheses around them to create the graph correctly.

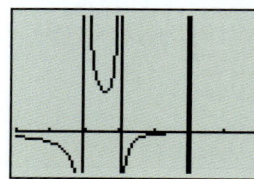 

e. The graph of this function shows a vertical asymptote at about $x = 3$, so the domain should be all real numbers except $x = 3$. We cannot see any holes in the given graph.

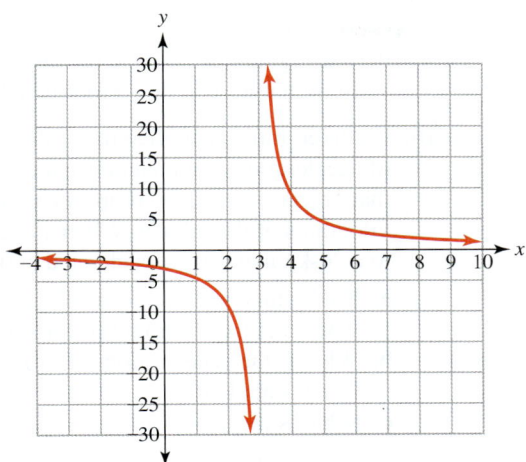

PRACTICE PROBLEM FOR EXAMPLE 6

Find the domain of the following rational functions.

a. $f(x) = \dfrac{3}{x}$ b. $g(x) = \dfrac{x+2}{x-7}$ c. $h(x) = \dfrac{x+2}{x^2 - 3x - 10}$

7.1 Exercises

1. Charity Poker, Inc. organizes poker tournaments to raise money for charitable organizations. A typical tournament for up to 100 players costs $600 to put together.
   a. Find a model for the per player cost for a charity poker tournament if p players participate.
   b. Find the cost per person for each player if 75 players participate.
   c. What are a reasonable domain and range for this model?

2. The Adventure Guides group at a local YMCA is planning a camping trip on the beach. It costs $1500 to rent the campsites for the weekend and pay for insurance for up to 90 campers.

   a. Find a model for the per camper cost for this camp out if p people attend.
   b. Find the cost per person for each camper if 60 people attend.
   c. What are a reasonable domain and range for this model?

3. The Fancy Affair catering company is catering an event for a local charity. Jan, the owner of Fancy Affair, is donating her time to the cause, but she needs to charge the charity for the food, decorations, and supplies used. If p people attend the charity event, Jan has figured her total cost to be modeled by

$$C(p) = 4.55p + 365.00$$

where $C(p)$ is the total cost in dollars for the food, decorations, and supplies when p people attend the charity event.
   a. Find the total cost for 100 people to attend the event.
   b. Find the cost per person for 100 people to attend the event.
   c. Find a new function that gives the cost per person for p people to attend the charity event.
   d. Use your new model to find the cost per person for 150 people to attend the event.
   e. If the location of the charity event can hold up to 250 people, what are a reasonable domain and range of the model you found in part c?

4. The math club is throwing a graduation party for five of its members. The club has decided to hire a band for $500 and to buy $300 worth of food and drinks for the party. Each member of the club who attends the party (except the five graduates) is to pay an equal amount of the costs.
   a. Find a model for the cost per person for this party if m members of the club attend the party. (m does include the five graduates.)
   b. Find the cost per person for each nongraduating member of the club who attends the party if 45 members attend.
   c. What are a reasonable domain and range for this model if the math club has 50 members?

5. The total cost for a hotel room varies directly with the number of nights you stay. At one hotel, three nights in the hotel cost a total of $450.
   a. Find a model for the total cost for a hotel room at this hotel if you stay for n nights.
   b. What is the total cost at this hotel for a seven-night stay?
   c. If you have a budget of $800 for your hotel, how many nights can you stay at this hotel?

6. Peter's weekly salary varies directly with the number of hours he works. In a week that Peter worked 30 hours he earned $382.50 before taxes.
   a. Find a model for Peter's weekly salary if he works h hours a week.
   b. What is Peter's salary if he works 40 hours a week?
   c. If Peter wants to earn $400 a week before taxes, how many hours does he need to work that week?

7. The volume in liters needed to store 0.05 mole of helium is directly proportional to the temperature in Kelvin. It takes 0.821 liter to store 0.05 mole of helium at a temperature of 400 K.
   a. Find a model for the volume it takes to store this helium if the temperature is t Kelvin.
   b. What volume is needed to store this helium if the temperature is 250 K?
   c. If you have 0.75 liter to store this helium, what temperature do you need to store it at?

8. The circumference of a circle varies directly with its radius. A circle with a radius of 5 inches has an approximate circumference of 31.4 inches.
   a. Find a model for the circumference of a circle depending on its radius.
   b. Find the circumference of a circle with a radius of 12 inches?
   c. What is the radius of a circle with a circumference of 8 inches? (Round to two decimal places.)

9. A light has an illumination of 25.5 foot-candles at a distance of 10 feet from the light source. (See Example 3.)
   a. Find a model for the illumination of this light.
   b. What is the illumination of this light at a distance of 5 feet from the light source?
   c. What is the illumination of this light at a distance of 30 feet from the light source?
   d. Use a graph to estimate the distance at which the illumination will be 50 foot-candles.

10. If a fixed force is applied to an object, the amount of acceleration a in meters per second squared of that object is inversely proportional to the mass in kilograms of that object. When a fixed force is applied to a 20-kg mass, it is accelerated at 0.5 m/s^2.
    a. Find a function for the acceleration a of a mass m when this fixed amount of force is applied.
    b. What is the acceleration of a 10-kg mass when this force is applied?
    c. What is the acceleration of a 5-kg mass when this force is applied?
    d. Estimate numerically what size mass this force can accelerate at 0.7 m/s^2.

11. The weight of a body varies inversely as the square of its distance from the center of the earth. If the radius of the earth is 4000 miles, how much would a 220-pound man weigh 2000 miles above the surface of the earth? (*Hint:* The person weighs 220 pounds when 4000 miles from the center of the earth.)

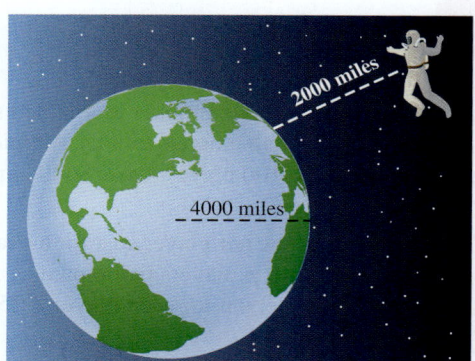

12. The pressure P of a certain amount of gas in a balloon is inversely proportional to the volume of the balloon. If the pressure in a balloon is 5 pounds per square inch when the volume of the balloon is 4 cubic inches, what is the pressure in the balloon if the volume is only 2 cubic inches?

13. The current that flows through an electrical circuit is inversely proportional to the resistance of that circuit. When the resistance R is 200 ohms, the current I is 1.2 amperes. Find the current when the resistance is 130 ohms.

14. The force needed to balance a 200-g weight on a fulcrum is inversely proportional to the distance from the fulcrum at which the force is applied. 2.25 newtons of force applied 4 units away from the fulcrum is

needed to balance the weight. (A newton is the unit of measure used with force.)

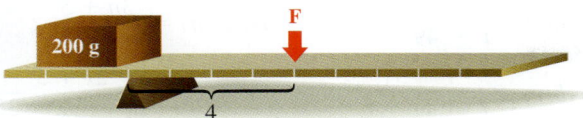

a. Find the force needed to balance the 200-g weight if it is applied 1 unit away from the fulcrum.
b. Find the force needed to balance the 200-g weight if it is applied 3 units away from the fulcrum.
c. Find the force needed to balance the 200-g weight if it is applied 10 units away from the fulcrum.

15. The pressure of a certain amount of gas in a balloon is inversely proportional to the volume of the balloon. For one such balloon, we have the function

$$P(v) = \frac{30}{v}$$

where $P(v)$ is the pressure of the gas in the balloon in pounds per square inch when the volume of the balloon is v cubic inches.

a. Find $P(2)$ and explain its meaning in this situation.
b. Find $P(10)$ and explain its meaning in this situation.

16. The force needed to lift a boulder using a pry-bar and fulcrum can be modeled by the function

$$F(d) = \frac{1500}{d}$$

where $F(d)$ is the force in newtons needed to lift the boulder when it is applied d inches away from the fulcrum.

a. Find $F(6)$ and explain its meaning in this situation.
b. Find $F(120)$ and explain its meaning in this situation.

17. a. If y varies directly with x^3 and $y = 150.4$ when $x = 4$, find an equation to represent this relationship.
b. Find y if $x = 8$.

18. a. If y varies directly with x^2 and $y = 648$ when $x = 9$, find an equation to represent this relationship.
b. Find y if $x = 3$.

19. a. If y varies inversely with x^3 and $y = 405$ when $x = 3$, find an equation to represent this relationship.
b. Find y if $x = 5$.

20. a. If y varies inversely with x^7 and $y = 114$ when $x = 4$, find an equation to represent this relationship.
b. Find y if $x = 6$.

21. a. If M is inversely proportional to $5\sqrt{t}$ and $M = 115$ when $t = 9$, find an equation to represent this relationship.
b. Find M if $t = 25$.

22. a. If y is inversely proportional to $8\sqrt{x}$ and $y = 25$ when $x = 6$, find an equation to represent this relationship.
b. Find y if $x = 4$.

23. The average amount of benefits received by people in the U.S. food stamp program can be modeled by

$$B(t) = \frac{-470001t^2 + 4110992t + 14032612}{-469.4t^2 + 3745t + 19774}$$

where $B(t)$ is the average benefit in dollars per person for people in the U.S. food stamp program t years since 1990.
Source: Model derived from data from Statistical Abstract of the United States, 2001.

a. Find the average benefit for a person participating in the U.S. food stamp program in 1995.
b. Use a graph to estimate when the average benefit for a person participating in the U.S. food stamp program was $800.

24. The average amount of benefits received by people in the U.S. food stamp program can be modeled by

$$B(t) = \frac{-470001t^2 + 4110992t + 14032612}{-469.4t^2 + 3745t + 19774}$$

where $B(t)$ is the average benefit in dollars per person for people in the U.S. food stamp program t years since 1990.
Source: Model derived from data from Statistical Abstract of the United States, 2001.

a. Estimate the average benefit in 2000.
b. Use a graph to estimate when the average benefit for a person participating in the U.S. food stamp program was $700.

25. The population of the United States since 1970 can be modeled by

$$P(t) = 0.226t^2 + 1.885t + 204.72$$

where $P(t)$ represents the population of the United States in millions t years since 1970. The national debt of the United States in millions of dollars t years since 1970 can be modeled by

$$D(t) = 8215.1t^2 - 23035.4t + 413525.6$$

where $D(t)$ represents the national debt of the United States in millions of dollars t years since 1970.
Source: Models derived from data from the U.S. Department of Commerce Bureau of Economic Analysis.

a. Find a new model for the average amount of national debt per person in the United States t years since 1970.
b. Find the average amount of national debt per person in 2000.
c. Estimate numerically the year in which the average amount of national debt per person was $10,000.

26. The state of California's spending per person has increased dramatically since 1950. The state's population from 1950 to 2000 can be modeled by

$$P(t) = 0.464t - 12.47$$

where $P(t)$ is California's population in millions of people t years since 1900. The amount that California has spent in millions of dollars can be modeled by

$$S(t) = 55.125t^2 - 6435.607t + 186914.286$$

where $S(t)$ is the amount California spent in millions of dollars t years since 1900.
Source: Models derived from data in the Governor's Budget Summary as printed in the North County Times, Feb. 9, 2003.

a. Estimate the population of California in 1980.
b. Estimate the amount California spent in 1990.
c. Find a new function that gives the spending per capita (per person) t years since 1900.
d. Estimate the per capita spending in California in 1980.
e. Estimate numerically when the per capita spending in California reached $2500.

For Exercises 27 through 50, give the domain of the rational functions.

27. $g(x) = \dfrac{25}{x}$

28. $f(x) = -\dfrac{20}{x}$

29. $f(x) = \dfrac{x + 5}{x - 3}$

30. $h(t) = \dfrac{t - 6}{t + 7}$

31. $m(b) = \dfrac{b + 9}{b + 11}$

32. $p(n) = \dfrac{n + 8}{n - 16}$

33. $p(t) = \dfrac{3t - 7}{4t + 5}$

34. $f(x) = \dfrac{2x - 7}{6x - 17}$

35. $h(a) = \dfrac{3a - 1}{(2a + 7)(a - 3)}$

36. $g(x) = \dfrac{2x - 7}{(x + 4)(x - 9)}$

37. $C(n) = \dfrac{n + 10}{(n + 4)(n + 10)}$

38. $h(t) = \dfrac{t - 3}{(t - 3)(t + 7)}$

39. $h(m) = \dfrac{4m^2 + 2m - 9}{m^2 + 7m + 12}$

40. $f(x) = \dfrac{3x + 7}{x^2 - 9x + 8}$

41. $f(x) = \dfrac{2x + 1}{x^2 + 3x + 19}$

42. $f(x) = \dfrac{x + 8}{x^2 + 5x + 21}$

43. $h(t) = \dfrac{t + 7}{t^2 - 49}$

44. $R(n) = \dfrac{3n + 5}{9n^2 + 25}$

45. $b(r) = \dfrac{r + 3}{r^2 + 5r + 6}$

46. $g(x) = \dfrac{2x + 5}{2x^2 - 9x - 35}$

47. $h(t) = \dfrac{t^2 - 3t + 5}{t^2 - 7t + 10}$

48. $f(x) = \dfrac{4x^2 + 3x - 8}{6x^2 - 7x - 20}$

49. $P(a) = \dfrac{a^2 + 2a + 1}{a^2 + 9}$

50. $g(x) = \dfrac{x - 8}{x^2 + 16}$

For Exercises 51 through 54, use the graph to answer the questions.

51. Given the graph of $f(x)$,

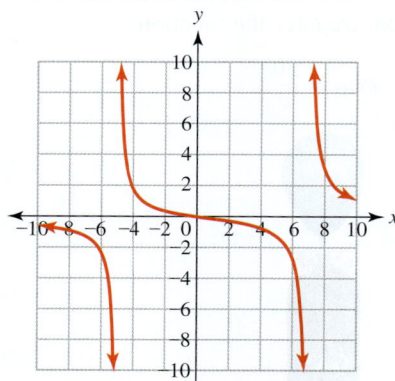

a. Find the domain of $f(x)$.
b. Estimate $f(0)$.
c. Estimate x such that $f(x) = 5$.

52. Given the graph of $f(x)$,

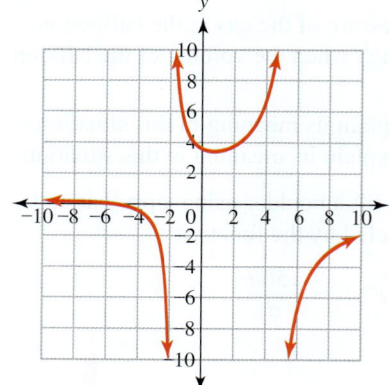

a. Find the domain of $f(x)$.
b. Estimate $f(0)$.
c. Estimate x such that $f(x) = 6$.

53. Given the graph of $h(x)$,

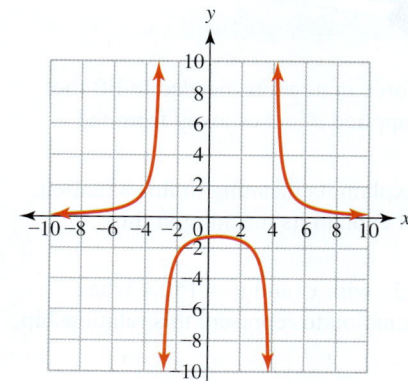

a. Find the domain of $h(x)$.
b. Estimate $h(1)$.
c. Estimate $h(-4)$.
d. Estimate x such that $h(x) = -2$.

54. Given the graph of $h(x)$,

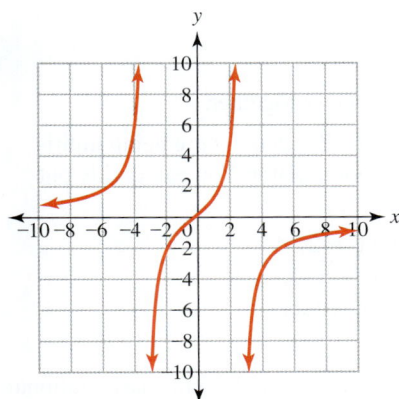

a. Find the domain of $h(x)$.
b. Estimate $h(8)$.
c. Estimate $h(-6)$.
d. Estimate x such that $h(x) = -8$.

59.

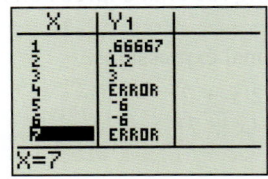

60.

For Exercises 61 through 76, find the domain of the given rational functions. Determine whether the values excluded from the domain represent where a vertical asymptote or a hole appears in the graph. Check your answer using the graphing calculator.

61. $f(x) = \dfrac{8}{x+3}$

62. $f(x) = \dfrac{7}{x-6}$

63. $g(x) = \dfrac{x+7}{(x+3)(x+1)}$

64. $h(x) = \dfrac{x-5}{(x-2)(x+8)}$

65. $g(x) = \dfrac{x+2}{(x+5)(x+2)}$

66. $h(x) = \dfrac{x-4}{(x-4)(x+9)}$

67. $g(x) = \dfrac{3x+9}{6x^2+7x-20}$

68. $h(x) = \dfrac{5x-11}{8x^2-26x+21}$

69. $f(x) = \dfrac{x+2}{x^2+3x+2}$

70. $g(x) = \dfrac{x-7}{x^2-2x-35}$

71. $h(x) = \dfrac{5x}{x^3-2x^2-8x}$

72. $f(x) = \dfrac{2x}{4x^3-32x^2+60x}$

73. $g(x) = \dfrac{3x+12}{x^2-16}$

74. $h(x) = \dfrac{5x+10}{x^2-4}$

75. $g(x) = \dfrac{x+5}{x^2+25}$

76. $h(x) = \dfrac{x-3}{x^2+3x+10}$

For Exercises 55 through 60, use the table to give values that are not included in the function's domain.

55.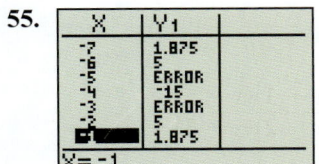

56.

57.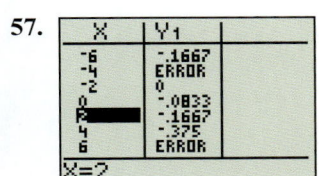

58.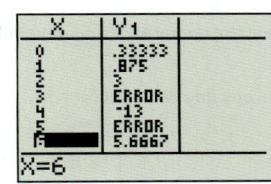

7.2 Simplifying Rational Expressions

LEARNING OBJECTIVES
- Simplify rational expressions using factoring.
- Divide polynomials using long division.
- Divide polynomials using synthetic division.

To prepare us to solve equations involving rational functions, we need to develop some basic skills to handle rational expressions. We are going to start by learning to simplify rational expressions. When the denominator of a rational expression has only one term, we can divide each term in the numerator by the denominator. If the denominator has more than one term, we will either use factoring or long division of polynomials.

What's That Mean?

Meanings for "cancel out"

Students often use "cancel out" in place of more specific phrases such as:
- Adds to zero
- Divides to 1
- Is the inverse of

This is usually done to simplify the language that is being used to make it easier to understand or read. Pay attention to the operation you are using to "cancel" something out, or else you might simplify at the wrong time.

Simplifying Rational Expressions

Simplifying rational expressions is the same as simplifying numeric fractions. The numerator and denominator must both include multiplication for a factor to divide out and thus reduce the fraction. When working with numeric fractions, we often take this for granted and do the process without thinking about all the steps. To work with rational expressions, we will need to focus on each step of the process to properly reduce the expression.

CHAPTER 7 Rational Functions

Skill Connection

Simplifying Fractions

Simplifying rational expressions is very similar to simplifying numerical fractions. Factor the numerator and denominator and divide out any common factors.

Example:

Simplify $\dfrac{420}{770}$

Solution:

Factor the numerator and denominator.

$$\dfrac{420}{770} = \dfrac{2 \cdot 2 \cdot 3 \cdot 5 \cdot 7}{2 \cdot 5 \cdot 7 \cdot 11}$$

Divide out the common factors.

$$= \dfrac{2 \cdot \cancel{2} \cdot 3 \cdot \cancel{5} \cdot \cancel{7}}{\cancel{2} \cdot \cancel{5} \cdot \cancel{7} \cdot 11}$$

$$= \dfrac{2 \cdot 3}{11}$$

Multiply the remaining factors.

$$= \dfrac{6}{11}$$

Simplifying Rational Expressions

1. Factor the numerator and denominator (if possible).
2. Divide out any common factors.
3. Leave in factored form or multiply remaining factors together.

Caution: Do not reduce unless the expressions to be divided out are being multiplied by the remaining parts of the numerator or denominator. You can divide out factors, but you cannot divide out terms.

$$\dfrac{5\cancel{(x+2)}}{(x+3)\cancel{(x+2)}} \qquad \dfrac{5+(x+2)}{(x+3)(x+2)}$$

Reduces Does not reduce.

Let's start with some examples with numeric fractions and some basic rational expressions.

Example 1 Simplifying rational expressions

Simplify the following rational expressions.

a. $\dfrac{5x^2}{10x}$ b. $\dfrac{2(x+7)}{(x+7)}$ c. $\dfrac{(x+3)(x-5)}{(x-2)(x+3)}$

SOLUTION

a. $\dfrac{5x^2}{10x} = \dfrac{5 \cdot x \cdot x}{2 \cdot 5 \cdot x}$ Factor the numerator and denominator.

$= \dfrac{\cancel{5} \cdot \cancel{x} \cdot x}{2 \cdot \cancel{5} \cdot \cancel{x}}$ Divide out any common factors.

$= \dfrac{x}{2}$

b. $\dfrac{2(x+7)}{(x+7)} = \dfrac{2\cancel{(x+7)}}{\cancel{(x+7)}}$ Already factored, so divide out common factors.

$= \dfrac{2}{1}$ When all the factors in the numerator or denominator divide out, a 1 is left.

$= 2$ Reduce.

c. $\dfrac{(x+3)(x-5)}{(x-2)(x+3)} = \dfrac{\cancel{(x+3)}(x-5)}{(x-2)\cancel{(x+3)}}$ Divide out common factors.

$= \dfrac{x-5}{x-2}$ Note that the remaining x does not reduce because it is not multiplied, but subtracted.

Note that an expression involving addition or subtraction can be divided out, but only if that expression is being multiplied by the remaining factors. Any addition or subtraction that remains separate from the multiplication in the numerator or denominator will stop us from dividing out the like expressions. The next example will work with some more complicated rational expressions. We will also consider a couple of situations in which we are not able to reduce the expression.

Connecting the Concepts

When do we divide out?

The most common place to reduce when you should not is in division. We cannot reduce when there is addition involved rather than multiplication.

$$\dfrac{x+5}{x+2} \neq \dfrac{5}{2}$$

Here, the x's cannot reduce using division because of the addition that is involved in the numerator and denominator.

$$\dfrac{5x}{2x} = \dfrac{5}{2}$$

In this fraction, the x's can divide out because only multiplication is involved, making the x's common factors that do divide to 1.

Example 2 Simplifying rational expressions

Simplify the following rational expressions.

a. $\dfrac{x^2 + 8x + 12}{x^2 + 7x + 10}$ b. $\dfrac{6x^2 - 29x + 28}{15x^2 - 41x + 28}$

c. $\dfrac{(x+4)(x-3)}{(x-3)+5}$ d. $\dfrac{18x^2 - 3x - 10}{x^3 - 2x^2 - 3x}$

SECTION 7.2 Simplifying Rational Expressions

SOLUTION

a. $\dfrac{x^2 + 8x + 12}{x^2 + 7x + 10} = \dfrac{(x+6)(x+2)}{(x+5)(x+2)}$ Factor the numerator and denominator.

$= \dfrac{(x+6)\cancel{(x+2)}}{(x+5)\cancel{(x+2)}}$ Divide out any common factors.

$= \dfrac{x+6}{x+5}$

b. $\dfrac{6x^2 - 29x + 28}{15x^2 - 41x + 28} = \dfrac{(2x-7)(3x-4)}{(3x-4)(5x-7)}$ Factor.

$= \dfrac{(2x-7)\cancel{(3x-4)}}{\cancel{(3x-4)}(5x-7)}$ Divide out any common factors.

$= \dfrac{2x-7}{5x-7}$

c. $\dfrac{(x+4)(x-3)}{(x-3)+5} = \dfrac{(x+4)(x-3)}{x+2}$ The x − 3 cannot reduce because of the addition present in the denominator.

d. $\dfrac{18x^2 - 3x - 10}{x^3 - 2x^2 - 3x} = \dfrac{(3x+2)(6x-5)}{x(x-3)(x+1)}$ There are no common factors, so the expression is already simplified.

PRACTICE PROBLEM FOR EXAMPLE 2

Simplify the following rational expressions.

a. $\dfrac{(x+5)(x+7)}{(x+7)(x-9)}$ b. $\dfrac{x^2 - 5x + 6}{x^2 - 13x + 30}$

c. $\dfrac{x^3 - 5x^2 + 6x}{x^2 + 5x - 14}$ d. $\dfrac{5(x+7)}{10(x+2) + 3}$

In parts c and d of Example 2, note that the lack of common factors keeps us from simplifying the rational expression further. In all of these expressions, we can multiply the factors back together as we would with numeric fractions, or we can leave them in factored form.

One special case that can occur is when an expression in the numerator is almost the same as an expression in the denominator. Most of the time, the expression cannot be reduced. Sometimes, we will be able to take out a common factor and make the two expressions the same. The common factor that most often needs to be taken out is −1.

Example 3 Special cases for simplifying

Simplify the following rational expressions.

a. $\dfrac{7(x-5)}{(5-x)}$ b. $\dfrac{2m+6}{5(m+3)}$

SOLUTION

a. $\dfrac{7(x-5)}{(5-x)} = \dfrac{7(x-5)}{-1(x-5)}$ Factor −1 out of the denominator to make the expressions the same.

$= \dfrac{7\cancel{(x-5)}}{-1\cancel{(x-5)}}$ Divide out common factors.

$= -\dfrac{7}{1} = -7$ The negative is typically put in front of the fraction or in the numerator.

b. $\dfrac{2m+6}{5(m+3)} = \dfrac{2(m+3)}{5(m+3)}$ Factor completely.

$= \dfrac{2\cancel{(m+3)}}{5\cancel{(m+3)}}$ Divide out common factors.

$= \dfrac{2}{5}$

Skill Connection

Factoring

The process for simplifying many rational expressions will require you to factor the numerator and denominator. Let's review the steps to factoring we went over in Section 3.4.

AC Method of Factoring:

1. Take out the greatest common factor.
2. Multiply *a* and *c* together.
3. Find factors of *ac* that add up to *b*.
4. Rewrite the middle (*bx*) term using the factors from step 3.
5. Group and factor out what is in common.

Example: Factor $3x^2 - 13x - 10$.

Solution:

Step 1: Nothing in common.
Step 2:
$a = 3 \quad c = -10 \quad ac = -30$
Step 3: $ac = -30$

1	−30	−1	30
2	−15	−2	15
3	−10	−3	10
5	−6	−5	6

$2 - 15$ gives us -13.

Step 4:
$3x^2 - 15x + 2x - 10$
Step 5:
$3x^2 - 15x + 2x - 10$
$(3x^2 - 15x) + (2x - 10)$
$3x(x-5) + 2(x-5)$
$(3x+2)(x-5)$

PRACTICE PROBLEM FOR EXAMPLE 3

Simplify the following.

a. $\dfrac{4(9-x)}{7(x-9)}$
b. $\dfrac{x-7}{21-3x}$

When dividing polynomials by a monomial, we can divide each term of the numerator by the monomial and simplify.

Skill Connection

Quotient rule for exponents

Recall from Section 3.1 that we used the quotient rule for exponents to simplify fractions that contained variables raised to powers.

$$\dfrac{x^m}{x^n} = x^{m-n} \qquad x \neq 0$$

Example 4 Dividing a polynomial by a monomial

Simplify

a. $\dfrac{5x^3 + 20x^2 - 8x}{5x^2}$
b. $\dfrac{x^5 y + 4x^4 y^3 - 15x^2 y^4}{x^2 y}$

SOLUTION

a. We will divide each term separately by the monomial in the denominator.

$$\dfrac{5x^3 + 20x^2 - 8x}{5x^2} = \dfrac{5x^3}{5x^2} + \dfrac{20x^2}{5x^2} - \dfrac{8x}{5x^2} \qquad \text{Separate each term and simplify.}$$

$$= x + 4 - \dfrac{8}{5x}$$

b. We will divide each term separately by the monomial in the denominator.

$$\dfrac{x^5 y + 4x^4 y^3 - 15x^2 y^4}{x^2 y} = \dfrac{x^5 y}{x^2 y} + \dfrac{4x^4 y^3}{x^2 y} - \dfrac{15x^2 y^4}{x^2 y}$$

$$= x^3 + 4x^2 y^2 - 15 y^3$$

PRACTICE PROBLEM FOR EXAMPLE 4

Simplify $\dfrac{12x^6 y^2 + 20x^4 y^3 - 18x^2 y^4}{3x^2 y^3}$

Long Division of Polynomials

When we divide a polynomial by another polynomial, we can use long division. Long division of polynomials is basically the same process as long division with numbers. We will start by using long division to divide 458 by 6.

```
      76
   6)458
    -42
     38
    -36
      2
```

6 does not divide into 4, so we divide into 45 first. 6 divides into 45 seven times, so 7 goes above the 5, and we multiply 6(7) = 42.

Subtract and bring down the next digit (8).

6 divides into 38 six times, so the 6 goes above the 8, and we multiply 6(6) = 36.

Subtract. The remainder is 2.

From this long division, we get $76\dfrac{2}{6} = 76\dfrac{1}{3}$. We can check this answer by multiplying it by 6.

This same division process can be used with polynomials. When the divisor has a higher degree than the remainder, we can stop the process. The remainder will be written over the original divisor.

Example 5 Dividing a polynomial by a polynomial

a. Divide $3x^2 + 17x + 20$ by $x + 4$.

b. Divide $10x^2 + 7x - 19$ by $2x + 3$.

SECTION 7.2 Simplifying Rational Expressions

SOLUTION

a. Divide the first term $3x^2$ by x. Multiply $x + 4$ by $3x$ and subtract.

$$\begin{array}{r} 3x \\ x + 4 \overline{) 3x^2 + 17x + 20} \\ -(3x^2 + 12x) \\ \hline 5x + 20 \end{array} \qquad \begin{array}{r} 3x \\ x + 4 \overline{) 3x^2 + 17x + 20} \\ -3x^2 - 12x \\ \hline 5x + 20 \end{array}$$

To subtract, distribute the negative sign and combine like terms. Bring down the next term, 20.

Continue dividing $5x$ by x. Multiply $x + 4$ by 5 and subtract.

$$\begin{array}{r} 3x + 5 \\ x + 4 \overline{) 3x^2 + 17x + 20} \\ -3x^2 - 12x \\ \hline 5x + 20 \\ -(5x + 20) \\ \hline 0 \end{array} \qquad \begin{array}{r} 3x + 5 \\ x + 4 \overline{) 3x^2 + 17x + 20} \\ -3x^2 - 12x \\ \hline 5x + 20 \\ -5x - 20 \\ \hline 0 \end{array}$$

Distribute the negative sign and combine like terms.

The remainder is zero, so we are done.

Therefore,

$$\frac{3x^2 + 17x + 20}{(x + 4)} = 3x + 5$$

b.

$$\begin{array}{r} 5x \\ 2x + 3 \overline{) 10x^2 + 7x - 19} \\ -(10x^2 + 15x) \\ \hline -8x - 19 \end{array}$$

Divide the first term $10x^2$ by $2x$. Multiply $2x + 3$ by $5x$ and subtract.

Bring down the next term.

$$\begin{array}{r} 5x - 4 \\ 2x + 3 \overline{) 10x^2 + 7x - 19} \\ -10x^2 - 15x \\ \hline -8x - 19 \\ -(-8x - 12) \\ \hline -7 \end{array}$$

Divide the $-8x$ by $2x$. Multiply $2x + 3$ by -4 and subtract.

The divisor, $2x + 3$ has a higher degree than the the remainder, -7, so we can stop here.

Therefore,

$$\frac{10x^2 + 7x - 19}{2x + 3} = 5x - 4 - \frac{7}{2x + 3}$$

The remainder remains over the divisor.

PRACTICE PROBLEM FOR EXAMPLE 5

a. Divide $4x^2 + 3x - 10$ by $x + 2$.

b. Divide $15x^2 - 13x + 10$ by $3x - 2$.

> **What's That Mean?**
>
> **Division Terms**
>
> Dividend: The number or polynomial being divided into.
> Numerator of a fraction.
> Inside the division symbol.
>
> Divisor: The number or polynomial that is dividing into the dividend.
> Denominator of a fraction.
> Outside the division symbol.
>
> $$\text{Divisor} \underbrace{x + 5 \overline{) x^2 + 12x + 35}}_{\text{Division Symbol}} \text{—Dividend}$$
>
> $$\text{Divisor} \underbrace{\frac{x^2 + 12x + 35}{x + 5}}_{\text{Division Symbol}} \text{—Dividend}$$

> **Skill Connection**
>
> **Checking a division problem**
>
> We can check a long division problem by multiplying the result by the original divisor.
> In Example 5b, we can check the result by multiplying the answer by $2x + 3$.
>
> $$(2x + 3) \cdot \left(5x - 4 - \frac{7}{2x + 3}\right)$$
>
> $$5x(2x + 3) - 4(2x + 3) - 7$$
>
> $$10x^2 + 15x - 8x - 12 - 7$$
>
> $$10x^2 + 7x - 19$$
>
> The multiplication results in the original dividend, so the long division is correct.

Whenever either of the polynomials is missing a term, we will replace it with a placeholder that has zero as the coefficient. In the polynomial $x^2 + 5$, we will put $0x$ in the place of the missing first degree term to get $x^2 + 0x + 5$. If more than one term is missing, we can put placeholders in for each missing term.

Example 6 Dividing a polynomial by a polynomial

a. Divide $5x^3 + 2x^2 - 7$ by $x - 1$.

b. Divide $x^4 + 2x^3 + 2x^2 - 10x - 35$ by $x^2 - 5$.

SOLUTION

a. Since the dividend is missing a first degree term, we will replace that term with $0x$ to allow the terms to line up while we do the division process.

$$\begin{array}{r} 5x^2 \\ x - 1 \overline{) 5x^3 + 2x^2 + 0x - 7} \\ -(5x^3 - 5x^2) \\ \hline 7x^2 + 0x \end{array}$$

Add the $0x$ term and divide x into $5x^3$.
Multiply $5x^2$ by $x - 1$ and subtract.
Be careful with the signs when subtracting the negative $5x^2$.
Bring down the next term.

$$\begin{array}{r} 5x^2 + 7x \\ x - 1 \overline{) 5x^3 + 2x^2 + 0x - 7} \\ -5x^3 + 5x^2 \\ \hline 7x^2 + 0x \\ -(7x^2 - 7x) \\ \hline 7x - 7 \end{array}$$

Divide $7x^2$ by x. Multiply $x - 1$ by $7x$ and subtract.
Bring down the next term.

$$\begin{array}{r} 5x^2 + 7x + 7 \\ x - 1 \overline{) 5x^3 + 2x^2 + 0x - 7} \\ -5x^3 + 5x^2 \\ \hline 7x^2 + 0x \\ -7x^2 + 7x \\ \hline 7x - 7 \\ -(7x - 7) \\ \hline 0 \end{array}$$

Divide $7x^2$ by x. Multiply $x - 1$ by 7 and subtract.

No remainder.

Therefore,

$$\frac{5x^3 + 2x^2 - 7}{x - 1} = 5x^2 + 7x + 7$$

b. The divisor is missing the first-degree term, so we will replace it with $0x$ and divide using long division.

$$\begin{array}{r} x^2 \\ x^2 + 0x - 5 \overline{) x^4 + 2x^3 + 2x^2 - 10x - 35} \\ -(x^4 + 0x^3 - 5x^2) \\ \hline 2x^3 + 7x^2 - 10x \end{array}$$

Add the $0x$ term and divide x^2 into x^4.
Multiply x^2 by $x^2 + 0x - 5$ and subtract.
Be careful with the signs when subtracting the negative $5x^2$.
Bring down the next term.

$$\begin{array}{r} x^2 + 2x \\ x^2 + 0x - 5 \overline{) x^4 + 2x^3 + 2x^2 - 10x - 35} \\ -x^4 - 0x^3 + 5x^2 \\ \hline 2x^3 + 7x^2 - 10x \\ -(2x^3 + 0x^2 - 10x) \\ \hline 7x^2 + 0x - 35 \end{array}$$

Divide $2x^3$ by x^2. Multiply $x^2 + 0x - 5$ by $2x$ and subtract.
Bring down the next term.

$$\begin{array}{r} x^2 + 2x + 7 \\ x^2 + 0x - 5 \overline{) x^4 + 2x^3 + 2x^2 - 10x - 35} \\ -x^4 - 0x^3 + 5x^2 \\ \hline 2x^3 + 7x^2 - 10x \\ -2x^3 - 0x^2 + 10x \\ \hline 7x^2 + 0x - 35 \\ -(7x^2 + 0x - 35) \\ \hline 0 \end{array}$$

Divide $7x^2$ by x^2. Multiply $x^2 + 0x - 5$ by 7 and subtract.

No remainder.

Therefore,

$$\frac{x^4 + 2x^3 + 2x^2 - 10x - 35}{x^2 - 5} = x^2 + 2x + 7$$

PRACTICE PROBLEM FOR EXAMPLE 6
Divide $3x^3 - 4x + 16$ by $x + 2$.

Synthetic Division

Using long division requires that we keep track of a lot of details along the way. Writing out some of the steps used in long division sometimes is not necessary. Simplifying the procedure to keep track of only the most important steps results in a shorter method than long division. This method is called *synthetic division*. Synthetic division is a simplified method of long division that uses only the coefficients of each term and not the variables. This method works only for single-variable polynomials that are being divided by a binomial of the form $x - c$. Let's compare the long division in Example 5a with the synthetic division of the same polynomials.

Divide $\quad\quad\quad\quad 3x^2 + 17x + 20$ by $x + 4$

The basic setup is similar but without the variables. Synthetic division works for divisors of the from $x - c$, and $x + 4$ can be written in this form as $x - (-4)$. Therefore, we start with a -4 representing the divisor.

$$\begin{array}{c|c}
\text{Long Division} & \text{Synthetic Division} \\
x + 4 \overline{\smash{)}3x^2 + 17x + 20} & \underline{-4|\quad 3\quad 17\quad 20} \\
\end{array}$$

In synthetic division, the leading coefficient comes down to the bottom row in the first column. Multiply this by the divisor, -4. The result goes into the second row and next column. By changing the sign of 4 in the divisor, we accounted for the subtraction in the long division. We add the values in the second column to get the next value in the bottom row.

$$\begin{array}{c|c}
\begin{array}{r} 3x \\ x + 4 \overline{\smash{)}3x^2 + 17x + 20} \\ \underline{-(3x^2 + 12x)} \\ 5x + 20 \end{array} &
\begin{array}{r} -4|\quad 3\quad 17\quad 20 \\ \quad\quad\quad\; \downarrow -12 \\ \hline \quad\quad\quad 3\quad\; 5 \end{array}
\end{array}$$

Multiply the 5 in the bottom row by the divisor, -4, and the result goes in the second row and next column to the right. Add the values in the next column, and we get zero in the bottom row. When all the columns have been filled, we are done. The zero in the bottom row and last column means that there is no remainder. Any nonzero number in this position would be the remainder.

$$\begin{array}{c|c}
\begin{array}{r} 3x + 5 \\ x + 4 \overline{\smash{)}3x^2 + 17x + 20} \\ \underline{-(3x^2 + 12x)} \\ 5x + 20 \\ \underline{-(5x + 20)} \\ 0 \end{array} &
\begin{array}{r} -4|\quad 3\quad 17\quad\; 20 \\ \quad\quad\quad\;\; -12\; -20 \\ \hline \quad\quad\quad 3\quad\; 5\quad\; 0 \\ \quad\quad\quad \downarrow\quad \downarrow \\ \quad\quad\quad 3x + 5 \end{array}
\end{array}$$

Once we are done with the division we can write the result using the numbers in the last row. Each number in the last row is a coefficient for a term in the final answer. The first number will be the coefficient of the first term that will have degree one less than the degree of the dividend. The dividend had a degree of 2, so we start with a first degree term.

Example 7   Using synthetic division

a. Use synthetic division to divide $3x^3 - 10x^2 - 13x + 20$ by $x - 4$.

b. Use synthetic division to divide $5x^4 + 2x^3 - x - 50$ by $x + 2$.

SOLUTION

a. Since the divisor, $x - 4$ is of the form $x - c$, we know that c is 4. The dividend's coefficients are 3, -10, -13, and 20.

$\underline{4|}\ \ 3\ \ \ -10\ \ \ -13\ \ \ \ 20$ Write out the coefficients in order.
$\phantom{\underline{4|}\ \ }\overline{}$ Draw a line and bring down the first coefficient.
$\phantom{\underline{4|}\ \ }3$

$\underline{4|}\ \ 3\ \ \ -10\ \ \ -13\ \ \ \ 20$ Multiply 4 by 3 and write the product under
$\phantom{\underline{4|}\ \ \ \ \ \ \ }12$ the next coefficient.
$\phantom{\underline{4|}\ \ }\overline{}$
$\phantom{\underline{4|}\ \ }3$

$\underline{4|}\ \ 3\ \ \ -10\ \ \ -13\ \ \ \ 20$ Add the -10 and 12 and write the sum below.
$\phantom{\underline{4|}\ \ \ \ \ \ \ }12$
$\phantom{\underline{4|}\ \ }\overline{}$
$\phantom{\underline{4|}\ \ }3\ \ \ \ \ \ 2$

$\underline{4|}\ \ 3\ \ \ -10\ \ \ -13\ \ \ \ 20$ Multiply 4 by 2 and write the product under
$\phantom{\underline{4|}\ \ \ \ \ \ \ }12\ \ \ \ \ \ 8$ the next coefficient.
$\phantom{\underline{4|}\ \ }\overline{}$ Add the -13 and 8 and write the sum below.
$\phantom{\underline{4|}\ \ }3\ \ \ \ \ \ 2\ \ \ \ \ -5$

$\underline{4|}\ \ 3\ \ \ -10\ \ \ -13\ \ \ \ \ \ 20$ Multiply 4 by -5 and write the product under
$\phantom{\underline{4|}\ \ \ \ \ \ \ }12\ \ \ \ \ \ 8\ \ \ -20$ the next coefficient.
$\phantom{\underline{4|}\ \ }\overline{}$ Add the -20 and 20 and write the sum below.
$\phantom{\underline{4|}\ \ }3\ \ \ \ \ \ 2\ \ \ \ \ -5\ \ \ \ \ 0$

We have completed each column, so we have finished the synthetic division. Each number below the line is a coefficient for a term in the final answer. The first number will be the coefficient of the first term that will have degree one less than the degree of the dividend. The dividend had a degree of 3, so start with a term with degree 2. These coefficients result in the quotient.

$$3x^2 + 2x - 5$$

The zero on the end is the remainder, so this answer does not have a remainder.

b. The divisor $x + 2$ can be written in the form $x - c$ as $x - (-2)$, so $c = -2$. The dividend is missing a second-degree term, so we will write zero as its coefficient. Start by writing c and the coefficients in order.

$\underline{-2|}\ \ \ 5\ \ \ \ \ 2\ \ \ \ \ \ 0\ \ \ \ -1\ \ \ -50$
$\phantom{\underline{-2|}\ \ \ \ \ \ \ \ }-10\ \ \ 16\ \ -32\ \ \ \ \ 66$
$\phantom{\underline{-2|}\ \ }\overline{}$
$\phantom{\underline{-2|}\ \ \ }5\ \ \ -8\ \ \ 16\ \ -33\ \ \ \ 16$

The last number in the bottom row is 16, so that is our remainder. Therefore, we have the following quotient.

$$5x^3 - 8x^2 + 16x - 33 + \frac{16}{x + 2}$$

PRACTICE PROBLEM FOR EXAMPLE 7

a. Use synthetic division to divide $-2x^3 + 4x^2 - 9x + 45$ by $x - 3$.

b. Use synthetic division to divide $3x^5 + 4x^4 + x^2 + 6x + 600$ by $x + 5$.

7.2 Exercises

In Exercises 1 through 34, simplify the rational expression.

1. $\dfrac{20x^3}{14x}$

2. $\dfrac{36x^2 y}{15xy}$

3. $\dfrac{12(x+9)}{(x+5)(x+9)}$

4. $\dfrac{(x+5)}{(x-7)(x+5)}$

5. $\dfrac{x+3}{2x+6}$

6. $\dfrac{x-7}{3x-21}$

7. $\dfrac{8-x}{2x-16}$

8. $\dfrac{3-x}{7x-21}$

9. $\dfrac{4x-24}{18-3x}$

10. $\dfrac{5x-20}{8-2x}$

11. $\dfrac{(x+2)(x-5)}{(5-x)(x-7)}$

12. $\dfrac{(4x-3)(7-2x)}{(2x-7)(5x+1)}$

13. $\dfrac{(x+5)(x-3)}{(x-3)(x-2)}$

14. $\dfrac{(x+4)(x+6)}{(x+6)(x-3)}$

15. $\dfrac{x^2+6x+9}{(x+2)(x+3)}$

16. $\dfrac{p-8}{p^2-12p+32}$

17. $\dfrac{t^2+2t-15}{t^2-9t+18}$

18. $\dfrac{x^2+5x-14}{x^2-6x+8}$

19. $\dfrac{w^2-16}{w^2+w-12}$

20. $\dfrac{t^2+7t-8}{t^2-64}$

21. $\dfrac{5(x+7)}{(x+7)+10}$

22. $\dfrac{(x-5)}{2(x-5)+3}$

23. $\dfrac{(m+3)}{2(m+3)-5}$

24. $\dfrac{4(k-6)-7}{5(k-6)}$

25. $\dfrac{2x^2-23x-70}{x^2-21x+98}$

26. $\dfrac{2x^2+11x+15}{4x^2+17x+15}$

27. $\dfrac{10x^3+5x^2-4x}{5x}$

28. $\dfrac{15t^4+20t^2+6t}{5t}$

29. $\dfrac{3x^4-8x^3+x^2}{4x^3}$

30. $\dfrac{5b^3-4b^2-6b}{4b^2}$

31. $\dfrac{10a^3b^2+12a^2b^3-14ab^2}{2ab}$

32. $\dfrac{24m^4n^3-16m^2n+10mn^2}{8mn}$

33. $\dfrac{10g^5h^3-30g^4h^2+20g^2h}{5g^2h^2}$

34. $\dfrac{x^4y^2-2x^3y^4+5xy^3}{2x^2y^2}$

For Exercises 35 and 36, the students were asked to simplify the given expressions. Explain what the student did wrong. Give the correct simplification.

35. $\dfrac{x^2+6x+8}{x+4}$

> **Tom**
>
> $\dfrac{x^2+6x+8}{x+4} = \dfrac{x^2}{x+4} + \dfrac{6x}{x+4} + \dfrac{8}{x+4}$

36. $\dfrac{2(x+6)-5}{(x+6)}$

> **Shannon**
>
> $\dfrac{2(x+6)-5}{(x+6)} = \dfrac{2\cancel{(x+6)}-5}{\cancel{(x+6)}}$
>
> $= \dfrac{2-5}{1}$
>
> $= -3$

For Exercises 37 and 38, the students were asked to divide using long division or synthetic division. Explain what the student did wrong. Give the correct answer.

37. $(6x^2 - x - 35) \div (2x - 5)$

> **Amy**
>
> $$\begin{array}{r} 3x-8 \\ 2x-5\overline{)6x^2-x-35} \\ \underline{-6x^2-15x} \\ -16x-35 \\ \underline{-16x+40} \\ 5 \end{array}$$
>
> $(6x^2 - x - 35) \div (2x-5) = 3x - 8 + \dfrac{5}{2x-5}$

38. $(32x^3 - 16x^2 - 42x + 15) \div (4x - 5)$

> **Warrick**
>
> $\underline{5|}\ \ 32\ \ -16\ \ -42\ \ \ \ 15$
> $\phantom{\underline{5|}\ \ 32}\ \ \ \ 160\ \ \ \ 720\ \ 3390$
> $\phantom{\underline{5|}\ }\overline{32\ \ \ \ 144\ \ \ \ 678\ \ 3405}$
>
> $(32x^3 - 16x^2 - 42x + 15) \div (4x - 5)$
>
> $= 32x^2 + 144x + 678 + \dfrac{3405}{4x-5}$

39. Find an expression for the length of the missing side.

Area = $21x^2 - 37x - 120$, width = $3x + 5$

40. Find an expression for the length of the missing side.

Area = $40x^2 - 82x + 21$, width = $4x - 7$

41. Find an expression for the length of the missing side.

Area = $5x^3 + 27x^2 + 31x + 12$, width = $x + 4$

42. Find an expression for the length of the missing side.

Area = $6x^3 - 20x^2 + 11x - 15$, width = $x - 3$

In Exercises 43 through 52, divide using long division.

43. $(6x^2 + 29x + 20) \div (x + 4)$
44. $(4m^2 + 25m - 21) \div (m + 7)$
45. $(x^3 + 7x^2 + 6x - 8) \div (x + 2)$
46. $(r^3 + 2r^2 - 7r + 40) \div (r + 5)$
47. $(3x^3 + 5x^2 + 6x + 10) \div (x^2 + 2)$
48. $(4h^3 - 7h^2 - 12h + 21) \div (h^2 - 3)$
49. $(4b^4 - 3b^3 - 23b^2 + 21b - 35) \div (b^2 - 7)$
50. $(3x^4 + 2x^3 + 22x^2 + 12x + 24) \div (x^2 + 6)$
51. $(5n^4 + 19n^3 + 14n^2 - 16n - 16) \div (n^2 + 3n + 2)$
52. $(4a^4 - 11a^3 + 29a^2 - 26a + 28) \div (a^2 - 2a + 4)$

In Exercises 53 through 60, divide using synthetic division.

53. $(x^2 - 3x - 28) \div (x - 7)$
54. $(t^2 + 7t - 18) \div (t - 2)$
55. $(2m^2 - 5m - 42) \div (m - 6)$
56. $(3g^2 - 19g + 20) \div (g - 5)$
57. $(4x^2 + 17x + 15) \div (x + 3)$
58. $(2x^2 + 11x + 15) \div (x + 3)$
59. $(3x^2 + 17x + 25) \div (x + 4)$
60. $(7x^2 + 38x + 17) \div (x + 5)$

In Exercises 61 through 76, divide using long division or synthetic division. Remember that synthetic division is used when dividing by a binomial of the form $x - c$.

61. $(12x^2 + 32x + 21) \div (2x + 3)$
62. $(12x^2 + 23x + 10) \div (3x + 2)$
63. $(2x^3 + 11x^2 + 17x + 20) \div (x + 4)$
64. $(3x^3 + 14x^2 + 19x + 12) \div (x + 3)$
65. $(x^4 + 2x^3 - 10x - 25) \div (x^2 - 5)$
66. $(b^4 + 6b^3 - 18b - 9) \div (b^2 - 3)$
67. $(x^4 - 3x^3 + 8x^2 - 12x + 16) \div (x^2 + 4)$
68. $(x^4 - 7x^3 + 10x^2 - 35x + 25) \div (x^2 + 5)$
69. $(t^3 + 3t^2 - 12t + 16) \div (t + 3)$
70. $(m^3 - 4m^2 - 8m + 50) \div (m - 4)$
71. $(a^4 + 2a^3 + 8a^2 + 10a + 15) \div (a^2 + 5)$
72. $(w^4 - 5w^3 + w^2 + 30w - 42) \div (w^2 - 6)$
73. $(t^3 + 4t^2 - 24) \div (t - 2)$
74. $(x^3 + 5x^2 - 18) \div (x + 3)$
75. $(4a^3 + 2a + 40) \div (a + 2)$
76. $(6x^3 + 4x - 15) \div (x - 1)$

77. a. Divide 65 by 5.
 b. Write the prime factorization of 65.

78. a. Divide 119 by 7.
 b. Write the prime factorization of 119.

79. a. Divide $(4x^2 - 25x - 21) \div (x - 7)$.
 b. Write the factorization of $4x^2 - 25x - 21$.

80. a. Divide $(6x^2 - 47x - 8) \div (x - 8)$.
 b. Write the factorization of $6x^2 - 47x - 8$.

81. a. Divide $(4x^2 + 5x - 21) \div (x + 3)$.
 b. Write the factorization of $4x^2 + 5x - 21$.

82. a. Divide $(27x^2 - 6x - 8) \div (3x - 2)$.
 b. Write the factorization of $27x^2 - 6x - 8$.

83. a. Divide $(2x^3 + x^2 - 63x - 90) \div (x + 5)$.
 b. Write the factorization of $2x^3 + x^2 - 63x - 90$.

84. a. Divide $(8x^3 - 86x^2 + 237x - 189) \div (x - 7)$.
 b. Write the factorization of $8x^3 - 86x^2 + 237x - 189$.

85. a. Divide $(3t^4 - 17t^3 + 47t^2 - 153t + 180) \div (t^2 + 9)$.
 b. Write the factorization of $3t^4 - 17t^3 + 47t^2 - 153t + 180$.

86. a. Divide $(5m^4 - 44m^3 - 3m^2 + 308m - 224) \div (m^2 - 7)$.
 b. Write the factorization of $5m^4 - 44m^3 - 3m^2 + 308m - 224$.

# 7.3 Multiplying and Dividing Rational Expressions

LEARNING OBJECTIVES
- Multiply rational expressions.
- Divide rational expressions.

In this section, we will study how to multiply and divide rational expressions. In the next section, we will study how to add and subtract rational expressions. Notice that working with rational expressions is very much the same as working with fractions. The connection between working with rational expressions and working with fractions is important and can be a good basis for understanding these two sections of this chapter.

Simplifying is a key part of all operations involving rational expressions. When multiplying or dividing, simplify before the multiplication is performed or after. Many students find it easier to simplify the rational expressions first so that there is less multiplication to be done. Remember to always check the final product for any further simplification that can be done.

Multiplying Rational Expressions

When multiplying or dividing fractions, you do not need common denominators. This makes it easier to multiply or divide any two rational expressions. Recall that when multiplying fractions, you simply multiply the numerators together and multiply the denominators together. Don't forget to simplify your results.

Multiplying Rational Expressions
1. Factor the numerator and denominator of each fraction (if possible).
2. Divide out any common factors.
3. Rewrite as a single fraction.
4. Leave in factored form.

Skill Connection

Reducing before multiplying

In multiplying rational expressions, it is usually best to reduce any common factors before multiplying the numerators or denominators. Reducing first will keep the numbers or expressions from getting larger than necessary.

Reduce first

$$\frac{20}{35} \cdot \frac{49}{8} = \frac{\cancel{2} \cdot \cancel{2} \cdot 5}{5 \cdot \cancel{7}} \cdot \frac{\cancel{7} \cdot 7}{\cancel{2} \cdot \cancel{2} \cdot 2}$$
$$= \frac{7}{2}$$

Multiply first

$$\frac{20}{35} \cdot \frac{49}{8} = \frac{980}{280}$$
$$= \frac{\cancel{2} \cdot \cancel{2} \cdot \cancel{5} \cdot \cancel{7} \cdot 7}{\cancel{2} \cdot \cancel{2} \cdot 2 \cdot \cancel{5} \cdot \cancel{7}}$$
$$= \frac{7}{2}$$

Example 1 Multiplying rational expressions

Multiply the following rational expressions.

a. $\dfrac{3}{10} \cdot \dfrac{4}{7}$ b. $\dfrac{20x}{18y^2} \cdot \dfrac{6xy}{10}$ c. $\dfrac{x+5}{x-3} \cdot \dfrac{x+2}{x+7}$

SOLUTION

a. $\dfrac{3}{10} \cdot \dfrac{4}{7} = \dfrac{3}{2 \cdot 5} \cdot \dfrac{2 \cdot 2}{7}$ Factor the numerator and denominator.

$= \dfrac{3}{\cancel{2} \cdot 5} \cdot \dfrac{\cancel{2} \cdot 2}{7}$ Divide out the common factors.

$= \dfrac{6}{35}$ Multiply the numerators and denominators together.

b. $\dfrac{20x}{18y^2} \cdot \dfrac{6xy}{10} = \dfrac{2 \cdot 2 \cdot 5 \cdot x}{2 \cdot 3 \cdot 3 \cdot y \cdot y} \cdot \dfrac{2 \cdot 3 \cdot x \cdot y}{2 \cdot 5}$ Factor the numerators and denominators.

$= \dfrac{\cancel{2} \cdot 2 \cdot \cancel{5} \cdot x}{\cancel{2} \cdot \cancel{3} \cdot 3 \cdot \cancel{y} \cdot y} \cdot \dfrac{\cancel{2} \cdot \cancel{3} \cdot x \cdot \cancel{y}}{\cancel{2} \cdot \cancel{5}}$ Divide out the common factors.

$= \dfrac{2x^2}{3y}$ Multiply remaining factors together.

c. $\dfrac{x+5}{x-3} \cdot \dfrac{x+2}{x+7} = \dfrac{(x+5)(x+2)}{(x-3)(x+7)}$ There are no common factors, so rewrite as a single fraction.

PRACTICE PROBLEM FOR EXAMPLE 1

Multiply the following rational expressions.

a. $\dfrac{50a^2b}{20c^2} \cdot \dfrac{6ac}{9b}$

b. $\dfrac{x+8}{x-3} \cdot \dfrac{x+7}{x+8}$

If the rational expressions are more complicated, make sure that you focus on one part at a time. Do not forget any piece of the expressions you are multiplying.

Example 2 Multiplying rational expressions

Multiply the following rational expressions.

a. $\dfrac{(x+3)(x-4)}{(x-5)(x-4)} \cdot \dfrac{(x+2)(x+7)}{(x+3)(x+7)}$

b. $\dfrac{x^2-3x-18}{x^2+7x+10} \cdot \dfrac{x^2+3x-10}{x^2+7x+12}$

SOLUTION

a. $\dfrac{(x+3)(x-4)}{(x-5)(x-4)} \cdot \dfrac{(x+2)(x+7)}{(x+3)(x+7)} = \dfrac{\cancel{(x+3)}\cancel{(x-4)}}{(x-5)\cancel{(x-4)}} \cdot \dfrac{(x+2)\cancel{(x+7)}}{\cancel{(x+3)}\cancel{(x+7)}}$

$= \dfrac{x+2}{x-5}$ Divide out common factors and multiply.

b. $\dfrac{x^2-3x-18}{x^2+7x+10} \cdot \dfrac{x^2+3x-10}{x^2+7x+12} = \dfrac{(x+3)(x-6)}{(x+5)(x+2)} \cdot \dfrac{(x+5)(x-2)}{(x+3)(x+4)}$ Factor.

$= \dfrac{\cancel{(x+3)}(x-6)}{\cancel{(x+5)}(x+2)} \cdot \dfrac{\cancel{(x+5)}(x-2)}{\cancel{(x+3)}(x+4)}$ Reduce.

$= \dfrac{(x-6)(x-2)}{(x+2)(x+4)}$ Rewrite as a single fraction.

PRACTICE PROBLEM FOR EXAMPLE 2

Multiply the following rational expressions.

a. $\dfrac{(x+3)(x+4)}{(x-7)(x+3)} \cdot \dfrac{(x-2)(x-5)}{(x+4)(x-6)}$

b. $\dfrac{x^2-5x-14}{2x^2-x-15} \cdot \dfrac{x^2+2x-15}{3x^2+4x-4}$

Dividing Rational Expressions

Dividing rational expressions is the same as dividing numerical fractions in that you multiply by the reciprocal of the fraction you are dividing by. This means that division will simply be the same as multiplication after you have flipped over the second fraction.

> **Dividing Rational Expressions**
> 1. Multiply by the reciprocal of the second fraction. (Flip over the fraction you are dividing by and make the division into multiplication.)
> 2. Factor the numerator and denominator of each fraction (if possible).
> 3. Divide out any common factors.
> 4. Rewrite as a single fraction.
> 5. Leave in factored form.

Example 3 Dividing rational expressions

Divide the following.

a. $\dfrac{2}{3} \div \dfrac{5}{7}$

b. $\dfrac{\frac{7}{10}}{\frac{2}{5}}$

c. $\dfrac{x+3}{x-7} \div \dfrac{x-4}{x-7}$

SOLUTION

a. $\dfrac{2}{3} \div \dfrac{5}{7} = \dfrac{2}{3} \cdot \dfrac{7}{5}$ Multiply by the reciprocal of the second fraction.

$= \dfrac{14}{15}$

b. $\dfrac{\frac{7}{10}}{\frac{2}{5}} = \dfrac{7}{10} \cdot \dfrac{5}{2}$ Multiply by the reciprocal of the fraction you are dividing by.

$= \dfrac{7}{2 \cdot \cancel{5}} \cdot \dfrac{\cancel{5}}{2}$ Factor and divide out common factors.

$= \dfrac{7}{4}$ Multiply.

c. $\dfrac{x+3}{x-7} \div \dfrac{x-4}{x-7} = \dfrac{x+3}{x-7} \cdot \dfrac{x-7}{x-4}$ Multiply by the reciprocal of the second fraction.

$= \dfrac{x+3}{\cancel{x-7}} \cdot \dfrac{\cancel{x-7}}{x-4}$ Divide out common factors.

$= \dfrac{x+3}{x-4}$ Rewrite as a single fraction.

PRACTICE PROBLEM FOR EXAMPLE 3

Divide the following.

a. $\dfrac{4x^3}{10y^2} \div \dfrac{6x}{15y^3}$ **b.** $\dfrac{x+5}{x-9} \div \dfrac{x+5}{x+7}$

Example 4 Dividing rational expressions

Divide the following rational expressions.

a. $\dfrac{(x+2)(x+5)}{(x-3)(2x+7)} \div \dfrac{(x-3)(x+5)}{(2x+7)(x-9)}$

b. $\dfrac{2x^2+11x+12}{x^2-11x+30} \div \dfrac{2x^2+15x+18}{x^2+2x-35}$

SOLUTION

a. Multiply by the reciprocal of the fraction you are dividing by.

$\dfrac{(x+2)(x+5)}{(x-3)(2x+7)} \div \dfrac{(x-3)(x+5)}{(2x+7)(x-9)} = \dfrac{(x+2)(x+5)}{(x-3)(2x+7)} \cdot \dfrac{(2x+7)(x-9)}{(x-3)(x+5)}$

$= \dfrac{(x+2)\cancel{(x+5)}}{(x-3)\cancel{(2x+7)}} \cdot \dfrac{\cancel{(2x+7)}(x-9)}{(x-3)\cancel{(x+5)}}$ Divide out the common factors.

$= \dfrac{(x+2)(x-9)}{(x-3)(x-3)}$ Rewrite as a single fraction.

$= \dfrac{(x+2)(x-9)}{(x-3)^2}$

b. Multiply by the reciprocal of the fraction you are dividing by.

$\dfrac{2x^2+11x+12}{x^2-11x+30} \div \dfrac{2x^2+15x+18}{x^2+2x-35} = \dfrac{2x^2+11x+12}{x^2-11x+30} \cdot \dfrac{x^2+2x-35}{2x^2+15x+18}$

$= \dfrac{(x+4)(2x+3)}{(x-5)(x-6)} \cdot \dfrac{(x-5)(x+7)}{(2x+3)(x+6)}$ Factor each polynomial.

$= \dfrac{(x+4)\cancel{(2x+3)}}{\cancel{(x-5)}(x-6)} \cdot \dfrac{\cancel{(x-5)}(x+7)}{\cancel{(2x+3)}(x+6)}$ Divide out the common factors.

$= \dfrac{(x+4)(x+7)}{(x-6)(x+6)}$ Rewrite as a single fraction.

PRACTICE PROBLEM FOR EXAMPLE 4
Divide the following.

a. $\dfrac{(x+2)(x+4)}{(x-3)(x+7)} \div \dfrac{(x+4)(x-8)}{(x+7)(x-5)}$

b. $\dfrac{x^2-4x-21}{2x^2-9x-35} \div \dfrac{x^2+12x+27}{2x^2-3x-20}$

7.3 Exercises

For Exercises 1 through 20, multiply the rational expressions.

1. $\dfrac{15x^2}{9y^3} \cdot \dfrac{21y^5}{35x}$

2. $\dfrac{a^3b^4}{10c} \cdot \dfrac{4ac}{b}$

3. $\dfrac{35a^2}{12b^3c} \cdot \dfrac{40b}{a^5c}$

4. $\dfrac{3xy}{8z^2} \cdot \dfrac{6z^4}{5x}$

5. $\dfrac{x+3}{x+7} \cdot \dfrac{x+7}{x-5}$

6. $\dfrac{x-8}{x-4} \cdot \dfrac{x-5}{x-8}$

7. $\dfrac{7-x}{x+3} \cdot \dfrac{2x+3}{x-7}$

8. $\dfrac{x-8}{x-4} \cdot \dfrac{3x-4}{8-x}$

9. $\dfrac{2(x-4)}{3x-12} \cdot \dfrac{3(x+5)}{x-7}$

10. $\dfrac{5(w+7)}{2w+10} \cdot \dfrac{4(w+5)}{w-11}$

11. $\dfrac{2(x-4)}{4x+20} \cdot \dfrac{3(x+5)}{12-3x}$

12. $\dfrac{5(r+2)}{10r-70} \cdot \dfrac{4(7-r)}{8r+16}$

13. $\dfrac{(x+3)(x+7)}{(x-2)(x+3)} \cdot \dfrac{(x+7)(x-2)}{(x-3)(x-9)}$

14. $\dfrac{(c-3)(c+6)}{(c-7)(c+2)} \cdot \dfrac{(c-7)(c-11)}{(c+6)(c-3)}$

15. $\dfrac{(k+5)(7-k)}{(k-7)(k-3)} \cdot \dfrac{(k-3)(k+6)}{(k+9)(k+5)}$

16. $\dfrac{(t+7)(t-9)}{(t+3)(t+7)} \cdot \dfrac{(t+3)(t+13)}{(9-t)(t-13)}$

17. $\dfrac{m^2+8m+7}{m^2-2m-3} \cdot \dfrac{m^2-9}{m^2+9m+14}$

18. $\dfrac{t^2-16}{t^2+9t+14} \cdot \dfrac{t^2+4t-21}{t^2-t-20}$

19. $\dfrac{x^2-16x+55}{x^2-x-12} \cdot \dfrac{x^2+12x+27}{x^2-9x+20}$

20. $\dfrac{m^2+6m+9}{m^2+5m+6} \cdot \dfrac{m^2-6m-16}{m^2-9}$

For Exercises 21 and 22, the students were asked to multiply or divide the given rational expressions. Explain what the student did wrong. Give the correct answer.

21. $\dfrac{3x+5}{x-7} \cdot \dfrac{7-x}{x+2}$

Frank

$\dfrac{3x+5}{x-7} \cdot \dfrac{7-x}{x+2} = \dfrac{3x+5}{\cancel{x-7}} \cdot \dfrac{\cancel{7-x}}{x+2}$

$= \dfrac{3x+5}{x+2}$

22. $\dfrac{(h+4)(h-8)}{(h+2)(h+7)} \div \dfrac{(h-8)(h+2)}{(h+4)(h+5)}$

Shania

$\dfrac{(h+4)(h-8)}{(h+2)(h+7)} \div \dfrac{(h-8)(h+2)}{(h+4)(h+5)}$

$= \dfrac{\cancel{(h+4)}(h-8)}{\cancel{(h+2)}(h+7)} \div \dfrac{(h-8)\cancel{(h+2)}}{\cancel{(h+4)}(h+5)}$

$= \dfrac{(h-8)(h-8)}{(h+7)(h+5)}$

In Exercises 23 through 42, divide the rational expressions.

23. $\dfrac{x^2y}{z^3} \div \dfrac{x}{z}$

24. $\dfrac{10a^3}{6b^2c^4} \div \dfrac{15a}{20bc^2}$

25. $\dfrac{x+5}{x-3} \div \dfrac{x-5}{x-7}$

26. $\dfrac{x+15}{x-3} \div \dfrac{x+7}{x-3}$

27. $\dfrac{5x+20}{x-4} \div \dfrac{3x+12}{x+7}$

28. $\dfrac{w+2}{10w+15} \div \dfrac{7w+14}{2w+3}$

29. $\dfrac{5-x}{x+7} \div \dfrac{x-5}{x+9}$

30. $\dfrac{x-3}{x-6} \div \dfrac{x-2}{6-x}$

31. $\dfrac{8b-10}{3b+12} \div \dfrac{5-4b}{5b-20}$

32. $\dfrac{2n+12}{4-3n} \div \dfrac{4n-6}{12n-16}$

33. $\dfrac{(x+3)(x+2)}{(x-8)(x-7)} \div \dfrac{(x+2)(x-5)}{(x-7)(x-5)}$

34. $\dfrac{(m+7)(m-5)}{(m+2)(m+7)} \div \dfrac{(m-5)(m-12)}{(m+2)(m+6)}$

35. $\dfrac{w^2+8w+15}{w^2+12w+35} \div \dfrac{w^2-5w-24}{w^2+3w-28}$

36. $\dfrac{k^2+k-20}{k^2+10k+21} \div \dfrac{k^2-10k+24}{k^2-13k+42}$

37. $\dfrac{c^2-c-20}{c^2+6c+8} \div \dfrac{2c^2+11c+12}{3c^2+c-10}$

38. $\dfrac{3k^2+19k+20}{k^2+6k+5} \div \dfrac{2k^2+2k-40}{k^2-2k-3}$

39. $\dfrac{6t^2+t-35}{10t^2+17t-20} \div \dfrac{12t^2-t-63}{20t^2+29t-36}$

40. $\dfrac{2x^2-5x-12}{x^2-3x-28} \div \dfrac{6x^2-7x-24}{3x^2+13x-56}$

41. $\dfrac{x^2-25}{x^2+7x+10} \div \dfrac{x^2-4}{x^2+9x+8}$

42. $\dfrac{t^2+12t+20}{t^2-49} \div \dfrac{t^2-100}{t^2-9t+14}$

44. $\dfrac{t^2+8t+15}{t^2-9} \cdot \dfrac{2t^2-6t}{t^2+7t+10}$

45. $\dfrac{h^2-7h+12}{h^2+2h-8} \div \dfrac{h^2+4h-32}{6h^2-12h}$

46. $\dfrac{r^2-36}{r^2-11r+30} \div \dfrac{r^2+11r+30}{r^2-8r+15}$

47. $\dfrac{6x^2+x-35}{10x^2+39x+35} \div \dfrac{12x^2-37x+21}{15x^2+11x-14}$

48. $\dfrac{28p^2+41p+15}{6p^2+11p-72} \div \dfrac{4p^2+7p+3}{2p^2+11p+9}$

49. $\dfrac{10t^2-29t-21}{6t^2-5t-56} \cdot \dfrac{3t^2-4t-32}{8t^2-27t-20}$

50. $\dfrac{24n^2+2n-15}{12n^2+52n+35} \cdot \dfrac{10n^2+29n-21}{12n^2-25n+12}$

51. $\dfrac{x^3-8}{x^2+2x-8} \div \dfrac{x^2+2x+4}{x^2+7x+12}$

52. $\dfrac{m^3+27}{m^2-3m+9} \div \dfrac{5m^2+17m+6}{10m^2-31m-14}$

53. $\dfrac{x^4-81}{5x^3+45x} \cdot \dfrac{4x^2-15x+9}{4x^2+9x-9}$

54. $\dfrac{16r^4-625}{10r^2-9r-40} \cdot \dfrac{20r^2+17r-24}{6r^2+r-35}$

For Exercises 43 through 54, perform the indicated operation and simplify.

43. $\dfrac{a^2-25}{a^2+6a+5} \cdot \dfrac{a^2-1}{a^2+3a-4}$

7.4 Adding and Subtracting Rational Expressions

LEARNING OBJECTIVES

- Find the least common denominator of rational expressions.
- Add rational expressions.
- Subtract rational expressions.
- Simplify complex fractions.

Now that we have studied simplifying, multiplying, and dividing rational expressions, we are going to learn how to add and subtract them. Adding and subtracting fractions requires us to have common denominators. This has always been a condition for adding or subtracting fractions and requires a little more work than multiplying and dividing, in which we do not need common denominators. Being able to add or subtract rational expressions will allow us to simplify situations in which several rational expressions are used.

Least Common Denominator

Finding a common denominator for a rational expression will be the same process as finding one for a numeric fraction, but once again we will need to focus more carefully on each step of the process. When finding the least common denominator (LCD) of a fraction, we need to factor each denominator and compare them so that we can choose

Skill Connection

Finding a common denominator

When adding or subtracting fractions, we need a common denominator. If the fractions do not have a common denominator, we find the least common denominator (LCD). Then we rewrite the fractions in terms of the LCD.

Example:

Find the LCD for the given fractions and rewrite each fraction in terms of the LCD.

$$\frac{12}{35} \quad \frac{8}{495}$$

Solution

$$\frac{12}{35} \quad \frac{8}{495}$$

Factor both denominators.

$$\frac{12}{5 \cdot 7} \quad \frac{8}{3^2 \cdot 5 \cdot 11}$$

Take the highest power of each factor for the LCD.

$$LCD = 3^2 \cdot 5 \cdot 7 \cdot 11$$
$$LCD = 3465$$

Multiply each fraction by the factors that its denominator is missing.

$$\frac{3^2 \cdot 11}{3^2 \cdot 11} \cdot \frac{12}{5 \cdot 7} = \frac{1188}{3456}$$

$$\frac{7}{7} \cdot \frac{8}{3^2 \cdot 5 \cdot 11} = \frac{56}{3465}$$

---

the right combination of factors that will make up the least common denominator. In some cases, we might be able to figure out the LCD without doing all of these steps, but with rational expressions, it is going to take a little more patience.

> **Finding the Least Common Denominator (LCD) for Rational Expressions**
> 1. Factor the denominators (if possible).
> 2. Take the highest power of each factor for the LCD.
> 3. Leave in factored form or multiply the factors together.
>   We generally leave polynomials factored but multiply out single terms.

> **Writing Fractions in Terms of the LCD**
> 1. Find the least common denominator.
> 2. Determine what factors of the LCD the fraction's denominator is missing.
> 3. Multiply the numerator and denominator of the fraction by the missing factors.

Example 1 Finding the LCD with monomial denominators

Find the least common denominator for each of the following sets of fractions and rewrite each fraction in terms of the LCD.

a. $\dfrac{2}{5x} \quad \dfrac{3}{10x^2}$

b. $\dfrac{6ab}{25c^3d} \quad \dfrac{a}{14bcd^2}$

SOLUTION

a.
$$\frac{2}{5x} \quad \frac{3}{10x^2}$$

$$\frac{2}{5 \cdot x} \quad \frac{3}{2 \cdot 5 \cdot x^2} \quad \text{Factor both denominators.}$$

$$LCD = 2 \cdot 5 \cdot x^2 \quad \text{Take the highest power of each factor for the LCD.}$$
$$LCD = 10x^2$$

$$\frac{2 \cdot x}{2 \cdot x} \cdot \frac{2}{5 \cdot x} \quad \frac{3}{2 \cdot 5 \cdot x^2} \quad \text{Multiply each fraction by the factors that its denominator is missing.}$$

$$\frac{4x}{10x^2} \quad \frac{3}{10x^2}$$

b.
$$\frac{6ab}{25c^3d} \quad \frac{a}{14bcd^2}$$

$$\frac{2 \cdot 3 \cdot a \cdot b}{5^2 \cdot c^3 \cdot d} \quad \frac{a}{2 \cdot 7 \cdot b \cdot c \cdot d^2} \quad \text{Factor both denominators.}$$

$$LCD = 2 \cdot 5^2 \cdot 7 \cdot b \cdot c^3 \cdot d^2 \quad \text{Take the highest power of each factor for the LCD.}$$
$$LCD = 350bc^3d^2$$

$$\frac{2 \cdot 7 \cdot b \cdot d}{2 \cdot 7 \cdot b \cdot d} \cdot \frac{2 \cdot 3 \cdot a \cdot b}{5^2 \cdot c^3 \cdot d} \quad \frac{a}{2 \cdot 7 \cdot b \cdot c \cdot d^2} \cdot \frac{5^2 \cdot c^2}{5^2 \cdot c^2} \quad \text{Multiply each fraction by the factors that its denominator is missing.}$$

$$\frac{84ab^2d}{350bc^3d^2} \quad \frac{25ac^2}{350bc^3d^2}$$

PRACTICE PROBLEM FOR EXAMPLE 1

Find the least common denominator for the following two fractions and write each fraction in terms of the LCD.

$$\frac{7}{24m} \quad \frac{11}{45m^3n}$$

With denominators that have a single term, it is easier to find the factors you need to make up the LCD. Note that you want to take the highest power of each factor that is present to be a part of the LCD. When you are working with denominators with more than one term, you will find situations in which the denominators look very similar but cannot be changed easily. Let's find the LCD for the two fractions.

$$\frac{3}{x+2} \qquad \frac{5x}{x+4}$$

It seems as though you could either add 2 to $x + 2$ or multiply $x + 2$ by 2 and get $x + 4$. The problem with these ideas is that multiplying by 2 will give you $2x + 4$, not $x + 4$, and adding 2 to the denominator is not possible because that would change the value of the fraction you are working on. The only way to get a LCD from these two denominators is to use both of them as factors in the new denominator, giving you $(x + 4)(x + 2)$ as the LCD.

$$\frac{(x+4)}{(x+4)} \cdot \frac{3}{(x+2)} \qquad \frac{5x}{(x+4)} \cdot \frac{(x+2)}{(x+2)}$$

$$\frac{3(x+4)}{(x+4)(x+2)} \qquad \frac{5x(x+2)}{(x+4)(x+2)}$$

Example 2 — Finding the LCD with polynomial denominators

Find the least common denominator for each of the following sets of fractions and write each fraction in terms of the LCD.

a. $\dfrac{x+5}{x-6} \qquad \dfrac{x+2}{x+7}$

b. $\dfrac{x+3}{(x-5)(x+4)} \qquad \dfrac{x-7}{(x-5)(x+2)}$

c. $\dfrac{5x+7}{x^2+5x+6} \qquad \dfrac{3x-8}{2x^2+9x+9}$

SOLUTION

a. $\dfrac{x+5}{x-6} \qquad \dfrac{x+2}{x+7}$ The denominators are already factored, so we take the highest power of each factor for the LCD.

LCD $= (x-6)(x+7)$

$\dfrac{(x+7)}{(x+7)} \cdot \dfrac{(x+5)}{(x-6)} \qquad \dfrac{(x+2)}{(x+7)} \cdot \dfrac{(x-6)}{(x-6)}$

$\dfrac{(x+7)(x+5)}{(x-6)(x+7)} \qquad \dfrac{(x+2)(x-6)}{(x-6)(x+7)}$ Leave the fractions in this factored form.

b. $\dfrac{x+3}{(x-5)(x+4)} \qquad \dfrac{x-7}{(x-5)(x+2)}$ Because each denominator has only one copy of $(x-5)$, we will need only one copy in the LCD.

LCD $= (x-5)(x+4)(x+2)$

$\dfrac{(x+2)}{(x+2)} \cdot \dfrac{(x+3)}{(x-5)(x+4)} \qquad \dfrac{(x-7)}{(x-5)(x+2)} \cdot$

$\dfrac{(x+2)(x+3)}{(x-5)(x+4)(x+2)} \qquad \dfrac{(x-7)(x+4)}{(x-5)(x+4)(x+2)}$

c. $\dfrac{5x + 7}{x^2 + 5x + 6} \quad \dfrac{3x - 8}{2x^2 + 9x + 9}$ Factor the denominators.

$\dfrac{5x + 7}{(x + 3)(x + 2)} \quad \dfrac{3x - 8}{(2x + 3)(x + 3)}$

LCD $= (x + 3)(x + 2)(2x + 3)$
LCD $= 2x^3 + 13x^2 + 27x + 18$

$\dfrac{(2x + 3)}{(2x + 3)} \cdot \dfrac{(5x + 7)}{(x + 3)(x + 2)} \qquad \dfrac{(3x - 8)}{(2x + 3)(x + 3)} \cdot \dfrac{(x + 2)}{(x + 2)}$

$\dfrac{(2x + 3)(5x + 7)}{(x + 3)(x + 2)(2x + 3)} \qquad \dfrac{(3x - 8)(x + 2)}{(x + 3)(x + 2)(2x + 3)}$

PRACTICE PROBLEM FOR EXAMPLE 2

Find the least common denominator for each of the following sets of fractions and write each fraction in terms of the LCD.

a. $\dfrac{x + 3}{x + 8} \quad \dfrac{x + 5}{x + 9}$

b. $\dfrac{x + 2}{(x - 4)(x - 7)} \quad \dfrac{x + 4}{(x - 7)(x + 8)}$

c. $\dfrac{3x + 2}{6x^2 - 7x - 20} \quad \dfrac{4x - 9}{2x^2 - 19x + 35}$

Adding Rational Expressions

Getting the least common denominator is the hardest part of adding and subtracting rational expressions. Once the fractions have a common denominator, you can just add or subtract the numerators, and the LCD remains the denominator of the sum or difference.

> **Adding Rational Expressions**
> 1. Rewrite all fractions using the least common denominator.
> 2. Add the numerators and keep the denominator the same.
> 3. Factor the numerator and reduce (if possible).

Example 3 Adding rational expressions

Add the following rational expressions.

a. $\dfrac{5}{7x^2} + \dfrac{12}{5x}$ b. $\dfrac{x + 2}{x - 3} + \dfrac{x - 7}{x - 3}$ c. $x + \dfrac{7}{x + 3}$

SOLUTION

a. $\dfrac{5}{7x^2} + \dfrac{12}{5x} = \dfrac{5}{5} \cdot \dfrac{5}{7x^2} + \dfrac{12}{5x} \cdot \dfrac{7x}{7x}$ Determine the least common denominator. Rewrite each fraction over the LCD.

$= \dfrac{25}{35x^2} + \dfrac{84x}{35x^2}$ Add the numerators. In this case, they are not like terms, so you can write them with addition between them in one fraction.

$= \dfrac{84x + 25}{35x^2}$

b. $\dfrac{x + 2}{x - 3} + \dfrac{x - 7}{x - 3} = \dfrac{2x - 5}{x - 3}$ Because the denominators are already the same, simply add the numerators together.

c. $x + \dfrac{7}{x+3} = \dfrac{x}{1} + \dfrac{7}{x+3}$ The x has a denominator of 1, so we will need to get a common denominator.

$= \dfrac{(x+3)}{(x+3)} \cdot \dfrac{x}{1} + \dfrac{7}{x+3}$

$= \dfrac{(x+3)x}{x+3} + \dfrac{7}{x+3}$

$= \dfrac{x^2 + 3x}{x+3} + \dfrac{7}{x+3}$ We need to multiply out the numerators, so we can add them together.

$= \dfrac{x^2 + 3x + 7}{x+3}$

PRACTICE PROBLEM FOR EXAMPLE 3

Add the following rational expressions.

a. $\dfrac{16}{5x^3} + \dfrac{7}{2x}$ **b.** $x + \dfrac{4}{x-5}$

Example 4 Adding rational expressions

Add the following rational expressions.

a. $\dfrac{x-19}{(x+3)(x+5)} + \dfrac{x-7}{x+5}$

b. $\dfrac{x+5}{x^2+5x-24} + \dfrac{x-9}{x^2+16x+64}$

SOLUTION

a. $\dfrac{x-19}{(x+3)(x+5)} + \dfrac{x-7}{x+5}$

$= \dfrac{x-19}{(x+3)(x+5)} + \dfrac{x-7}{x+5} \cdot \dfrac{(x+3)}{(x+3)}$ Rewrite the fractions using the LCD.

$= \dfrac{x-19}{(x+3)(x+5)} + \dfrac{(x-7)(x+3)}{(x+5)(x+3)}$

$= \dfrac{x-19}{(x+3)(x+5)} + \dfrac{x^2 - 4x - 21}{(x+3)(x+5)}$ Multiply the numerator out.

$= \dfrac{x^2 - 3x - 40}{(x+3)(x+5)}$ Add the numerators.

$= \dfrac{\cancel{(x+5)}(x-8)}{(x+3)\cancel{(x+5)}}$ Factor the numerator and reduce.

$= \dfrac{x-8}{x+3}$

b. $\dfrac{x+5}{x^2+5x-24} + \dfrac{x-9}{x^2+16x+64} = \dfrac{x+5}{(x-3)(x+8)} + \dfrac{x-9}{(x+8)(x+8)}$

$= \dfrac{(x+8)}{(x+8)} \cdot \dfrac{x+5}{(x-3)(x+8)} + \dfrac{x-9}{(x+8)(x+8)} \cdot \dfrac{(x-3)}{(x-3)}$ Rewrite using the LCD.

$= \dfrac{(x+8)(x+5)}{(x-3)(x+8)(x+8)} + \dfrac{(x-9)(x-3)}{(x-3)(x+8)(x+8)}$

$= \dfrac{x^2 + 13x + 40}{(x-3)(x+8)(x+8)} + \dfrac{x^2 - 12x + 27}{(x-3)(x+8)(x+8)}$ Multiply the numerators out.

$= \dfrac{2x^2 + x + 67}{(x-3)(x+8)(x+8)}$ Add the numerators.

PRACTICE PROBLEM FOR EXAMPLE 4

Add the following rational expressions.

$$\frac{2x+5}{(x+3)(x-4)} + \frac{3x+7}{(x+3)(x+7)}$$

Subtracting Rational Expressions

Subtraction is the same process as addition but with one warning: Be sure to subtract the entire numerator of the second fraction. The most common error that students make is to forget to distribute the subtraction over the numerator of the second fraction.

> **Subtracting Rational Expressions**
> 1. Write all fractions with a common denominator.
> 2. Subtract the numerators and keep the denominator the same.
> 3. Factor the numerator and reduce (if possible).
>
> **Caution:** When subtracting rational expressions, be sure to distribute the subtraction to the entire numerator of the second fraction.

Example 5   Subtracting rational expressions

Subtract the following rational expressions.

a. $\dfrac{x+5}{x+8} - \dfrac{x-9}{x+8}$

b. $\dfrac{x-7}{(x+2)(x-4)} - \dfrac{x+5}{(x+6)(x+2)}$

c. $\dfrac{3x+2}{x^2-3x-10} - \dfrac{2x+7}{x^2-2x-15}$

SOLUTION

a. $\dfrac{x+5}{x+8} - \dfrac{x-9}{x+8} = \dfrac{x+5-(x-9)}{x+8}$

$= \dfrac{x+5-x+9}{x+8} = \dfrac{14}{x+8}$ Distribute the negative sign and combine like terms.

b. $\dfrac{x-7}{(x+2)(x-4)} - \dfrac{x+5}{(x+6)(x+2)}$ Determine the LCD.

$= \dfrac{(x+6)}{(x+6)} \cdot \dfrac{(x-7)}{(x+2)(x-4)} - \dfrac{(x+5)}{(x+6)(x+2)} \cdot \dfrac{(x-4)}{(x-4)}$ Rewrite using the LCD.

$= \dfrac{(x+6)(x-7)}{(x+6)(x+2)(x-4)} - \dfrac{(x+5)(x-4)}{(x+6)(x+2)(x-4)}$ Multiply out the numerators.

$= \dfrac{x^2-x-42}{(x+6)(x+2)(x-4)} - \dfrac{x^2+x-20}{(x+6)(x+2)(x-4)}$

$= \dfrac{x^2-x-42-(x^2+x-20)}{(x+6)(x+2)(x-4)}$ Subtract the numerators.

$= \dfrac{x^2-x-42-x^2-x+20}{(x+6)(x+2)(x-4)}$ Distribute the negative sign.

$= \dfrac{-2x-22}{(x+6)(x+2)(x-4)}$ Simplify by combining like terms.

SECTION 7.4 Adding and Subtracting Rational Expressions

c. $\dfrac{3x+2}{x^2-3x-10} - \dfrac{2x+7}{x^2-2x-15} = \dfrac{3x+2}{(x-5)(x+2)} - \dfrac{2x+7}{(x-5)(x+3)}$

$= \dfrac{(x+3)}{(x+3)} \cdot \dfrac{(3x+2)}{(x-5)(x+2)} - \dfrac{(2x+7)}{(x-5)(x+3)} \cdot \dfrac{(x+2)}{(x+2)}$ Rewrite using the LCD.

$= \dfrac{3x^2+11x+6}{(x+3)(x-5)(x+2)} - \dfrac{2x^2+11x+14}{(x+3)(x-5)(x+2)}$ Multiply out the numerators.

$= \dfrac{3x^2+11x+6-(2x^2+11x+14)}{(x+3)(x-5)(x+2)}$ Subtract the numerators.

$= \dfrac{3x^2+11x+6-2x^2-11x-14}{(x+3)(x-5)(x+2)}$ Distribute the negative sign.

$= \dfrac{x^2-8}{(x+3)(x-5)(x+2)}$ Simplify by combining like terms.

PRACTICE PROBLEM FOR EXAMPLE 5

Subtract the following rational expressions.

a. $\dfrac{5x-4}{(x+5)(x+6)} - \dfrac{3x-16}{(x+5)(x+6)}$

b. $\dfrac{3x-7}{x^2-4x-45} - \dfrac{2x+3}{x^2-5x-36}$

Simplifying Complex Fractions

In some circumstances, we will have rational expressions that contain fractions in their numerator, denominator, or both. We call these types of expressions *complex fractions*. Some examples of complex fractions are

$$\dfrac{\dfrac{2}{x}+5}{7x+\dfrac{3}{x+1}} \qquad \dfrac{\dfrac{5}{x+7}}{\dfrac{2}{x-8}}$$

Simplifying complex fractions uses many of the same skills we have learned in this section and the previous two sections. The least common denominator of the fractions plays a key role in the simplification process. Multiplying the numerator and denominator of the rational expression by the LCD of all the fractions involved will eliminate the fractions in both the numerator and denominator and simplify the rational expression.

> **Simplifying Complex Fractions**
> 1. Find the LCD of all fractions inside the rational expression.
> 2. Multiply the numerator and denominator of the rational expression by the LCD.
> 3. Factor and reduce (if possible).

Example 6   Simplifying complex fractions

Simplify the following complex fractions.

a. $\dfrac{5+\dfrac{2}{x+7}}{\dfrac{3}{x+2}}$

b. $\dfrac{\dfrac{2}{x}+\dfrac{7}{y^2}}{\dfrac{3}{x^3}-\dfrac{4}{y}}$

SOLUTION

a.
$$\frac{5 + \frac{2}{x+7}}{\frac{3}{x+2}} = \frac{5 + \frac{2}{x+7}}{\frac{3}{x+2}} \cdot \frac{(x+7)(x+2)}{(x+7)(x+2)}$$ Multiply the numerator and denominator by the LCD.

$$= \frac{5(x+7)(x+2) + \frac{2}{x+7}(x+7)(x+2)}{\frac{3}{x+2}(x+7)(x+2)}$$ Distribute the LCD to every term.

$$= \frac{5(x^2 + 9x + 14) + 2(x+2)}{3(x+7)}$$ Multiply out and combine like terms.

$$= \frac{5x^2 + 45x + 70 + 2x + 4}{3x + 21}$$

$$= \frac{5x^2 + 47x + 74}{3x + 21}$$

$$= \frac{(5x + 37)(x + 2)}{3(x + 7)}$$ Factor and simplify if possible.

b.
$$\frac{\frac{2}{x} + \frac{7}{y^2}}{\frac{3}{x^3} - \frac{4}{y}} = \frac{\frac{2}{x} + \frac{7}{y^2}}{\frac{3}{x^3} - \frac{4}{y}} \cdot \frac{x^3 y^2}{x^3 y^2}$$ Multiply the numerator and denominator by the LCD.

$$= \frac{\frac{2}{x}(x^3 y^2) + \frac{7}{y^2}(x^3 y^2)}{\frac{3}{x^3}(x^3 y^2) - \frac{4}{y}(x^3 y^2)}$$ Distribute the LCD to every term.

$$= \frac{2x^2 y^2 + 7x^3}{3y^2 - 4x^3 y}$$

$$= \frac{x^2(2y^2 + 7x)}{y(3y - 4x^3)}$$ Factor and simplify if possible.

PRACTICE PROBLEM FOR EXAMPLE 6

Simplify the following complex fractions.

a. $\dfrac{\frac{7}{x+8} + 4}{\frac{6}{x-1}}$ **b.** $\dfrac{\frac{4}{x} + \frac{3}{y}}{\frac{2}{x^2} - \frac{6}{y^2}}$

7.4 Exercises

For Exercises 1 through 20, find the least common denominator for each set of rational expressions. Write each expression in terms of the LCD.

1. $\dfrac{7}{12x^2 y}$ $\dfrac{3}{28xy}$

2. $\dfrac{11a}{15b^2 c}$ $\dfrac{14b}{25ac^3}$

3. $\dfrac{7m}{180n^2 p}$ $\dfrac{4n}{150p}$

4. $\dfrac{16xz}{300y^4}$ $\dfrac{9}{84x^2 y^2 z}$

5. $\dfrac{x+2}{x+8}$ $\dfrac{x-4}{x+7}$

6. $\dfrac{x+9}{x-4}$ $\dfrac{x+7}{x+3}$

7. $\dfrac{x-5}{x+6}$ $\dfrac{x-5}{x+5}$

8. $\dfrac{x+1}{x-4}$ $\dfrac{x-3}{x-7}$

9. $\dfrac{2x}{5-x}$ $\dfrac{3x}{x-5}$

10. $\dfrac{2m}{m-6}$ $\dfrac{7m}{6-m}$

SECTION 7.4 Adding and Subtracting Rational Expressions

11. $2x \quad \dfrac{5}{x+1}$

12. $7y \quad \dfrac{x}{3y}$

13. $\dfrac{h+2}{(h+3)(h+7)} \quad \dfrac{h-4}{(h+3)(h+5)}$

14. $\dfrac{c-2}{(c+9)(c-3)} \quad \dfrac{c+7}{(c+4)(c-3)}$

15. $\dfrac{n+1}{(n-7)(n-2)} \quad \dfrac{n+2}{(n-8)(n-2)}$

16. $\dfrac{a+5}{(a-1)(a-6)} \quad \dfrac{a+3}{(a-1)(a-2)}$

17. $\dfrac{x+1}{x^2+5x+6} \quad \dfrac{x-3}{x^2-2x-8}$

18. $\dfrac{p+5}{p^2+8p+7} \quad \dfrac{p-6}{p^2+3p-28}$

19. $\dfrac{t+3}{t^2-2t-35} \quad \dfrac{t-7}{t^2+t-20}$

20. $\dfrac{x-5}{2x^2-9x-35} \quad \dfrac{x+6}{2x^2+11x+15}$

For Exercises 21 through 36, add or subtract the rational expressions.

21. $\dfrac{x+2}{x+7} + \dfrac{x+8}{x+7}$

22. $\dfrac{b-3}{b-9} + \dfrac{b-15}{b-9}$

23. $\dfrac{m+7}{m+5} - \dfrac{4m+22}{m+5}$

24. $\dfrac{k+7}{k-2} - \dfrac{k-6}{k-2}$

25. $\dfrac{d+5}{d+3} + \dfrac{d-2}{d+7}$

26. $\dfrac{x-4}{x+6} + \dfrac{x+3}{x-5}$

27. $\dfrac{c+2}{c+3} - \dfrac{c+4}{c+7}$

28. $\dfrac{y-7}{y+2} - \dfrac{y+3}{y+4}$

29. $\dfrac{n+6}{n-2} + \dfrac{n-5}{2-n}$

30. $\dfrac{x-7}{5-x} + \dfrac{x+4}{x-5}$

31. $\dfrac{x+7}{x-9} + \dfrac{x-8}{9-x}$

32. $\dfrac{x-1}{7-x} + \dfrac{x+1}{x-7}$

33. $5 + \dfrac{2}{x+7}$

34. $8 - \dfrac{3x}{x+4}$

35. $2x + \dfrac{7}{xy^2}$

36. $5y - \dfrac{6}{x^2}$

For Exercises 37 through 40, the students were asked to add or subtract the given rational expressions. Explain what the student did wrong. Give the correct answer.

37. $\dfrac{x+5}{x-7} + \dfrac{x+3}{x+2}$

Marie

$\dfrac{x+5}{x-7} + \dfrac{x+3}{x+2} = \dfrac{2x+8}{2x-5}$

38. $\dfrac{b+3}{b+5} - \dfrac{5b+2}{b+5}$

Hung

$\dfrac{b+3}{b+5} - \dfrac{5b+2}{b+5} = \dfrac{b+3-5b+2}{b+5}$

$= \dfrac{-4b+5}{b+5}$

39. $\dfrac{x+1}{x-7} + \dfrac{x-7}{x+4}$

Matt

$\dfrac{x+1}{x-7} + \dfrac{x-7}{x+4} = \dfrac{x+1}{\cancel{x-7}} + \dfrac{\cancel{x-7}}{x+4}$

$= \dfrac{x+1}{x+4}$

40. $\dfrac{h-3}{h+2} - \dfrac{h+5}{h^2+5h+6}$

Michelle

$\dfrac{h-3}{h+2} - \dfrac{h+5}{h^2+5h+6}$

$= \dfrac{h^2+5h+6}{h^2+5h+6} \cdot \dfrac{h-3}{h+2} - \dfrac{h+5}{h^2+5h+6} \cdot \dfrac{h+2}{h+2}$

$= \dfrac{(h^3+2h^2-9h-18)-(h^2+7h+10)}{(h^2+5h+6)(h+2)}$

$= \dfrac{(h^3+h^2-16h-28)}{(h^2+5h+6)(h+2)}$

41. The illumination in foot-candles provided by a particular light that is d feet away can be represented by

$$\frac{50}{d^2}$$

Another light from the same distance d provides an illumination of

$$\frac{75}{d^2}$$

a. Find an expression for the illumination if these two lights are both used from a distance of d feet.
b. If both of these lights are placed 5 feet away, how much illumination will be provided?

42. The illumination in foot-candles provided by a particular light that is d feet away can be represented by

$$\frac{100}{d^2}$$

Another light from the same distance d provides an illumination of

$$\frac{150}{d^2}$$

a. Find an expression for the illumination if these two lights are both used from a distance of d feet.
b. If both of these lights are placed 10 feet away, how much illumination will be provided?

43. The average cost in dollars per person to feed p people at a charity event can be represented by

$$\frac{3.5p + 250}{p}$$

The average cost in dollars per person for entertainment, brochures, and other supplies can be represented by

$$\frac{2p + 600}{p}$$

a. Find an expression for the average cost of all these expenses for this charity event.
b. If 100 people attend this event, what will the average cost be for all of these expenses?

44. A basketball team on a road trip calculates the average cost in dollars per person to feed the team and coaches. If p players and the three coaches go on the road trip, the average cost for food can be represented by

$$\frac{75p + 500}{p + 3}$$

The average cost in dollars per person for housing and transportation can be represented by

$$\frac{80p + 600}{p + 3}$$

a. Find an expression for the average cost of all these expenses for this team's road trip.
b. If 20 players on the team go on the road trip, what will the average cost be for all these expenses?

For Exercises 45 through 60, add or subtract the rational expressions.

45. $\dfrac{r + 2}{(r + 5)(r + 3)} + \dfrac{r + 4}{(r + 5)(r + 3)}$

46. $\dfrac{3a - 6}{(2a + 7)(a + 3)} - \dfrac{7a + 8}{(2a + 7)(a + 3)}$

47. $\dfrac{x + 1}{(x - 7)(x - 2)} + \dfrac{x + 2}{(x - 8)(x - 2)}$

48. $\dfrac{2b + 5}{(b - 11)(b + 3)} + \dfrac{b - 3}{(b - 11)(b - 4)}$

49. $\dfrac{w - 2}{(w + 4)(w - 6)} - \dfrac{3w + 1}{(w + 3)(w - 6)}$

50. $\dfrac{c - 7}{(3c + 2)(2c + 5)} - \dfrac{c + 4}{(2c + 5)(c - 9)}$

51. $\dfrac{h + 3}{h^2 - 9h + 14} - \dfrac{h - 4}{h^2 - 10h + 16}$

52. $\dfrac{x + 2}{x^2 + 10x + 21} - \dfrac{x - 4}{x^2 + 8x + 15}$

53. $\dfrac{t - 2}{t^2 + 6t - 27} + \dfrac{t + 7}{t^2 + t - 12}$

54. $\dfrac{x - 5}{2x^2 - 9x - 35} + \dfrac{x + 6}{2x^2 + 11x + 15}$

55. $\dfrac{k + 2}{2k^2 + 7k + 3} + \dfrac{k - 4}{4k^2 + 7k - 15}$

56. $\dfrac{3x}{3x^2 + 24x + 36} + \dfrac{x^2}{2x^2 + 16x + 24}$

57. $\dfrac{r^2 + 10r}{r^2 + 16r + 63} - \dfrac{4r + 27}{r^2 + 16r + 63}$

58. $\dfrac{5x + 43}{x^2 + 6x - 55} - \dfrac{2x + 10}{x^2 + 6x - 55}$

59. $\dfrac{d + 4}{d + 2} - \dfrac{2}{d^2 + 5d + 6}$

60. $\dfrac{4g - 8}{g^2 - 4g - 5} - \dfrac{2}{g - 5}$

SECTION 7.4 Adding and Subtracting Rational Expressions

For Exercises 61 through 76, simplify the complex fractions.

61. $\dfrac{3 + \dfrac{6}{x}}{2 - \dfrac{5}{x}}$

62. $\dfrac{8 - \dfrac{2}{t}}{3 - \dfrac{5}{t}}$

63. $\dfrac{7 + \dfrac{4}{a^2}}{5 + \dfrac{2}{a}}$

64. $\dfrac{4 + \dfrac{1}{n}}{2 - \dfrac{7}{n^2}}$

65. $\dfrac{3 + \dfrac{75}{r^2}}{2 - \dfrac{10}{r}}$

66. $\dfrac{3 + \dfrac{6}{n}}{2 - \dfrac{8}{n^2}}$

67. $\dfrac{\dfrac{3}{x} + \dfrac{5}{y^2}}{\dfrac{7}{y} - \dfrac{6}{x^4}}$

68. $\dfrac{\dfrac{7}{x^3} - 4}{\dfrac{x}{y} + \dfrac{2}{x}}$

69. $\dfrac{\dfrac{2x}{5y^2} + \dfrac{6}{x}}{\dfrac{8}{x} + \dfrac{2}{3y}}$

70. $\dfrac{\dfrac{x}{y} - \dfrac{y}{x}}{\dfrac{x}{y^3} + \dfrac{y}{x^2}}$

71. $\dfrac{\dfrac{3}{x+5}}{\dfrac{2}{x+1} + 4x}$

72. $\dfrac{\dfrac{7}{x+2}}{\dfrac{3}{x+4} - 5x}$

73. $\dfrac{7 + \dfrac{2}{x+8}}{\dfrac{1}{x-3}}$

74. $\dfrac{10 - \dfrac{3}{2x+5}}{\dfrac{7}{3x-4}}$

75. $\dfrac{\dfrac{2}{x+3} - \dfrac{5x}{x+2}}{\dfrac{4}{x+2} + \dfrac{7x}{x-3}}$

76. $\dfrac{\dfrac{7}{x+4} - \dfrac{6x}{x+5}}{\dfrac{3}{x-4} - \dfrac{2x}{x-3}}$

77. In electronics, if three resistors are connected in parallel, their combined resistance can be represented by

$$R = \dfrac{1}{\dfrac{1}{R_1} + \dfrac{1}{R_2} + \dfrac{1}{R_3}}$$

where R is the total resistance and R_1, R_2, and R_3 are the resistance of the three resistors.

a. Simplify the complex fraction on the right side of the equation.
b. If R_1 is 3 ohms, R_2 is 6 ohms, and R_3 is 9 ohms, then find R.

78. Use the simplified complex fraction you found in Exercise 77 part a to find R if R_1 is 2 ohms, R_2 is 4 ohms, and R_3 is 8 ohms.

79. In photography, the focal length inside a camera needed to focus an object in front of the lens is related to the distance between the object and the lens and the distance between the lens and the film in the camera. This relationship can be represented by

$$f = \dfrac{1}{\dfrac{1}{d} + \dfrac{1}{c}}$$

where f is the focal length, d is the distance between the object being photographed and the lens, and c is the distance between the lens and the film inside the camera.

a. Simplify the complex fraction on the right side of the equation.
b. If the distance from the lens to the object being photographed is 5 feet, and the distance from the lens to the film is 0.75 foot, find the focal length.

80. Use the simplified complex fraction you found in part a of Exercise 79 to find the focal length if the distance from the lens to the object being photographed is 16 feet and the distance from the lens to the film is 0.25 foot.

For Exercises 81 through 90, perform the indicated operation and simplify.

81. $\dfrac{a+2}{a^2+6a+9} \cdot \dfrac{a+3}{a^2-4a+3}$

82. $\dfrac{\dfrac{2}{x+3}}{\dfrac{5}{x-6}}$

83. $\dfrac{m^2-25}{m^2+8m+7} - \dfrac{m+7}{m+1}$

84. $\dfrac{x^3+8}{x^2+3x-10} \div \dfrac{x^2-2x+4}{3x^2+16x+5}$

85. $\dfrac{6-x}{x+2} \cdot \dfrac{x+7}{x-6}$

86. $\dfrac{2r-5}{r^2+7r-30} + \dfrac{8r-3}{r^2+11r+10}$

87. $\dfrac{7x+8}{x^2+5x+6} - \dfrac{x-4}{x^2+5x+6}$

88. $\dfrac{45a^2b^3}{12ab^5c^2} \cdot \dfrac{6b^4c}{10ab}$

89. $\dfrac{5x}{x^2-16} - \dfrac{7}{x^2+7x+12}$

90. $\dfrac{d+7}{d^2-8d+12} + \dfrac{3d+2}{d^2+5d+4}$

7.5 Solving Rational Equations

LEARNING OBJECTIVES
- Solve rational equations.
- Solve applications that contain rational equations.

In this section, we will learn to solve rational equations. Most often, we will want to eliminate all the fractions by multiplying both sides of the equation by the common denominator. This will allow us to solve the remaining equation using the techniques we learned in earlier chapters. When solving rational equations, keep in mind that some solutions will not be valid. When variables are in the denominator of a fraction, we must be very careful not to allow division by zero to take place. To avoid this, always check solutions in the original equation to be sure that no excluded values are kept.

> **Solving Rational Equations**
> 1. State what values should be excluded.
> 2. Multiply both sides of the equation by the least common denominator.
> 3. Solve the remaining equation.
> 4. Check the answer(s) in the original equation. (Watch for division by zero.)

Example 1 Solving rational equations

Solve the following rational equation. Check your answer in the original equation.

$$\frac{10}{x-8} = 5$$

SOLUTION

$$\frac{10}{x-8} = 5 \qquad \text{The denominator equals zero when } x = 8, \text{ so it will be an excluded value.}$$

$$(x-8)\left(\frac{10}{x-8}\right) = 5(x-8) \qquad \text{Multiply both sides of the equation by the denominator.}$$

$$10 = 5x - 40 \qquad \text{Solve for } x.$$

$$50 = 5x$$

$$10 = x$$

$$\frac{10}{(10)-8} \stackrel{?}{=} 5 \qquad \text{Check the answer using the original equation.}$$

$$\frac{10}{2} \stackrel{?}{=} 5$$

$$5 = 5 \qquad \text{The answer works.}$$

PRACTICE PROBLEM FOR EXAMPLE 1

Solve the following rational equation.

$$\frac{7}{x-4} = 14$$

Example 2 Solving an average cost equation

The Green Club is planning a ski trip to a local mountain resort. The students want to keep the costs for each person as low as possible, so they plan to rent a van and split the costs evenly between them. They have decided that the person driving the van should not have to pay. If the van is going to cost $130 to rent and can hold up to 15 people, how many people need to go in the van to keep the transportation costs per person at $12?

SOLUTION

Let p be the number of people going on the ski trip in the rented van, and let $C(p)$ be the cost per person going on the trip in dollars per person. Because the driver will not be paying for the van expenses, we need to subtract the driver from the total number of people p. If we take the total cost for the van and divide it by the number of people paying, we get the cost per person. This leads us to the following equation.

$$C(p) = \frac{130}{p-1}$$

Because we want the average cost to be $12 per person, we can replace $C(p)$ with 12 and solve. To solve this equation, we need to get the variable p out of the denominator of the fraction. We can do this by multiplying both sides of the equation by the denominator $p - 1$.

$\frac{130}{p-1} = 12$ The denominator equals zero when $p = 1$, so it will be an excluded value.

$(p-1)\left(\frac{130}{p-1}\right) = 12(p-1)$ Multiply both sides by the denominator.

$130 = 12p - 12$ Solve for p.

$142 = 12p$

$\frac{142}{12} = p$

$\frac{71}{6} = p$

$11.83 \approx p$

Therefore, the Green Club needs to have at least 12 people go in the van to keep the transportation cost at $12 per person or less. Note that we had to round this answer up, or the cost would have been more than $12 per person. In this type of situation, we always want to round up to the next whole person.

PRACTICE PROBLEM FOR EXAMPLE 2

In Section 7.1 we found the model $a(m) = \frac{50}{m}$ for the acceleration a in meters per second squared that a fixed amount of force would give when applied to a mass of m kg.

a. If an object has an acceleration of 4 m/s² when this force is applied, what is the object's mass?

b. What mass can be accelerated at 0.5 m/s² using the given fixed amount of force?

Example 3 Solving rational equations

Solve the following rational equation. Check your answer in the original equation.

a. $\frac{5}{x+1} = 3 - \frac{7}{x+1}$

SOLUTION

a.
$$\frac{5}{x+1} = 3 - \frac{7}{x+1}$$ The denominator equals zero when $x = -1$, so it will be an excluded value.

$$(x+1)\left(\frac{5}{x+1}\right) = \left(3 - \frac{7}{x+1}\right)(x+1)$$ Multiply both sides by the common denominator.

$$5 = 3(x+1) - \left(\frac{7}{x+1}\right)(x+1)$$ Use the distributive property.

$$5 = 3x + 3 - 7$$ Solve for x.
$$5 = 3x - 4$$
$$9 = 3x$$
$$x = 3$$

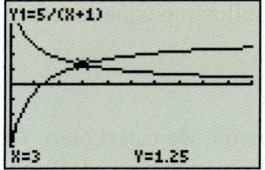

Check the answer.
The answer works.

PRACTICE PROBLEM FOR EXAMPLE 3

Solve the following rational equation.

$$\frac{35}{x+3} = 4 + \frac{15}{x+3}$$

For rational functions that arise from dividing two functions, the technique of multiplying both sides of the equation by the denominator quickly clears the fraction and allows us to solve the remaining equation. Although some of these equations might look very complicated, the use of a calculator can help to manage the solution process.

Example 4   Finding average disbursements

In Section 7.1, we found a model for the average disbursements made by the U.S. Medicare program.

$$P(t) = \frac{-3160.64t^2 + 52536.5t - 4548.31}{0.39t + 35.65}$$

where $P(t)$ represents the average dollars disbursed per person by Medicare t years since 1990. In what year did the average dollars disbursed reach $5500?

SOLUTION

We are given an average disbursement, so we can substitute $P(t) = 5500$.

$$5500 = \frac{-3160.64t^2 + 52536.5t - 4548.31}{0.39t + 35.65}$$

$$(0.39t + 35.65)(5500) = \frac{-3160.64t^2 + 52536.5t - 4548.31}{0.39t + 35.65} \cdot (0.39t + 35.65)$$

$$2145t + 196{,}075 = -3160.64t^2 + 52536.5t - 4548.31$$

$$0 = -3160.64t^2 + 50391.5t - 200{,}623.31$$

$$t \approx 7.7 \quad t \approx 8.24 \quad \text{By the quadratic formula.}$$

Because both $t = 7.7$ and $t = 8.24$ represent the year 1998, we have that in 1998, the average disbursement per person made by Medicare was about $5500. We can check these solutions using the graph or table.

Once you have multiplied by the denominator, you can combine all the like terms. In the case of Example 4, we were left with a quadratic equation and therefore, can use the quadratic formula or factoring to solve. In general, multiplying by the common denominator will allow us to solve most equations involving rational expressions.

This solving technique can be used with more complicated rational equations. When there is more than one denominator, it is quickest if you first find the least common denominator for all fractions in the equation. When you multiply by the least common denominator, you will eliminate all of the fractions, and you should then be able to solve the remaining equation. Remember to check your answers for values that are not valid.

Example 5 Solving rational equations

Solve the following rational equations. Check your answer(s) in the original equation.

a. $\dfrac{12}{x+3} = \dfrac{7}{x-5}$

b. $\dfrac{4}{x-4} = \dfrac{3}{x+7}$

c. $\dfrac{7}{x+5} + \dfrac{3x}{x-2} = \dfrac{42}{x^2+3x-10}$

d. $\dfrac{2}{x+3} + \dfrac{7}{x+4} = \dfrac{4x-11}{x^2+7x+12}$

SOLUTION

a.
$$\dfrac{12}{x+3} = \dfrac{7}{x-5}$$ The denominators equal zero when $x = -3$ or 5, so these will be excluded values.

$$(x+3)(x-5)\left(\dfrac{12}{x+3}\right) = \left(\dfrac{7}{x-5}\right)(x+3)(x-5)$$ Multiply both sides by the common denominator.

$$(x-5)12 = 7(x+3)$$ Reduce the fractions and distribute the multiplication.
$$12x - 60 = 7x + 21$$
$$5x = 81$$ Solve for x.
$$x = \dfrac{81}{5}$$
$$x = 16.2$$

$$\dfrac{12}{(16.2)+3} \stackrel{?}{=} \dfrac{7}{(16.2)-5}$$ Check the answer.
$$0.625 = 0.625$$ The answer works.

b.
$$\dfrac{4}{x-4} = \dfrac{3}{x+7}$$ The denominators equal zero when $x = -7$ or 4, so these will be excluded values.

$$(x-4)(x+7)\left(\dfrac{4}{x-4}\right) = \left(\dfrac{3}{x+7}\right)(x-4)(x+7)$$ Multiply both sides by the common denominator.

$$(x+7)(4) = (3)(x-4)$$
$$4x + 28 = 3x - 12$$ Solve for x.
$$x = -40$$

$$\dfrac{4}{(-40)-4} \stackrel{?}{=} \dfrac{3}{(-40)+7}$$ Check the answer.
$$-0.091 = -0.091$$ The answer works.

c.
$$\frac{7}{x+5} + \frac{3x}{x-2} = \frac{42}{x^2+3x-10}$$

$$\frac{7}{x+5} + \frac{3x}{x-2} = \frac{42}{(x+5)(x-2)}$$

The denominators equal zero when $x = -5$ or 2, so these will be excluded values.

$$(x+5)(x-2)\left(\frac{7}{x+5} + \frac{3x}{x-2}\right) = \left(\frac{42}{(x+5)(x-2)}\right) \cdot (x+5)(x-2)$$

Note that we cannot reduce the left side until we distribute.

$$(x+5)(x-2)\left(\frac{7}{x+5}\right) + (x+5)(x-2)\left(\frac{3x}{x-2}\right) = 42$$

$$7(x-2) + 3x(x+5) = 42$$

$$7x - 14 + 3x^2 + 15x = 42 \quad \text{Solve for } x.$$

$$3x^2 + 22x - 56 = 0$$

$$x = 2 \qquad x = -\frac{28}{3}$$

Check the answers.

Because 2 results in division by zero, it is not a valid answer.

Because $x = 2$ causes some of the fractions to have division by zero, we must eliminate that answer; thus we have only $x = -\frac{28}{3}$ as an answer.

d.
$$\frac{2}{x+3} + \frac{7}{x+4} = \frac{4x-11}{x^2+7x+12}$$

Factor to find the common denominator.

$$\frac{2}{x+3} + \frac{7}{x+4} = \frac{4x-11}{(x+3)(x+4)}$$

The denominators equal zero when $x = -3$ or -4, so these will be excluded values.

$$(x+3)(x+4)\left(\frac{2}{x+3} + \frac{7}{x+4}\right) = \left(\frac{4x-11}{(x+3)(x+4)}\right)(x+3)(x+4)$$

$$(x+3)(x+4)\left(\frac{2}{x+3}\right) + (x+3)(x+4)\left(\frac{7}{x+4}\right) = 4x - 11$$

$$(x+4)(2) + (x+3)(7) = 4x - 11$$

$$2x + 8 + 7x + 21 = 4x - 11$$

$$9x + 29 = 4x - 11$$

$$5x = -40$$

$$x = -8 \qquad \text{Check the answer.}$$

$$\frac{2}{(-8)+3} + \frac{7}{(-8)+4} \stackrel{?}{=} \frac{4(-8)-11}{(-8)^2+7(-8)+12}$$

$$-2.15 = -2.15 \qquad \text{The answer works.}$$

PRACTICE PROBLEM FOR EXAMPLE 5
Solve the following rational equations.

a. $\dfrac{3}{x+2} = \dfrac{6}{x+9}$

b. $\dfrac{5x}{x-4} + \dfrac{3}{x+2} = \dfrac{-18}{x^2-2x-8}$

c. $\dfrac{5}{x+2} = 8 - \dfrac{52}{x+7}$

Example 6 Solving a work problem

Jim working alone can paint a house in 30 hours. Mary working alone can paint a house in 24 hours. How long will it take if Jim and Mary work together to paint that house?

SOLUTION

Let t be the time it takes Jim and Mary working together to paint the house.

Since Jim can paint the house in 30 hours, he gets $\frac{1}{30}$ of the house painted in 1 hour.

Since Mary can paint the house in 24 hours, she gets $\frac{1}{24}$ of the house painted in 1 hour.

Together, they can paint $\frac{1}{t}$ of the house in 1 hour.

If we add how much each can do in 1 hour, we get the amount of work that they can do together in 1 hour. Therefore, we get the following equation.

$$\frac{1}{30} + \frac{1}{24} = \frac{1}{t}$$

Solve the equation for t.

$$\frac{1}{30} + \frac{1}{24} = \frac{1}{t}$$

$$120t\left(\frac{1}{30} + \frac{1}{24}\right) = \left(\frac{1}{t}\right) \cdot 120t \quad \text{Multiply both sides of the equation by the LCD.}$$

$$\overset{4}{\cancel{120t}} \cdot \frac{1}{\cancel{30}} + \overset{5}{\cancel{120t}} \cdot \frac{1}{\cancel{24}} = \left(\frac{1}{\cancel{t}}\right) \cdot 120\cancel{t} \quad \text{Distribute and simplify.}$$

$$4t + 5t = 120$$

$$9t = 120 \quad \text{Solve for } t.$$

$$t = \frac{120}{9}$$

$$t = \frac{40}{3}$$

$$t = 13\frac{1}{3}$$

$$\frac{1}{30} + \frac{1}{24} \stackrel{?}{=} \frac{1}{\left(\frac{40}{3}\right)} \quad \text{Check the answer.}$$

$$\frac{1}{30} + \frac{1}{24} \stackrel{?}{=} \frac{3}{40}$$

$$\frac{4}{120} + \frac{5}{120} \stackrel{?}{=} \frac{9}{120} \quad \text{Get a common denominator.}$$

$$\frac{9}{120} = \frac{9}{120} \quad \text{The answer works.}$$

Therefore, working together Jim and Mary can paint the house in $13\frac{1}{3}$ hours.

PRACTICE PROBLEM FOR EXAMPLE 6

It takes a garden hose 48 hours to fill a backyard pool. A fire hydrant can fill the same pool in 20 hours. How long will it take to fill the pool if both the garden hose and the fire hydrant are used?

7.5 Exercises

For Exercises 1 through 8, solve the following rational equations. Check the answer(s) using the original equation.

1. $6 = \dfrac{30}{x+2}$
2. $4 = \dfrac{24}{x-7}$
3. $5 = \dfrac{4}{2w+3}$
4. $10 = \dfrac{6}{3p+4}$
5. $2t = \dfrac{8}{t-3}$
6. $5x = \dfrac{-10}{3x-7}$
7. $3m = \dfrac{-9}{m+4}$
8. $2x = \dfrac{-7}{x+6}$

9. In Exercise 1 of Section 7.1, we found the model
$$C(p) = \dfrac{600}{p}$$
where $C(p)$ is the per player cost in dollars for a charity poker tournament if p players participate.
    a. How many players must participate if the per player cost is $20?
    b. How many players must participate if the per player cost is $7?

10. In Exercise 2 of Section 7.1, we found the model
$$C(p) = \dfrac{1500}{p}$$
where $C(p)$ is the per camper cost in dollars for an Adventure Guides camp out if p people attend. Up to 90 people can attend.
    a. How many people must attend if the per camper cost is $30?
    b. How many people must attend if the per camper cost is $10?

11. The Vermont History Society is hosting a charity dinner to raise funds. The dinner will cost a total of $3400 for up to 100 people to attend.
    a. Find a function that gives the cost per person for p people to attend the charity event.
    b. Use your new model to find how many people need to attend to keep the cost per person down to $40 each.

12. The Fancy Affair catering company is catering an event for a local charity. Jan, the owner of Fancy Affair, is donating her time to the cause but she needs to charge the charity for the food, decorations, and supplies used. If p people attend the charity event, Jan has figured her total cost to be modeled by
$$C(p) = 4.55p + 365.00$$
where $C(p)$ is the total cost in dollars for the food, decorations, and supplies when p people attend the charity event.
    a. Find a new function that gives the cost per person for p people to attend the charity event.
    b. Use your new model to find how many people need to attend to keep the cost per person down to $7.50 each.

For Exercises 13 through 20, solve the following rational equations. Check the answer(s) using the original equation.

13. $2 - \dfrac{6}{x} = \dfrac{4}{x}$
14. $5 + \dfrac{7}{a} = \dfrac{11}{a}$
15. $6 + \dfrac{2}{5m} = \dfrac{3}{m}$
16. $3 + \dfrac{5}{2p} = \dfrac{4}{p}$
17. $\dfrac{4}{x+2} = -1 + \dfrac{8}{x+2}$
18. $\dfrac{10}{t-4} = 3 - \dfrac{5}{t-4}$
19. $4 + \dfrac{6}{h-3} = 7 + \dfrac{12}{h-3}$
20. $-2 - \dfrac{15}{c+6} = 11 - \dfrac{145}{c+6}$

21. The percentage of housing units in Colorado that were vacant in the 1990s can be modeled by
$$P(t) = \dfrac{216.89t^2 - 2129.2t + 19114.9}{33.67t + 1444.29}$$
where $P(t)$ is the percentage of housing units in Colorado that were vacant t years since 1990.
Source: Model derived from data from the U.S. Department of Commerce Bureau of Economic Analysis.
    a. Find the percentage of housing units vacant in 1995.
    b. Find when there was an 11.5% vacancy rate for housing units in Colorado.

22. The per capita personal spending in the United States can be modeled by
$$s(t) = \dfrac{248158.9t + 186907.6}{0.0226t^2 + 1.885t + 204.72}$$
where $s(t)$ represents the per capita personal spending in dollars in the United States t years since 1970.
Source: Model derived from data from the U.S. Department of Commerce Bureau of Economic Analysis.
    a. Find the per capita personal spending in the United States in 1995.
    b. In what year did the per capita personal spending in the United States reach $25,000?

23. The state of California's spending per person has increased dramatically since 1950. The state's population from 1950 can be modeled by
$$P(t) = 0.464t - 12.47$$
where $P(t)$ is California's population in millions of people t years since 1900. The amount that California

spent in millions of dollars can be modeled by

$$S(t) = 55.125t^2 - 6435.607t + 186{,}914.286$$

where $S(t)$ is the amount California spent in millions of dollars t years since 1900.
Source: Models derived from data in the Governor's Budget Summary as printed in the North County Times, *Feb. 9, 2003.*

a. Find a new function that gives the spending per capita (per person) t years since 1900.
b. Find the per capita spending in California in 1995.
c. Find the year in which the per capita spending in California reached $3000 per person.

24. The population of the United States since 2000 can be modeled by

$$P(t) = 2.74t + 282.63$$

where $P(t)$ represents the population of the United States in millions t years since 2000. The national debt of the United States in millions of dollars t years since 2000 can be modeled by

$$D(t) = 540{,}872.6t + 5{,}109{,}943.3$$

where $D(t)$ represents the national debt of the United States in millions of dollars t years since 2000.

National debt as of July 21, 2008, 8:31:30 PM GMT

Source: Models derived from data from the U.S. Department of Commerce Bureau of Economic Analysis.

a. Find a new model for the average amount of national debt per person in the United States t years since 2000.
b. Find the average amount of national debt per person in 2015.
c. Find when the average amount of national debt per person will reach $50,000 per person.

25. The per person spending by the state of Connecticut can be modeled by

$$S(t) = \frac{1100t + 10600}{0.007t + 3.453}$$

where $S(t)$ is Connecticut's spending per person in dollars t years since 2000.
Source: Model derived from data found at www.USgovernmentspending.com.

a. Find the per capita spending in Connecticut in 2008.
b. Find the year in which the per capita spending in Connecticut reached $4000 per person.

26. The average amount of benefits received by people in the U.S. food stamp program can be modeled by

$$B(t) = \frac{-470{,}001t^2 + 411{,}0992t + 1{,}4032{,}612}{-469.4t^2 + 3745t + 19774}$$

where $B(t)$ is the average benefit in dollars per person for people in the U.S. food stamp program t years since 1990.
Source: Model derived from data from Statistical Abstract of the United States, 2001.

Find when the average benefit was $800 per person.

27. In electronics, Ohm's Law states that if the voltage in a circuit is constant, the current that flows through the circuit is inversely proportional to the resistance of that circuit. When the resistance R is 200 ohms, the current I is 1.2 amp. Find the resistance when the current is 0.9 amp.
(*Note:* The proportionality constant in this problem is actually the voltage supplied to the circuit.)

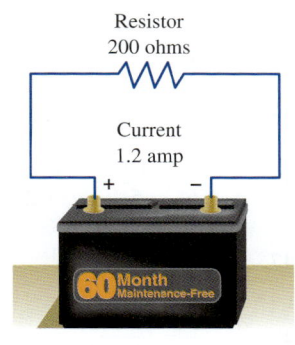

28. The illumination of a light source is inversely proportional to the square of the distance from the light source. If a light has an illumination of 38 foot-candles at a distance of 10 feet from the light source:

a. Find a model for the illumination of this light.
b. What is the illumination of this light at a distance of 5 feet from the light source?
c. How far away from this light source would you need to be for the illumination to be 20 foot-candles?

29. a. If y varies inversely with the cube of x and $y = 11$ when $x = 5$, find an equation to represent this relationship.
b. Find x if $y = 4$.

30. a. If m varies inversely with n^5 and $m = 25$ when $n = 2$, find an equation to represent this relationship.
b. Find m if $n = 4$.
c. Find n if $m = 50$.

31. a. If W varies inversely with p^4 and $W = 135$ when $p = 3$, find an equation to represent this relationship.
b. Find W when $p = 10$.
c. Find p when $W = 400$.

32. a. If y varies inversely with x^3 and $y = 310$ when $x = 2$, find an equation to represent this relationship.
b. Find y when $x = 5$.
c. Find x when $y = 38.75$.

For Exercises 33 through 60, solve and check the answer(s) in the original equation.

33. $\dfrac{x+5}{3x-6} = \dfrac{2}{3}$

34. $\dfrac{2t-3}{5t+6} = \dfrac{11}{17}$

35. $\dfrac{x^2}{x+5} = \dfrac{25}{x+5}$

36. $\dfrac{x^2}{x+6} = \dfrac{36}{x+6}$

37. $\dfrac{w^2}{w-3} = \dfrac{9}{w-3}$

38. $\dfrac{x^2}{x-7} = \dfrac{49}{x-7}$

39. $\dfrac{4}{x+5} = \dfrac{6}{x+8}$

40. $\dfrac{12}{p-4} = \dfrac{2}{p+6}$

41. $\dfrac{2r}{r-3} = \dfrac{9}{r-7}$

42. $\dfrac{b}{b+4} = \dfrac{4b}{b+6}$

43. $\dfrac{12}{x+5} - 2 = \dfrac{-10}{x+3}$

44. $\dfrac{8}{k+3} + 5 = \dfrac{-21}{k-4}$

45. $\dfrac{2}{a+6} + \dfrac{19}{5} = \dfrac{8}{a-2}$

46. $\dfrac{9}{t+3} - \dfrac{1}{8} = \dfrac{4}{t-1}$

47. $\dfrac{v+6}{v-2} = \dfrac{v-1}{v-8}$

48. $\dfrac{x+9}{x-3} = \dfrac{x+4}{x-6}$

49. $\dfrac{d+3}{d+5} = \dfrac{d+8}{d+2}$

50. $\dfrac{n-8}{n-5} = \dfrac{n+9}{n-2}$

51. $\dfrac{2x+5}{x-7} = \dfrac{x-2}{x+3}$

52. $\dfrac{3c-7}{c+5} = \dfrac{2c+3}{c-8}$

53. $\dfrac{x}{x-14} + \dfrac{9}{x-7} = \dfrac{x^2}{x^2-21x+98}$

54. $\dfrac{p}{p-8} + \dfrac{6}{p-4} = \dfrac{p^2}{p^2-12p+32}$

55. $\dfrac{k-2}{k+1} = \dfrac{7k+22}{k^2+6k+5}$

56. $\dfrac{m+4}{m+1} = \dfrac{30}{m^2-2m-3}$

57. $\dfrac{t}{t-6} + \dfrac{5}{t-3} = \dfrac{t^2}{t^2-9t+18}$

58. $\dfrac{x}{x-4} + \dfrac{4}{x-2} = \dfrac{x^2}{x^2-6x+8}$

59. $\dfrac{25}{w^2+w-12} = \dfrac{w+4}{w-3}$

60. $\dfrac{c-3}{c+4} = \dfrac{14}{c^2+6c+8}$

61. Gina can solve all the problems in a chapter in 12 hours. Karen can solve all the problems in a chapter in 16 hours. How long would it take Gina and Karen working together to solve all the problems in the chapter?

62. Fred can stock the shelves of the canned vegetable aisle in 3 hours. Wanda can stock the same shelves in 2 hours. How long would it take Fred and Wanda working together to stock the shelves?

63. Rosemary can mow a large lawn using a riding lawn mower in 4 hours. Will can mow the same lawn using a push mower in 12 hours. How long would it take Rosemary and Will working together to mow the lawn?

64. At a local gravel yard, a large skip loader can move a large pile of gravel in 45 minutes. A smaller skip loader can move the same pile of gravel in 1.25 hours. How long would it take the two skip loaders working together to move the gravel pile?

For Exercises 65 through 74, solve each equation or simplify each expression by performing the indicated operation.

65. $\dfrac{x+3}{x+2} + \dfrac{x+7}{(x+2)(x+5)}$

66. $\dfrac{a}{a+5} + \dfrac{7}{a+2} = \dfrac{-3a}{a^2+7a+10}$

67. $\dfrac{4m+2}{2m^2-5m-3} \cdot \dfrac{5m-15}{m+6}$

68. $\dfrac{h-4}{h^2+4h+3} - \dfrac{h+2}{h^2+5h+6}$

69. $\dfrac{x+8}{x-7} = \dfrac{x+2}{x-6}$

70. $\dfrac{\dfrac{2}{x}+5}{\dfrac{3}{x}-7}$

71. $\dfrac{c+5}{c+4} = \dfrac{12}{c^2+6c+8}$

72. $5 - \dfrac{x+2}{x+7}$

73. $\dfrac{4x}{x+3} = 5 + \dfrac{x}{x+3}$

74. $\dfrac{\dfrac{5}{x+2}-4}{6+\dfrac{2}{x-3}}$

For Exercises 75 through 84, determine whether the equation is linear, quadratic, exponential, logarithmic, rational, or other. Solve the equation. Check your answers in the original equation.

75. $3x^2 + 5x - 10 = 5$

76. $2(1.6)^x = 86$

77. $\dfrac{2}{3}(x+5) = \dfrac{5}{6} + \dfrac{2}{5}x$

78. $2(b-4)^2 + 7 = 15$

79. $\ln(x+3) = 4$

80. $4c^5 - 2400 = 10100$

81. $\dfrac{3x}{x+2} - \dfrac{5}{x-7} = \dfrac{x^2+2}{x^2-5x-14}$

82. $6x^3 + 34x^2 = 56x$

83. $\log(x+2) + \log(x-5) = 1$

84. $\dfrac{5}{x+2} = \dfrac{x+3}{x-7}$

Chapter Summary

Section 7.1 Rational Functions and Variation

- A **rational expression** is anything of the form

$$\frac{P(x)}{Q(x)}$$

  where $P(x)$ and $Q(x)$ are polynomials and $Q(x) \neq 0$.
- Many rational functions come from the division necessary to calculate averages.
- The domain of a rational function that is not restricted by an application is found by setting the denominator equal to zero. The domain will be all real numbers except where the denominator equals zero.
- The variable y varies directly with x if

$$y = kx$$

- The variable y varies inversely with x if

$$y = \frac{k}{x}$$

- For direct and inverse variation, k is called the variation constant and can be any nonzero real number.

Example 1

a. If y varies inversely with x^5 and $y = 7$ when $x = 4$, find an equation to represent this relationship.

b. Find y if $x = 6$.

SOLUTION

a. $y = \dfrac{k}{x^5}$

$7 = \dfrac{k}{4^5}$

$7 = \dfrac{k}{1024}$ Substitute the values for x and y and find k.

$7168 = k$

$y = \dfrac{7168}{x^5}$ Write the equation.

b. $y = \dfrac{7168}{x^5}$

$y = \dfrac{7168}{6^5}$

$y \approx 0.922$

Example 2 Give the domain of the rational function

$$f(x) = \frac{5x + 7}{(x + 3)(x - 7)}$$

SOLUTION The function $f(x) = \frac{5x + 7}{(x + 3)(x - 7)}$ will be undefined if the denominator is equal to zero, so we set the denominator equal to zero and solve. This will give us the values of x that should not be in the domain.

$(x + 3)(x - 7) = 0$ *Set the denominator equal to zero and solve.*

$x + 3 = 0 \quad\quad x - 7 = 0$

$x = -3 \quad\quad\quad x = 7$

The domain of $f(x)$ is all real numbers except $x = -3$ or 7.

Section 7.2 Simplifying Rational Expressions

- To simplify rational expressions, factor the numerator(s) and denominator(s) so that any common factors can be divided out.
- Always use caution in reducing rational expressions. You can reduce only if the entire numerator and denominator is factored (written as a multiplication). If there are any separate terms, you cannot reduce.
- Long division of polynomials is done in the same way you divide large numbers using long division.
- If terms are missing in a divisor or dividend, replace them with a zero term so that the process will have a better flow.
- Synthetic division can be used to simplify long division problems when the divisor can be written in the form $x - c$.

Example 3 Simplify the rational expression $\frac{5x + 20}{x^2 + 5x + 4}$.

SOLUTION

$\dfrac{5x + 20}{x^2 + 5x + 4} = \dfrac{5(x + 4)}{(x + 1)(x + 4)}$ *Factor the numerator and denominator to find common factors that can be divided out.*

$= \dfrac{5\cancel{(x + 4)}}{(x + 1)\cancel{(x + 4)}}$

$= \dfrac{5}{x + 1}$

Example 4 Use long division to divide $(3x^3 - 12x + 144) \div (x + 4)$.

SOLUTION

$$\begin{array}{r}
3x^2 - 12x + 36 \\
x + 4 \overline{\smash{)}\,3x^3 + 0x^2 - 12x + 144} \\
\underline{-(3x^3 + 12x^2)} \\
-12x^2 - 12x \\
\underline{-(12x^2 - 48x)} \\
36x + 144 \\
\underline{-(36x + 144)} \\
0
\end{array}$$

Include a zero term for the missing squared term in the dividend.

Watch the signs when you subtract.

No remainder.

Therefore, $(3x^3 - 12x + 144) \div (x + 4) = 3x^2 - 12x + 36$.

Example 5 Use synthetic division to divide $(6x^3 - 11x^2 - 80x + 112) \div (x - 4)$.

SOLUTION Since the divisor is of the form $x - c$, we know that c is 4.

$$\begin{array}{r|rrrr} 4 & 6 & -11 & -80 & 112 \\ & & 24 & 52 & -112 \\ \hline & 6 & 13 & -28 & 0 \end{array}$$

Therefore, $(6x^3 - 11x^2 - 80x + 112) \div (x - 4) = 6x^2 + 13x - 28$.

Section 7.3 Multiplying and Dividing Rational Expressions

- When multiplying or dividing rational expressions, factor the numerator(s) and denominator(s) so that any common factors can be divided out as the first step.
- Multiply rational expressions in the same way that you multiply fractions. Simplify if possible. Then multiply the numerators together and multiply the denominators together.
- Divide rational expressions in the same way that you divide fractions. Multiply by the reciprocal of the divisor.
- The result from the multiplication or division can be left in factored form.

Example 6 Perform the indicated operation and simplify.

a. $\dfrac{x^2 + 7x + 10}{x + 3} \cdot \dfrac{x - 7}{x^2 + x - 20}$

b. $\dfrac{x^2 + 2x - 15}{2x^2 + 13x + 15} \div \dfrac{x^2 + 4x - 21}{2x^2 + 23x + 63}$

SOLUTION a. $\dfrac{x^2 + 7x + 10}{x + 3} \cdot \dfrac{x - 7}{x^2 + x - 20}$

$\dfrac{(x + 5)(x + 2)}{x + 3} \cdot \dfrac{x - 7}{(x + 5)(x - 4)}$ Factor the numerators and denominators.

$\dfrac{\cancel{(x + 5)}(x + 2)}{x + 3} \cdot \dfrac{x - 7}{\cancel{(x + 5)}(x - 4)}$ Divide out common factors.

$\dfrac{(x + 2)(x - 7)}{(x + 3)(x - 4)}$

b. $\dfrac{x^2 + 2x - 15}{2x^2 + 13x + 15} \div \dfrac{x^2 + 4x - 21}{2x^2 + 23x + 63}$ Multiply by the reciprocal of the divisor.

$\dfrac{x^2 + 2x - 15}{2x^2 + 13x + 15} \cdot \dfrac{2x^2 + 23x + 63}{x^2 + 4x - 21}$ Factor the numerators and denominators.

$\dfrac{(x + 5)(x - 3)}{(2x + 3)(x + 5)} \cdot \dfrac{(2x + 9)(x + 7)}{(x - 3)(x + 7)}$ Divide out the common factors.

$\dfrac{\cancel{(x + 5)}\cancel{(x - 3)}}{(2x + 3)\cancel{(x + 5)}} \cdot \dfrac{(2x + 9)\cancel{(x + 7)}}{\cancel{(x - 3)}\cancel{(x + 7)}}$

$\dfrac{2x + 9}{2x + 3}$ Note that the 2x's do not reduce because they are terms and not factors.

Section 7.4 Adding and Subtracting Rational Expressions

- To add or subtract rational expressions, you need a common denominator.
- The least common denominator (LCD) can be found by factoring all the denominators and taking the highest power of each factor for the LCD.
- Once a common denominator is found, rewrite each fraction in terms of the common denominator by multiplying both the numerator and the denominator by any missing factors of the common denominator.
- To add or subtract rational expressions that have a common denominator, add or subtract the numerators and keep the denominator the same.
- Simplify complex fractions using the LCD of all the fractions in the complex rational expression.

Example 7 Perform the indicated operation and simplify.

a. $\dfrac{x+2}{x^2-3x-28} + \dfrac{x-3}{x^2-9x+14}$

b. $\dfrac{x+7}{x^2+8x+15} - \dfrac{x+4}{2x^2+6x-20}$

SOLUTION

a.
$$\dfrac{x+2}{x^2-3x-28} + \dfrac{x-3}{x^2-9x+14}$$

$$\dfrac{x+2}{(x+4)(x-7)} + \dfrac{x-3}{(x-2)(x-7)} \quad \text{Factor the denominators to find the LCD.}$$

$$\dfrac{x-2}{x-2} \cdot \dfrac{x+2}{(x+4)(x-7)} + \dfrac{x-3}{(x-2)(x-7)} \cdot \dfrac{x+4}{x+4} \quad \text{Multiply by the missing factors.}$$

$$\dfrac{x^2-4}{(x-2)(x+4)(x-7)} + \dfrac{x^2+x-12}{(x-2)(x+4)(x-7)}$$

$$\dfrac{2x^2+x-16}{(x-2)(x+4)(x-7)} \quad \text{Add the numerators.}$$

b.
$$\dfrac{x+7}{x^2+8x+15} - \dfrac{x+4}{x^2+3x-10}$$

$$\dfrac{x+7}{(x+3)(x+5)} - \dfrac{x+4}{(x-2)(x+5)} \quad \text{Factor the denominators to find the LCD.}$$

$$\dfrac{x-2}{x-2} \cdot \dfrac{x+7}{(x+3)(x+5)} - \dfrac{x+4}{(x-2)(x+5)} \cdot \dfrac{x+3}{x+3} \quad \text{Multiply by the missing factors.}$$

$$\dfrac{x^2+5x-14}{(x-2)(x+3)(x+5)} - \dfrac{x^2+7x+12}{(x-2)(x+3)(x+5)}$$

$$\dfrac{-2x-26}{(x-2)(x+3)(x+5)} \quad \text{Subtract the numerators.}$$

$$\dfrac{-2(x+13)}{(x-2)(x+3)(x+5)}$$

Section 7.5 Solving Rational Equations

- To solve an equation that contains rational expressions, follow these steps:
  1. State what values should be excluded.
  2. Multiply both sides of the equation by the least common denominator.
  3. Solve the remaining equation.
  4. Check the answer(s) in the original equation.
- Keep in mind when solving rational equations that some potential answers will not work. Always check for division by zero in the original equation.

Example 8 Solve the following equations.

a. $\dfrac{3}{x+5} = 4 - \dfrac{7}{x+5}$

b. $\dfrac{2x}{x+3} + \dfrac{5}{x-7} = \dfrac{8x-6}{x^2-4x-21}$

SOLUTION

a.
$\dfrac{3}{x+5} = 4 - \dfrac{7}{x+5}$ The denominator equals zero when $x = -5$, so it will be an excluded value.

$(x+5) \cdot \dfrac{3}{x+5} = \left(4 - \dfrac{7}{x+5}\right) \cdot (x+5)$ Multiply by the common denominator.

$(x+5) \cdot \dfrac{3}{x+5} = 4(x+5) - \dfrac{7}{x+5} \cdot (x+5)$ Distribute and divide out common factors.

$\cancel{(x+5)} \cdot \dfrac{3}{\cancel{x+5}} = 4(x+5) - \dfrac{7}{\cancel{x+5}} \cdot \cancel{(x+5)}$

$3 = 4x + 20 - 7$ Solve the remaining equation.

$3 = 4x + 13$

$-10 = 4x$

$-2.5 = x$

X	Y₁	Y₂
-2.5	1.2	1.2

X=

Check your answer.
The answer works.

b.
$\dfrac{2x}{x+3} + \dfrac{5}{x-7} = \dfrac{8x-6}{x^2-4x-21}$ The denominator equals zero when $x = -3$ or 7, so these will be excluded values.

$\dfrac{2x}{x+3} + \dfrac{5}{x-7} = \dfrac{8x-6}{(x+3)(x-7)}$ Factor to find the LCD. Multiply both sides by the LCD.

$(x+3)(x-7) \cdot \left(\dfrac{2x}{x+3} + \dfrac{5}{x-7}\right) = \dfrac{8x-6}{(x+3)(x-7)} \cdot (x+3)(x-7)$

$(x+3)(x-7) \cdot \dfrac{2x}{x+3} + (x+3)(x-7) \cdot \dfrac{5}{x-7}$

$= \dfrac{8x-6}{(x+3)(x-7)} \cdot (x+3)(x-7)$ Distribute.

$\cancel{(x+3)}(x-7) \cdot \dfrac{2x}{\cancel{x+3}} + (x+3)\cancel{(x-7)} \cdot \dfrac{5}{\cancel{x-7}}$

$= \dfrac{8x-6}{\cancel{(x+3)(x-7)}} \cdot \cancel{(x+3)(x-7)}$ Simplify.

$2x(x-7) + 5(x+3) = 8x - 6$ Solve the remaining equation.

$2x^2 - 14x + 5x + 15 = 8x - 6$

$2x^2 - 17x + 21 = 0$

$(2x - 3)(x - 7) = 0$

$2x - 3 = 0 \quad\quad x - 7 = 0$

$x = \dfrac{3}{2} \quad\quad x = 7$ The 7 causes division by zero.

$x = 1.5$

X	Y₁	Y₂
1.5	-.2424	-.2424
7	ERROR	ERROR

X=

Check your answers.
Only one answer works.

Chapter Review Exercises

1. The illumination of a light source is inversely proportional to the square of the distance from the light source. A certain light has an illumination of 40 foot-candles at a distance of 30 feet from the light source.
   a. Find a model for the illumination of this light.
   b. What is the illumination of this light at a distance of 40 feet from the light source?
   c. At what distance would the illumination of this light be 50 foot-candles? [7.1]

2. The cost for making an international phone call using a prepaid calling card is directly proportional to the number of minutes the call lasts. A certain call that lasted 7 minutes cost $1.89.
   a. Find a model for the cost of an international call using this calling card.
   b. What is the cost of a 30-minute international call using this calling card?
   c. How long of an international call could you make using a calling card that cost $20? [7.1]

3. a. If y varies directly with x^2 and $y = 144$ when $x = 6$, find an equation to represent this relationship.
   b. Find y if $x = 2$. [7.1]

4. a. If y varies inversely with x^5 and $y = 7$ when $x = 2$, find an equation to represent this relationship.
   b. Find y if $x = 4$. [7.1]

5. Math R Us (MRU) is having a math competition and wants to have a total of $500 in prize money. To run the competition, MRU will need to have the prize money plus $7.50 per person competing in the competition. They want to set an entrance fee for each competitor that will cover their costs. MRU plans to give 10 scholarships to people who cannot afford to pay.
   a. Find a model that gives the entrance fee MRU should charge if they have p people compete. (p includes the 10 people who are on scholarship and, therefore, will not pay an entrance fee.) [7.1]
   b. Use your model to estimate the entrance fee MRU should charge if 100 people compete. [7.1]
   c. If MRU wants to have an entrance fee of $10, how many people do they need to participate in this competition? [7.5]

6. The per capita amount of cheese consumed by Americans can be modeled by
$$C(t) = \frac{637.325t + 7428.649}{0.226t^2 + 10.925t + 332.82}$$
where $C(t)$ represents the per capita amount of cheese consumed by Americans in pounds per person t years since 1990.

Source: Models derived from data in Statistical Abstract of the United States, 2001.
   a. Estimate the per capita amount of cheese consumed by Americans in 1995. [7.1]
   b. Estimate when the per capita amount of cheese consumed by Americans will reach 30 pounds per person. [7.5]

For Exercises 7 through 12, find the domain of the given rational function.

7. $f(x) = \dfrac{3x + 2}{x - 9}$ [7.1]

8. $g(x) = \dfrac{x + 5}{(x + 5)(x - 3)}$ [7.1]

9. $h(x) = \dfrac{5x + 2}{2x^2 - 7x - 15}$ [7.1]

10. $f(x) = \dfrac{x + 3}{7x^2 + 5x + 25}$ [7.1]

11. [7.1]

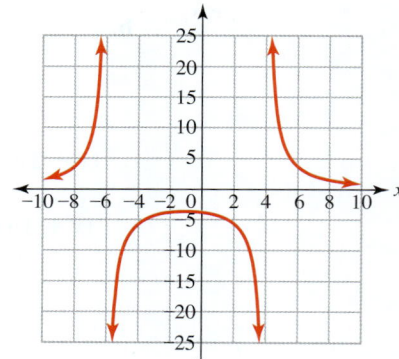

12. [7.1]

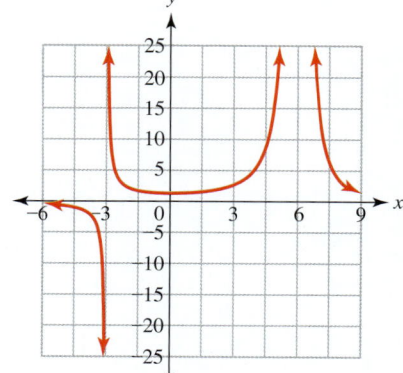

For Exercises 13 through 16, simplify the rational expressions.

13. $\dfrac{24x^8 + 16x^5 - 30x^2}{4x^3}$ [7.2]

14. $\dfrac{4-m}{(m+5)(m-4)}$ [7.2]

15. $\dfrac{4a+8}{a^2+6a+8}$ [7.2]

16. $\dfrac{6p^2+25p+14}{2p^2+25p+63}$ [7.2]

For Exercises 17 through 20, use long division to divide.

17. $(4h^2+17h-15) \div (h+5)$ [7.2]

18. $(b^2+3b+5) \div (b-4)$ [7.2]

19. $(3x^3+20x^2-3x-10) \div (3x+2)$ [7.2]

20. $(c^4+3c^3+c^2+15c-20) \div (c^2+5)$ [7.2]

For Exercises 21 through 24, use synthetic division to divide.

21. $(2t^2-t-15) \div (t-3)$ [7.2]

22. $(5r^2+17r-10) \div (r+4)$ [7.2]

23. $(4x^3+15x^2+12x-4) \div (x+2)$ [7.2]

24. $(5y^3-20y-75) \div (y-3)$ [7.2]

For Exercises 25 through 38, perform the indicated operation and simplify the rational expressions.

25. $\dfrac{x+3}{x-15} \cdot \dfrac{x-15}{x-2}$ [7.3]

26. $\dfrac{4}{m+3} \cdot \dfrac{m+9}{m-7}$ [7.3]

27. $\dfrac{x+2}{x-3} \div \dfrac{x+7}{x-3}$ [7.3]

28. $\dfrac{h-5}{h+2} \div \dfrac{5-h}{h-4}$ [7.3]

29. $\dfrac{(d+3)(d-4)}{(d-4)(d+2)} \cdot \dfrac{(d+2)(d-1)}{(d+2)(d-3)}$ [7.3]

30. $\dfrac{(x+4)(x+5)}{(x-5)(x+2)} \div \dfrac{(x+5)(x-7)}{(x+2)(x-7)}$ [7.3]

31. $\dfrac{2n+6}{n^2+8n+15} \cdot \dfrac{3n+15}{n^2+11n+28}$ [7.3]

32. $\dfrac{v+5}{v^2-6v-55} \div \dfrac{v-8}{v^2-4v-77}$ [7.3]

33. $\dfrac{12}{x+6} + \dfrac{4x+3}{x+6}$ [7.4]

34. $\dfrac{3}{h-5} - \dfrac{7}{h+2}$ [7.4]

35. $\dfrac{x+2}{(x+3)(x+5)} + \dfrac{6}{(x+3)(x-7)}$ [7.4]

36. $\dfrac{2a+5}{(a+1)(a-4)} - \dfrac{3}{(a+3)(a+1)}$ [7.4]

37. $\dfrac{2t}{3t^2-13t-10} - \dfrac{t+1}{3t^2+14t+8}$ [7.4]

38. $\dfrac{2x-7}{x^2-5x-14} + \dfrac{x+5}{x^2-4x-21}$ [7.4]

For Exercises 39 through 42, simplify the complex fractions.

39. $\dfrac{3+\dfrac{4}{a}}{4-\dfrac{7}{a}}$ [7.4]

40. $\dfrac{5-\dfrac{7}{4n}}{2+\dfrac{6}{n}}$ [7.4]

41. $\dfrac{2+\dfrac{4}{x+3}}{\dfrac{5}{x-2}}$ [7.4]

42. $\dfrac{\dfrac{5}{t+2}-\dfrac{3}{t+5}}{\dfrac{1}{t+5}+\dfrac{3}{t-2}}$ [7.4]

For Exercises 43 through 54, solve the rational equations. Check your answers with a table or graph.

43. $\dfrac{6}{m+2} = \dfrac{2m}{m+2}$ [7.5]

44. $\dfrac{5}{c-3} = \dfrac{2}{c+7}$ [7.5]

45. $\dfrac{4x}{x+3} = \dfrac{2x}{x-9}$ [7.5]

46. $\dfrac{2}{b} = 3 + \dfrac{7}{4b}$ [7.5]

47. $\dfrac{2}{x+5} + 6 = \dfrac{-5}{x-3}$ [7.5]

48. $\dfrac{3}{h+2} + 4 = \dfrac{-20}{h-5}$ [7.5]

49. $\dfrac{x}{x+5} + \dfrac{7}{x-3} = \dfrac{56}{(x+5)(x-3)}$ [7.5]

50. $\dfrac{w}{w-7} - \dfrac{3}{w-4} = \dfrac{9}{(w-7)(w-4)}$ [7.5]

51. $\dfrac{5x+3}{x^2+7x-9} = \dfrac{10x-12}{x^2+7x-9}$ [7.5]

52. $\dfrac{3k}{k+2} + \dfrac{2}{k+5} = \dfrac{50}{k^2+7k+10}$ [7.5]

53. $\dfrac{5}{x^2+3x-28} = \dfrac{3x}{x^2-8x+16}$ [7.5]

54. $\dfrac{b}{2b^2+9b+10} = \dfrac{1}{b^2+b-2}$ [7.5]

55. Sam can put up a fence in 20 hours. Craig can put up the same fence in only 16 hours. How long would it take Sam and Craig working together to put up the fence? [7.5]

56. Mark can write a book in 6 years. Cindy can write a book in 4 years. How long would it take Mark and Cindy working together to write the book? [7.5]

Chapter Test

1. The illumination of a light source is inversely proportional to the square of the distance from the light source. A certain light has an illumination of 30 foot-candles at a distance of 12 feet from the light source.
   a. Find a model for the illumination of this light.
   b. What is the illumination of this light at a distance of 7 feet from the light source?
   c. At what distance would the illumination of this light be 20 foot-candles?

2. The length of a wall is directly proportional to the drawing of that wall on a blueprint. A 7.5-foot wall is represented by a 3-inch drawing on the blueprint.
   a. Find a model for the length of a wall if you know the length of the drawing on the blueprint.
   b. What is the length of a wall that is represented by a 5-inch drawing on the blueprint?
   c. How long should the drawing be if the wall is supposed to be 20-feet long?

For Exercises 3 through 9, perform the indicated operation and simplify the rational expressions.

3. $\dfrac{2x}{x+7} \cdot \dfrac{x+3}{x-4}$

4. $\dfrac{(x+3)(x+5)}{(x-7)(x+3)} \div \dfrac{(x+5)(x+2)}{(x+4)(x-7)}$

5. $\dfrac{5w+3}{w^2-4w-21} \cdot \dfrac{2w-14}{5w^2-17w-12}$

6. $\dfrac{2m+4}{m^2-m-20} \div \dfrac{m+2}{m^2-4m-5}$

7. $\dfrac{5x}{2x-7} + \dfrac{3x-8}{2x-7}$

8. $\dfrac{x+5}{x-3} - \dfrac{x-2}{x+4}$

9. $\dfrac{5x+2}{x^2+5x+6} + \dfrac{x-4}{x^2+9x+14}$

For Exercises 10 and 11, simplify the complex fractions.

10. $\dfrac{2 + \dfrac{5}{6d}}{4 - \dfrac{8}{d}}$

11. $\dfrac{\dfrac{2}{x+1} + 3}{\dfrac{5}{x+1} - \dfrac{4}{x-1}}$

12. Find the domain for the rational functions.
   a. $h(x) = \dfrac{3x-7}{x^2+6x-27}$

   b.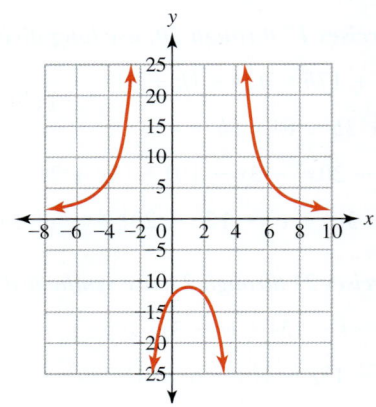

13. Divide $(5x^3 + 22x^2 - 17x - 10) \div (5x + 2)$

14. Use synthetic division to divide $(x^4 + 5x^3 + 3x^2 + 19x + 20) \div (x + 5)$

For Exercises 15 through 19, solve the rational equations. Check your answer(s) in the original equation.

15. $\dfrac{2x+5}{x-7} = \dfrac{3x-12}{x-7}$

16. $\dfrac{5}{x+6} = \dfrac{3}{x-4}$

17. $\dfrac{5}{x+2} + 7 = \dfrac{-16}{x-5}$

18. $\dfrac{x^2}{(x+2)(x+3)} + \dfrac{5}{x+3} = \dfrac{4}{x+2}$

19. $\dfrac{5.6}{x^2-3x-10} = \dfrac{2x}{x^2+9x+14}$

20. The total labor force in the state of Florida can be modeled by
$$L(t) = 141.5t + 7826.8$$
where $L(t)$ represents the labor force in thousands in the state of Florida t years since 2000. The number of people unemployed in Florida can be modeled by
$$E(t) = 80.5t + 287.5$$
where $E(t)$ represents the number of people in the workforce who are unemployed in thousands in the state of Florida. The unemployment rate is the percentage of people in the labor force who are not employed.
Source: Models derived from data obtained from the Florida Research and Economic Database 2003.

   a. Find a model for the unemployment rate for the state of Florida.
   b. Estimate the unemployment rate in Florida during 2003.
   c. Estimate when the unemployment rate in Florida will reach 8%.

Chapter 7 Projects

How Bright Is That Light?

Group Experiment
Three or more people

What you will need
- A tape measure
- A flashlight
- A Texas Instruments CBL unit with a light sensor probe

In this project, you will investigate the inverse relationship between the illumination of a light and the square of the distance from the light source. Fix the position of the flashlight and place the tape measure out from the front of the flashlight. Using the tape measure, collect eight or more data values for the light's illumination, by placing the CBL light sensor probe at different distances. Measure your distances in feet, but remember that you can use fractions of feet in your measurements. Be sure to turn off any room lights you can, and try not to allow other light sources interfere with your measurements.

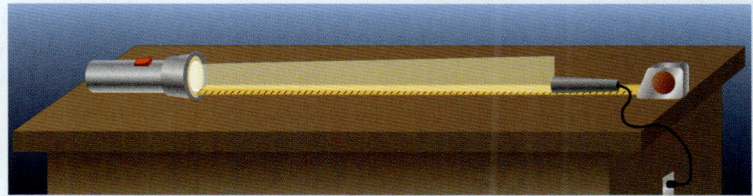

Write up
a. Describe how you collected the data. What was the most important thing to do in order to get good data? Explain.
b. Create a scattergram of the data on the calculator or computer and print it out or draw it neatly by hand on graph paper.
c. Use the inverse relationship to find a model to fit the data.
d. What is the variation constant for the light source in your experiment? If someone else in your class also did this experiment, how do the variation constants compare?
e. What is the vertical intercept of your model and what does it represent in this experiment?
f. What is a reasonable domain and range for your model? (Remember to consider any restrictions on the experiment.)
g. Describe where this may be relevant in the real world.

Drain That Bucket

Group Experiment
Three or more people

What you will need
- A 1-gallon milk container or 2-liter bottle.
- Something to make holes in the container
- Two plugs for the holes
- A stopwatch or timer

In this project, you will explore the value of things working together. Using a 1-gallon container, you will need to put two different-sized holes in the bottom of the container and plug them closed to start. Fill the container with water and record the time for how long it takes for one of the holes to empty the container. Repeat the experiment with only the second hole open. Finally, repeat the experiment a third time with both holes open at the same time.

Write up
a. Describe the results from each of the three runs of the experiment.
b. Which hole took the least amount of time to drain the container?
c. Did having both holes open make the container drain more quickly? Is this what you expected to happen?

d. Let A be the time it takes hole one to drain the container.
Let B be the time it takes hole two to drain the container.
Using the results from your experiments, fill in the values for the equation

$$\frac{1}{A} + \frac{1}{B} = \frac{1}{t}$$

e. Solve the equation from part d for t. Here, t is the time it should have taken both holes working together to drain the container. How does the value you found for t compare to what you found when you ran the experiment the third time? Explain.

f. Describe where this application would be relevant in the real world.

Explore Your Own Rational Function

Research Project
One or more people

What you will need
- Find a real-world situation that can be modeled with a rational function.
- You might want to use the Internet or library. Statistical abstracts and some journals and scientific articles are good resources for real-world data.
- Follow the MLA style guide for all citations. If your school recommends another style guide, use that.

In this project, you are given the task of finding and exploring a real-world situation that you can model using a rational function. You may use the problems in this chapter to get ideas of things to investigate, but your data should not be discussed in this textbook. Some items that you might wish to investigate would be formulas from physics and other sciences. Remember that you can look for inversely proportional relationships as one example.

Write up

a. Describe the real-world situation you found and where you found it. Cite any sources you used following the MLA style guide or the style guide that your school recommends.

b. Either run your own experiment or find data that can be used to verify the relationship you are investigating.

c. Create a scatterplot of the data on the calculator or computer and print it out or draw it neatly by hand on graph paper.

d. Find a model to fit the data.

e. Where are the vertical asymptotes of your model?

f. What is a reasonable domain for your model?

g. Use your model to estimate an output value of your model for an input value that you did not collect in your original data.

h. Use your model to estimate for what input value your model will give you a specific output value that you did not collect in your original data.

Review Presentation

Research Project
One or more people

What you will need
- This presentation may be done in class or as a video.
- If it is done as a video, you might want to post the video on YouTube or another site.
- You might want to use problems from the homework set or review material in this book.
- Be creative, and make the presentation fun for students to watch and learn from.

Create a 5-minute presentation or video that reviews a section of Chapter 7. The presentation should include the following:

- Examples of the important skills presented in the section
- Explanation of any important terminology and formulas
- Common mistakes made and how to recognize them

Radical Functions

8

- **8.1** Radical Functions
- **8.2** Simplifying, Adding, and Subtracting Radicals
- **8.3** Multiplying and Dividing Radicals
- **8.4** Solving Radical Equations
- **8.5** Complex Numbers

When riding a bike or driving a car, how fast you are going will determine how sharp a turn you can make. Cyclists in the Tour de France and other top races can reach downhill speeds in excess of 50 miles per hour. When navigating their way down the back side of a climb, cyclists must be aware of the speed at which they can enter a corner and still maintain control of the bike. In this chapter, we will discuss how radical functions relate to a variety of life applications such as skid marks made by a car. Also, in one of the chapter projects, you will investigate how radical functions relate to the turning radius at different speeds.

8.1 Radical Functions

LEARNING OBJECTIVES

- Use radicals that model applications.
- Find the domain and range of a radical function in a context.
- Recognize the graphs of radicals with odd and even indexes.
- Find the domain and range of a radical function.
- Graph radical functions.

In this chapter, we will investigate and work with a new type of function that involves using square roots and other higher roots. We will start by looking at some situations and data that are best modeled by using square roots or other powers that result in radical expressions. The name **radical** is given to the symbol $\sqrt{}$, and the expression inside of the radical is called the **radicand**. The word *radical* is also used to describe any function that uses a radical with variables in the radicand. When we want to represent a root other than a square root, we indicate that by using an **index** in the nook of the radical symbol. Whenever the index is higher than 2, the radical is considered a higher root.

> **DEFINITION**
>
> **Radical Expression**
>
> $$\sqrt{x} \qquad \sqrt[n]{x}$$
>
> A square root or *n*th root is called a radical expression. In these examples, *x* is called the radicand, and *n* is the index. Square roots have an index of 2, but the 2 is not written in the nook of the radical.
>
>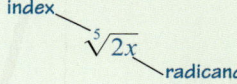

We will give the formal definitions of square roots and higher roots in Section 8.2. To start our investigation of radical functions, let's consider the following data.

Radical Functions That Model Data

■ CONCEPT INVESTIGATION

What shape is that?

1. Fill in the missing parts to this table of data.

x	0	1	4	9	16	25	36	49	100	144	196
$f(x)$	0	1	2	3	4	5					

2. Describe the pattern in the table above.

3. Create a scatterplot for these data and describe its shape compared to the shapes of other model types you have learned.

4. Graph the following models and describe their fit to these data.

$$y = 0.0666x + 2.305$$
$$y = -0.0003139x^2 + 0.1233x + 1.4236$$

5. Graph $f(x) = \sqrt{x}$ with the data. How well does this fit the data?

6. Change the window to Xmin = −50, Xmax = 250, Ymin = −5, and Ymax = 20, Xscl = 25, Yscl = 5.
   Does this model exist for negative values of x? Why or why not?

7. Does this model have negative output values? Why or why not?

Now that we have explored a radical model, let's consider the following applications of some different types of radicals.

Example 1 A radical model from zoology

Allometry is the study of how some aspect of the physiology of a certain species of animal changes in proportion with a change in body size. For example, a study of neotropical butterflies at the Department of Zoology, University of Texas–Austin, studied the relationship of the airspeeds during natural free flight and several characteristics of the butterflies. One relationship that was studied was between the body mass (in grams) of the butterfly and its mean forward airspeed, velocity (in meters per second). A sample of the data collected is given in the table.

Body Mass (g)	Velocity (m/s)
0.1	2.75
0.13	3.14
0.16	3.46
0.21	3.99
0.25	4.36
0.3	4.77
0.4	5.49
0.63	6.91

Source: The Journal of Experimental Biology, 191, 125–139 (1994) as found at jeb.biologist.org.

a. Create a scatterplot of the data in the table.

b. An estimated model for these data is given by the following. Let $V(m)$ be the mean forward airspeed in meters per second for neotropical butterflies and let m be the body mass of the butterfly in grams.

$$V(m) = 8.7\sqrt{m}$$

   i. Graph this model with the given data.
   ii. How well does it fit the data?
   iii. Describe the general shape of the graph.

c. Estimate the airspeed of a butterfly with a body mass of 0.5 gram.

d. Give a reasonable domain and range for this model.

e. Use the graph to estimate the body mass of a butterfly that has an airspeed of 5 meters per second.

SOLUTION

a.

b. i.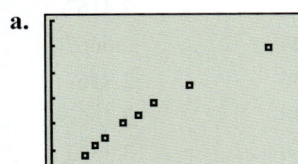

   ii. The graph seems to follow the pattern the data are distributed in.
   iii. This graph is increasing and is slightly curved and, thus, is not linear.

c. We are given the body mass of the butterfly, so we can let $m = 0.5$, and we get

$$V(0.5) = 8.7\sqrt{0.5}$$
$$V(0.5) \approx 6.15$$

Therefore, a butterfly with a body mass of 0.5 gram would have a mean forward airspeed of approximately 6.15 meters per second.

d. Because the input variable represents the body mass of a butterfly, we should choose only small positive numbers. Therefore, one possible domain could be [0.08, 0.75]. Because the range will be the lowest point to highest point within the domain, we would get a range of [2.46, 7.53].

e. Using the graph and trace, we get the following.

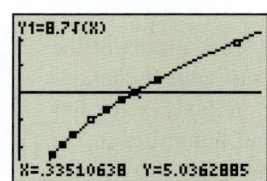

Therefore, a butterfly with a body mass of about 0.34 gram will have an airspeed of about 5 meters per second.

The model in Example 1 is a radical function that contains a square root and can be easily evaluated by using your calculator. The basic shape of this type of data is an increasing function that is slightly curved. We can see that the graph is shaped very similarly to a logarithm's graph. The main difference is that it grows more quickly than a log would and does not have a vertical asymptote as a log does. In the situation of the butterfly's velocity (airspeed), the outputs would be only positive.

The domain of a radical function that is used in a context can be found in the same way that we have done throughout the text. Start by trying to expand the domain beyond the given data, avoiding model breakdown and any restrictions that the problem states. The range will then be the lowest to highest output values within that domain. The range of a radical function will again have its output values come from the endpoints of the domain. Let's look at another example that uses a higher root as part of the radical function.

■ Using Your TI Graphing Calculator

When entering an equation with a square root into your Y= screen, you should be careful to use parentheses around the expression inside the radical.

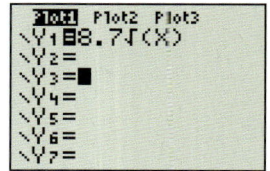

In entering higher roots such as cube roots, it is easiest to use rational exponents, as we did in Section 3.1, to represent the radical.

$$\sqrt[3]{x} = x^{\frac{1}{3}}$$
$$\sqrt[4]{m} = m^{\frac{1}{4}}$$

Use parentheses around any expressions you want to have in the radical and parentheses around the rational exponent.

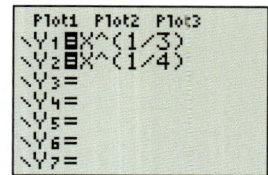

Note that higher roots can also be put in the calculator using the math menu and using the

$$\sqrt[3]{(} \quad \sqrt[x]{}$$

options. You need to enter the index of the radical you want before using the $\sqrt[x]{}$ option.

Example 2 Growth rate application

Pediatricians often use growth charts to compare a child's height and weight to those of the average child of the same age. This helps doctors gauge the growth and development of their patients compared to what is expected in general. The Internet provides parents with several samples of growth charts with heights and weights for girls and boys. Data from one of these charts for girls are given in the table.

Age (months)	Height (feet)
4.5	2.0
10.5	2.1667
12	2.4167
24	2.6667
36	2.9167
48	3.1667
60	3.3333

Source: "The Wellness Site," aarogya.com.

a. Create a scatterplot for these data.

b. Let $H(a)$ be the average height of a girl in feet at age a months old.

  i. Graph the function $H(a) = 1.44\sqrt[5]{a}$ with your data.

  ii. How well does this function fit these data?

c. Estimate the average height of girls who are 18 months old.

d. Estimate the average height of girls who are $2\frac{1}{2}$ years old.

e. What would be a reasonable starting value for the domain of this function?

f. Estimate numerically at what age the average height of a girl would be 2.5 feet.

SOLUTION

a.
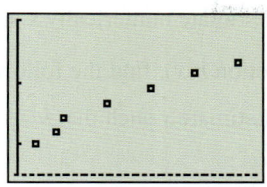

b. i. Note the radical was put into the equation editor using fraction exponents.

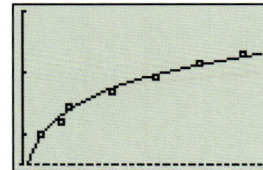 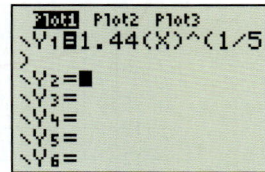

ii. This function follows the pattern in the data.

c. We are given the age of the girls, so we can substitute $a = 18$ and get

$$H(18) = 1.44\sqrt[5]{18}$$
$$H(18) \approx 2.5669$$

If we convert the 0.5669 to inches by multiplying by 12, we get $0.5669 \cdot 12 = 6.8$ inches. Therefore, the average height of 18-month-old girls is about 2 feet 7 inches.

d. We are told that the girls are $2\frac{1}{2}$ years old, so we can substitute $a = 30$ months and get

$$H(30) = 1.44\sqrt[5]{30}$$
$$H(30) \approx 2.8431$$

If we convert the 0.8431 into inches, we get $0.8431 \cdot 12 = 10.1$ inches. Therefore, the average height of $2\frac{1}{2}$ year old girls is about 2 feet 10 inches.

e. Because we are talking about the average height of girls at age a, the domain will be the ages of girls in months. Therefore, negative values would not make sense, and $a = 0$ would result in an average height of zero feet, so that also is model breakdown. We should probably start this domain at about 3 months of age. This gives us a reasonable height of about 21.5 inches.

f. Using the table and starting at 12 months, we get the following.

X	Y1
12	2.367
13	2.4052
14	2.4411
15	2.475
16	2.5072
17	2.5378

Therefore, the average height of 16-month-old girls is 2.5 feet.

Both of these examples require us to evaluate the function for a given value of the input variable, which is easily done on the calculator. We have also solved these functions using a graph or table. To solve these types of models algebraically for a missing value of the input variable, we will need several other skills that we will be learning in the next few sections. In Section 8.4, we will come back to some of these examples and solve them algebraically for the input variables.

Example 3 — Working with radical functions

a. Given the function $f(x) = 4.6\sqrt[5]{x}$, find the following.

  i. $f(84)$
  ii. Estimate numerically x such that $f(x) = 14$.

b. Given the graph of the function $h(x)$, find the following.

  i. Estimate $h(0)$.
  ii. Estimate x such that $h(x) = -6$.

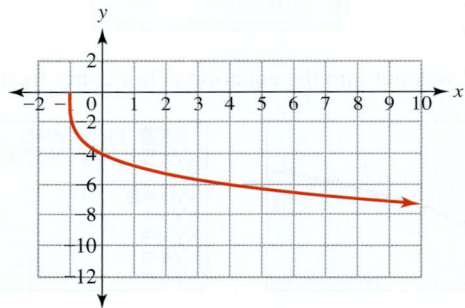

SOLUTION

a. i. Enter the equation into the calculator using a fraction exponent for the radical. Evaluate the function at $x = 84$.

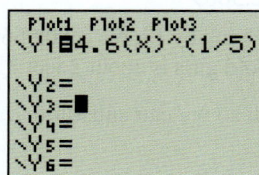

Therefore, $f(84) \approx 11.159$.

ii. Use the table to try values of x. We know that it is more than 84, since that gave us approximately 11.

Therefore, $f(x) = 14$ when $x \approx 261$.

b. Reading the graph, we can find the values we want.

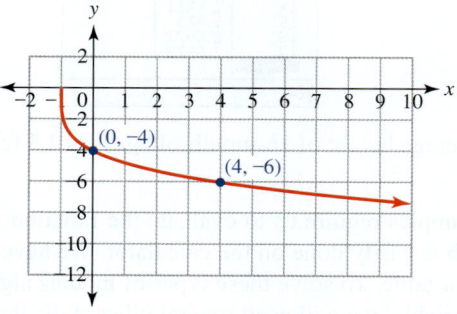

  i. $h(0) = -4$.
  ii. $h(x) = -6$ when $x = 4$.

PRACTICE PROBLEM FOR EXAMPLE 3

a. Given the function $f(x) = 7.2\sqrt[3]{x}$, find the following.

   i. $f(20)$ ii. Estimate numerically when $f(x) = 34$.

b. Given the graph of the function $h(x)$, find the following.

   i. Estimate $h(2)$. ii. Estimate $h(x) = -8$.

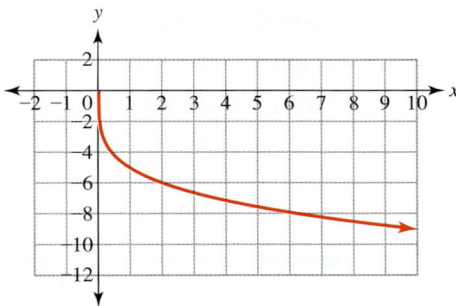

Domain and Range of Radical Functions

Let's look at the graphs of several radical functions and consider the domain and range of those functions in problems that are not restricted by an application.

CONCEPT INVESTIGATION

The index is odd or even—does it matter?

Set your graphing calculator's window to Xmin = −10, Xmax = 10, Ymin = −3.5, and Ymax = 3.5.

1. Graph the following functions on your graphing calculator.

   a. $f(x) = \sqrt{x}$

   b. $g(x) = \sqrt[4]{x}$ Remember to enter the index with fraction exponents.

   c. $h(x) = \sqrt[6]{x}$

2. Describe the shape of these graphs.

3. How does the graph change as you take higher roots?

4. What appear to be the domain and range for these functions?

5. Graph the following functions on your graphing calculator.

   a. $f(x) = \sqrt[3]{x}$ b. $g(x) = \sqrt[5]{x}$ c. $h(x) = \sqrt[7]{x}$

6. Describe the shape of these graphs.

7. How does the graph change as you get higher roots?

8. What appear to be the domain and range for these functions?

9. What is the difference between domain and range of the radical functions in part 1 and the radical functions in part 5?

10. Explain why the difference you described in part 9 occurs.

As you can see from this concept investigation, whether the root is odd or even makes a difference in the shape of the graph as well as what its domain and range are. You should note that in the real number system, even roots cannot have negative numbers under the radical and have only values that make the radicand nonnegative as part of their domain. Odd roots do not have this restriction because negatives are possible under an odd root and typically have all real numbers as a domain. We will study what happens when negatives are under a square root in the last section of this chapter.

We can find the domain of a radical function by first determining whether the radical is an even or odd root. If the root is even we set the radicand greater than or equal to zero and solve. If the root is odd, then we know the domain is all real numbers. It is easiest to determine the range of a radical function by using the graph.

> **Finding the Domain and Range of Radical Functions**
>
> For **even roots**, set the radicand greater than or equal to zero.
>
> $$f(x) = \sqrt{x + 10}$$
>
> Domain
>
> $$x + 10 \geq 0$$
> $$x \geq -10$$
>
> Look at the graph of the function to determine the lowest and highest output values of the function for the range.
>
> For **odd roots**, the domain and range are both all real numbers.

What's That Mean?

Nonnegative

We say that something must be nonnegative when we want to include zero with all the positive numbers. If we say *positive*, it does not include the number zero. In an even root, the radicand can be zero.

$$0^2 = 0$$
$$\sqrt{0} = 0$$

Therefore, we say that the radicand must be nonnegative or greater than or equal to zero. Both of these statements include zero with the positive numbers.

Example 4   Domain and range of radical functions

Give the domain and range of the following radical functions.

a. $f(x) = \sqrt{x + 3}$
b. $g(x) = -\sqrt{x - 9}$
c. $f(x) = \sqrt[3]{x - 7}$
d. $h(x) = \sqrt{-x + 5}$

SOLUTION

a. The function $f(x) = \sqrt{x + 3}$ is an even root, so the domain can include only values that make the radicand nonnegative. Solve the following inequality.

$$x + 3 \geq 0$$
$$x \geq -3$$

Therefore, the domain of this function is $x \geq -3$. The lowest point on the graph is zero. The graph continues upward to infinity, so we get a range of $[0, \infty)$. We can confirm this domain and range by looking at its graph on the calculator.

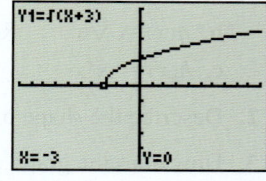

b. The function $g(x) = -\sqrt{x - 9}$ is an even root, so the domain can include only values that make the radicand nonnegative. Solve the following inequality.

$$x - 9 \geq 0$$
$$x \geq 9$$

Therefore, the domain of this function is $x \geq 9$. The graph increases from negative infinity up to the highest point on the graph (9, 0), so we get a range of $(-\infty, 0]$. We can confirm this domain and range by looking at its graph on the calculator.

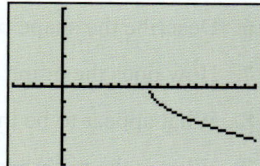

c. The function $f(x) = \sqrt[3]{x - 7}$ is a cube root, which means that it is an odd root. The radicand may be either positive or negative, so the domain would include all real numbers. The range will also be all real numbers because the graph goes down to negative infinity and up to positive infinity. Looking at the graph of this function confirms this.

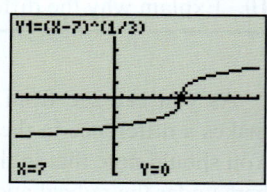

d. Because the function $h(x) = \sqrt{-x + 5}$ is an even root, the radicand must be nonnegative. Although the radicand looks negative, if x is a negative number, then $-x$ will be a positive number. We solve the following inequality,

$$-x + 5 \geq 0$$
$$-x \geq -5$$
$$\frac{-x}{-1} \leq \frac{-5}{-1}$$
To isolate x, we divide both sides by -1. Remember that we need to reverse the inequality symbol whenever we multiply or divide both sides by a negative number.
$$x \leq 5$$

Therefore, the domain of this function is $x \leq 5$. The lowest point on the graph of this function is $(5, 0)$, so the range will start at zero. The function continues up to infinity, so we get a range of $[0, \infty)$. We can confirm this domain and range by looking at its graph on the calculator.

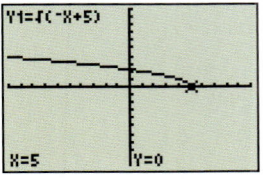

PRACTICE PROBLEM FOR EXAMPLE 4

Give the domain and range of the following radical functions.

a. $f(x) = \sqrt{x + 8}$ **b.** $g(x) = \sqrt{-x + 3}$ **c.** $h(x) = \sqrt[5]{x - 9}$

If you are given the graph of a radical function, you can determine the domain by looking for which values of the input (x) are represented by the graph. The range continues to be the lowest output value to the highest output value within the domain.

Using the Graphs of Radical Functions to Determine Domain and Range

Even roots

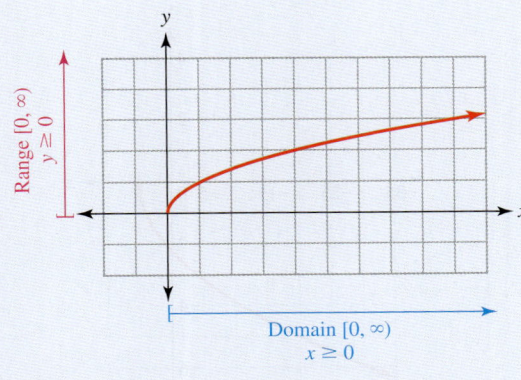

Odd roots

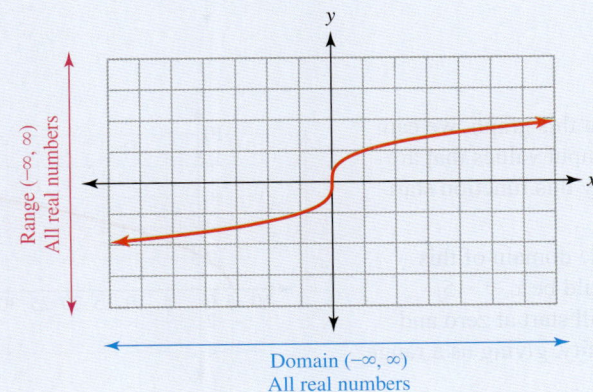

Example 5 — Domain and range of radical functions

Give the domain and range of the following radical functions.

a.

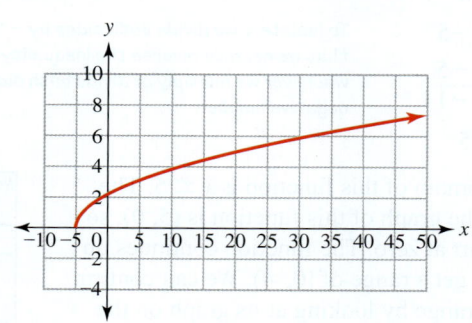

b.

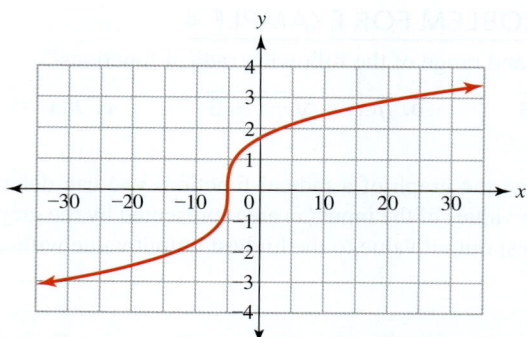

c.

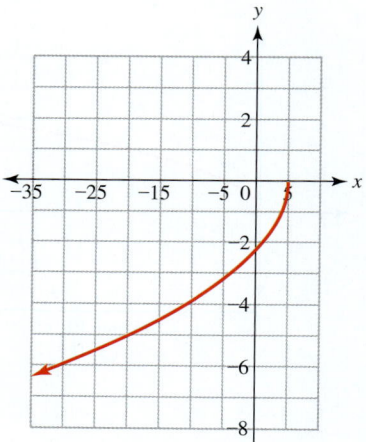

SOLUTION

a. By looking at this graph, we can see that the input values that are being used by this function start at $x = -5$.

Therefore, the domain of this function should be $x \geq -5$. The range will start at zero and go up to infinity, giving us a range of $[0, \infty)$.

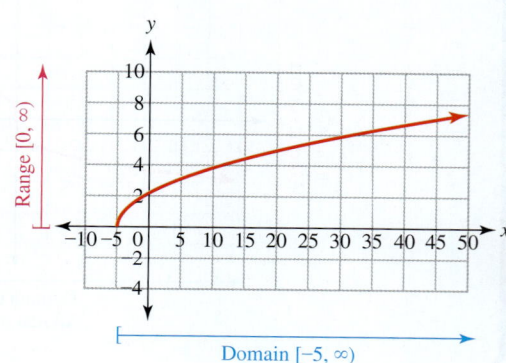

Range $[0, \infty)$

Domain $[-5, \infty)$

b. This graph uses all the input values and appears to be an odd root, so the domain of this function will be all real numbers. The range is also all real numbers because the graph goes down to negative infinity and up to positive infinity.

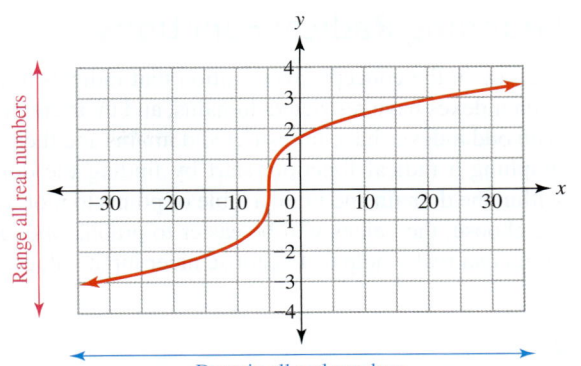

c. From this graph, we can see that only input values less than or equal to 5 are being used. The domain will be $x \leq 5$. The graph increases from negative infinity up to zero, so the range will be $(-\infty, 0]$.

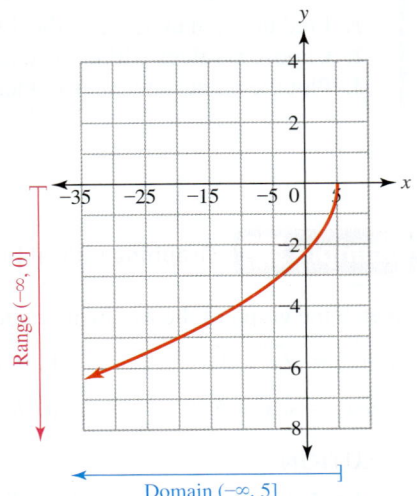

PRACTICE PROBLEM FOR EXAMPLE 5

Give the domain and range of the following radical functions.

a.

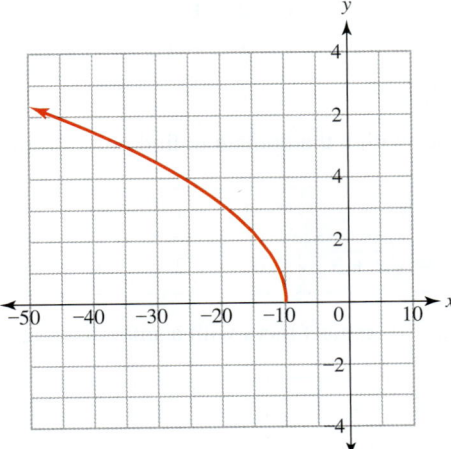

b.

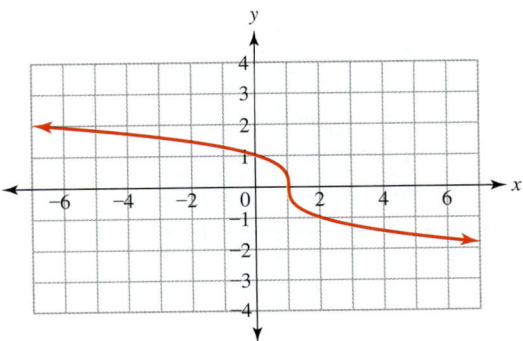

Graphing Radical Functions

We saw in the concept investigation that radicals have two basic graphs. Radicals with even indexes have restricted domains, and their graphs only go in one direction. Radicals with odd indexes have unrestricted domains and their graphs go in both directions. When graphing a radical function, start by finding the domain and then choose values of x within the domain and build a table of points to plot. If we are careful with the points that we choose, the values will be easier to graph. Knowing the perfect squares and perfect cubes can really help us to choose nicer input values.

> **Steps for Graphing Radical Functions**
> 1. Find the domain of the radical function.
> 2. Choose x-values within the domain to find points to plot.
> 3. Plot the points and connect them with a smooth curve.

Example 6 Graphing radical functions

Sketch the graph of the following radical functions.

a. $f(x) = \sqrt{x+3}$ **b.** $f(x) = -\sqrt{x-8}$

c. $g(x) = \sqrt[3]{x+5}$ **d.** $h(x) = \sqrt{4-x}$

SOLUTION

a. Step 1 Find the domain of the radical function.

The function $f(x) = \sqrt{x+3}$ is an even root, so the radicand must be nonnegative.

$$x + 3 \geq 0$$
$$x \geq -3$$

The domain is $x \geq -3$.

Step 2 Choose x-values within the domain to find points to plot.

x	-3	1	6	13
$f(x) = \sqrt{x+3}$	0	2	3	4

Step 3 Plot the points and connect them with a smooth curve.

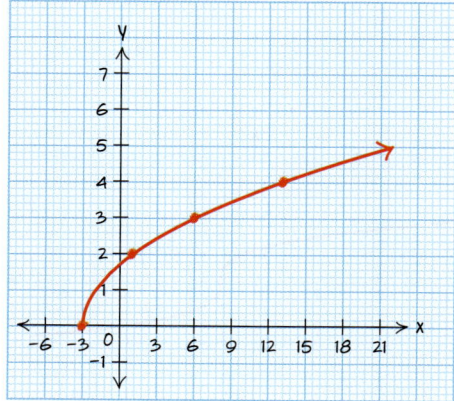

b. **Step 1** Find the domain of the radical function.

The function $f(x) = -\sqrt{x - 8}$ is an even root, so the radicand must be nonnegative.
$$x - 8 \geq 0$$
$$x \geq 8$$

The domain is $x \geq 8$.

Step 2 Choose x-values within the domain to find points to plot.

x	8	12	33	44
$f(x) = -\sqrt{x - 8}$	0	-2	-5	-6

Step 3 Plot the points and connect them with a smooth curve.

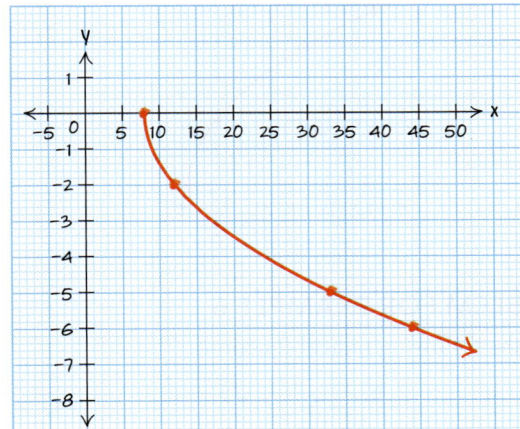

c. **Step 1** Find the domain of the radical function.

The function $g(x) = \sqrt[3]{x + 5}$ is an odd root, so the domain is all real numbers.

Step 2 Choose x-values within the domain to find points to plot.

x	-69	-13	-5	3	59
$g(x) = \sqrt[3]{x + 5}$	-4	-2	0	2	4

Step 3 Plot the points and connect them with a smooth curve.

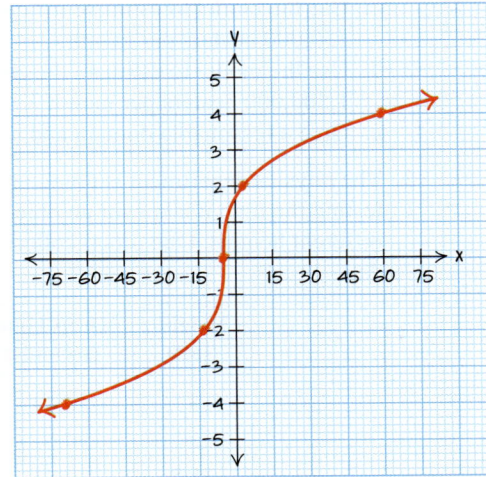

d. **Step 1** Find the domain of the radical function.

The function $h(x) = \sqrt{4 - x}$ is an even root, so the radicand must be nonnegative.
$$4 - x \geq 0$$
$$-x \geq -4$$
$$x \leq 4$$

The domain is $x \leq 4$.

Step 2 Choose x-values within the domain to find points to plot.

x	−21	−12	0	4
$h(x) = \sqrt{4-x}$	5	4	2	0

Step 3 Plot the points and connect them with a smooth curve.

PRACTICE PROBLEM FOR EXAMPLE 6
Sketch the graph of the following radical functions.

a. $f(x) = \sqrt{x+6}$
b. $f(x) = -\sqrt{5-x}$
c. $g(x) = -\sqrt[3]{x-4}$
d. $h(x) = -\sqrt{x+1}$

8.1 Exercises

1. Jim Bob's Heat Source manufactures h hundred thousand space heaters per month. Jim Bob has $3.2 million in monthly fixed costs to run the plant. The following data have been collected on the cost of various numbers of heaters produced per month.

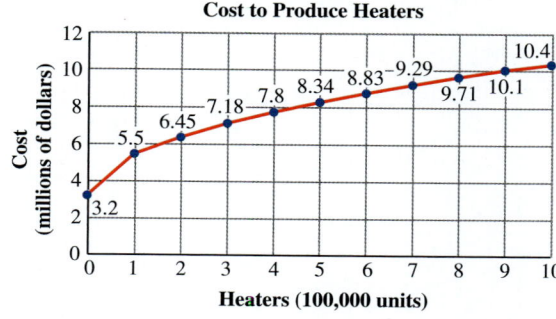

a. Create a scatterplot for these data. Does a radical function seem appropriate for this model? Explain.
b. Let $C(h)$ be the cost in millions of dollars to produce h hundred thousand heaters per month. Graph the following function with the data.
$$C(h) = 2.29\sqrt{h} + 3.2$$

How well does this model fit the data? Explain.
c. Use the model to estimate the cost to produce 550,000 heaters per month.
d. Use the model to estimate the cost to produce 1,500,000 heaters per month.
e. Give a reasonable domain and range for this model.

2. The physics department is doing an experiment testing the time it takes for an object to fall from different heights. One group collected the following data.

Height (feet)	Drop Time (seconds)
4	0.49
6	0.6
11.5	0.82
18.5	1.05
23	1.17
37	1.48
54.5	1.8

SECTION 8.1 Radical Functions

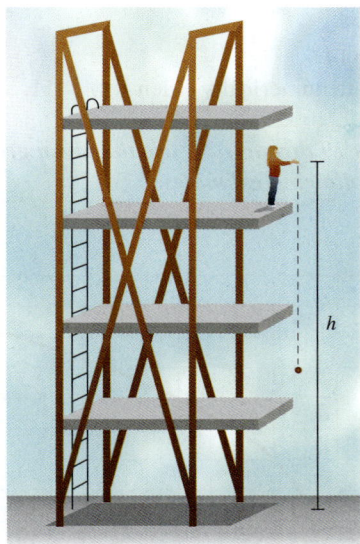

a. Let $T(h)$ be the drop time in seconds for an object that is dropped from h feet. Plot the data and the model
$$T(h) = 0.243\sqrt{h}$$
b. Use the model to determine the drop time for an object dropped from 15 feet.
c. Use the model to determine the drop time for an object that is dropped from 100 feet.
d. Determine a reasonable domain and range for this model.
e. Estimate numerically what height an object must be dropped from to have a drop time of 2 seconds.

3. In a study of basilisk lizards, a biologist found that there were several allometric relationships between the body mass of a lizard and different characteristics of its body. One such relationship is demonstrated by the following data and model.

Body Mass (grams)	Leg Length (meters)
2.5	0.022
15	0.041
34.9	0.054
41.7	0.057
72.1	0.069
90.3	0.074
140.6	0.086
164.2	0.09

Source: Data and model derived from *Size-Dependence of Water Running Ability in Basilisk Lizards* by Blasheen and McMahon. *The Journal of Experimental Biology* 199, 2611–2618, 1996.

Let $L(M)$ represent the leg length in meters of a basilisk lizard with a body mass of M grams.
$$L(M) = 0.0165\sqrt[3]{M}$$

a. How well does this model fit the data given? Explain.
b. Estimate the leg length of a basilisk lizard that has a body mass of 50 grams.
c. Estimate the leg length of a basilisk lizard that has a body mass of 100 grams.
d. Estimate graphically the body mass of a basilisk lizard that has a leg length of 0.1 meter.

4. Dr. Marina Silva, in the *Journal of Mammalogy* (1998), gives the following formula for the approximate body length of mammals based on the mammal's body mass.
$$L(M) = 0.330\sqrt[3]{M}$$
where $L(M)$ represents the body length in meters of a mammal with a body mass of M kilograms.

a. Use this model to estimate the body length of a mammal with a body mass of 4.6 kilograms.
b. Use this model to estimate the body length of a mammal with a body mass of 25 kilograms.
c. If this model is valid for mammals with body masses between 0.01 and 250 kilograms, what is this model's range?
d. Estimate numerically the body mass of a mammal with a body length of 1 meter.

5. The butterfly study from Example 1 on page 625 also developed the following formula for the relationship between the thoracic (middle body portion) mass of the butterfly and its mean forward airspeed.
$$V(m) = 17.6\sqrt[5]{m^3}$$
where $V(m)$ represents the mean forward airspeed in meters per second of a butterfly with a thoracic mass of m grams.

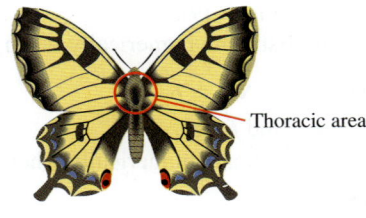
Thoracic area

a. Use the model to estimate the airspeed of a butterfly with a thoracic mass of 0.05 gram.
b. Use the model to estimate the airspeed of a butterfly with a thoracic mass of 0.12 gram.
c. If this model is valid for butterflies with a thoracic mass between 0.001 and 0.6 gram, determine the range of this model.

6. The average height of boys of various ages are given in the table.

Age (months)	Average Height of Boys (feet)
4.5	2.05
10.5	2.3
12	2.45
24	2.75
36	2.95
48	3.2

Source: "The Wellness Site," aarogya.com.

626 CHAPTER 8 Radical Functions

a. Let $H(a)$ be the average height in feet of boys at a months old. The height can be modeled by

$$H(a) = 1.46\sqrt[5]{a}$$

Graph the data and the model together. How well does this model fit the data?

b. Estimate the average height for boys who are 18 months old.

c. Estimate the average height for boys who are $2\frac{1}{2}$ years old.

d. How do your answers in parts b and c compare to the average heights of girls found in Example 2? Explain.

e. Give a reasonable domain and range for this model.

For Exercises 7 through 18, use the given functions to answer the questions. Round answers to three decimal places.

7. $f(x) = 2.5\sqrt{x}$
   a. Find $f(81)$.
   b. Estimate numerically when $f(x) = 12.5$.

8. $f(x) = 7.2\sqrt{x}$
   a. Find $f(49)$.
   b. Estimate numerically when $f(x) = 86.4$.

9. $f(x) = 6\sqrt[5]{x}$
   a. Find $f(45)$.
   b. Estimate numerically when $f(x) = 12$.

10. $f(x) = 8\sqrt[4]{x}$
    a. Find $f(49)$.
    b. Estimate numerically when $f(x) = 24$.

11. $f(x) = \sqrt{x + 8}$
    a. Find $f(28)$.
    b. Estimate numerically when $f(x) = 11$.

12. $f(x) = \sqrt{x + 5}$
    a. Find $f(11)$.
    b. Estimate numerically when $f(x) = 8$.

13. $f(x) = \sqrt{x - 10}$
    a. Find $f(50)$.
    b. Estimate numerically when $f(x) = 7$.

14. $f(x) = \sqrt{x - 6}$
    a. Find $f(40)$.
    b. Estimate numerically when $f(x) = 5$.

15. $f(x) = \sqrt[7]{x + 20}$
    a. Find $f(-60)$.
    b. Estimate numerically when $f(x) = -3$.

16. $f(x) = \sqrt[5]{x - 14}$
    a. Find $f(-12)$.
    b. Estimate numerically when $f(x) = -9.5$.

17. $f(x) = \sqrt[4]{x^3}$
    a. Find $f(5)$.
    b. Estimate numerically when $f(x) = 12$.

18. $f(x) = \sqrt[5]{x^2}$
    a. Find $f(14)$.
    b. Estimate numerically when $f(x) = 11$.

For Exercises 19 through 22, use the given graphs of $f(x)$ to make the following estimates.

19.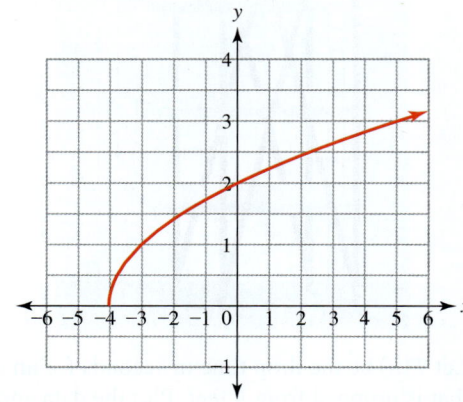

a. Estimate $f(5)$.
b. Estimate when $f(x) = 1$.

20.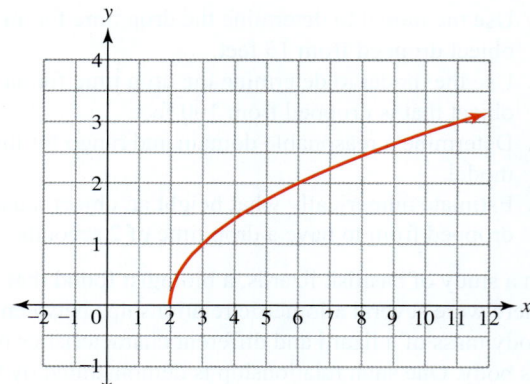

a. Estimate $f(3)$.
b. Estimate when $f(x) = 2.5$.

21.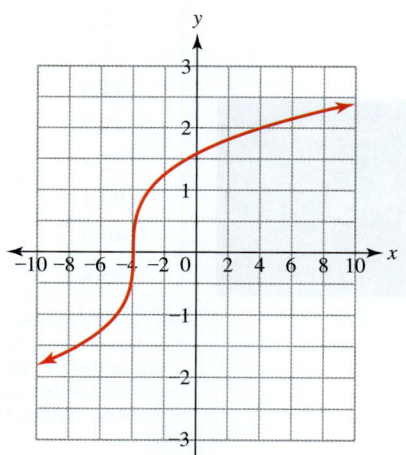

a. Estimate $f(4)$.
b. Estimate when $f(x) = -1.5$.

22.

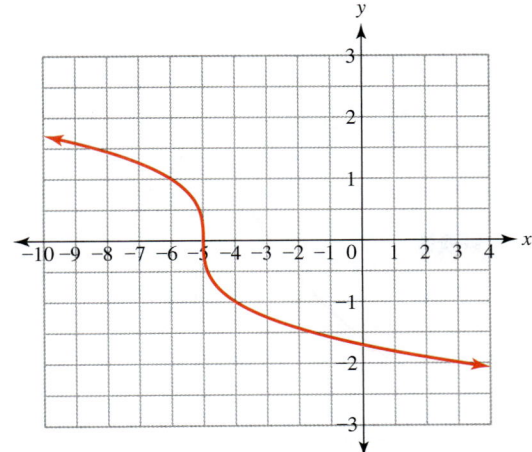

a. Estimate $f(-2)$.
b. Estimate when $f(x) = 1$.

23.

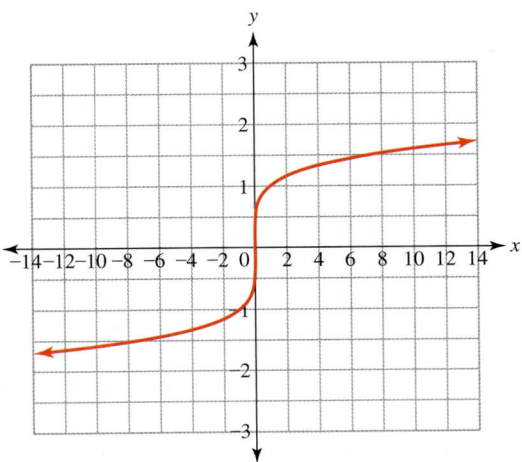

a. Is this the graph of an odd or even root?
b. Give the domain of this function.
c. Give the range of this function.

24.

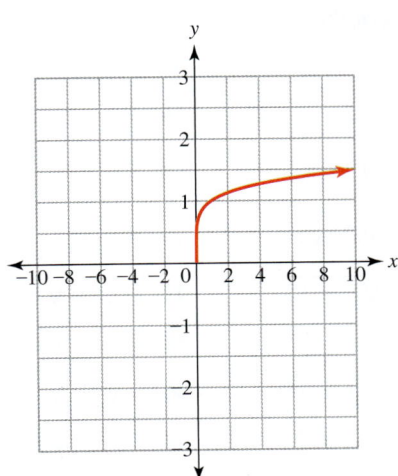

a. Is this the graph of an odd or even root?
b. Give the domain of this function.
c. Give the range of this function.

25.

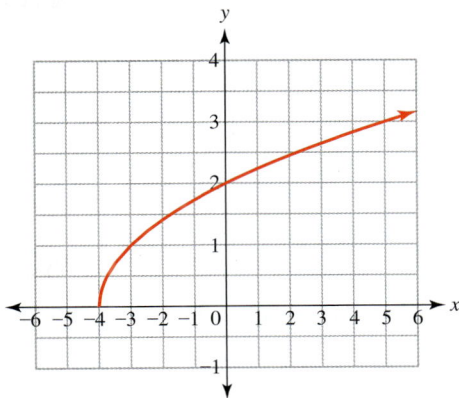

a. Is this the graph of an odd or even root?
b. Give the domain of this function.
c. Give the range of this function.

26.

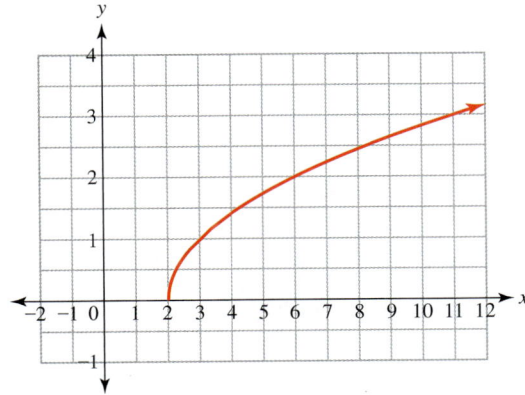

a. Is this the graph of an odd or even root?
b. Give the domain of this function.
c. Give the range of this function.

27.

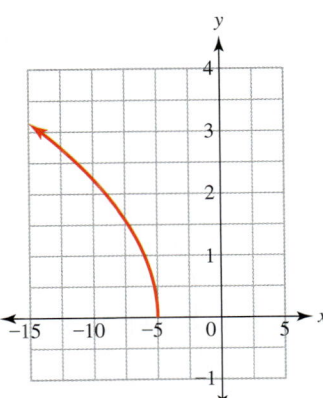

a. Is this the graph of an odd or even root?
b. Give the domain of this function.
c. Give the range of this function.

628 CHAPTER 8 Radical Functions

28.

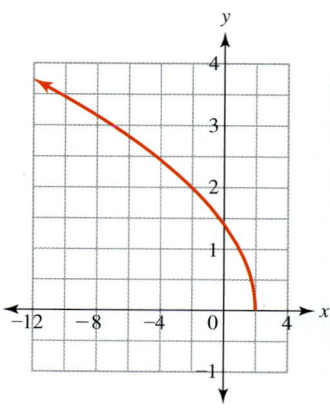

a. Is this the graph of an odd or even root?
b. Give the domain of this function.
c. Give the range of this function.

29.

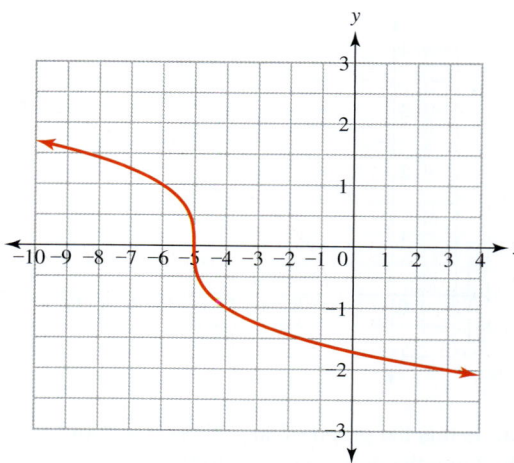

a. Is this the graph of an odd or even root?
b. Give the domain of this function.
c. Give the range of this function.

30.

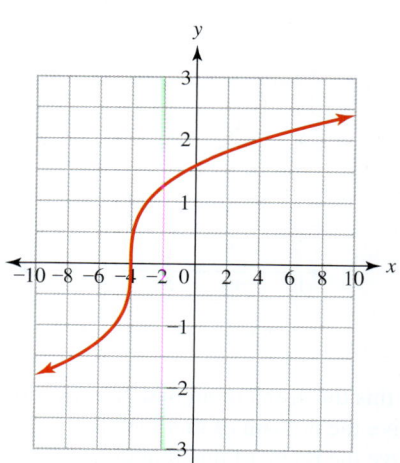

a. Is this the graph of an odd or even root?
b. Give the domain of this function.
c. Give the range of this function.

31.

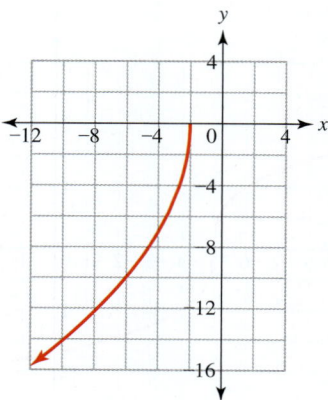

a. Is this the graph of an odd or even root?
b. Give the domain of this function.
c. Give the range of this function.

32.

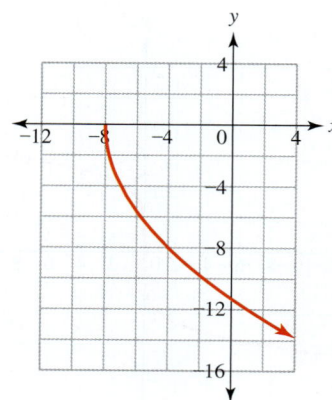

a. Is this the graph of an odd or even root?
b. Give the domain of this function.
c. Give the range of this function.

For Exercises 33 through 60, find the domain and range of the given function algebraically. Verify your answer using the graph.

33. $f(x) = 4\sqrt[3]{x}$ **34.** $g(m) = -3\sqrt[5]{m}$

35. $f(x) = 5\sqrt{x}$ **36.** $g(x) = -7\sqrt{x}$

37. $h(x) = 0.7\sqrt[9]{x}$ **38.** $f(x) = -\dfrac{5}{8}\sqrt[7]{x}$

39. $h(x) = 0.24\sqrt[4]{x}$ **40.** $f(x) = -\dfrac{2}{3}\sqrt[8]{x}$

41. $f(x) = \sqrt{x + 12}$ **42.** $g(x) = \sqrt{x + 3}$

43. $f(x) = \sqrt{x + 7}$ **44.** $g(x) = \sqrt{x + 4}$

45. $f(x) = \sqrt[3]{x + 15}$ **46.** $g(x) = \sqrt[3]{x + 12}$

47. $h(x) = \sqrt[4]{x + 2}$ **48.** $f(x) = \sqrt[6]{x + 9}$

49. $g(x) = \sqrt{x - 23}$ **50.** $h(x) = \sqrt{x - 35}$

51. $h(x) = \sqrt{-x + 3}$ **52.** $f(x) = \sqrt{-x + 8}$

53. $g(x) = \sqrt{-x - 10}$ **54.** $h(x) = \sqrt{-x - 11}$

55. $f(x) = \sqrt[3]{-x+9}$
56. $g(x) = \sqrt[3]{-x+12}$
57. $h(x) = \sqrt[4]{3x+5}$
58. $f(x) = \sqrt[6]{2x-6}$
59. $g(x) = \sqrt{-4x+9}$
60. $h(x) = \sqrt{-5x-12}$

For Exercises 61 through 80, sketch the graph of the given radical functions.

61. $f(x) = \sqrt{x+7}$
62. $f(x) = \sqrt{x+5}$
63. $g(x) = \sqrt{x-8}$
64. $h(x) = \sqrt{x-2}$
65. $f(x) = -\sqrt{x+3}$
66. $g(x) = -\sqrt{x+9}$
67. $f(x) = -\sqrt{x-6}$
68. $h(x) = -\sqrt{x-3}$
69. $f(x) = \sqrt[3]{x+2}$
70. $g(x) = \sqrt[3]{x+7}$
71. $h(x) = \sqrt[3]{x-11}$
72. $f(x) = \sqrt[3]{x-8}$
73. $f(x) = \sqrt{9-x}$
74. $h(x) = \sqrt{2-x}$
75. $g(x) = -\sqrt{x-3}$
76. $f(x) = -\sqrt{x-12}$
77. $h(x) = \sqrt[5]{2x}$
78. $f(x) = \sqrt[5]{5x}$
79. $g(x) = 2\sqrt{x+3}$
80. $f(x) = -5\sqrt{x+4}$

8.2 Simplifying, Adding, and Subtracting Radicals

LEARNING OBJECTIVES

- Calculate square roots and higher roots.
- Simplify radical expressions.
- Add and subtract radical expressions.

Square Roots and Higher Roots

Radicals include square roots and other higher roots such as cube roots and fourth roots. To solve problems that involve radicals, we should know some basic rules for doing arithmetic operations. In this section, we will simplify radical expressions and perform basic operations with radicals.

Let's start with a basic definition of a square root.

> **DEFINITION**
>
> **Square root**
> b is the square root of a if $b^2 = a$ and $a \geq 0$.
> $$\sqrt{a} = b$$

Although we are probably more familiar with square roots, higher roots have a very similar definition.

> **DEFINITION**
>
> **nth root**
> b is the nth root of a if $b^n = a$ and n is a natural number $n \geq 2$.
> $$\sqrt[n]{a} = b$$
>
> n is called the index of the radical, a is called the radicand, and b is called the root.

When working with radicals, we might notice that the outputs of square roots always seem to be positive. This is true only by the choice of mathematicians. In general, the square root would have both a positive and a negative answer.

$$\sqrt{25} = \begin{cases} \sqrt{5^2} = 5 \\ \sqrt{(-5)^2} = -5 \end{cases}$$

Because having two possible answers will cause many problems with consistency, mathematicians agree to use the **principal root,** which is the positive root only. If a negative root is needed, then we use a negative sign outside of the square root to designate the negative root.

$$-\sqrt{100} = -10 \quad \text{but} \quad \sqrt{100} = 10$$

Principal roots are also commonly used for higher roots.

> **DEFINITION**
>
> **Principal nth root**
>
> If a is a real number and $n \geq 2$ is a natural number, then
>
> $$\sqrt[n]{a^n} = a \quad \text{If } n \text{ is odd.}$$
>
> $$\sqrt[n]{a^n} = |a| \quad \text{If } n \text{ is even.}$$

Note that a root with an odd index can have a negative radicand and will have a negative root. If a root with an even index has a negative radicand, the root will not be a real number. Square roots with negative radicands will be introduced in Section 8.5.

Example 1 Evaluating radicals

Evaluate the following radicals without a calculator.

a. $\sqrt{36}$ b. $\sqrt[3]{8}$ c. $\sqrt[3]{-27}$ d. $\sqrt[4]{625}$

SOLUTION

a. Rewrite the radicand using perfect squares and then simplify the square root.

$$\sqrt{36} = \sqrt{6^2} = 6$$

b. Since this is a cube root, rewrite the radicand using perfect cubes. Then simplify the cube root.

$$\sqrt[3]{8} = \sqrt[3]{2^3} = 2$$

c. Since this is a cube root, rewrite the radicand using perfect cubes. Then simplify the cube root.

$$\sqrt[3]{-27} = \sqrt[3]{(-3)^3} = -3$$

d. Since this is a fourth root, rewrite the radicand as something to the fourth power. If you factor 625, you can see that $625 = 5^4$. Then simplify the radical.

$$\sqrt[4]{625} = \sqrt[4]{5^4} = 5$$

PRACTICE PROBLEM FOR EXAMPLE 1

Evaluate the following radicals without a calculator.

a. $\sqrt{49}$ b. $\sqrt{144}$ c. $\sqrt[3]{-512}$ d. $\sqrt[4]{10{,}000}$

Simplifying Radicals

When simplifying radical expressions, we are looking for any factors of the radicand that can be pulled out of the radical. Anything that cannot come out of the radical can be left inside as a new simplified radicand. Whenever we are given a radical with a variable in the radicand, we will assume that the values of the variables are nonnegative.

Working with radicals is much like working with the rules for exponents from Section 3.1. Radicals have properties that are derived from the rules for exponents. The product property of radicals is derived from the powers of products rule for exponents. We will use this property to simplify radical expressions throughout this chapter.

> **What's That Mean?**
>
> **Perfect Powers**
>
> Perfect powers are numbers that are the result of a natural number being raised to a certain power. The most common examples are perfect squares and perfect cubes.
>
> Perfect squares are numbers that are the result of a natural number being squared.
>
> Perfect cubes are numbers that are the result of a natural number being cubed.
>
> When simplifying square roots and cube roots, knowing some common perfect squares and perfect cubes will help you to simplify faster.
>
> | Perfect Squares | Perfect Cubes |
> |---|---|
> | $2^2 = 4$ | $2^3 = 8$ |
> | $3^2 = 9$ | $3^3 = 27$ |
> | $4^2 = 16$ | $4^3 = 64$ |
> | $5^2 = 25$ | $5^3 = 125$ |
> | $6^2 = 36$ | $6^3 = 216$ |
> | $7^2 = 49$ | $7^3 = 343$ |
> | $8^2 = 64$ | $8^3 = 512$ |
> | $9^2 = 81$ | $9^3 = 729$ |
> | $10^2 = 100$ | $10^3 = 1000$ |

SECTION 8.2 Simplifying, Adding, and Subtracting Radicals

Product Property of Radicals

If $\sqrt[n]{a}$ and $\sqrt[n]{b}$ are real numbers and $n \geq 2$ is a natural number, then
$$\sqrt[n]{a \cdot b} = \sqrt[n]{a} \cdot \sqrt[n]{b}$$

If a factor has an exponent higher than the index of the radical, we will break up the factor into a factor that is a multiple of the index. For example,
$$\sqrt[5]{x^{13}} = \sqrt[5]{x^{10} \cdot x^3}$$

Breaking this up allows us to simplify the radical using the product property of radicals and the definition of an nth root.
$$\sqrt[5]{x^{13}} = \sqrt[5]{x^{10} \cdot x^3} = \sqrt[5]{x^{10}} \cdot \sqrt[5]{x^3} = x^{\frac{10}{5}} \sqrt[5]{x^3} = x^2 \sqrt[5]{x^3}$$

Steps to Simplify a Radical Expression

1. Factor the radicand into factors that are raised to powers that are a multiple of the index of the radical.
2. Simplify the radicals.
3. Simplify by multiplying any remaining radicands together and multiplying anything that has been taken out of the radicals.

A Radical Is Considered Simplified When

1. The radicand does not contain any factors that are perfect powers of the index.
2. All exponents in the radicand are less than the index.
3. There are no fractions in the radicand.
4. There are no radicals in the denominator of a fraction.

Connecting the Concepts

Do exponent rules apply to radicals?

Recall from Section 3.1 that a rational exponent is the same as a radical.
$$x^{\frac{1}{2}} = \sqrt{x} \qquad x^{\frac{1}{n}} = \sqrt[n]{x}$$
and
$$\sqrt[3]{x^5} = x^{\frac{5}{3}} \qquad \sqrt[n]{x^m} = x^{\frac{m}{n}}$$

Rewriting radicals as rational exponents will help to simplify radicals.

If you think of radicals as rational exponents, you can use some of the rules for exponents to derive important properties of radicals. The following rules for exponents are some of the most helpful when simplifying radicals.

Powers of Products and Quotients:

$$(xy)^m = x^m y^m \qquad \left(\frac{x}{y}\right)^m = \frac{x^m}{y^m}$$

$$(xy)^5 = x^5 y^5 \qquad \left(\frac{x}{y}\right)^4 = \frac{x^4}{y^4}$$

Example 2 Simplifying radicals

Simplify the following radicals. Assume that all variables are nonnegative.

a. $\sqrt{25x^2}$ b. $\sqrt[3]{8m^3 n^6}$ c. $\sqrt{15x^3 y^6}$ d. $\sqrt[5]{96 a^2 b^8 c^{10}}$

SOLUTION

a. $\sqrt{25x^2} = \sqrt{5^2} \sqrt{x^2}$ Break the radicand into factors that are squared.

$\phantom{\sqrt{25x^2}} = 5x$ Simplify.

b. Because this is a cube root, look for factors that are cubed.

$\sqrt[3]{8m^3 n^6} = \sqrt[3]{2^3} \sqrt[3]{m^3} \sqrt[3]{n^6}$ Break the radicand into factors that have exponents that are multiples of the index 3.

$\phantom{\sqrt[3]{8m^3 n^6}} = 2^{\frac{3}{3}} m^{\frac{3}{3}} n^{\frac{6}{3}}$ Divide each exponent by the index 3.

$\phantom{\sqrt[3]{8m^3 n^6}} = 2mn^2$ Simplify.

c. $\sqrt{15x^3 y^6} = \sqrt{x^2 y^6} \sqrt{3 \cdot 5 \cdot x}$ Break the radicand into factors that have exponents that are multiples of the index 2.

$\phantom{\sqrt{15x^3 y^6}} = x^{\frac{2}{2}} y^{\frac{6}{2}} \sqrt{3 \cdot 5 \cdot x}$ Divide each exponent by the index 2.

$\phantom{\sqrt{15x^3 y^6}} = xy^3 \sqrt{3 \cdot 5 \cdot x}$

$\phantom{\sqrt{15x^3 y^6}} = xy^3 \sqrt{15x}$ Simplify the remaining radicand.

d. $\sqrt[5]{96 a^2 b^8 c^{10}} = \sqrt[5]{2^5 b^5 c^{10}} \sqrt[5]{3 a^2 b^3}$ Break the radicand into factors that have exponents that are multiples of the index 5.

$\phantom{\sqrt[5]{96 a^2 b^8 c^{10}}} = 2bc^2 \sqrt[5]{3 a^2 b^3}$ Simplify.

PRACTICE PROBLEM FOR EXAMPLE 2
Simplify the following radicals. Assume that all variables are nonnegative.

a. $\sqrt{16a^2b^4}$ **b.** $\sqrt{180x^2y^3}$ **c.** $\sqrt[3]{108m^5n^6}$ **d.** $\sqrt[9]{x^{36}y^{18}z^{11}}$

Adding and Subtracting Radicals

Adding and subtracting radical expressions is basically the same as adding and subtracting like terms in a polynomial. You can add or subtract only two radical expressions that have the same index, the same variables and exponents outside the radical and the same radicand inside. This can also be looked at as a form of factoring.

$$5\sqrt{3x} + 7\sqrt{3x}$$
$$\sqrt{3x}(5 + 7) \quad \text{Factor out the } \sqrt{3x} \text{ from both terms.}$$
$$12\sqrt{3x} \quad \text{Combine the constants, and you have combined like terms.}$$

> **DEFINITION**
>
> **Like Radical Expression**
> Radical expressions are like if they have the same index and the same radicand.
>
> | Like Radicals | Unlike Radicals |
> |---|---|
> | $\sqrt{3x} \quad 8\sqrt{3x}$ | $\sqrt[3]{x-7} \quad \sqrt[3]{x+4}$ |

> **Steps to Adding or Subtracting Radical Expressions**
> 1. Simplify each radical.
> 2. Add or subtract any like radical expressions by adding or subtracting the coefficients in front of the radical and keeping the radicand the same.

Example 3 Adding and subtracting radicals

Add or subtract the following expressions. Assume that all variables are nonnegative.

a. $3\sqrt{5} + 7\sqrt{5}$ **b.** $2x\sqrt{10} - 5x\sqrt{10}$

c. $\sqrt{25x^3yz^5} + 2x\sqrt{xyz^5}$ **d.** $5\sqrt{x} + 2\sqrt{3x} + 4\sqrt{x} - 7\sqrt{3x}$

SOLUTION

a. $3\sqrt{5} + 7\sqrt{5} = 10\sqrt{5}$ Combine like terms.

b. $2x\sqrt{10} - 5x\sqrt{10} = -3x\sqrt{10}$ Combine like terms.

c. $\sqrt{25x^3yz^5} + 2x\sqrt{xyz^5} = 5xz^2\sqrt{xyz} + 2xz^2\sqrt{xyz}$ Simplify each radical.
$$= 7xz^2\sqrt{xyz} \quad \text{Combine like terms.}$$

d. $5\sqrt{x} + 2\sqrt{3x} + 4\sqrt{x} - 7\sqrt{3x} =$ Combine like terms.
$$= (5\sqrt{x} + 4\sqrt{x}) + (2\sqrt{3x} - 7\sqrt{3x})$$
$$= 9\sqrt{x} - 5\sqrt{3x}$$

PRACTICE PROBLEM FOR EXAMPLE 3
Add or subtract the following expressions. Assume that all variables are nonnegative.

a. $2\sqrt{3} - 7\sqrt{3}$ **b.** $4\sqrt{50} + 7\sqrt{2}$

c. $3x\sqrt{5y^3} + 2xy\sqrt{5y}$ **d.** $\sqrt{5x} + 2\sqrt{2x} - 6\sqrt{2x} + 4\sqrt{5x}$

When combining radical expressions, you combine only like expressions. This will include only combining similar radicals. For instance, you should not add or subtract a square root with a cube root or any other higher root.

Example 4 Adding and subtracting radicals

Perform the indicated operation and simplify. Assume that all variables are nonnegative.

a. $5\sqrt{x} + 2\sqrt[3]{x} - 2\sqrt{x} + 3\sqrt[3]{x}$
b. $\sqrt[3]{8x^5y^2} + 5x\sqrt[4]{x^2y^2} + 7x\sqrt[3]{x^2y^2}$

SOLUTION

a. $5\sqrt{x} + 2\sqrt[3]{x} - 2\sqrt{x} + 3\sqrt[3]{x} = 3\sqrt{x} + 5\sqrt[3]{x}$ Combine like expressions.

b. $\sqrt[3]{8x^5y^2} + 5x\sqrt[4]{x^2y^2} + 7x\sqrt[3]{x^2y^2} = 2x\sqrt[3]{x^2y^2} + 5x\sqrt[4]{x^2y^2} + 7x\sqrt[3]{x^2y^2}$
$= 9x\sqrt[3]{x^2y^2} + 5x\sqrt[4]{x^2y^2}$

PRACTICE PROBLEM FOR EXAMPLE 4

Perform the indicated operation and simplify. Assume that all variables are nonnegative.

a. $12\sqrt{2m} + 5\sqrt[3]{7m} - 4\sqrt{2m} + 3\sqrt[3]{7m}$
b. $\sqrt[5]{32x^3y} + 5\sqrt[4]{xy^2} + 6\sqrt[5]{x^3y}$

8.2 Exercises

Note: Assume that all variables are positive.

For Exercises 1 through 22, simplify the given radical expression. Give exact answers. Assume that all variables are nonnegative.

1. $\sqrt{100}$
2. $\sqrt{16}$
3. $\sqrt{121}$
4. $\sqrt{64}$
5. $\sqrt[3]{64}$
6. $\sqrt[3]{125}$
7. $\sqrt[3]{-8}$
8. $\sqrt[3]{-216}$
9. $\sqrt{-9}$
10. $\sqrt{-4}$
11. $\sqrt{50}$
12. $\sqrt{48}$
13. $\sqrt{180}$
14. $\sqrt{192}$
15. $\sqrt[3]{40}$
16. $\sqrt[3]{375}$
17. $\sqrt[3]{-16}$
18. $\sqrt[3]{-640}$
19. $\sqrt{49x^2}$
20. $\sqrt{36a^2}$
21. $\sqrt[3]{125y^6}$
22. $\sqrt[3]{27m^3}$

For Exercises 23 and 24, the students were asked to simplify the given radical expressions. Explain what the student did wrong. Give the correct answer.

23. $\sqrt{10x^2y^8}$

> **Mark**
> $\sqrt{10x^2y^8} = 5xy^4$

24. $\sqrt[3]{16a^6b^{10}}$

> **Shannon**
> $\sqrt[3]{16a^6b^{10}} = 4a^3b^5$

For Exercises 25 through 40, simplify the given radical expression. Assume that all variables are nonnegative.

25. $\sqrt{196m^4n^2}$
26. $\sqrt{50x^2y^6}$
27. $\sqrt{36a^3}$
28. $\sqrt{24w^4z^5}$
29. $\sqrt{1296a^5b^8c^{15}}$
30. $\sqrt{6480x^3y^5z^8}$
31. $\sqrt[3]{8x^3y^6}$
32. $\sqrt[3]{40a^3b^5}$
33. $\sqrt[3]{3888x^3y^5z^8}$
34. $\sqrt[3]{-27a^3b^6}$
35. $\sqrt[4]{960m^3n^4p^5}$
36. $\sqrt[4]{16a^4b^{12}}$
37. $\sqrt[5]{-32c^5d^{10}}$
38. $\sqrt[4]{48a^3b^8c^{21}}$
39. $\sqrt[5]{32m^5n^{10}}$
40. $\sqrt[5]{243s^3t^5u^{10}}$

For Exercises 41 through 54, perform the indicated operation and simplify. Assume that all variables are nonnegative.

41. $\sqrt{5x} + 3\sqrt{5x}$
42. $4\sqrt{2a} + 9\sqrt{2a}$
43. $5\sqrt{t} - 3\sqrt{t}$
44. $14\sqrt{7p} - 5\sqrt{7p}$
45. $10\sqrt[3]{5b} + 4\sqrt[3]{5b}$
46. $7\sqrt[3]{3x} + 3\sqrt[3]{3x}$

47. $3n\sqrt{6} + 2n\sqrt{6} + 8n\sqrt{6}$

48. $2x\sqrt{3} + 7x\sqrt{3} + 4x\sqrt{3}$

49. $5t\sqrt{11r} - 8t\sqrt{11r}$

50. $3x\sqrt{2y} - 7x\sqrt{2y}$

51. $2\sqrt{4x^2y} - 3x\sqrt{y}$

52. $5\sqrt{49m^2n^5} - 3m\sqrt{16n^5}$

53. $\sqrt[3]{8a} + 4\sqrt[3]{a}$

54. $\sqrt[3]{125h^2} + 9\sqrt[3]{h^2}$

For Exercises 55 through 58, the students were asked to perform the indicated operation and simplify. Explain what the student did wrong. Give the correct answer.

55. $8\sqrt{5} + 3\sqrt{5}$

> **Matt**
> $8\sqrt{5} + 3\sqrt{5} = 11\sqrt{10}$

56. $8\sqrt{20} - 7\sqrt{5}$

> **Shania**
> $8\sqrt{20} - 7\sqrt{5} = 10\sqrt{5} - 7\sqrt{5}$
> $= 3\sqrt{5}$

57. $4\sqrt{7} + 10\sqrt[3]{7}$

> **Amy**
> $4\sqrt{7} + 10\sqrt[3]{7} = 14\sqrt{7}$

58. $5\sqrt[3]{8x^4y} + 14x\sqrt[3]{27xy}$

> **Frank**
> $5\sqrt[3]{8x^4y} + 14x\sqrt[3]{27xy} = 10x\sqrt[3]{xy} + 42x\sqrt[3]{xy}$
> $= 52x\sqrt[6]{2xy}$

For Exercises 59 through 76, perform the indicated operation and simplify. Assume that all variables are nonnegative.

59. $7\sqrt{50a^3b^4c} - 2ab\sqrt{162ab^2c}$

60. $4m^3\sqrt{150m^5n^6p^3} - 5n^2\sqrt{600m^{11}n^2p^3}$

61. $7\sqrt[3]{125x^6y^9z^2} - 9x^2y\sqrt[3]{216y^3z^2}$

62. $2\sqrt[3]{1080a^3b^6c^9} - 8abc\sqrt[3]{625b^3c^6}$

63. $5\sqrt{13x} + 7x\sqrt{2} - 12\sqrt{13x} + 3x\sqrt{2}$

64. $\sqrt{11b} + 3a\sqrt{5} - 15a\sqrt{5} + 3\sqrt{11b}$

65. $4\sqrt{18} + 5\sqrt{2} - 8\sqrt{75} + 3\sqrt{48}$

66. $8\sqrt{12} + 9\sqrt{7} - 4\sqrt{112} - 2\sqrt{108}$

67. $3xy\sqrt{9z^5} + 7xz^2\sqrt{yz} - 2xyz^2\sqrt{z}$

68. $5m^2n\sqrt{12mn^2} + 7mn^2\sqrt{3mn} - 10\sqrt{243m^5n^4}$

69. $5\sqrt{2x} + 4\sqrt[3]{2x} + 7\sqrt{2x}$

70. $2\sqrt{10gh} + 6\sqrt[3]{10gh} + 4\sqrt{10gh}$

71. $3\sqrt[5]{64a^5b^{10}c^3} - 10ab\sqrt[5]{2b^5c^3}$

72. $\sqrt[5]{10m^3n^5} + 4n\sqrt[5]{10m^3}$

73. $\sqrt[4]{16x^5y^8} + 7xy^2\sqrt[4]{81x}$

74. $2p\sqrt[4]{405m^6n} - 7m\sqrt[4]{5m^2np^4}$

75. $\sqrt[3]{7xy} + \sqrt[5]{7xy} - \sqrt[3]{448xy} + \sqrt[5]{224xy}$

76. $\sqrt{27x^5y^3} + \sqrt[3]{27x^5y^3} - 3xy\sqrt[3]{x^2} + xy\sqrt{3x^3y}$

For Exercises 77 and 78, find the exact perimeter of the given triangle.

77.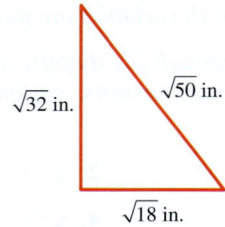
Triangle with sides $\sqrt{32}$ in., $\sqrt{50}$ in., $\sqrt{18}$ in.

78.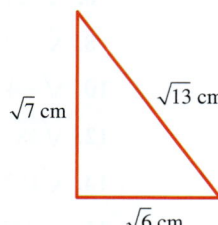
Triangle with sides $\sqrt{7}$ cm, $\sqrt{13}$ cm, $\sqrt{6}$ cm.

79. $\sqrt{64} = 8$ and $\sqrt[3]{64} = 4$. Does 64 have any other roots that are whole numbers?

80. $\sqrt{729} = 27$ and $\sqrt[3]{729} = 9$. Does 729 have any other roots that are whole numbers?

81. Find a number that has a square root and a cube root that are both whole numbers. Answers will vary.

82. Find a number that has a cube root and a fourth root that are both whole numbers. Answers will vary.

8.3 Multiplying and Dividing Radicals

LEARNING OBJECTIVES
- Multiply radical expressions.
- Divide radical expressions.
- Use conjugates to rationalize denominators.

Multiplying Radicals

Since we have learned the product property of radicals, we can now use it to multiply radicals. To use the product property of radicals, the two radical expressions must have the same index. To multiply two radical expressions with the same index, multiply the radicands (insides) together. After multiplying the radicands, simplify the result if possible. Remember that the radical expressions do not need to be like to multiply, which is not the same as the operations of addition and subtraction.

> **Steps to Multiplying Radicals**
> 1. Confirm that the indices of the radicals are the same.
> 2. Multiply the radicands together using the product property of radicals.
> 3. Simplify the result if possible.
>
> $$\sqrt{3} \cdot \sqrt{7} = \sqrt{21}$$
> $$\sqrt[4]{5x} \cdot \sqrt[4]{6xy} = \sqrt[4]{30x^2y}$$

Example 1 Multiplying radicals

Multiply the following and simplify the result.

a. $\sqrt{5} \cdot \sqrt{2}$ b. $\sqrt{2x} \cdot \sqrt{10y}$ c. $\sqrt{2a} \cdot \sqrt{18a}$ d. $\sqrt[3]{4m^2} \cdot \sqrt[3]{5m^2}$

SOLUTION

a. $\sqrt{5} \cdot \sqrt{2} = \sqrt{5 \cdot 2}$
$\phantom{\sqrt{5} \cdot \sqrt{2}} = \sqrt{10}$ There are no square factors, so it is simplified.

b. $\sqrt{2x} \cdot \sqrt{10y} = \sqrt{20xy}$ 20 has a perfect square factor, so simplify.
$\phantom{\sqrt{2x} \cdot \sqrt{10y}} = \sqrt{4}\sqrt{5xy}$
$\phantom{\sqrt{2x} \cdot \sqrt{10y}} = 2\sqrt{5xy}$

c. $\sqrt{2a} \cdot \sqrt{18a} = \sqrt{36a^2}$ 36 and a^2 are perfect squares, so simplify.
$\phantom{\sqrt{2a} \cdot \sqrt{18a}} = 6a$

d. $\sqrt[3]{4m^2} \cdot \sqrt[3]{5m^2} = \sqrt[3]{20m^4}$ 20 has no perfect cube factors, but the m^4 can
$\phantom{\sqrt[3]{4m^2} \cdot \sqrt[3]{5m^2}} = m\sqrt[3]{20m}$ be simplified.

PRACTICE PROBLEM FOR EXAMPLE 1

Multiply the following and simplify the result.

a. $\sqrt[3]{3a} \cdot \sqrt[3]{5bc}$ b. $\sqrt{5m} \cdot \sqrt{20m^3}$ c. $\sqrt{3y} \cdot \sqrt{12y}$

If the expressions are more complicated, you will multiply the coefficients of the radicals together and multiply the radicands together. Once you have multiplied each of these together, you should simplify the result. Some students find it easiest to simplify each radical first, then multiply them together, and simplify again if necessary.

Example 2 Multiplying radicals

Multiply the following and simplify the result.

a. $5\sqrt{2x} \cdot 3\sqrt{7}$
b. $2x\sqrt{3y} \cdot 5\sqrt{12y}$
c. $5a^2b\sqrt{3ab^3} \cdot 2ab^2\sqrt{21a^2b}$
d. $2mn^2\sqrt[4]{m^2n} \cdot 5mn^3\sqrt[4]{mn^9}$

SOLUTION

a. $5\sqrt{2x} \cdot 3\sqrt{7} = 3 \cdot 5\sqrt{2x \cdot 7}$ Multiply the coefficients and the **radicands**.
$= 15\sqrt{14x}$

b. $2x\sqrt{3y} \cdot 5\sqrt{12y} = 10x\sqrt{36y^2}$ Multiply the coefficients and the radicands.
$= 10x \cdot 6y$ Simplify the radical.
$= 60xy$

c. $5a^2b\sqrt{3ab^3} \cdot 2ab^2\sqrt{21a^2b} = 10a^3b^3\sqrt{63a^3b^4}$
$= 10a^3b^3\sqrt{3^2 \cdot 7a^3b^4}$ Break the radicand into factors whose exponents are multiples of 2.
$= 10a^3b^3\sqrt{3^2a^2b^4}\sqrt{7a}$
$= 10a^3b^3(3ab^2)\sqrt{7a}$ Simplify.
$= 30a^4b^5\sqrt{7a}$

d. $2mn^2\sqrt[4]{m^2n} \cdot 5mn^3\sqrt[4]{mn^9} = 10m^2n^5\sqrt[4]{m^3n^{10}}$
$= 10m^2n^5\sqrt[4]{m^3n^{10}}$ Break the radicand into factors whose exponents are multiples of 4.
$= 10m^2n^5\sqrt[4]{n^8}\sqrt[4]{m^3n^2}$
$= 10m^2n^5(n^2)\sqrt[4]{m^3n^2}$ Simplify.
$= 10m^2n^7\sqrt[4]{m^3n^2}$

PRACTICE PROBLEM FOR EXAMPLE 2

Multiply the following and simplify the result.

a. $4\sqrt{3a} \cdot 6\sqrt{5b}$
b. $3m\sqrt{14n} \cdot 5\sqrt{21mn}$
c. $7x^2y^3\sqrt[3]{x^2y^5} \cdot 4xy\sqrt[3]{xy^2}$

Example 3 Multiplying radicals

Multiply the following and simplify the result.

a. $(3 + \sqrt{5})(4 - \sqrt{6})$
b. $(5 + \sqrt{2})^2$

SOLUTION

a. Since the expressions being multiplied include addition or subtraction, use the distributive property (FOIL) to multiply them together.

$(3 + \sqrt{5})(4 - \sqrt{6}) = 12 - 3\sqrt{6} + 4\sqrt{5} - \sqrt{30}$

b. This expression is being squared, so multiply it by itself.

$(5 + \sqrt{2})(5 + \sqrt{2}) = 25 + 5\sqrt{2} + 5\sqrt{2} + \sqrt{4}$ Use the distributive property.
$= 25 + 10\sqrt{2} + 2$ Combine like terms and simplify.
$= 27 + 10\sqrt{2}$

PRACTICE PROBLEM FOR EXAMPLE 3

Multiply the following and simplify the result.

a. $(7 + \sqrt{3})(2 - \sqrt{5})$
b. $(6 + \sqrt{7})^2$

SECTION 8.3 Multiplying and Dividing Radicals

Dividing Radicals and Rationalizing the Denominator

Division inside a radical can be simplified in the same way that a fraction would be reduced if it were by itself. This follows from the powers of quotients rule for exponents.

$$\sqrt{\frac{50}{2}} = \sqrt{25} = 5 \quad \text{or} \quad \sqrt{\frac{50}{2}} = \left(\frac{50}{2}\right)^{\frac{1}{2}} = 25^{\frac{1}{2}} = 5$$

You can use this rule to simplify some radical expressions that have fractions in them. Please note that you can simplify only fractions that are either both inside a radical or both outside the radical. That is, you cannot divide out something that is inside the radical with something that is outside of a radical.

> **The Quotient Property of Radicals**
>
> If $\sqrt[n]{a}$ and $\sqrt[n]{b}$ are real numbers, $b \neq 0$, and $n \geq 2$ is an integer, then
>
> $$\sqrt[n]{\frac{a}{b}} = \frac{\sqrt[n]{a}}{\sqrt[n]{b}}$$
>
> A fraction inside a radical can be made into a fraction with separate radicals in the numerator and denominator of the fraction. This rule is often used in both directions.

Example 4 Simplifying radicals with division

Simplify the following radicals.

a. $\sqrt{\dfrac{100}{25}}$ b. $\sqrt{\dfrac{49}{36}}$ c. $\sqrt{\dfrac{400x^3 y}{4xy^5}}$ d. $\sqrt[3]{\dfrac{a^5 b^2}{a^2 b}}$

SOLUTION

a. $\sqrt{\dfrac{100}{25}} = \sqrt{4}$ Reduce the fraction and then simplify the remaining radical.

$= 2$

b. $\sqrt{\dfrac{49}{36}} = \dfrac{\sqrt{49}}{\sqrt{36}}$ The fraction does not reduce, so separate the radical and then simplify each remaining radical.

$= \dfrac{7}{6}$

c. $\sqrt{\dfrac{400x^3 y}{4xy^5}} = \sqrt{\dfrac{100x^2}{y^4}}$ Reduce the fraction.

$= \dfrac{\sqrt{100x^2}}{\sqrt{y^4}}$ Since the fraction does not reduce further, separate the radical and simplify.

$= \dfrac{10x}{y^2}$

d. $\sqrt[3]{\dfrac{a^5 b^2}{a^2 b}} = \sqrt[3]{a^3 b}$ Reduce the fraction.

$= a\sqrt[3]{b}$ Simplify the radical.

PRACTICE PROBLEM FOR EXAMPLE 4

Simplify the following radicals.

a. $\sqrt{\dfrac{63}{7}}$ b. $\sqrt{\dfrac{50}{18}}$ c. $\sqrt[4]{\dfrac{144 m^5 n^3}{mn^7}}$

In the previous example and practice problem, all of the denominators simplified to the point at which no radicals remain. This will not always happen, but mathematicians often like to have no radicals in the denominator of a fraction. Having no radicals in a denominator makes some operations easier and is considered a standard way to write a simplified fraction. Clearing any remaining radicals from the denominator of a fraction is called **rationalizing the denominator.** This process uses multiplication on the top and bottom of the fraction to force any radicals in the denominator to simplify completely.

The key to rationalizing the denominator of a fraction is to multiply both the numerator and the denominator of the fraction by the right radical expression. This will allow the resulting denominator to simplify and be without any remaining radicals. Use the concept investigation to consider what makes a radical simplify completely.

CONCEPT INVESTIGATION
What can make radicals disappear?

1. Multiply the following expressions together and simplify.
   a. $\sqrt{7} \cdot \sqrt{7}$
   b. $\sqrt{3x} \cdot \sqrt{3x}$
   c. $\sqrt{x+2} \cdot \sqrt{x+2}$

2. Describe what happens when you multiply these expressions together.

3. Multiply the following expressions together and simplify.
   a. $\sqrt[3]{2} \cdot \sqrt[3]{5}$
   b. $\sqrt[4]{7x} \cdot \sqrt[4]{7x}$

4. Do the radicals simplify so that no radicals remain when you multiply these expressions together?

5. Multiply the following expressions together and simplify.
   a. $\sqrt[3]{2} \cdot \sqrt[3]{2^2}$
   b. $\sqrt[4]{8x} \cdot \sqrt[4]{2x^3}$

6. Do the radicals simplify so that no radicals remain when you multiply these expressions together?

7. What radical would you have to multiply the given radicals by to have the product simplify completely?
   a. $\sqrt{3a}$ b. $\sqrt[3]{5}$ c. $\sqrt[4]{7x}$

With a single square root, rationalizing the denominator is usually accomplished by multiplying the numerator and denominator by the radical factor of the denominator.

> **Steps to Rationalizing Denominators with One Term Containing a Square Root**
> 1. Separate radicals for the numerator and denominator. Simplify if possible.
> 2. Multiply the numerator and denominator by the radical factor of the denominator.
> 3. Simplify the result if possible.

Example 5 Rationalizing the denominator

Rationalize the denominator and simplify the following radical expressions.

a. $\sqrt{\dfrac{3}{5}}$ b. $\sqrt{\dfrac{5y}{2x}}$ c. $\dfrac{2\sqrt{3n}}{5\sqrt{6m}}$

SECTION 8.3 Multiplying and Dividing Radicals

SOLUTION

a. $\sqrt{\dfrac{3}{5}} = \dfrac{\sqrt{3}}{\sqrt{5}}$ Separate into two radicals.

$= \dfrac{\sqrt{3}}{\sqrt{5}} \cdot \dfrac{\sqrt{5}}{\sqrt{5}}$ Multiply the numerator and denominator by the denominator. This is the same as multiplying by 1.

$= \dfrac{\sqrt{15}}{\sqrt{25}}$

$= \dfrac{\sqrt{15}}{5}$ Simplify the radicals.

b. $\sqrt{\dfrac{5y}{2x}} = \dfrac{\sqrt{5y}}{\sqrt{2x}}$

$= \dfrac{\sqrt{5y}}{\sqrt{2x}} \cdot \dfrac{\sqrt{2x}}{\sqrt{2x}}$ Multiply the numerator and denominator by the denominator.

$= \dfrac{\sqrt{10xy}}{\sqrt{4x^2}}$

$= \dfrac{\sqrt{10xy}}{2x}$

c. $\dfrac{2\sqrt{3n}}{5\sqrt{6m}} = \dfrac{2\sqrt{n}}{5\sqrt{2m}}$ Reduce the radicands.

$= \dfrac{2\sqrt{n}}{5\sqrt{2m}} \cdot \dfrac{\sqrt{2m}}{\sqrt{2m}}$ Multiply the numerator and denominator by the radical factor of the denominator.

$= \dfrac{2\sqrt{2mn}}{5\sqrt{4m^2}}$

$= \dfrac{2\sqrt{2mn}}{10m}$ Simplify the radicals.

$= \dfrac{\sqrt{2mn}}{5m}$ Reduce the fraction.

PRACTICE PROBLEM FOR EXAMPLE 5

Simplify the following radical expressions.

a. $\sqrt{\dfrac{2n}{5m}}$ b. $\dfrac{7x\sqrt{3xy}}{2y\sqrt{5x}}$

If the denominator contains a higher root, it will take more thought to choose an appropriate expression to multiply by. If you factor the radicand in the denominator, you can then determine what factors are needed to allow the radical to simplify completely. Remember that you want each factor's exponent to be a multiple of the index. If you are working with a cube root, you will want each factor to be a perfect cube. This will allow the cube root to simplify completely, thus eliminating the radical from the denominator.

$\dfrac{1}{\sqrt[3]{x}} = \dfrac{1}{\sqrt[3]{x}} \cdot \dfrac{\sqrt[3]{x^2}}{\sqrt[3]{x^2}}$ Multiplying by $\sqrt[3]{x^2}$ gives you $\sqrt[3]{x^3}$, so the cube root in the denominator will reduce.

$= \dfrac{\sqrt[3]{x^2}}{\sqrt[3]{x^3}}$

$= \dfrac{\sqrt[3]{x^2}}{x}$ The denominator is clear of radicals and rationalized.

> **Steps to Rationalizing Denominators with One Term Containing a Higher Root**
> 1. Separate radicals for the numerator and denominator. Simplify if possible.
> 2. Multiply the numerator and denominator by the radical factor that will allow the radical in the denominator to simplify completely, leaving no radical part. Each factor inside the radical will need to have an exponent that is a multiple of the index.
> 3. Simplify the result if possible.

Example 6 — Rationalizing denominators with higher roots

Simplify the following radical expressions.

a. $\sqrt[3]{\dfrac{5xy}{10x^2}}$ b. $\dfrac{2a}{\sqrt[4]{9a^3b^6}}$ c. $\dfrac{\sqrt[3]{2m}}{\sqrt[3]{8np^2}}$

SOLUTION

a. $\sqrt[3]{\dfrac{5xy}{10x^2}} = \sqrt[3]{\dfrac{y}{2x}}$ Reduce the fraction.

$= \dfrac{\sqrt[3]{y}}{\sqrt[3]{2x}}$ Separate the radical.

$= \dfrac{\sqrt[3]{y}}{\sqrt[3]{2x}} \cdot \dfrac{\sqrt[3]{2^2x^2}}{\sqrt[3]{2^2x^2}}$ Multiply by the necessary number of factors to clear the denominator. In this case, two more 2's and two more x's are needed to get the radical to simplify. Note that a cube root will undo a perfect cube.

$= \dfrac{\sqrt[3]{4x^2y}}{\sqrt[3]{2^3x^3}}$

$= \dfrac{\sqrt[3]{4x^2y}}{2x}$ Simplify the radicals.

b. $\dfrac{2a}{\sqrt[4]{9a^3b^6}} = \dfrac{2a}{\sqrt[4]{3^2a^3b^6}}$ Factor the radicand.

$= \dfrac{2a}{\sqrt[4]{3^2a^3b^6}} \cdot \dfrac{\sqrt[4]{3^2ab^2}}{\sqrt[4]{3^2ab^2}}$ Multiply by the needed factors. Two more 3's, one more a, and two more b's are needed. Note that eight b's are necessary to get an exponent on b that is divisible by 4 (the root's index).

$= \dfrac{2a\sqrt[4]{3^2ab^2}}{\sqrt[4]{3^4a^4b^8}}$

$= \dfrac{2a\sqrt[4]{9ab^2}}{3ab^2}$ Simplify the radical.

$= \dfrac{2\sqrt[4]{9ab^2}}{3b^2}$ Reduce the fraction.

c. $\dfrac{\sqrt[3]{2m}}{\sqrt[3]{8np^2}} = \dfrac{\sqrt[3]{m}}{\sqrt[3]{4np^2}}$ Reduce the fraction.

$= \dfrac{\sqrt[3]{m}}{\sqrt[3]{2^2np^2}}$ Factor the radicand.

$= \dfrac{\sqrt[3]{m}}{\sqrt[3]{2^2np^2}} \cdot \dfrac{\sqrt[3]{2n^2p}}{\sqrt[3]{2n^2p}}$ Multiply by the needed factors.

$= \dfrac{\sqrt[3]{2mn^2p}}{\sqrt[3]{2^3n^3p^3}}$

$= \dfrac{\sqrt[3]{2mn^2p}}{2np}$ Simplify the radical.

SECTION 8.3 Multiplying and Dividing Radicals   641

PRACTICE PROBLEM FOR EXAMPLE 6

Simplify the following radical expressions.

a. $\sqrt[3]{\dfrac{24m^2}{15mn^2}}$ b. $\dfrac{7x^2}{\sqrt[5]{27xy^7}}$ c. $\dfrac{\sqrt[3]{5a^2}}{\sqrt[3]{20ab}}$

Conjugates

> **CONCEPT INVESTIGATION**
>
> *What happened to the radical?*
>
> 1. Multiply the following expressions together and simplify.
>   a. $(x + 2)(x - 2)$
>   b. $(5 + \sqrt{7})(5 - \sqrt{7})$
>   c. $(6 + \sqrt{2a})(6 - \sqrt{2a})$
>
> 2. Describe what happens when you multiply these expressions together.

You should be seeing a pattern: When you multiply the expressions from the concept investigation together, the result has no radicals remaining. These expressions are special in that they are basically the same except for the sign between the two terms. These expressions are called **conjugates** of one another. Although the term conjugates was not used, conjugates were introduced in Chapter 4 with the quadratic formula. The plus/minus symbol in the quadratic formula resulted in the solutions being conjugates.

> **DEFINITION**
>
> **Conjugates**
>
> $$a + b \quad \text{and} \quad a - b$$
>
> These expressions are called conjugates of one another. They are simply the sum and difference of the same two terms.
>
> $$5 + x \qquad 5 - x$$
> $$2 + \sqrt{3x} \qquad 2 - \sqrt{3x}$$
> $$-17 + 4\sqrt{5m} \qquad -17 - 4\sqrt{5m}$$

Often fractions will have square roots in the denominator with another term. This requires you to use a different approach to rationalizing the denominator. If there are two terms, you will multiply by the conjugate of the denominator to clear all of the radicals in the denominator.

> **Steps to Rationalizing Denominators with Two Terms**
>
> 1. Multiply the numerator and denominator by the conjugate of the denominator.
> 2. Simplify the result if possible.

Example 7 Using conjugates to rationalize denominators

Rationalize the denominator of the following fractions.

a. $\dfrac{5}{2 + \sqrt{7}}$ b. $\dfrac{2 + 3\sqrt{5}}{8 + \sqrt{10}}$ c. $\dfrac{2 + \sqrt{x}}{3 - 5\sqrt{x}}$

SOLUTION

a. $\dfrac{5}{2+\sqrt{7}} = \dfrac{5}{(2+\sqrt{7})} \cdot \dfrac{(2-\sqrt{7})}{(2-\sqrt{7})}$ Multiply the numerator and denominator by the conjugate of the denominator.

$= \dfrac{10 - 5\sqrt{7}}{4 - 2\sqrt{7} + 2\sqrt{7} - 7}$

$= \dfrac{10 - 5\sqrt{7}}{-3}$

b. $\dfrac{2+3\sqrt{5}}{8+\sqrt{10}} = \dfrac{(2+3\sqrt{5})}{(8+\sqrt{10})} \cdot \dfrac{(8-\sqrt{10})}{(8-\sqrt{10})}$ Multiply the numerator and denominator by the conjugate of the denominator.

$= \dfrac{16 - 2\sqrt{10} + 24\sqrt{5} - 3\sqrt{50}}{64 - 10}$ Use the distributive property.

$= \dfrac{16 - 2\sqrt{10} + 24\sqrt{5} - 15\sqrt{2}}{54}$ Simplify the radicals.

c. $\dfrac{2+\sqrt{x}}{3-5\sqrt{x}} = \dfrac{(2+\sqrt{x})}{(3-5\sqrt{x})} \cdot \dfrac{(3+5\sqrt{x})}{(3+5\sqrt{x})}$ Multiply the numerator and denominator by the conjugate of the denominator.

$= \dfrac{6 + 10\sqrt{x} + 3\sqrt{x} + 5\sqrt{x^2}}{9 - 25x}$ Use the distributive property.

$= \dfrac{6 + 5x + 13\sqrt{x}}{9 - 25x}$ Simplify the radicals and combine like terms.

PRACTICE PROBLEM FOR EXAMPLE 7

Rationalize the denominator of the following fractions.

a. $\dfrac{7}{4+\sqrt{13}}$ b. $\dfrac{4+5\sqrt{3}}{7+\sqrt{15}}$ c. $\dfrac{3+\sqrt{a}}{5-7\sqrt{ab}}$

Simplifying radical expressions completely requires you to rationalize all denominators and simplify each radical. There should be no fractions remaining inside a radical and no factors left inside a radical that can be pulled out of the radical.

8.3 Exercises

Note: Assume that all variables are positive.

For Exercises 1 through 22, multiply the given radical expressions and simplify the result.

1. $\sqrt{7} \cdot \sqrt{3}$
2. $\sqrt{5} \cdot \sqrt{11}$
3. $\sqrt{6} \cdot \sqrt{10}$
4. $\sqrt{15} \cdot \sqrt{35}$
5. $\sqrt{5x} \cdot \sqrt{10x}$
6. $\sqrt{8a} \cdot \sqrt{50a}$
7. $\sqrt{3m} \cdot \sqrt{12n}$
8. $\sqrt{7r} \cdot \sqrt{21t}$
9. $\sqrt[3]{4x} \cdot \sqrt[3]{2x^2}$
10. $\sqrt[3]{24b^2} \cdot \sqrt[3]{9b}$
11. $\sqrt{7m} \cdot \sqrt{14mn}$
12. $\sqrt[3]{3xy} \cdot \sqrt[3]{4y}$
13. $\sqrt{12a^3b} \cdot \sqrt{15a^2b^5}$
14. $\sqrt[5]{18m^5} \cdot \sqrt{2m}$
15. $5\sqrt{3xy} \cdot 7x\sqrt{2y}$
16. $8m\sqrt{14mn} \cdot 2p\sqrt{50m}$
17. $5n^3 \sqrt[4]{m^2n^2} \cdot 7m^2n\sqrt[4]{m^3n^7}$
18. $8a^5 \sqrt[3]{a^4b^2} \cdot 6a^2b^5 \sqrt[3]{a^2b^5}$
19. $2x^2 \sqrt[3]{5x^2y} \cdot 7y\sqrt[3]{4xy}$
20. $4ab\sqrt[3]{a^5b^2} \cdot 9a\sqrt[3]{ab^2}$
21. $12x^2y\sqrt[5]{7x^3y^4} \cdot 3xy^3\sqrt[5]{98x^4y^6}$
22. $8mn\sqrt[5]{2mn^2p} \cdot 5mp\sqrt[5]{16m^3n^3}$

For Exercises 23 through 38, perform the indicated operation and simplify the result.

23. $(4+\sqrt{5})(3+\sqrt{2})$
24. $(2-\sqrt{3})(5-\sqrt{7})$
25. $(5+\sqrt{6})(3-\sqrt{15})$
26. $(8-\sqrt{18})(4+\sqrt{14})$
27. $(3+\sqrt{2x})(6+\sqrt{6x})$
28. $(9-\sqrt{10a})(7-\sqrt{35a})$
29. $(\sqrt{3}-\sqrt{c})(\sqrt{6}-\sqrt{3c})$
30. $(\sqrt{5}+\sqrt{7m})(\sqrt{2}+\sqrt{14m})$

31. $(2+\sqrt{5})^2$
32. $(7-\sqrt{6})^2$
33. $(4+\sqrt{3m})^2$
34. $(15-\sqrt{5b})^2$
35. $(3+\sqrt{7})(3-\sqrt{7})$
36. $(4-\sqrt{2x})(4+\sqrt{2x})$
37. $(8+5\sqrt{7ab})(8-5\sqrt{7ab})$
38. $(2\sqrt{3}+5\sqrt{7})(2\sqrt{3}-5\sqrt{7})$

For Exercises 39 and 40, the students were asked to perform the indicated operation and simplify the result. Explain what the student did wrong. Give the correct answer.

39. $(5+\sqrt{3})^2$

Tom
$(5+\sqrt{3})^2 = 25 + 3$
$= 28$

40. $(3+\sqrt{2x})(5+\sqrt{7y})$

Warrick
$(3+\sqrt{2x})(5+\sqrt{7y}) = 15 + \sqrt{14xy}$

For Exercises 41 through 48, simplify the given radical expressions.

41. $\sqrt{\dfrac{20}{5}}$
42. $\sqrt{\dfrac{63}{7}}$
43. $\sqrt{\dfrac{72}{8}}$
44. $\sqrt{\dfrac{432}{75}}$
45. $\sqrt{\dfrac{7x^5}{25x}}$
46. $\sqrt{\dfrac{3a}{4a^3}}$
47. $\sqrt{\dfrac{5x^3}{45xy^2}}$
48. $\sqrt{\dfrac{6m^7}{75m^3n^2}}$

For Exercises 49 through 62, simplify the given radical expressions. Rationalize the denominator.

49. $\dfrac{7}{\sqrt{5n}}$
50. $\dfrac{5}{\sqrt{3x}}$
51. $\sqrt{\dfrac{7cd^2}{3c}}$
52. $\sqrt{\dfrac{11mn}{5n}}$
53. $\dfrac{5\sqrt{2x}}{3\sqrt{7y}}$
54. $\dfrac{12\sqrt{3ab}}{\sqrt{6a}}$
55. $\sqrt[3]{\dfrac{5x}{7x^2y^2}}$
56. $\dfrac{5m^2}{\sqrt[3]{2m^4n}}$
57. $\dfrac{\sqrt[3]{9x^2y^2}}{7x\sqrt[3]{2xy^3}}$
58. $\dfrac{\sqrt[3]{4ab^2}}{\sqrt[3]{6a}}$
59. $\sqrt[4]{\dfrac{5ab}{8a^3b^6}}$
60. $\dfrac{7g}{\sqrt[4]{2g^3h^7}}$
61. $\dfrac{8\sqrt[4]{4xy}}{\sqrt[4]{144xy^3z^2}}$
62. $\dfrac{3\sqrt[4]{9xy}}{\sqrt[4]{18x^2y^3}}$

For Exercises 63 and 64, the students were asked to simplify and rationalize the denominator. Explain what the student did wrong. Give the correct answer.

63. $\dfrac{8\sqrt{7xy}}{\sqrt{2x}}$

Hung
$\dfrac{8\sqrt{7xy}}{\sqrt{2x}} = \dfrac{8\sqrt{7xy}}{\sqrt{2x}} \cdot \dfrac{\sqrt{2x}}{\sqrt{7xy}}$
$= 8$

64. $\dfrac{12}{4+\sqrt{5x}}$

Michelle
$\dfrac{12}{4+\sqrt{5x}} = \dfrac{12}{4+\sqrt{5x}} \cdot \dfrac{4+\sqrt{5x}}{4+\sqrt{5x}}$
$= \dfrac{48 + 12\sqrt{5x}}{16 + 5x}$

In Exercises 65 through 80, rationalize the denominator of each fraction.

65. $\dfrac{2+\sqrt{6}}{\sqrt{3}}$
66. $\dfrac{7-\sqrt{5}}{\sqrt{6}}$
67. $\dfrac{2+\sqrt{3}}{8\sqrt{15}}$
68. $\dfrac{4+\sqrt{5}}{7\sqrt{10}}$
69. $\dfrac{5}{2+\sqrt{3}}$
70. $\dfrac{7}{8-\sqrt{6}}$
71. $\dfrac{2+\sqrt{3}}{5-\sqrt{7}}$
72. $\dfrac{4+\sqrt{5}}{7-5\sqrt{3}}$
73. $\dfrac{9}{3+\sqrt{2x}}$
74. $\dfrac{6}{4+\sqrt{3a}}$
75. $\dfrac{4+2\sqrt{7x}}{2-3\sqrt{12x}}$
76. $\dfrac{2+\sqrt{15}}{3+4\sqrt{30m}}$
77. $\dfrac{2+5\sqrt{m}}{3-\sqrt{mn}}$
78. $\dfrac{9+4\sqrt{8g}}{10-\sqrt{3gh}}$
79. $\dfrac{4+8\sqrt{5x}}{2+\sqrt{30x}}$
80. $\dfrac{7-6\sqrt{3ab}}{3+5\sqrt{15ab}}$

For Exercises 81 through 90, perform the indicated operation and simplify the result. Rationalize all denominators.

81. $\sqrt{5x} + 7\sqrt{2} - 8\sqrt{5x}$
82. $8\sqrt{10} - 4\sqrt{6b} + \sqrt{40} + 9\sqrt{24b}$
83. $(\sqrt{3}+\sqrt{5})^2$
84. $(2\sqrt{x}-\sqrt{6})^2$
85. $\dfrac{8}{3+\sqrt{5x}}$
86. $\dfrac{3m}{4-\sqrt{2m}}$
87. $\sqrt{3x} + (4+\sqrt{2x})(7+\sqrt{6})$
88. $\sqrt{14a} - (10+\sqrt{8})(4-\sqrt{7a})$
89. $\dfrac{4\sqrt[3]{6m^2n}}{55\sqrt[3]{4m^4n^2}}$
90. $\dfrac{2\sqrt[3]{gh}}{15\sqrt[3]{5g^2h}}$

8.4 Solving Radical Equations

LEARNING OBJECTIVES

- Solve radical equations involving one or more square roots.
- Use radical equations to solve application problems.
- Solve radical equations involving higher roots.

Solving Radical Equations

So far, we have simplified radical expressions and done arithmetic operations involving radicals. We have also evaluated radicals for different values. Recall from Section 3.1 that a radical can be written using fraction exponents.

$$\sqrt{x} = x^{1/2} \qquad \sqrt[n]{x} = x^{1/n}$$

Using the fraction exponent representation of radicals, we can solve radical equations using the same basic steps we used in Section 5.2 to solve power equations. In the case of a square root, isolating the square root and then squaring both sides will eliminate the radical and allow us to solve for the variable. This is the same technique as raising both sides to the reciprocal power that we used in Section 5.2.

> **Steps to Solving Radical Equations Involving Square Roots**
> 1. Isolate a radical on one side of the equation.
> 2. Square both sides of the equation to eliminate the radical.
> 3. Isolate any remaining radical if needed and square both sides again.
> 4. Solve the remaining equation.
> 5. Check your answer in the original equation. Some answers might not check in the original equation.

Example 1 Solving radical equations involving one square root

Solve the following equations. Check the answers in the original equation.

a. $\sqrt{x} = 5$ **b.** $\sqrt{x+2} - 3 = 0$

SOLUTION

a. The radical is already isolated on one side, so square both sides of the equation to eliminate the radical.

$$\sqrt{x} = 5$$
$$(\sqrt{x})^2 = 5^2 \qquad \text{Square both sides of the equation.}$$
$$x = 25$$

The square root of 25 is 5, so this is the correct answer.

b. Start by adding 3 to both sides to isolate the radical on one side of the equation. Then we square both sides to eliminate the radical.

$$\sqrt{x+2} - 3 = 0 \qquad \text{Add 3 to both sides of the equation.}$$
$$\sqrt{x+2} = 3$$
$$(\sqrt{x+2})^2 = 3^2 \qquad \text{Square both sides of the equation.}$$
$$x + 2 = 9 \qquad \text{Solve the remaining equation by}$$
$$x = 7 \qquad \text{subtracting 2 from both sides.}$$

Check this answer by substituting $x = 7$ into the original equation.

$$\sqrt{7+2} \stackrel{?}{=} 3$$
$$\sqrt{9} \stackrel{?}{=} 3$$
$$3 = 3 \quad \text{The answer works.}$$

PRACTICE PROBLEM FOR EXAMPLE 1

Solve the following equations.

a. $\sqrt{x+5} = 7$ **b.** $2\sqrt{3x+4} - 15 = -7$

Many applications in physics and other sciences use formulas that contain radicals. Often these formulas need to be solved for a variable that is inside the radical itself. In the next two examples, we will see two such formulas.

Example 2 Period of a pendulum

Traditionally, clocks have used pendulums to keep time. The period (the time to go back and forth) of a simple pendulum for a small amplitude is given by the function

$$T(L) = 2\pi\sqrt{\frac{L}{32}}$$

where $T(L)$ is the period in seconds and L is the length of the pendulum in feet. Use this formula to find the following:

a. The period of a pendulum if its length is 2 feet.

b. How long must a pendulum be if we want the period to be 3 seconds?

SOLUTION

a. Because the length of the pendulum is 2 feet, we know that $L = 2$, so we get

$$T(L) = 2\pi\sqrt{\frac{L}{32}}$$
$$T(2) = 2\pi\sqrt{\frac{2}{32}}$$
$$T(2) \approx 1.57$$

Therefore, a 2-foot pendulum will have a period of about 1.57 seconds.

b. We are given that $T(L) = 3$, so we get

$$T(L) = 2\pi\sqrt{\frac{L}{32}}$$
$$3 = 2\pi\sqrt{\frac{L}{32}} \quad \text{Isolate the radical on one side of the equation.}$$
$$\frac{3}{2\pi} = \frac{2\pi\sqrt{\frac{L}{32}}}{2\pi}$$
$$\frac{3}{2\pi} = \sqrt{\frac{L}{32}}$$
$$\left(\frac{3}{2\pi}\right)^2 = \left(\sqrt{\frac{L}{32}}\right)^2 \quad \text{Square both sides of the equation to eliminate the radical.}$$
$$\frac{9}{4\pi^2} = \frac{L}{32}$$
$$32\left(\frac{9}{4\pi^2}\right) = 32\left(\frac{L}{32}\right) \quad \text{Solve for } L.$$
$$\frac{72}{\pi^2} = L$$
$$7.295 \approx L$$

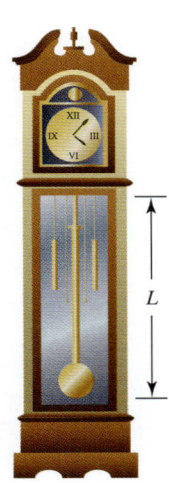

We can check this answer with the table.

To have a period of 3 seconds, we will need a pendulum that is about 7.3 feet long.

Example 3 The speed of sound

At sea level, the speed of sound through air can be calculated by using the following formula.

$$c = 340.3\sqrt{\frac{T + 273.15}{288.15}}$$

where c is the speed of sound in meters per second and T is the temperature in degrees Celsius. Use this formula to find the following.

a. The speed of sound when the temperature is 30°C.

b. The temperature if the speed of sound is 330 meters per second.

SOLUTION

a. $T = 30$, so we substitute and simplify the expression on the right side of the equation.

$$c = 340.3\sqrt{\frac{T + 273.15}{288.15}}$$

$$c = 340.3\sqrt{\frac{30 + 273.15}{288.15}} \quad \text{Simplify inside the radical.}$$

$$c \approx 340.3\sqrt{1.052} \quad \text{Calculate the radical and multiply.}$$

$$c \approx 349.05$$

At sea level, if the temperature is 30°C, then the speed of sound will be about 349.05 meters per second.

b. $c = 330$, so we get

$$c = 340.3\sqrt{\frac{T + 273.15}{288.15}}$$

$$330 = 340.3\sqrt{\frac{T + 273.15}{288.15}}$$

$$\frac{330}{340.3} = \frac{340.3\sqrt{\frac{T + 273.15}{288.15}}}{340.3} \quad \text{Isolate the square root.}$$

$$0.9697 \approx \sqrt{\frac{T + 273.15}{288.15}}$$

$$(0.9697)^2 \approx \left(\sqrt{\frac{T + 273.15}{288.15}}\right)^2 \quad \text{Square both sides to eliminate the square root.}$$

$$0.94038 \approx \frac{T + 273.15}{288.15}$$

$$288.15(0.94038) \approx 288.15\left(\frac{T + 273.15}{288.15}\right) \quad \text{Solve for } T.$$

$$270.9709 \approx T + 273.15$$

$$-2.179 \approx T$$

We can check this answer using the graph and trace.

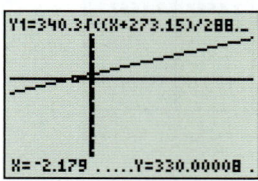

At sea level, if the speed of sound through air is 330 meters per second, then the temperature must be about $-2.18°C$.

Solving Radical Equations Involving More Than One Square Root

Some equations may have more than one radical involved and will require more work to solve. You will want to first isolate one of the radicals so that squaring both sides will eliminate that radical and then work on the remaining problem.

Example 4 Solving radical equations with more than one square root

Solve the following.

a. $\sqrt{x+1} = \sqrt{2x}$ b. $\sqrt{x+10} = 3 + \sqrt{x}$ c. $\sqrt{3x+5} = 2 - \sqrt{3x+2}$

SOLUTION

a.
$\sqrt{x+1} = \sqrt{2x}$
$(\sqrt{x+1})^2 = (\sqrt{2x})^2$ Square both sides to eliminate the radical.
$x + 1 = 2x$
$x + 1 = 2x$
$\underline{-x \quad\quad -x}$ Solve for x.
$1 = x$
$\sqrt{(1)+1} \stackrel{?}{=} \sqrt{2(1)}$ Check the answer.
$\sqrt{2} = \sqrt{2}$ The answer works.

b.
$\sqrt{x+10} = 3 + \sqrt{x}$ One radical is isolated.
$(\sqrt{x+10})^2 = (3 + \sqrt{x})^2$ Square both sides to eliminate the isolated radical.
$x + 10 = (3 + \sqrt{x})(3 + \sqrt{x})$
$x + 10 = 9 + 3\sqrt{x} + 3\sqrt{x} + x$ The right side will need the distributive property because of the addition involved.
$x + 10 = 9 + 6\sqrt{x} + x$
$\underline{-x \quad\quad\quad\quad\quad -x}$ Isolate the remaining radical.
$10 = 9 + 6\sqrt{x}$
$10 = 9 + 6\sqrt{x}$
$\underline{-9 \quad -9}$
$1 = 6\sqrt{x}$
$\dfrac{1}{6} = \dfrac{6\sqrt{x}}{6}$
$\dfrac{1}{6} = \sqrt{x}$
$\left(\dfrac{1}{6}\right)^2 = (\sqrt{x})^2$ Square both sides again to eliminate the remaining radical.
$\dfrac{1}{36} = x$

Check the answer.
The answer works.

c.
$$\sqrt{3x+5} = 2 - \sqrt{3x+2}$$
$$(\sqrt{3x+5})^2 = (2-\sqrt{3x+2})^2 \qquad \text{Square both sides.}$$
$$3x+5 = (2-\sqrt{3x+2})(2-\sqrt{3x+2}) \qquad \text{Use the distributive property.}$$
$$3x+5 = 4 - 2\sqrt{3x+2} - 2\sqrt{3x+2} + (3x+2)$$
$$3x+5 = 4 - 4\sqrt{3x+2} + 3x + 2 \qquad \text{Simplify.}$$
$$3x+5 = 6 + 3x - 4\sqrt{3x+2}$$
$$-1 = -4\sqrt{3x+2} \qquad \text{Isolate the radical.}$$
$$\frac{1}{4} = \sqrt{3x+2}$$
$$\left(\frac{1}{4}\right)^2 = (\sqrt{3x+2})^2 \qquad \text{Square both sides.}$$
$$\frac{1}{16} = 3x + 2$$
$$\frac{1}{16} - 2 = 3x$$
$$-\frac{31}{16} = 3x$$
$$-\frac{31}{48} = x$$

Check the answer.
The answer works.

PRACTICE PROBLEM FOR EXAMPLE 4

Solve the following.

a. $\sqrt{x+3} = \sqrt{2x-8}$

b. $\sqrt{2x+3} = 8 - \sqrt{2x+1}$

Equations involving radicals can be more complicated and require you to do more solving after eliminating the radicals. After you have eliminated the radicals, use the tools you have learned in other chapters to solve the remaining equation.

Example 5 Solving more complicated radical equations

Solve the following.

a. $\sqrt{3x+1} = 3 + \sqrt{x-4}$

b. $\sqrt{7x+9} = 5 + \sqrt{2x-1}$

SOLUTION

a.
$$\sqrt{3x+1} = 3 + \sqrt{x-4}$$
$$(\sqrt{3x+1})^2 = (3 + \sqrt{x-4})^2 \quad \text{Square both sides.}$$
$$3x + 1 = (3 + \sqrt{x-4})(3 + \sqrt{x-4}) \quad \text{Use the distributive property.}$$
$$1 = 9 + 3\sqrt{x-4} + 3\sqrt{x-4} + (x-4)$$
$$3x + 1 = 5 + x + 6\sqrt{x-4} \quad \text{Isolate the radical.}$$
$$2x - 4 = 6\sqrt{x-4}$$
$$(2x-4)^2 = (6\sqrt{x-4})^2 \quad \text{Square both sides. Use the distributive property again.}$$
$$(2x-4)(2x-4) = 36(x-4)$$
$$4x^2 - 16x + 16 = 36x - 144$$
$$4x^2 - 52x + 160 = 0 \quad \text{Solve the remaining quadratic equation using the quadratic formula.}$$
$$x = 5 \quad x = 8$$
$$\sqrt{3(5)+1} \stackrel{?}{=} 3 + \sqrt{(5)-4} \quad \text{Check the answers.}$$
$$4 = 4$$
$$\sqrt{3(8)+1} \stackrel{?}{=} 3 + \sqrt{(8)-4}$$
$$5 = 5 \quad \text{The answers work.}$$

b.
$$\sqrt{7x+9} = 5 + \sqrt{2x-1}$$
$$(\sqrt{7x+9})^2 = (5 + \sqrt{2x-1})^2 \quad \text{Square both sides. Use the distributive property.}$$
$$7x + 9 = (5 + \sqrt{2x-1})(5 + \sqrt{2x-1})$$
$$7x + 9 = 25 + 5\sqrt{2x-1} + 5\sqrt{2x-1} + (2x-1)$$
$$7x + 9 = 24 + 10\sqrt{2x-1} + 2x \quad \text{Isolate the radical.}$$
$$5x - 15 = 10\sqrt{2x-1}$$
$$(5x-15)^2 = (10\sqrt{2x-1})^2 \quad \text{Square both sides.}$$
$$(5x-15)(5x-15) = 100(2x-1)$$
$$25x^2 - 150x + 225 = 200x - 100$$
$$25x^2 - 350x + 325 = 0 \quad \text{Solve the remaining quadratic equation using the quadratic formula.}$$
$$x = 13 \quad x = 1$$

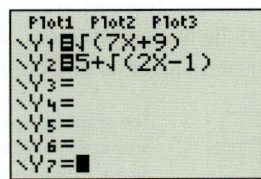

Check the answers. Only $x = 13$ works. $x = 1$ does not work, so it is not a solution to the equation.

Therefore, $x = 13$ is the only solution to this equation.

PRACTICE PROBLEM FOR EXAMPLE 5

Solve $\sqrt{4x+1} = 3 + \sqrt{x-2}$.

Example 6 Identifying extraneous solutions

Solve the following.

a. $\sqrt{x-5} + 7 = 4$

b. $-5\sqrt{x+4} - 10 = 15$

SOLUTION

a.
$$\sqrt{x-5} + 7 = 4$$
$$\sqrt{x-5} = -3 \quad \text{Isolate the radical.}$$
$$(\sqrt{x-5})^2 = (-3)^2 \quad \text{Square both sides.}$$
$$x - 5 = 9$$
$$x = 14$$
$$\sqrt{(14)-5} + 7 \stackrel{?}{=} 4 \quad \text{Check the answer.}$$
$$10 \neq 4 \quad \text{This answer does not work.}$$

This solution does not work. There are no real solutions to this equation, and $x = 14$ is an extraneous solution. Notice that in the second step, the square root was equal to a negative number. Because this is not possible, there will be no solution. We can use the graph to confirm that there are no real solutions. The graphs of the two sides do not cross, so there is no real solution.

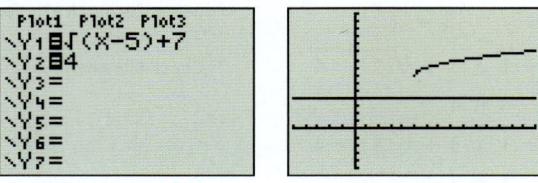

b.
$$-5\sqrt{x+4} - 10 = 15$$
$$-5\sqrt{x+4} = 25 \quad \text{Isolate the radical.}$$
$$\sqrt{x+4} = -5$$
$$(\sqrt{x+4})^2 = (-5)^2 \quad \text{Square both sides.}$$
$$x + 4 = 25 \quad \text{Solve for x.}$$
$$x = 21$$

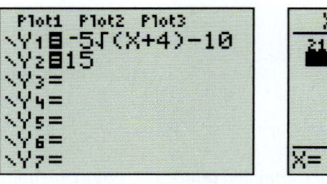 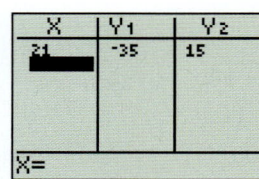

Check the answer.
$x = 21$ does not work, so it is not a solution.

The solution that we found does not work, so we believe that there are no solutions to this equation. Notice that in the second step, the square root was equal to a negative number. Because this is not possible, there will be no solution. To check that there are no solutions, we can graph both sides of the equation. If the two graphs cross, there would be a solution and we would have to solve the equation again to find our mistake.

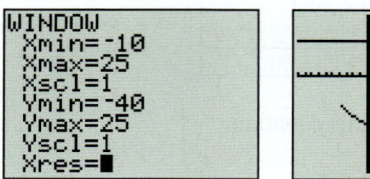 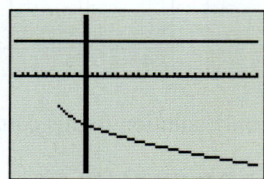

The graphs do not cross, so there are no solutions to this equation.

PRACTICE PROBLEM FOR EXAMPLE 6

Solve $4\sqrt{2x+8} + 20 = 4$.

Solving Radical Equations Involving Higher-Order Roots

When an equation has a higher-order root, the solving process will be basically the same as with square roots. After isolating the radical, raise both sides of the equation to the power of the roots index. If the root has an index of n, then we raise both sides of the equation to the nth power.

Another way to think of this process is to consider the root as a fraction exponent.

$$\sqrt[n]{x} = x^{\frac{1}{n}}$$

If an equation has a higher-order root we can undo the fraction exponent by raising both sides to the reciprocal exponent.

$$\sqrt[7]{x} = 3$$

$$x^{\frac{1}{7}} = 3 \qquad \text{Rewrite the radical using a fraction exponent.}$$

$$\left(x^{\frac{1}{7}}\right)^7 = 3^7 \qquad \text{Raise both sides to the reciprocal exponent.}$$

$$x = 2187$$

> **Steps to Solving Radical Equations Involving Higher-Order Roots**
> 1. Isolate a radical on one side of the equation.
> 2. Raise both sides of the equation to the reciprocal exponent. This will eliminate the isolated radical.
> 3. Isolate any remaining radical if needed and raise both sides to the reciprocal exponent again.
> 4. Solve the remaining equation.
> 5. Check your answer in the original equation.

Example 7 Solving higher-order radical equations

Solve the following equations.

a. $\sqrt[3]{2x} = 5$ b. $\sqrt[5]{4x - 9} = 2$

c. $5\sqrt[3]{x - 7} = 40$

SOLUTION

a.
$$\sqrt[3]{2x} = 5$$
$$\left(\sqrt[3]{2x}\right)^3 = 5^3 \qquad \text{Because this is a cube root, raise both sides to the third power.}$$
$$2x = 125$$
$$\frac{2x}{2} = \frac{125}{2}$$
$$x = 62.5$$

$$\sqrt[3]{2(62.5)} \stackrel{?}{=} 5 \qquad \text{Check the answer.}$$
$$5 = 5 \qquad \text{The answer works.}$$

b.
$$\sqrt[5]{4x - 9} = 2$$
$$\left(\sqrt[5]{4x - 9}\right)^5 = 2^5 \qquad \text{Because this is a fifth root, raise both sides to the power of 5.}$$
$$4x - 9 = 32$$
$$4x = 41$$
$$x = 10.25$$

$$\sqrt[5]{4(10.25) - 9} \stackrel{?}{=} 2 \qquad \text{Check the answer.}$$
$$2 = 2 \qquad \text{The answer works.}$$

c. Again, we will isolate the radical. Since this is a cube root, we will raise both sides to the power of 3, the reciprocal power.

$$5\sqrt[3]{x-7} = 40$$

$$\frac{5\sqrt[3]{x-7}}{5} = \frac{40}{5}$$ Isolate the radical by dividing both sides of the equation by 5.

$$\sqrt[3]{x-7} = 8$$ Raise both sides of the equation to the reciprocal exponent, 3.

$$(\sqrt[3]{x-7})^3 = 8^3$$

$$x - 7 = 512$$ Solve the remaining equation by adding 7 to both sides.

$$x = 519$$

$$5\sqrt[3]{(519)-7} \stackrel{?}{=} 40$$ Check the answer.

$$40 = 40$$ The answer works.

PRACTICE PROBLEM FOR EXAMPLE 7

Solve the following equations.

a. $\sqrt[4]{2x+9} = 3$

b. $4\sqrt[3]{7x-6} - 20 = -68$

8.4 Exercises

For Exercises 1 through 8, solve each equation. Check your answers in the original example.

1. $\sqrt{x+2} = 5$
2. $\sqrt{w-7} = 10$
3. $\sqrt{3x+4} = 15$
4. $\sqrt{7x+1} = 13$
5. $\sqrt{-2t+4} = 7$
6. $\sqrt{-4m-9} = 6$
7. $3\sqrt{2x+5} + 12 = 24$
8. $-2\sqrt{3x-7} - 11 = -19$

For Exercises 9 through 12, use the following information: Police investigating traffic accidents use the fact that the speed, in miles per hour, of a car traveling on an asphalt road can be determined by the length of the skid marks left by the car after sudden braking. The speed can be modeled by

$$s = \sqrt{30fd}$$

where s is the speed of the car in miles per hour, f is the coefficient of friction for the road, and d is the length in feet of the skid marks.

9. Find the speed of a car if it leaves skid marks 150 feet long on a dry asphalt road that has a coefficient of friction equal to 1.

10. Find the speed of a car if it leaves skid marks 120 feet long on an asphalt road with a coefficient of friction equal to 0.444.

11. Find the coefficient of friction of a road if a car traveling 55 mph leaves skid marks 115 feet long after sudden braking. (Round to three decimal places.)

12. Find the length of skid marks a police officer would expect to find if a person traveling 25 mph suddenly brakes on a road with a coefficient of friction equal to 1.

For Exercises 13 through 16, recall the formula for the period of a pendulum from Example 2.

$$T(L) = 2\pi\sqrt{\frac{L}{32}}$$

13. Find the period of a pendulum if its length is 3 feet. See Example 1.

14. Find the period of a pendulum if its length is 10 feet.

15. If you want a clock's pendulum to have a period of 1 second, what must its length be?

16. If you want a clock's pendulum to have a period of 2 seconds, what must its length be?

For Exercises 17 through 20, use the following information: When an object is dropped from a height of h feet, the time it will take to hit the ground can be approximated by the formula

$$t = \sqrt{\frac{2h}{32}}$$

where t is the time in seconds for the object to fall h feet.

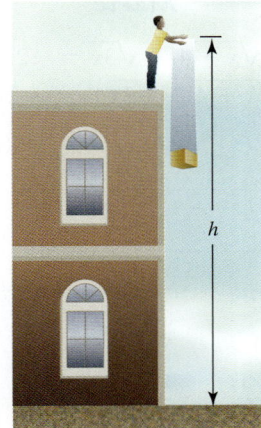

17. Find how long it will take an object to fall 100 feet.
18. Find how long it will take an object to fall 50 feet.
19. If you want an object to fall for 10 seconds, from what height should the object be dropped?
20. If you want an object to fall for 20 seconds, from what height should the object be dropped?

For Exercises 21 through 24, use the following information: The distance a person can see to the horizon depends on their height above the ground. The distance can be found using the formula

$$d = \sqrt{1.5h}$$

where d is the distance in miles when the person's eye is h feet above ground.

21. Find the distance someone can see to the horizon if their eye is 6 feet off the ground.
22. Find the distance someone can see to the horizon if their eye is 24 feet off the ground.
23. A person in an airplane can see 175 miles to the horizon, what altitude is the plane?
24. How high off the ground must a person's eye be to see 15 miles to the horizon?

For Exercises 25 through 32, solve each equation. Check your answers in the original equation.

25. $5\sqrt{2x} = 40$
26. $-0.25\sqrt{2.3w} = -14$
27. $2.4\sqrt{2.5g + 4} - 7.5 = 4.6$
28. $3.5\sqrt{1.7x + 6} + 2.8 = 14.3$
29. $\sqrt{2x + 5} = \sqrt{3x - 4}$
30. $\sqrt{p + 6} = \sqrt{3p + 2}$
31. $\sqrt{4 + 2z} = \sqrt{3 - 7z}$
32. $\sqrt{2 + 5x} = \sqrt{3x + 10}$

For Exercises 33 and 34, recall the formula for the speed of sound at sea level from Example 3.

$$c = 340.3\sqrt{\frac{T + 273.15}{288.15}}$$

33. Find the speed of sound at sea level when the temperature is 28°C. See Example 2.
34. Find the temperature at sea level if the speed of sound is 345 m/s.
35. In Exercise 1 of Section 8.1 on page 624, we used the function

$$C(h) = 2.29\sqrt{h} + 3.2$$

to calculate the monthly cost in millions of dollars to produce h hundred thousand space heaters at Jim Bob's Heater Source.

a. How many heaters can Jim Bob's make in a month if they have a budget of $15 million?
b. How many heaters can Jim Bob's make if they have a budget cut and can spend only $12 million?

36. In Exercise 3 of Section 8.1 on page 625, we used the function

$$L(M) = 0.0165\sqrt[3]{M}$$

to determine the leg length in meters of a basilisk lizard with a body mass of M grams.

a. Use this model to estimate the live body mass of a museum specimen that has a leg length of 0.1 meter.
b. Use the model to estimate the live body mass of a specimen that has a leg length of 0.075 meter.

37. In Exercise 4 of Section 8.1 on page 625, we used the function

$$L(M) = 0.330\sqrt[3]{M}$$

where L(M) represents the body length in meters of a mammal with a body mass of M kilograms.

a. Use this model to estimate the body mass of a mammal whose body length is 1 meter.
b. Use this model to estimate the body mass of a mammal whose body length is 2.4 meters.

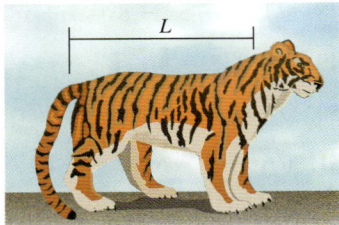

38. In Exercise 5 of Section 8.1 on page 625, we were given the following formula for the relationship between the thoracic mass of the butterfly and its mean forward airspeed.

$$V(m) = 17.6\sqrt[5]{m^3}$$

where V(m) represents the mean forward airspeed in meters per second of a butterfly with a thoracic mass of m grams.

a. Use the model to estimate the thoracic mass of a butterfly that can fly at 5 meters per second.

b. Use the model to estimate the thoracic mass of a butterfly that can fly at 6.2 meters per second.

For Exercises 39 through 42, use the following information: according to Kepler's Third Law of Planetary Motion, the average distance of an object from the sun can be estimated using the formula

$$a = \sqrt[3]{P^2}$$

where a is the average distance from the sun in astronomical units (AU) when the object's orbit is P years. (An astronomical unit is the distance from the earth to the sun.)

39. Find the distance from the sun to Uranus if its orbit is approximately 84 years.

40. Find the distance from the sun to Neptune if its orbit is approximately 165 years.

41. Find the number of years it takes Pluto to orbit the sun if its average distance from the sun is 39.5 AU.

42. Find the number of years it takes Jupiter to orbit the sun if its average distance from the sun is 5.2 AU.

For Exercises 43 through 58, solve each equation. Check your answers in the original equation.

43. $\sqrt{x-7} - \sqrt{x} = -1$
44. $\sqrt{x-11} - \sqrt{x} = -1$
45. $\sqrt{2x+3} = 1 - \sqrt{x+5}$
46. $\sqrt{6x-5} = 4 + \sqrt{5x+2}$
47. $\sqrt{4x^2+9x+5} = 5 + x$
48. $\sqrt{12x^2+5x-8} = 7 + x$
49. $\sqrt{w+2} = 4 - w$
50. $\sqrt{-m+2} = m - 2$
51. $\sqrt{6x+1} - \sqrt{9x} = -1$
52. $\sqrt{2x+11} + \sqrt{2x+7} + 4 = 0$
53. $\sqrt{6x+5} + \sqrt{3x+2} = 5$
54. $\sqrt{3x-5} - \sqrt{3x+3} = 4$
55. $\sqrt{8x+8} - \sqrt{8x-4} = -2$
56. $\sqrt{5x+4} - \sqrt{5x-7} = -9$
57. $\sqrt{2x+3} = 1 + \sqrt{x+1}$
58. $\sqrt{2x+5} - 3 = \sqrt{x-2}$

For Exercises 59 and 60, the students were asked to solve the given radical equation. Explain what the student did wrong. Give the correct answer.

59. $\sqrt{x+4} - \sqrt{3x} = -2$

Marie
$\sqrt{x+4} - \sqrt{3x} = -2$
$(\sqrt{x+4} - \sqrt{3x})^2 = (-2)^2$
$x + 4 - 3x = 4$
$-2x + 4 = 4$
$-2x = 0$
$x = 0$

60. $\sqrt{x^2+1} = 10$

John
$\sqrt{x^2+1} = 10$
$x + 1 = 10$
$x = 9$

For Exercises 61 through 68, solve each equation. Check your answers in the original equation.

61. $\sqrt[3]{2x+5} = 6$
62. $\sqrt[5]{x-8} = 4$
63. $\sqrt[4]{5x-9} = 2$
64. $\sqrt[7]{145x} = 6$
65. $\sqrt[3]{5x+2} + 7 = 12$
66. $\sqrt[5]{2x-15} + 9 = 4$
67. $\sqrt[3]{-4x+3} = -5$
68. $\sqrt[5]{-2x+20} = -4$

For Exercises 69 through 80, solve each equation or simplify each expression by performing the indicated operation.

69. $\sqrt{7x+2} = 3$
70. $\sqrt{x+2} + 3\sqrt{x+2}$
71. $\sqrt[3]{2x} \cdot \sqrt[3]{20x}$
72. $\sqrt[4]{3x+5} - 7 = -4$
73. $(6 + \sqrt{2x})(6 - \sqrt{2x})$
74. $\sqrt{\dfrac{10a^3b}{8x}}$
75. $5\sqrt{x-8} = \sqrt{x+4} + 6$
76. $\dfrac{10}{3 + \sqrt{5}}$
77. $\sqrt[6]{4ab^2} + 5b\sqrt{32a} + \sqrt[6]{4ab^2} + 7\sqrt{2ab^2}$
78. $\sqrt[3]{4m+3} = -5$
79. $\sqrt{6a+1} = 1 + \sqrt{5a-4}$
80. $\sqrt[3]{108m^5n^6p}$

For Exercises 81 through 90, determine whether the equation is linear, quadratic, exponential, logarithmic, radical, rational, or other. Solve the equation. Check your answers in the original equation.

81. $4t^2 + 3t = 232$
82. $40(0.8)^n = 20.48$
83. $6g + 8 = 3(g - 4)$
84. $\sqrt{m + 8} = 30$
85. $2x^3 + 8x^2 = 24x$
86. $\ln(a + 5) = 4$
87. $\sqrt{2n + 6} + \sqrt{n - 4} = 5$
88. $\dfrac{5}{x + 2} + \dfrac{2x}{x - 7} = \dfrac{10}{x^2 - 5x - 14}$
89. $7d^4 = 45927$
90. $\sqrt[3]{7d - 8} = -4$

8.5 Complex Numbers

LEARNING OBJECTIVES
- Identify complex numbers.
- Perform arithmetic operations with complex numbers.
- Find complex solutions to equations.

Definition of Imaginary and Complex Numbers

Throughout our work with radicals, we have noted that the square root of a negative number is not a real number. We saw this in the chapter on quadratics (Chapter 4) when we found no real solutions to some problems. This was a problem that faced mathematicians up until the 1500s. During the early 1500s, negative numbers were just starting to be accepted by mathematicians, and that led two Italian mathematicians, Cardano and Tartaglia, to negative numbers under square roots while trying to solve cubic equations. This was later expanded on by mathematicians such as Bombelli, Euler, and Gauss. Today, something like $\sqrt{-4}$ is considered a nonreal number and is called an **imaginary number.** Because these numbers were not believed to really exist, they were called imaginary numbers. Later they were proven to exist, and they have been shown to be applicable in different fields of mathematics and science. Although imaginary numbers were proven to exist, the name had stuck by then, so we must remember that the name *imaginary* does not mean that these numbers do not exist. In mathematics, the number $\sqrt{-1}$ is the **imaginary unit** and is usually represented by the letter i. Using the letter i, we can represent imaginary numbers without showing a negative under a square root.

> **DEFINITION**
>
> **Imaginary unit**
>
> $$\sqrt{-1} = i$$

Example 1 Simplifying radicals into complex numbers

Simplify the following, using the imaginary number i. Round any approximations to three decimal places.

a. $\sqrt{-25}$
b. $\sqrt{-100}$
c. $\sqrt{-30}$
d. $-\sqrt{16}$

SOLUTION

a. $\sqrt{-25} = \sqrt{-1 \cdot 25}$ Factor out the negative 1.

$\phantom{\sqrt{-25}} = \sqrt{-1} \cdot \sqrt{25}$

$\phantom{\sqrt{-25}} = i \cdot 5$

$\phantom{\sqrt{-25}} = 5i$

b. $10i$

c. $i\sqrt{30} \approx 5.477i$

d. $-\sqrt{16} = -4$. Note that this is not an imaginary number because the negative is not inside the square root.

Imaginary numbers can be combined with the real numbers into what are called **complex numbers.** Any number that can be written in the form $a + bi$ is a complex number; a is considered the real part of a complex number, and b is considered the imaginary part. All real numbers are considered complex numbers whose imaginary part is equal to zero. All imaginary numbers are also considered complex numbers whose real part is equal to zero.

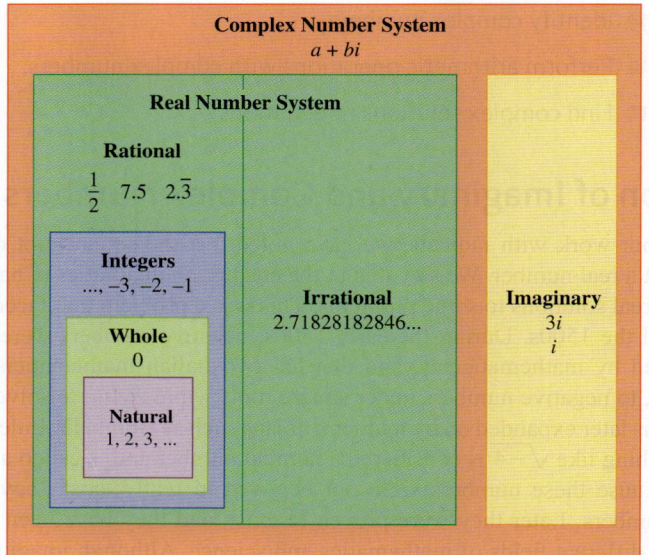

Complex numbers are used in many areas of mathematics and many physics and engineering fields. Electronics uses complex numbers to work with voltage calculations. The shapes of airplane wings are developed and studied by using complex numbers. Many areas of algebra, such as fractals and chaos theory, can be studied by working in the complex number system.

DEFINITION

Complex number   Any number that can be written in the form

$$a + bi$$

where a and b are real numbers.

Example 2 Name the parts of complex numbers

For each complex number, name the real part and the imaginary part.

a. $5 + 4i$ **b.** $-3 + 7i$ **c.** $5i$ **d.** 9

SOLUTION

a. $5 + 4i$; Real part = 5; imaginary part = 4.

b. $-3 + 7i$; Real part = -3; imaginary part = 7.

c. In standard complex number form, $5i = 0 + 5i$.
Real part = 0; imaginary part = 5.

d. In standard complex number form, $9 = 9 + 0i$.
Real part = 9; imaginary part = 0.

Operations with Complex Numbers

Complex numbers can be added and subtracted easily by adding or subtracting the real parts together and then adding or subtracting the imaginary parts together. This is similar to combining like terms with variables.

> **Adding Complex Numbers**
> $$(a + bi) + (c + di)$$
> $$(a + c) + (b + d)i$$
> When adding complex numbers, add the real parts together and then add the imaginary parts.
>
> **Subtracting Complex Numbers**
> $$(a + bi) - (c + di)$$
> $$(a - c) + (b - d)i$$
> When subtracting complex numbers, subtract the real parts and then subtract the imaginary parts. Be careful with the positive and negative signs.

Example 3 Add or subtract complex numbers

Add or subtract the following complex numbers.

a. $(2 + 8i) + (6 + 7i)$ **b.** $(5 - 4i) + (7 + 6i)$

c. $(6 + 3i) - (4 + 8i)$ **d.** $(2.5 + 3.8i) - (4.6 - 7.2i)$

SOLUTION

a. $(2 + 8i) + (6 + 7i) = 2 + 6 + 8i + 7i$ Add the real parts.
$= (2 + 6) + (8i + 7i)$ Add the imaginary parts.
$= 8 + 15i$

b. $12 + 2i$

c. $(6 + 3i) - (4 + 8i) = 6 + 3i - 4 - 8i$ Distribute the negative sign.
$= (6 - 4) + (3i - 8i)$ Subtract the real parts.
$= 2 - 5i$ Subtract the imaginary parts.

d. $(2.5 + 3.8i) - (4.6 - 7.2i) = 2.5 + 3.8i - 4.6 + 7.2i$ Distribute the negative sign.
$= (2.5 - 4.6) + (3.8i + 7.2i)$ Combine the real parts.
$= -2.1 + 11i$ Combine the imaginary parts.

PRACTICE PROBLEM FOR EXAMPLE 3

Add or subtract the following complex numbers.

a. $(2 + 6i) + (-3 + 5i)$ **b.** $(3.4 - 7.5i) - (-4.2 - 5.9i)$

c. $(-7 + 6i) + (-3 - 4i)$ **d.** $(-3 + 5i) - (4 + 9i)$

Because $i = \sqrt{-1}$, other powers of i can be calculated by considering the following pattern,

$$i = \sqrt{-1} \qquad\qquad i^5 = 1i = i$$
$$i^2 = (\sqrt{-1})^2 = -1 \qquad\qquad i^6 = -1$$
$$i^3 = -1(\sqrt{-1}) = -\sqrt{-1} = -i \qquad\qquad i^7 = -i$$
$$i^4 = i^2 i^2 = (-1)(-1) = 1 \qquad\qquad i^8 = 1$$

This pattern will continue to repeat as you have higher and higher powers of i. If you note that $i^4 = 1$, then you can basically reduce any powers of i by a multiple of 4 without changing the resulting value. For example,

$$i^5 = i^{5-4} = i \qquad\qquad \text{Reduce the exponent by 4.}$$
$$i^{10} = i^{10-8} = i^2 = -1 \qquad\qquad \text{Reduce the exponent by 8 and simplify.}$$

In most problems we will deal with, the most important power of i that you should know and use is $i^2 = -1$. When you multiply complex numbers, using this fact will help you to reduce the answers to complex form. Whenever you see i^2 in a calculation, you should replace it with a -1 and continue to combine like terms and simplify.

Multiplying Complex Numbers

$$(a + bi)(c + di)$$

When multiplying complex numbers, use the distributive property, and change any i^2 to -1. Simplify the result by combining like terms and writing the result in the standard form of a complex number.

Example 4 Multiplying complex numbers

Multiply the following complex numbers.

a. $3(4 + 9i)$ 	**b.** $2i(7 - 3i)$
c. $(2 + 5i)(4 + 8i)$ 	**d.** $(3 + 2i)(3 - 2i)$

SOLUTION

In all of these problems, we will use the distributive property and simplify where possible.

a. $3(4 + 9i) = 12 + 27i$ 	Distribute the 3.

b. $2i(7 - 3i) = 14i - 6i^2$ 	Distribute the $2i$.
$ = 14i - 6(-1)$ 	Replace i^2 with -1.
$ = 14i + 6$ 	Simplify.
$ = 6 + 14i$ 	Put in standard complex form $a + bi$.

c. $(2 + 5i)(4 + 8i) = 8 + 16i + 20i + 40i^2$ 	Use the distributive property (FOIL).
$ = 8 + 36i + 40i^2$ 	Combine like terms.
$ = 8 + 36i + 40(-1)$ 	Replace i^2 with -1.
$ = 8 + 36i - 40$ 	Simplify.
$ = -32 + 36i$

d. $(3 + 2i)(3 - 2i) = 9 - 6i + 6i - 4i^2$ Use the distributive property (FOIL).
$= 9 - 4i^2$ Combine like terms.
$= 9 - 4(-1)$ Replace i^2 with -1.
$= 9 + 4$ Simplify.
$= 13$

PRACTICE PROBLEM FOR EXAMPLE 4

Multiply the following complex numbers.

a. $3i(4 + 8i)$
b. $(2 + 7i)(3 + 5i)$
c. $(4 - 2i)(7 + 9i)$
d. $(2 + 5i)(2 - 5i)$

Part d of Example 4 is an example of multiplying two complex numbers that have a very special relationship. These two complex numbers are what are called **complex conjugates** of one another. When complex conjugates are multiplied together, note that the product is a real number. Therefore, it has no imaginary part remaining.

DEFINITION

Complex conjugates

$$a + bi \quad \text{and} \quad a - bi$$

are conjugates of one another. The conjugate of a complex number is that complex number with the sign of the imaginary number changed.

$$5 - 3i \quad \text{and} \quad 5 + 3i$$

are complex conjugates of one another.

$$(a + bi)(a - bi) = a^2 + b^2$$

Connecting the Concepts

How do we use the word conjugates?

The word *conjugate* was used earlier in this chapter when we discussed radical expressions. Now we are using it with complex numbers. Is there a connection?

- **Radical expressions**

    $2 + 5\sqrt{x}$ and $2 - 5\sqrt{x}$

    are conjugates.

- **Complex numbers**

    $3 + 7i$ and $3 - 7i$

    are conjugates.

These two situations are related because we know that $i = \sqrt{-1}$. This makes complex conjugates similar to the conjugates in radical expressions.

$3 + 7\sqrt{-1}$ and $3 - 7\sqrt{-1}$

are conjugates.

Example 5 Writing complex conjugates

Write the complex conjugate of the following.

a. $2 + 9i$
b. $4 - 6i$
c. $3.4i$
d. 10

SOLUTION

a. $2 - 9i$

b. $4 + 6i$

c. $-3.4i$

d. 10 Because the imaginary part is zero, there is no other complex conjugate.

Example 6 Multiplying by complex conjugates

Multiply the following complex numbers by their conjugates.

a. $2 + 9i$
b. $4 - 6i$
c. $3.4i$

SOLUTION

a. $(2 + 9i)(2 - 9i) = 4 - 18i + 18i - 81i^2$ Use the distributive property.

$ = 4 - 81i^2$ Combine like terms.

$ = 4 - 81(-1)$ Replace i^2 with -1.

$ = 4 + 81$ Simplify.

$ = 85$

b. $(4 - 6i)(4 + 6i) = 16 + 24i - 24i - 36i^2$ Use the distributive property.

$ = 16 - 36i^2$ Combine like terms.

$ = 16 - 36(-1)$ Replace i^2 with -1.

$ = 16 + 36$ Simplify.

$ = 52$

c. $3.4i(-3.4i) = -11.56i^2$ Distribute.

$ = -11.56(-1)$ Replace i^2 with -1.

$ = 11.56$ Simplify.

When complex conjugates are multiplied together, the sign change in the conjugate allows the imaginary parts to add to zero, and we are left with a real number. This is a very helpful attribute of conjugates when complex numbers are involved in a division problem. In dealing with division or fractions, it is standard practice not to have complex numbers in denominators. Imaginary numbers should be eliminated from any denominator. This can be done by multiplying both the numerator and the denominator by the conjugate of the denominator (similar to what we did to rationalize denominators). This will not change the value of the fraction, but it will change the denominator into a real number. The resulting fraction can then be written in standard complex form.

■ Connecting the Concepts

What do we rationalize?

In Section 8.3, we rationalized the denominator of fractions that had radicals. Rationalizing fractions with complex numbers is related to rationalizing fractions with radicals because we know that $i = \sqrt{-1}$. This makes complex numbers similar to the radical expressions.

Just as we used conjugates to rationalize fractions with radicals, we use conjugates to rationalize fractions that contain complex numbers in the denominator.

Fraction with radicals

$\dfrac{2 + \sqrt{3}}{3 + \sqrt{5}} =$

$= \dfrac{2 + \sqrt{3}}{3 + \sqrt{5}} \cdot \dfrac{3 - \sqrt{5}}{3 - \sqrt{5}}$

$= \dfrac{6 - 2\sqrt{5} + 3\sqrt{3} - \sqrt{15}}{4}$

Fraction with complex numbers

$\dfrac{4 + 5i}{7 + 8i} =$

$= \dfrac{4 + 5i}{7 + 8i} \cdot \dfrac{7 - 8i}{7 - 8i}$

$= \dfrac{68 + 3i}{113}$

$= \dfrac{68}{113} + \dfrac{3}{113}i$

> **Dividing Complex Numbers**
>
> $$\dfrac{a + bi}{c + di}$$
>
> $$\dfrac{(a + bi)}{(c + di)} \cdot \dfrac{(c - di)}{(c - di)}$$
>
> In dividing complex numbers, you should clear all denominators of imaginary numbers. This is done by multiplying both the denominator and the numerator by the conjugate of the denominator. Remember to reduce the final answer and write it in the standard form for complex numbers.

Example 7   Dividing complex numbers

Divide the following. Give answers in the standard form for a complex number.

a. $\dfrac{10 + 8i}{2}$ **b.** $\dfrac{4 + 7i}{5 + 3i}$ **c.** $\dfrac{2 - 9i}{4 + 3i}$ **d.** $\dfrac{6 - 7i}{5i}$

SOLUTION

a. Since the denominator does not have an imaginary part, we use the standard division algorithm.

$\dfrac{10 + 8i}{2} = \dfrac{10}{2} + \dfrac{8}{2}i$ Reduce the fraction and put it into the standard form of a complex number.

$\phantom{\dfrac{10 + 8i}{2}} = 5 + 4i$ Simplify.

b.

$$\frac{4+7i}{5+3i}$$

$$\frac{(4+7i)}{(5+3i)} \cdot \frac{(5-3i)}{(5-3i)}$$ Multiply the numerator and denominator by the conjugate of the denominator. Use the distributive property (FOIL).

$$\frac{20-12i+35i-21i^2}{25-15i+15i-9i^2}$$

$$\frac{20+23i-21(-1)}{25-9(-1)}$$

$$\frac{20+23i+21}{25+9}$$

$$\frac{41+23i}{34}$$

$$\frac{41}{34}+\frac{23}{34}i$$ Write in the standard form of a complex number.

c.

$$\frac{2-9i}{4+3i}$$

$$\frac{(2-9i)}{(4+3i)} \cdot \frac{(4-3i)}{(4-3i)}$$ Multiply the numerator and denominator by the conjugate of the denominator. Use the distributive property (FOIL).

$$\frac{8-6i-36i+27^2}{16-12i+12i-9i^2}$$

$$\frac{8-42i-27}{16+9}$$

$$\frac{-19-42i}{25}$$

$$-\frac{19}{25}-\frac{42}{25}i$$ Write in the standard form of a complex number.

d.

$$\frac{6-7i}{5i}$$

$$\frac{(6-7i)}{5i} \cdot \frac{i}{i}$$ Multiply the numerator and denominator by i. Because the denominator does not have a real part, multiply by i to rationalize the denominator.

$$\frac{6i-7i^2}{5i^2}$$

$$\frac{6i+7}{-5}$$

$$-\frac{7}{5}-\frac{6}{5}i$$ Write in the standard form of a complex number.

PRACTICE PROBLEM FOR EXAMPLE 7

Divide the following. Give answers in the standard form for a complex number.

a. $\dfrac{14+8i}{2}$ **b.** $\dfrac{4-7i}{3i}$ **c.** $\dfrac{4+2i}{3-7i}$ **d.** $\dfrac{7.2+3.4i}{1.4+5.6i}$

Solving Equations with Complex Solutions

Some equations will have complex solutions. The most common place in which we will see these types of solutions is in working with quadratics. The quadratic formula is a great tool to find both real and complex solutions to any quadratic equation. Now when a discriminant ($b^2 - 4ac$) is a negative number, we can write our solutions using complex numbers instead of just saying that there are no real solutions. This results in a more complete answer to the equation. Notice that if a complex number is a solution to a polynomial, the complex conjugate will also be a solution to that equation. In most applications, we are interested only in the real solutions to an equation, but in some contexts, such as electrical engineering, complex solutions are of interest.

> **Example 8** Solving equations with complex solutions
>
> Solve the following equations. Give answers in the standard form for a complex number.
>
> **a.** $t^2 + 2t + 5 = 0$ **b.** $x^2 = -25$
>
> **c.** $x^2 + 4x = -30$ **d.** $x^3 - 10x^2 + 29x = 0$
>
> **SOLUTION**
>
> **a.**
> $$t^2 + 2t + 5 = 0$$
> $$t = \frac{-(2) \pm \sqrt{(2)^2 - 4(1)(5)}}{2(1)} \quad \text{Use the quadratic formula.}$$
> $$t = \frac{-2 \pm \sqrt{-16}}{2} \quad \text{The discriminant is } -16, \text{ so the answer will be a complex number.}$$
> $$t = \frac{-2 \pm 4i}{2}$$
> $$t = -1 + 2i \quad t = -1 - 2i \quad \text{Write in standard form.}$$
>
> **b.**
> $$x^2 = -25$$
> $$x = \pm\sqrt{-25} \quad \text{Use the square root property.}$$
> $$x = \pm 5i$$
>
> **c.**
> $$x^2 + 4x = -30$$
> $$x^2 + 4x + 30 = 0$$
> $$x = \frac{-(4) \pm \sqrt{(4)^2 - 4(1)(30)}}{2(1)} \quad \text{Use the quadratic formula.}$$
> $$x = \frac{-4 \pm \sqrt{-104}}{2} \quad \text{The discriminant is } -104, \text{ so the answer will be a complex number.}$$
> $$x = \frac{-4 \pm 2i\sqrt{26}}{2}$$
> $$x = -2 + i\sqrt{26} \quad x = -2 - i\sqrt{26} \quad \text{Write in standard form.}$$
> $$x \approx -2 + 5.1i \quad x \approx -2 - 5.1i$$
>
> **d.**
> $$x^3 - 10x^2 + 29x = 0 \quad \text{Factor the common term out.}$$
> $$x(x^2 - 10x + 29) = 0 \quad \text{Set each factor equal to zero and continue to solve.}$$
> $$x = 0 \quad x^2 - 10x + 29 = 0$$
> $$x = 0 \quad x = \frac{-(-10) \pm \sqrt{(-10)^2 - 4(1)(29)}}{2(1)} \quad \text{Use the quadratic formula.}$$
> $$x = 0 \quad x = \frac{10 \pm \sqrt{100 - 116}}{2}$$
> $$x = 0 \quad x = \frac{10 \pm \sqrt{-16}}{2} \quad \text{The discriminant is } -16, \text{ so the answer will be a complex number.}$$
> $$x = 0 \quad x = \frac{10 \pm 4i}{2}$$
> $$x = 0 \quad x = 5 + 2i \quad x = 5 - 2i \quad \text{Write in standard form.}$$

■ **Connecting the Concepts**

When should you give a nonreal answer?

Because imaginary numbers are not commonly used in applications, we rarely give a nonreal answer to an application problem. Some areas of science use imaginary numbers. In those applications, it should be made clear that an imaginary number answer makes sense.

In a math problem, giving the real and nonreal answers, provides the most complete answer. If the instructions specifically ask for all real solutions, then a nonreal answer should not be given.

Whenever giving a complex number as an answer, be sure to put it in the standard form for a complex number: $(a + bi)$.

PRACTICE PROBLEM FOR EXAMPLE 8

Solve the following equations. Give answers in the standard form for a complex number.

a. $w^2 + 6w + 18 = 0$ **b.** $x^2 = -81$

c. $x^2 + 8x = -18$ **d.** $x^3 - 4x^2 + 13x = 0$

8.5 Exercises

In Exercises 1 through 20, simplify the given expression and write your answer in terms of i. Give exact answers or round approximations to three decimal places.

1. $\sqrt{-64}$
2. $\sqrt{-144}$
3. $\sqrt{-81}$
4. $\sqrt{-16}$
5. $-\sqrt{36}$
6. $-\sqrt{25}$
7. $\sqrt{-20}$
8. $\sqrt{-75}$
9. $\sqrt{-200}$
10. $\sqrt{-60}$
11. $12 - \sqrt{-100}$
12. $5 + \sqrt{-36}$
13. $\sqrt{-4} + \sqrt{-36}$
14. $\sqrt{-49} + \sqrt{-25}$
15. $\sqrt{9} + \sqrt{-16}$
16. $\sqrt{4} - \sqrt{-64}$
17. $\sqrt{-3} + \sqrt{-5.6}$
18. $\sqrt{-5} + \sqrt{-24}$
19. $-4(3 + \sqrt{-63})$
20. $3(5 + \sqrt{-9})$

In Exercises 21 through 28, name the real part and the imaginary part of each complex number.

21. $2 + 5i$
22. $3 - 8i$
23. $2.3 - 4.9i$
24. $-5 + 8i$
25. 3
26. 2
27. $7i$
28. $-5i$

In Exercises 29 through 42, add or subtract the given complex numbers. Write your answer in the standard form for a complex number.

29. $(2 + 5i) + (6 + 4i)$
30. $(3 - 8i) + (2 + 6i)$
31. $(5.6 + 3.2i) + (2.3 - 4.9i)$
32. $(3 + 2i) - (5 + 8i)$
33. $(5i) + (3 - 9i)$
34. $(2 - 7i) - (5 - 6i)$
35. $(5 + 7i) - 10$
36. $(3 - 8i) - (10 - 12i)$
37. $(3 + 5i) - (7 + 8i)$
38. $(-4 - 6i) - (3 - 9i)$
39. $(4.7 - 3.5i) + (1.8 - 5.7i)$
40. $(2.4 - 9.6i) - (3.5 + 8.6i)$
41. $(2 + 8i) + (3 - 8i)$
42. $(4 - 7i) - (4 - 7i)$

In Exercises 43 through 54, multiply the given complex numbers. Write your answer in the standard form for a complex number.

43. $(2 + 3i)(4 + 7i)$
44. $(5 + 4i)(3 + 2i)$
45. $(2 - 4i)(3 - 5i)$
46. $(4 - 5i)(1 - 6i)$
47. $(7.5 + 3i)(4.5 - 9i)$
48. $(2 + 7.6i)(6 - 5.4i)$
49. $(2 + 8i)(2 + 8i)$
50. $(5 + 2i)(5 + 2i)$
51. $(3 - 7i)(3 + 7i)$
52. $(2 + 5i)(2 - 5i)$
53. $(7 + 6i)^2$
54. $(3 - 5i)^2$

In Exercises 55 through 68, divide the given complex numbers. Write your answer in the standard form for a complex number.

55. $\dfrac{12 + 9i}{3}$
56. $\dfrac{45 - 15i}{5}$
57. $\dfrac{4 + 7i}{2i}$
58. $\dfrac{5 + 8i}{3i}$
59. $\dfrac{2.4 + 3.7i}{7.2i}$
60. $\dfrac{3.5 + 8.3i}{2.4i}$
61. $\dfrac{5 + 2i}{2 + 3i}$
62. $\dfrac{4 + 6i}{2 - 7i}$
63. $\dfrac{2 - 4i}{3 - 7i}$
64. $\dfrac{3 + 7i}{2 + 5i}$
65. $\dfrac{-8 - 5i}{-2 + 9i}$
66. $\dfrac{-7 - 8i}{-3 - 5i}$
67. $\dfrac{-12 - 73i}{-5 - 14i}$
68. $\dfrac{-8 - 14i}{-11 - 9i}$

In Exercises 69 through 82, solve each equation. Write any complex solutions in standard form for a complex number.

69. $x^2 + 9 = 0$
70. $w^2 + 25 = 0$
71. $3t^2 + 45 = -3$
72. $2x^2 - 18 = 0$
73. $3(x - 8)^2 + 12 = 4$
74. $-0.25(t + 3.4)^2 + 9 = 15$
75. $2x^2 + 5x + 4 = -4$
76. $3x^2 + 2x + 7 = -8$
77. $5x^3 + 2x^2 - 7x = 0$
78. $8x^3 + 4x^2 - 14x = 0$
79. $0.25x^3 - 3.5x^2 + 16.25x = 0$
80. $3x^3 + 24x^2 + 53x = 0$
81. $2x^2 - 28x + 45 = -54$
82. $-3w^2 + 12w - 26 = 0$

Chapter Summary

Section 8.1 Radical Functions

- A **radical** is a square root $\sqrt{x}$ or higher root $\sqrt[n]{x}$.
- **Radical functions** can be used to model some real-life situations.
- **Higher roots** can be calculated on the calculator by using rational exponents. Use parentheses around all rational exponents.
- When not in a context, the domain of an even root is restricted to values that keep the radicand nonnegative.
- When not in a context, the domain and range of odd roots are typically all real numbers.

Example 1 The distance to the horizon at a particular altitude can be modeled by the function $H(a) = \sqrt{1.5a}$, where $H(a)$ is the distance to the horizon in miles at an altitude a feet above sea level. Use this equation to determine the distance to the horizon if you are standing on top of the Empire State Building at 1250 feet.

SOLUTION $H(1250) \approx 43.3$. The horizon will be 43.3 miles away when you are on the top of the Empire State Building.

Example 2 Give the domain and range of the function. Sketch a graph of the function.

  a. $f(x) = -\sqrt{x+6}$ **b.** $g(x) = \sqrt[5]{3x}$

SOLUTION **a.** For the domain the radicand of an even root must be nonnegative, so

$$x + 6 \geq 0$$
$$x \geq -6$$

Domain: $[-6, \infty)$ Range: $(-\infty, 0]$

x	$f(x)$
-6	0
3	-3
10	-4
30	-6

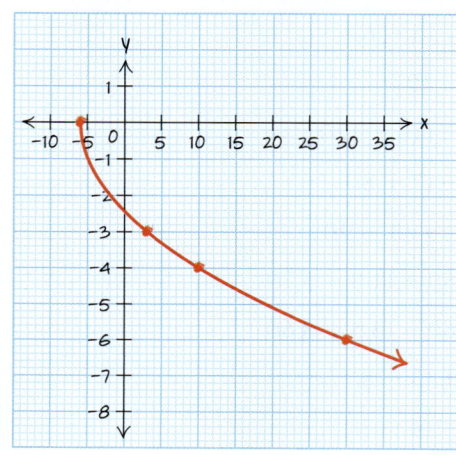

b. The function $g(x) = \sqrt[5]{3x}$ is an odd root.
Domain: All real numbers   Range: All real numbers

x	$f(x)$
-81	-3
$-\frac{1}{3}$	-1
$\frac{1}{3}$	1
81	3

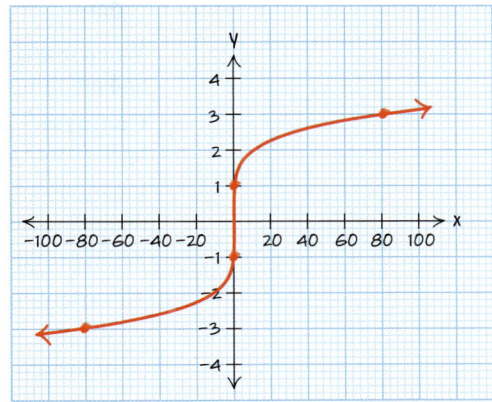

Section 8.2 Simplifying, Adding, and Subtracting Radicals

- The **principal root** is the positive root of a square root.
- To get a **negative root**, put a negative sign outside the square root.
- **Negative radicands** result in nonreal solutions when in radicals with an even index.
- The **product property of radicals** states that if $\sqrt[n]{a}$ and $\sqrt[n]{b}$ are real numbers and $n \geq 2$ is a natural number, then
$$\sqrt[n]{a \cdot b} = \sqrt[n]{a} \cdot \sqrt[n]{b}$$
- **Simplified radicals** have the following characteristics.
  1. The radicand does not contain any factors that are perfect powers of the index.
  2. All exponents in the radicand are less than the index.
  3. There are no fractions in the radicand.
  4. There are no radicals in the denominator of a fraction.
- To **add or subtract radicals,** they need to have the same radicand and index.
- Simplify radicals before adding or subtracting. The radicands do not change in adding or subtracting radicals only the coefficients will be added or subtracted.

Example 3 Simplify the following radicals.

a. $\sqrt{12x^3y^8}$ **b.** $\sqrt[5]{32a^4b^{10}c^7}$

SOLUTION
a. $\sqrt{12x^3y^8} = \sqrt{4x^2y^8}\ \sqrt{3x}$
$= 2xy^4\sqrt{3x}$

b. $\sqrt[5]{32a^4b^{10}c^7} = \sqrt[5]{2^5b^{10}c^5}\ \sqrt[5]{a^4c^2}$
$= 2b^2c\ \sqrt[5]{a^4c^2}$

Example 4 Perform the indicated operation and simplify.

a. $5\sqrt{x} + 3\sqrt{x}$ **b.** $\sqrt{8x^3y} - 7x\sqrt{2xy}$

SOLUTION
a. $5\sqrt{x} + 3\sqrt{x} = 8\sqrt{x}$

b. $\sqrt{8x^3y} - 7x\sqrt{2xy} = 2x\sqrt{2xy} - 7x\sqrt{2xy}$
$= -5x\sqrt{2xy}$

Section 8.3 Multiplying and Dividing Radicals

- When **multiplying radicals** with the same index, multiply the radicands together, and anything outside each radical is also multiplied together.
- **Conjugates** are expressions that are the sum and difference of the same two terms: $(a + b)$ and $(a - b)$.
- Fractions inside radicals can either be reduced inside the radical or separated into distinct radicals and then reduced.
- Fractions that have a radical in the denominator are usually rationalized.
- To **rationalize** a denominator of a fraction, use one of these techniques:
    1. When a single square root is in the denominator, you simply multiply both the numerator and the denominator by the original radical in the denominator.
    2. When higher roots are in the denominator, you multiply by the root with the same index with the needed factors to make the radical simplify completely.
    3. If two terms with a radical are in the denominator, you will need to use conjugates to eliminate the radical.

Example 5 Perform the indicated operations and simplify. Rationalize all denominators.

a. $2w\sqrt{3r} \cdot 5\sqrt{7wr}$ b. $\sqrt{\dfrac{4}{7x}}$

c. $\dfrac{7}{\sqrt[3]{5x^2}}$ d. $\dfrac{2 + \sqrt{3}}{5 - \sqrt{7}}$

SOLUTION

a. $2w\sqrt{3r} \cdot 5\sqrt{7wr} = 10w\sqrt{21wr^2}$
$= 10wr\sqrt{21w}$

b. $\sqrt{\dfrac{4}{7x}} \cdot \dfrac{\sqrt{7x}}{\sqrt{7x}} = \dfrac{\sqrt{28x}}{7x} = \dfrac{2\sqrt{7x}}{7x}$

c. $\dfrac{7}{\sqrt[3]{5x^2}} \cdot \dfrac{\sqrt[3]{25x}}{\sqrt[3]{25x}} = \dfrac{7\sqrt[3]{25x}}{5x}$

d. $\dfrac{2 + \sqrt{3}}{5 - \sqrt{7}} \cdot \dfrac{5 + \sqrt{7}}{5 + \sqrt{7}} = \dfrac{10 + 2\sqrt{7} + 5\sqrt{3} + \sqrt{21}}{25 + 5\sqrt{7} - 5\sqrt{7} - 7}$
$= \dfrac{10 + 2\sqrt{7} + 5\sqrt{3} + \sqrt{21}}{18}$

Section 8.4 Solving Radical Equations

- When **solving an equation with a radical,** isolate the radical on one side of the equation and then square both sides of the equation.
- If there is more than one radical, you might need to square both sides of the equation twice. Remember to always isolate a radical before you square both sides.
- If a radical is not a square root, then raise both sides of the equation to the same power as the index of the radical.
- Be sure to check your solutions in the original equation.

Example 6

The distance to the horizon at a particular altitude can be modeled by the function $H(a) = \sqrt{1.5a}$, where $H(a)$ is the distance to the horizon in miles at an altitude a feet above sea level. Use this function to find the altitude you must be at for the horizon to be 100 miles away.

SOLUTION

$$100 = \sqrt{1.5a}$$
$$100^2 = (\sqrt{1.5a})^2$$
$$1000 = 1.5a$$
$$6666.67 \approx a$$

For the horizon to be 100 miles away, you must be about 6666.67 feet above sea level.

Example 7

Solve the following equations.

a. $\sqrt{5x - 6} = 7 + \sqrt{x}$

b. $\sqrt[4]{2x + 8} + 9 = 20$

SOLUTION

a.
$$\sqrt{5x - 6} = 7 + \sqrt{x}$$
$$(\sqrt{5x - 6})^2 = (7 + \sqrt{x})^2$$
$$5x - 6 = (7 + \sqrt{x})(7 + \sqrt{x})$$
$$5x - 6 = 49 + 7\sqrt{x} + 7\sqrt{x} + x$$
$$5x - 6 = 49 + 14\sqrt{x} + x$$
$$4x - 55 = 14\sqrt{x}$$
$$(4x - 55)^2 = (14\sqrt{x})^2$$
$$16x^2 - 440x + 3025 = 196x$$
$$16x^2 - 636x + 3025 = 0$$
$$x \approx 34.23 \qquad x \approx 5.52 \qquad \text{Solve using the quadratic formula.}$$
$$x \approx 34.23$$

X	Y₁	Y₂
34.23	12.851	12.851
5.52	4.6476	9.3495

X=

Check your answers.

Only one of these answers is valid.

b.
$$\sqrt[4]{2x + 8} + 9 = 20$$
$$\sqrt[4]{2x + 8} = 11$$
$$(\sqrt[4]{2x + 8})^4 = (11)^4$$
$$2x + 8 = 14641$$
$$x = 7316.5$$

X	Y₁	Y₂
7316.5	20	20

X=

Section 8.5 Complex Numbers

- The imaginary unit is $i = \sqrt{-1}$.
- When a negative number is the radicand of a square root, it results in an imaginary number.
- A standard complex number is $a + bi$, where a is the real part and b is the imaginary part.
- Complex conjugates are $a + bi$ and $a - bi$.
- To add or subtract complex numbers, combine the real parts and combine the imaginary parts.
- To multiply complex numbers, use the distributive property if necessary, and use the fact that $i^2 = -1$ to reduce any higher powers of the imaginary unit.
- When dividing complex numbers, rationalize the denominator by multiplying both the numerator and denominator by the complex conjugate.
- When solving a quadratic equation, the quadratic formula can be used to find any complex solutions.

Example 8

Perform the indicated operation and answer in standard complex form.

a. $(2 + 5i) + (7 - 8i)$
b. $(3 - 7i)(4 + 2i)$
c. $\dfrac{5 + 3i}{2i}$
d. $\dfrac{4 - 3i}{2 + 5i}$

SOLUTION

a. $(2 + 5i) + (7 - 8i) = 9 - 3i$

b. $(3 - 7i)(4 + 2i) = 12 + 6i - 28i - 14i^2$
$= 12 - 22i + 14$
$= 26 - 22i$

c. $\dfrac{5 + 3i}{2i} \cdot \dfrac{i}{i} = \dfrac{5i + 3i^2}{2i^2}$

$= \dfrac{-3 + 5i}{-2}$

$= \dfrac{3}{2} - \dfrac{5}{2}i$

d. $\dfrac{4 - 3i}{2 + 5i} \cdot \dfrac{2 - 5i}{2 - 5i} = \dfrac{8 - 20i - 6i + 15i^2}{4 + 25}$

$= -\dfrac{7}{29} - \dfrac{26}{29}i$

Example 9

Solve $2x^2 + 4x + 20 = 0$ over the complex number system. Write your answers in standard complex form.

SOLUTION Using the quadratic formula, we get.

$x = \dfrac{-4 \pm \sqrt{(4)^2 - 4(2)(20)}}{2(2)}$

$x = \dfrac{-4 \pm \sqrt{-144}}{4}$

$x = \dfrac{-4 \pm 12i}{4}$

$x = -1 \pm 3i$

Chapter Review Exercises

1. The profit at Big Jim's Mart the first year it was open can be modeled by
   $$P(m) = 10.3 + 2\sqrt{m}$$
   where $P(m)$ represents the profit in thousands of dollars for Big Jim's Mart during the mth month of the year.
   a. Find the profit for Big Jim's Mart in the third month of the year.
   b. Find the profit for Big Jim's Mart during August. [8.1]

2. The weight of an alpaca, a grazing animal similar to a llama that is found mostly in the Andes, can be modeled by
   $$W(a) = 22.37\sqrt[5]{a^3}$$
   where $W(a)$ represents the body weight in kilograms of an alpaca that is a years old.
   Source: Model derived from data given in the Alpaca Registry Journal, *Volume II, Summer–Fall 1997.*

   Find the body weight of a four-year-old alpaca. [8.1]

In Exercises 3 through 16, give the domain and range of the given functions.

3. $f(x) = -5\sqrt{x}$ [8.1]
4. $h(x) = 5\sqrt{x}$ [8.1]
5. $f(x) = \sqrt{3x + 5}$ [8.1]
6. $g(x) = \sqrt{2x - 7}$ [8.1]
7. $f(x) = \sqrt{4 - x}$ [8.1]
8. $h(x) = \sqrt{7 - x}$ [8.1]
9. $f(x) = -2\sqrt[3]{x}$ [8.1]
10. $f(x) = 4\sqrt[3]{x + 3}$ [8.1]
11. $k(a) = 3\sqrt[4]{a}$ [8.1]
12. $M(p) = -2.3\sqrt[6]{p}$ [8.1]
13. $g(x) = 3.4\sqrt[5]{x}$ [8.1]
14. $f(x) = 8\sqrt[9]{x - 7}$ [8.1]

15. 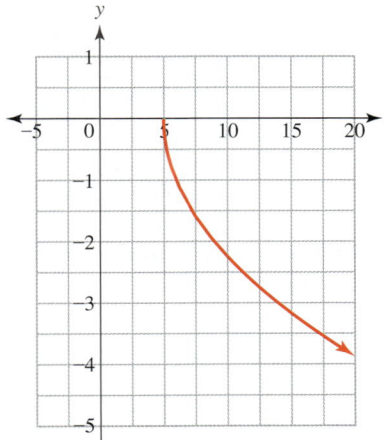 [8.1]

16. 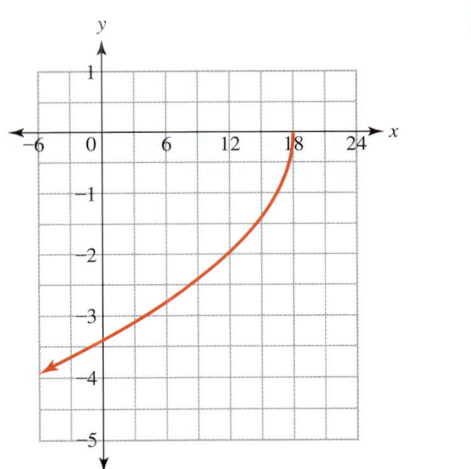 [8.1]

In Exercises 17 through 24, sketch the graph of the given radical functions.

17. $f(x) = \sqrt{x + 2}$ [8.1]
18. $h(x) = -\sqrt{x - 4}$ [8.1]
19. $f(x) = \sqrt{x - 6}$ [8.1]
20. $g(x) = -\sqrt{x + 12}$ [8.1]
21. $f(x) = \sqrt[3]{x + 6}$ [8.1]
22. $h(x) = \sqrt[3]{x - 10}$ [8.1]
23. $f(x) = -\sqrt[3]{x + 4}$ [8.1]
24. $g(x) = \sqrt[5]{6x}$ [8.1]

In Exercises 25 through 38, simplify the following radical expressions.

25. $\sqrt{121}$
26. $\sqrt{49}$ [8.2]
27. $\sqrt{24x}$
28. $\sqrt{72m}$ [8.2]
29. $\sqrt{144x^2y^4}$
30. $\sqrt{100a^3b}$ [8.2]
31. $5mn^3\sqrt{180m^4n^5}$
32. $2c^2d\sqrt{220c^7d^6}$ [8.2]
33. $\sqrt[3]{-8x^3y^7}$
34. $\sqrt[3]{64x^3y^5}$ [8.2]
35. $\sqrt[4]{56a^2b^{10}}$
36. $\sqrt[4]{25m^3n^{11}}$ [8.2]
37. $\sqrt[5]{32a^5b^{10}c^{30}}$
38. $\sqrt[5]{-40a^7b^3c^5}$ [8.2]

In Exercises 39 through 44 perform the indicated operation and simplify.

39. $5\sqrt{3x} + 2\sqrt{3x}$ [8.2]
40. $3x\sqrt{5y} - 6\sqrt{5x^2y}$ [8.2]
41. $5x\sqrt{3xy^3} + 7\sqrt{27x^3y^3} + 4\sqrt{3y}$ [8.2]
42. $4\sqrt{3a} + 7\sqrt[3]{3a} + 8\sqrt{3a}$ [8.2]
43. $\sqrt[3]{27a^3b^2} - 5a\sqrt[3]{b^2}$ [8.2]
44. $\sqrt[5]{-64a^7b^3c^{10}} + 4ac\sqrt[5]{2a^2b^3c^5} + 19ac^2\sqrt[5]{2a^2b^3}$ [8.2]

In Exercises 45 through 52, multiply the following radical expressions. Simplify the result.

45. $\sqrt{15} \cdot \sqrt{20}$ [8.3]
46. $\sqrt{3xy} \cdot \sqrt{6x}$ [8.3]
47. $4\sqrt{5a^3b} \cdot 3\sqrt{2ab}$ [8.3]
48. $\sqrt[3]{3a^3b^2} \cdot \sqrt[3]{18b^2}$ [8.3]
49. $3x^2y\sqrt[3]{15xy^2z} \cdot 2xy^3\sqrt[3]{18xyz^5}$ [8.3]
50. $(9 + 2\sqrt{5})^2$ [8.3]
51. $(3 + 5\sqrt{7})(2 - 3\sqrt{7})$ [8.3]
52. $(3 + 5\sqrt{6})(3 - 5\sqrt{6})$ [8.3]

In Exercises 53 through 60, simplify the following radical expressions. Rationalize the denominator if necessary.

53. $\sqrt{\dfrac{36}{2}}$
54. $\dfrac{\sqrt{24}}{\sqrt{3}}$ [8.3]
55. $\dfrac{\sqrt{5x}}{\sqrt{3x}}$
56. $\sqrt{\dfrac{7ab}{3ac}}$ [8.3]
57. $\dfrac{5x}{\sqrt[3]{2x^2y}}$
58. $\dfrac{5}{2+\sqrt{3}}$ [8.3]
59. $\dfrac{5+2\sqrt{x}}{3-4\sqrt{x}}$
60. $\dfrac{4-\sqrt{7}}{2+\sqrt{6}}$ [8.3]

For Exercises 61 and 62, recall the formula for the period of a pendulum from Example 2 in Section 8.4.

$$T(L) = 2\pi\sqrt{\dfrac{L}{32}}$$

61. Find the period of a pendulum if its length is 2 feet. [8.4]

62. If you want a pendulum to have a period of 2.5 seconds, what must its length be? [8.4]

63. Use the model from Exercise 1 to estimate the month in which the profit for Big Jim's Mart will be $17,000. [8.4]

64. Use the model from Exercise 2 to estimate the age of an alpaca that weighs 60 kilograms. [8.4]

In Exercises 65 through 74, solve the following radical equations. Check your answers in the original equation.

65. $\sqrt{5+x} = 4$ [8.4]
66. $\sqrt{3x-7} = 2$ [8.4]
67. $-3\sqrt{2x-7} + 14 = -1$ [8.4]
68. $\sqrt{x-4} = \sqrt{2x+8}$ [8.4]
69. $\sqrt{x+2} = \sqrt{3x-7}$ [8.4]
70. $\sqrt{x-5} - \sqrt{x} = -1$ [8.4]
71. $\sqrt{x-4} = 5 + \sqrt{3x}$ [8.4]
72. $\sqrt{x+5} + \sqrt{3x+4} = 13$ [8.4]
73. $\sqrt[3]{2x} = 3$ [8.4]
74. $\sqrt[4]{5x+2} = 2$ [8.4]

In Exercises 75 through 78, simplify the given expressions and write your answer in terms of i.

75. $\sqrt{-25}$ [8.5]
76. $\sqrt{-32}$ [8.5]
77. $\sqrt{-4} + \sqrt{-25}$ [8.5]
78. $(2 + \sqrt{-5})(3 - \sqrt{-10})$ [8.5]

In Exercises 79 through 94, perform the indicated operation. Write your answer in the standard form for a complex number.

79. $(2 + 3i) + (5 + 7i)$ [8.5]
80. $(3.5 + 1.2i) + (2.4 - 3.6i)$ [8.5]
81. $(4 + 15i) - (3 + 11i)$ [8.5]
82. $(4.5 + 2.9i) - (1.6 - 4.2i)$ [8.5]
83. $(3 + 4i)(2 - 7i)$ [8.5]
84. $(6 + 5i)(2 + 7i)$ [8.5]
85. $(2 - 3i)(2 + 3i)$ [8.5]

86. $(2.3 + 4.1i)(3.7 - 9.2i)$ [8.5]

87. $\dfrac{12 + 7i}{3}$ [8.5]

88. $\dfrac{12 + 9i}{3}$ [8.5]

89. $\dfrac{4 + 7i}{5i}$ [8.5]

90. $\dfrac{7}{2 + 3i}$ [8.5]

91. $\dfrac{5 + 7i}{3 - 4i}$ [8.5]

92. $\dfrac{2 - 9i}{4 + 3i}$ [8.5]

93. $\dfrac{2.5 + 6.4i}{3.3 + 8.2i}$ [8.5]

94. $\dfrac{1.5 + 7.25i}{3.25 - 4.5i}$ [8.5]

97. $2(a - 7)^2 + 11 = 5$ [8.5]

98. $5(g - 9)^2 + 25 = 5$ [8.5]

99. $3x^2 + 12x + 15 = 2$ [8.5]

100. $-b^2 - 4b + 6 = 15$ [8.5]

101. $t^3 - 6t^2 = -13t$ [8.5]

102. $h^3 + 8h^2 + 65h = 0$ [8.5]

For Exercises 95 through 102, solve each equation. Write any complex solutions in standard form.

95. $x^2 + 25 = 0$ [8.5]

96. $m^2 - 4 = -20$ [8.5]

Chapter Test

8

1. The speed of a sound wave in air can be modeled by
$$v(T) = 20.1\sqrt{273 + T}$$
where $v(T)$ represents the speed of sound in air in meters per second when the temperature is T degrees Celsius. Find the speed of a sound wave at room temperature (20°C).

2. A daredevil wants to free-fall from a plane for 1.5 minutes before he needs to pull his ripcord and release his parachute at 1500 feet. How high should the plane be, to allow for this long of a free-fall? Use the function
$$T(h) = 0.243\sqrt{h}$$
where $T(h)$ is the drop time in seconds for an object that is dropped from h feet.

© iStockphoto.com/2happy

In Exercises 3 through 5, give the domain and range of the following radical functions.

3. $f(x) = -4.5\sqrt{x + 7}$

4. $f(x) = 4\sqrt[3]{x}$

5.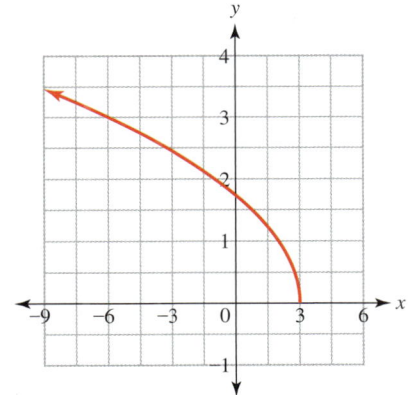

In Exercises 6 and 7, sketch a graph of the given function.

6. $f(x) = -\sqrt{10 - x}$

7. $g(x) = \sqrt[3]{x - 7}$

In Exercises 8 through 14, simplify the following radical expressions. Rationalize the denominator if necessary.

8. $\dfrac{\sqrt{6}}{\sqrt{5x}}$

9. $\sqrt{\dfrac{8b}{10a}}$

10. $\dfrac{5m}{\sqrt[3]{3mn^2}}$

11. $\dfrac{3 - \sqrt{2}}{2 + \sqrt{5}}$

12. $\sqrt{36xy^2}$

13. $7\sqrt{120a^2b^3}$

14. $\sqrt[5]{-32m^3n^7p^{10}}$

In Exercises 15 and 16, solve each equation. Write any complex solutions in standard form.

15. $-4(x+5)^2 + 7 = 9$

16. $2.3x^2 + 4.6x + 9 = 5$

17. How high off the ground must a person's eye be to see 20 miles to the horizon? Use the formula
$$d = \sqrt{1.5h}$$
where d is the distance in miles a person can see to the horizon when the person's eye is h feet above ground.

In Exercises 18 through 23, perform the indicated operation and simplify.

18. $5n\sqrt{6m} - 2\sqrt{24mn^2}$

19. $2a\sqrt{7yab^5} + 7b\sqrt{28a^3b^3} + 4b\sqrt{7a}$

20. $\sqrt{3ab} \cdot 2\sqrt{15ac}$

21. $\sqrt[3]{4a^4b} \cdot \sqrt[3]{18b^2}$

22. $2xy\sqrt[3]{18xz^2} \cdot 5xz^2\sqrt[3]{6x^7yz^4}$

23. $(5 + 2\sqrt{3})(2 - 4\sqrt{3})$

In Exercises 24 through 27, solve the following radical equations.

24. $2\sqrt{5x+4} - 11 = -3$

25. $\sqrt{x-13} - \sqrt{x} = -1$

26. $\sqrt{x-7} + \sqrt{4x-11} = 13$

27. $\sqrt[3]{x+5} = 4$

In Exercises 28 through 33, perform the indicated operation. Write your answer in the standard form for a complex number.

28. $(2.7 + 3.4i) + (1.4 - 4.8i)$

29. $(5 + 11i) - (4 - 7i)$

30. $(7 + 2i)(3 - 5i)$

31. $(1.5 - 4.5i)(2.25 - 6.5i)$

32. $\dfrac{8}{4 + 5i}$

33. $\dfrac{3 + 2i}{6 - 7i}$

34. In helium, the speed of sound can be modeled by
$$v(T) = 58.8\sqrt{273 + T}$$
where $v(T)$ represents the speed of sound in helium in meters per second when the temperature is T degrees Celsius.

   a. Find the speed of a sound wave at room temperature (20 degrees Celsius).

   b. How does the speed of sound in air compare to the speed of sound in helium? Explain.

Chapter 8 Projects

How Tight Can That Turn Be?

Written Project

One or more people

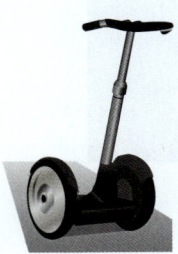

In this project, you will investigate the relationship between the speed of a Segway Human Transporter and the radius of a curve on which it is traveling. Although a Segway can "turn on a dime" when stopped, if it is traveling at any speed, it will need a certain turning radius to turn safely. According to the manufacturer, a Segway travels at top speeds ranging from 6 to 12.5 miles per hour depending on the chosen settings. Use this information and the formula

$$v = \sqrt{4.6r}$$

where v is the velocity in miles per hour and r is the minimum turning radius in feet, to answer the following questions.

Write up

a. What is the maximum speed of a Segway on a turn with radius of 25 feet?

b. What turning radius must be available to turn at the maximum speed of 12.5 miles per hour?

c. What turning radius must be available to turn at 6 miles per hour?

d. The turning radius of a bicycle is similar to that of a Segway, but a bicycle can travel at a much higher speed. Find the turning radius of a bicycle traveling at 20 miles per hour.

e. The Tour de France is a 23-day bicycle race that takes place in and around France each year during the month of July. In 2004, the race was about 2100 miles long, and Lance Armstrong finished the 23-day race in about 83 hours and 36 minutes. Use this information to determine his average speed during the race and what turning radius would be needed to turn at that average speed.

f. The next-to-last stage of the Tour de France in 2004 was approximately 34.4 miles long, and Armstrong finished that stage in just under 1 hour and 7 minutes. Find Armstrong's average speed and what turning radius would be needed for him to make a turn at that speed. Do you think he ever made a turn at this speed? Why or why not?

g. If this formula is used to represent the velocity and turning radius of a Segway, what are a reasonable domain and range for your model?

h. If this formula is used to represent the velocity and turning radius of a bicycle, what are a reasonable domain and range for your model?

i. If you gave different domains and ranges in parts g and h, explain why.

What Is That Planet's Period?

Research Project

One or more people

What you will need

- Find information about the orbital radius and period of different planets and moons.

In this project, you will explore the time it takes different planets to go around the sun. The earth has a period of 1 year, or 365.25 days, and has an orbital radius of about 150 million kilometers. In 1610, the German mathematician and astronomer Johannes Kepler formulated what became known as Kepler's Third Law:

$$T^2 = R^3$$

where T is the period of a planet with an average orbital radius of R. We will use an earth year, 365.25 days, to be a standard unit of time and millions of kilometers for the units for radius. Making these the standard units will give us a variation constant that will work for all planets that orbit our sun. Solving for T, we get

$$T = 0.000544\sqrt{R^3}$$

Use this formula to answer the following questions.

Write up

a. Mercury has an orbital radius of about 58 million kilometers. Find how long it takes Mercury to orbit the sun.

b. Mars has an orbital radius of about 228 million kilometers. Find how long it takes Mars to orbit the sun.

c. Find the average orbital radius of the other planets. Give the source of your information.

d. Use the values that you found for the other planets to calculate their periods.

e. If you wanted to have a satellite orbit the sun every $\frac{1}{2}$ year, how far from the sun must it orbit?

Use the basic formula $T = k\sqrt{R^3}$ to answer the following questions.

f. Saturn has over 31 known moons. One of Saturn's moons, Hyperion, has an orbital radius of 1.48 million kilometers and a period of about 21.3 days. Use this information to find a model for the orbits of Saturn's moons.

g. Another of Saturn's moons, Titan, has an orbital radius of about 1.22 million kilometers. Find the period of Titan's orbit.

h. Epimetheus orbits Saturn in only 0.694 day. Find its orbital radius.

i. Find the orbital radius of two more of Saturn's moons and calculate their periods.

Explore Your Own Radical Function

Research Project
One or more people

In this project, you are given the task of finding and exploring a real-world situation that you can model using a radical function. You may use the problems in this chapter to get ideas of things to investigate, but your application should not be discussed in this textbook. Some items that you might wish to investigate could be formulas from physics and other sciences. Remember that any kind of root can be used, not just square roots.

Write up

a. Describe the real-world situation that you found and where you found it. Cite any sources you used, following the MLA style guide or your school's recommended style guide.

b. Either run your own experiment or find data that can be used to verify the relationship that you are investigating.

What you will need

- Find a real-world situation that can be modeled with a radical function.
- You might want to use the Internet or library.
- Follow the MLA style guide for all citations. If your school recommends another style guide, use that.

c. Create a scatterplot of the data on the calculator or computer and print it out or draw it neatly by hand on graph paper.
d. Find a model to fit the data.
e. What is a reasonable domain for your model?
f. Use your model to estimate an output value of your model for an input value that you did not collect in your original data.
g. Use your model to estimate the input value for which your model will give you a specific output value that you did not collect in your original data.

Review Presentation

Research Project
One or more people

What you will need

- This presentation may be done in class or as a video.
- If it is done as a video, you may want to post the video on YouTube or another site.
- You might want to use problems from the homework set or review material in this book.
- Be creative, and make the presentation fun for students to watch and learn from.

Create a 5-minute presentation or video that reviews a section of Chapter 8. The presentation should include the following:

- Examples of the important skills presented in the section
- Explanation of any important terminology and formulas
- Common mistakes made and how to recognize them

Equation-Solving Toolbox — Chapters 1-8

Linear Equations

Name	Uses	Example	Where to Find It
Distributive Property	Use when an equation contains grouping symbols.	$2(x + 4) = 3x - 5$	Section 1.1 and Appendix A
Addition Property of Equality	Use to isolate a term.	$x + 7 = 20$	Section 1.1 and Appendix A
Multiplication Property of Equality	Use to isolate a variable.	$5x = 13$	Section 1.1 and Appendix A

Systems of Equations

Name	Uses	Example	Where to Find It
Substitution Method	Use when a variable is isolated or can be easily isolated.	$y = 6x + 8$ $2x + 4y = 30$	Section 2.2
Elimination Method	Use when the equations are in general form.	$5x - 7y = 42$ $3x + 20y = 103$	Section 2.3

Quadratic Equations

Name	Uses	Example	Where to Find It
Square Root Property	Use when there is a squared term but no first-degree term.	$3(x - 4)^2 + 5 = 0$	Section 4.4
Completing the Square	Use if the vertex form is required.	$x^2 + 6x + 4 = 0$	Section 4.4
Factoring	Use when the quadratic has small coefficients that factor easily.	$x^2 + 7x + 10 = 0$	Sections 3.4, 3.5, and 4.5
Quadratic Formula $$x = \frac{-b \pm \sqrt{b^2 - 4ac}}{2a}$$	Use when there are fractions, decimals, or large numbers. The quadratic formula will always give you the answers.	$11x^2 + 42x - 8 = 0$	Section 4.6

Exponential Equations

Name	Uses	Example	Where to Find It
Rewrite in logarithmic form	Use when the equation has a variable in the exponent. Isolate the base and exponent part first.	$5^x + 20 = 300$	Sections 6.2 and 6.5

Logarithmic Equations

Name	Uses	Example	Where to Find It
Rewrite in exponential form	Use when the equation contains a logarithm. Isolate the logarithm first.	$\log(x + 2) = 4$	Sections 6.2 and 6.6

Rational Equations

Name	Uses	Example	Where to Find It
Multiply both sides by the least common denominator	Use when the equation contains a rational expression. Watch for answers that cause division by zero and are therefore extraneous.	$\frac{5}{x + 2} = 4 - \frac{2}{x - 1}$	Section 7.5

Radical Equations

Isolate the radical and raise both sides to the reciprocal power.	Use when the equation contains a radical. If the radical is a square root, square both sides. Some answers might not work and are extraneous.	$\sqrt{3x+2} = 4$	Section 8.4

General Solving Techniques

Graphing	Use when you cannot solve the equation algebraically. Graph each side of the equation and find the intersection(s).		All chapters
Numerically	Use when you cannot solve the equation algebraically. Estimate a solution and check it. Best used to check solutions.		All chapters

Modeling Processes   Chapters 1-8

Linear	Quadratic	Exponential
1. Define the variables and adjust the data (if needed).	1. Define the variables and adjust the data (if needed).	1. Define the variables and adjust the data (if needed).
2. Create a scatterplot.	2. Create a scatterplot.	2. Create a scatterplot.
3. Select a model type.	3. Select a model type.	3. Select a model type.
4. **Linear model:** Pick two points and find the equation of the line. $y = mx + b$	4. **Quadratic model:** Pick a vertex and substitute it for h and k in the vertex form $f(x) = a(x - h)^2 + k$.	4. **Exponential model:** Pick two points, write two equations using the form $f(x) = a \cdot b^x$ and solve for b.
5. Write the equation of the model using function notation.	5. Pick another point and use it to find a.	5. Use b and one equation to find a.
6. Check the model by graphing it with the scatterplot.	6. Write the equation of the model using function notation.	6. Write the equation of the model using function notation.
	7. Check the model by graphing it with the scatterplot.	7. Check the model by graphing it with the scatterplot.

Cumulative Review — Chapters 1-8

For Exercises 1 through 20, solve the given equation using any method you have learned. Check your answers in the original equation.

1. $5(2)^x - 4 = 76$
2. $5h + 20 = 3h - 8$
3. $n^2 - 56 = 8$
4. $\dfrac{4}{a-6} = \dfrac{7}{a+5}$
5. $3t^2 - 17t = 56$
6. $\ln(4x - 9) = 3$
7. $2\sqrt{x+5} - 12 = 4$
8. $\dfrac{6}{x+2} + 4 = \dfrac{4x}{x-3}$
9. $7c^3 - 120 = -1632$
10. $\log_5(x+8) + \log_5(x+12) = 1$
11. $\sqrt{x+5} + \sqrt{2x-3} = 3$
12. $|2x + 7| = 31$
13. $3.4g^2 - 1.6g - 8.4 = 147$
14. $\dfrac{4}{w^2 + 2w - 24} = \dfrac{-0.5w}{w^2 - 13w + 36}$
15. $\dfrac{2}{3}(b+5) = \dfrac{5}{6}b - \dfrac{2}{9}$
16. $\sqrt[5]{4x - 9} = 4$
17. $4r^2 - 12r + 36 = 0$
18. $-3|2t - 5| = -24$
19. $1.5^{x^2 - 8} = 2.25$
20. $2d^3 - 4d^2 - 63d + 124 = 63d + 124$

For Exercises 21 through 24, solve the given systems. Check your answer.

21. $3x + 6y = 20$
    $4y = -2x + 15$
22. $x = 4y - 6$
    $\dfrac{1}{2}x + 6y = 13$
23. $y = 3x^2 + 5x - 10$
    $y = \dfrac{1}{4}x - \dfrac{9}{4}$
24. $y = 2x^2 + 5x - 20$
    $y = -x^2 + 3x + 36$

25. The number of trademark applications for items with a stars-and-stripes motif is given in the chart.

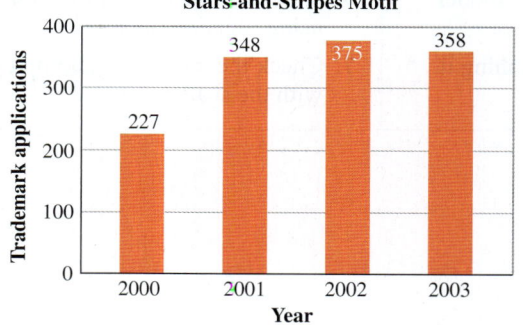

Source: USAToday.com.

   a. Find an equation for a model for these data.
   b. Give a reasonable domain and range for the model.
   c. Use the model to estimate the number of trademark applications in 2005.
   d. Give the vertex of your model and what it means in this context.

26. The net sales for Odyssey HealthCare, Inc., a leading hospice care provider, is given in the table.

Year	Net Sales (millions of $)
2000	85.3
2001	130.2
2002	194.5
2003	274.3

Source: CBSMarketwatch.com.

   a. Find an equation for an exponential model for these data.
   b. Give a reasonable domain and range.
   c. Use your model to estimate the net sales for Odyssey HealthCare, Inc. in 1998.
   d. According to your model, when will Odyssey HealthCare, Inc. reach $600 million in net sales?

27. The average number of hours per year Americans spent listening to recorded music are given in the graph.

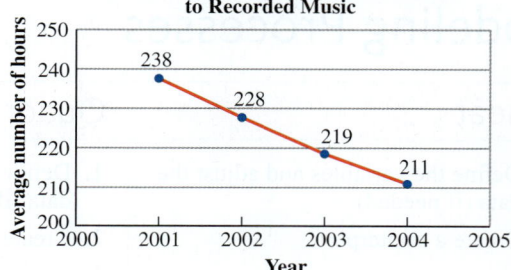

Source: Statistical Abstract of the United States, 2005.

   a. Find an equation for a model for these data.
   b. Use your model to estimate the average number of hours per year Americans spent listening to recorded music in 1995.
   c. Find when Americans will spend an average of only 150 hours a year listening to recorded music.
   d. What is the slope of your model and what does it mean in this context?
   e. Give a reasonable domain and range for your model.

28. Find the domain for the rational function
$$f(x) = \dfrac{x+2}{x^2 + 6x + 8}$$

29. Graph the system of inequalities.
$$4x - 6y > 3$$
$$y > -2x + 5$$

For Exercises 30 through 34, solve the given inequalities.

30. $4x + 20 > 10x - 10$
31. $\dfrac{1}{4}n + 3 \le \dfrac{2}{5}n + \dfrac{3}{10}$

32. $|2w - 13| \geq 3$ 33. $-2|d + 5| < -10$

34. $|2x + 4| \leq -8$

35. The number of visitors to U.S. amusement parks can be modeled by
$$V(t) = 0.275t^2 - 2.77t + 85.73$$
where $V(t)$ represents the number of visitors in millions to U.S. amusement parks t years since 1990.
   a. Use the model to estimate the number of visitors in 2005.
   b. Give the vertex of this model and what it means in this context.
   c. Estimate the year(s) in which U.S. amusement parks had 90 million visitors.

36. A spill of juice from a package of raw chicken is sitting on a kitchen counter unnoticed. At 2:00 P.M., there were 4000 salmonella bacteria on the counter, and the number of bacteria is doubling every 10 minutes.
   a. Find a model for the number of salmonella bacteria m minutes after 2:00 P.M.
   b. Use your model to estimate when there will be 1 million bacteria on the counter.
   c. If the spill stays unnoticed and the number of bacteria continues to grow, how many bacteria will be present when dinner is prepared at 6:00 P.M.?

For Exercises 37 and 38, find the inverse of the given function.

37. $f(x) = \dfrac{1}{2}x - 5$ 38. $g(x) = -3(5)^x$

39. $f(x) = -7$
   a. Find $f(9)$.
   b. Give the domain and range of $f(x)$.

40. Let $f(x) = 5x + 9$ and $g(x) = -1.4x + 3.2$.
   a. Find $f(x) + g(x)$.   b. Find $f(g(x))$.
   c. Find $f(x)g(x)$.

41. Let $f(x) = \dfrac{1}{2}x - 1$ and $g(x) = -8x + 5$.
   a. Find $f(6) - g(6)$.   b. Find $f(g(7))$.
   c. Find $f(3)g(3)$.

42. Let $f(x) = 2x^2 + 7x - 8$ and $g(x) = 2x - 5$.
   a. Find $f(x) + g(x)$.   b. Find $f(g(x))$.
   c. Find $f(x)g(x)$.

43. The 1906 San Francisco earthquake had a magnitude of 7.8. Find its intensity. Use the formula
$$M = \log\left(\dfrac{I}{10^{-4}}\right)$$

44. Write the equation of the line passing through the point $(-8, 7)$ and perpendicular to the line $2x + 5y = 11$. Write the equation in slope-intercept form.

45. Write the inequality for the following graph.

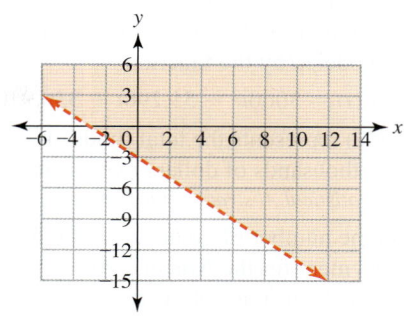

For Exercises 46 through 60, sketch the graph of the given equation or inequality. Label the vertical and horizontal intercepts. If quadratic label the vertex.

46. $3x - 7y = 28$ 47. $y = \sqrt{x + 1}$

48. $y = 2x^2 - 4x - 18$ 49. $y = 45(0.7)^x$

50. $y = -0.3(x - 7)^2 + 30$ 51. $y = \sqrt[3]{5 - x}$

52. $y = 4x - 22$ 53. $y = 3(4)^x$

54. $y = \log_7 x$ 55. $y = -2.5(2)^x - 30$

56. $y = -\sqrt{x - 15}$ 57. $y < x^2 - 3x - 10$

58. $y \geq 0.5x + 6$ 59. $4x - 3y < 12$

60. $y \geq -\dfrac{1}{2}(x - 8)^2 + 4$

61. The pressure of a certain amount of gas in a syringe is inversely proportional to the volume of the syringe. The pressure in the syringe is 4 pounds per square inch when the volume is 2 cubic inches.
   a. Find a model for the pressure in the syringe.
   b. Use your model to find the pressure in the syringe when the volume is 1 cubic inch.

For Exercises 62 through 69, perform the indicated operation and simplify.

62. $5\sqrt{x} + 7\sqrt{4x} - 10$

63. $(6x^2 + 8x - 14) - (8x^2 - 3x + 5)$

64. $\dfrac{x + 2}{x - 8} \cdot \dfrac{x + 5}{x + 3}$

65. $(3 + \sqrt{2c})(4 + \sqrt{18c})$

66. $(16b^3 - 12b^2 - 130b + 150) \div (2b - 5)$

67. $(3x - 8)^2$

68. $\dfrac{4h + 7}{h^2 + 10h + 21} \div \dfrac{h - 6}{h^2 + 3h - 28}$

69. $\dfrac{2x}{x + 4} + \dfrac{3}{x - 6}$

70. The National Endowment for the Arts grants funds to promote performing and visual arts. The number of grants that are awarded can be modeled by the function
$$G(t) = 210.9t - 226.3$$

where $G(t)$ represents the number of grants given by the National Endowment for the Arts t years since 1990. The total funds that are obligated for these grants can be modeled by the function

$$F(t) = 3060t^2 - 55{,}290t + 330{,}910$$

where $F(t)$ represents the total funds obligated for grants in thousands of dollars t years since 1990.
Source: Statistical Abstract of the United States, 2004.

   a. Find the number of grants given by the National Endowment for the Arts in 2000.
   b. Find the total funds obligated by the National Endowment for the Arts in 2000.
   c. What was the average size of a grant given by the National Endowment for the Arts in 2000?
   d. Use the given functions to find a function that will give the average size of a grant given by the National Endowment for the Arts t years since 1990.
   e. Use your new model to estimate when the average size of a grant given was $50,000.

For Exercises 71 through 75, factor the given expression using any method.

71. $8r^2 - 128$
72. $c^2 + 10c + 25$
73. $8t^2 + 14t - 30$
74. $t^3 - 64$
75. $21x^3 - 68x^2 + 32x$

76. What is the degree of the polynomial given in Exercise 72?

77. Complete the square to convert the function $f(x) = -2x^2 + 20x - 44$ to vertex form.

78. The average weight in pounds of girls who are m months old can be modeled by
$$W(m) = 9.3\sqrt[3]{m}$$
Source: Model derived from data from babybag.com.
   a. Use the model to estimate the average weight of girls who are 10 months old.
   b. What age girls should have an average weight of 25 pounds?

For Exercises 79 and 80, use the graph of the radical function $f(x)$ to answer the following.

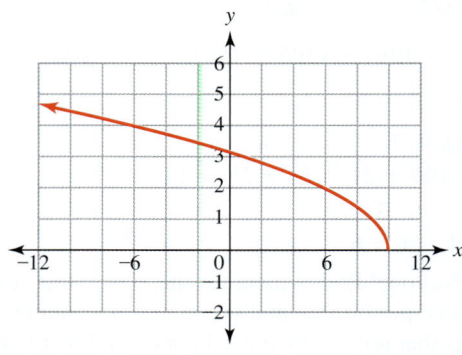

79. Find x such that $f(x) = 3$.
80. Give the domain and range of $f(x)$.

For Exercises 81 through 83, use the graph of the exponential $f(x) = a \cdot b^x$ to answer the following.

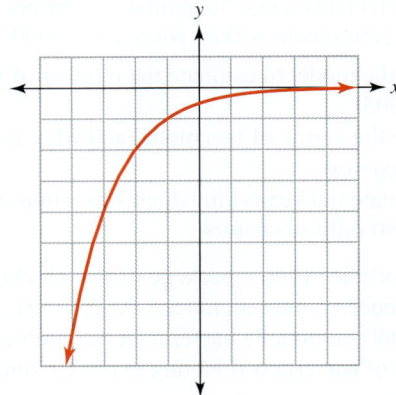

81. Is the value of a positive or negative? Explain your reasoning.
82. Is the value of b less than or greater than 1? Explain your reasoning.
83. Give the domain and range of $f(x)$.

For Exercises 84 through 88, use the graph of the quadratic $f(x) = a(x - h)^2 + k$ to answer the following.

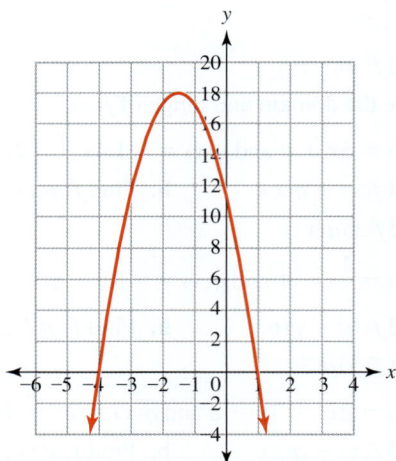

84. Is the value of a positive or negative? Explain your reasoning.
85. Estimate the values of h and k? Explain your reasoning.
86. $f(-3) = $?
87. Find x such that $f(x) = 0$.
88. Give the domain and range of $f(x)$.

For Exercises 89 through 93, use the graph of the line $f(x) = mx + b$ to answer the following.

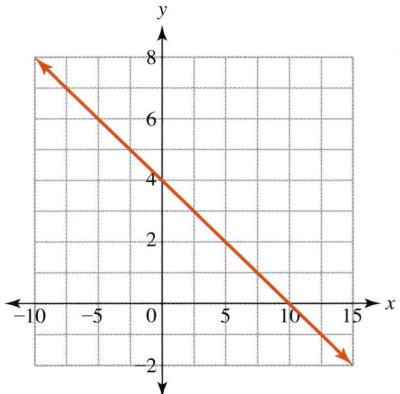

89. Is the value of m positive or negative? Explain your reasoning.

90. Estimate the value of b? Explain your reasoning.

91. Find the slope of the line.

92. Find x such that $f(x) = 1$

93. Give the domain and range of $f(x)$.

For Exercises 94 through 97, simplify the following radical expressions. Rationalize the denominator if necessary.

94. $\sqrt{100x^4y^8z^2}$

95. $\sqrt[3]{32a^4b^8c^6}$

96. $\dfrac{\sqrt{2x}}{\sqrt{5xy}}$

97. $\dfrac{5 + \sqrt{3}}{4 - \sqrt{7}}$

For Exercises 98 through 104, perform the indicated operation and write your answer in the standard form of a complex number.

98. $(3 + 2i) + (8 - 7i)$

99. $(4 + 3i) - (7 - 5i)$

100. $(2 + 6i)(2 - 6i)$

101. $(2 - 6i)(4 + 7i)$

102. $\dfrac{2 + 3i}{7i}$

103. $\dfrac{2 - 5i}{3 - 8i}$

104. $\dfrac{5}{4 - 9i}$

For Exercises 105 through 108, solve the equation and write any complex solutions in standard form.

105. $-x^2 + 4x = 20$

106. $2x^3 - 20x^2 + 68x = 0$

107. $x^2 + 9 = 0$

108. $x^2 - 6x + 30 = -4$

Conic Sections, Sequences, and Series

9

- **9.1** Parabolas and Circles
- **9.2** Ellipses and Hyperbolas
- **9.3** Arithmetic Sequences
- **9.4** Geometric Sequences
- **9.5** Series

The designs of car headlights, satellite dishes, and many other products are influenced by the properties of parabolas, ellipses, and other geometric shapes. In this chapter, we will discuss conic sections and some of the properties that make them an important part of engineering. We will also investigate some models that make sense only for whole number inputs. Sequences and series can model annual salaries, epidemics, and many other situations.

9.1 Parabolas and Circles

LEARNING OBJECTIVES

- Graph a parabola.
- Identify the focus and directrix of a parabola.
- Identify the equation for a parabola.
- Graph a circle.
- Identify the center and radius of a circle.
- Write the equation of a circle in standard form.

Introduction to Conic Sections

We will now study four shapes that we call **conic sections.** These shapes are formed when two cones connected at their **vertices** are sliced by a **plane.** The four types of conic sections are illustrated below.

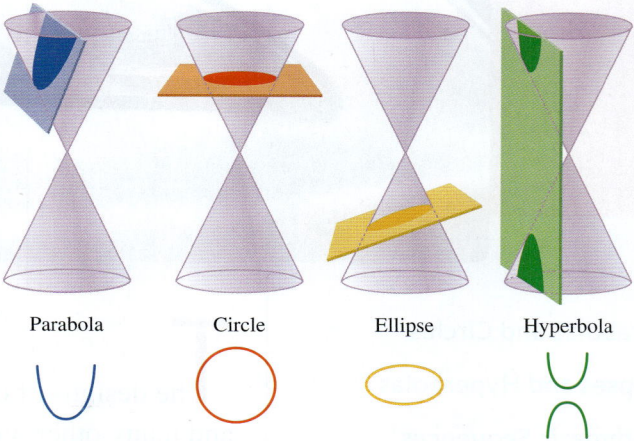

Source: www.BrooksCole.com WebQuest.

Conic sections are some of the oldest curves studied. A Greek named Menaechmus, a tutor to Alexander the Great, discovered conics in an attempt to solve several famous geometry problems. Conics were later expanded on by Appollonius, known as the Great Geometer, and by many other prominent mathematicians over the years. Johannes Kepler used conic sections to describe the motion of the planets in Kepler's Laws of Planetary Motion.

In the next two sections, we will study the four conic sections and more applications of each. We will look at the equations that describe these conic sections as well as their graphs. In this section, we will review and expand on parabolas, and we will study circles. Ellipses and hyperbolas will be introduced in the next section.

We will start our study of conic sections by investigating whether or not the conic sections represent functions.

■ CONCEPT INVESTIGATION
Are conic sections functions?

Use the vertical line test to determine whether or not the graphs of the different conic sections represent functions.

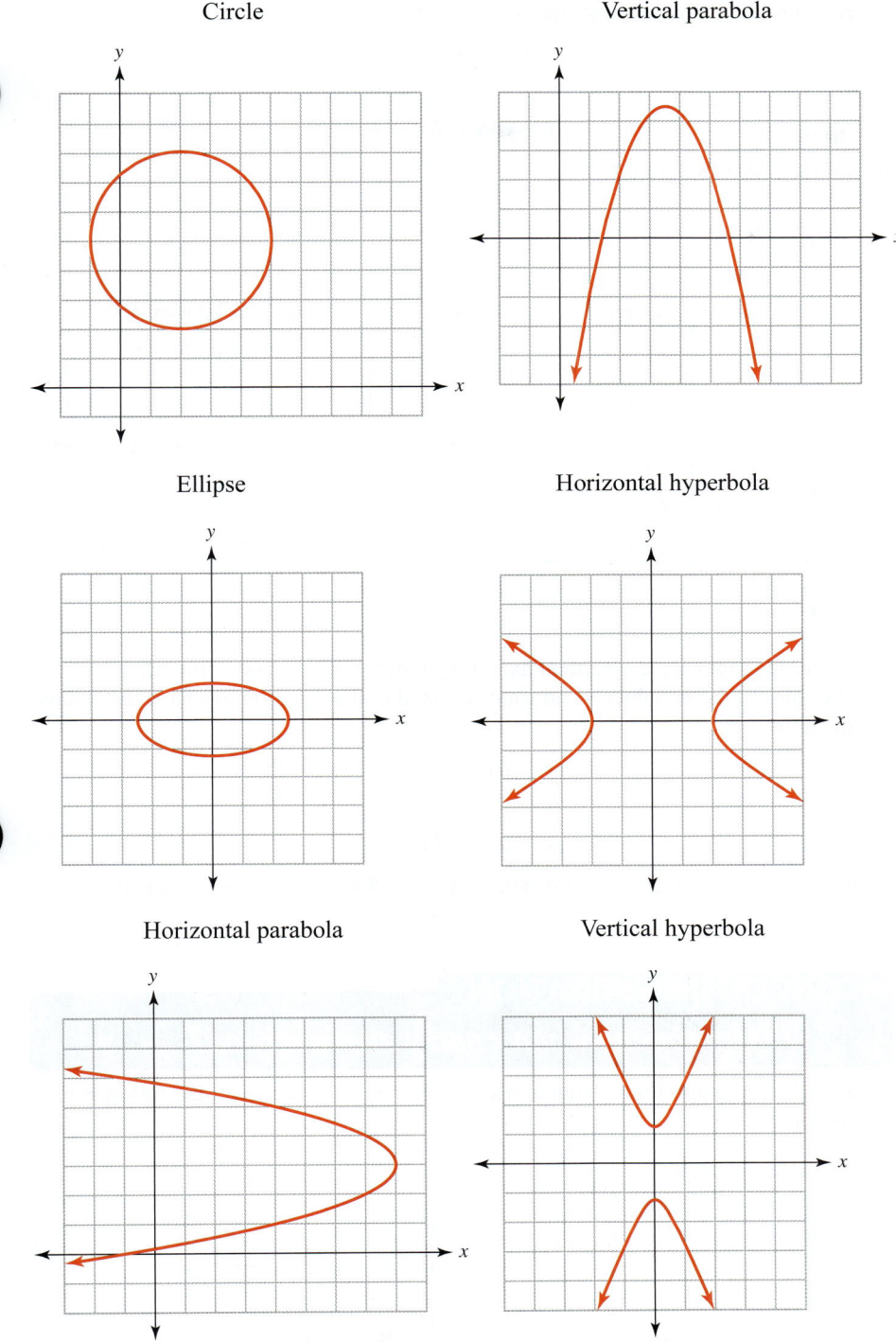

Although the parabolas that we studied in Chapter 4 were all functions, parabolas that open left or right are not functions. They fail the vertical line test. Only parabolas that open up or down are functions. None of the other conic sections pass the vertical line test, so they are not functions.

Revisiting Parabolas

In Chapter 4, we studied the parabola. Parabolas are formed when a plane that is parallel to one side of the cone slices through the cone. Although we have looked at parabolas that open up or down, we can also consider parabolas that open left and right.

The equation for a parabola that opens up or down is of the form:

$$y = ax^2 + bx + c$$

or

$$y = a(x - h)^2 + k$$

and has the following graphs.

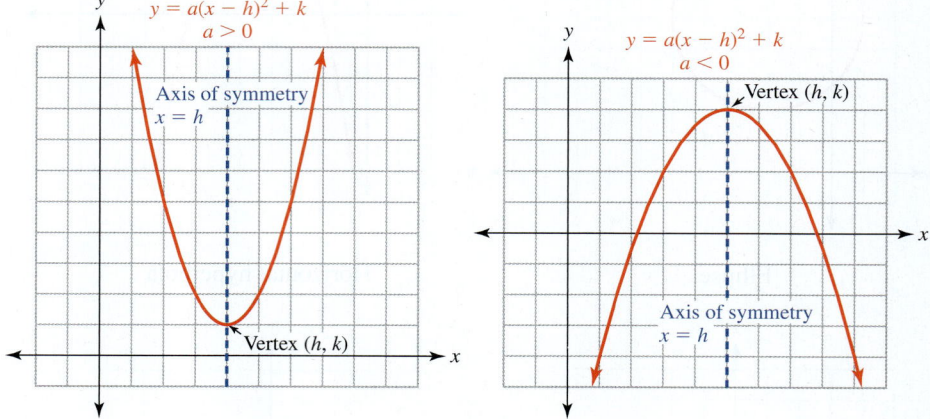

Both of these graphs have a vertical axis of symmetry, $x = h$, and vertex (h, k).

If the roles of the variables x and y are reversed in the equation, we have the following equations.

$$x = ay^2 + by + c$$

or

$$x = a(y - k)^2 + h$$

Use the concept investigation to investigate the graphs of these types of equations.

CONCEPT INVESTIGATION

What happens when x and y are reversed?

1. Use the given graph for the equation $x = 2(y - 3)^2 + 5$ to answer the following questions.

x	y
23	0
13	1
7	2
5	3
7	4
13	5
23	6

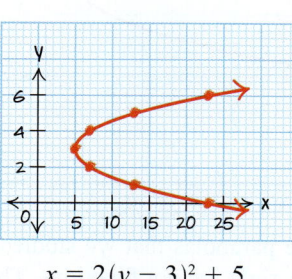

$x = 2(y - 3)^2 + 5$

   a. What is the vertex of the parabola?
   b. How is the vertex related to the equation?
   c. Is the axis of symmetry vertical or horizontal?
   d. What is the equation for the axis of symmetry?
   e. In what direction does the parabola open?
   f. What is the value of a in this equation?

2. Use the given graph for the equation $x = -0.5(y - 2)^2 + 18$ to answer the following questions.

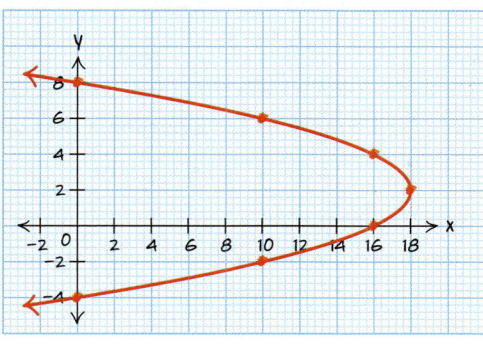

x	y
0	−4
10	−2
16	0
18	2
16	4
10	6
0	8

$x = -0.5(y - 2)^2 + 18$

a. What is the vertex of the parabola?
b. How is the vertex related to the equation?
c. Is the axis of symmetry vertical or horizontal?
d. What is the equation for the axis of symmetry?
e. In what direction does the parabola open?
f. What is the value of a in this equation?

From the concept investigation, we can summarize the characteristics of left- and right- facing parabolas using the following graphs.

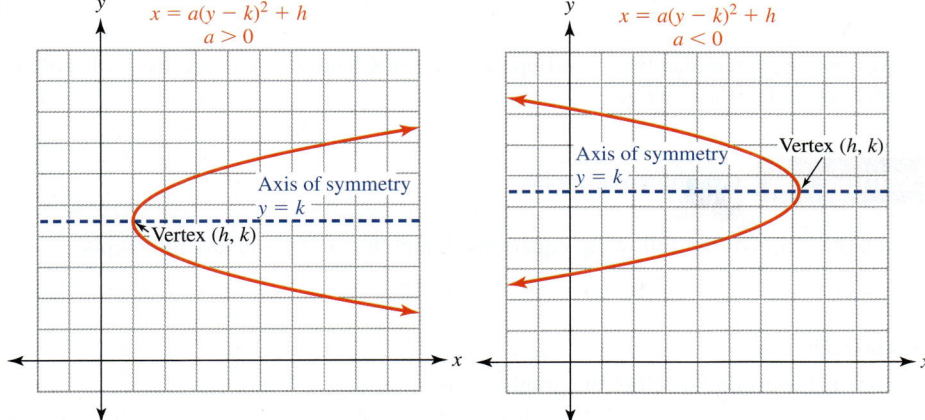

Both of these graphs have a horizontal axis of symmetry, $y = k$, and a vertex (h, k).

When considering the different equations of parabolas, pay attention to a couple of details that are similar to the original equations studied in Chapter 4. When the equation for a horizontal parabola is written in the vertex form, $x = a(y - k)^2 + h$, the vertex is still (h, k). The constant h still moves the graph left and right. The constant k still moves the graph up and down. Notice, though, that they are in different places in the equations. When x and y traded places in the equation, h and k also traded places. The constant k is now inside the parentheses, with y being squared, and h is outside.

The sign of a still affects the direction of the parabola. When a is positive, the parabola opens the positive direction, but on the x-axis. When a is negative, the parabola opens the negative direction on the x-axis. When the equation is written in general form $x = ay^2 + by + c$, the value $\frac{-b}{2a}$ is now the y-value of the vertex instead of the x-value.

What's That Mean?

Axes

Throughout the book, different terms have been used for the vertical and horizontal axes. Using the terms *vertical* and *horizontal* to describe the axes or intercepts allows for different variables to be used.

The standard equations for conic sections use the variables x and y. Because of this, the terms x- and y-axis as well as x- and y-intercepts will be used in this chapter.

DEFINITIONS

The standard equations for a parabola that opens up or down

$$y = ax^2 + bx + c \quad \text{or} \quad y = a(x - h)^2 + k$$

The parabola opens up when $a > 0$ and opens down when $a < 0$.

The standard equations for a parabola that opens right or left

$$x = ay^2 + by + c \quad \text{or} \quad x = a(y - k)^2 + h$$

The parabola opens right when $a > 0$ and opens left when $a < 0$.

In both sets of equations, a, b, c, h, and k are real numbers, and $a \neq 0$. The vertex is (h, k) in both vertex forms.

The process used in Chapter 4 to graph parabolas, will remain the same, except that the direction in which the parabola opens can now be up, down, left, or right. As in Chapter 4, we will continue to label the points of interest. For a parabola, these will include the intercepts and vertex.

Steps to Graphing a Parabola

1. Determine whether the graph opens up or down, left or right.
2. Find the vertex.
3. Find the axis of symmetry.
4. Find the y-intercepts (if any).
5. Find the x-intercepts (if any).
6. Plot the points, plot their symmetric pairs, and sketch the graph. Find an additional symmetric pair if needed.

We graphed parabolas that faced up or down in Chapter 4, so here we will look at parabolas that face left or right.

Example 1   Graphing parabolas

Sketch the graph of the following equations. Label the vertex and any intercepts.

a. $x = (y - 4)^2 + 3$ **b.** $x = -y^2 + 6y + 12$

SOLUTION

a. Step 1 Determine whether the graph opens up or down, left or right.
The equation $x = (y - 4)^2 + 3$ is in vertex form $x = a(y - k)^2 + h$. Because $a = 1$ is positive, the parabola will open to the right.

Step 2 Find the vertex.
This equation is in vertex form with $h = 3$ and $k = 4$, so the vertex is $(3, 4)$.

Step 3 Find the axis of symmetry.
This graph opens sideways and the vertex is $(3, 4)$, so the axis of symmetry will be the horizontal line $y = 4$.

Step 4 Find the y-intercepts (if any).
This vertex is to the right of the y-axis. The parabola opens right, so it will not intersect the y-axis. For this reason, the parabola has no y-intercepts. If we set $x = 0$ and try to solve for the y-intercept, we will get no real solutions.

$$0 = (y - 4)^2 + 3$$
$$-3 = (y - 4)^2 \quad \text{Use the square root property.}$$
$$\pm\sqrt{-3} = y - 4 \quad \text{There are no real solutions.}$$

Step 5 Find the *x*-intercepts (if any).
Find the *x*-intercept by making $y = 0$ and calculating x.

$$x = (0 - 4)^2 + 3$$
$$x = (-4)^2 + 3$$
$$x = 16 + 3$$
$$x = 19$$

Therefore, the *x*-intercept is $(19, 0)$.

Step 6 Plot the points, plot their symmetric pairs, and sketch the graph.
(Find an additional symmetric pair if needed.)
We have found the following points.

$$\text{Vertex} = (3, 4)$$
$$x\text{-intercept} = (19, 0)$$

The symmetric points are now vertical. Therefore, there is a symmetric pair to the *x*-intercept. We need to find an additional symmetric pair, so pick a value of *y* and find *x*.

The axis of symmetry is $y = 4$. Therefore, choose any *y*-value less than or greater than 4. We will arbitrarily choose $y = 2$. Substitute 2 into *y* and find *x*.

$$x = (2 - 4)^2 + 3$$
$$x = (-2)^2 + 3$$
$$x = 4 + 3$$
$$x = 7$$

Therefore, the point is $(7, 2)$. Plot the points and then plot their symmetric pairs.

x	*y*
3	4
19	0
7	2

Symmetric Pairs

x	*y*
19	8
7	6

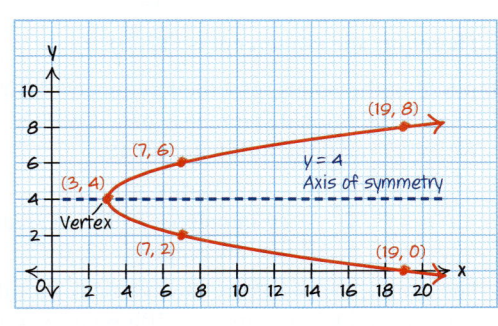

b. Step 1 Determine whether the graph opens up or down, left or right.
The equation $x = -y^2 + 6y + 12$ is in the general form $x = ay^2 + by + c$.
Because $a = -1$ is negative, the parabola will open to the left.

Step 2 Find the vertex.
This equation is in the general form. Use the formula $y = \dfrac{-b}{2a}$ to find the *y*-coordinate of the vertex.

$$x = -y^2 + 6y + 12$$
$$y = \frac{-(6)}{2(-1)} = 3$$

Substitute the *y*-value into the equation to find *x*.

$$x = -(3)^2 + 6(3) + 12$$
$$x = -(9) + 18 + 12$$
$$x = 21$$

The vertex is $(21, 3)$.

Step 3 Find the axis of symmetry.
This graph opens sideways. The axis of symmetry will be the horizontal line $y = 3$.

Step 4 Find the y-intercepts (if any).
This vertex is to the right of the y-axis. The parabola opens left, so it will have y-intercepts. Set $x = 0$ and solve for y.

$$0 = -y^2 + 6y + 12$$

$$y = \frac{-b \pm \sqrt{b^2 - 4ac}}{2a} \quad \text{Use the quadratic formula.}$$

$$y = \frac{-(6) \pm \sqrt{(6)^2 - 4(-1)(12)}}{2(-1)}$$

$$y = \frac{-6 \pm \sqrt{36 + 48}}{-2}$$

$$y = \frac{-6 \pm \sqrt{84}}{-2}$$

$$y \approx \frac{-6 \pm 9.165}{-2}$$

$$y \approx \frac{-6 + 9.165}{-2} \qquad y \approx \frac{-6 - 9.165}{-2}$$

$$y \approx 1.5825 \qquad y \approx 7.5825$$

The y-intercepts are $(0, -1.6)$ and $(0, 7.6)$.

Step 5 Find the x-intercepts (if any).
Find the x-intercept by making $y = 0$ and calculating x.

$$x = -(0)^2 + 6(0) + 12$$

$$x = 12$$

Therefore, the x-intercept is $(12, 0)$.

Step 6 Plot the points, plot their symmetric pairs, and sketch the graph. (Find an additional symmetric pair if needed.)
We have found the following points.

$$\text{Vertex} = (21, 3)$$
$$y\text{-intercepts} = (0, -1.6) \text{ and } (0, 7.6)$$
$$x\text{-intercept} = (12, 0)$$

The symmetric points are now vertical. Plot the symmetric pair to the x-intercept. Doing so gives a fifth point to work with, and there is no reason to find another symmetric pair. Plot the points and then plot their symmetric pairs.

x	y
21	3
12	0
0	-1.6
0	7.6

Symmetric Pairs

x	y
12	6

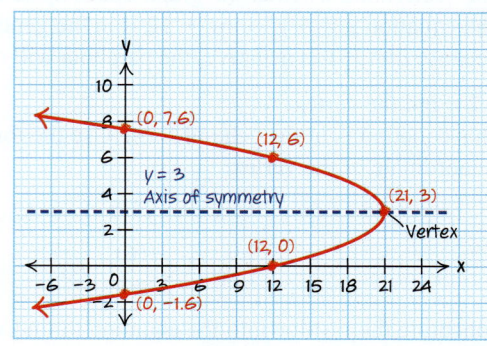

PRACTICE PROBLEM FOR EXAMPLE 1
Sketch the graphs of the following equations. Label the vertex and any intercepts.

a. $x = (y - 2)^2 - 8$ **b.** $x = -2y^2 + 16y + 5$

A Geometric Approach to Parabolas

Another way to study conic sections is through geometry. In geometry, a parabola would be defined as the set of all points on a plane equidistant from a fixed point, called the **focus**, and a fixed line, called the **directrix**.

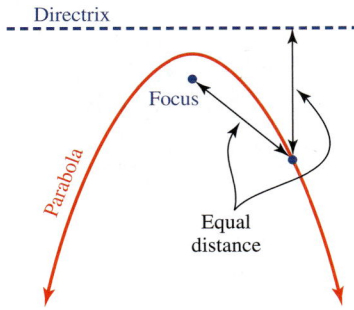

Connecting the Concepts

Where is the focus?

The focus of a parabola is always inside the parabola on the axis of symmetry.

The directrix is outside the parabola and perpendicular to the axis of symmetry.

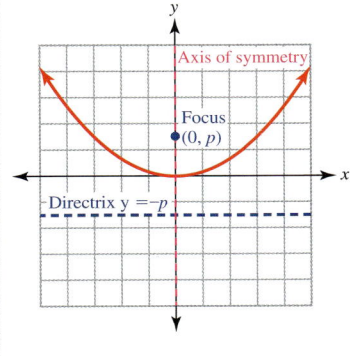

In other words, the distance from each point on the parabola to the focus must be the same as its distance to the directrix.

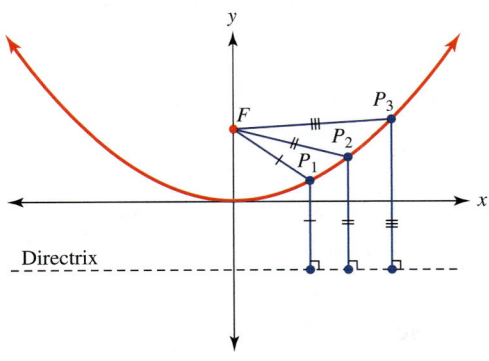

DEFINITION

Geometric Definition of a Parabola The set of all points in a plane that are equidistant from a fixed point (the focus) and a fixed line (the directrix) is called a **parabola**.

The focus and directrix were studied by Pappus of Alexandria, a Greek geometer. Later, in the 17th and 18th centuries, James Gregory and Sir Isaac Newton studied the reflective properties of a parabola that bring parallel rays of light to a focus. There are many applications of parabolas that take advantage of the concept of a focus. Many flashlights are designed to position the bulb at the focus of a reflector shaped like a parabola. This reflects the light out of the flashlight most efficiently. A cross section of a parabolic reflector is a parabola.

When a satellite dish transmits a signal, the transmitter is positioned at the focus of the parabolic dish. This allows most of the signal to be directed straight out in the direction of the receiver. The receiver is also at the focus point of a parabolic dish so that the incoming signal will be directed to the receiver.

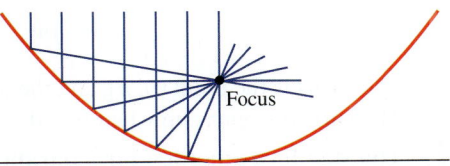

Satellite dish transmitter

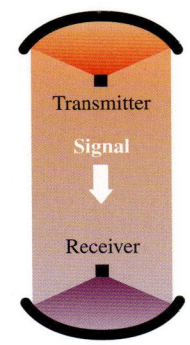

Satellite dish receiver

When a parabola has its vertex at the origin, you can find the equation of the parabola using the distance formula and the geometric definition of a parabola. The distance d between two points (x_1, y_1) and (x_2, y_2) can be found by using the equation

$$d = \sqrt{(x_1 - x_2)^2 + (y_1 - y_2)^2}$$

Assume that the point $P(x, y)$ is on the parabola with a vertex at the origin $(0, 0)$, focus at the point $(0, p)$, and directrix $y = -p$.

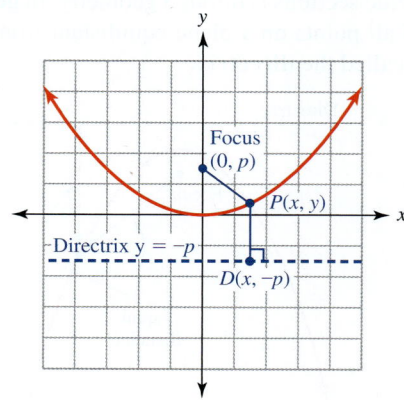

The distance from the focus to point P must equal the distance from the directrix to point P. Using the distance formula, we get the following equation for the parabola.

Distance from $(0, p)$ to (x, y) Distance from (x, y) to $(x, -p)$

$$\sqrt{(x - 0)^2 + (y - p)^2} = \sqrt{(x - x)^2 + (y - (-p))^2} \quad \text{Set the two distances equal to each other and simplify.}$$

$$\sqrt{x^2 + (y - p)^2} = \sqrt{0 + (y + p)^2}$$

$$\sqrt{x^2 + (y - p)^2} = \sqrt{(y + p)^2}$$

$$\left(\sqrt{x^2 + (y - p)^2}\right)^2 = \left(\sqrt{(y + p)^2}\right)^2 \quad \text{Square both sides of the equation.}$$

$$x^2 + (y - p)^2 = (y + p)^2$$

$$\underline{x^2 + y^2 - 2py + p^2 = y^2 + 2py + p^2} \quad \text{Simplify.}$$
$$-y^2 -y^2$$

$$\underline{x^2 - 2py + p^2 = 2py + p^2}$$
$$-p^2 -p^2$$

$$\underline{x^2 - 2py = 2py}$$
$$+2py +2py$$

$$x^2 = 4py$$

Using a similar approach, we could find the standard equation for a parabola that has a vertex at the origin and opens left or right. With an equation in one of these forms, we can tell where the focus and directrix are.

> ### DEFINITIONS
>
> **The standard equation for a parabola with vertex $(0, 0)$ that opens up or down using a geometric approach**
>
> $$x^2 = 4py \quad \text{or} \quad y = \frac{x^2}{4p}$$
>
> The focus is the point $(0, p)$, and the directrix is the horizontal line $y = -p$. The parabola opens up when $p > 0$ and opens down when $p < 0$.
>
> **The standard equation for a parabola with vertex $(0, 0)$ that opens right or left using a geometric approach**
>
> $$y^2 = 4px \quad \text{or} \quad x = \frac{y^2}{4p}$$
>
> The focus is the point $(p, 0)$, and the directrix is the vertical line $x = -p$. The parabola opens right when $p > 0$ and opens left when $p < 0$.

SECTION 9.1 Parabolas and Circles   693

Example 2 Using the focus and directrix to graph a parabola

Find the directrix and focus for each parabola. Sketch the graph and label the focus and the directrix.

a. $y = \dfrac{x^2}{20}$ **b.** $y = -\dfrac{1}{12}x^2$ **c.** $x = \dfrac{y^2}{11}$

SOLUTION

a. The equation $y = \dfrac{x^2}{20}$ is in the standard form for a parabola with a vertex at the origin. Therefore,

$$y = \dfrac{x^2}{20} \longleftrightarrow y = \dfrac{x^2}{4p}$$

Therefore,

$$4p = 20$$
$$\dfrac{4p}{4} = \dfrac{20}{4}$$
$$p = 5$$

This parabola has a focus at the point $(0, 5)$, and the directrix is the line $y = -5$. To get a couple of points on the parabola, choose a value for x and solve for y.

$$x = 10 \qquad x = 20$$
$$y = \dfrac{10^2}{20} \qquad y = \dfrac{20^2}{20}$$
$$y = 5 \qquad y = 20$$

Therefore, we have the points $(10, 5)$ and $(20, 20)$. Using this information and symmetry, we can sketch the following graph.

x	y
0	0
10	5
20	20

Symmetric Pairs

x	y
−10	5
−20	20

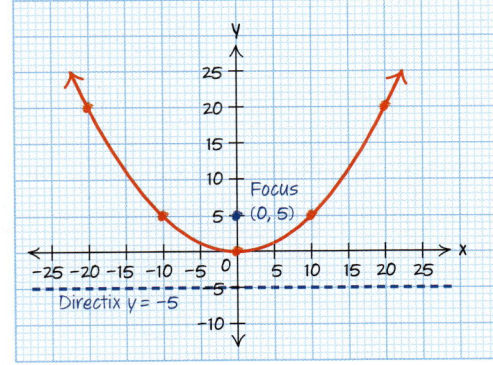

b. The equation $y = -\dfrac{1}{12}x^2$ is in the vertex form for a parabola with a vertex at the origin. To find the focus and directrix, we rewrite it and solve for p.

$$y = -\dfrac{1}{12}x^2$$
$$y = \dfrac{x^2}{-12}$$
$$y = \dfrac{x^2}{-12} \qquad y = \dfrac{x^2}{4p}$$

Therefore,

$$4p = -12$$
$$\dfrac{4p}{4} = \dfrac{-12}{4}$$
$$p = -3$$

This parabola has a focus at the point $(0, -3)$, and the directrix is the line

$$y = -(-3)$$
$$y = 3$$

To get a few points on the parabola, choose a couple of values for x and solve for y.

$$x = 6 \qquad\qquad x = 12$$
$$y = -\frac{1}{12}(6)^2 \qquad y = -\frac{1}{12}(12)^2$$
$$y = -3 \qquad\qquad y = -12$$

Therefore, we have the points $(6, -3)$ and $(12, -12)$. Using this information and symmetry, we can sketch the following graph.

x	y
0	0
6	-3
12	-12

Symmetric Pairs

x	y
-6	-3
-12	-12

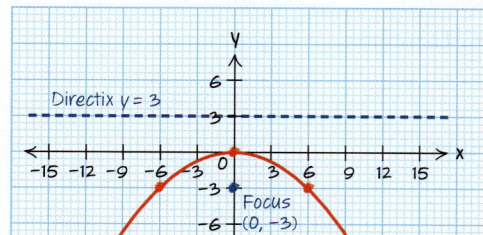

c. The equation $x = \dfrac{y^2}{11}$ is in the standard form of a parabola. Since y is being squared, it will open left or right.

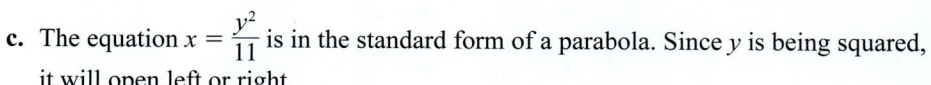

Therefore,

$$4p = 11$$
$$\frac{4p}{4} = \frac{11}{4}$$
$$p = \frac{11}{4} = 2.75$$

Since p is positive, the parabola will open to the right. This parabola has a focus at the point $\left(\dfrac{11}{4}, 0\right)$. The directrix is the line $x = -\dfrac{11}{4}$. To get a couple of points on the parabola, choose a value for y and solve for x.

$$y = 4 \qquad\qquad y = 8$$
$$x = \frac{(4)^2}{11} \qquad x = \frac{(8)^2}{11}$$
$$x \approx 1.45 \qquad\quad x \approx 5.82$$

Therefore, we have the points $(1.45, 4)$ and $(5.82, 8)$. Using this information and symmetry, we can sketch the following graph.

x	y
0	0
1.45	4
5.82	8

Symmetric Pairs

x	y
1.45	-4
5.82	-8

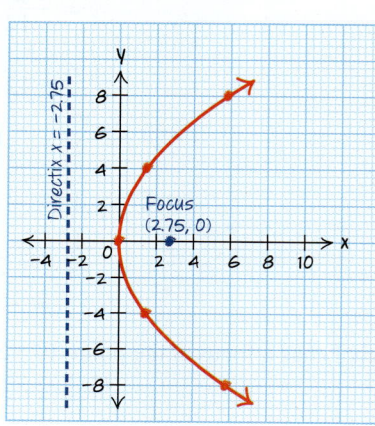

SECTION 9.1 Parabolas and Circles   695

PRACTICE PROBLEM FOR EXAMPLE 2

Find the directrix and focus for the given parabolas. Sketch the graph and label the focus and the directrix.

a. $y = \dfrac{x^2}{6}$ **b.** $-24x = y^2$

Example 3 Finding the equation of a parabola from its graph

Find the equation of the parabola.

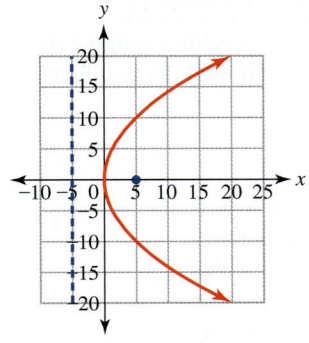

SOLUTION

The graph is facing right and has a focus at the point $(5, 0)$, so $p = 5$. Therefore, the equation is

$$x = \frac{y^2}{4(5)}$$

$$x = \frac{y^2}{20}$$

or

$$y^2 = 20x$$

PRACTICE PROBLEM FOR EXAMPLE 3

Find the equation of the parabola.

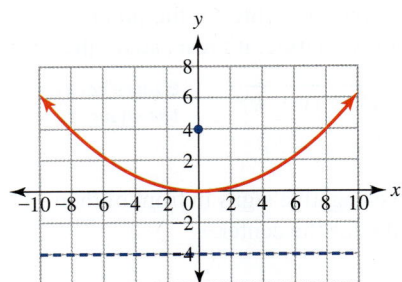

Example 4 Finding the best location for the light bulb

The optimal shape for the reflector in a flashlight is a parabola with the bulb at the focus. If a flashlight has a 2-inch-wide reflector that is 0.5 inch deep, find the best location for the bulb.

SOLUTION

We want to know the best location for the bulb, so we want to find the focus of the parabola. To find the focus, we will draw a picture of the parabola and the points on the parabola that we are given.

Allowing the base of the flashlight to be the vertex, we have a parabola that has an equation of the form $x^2 = 4py$. Using the point $(1, 0.5)$, we can solve for p.

$$x^2 = 4py$$
$$1^2 = 4p(0.5)$$
$$1 = 2p$$
$$0.5 = p$$

The focus is at the point $(0, 0.5)$, so the bulb should be 0.5 inch above the base of the reflector on the axis of symmetry.

PRACTICE PROBLEM FOR EXAMPLE 4

A satellite dish is in the shape of a parabola with the receiver at the focus. If the satellite dish is 10 feet wide and 1 foot deep, find the best location for the receiver.

Circle

Circles

Circles are all around us. A circle is formed when a plane parallel to the bottom of the cone slices through the cone. A circle can also be described as the set of all points in a plane that are equally distant from a fixed point. The fixed point is the **center** (h, k) of the circle, and the distance from the center to the points on the circle is called the **radius** (r).

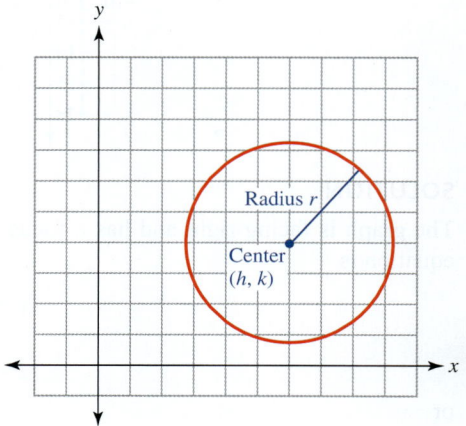

Since a circle is defined by using distance, the distance formula can be used to discover its general equation. The distance d between two points (x_1, y_1) and (x_2, y_2) can be found by using the equation

$$d = \sqrt{(x_1 - x_2)^2 + (y_1 - y_2)^2}$$

We will call the center of the circle (h, k) and put it in the place of the point (x_2, y_2). Since the radius is the distance from the center to the points on the circle, we will replace d with r. If (x, y) is a point on the circle, it must satisfy the equation

$$r = \sqrt{(x - h)^2 + (y - k)^2}$$ *Square both sides to remove the square root.*
$$r^2 = (x - h)^2 + (y - k)^2$$

Note that both h and k have negative signs in front of them. Therefore, we must be careful with the signs of h and k for the center.

DEFINITION

Geometric Definition of a Circle   The set of all points in a plane equidistant from a fixed point.

The standard equation for a circle with center (h, k) and radius r

$$(x - h)^2 + (y - k)^2 = r^2$$

h, k, and r are real numbers, and $r > 0$.

If a circle has its center at the origin $(0, 0)$, its equation simplifies to

$$x^2 + y^2 = r^2$$

SECTION 9.1 Parabolas and Circles **697**

Example 5 — Writing the equation of a circle

Write the equation of a circle with the given center and radius.

a. Center = (8, 6), and radius = 7.

b. Center = (4, −1), and radius = 2.

SOLUTION

a. The center is (8, 6), so $h = 8$ and $k = 6$. The radius is 7, so $r = 7$. Substitute these values into the standard form for the equation of a circle.

$$(x - 8)^2 + (y - 6)^2 = 7^2$$
$$(x - 8)^2 + (y - 6)^2 = 49$$

b. The center is (4, −1), so $h = 4$ and $k = -1$. The radius is 2, so $r = 2$. Substitute these values into the standard form for the equation of a circle.

$$(x - 4)^2 + (y - (-1))^2 = 2^2 \qquad \text{Be careful with the signs.}$$
$$(x - 4)^2 + (y + 1)^2 = 4$$

PRACTICE PROBLEM FOR EXAMPLE 5

Write the equation of a circle with the given center and radius.

a. Center = (5, 7), and radius = 10.

b. Center = (−5, −4), and radius = 9.

Example 6 — Graphing circles

Sketch a graph of the following circles. Find the radius and the center of the circle.

a. $x^2 + y^2 = 25$ **b.** $(x - 4)^2 + (y + 2)^2 = 9$

SOLUTION

a. This equation is in the standard form of a circle with center at the origin (0, 0). The radius is 5 because

$$r^2 = 25$$
$$r = \sqrt{25} = 5$$

Therefore, we have the following graph.

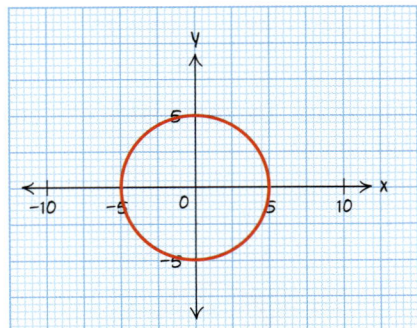

b. The equation $(x - 4)^2 + (y + 2)^2 = 9$ is in the standard form of circle with a center of (4, −2). The radius is 3 because

$$r^2 = 9$$
$$r = \sqrt{9} = 3$$

Therefore, we have the following graph.

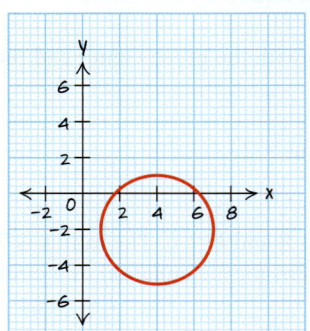

PRACTICE PROBLEM FOR EXAMPLE 6

Sketch a graph of the following circles. Find the radius and the center of the circle.

a. $x^2 + y^2 = 16$ **b.** $(x - 2)^2 + (y + 3)^2 = 4$

Some equations of circles will not be presented in standard form. An equation of the form

$$Ax^2 + By^2 + Cx + Dy + E = 0$$

where $A = B$ and A and B are not zero, can be rewritten into the standard form for a circle. We will use the process of completing the square from Chapter 4 to change these equations into standard form.

Skill Connection

Steps to Completing the Square (Section 4.4)

1. Isolate the terms on one side of the equation.
2. If the coefficient of x^2 is not 1, then divide both sides of the equation by the coefficient of x^2.
3. Take half the coefficient of x and square it. Add this number to both sides of the equation. For circles, repeat for coefficient of y.
4. Factor the quadratics into the squares of a binomial.

Example 7 Putting the equation of a circle into standard form

Write the equation in the standard form of a circle. Sketch a graph of the circle and state its radius and center.

a. $x^2 + y^2 - 10x + 14y + 10 = 0$ **b.** $2x^2 + 2y^2 + 12x - 16y - 22 = 0$

SOLUTION

a. The coefficients of the squared terms are the same. Rewrite this equation into the standard form of a circle using the method of completing the square.

$$x^2 + y^2 - 10x + 14y + 10 = 0$$
$$x^2 - 10x + y^2 + 14y = -10$$
$$(x^2 - 10x + \underline{}) + (y^2 + 14y + \underline{}) = -10$$
$$(x^2 - 10x + 25) + (y^2 + 14y + 49) = -10 + 25 + 49$$
$$(x - 5)^2 + (y + 7)^2 = 64$$

Regroup like variable terms and move the constant to right side.

$\left(\dfrac{-10}{2}\right)^2 = 25 \quad \left(\dfrac{14}{2}\right)^2 = 49$

Add constants to complete the square. Factor.

The equation is in standard form. The center is $(5, -7)$, and the radius is 8. The graph can be sketched as follows.

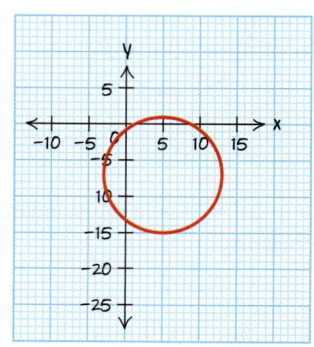

SECTION 9.1 Parabolas and Circles

b. Since the coefficients of the squared terms are both 2, rewrite this equation into the standard form of a circle using the method of completing the square.

$$2x^2 + 2y^2 + 12x - 16y - 22 = 0 \quad \text{Regroup like variable terms and move the constant to right side.}$$

$$2x^2 + 12x + 2y^2 - 16y = 22$$

$$\frac{2x^2 + 12x + 2y^2 - 16y}{2} = \frac{22}{2} \quad \text{Divide both sides by the coefficient of the squared terms.}$$

$$x^2 + 6x + y^2 - 8y = 11 \quad \left(\frac{6}{2}\right)^2 = 9 \quad \left(\frac{-8}{2}\right)^2 = 16$$

$$(x^2 + 6x + \underline{}) + (y^2 - 8y + \underline{}) = 11$$

$$(x^2 + 6x + 9) + (y^2 - 8y + 16) = 11 + 9 + 16 \quad \text{Add constants to complete the square.}$$

$$(x + 3)^2 + (y - 4)^2 = 36 \quad \text{Factor.}$$

Now that the equation is in standard form, the center is $(-3, 4)$ and the radius is 6. The graph can be sketched as follows.

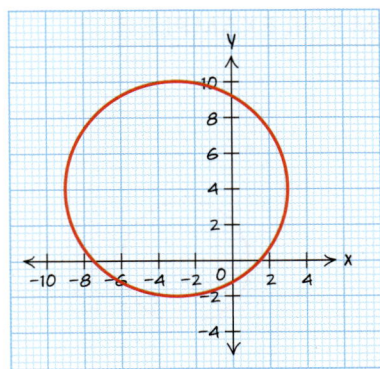

PRACTICE PROBLEM FOR EXAMPLE 7

Write the equation in the standard form of a circle. Sketch a graph of the circle and state its radius and center.

$$x^2 + y^2 - 4x + 10y + 20 = 0$$

By determining the center and radius of a circle's graph, we can find the equation of the circle. Remember that h and k will be subtracted in each squared term.

Example 8 Finding the equation of a circle from its graph

Find the equation of the circle.

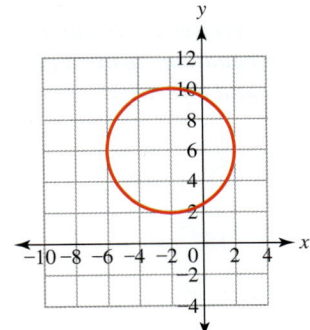

SOLUTION

This circle has its center at the point $(-2, 6)$, so $h = -2$ and $k = 6$. Choose any point on the circle to find the radius using the distance formula. Picking the point $(2, 6)$, we have

$$r = \sqrt{(-2 - 2)^2 + (6 - 6)^2}$$
$$r = \sqrt{(-4)^2 + 0}$$
$$r = \sqrt{16}$$
$$r = 4$$

Therefore, the radius is $r = 4$, and the equation of the circle is

$$(x - h)^2 + (y - k)^2 = r^2$$
$$(x - (-2))^2 + (y - 6)^2 = (4)^2$$
$$(x + 2)^2 + (y - 6)^2 = 16$$

PRACTICE PROBLEM FOR EXAMPLE 8

Find the equation of the circle.

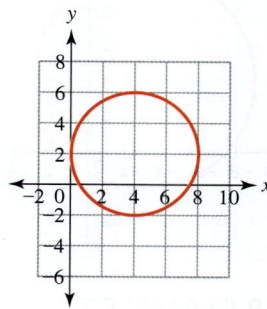

9.1 Exercises

For Exercises 1 through 10, determine whether the parabola will open up, down, right, or left. Find its vertex and axis of symmetry.

1. $y = 3x^2 + 24x - 9$
2. $y = 2x^2 - 16x + 3$
3. $x = 3y^2 + 12y - 7$
4. $x = y^2 + 8y + 9$
5. $x = -\frac{1}{2}y^2 + 4$
6. $x = -\frac{1}{4}y^2 + 5y - 2$
7. $y = -0.2x^2 + 5x - 3$
8. $x = -0.4y^2 + 3y - 5$
9. $x = \frac{y^2}{32}$
10. $y = -\frac{x^2}{28}$

For Exercises 11 through 20, sketch the graph of the given equations. Label the vertex and any intercepts.

11. $x = y^2 + 10y + 21$
12. $x = y^2 - 2y - 15$
13. $x = -2y^2 + 15y + 8$
14. $x = -3y^2 - y + 10$
15. $x = \frac{1}{2}(y - 4)^2 + 5$
16. $x = \frac{2}{5}(y + 3)^2 + 2$
17. $x = (y + 2)^2 - 6$
18. $x = (y - 7)^2 - 3$
19. $x = -2(y + 5)^2 + 4$
20. $x = -(y - 6)^2 + 2$

For Exercises 21 through 26, sketch the graph of the given equations using a geometric approach. Label the focus and directrix.

21. $x = \frac{y^2}{36}$
22. $x = \frac{y^2}{44}$
23. $y = \frac{1}{15}x^2$
24. $y = -\frac{1}{18}x^2$
25. $28x = y^2$
26. $-32x = y^2$

SECTION 9.1 Parabolas and Circles   701

For Exercises 27 through 32, use the geometric definition of a parabola to find the equation of the parabola from the given graph.

27.
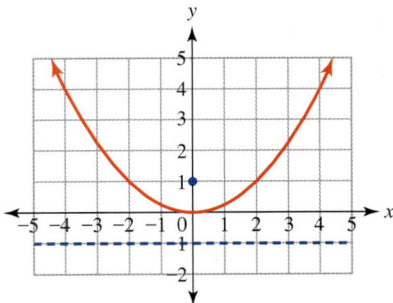

28.
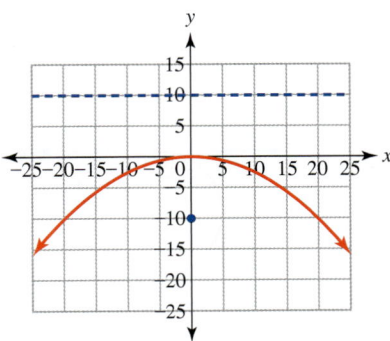

29.
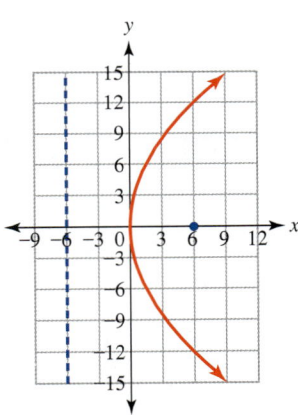

30.
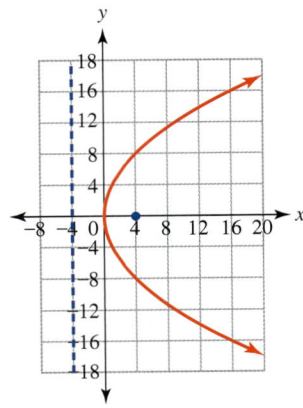

31.
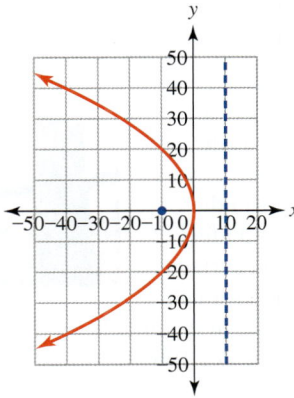

32.
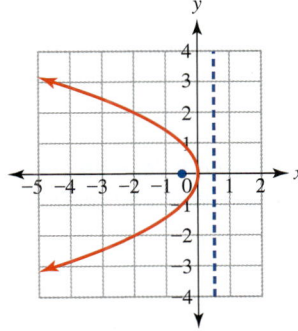

33. A satellite dish is in the shape of a parabola with the receiver at the focus. If the satellite dish is 6 feet wide and 2 feet deep, find the best location for the receiver.

34. A satellite dish is in the shape of a parabola with the receiver at the focus. If the satellite dish is 20 meters wide and 2.5 meters deep, find the best location for the receiver.

35. An arch over a bridge spans a 200-foot overpass. The maximum height of the arch is 50 feet above the road surface. Find the height of an arch 40 feet from the center of the bridge.

36. The Tyne Bridge in England crosses the Tyne River. The bridge's arch has a span of approximately 162 meters and a maximum height of

55 meters. Find the height of the arch 30 meters from the center of the bridge.

37. A suspension bridge's main cables form a parabola. The span between the two towers is 2500 feet, and the towers stand 300 feet above the road surface. The main cables touch the road surface in the middle of the span. Find the height of the cables 1000 feet from the center of the bridge.

38. A suspension bridge's main cables form a parabola. The span between the two towers is 1200 feet, and the towers stand 200 feet above the road surface. The main cables touch the road surface in the middle of the span. Find the height of the cables 400 feet from the center of the bridge.

For Exercises 39 through 46, write the equation of a circle with the given center and radius.

39. Center = (0, 0), radius = 6
40. Center = (0, 0), radius = 1.5
41. Center = (4, 9), radius = 4
42. Center = (7, 1), radius = 8
43. Center = (−2, 3), radius = 12
44. Center = (5, −8), radius = 2
45. Center = (−1, −4), radius = 0.5
46. Center = (−12, −10), radius = 9

For Exercises 47 through 54, sketch a graph of the given circle. Find the radius and the center of the circle.

47. $x^2 + y^2 = 9$
48. $x^2 + y^2 = 0.36$
49. $(x - 5)^2 + (y - 4)^2 = 25$
50. $(x - 2)^2 + (y - 7)^2 = 49$
51. $(x + 4)^2 + (y + 1)^2 = 1$
52. $(x + 5)^2 + (y + 5)^2 = 16$
53. $(x - 8)^2 + (y + 2)^2 = 36$
54. $(x + 3)^2 + (y - 6)^2 = 91$

For Exercises 55 through 60, write the equation in the standard form of a circle. Sketch a graph of the circle and find its radius and center.

55. $x^2 + y^2 + 2x - 6y + 9 = 0$
56. $x^2 + y^2 - 8x + 10y - 23 = 0$
57. $3x^2 + 3y^2 + 18x - 36y + 27 = 0$
58. $-2x^2 - 2y^2 + 4x + 8y + 40 = 0$
59. $4x^2 + 4y^2 + 4x - 4y = 34$
60. $9x^2 + 9y^2 - 6x - 6y + 1 = 0$

For Exercises 61 through 66, find the equation of the circle from the given graph.

61.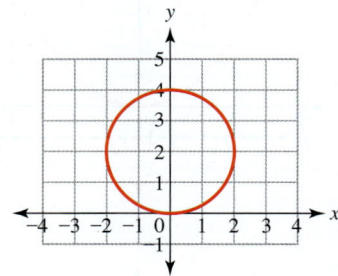

62.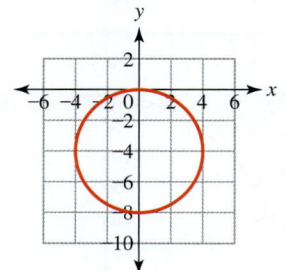

63.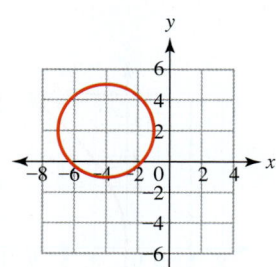

64.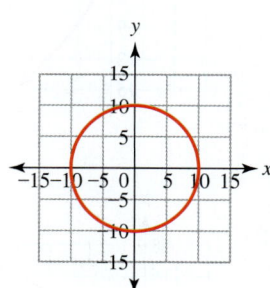

65.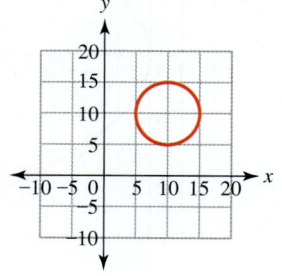

66.

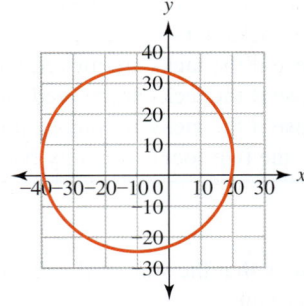

For Exercises 67 through 76, determine whether the equation represents a parabola or circle. Sketch a graph. If the graph is a parabola, label the vertex. If the graph is a circle, label the center and find the radius.

67. $2x^2 + 2y^2 + 4 = 22$

68. $x + 2y^2 - 3y + 10 = 0$

69. $x^2 - 4x - y - 12 = 0$

70. $x^2 + 2x + y^2 + 6x - 4 = 0$

71. $x - y^2 - 8y = 15$

72. $x^2 + 10 = 26 - y^2$

73. $x^2 = -(y-3)^2 + 49$

74. $x + (y+4)^2 - 8 = 0$

75. $x = \dfrac{y^2}{32}$

76. $\dfrac{x^2}{4} + 5 = 13 - \dfrac{y^2}{4}$

9.2 Ellipses and Hyperbolas

LEARNING OBJECTIVES

- Graph an ellipse centered at the origin.
- Find the foci of an ellipse.
- Write the equation for an ellipse.
- Graph a hyperbola centered at the origin.
- Write the equation for a hyperbola.
- Recognize the equations for the different conic sections.

Ellipses

In the first part of this section we will study the **ellipse**. An ellipse is formed when a plane that is not parallel to the bottom or side of the cone slices through the cone.

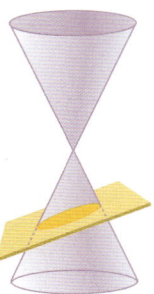

Ellipse

■ CONCEPT INVESTIGATION

How can I draw an ellipse?

One way to draw an ellipse is to pin two tacks to a piece of paper on a drawing board. Tie a string connecting both tacks. Keeping the string taut with a pencil, draw a curve around the two tacks. The resulting curve will be an ellipse with the two tacks being the **foci** (plural of focus). *Source: www.maa.org*

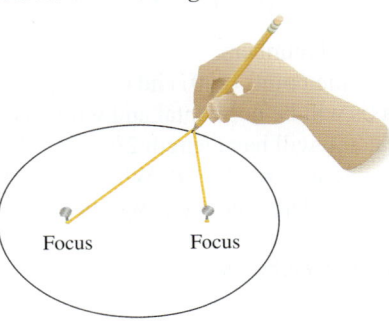

Focus   Focus

a. Pin two tacks to a piece of paper 3 inches apart and tie a piece of string approximately 5 inches long to the two tacks. Draw an ellipse.
b. What happens to the shape of the ellipse if the tacks are moved farther apart?
c. What happens to the shape of the ellipse if the tacks are moved closer together?
d. What happens to the shape of the ellipse if the tacks are put right next to each other? (When right next to each other, the two foci essentially become one focus.)

The geometric description of an ellipse is that the sum of the distances from the two foci to any point on the curve must be constant.

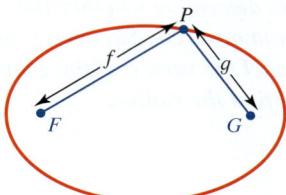

Source: www.mathisfun.com

The sum $f + g$ is the same for every point P on the ellipse. This procedure leads to a way of describing an ellipse. An ellipse is the set of all points in a plane such that the sum of the distances from two fixed points, the foci, is constant.

An ellipse has several characteristics that can be used to describe its graph. In the following illustrations, the **major axis** and the **minor axis** are labeled as well as the **vertices**. The major axis is the longer of the two axes that go through the center of the ellipse. The vertices are the endpoints of the major axis.

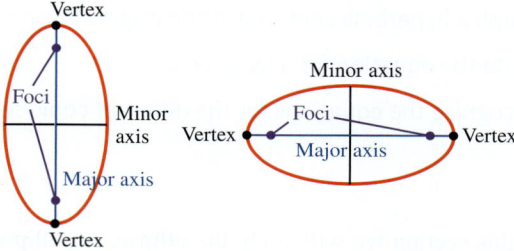

In this text, we will work with ellipses that are centered on the origin.

DEFINITION

Geometric Definition of an Ellipse   The set of all points in a plane such that the sum of the distances from two fixed points is constant is called an **ellipse.**

The standard equation for an ellipse centered at the origin

$$\frac{x^2}{a^2} + \frac{y^2}{b^2} = 1$$

where a and b are positive real numbers.

When $a > b$, the foci are the points $(c, 0)$ and $(-c, 0)$, where $c^2 = a^2 - b^2$. When $a > b$, the major axis will be horizontal and will have length $2a$. The minor axis will be vertical and will have length $2b$. When $b > a$, the foci are the points $(0, c)$ and $(0, -c)$, where $c^2 = b^2 - a^2$. When $a < b$, the major axis will be vertical and have length $2b$. The minor axis will be horizontal and will have length $2a$.

The foci are always on the major axis.

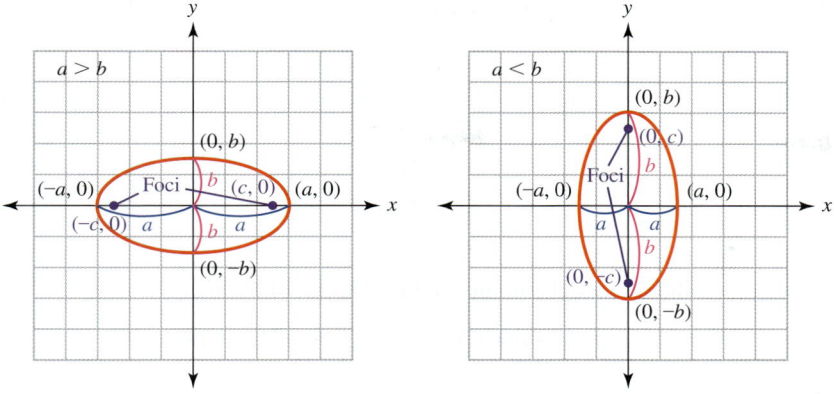

Using the geometric definition of an ellipse and the Pythagorean Theorem we can find the formula for the foci. Assuming that $a > b$, we have

$$(a - c) + (a + c) = 2\sqrt{b^2 + c^2}$$
$$2a = 2\sqrt{b^2 + c^2}$$
$$a = \sqrt{b^2 + c^2}$$
$$a^2 = b^2 + c^2$$
$$a^2 - b^2 = c^2$$

This same calculation can be done with $b > a$, resulting in $b^2 - a^2 = c^2$. These two formulas are used to find the value of c for the foci of ellipses.

> **Steps to Graphing an Ellipse Centered at the Origin**
> 1. Find the values of a and b.
> 2. Determine the direction of the major axis.
> 3. Find the foci.
> 4. Plot the vertices at $(a, 0), (-a, 0), (0, b),$ and $(0, -b)$.
> 5. Draw a smooth curve through the vertices.

Example 1 Graphing ellipses

Sketch the graph of the following equations.

a. $\dfrac{x^2}{25} + \dfrac{y^2}{4} = 1$ **b.** $\dfrac{x^2}{9} + \dfrac{y^2}{16} = 1$

SOLUTION

a. Step 1 This equation is in the standard form of an ellipse centered at the origin. $a^2 = 25$, so $a = 5$, and $b^2 = 4$, so $b = 2$.

Step 2 The bigger number is under x^2, so $a > b$. Therefore, the major axis will be horizontal.

Step 3 $a > b$, so to find the foci, use the equation $c^2 = a^2 - b^2$ and solve for c.

$$c^2 = a^2 - b^2$$
$$c^2 = 5^2 - 2^2$$
$$c^2 = 25 - 4$$
$$c^2 = 21$$
$$c = \pm\sqrt{21}$$
$$c \approx \pm 4.58$$

Therefore, the foci are the points $(4.58, 0)$ and $(-4.58, 0)$.

Steps 4 and 5 Using these values, we get the following graph.

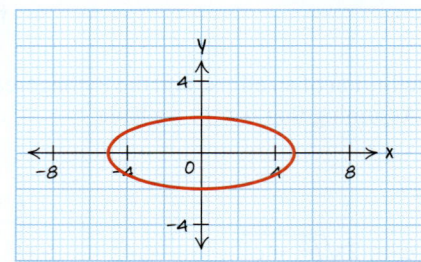

b. Step 1 This equation is in the standard form of an ellipse centered at the origin. $a^2 = 9$, so $a = 3$, $b^2 = 16$, so $b = 4$.

Step 2 $b > a$, so the major axis will be vertical.

Step 3 $b > a$, so to find the foci, use the equation $c^2 = b^2 - a^2$ and solve for c.

$$c^2 = b^2 - a^2$$
$$c^2 = 4^2 - 3^2$$
$$c^2 = 16 - 9$$
$$c^2 = 7$$
$$c = \pm\sqrt{7}$$
$$c \approx \pm 2.65$$

Therefore, the foci are the points $(0, 2.65)$ and $(0, -2.65)$.

Steps 4 and 5 Using these values, we get the following graph.

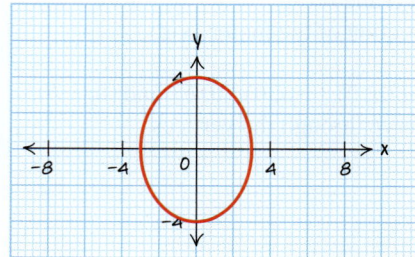

PRACTICE PROBLEM FOR EXAMPLE 1

Sketch the graph of the following equations.

a. $\dfrac{x^2}{4} + \dfrac{y^2}{36} = 1$ **b.** $\dfrac{x^2}{100} + \dfrac{y^2}{16} = 1$

Example 2 Converting an equation to standard form for an ellipse

Convert the equation $25x^2 + 49y^2 = 1225$ to the standard form for an ellipse centered at the origin. Sketch a graph of the ellipse.

SOLUTION

To be in standard form, the equation must equal 1. Divide both sides of the equation by 1225.

$$25x^2 + 49y^2 = 1225$$

$$\frac{25x^2 + 49y^2}{1225} = \frac{1225}{1225} \quad \text{Divide both sides of the equation by 1225.}$$

$$\frac{25x^2}{1225} + \frac{49y^2}{1225} = 1 \quad \text{Divide each term by 1225.}$$

$$\frac{x^2}{49} + \frac{y^2}{25} = 1 \quad \text{Reduce.}$$

From the new equation, $a^2 = 49$, so $a = 7$. Then $b^2 = 25$, so $b = 5$. Because $a > b$, use the equation $c^2 = a^2 - b^2$ to find the value of c.

$$c^2 = 7^2 - 5^2$$
$$c^2 = 49 - 25$$
$$c^2 = 24$$
$$c = \pm\sqrt{24}$$
$$c \approx \pm 4.90$$

Therefore, the foci are the points $(4.90, 0)$ and $(-4.90, 0)$. Using these values, we get the following graph.

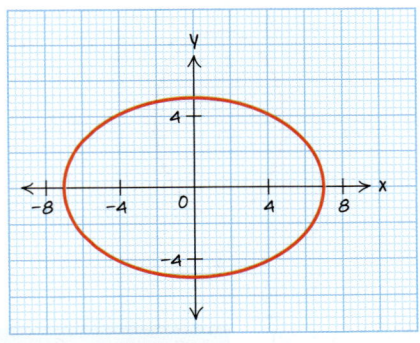

PRACTICE PROBLEM FOR EXAMPLE 2

Convert the equation $4x^2 + y^2 = 16$ to the standard form for an ellipse centered at the origin. Sketch a graph of the ellipse.

Example 3 Finding the equation of an ellipse from the graph

Find the equation of the ellipse. Find the foci.

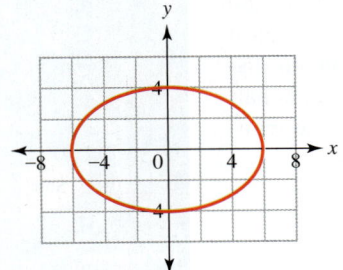

SOLUTION

The vertices of the ellipse are at $(6, 0)$, $(-6, 0)$, $(0, 4)$, and $(0, -4)$. Therefore, $a = 6$ and $b = 4$. The major axis is horizontal, which confirms that a should be greater than b. Using the values for a and b gives the following equation.

$$\frac{x^2}{6^2} + \frac{y^2}{4^2} = 1$$

$$\frac{x^2}{36} + \frac{y^2}{16} = 1$$

The foci can be found by using the equation $c^2 = a^2 - b^2$.

$$c^2 = a^2 - b^2$$
$$c^2 = 6^2 - 4^2$$
$$c^2 = 36 - 16$$
$$c^2 = 20$$
$$c = \pm\sqrt{20}$$
$$c \approx \pm 4.47$$

Therefore, the foci are the points $(4.47, 0)$ and $(-4.47, 0)$.

PRACTICE PROBLEM FOR EXAMPLE 3

Find the equation of the ellipse. Find the foci.

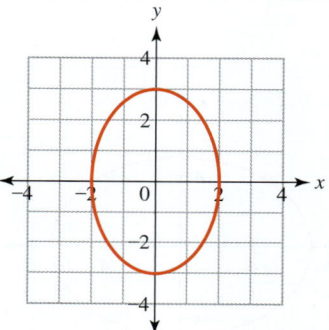

According to Kepler's Laws of Planetary Motion, the planets follow an elliptical orbit, with the sun being at one of the focal points. There isn't anything at the other focal point.

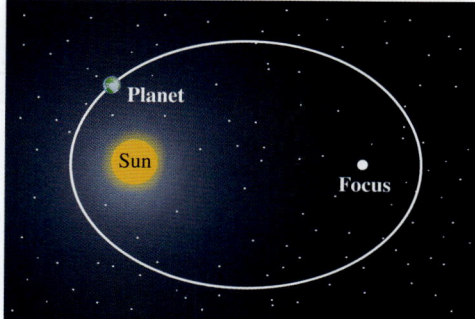

The **perihelion** of a planet is its shortest distance from the sun, and the **aphelion** is the planet's greatest distance from the sun. The **mean distance** of a planet from the sun is half the length of the major axis (the semi-major axis).

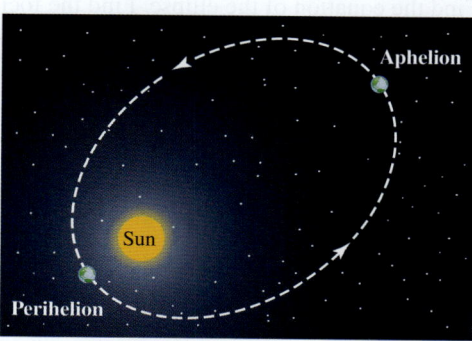

Example 4 Finding the equation of the earth's elliptical orbit

The earth has a mean distance from the sun of about 93 million miles. The earth's aphelion is about 94.5 million miles, and its perihelion is about 91.5 million miles. Write an equation for the orbit of the earth around the sun.

SOLUTION

The sun is at one of the focal points, so we can start a drawing of the orbit with the information we have been given.

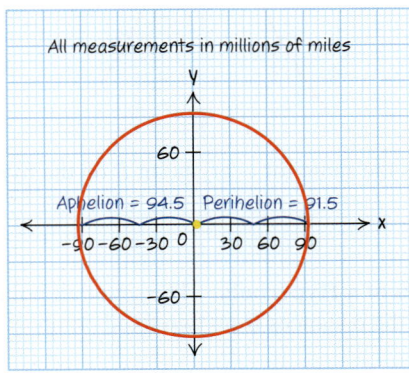

The major axis is twice the earth's mean distance from the sun, so a = the mean distance, $a = 93$. The sun is at one of the focal points. Therefore, c is the difference between the mean distance and perihelion.

$$c = 93 - 91.5 = 1.5$$

Use the values for a and c to find the value of b.

$$c^2 = a^2 - b^2$$
$$(1.5)^2 = (93)^2 - b^2$$
$$2.25 = 8649 - b^2$$
$$-8646.75 = -b^2$$
$$8646.75 = b^2$$
$$\sqrt{8646.75} = b$$
$$92.9879 \approx b$$

Using the values for a and b, write the equation for the earth's orbit.

$$\frac{x^2}{8649} + \frac{y^2}{8646.75} = 1$$

PRACTICE PROBLEM FOR EXAMPLE 4

Mars has a mean distance from the sun of about 141.5 million miles. Mars's aphelion is about 155 million miles, and its perihelion is about 128 million miles. Write an equation for the orbit of Mars around the sun.

Hyperbolas

The last type of conic section we will study is the **hyperbola.** Hyperbolas occur when a plane slices through the upper and lower parts of joined cones. This occurs when the plane is parallel to the axis of the cones.

The equation for a hyperbola is very similar to that for an ellipse. The distinction is that instead of a sum of two distances, there is a difference of two distances. A **hyperbola**

Hyperbola

is defined as the set of all points in a plane such that the difference of the distances from two fixed points (the foci) is a constant.

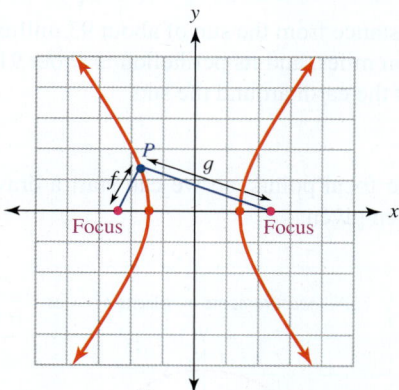

The difference $f - g$ is the same for every point P on the hyperbola.

> **DEFINITION**
>
> **Geometric Definition of a Hyperbola** The set of all points in a plane such that the difference of the distances from two fixed points is a constant is called a **hyperbola**.
>
> **The standard equation for a hyperbola centered at the origin that opens right and left**
>
> $$\frac{x^2}{a^2} - \frac{y^2}{b^2} = 1$$
>
> where a and b are positive real numbers.
>   The foci are the points $(c, 0)$ and $(-c, 0)$, where $c^2 = a^2 + b^2$, and the vertices are $(a, 0)$ and $(-a, 0)$.
>
> **The standard equation for a hyperbola centered at the origin that opens up and down**
>
> $$\frac{y^2}{b^2} - \frac{x^2}{a^2} = 1$$
>
> where a and b are positive real numbers.
>   The foci are the points $(0, c)$ and $(0, -c)$, where $c^2 = a^2 + b^2$, and the vertices are $(0, b)$ and $(0, -b)$.

The direction in which the hyperbola opens is determined by which variable term is positive. When the $\frac{x^2}{a^2}$ term is first, the hyperbola will open left and right and will have x-intercepts at $(a, 0)$ and $(-a, 0)$. When the term $\frac{y^2}{b^2}$ is first, the hyperbola will open up and down and will have y-intercepts at $(0, b)$ and $(0, -b)$. The following two graphs demonstrate the basic graphs of hyperbolas.

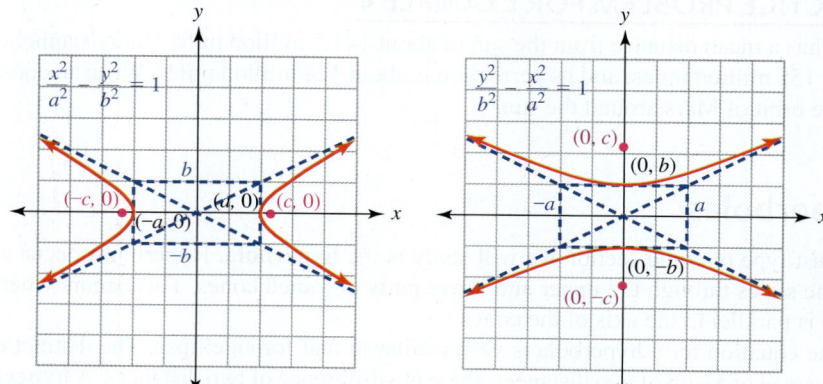

When graphing hyperbolas, we will use the **fundamental rectangle** to help us sketch the graph. The fundamental rectangle has vertical sides at $x = a$ and $x = -a$ and horizontal sides at $y = b$ and $y = -b$. When the diagonals of the fundamental rectangle are extended, they form asymptotes for the hyperbola. Therefore, the ends of the hyperbola will come closer and closer to these asymptotes but never touch them.

An example of hyperbolas is the path a comet's orbit may follow. While the planets follow an elliptical orbit, a comet can follow an elliptical, parabolic, or hyperbolic orbit. If a comet's speed around the sun is fast enough to break the sun's gravitational pull, its orbit will be either parabolic or hyperbolic.

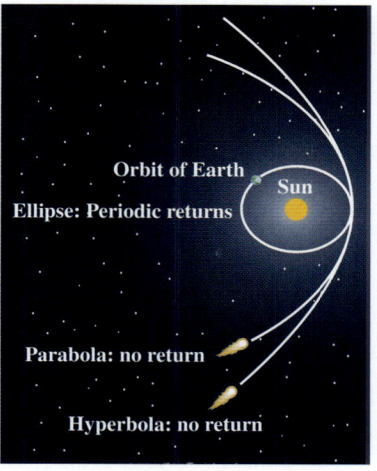

■ **Connecting the Concepts**

What is an asymptote?

Recall from working with exponential functions in Chapter 5, and rational functions in Chapter 7 that an asymptote is a line that a curve will get closer and closer to.

In the case of hyperbolas, the curve will approach the asymptotes but will not cross them.

Steps to Graphing a Hyperbola Centered at the Origin

1. Determine in which directions the hyperbola will open. (up/down or left/right)
2. Find the values of a and b and list the vertices.
3. Find the foci.
4. Plot the vertices, plot the fundamental rectangle, and draw the diagonals extended out in both directions.
5. Sketch the graph.

Example 5 Sketching the graph of a hyperbola

Sketch the graph of the given hyperbolas. Label the vertices and foci.

a. $\dfrac{x^2}{9} - \dfrac{y^2}{16} = 1$ **b.** $4y^2 - 25x^2 = 100$

SOLUTION

a. $\dfrac{x^2}{9} - \dfrac{y^2}{16} = 1$

Step 1 This equation is in the standard form for a hyperbola. The $\dfrac{x^2}{a^2}$ term is first, so the hyperbola will open left and right.

Step 2 $a^2 = 9$, so $a = 3$, and $b^2 = 16$, so $b = 4$. Therefore, the vertices are $(3, 0)$ and $(-3, 0)$.

Step 3 To find the foci, use the equation $c^2 = a^2 + b^2$ and solve for c.

$$c^2 = a^2 + b^2$$
$$c^2 = 3^2 + 4^2$$
$$c^2 = 9 + 16$$
$$c^2 = 25$$
$$c = \pm\sqrt{25}$$
$$c = \pm 5$$

Therefore, the foci are the points $(5, 0)$ and $(-5, 0)$.

Steps 4 and 5 The fundamental rectangle will have vertical sides at $x = 3$ and $x = -3$ and horizontal sides at $y = 4$ and $y = -4$.

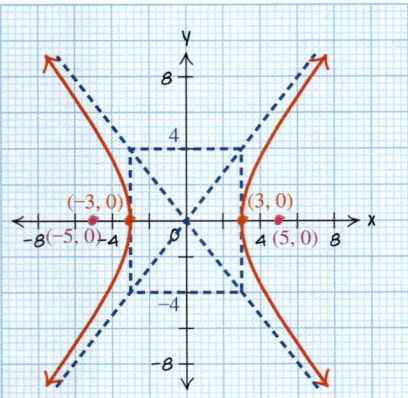

b. $4y^2 - 25x^2 = 100$

Step 1 This equation is not in the standard form for a hyperbola. Start by dividing by 100 to make the equation equal 1.

$$4y^2 - 25x^2 = 100$$

$$\frac{4y^2}{100} - \frac{25x^2}{100} = \frac{100}{100}$$

$$\frac{y^2}{25} - \frac{x^2}{4} = 1$$

The $\frac{y^2}{b^2}$ term is first, so the hyperbola will open up and down.

Step 2 $a^2 = 4$, so $a = 2$, and $b^2 = 25$, so $b = 5$. Therefore, the vertices are $(5, 0)$ and $(-5, 0)$.

Step 3 To find the foci, use the equation $c^2 = a^2 + b^2$ and solve for c.

$$c^2 = a^2 + b^2$$
$$c^2 = 2^2 + 5^2$$
$$c^2 = 4 + 25$$
$$c^2 = 29$$
$$c = \pm\sqrt{29}$$
$$c \approx \pm 5.39$$

Therefore, the foci are the points $(0, 5.39)$ and $(0, -5.39)$.

Steps 4 and 5

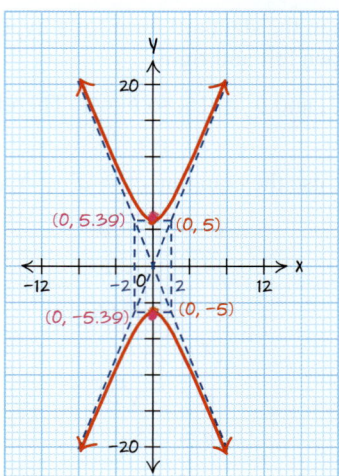

PRACTICE PROBLEM FOR EXAMPLE 5

Sketch the graph of the given hyperbolas. Label the vertices and foci.

a. $\dfrac{x^2}{36} - \dfrac{y^2}{64} = 1$ **b.** $16y^2 - 9x^2 = 144$

Example 6 Finding the equation of a hyperbola from a graph

Find the equation for the hyperbola. Find the foci.

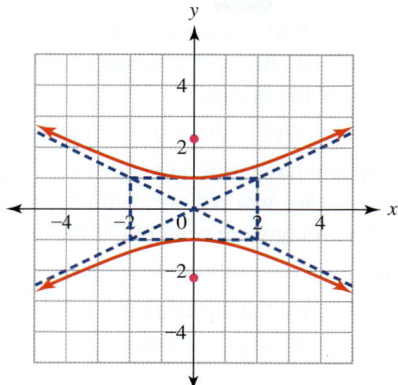

SOLUTION

The vertices of the hyperbola are $(0, 1)$ and $(0, -1)$. The fundamental rectangle intercepts the x-axis at $(2, 0)$ and $(-2, 0)$. From these values, we find that $a = 2$ and $b = 1$. Since the hyperbola is facing up and down, the $\frac{y^2}{b^2}$ term must be positive. These values give the equation

$$\frac{y^2}{1^2} - \frac{x^2}{2^2} = 1$$

$$\frac{y^2}{1} - \frac{x^2}{4} = 1$$

We can find the foci for this hyperbola using the equation $c^2 = a^2 + b^2$.

$$c^2 = 2^2 + 1^2$$
$$c^2 = 4 + 1$$
$$c^2 = 5$$
$$c = \pm\sqrt{5}$$
$$c \approx \pm 2.24$$

Therefore, the foci are the points $(0, 2.24)$ and $(0, -2.24)$.

PRACTICE PROBLEM FOR EXAMPLE 6

Find the equation for the hyperbola. Find the foci.

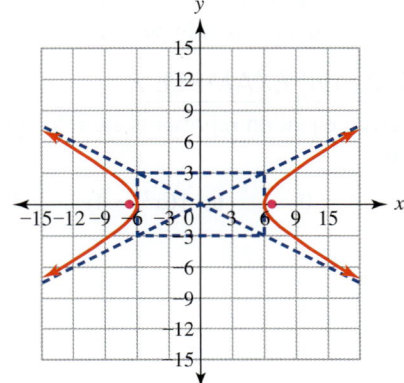

Recognizing the Equations for Conic Sections

All of the conic sections that we have studied can be described by the equation

$$Ax^2 + Bxy + Cy^2 + Dx + Ey + F = 0$$

in which A, B, and C cannot all be zero. If we assume that $B = 0$, then we can determine which conic section we have. Four different rules follow.

- If $A = C$, the equation represents a circle.
$$6x^2 + 6y^2 + 4x - 3y - 8 = 0 \quad \text{Circle}$$

- If A and C are different but both are positive, then the equation represents an ellipse. Note that we worked only with ellipses where $D = 0$ and $E = 0$ and, therefore, were centered at the origin.
$$4x^2 + 9y^2 - 36 = 0 \quad \text{Ellipse}$$

- If A and C are different and have different signs, then the equation represents a hyperbola. Again note that we worked only with hyperbolas where $D = 0$ and $E = 0$ and that, therefore, they were centered at the origin.
$$16x^2 - 25y^2 - 400 = 0 \quad \text{Hyperbola}$$

- If only one of A and C is zero, the equation is a parabola.
$$4x^2 + 5x + y - 8 = 0$$
or $\quad\quad\quad\quad\quad\quad\quad\quad\quad\quad\quad\quad\quad\quad\quad\quad\quad$ Parabola
$$2y^2 + 3x + 4y - 10 = 0$$

These descriptions give an idea of what type of equation makes each conic section. These descriptions can be used to determine the type of conic section that an equation represents.

Example 7 Determining the type of conic section

Determine whether the given equation represents a circle, ellipse, hyperbola, or parabola. Explain your answer.

a. $49x^2 + 36y^2 - 1764 = 0$ b. $3y^2 + x + 4y - 15 = 0$

c. $4x^2 - 9y^2 - 36 = 0$

SOLUTION

a. $49x^2 + 36y^2 - 1764 = 0$
In this equation, the coefficients of the two squared terms are different and both positive. This is the equation of an ellipse.

b. $3y^2 + x + 4y - 15 = 0$
In this equation, the x^2 term is not there. This is the equation of a parabola.

c. $4x^2 - 9y^2 - 36 = 0$
In this equation, the coefficients of the two squared terms are different and have different signs. This is the equation of a hyperbola.

PRACTICE PROBLEM FOR EXAMPLE 7

Determine whether the given equation represents a circle, ellipse, hyperbola, or parabola. Explain your answer.

a. $9x^2 + 9y^2 + 2x + 7y - 4 = 0$ b. $16x^2 - 4y^2 - 64 = 0$

c. $4x^2 + y^2 - 4 = 0$

9.2 Exercises

For Exercises 1 through 8, sketch the graph of the given equations.

1. $\dfrac{x^2}{4} + \dfrac{y^2}{16} = 1$ 2. $\dfrac{x^2}{9} + \dfrac{y^2}{36} = 1$ 3. $\dfrac{x^2}{25} + \dfrac{y^2}{9} = 1$ 4. $\dfrac{x^2}{100} + \dfrac{y^2}{64} = 1$ 5. $x^2 + \dfrac{y^2}{9} = 1$ 6. $\dfrac{x^2}{25} + y^2 = 1$

7. $\dfrac{x^2}{10} + \dfrac{y^2}{49} = 1$ 8. $\dfrac{x^2}{81} + \dfrac{y^2}{50} = 1$

For Exercises 9 through 14, convert the given equation to the standard form for an ellipse centered at the origin. Sketch a graph of the ellipse.

9. $4x^2 + 9y^2 = 36$
10. $16x^2 + 25y^2 = 400$
11. $x^2 + 36y^2 = 36$
12. $49x^2 + y^2 = 49$
13. $20x^2 + 9y^2 = 180$
14. $16x^2 + 8y^2 = 128$

For Exercises 15 through 20, find the equation of the ellipse shown. Find the foci.

15.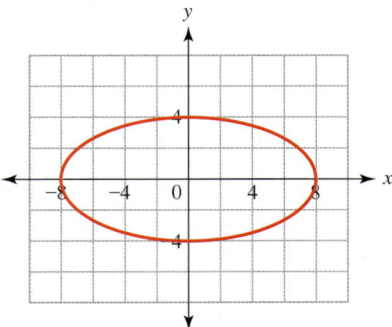

16.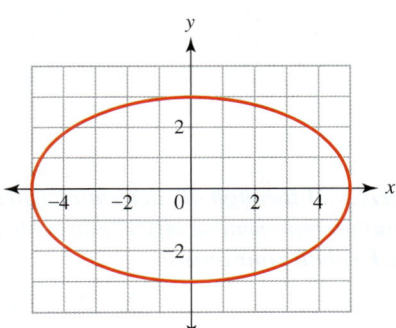

17.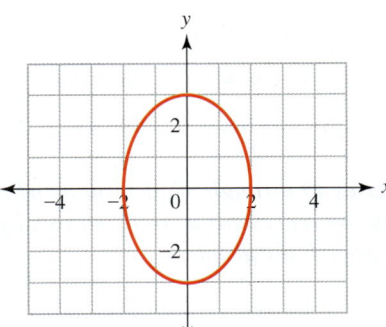

18.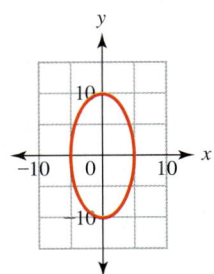

19.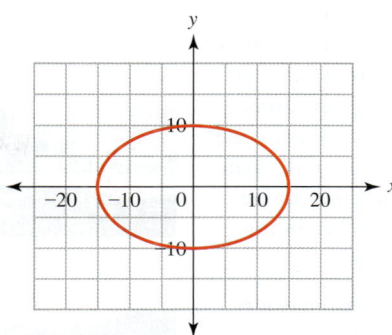

20.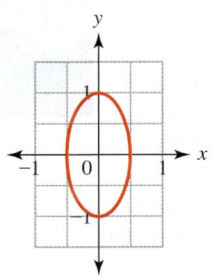

21. Pluto has a mean distance from the sun of about 3671.5 million miles. Pluto's aphelion is about 4583 million miles, and its perihelion is about 2760 million miles. Write an equation for the orbit of Pluto around the sun.

22. Neptune has a mean distance from the sun of about 2796.5 million miles. Neptune's aphelion is about 2823 million miles, and its perihelion is about 2770 million miles. Write an equation for the orbit of Neptune around the sun.

23. Saturn's mean distance from the sun is about 886 million miles. If Saturn has a perihelion of 840 million miles, find its aphelion. Write an equation for the orbit of Saturn around the sun.

24. Venus' mean distance from the sun is about 67 million miles. If Venus has an aphelion of 68 million miles, find its perihelion. Write an equation for the orbit of Venus around the sun.

25. The equation for the top of a semi-elliptical desk is

$$\frac{w^2}{4556.25} + \frac{d^2}{945.5} = 1$$

where w is half the width in centimeters and d is the depth in centimeters. Find the width and depth of the table.

26. The equation for the top of an elliptical table is

$$\frac{l^2}{588} + \frac{w^2}{315} = 1$$

where l is half the length of the table in inches and w is half the width of the table in inches. Find the length and width of the table.

27. The Waterloo Bridge over the Thames River in Britain is made of nine elliptical arches that are 120 feet wide and 35 feet tall. Write an equation for one of the arches in this bridge.

28. The largest masonry arch bridge in the early 1900s was the Adolphe Bridge in Luxembourg. The arch has a span of 278 feet and a rise of 138 feet. Write an equation for the arch.

For Exercises 29 through 36, sketch the graph of the given hyperbolas. Label the vertices and foci.

29. $\dfrac{x^2}{4} - \dfrac{y^2}{9} = 1$

30. $\dfrac{x^2}{25} - \dfrac{y^2}{16} = 1$

31. $\dfrac{y^2}{36} - \dfrac{x^2}{4} = 1$

32. $\dfrac{y^2}{9} - \dfrac{x^2}{49} = 1$

33. $x^2 - 25y^2 = 25$

34. $49y^2 - 4x^2 = 196$

35. $9y^2 - 81x^2 = 81$

36. $16x^2 - 36y^2 = 576$

For Exercises 37 through 40, find the equation for the hyperbola shown. Find the foci.

37.

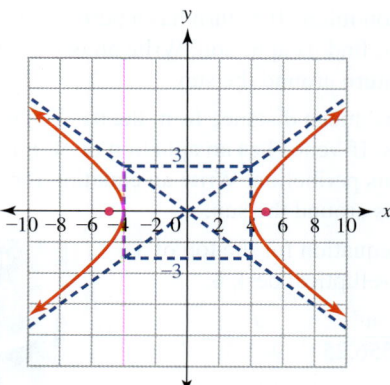

38.

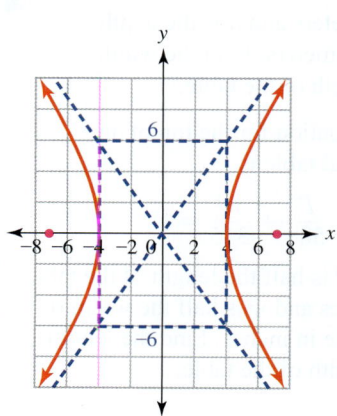

39.

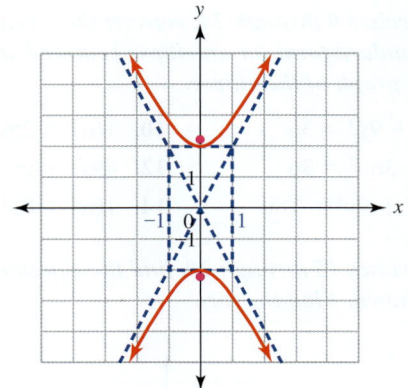

40.

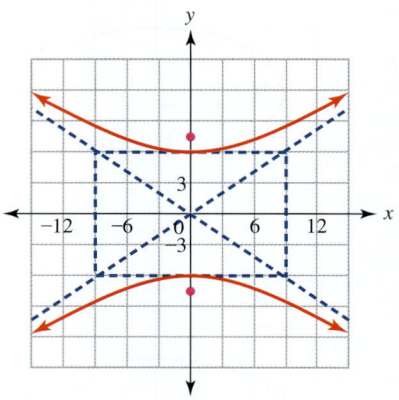

For Exercises 41 through 48, determine whether the given equation represents a circle, ellipse, hyperbola, or parabola. Explain your answer.

41. $3x^2 + 3y^2 = 27$

42. $12x^2 + 108y^2 = 108$

43. $4x^2 - 9y^2 = 36$

44. $5x^2 + 5y^2 + 10x - 20y - 10 = 0$

45. $6y^2 + 2y + x - 9 = 0$

46. $100y^2 - 9x^2 = 900$

47. $64x^2 + 36y^2 = 576$

48. $4x^2 + 3x - y = 0$

For Exercises 49 through 58, determine whether the given equation represents a circle, ellipse, hyperbola, or parabola. Sketch the graph.

49. $\dfrac{x^2}{16} + \dfrac{y^2}{36} = 1$

50. $(x + 4)^2 + (y - 7)^2 = 4$

51. $\dfrac{x^2}{49} - \dfrac{y^2}{100} = 1$

52. $x^2 - 4x + y - 12 = 0$

53. $18x^2 + 9y^2 = 36$

54. $2x - 6y^2 + 8y = 20$

55. $8x^2 + 8y^2 = 2$

56. $20y^2 - 45x^2 = 180$

57. $2x^2 + 6x + 2y^2 - 10y - 5 = 0$

58. $-24x = y^2$

9.3 Arithmetic Sequences

LEARNING OBJECTIVES

- Compute and graph terms of a sequence.
- Identify an arithmetic sequence.
- Find the general term of an arithmetic sequence.
- Solve applied problems that involve arithmetic sequences.

Introduction to Sequences

We are now going to examine a new concept called a **sequence**. A sequence is an ordered list. An alphabetical class roster and a list of the number of hours worked each week of a month are both sequences. To investigate the idea of a sequence further, we will consider the following scenario.

CONCEPT INVESTIGATION

How much will I get paid?

Suppose you just got hired by a new company and you are earning a salary of $38,500 for the first year. Your employer tells you that you have two options to choose from: a $1000 raise each year or a 2.5% raise each year.

Option 1

- $1000 raise each year.

To calculate the next year's salary, add $1000.

$$38,500 + 1000 = 39,500$$

Option 2

- 2.5% raise each year.

To calculate the next year's salary:

1. To find the amount of the raise, multiply the current salary by 2.5%.

$$0.025 \cdot 38,500 = 962.50$$

2. Add the raise to the current salary.

$$38,500 + 962.50 = 39,462.50$$

1. List the first five years' salaries for each option in the table.

Year	Salary with Option 1
1	$38,500.00
2	$39,500.00
3	$
4	$
5	$

Year	Salary with Option 2
1	$38,500.00
2	$39,462.50
3	$
4	$
5	$

2. Which salary option is the best choice if you work for the company for only 5 years?

3. Will the salary from option 2 ever be larger than the salary from option 1? Explain.

In both salary options, the inputs are the numbers (years) 1, 2, 3, 4, or 5, and the outputs are your salary {$38,500.00, $39,462.50, $40,449.06, $41,460.29, $42,496.80}. This set of outputs is an ordered list and is a sequence of numbers. In this scenario, we computed your salary only for the first five years. Our set of salaries is a finite set, which is also known as a **finite sequence**. If we had continued computing your salary forever, we would have obtained an **infinite sequence**.

This ordered list can be thought of as a function whose outputs are the values in the ordered list and whose domain (inputs) is the set of natural numbers. The inputs represent the position of the output in the sequence.

■ **Connecting the Concepts**

Which set of numbers?

Recall that the natural numbers are the positive whole numbers. They are sometimes called the counting numbers.

$$1, 2, 3, 4, 5, \ldots$$

DEFINITIONS

Sequence A **sequence** is an ordered list.

Finite Sequence A function whose domain is the set $\{1, 2, 3, \ldots, n\}$, where n is a natural number, is called a **finite sequence**. A finite sequence stops and has a fixed number of elements in the domain.

Infinite Sequence A function whose domain is the set of natural numbers is called an **infinite sequence**. An infinite sequence goes on forever.

What is very important about this definition is that the domain is the set of natural numbers only. As we noted before, sequences are functions. Since their inputs come from the set of natural numbers, we use n as the input variable, instead of x ("natural" numbers starts with an n). To designate the outputs, we can use either the function notation $f(n)$ or, more traditionally, a_n. Sequences are listed either in a table or as a list of numbers. The outputs of a sequence are called **terms**. In the sequence, 3, 5, 7, 9, 11, 13, ..., a_n:

3 is the first term. We also write $a_1 = 3$.

5 is the second term. $a_2 = 5$.

7 is the third term and so on. $a_3 = 7$.

a_n is called the nth term or the **general term.**

For salary option 1 in the scenario we looked at earlier, the first term is $38,500.00, the second term is $39,500.00, and so on.

DEFINITIONS

Term of a Sequence The outputs of a sequence are called the **terms of the sequence.**

General Term The nth term, a_n, is called the **general term** of a sequence, where n is a natural number.

Example 1 Computing terms of a sequence

a. Let $f(n) = 5 + (n - 1) \cdot 4$. Compute and list the first five terms of this sequence in a table.

b. Let $a_n = 2^n + 3$. Compute and list the first four terms of this sequence.

c. Let $a_n = \dfrac{(-1)^n}{2n}$. Compute and list the first six terms of this sequence.

SOLUTION

a. Recall that the first five inputs are $n = 1, 2, 3, 4, 5$.

n	$f(n)$
1	$f(1) = 5 + (1 - 1) \cdot 4 = 5$
2	$f(2) = 5 + (2 - 1) \cdot 4 = 9$
3	$f(3) = 5 + (3 - 1) \cdot 4 = 13$
4	$f(4) = 5 + (4 - 1) \cdot 4 = 17$
5	$f(5) = 5 + (5 - 1) \cdot 4 = 21$

The first five terms of this sequence are 5, 9, 13, 17, 21.

b. Recall that the first four inputs are $n = 1, 2, 3, 4$. When $n = 1$, we compute the first term as follows.

$$n = 1$$
$$a_1 = 2^1 + 3$$
$$a_1 = 5$$

Computing the next three terms in a similar way yields the following.

$n = 2$ | $n = 3$ | $n = 4$
$a_2 = 2^2 + 3$ | $a_3 = 2^3 + 3$ | $a_4 = 2^4 + 3$
$a_2 = 7$ | $a_3 = 11$ | $a_4 = 19$

The first four terms of this sequence are 5, 7, 11, 19.

c. Computing the first six terms yields the following.

$n = 1$ | $n = 2$ | $n = 3$
$a_1 = \dfrac{(-1)^1}{2(1)}$ | $a_2 = \dfrac{(-1)^2}{2(2)}$ | $a_3 = \dfrac{(-1)^3}{2(3)}$
$a_1 = -\dfrac{1}{2}$ | $a_2 = \dfrac{1}{4}$ | $a_3 = -\dfrac{1}{6}$

$n = 4$ | $n = 5$ | $n = 6$
$a_4 = \dfrac{(-1)^4}{2(4)}$ | $a_5 = \dfrac{(-1)^5}{2(5)}$ | $a_6 = \dfrac{(-1)^6}{2(6)}$
$a_4 = \dfrac{1}{8}$ | $a_5 = -\dfrac{1}{10}$ | $a_6 = \dfrac{1}{12}$

The first six terms of this sequence are $-\dfrac{1}{2}, \dfrac{1}{4}, -\dfrac{1}{6}, \dfrac{1}{8}, -\dfrac{1}{10}, \dfrac{1}{12}$. Notice that the factor $(-1)^n$ made the signs of the terms alternate.

PRACTICE PROBLEM FOR EXAMPLE 1

a. List the first five terms of the sequence whose general term is given by

$$f(n) = 5n + 7$$

b. List the first four terms of the sequence whose general term is given by

$$a_n = \dfrac{1}{3^n}$$

If we know the value of a term, we can figure out what term in the sequence it is by setting the general term equal to the given value and solving for n.

Example 2 Finding a term number

Find n for the given value of a_n.

a. $a_n = 2n - 30$ when $a_n = -4$

b. $a_n = \dfrac{5n}{n+4}$ when $a_n = 3.75$

SOLUTION

a. Set the general term equal to -4 and solve for n.

$$-4 = 2n - 30$$
$$26 = 2n$$
$$\frac{26}{2} = \frac{2n}{2}$$
$$13 = n$$

-4 is the 13th term in the sequence.

b. Set the general term equal to 3.75 and solve.

$$3.75 = \frac{5n}{n+4}$$
$$(n+4) \cdot 3.75 = \frac{5n}{n+4} \cdot (n+4) \quad \text{Rational equation, so multiply by the LCD.}$$
$$3.75n + 15 = 5n$$
$$15 = 1.25n$$
$$\frac{15}{1.25} = \frac{1.25n}{1.25}$$
$$12 = n$$

3.75 is the 12th term of the sequence.

PRACTICE PROBLEM FOR EXAMPLE 2

Find n for the given value of a_n.

a. $a_n = -7n + 24$ when $a_n = -11$

b. $a_n = n^2 + 8$ when $a_n = 204$. Remember that n must be a natural number.

Graphing Sequences

Another way to represent a sequence is to graph its terms. Recall that the domain of a sequence is the set of natural numbers. We can never substitute into a sequence an input that is not a natural number. Therefore, we cannot let $n = 1.5$, as it is not a natural number. When we graph a sequence, we graph a discrete set of points only. We *do not connect the points* because the function does not exist except for inputs that are the natural numbers.

The sequence for the Option 1 salary from the Concept Investigation can be graphed as shown.

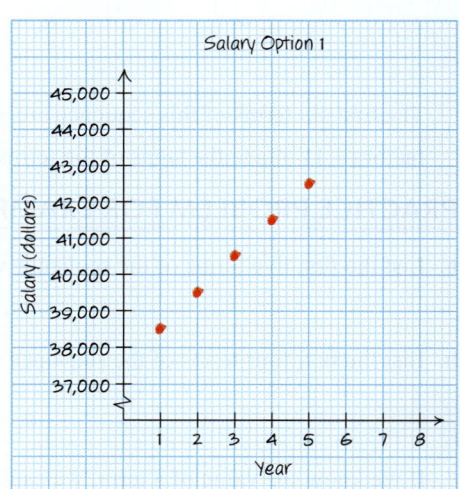

Example 3 Graphing sequences

Graph the first six terms of each sequence. Clearly label and scale the axes. Do not connect the points.

a. $a_n = -3 + (n-1) \cdot 7$ **b.** $f(n) = (-1)^n \cdot n^2$

SOLUTION

a. Start by determining the first six terms. To do this, substitute in the values $n = 1, 2, 3, 4, 5,$ and 6 into the general term and simplify. The results are listed in the following table.

n	a_n
1	$a_1 = -3 + (1-1) \cdot 7 = -3$
2	$a_2 = -3 + (2-1) \cdot 7 = 4$
3	$a_3 = -3 + (3-1) \cdot 7 = 11$
4	$a_4 = -3 + (4-1) \cdot 7 = 18$
5	$a_5 = -3 + (5-1) \cdot 7 = 25$
6	$a_6 = -3 + (6-1) \cdot 7 = 32$

The graph is shown below. Remember that you do NOT connect the points. The graph of a sequence is a discrete (separated) set of points.

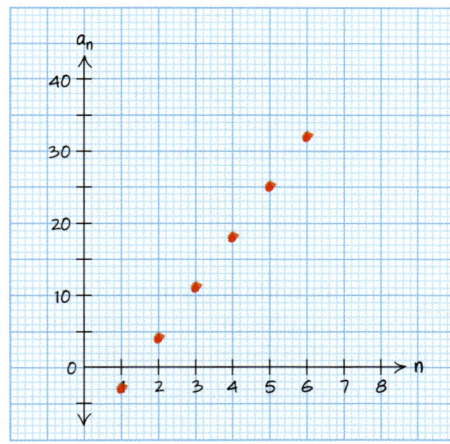

b. Begin by finding the first six terms of the sequence. Then plot the terms.

n	$f(n)$
1	$f(1) = (-1)^1 \cdot (1)^2 = -1$
2	$f(2) = (-1)^2 \cdot (2)^2 = 4$
3	$f(3) = (-1)^3 \cdot (3)^2 = -9$
4	$f(4) = (-1)^4 \cdot (4)^2 = 16$
5	$f(5) = (-1)^5 \cdot (5)^2 = -25$
6	$f(6) = (-1)^6 \cdot (6)^2 = 36$

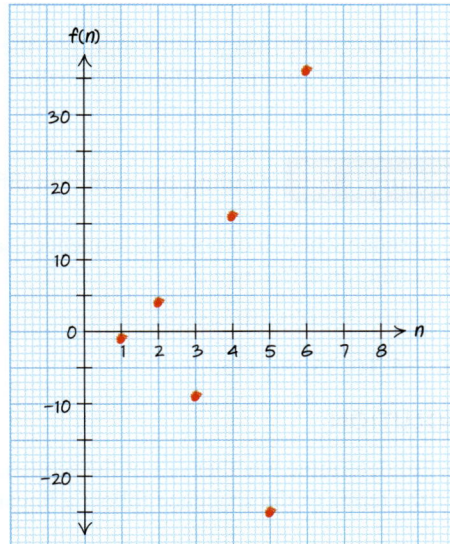

PRACTICE PROBLEM FOR EXAMPLE 3

Graph the first six terms of each sequence. Clearly label and scale your axes.

a. $a_n = 3 + (n-1) \cdot 5$ **b.** $f(n) = (-1)^n$

Arithmetic Sequences

In part a of Example 3, the graph of the sequence of points lies in a straight line. If you look carefully at the general term, you might notice that it has the form of a linear function. In discussing sequences, linear functions are called arithmetic sequences. Salary option 1 from the concept investigation at the beginning of this section is another example of an arithmetic sequence.

Let's reexamine the table of values from part a of Example 3. Remember, when we graphed the terms we saw that the points followed a linear pattern.

n	a_n	Difference
1	-3	$a_2 - a_1 = 4 - (-3) = 7$
2	4	$a_3 - a_2 = 11 - 4 = 7$
3	11	$a_4 - a_3 = 18 - 11 = 7$
4	18	$a_5 - a_4 = 25 - 18 = 7$
5	25	$a_6 - a_5 = 32 - 25 = 7$
6	32	

Looking at the outputs, we see that to get from one term to the next we add 7. Another way to look at this situation is that each term differs from the term before it by 7. The number 7 is called the **common difference** d for this sequence.

> **DEFINITIONS**
>
> **Arithmetic Sequence** An **arithmetic sequence** is a sequence in which successive terms have a common difference. An arithmetic sequence is a linear function with a domain restricted to the natural numbers.
>
> **Common Difference** The **common difference** d for an arithmetic sequence is the constant (same) amount that each term differs from the previous term.
>
> $$d = a_n - a_{n-1} \quad \text{or} \quad d = a_{n+1} - a_n$$

For each unit change in the input (n), the output changes by d. Recall that for a linear function, this common difference is called the slope.

Example 4 Finding the common difference

Determine whether or not each sequence is arithmetic. If the sequence is arithmetic, find the common difference d.

a. $-10, -5, 0, 5, 10, 15, \ldots$

b. $-5, 10, 35, 50, 115, \ldots$

SOLUTION

a. To determine whether or not the sequence is arithmetic, first determine whether the terms of the sequence follow a linear pattern. Compute the difference between each term and the preceding term. If the difference is common (the same), then the sequence is arithmetic (linear).

$$a_2 - a_1 = (-5) - (-10) = 5$$
$$a_3 - a_2 = 0 - (-5) = 5$$
$$a_4 - a_3 = 5 - 0 = 5$$
$$a_5 - a_4 = 10 - 5 = 5$$
$$a_6 - a_5 = 15 - 10 = 5$$

Find the difference between each term and the preceding term.

The common difference $d = 5$ is constant, and therefore, the sequence is arithmetic. Another method to determine whether or not the sequence is arithmetic is to graph the terms as in Example 3. If the points follow a straight line, the sequence is arithmetic.

b. To determine whether or not the sequence is arithmetic, compute the difference between each term and the preceding term and see whether that difference is common (the same) for the entire sequence.

$$a_2 - a_1 = 10 - (-5) = 15$$
$$a_3 - a_2 = 35 - 10 = 25$$
$$15 \neq 25$$

Find the difference between each term and the preceding term.

The differences are not equal.

Since the difference between the first two terms is 15 and the difference between the second two terms is 25, we see that the difference is not the same for the entire sequence. Thus, the sequence is not arithmetic.

PRACTICE PROBLEM FOR EXAMPLE 4

Determine whether or not each sequence is arithmetic. If the sequence is arithmetic, find the common difference d.

a. $-2, -8, -14, -20, -26, \ldots$

b. $-4, -2, 0, 2, 4, \ldots$

c. $0, -3, -8, -15, -24, \ldots$

Now we are going to use our skills from previous chapters to write the general term of an arithmetic sequence given a finite number of terms in the sequence. Because an arithmetic sequence is a linear function, we can find the general term using the point-slope formula or the slope-intercept form. We can also write a formula for the general term using the notation of sequences.

Starting with the point-slope formula, substitute the sequence notation for the different pieces and the result is the general term a_n.

$$y - y_1 = m(x - x_1) \quad \text{Point-slope formula.}$$
$$y - y_1 = d(x - x_1) \quad \text{Replace the slope with the common difference } d.$$
$$y - a_1 = d(x - 1) \quad \text{Replace the point } (x_1, y_1) \text{ with the term } (1, a_1).$$
$$y = a_1 + d(x - 1) \quad \text{Isolate } y \text{ and replace the variables } x \text{ and } y \text{ with the notation for the general term } n \text{ and } a_n.$$
$$a_n = a_1 + (n - 1)d$$

Skill Connection

Point-Slope Formula

In Section 1.5, we first used the point-slope formula to find the equation of a line.

$$y - y_1 = m(x - x_1)$$

DEFINITION

General Term, a_n, of an Arithmetic Sequence The **general term, a_n, of an arithmetic sequence** is given by $a_n = a_1 + (n - 1)d$, where a_1 is the first term, d is the common difference, and n is a natural number.

> **Example 5** Finding the general term of an arithmetic sequence

Determine whether or not each sequence is arithmetic. If the sequence is arithmetic, find the common difference d and the general term a_n.

a. 7, 10, 13, 16, 19, ...

b. $-\frac{1}{2}, 0, \frac{1}{2}, 1, \frac{3}{2}, 2, \ldots$

c. 6, 12, 24, 48, ...

SOLUTION

a. To determine whether the sequence is arithmetic, compute the difference between each term and the preceding term. See whether that difference is common (the same) for the entire sequence.

$$a_2 - a_1 = 10 - 7 = 3$$
$$a_3 - a_2 = 13 - 10 = 3$$
$$a_4 - a_3 = 16 - 13 = 3$$
$$a_5 - a_4 = 19 - 16 = 3$$

The difference is the same between each consecutive term, so the common difference is $d = 3$. To find the general term a_n, use the formula for the general term of an arithmetic sequence. The first term is $a_1 = 7$ and the common difference is $d = 3$, so substitute these values into the form and simplify.

$a_n = a_1 + (n - 1)d$ Formula for the general term.
$a_n = 7 + (n - 1)3$ Substitute $d = 3$ and $a_1 = 7$.
$a_n = 7 + 3n - 3$ Combine like terms to simplify.
$a_n = 3n + 4$

The general term for this arithmetic sequence is $a_n = 3n + 4$.

b. Check the differences to determine whether the sequence is arithmetic.

$$a_2 - a_1 = 0 - \left(-\frac{1}{2}\right) = \frac{1}{2}$$
$$a_3 - a_2 = \frac{1}{2} - 0 = \frac{1}{2}$$
$$a_4 - a_3 = 1 - \frac{1}{2} = \frac{1}{2}$$
$$a_5 - a_4 = \frac{3}{2} - 1 = \frac{1}{2}$$

The difference is the same between each consecutive term, so the common difference is $d = \frac{1}{2}$. To find the general term a_n, use the formula for the general term of an arithmetic sequence. The first term is $a_1 = -\frac{1}{2}$ and the common difference is $d = \frac{1}{2}$, so substitute these values into the form and simplify.

$a_n = a_1 + (n - 1)d$ Formula for the general term.
$a_n = -\frac{1}{2} + (n - 1)\frac{1}{2}$ Substitute $d = \frac{1}{2}$ and $a_1 = -\frac{1}{2}$.
$a_n = -\frac{1}{2} + \frac{1}{2}n - \frac{1}{2}$ Combine like terms to simplify.
$a_n = \frac{1}{2}n - 1$

The general term for this arithmetic sequence is $a_n = \frac{1}{2}n - 1$.

c. Check the differences to determine whether the sequence is arithmetic.

$$a_2 - a_1 = 12 - 6 = 6$$
$$a_3 - a_2 = 24 - 12 = 12$$
$$6 \neq 12$$

The differences between consecutive terms are not the same, so this is not an arithmetic sequence, and the formula does not apply.

PRACTICE PROBLEM FOR EXAMPLE 5

Determine whether or not each sequence is arithmetic. If the sequence is arithmetic, find the common difference d and the general term a_n.

a. 9, 7, 5, 3, 1, ...   b. 4, 6, 9, 13, ...   c. $-\frac{5}{3}, -\frac{4}{3}, -1, -\frac{2}{3}, -\frac{1}{3}, \ldots$

Example 6 Finding a car's value

A one-year-old car is valued at $34,000 and is depreciating by $3000 per year.

a. Write a formula for the value of the car a_n in dollars when it is n years old.

b. What will the value of the car be in the fifth year?

c. In what year will the car's value be $10,000?

SOLUTION

a. Since the car is valued at $34,000 when it is a year old, we have $a_1 = 34,000$. It is depreciating, so the value is going down $3000 per year. Therefore, $d = -3000$. With this information, we can find the general term a_n.

$$a_n = a_1 + (n-1)d$$
$$a_n = 34,000 + (n-1)(-3000)$$
$$a_n = 34,000 - 3000n + 3000$$
$$a_n = -3000n + 37,000$$

b. The fifth year is represented by $n = 5$, so substitute 5 for n and solve for a_5.

$$a_5 = -3000(5) + 37,000$$
$$a_5 = -15,000 + 37,000$$
$$a_5 = 22,000$$

The car will be worth $22,000 in the fifth year.

c. We are given a value, so we will substitute 10,000 for a_n and solve for n.

$$10,000 = -3000n + 37,000$$
$$-27,000 = -3000n$$
$$\frac{-27,000}{-3000} = \frac{-3000n}{-3000}$$
$$9 = n$$

The car will be worth $10,000 during the ninth year.

> **Connecting the Concepts**
>
> **How does value change?**
>
> When the value of an object is decreasing over time, the object is said to be **depreciating**. Depreciation of a car or other business assets is something that can be deducted in calculating taxes. Depreciation is often assumed to be a straight line (linear) depreciation over a certain period of time set by the government. When that time is over, the object is assumed to have no value remaining.

PRACTICE PROBLEM FOR EXAMPLE 6

Your starting salary is $44,000.00, and each year you are getting a $1200 raise.

a. Write a formula for your salary a_n in dollars in the nth year.

b. What will your salary be in the 14th year?

c. In what year will your salary be $68,000?

If we don't know the first term of an arithmetic sequence, we can use the point-slope formula or the slope-intercept form of a line to find the general term. Any two terms can be used to find the common difference the same way we found slope in Chapter 1.

Example 7 Using two terms to find the general term of an arithmetic sequence

Find the general term a_n for the arithmetic sequences with the given terms.

a. $a_4 = 5$ and $a_7 = 20$

b. $a_3 = -6$ and $a_8 = -26$

SOLUTION

a. We are told that the sequences are arithmetic, so we can use the two terms given to find the general term. First we will find the common difference d in the same way we found slope in Chapter 1.

$$a_4 = 5 \qquad a_7 = 20$$
$$(4, 5) \qquad (7, 20)$$
$$d = \frac{20 - 5}{7 - 4}$$
$$d = \frac{15}{3} = 5$$

Now we can use the point-slope formula and the common difference as the slope to find the equation.

$$y - y_1 = m(x - x_1) \qquad \text{Point-slope formula.}$$
$$y - 20 = 5(x - 7) \qquad \text{Substitute common difference and one point.}$$
$$y - 20 = 5x - 35 \qquad \text{Solve for y.}$$
$$y = 5x - 15$$
$$a_n = 5n - 15 \qquad \text{Write the general term using sequence notation.}$$

The general term for this arithmetic sequence is $a_n = 5n - 15$.

b. Again this is an arithmetic sequence, so we will find the common difference. Then we will use the slope-intercept form to find the equation of the general term.

$$a_3 = -6 \qquad a_8 = -26$$
$$(3, -6) \qquad (8, -26)$$
$$d = \frac{(-26) - (-6)}{8 - 3} = \frac{-20}{5} = -4 \qquad \text{Find the common difference.}$$
$$y = mx + b$$
$$y = -4x + b \qquad \text{Substitute common difference into the slope-intercept form.}$$
$$-6 = -4(3) + b \qquad \text{Substitute a point and solve for b.}$$
$$-6 = -12 + b$$
$$6 = b$$
$$y = -4x + 6 \qquad \text{Write the general term in sequence notation.}$$
$$a_n = -4n + 6$$

The general term for this arithmetic sequence is $a_n = -4n + 6$.

PRACTICE PROBLEM FOR EXAMPLE 7

Find the general term a_n for the arithmetic sequences with the given terms.

a. $a_5 = -2$ and $a_{20} = 4$

b. $a_3 = 1$ and $a_{10} = 22$

9.3 Exercises

For Exercises 1 through 6, compute the terms of the given sequence to complete the table.

1.

n	$a_n = 4 + (n-1)6$
1	
2	
3	
4	

2.

n	$a_n = -7 + (n-1)3$
1	
2	
3	
4	

3.

n	$f(n) = 2n^2$
1	
2	
3	
4	

4.

n	$f(n) = 3(2)^n$
1	
2	
3	
4	

5.

n	$a_n = 4(-1)^n$
1	
2	
3	
4	
5	

6.

n	$a_n = 2n(-1)^n$
1	
2	
3	
4	
5	

For Exercises 7 through 14, compute and list the first five terms of the given sequence.

7. $f(n) = 3^n - 20$

8. $f(n) = 1.5^n - 18$

9. $a_n = \dfrac{4n}{n+7}$

10. $a_n = \dfrac{-3n}{n-15}$

11. $a_n = \dfrac{1}{4^n}$

12. $a_n = \dfrac{2}{3^n}$

13. $a_n = \dfrac{(-1)^n}{n^2}$

14. $a_n = \dfrac{(-1)^n}{5n}$

For Exercises 15 through 24, find n for the given value of a_n.

15. $a_n = 4 + (n-1)6$ when $a_n = 118$

16. $a_n = -7 + (n-1)3$ when $a_n = 95$

17. $a_n = 2n^2 + 5$ when $a_n = 397$

18. $a_n = -5n^2 + 40$ when $a_n = -5405$

19. $a_n = 2(3)^n - 24$ when $a_n = 354270$

20. $a_n = 2048\left(\dfrac{1}{2}\right)^n - 80$ when $a_n = -79.5$

21. $a_n = \dfrac{2n}{n+8}$ when $a_n = 1.2$

22. $a_n = \dfrac{-5n}{n+8}$ when $a_n = -\dfrac{10}{3}$

23. $a_n = 50 + \sqrt{2n}$ when $a_n = 68$

24. $a_n = 2 + \sqrt{n+3}$ when $a_n = 7$

For Exercises 25 through 34, graph the first six terms of each sequence. Clearly label and scale the axes. Do not connect the points.

25. $a_n = 4n - 5$

26. $a_n = -2n + 8$

27. $a_n = 4 + (n-1)6$

28. $a_n = -7 + (n-1)3$

29. $f(n) = 3^n - 20$

30. $f(n) = 1.5^n - 18$

31. $a_n = 4n(-1)^n$

32. $a_n = 3n^2(-1)^n$

33. $a_n = \dfrac{4n}{n+7}$

34. $a_n = \dfrac{-3n}{n-15}$

For Exercises 35 through 44, determine whether or not each sequence is arithmetic. If the sequence is arithmetic, find the common difference d.

35. 12, 16, 20, 24, 28, 32, ...

36. 14, 17, 20, 23, 26, 29, ...

37. $-6.65, -4.3, -1.95, 0.4, 2.75, 5.1, \ldots$
38. $-10.05, -8.6, -7.15, -5.7, -4.25, -2.8, \ldots$
39. $100, 80, 64, 51.2, 40.96, 32.768, \ldots$
40. $12, 18, 27, 40.5, 60.75, 91.125, \ldots$
41. $14, 10.55, 7.1, 3.65, 0.2, -3.25, \ldots$
42. $20, 18.25, 16.5, 14.75, 13, 11.25, \ldots$
43. $4, \frac{7}{2}, 3, \frac{5}{2}, 2, \frac{3}{2}, 1, \ldots$
44. $-2, -\frac{5}{3}, -\frac{4}{3}, -1, -\frac{2}{3}, -\frac{1}{3}, \ldots$

For Exercises 45 through 52, determine whether or not each sequence is arithmetic. If the sequence is arithmetic, find the common difference d and the general term a_n.

45. $4, 10, 16, 22, 28, 34, \ldots$
46. $9, 16, 23, 30, 37, 44, \ldots$
47. $4, 5.25, 6.5, 7.75, 9, 10.25, \ldots$
48. $-5, -1.5, 2, 5.5, 9, 12.5, \ldots$
49. $3, 9, -1, 13, -5, 17, \ldots$
50. $-4, -6, -6, -4, 0, 6, \ldots$
51. $4, \frac{10}{3}, \frac{8}{3}, 2, \frac{4}{3}, \frac{2}{3}, \ldots$
52. $8, \frac{37}{5}, \frac{34}{5}, \frac{31}{5}, \frac{28}{5}, 5, \ldots$

53. Your starting salary is $36,000.00, and each year you receive a $900 raise.
    a. Write a formula for your salary a_n in dollars in the nth year.
    b. What will your salary be in the 12th year?
    c. In what year will your salary be $60,300?

54. Your starting salary is $52,000.00, and each year you receive a $1400 raise.
    a. Write a formula for your salary a_n in dollars in the nth year.
    b. What will your salary be in the 16th year?
    c. In what year will your salary be $101,000?

55. A one-year-old car is valued at $48,000, and is depreciating by $4500 per year.
    a. Write a formula for the value of the car a_n in dollars when it is n years old.
    b. What will the value of the car be in the 11th year?
    c. At what age will the car's value be $21,000?

56. A one-year-old factory machine is valued at $180,000 and is depreciating by $15,000 per year.
    a. Write a formula for the value of the machine a_n in dollars when it is n years old.
    b. What will the value of the machine be in the seventh year?
    c. At what age will the machine's value be zero dollars?

57. You start saving money. You begin by saving $10 one week. The next week, you save $11. Each following week, you save $1 more than the previous week.
    a. Write a formula for the amount you saved a_n in dollars during the nth week.
    b. How much money will you save during the 32nd week?
    c. In what week will you save $61?

58. Suppose you are going to start saving money. You begin by saving $50 one month. The next month, you save $55. Each following month, you save $5 more than the previous month.
    a. Write a formula for the amount you saved a_n in dollars during the nth month.
    b. How much money will you save during the eighth month?
    c. In what month will you save $100?

59. The cost in dollars to produce n candy bars is represented by the sequence
$$a_n = 400 + 0.15n$$
    a. Use the sequence to find the cost to produce 100 candy bars.
    b. How many candy bars can be produced for a cost of $550?
    c. Find the common difference for this sequence and explain its meaning.

60. The profit in dollars from selling n concert tickets is represented by the sequence
$$a_n = 4.5n - 240$$
    a. Use the sequence to find the profit from selling 40 concert tickets.
    b. How many concert tickets must be sold to break even?
    c. Find the common difference for this sequence and explain its meaning.

61. A child stacks cups to form a pyramid. The top row has one cup. Each row down has one more cup than the row above it.
    a. Write a formula for the number of cups a_n in the nth row.
    b. How many cups are in the 10th row?

62. A retaining wall is put into a backyard. The top row has 11 blocks in it. Each row after that has one more block than the row above it.
    a. Write a formula for the number of blocks a_n in the nth row.
    b. How many blocks are in the eighth row?

For Exercises 63 through 70, find the general term a_n for the arithmetic sequences with the given terms.

63. $a_4 = 9$ and $a_{15} = -13$
64. $a_3 = 12$ and $a_{18} = -78$
65. $a_7 = 28$ and $a_{21} = 77$
66. $a_5 = 4$ and $a_{33} = 46$
67. $a_{10} = 56.75$ and $a_{32} = 161.25$
68. $a_4 = -21.1$ and $a_{22} = 92.3$
69. $a_6 = -36$ and $a_{25} = -32.2$
70. $a_9 = -18.4$ and $a_{41} = 4$

9.4 Geometric Sequences

LEARNING OBJECTIVES

- Identify a geometric sequence.
- Find the general term of a geometric sequence.
- Solve application problems that involve geometric sequences.

In Section 9.3, we studied arithmetic sequences which were similar to linear functions. In this section, we will consider a different type of sequence that is similar to exponential functions. This type of sequence is called a **geometric sequence**.

With arithmetic sequences, there must be a common difference between consecutive terms. In a geometric sequence, there must be a **common ratio** between consecutive terms. The common ratio can be found by dividing each term by the preceding term. In Section 9.3, we looked at two options for a salary increase. Option 2 was described as having a starting salary of $38,500 and getting a 2.5% raise each year. The salaries for each year represent a geometric sequence because each year's salary was 1.025 times the salary of the previous year.

Use the following Concept Investigation to compare the difference between arithmetic and geometric sequences.

■ CONCEPT INVESTIGATION

How do I get to the next term?

1. Fill in the following tables with the terms of the sequences given.

n	$a_n = 5 + (n-1)3$
1	
2	
3	
4	

n	$a_n = 4 \cdot 3^{(n-1)}$
1	
2	
3	
4	

2. Looking at the first table, what operation ($+, -, \cdot, \div$) can you use to get from one term to the next without using the formula?

3. Looking at the second table, how can you get from one term to the next without using the formula?

4. Which table represents an arithmetic sequence? Explain why.

If we look at the two sequences from the Concept Investigation, we can see that the arithmetic sequence has a common difference and the geometric sequence has a common ratio.

Arithmetic Sequence	**Geometric Sequence**
5, 8, 11, 14, …	3, 12, 48, 192, …
$a_2 - a_1 = 8 - 5 = 3$	$\dfrac{a_2}{a_1} = \dfrac{12}{3} = 4$
$a_3 - a_2 = 11 - 8 = 3$	$\dfrac{a_3}{a_2} = \dfrac{48}{12} = 4$
$a_4 - a_3 = 14 - 11 = 3$	$\dfrac{a_4}{a_3} = \dfrac{192}{48} = 4$
Common difference = 3	Common ratio = 4

Each term of a geometric sequence is the previous term multiplied by the common ratio. Starting with the first term, we get the following pattern.

n	a_n
1	$a_1 = a_1$
2	$a_2 = a_1 r$
3	$a_3 = a_1 rr$
4	$a_4 = a_1 rrr$

Using exponents, we have the following list of terms.

n	a_n
1	$a_1 = a_1$
2	$a_2 = a_1 r$
3	$a_3 = a_1 r^2$
4	$a_4 = a_1 r^3$

If we look at the pattern connecting the number of the term n and the exponent for r, we have the general term $a_n = a_1 r^{(n-1)}$.

> **DEFINITIONS**
>
> **Geometric Sequence** A **geometric sequence** is a sequence whose consecutive terms have a common ratio, r.
>
> $$r = \frac{a_{n+1}}{a_n}$$
>
> **General Term, a_n, of a Geometric Sequence** The **general term, a_n, of a geometric sequence** is given by $a_n = a_1 \cdot r^{(n-1)}$, where a_1 is the first term of the sequence, r is the common ratio, and n is a natural number.

If we consider the general form of an exponential function and that of a geometric sequence, we will see some similarities and some differences.

Exponential Function	**Geometric Sequence**
General form	General term
$f(x) = ab^x$	$a_n = a_1 \cdot r^{(n-1)}$
where $b > 0$ and $b \neq 1$.	where $r \neq 0$ or 1.
Domain: All real numbers	Domain: Natural numbers

The domains of the two are different. An exponential function has a domain of all real numbers, so it will be a smooth curve when graphed. A geometric sequence has a domain of the natural numbers, so it will be a discrete set of points that are not connected.

Exponential Function

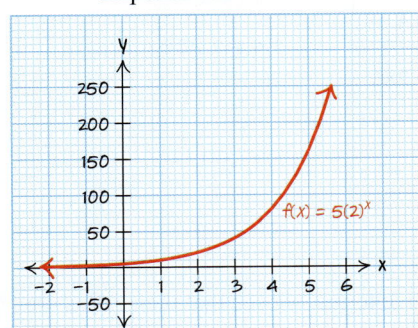

Geometric Sequence

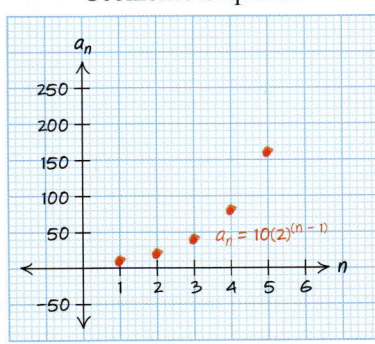

The difference in domains also requires different restrictions on the base for the exponential. Recall from Chapter 5 that the base b of an exponential function must be a positive number. This was to avoid any nonreal numbers when raising the base to fractional exponents. The base r in a geometric sequence can be negative, since we are raising it only to natural number powers, not fractional powers, and, therefore, we will have only real number answers.

To determine whether a sequence is geometric, we need to confirm that each set of consecutive terms has a common ratio.

Example 1 Finding the common ratio

Determine whether or not each of the following sequences is geometric. If the sequence is geometric, find the common ratio r.

a. 6, 18, 54, 162, 486, ...

b. 5, 7, 9, 11, 13, 15, ...

c. 2.5, 6.25, 15.625, 39.0625, 97.65625, ...

SOLUTION

a. Put the terms of the sequence in a table. Looking at the second column of the sequence values, a_n, we see that the sequence is not arithmetic, as we do not add or subtract the same number each time to get the next term. Dividing each term by the previous term yields the third column of ratios in the table.

n	a_n	Ratio
1	6	$\frac{a_2}{a_1} = \frac{18}{6} = 3$
2	18	
3	54	$\frac{a_3}{a_2} = \frac{54}{18} = 3$
4	162	$\frac{a_4}{a_3} = \frac{162}{54} = 3$
5	486	$\frac{a_5}{a_4} = \frac{486}{162} = 3$

The result in the third column is always 3; therefore, the sequence is geometric, and the common ratio is $r = 3$.

b. As in part a, put the terms of the sequence in a table and check the ratio of consecutive terms.

n	a_n	Ratios
1	5	$\frac{a_2}{a_1} = \frac{7}{5} = 1.4$
2	7	$\frac{a_3}{a_2} = \frac{9}{7} \approx 1.286$
3	9	
4	11	
5	13	
6	15	

$1.4 \neq 1.286$

By comparing the first two sets of consecutive terms, we see that their ratios are not the same. Therefore, the sequence is not geometric. To get from one term of the sequence to the next, add 2. Therefore, the sequence is arithmetic.

c. Looking at the terms of the sequence, it is not arithmetic as we do not add (or subtract) the same number each time to get the next term. Therefore, check to see whether the sequence is geometric. Placing the terms of the sequence in a table yields the following.

n	a_n	Ratio
1	2.5	$\frac{a_2}{a_1} = \frac{6.25}{2.5} = 2.5$
2	6.25	$\frac{a_3}{a_2} = \frac{15.625}{6.25} = 2.5$
3	15.625	$\frac{a_4}{a_3} = \frac{39.0625}{15.625} = 2.5$
4	39.0625	$\frac{a_5}{a_4} = \frac{97.65625}{39.0625} = 2.5$
5	97.65625	

Dividing each term by the previous term results in the third column ratios. The result in the third column is always 2.5, so the sequence is geometric with a common ratio $r = 2.5$.

PRACTICE PROBLEM FOR EXAMPLE 1

Determine whether or not each of the following sequences is geometric. If the sequence is geometric, find the common ratio r.

a. $0.75, 0.1875, 0.046875, 0.01171875, \ldots$

b. $17, 22, 27, 32, 37, \ldots$

Now we will use the formula for the general term of a geometric sequence, $a_n = a_1 \cdot r^{(n-1)}$, to find the general term for a given geometric sequence.

Example 2 Finding a general term for a geometric sequence

Determine whether or not each sequence is geometric. If the sequence is geometric, find the general term a_n.

a. $6, 9, 13.5, 20.25, 30.375, \ldots$

b. $6, -18, 54, -162, 486, \ldots$

SOLUTION

a. First we will check the ratios of consecutive terms to see whether the sequence is a geometric sequence.

n	a_n	Ratio
1	6	$\frac{a_2}{a_1} = \frac{9}{6} = \frac{3}{2}$
2	9	$\frac{a_3}{a_2} = \frac{13.5}{9} = 1.5$ or $\frac{3}{2}$
3	13.5	$\frac{a_4}{a_3} = \frac{20.25}{13.5} = 1.5$ or $\frac{3}{2}$
4	20.25	$\frac{a_5}{a_4} = \frac{30.375}{20.25} = 1.5$ or $\frac{3}{2}$
5	30.375	

The sequence has a common ratio of $r = \frac{3}{2}$, so it is a geometric sequence. Use the formula for the general form of a geometric sequence to find a_n. $r = \frac{3}{2}$ and $a_1 = 6$, so

$$a_n = a_1 \cdot r^{(n-1)}$$
$$a_n = 6\left(\frac{3}{2}\right)^{(n-1)}$$

The general term of this geometric sequence is $a_n = 6\left(\frac{3}{2}\right)^{(n-1)}$.

b. First check whether there is a common ratio.

n	a_n	Ratio
1	6	$\frac{a_2}{a_1} = \frac{-18}{6} = -3$
2	-18	$\frac{a_3}{a_2} = \frac{54}{-18} = -3$
3	54	$\frac{a_4}{a_3} = \frac{-162}{54} = -3$
4	-162	$\frac{a_5}{a_4} = \frac{486}{-162} = -3$
5	486	

The sequence has a common ratio of $r = -3$, so it is a geometric sequence. Now use the formula for the general form of a geometric sequence to find a_n. $r = -3$ and $a_1 = 6$, so

$$a_n = a_1 \cdot r^{(n-1)}$$
$$a_n = 6(-3)^{(n-1)}$$

The general term of this geometric sequence is $a_n = 6(-3)^{(n-1)}$.

■ Skill Connection

Oversimplifying

When you work with exponential functions or geometric sequences, it is often tempting to try to simplify the equation too much. In the equation

$$a_n = 6\left(\frac{3}{2}\right)^{(n-1)}$$

we cannot simplify the fraction by reducing the 6 with the 2 in the denominator because the fraction $\frac{3}{2}$ is being raised to a power. Recall that the order of operations says that we must calculate the exponents before multiplication or division.

PRACTICE PROBLEM FOR EXAMPLE 2

Determine whether or not each sequence is geometric. If the sequence is geometric, find the general term a_n.

a. 25, 30, 36, 43.2, 51.84, . . .

b. 200, -80, 32, -12.8, 5.12, . . .

Example 3 Modeling white-tailed deer population

The white-tailed deer population has been increasing by about 8% per year. In 1980, there were approximately 1.19 million white-tailed deer in the Northeast.

a. Find the general term of a sequence that will model the population of white-tailed deer in millions n years since 1979. Using 1979 will allow 1980 to be represented with $n = 1$, and thus, 1.19 will be the first term in the sequence.

b. Find a_{10} and explain its meaning.

c. If this population growth continued, when did the white-tailed deer population reach 3 million?

SOLUTION

a. Since we are given the growth rate of 8%, we can find the base of the exponential by adding 1 to the growth rate, as we did in Section 5.5.

$$r = 1 + 0.08 = 1.08$$

Since $n = 1$ represents 1980, $a_1 = 1.19$. Therefore, the general term is

$$a_n = 1.19(1.08)^{(n-1)}$$

b. Using the general term, calculate a_{10}.

$$a_{10} = 1.19(1.08)^{(10-1)}$$
$$a_{10} \approx 2.379$$

In 1989, there were approximately 2.38 million white-tailed deer in the Northeast.

c. Since we want the population to be 3 million, we can set the general term equal to 3 and solve for n.

$$3 = 1.19(1.08)^{(n-1)}$$
$$\frac{3}{1.19} = \frac{1.19(1.08)^{(n-1)}}{1.19} \quad \text{Isolate the base and exponent.}$$
$$2.521 \approx (1.08)^{(n-1)}$$
$$\log_{1.08} 2.521 \approx n - 1 \quad \text{Rewrite in log form.}$$
$$\frac{\ln 2.521}{\ln 1.08} \approx n - 1 \quad \text{Use the change of base formula.}$$
$$\frac{\ln 2.521}{\ln 1.08} + 1 \approx n$$
$$13.01 \approx n$$

Since n must be a natural number, we will round to $n = 13$. In 1992, there were approximately 3 million white-tailed deer in the Northeast.

PRACTICE PROBLEM FOR EXAMPLE 3

Your new employer offers you a starting salary of $40,000 with a 2% raise each year.

a. Find the general term for the sequence that will model your salary during your nth year of employment.

b. Find a_5 and explain its meaning.

c. After how many years will you earn $100,000?

If we are not given the common ratio or consecutive terms, we will use the skills we learned in Chapter 5 to find the general term. Since a geometric sequence is of the same form as an exponential function, we can find the general term in the same way that we found an exponential model.

■ **Skill Connection**

Exponential Equations

We solved exponential equations in Section 6.5 using the following steps.

1. Isolate the base and exponent on one side of the equation.
2. Rewrite the exponential in logarithm form.
3. Use the change of base formula to write the log in a convenient base.
4. Isolate the variable.

From Section 6.2, recall that rewriting in log form uses the relationship between an exponential and logarithm.

$$\log_b n = m$$
$$n = b^m$$

The change of base formula is

$$\log_b x = \frac{\log_c x}{\log_c b}$$

Example 4 — Using two terms to find the general term of a geometric sequence

Find the general term a_n for the geometric sequences with the given terms.

a. $a_4 = 250$ and $a_7 = 31{,}250$

b. $a_2 = -60$ and $a_4 = -15$, $r < 0$

SOLUTION

a. The sequence is geometric, so use the two terms to find the common ratio.

$$250 = a_1(r)^{4-1} \qquad 31250 = a_1(r)^{7-1} \quad \text{Write as two equations}$$
$$250 = a_1(r)^3 \qquad 31250 = a_1(r)^6 \quad \text{and simplify.}$$

$$\frac{31250}{250} = \frac{a_1(r)^6}{a_1(r)^3} \quad \text{Divide the two equations to find } r.$$

$$125 = r^{6-3} \quad \text{Reduce and subtract exponents.}$$

$$125 = r^3$$

$$(125)^{\frac{1}{3}} = (r^3)^{\frac{1}{3}} \quad \text{Raise to the reciprocal power.}$$

$$5 = r$$

The common ratio is $r = 5$. Use either equation to find a_1.

$$31250 = a_1(5)^{7-1} \quad \text{Substitute in the common ratio.}$$
$$31250 = a_1(5)^6$$
$$31250 = a_1(15625) \quad \text{Solve.}$$
$$\frac{31250}{15625} = \frac{a_1(15625)}{15625}$$
$$2 = a_1$$

The common ratio is $r = 5$ and the first term is $a_1 = 2$, so write the general term for the sequence

$$a_n = 2(5)^{(n-1)}$$

Check this solution by confirming that both terms come from the general term.

$$a_4 = 250 \qquad\qquad a_7 = 31{,}250$$
$$250 \stackrel{?}{=} 2(5)^{(4-1)} \qquad 31250 \stackrel{?}{=} 2(5)^{(7-1)}$$
$$250 \stackrel{?}{=} 2(5)^3 \qquad\qquad 31250 \stackrel{?}{=} 2(5)^6$$
$$250 = 250 \qquad\qquad 31250 = 31250$$

Both terms check, so the general term is correct.

b. Using the two points, find the common ratio.

$$-60 = a_1(r)^{2-1} \qquad -15 = a_1(r)^{4-1} \quad \text{Write as two equations}$$
$$-60 = a_1(r)^1 \qquad -15 = a_1(r)^3 \quad \text{and simplify.}$$

$$\frac{-15}{-60} = \frac{a_1(r)^3}{a_1(r)^1} \quad \text{Divide the two equations.}$$

$$\frac{1}{4} = r^{3-1} \quad \text{Reduce and subtract exponents.}$$

$$\frac{1}{4} = r^2$$

$$\pm\sqrt{\frac{1}{4}} = r \quad \text{Use the square root property.}$$

$$\pm\frac{1}{2} = r$$

Since $r < 0$, $r = -\frac{1}{2}$. Use the first equation to find a_1.

$$-60 = a_1\left(-\frac{1}{2}\right)^{2-1} \quad \text{Substitute in the common ratio.}$$

$$-60 = a_1\left(-\frac{1}{2}\right)^{1} \quad \text{Solve.}$$

$$\frac{-60}{-\frac{1}{2}} = \frac{a_1\left(-\frac{1}{2}\right)}{-\frac{1}{2}}$$

$$120 = a_1$$

The common ratio is $r = -\frac{1}{2}$ and the first term is $a_1 = 120$, so write the general term for the sequence $a_n = 120\left(-\frac{1}{2}\right)^{(n-1)}$. Check this general term using the calculator table.

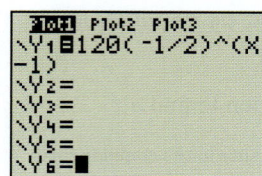 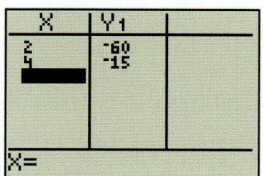

Both points work, so the general term is correct.

PRACTICE PROBLEM FOR EXAMPLE 4

Find the general term a_n for the geometric sequences with the given terms.

a. $a_4 = -270$ and $a_9 = 65{,}610$

b. $a_3 = 12.8$ and $a_5 = 0.512$, $r > 0$

9.4 Exercises

For Exercises 1 through 10, determine whether or not each of the following sequences is geometric. If the sequence is geometric, find the common ratio r.

1. $2, 3, 4.5, 6.75, 10.125, 15.1875, \ldots$
2. $4, 10, 25, 62.5, 156.25, 390.625, \ldots$
3. $270, 90, 30, 10, \frac{10}{3}, \frac{10}{9}, \ldots$
4. $125, 25, 5, 1, \frac{1}{5}, \frac{1}{25}, \ldots$
5. $-4, -1, 4, 11, 20, 31, \ldots$
6. $-7.75, -1.5, 4.75, 11, 17.25, 23.5, \ldots$
7. $-6, -12, -24, -48, -96, -192, \ldots$
8. $-7, -21, -63, -189, -567, -1701, \ldots$
9. $4, -12, 36, -108, 324, -972, \ldots$
10. $8, -16, 32, -64, 128, -256, \ldots$

For Exercises 11 through 20, determine whether or not each of the following sequences is geometric. If the sequence is geometric, find the general term a_n.

11. $3, 6, 12, 24, 48, 96, \ldots$
12. $8, 24, 72, 216, 648, 1944, \ldots$
13. $2, -8, 32, -128, 512, -2048, \ldots$

14. 5, −10, 20, −40, 80, −160, ...

15. 625, 125, 25, 5, 1, 0.2, ...

16. 81, 27, 9, 3, 1, $\frac{1}{3}$, ...

17. −4.5, 0, 4.5, 9, 13.5, 18, ...

18. 4.75, 2.5, 0.25, −2, −4.25, −6.5, ...

19. −4, −2, −1, $-\frac{1}{2}, -\frac{1}{4}, -\frac{1}{8}$, ...

20. −80, −20, −5, $-\frac{5}{4}, -\frac{5}{16}, -\frac{5}{64}$, ...

21. Your new employer offers you a starting salary of $56,000 with a 1.5% raise each year.
    a. Find the general term for the sequence that will model your salary during your nth year of employment.
    b. Find a_7 and explain its meaning in this situation.
    c. After how many years will you earn $80,052?

22. Your new employer offers you a starting salary of $36,000 with a 3% raise each year.
    a. Find the general term for the sequence that will model your salary during your nth year of employment.
    b. Find a_7 and explain its meaning in this context.
    c. After how many years will you earn $90,003?

23. A rubber ball dropped on a hard surface will take a sequence of bounces. Each bounce will be half as tall as the previous bounce. If the ball is dropped from a height of 16 feet, find the following.
    a. Find the general term for the sequence that will model the height of the ball after n bounces.
    b. Find a_3 and explain its meaning in this situation.

24. A radioactive substance has a half-life of one day. (Recall half-life from Section 6.1.) If there is 400 mg of the radioactive material on day 1, find the following:
    a. Find the general term for the sequence that will model the amount of radioactive material left on the nth day.
    b. Find a_5 and explain its meaning in this situation.

25. A very contagious flu is spreading quickly. The first day, five people are diagnosed with this flu. Each day after that, twice as many people are diagnosed.
    a. Find the general term for the sequence that will model the number of flu diagnoses on the nth day.
    b. Find a_7 and explain its meaning in this situation.

26. A store having a grand opening advertises that they will give away prizes every hour they are open the first day. They plan to give out three prizes the first hour. Every hour later they plan to give out twice as many prizes as the hour before.
    a. Find the general term for the sequence that will model the number of prizes given out n hours after the store opens.

    b. How many prizes will the store give out during the ninth hour of being open?

27. A one-year-old Honda Civic Hybrid is worth $21,000 and is depreciating 7% per year.
    a. Find the general term for the sequence that will model the value of the car when it is n years old.
    b. Find the value of this Honda Civic Hybrid when it is four years old.

28. A one-year-old Ford Mustang is worth $15,000 and is depreciating 10% per year.
    a. Find the general term for the sequence that will model the value of the car when it is n years old.
    b. Find the value of this Ford Mustang when it is 6 years old.

29. Morphine has a half-life of two hours when injected into a patient. A patient is given 10 mg of morphine.
    a. Find the general term for the sequence that will model the amount of morphine remaining after n hours. (Recall half-life from Section 6.1.)
    b. Find the amount of morphine remaining after 6 hours.

30. Warfarin, an anticoagulant, has a half-life of 37 hours when given to a patient. A patient is given a 5-mg dose of warfarin.
    a. Find the general term for the sequence that will model the amount of warfarin remaining after n hours. (Recall half-life from Section 6.1.)
    b. Find the amount of warfarin remaining after 48 hours.

For Exercises 31 through 38, find the general term a_n for the geometric sequences with the given terms.

31. $a_4 = 250$ and $a_9 = 781{,}250$

32. $a_6 = 288$ and $a_{13} = 36{,}864$

33. $a_4 = 256$ and $a_{10} = \frac{1}{16}$

34. $a_6 = 27$ and $a_{12} = \frac{1}{27}$

35. $a_4 = -24$ and $a_{11} = 3072$

36. $a_6 = -1944$ and $a_9 = 52{,}488$

37. $a_5 = 16$ and $a_{12} = -\dfrac{1}{8}$

38. $a_7 = 1600$ and $a_{12} = -16.384$

For Exercises 39 through 46, determine whether each of the given sequences is geometric or arithmetic. Find the general term a_n.

39. $11, 22, 44, 88, 176, 352, \ldots$

40. $9, 27, 81, 243, 729, 2187, \ldots$

41. $14, 20, 26, 32, 38, 44, \ldots$

42. $27, 42, 57, 72, 87, 102, \ldots$

43. $7776, 3888, 1944, 972, 486, 243, \ldots$

44. $7776, 2592, 864, 288, 96, 32, \ldots$

45. $-38, -34.75, -31.5, -28.25, -25, -21.75, \ldots$

46. $40, -80, 160, -320, 640, -1280, \ldots$

47. The sum of the interior angles of a triangle is 180°, that of a quadrilateral is 360°, and that of a pentagon is 540°.
    a. Find the general term for the sequence that will give the sum of the interior angles of an $n + 2$ sided polygon.
    b. Find a_{10} and explain its meaning in this situation.

48. A $200,000 production machine is depreciating 9% per year.
    a. Find the general term for the sequence that will model the value of the machine when it is n years old.
    b. Find the value of this machine when it is 10 years old.

49. A $200,000 production machine is depreciating $8000 per year.
    a. Find the general term for the sequence that will model the value of the machine when it is n years old.
    b. Find the value of this machine when it is 10 years old.

50. After surgery, your doctor wants you to slowly return to your jogging routine. The first week, she wants you to jog for only 15 minutes each day. Each week later, you can add 20% more time to your daily jog.
    a. Find the general term for the sequence that will give the amount of time you should jog each day during the nth week after surgery.
    b. Find a_5 and explain its meaning in this context.

51. After surgery, your doctor wants you to slowly return to your jogging routine. The first week, she wants you to jog for only 15 minutes each day. Each week later, you can jog an additional 5 minutes a day.
    a. Find the general term for the sequence that will give the amount of time you should jog each day during the nth week after surgery.
    b. Find a_5 and explain its meaning in this situation.

52. The cost of a product will go up at the rate of inflation. A product costs $20,000, and the cost is increasing at a rate of 3% per year.
    a. Find the general term for the sequence that will give the cost of the product after n years.
    b. Find a_7 and explain its meaning in this situation.

9.5 Series

LEARNING OBJECTIVES

- Use sigma notation to find a sum.
- Identify geometric and arithmetic series.
- Find finite sums of an arithmetic series.
- Find finite sums of a geometric series.
- Solve applied problems that involve geometric and arithmetic series.

Introduction to Series

In the previous two sections, we learned that a sequence is a function whose domain is the set of natural numbers. There are situations in which we are interested in adding up all the terms of a sequence.

For example, Ali's monthly salary for one calendar year is given in the following table. We will denote the month of January by 1, February by 2, and so on, until

December, which is denoted by 12. Ali's monthly salary forms a sequence as the inputs are from the set of natural numbers.

Month	Ali's Salary
1	$2100
2	$2400
3	$2000
4	$1900
5	$2300
6	$2300
7	$2100
8	$1800
9	$2000
10	$2100
11	$2200
12	$2400

For tax purposes, Ali has to compute his yearly income. He adds up his monthly salaries (the terms of the sequence) to find his yearly income. Adding up his salary, he determines that his yearly income is $25,600. Ali's total yearly income is the sum of the first 12 terms on a sequence.

$$2100 + 2400 + 2000 + \cdots + 2100 + 2200 + 2400 = 25{,}600$$
$$= a_1 + a_2 + a_3 + \cdots + a_{10} + a_{11} + a_{12}$$

The sum of the terms of a sequence is called a **series**. The yearly income above is an example of a **finite series**. We say that the sum of the series is $25,600.

DEFINITIONS

Series A **series** is the sum of a sequence.

Finite Series A **finite series** sums up a finite sequence.

Example 1 Total deaths from swine flu

The first death from the H1N1 virus (swine flu) was reported by the World Health Organization on April 24, 2009. The number of deaths reported each day from April 24 to May 7, 2009 is listed in the tables.

Date	4-24-09	4-25-09	4-26-09	4-27-09	4-28-09	4-29-09	4-30-09
Deaths Reported	3	0	4	0	0	0	1

Date	5-1-09	5-2-09	5-3-09	5-4-09	5-5-09	5-6-09	5-7-09
Deaths Reported	2	7	3	5	4	2	13

Source: World Health Organization. http://www.who.int/csr/disease/swineflu/en/

a. Write a series that sums the total deaths reported from April 24 to May 7, 2009.

b. Find the sum.

SOLUTION

a. The series will be each day's reported death toll added together.

$$3 + 0 + 4 + 0 + 0 + 0 + 1 + 2 + 7 + 3 + 5 + 4 + 2 + 13$$

b.

$$3 + 0 + 4 + 0 + 0 + 0 + 1 + 2 + 7 + 3 + 5 + 4 + 2 + 13 = 44$$

During the first 14 days of reporting for the H1N1 virus, a total of 44 deaths occurred.

Notice that in Example 1, we had to add up 14 numbers to determine the total number of deaths. Since writing long sums can get a bit tedious, mathematicians have a notation that tells us to sum up a sequence. This is called **summation notation.**

Summation Notation (finite sums)

$$\sum_{n=1}^{k} a_n = a_1 + a_2 + \cdots + a_k$$

- The Greek letter sigma $\sum$ means to sum in mathematics.
- The subscript below the sigma $\sum_{n=1}$ means to start the sum with $n = 1$ or the first term, a_1.
- The superscript above the sigma $\sum^{k}$ means to stop the sum with $n = k$ or with the kth term, a_k.

Example 2 Using summation notation to find a sum

Write out each series in expanded form and find the sum.

a. $\sum_{n=1}^{6}(2n)$ **b.** $\sum_{n=1}^{4}(-1)^n$ **c.** $\sum_{n=1}^{5}(2 \cdot 3^n)$

SOLUTION

a. First, expand the sigma notation. Then sum the first six terms, starting with the first term (a_1) and ending with the sixth term (a_6). The general term is $a_n = 2n$. Doing so yields

$$\sum_{n=1}^{6}(2n) = a_1 + a_2 + a_3 + a_4 + a_5 + a_6 \qquad \text{Add the first six terms.}$$

$$\sum_{n=1}^{6}(2n) = (2 \cdot 1) + (2 \cdot 2) + (2 \cdot 3) + (2 \cdot 4) + (2 \cdot 5) + (2 \cdot 6) \qquad \text{Substitute the values of } n.$$

$$\sum_{n=1}^{6}(2n) = 2 + 4 + 6 + 8 + 10 + 12$$

$$\sum_{n=1}^{6}(2n) = 42 \qquad \text{Simplify.}$$

b. Sum the first four terms of the sequence whose general term is $a_n = (-1)^n$.

This yields

$$\sum_{n=1}^{4}(-1)^n = (-1)^1 + (-1)^2 + (-1)^3 + (-1)^4 \quad \text{Substitute the values of } n.$$

$$\sum_{n=1}^{4}(-1)^n = (-1) + 1 + (-1) + 1 \quad \text{Simplify.}$$

$$\sum_{n=1}^{4}(-1)^n = 0$$

c. Sum the first five terms of the sequence whose general term is $a_n = 2 \cdot 3^n$. This yields

$$\sum_{n=1}^{5}(2 \cdot 3^n) = (2 \cdot 3^1) + (2 \cdot 3^2) + (2 \cdot 3^3) + (2 \cdot 3^4) + (2 \cdot 3^5)$$

$$\sum_{n=1}^{5}(2 \cdot 3^n) = 6 + 18 + 54 + 162 + 486$$

$$\sum_{n=1}^{5}(2 \cdot 3^n) = 726$$

PRACTICE PROBLEM FOR EXAMPLE 2
Write out each series in expanded form. Find the sum.

a. $\sum_{n=1}^{4}(4n - 20)$ **b.** $\sum_{n=1}^{5}(2(-3)^n)$

Arithmetic Series

When the terms of an arithmetic sequence are summed, it is called an **arithmetic series**. When the terms of a finite arithmetic series are added, an interesting pattern appears. Use the following concept investigation to help discover this pattern.

■ CONCEPT INVESTIGATION
How much will I get paid?

1. You start a job with an annual salary of $54,000. You receive a $3000 raise each year. Write a series for the total income you will make in the first five years of the job. Find the sum.

2. The series for the total income in thousands you will make in the first eight years is given by

$$= a_1 + a_2 + a_3 + a_4 + a_5 + a_6 + a_7 + a_8$$
$$54 + 57 + 60 + 63 + 66 + 69 + 72 + 75$$

Take this series and add it to the same series in the reverse order.

$54 + 57 + 60 + 63 + 66 + 69 + 72 + 75$ Original series.
$75 + 72 + 69 + 66 + 63 + 60 + 57 + 54$ Same series in reverse order.
$129 +$

How many 129's do you have?
Use multiplication to find the total of the 129's. This is twice the sum of the original series.
Since we added the series to itself, we can find the sum of the original series by dividing the total by 2. Divide the result of your multiplication by 2.
This is the total income in thousands of dollars for the first eight years on the job.

This concept investigation goes through the process of finding a finite sum for an arithmetic series. If we want to add more terms, this process can be long and tedious. Notice, too, that when we add the series with the terms in reverse order, we get the same result for each pair of terms. Therefore, the sum of each pair of terms is the same as the sum of the first term a_1 and the last term a_n.

$$(a_1 + a_n)$$

Since there are n pairs of terms, we can multiply this sum by n to get the total of all the pairs of terms.

$$n(a_1 + a_n)$$

This product is twice the sum of the original series. If we divide by 2, we will have the sum of the series.

$$\frac{n(a_1 + a_n)}{2}$$

This formula will find the sum of the first n terms for any arithmetic sequence or series.

> **Sum of the First n Terms of an Arithmetic Sequence**
>
> The finite sum S_n, of the first n terms of an arithmetic sequence a_n, can be found by using the formula
>
> $$S_n = a_1 + a_2 + \cdots + a_n = \frac{n(a_1 + a_n)}{2}$$

Example 3 Total income with fixed raises

Find the total income from a 30-year career at a job with a starting salary of $48,000 and an annual raise of $1500.

SOLUTION

Since there is a constant change in the salary each year, we can represent the annual salary with an arithmetic sequence. We will start by finding the general term.

$$a_n = 48{,}000 + 1500(n - 1)$$

Now we want to sum the first 30 terms, so we can use the formula for a finite sum.

$$a_1 = 48{,}000 \qquad a_{30} = 48{,}000 + 1500(30 - 1) = 91{,}500$$

$$S_{30} = \frac{30(48{,}000 + 91{,}500)}{2}$$

$$S_{30} = 15(139500)$$

$$S_{30} = 2{,}292{,}500$$

Over a 30-year career, the total income will be $2,292,500.

PRACTICE PROBLEM FOR EXAMPLE 3

The first row of an amphitheater has 13 seats. Each row adds an additional 3 seats. If the amphitheater has a total of 8 rows, how many seats are there in all?

Example 4 Finding the sum of an arithmetic sequence

Find the indicated sum.

a. S_{20} if $a_n = 3n + 10$ **b.** S_{55} if $a_n = -4n + 8$

SOLUTION

a. Since the given sequence is arithmetic, the sum will be an arithmetic series. First, find the first and 20th terms.

$$a_1 = 3(1) + 10 = 13 \qquad a_{20} = 3(20) + 10 = 70$$

Using the formula for the finite sum of the arithmetic sequence,

$$S_n = \frac{n(a_1 + a_n)}{2}$$

$$S_{20} = \frac{20(13 + 70)}{2} \qquad \text{Substitute } n = 20 \text{ and the values for the first and 20th terms.}$$

$$S_{20} = \frac{\overset{10}{\cancel{20}}(83)}{\cancel{2}} \qquad \text{Simplify.}$$

$$S_{20} = 10(83)$$

$$S_{20} = 830$$

b. The given sequence is arithmetic. Therefore, we can use the formula for the sum of an arithmetic sequence.

$$a_1 = -4(1) + 8 = 4 \qquad a_{55} = -4(55) + 8 = -212$$

$$S_n = \frac{n(a_1 + a_n)}{2}$$

$$S_{55} = \frac{55(4 + (-212))}{2}$$

$$S_{55} = \frac{55(-208)}{2}$$

$$S_{55} = \frac{-11{,}440}{2}$$

$$S_{55} = -5720$$

PRACTICE PROBLEM FOR EXAMPLE 4

Find S_{44} if $a_n = 2n + 15$.

Geometric Series

When the terms of a geometric sequence are summed, it is called a **geometric series.** As with arithmetic sequences, there is a formula to find the finite sum of a geometric sequence. To find the formula, we again work with the basic idea of a series being the sum of the terms of a sequence. Because geometric series have a common ratio, we must work with the exponents of the common ratio to develop the formula for the finite sum. In this case, we will multiply the sum by the common ratio r and then by -1 in order to get most of the terms in the series to cancel when we add it to the original series. Once most of the terms have canceled, we can solve for the finite sum.

We will start with the definition of a series.

$$S_n = a_1 + a_2 + a_3 + a_4 + \cdots + a_n$$

Substitute the general form of each geometric sequence term.

$$S_n = a_1 + a_1 r + a_1 r^2 + a_1 r^3 + \cdots + a_1 r^{(n-3)} + a_1 r^{(n-2)} + a_1 r^{(n-1)}$$

Multiply both sides by r.

$$rS_n = a_1 r + a_1 r^2 + a_1 r^3 + a_1 r^4 + \cdots + a_1 r^{(n-2)} + a_1 r^{(n-1)} + a_1 r^n$$

Multiply both sides by -1.

$$-rS_n = -a_1 r - a_1 r^2 - a_1 r^3 - a_1 r^4 - \cdots - a_1 r^{(n-2)} - a_1 r^{(n-1)} - a_1 r^n$$

Add this equation to the original series. All the middle terms will cancel.

$$S_n = a_1 + a_1 r + a_1 r^2 + a_1 r^3 + \cdots + a_1 r^{(n-3)} + a_1 r^{(n-2)} + a_1 r^{(n-1)}$$
$$-rS_n = - a_1 r - a_1 r^2 - a_1 r^3 - a_1 r^4 - \cdots - a_1 r^{(n-2)} - a_1 r^{(n-1)} - a_1 r^n$$
$$S_n - rS_n = a_1 - a_1 r^n$$

Factor out S_n from the left side and a_1 from the right side.

$$S_n - rS_n = a_1 - a_1 r^n$$
$$S_n(1 - r) = a_1(1 - r^n)$$

Divide by $1 - r$, to solve for S_n.

$$S_n = \frac{a_1(1 - r^n)}{1 - r}$$

> **Sum of the First n Terms of a Geometric Sequence**
>
> The finite sum S_n of the first n terms of a geometric sequence a_n with common ratio $r \neq 1$ can be found by using the formula
>
> $$S_n = a_1 + a_2 + \cdots + a_n = \frac{a_1(1 - r^n)}{1 - r}$$

Example 5 — Total income with percentage raises

Find the total income from a 50-year career at a job that has a starting salary of $48,000 and an annual raise of 2.5%.

SOLUTION

Since the salary increases by a set percentage each year, we have a common ratio and, therefore, a geometric sequence. Since this is a growth rate problem, we know the common ratio will be 1 plus the rate $r = 1.025$. The starting salary is the first term of the sequence, so we know that $a_1 = 48,000$. With this information, use the formula for the finite sum of a geometric sequence.

$$S_n = \frac{a_1(1 - r^n)}{1 - r}$$

$$S_{50} = \frac{48,000(1 - (1.025)^{50})}{1 - 1.025}$$

Use your calculator memory or keep as many decimals as you can to avoid rounding error.

$$S_{50} \approx \frac{48,000(1 - 3.43710872)}{-0.025}$$

$$S_{50} \approx \frac{48,000(-2.43710872)}{-0.025}$$

$$S_{50} \approx \frac{-116,981.2185}{-0.025}$$

$$S_{50} \approx 4,679,248.74$$

Over a 50-year career, the total income will be $4,679,248.74.

■ **Using Your TI Graphing Calculator**

When working with large exponents, you want to avoid rounding during your calculations, or you will get more error. Using the ANS feature on your calculator allows you to keep all the decimals from a calculation. The ANS feature is above the $(-)$ key.

PRACTICE PROBLEM FOR EXAMPLE 5

A family uses 11,000 gallons of water in January. During the first nine months of the year, their water usage increases by 1.5% each month. Find the total amount of water used during the first nine months of the year.

Example 6 Finding the sum of a geometric sequence

Find the indicated sum.

a. S_8 if $a_n = 2(-3)^n$

b. S_{10} if $a_n = 360\left(\dfrac{1}{2}\right)^n$

SOLUTION

a. We want to sum the first eight terms of a geometric sequence, so we can use the formula for the finite sum of a geometric sequence. First we find the first term,

$$a_1 = 2(-3)^1 = -6$$

Now, using the formula, we get

$$S_n = \dfrac{a_1(1 - r^n)}{1 - r}$$

$$S_8 = \dfrac{-6(1 - (-3)^8)}{1 - (-3)}$$

$$S_8 = \dfrac{-6(1 - 6561)}{4}$$

$$S_8 = \dfrac{-6(-6560)}{4}$$

$$S_8 = \dfrac{39360}{4}$$

$$S_8 = 9840$$

b. Again this is a geometric sequence. Therefore, we use the formula for a finite sum of a geometric sequence.

$$a_1 = 360\left(\dfrac{1}{2}\right)^1 = 180$$

$$S_n = \dfrac{a_1(1 - r^n)}{1 - r}$$

$$S_{10} = \dfrac{180\left(1 - \left(\dfrac{1}{2}\right)^{10}\right)}{1 - \left(\dfrac{1}{2}\right)}$$

$$S_{10} = \dfrac{180\left(1 - \dfrac{1}{1024}\right)}{\dfrac{1}{2}}$$

$$S_{10} = \dfrac{180\left(\dfrac{1024}{1024} - \dfrac{1}{1024}\right)}{\dfrac{1}{2}}$$

$$S_{10} = \dfrac{180\left(\dfrac{1023}{1024}\right)}{\dfrac{1}{2}}$$

$$S_{10} = \dfrac{\dfrac{46,035}{256}}{\dfrac{1}{2}}$$

$$S_{10} = \dfrac{46,035}{128} \approx 359.65$$

PRACTICE PROBLEM FOR EXAMPLE 6
Find the indicated sum.

a. S_{12} if $a_n = 5(-2)^n$

b. S_{15} if $a_n = 600(0.4)^n$

9.5 Exercises

1. The 2009 Giro d'Italia, a professional bicycle race, had 21 stages with the following lengths.

Stage	1	2	3	4	5	6	7
Length (km)	20.5	156	198	162	125	248	244

Stage	8	9	10	11	12	13	14
Length (km)	209	165	262	214	60.6	176	172

Stage	15	16	17	18	19	20	21
Length (km)	161	237	83	182	164	203	14.4

Source: www.cyclingnews.com.

a. Write a series that sums the total distance traveled at the 2009 Giro d'Italia.
b. Find the sum.

2. The 2009 Tour de France, a professional bicycle race, had 21 stages with the following lengths.

Stage	1	2	3	4	5	6	7
Length (km)	15	182	196	38	197	175	224

Stage	8	9	10	11	12	13	14
Length (km)	176	160	193	192	200	200	199

Stage	15	16	17	18	19	20	21
Length (km)	207	160	169	40	195	167	160

Source: www.cyclingnews.com.

a. Write a series that sums the total distance traveled at the 2009 Tour de France.
b. Find the sum.

For Exercises 3 through 14, write out each series in expanded form. Find the sum.

3. $\sum_{n=1}^{5}(2n+7)$

4. $\sum_{n=1}^{5}(-3n+12)$

5. $\sum_{n=1}^{4}(4^n)$

6. $\sum_{n=1}^{4}(3^n)$

7. $\sum_{n=1}^{6}(5n(-1)^n)$

8. $\sum_{n=1}^{7}(2n(-1)^n)$

9. $\sum_{n=1}^{5}\left(\frac{1}{n}\right)$

10. $\sum_{n=1}^{5}\left(\frac{1}{2n}\right)$

11. $\sum_{n=1}^{8}(n^2+5n)$

12. $\sum_{n=1}^{6}(-2n^2+12n)$

13. $\sum_{n=1}^{8}(3(-2)^n)$

14. $\sum_{n=1}^{6}(4(-3)^n)$

15. Your starting salary is $42,000.00, and each year you receive an $800 raise.
    a. Write a formula for your salary a_n in dollars during the nth year.
    b. What will your salary be during the 15th year?
    c. How much total salary will you have earned during the first 15 years?

16. Your starting salary is $58,000.00, and each year you receive an $1,800 raise.
    a. Write a formula for your salary a_n in dollars during the nth year.
    b. What will your salary be during the 25th year?
    c. How much total salary will you have earned during the first 25 years?

17. You decide to start saving money. You begin by saving $8 one week. The next week, you save $10. Each following week, you save $2 more than the previous week.
    a. Write a formula for the amount you saved a_n in dollars during the nth week.
    b. How much money will you save during the eighth week?
    c. Find the total amount you saved during the first 26 weeks.

18. You start saving money. You begin by saving $20 one month. The next month, you save $25. Each following month, you save $5 more than the previous month.

a. Write a formula for the amount you saved a_n in dollars during the nth month.
b. How much money will you save during the 12th month?
c. Find the total amount you will save during the first two years.

19. A retaining wall is put into a backyard. The top row has 15 blocks in it. Each row below has one more block than the row above it.

a. Write a formula the number of blocks a_n in the nth row.
b. If the wall is 12 rows tall, how many blocks are needed to build the entire wall?

20. A child stacks cups to form a pyramid. The top row has one cup. Each row down has one more cup than the row above it.

a. Write a formula for the number of cups a_n in the nth row.
b. If the pyramid has 16 rows in it, how many cups are in the pyramid altogether?

For Exercises 21 through 30, find the indicated sum.

21. S_{30} if $a_n = 4n + 1$ **22.** S_{25} if $a_n = 6n - 15$

23. S_{15} if $a_n = 20 + (n-1)5$

24. S_{50} if $a_n = 14 + (n-1)8$

25. S_{40} if $a_n = -9n - 2$ **26.** S_{26} if $a_n = -8n - 15$

27. S_{52} if $a_n = 2.4n + 6.2$

28. S_{24} if $a_n = 3.6n - 18$

29. S_{27} if $a_n = \frac{1}{3}n + 4$ **30.** S_{25} if $a_n = \frac{2}{5}n - 8$

31. Your starting salary is $42,000.00, and each year you receive a 2% raise.

a. Write a formula for your salary a_n in dollars during the nth year.
b. What will your salary be during the 15th year?
c. How much total salary will you have earned during the first 15 years?

32. Your starting salary is $58,000.00, and each year you receive a 3% raise.

a. Write a formula for your salary a_n in dollars during the nth year.
b. What will your salary be during the 25th year?
c. How much total salary will you have earned during the first 25 years?

33. You make up your mind to start saving money. You begin by saving $10 one week. Each following week, you save 10% more than the previous week.

a. Write a formula for the amount you saved a_n in dollars during the nth week.
b. How much money will you save during the eighth week?
c. Find the total amount you saved during the first 26 weeks.

34. You decide to start saving money. You begin by saving $50 one month. Each following month, you save 5% more than the previous month.

a. Write a formula for the amount you saved a_n in dollars during the nth month.
b. How much money will you save during the 12th month?
c. Find the total amount you will save during the first two years.

35. A very contagious flu is spreading quickly. The first day, three people are diagnosed with this flu. Each day after that, twice as many people are diagnosed.

a. Find the general term for the sequence that will model the number of flu diagnoses on the nth day.
b. Find the number of people diagnosed with this flu on the fifth day.
c. How many people in all were diagnosed with this flu during the first two weeks?

36. A store having a grand opening advertises that it will give away prizes every hour the store is open the first day. The store plans to give out seven prizes the first hour. Every hour later, the store plans to give out twice as many prizes as the hour before.

a. Find the general term for the sequence that will model the number of prizes given out n hours after the store opens.
b. How many prizes will the store give out during the seventh hour of being open?
c. If the store is open for 12 hours the first day, how many prizes in all will they give out?

For Exercises 37 through 46, find the indicated sum.

37. S_9 if $a_n = 3^n$ **38.** S_8 if $a_n = 4^n$

39. S_6 if $a_n = 2(7)^n$ **40.** S_8 if $a_n = 4(5)^n$

41. S_{11} if $a_n = 2(-6)^n$ **42.** S_{15} if $a_n = 7(-4)^n$

43. S_{12} if $a_n = 900\left(\frac{1}{3}\right)^n$ **44.** S_{15} if $a_n = 800\left(\frac{1}{2}\right)^n$

45. S_{10} if $a_n = -1000\left(\frac{2}{5}\right)^n$ **46.** S_{13} if $a_n = -2500\left(\frac{3}{5}\right)^n$

For Exercises 47 through 56, determine whether the general term represents an arithmetic or geometric sequence. Find the indicated sum.

47. S_{10} if $a_n = 5(2)^n$

48. S_{11} if $a_n = 300 + 4(n-1)$

49. S_6 if $a_n = 2(-3)^n$ **50.** S_8 if $a_n = 4000\left(\frac{3}{4}\right)^n$

51. S_{23} if $a_n = 8n + 25$

52. S_{30} if $a_n = 700 - 2(n-1)$

53. S_{12} if $a_n = 900(0.94)^n$ **54.** S_6 if $a_n = 800(1.02)^n$

55. S_{10} if $a_n = -\frac{1}{2}n + 30$ **56.** S_9 if $a_n = -3000\left(\frac{2}{3}\right)^n$

Chapter Summary

Section 9.1 Conic Sections

- **The standard equations for a parabola that opens up or down** are the following:

$$y = ax^2 + bx + c \quad \text{or} \quad y = a(x - h)^2 + k$$

The parabola opens up when $a > 0$ and opens down when $a < 0$. The vertex is (h, k).

- **The standard equations for a parabola that opens right or left** are the following:

$$x = ay^2 + by + c \quad \text{or} \quad x = a(y - k)^2 + h$$

The parabola opens right when $a > 0$ and opens left when $a < 0$. The vertex is (h, k).

- Use the following steps to **graph a parabola.**

  1. Determine whether the graph opens up or down, left or right.
  2. Find the vertex.
  3. Find the axis of symmetry.
  4. Find the y-intercepts (if any).
  5. Find the x-intercepts (if any).
  6. Plot the points, plot their symmetric pairs, and sketch the graph. Find an additional symmetric pair if needed.

- **Geometric Definition of a Parabola:** The set of all points in a plane that are equidistant from a fixed point (the focus) and a fixed line (the directrix).

- **The standard equations for a parabola with vertex at the origin that opens up or down are the following:**

$$x^2 = 4py \quad \text{or} \quad y = \frac{x^2}{4p}$$

The focus is the point $(0, p)$ and the directrix is the horizontal line $y = -p$. The parabola opens up when $p > 0$ and opens down when $p < 0$.

- **The standard equations for a parabola with vertex at the origin that opens right or left are the following:**

$$y^2 = 4px \quad \text{or} \quad x = \frac{y^2}{4p}$$

The focus is the point $(p, 0)$, and the directrix is the vertical line $x = -p$. The parabola opens right when $p > 0$ and opens left when $p < 0$.

- **Geometric Definition of a Circle:** The set of all points in a plane equidistant from a fixed point (center).

- **The standard equation for a circle with center (h, k) and radius r.**

$$(x - h)^2 + (y - k)^2 = r^2$$

- You can use the process of completing the square to put equations of circles into standard form.

CHAPTER 9 Summary

Example 1 Sketch the graph of $x = (y + 2)^2 - 16$.

SOLUTION This equation is in the standard form for a parabola that opens to the right.
The vertex is the point $(-16, -2)$.
The axis of symmetry is the horizontal line $y = -2$.
Make $y = 0$ to find the x-intercept.

$$x = (0 + 2)^2 - 16$$
$$x = -12$$

The x-intercept is the point $(-12, 0)$.
Make $x = 0$ to find the y-intercepts.

$$0 = (y + 2)^2 - 16$$
$$16 = (y + 2)^2$$
$$\pm\sqrt{16} = y + 2$$
$$\pm 4 = y + 2$$
$$4 = y + 2 \qquad -4 = y + 2$$
$$2 = y \qquad -6 = y$$

The y-intercepts are the points $(0, 2)$ and $(0, -6)$.
Using these points and symmetry, we get the following graph.

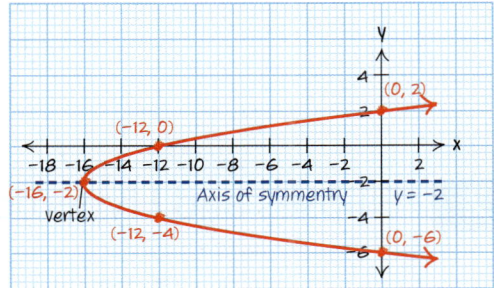

Example 2 Determine the equation from the given graph.

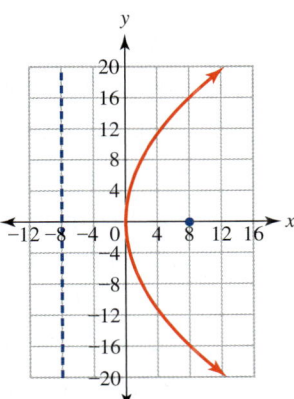

SOLUTION This parabola opens right and has a focus at the point $(6, 0)$. Therefore, $p = 6$, and the equation is

$$x = \frac{y^2}{24}$$

750 CHAPTER 9 Conic Sections, Sequences, and Series

Example 3 Sketch the graph of $(x - 4)^2 + (y + 3)^2 = 36$.

SOLUTION This equation is in the standard form for a circle. The center is $(4, -3)$, and the radius is $r = 6$. Using this information, we get the following graph.

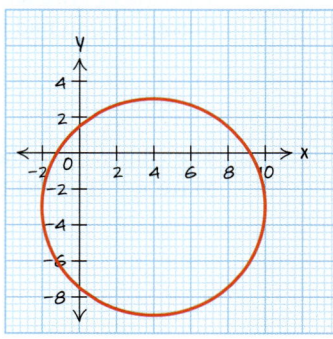

Example 4 Write the equation $x^2 + y^2 + 3x - 8y + 9 = 0$ in the standard form of a circle.

SOLUTION Use completing the square to put the equation into the standard form for a circle.

$$x^2 + y^2 + 6x - 8y + 9 = 0$$
$$x^2 + 6x + 9 + y^2 - 8y + 16 = -9 + 9 + 16$$
$$(x + 3)^2 + (y - 4)^2 = 16$$

Section 9.2 Ellipses and Hyperbolas

- **Geometric Definition of an Ellipse:** The set of all points in a plane that the sum of the distances from two fixed points (foci) is constant.
- **The standard equation for an ellipse centered at the origin:**

$$\frac{x^2}{a^2} + \frac{y^2}{b^2} = 1$$

When $a > b$, the foci are the points $(c, 0)$ and $(-c, 0)$, where $c^2 = a^2 - b^2$.
When $b > a$, the foci are the points $(0, c)$ and $(0, -c)$, where $c^2 = b^2 - a^2$.

- Use the following steps to graph an ellipse centered at the origin.
  1. Find the values of a and b.
  2. Determine the direction of the major axis.
  3. Find the foci.
  4. Plot the vertices at $(a, 0), (-a, 0), (b, 0),$ and $(-b, 0)$.
  5. Draw a smooth curve through the vertices.

- **Geometric Definition of a Hyperbola:** The set of all points in a plane such that the difference of the distances from two fixed points is a constant.
- **The standard equation for a hyperbola centered at the origin that opens right and left:**

$$\frac{x^2}{a^2} - \frac{y^2}{b^2} = 1$$

- The foci are the points $(c, 0)$ and $(-c, 0)$, where $c^2 = a^2 + b^2$, and the vertices are $(a, 0)$ and $(-a, 0)$.

- **The standard equation for a hyperbola centered at the origin that opens up and down:**

$$\frac{y^2}{b^2} - \frac{x^2}{a^2} = 1$$

The foci are the points $(0, c)$ and $(0, -c)$, where $c^2 = a^2 + b^2$, and the vertices are $(0, b)$ and $(0, -b)$.

- **Use the following steps to graph a hyperbola centered at the origin.**
  1. Determine in which directions the hyperbola will open.
  2. Find the values of a and b and list the vertices.
  3. Find the foci.
  4. Plot the vertices, plot the fundamental rectangle, and draw the diagonals extended out in both directions.
  5. Sketch the graph.

- **All the conic sections we have studied can be described by using the equation**

$$Ax^2 + Bxy + Cy^2 + Dx + Ey + F = 0$$

Assume $B = 0$

- If $A = C$, the equation represents a circle.
- If A and C are different but both positive, the equation represents an ellipse.
- If A and C are different and have different signs, the equation represents a hyperbola.
- If only one of A and C is zero, the equation is a parabola.

Example 5 Sketch the graph of $\frac{x^2}{16} + \frac{y^2}{9} = 1$.

SOLUTION This is the equation of an ellipse, where $a = 4$ and $b = 3$. Find c using the equation $c^2 = a^2 - b^2$.

$$c^2 = a^2 - b^2$$
$$c^2 = 4^2 - 3^2$$
$$c^2 = 16 - 9$$
$$c^2 = 7$$
$$c = \pm\sqrt{7}$$
$$c \approx \pm 2.646$$

Therefore, the foci are the points $(2.646, 0)$ and $(-2.646, 0)$. Using this information, we get the following graph.

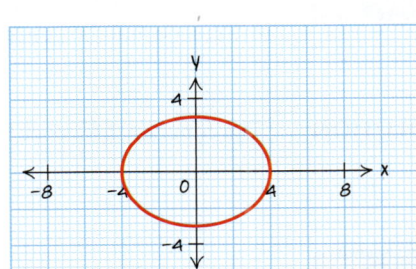

Example 6 Sketch the graph of $\frac{x^2}{16} - \frac{y^2}{9} = 1$.

SOLUTION This is the equation of a hyperbola, where $a = 4$ and $b = 3$. Therefore, the vertices are $(4, 0)$ and $(-4, 0)$. Find c using the equation $c^2 = a^2 + b^2$.

$$c^2 = a^2 + b^2$$
$$c^2 = 4^2 + 3^2$$
$$c^2 = 16 + 9$$
$$c^2 = 25$$
$$c = \pm\sqrt{25}$$
$$c = \pm 5$$

Therefore, the foci are the points (5, 0) and (−5, 0). Using this information, we get the following graph.

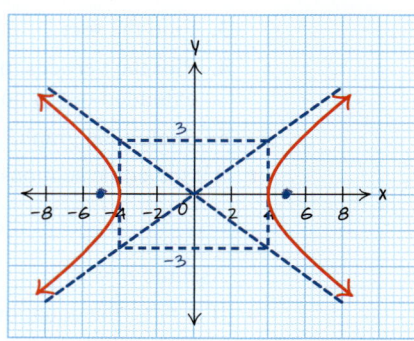

Section 9.3 Arithmetic Sequences

- A **sequence** is an ordered list.
- A **finite sequence** is a function whose domain is the set {1, 2, 3, ..., n}, where n is a natural number.
- To graph a sequence, plot the terms of the sequence as points. Do not connect the points.
- An **arithmetic sequence** is a linear function whose domain is the set of natural numbers.
- An arithmetic sequence will have a **common difference** d. This is the same as the slope of a linear function.
- The **general term of an arithmetic sequence** is given by $a_n = a_1 + (n-1)d$, where a_1 is the first term and d is the common difference.

Example 7 Determine whether or not the following sequence is arithmetic. If it is, find the common difference and the general term.

$$12, 15, 18, 21, 24, 27, \ldots$$

SOLUTION The common difference between terms is 3. The first term is 12. Using these values, we get

$$a_n = 12 + (n-1)3$$

Example 8 A car collector starts his collection with four cars. Each year after that he adds two more cars to his collection.

a. Write a formula for the number of cars a_n in the collection after n years.

b. How many cars will be in the collection after 10 years?

SOLUTION a. The collection started with four cars, so the first term is $a_1 = 4$. The collector adds two more cars per year, so the common difference is $d = 2$. Therefore, we get

$$a_n = 4 + (n-1)2$$

b. 10 years is represented by the 10th term, so we get

$$a_{10} = 4 + (10 - 1)2$$
$$a_{10} = 22$$

Therefore, there will be 22 cars in the collection after 10 years.

Section 9.4 Geometric Sequences

- A **geometric sequence** is a sequence whose consecutive terms have a common ratio.
- The **general term of a geometric sequence** is given by $a_n = a_1 \cdot r^{(n-1)}$, where a_1 is the first term and r is the common ratio.
- A **geometric sequence** is an exponential function with a domain of the set of natural numbers.

Example 9 Determine whether or not the following sequence is geometric. If it is, find the common ratio and the general term.

$$5.75, 11.5, 23, 46, 92, \ldots$$

SOLUTION Divide consecutive terms and check for a common ratio.

$$\frac{11.5}{5.75} = 2 \qquad \frac{23}{11.5} = 2$$

$$\frac{46}{23} = 2 \qquad \frac{92}{46} = 2$$

The common ratio is $r = 2$. The first term is $a_1 = 5.75$. Therefore, we get $a_n = 5.75 \cdot 2^{(n-1)}$.

Example 10 A one-year-old car is valued at $40,000 and is depreciating 10% per year.

a. Write a formula for the value of the car a_n in dollars when it is n years old.

b. How much will the car be worth when it is eight years old?

SOLUTION **a.** The first term in the sequence is the value of the one-year-old car $a_1 = 40,000$. The car is depreciating 10% per year, so the common ratio is $r = 0.90$. Using these values, we get

$$a_n = 40000 \cdot 0.90^{(n-1)}$$

b. The value of the car at eight years old is the eighth term of the sequence.

$$a_8 = 40000 \cdot 0.90^{(8-1)}$$
$$a_8 \approx 19132$$

The car will be worth $19,132 when it is eight years old.

Section 9.5 Series

- A **series** is the sum of a sequence.
- Summation notation is used to represent the sum of terms of a sequence

$$\sum_{n=1}^{k} a_n = a_1 + a_2 + \cdots + a_k$$

- An **arithmetic series** is the sum of an arithmetic sequence.
- The finite sum of an arithmetic sequence can be found by using the formula

$$S_n = a_1 + a_2 + \cdots + a_n = \frac{n(a_1 + a_n)}{2}$$

754 CHAPTER 9 Conic Sections, Sequences, and Series

- A **geometric series** is the sum of a geometric sequence.
- The finite sum of a geometric sequence can be found by using the formula

$$S_n = a_1 + a_2 + \cdots + a_n = \frac{a_1(1 - r^n)}{1 - r}$$

Example 11 Write out the series in expanded form. Find the sum.

$$\sum_{n=1}^{4} (2(-3)^n)$$

SOLUTION First write out the terms and then add.

$$\sum_{n=1}^{4} (2(-3)^n) = 2(-3)^1 + 2(-3)^2 + 2(-3)^3 + 2(-3)^4$$

$$\sum_{n=1}^{4} (2(-3)^n) = -6 + 18 - 54 + 162$$

$$\sum_{n=1}^{4} (2(-3)^n) = 120$$

Example 12 Find the indicated sums.

a. S_{13} if $a_n = 10 + (n - 1)4$ **b.** S_8 if $a_n = -40\left(\frac{1}{2}\right)^n$

SOLUTION **a.** The general term $a_n = 10 + (n - 1)4$ represents an arithmetic sequence. Therefore, use the formula

$$S_n = \frac{n(a_1 + a_n)}{2}$$

to find the sum of the first 13 terms.

$$a_1 = 10 + (1 - 1)4 = 10$$
$$a_{13} = 10 + (13 - 1)4 = 58$$
$$S_{13} = \frac{13(10 + 58)}{2}$$
$$S_{13} = 442$$

b. The general term $a_n = -40\left(\frac{1}{2}\right)^n$ represents a geometric sequence. Therefore, use the formula

$$S_n = \frac{a_1(1 - r^n)}{1 - r}$$

to find the sum of the first eight terms.

$$a_1 = -20$$
$$r = \frac{1}{2}$$
$$S_8 = \frac{-20\left(1 - \frac{1}{2}^8\right)}{1 - \frac{1}{2}}$$
$$S_8 \approx \frac{-20(0.9961)}{\frac{1}{2}}$$
$$S_8 = -39.84375$$

Chapter Review Exercises

For Exercises 1 through 4, sketch the graph of the given equations. Label the vertex and any intercepts.

1. $x = y^2 + 6y - 16$ [9.1]
2. $y = 2x^2 - 24x - 30$ [9.1]
3. $x = -1.5(y + 2)^2 + 6$ [9.1]
4. $x = -\frac{1}{2}y^2 - 3y - 8$ [9.1]

For Exercises 5 through 8, using the geometric definition of a parabola, sketch the graph of the given equations. Label the focus and directrix.

5. $x = \dfrac{y^2}{48}$ 6. $y = \dfrac{x^2}{-20}$ [9.1]

7. $-8x = y^2$ 8. $14y = x^2$ [9.1]

9. Write the equation of the circle with center $(2, 8)$ and radius 3.   [9.1]

10. Write the equation of the circle with center $(5, -2)$ and radius 2.5.   [9.1]

For Exercises 11 through 14, sketch the graph of the following circles. Find the radius and the center of the circle.

11. $x^2 + y^2 = 16$ [9.1]
12. $(x - 5)^2 + (y - 2)^2 = 4$ [9.1]
13. $(x + 4)^2 + (y - 6)^2 = 1$ [9.1]
14. $(x + 3)^2 + (y + 7)^2 = 0.25$ [9.1]

For Exercises 15 through 16, write the equation in the standard form of a circle. Sketch the graph of the circle and find the radius and the center.

15. $x^2 + y^2 + 12x - 10y - 39 = 0$ [9.1]
16. $3x^2 + 3y^2 - 24x + 12y - 15 = 0$ [9.1]

For Exercises 17 through 20, find the equation of the parabola or circle from the given graph.

17. [9.1]
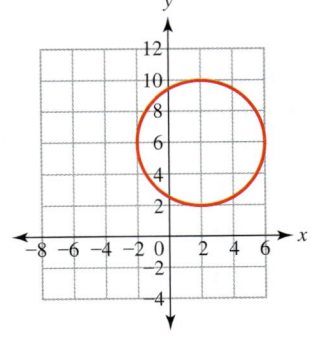

18. [9.1]
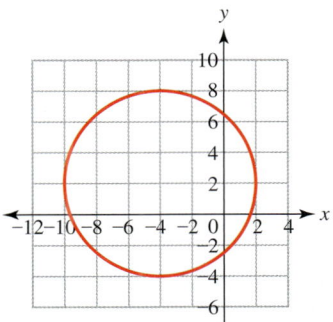

19. [9.1]
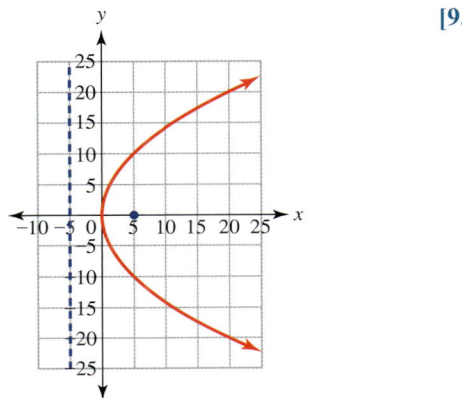

20. [9.1]
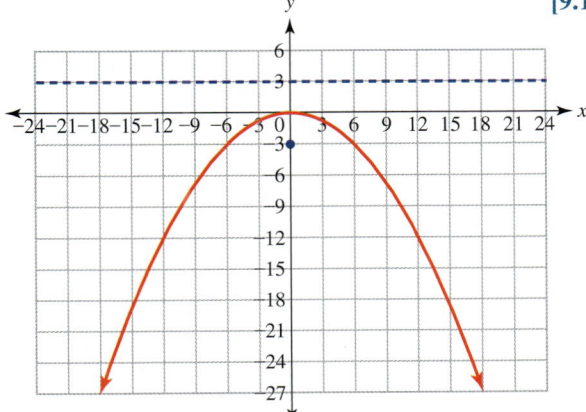

For Exercises 21 through 24, convert the given equation to the standard form for an ellipse centered at the origin. Sketch a graph of the ellipse.

21. $\dfrac{x^2}{16} + \dfrac{y^2}{25} = 1$ [9.2]

22. $\dfrac{x^2}{0.25} + y^2 = 1$ [9.2]

23. $100x^2 + 4y^2 = 400$ [9.2]

24. $10x^2 + 2y^2 = 40$ [9.2]

For Exercises 25 through 28, sketch a graph of the given hyperbolas. Label the vertices and foci.

25. $\dfrac{x^2}{16} - \dfrac{y^2}{25} = 1$ [9.2]

26. $\dfrac{y^2}{100} - \dfrac{x^2}{4} = 1$ [9.2]

27. $x^2 - 16y^2 = 16$ [9.2]

28. $9y^2 - 25x^2 = 225$ [9.2]

For Exercises 29 through 32, find the equation of the ellipse or hyperbola from the given graph.

29. [9.2]

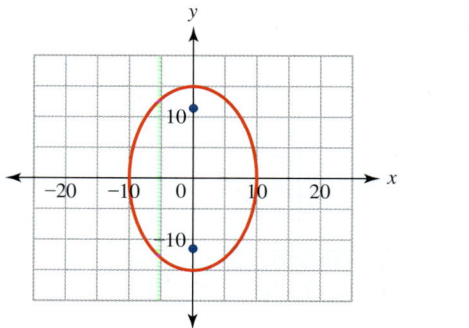

30. [9.2]

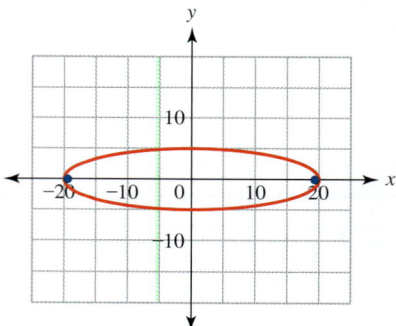

31. [9.2]

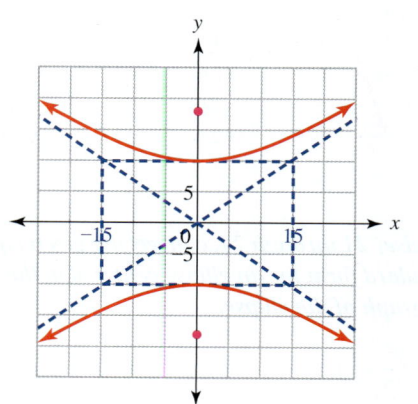

32. [9.2]

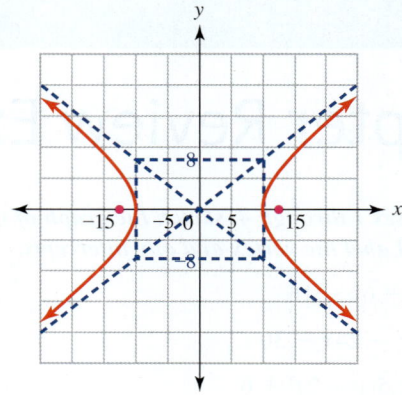

For Exercises 33 through 40, determine whether the given equation represents a circle, ellipse, hyperbola, or parabola. Sketch the graph.

33. $4x^2 + 36y^2 = 36$ [9.2]

34. $(x - 9)^2 + (y - 6)^2 = 9$ [9.2]

35. $2y^2 - 50x^2 = 50$ [9.2]

36. $-2y^2 + 3x + 6y - 21 = 0$ [9.2]

37. $4x^2 + 4y^2 + 24x = 0$ [9.2]

38. $9x^2 - 36y^2 = 324$ [9.2]

39. $0.5x^2 + 4x - y = 6$ [9.2]

40. $x^2 + 49y^2 = 49$ [9.2]

For Exercises 41 through 44, compute and list the first five terms of the given sequence. Graph the terms of each sequence.

41. $a_n = 7n - 20$ [9.3]

42. $a_n = 4(1.5)^n$ [9.3]

43. $a_n = \dfrac{(-1)^n}{8n}$ [9.3]

44. $a_n = \dfrac{2n}{n + 8}$ [9.3]

For Exercises 45 through 48, find n for the given value of a_n.

45. $a_n = 50 + 2(n - 1)$ when $a_n = 78$ [9.3]

46. $a_n = 4n^2 + 3n$ when $a_n = 351$ [9.3]

47. $a_n = 0.25(2)^n$ when $a_n = 1{,}048{,}576$ [9.3]

48. $a_n = \sqrt{6x + 4} + 48$ when $a_n = 68$ [9.3]

For Exercises 49 through 52, determine whether or not each sequence is arithmetic. If the sequence is arithmetic, find the common difference d and the general term a_n.

49. 82, 79, 76, 73, 70, 67, ... [9.3]

50. $-4, -1.5, 1, 3.5, 6, 8.5, \ldots$ [9.3]

51. 48, 72, 108, 162, 243, 364.5, ... [9.3]

52. $6, \frac{7}{2}, 1, -\frac{3}{2}, -4, -\frac{13}{2}, \ldots$ [9.3]

53. In 2001, 4000 people attended North Coast Church on Easter Sunday. Each year, after that the number of people in attendance rose on average 200. [9.3]

    a. Write a formula for the attendance a_n on Easter Sunday at North Coast Church n years after 2000.

    b. If this trend continues, what will the attendance be in 2015?

    c. In what year will the attendance be 6000?

54. A one-year-old car is valued at $27,000 and is depreciating by $2500 per year. [9.3]

    a. Write a formula for the value of the car a_n in dollars when it is n years old.

    b. What will the value of the car be in the sixth year?

    c. At what age will the car be worth $2000?

For Exercises 55 and 56, find the general term a_n for the arithmetic sequence with the given terms.

55. $a_5 = 66$ and $a_{17} = 144$ [9.3]

56. $a_4 = -12$ and $a_{16} = -108$ [9.3]

For Exercises 57 through 60, determine whether or not each sequence is geometric. If the sequence is geometric, find the general term a_n.

57. $3, 12, 48, 192, 768, 3072, \ldots$ [9.4]

58. $36, 12, 4, \frac{4}{3}, \frac{4}{9}, \frac{4}{27}, \ldots$ [9.4]

59. $0.5, -1, 2, -4, 8, -16, \ldots$ [9.4]

60. $6, 34, 64, 106, 160, 226$ [9.4]

61. The cost of a product will go up at the rate of inflation. A product costs $4000 and the cost is increasing at a rate of 2% per year. [9.4]

    a. Find the general term for the sequence that will give the cost of the product after n years.

    b. Find a_8 and explain its meaning in this situation.

62. Your new employer offers you a starting salary of $42,000 with a 1.25% raise each year. [9.4]

    a. Find the general term for the sequence that will model your salary during your nth year of employment.

    b. Find a_5 and explain its meaning in this situation.

    c. After how many years will you earn $50,603?

For Exercises 63 and 64, find the general term a_n for the geometric sequence with the given terms.

63. $a_3 = 72$ and $a_6 = 243$ [9.4]

64. $a_2 = 27$ and $a_5 = 1$ [9.4]

For Exercises 65 and 66, write out each series in expanded form. Find the sum.

65. $\sum_{n=1}^{7}(4n-20)$ [9.5]

66. $\sum_{n=1}^{6}(5(-2)^n)$ [9.5]

67. Your starting salary is $38,000, and each year you receive a $900 raise. [9.5]

    a. Find the general term for the sequence that will give your salary in dollars during the nth year.

    b. What will your salary be during the 15th year?

    c. How much total salary will you have earned during the first 15 years?

68. One day you pick five ripe apricots from a tree. Each day after that, you pick three more ripe apricots than you did on the previous day. [9.5]

    a. Find the general term for the sequence that will give the number of ripe apricots you picked on the nth day.

    b. How many apricots will you pick on the fifth day?

    c. How many total ripe apricots will you pick during the first two weeks?

69. Using the sequence you found in Exercise 62, find the total salary you will have earned during a 15-year career. [9.5]

70. The attendance the first day at a car show was 10,000. The attendance at the car show increased by 20% each day. [9.5]

    a. Find the general term for the sequence that will give the attendance at the car show on the nth day.

    b. How many total people attended the car show during the seven-day car show?

For Exercises 71 through 74, find the indicated sum.

71. S_{14} if $a_n = 4n + 7$ [9.5]

72. S_{23} if $a_n = 300 - 2.5(n-1)$ [9.5]

73. S_{11} if $a_n = 7(1.8)^n$ [9.5]

74. S_{20} if $a_n = 48\left(-\frac{1}{2}\right)^n$ [9.5]

Chapter Test

1. Write the equation of a circle with center $(-7, 4)$ and radius 8.

2. Find the focus and directrix for the parabola
$$x = \frac{y^2}{36}$$

For Exercises 3 through 6, determine whether the given equation represents a circle, ellipse, hyperbola, or parabola. Sketch the graph.

3. $\dfrac{x^2}{100} - \dfrac{y^2}{36} = 1$

4. $y^2 + 8 + x - 20 = 0$

5. $49x^2 + 4y^2 = 196$

6. $(x - 3)^2 + (y - 7)^2 = 81$

7. Find the equation of the parabola from the graph.

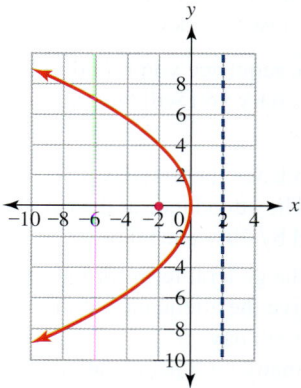

8. Find the equation of the ellipse from the graph.

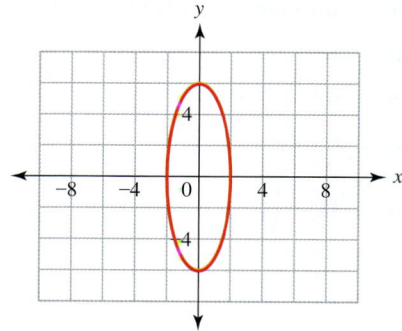

9. Describe the difference between arithmetic and geometric sequences.

10. Compute and list the first five terms of $a_n = 4(-3)^n$.

11. Compute and list the first five terms of $a_n = \dfrac{6n}{n+2}$.

12. If $a_n = 400 - 11(n - 1)$, find n when $a_n = 158$.

13. If $a_n = 20 + \sqrt{n + 8}$, find n when $a_n = 38$.

For Exercises 14 and 15, determine whether each sequence is arithmetic or geometric. Find the general term a_n.

14. 250, 50, 10, 2, 0.4, 0.08, . . .

15. 24, 16.5, 9, 1.5, −6, −13.5, . . .

16. After surgery, your doctor instructs you to stretch your injured knee five minutes at a time for the first week. Each week after that, you are to increase the stretching time by three minutes.

   a. Find the general term for the sequence that will give the number of minutes you are to stretch at one time during the nth week after surgery.
   b. Find a_3 and explain its meaning in this context.
   c. In what week will you be stretching for half an hour at one time?

17. Your starting salary is $60,000.00, and each year you receive a 2.5% raise.

   a. Write a formula for your salary a_n in dollars during the nth year.
   b. What will your salary be during the 20th year?
   c. How much total salary will you have earned during the first 20 years?

18. Write the series $\sum_{n=1}^{6} (2n^2)$ in expanded form and find the sum.

19. Find S_{15} if $a_n = 600 - 18(n - 1)$.

20. Find S_{12} if $a_n = 20(1.75)^n$.

Chapter 9 Projects

How Are Conics Used?

Research Project

One or more people

What you will need

- Find an application for a conic section using the Internet or library.
- Follow the MLA style guide for all citations. If your school recommends another style guide, use that.

Research and describe an application of a conic section. Cite any sources you used following the MLA style guide or your school's recommended style guide.

Write up

a. Describe the application and how the conic section works in that application.
b. Provide a historical background of the application and how conics have been used in the situation.
c. Give a specific example of a conic section used in the application and find an equation for that conic section.
d. Provide a graphic or drawing of the conic used in the application.

Which Conic Does That Comet Follow?

Research Project

One or more people

What you will need

- Find information about two comets using the Internet or library.
- Follow the MLA style guide for all citations. If your school recommends another style guide, use that.

Research and describe the path that at least two comets follow. Find a comet that follows an elliptical path and one that follows a parabolic or hyperbolic path. Cite any sources you used following the MLA style guide or your school's recommended style guide.

Write up

a. Describe the path that each of the comets you chose follows.
b. Describe what determines the type of conic that the comet's path will follow.
c. Provide a historical background of the comet you chose.
d. Find an equation for the conic sections that represent the two comets' paths.

Fibonacci Who?

Research Project

One or more people

What you will need

- Find information about the Fibonacci sequence using the Internet or library.
- Follow the MLA style guide for all citations. If your school recommends another style guide, use that.

Research and describe the Fibonacci sequence. Cite any sources you used following the MLA style guide or your school's recommended style guide.

Write up

a. Describe the Fibonacci sequence and how to determine the terms in the sequence.
b. List the first 30 terms of the Fibonacci sequence.
c. Provide a historical background of the Fibonacci sequence.
d. Describe two or more applications of the Fibonacci sequence in nature.

Review Presentation

Research Project
One or more people

What you will need
- This presentation may be done in class or as a video.
- If it is done as a video, you might want to post the video on YouTube or another Website.
- You might want to use problems from the homework set or review material in this book.
- Be creative, and make the presentation fun for students to watch and learn from.

Create a 5- to 10-minute presentation or video that reviews a section of Chapter 9. The presentation should include the following:

- Examples of the important skills presented in the section
- Explanation of any important terminology and formulas
- Common mistakes made and how to recognize them

Cumulative Review — Chapters 1–9

For Exercises 1 through 20, solve the given equation using any method you have learned. Check your answers in the original equation.

1. $7a^3 + 18 = 1530$
2. $-2(t + 3)^2 - 8 = -80$
3. $6x + 5 = 2(4x - 9)$
4. $\dfrac{2}{m - 3} = \dfrac{5}{m + 8}$
5. $-2\sqrt{3h + 7} + 10 = -4$
6. $40\left(\dfrac{4}{5}\right)^n + 300 = 320.48$
7. $\ln(2x - 5) = 4$
8. $\dfrac{4}{b + 3} + 2 = \dfrac{-5}{b - 7}$
9. $|5h - 8| = 32$
10. $\log_2(x + 3) + \log_2(x + 5) = 3$
11. $\sqrt{t + 12} + \sqrt{2t + 1} = 7$
12. $30(2^n) - 8400 = 6960$
13. $3.2g^2 + 2.4g - 16.2 = 59.4$
14. $\dfrac{44}{a^2 + 3a - 28} = \dfrac{7a}{a^2 - a - 12}$
15. $\dfrac{1}{5}(x + 4) = \dfrac{3}{10}x - \dfrac{5}{4}$
16. $\sqrt[5]{7 - 3m} = 3$
17. $3r^3 + 9r^2 - 120r = 0$
18. $-5|6h + 4| = -30$
19. $2^{4x+5} = 0.125$
20. $5t^4 + 16 = 1269$

For Exercises 21 through 24, solve the given systems. Check your answer.

21. $6x + 8y = 15$
    $4y = -3x + 15$
22. $2x + 7y = 29$
    $10x - 3y = -7$
23. $y = 4x^2 - 6x + 5$
    $y = \dfrac{1}{2}x + 8$
24. $y = -2x^2 + 3x + 10$
    $y = x^2 - 5x - 25$

25. The profit from manufacturing and selling custom MP3 players is given in the table.

Number of MP3 Players	Profit ($)
200	800
500	2500
750	4125
900	4450
1200	4800
1500	4420

   a. Find an equation for a model for these data.
   b. Give a reasonable domain and range for the model.
   c. Use the model to estimate the profit from manufacturing and selling 1000 custom MP3 players.
   d. Give the vertex of your model and what it means in this context.

26. The Wikimedia Foundation supports the development and availability of free wiki-based content on the Internet. The annual primary revenue from donations for the Wikimedia Foundation is given in the table.

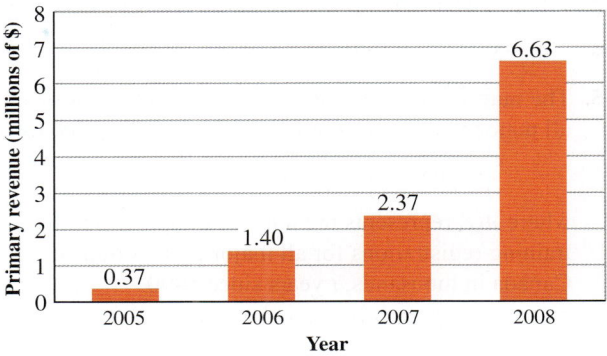

Source: www.charitynavigator.org.

   a. Find an equation for an exponential model for these data.
   b. Give a reasonable domain and range.
   c. Use your model to estimate the primary revenue for the Wikimedia Foundation in 2010.
   d. According to your model, when will the Wikimedia Foundation reach $50 million in annual primary revenue?

27. The total circulation of U.S. Sunday newspapers are given in the graph.

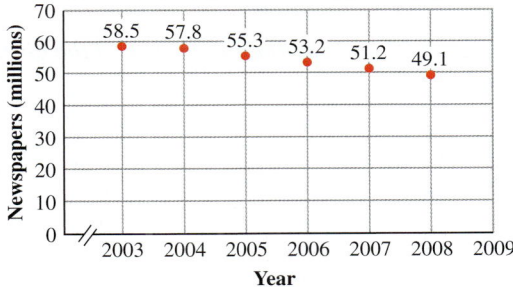

Source: Newspaper Association of America, www.naa.org.

   a. Find an equation for a model for these data.
   b. Use your model to estimate the total circulation of U.S. Sunday newspapers in 2010.
   c. Find when your model predicts that the total circulation will be only 25 million.
   d. What is the slope of your model and what does it mean in this situation?
   e. Give a reasonable domain and range for your model.

28. Find the domain for the rational function
$$f(x) = \dfrac{x - 2}{x^2 + 7x - 18}$$

29. Graph the system of inequalities.
$$2x - 3y > 6$$
$$y < -x + 4$$

For Exercises 30 through 34, solve the given inequalities.

30. $6a - 15 < 11a + 20$
31. $\frac{1}{2}n + 7 \leq \frac{3}{4}n + \frac{5}{12}$
32. $|4h - 9| \geq 7$
33. $-6|g - 4| < 30$
34. $-4|-5x + 7| \leq -10$
35. The number of apprenticeship training registrations for all major trade groups in Canada can be modeled by

    $$A(t) = 1.4t^2 - 14.3t + 202.5$$

    where $A(t)$ represents the number of apprenticeship training registrations for all major trade groups in Canada in thousands, t years since 1990.
    Source: Model derived from Statistics Canada data.
    a. Use the model to estimate the number of registrations in 2008.
    b. Give the vertex of this model and what it means in this context.
    c. Estimate the year(s) in which Canada had 300,000 apprenticeship registrations for all major trade groups.
36. A bacteria is doubling every 20 minutes. At the start of the experiment there are 2000 of this bacteria.
    a. Find an equation for a model for the number of bacteria after h hours.
    b. Use your model to estimate when there will be 1 million bacteria.
    c. How many bacteria will be present after five hours?

For Exercises 37 and 38, find the inverse of the given function.

37. $f(x) = \frac{2}{3}x - 8$
38. $g(x) = 4(3)^x$
39. $f(x) = 2.5$
    a. Find $f(3)$.
    b. Give the domain and range of $f(x)$.
40. Let $f(x) = 2x - 7$ and $g(x) = 3.4x + 2.7$.
    a. Find $f(x) + g(x)$.
    b. Find $f(g(x))$.
    c. Find $f(x)g(x)$.
41. Let $f(x) = \frac{3}{4}x - 10$ and $g(x) = -12x + 5$.
    a. Find $f(2) - g(2)$.
    b. Find $(f \circ g)(7)$.
    c. Find $f(2)g(2)$.
42. Let $f(x) = 3x^2 + 4x - 6$ and $g(x) = 3x - 4$.
    a. Find $f(x) + g(x)$.
    b. Find $f(g(x))$.
    c. Find $f(x)g(x)$.

43. In 2010 an earthquake off the coast of California near Eureka had a magnitude of 6.5. Find its intensity. Use the formula

    $$M = \log\left(\frac{I}{10^{-4}}\right)$$

44. Write the equation of the line passing through the point $(-5, 2)$ and perpendicular to the line $-3x + 7y = 4$. Write the equation in slope-intercept form.
45. Write the inequality for the following graph.

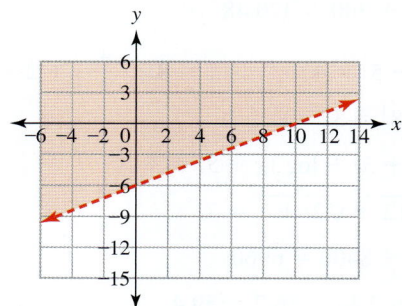

For Exercises 46 through 59, sketch the graph of the given equation or inequality. Label the vertical and horizontal intercepts. If the graph is quadratic, label the vertex.

46. $y = \frac{1}{4}x - 6$
47. $y = -3x^2 - 24x + 2$
48. $y = -1.5(2.5)^x$
49. $y = \sqrt{x + 17}$
50. $y = \frac{1}{3}(x + 4)^2 - 10$
51. $5x - 30y = 180$
52. $y = 800(0.2)^x$
53. $y = \log_9 x$
54. $y = \sqrt[3]{x + 9}$
55. $y = 3(1.6)^x + 20$
56. $y < x^2 - 5x - 14$
57. $y \geq 2x + 3$
58. $5x - 6y < 6$
59. $y \geq -3(x - 1)^2 + 8$

60. The profit for a manufacturing company is inversely proportional to the time it takes to produce a product. The profit is $1.2 million when the time to produce the product is 12 seconds each.
    a. Write an equation for a model for the profit.
    b. Use your model to find the profit if it takes 10 seconds each to produce the product.

For Exercises 61 through 68, perform the indicated operation and simplify.

61. $7\sqrt{12a} + 8\sqrt{3a} - 10\sqrt{a}$
62. $(14m^3 + 3m^2 - 2m + 15) - (5m^3 - 8m^2 + 6)$
63. $\dfrac{x^2 - 16}{x - 3} \cdot \dfrac{x + 2}{x + 4}$
64. $(7 + \sqrt{3w})(-2 + \sqrt{12w})$
65. $(12a^3 + 65a^2 - 3a - 140) \div (3a - 4)$
66. $(5x - 6)^2$
67. $\dfrac{3h - 4}{h^2 - 5h + 6} \div \dfrac{h + 2}{h^2 + 4h - 21}$
68. $\dfrac{5m}{m - 3} + \dfrac{4}{m + 7}$

69. The number of graduates from a certain college can be modeled by the function
$$G(t) = 3400 + 155t$$
where $G(t)$ represents the number of graduating students from the college t years since 2000. The percentage of graduating students who attend the college's graduation ceremonies can be modeled by the function
$$P(t) = 0.86 - 0.03t$$
where $P(t)$ represents the percentage, as a decimal, of graduating students who attend the college's graduation ceremonies t years since 2000.
   a. Find the number of graduating students at this college in 2012.
   b. Find the percentage of graduating students who will attend graduation ceremonies in 2012.
   c. Find the number of graduating students who will attend graduation ceremonies in 2012.
   d. Use the given functions to find a function that will give the number of students who will attend graduation ceremonies t years since 2000.
   e. Use your new model to estimate the number of graduating students who will attend graduation ceremonies in 2015.

For Exercises 70 through 74, factor the given expression using any method.

70. $2x^2 - x - 28$
71. $20a^2 - 45b^2$
72. $h^3 + 125$
73. $21m^3 - 162m^2 - 48m$
74. $60x^2 - 265x + 280$

75. What is the degree of the polynomial given in Exercise 73?

76. Complete the square to convert the function $f(x) = 4x^2 - 48x + 126$ to vertex form.

77. The monthly profit for Beach Shack Rentals can be modeled by
$$P(m) = 6.5 + 3\sqrt{m}$$
where $P(m)$ is the monthly profit in thousands of dollars m months after opening.
   a. Use the model to estimate the profit one year after Beach Shack Rentals opened.
   b. After how many months will Beach Shack Rentals have a monthly profit of $25,000?

For Exercises 78 and 79, use the graph of the radical function $f(x)$ to answer the following.

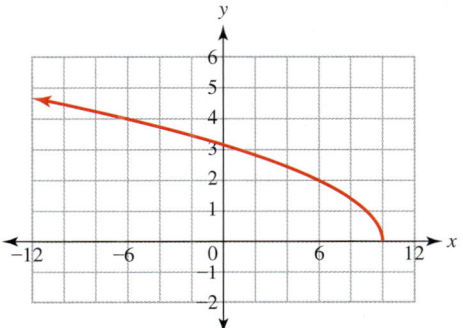

78. Find x such that $f(x) = 4$.
79. Give the domain and range of $f(x)$.

For Exercises 80 through 82, use the graph of the exponential $f(x) = a \cdot b^x$ to answer the following.

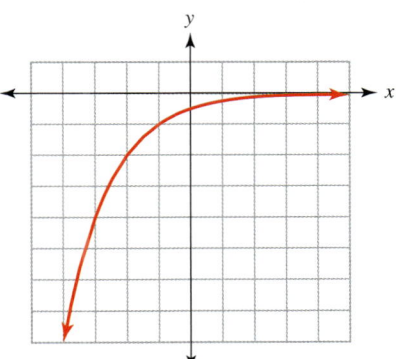

80. Is the value of a positive or negative? Explain your reasoning.
81. Is the value of b less than or greater than 1? Explain your reasoning.
82. Give the domain and range of $f(x)$.

For Exercises 83 through 87, use the graph of the quadratic $f(x) = a(x-h)^2 + k$ to answer the following.

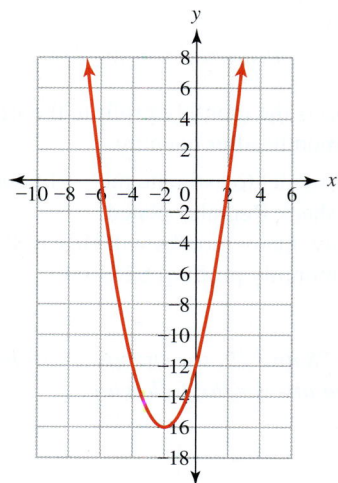

83. Is the value of a positive or negative? Explain your reasoning.
84. Estimate the values of h and k. Explain your reasoning.
85. $f(-4) = $?
86. Find x such that $f(x) = 0$.
87. Give the domain and range of $f(x)$.

For Exercises 88 through 92, use the graph of the line $f(x) = mx + b$ to answer the following.

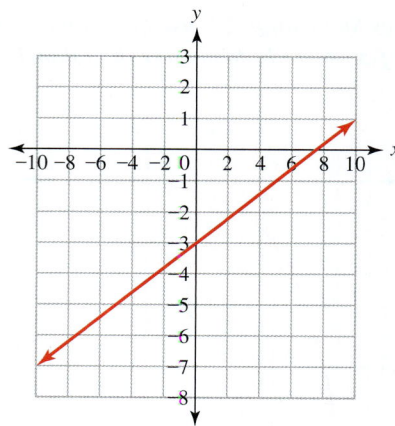

88. Is the value of m positive or negative? Explain your reasoning.
89. Estimate the value of b.
90. Find the slope of the line.
91. Find x such that $f(x) = -5$.
92. Give the domain and range of $f(x)$.

For Exercises 93 through 96, simplify the following radical expressions. Rationalize the denominator if necessary.

93. $\sqrt{36a^4b^6c^2}$ 94. $\sqrt[3]{54x^3y^5}$

95. $\dfrac{\sqrt{7a}}{\sqrt{2b}}$ 96. $\dfrac{6+\sqrt{5}}{2-\sqrt{3}}$

For Exercises 97 through 103, perform the indicated operation and write your answer in the standard form of a complex number.

97. $(7+4i)+(2-7i)$ 98. $(6+2i)-(10-6i)$

99. $(4+9i)(4-9i)$ 100. $(11-3i)(8+2i)$

101. $\dfrac{6+5i}{4i}$ 102. $\dfrac{8-6i}{2-5i}$

103. $\dfrac{10}{3+4i}$

For Exercises 104 through 107, solve the equation and write any complex solutions in standard form.

104. $x^2 + 4x + 29 = 0$ 105. $x^3 + 40x = -12x^2$

106. $x^2 + 70 = 0$ 107. $-3(x-6)^2 + 20 = 32$

108. Write the equation of a circle with center $(3, -9)$ and radius 12.

109. Find the focus and directrix for the parabola
$$x = \dfrac{y^2}{-44}$$

For Exercises 110 through 113, determine whether the given equation represents a circle, ellipse, hyperbola, or parabola. Sketch the graph.

110. $\dfrac{x^2}{144} - \dfrac{y^2}{64} = 1$

111. $121x^2 + 4y^2 = 484$

112. $y^2 + 2y + x - 10 = 0$

113. $(x+8)^2 + (y-3)^2 = 36$

114. Find the equation of the parabola from the graph.

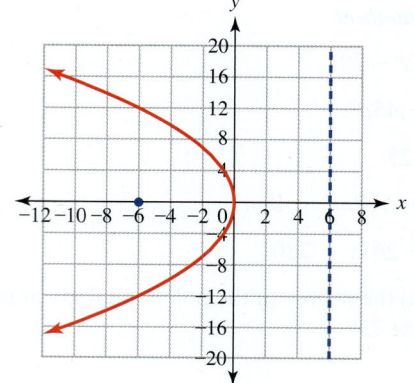

115. Find the equation of the ellipse from the graph.

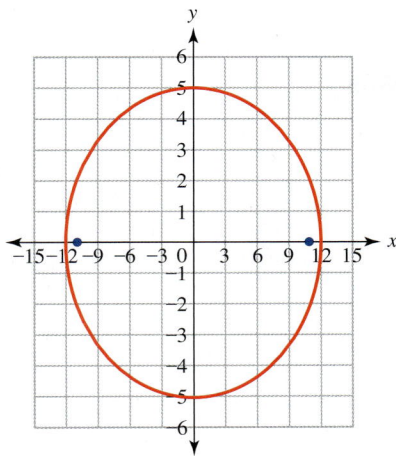

116. Write an arithmetic sequence with common difference $d = 11$ and first term $a_1 = 34$.

117. Compute and list the first five terms of $a_n = 5(-2)^n$.

118. Compute and list the first five terms of $a_n = \dfrac{-12n}{3n+5}$.

119. If $a_n = 26 + 7(n-1)$, find n when $a_n = 607$.

120. If $a_n = -45 + \sqrt{n+20}$, find n when $a_n = -37$.

For Exercises 121 and 122, determine whether the each sequence is arithmetic or geometric. Find the general term a_n.

121. 48, 45, 42, 39, 36, 33, . . .

122. 96, 384, 1536, 6144, 24576, 98304, . . .

123. Your starting salary is $55,000.00, and each year you receive an $825 raise.
   a. Write a formula for your salary a_n in dollars during the nth year.
   b. What will your salary be during the 10th year?
   c. How much total salary will you have earned during the first 10 years?

124. Your starting salary is $55,000.00, and each year you receive a 1.5% raise.
   a. Write a formula for your salary a_n in dollars during the nth year.
   b. What will your salary be during the 10th year?
   c. How much total salary will you have earned during the first 10 years?

125. Write the series $\sum\limits_{n=1}^{6}(2n^3)$ in expanded form and find the sum.

126. Find S_{15} if $a_n = 75 - 6(n-1)$.

127. Find S_{12} if $a_n = 300(1.3)^n$.

Basic Algebra Review

APPENDIX A

APPENDIX TOPICS

- Number systems
- Rectangular coordinate system
- Operations with integers
- Operations with rational numbers
- Order of operations
- Basic solving techniques
- Scientific notation
- Interval notation

Number Systems

In this appendix, we will review some of the basic material from a beginning algebra course. You might want to review this material before you begin studying from this book.

One of the ways in which people keep track of things is by classifying them into categories. In mathematics, we classify different types of numbers into categories called *number systems*. Most of these systems are actually built on a previously defined number system. In this textbook, we will use one of the most general number systems, called the **real number system.** This system includes many types of numbers that we use everyday. To get a complete picture of the real number system, we will define several other number systems that are included in the real numbers. The natural numbers and whole numbers are the most common numbers we use, since they do not include any decimals or fractions and are not negative. The integers include all the numbers from the natural and whole numbers but in addition also include the whole negative numbers. The rational numbers again include all the previous sets but now also include any fraction made of two integers. The rational numbers also include any terminating or repeating decimals, since they can be represented by a fraction of two integers. The final set included in the real number system is the irrational numbers, which include

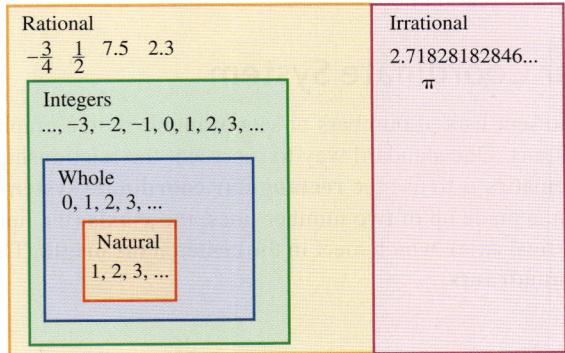

numbers that cannot be written as a fraction. These include numbers such as pi and $\sqrt{2}$. All of these sets together make up what we call the real number system.

> **DEFINITIONS**
>
> **Natural numbers**   The set $\{1, 2, 3, \ldots\}$.
> **Whole numbers**   The set $\{0, 1, 2, 3, \ldots\}$.
> **Integers**   The set $\{\ldots, -3, -2, -1, 0, 1, 2, 3, \ldots\}$.
> **Rational numbers**   The set of numbers that are ratios of integers.
> **Irrational numbers**   The set of numbers that are infinite, nonrepeating decimals. These numbers cannot be written as a ratio of integers.
> **Real numbers**   The set of all rational and irrational numbers.
> - *Note:* The symbols { } in mathematics indicate a set.
> - *Note:* The symbols . . . indicate that the set continues in this way indefinitely.

The real numbers can be represented by the following phrases or symbols:

All real numbers

$(-\infty, \infty)$ (*Note:* The symbol ∞ represents infinity.)

$\mathbb{R}$ (*Note:* This symbol has a double stroke on the R.)

Throughout this text, we will use these different notations for all real numbers.

Example 1   Determining the set of numbers

Determine which set of numbers each of the following belong to. Remember that a number can belong to more than one set.

a. 5   b. $\frac{2}{3}$ c. -7 d. $\sqrt{5}$

SOLUTION

a. The number 5 is a natural number, so it also belongs to the whole numbers, the integers, the rational numbers, and the real numbers. It is considered a rational number because it can be written as a fraction such as $\frac{5}{1}$.

b. The number $\frac{2}{3}$ is a rational number because it is the ratio of two integers. It is also a real number.

c. -7 is an integer and thus is also a rational and real number.

d. $\sqrt{5}$ is an irrational number because it can be represented by an infinite nonrepeating decimal. It is also a real number.

Rectangular Coordinate System

Being able to represent lists of numbers or sets of numbers in a graphical way is helpful in many situations. One standard way to represent the relationship between sets of numbers in mathematics is to use the **rectangular coordinate system**. The rectangular coordinate system is made up of two number lines, the *x*-axis (the horizontal axis) and the *y*-axis (the vertical axis), which meet in the center at the **origin**. The axes divide the system into four **quadrants**.

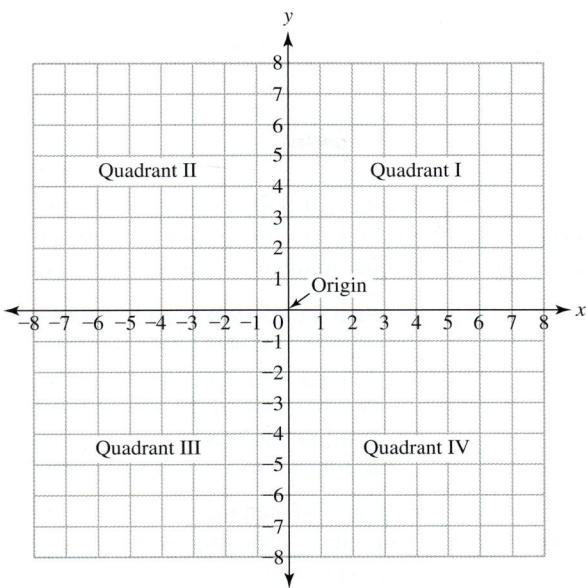

The rectangular coordinate system helps us to graph points that represent ordered pairs of numbers. In life, we often organize things as ordered pairs. For example, we might list the years and the population of a state. Each year would have a population associated with it. The order of the numbers is important because it tells us which number represents what type of value. In mathematics, we typically write ordered pairs in parentheses, such as $(2, 5)$. This ordered pair has two parts: the x-coordinate, equal to 2, and the y-coordinate, equal to 5. These values help us to know where to plot this point on a graph. The 2 tells us to move from the origin, $(0, 0)$, two units to the right, and the 5 tells us to move five units up. This will put our point at the location $(2, 5)$. In the ordered pair (x, y), the x-value moves horizontally with the x-axis or horizontal axis, and the y-value moves vertically with the y-axis or vertical axis. The point $(2, 5)$ is plotted on the graph below, along with several other points.

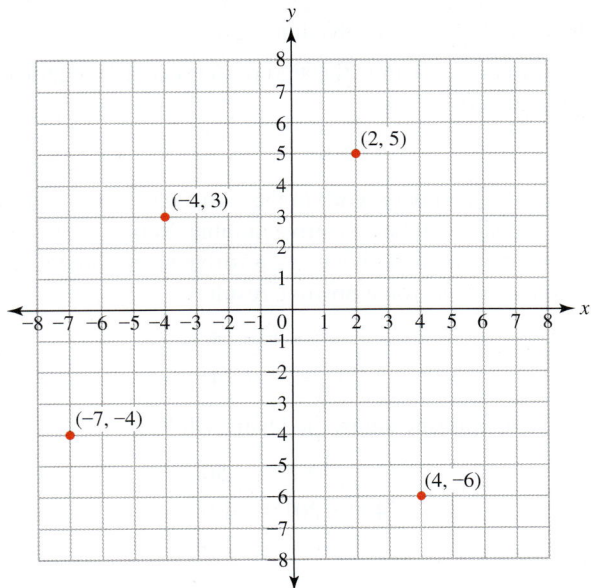

Operations with Integers

Because the real number system includes both signed numbers and fractions, we will review how to add, subtract, multiply, and divide these types of numbers.

When working with signed numbers, you want to remember some basic properties. When adding, you already know how to add two positive numbers, and the result becomes a larger positive number. For example, $5 + 7 = 12$. The same basic thing happens with two negative numbers. The two numbers will add together, and you will get a larger negative number.

$$-4 + (-5) = -9$$
$$-7 + (-21) = -28$$

When you add two numbers that have different signs, the positive and negative numbers start to cancel each other out, so you actually subtract the two numbers from one another, and then the sign of the larger number will be the sign of the sum.

$-13 + 8 = -5$ — Subtract 13 and 8 to get 5, and the negative sign stays because 13 is bigger than 8.

$4 + (-17) = -13$ — Subtract 17 and 4 to get 13, and the negative sign stays because 17 is bigger than 4.

$14 + (-6) = 8$ — Subtract 14 and 6 to get 8, and the positive sign stays because 14 is bigger than 6.

The basic properties for subtraction of signed numbers can be thought of as the same as those for addition except that you start by changing the subtraction problem into addition and changing the sign of the second number.

$5 - (-8) = 5 + (8)$ — Change the subtraction into an addition problem, and change the sign of the -8 to positive.
$ = 13$ — Add the two numbers together.

$3 - 12 = 3 + (-12)$ — Change the subtraction into an addition problem, and change the sign of the 12 into negative.
$ = -9$ — Subtract the 12 and 3 to get 9, and the negative sign stays because 12 is bigger than 3.

> ### Steps to Add or Subtract Integers
> - **To add integers with the same sign:** Add the absolute value of the numbers. Attach the same sign of the numbers to the sum.
> - **To add integers with different signs:** Take the absolute value of each number. Subtract the smaller absolute value from the larger. Attach the sign of the number that is larger in absolute value.
> - **To subtract integers:** Change the sign of the second integer (reading from left to right) and add.

Multiplication and division both work in the same way. If the signs of the two numbers are the same, the result will be a positive number. You already know that two positive numbers multiplied or divided give you a positive result, but two negative numbers multiplied or divided also give you a positive result.

$$5 \cdot 3 = 15$$
$$(-7)(-2) = 14$$

If the two numbers have opposite signs, the result will be negative.

$$-7 \cdot 8 = -56$$
$$4 \cdot (-20) = -80$$

> ### Steps to Multiply or Divide Integers
> - **To multiply or divide two integers with the same sign:** Multiply or divide the absolute values of the numbers. The solution is always positive.
> - **To multiply or divide two integers with different signs:** Multiply or divide the absolute values of the numbers. The solution is always negative.

Example 2 — Performing operations with integers

Perform the indicated operation.

a. $5 + (-8)$ **b.** $-7 + (-12)$ **c.** $-2 - 15$
d. $-3(4)$ **e.** $\dfrac{-20}{4}$ **f.** $(-5)(-6)$

SOLUTION

a. $5 + (-8) = -3$ Subtract the 8 and 5, and the negative sign stays because the 8 is larger.

b. $-7 + (-12) = -19$ Add the two numbers together, and the sign stays negative.

c. $-2 - 15$
$-2 + (-15)$
-17
Change the subtraction to an addition problem, and change the sign of the 15. Add the two numbers together, and the sign stays negative.

d. $-3(4) = -12$ The factors have opposite signs, so the product is negative.

e. $\dfrac{-20}{4} = -5$ The numbers have opposite signs, so the quotient is negative.

f. $(-5)(-6) = 30$ The factors have the same sign, so the product is positive.

Operations with Rational Numbers

When working with fractions, we again will need to remember some basic rules about how to add, subtract, multiply, and divide. When adding or subtracting fractions, you will need to find common denominators and then add or subtract the numerators.

$$\dfrac{1}{2} + \dfrac{2}{5}$$ The least common denominator will be 10.

$$= \dfrac{5}{5} \cdot \dfrac{1}{2} + \dfrac{2}{5} \cdot \dfrac{2}{2}$$ Multiply each fraction by a version of 1 to get the common denominator.

$$= \dfrac{5}{10} + \dfrac{4}{10}$$ Add the numerators together.

$$= \dfrac{9}{10}$$

Steps to Add or Subtract Fractions

1. Rewrite each fraction over the least common denominator.
2. Add or subtract the numerators. The denominator will stay the same.
3. Reduce the final answer to lowest terms.

Multiplying and dividing fractions does not require you to have common denominators. When multiplying, you simply multiply the numerators together and multiply the denominators together. When dividing fractions, you multiply by the reciprocal of the second fraction.

$$\dfrac{3}{5} \cdot \dfrac{2}{7} = \dfrac{6}{35}$$ Multiply the numerators and multiply the denominators.

$$\dfrac{2}{5} \div \dfrac{3}{4} = \dfrac{2}{5} \cdot \dfrac{4}{3}$$ First change the division into multiplication and change the second fraction into its reciprocal.

$$= \dfrac{8}{15}$$ Multiply the numerators and multiply the denominators.

Skill Connection

When finding common denominators for fractions, find a number into which all the denominators will divide equally. This can be done in many ways, but sometimes just looking at the denominators is enough. When that is not enough, it is best to factor each denominator into primes. Then take all the prime numbers and any primes that are repeated in the same number, and multiply them all together to get the least common denominator.

Example 1:
Find the common denominator of the given fractions.

a. $\dfrac{2}{5} \quad \dfrac{3}{8}$ **b.** $\dfrac{3}{10} \quad \dfrac{11}{14}$

Solution:

a. Because 5 and 8 both divide into 40 evenly, that will be a common denominator.

$$\dfrac{2}{5} \cdot \dfrac{8}{8} = \dfrac{16}{40} \qquad \dfrac{3}{8} \cdot \dfrac{5}{5} = \dfrac{15}{40}$$

b. 140 could be a common denominator for these two fractions, but a smaller number might also work. First we can factor the two denominators and find the smallest common denominator.

$$10 = 2 \cdot 5 \qquad 14 = 2 \cdot 7$$

The factors include 2, 5, and 7, so $2 \cdot 5 \cdot 7 = 70$ is the least common denominator.

$$\dfrac{3}{10} \cdot \dfrac{7}{7} = \dfrac{21}{70} \qquad \dfrac{11}{14} \cdot \dfrac{5}{5} = \dfrac{55}{70}$$

> **Steps to Multiply or Divide Fractions**
>
> - **To multiply fractions:** First multiply the numerators together and then multiply the denominators together. Reduce the final answer to lowest terms.
> - **To divide fractions:** First find the reciprocal of the second fraction. Then multiply the numerators together and the denominators together. Reduce the final answer to lowest terms.

Example 3 Performing operations with fractions

Perform the indicated operation.

a. $\dfrac{2}{7} + \dfrac{3}{7}$ b. $\dfrac{4}{9} - \dfrac{2}{6}$ c. $\dfrac{7}{10} \cdot \dfrac{3}{8}$ d. $\dfrac{4}{7} \div \dfrac{5}{9}$

SOLUTION

a. $\dfrac{2}{7} + \dfrac{3}{7} = \dfrac{5}{7}$ The denominators are the same, so add the numerators.

b. $\dfrac{4}{9} - \dfrac{2}{6}$ These denominators are not the same, so we need a common denominator.

$= \dfrac{2}{2} \cdot \dfrac{4}{9} - \dfrac{2}{6} \cdot \dfrac{3}{3}$ Get a common denominator.

$= \dfrac{8}{18} - \dfrac{6}{18}$ Subtract the numerators.

$= \dfrac{2}{18}$ Reduce the fraction.

$= \dfrac{1}{9}$

c. $\dfrac{7}{10} \cdot \dfrac{3}{8} = \dfrac{21}{80}$ Multiply the numerators together, and multiply the denominators together.

d. $\dfrac{4}{7} \div \dfrac{5}{9} = \dfrac{4}{7} \cdot \dfrac{9}{5}$ Change the division into multiplication, using the reciprocal of the second fraction.

$= \dfrac{36}{35}$ Multiply the numerators together and the denominators together.

Order of Operations

One of the underlying themes in mathematics is the order in which operations are to be done. Mathematicians needed to agree on the order in which certain operations would be done so that anyone trying to repeat a calculation would obtain the same result. The order of operations is a vital part of all mathematics. Many students remember the order of operations by using an acronym such as PEMDAS (Please Excuse My Dear Aunt Sally). This acronym is meant to help you remember the order in which the different operations are made.

P—Please: Parenthesis, or grouping symbols such as brackets, absolute value, fraction bars, square roots, etc.
E—Excuse: Exponents
M—My: Multiplication
D—Dear: Division
A—Aunt: Addition
S—Sally: Subtraction

In using this acronym, it is very important that you pay attention to the order of the items and that you understand that operations are all done from left to right. Multiplication and division are actually the same in order as long as you perform them from left to right. Addition and subtraction are also the same in order from left to right—and please note that they are the last operations you should be doing. Many students will do an addition or subtraction before a multiplication or exponent and get the problem wrong.

Example 4 Using the order of operations

Perform the indicated operations following the order of operations.

a. $5 + 2(7 - 3)^2$

b. $\dfrac{31 - 9}{14 - 3} + 17 \cdot 2$

c. $5 + 2[3^2 - 4(6 - 8)]$

d. $\dfrac{1}{2} + \dfrac{2}{3}(6 + 5) - \left(\dfrac{3}{5}\right)^2$

SOLUTION

a. $5 + 2(7 - 3)^2$ Perform the operations in the parentheses first.
$= 5 + 2(4)^2$ Calculate the exponent.
$= 5 + 2(16)$ Multiply.
$= 5 + 32$ Add.
$= 37$

b. $\dfrac{31 - 9}{14 - 3} + 17 \cdot 2$ The fraction bar is a grouping symbol, so the subtractions need to be done first.

$= \dfrac{22}{11} + 17 \cdot 2$ Divide because it is first when looking left to right.

$= 2 + 17 \cdot 2$ Multiply.

$= 2 + 34$ Add.

$= 36$

c. $5 + 2[3^2 - 4(6 - 8)]$ The inside parentheses go first, so perform the subtraction.
$= 5 + 2[3^2 - 4(-2)]$ Calculate the exponent inside the brackets.
$= 5 + 2[9 - 4(-2)]$ Multiply inside the brackets.
$= 5 + 2[9 + 8]$ Add inside the brackets.
$= 5 + 2[17]$ Multiply.
$= 5 + 34$ Add.
$= 39$

d. $\dfrac{1}{2} + \dfrac{2}{3}(6 + 5) - \left(\dfrac{3}{5}\right)^2$ Perform the addition inside the parentheses.

$= \dfrac{1}{2} + \dfrac{2}{3}(11) - \left(\dfrac{3}{5}\right)^2$ Calculate the exponent.

$= \dfrac{1}{2} + \dfrac{2}{3}(11) - \dfrac{9}{25}$ Multiply.

$= \dfrac{1}{2} + \dfrac{22}{3} - \dfrac{9}{25}$ Get a common denominator.

$= \dfrac{75}{150} + \dfrac{1100}{150} - \dfrac{54}{150}$ Add and subtract.

$= \dfrac{1121}{150}$

Basic Solving Techniques

The basic solving techniques that are used to solve equations are necessary from the start of this textbook. In general, you will want to undo anything that is happening to a variable. To solve for a variable, you will need to isolate that variable on one side of the equal sign. We start to do this by simplifying each side and then slowly isolating the variable on one side. To isolate the variable, we use the inverse operation of what is happening to the variable. One way to remember this is to ask the following two questions.

"What is happening to the variable?"

"What is the inverse operation?"

Be sure always to do the inverse operation to *both sides* of the equation. As long as you do this, you will typically keep the equation true. When trying to decide what to do first, you generally want to follow the order of operations in reverse. This means that you want to "undo" the addition and subtraction followed by the multiplication and division and finally any exponents and grouping symbols.

Example 5 Solving equations

Solve the following equations.

a. $2x + 5 = 17$

b. $3(4 + x) - 6 = 5 + 10$

c. $\dfrac{m + 7}{4} - 9 = 2$

d. $\dfrac{1}{2}w + 5 = \dfrac{2}{3}(7) - \dfrac{1}{2}$

SOLUTION

a. $2x + 5 = 17$
$\underline{-5 \quad -5}$ 5 is being added to the 2x, so subtract 5 from both sides of the equation.
$ 2x = 12$
$ \dfrac{2x}{2} = \dfrac{12}{2}$ 2 is being multiplied by the x, so divide both sides by 2.
$ x = 6$

$2(6) + 5 \stackrel{?}{=} 17$ Chek your answer by plugging 6 into the equation.
$12 + 5 \stackrel{?}{=} 17$
$17 = 17$ Both sides are the same, so 6 is the answer.

b. $3(4 + x) - 6 = 5 + 10$ Simplify both sides of the equation.
$3(4 + x) - 6 = 15$ Add 6 to both sides of the equation.
$\underline{+6 \quad +6}$
$3(4 + x) = 21$
$\dfrac{3(4 + x)}{3} = \dfrac{21}{3}$ Divide both sides of the equation by 3.
$4 + x = 7$ Subtract 4 from both sides of the equation.
$\underline{-4 -4}$
$x = 3$

$3(4 + 3) - 6 \stackrel{?}{=} 5 + 10$ Check your answer.
$3(7) - 6 \stackrel{?}{=} 15$
$21 - 6 \stackrel{?}{=} 15$
$15 = 15$ Both sides are equal, so 3 is the answer.

Basic Solving Techniques   A-9

c. $\dfrac{m+7}{4} - 9 = 2$ Add 9 to both sides of the equation.

$\underline{+9 +9}$

$\dfrac{m+7}{4} = 11$ Multiply both sides by 4.

$4\left(\dfrac{m+7}{4}\right) = 4(11)$

$m + 7 = 44$ Subtract 7 from both sides of the equation.

$\underline{-7 -7}$

$m = 37$

$\dfrac{37+7}{4} - 9 \stackrel{?}{=} 2$ Check your answer.

$\dfrac{44}{4} - 9 \stackrel{?}{=} 2$

$11 - 9 \stackrel{?}{=} 2$

$2 = 2$ Both sides are equal, so 37 is the answer.

d. $\dfrac{1}{2}w + 5 = \dfrac{2}{3}(7) - \dfrac{1}{2}$ Simplify both sides of the equation.

$\dfrac{1}{2}w + 5 = \dfrac{14}{3} - \dfrac{1}{2}$ Get like denominators before subtracting.

$\dfrac{1}{2}w + 5 = \dfrac{28}{6} - \dfrac{3}{6}$

$\dfrac{1}{2}w + 5 = \dfrac{25}{6}$ Subtract 5 from both sides of the equal sign.

$\underline{-5 -5}$

$\dfrac{1}{2}w = -\dfrac{5}{6}$ Multiply both sides by 2.

$2\left(\dfrac{1}{2}w\right) = 2\left(-\dfrac{5}{6}\right)$

$w = -\dfrac{5}{3}$

$\dfrac{1}{2}\left(-\dfrac{5}{3}\right) + 5 \stackrel{?}{=} \dfrac{2}{3}(7) - \dfrac{1}{2}$ Check your answer.

$-\dfrac{5}{6} + 5 \stackrel{?}{=} \dfrac{14}{3} - \dfrac{1}{2}$

$\dfrac{25}{6} = \dfrac{25}{6}$ Both sides are equal, so $-\dfrac{5}{3}$ is a answer.

In solving an equation that contains fractions, it may be easier to multiply both sides of the equation by the least common denominator of all the fractions to eliminate the fractions. Then you can continue to solve the equation without fractions.

Example 6 Solving equations with fractions

Solve the following equations.

a. $\dfrac{1}{2}w + 5 = \dfrac{2}{3}(7) - \dfrac{1}{2}$

b. $\dfrac{1}{5}x + 7 = \dfrac{3}{4}(x + 10)$

SOLUTION

a. This is the same equation as in part c of Example 5 above. This time, we will clear the fractions by multiplying by the least common denominator of the fractions.

$$\frac{1}{2}w + 5 = \frac{2}{3}(7) - \frac{1}{2} \qquad \text{Simplify both sides.}$$

$$\frac{1}{2}w + 5 = \frac{14}{3} - \frac{1}{2}$$

$$6\left(\frac{1}{2}w + 5\right) = \left(\frac{14}{3} - \frac{1}{2}\right)6 \qquad \text{Multiply by the LCD} = 6.$$

$$6\left(\frac{1}{2}w\right) + 6(5) = 6\left(\frac{14}{3}\right) - 6\left(\frac{1}{2}\right) \qquad \text{Distribute to each term.}$$

$$\overset{3}{6}\left(\frac{1}{\cancel{2}}w\right) + 6(5) = \overset{2}{6}\left(\frac{14}{\cancel{3}}\right) - \overset{3}{6}\left(\frac{1}{\cancel{2}}\right) \qquad \text{Reduce. All denominators should cancel.}$$

$$3w + 30 = 28 - 3$$

$$3w + 30 = 25 \qquad \text{Solve for } w.$$

$$3w = -5$$

$$w = -\frac{5}{3} \qquad \text{We already saw that this was the answer, so we will not check this one.}$$

b.
$$\frac{1}{5}x + 7 = \frac{3}{4}(x + 10)$$

$$\frac{1}{5}x + 7 = \frac{3}{4}x + \frac{15}{2} \qquad \text{Simplify both sides.}$$

$$20\left(\frac{1}{5}x + 7\right) = \left(\frac{3}{4}x + \frac{15}{2}\right)20 \qquad \text{Multiply both sides by the LCD} = 20.$$

$$20\left(\frac{1}{5}x\right) + 20(7) = 20\left(\frac{3}{4}x\right) + 20\left(\frac{15}{2}\right) \qquad \text{Distribute to every term.}$$

$$4x + 140 = 15x + 150 \qquad \text{Solve for } x.$$

$$-10 = 11x$$

$$-\frac{10}{11} = x$$

$$\frac{1}{5}\left(-\frac{10}{11}\right) + 7 \stackrel{?}{=} \frac{3}{4}\left(\left(-\frac{10}{11}\right) + 10\right) \qquad \text{Check your answer.}$$

$$-\frac{2}{11} + 7 \stackrel{?}{=} \frac{3}{4}\left(-\frac{10}{11} + \frac{110}{11}\right)$$

$$-\frac{2}{11} + \frac{77}{11} \stackrel{?}{=} \frac{3}{4}\left(\frac{100}{11}\right)$$

$$\frac{75}{11} = \frac{75}{11} \qquad \text{The answer works.}$$

Scientific Notation

In mathematics and many areas of science, very large or very small numbers can be involved in many calculations. In working with very large or very small numbers, it is often best to write the number in scientific notation.

Remember that we work in a base 10 number system. This means that each place value is a power of 10. When we multiply by 10, the number simply increases one place value. We see this when the decimal point moves one place to the right. If we multiply by a power of 10, the decimal point will move the same number of places to the right as the power. When we divide by a power of 10, the decimal point simply moves to the left. Because multiplying or dividing by 10 changes only the place value, it becomes an easy way to shorten a very large number.

$$275{,}000{,}000{,}000 = 2.75 \times 10^{11}$$

In this case, the decimal point was moved 11 times, so the reduced number is multiplied by 10 to the 11th power. A number written in the form 2.75×10^{11} is said to be written in scientific notation.

> **DEFINITION**
>
> **Scientific Notation** A number in the form $a \times 10^n$, where $1 \leq |a| < 10$ and n is an integer, is said to be in **scientific notation.**

In working with very small numbers, the exponent of 10 will be negative to indicate that you are dividing by powers of 10 instead of multiplying. Negative exponents are covered in more detail in Section 3.1.

$$0.0000000047 = 4.7 \times 10^{-9}$$

> **Steps to Convert a Number in Scientific Notation to Standard Form**
> - If the exponent on 10 is positive n, multiply by 10^n. This is equivalent to moving the decimal point n places to the right.
> - If the exponent on 10 is negative n, divide by $10^{|n|}$. This is equivalent to moving the decimal point n places to the left.

Example 7   Converting scientific notation to standard form

Write the following numbers in standard form.

a. 8.46×10^9 b. -4.3×10^{14}

c. 5.2×10^{-7} d. -1.47×10^{-11}

SOLUTION

a. The exponent on 10 is positive, so multiply 8.46 by 10^9. This moves the decimal point 9 places to the right.

$$8.46 \times 10^9 = 8{,}460{,}000{,}000$$

b. The exponent on 10 is positive, so multiply -4.3 by 10^{14}. This moves the decimal point 14 places to the right.

$$-4.3 \times 10^{14} = -430{,}000{,}000{,}000{,}000$$

c. The exponent on 10 is negative, so divide 5.2 by 10^7. This moves the decimal point 7 places to the left.

$$5.2 \times 10^{-7} = 0.00000052$$

d. The exponent on 10 is negative, so divide -1.47 by 10^{11}. This moves the decimal point 11 places to the left.

$$-1.47 \times 10^{-11} = -0.0000000000147$$

In rewriting a number into scientific notation, remember that the exponent will be positive if the original number is very large and negative if the the original number is very small. In scientific notation, there should always be only one nonzero digit in front of the decimal point.

Example 8 Converting to scientific notation

Write the following numbers in scientific notation.

a. 32,100,000,000   **b.** −486,000,000,000,000

c. 0.00000000248   **d.** −0.000000000000983

SOLUTION

a. This is a very large number, so the exponent on 10 will be positive. The decimal point must be moved 10 times to get only one nonzero digit in front of the decimal point, so the exponent of 10 will be 10.

$$32{,}100{,}000{,}000 = 3.21 \times 10^{10}$$

b. This is a very large number, so the exponent on 10 will be positive. The decimal point must be moved 14 times to get only one nonzero digit in front of the decimal point, so the exponent of 10 will be 14.

$$-486{,}000{,}000{,}000{,}000 = -4.86 \times 10^{14}$$

c. This is a very small number, so the exponent on 10 will be negative. The decimal point must be moved 9 times to get only one nonzero digit in front of the decimal point, so the exponent of 10 will be −9.

$$0.00000000248 = 2.48 \times 10^{-9}$$

d. This is a very small number, so the exponent on 10 will be negative. The decimal point must be moved 13 times to get only one nonzero digit in front of the decimal point, so the exponent of 10 will be −13.

$$-0.000000000000983 = -9.83 \times 10^{-13}$$

Interval Notation

When working with inequalities, we often have a set of numbers that represents the solution. Often, this set can be represented by an interval on a number line.

$$-9 \leq x \leq 7$$

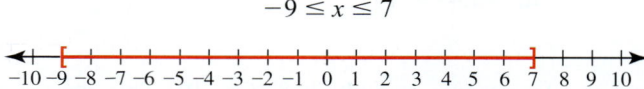

Drawing the interval on a number line is not always convenient, so interval notation or inequalities is used. Interval notation uses parentheses and brackets to indicate the endpoints of the intervals. A parenthesis is used to indicate that the value at the end of the interval is not included in the set, and a bracket is used when the value at the end of the interval is included in the set.

The interval drawn on the number line above can be represented by using interval notation as

$$[-9, 7]$$

The interval is always written from lowest value to highest, or from left to right. The "equal to" part of the inequality and the brackets on the number line indicate that the endpoints of the interval are included. Therefore, we use the brackets in the interval notation to indicate the inclusion of the values −9 and 7.

If the endpoints are not included, there will not be an "equal to" part in the inequality symbol, and the number line will use parentheses at the ends of the interval.

$$-4 < x < 6$$

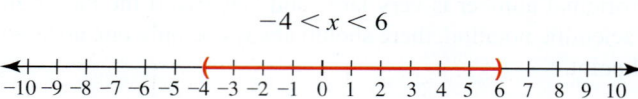

In this case, the interval notation will use parentheses to indicate that the values −4 and 6 are not included.

$$(-4, 6)$$

When an interval extends in a direction forever, the infinity symbol ∞ is used. In interval notation, a parenthesis is always placed next to the infinity symbol because we can never reach infinity as a specific number.

$$x \geq -9$$

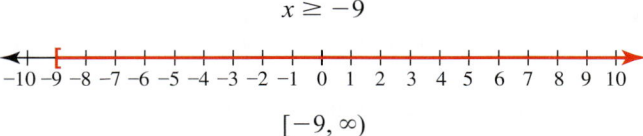

$$[-9, \infty)$$

Notice that the −9 is included in the interval and has a bracket next to it, while the infinity symbol has a parenthesis. If the interval goes toward negative infinity we start the interval with −∞.

$$x < 7$$

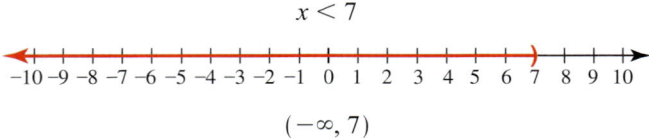

$$(-\infty, 7)$$

If all real numbers are included in the interval, we use the interval $(-\infty, \infty)$ to indicate that all real numbered values are included.

Example 9 Writing interval notation

Write the given intervals in interval notation.

a.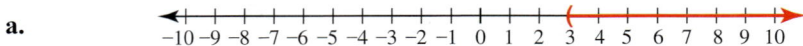

b.

c.

d.

e.

SOLUTION

a.

This interval does not include the starting value 3 and goes out toward infinity. Therefore, use parentheses on both ends of the interval.

$$(3, \infty)$$

b.

This interval does not include either endpoint value. Therefore, use parentheses on both ends of the interval.

$$(-5, 3)$$

c.

This interval goes out to negative infinity and ends at 2. The 2 is included in the interval, so a bracket will be used in the interval notation. The infinity symbol always has a parenthesis next to it.

$$(-\infty, 2]$$

d.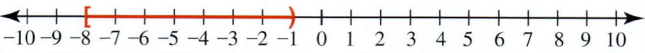

This interval includes the -8 but not the -1, so we will use one bracket and one parenthesis.

$$[-8, -1)$$

e.

This interval goes from negative infinity to infinity, so all real numbers are included.

$$(-\infty, \infty)$$

A Exercises

For Exercises 1 through 10, determine which set(s) of numbers each of the numbers belongs to. Remember that a number can belong to more than one set.

1. 12
2. $\sqrt{9}$
3. 0
4. -8
5. $\sqrt{17}$
6. $\frac{2}{5}$
7. -0.568
8. π
9. 1,522,658,235,120
10. $-\frac{3}{8}$

For Exercises 11 through 20, graph the given points on a rectangular coordinate plane.

11. $(2, 9)$
12. $(5, -7)$
13. $(4, -8.5)$
14. $(-2, -5)$
15. $(5, 3)$
16. $(-3, 7)$
17. $(-8, 6)$
18. $(-7, -4)$
19. $(-8, 0)$
20. $(0, 5)$

For Exercises 21 through 38, perform the indicated operation.

21. $7 + (-25)$
22. $-6 - 18$
23. $-5 + (-9)$
24. $17 - (-6)$
25. $-14 - (-12)$
26. $-8 - (-15)$
27. $2(-8)$
28. $-6 \cdot 10$
29. $-5(-4)$
30. $-6(-7)$
31. $\frac{30}{-5}$
32. $\frac{-40}{8}$
33. $\frac{-100}{-25}$
34. $\frac{-36}{-3}$
35. $\frac{144}{6}$
36. $\frac{120}{8}$
37. $\frac{-136}{8}$
38. $\frac{250}{-10}$

For Exercises 39 through 54, perform the indicated operation.

39. $\frac{2}{3} + \frac{7}{3}$
40. $\frac{4}{9} + \frac{1}{9}$
41. $\frac{15}{34} - \frac{8}{34}$
42. $\frac{7}{18} - \frac{1}{18}$
43. $\frac{2}{3} + \frac{5}{6}$
44. $\frac{3}{5} + \frac{3}{10}$
45. $\frac{2}{5} \cdot \frac{3}{7}$
46. $\frac{4}{9} \cdot \frac{2}{7}$
47. $\frac{4}{15} + \frac{7}{25}$
48. $\frac{8}{15} - \frac{4}{7}$
49. $\frac{3}{4} \div \frac{7}{5}$
50. $\frac{2}{9} \div \frac{4}{11}$

51. $\dfrac{3}{20} \div \dfrac{6}{7}$

52. $\dfrac{4}{25} \div \dfrac{6}{15}$

53. $\dfrac{8}{26} \cdot \dfrac{3}{15}$

54. $\dfrac{16}{32} \cdot \dfrac{10}{15}$

For Exercises 55 through 70, perform the indicated operations following the order of operations.

55. $7 - 2(4 + 5)$

56. $8 + 3(14 - 10)$

57. $2 + 3^2(5 + 7)$

58. $5^2 - 14 + 2(5 - 1)^2$

59. $3^2 + 6(13 - 8)^2$

60. $\dfrac{1}{2} + \dfrac{3}{4}(7 + 15)$

61. $\dfrac{2}{3}\left(\dfrac{5}{2} + \dfrac{7}{3}\right) - \left(\dfrac{2}{3}\right)^2$

62. $\dfrac{8}{9}\left(\dfrac{2}{5} - \dfrac{7}{10}\right) + \left(\dfrac{3}{5}\right)^2$

63. $5 + 7[3 - 5(6 + 2)^2]$

64. $4^2 - 3[2 - 7(6 - 9)]$

65. $2^3 + 5[6^2 - 4(2 + 7)]$

66. $\dfrac{12 - 8}{3 + 2} + 6 \cdot 5$

67. $\dfrac{14 + 10}{5} - \dfrac{6}{3} + 4$

68. $\dfrac{25 - 13}{6 + 5} + 2 \cdot \dfrac{7}{11}$

69. $5 + 3^2\left[6 - \left(4 + \dfrac{12 + 3}{6 - 3}\right)\right]$

70. $2^3 + 4\left[7^2 + \dfrac{15 + 6}{3 - 4}\right]$

For Exercises 71 through 88, solve the given equations.

71. $3x + 8 = 20$

72. $-4p + 21 = 13$

73. $\dfrac{2}{3}x - 5 = 4$

74. $\dfrac{4}{5}w + 9 = -7$

75. $2(m - 6) + 11 = 61$

76. $3(x + 8) - 15 = -11$

77. $\dfrac{h + 6}{5} = 17$

78. $\dfrac{d - 9}{3} = 11$

79. $\dfrac{k - 5}{4} + 9 = 21$

80. $\dfrac{8 + g}{2} - 7 = -4$

81. $\dfrac{4}{5}b - 5 = 6$

82. $\dfrac{3}{8}h + 7 = 13$

83. $\dfrac{2}{9}w - 3 = \dfrac{4}{7}$

84. $\dfrac{3}{8}r + \dfrac{5}{6} = \dfrac{2}{3}$

85. $\dfrac{2}{7}m + 4 = \dfrac{3}{7}(5) + 7$

86. $\dfrac{1}{5}x + 6 = \dfrac{2}{5}(3) - 8$

87. $\dfrac{7}{10}x - \dfrac{4}{5} = \dfrac{5}{6}(x + 2)$

88. $\dfrac{2}{3}(x + 5) = \dfrac{1}{4}x + 3$

For Exercises 89 through 96, write the numbers in standard form.

89. 2.74×10^8

90. 3.27×10^{13}

91. -4.63×10^{12}

92. -1.5×10^7

93. 1.28×10^{-5}

94. 5.9×10^{-14}

95. -9.6×10^{-13}

96. -4.81×10^{-6}

For Exercises 97 through 104, write the numbers in scientific notation.

97. 14,000,000,000

98. 6,780,000,000,000

99. $-547{,}000{,}000$

100. $-2{,}000{,}000{,}000{,}000{,}000$

101. 0.000000078

102. 0.000000000000236

103. -0.00000000000751

104. -0.00000000141

For Exercises 105 through 112, write the given intervals in interval notation.

105.

106.

107.
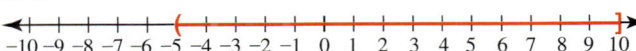

108.
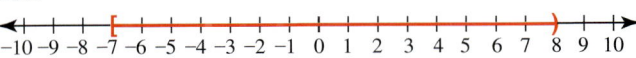

109.
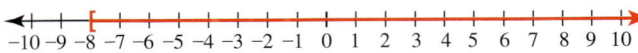

110.
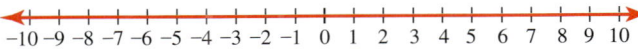

111.
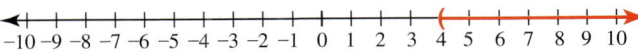

112.

APPENDIX B

Matrices

APPENDIX TOPICS

- Solve systems of three equations
- Work with matrices
- Matrix row reduction
- Solve a system using matrices
- Finding a quadratic function using matrices

Solving Systems of Three Equations

In Section 2.3, we learned how to solve a system of two equations in two unknowns using the elimination method. We are going to expand this method to solve a system of three equations in three unknowns. Such a system looks like the following.

$$x + y + z = 2$$
$$x - y - z = 1$$
$$3x + 2y + z = 5$$

The three unknowns are the variables x, y, and z. A solution of this system will consist of values of x, y, and z that make all three equations simultaneously true. Be sure to write the point with x, y, and z in the correct order as (x, y, z). Points in the form (x, y, z) are sometimes called *ordered triples*.

> **DEFINITION**
>
> **Solution to a system of three equations in three unknowns** A **solution** is one or more points (x, y, z) that satisfy all three equations.

Example 1 Checking solutions

Consider the following system.

$$x + y + z = 4$$
$$x - y + z = 2$$
$$2x + y - z = 7$$

a. Is $(3, 1, 0)$ a solution of the system?

b. Is $(4, 0, 1)$ a solution of the system?

SOLUTION

a. Substitute in the values $(x, y, z) = (3, 1, 0)$ into all three equations and simplify.

Substitute in $(x, y, z) = (3, 1, 0)$ **Simplify and check**

$$x + y + z = 4 \qquad 3 + 1 + 0 = 4$$
$$x - y + z = 2 \qquad 3 - 1 + 0 = 2$$
$$2x + y - z = 7 \qquad 2(3) + 1 - 0 = 7$$

Since all three equations are true, the point $(x, y, z) = (3, 1, 0)$ is a solution of the system.

b. Substitute in the values $(x, y, z) = (4, 0, 1)$ into all three equations and simplify.

Substitute in $(x, y, z) = (4, 0, 1)$ **Simplify and check**

$$x + y + z = 4 \qquad 4 + 0 + 1 = 5 \neq 4$$
$$x - y + z = 2 \qquad 4 - 0 + 1 = 5 \neq 2$$
$$2x + y - z = 7 \qquad 2(4) + 0 - 1 = 7$$

Because the first two equations are not true, the point $(x, y, z) = (4, 0, 1)$ is not a solution of the system.

To solve a system of three equations in three unknowns, we can use the elimination method that we discussed in Section 2.3. It can help to number the equations.

Example 2 Solving a system of three equations in three unknowns using elimination

Solve the system. Write the solution as a point (x, y, z).

$$x + y + z = 4$$
$$x - y + z = 2$$
$$2x + y - z = 7$$

SOLUTION

First, number the equations. It will help us to keep track of them during the elimination method.

1. $x + y + z = 4$
2. $x - y + z = 2$
3. $2x + y - z = 7$

Add equations 1 and 2 together to eliminate the variable y. Call the result equation 4.

1. $x + y + z = 4$
2. $x - y + z = 2$
4. $\quad 2x + 2z = 6$

Add equations 2 and 3 together to eliminate the variable y. Call the result equation 5.

2. $x - y + z = 2$
3. $2x + y - z = 7$
5. $3x \qquad\quad = 9$

From equation 5, we have that

$$3x = 9 \qquad \text{This is equation 5.}$$
$$\frac{3x}{3} = \frac{9}{3} \qquad \text{Divide by 3 to solve for } x.$$
$$x = 3$$

Equation 4 only has two variables, x and z. Substitute this value of x into equation 4 to find the value of z.

$$2x + 2z = 6 \quad \text{Substitute in } x = 3.$$
$$2(3) + 2z = 6$$
$$6 + 2z = 6 \quad \text{Solve for } z.$$
$$\underline{-6 \qquad\quad -6}$$
$$2z = 0$$
$$z = 0$$

Now that we have the values of x and z, we find the value of y. Any of the three original equations involves all three variables. Therefore, we can substitute x and z into any of the original equations. Just picking equation 1, we have that

$$x + y + z = 4 \quad \text{Substitute in } x = 3 \text{ and } z = 0.$$
$$3 + y + 0 = 4 \quad \text{Simplify.}$$
$$3 + y = 4 \quad \text{Solve for } y.$$
$$\underline{-3 \qquad -3}$$
$$y = 1$$

The solution to this system is the point $(x, y, z) = (3, 1, 0)$. The check was done in Example 1.

As when we solved systems in Chapter 2, sometimes we must multiply one or both equations by a constant before eliminating a variable.

Example 3 Multiplying by a constant first

Solve the system. Write the solution as a point (x, y, z). Check the solution.

$$x + y + z = 2$$
$$x + 2y - z = 2$$
$$-2x + y + z = 5$$

SOLUTION

Once again, we number the equations first to keep track of them.

1. $x + y + z = 2$
2. $x + 2y - z = 2$
3. $-2x + y + z = 5$

Adding equations 1 and 2 will eliminate the variable z. Call the result equation 4.

1. $x + y + z = 2$
2. $\underline{x + 2y - z = 2}$
4. $\quad 2x + 3y = 4$

Likewise, adding equations 2 and 3 will eliminate z in those equations. Call this result equation 5.

2. $x + 2y - z = 2$
3. $\underline{-2x + y + z = 5}$
5. $\quad -x + 3y = 7$

Adding equations 4 and 5 will eliminate neither variable x nor y. If we first multiply equation 5 by 2 and then add it to equation 4, we can eliminate x.

$$5. \quad -x + 3y = 7$$
$$2(5.) \quad 2(-x + 3y) = 2(7) \quad \text{Multiply both sides by 2.}$$
$$6. \quad -2x + 6y = 14 \quad \text{Distribute 2 on the left side.}$$

Now add equation 6 to equation 4.

$$4. \quad 2x + 3y = 4$$
$$6. \quad \underline{-2x + 6y = 14}$$
$$9y = 18$$
$$\frac{9y}{9} = \frac{18}{9}$$
$$y = 2$$

Substitute this into equation 5 (or equation 6) to find x.

$$-x + 3y = 7$$
$$-x + 3(2) = 7 \quad \text{Substitute in } y = 2.$$
$$-x + 6 = 7 \quad \text{Isolate } x.$$
$$\underline{-6 \quad\quad -6}$$
$$-x = 1$$
$$(-1)(-x) = (-1)(1) \quad \text{Multiply by } -1 \text{ to change the signs.}$$
$$x = -1$$

Substitute the values of x and y into any of the original equations to find z. Using equation 1, we see that

$$(-1) + (2) + z = 2 \quad \text{Substitute in } x = -1 \text{ and } y = 2.$$
$$1 + z = 2 \quad \text{Solve for } z.$$
$$\underline{-1 \quad\quad -1}$$
$$z = 1$$

The solution to this system is the point $(x, y, z) = (-1, 2, 1)$.
To check the solution, substitute the point into all three equations.

	Equation	Substitute in $(-1, 2, 1)$	Simplify
1.	$x + y + z = 2$	$(-1) + (2) + (1) \stackrel{?}{=} 2$	$2 = 2$
2.	$x + 2y - z = 2$	$(-1) + 2(2) - (1) \stackrel{?}{=} 2$	$2 = 2$
3.	$-2x + y + z = 5$	$-2(-1) + (2) + (1) \stackrel{?}{=} 5$	$5 = 5$

The solution checks.

Matrices

Using the elimination method requires a lot of manipulation with several equations and can get confusing. To reduce the confusion, we can simplify the work using **matrices.** Using a matrix to solve a system of equations focuses on the coefficients of each term and not on the variables themselves.

Think of a matrix as being like a spreadsheet. When we create a spreadsheet or a table on a computer, we must first specify how many rows and columns it should have. With a matrix, we also specify how many rows and columns it has. The following matrix has two rows and three columns. We write this as 2×3, and say, "2 by 3." The size of a matrix is always given as rows $\times$ columns, in that order.

$$\begin{bmatrix} 3 & 4 & -1 \\ 0 & 5 & 7 \end{bmatrix}$$

Matrices are usually written with square brackets around the elements.

We discuss individual elements of a matrix by their position in the matrix. In the matrix $\begin{bmatrix} 3 & 4 & -1 \\ 0 & 5 & 7 \end{bmatrix}$, the number 3 is in row 1, column 1, so we call it the (1, 1) element. The number 7 is in row 2, column 3, so we call it the (2, 3) element.

> **DEFINITION**
>
> **Matrix** A **matrix** is a rectangular array of numbers. The *size* of a matrix is given as rows × columns. An individual element in a matrix is located by its position: (row, column).
>
> $\begin{bmatrix} 4 & 5 \\ -1 & 3 \\ 2 & 0 \end{bmatrix}$ is a 3 × 2 matrix. The number −1 is in the (2, 1) position.

In Examples 1 and 2, we saw that the elimination method can be used to solve systems of equations. Now we will consider another way to write a system of equations without the variables: using a matrix.

The following system consists of two equations in two unknowns, x and y. To write this system in matrix form, we put the coefficients of x in the first column. The coefficients of y are written in the second column. The right-hand side is put in the third column.

$$\begin{array}{cc} \textbf{System} & \textbf{Matrix form} \\ \begin{array}{c} 3x + 2y = 6 \\ -x + 4y = 10 \end{array} & \begin{bmatrix} 3 & 2 & 6 \\ -1 & 4 & 10 \end{bmatrix} \end{array}$$

Think of the first column as containing the coefficients of x, the second column as containing the coefficients of y, and the third column as containing the right-hand side. The rows correspond to the two equations.

$$\begin{array}{ccc} x & y & \\ \begin{bmatrix} 3 & 2 & 6 \\ -1 & 4 & 10 \end{bmatrix} \end{array}$$

This type of matrix that represents a system of equations, where the last column contains the right-hand side, has a special name. It is called an *augmented matrix*.

Example 4 Writing an augmented matrix

Write the augmented matrix for each of the following linear systems.

a. $\begin{cases} -3x + 5y = 9 \\ 2x + y = -7 \end{cases}$

b. $\begin{cases} x + y + z = 2 \\ -x + 2y + z = 5 \\ x - y = 0 \end{cases}$

SOLUTION

a. The corresponding augmented matrix is

$$\begin{bmatrix} -3 & 5 & 9 \\ 2 & 1 & -7 \end{bmatrix}$$

b. Notice that the third equation does not contain the variable z. If we rewrite the equation so that the missing variable has a coefficient of 0, $x - y = 0$ becomes $x - y + 0z = 0$. The corresponding augmented matrix is

$$\begin{bmatrix} 1 & 1 & 1 & 2 \\ -1 & 2 & 1 & 5 \\ 1 & -1 & 0 & 0 \end{bmatrix}$$

Matrix Row Reduction

Now that we can write a linear system as a matrix, we want to solve the system in matrix form. This is because we won't have to write all of the variables down constantly. The coefficients of the variables are stored in the matrix as we saw in Example 3.

We first look at an augmented matrix that is in a special form. This form is called *upper triangular form*. Below are two matrices in upper triangular form.

$$\begin{bmatrix} 1 & 1 & 6 \\ 0 & 2 & -4 \end{bmatrix} \qquad \begin{bmatrix} -1 & 1 & 1 & 1 \\ 0 & -3 & 4 & 3 \\ 0 & 0 & -1 & 4 \end{bmatrix}$$

The *main diagonal* of the matrix runs from the (1, 1) element, through the (2, 2) element, through the (3, 3) element, and so on. It has been drawn on the matrices below.

Main diagonal   **Main diagonal**

$$\begin{bmatrix} 1 & 1 & 6 \\ 0 & 2 & -4 \end{bmatrix} \qquad \begin{bmatrix} -1 & 1 & 1 & 1 \\ 0 & -3 & 4 & 3 \\ 0 & 0 & -1 & 4 \end{bmatrix}$$

The key to recognizing whether a matrix is in upper triangular form is that all the elements below the main diagonal are 0.

When an augmented matrix is in upper triangular form, we can solve the system using a technique called *back-solving*.

> **Steps to Solving an Augmented Matrix Using Back-Solving**
> 1. Write the last row as an equation in variable form. It will have only one variable. Solve for that variable.
> 2. Working backwards, write the next to last row as an equation in variable form. It will have only two variables. Substitute in the value of the variable from step 1 to find the value of the unknown variable.
> 3. If necessary, work backwards once again to write the first row as an equation in variable form. Substitute the values of the variables found in steps 1 and 2 into the equation and solve for the unknown.
> 4. Write the solution as an ordered pair (x, y) or as an ordered triple (x, y, z).
>
> *Note:* The matrix must be in upper triangular form.

Example 5 Using back-solving

Solve each augmented matrix using back-solving. Write the solution as an ordered pair or ordered triple.

a. $\begin{bmatrix} 1 & 1 & 6 \\ 0 & 2 & -4 \end{bmatrix}$ **b.** $\begin{bmatrix} -1 & 1 & 1 & 1 \\ 0 & -3 & 4 & -1 \\ 0 & 0 & -1 & 4 \end{bmatrix}$

SOLUTION

a. $\begin{bmatrix} 1 & 1 & 6 \\ 0 & 2 & -4 \end{bmatrix}$

Rewriting the last row in equation form yields

$$0x + 2y = -4 \quad \text{or} \quad 2y = -4$$

Solving this equation for the variable y yields

$$\frac{2y}{2} = \frac{-4}{2}$$

$$y = -2$$

Rewriting the first row in equation form and substituting in $y = -2$ yields

$$x + y = 6 \quad \text{Rewrite the first row as an equation.}$$
$$x + (-2) = 6 \quad \text{Substitute in } y = -2.$$
$$\underline{+2 \quad +2} \quad \text{Solve for x.}$$
$$x = 8$$

The solution to the system is $(x, y) = (8, -2)$.

b. $\begin{bmatrix} -1 & 1 & 1 & 1 \\ 0 & -3 & 4 & -1 \\ 0 & 0 & -1 & 4 \end{bmatrix}$

Rewriting the last row in equation form yields

$$0x + 0y - 1z = 4 \quad \text{or} \quad -z = 4$$

Solving this equation for the variable z yields

$$-1z = 4$$
$$z = -4$$

Working backwards, write the second row in equation form.

$$0x - 3y + 4z = -1$$

Solve this equation for y.

$$-3y + 4z = -1$$
$$-3y + 4(-4) = -1 \quad \text{Substitute in } z = -4.$$
$$-3y - 16 = -1 \quad \text{Solve for y.}$$
$$\underline{+16 \quad +16}$$
$$-3y = 15$$
$$\frac{-3y}{-3} = \frac{15}{-3}$$
$$y = -5$$

Rewrite the first row in equation form. Substitute in $y = -5$ and $z = -4$ and solve for x.

$$-x + y + z = 1 \quad \text{Rewrite the first row as an equation.}$$
$$-x + (-5) + (-4) = 1 \quad \text{Substitute in } y = -5 \text{ and } z = -4.$$
$$-x + (-9) = 1 \quad \text{Solve for x.}$$
$$\underline{+9 \quad +9}$$
$$-x = 10$$
$$x = -10$$

The solution is the ordered triple $(x, y, z) = (-10, -5, -4)$.

Putting an augmented matrix in upper triangular form is done by using the elementary row operations. There are three operations that are allowable on matrix rows. This means that if we use these three operations, they will not change the solution to the system.

> **DEFINITION**
>
> **Elementary Row Operations** The three elementary row operations are as follows
>
> - Two rows may be interchanged (row swap).
> - A row may be multiplied by a nonzero constant.
> - A multiple of one row may be added to another row. Use this to overwrite the row we are trying to change.

Let's see how to use these three row operations to put a matrix in upper triangular form. We will work in a systematic fashion, beginning with the first row. We will use the first row and the elementary row operations to put 0's below the (1, 1) element (the first main diagonal element). We will then move on to the second row. Using the (2, 2) element and the elementary row operations, we will put 0's below the (2, 2) element. We will continue on in this fashion until the matrix is in upper triangular form.

Example 6 Putting a matrix in upper triangular form

Put each matrix in upper triangular form.

a. $\begin{bmatrix} 1 & 1 & -1 \\ -2 & 3 & 7 \end{bmatrix}$

b. $\begin{bmatrix} 1 & 1 & 1 & 5 \\ -1 & 2 & 1 & 4 \\ 3 & -1 & -1 & -5 \end{bmatrix}$

SOLUTION

a. $\begin{bmatrix} 1 & 1 & -1 \\ -2 & 3 & 7 \end{bmatrix}$

Begin with the (1, 1) element on the main diagonal. This element is the number 1. We want to put a 0 in the (2, 1) position. This is the first element of the second row. The number -2 is currently in that position. The elementary row operations say that it is okay to add a multiple of one row to another. We multiply row 1 by 2 and add it to row 2. We are trying to change row 2, so we overwrite row 2 with the result.

$\begin{bmatrix} 1 & 1 & -1 \\ -2 & 3 & 7 \end{bmatrix}$

$= \begin{bmatrix} 1 & 1 & -1 \\ 2(1) + -2 & 2(1) + 3 & 2(-1) + 7 \end{bmatrix}$ 2 · row 1 + row 2. Overwrite row 2.

$= \begin{bmatrix} 1 & 1 & -1 \\ 0 & 5 & 5 \end{bmatrix}$ This matrix is in upper triangular form.

b. $\begin{bmatrix} 1 & 1 & 1 & 5 \\ -1 & 2 & 1 & 4 \\ 3 & -1 & -1 & -5 \end{bmatrix}$

Begin with the (1, 1) element on the main diagonal. This element is the number 1. We want to put a 0 in the (2, 1) position: This is the first element of the second row. The number -1 is currently in that position. The elementary row operations say that it is okay to add a multiple of one row to another. We add rows 1 and 2. We are trying to change row 2, so overwrite row 2 with the result.

$\begin{bmatrix} 1 & 1 & 1 & 5 \\ -1 & 2 & 1 & 4 \\ 3 & -1 & -1 & -5 \end{bmatrix}$

$= \begin{bmatrix} 1 & 1 & 1 & 5 \\ 1 + (-1) & 1 + 2 & 1 + 1 & 5 + 4 \\ 3 & -1 & -1 & -5 \end{bmatrix}$ row 1 + row 2. Overwrite row 2.

$= \begin{bmatrix} 1 & 1 & 1 & 5 \\ 0 & 3 & 2 & 9 \\ 3 & -1 & -1 & -5 \end{bmatrix}$

We want to next put a 0 in the (3, 1) position. The number 3 is currently in that position. Multiply row 1 by -3, add it to row 3, and overwrite row 3.

$$\begin{bmatrix} 1 & 1 & 1 & 5 \\ 0 & 3 & 2 & 9 \\ 3 & -1 & -1 & -5 \end{bmatrix}$$

$$= \begin{bmatrix} 1 & 1 & 1 & 5 \\ 0 & 3 & 2 & 9 \\ -3(1)+3 & -3(1)+(-1) & -3(1)+(-1) & -3(5)+(-5) \end{bmatrix} \quad \begin{array}{l} -3 \cdot \text{row 1 +} \\ \text{row 3.} \\ \text{Overwrite row 3.} \end{array}$$

$$= \begin{bmatrix} 1 & 1 & 1 & 5 \\ 0 & 3 & 2 & 9 \\ 0 & -4 & -4 & -20 \end{bmatrix}$$

There are 0's in the column below the (1, 1) element on the main diagonal. We move on to the (2, 2) element on the main diagonal. We want to put 0's below this element (in the same column). Looking at the matrix, we see that the (2, 2) element is 3 and the element below it is -4, in the (3, 2) position.

$$\begin{bmatrix} 1 & 1 & 1 & 5 \\ 0 & 3 & 2 & 9 \\ 0 & -4 & -4 & -20 \end{bmatrix}$$

The least common multiple for 3 and -4 is -12. We multiply row 2 by 4 and row 3 by 3.

$$\begin{bmatrix} 1 & 1 & 1 & 5 \\ 4(0) & 4(3) & 4(2) & 4(9) \\ 3(0) & 3(-4) & 3(-4) & 3(-20) \end{bmatrix} \quad \text{Multiply row 2 by 4 and row 3 by 3.}$$

$$= \begin{bmatrix} 1 & 1 & 1 & 5 \\ 0 & 12 & 8 & 36 \\ 0 & -12 & -12 & -60 \end{bmatrix}$$

If we now add rows 2 and 3, overwriting row 3, the (3, 2) element will become 0.

$$\begin{bmatrix} 1 & 1 & 1 & 5 \\ 0 & 12 & 8 & 36 \\ 0+0 & 12+(-12) & 8+(-12) & 36+(-60) \end{bmatrix} \quad \begin{array}{l} \text{Add row 2 to row 3,} \\ \text{overwriting row 3.} \end{array}$$

$$\begin{bmatrix} 1 & 1 & 1 & 5 \\ 0 & 12 & 8 & 36 \\ 0 & 0 & -4 & -24 \end{bmatrix} \quad \begin{array}{l} \text{This matrix is in upper} \\ \text{triangular form.} \end{array}$$

Solving Systems with Matrices

We now have the skills to solve a system of equations using augmented matrices. Pulling this all together leads to the following steps.

Steps to Solving a System of Equations Using Augmented Matrices

1. Write the system as an augmented matrix.
2. Using the elementary row operations, put the augmented matrix in upper triangular form.
3. Back-solve to find the values of the variables.
4. Write the solution as an ordered pair (x, y) or an ordered triple (x, y, z).
5. Check the solution in the original equations.

Example 7 Solving a system using augmented matrices

Solve the system using augmented matrices. Write the solution as an ordered pair (x, y) or an ordered triple (x, y, z). Check the solution in the original equation.

a. $\begin{cases} x - 3y = 11 \\ -5x + y = -13 \end{cases}$

b. $\begin{cases} x - y + 2z = 7 \\ 2x + y - z = -7 \\ 3y + z = -3 \end{cases}$

SOLUTION

a. $\begin{cases} x - 3y = 11 \\ -5x + y = -13 \end{cases}$

First, rewrite this system as an augmented matrix.

$$\begin{bmatrix} 1 & -3 & 11 \\ -5 & 1 & -13 \end{bmatrix}$$

Next, we put the matrix in upper triangular form. Using the $(1, 1)$ element to put a 0 below it yields

$$\begin{bmatrix} 1 & -3 & 11 \\ -5 & 1 & -13 \end{bmatrix}$$

$$= \begin{bmatrix} 1 & -3 & 11 \\ 5 + -5 & -15 + 1 & 55 + -13 \end{bmatrix} \quad \text{5 · row 1 + row 2. Overwrite row 2.}$$

$$= \begin{bmatrix} 1 & -3 & 11 \\ 0 & -14 & 42 \end{bmatrix}$$

This matrix is now in upper triangular form. Back-solving yields

$$\begin{bmatrix} 1 & -3 & 11 \\ 0 & -14 & 42 \end{bmatrix}$$

$-14y = 42$ Rewrite row 2 in equation form.

$\dfrac{-14y}{-14} = \dfrac{42}{-14}$ Solve for y.

$y = -3$

Rewrite row 1 in equation form and substitute in $y = -3$.

$x - 3y = 11$ Write the first row in equation form.
$x - 3(-3) = 11$ Substitute in $y = -3$.
$x + 9 = 11$ Solve for x.
$\underline{-9 \quad -9}$
$x = 2$

The solution is the point $(2, -3)$.
Check the solution in both original equations.

First equation	Second equation
$x - 3y = 11$	$-5x + y = -13$
$2 - 3(-3) \stackrel{?}{=} 11$	$-5(2) + (-3) \stackrel{?}{=} -13$
$11 = 11$	$-13 = -13$

b. $\begin{cases} x - y + 2z = 7 \\ 2x + y - z = -7 \\ 3y + z = -3 \end{cases}$

Rewrite the system as an augmented matrix.

$\begin{cases} x - y + 2z = 7 \\ 2x + y - z = -7 \\ 0x + 3y + z = -3 \end{cases}$ Rewrite equation 3 with a term of 0x.

$\begin{bmatrix} 1 & -1 & 2 & 7 \\ 2 & 1 & -1 & -7 \\ 0 & 3 & 1 & -3 \end{bmatrix}$

To put this augmented matrix in upper triangular form, we use the (1, 1) element to introduce 0's in the column below it.

$\begin{bmatrix} 1 & -1 & 2 & 7 \\ 2 & 1 & -1 & -7 \\ 0 & 3 & 1 & -3 \end{bmatrix}$ $-2 \cdot$ row 1 + row 2. Overwrite row 2.

$= \begin{bmatrix} 1 & -1 & 2 & 7 \\ -2(1)+2 & -2(-1)+1 & -2(2)+-1 & -2(7)+-7 \\ 0 & 3 & 1 & -3 \end{bmatrix}$

$= \begin{bmatrix} 1 & -1 & 2 & 7 \\ 0 & 3 & -5 & -21 \\ 0 & 3 & 1 & -3 \end{bmatrix}$ Use the (2, 2) element to put a 0 below it.

$= \begin{bmatrix} 1 & -1 & 2 & 7 \\ 0 & 3 & -5 & -21 \\ 0 & 3 & 1 & -3 \end{bmatrix}$ $-1 \cdot$ row 2 + row 3. Overwrite row 3

$= \begin{bmatrix} 1 & -1 & 2 & 7 \\ 0 & 3 & -5 & -21 \\ 0 & -1(3)+3 & -1(-5)+1 & -1(-21)+-3 \end{bmatrix}$

$= \begin{bmatrix} 1 & -1 & 2 & 7 \\ 0 & 3 & -5 & -21 \\ 0 & 0 & 6 & 18 \end{bmatrix}$ This matrix is in upper triangular form.

Back-solving yields the following.

$6z = 18$ Rewrite row 3 in equation form.

$\dfrac{6z}{6} = \dfrac{18}{6}$ Solve for z.

$z = 3$

Rewrite row 2 in equation form and solve for y.

$3y - 5z = -21$ Rewrite row 2 in equation form.

$3y - 5(3) = -21$ Substitute in z = 3.

$3y - 15 = -21$ Solve for y.

$\underline{+15 \quad +15}$

$3y = -6$

$y = -2$

Substitute $y = -2$ and $z = 3$ into row 1 to solve for x.

$$x - y + 2z = 7 \quad \text{Rewrite row 1 as an equation.}$$
$$x - (-2) + 2(3) = 7 \quad \text{Substitute in } y = -2 \text{ and } z = 3.$$
$$x + 2 + 6 = 7 \quad \text{Solve for } x.$$
$$x + 8 = 7$$
$$\underline{-8 \quad -8}$$
$$x = -1$$

The solution is the point $(-1, -2, 3)$.
To check, substitute the point into all three equations.

First equation	Second equation	Third equation
$x - y + 2z = 7$	$2x + y - z = -7$	$3y + z = -3$
$(-1) - (-2) + 2(3) \stackrel{?}{=} 7$	$2(-1) + (-2) - 3 \stackrel{?}{=} -7$	$3(-2) + (3) \stackrel{?}{=} -3$
$-1 + 2 + 6 \stackrel{?}{=} 7$	$-2 - 2 - 3 \stackrel{?}{=} -7$	$-6 + 3 \stackrel{?}{=} -3$
$7 = 7$	$-7 = -7$	$-3 = -3$

The solution checks.

Solving Systems of Three Equations to Model Quadratics

Matrices can be used to find the equation of a quadratic function in standard form. Recall that a quadratic function in standard form looks like $f(x) = ax^2 + bx + c$. The three unknowns are the coefficients a, b, and c. If we are given three points the function passes through, then we can substitute in those three points and find the values of a, b, and c. To find the quadratic function that passes through three points, use the following steps.

> **Steps to Find a Quadratic Function Given Three Points**
> 1. Write the function $f(x) = ax^2 + bx + c$ in the form $y = ax^2 + bx + c$.
> 2. Substitute each of the three given points in for x and y. This will result in three equations with the variables a, b, and c.
> 3. Write the equations as an augmented matrix.
> 4. Put the matrix in upper triangular form.
> 5. Back-solve to find the values of a, b, and c.
> 6. Write the quadratic function using the values from step 5.

Example 8 Finding a quadratic function using three points

Find the equation of the quadratic function that passes through the points $(1, -18)$, $(2, -14)$, and $(3, -8)$.

SOLUTION

First, substitute in the equation $y = ax^2 + bx + c$ the given points for x and y.

$$y = ax^2 + bx + c$$

Point	Substitute in the point	Simplify
$(1, -18)$	$-18 = a(1)^2 + b(1) + c$	$a + b + c = -18$
$(2, -14)$	$-14 = a(2)^2 + b(2) + c$	$4a + 2b + c = -14$
$(3, -8)$	$-8 = a(3)^2 + b(3) + c$	$9a + 3b + c = -8$

The system is

$$\begin{cases} a + b + c = -18 \\ 4a + 2b + c = -14 \\ 9a + 3b + c = -8 \end{cases}$$

Rewrite the system as an augmented matrix.

$$\begin{bmatrix} 1 & 1 & 1 & -18 \\ 4 & 2 & 1 & -14 \\ 9 & 3 & 1 & -8 \end{bmatrix}$$

Reduce the matrix to upper triangular form.

$$\begin{bmatrix} 1 & 1 & 1 & -18 \\ 4 & 2 & 1 & -14 \\ 9 & 3 & 1 & -8 \end{bmatrix}$$ Use the (1, 1) element to put 0's below it.

$$= \begin{bmatrix} 1 & 1 & 1 & -18 \\ -4(1)+4 & -4(1)+2 & -4(1)+1 & -4(-18)+-14 \\ 9 & 3 & 1 & -8 \end{bmatrix}$$ Multiply $-4 \cdot$ row 1, add to row 2, and overwrite row 2.

$$= \begin{bmatrix} 1 & 1 & 1 & -18 \\ 0 & -2 & -3 & 58 \\ 9 & 3 & 1 & -8 \end{bmatrix}$$ Multiply $-9 \cdot$ row 1, add to row 3, and overwrite row 3.

$$= \begin{bmatrix} 1 & 1 & 1 & -18 \\ 0 & -2 & -3 & 58 \\ -9(1)+9 & -9(1)+3 & -9(1)+1 & -9(-18)+-8 \end{bmatrix}$$

$$= \begin{bmatrix} 1 & 1 & 1 & -18 \\ 0 & -2 & -3 & 58 \\ 0 & -6 & -8 & 154 \end{bmatrix}$$ Use the (2, 2) element to put a 0 below it.

$$= \begin{bmatrix} 1 & 1 & 1 & -18 \\ 0 & -2 & -3 & 58 \\ 0 & -6 & -8 & 154 \end{bmatrix}$$ Multiply $-3 \cdot$ row 2, add to row 3, and overwrite row 3.

$$= \begin{bmatrix} 1 & 1 & 1 & -18 \\ 0 & -2 & -3 & 58 \\ 0 & -3(-2)+-6 & -3(-3)+-8 & -3(58)+154 \end{bmatrix}$$

$$= \begin{bmatrix} 1 & 1 & 1 & -18 \\ 0 & -2 & -3 & 58 \\ 0 & 0 & 1 & -20 \end{bmatrix}$$ This matrix is in upper triangular form.

Back-solving yields
$c = -20$
Rewrite the second row in equation form and substitute in $c = -20$.

$$-2b - 3c = 58$$
$$-2b - 3(-20) = 58$$
$$-2b + 60 = 58$$
$$\underline{ -60 \quad -60}$$
$$-2b = -2$$
$$b = 1$$

Rewrite the first row in equation form and substitute in $b = 1$ and $c = -20$.

$$a + b + c = -18$$
$$a + (1) + (-20) = -18$$
$$a - 19 = -18$$
$$\underline{+19 \quad +19}$$
$$a = 1$$

The quadratic function that passes through these three points is

$$y = ax^2 + bx + c$$
$$y = (1)x^2 + (1)x + (-20)$$
$$y = x^2 + x - 20$$
$$f(x) = x^2 + x - 20$$

This technique can be used to find a quadratic model from standard form. Substitute in the three data points into standard form, $y = ax^2 + bx + c$, and solve for a, b, and c as above.

B Exercises

For Exercises 1 through 6, check the solution to the system of three equations in three unknowns.

1. $\begin{cases} x + y + z = -2 \\ 2x - y + 4z = -17 \\ x + 3y - z = 18 \end{cases}$

   a. Is $(x, y, z) = (0, 5, 1)$ a solution of the system?
   b. Is $(x, y, z) = (0, 5, -3)$ a solution of the system?

2. $\begin{cases} -x + y + z = 1 \\ x + 2y - z = -7 \\ 3x - y + z = 9 \end{cases}$

   a. Is $(x, y, z) = (1, -2, 4)$ a solution of the system?
   b. Is $(x, y, z) = (0, 0, 4)$ a solution of the system?

3. $\begin{cases} x + y + z = 7 \\ -2x + 3y - 2z = -14 \\ -x - y + 3z = -19 \end{cases}$

   a. Is $(x, y, z) = (-5, 0, 1)$ a solution of the system?
   b. Is $(x, y, z) = (10, 0, -3)$ a solution of the system?

4. $\begin{cases} x + y - z = 4 \\ 2x + 3y + z = 5 \\ x + z = 2 \end{cases}$

   a. Is $(x, y, z) = (4, 1, 0)$ a solution of the system?
   b. Is $(x, y, z) = (3, 0, -1)$ a solution of the system?

5. $\begin{cases} x - y + z = -15 \\ y + z = -1 \\ 3x + 2y = 14 \end{cases}$

   a. Is $x = 16, y = 1, z = 0$ a solution of the system?
   b. Is $x = 0, y = 7, z = -8$ a solution of the system?

6. $\begin{cases} x - y + 2z = 4 \\ 3x + z = -13 \\ 5y - 6z = -30 \end{cases}$

   a. Is $x = 10, y = 6, z = 4$ a solution of the system?
   b. Is $x = -6, y = 0, z = 5$ a solution of the system?

For Exercises 7 through 14, solve the system using the elimination method without using matrices. Check the solution in the original equations.

7. $\begin{cases} x + y - z = 3 \\ x - y + 2z = 6 \\ -x - 2y + 3z = -1 \end{cases}$

8. $\begin{cases} x + 2y - z = -3 \\ -x - y + 4z = 1 \\ x - y + 3z = 3 \end{cases}$

9. $\begin{cases} x + y - 2z = -6 \\ -x + y + z = 7 \\ x - 2y + z = -3 \end{cases}$

10. $\begin{cases} x + y - z = -5 \\ -x + 2y + z = -1 \\ x - 2y + z = 7 \end{cases}$

11. $\begin{cases} x - y + 2z = 6 \\ 2x + y - z = -1 \\ -x - 2y + z = 3 \end{cases}$

12. $\begin{cases} x + y - 2z = -1 \\ -3x + 2z = 8 \\ 2x - 2y + z = -9 \end{cases}$

13. $\begin{cases} 2x + 2y - z = -5 \\ -x - y + z = 4 \\ 3x - y - 2z = -1 \end{cases}$

14. $\begin{cases} 2x + 5y - z = -11 \\ -x + 3y + z = 7 \\ -4x - y + 2z = 26 \end{cases}$

For Exercises 15 through 20, answer the questions about the given matrix.

15. $\begin{bmatrix} 3 & -1 \\ 4 & 7 \end{bmatrix}$

    a. What is the size of this matrix?
    b. What element is in the (1, 2) position?

16. $\begin{bmatrix} 5 & 9 \\ 0 & -3 \end{bmatrix}$

   a. What is the size of this matrix?
   b. What element is in the (2, 1) position?

17. $\begin{bmatrix} 4 & 0 & -2 \\ 3 & -1 & 5 \end{bmatrix}$

   a. What is the size of this matrix?
   b. What element is in the (1, 3) position?

18. $\begin{bmatrix} 9 & -3 & 7 & 1 \\ 0 & -2 & 6 & -3 \\ 1 & 1 & 5 & -4 \end{bmatrix}$

   a. What is the size of this matrix?
   b. What element is in the (3, 3) position?

19. $\begin{bmatrix} 1 & -2 & 6 \\ 0 & -3 & 2 \\ 0 & 1 & 1 \\ -3 & 2 & 5 \end{bmatrix}$

   a. What is the size of this matrix?
   b. What element is in the (4, 1) position?

20. $\begin{bmatrix} 6 & 3 & 2 \\ 1 & 0 & 0 \\ -3 & 0 & 6 \end{bmatrix}$

   a. What is the size of this matrix?
   b. What element is in the (3, 3) position?

For Exercises 21 through 28, write the augmented matrix for the given system. Do not solve the system.

21. $\begin{cases} 2x - y = -5 \\ 4x + 6y = -1 \end{cases}$

22. $\begin{cases} 5x - y = -1 \\ -x + 2y = 0 \end{cases}$

23. $\begin{cases} x - y = 0 \\ -3x + 4y = -1 \end{cases}$

24. $\begin{cases} -x - y = 4 \\ 2x - 7 = -1 \end{cases}$

25. $\begin{cases} -2x + 3y - z = 0 \\ x + y - z = 5 \\ x - y + z = 2 \end{cases}$

26. $\begin{cases} -3x + 5y - 2z = 0 \\ 4x - y - 3z = -7 \\ x - 6y + 3z = 12 \end{cases}$

27. $\begin{cases} -x + y + z = 3 \\ 3y - 2z = 1 \\ -x + 2z = 0 \end{cases}$

28. $\begin{cases} x + 3z = -1 \\ -x - 5y = 6 \\ 2y - z = -3 \end{cases}$

For Exercises 29 through 36, back-solve the given augmented matrix. Assume that the variables are x and y or x, y, and z.

29. $\begin{bmatrix} 3 & 1 & 5 \\ 0 & 2 & -4 \end{bmatrix}$

30. $\begin{bmatrix} 1 & -2 & 0 \\ 0 & 3 & 9 \end{bmatrix}$

31. $\begin{bmatrix} 4 & -3 & 7 \\ 0 & -1 & 5 \end{bmatrix}$

32. $\begin{bmatrix} 5 & -1 & 0 \\ 0 & 2 & 0 \end{bmatrix}$

33. $\begin{bmatrix} 1 & -1 & 0 & 2 \\ 0 & 2 & 1 & 3 \\ 0 & 0 & 1 & -2 \end{bmatrix}$

34. $\begin{bmatrix} 1 & 0 & -3 & 4 \\ 0 & 1 & -1 & 2 \\ 0 & 0 & 1 & 5 \end{bmatrix}$

35. $\begin{bmatrix} 2 & -1 & 1 & 7 \\ 0 & -2 & 1 & -4 \\ 0 & 0 & -3 & 15 \end{bmatrix}$

36. $\begin{bmatrix} 1 & 0 & 5 & -2 \\ 0 & -2 & 1 & -3 \\ 0 & 0 & -4 & 12 \end{bmatrix}$

For Exercises 37 through 44, put the given augmented matrix in upper triangular form.

37. $\begin{bmatrix} 1 & -1 & -5 \\ -2 & 1 & 7 \end{bmatrix}$

38. $\begin{bmatrix} 1 & -1 & 2 \\ -2 & 1 & -5 \end{bmatrix}$

39. $\begin{bmatrix} 4 & 3 & 7 \\ 4 & 11 & 24 \end{bmatrix}$

40. $\begin{bmatrix} 1 & -3 & -5 \\ 3 & -10 & 16 \end{bmatrix}$

41. $\begin{bmatrix} 1 & 1 & 1 & 2 \\ 2 & 3 & 2 & 3 \\ 0 & -2 & 2 & 6 \end{bmatrix}$

42. $\begin{bmatrix} 1 & 2 & 1 & -7 \\ 1 & 3 & 1 & -10 \\ -2 & -3 & 0 & 13 \end{bmatrix}$

43. $\begin{bmatrix} 3 & 0 & 2 & 9 \\ 4 & 1 & 2 & 8 \\ 3 & 2 & 1 & 2 \end{bmatrix}$

44. $\begin{bmatrix} 5 & 2 & 1 & 1 \\ 3 & 0 & 1 & -4 \\ 5 & 1 & 1 & -3 \end{bmatrix}$

For Exercises 45 through 54, solve each system using augmented matrices. Check the solution in the original equations.

45. $\begin{cases} 2x + y = -1 \\ 4x + y = -7 \end{cases}$

46. $\begin{cases} x - 5y = 22 \\ 3x + y = 2 \end{cases}$

47. $\begin{cases} -2x + 3y = 24 \\ 6x + y = -12 \end{cases}$

48. $\begin{cases} 3x + 6y = 21 \\ -2x + 5y = -14 \end{cases}$

49. $\begin{cases} x - 4y + 6z = -3 \\ -x + 5y - 2z = -1 \\ 2x + y - z = 7 \end{cases}$

50. $\begin{cases} -x + 2y + z = -4 \\ 3x - y + 2z = 17 \\ x + 3y + 2z = -3 \end{cases}$

51. $\begin{cases} 4x + y - z = -1 \\ -2x - y + z = -1 \\ -x + 7y - 2z = 7 \end{cases}$

52. $\begin{cases} 2x - y + 3z = 5 \\ -3x + y - 3z = -11 \\ -4x + 3y - z = -3 \end{cases}$

53. $\begin{cases} 2x - y - 2z = 4 \\ -x + 5y = 26 \\ -y + 4z = -10 \end{cases}$

54. $\begin{cases} x - 3y + z = -12 \\ 3x + 4y = -3 \\ 2x - 3z = -16 \end{cases}$

For Exercises 55 through 62, find the equation of the quadratic function in standard form that passes through the given points.

55. The three points are $(-1, -6)$, $(1, 2)$, and $(2, 3)$.

56. The three points are $(-2, -5)$, $(-1, 0)$, and $(0, 3)$.

57. The three points are $(1, 10)$, $(2, 18)$, and $(3, 28)$.

58. The three points are $(-2, -3)$, $(-1, 0)$, and $(1, 12)$.

59. The three points are $(1, 4)$, $(2, 15)$, and $(3, 30)$.

60. The three points are $(-2, -6)$, $(-1, -15)$, and $(1, -21)$.

61. The three points are $(-1, -5)$, $(1, 15)$, and $(2, 16)$.

62. The three points are $(-1, -40)$, $(1, 18)$, and $(2, 35)$.

Using the Graphing Calculator

APPENDIX C

These instructions are based on the Texas Instruments 83/84 family of calculators. Most of the instructions given here will also work with the TI-82 calculator.

Basic Keys and Calculations

When you are doing basic calculations, a graphing calculator works much the same as a scientific calculator. The home screen is a blank screen where calculations are performed. The following table introduces some of the important keys used for calculations in this text.

Key	What It Does
ENTER	Tells the calculator to do the calculation. Acts like an equals key on most scientific calculators.
÷ × − +	Performs the basic operations of division, multiplication, subtraction and addition.
(−)	Used for negative numbers. Using the subtraction key for a negative number will result in a syntax error. See the error messages later in this appendix.
x^2	Squares the expression immediately before it. To square more than one thing, parentheses must be used. *Note:* -5^2 is different from $(-5)^2$. $$-5^2$$ (−) 5 x^2 ENTER $$(-5)^2$$ ((−) 5) x^2 ENTER Display: $-5^2 = -25$; $(-5)^2 = 25$
^	Raises the expression immediately before it to the power you choose. To raise more than one thing to a power, parentheses must be used. $$3^7$$ 3 ^ 7 ENTER Display: $3^7 = 2187$
2nd	Acts like a shift key to access things typed above the keys in blue or yellow.
2nd $\sqrt{}$	Using 2nd and then x^2 results in a square root. The calculator will automatically give a left parenthesis. Close with a right parenthesis when you are done entering the radicand. $$\sqrt{25}$$ 2nd $\sqrt{}$ 2 5) ENTER Display: $\sqrt{(25)} = 5$

Continued

Key	What It Does
2nd (-) [ANS ?]	Using 2nd and then (-) results in the answer feature. The answer feature uses the previous result in the next calculation. See the long calculations in the next section.
DEL	Deletes *one piece* of an entry at a time.
CLEAR	Clears (deletes) an *entire* entry at a time. Pressing CLEAR twice on the home screen will clear all previous entries on the screen.
2nd MODE [QUIT]	Using 2nd and then MODE will exit from any screen and return to the home screen where calculations are done.
MATH	Brings up the math menu that can be used for many operations. 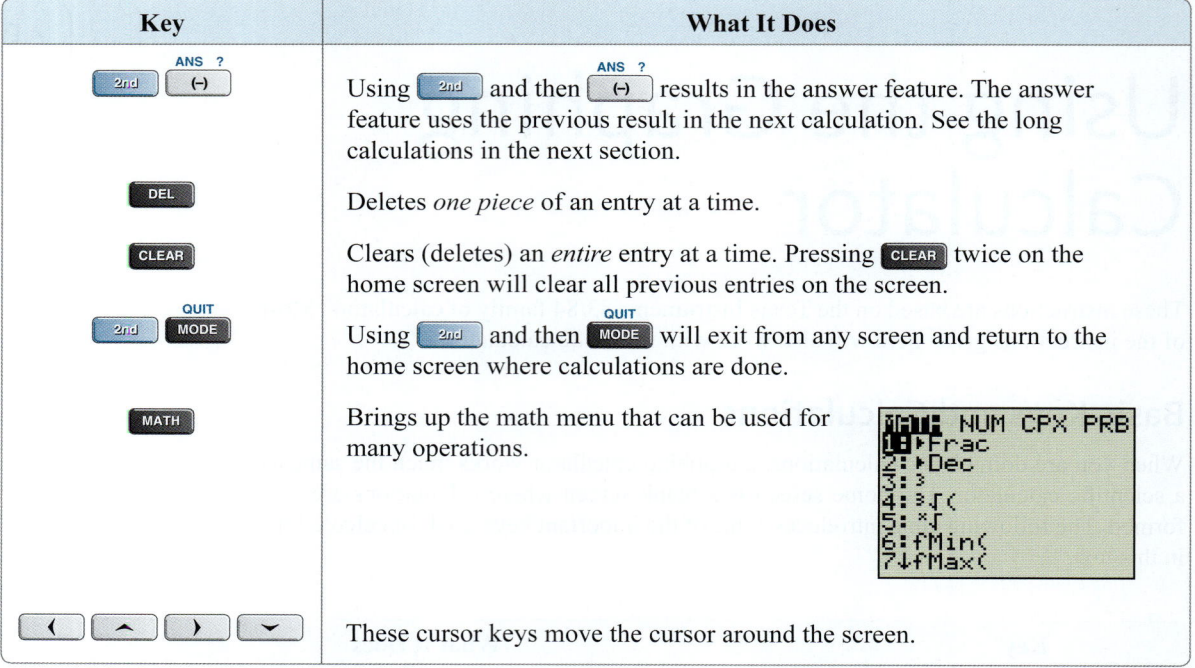
◄ ▲ ► ▼	These cursor keys move the cursor around the screen.

Additional keys and menus will be introduced throughout this appendix.

Long Calculations

When entering a long expression, use parentheses to force the calculator to correctly use the order of operations. Another option is to do calculations in parts, while using the ANS key to keep values that have already been calculated.

Option 1: All at once.

$$2000\left(1 + \frac{0.05}{12}\right)^{12(5)}$$

[2] [0] [0] [0] [( ] [1] [+] [0] [.] [0] [5] [÷]
[1] [2] [ )] [^] [( ] [1] [2] [( ] [5] [ )] [ )] [ENTER]

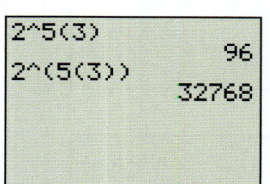

Notice the extra parentheses around the exponent (12(5)). Without the extra set of parentheses, the calculator would raise the base to the 12th power and then multiply the result by 5. This would give an incorrect result. Compare these two calculations: one with parentheses and the other without.

$$2^{5(3)} = 32768$$

Option 2: Using [2nd] [(-)] [ANS ?]. Here, the [(-)] [ANS ?] will keep the value of 60 from the first calculation and use it in the second. It will then take the result from the second calculation to multiply by 2000 in the final step.

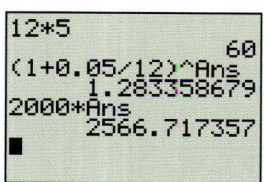

[2nd] [(-)] [ANS ?]

$$2000\left(1 + \frac{0.05}{12}\right)^{12(5)}$$

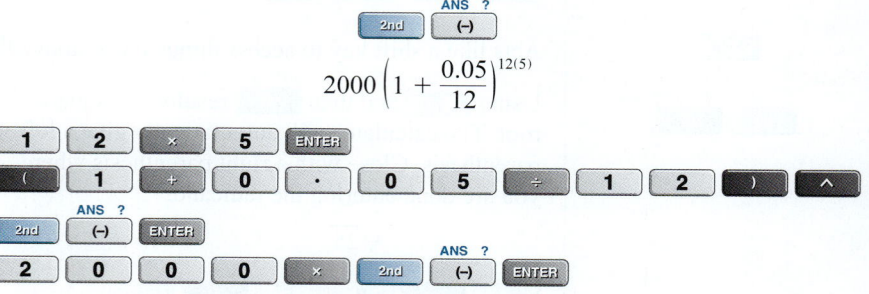

The ANS key recalls the previous numerical answer from the last line.

Complex Number Calculations   C-3

Converting Decimals to Fractions

To convert a decimal result to a reduced fraction, use the math menu and the first option is ▶Frac.

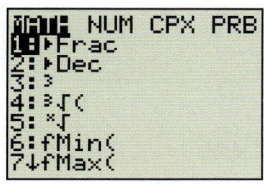

Without ▶Frac, the result is a decimal approximation.
With ▶Frac, the result is given as a fraction.

$$\frac{1}{5} + \frac{7}{3}$$

Convert 0.2685 to a fraction.

Absolute Values

To use an absolute value in a calculation or equation, use the NUM menu found within the math menu.

Use the cursor to highlight the NUM option in the math menu. **abs(** is the first option, so press [1]. Be sure to finish off the parentheses in the calculation.

$$|5 - 24|$$

Complex Number Calculations

In Chapter 8, complex numbers will be introduced. To perform arithmetic with complex numbers, use [2nd] [.]. The i on top of the [.] key represents the imaginary number.

Arithmetic with complex numbers.

$$(2 + 5i) + (8 - 9i)$$

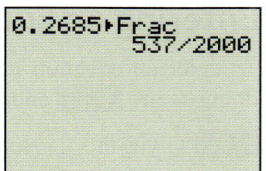

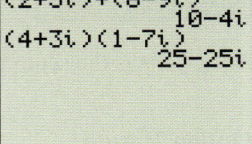

$$(4 + 3i)(1 - 7i)$$

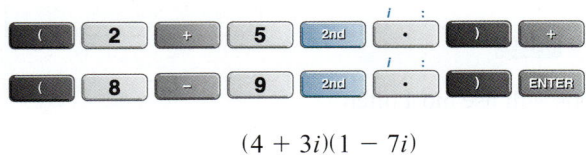

To get a complex number result from taking the square root of a negative number, the calculator must be in complex mode. Change the calculator into complex mode using the mode menu [MODE].

Calculations in complex mode.

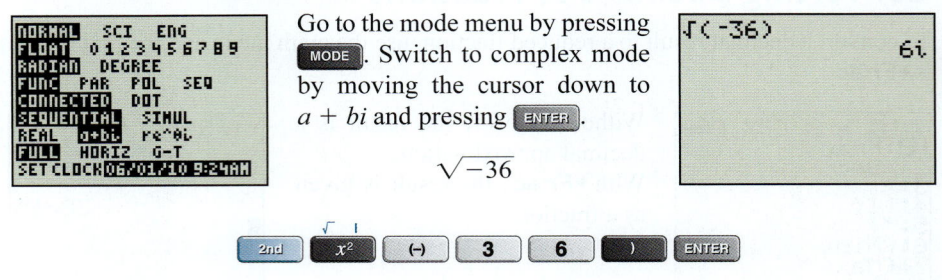

Go to the mode menu by pressing [MODE]. Switch to complex mode by moving the cursor down to $a + bi$ and pressing [ENTER].

$$\sqrt{-36}$$

Entering an Equation

To enter an equation into the calculator, use the Y= screen. The important keys for doing so are [Y=] and [X,T,θ,n].

[Y=] Gives the Y= screen where equations are entered.

[X,T,θ,n] Enters the variable x into the equation. The other variables listed on this key will not be used in this textbook.

Enter the equation $y = 2x + 5$:

[Y=] [2] [X,T,θ,n] [+] [5]

To enter more than one equation, use the [▼] key to move to the $Y_2 =$ line, and enter the second equation.

Note: The equations must have the y-variable isolated before they are entered into the calculator.

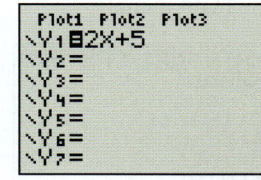

Using the Table Feature

The table feature can be used to evaluate an equation for several x-values. This can be useful to check answers, look for patterns, or solve an equation numerically.

Set up the table feature using [2nd] [WINDOW] (TBLSET F2).

Option 1: Ask
Have the calculator ask for x-values and give the y-values for the equation entered. Select the following by highlighting them with a black box.

Indpnt: Ask   Depend: Auto

Note: This is the setup the textbook will use most often.

Option 2: Auto
Have the calculator automatically fill up the table with equally spaced x-values and the resulting y-values. Select the following by highlighting them with a black box.

Indpnt: Auto   Depend: Auto

Note: This table with start at $x = 0$ and increase by 1's, giving 0, 1, 2, 3, ...

To start at a different x-value, change TblStart to a different number. To change the increments, change ΔTbl.

TblStart = 0   ΔTbl = 1   TblStart = 15   ΔTbl = 0.2

Access the table feature using [2nd] [GRAPH] (TABLE F5).

Enter the equation $y = 2x + 5$:

[Y=] [2] [X,T,Θ,n] [+] [5]

Open the table: [2nd] [GRAPH] (TABLE F5)

With table set for Ask, enter values for x, and press [ENTER].

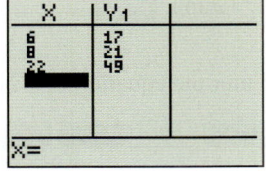

[6] [ENTER] [8] [ENTER] [2] [2] [ENTER]

Setting the Window

When setting the window, determine what region of the rectangular coordinate system will be displayed. To open the window settings screen, press [WINDOW].

```
WINDOW
 Xmin=-10
 Xmax=10
 Xscl=1
 Ymin=-10
 Ymax=10
 Yscl=1
 Xres=1
```

The standard window settings view from −10 to 10 on both the x- and y-axes.

Xmin   Sets the left-hand side (minimum x-value) of the viewing window.
Xmax   Sets the right-hand side (maximum x-value) of the viewing window.
Xscl   Sets the scale (space between tick marks) for the x-axis.
Ymin   Sets the bottom (minimum y-value) of the viewing window.
Ymax   Sets the top (maximum y-value) of the viewing window.
Yscl   Sets the scale (space between tick marks) for the y-axis.

Use the cursor keys to move the cursor to each setting, and type in the new value for that setting.

To set a window for given data, choose values as follows:

Xmin   A slightly smaller value than the smallest input (x-value)
Xmax   A slightly larger value than the largest input (x-value)
Xscl   A scale for the x-axis. One possibility is to use $\text{Xscl} \approx \frac{|\text{Xmax} - \text{Xmin}|}{20}$. The scale can typically be left as 1.
Ymin   A slightly smaller value than the smallest output (y-value)
Ymax   A slightly larger value than the largest output (y-value)
Yscl   A scale for the y-axis. One possibility is to use $\text{Yscl} \approx \frac{|\text{Ymax} - \text{Ymin}|}{20}$. The scale can typically be left as 1.

If you are given an equation and you do not know what window to set, use the table feature to evaluate the equation for several values of x, and use the resulting y-values to set an approximate window. Adjust the window if needed, once you see the portion of the graph shown. The standard window can be set by using the Zoom features of the calculator. These features are discussed later in this appendix.

Graphing a Function

Graphing an equation requires three steps:

1. Enter the equation in the Y= screen: [Y=]
2. Set an appropriate window: [WINDOW]
3. Graph: [GRAPH]

Enter the equation $y = 2x + 5$:

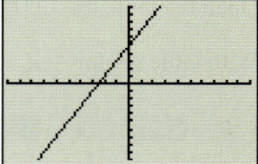

Set an appropriate window [WINDOW]. A standard window should work well for this equation.

Graph the function by pressing [GRAPH].

Tracing a Graph

The TRACE feature [TRACE], follows the curve and gives the coordinates of points along the curve.

Using TRACE [TRACE]

Graph the equation $y = 3x - 6$:

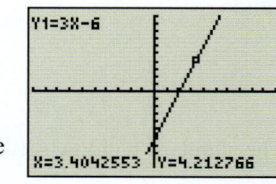

Press [TRACE], and use [◄] or [►] to move along the curve left or right, respectively.

The values of points along the curve will be displayed at the bottom on the window.

You can also press [TRACE] then enter a value for x, and the calculator will jump to that point on the graph and display the x- and y-values. Entering a value directly works only if the value is between Xmin and Xmax (i.e., can be seen in the window).

Graphing a Scatterplot

To create a scatterplot requires four steps:

1. Enter the data: [STAT]
2. Set an appropriate window: [WINDOW]
3. Turn on the STAT PLOT: [Y=] (STAT PLOT F1)
4. Graph: [GRAPH]

Since a scatterplot is a statistical plot of data, use the STAT menu to work with data.

Pressing the [STAT] key will give the STAT menu. The features of this menu that will be used most often are as follows:

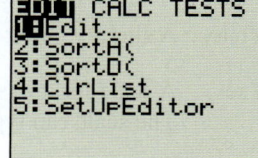

1: Edit   To input data into lists.
5: SetUpEditor   To restore or bring back any accidentally deleted lists.

Step 1: Enter the data into lists

Open the lists screen by selecting 1: Edit:

STAT 1

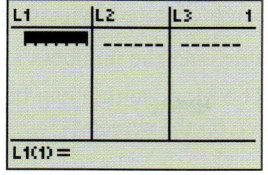

To empty old data from a list, use the ▲ to move the cursor onto the title of the list L1 and press CLEAR ENTER. Repeat this for each list that needs to be emptied of old data. Avoid using DEL when clearing lists. If you press DEL, the entire list will disappear. The list can be brought back by using SetUpEditor from the main STAT menu.

To enter the given data, move the cursor to the list where the data should be entered, and type the coordinates of the data values one at a time, pressing ENTER after each one.

x	y
-2	5
4	9
10	-3
17	-15

With the cursor in L1,

(−) 2 ENTER 4 ENTER 1
0 ENTER 1 7 ENTER

Move the cursor over to L2, using ▶.

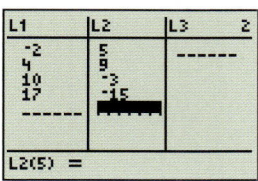

5 ENTER 9 ENTER (−) 3 ENTER (−) 1 5 ENTER

Note: Use the (−) key, not the subtraction key, for negatives.

It is important that each list have the same number of data entered. If the numbers of data in each list are not equal, an error message will be given when you are creating a scatterplot of the data.

Step 2: Setting the window

Open the window settings screen: WINDOW

Xmin The smallest x-value is -2. Something slightly smaller is $x = -4$.

Xmax The largest x-value is 17. Something slightly larger is $x = 20$.

Xscl 1 seems like a reasonable scale for the x-axis.

Ymin The smallest y-value is -15. Something slightly smaller is $y = -17$.

Ymax The largest y-value is 9. Something slightly larger is $y = 11$.

Yscl 1 seems like a reasonable scale for the y-axis.

Step 3: Set up and turn on STAT PLOT

Statplots are set up and turned on by using the STAT PLOT menu: 2nd Y=.

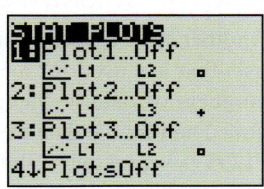

Here, you can tell that there are three plots. Any number of these can be turned on at a time, but only one is needed for this scatterplot.

Enter the setup screen for Plot1 by pressing 1.

To be on, the word "On" must have the black box around it. If it does not have this black box, move the cursor over the word On and press ENTER.

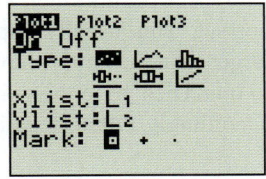

For a scatterplot, be sure that the first option after the word Type is selected with a black box. The Xlist and Ylist should be the same as the lists in which the data were entered.

If these lists are not correct, change them. The default list names are L1 and L2, which can be entered by using 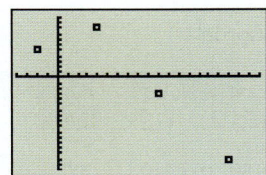 and . The other lists are also available over the numbers 3, 4, 5, and 6.

The Mark setting can be set to any of the three options, but the single dot is harder to see on the screen.

Step 4: Graph the scatterplot

Now that the data have been entered, an appropriate window has been set, and the STAT PLOT has been turned on, simply press GRAPH.

Graphing an Inequality with Shading

To graph inequalities, the calculator can be set to shade above or below the curve.

Enter the equation $y \leq x^2 - 5x - 3$:

Move the cursor to the left of the Y1, and press ENTER until the shading desired is shown. In this case, shading is below the curve.

Set an appropriate window WINDOW. A standard window should be all right for this equation.

Graph the function by pressing GRAPH.

Error Messages

ERR: INVALID DIM

This error occurs when a STAT PLOT is turned on but the stat lists are empty. To fix this error, press 1 to quit. Then either enter data into the lists or turn off the STAT PLOT if you are not graphing a scatterplot.

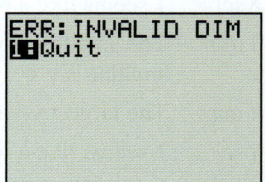

ERR: DIM MISMATCH

Another error related to stat plots. This error will occur if the lists that are used for a STAT PLOT do not have the same numbers of data in them. To fix this error, press 1 to quit. Then fix the data in the stat lists. Be sure that the STAT PLOT is set up to use the same lists that the data are in.

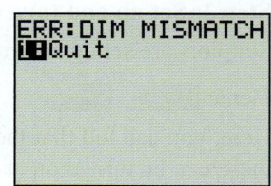

ERR: WINDOW RANGE

This error occurs when a setting in the window set up is not correct. This usually happens when the Xmin or Ymin is larger than the Xmax or Ymax. To fix this error, press 1 to quit. Then go to the window screen and adjust the settings accordingly. Often, negative numbers get entered incorrectly.

ERR: SYNTAX

This error occurs when you have entered something into the calculator incorrectly. The most common mistake is to use (-) instead of - or vice versa. Other reasons would be misplacement of commas or parentheses.

To fix syntax errors, press 2 to go to the error. The error will be highlighted with the cursor. In this case, the cursor is flashing on the – because the (-) key should have been used instead of -.

ERR: DIVIDE BY 0

This error occurs whenever a calculation results in division by zero. To fix this error, press 2 to go to the error. The error will be highlighted with the cursor.

ERR: NONREAL ANS

This error occurs when a calculation results in a nonreal number. This happens most often when the square root of a negative number is taken. To fix this error, press 2 to go to the error. The error will be highlighted with the cursor.

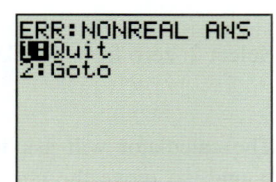

Additional Features (check with your instructor if you are allowed to use these in your class)

Zooming to an Appropriate Window

The ZOOM menu helps find an appropriate window for graphs and scatterplots. ZOOM
The main zoom options that are useful in this textbook are as follows:

5: ZSquare	Square window so the graph does not appear stretched.	
6: ZStandard	Standard window from -10 to 10 on both axes.	
9: ZoomStat	Appropriate window for any stat plot that is turned on.	

Using ZoomStat

Enter the data.	Turn on STAT PLOT	ZoomStat
Step 1: STAT 1	Step 2: 2nd Y= 1 (STAT PLOT F1)	Step 3: ZOOM 9 (ZoomStat)

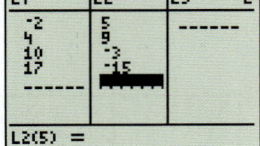

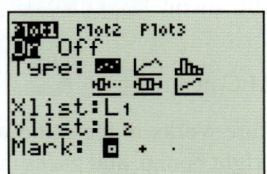

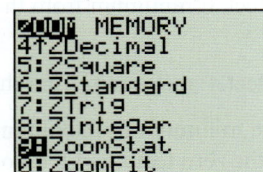

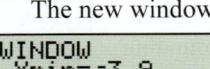

The new window   The final graph

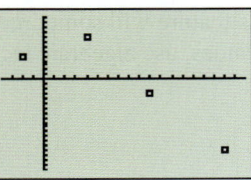

Zero, Minimum, Maximum, and Intersect Features

When you are graphing, several features in the CALC menu help you to find special points on the graph. [2nd] [TRACE] (CALC F4)

2: zero   Finds the nearest horizontal intercept (zero) of the curve.

3: minimum   Finds the minimum value of the function within an interval.

4: maximum   Finds the maximum value of the function within an interval.

5: intersect   Finds the intersection point of two curves.

Using the Zero Feature

Select an appropriate window, and graph the equation $y = 0.25x^2 + 0.5x - 6$:

Select 2: zero from the CALC menu. [2nd] [TRACE] [2] (CALC F4)

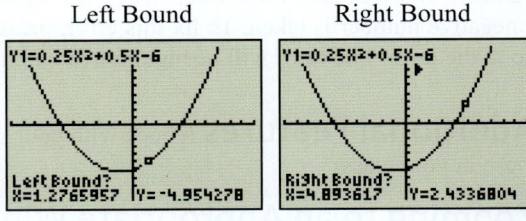

Left Bound | Right Bound

The calculator will ask for a left bound, so move the cursor to the left of the desired zero, and press [ENTER]. The calculator will ask for a right bound, so move the cursor to the right of the desired zero, and press [ENTER].

Guess | Zero

The calculator will ask for a guess, so move the cursor close to the desired zero, and press [ENTER]. The zero is shown at the bottom of the screen.

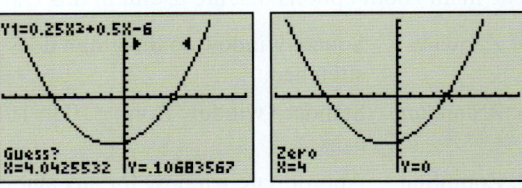

Repeat this process for any other zeros. This will not always work. If a zero only touches the x-axis but does not cross it, the calculator will give back a "no sign change" error message.

Using the Minimum or Maximum Feature

Select an appropriate window, and graph the equation $y = 0.25x^2 + 0.5x - 6$.

Select 3: minimum from the CALC menu. [2nd] [TRACE] [3] (CALC F4)

or

Select 4: maximum from the CALC menu. [2nd] [TRACE] [4] (CALC F4)

The minimum or maximum feature works in the same way as the zero feature. A left bound, right bound, and guess must be provided, as with the process for finding the zeros. The minimum or maximum point will then display at the bottom of the screen. This feature will sometimes result in rounding error. For exact values, use algebraic techniques.

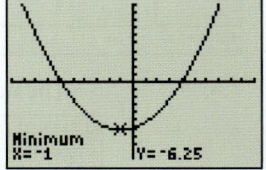

Using the Intersect Feature

Select an appropriate window, and graph the two equations. Enter one in Y1 and the other in Y2.

$$y = 0.25x^2 + 0.5x - 6 \qquad y = 2x - 3$$

Select 5: intersect from the CALC menu: [2nd] [TRACE] [5]

The calculator will ask for a first curve, so use the up or down cursor to select one of the curves, and press [ENTER]. The calculator will ask for a second curve, so select the other curve, and press [ENTER]. The calculator will ask for a guess, so move the cursor close to the intersection, and press [ENTER]. The intersection point is shown at the bottom of the screen.

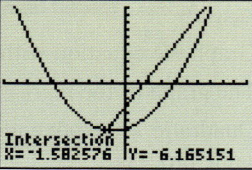

Regression

The statistical process for finding a model for data is called *regression*. Regression finds the regression (best fit) curve for a set of data. The calculations that are involved take into account the differences between each point and the regression curve and minimize these differences over all the data. Although these calculations can be tedious and long, the calculator has a regression feature that will quickly calculate the regression curve for data entered into the stat lists.

The regression feature is located in the CALC portion of the STAT menu that we used earlier when creating scatterplots. [STAT] [▶]

The regression features used in this textbook include the following:

    4: LinReg(ax+b)   Finds a linear regression equation.

    5: QuadRed   Finds a quadratic regression equation.

    0: ExpReg   Finds an exponential regression equation.

Calculating a Regression Equation

Enter the data into the STAT lists: [STAT] [1] (Edit)

Turn on a STAT PLOT: [2nd] [Y=] [1] (Plot1)

Graph the data using [ZOOM] [9] (ZoomStat).

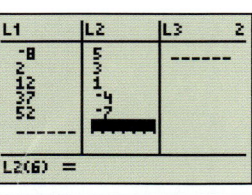

Decide the type of model that should fit the data. In this case, linear seems appropriate.

Choose the linear regression option in the STAT CALC menu:

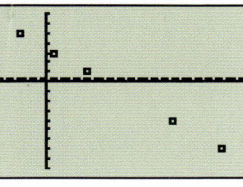

[STAT] [▶] [4] (LinReg(ax+b)).

If the data are in lists L1 and L2, simply press [ENTER].

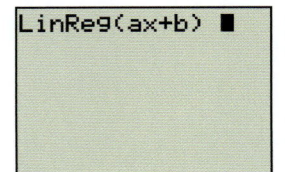 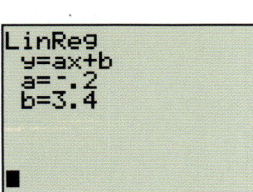

To enter specific lists in which the data are located and to put the equation into the Y= screen so that it can be graphed, enter the list names and the equation name following the LinReg(ax+b):

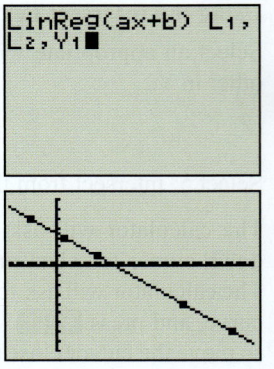

STAT ▶ 4 2nd 1 (L1) , 2nd 2 (L2) , VARS ▶ 1 1 ENTER

Graph the equation with the data: GRAPH

The graph of the regression curve fits the points nicely.

Quadratic or exponential regressions both work with the same process.

Answers to Practice Problems

APPENDIX D

CHAPTER 1

Section 1.1

*PP for Example 1.

a. An 8-mile. Taxi ride costs $21.30.

b. $m \approx 13.708$. For $35, you can take a taxi ride of up to 13 miles.

PP for Example 2.

a. $v =$ Number of patient visits in a month at this chiropractor's office
$C =$ Monthly cost, in dollars, for v patient visits at this chiropractor's office
$$C = 15v + 8000$$

b. $C = 9500$. This chiropractor's office has a total monthly cost of $9500 if they have 100 patient visits during the month.

c. $R =$ Monthly revenue, in dollars, for this chiropractor's office
$$R = 80v$$

d. $P =$ Monthly profit, in dollars, for this chiropractor's office
$$P = 65v - 8000$$

e. $P = 1750$. This chiropractor's office has a profit of $1750 in a month with 150 patient visits.

f. $200 = v$. For this chiropractor's office to have a monthly profit of $5000, they would need to have 200 patient visits during the month.

PP for Example 3.

a. $x = 116$

b. $x \approx -3.652$

PP for Example 4.

a. $\dfrac{v}{G} = t$

b. $\dfrac{v - v_o}{t} = a$

*PP = Practice Problem

Section 1.2

PP for Example 1.

There are many possible correct answers; only one of these possibilities is given here.

a. $t =$ Time in years since 2000
$P =$ Population of Arizona (in millions)

b.

t	Population of Arizona (millions)
0	5.17
1	5.30
2	5.44
3	5.58
4	5.74
5	5.94
6	6.17

c.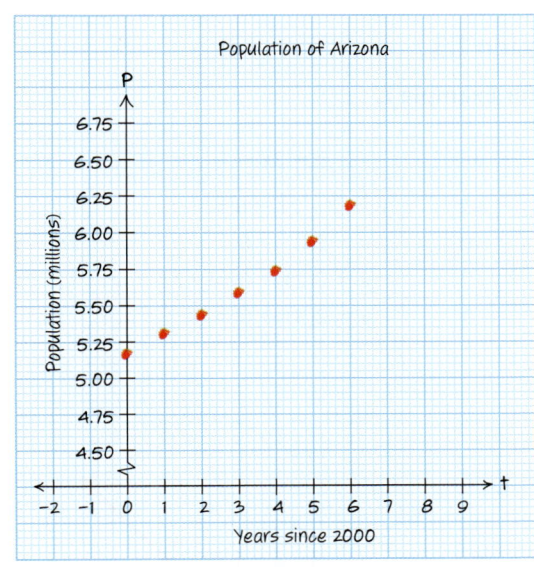

D-1

PP for Example 2.

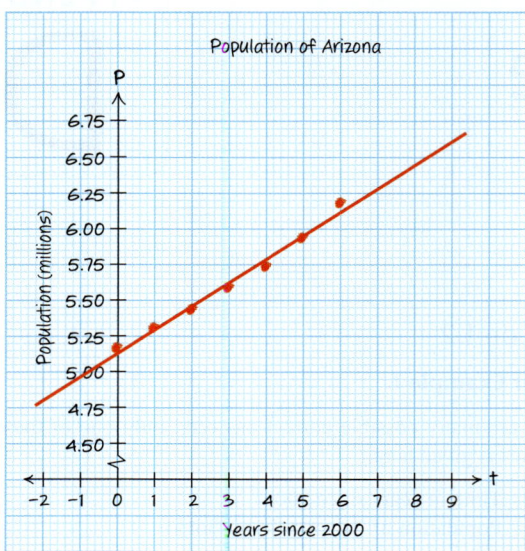

PP for Example 3.

a. y-intercept: $(0, -4)$.

b. x-intercept: $(-7, 0)$.

c. The line has a y value of -10 when $x = 10$.

d. The line has a y value of 5 when $x = -15$.

PP for Example 4.

Domain $[-2, 8]$ or $-2 \leq t \leq 8$, range $[4.76, 6.45]$ or $4.76 \leq P \leq 6.45$.

Section 1.3

PP for Example 1.

a.

x	$y = 2x - 6$	(x, y)
-2	$y = 2(-2) - 6 = -10$	$(-2, -10)$
0	$y = 2(0) - 6 = -6$	$(0, -6)$
2	$y = 2(2) - 6 = -2$	$(2, -2)$
3	$y = 2(3) - 6 = 0$	$(3, 0)$

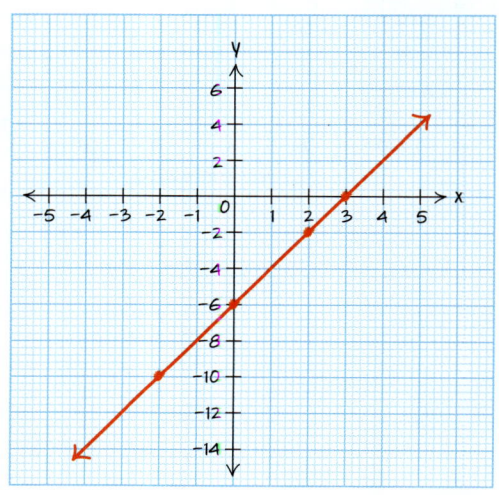

b.

x	$y = x^2 - 8$	(x, y)
-2	$y = (-2)^2 - 8 = -4$	$(-2, -4)$
-1	$y = (-1)^2 - 8 = -7$	$(-1, -7)$
0	$y = (0)^2 - 8 = -8$	$(0, -8)$
1	$y = (1)^2 - 8 = -7$	$(1, -7)$
2	$y = (2)^2 - 8 = -4$	$(2, -4)$

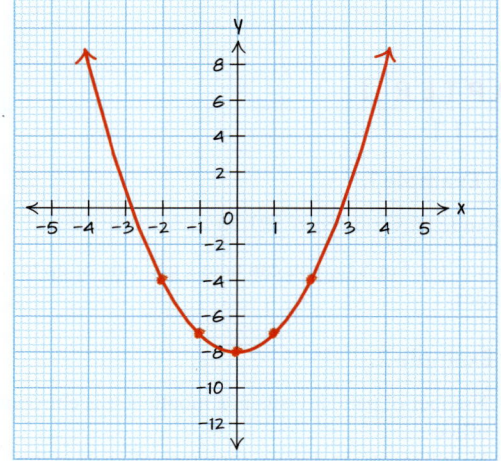

PP for Example 3.

Slope $= -\dfrac{5}{2}$. The graph is decreasing.

PP for Example 4.

Slope $= -8$

PP for Example 5.

a. The slope between each pair of points is 2, so these points all lie on the same line.

b. The slope between each pair of points is not always the same, so these points will not all lie on the same line.

PP for Example 6.

a. Slope $= -3$, y-intercept: $(0, 7)$.

b. Slope $= \dfrac{2}{7}$, y-intercept: $(0, -8)$.

c. Slope is $\dfrac{3}{4}$, y-intercept: $\left(0, -\dfrac{15}{4}\right)$

PP for Example 7.

a. Slope $= -0.07 = \dfrac{-0.07}{1} = \dfrac{-0.07 \text{ psi}}{1 \text{ second}}$. This means that the pressure inside the vacuum chamber is decreasing by 0.07 psi per second.

b. Slope $= 0.55 = \dfrac{0.55}{1} = \dfrac{\$0.55}{1 \text{ taco}}$. Therefore, the cost for making tacos at a local street stand is increasing by $0.55 (55 cents) per taco made.

PP for Example 8.

a.
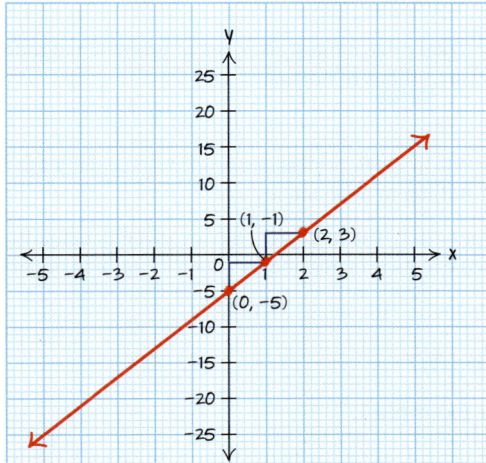

b.
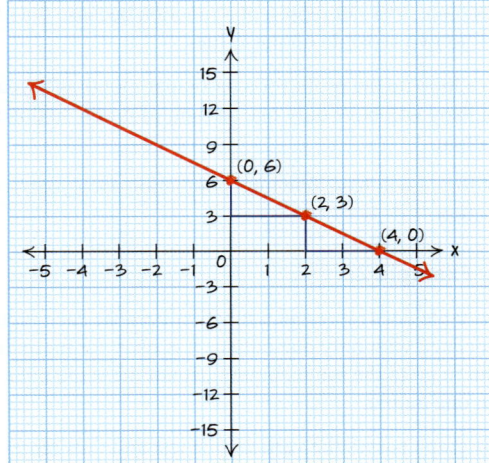

Section 1.4

PP for Example 1.

a. $2x - y = -10$ b. $7x - 14y = 6$

PP for Example 2.

a. The vertical intercept is $(0, 35)$. This means that after zero seconds, the vacuum chamber has a pressure of 35 psi.

The horizontal intercept is $(500, 0)$. This intercept means that after 500 seconds, the vacuum chamber will have a pressure of zero psi.

b. The vertical intercept is $(0, 140.00)$. This intercept means that if the local stand makes no tacos, their cost is still $140.00.

The horizontal intercept is $(-254.55, 0)$. This intercept means that if the neighborhood stand makes negative 255 tacos, their cost will be zero dollars. This is model breakdown, because they cannot make a negative number of tacos.

PP for Example 3.

The horizontal intercept is $(5, 0)$.
The vertical intercept is $(0, 20)$.

PP for Example 4.

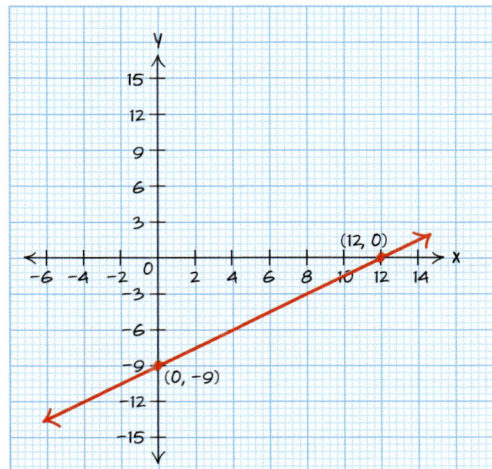

PP for Example 5.

a.
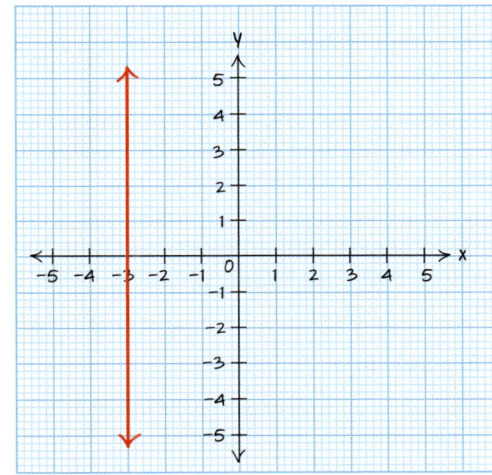

b.
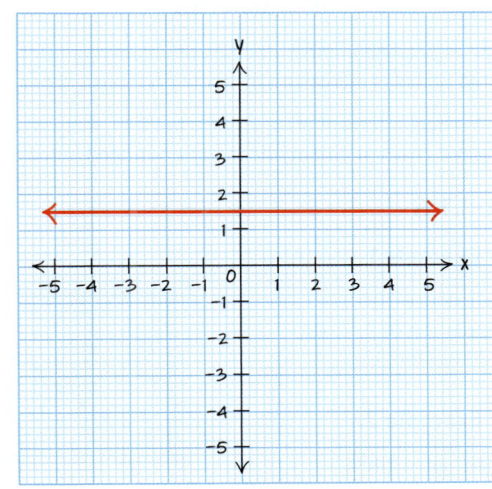

Section 1.5

PP for Example 1.

$$y = \frac{1}{5}x + 3$$

APPENDIX D Answers to Practice Problems

PP for Example 3.

H = Number of sports cards hobby stores
t = Years since 1995

$$H = -300t + 4500$$

PP for Example 4.

$$y = 2x + 2$$

PP for Example 5.

$$y = -\frac{5}{3}x + \frac{23}{3}$$

PP for Example 6.

a. $y = 4x - 11$ b. $y = \frac{5}{2}x - 13$

PP for Example 8.

a. Slope $= m = -300 = \dfrac{-300}{1} = \dfrac{-300 \text{ stores}}{1 \text{ year}}$. The slope means that the number of sports cards hobby stores is decreasing by about 300 per year.

b. The vertical intercept is $(0, 4500)$. This intercept means that in 1995, year zero, there were about 4500 sports cards hobby stores.

c. The horizontal intercept is $(15, 0)$, which implies that there were no more sports cards hobby stores in 2010. This is model breakdown, since there are still sports cards hobby stores today.

PP for Example 9.

a. C = Thousands of cases of chlamydia infection in the United States
t = Years since 2000

$$C = 56t + 709.5$$

b. Slope $= 56$. The number of cases of chlamydia infection in the United States is increasing by about 56 thousand per year.

c. The vertical intercept is $(0, 709.5)$. In 2000, there were 709.5 thousand cases of chlamydia infection in the United States.

Section 1.6

PP for Example 1.

t = Years since 2000.
P = Population of Arizona (in thousands)

PP for Example 2.

t = Years since 2000.
P = Population of Arizona (in thousands)

$$P = 196.14t + 5046.72$$

PP for Example 3.

a. t = Years since 2000

P = Total prize money given out at professional rodeo events in millions of dollars

$$P = 1.1t + 31.1$$

b. $P = 41$. In 2009, professional rodeos gave out a total of about $41 million in prize money.

c. Slope $= 1.1$. The total prize money given out at profession rodeo events is increasing by about 1.1 million dollars per year.

d. Domain $[0, 10]$, range $[31.1, 42.1]$.

Section 1.7

PP for Example 1.

a. The table does represent a function, since each input relates to a single output.

b. This table does not represent a function because the name John goes to two separate output values.

c. If we look up the profit for a particular month of a year, the company will have only one amount of profit, so this is a function.

d. California will have one total population each year, so this is a function.

PP for Example 2.

a. This table does represent a function because each input is associated with only one output. It is okay for more than one input to go to the same output, just not the same input to more than one output.

b. This graph passes the vertical line test and thus represents a function.

c. This graph does not pass the vertical line test, since almost every vertical line drawn would hit the graph in more than one place.

d. This equation is a function, since any input value u will result in only one output value C.

e. This equation does not represent a function, since the plus/minus symbol ($\pm$) will give two outputs for almost any input given.

PP for Example 3.

a. 2500 is the value of the input variable, and thus the number of miracle mops produced, and 189 is the cost in hundreds of dollars. Therefore, a good interpretation of this notation is "The cost of producing 2500 miracle mops is $18,900."

b. In 2010, the population of Michigan was 10.4 million people.

PP for Example 5.

a. $f(3) = -7$ b. $x = 3$

PP for Example 7.

a. Domain: all real numbers or $(-\infty, \infty)$, range: all real numbers or $(-\infty, \infty)$.

b. Domain: all real numbers or $(-\infty, \infty)$, range is $\{-8\}$.

c. Domain: all real numbers or $(-\infty, \infty)$, range: all real numbers or $(-\infty, \infty)$.

CHAPTER 2

Section 2.1

PP for Example 1.

a.

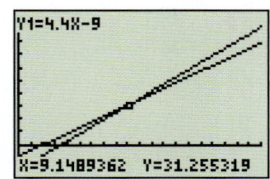

b. $(9, 31.26)$. American male smokers have the same risk of dying from a heart attack as they do of dying from lung cancer at about age 49. At age 49, about 31 male smokers out of every 1000 male smokers are expected to die in the next 10 years from a heart attack, and about 31 of every 1000 are expected to die from lung cancer.

PP for Example 2.

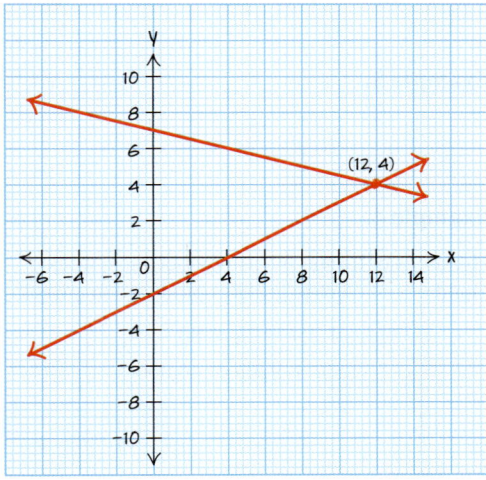

The solution to the system is the point $(12, 4)$.

PP for Example 4.

a. $(6, 4)$ b. $(-3, 10.8)$

PP for Example 5.

a. This is an inconsistent system and has no solutions.

b. This is a dependent system and has an infinite number of solutions. Any solution to the equation $y = \frac{1}{4}x + 7$ is a solution to the system.

Section 2.2

PP for Example 2.

$t = 31.6$. New Zealand will spend about 41.6 million New Zealand dollars both on diabetes research and on diabetes treatment in 2021. This is probably not reasonable because the prediction is so far into the future.

PP for Example 3.

a. $(10, 42)$

b. The values $n = 25$ and $m = 5.5$ are the answer to the system.

PP for Example 5.

George needs to deposit \$125,000 in the 3.6% account and \$225,000 into the account paying 6% interest.

PP for Example 6.

a. Inconsistent system with no solutions.

b. Dependent system with an infinite number of solutions. Any solution to the equation $x = -\frac{7}{6}y + \frac{2}{3}$ will be a solution to the system.

Section 2.3

PP for Example 2.

Jim should use 150 pounds of Diamond Pet Foods Maintenance Formula and 75 pounds of Diamond Pet Foods Professional Formula to make the 225 pounds of dog food with 24% protein.

PP for Example 3.

a. $(-3, -4)$ b. $d = -6$ and $g = 9$

PP for Example 4.

a. $\left(\frac{2}{3}, \frac{3}{7}\right)$

b. Inconsistent system with no solutions.

Section 2.4

PP for Example 1.

a. $x > -\frac{1}{2}$ b. $h \geq 4$ c. $m > -54$

PP for Example 2.

Shannon made a mistake on line 5 when she reversed the symbol while dividing by a positive number. The inequality symbol should be reversed only when multiplying or dividing by a negative number. Tom did the problem correctly.

PP for Example 4.

a. t = Time in years since 1980
$M(t)$ = Percentage of college freshmen who are male
$F(t)$ = Percentage of college freshmen who are female

$$M(t) = -0.152t + 48.6$$
$$F(t) = 0.152t + 51.4$$

b. $M(t) \geq F(t)$ where $t \leq -9.21$. This answer means that in about 1971 and before, the percentage of freshmen who were male was greater than or equal to the percentage of freshmen who were female.

PP for Example 5.
$$x < 3$$

PP for Example 6.
$$x \leq 4$$

Section 2.5

PP for Example 1.

a. $x = 9$ and $x = -5$

b. There is no real solution to this equation.

c. $x = 4$ and $x = -6$

PP for Example 2.

Svetlana will be about 25 miles from Columbus, Texas, after driving 1.5 hours and again after 2.3 hours. The second time is after she has passed Columbus and is 25 miles beyond the town.

PP for Example 3.
$$-6 < x < 9$$

PP for Example 4.

a. $6 < x < 10$

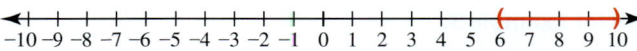

b. $-9 \leq d \leq 7$

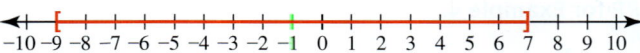

PP for Example 5.

$4.1245 \leq L \leq 4.1255$. The rod's length must be between 4.1245 inches and 4.1255 inches.

PP for Example 6.
$$x < 1 \quad \text{or} \quad x > 5$$

PP for Example 7.

a. $x < -9$ or $x > 5$

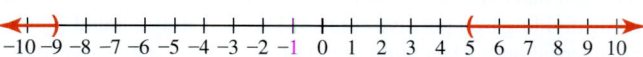

b. $g \leq -2$ or $g \geq 8$

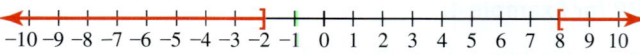

PP for Example 8.

$h < 23.18$ or $h > 40.82$. Watching TV less than 23.18 hours a week or more than 40.82 hours a week would be considered unusual.

Section 2.6

PP for Example 2.

a.

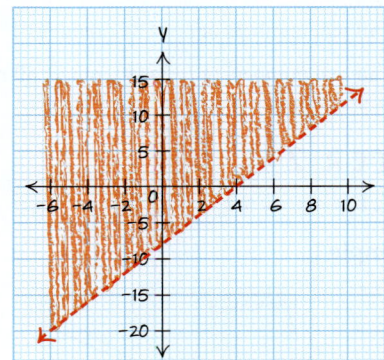

b.

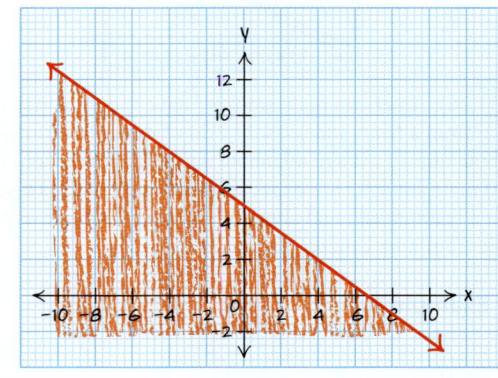

PP for Example 3.
$$y > \frac{5}{4}x - 2$$

PP for Example 4.

a. $t =$ Hundreds of toaster ovens produced in a week
$g =$ Hundreds of electric griddles produced in a week

$$1.25t + 2g \leq 80$$
$$850t + 920g \leq 40{,}000$$

b.

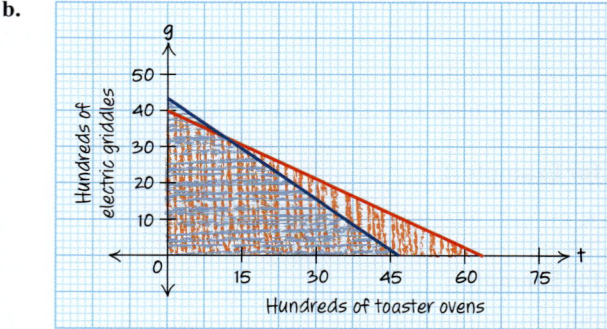

c. The plant can produce 1600 toaster ovens and 2000 electric griddles in a week.

d. Under these constraints, the plant cannot produce 2000 toaster ovens and 3000 electric griddles in a week.

PP for Example 5.

A = Amount in the account paying 6.5% annual interest
B = Amount in the account paying 4.5% annual interest

$$A + B \leq 800{,}000$$
$$0.065A + 0.045B \geq 40{,}500$$

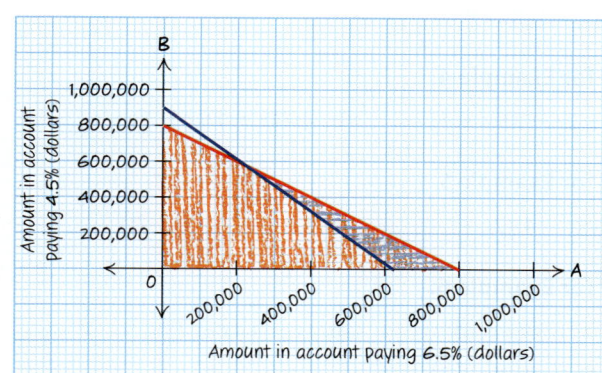

Any combination of investments from the overlapping section of the graph can be used to meet Don's investment goal.

PP for Example 6.

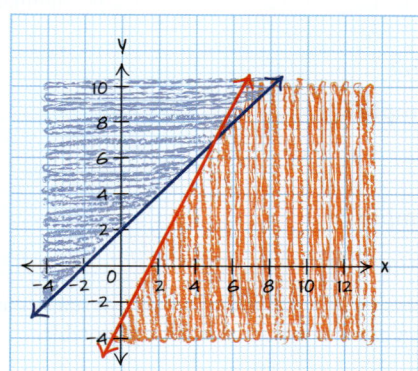

PP for Example 7.

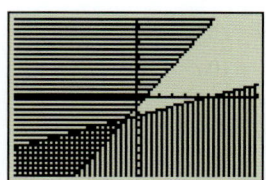

CHAPTER 3

Section 3.1

PP for Example 1.
a. m^9 b. $32w^7 x^{11}$

PP for Example 2.
a. $ab^6 c$ b. $8t^8 w^5$ c. $\dfrac{12b^8 c}{7} = \dfrac{12}{7} b^8 c$

PP for Example 3.
a. $27x^{15} y^6 z^3$ b. $\dfrac{16 m^2 p^6}{25} = \dfrac{16}{25} m^2 p^6$

PP for Example 4.
a. $24 x^{11} y^8$ b. $\dfrac{8 m^3 n^2}{5} = \dfrac{8}{5} m^3 n^2$ c. $m^9 n^2$

PP for Example 5.
a. $\dfrac{5 d^4}{c^3}$ b. $\dfrac{6 x^3 z^6}{5 y^2}$ c. $\dfrac{-2 h^7}{g^7}$

PP for Example 6.
a. $\dfrac{49 x^4 y}{125}$ b. $\dfrac{g^{18}}{8 h^{30}}$ c. 1

PP for Example 7.
a. $\sqrt[9]{g^4}$ b. $\sqrt[10]{m^7}$

PP for Example 8.
a. $(7a)^{\frac{1}{2}}$ b. $t^{\frac{3}{4}}$

PP for Example 9.
a. $2ab^2$ b. $\dfrac{4n}{m}$

Section 3.2

PP for Example 2.
a. The expression $4^w + 5$ is not a polynomial because the variable is in the exponent.

b. The expression $4\sqrt{t} + 20t$ is not a polynomial because the square root of the variable t is the same as a fraction exponent $\dfrac{1}{2}$.

c. The expression $8x - 9$ is a polynomial.

PP for Example 3.
a. $-0.23x^2$ has degree 2, $4x$ has degree 1, and -3 has degree zero. The polynomial expression has degree 2.

b. $8m^3 n$ has degree 4, $6mn^2$ has degree 3, and $-2n^3$ has degree 3. The polynomial expression has degree 4.

c. $-9s^5 t^4 u^2$ has degree 11, and $-7s^3 t^2 u^2$ has degree 7. The polynomial expression has degree 11.

PP for Example 4.
a. $R(10) = 5988.8$

$P(10) = 2247.2$

Thus, in 2010, the revenue for the California lottery was about $5988.8 million, and the profits were about $2247.2 million.

b. Let $C(t)$ be the costs for the California lottery in millions of dollars t years since 2000. $C(t) = 36.6 t^2 - 178.4 t + 1865.6$

c. $C(15) = 7424.6$. In 2015, the cost for the California lottery is approximately $7424.6 million.

PP for Example 7.
a. $f(x) + g(x) = 11x - 3$ b. $g(x) - f(x) = -3x - 13$

PP for Example 8.

a. $F(t) \cdot I(t)$ = Total amount spent on health insurance in dollars for Ford Motor Company employees in year t

b. $F(t) - M(t)$ = Number of nonmanagement employees at Ford Motor Company in year t

c. $\dfrac{V(t)}{(F(t) - M(t))}$ = Average cost for vacations in dollars per nonmanagement employee at Ford Motor Company in year t

PP for Example 9.

a. $S(45) = 23.7$. At a selling price of $45,000, this car manufacturer is willing to supply 23,700 SUVs.

b. Let $R(p)$ represent the revenue in millions of dollars from selling SUVs at a price of p thousand dollars. $R(p) = 0.008p^3 - 0.5p^2 + 30p$

c. $R(40) = 912$. At a price of $40,000, the car manufacturer can expect a revenue of about $912 million.

PP for Example 10.

a. $f(x)g(x) = 28x^2 - 23x - 15$
b. $\dfrac{g(x)}{h(x)} = \dfrac{4x - 5}{9x + 2}$

PP for Example 11.

a. $7x^2y + 3xy - 9y + 10$

b. $3m^4n^2 - 13m^2n + 5mn - 4n^2$

c. $63x^2 - 78x - 45$

d. $8x^4 - 2x^3 + 13x^2 + 35x$

Section 3.3

PP for Example 2.

a. $P(5) = 9.4$. This urban area had a population of 9.4 million people in 2005.

b. $A(9.4) \approx 41.51$. This urban area had air pollution levels of 41.51 parts per million in 2005.

c. $a(t) = A(P(t))$ = Amount of air pollution in parts per million t years since 2000. $a(t) = A(P(t)) = 18.576t - 51.368$

d. $a(5) \approx 41.5$, $a(7) \approx 78.7$, $a(10) \approx 134.4$. The amount of air pollution in this urban area was 41.5 ppm in 2005, 78.7 ppm in 2007, and 134.4 ppm in 2010.

PP for Example 3.

a. $f(g(x)) = 20x - 27$
b. $(h \circ g)(x) = 50x^2 - 140x + 95$

PP for Example 4.

a. $C(S(h)) = -0.0272h^2 + 0.7548h - 3.8168$

b. $C(S(15)) = 1.3852$. At 3:00 P.M., the CO concentration at Old Faithful was approximately 1.39 parts per million.

Section 3.4

PP for Example 1.

a. $4t(3t^2 - 5)$
b. $2(3x^2 + 4x - 7)$

c. $(x + 4)(5x - 3)$

PP for Example 2.

a. $(4x - 3)(3x - 5)$
b. $(2m + 5n)(4m - 7)$

PP for Example 3.

a. $(x - 8)(x + 6)$
b. $(5x + 4)(2x + 3)$

c. $(x + 5y)(2x + y)$

PP for Example 4.

a. $3(2x - 3)(x + 4)$
b. $4x(4x + 1)(x + 3)$

PP for Example 5.

a. $(5x - 4)(2x - 7)$
b. $5x(6x - 1)(3x - 2)$

PP for Example 6.

a. $(x + 3)(x + 4)$
b. $(x - 4)(x - 7)$

c. $(3x - 4)(5x + 2)$

PP for Example 7.

a. $2(2x^2 - 10x + 3)$

b. The polynomial $g^2 + 52g - 4$ is a prime polynomial.

Section 3.5

PP for Example 1.

a. $(x + 4)^2$
b. $4(3b - 1)^2$
c. $(r - 5t)^2$

PP for Example 2.

a. $(x + 6)(x - 6)$
b. $(4a + 9b)(4a - 9b)$

c. The expression $m^2 + 25$ is the sum of two squares and is therefore a prime polynomial and not factorable.

PP for Example 3.

a. $(x + 2)(x^2 - 2x + 4)$

b. $(x - 3y)(x^2 + 3xy + 9y^2)$

c. $2(5a - 2b)(25a^2 + 10ab + 4b^2)$

PP for Example 4.

$(t + 2)(t^2 - 2t + 4)(t - 2)(t^2 + 2t + 4)$

PP for Example 5.

$(3a^5 + 7)(a^5 - 4)$

CHAPTER 4

Section 4.1

PP for Example 2.

a. This curve increases for $x < 1$ and decreases for $x > 1$.

b. Vertex = $(1, 8)$

c. x-intercept(s): $(-1, 0)$ and $(3, 0)$

d. y-intercept: $(0, 6)$

e. $f(2) = 6$

f. The output of the function is $y = -10$ when the input is about $x = -2$ and $x = 4$.

PP for Example 4.

a.
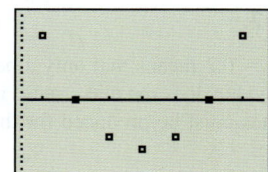

A quadratic function would be used to model the distribution. The vertex is the minimum point at approximately $(5, -8)$.

b.
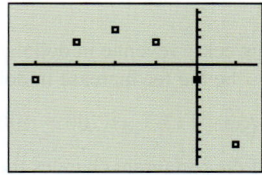

A quadratic function would be used to model it. The vertex is the maximum point at $(-2, 4)$.

Section 4.2

PP for Example 1.

$$f(x) = (x - 4)^2 + 6$$

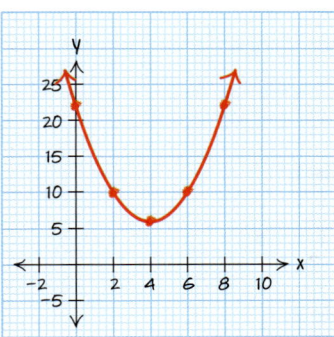

PP for Example 2.

$$f(x) = -0.5(x + 2)^2 + 4$$

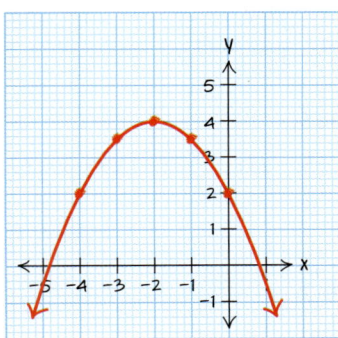

PP for Example 3.

$$f(x) = -x^2 + 8$$

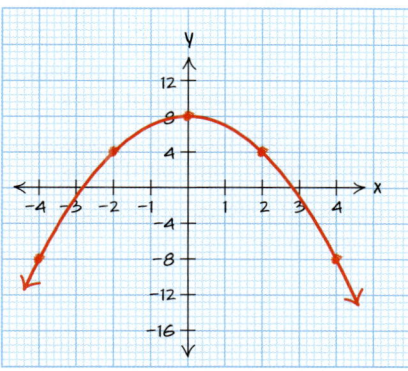

PP for Example 4.

$$C(t) = 833.53(t + 1)^2 + 3200$$

a. $C(10) \approx 104{,}057$. There were approximately 104,057 thousand cellular telephone subscribers in 2000. This can also be stated as 104 million cellular telephone subscribers.

b.
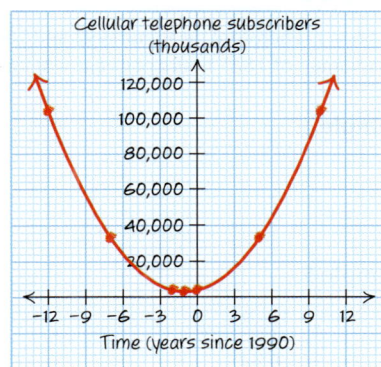

c. The vertex is the lowest point, so the number of cellular telephone subscribers was the least in 1989.

d. $C(-5) \approx 16{,}536$. This answer does not make sense, since there were only 3200 thousand in 1989, so there would not be more than that in 1985.

e. Domain: $[-1, 10]$, range: $[3200, 104{,}057]$.

PP for Example 5.

a. Domain: All real numbers or $(-\infty, \infty)$, range: $y \leq 15$ or $(-\infty, 15]$.

b. Domain: All real numbers or $(-\infty, \infty)$, range: $y \geq 4$ or $[4, \infty)$.

c. Domain: All real numbers or $(-\infty, \infty)$, range: $y \leq -8.6$ or $(-\infty, -8.6]$.

Section 4.3

PP for Example 1.

a. $C(t) =$ Number of cable television systems in the United States
$t =$ Time in years since 1990

$$C(t) = -33.3(t - 4)^2 + 11250$$

b. The vertex of (4, 11,250) represents when the number of cable television systems reached a maximum. Specifically, in 1994, there were the most cable television systems, with a total of 11,250 systems.

c. $C(9) = 10417.5$. In 1999, there were approximately 10,418 cable television systems in the United States.

d. $C(0) = 10717.2$. In 1990, there were approximately 10,717 cable television systems in the United States.

e. In the years 1989 and 1999, there were about 10,500 cable television systems in the United States.

PP for Example 2.

The parabola should be facing downward instead of upward so make the *a*-value negative. $f(x) = -0.5(x - 3)^2 + 6$

PP for Example 4.

$$f(x) = (x + 2)^2 - 49$$

Section 4.4

PP for Example 2.

$12.44 = t$, $1.56 = t$. Therefore, in about 1992 and 2002, people were spending an average of 1825 hours per year watching television.

PP for Example 3.

a. $x = \pm 4$

b. $x = 3, \quad x = \dfrac{1}{2}$

PP for Example 5.

a. $x = 2, \quad x = -10$

b. $x \approx 4.56, \quad x \approx 0.44$

c. No real answers.

PP for Example 6.

a. $f(x) = 0.25(x + 20)^2 - 103$

b. $h(x) = -3\left(x - \dfrac{2}{3}\right)^2 + \dfrac{34}{3}$

PP for Example 7.

$$f(x) = 2(x + 1)^2 - 32$$

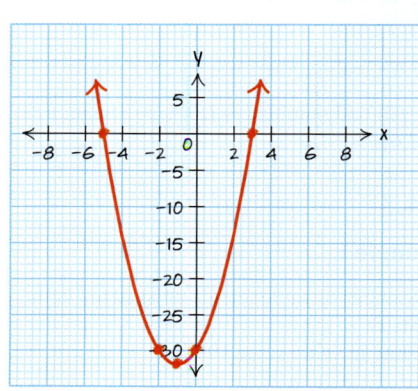

Domain: all real numbers, range: $y \geq -32$ or $[-32, \infty)$.

Section 4.5

PP for Example 1.

a. $x = -9 \quad x = -4$ **b.** $x = 0 \quad x = -3 \quad x = \dfrac{-5}{2}$

PP for Example 2.

$p = 0.2$, $p = 6$. $p = 0.2$ means that only a part of a plane should be produced. This value does not make sense in this context, so we know that six planes must be produced for the average cost to be $200,000 per plane.

PP for Example 3.

a. The weekly revenue from valet parking will be $2400 if the charge is $4.

b. Let $R(x)$ be the weekly revenue in dollars from the valet service, and let x be the fee increase in dollars.

$$R(x) = (3 + x)(700 - 100x)$$

c. $R(2) = 2500$. If the restaurant changes the fee for valet parking to $5, weekly revenue will be about $2500.

d. $R(3) = 2400$. If the restaurant changes the fee for valet parking to $6, weekly revenue will be about $2400.

e. From the answers to parts (a) to (d), it seems that a $5 fee brings in the highest weekly revenue from valet parking.

PP for Example 4.

$$y = \dfrac{5}{3}x^2 - \dfrac{5}{3}x - 10$$

Section 4.6

PP for Example 1.

a. $x = \dfrac{-3}{2} \quad x = -5$ **b.** $x = \dfrac{-1}{3} \quad x = -4$

PP for Example 2.

$t \approx 9.87$, $t \approx 3.66$. In about 1994 and 2000, there were 30,000 people in Hawaii who were unemployed.

PP for Example 3.

a. $x = 0 \quad x = -\dfrac{4}{3} \quad x = \dfrac{5}{2}$ **b.** $x = \pm 4$

PP for Example 5.

$(-5, -24) \quad (-13.33, 392.67)$

Section 4.7

PP for Example 1.

a. $f(x) = 4x^2 + 40x + 7$
Vertical intercept = $(0, 7)$.
Vertex = $(-5, -93)$. The *a*-value of this quadratic is positive, so this parabola faces upward. This vertex is a minimum point.

b. $k(x) = -2(x-3)^2 + 12$
Vertical intercept = $(0, -6)$.
Vertex = $(3, 12)$. The *a*-value of this quadratic is negative, so this parabola faces downward. This vertex is a maximum point.

c. $h(t) = -7t^2 - 15t - 6$
Vertical intercept = $(0, -6)$.
Vertex = $(-1.07, 2.04)$. The *a*-value of this quadratic is negative, so this parabola faces downward. This vertex is a maximum point.

PP for Example 2.

a. $f(x) = 0.25x^2 + 6x - 15$

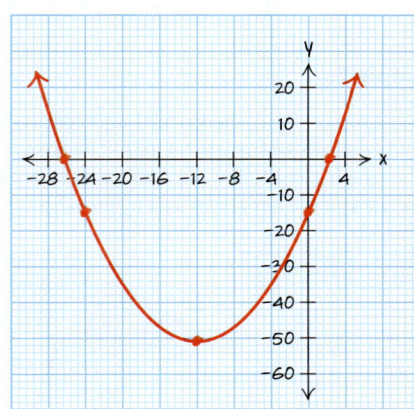

b. $f(x) = 3x^2 + 6x + 7$

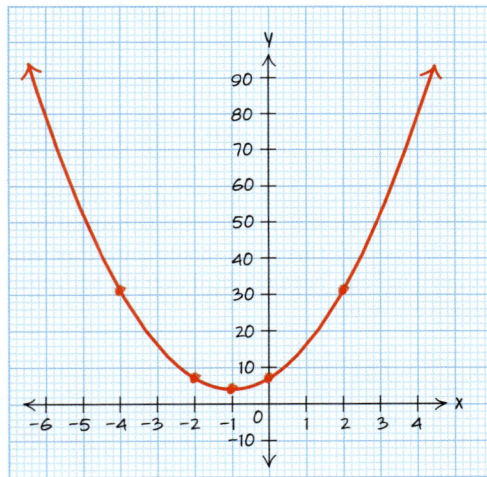

PP for Example 3.

a. $S(15) = 3728.9$. Therefore, Gateway's net sales in 1995 were approximately $3728.9 million.

b. $t \approx 18.86$, $t \approx -2.99$. Therefore, in 1999, Gateway had net sales of about $9000 million. At $t = -2.99$, model breakdown occurs because in 1977, Gateway was a new company and could not have made that much in net sales.

c. The vertex is $(7.9, -53.7)$, which represents the lowest net sales for Gateway, Inc. The negative net sales represents model breakdown in this context.

PP for Example 4.

a. $y > 2x^2 - 12x - 8$

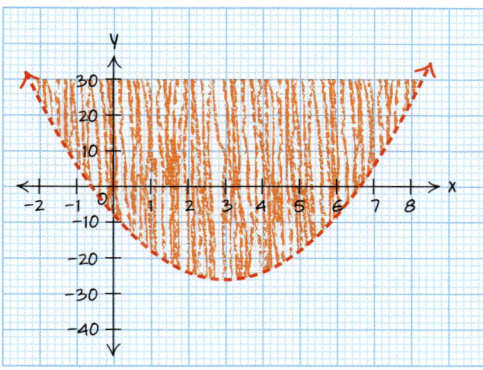

b. $y \leq -0.25x^2 - 3x + 5$

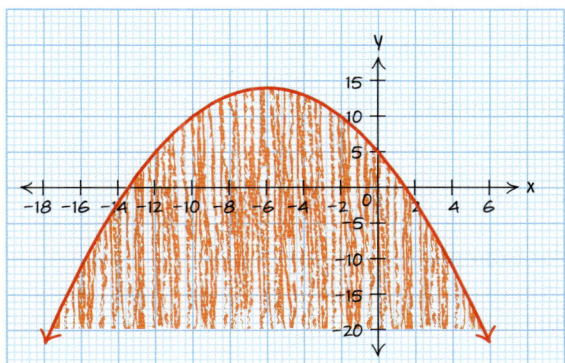

CHAPTER 5
Section 5.1

PP for Example 1.

a. $B(w)$ = Balance on the loan in dollars
w = Weeks after taking out the loan
$$B(w) = 5000(2)^w$$

b. $B(4) = 80000$. Thus, if you keep the $5000 for 4 weeks, you will owe this loan shark $80,000.

c. After keeping the money for 8 weeks, we would owe the loan shark about 1.28 million dollars.

PP for Example 2.

a. $P(t)$ = World population in billions of people
t = Time in years after 2000
$$P(t) = 6(2)^{\frac{t}{50}}$$

b. $P(500) = 6144$. This means that if the world population were to grow in ideal circumstances, the population in the year 2500 would be 6144 billion. This seems to be unrealistic and can be considered model breakdown.

c. If this exponential trend continues the world population will reach 10 billion in the year 2037.

PP for Example 3.

a. $P(d)$ = Percent of a thorium-234 sample remaining
   d = Time in days since the sample was taken
   $$P(d) = 100\left(\frac{1}{2}\right)^{\frac{d}{24.5}}$$

b. $P(180) = 0.614$. After 180 days, only 0.614% of the thorium-234 sample will remain.

c. It will take about 32 days for only 40% of the thorium-234 sample to remain.

PP for Example 5.

a. $f(x) = 10(3.1)^x$ b. $f(x) = 16(0.9)^{\frac{x}{3}}$

Section 5.2

PP for Example 1.

a. $x = 5$ b. $t = 6$ c. $x = -2$

PP for Example 2.

a. $x = 3$

b. $x = -3$

PP for Example 3.

$w = 5$. If you do not make any payments for 5 weeks, you will owe the loan shark $160,000.

PP for Example 4.

a. $x = 3$ b. $p \approx \pm 3.81$ c. $m = 3.4$

PP for Example 5.

a. The equation $5x^8 - 6 = 28$ is a power equation.

b. The equation $3(8^x) - 20 = 40$ is an exponential equation, because the variable is in the exponent.

Section 5.3

PP for Example 1.

a. $f(x) = 8(1.25)^x$. $a = 8$, the vertical intercept is $(0, 8)$, the graph will be above the horizontal axis. The base $b = 1.25$ is > 1, so this graph is increasing, showing exponential growth.

x	$f(x)$
-3	4.1
0	8
3	15.6
6	30.5
9	59.6

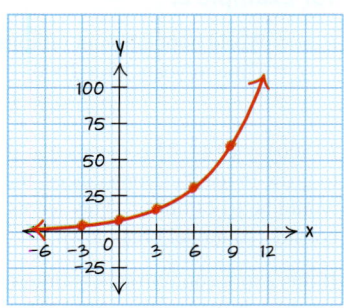

b. $g(x) = 300(0.5)^x$. $a = 300$, the vertical intercept is $(0, 300)$, the graph will be above the horizontal axis. The base $b = 0.5$ is < 1. The graph is decreasing, showing exponential decay.

x	$f(x)$
-1	600
0	300
1	150
3	37.5
6	4.7

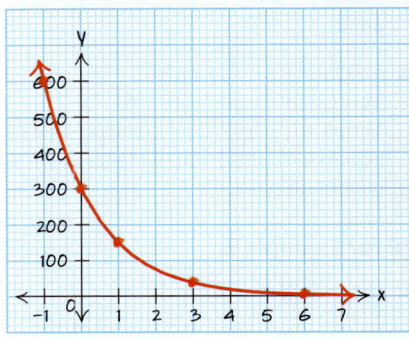

PP for Example 2.

a. $f(x) = -2(2.5)^x$. $a = -2$, the vertical intercept is $(0, -2)$, the graph will be below the horizontal axis. The base $b = 2.5$ is > 1. This graph should show exponential growth, but it has also been reflected over the horizontal axis by the negative a value.

x	$f(x)$
-1	-0.8
0	-2
2	-12.5
4	-78.13

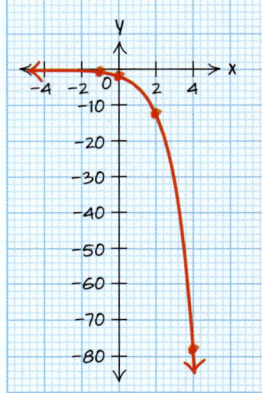

b. $h(x) = -145(0.5)^x$. $a = -145$, the vertical intercept is $(0, -145)$, the graph will be below the horizontal axis. The base $b = 0.5$ is < 1. This graph should show exponential decay, but it has also been reflected over the horizontal axis by the negative a value.

x	$f(x)$
-2	-580
-1	-290
0	-145
1	-72.5
2	-36.25
3	-18.13

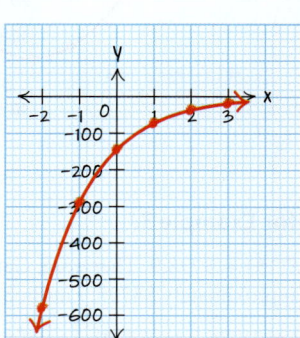

PP for Example 3.

a. Domain: All real numbers or $(-\infty, \infty)$, range: All negative real numbers or $(-\infty, 0)$.

b. Domain: All real numbers or $(-\infty, \infty)$, range: All positive real numbers or $(0, \infty)$.

PP for Example 4.

a. $f(x) = 100(0.2)^x + 20$
Horizontal asymptote $y = 20$.
Domain: All real numbers or $(-\infty, \infty)$,
range: $(20, \infty)$.

x	$f(x)$
-1	520
0	120
1	40
2	24
3	20.8
4	20.16

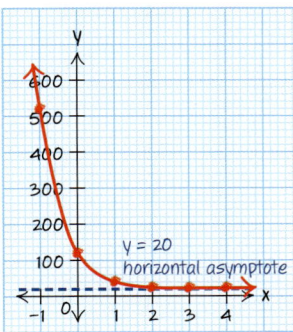

b. $g(x) = -4(3)^x - 2$
Horizontal asymptote $y = -2$.
Domain: All real numbers or $(-\infty, \infty)$,
range: $(-\infty, -2)$.

x	$g(x)$
-2	-2.44
-1	-3.33
0	-6
1	-14
2	-38
3	-110

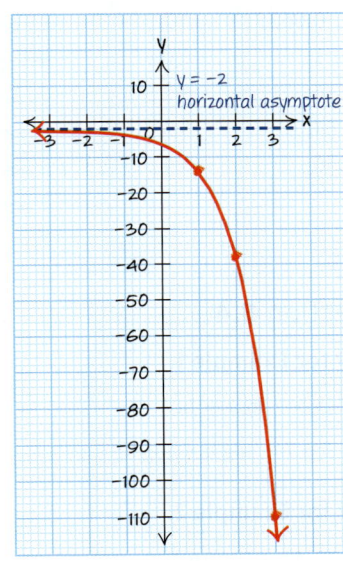

Section 5.4

PP for Example 1.

a. $y = 8000(0.75)^x$ **b.** $y = 9(4)^x$

PP for Example 2.

$$f(x) = 2.28(1.45)^x$$

PP for Example 4.

$$f(x) = 145(0.795)^x$$

Section 5.5

PP for Example 2.

a. $P(10) \approx 20.5759$. In 2010, the population of Australia would be approximately 20.6 million.

b. $r = 0.008$. The population of Australia is growing at a rate of 0.8% per year.

PP for Example 3.

a. $P(t)$ = Population of Boulder, Colorado
t = Time in years since 2000

$$P(t) = 96,000(1.10)^t$$

b. $P(2) = 116,160$. In 2002, Boulder, Colorado, would have had approximately 116,160 people if the population grew at 10% annually.

PP for Example 4.

a. $P(t)$ = Population of Trinidad and Tobago in millions
t = Time in years since 2008

$$P(t) = 1(0.991)^t$$

b. $P(12) \approx 0.897$. In 2020, Trinidad and Tobago will have approximately 897,000 people if the population decreases by 0.9% annually.

PP for Example 5.

a. $30,000 will grow to $66,589.21 in 10 years if earning 8% annual interest compounded monthly.

b. $30,000 will grow to $66,725.20 in 10 years if the account earns 8% annual interest compounded weekly.

c. $30,000 will grow to $66,760.38 in 10 years if the account earns 8% annual interest compounded daily.

PP for Example 6.

$A \approx 145769.504$. Therefore, after 12 years, this account will have $145,769.50.

CHAPTER 6

Section 6.1

PP for Example 1.

a. $s(P) = -14.29P + 500$

b. $s(5) = 428.55$. Therefore, it will take about 429 seconds to pump the vacuum chamber down to 5 psi.

c. Domain $0 \leq P \leq 35$, range: $0 \leq s \leq 500$

PP for Example 2.

a. $f^{-1}(x) = -0.25x + 2.25$

b. $g^{-1}(t) = 0.4t + 1.4$

PP for Example 3.

a.

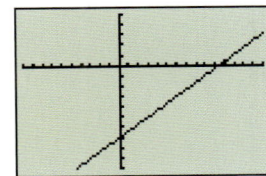

This graph passes the horizontal line test, so $f(t) = \frac{2}{3}t - 7$ is a one-to-one function.

b.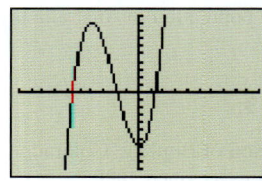

This graph fails the horizontal line test, so $g(w) = 0.5w^3 + 3w^2 - 7$ is not a one-to-one function.

c.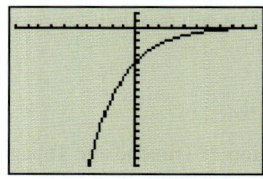

This graph passes the horizontal line test, so $g(x) = -5(0.7)^x$ is a one-to-one function.

PP for Example 4.

$$f \circ f^{-1}(x) = x \qquad f^{-1} \circ f(x) = x$$

Section 6.2

PP for Example 1.

a. $\log 10{,}000 = 4$
b. $\log_7 49 = 2$

PP for Example 2.

a. $\log_7 7^8 = 8$
b. $\log_4 1 = 0$

PP for Example 4.

a. $\log_3 278 \approx 5.122$
b. $\log_{17} 11 \approx 0.846$

PP for Example 5.

a. $\log_4 65536 = 8$
b. $6^4 = 1296$

PP for Example 6.

a. $f^{-1}(x) = \dfrac{\log x}{\log 7}$
b. $g^{-1}(t) = 2^t$
c. $h^{-1}(x) = \dfrac{\log\left(\dfrac{x}{3}\right)}{\log 4}$

PP for Example 7.

a. $x = e^4 \approx 54.598$
b. $x = 12.8$

Section 6.3

PP for Example 1.

x	$y = \log_5 x$
$\dfrac{1}{125}$	-3
$\dfrac{1}{25}$	-2
$\dfrac{1}{5}$	-1
1	0
5	1
25	2
125	3
625	4

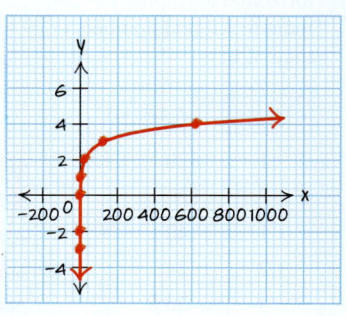

PP for Example 2.

x	$y = \log_{0.25} x$
0.0625	2
1	0
16	-2
64	-3
256	-4

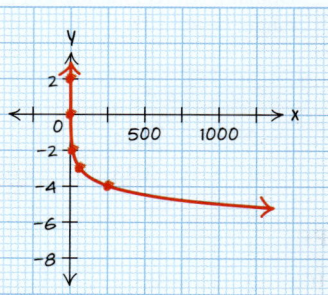

Section 6.4

PP for Example 1.

a. $\ln 5 + \ln a + \ln b$
b. $\log 7 + \log n - \log m$

PP for Example 2.

a. $\ln\left(\dfrac{18s}{3t}\right) = \ln\left(\dfrac{6s}{t}\right)$
b. $\log_7 (15mn)$
c. $\log(x^2 - 4x - 21)$

PP for Example 3.

a. $\log_4 5 + \log_4 x + 3 \log_4 y$
b. $\log 3 + 2 \log m - \log 2 - 3 \log n$
c. $\dfrac{1}{2}(\ln 2 + \ln x + \ln y)$

PP for Example 4.

a. $\log_5 \dfrac{7m^2 n^3}{p^2}$
b. $\log \dfrac{x\sqrt{y}}{z^3}$
c. $\ln\left(\dfrac{125 x^5 y^5 z^3}{32}\right)$

Section 6.5

PP for Example 3.

a. $x \approx 0.850$ b. $x \approx -0.371$

PP for Example 4.

$t \approx 24.587$. If the population of Boulder, Colorado, continues to grow at a rate of 10% per year, it will reach 1 million people around 2025.

PP for Example 5.

$r \approx 0.0463$. To have $60,000 double in 15 years, the interest rate would need to be about 4.63% compounded monthly.

PP for Example 6.

$t \approx 21.98$. Therefore, it will take about 22 years for $10,000 to triple if it is in an account that pays 5% compounded weekly.

PP for Example 7.

$t \approx 8.664$. It takes about 8.66 years for an investment to double if it is in an account paying 8% compounded continuously.

Section 6.6

PP for Example 1.

$$x = 9996$$

PP for Example 3.

$10000 = I$. Therefore, the intensity of a magnitude 8 earthquake would be 10,000 cm.

PP for Example 5.

$0.00501 \approx H^+$. Therefore, a solution with a pH value of 2.3 has a hydrogen ion concentration of 5.01×10^{-3} M.

PP for Example 6.

a. $x \approx 3162.28$ b. $x = 8$ c. $x = \dfrac{e^{-5}}{5} \approx 0.0013$

PP for Example 7.

a. $x \approx 12.91$ b. $x \approx 16.533$

c. $x = -2$ d. $x \approx 4.14$

CHAPTER 7

Section 7.1

PP for Example 2.

a. Let $c(m)$ be the number of calories burned by a 130-lb woman while rowing for m minutes. $c(m) = 11.8m$.

b. $c(m) = 165.2$. If Sheila rows for 14 minutes, she will burn about 165.2 calories.

PP for Example 3.

a. $a(m) = \dfrac{50}{m}$, where a is the acceleration in meters per second squared and m is the mass in kilograms.

b. $a(15) \approx 3.3$. When this force is applied to an object with a 15-kg mass, that object should accelerate at 3.3 m/s².

PP for Example 4.

a. $y = \dfrac{2300}{x^4}$ b. $y = 3.68$

PP for Example 5.

a. Let $P(t)$ be the average dollars disbursed per person by Medicare t years since 1990.

$$P(t) = \dfrac{-3160.64t^2 + 52536.5t - 4548.31}{0.39t + 35.65}$$

b. $P(9) \approx 5420.54$. Therefore, in 1999, the Medicare program disbursed an average of $5420.54 per person enrolled in Medicare.

c. $(8, 5505.87)$. In 1998, the average disbursements to Medicare recipients reached a maximum of about $5505.87.

PP for Example 6.

a. Domain: all real numbers except zero.

b. Domain: all real numbers except $x = 7$.

c. Domain: all real numbers except $x = 5$ or -2.

Section 7.2

PP for Example 2.

a. $\dfrac{x+5}{x-9}$ b. $\dfrac{x-2}{x-10}$

c. $\dfrac{x^2-3x}{x+7}$ d. $\dfrac{5x+35}{10x+23}$

PP for Example 3.

a. $-\dfrac{4}{7}$ b. $-\dfrac{1}{3}$

PP for Example 4.

$$\dfrac{4x^4}{y} + \dfrac{20x^2}{3} - 6y$$

PP for Example 5.

a. $4x - 5$ b. $5x - 1 + \dfrac{8}{3x-2}$

PP for Example 6.

$$3x^2 - 6x + 8$$

PP for Example 7.

a. $-2x^2 - 2x - 15$

b. $3x^4 - 11x^3 + 55x^2 - 274x + 1376 - \dfrac{6280}{x+5}$

Section 7.3

PP for Example 1.

a. $\dfrac{5a^3}{3c}$ b. $\dfrac{x+7}{x-3}$

APPENDIX D Answers to Practice Problems

PP for Example 2.

a. $\dfrac{(x-2)(x-5)}{(x-7)(x-6)}$ b. $\dfrac{(x-7)(x+5)}{(2x+5)(3x-2)}$

PP for Example 3.

a. $x^2 y$ b. $\dfrac{x+7}{x-9}$

PP for Example 4.

a. $\dfrac{(x+2)(x-5)}{(x-3)(x-8)}$ b. $\dfrac{x-4}{x+9}$

Section 7.4

PP for Example 1.

$$\text{LCD} = 360\,m^3 n \qquad \dfrac{105 m^2 n}{360 m^3 n} \qquad \dfrac{88}{360 m^3 n}$$

PP for Example 2.

a. LCD $= (x+9)(x+8)$ $\dfrac{(x+9)(x+3)}{(x+9)(x+8)}$ $\dfrac{(x+8)(x+5)}{(x+9)(x+8)}$

b. LCD $= (x-4)(x-7)(x+8)$

$\dfrac{(x+8)(x+2)}{(x-4)(x-7)(x+8)}$ $\dfrac{(x-4)(x+4)}{(x-4)(x-7)(x+8)}$

c. LCD $= (3x+4)(2x-5)(x-7)$

$\dfrac{(x-7)(3x+2)}{(3x+4)(2x-5)(x-7)}$ $\dfrac{(4x-9)(3x+4)}{(3x+4)(2x-5)(x-7)}$

PP for Example 3.

a. $\dfrac{35x^2 + 32}{10x^3}$ b. $\dfrac{x^2 - 5x + 4}{x - 5}$

PP for Example 4.

$\dfrac{5x^2 + 14x + 7}{(x+3)(x-4)(x+7)}$

PP for Example 5.

a. $\dfrac{2}{x+5}$ b. $\dfrac{x^2 - 8x - 43}{(x-9)(x+5)(x+4)}$

PP for Example 6.

a. $\dfrac{(4x+39)(x-1)}{6(x+8)}$ b. $\dfrac{4xy^2 + 3x^2 y}{2y^2 - 6x^2}$

Section 7.5

PP for Example 1.

$$x = 4.5$$

PP for Example 2.

a. $m = 12.5$. An object with a mass of 12.5 kg will accelerate at 4 m/s² when this force is applied.

b. $m = 100$. A 100-kg object will accelerate at only 0.5 m/s² when this force is applied.

PP for Example 3.

$$x = 2$$

PP for Example 5.

a. $x = 5$ b. $x = -\dfrac{3}{5}$ c. $x = -3$ $x = \dfrac{9}{8}$

PP for Example 6.

Let t be the time it takes to fill the pool when both the hose and the hydrant are used. $t \approx 14.12$. It will take about 14.12 hours using both the hose and hydrant to fill the pool.

CHAPTER 8

Section 8.1

PP for Example 3.

a. i. $f(20) = 19.544$
   ii. $f(x) = 34$ when $x \approx 105$

b. i. $h(2) = -6$
   ii. $h(x) = -8$ when $x = 6$

PP for Example 4.

a. Domain: $x \geq -8$, range: $[0, \infty)$

b. Domain: $x \leq 3$, range: $[0, \infty)$

c. Domain: all real numbers, range: all real numbers

PP for Example 5.

a. Domain: $x \leq -10$, range: $[0, \infty)$

b. Domain: all real numbers, range: all real numbers

PP for Example 6.

a. $f(x) = \sqrt{x+6}$

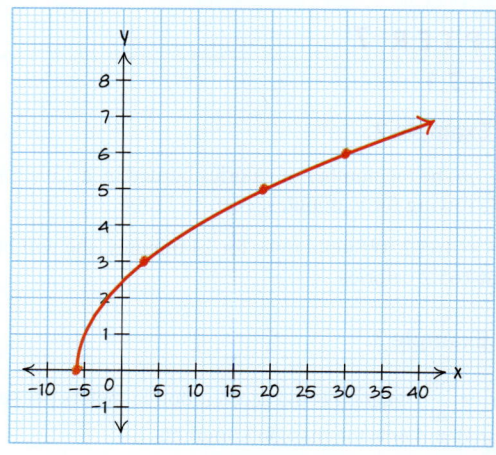

b. $f(x) = -\sqrt{5-x}$

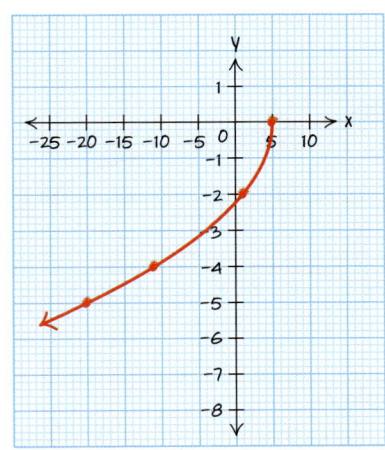

c. $g(x) = -\sqrt[3]{x} - 4$

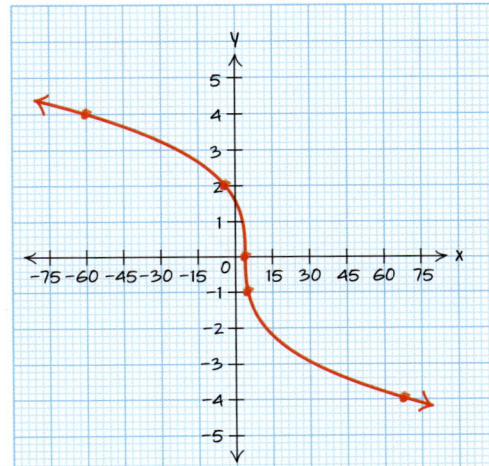

d. $h(x) = -\sqrt{x+1}$

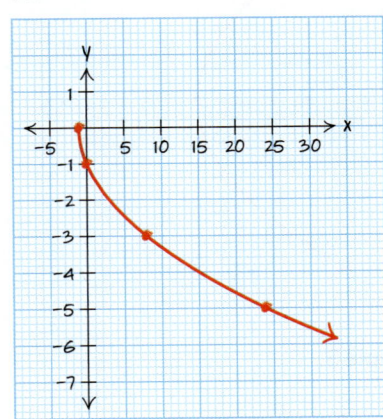

Section 8.2

PP for Example 1.
a. 7 **b.** 12 **c.** -8 **d.** 10

PP for Example 2.
a. $4ab^2$ **b.** $6xy\sqrt{5y}$
c. $3mn^2\sqrt[3]{4m^2}$ **d.** $x^4y^2z\sqrt[9]{z^2}$

PP for Example 3.
a. $-5\sqrt{3}$ **b.** $27\sqrt{2}$
c. $5xy\sqrt{5y}$ **d.** $5\sqrt{5x} - 4\sqrt{2x}$

PP for Example 4.
a. $8\sqrt{2m} + 8\sqrt[3]{7m}$ **b.** $8\sqrt[5]{x^3y} + 5\sqrt[4]{xy^2}$

Section 8.3

PP for Example 1.
a. $\sqrt[3]{15abc}$ **b.** $10m^2$ **c.** $6y$

PP for Example 2.
a. $24\sqrt{15ab}$ **b.** $105mn\sqrt{6m}$ **c.** $28x^4y^6\sqrt[3]{y}$

PP for Example 3.
a. $14 - 7\sqrt{5} + 2\sqrt{3} - \sqrt{15}$ **b.** $43 + 12\sqrt{7}$

PP for Example 4.
a. 3 **b.** $\dfrac{5}{3}$ **c.** $\dfrac{2m\sqrt[4]{9}}{n}$

PP for Example 5.
a. $\dfrac{\sqrt{10mn}}{5m}$ **b.** $\dfrac{7x\sqrt{15y}}{10y}$

PP for Example 6.
a. $\dfrac{2\sqrt[3]{25mn}}{5n}$ **b.** $\dfrac{7x\sqrt[5]{9x^4y^3}}{3y^2}$ **c.** $\dfrac{\sqrt[3]{2ab^2}}{2b}$

PP for Example 7.
a. $\dfrac{28 - 7\sqrt{13}}{3}$ **b.** $\dfrac{28 - 4\sqrt{15} + 35\sqrt{3} - 15\sqrt{5}}{34}$
c. $\dfrac{15 + 21\sqrt{ab} + 5\sqrt{a} + 7a\sqrt{b}}{25 - 49ab}$

Section 8.4

PP for Example 1.
a. $x = 44$ **b.** $x = 4$

PP for Example 4.
a. $x = 11$ **b.** $x \approx 7.01$

PP for Example 5.
$x = 6 \quad x = 2$

PP for Example 6.
No real solutions.

PP for Example 7.
a. $x = 36$ b. $x = -246$

Section 8.5
PP for Example 3.
a. $-1 + 11i$ b. $7.6 - 1.6i$
c. $-10 + 2i$ d. $-7 - 4i$

PP for Example 4.
a. $-24 + 12i$ b. $-29 + 31i$
c. $46 + 22i$ d. 29

PP for Example 7.
a. $7 + 4i$ b. $-\frac{7}{3} - \frac{4}{3}i$
c. $-\frac{1}{29} + \frac{17}{29}i$ d. $0.87395 - 1.0672i$

PP for Example 8.
a. $w = -3 + 3i$ $w = -3 - 3i$
b. $x = \pm 9i$
c. $x = -4 + i\sqrt{2}$ $x = -4 - i\sqrt{2}$
d. $x = 0$ $x = 2 + 3i$ $x = 2 - 3i$

CHAPTER 9
Section 9.1
PP for Example 1.
a. $x = (y - 2)^2 - 8$

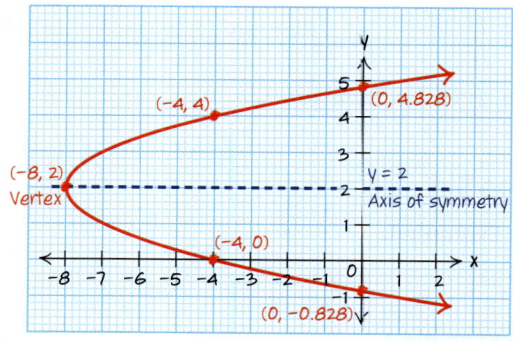

b. $x = -2y^2 + 16y + 5$

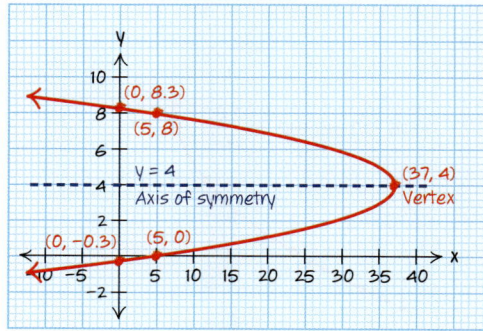

PP for Example 2.
a. $y = \frac{x^2}{6}$ Focus: $\left(0, \frac{3}{2}\right)$, directrix: $y = -\frac{3}{2}$.

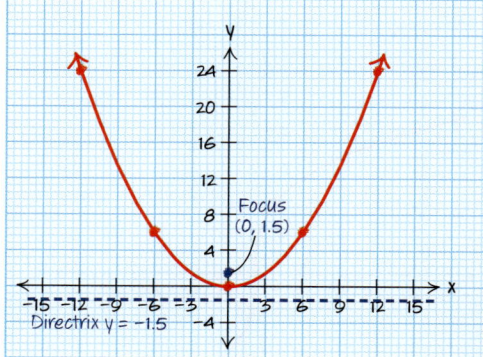

b. $-24x = y^2$ Focus: $(-6, 0)$, directrix: $x = 6$.

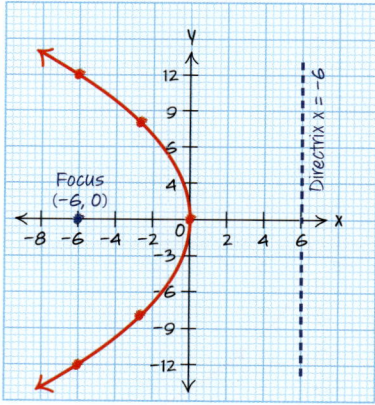

PP for Example 3.

$$y = \frac{x^2}{16} \quad \text{or} \quad 16y = x^2$$

PP for Example 4.

The focus should be at the point (0, 6.25), so the receiver should be 6.25 feet above the base of the satellite on the axis of symmetry.

PP for Example 5.

a. $(x-5)^2 + (y-7)^2 = 100$ **b.** $(x+5)^2 + (y+4)^2 = 81$

PP for Example 6.

a. $x^2 + y^2 = 16$

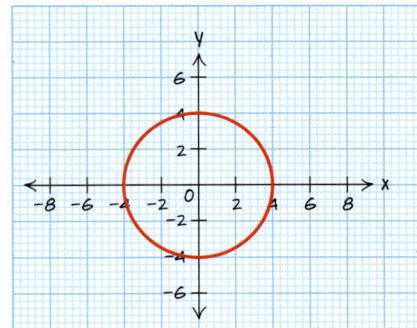

b. $(x-2)^2 + (y+3)^2 = 4$

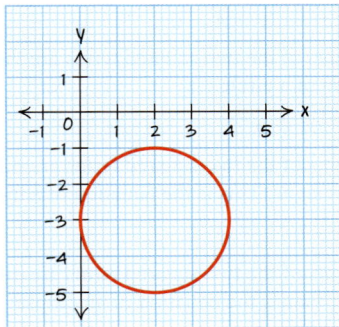

PP for Example 7.

$(x-2)^2 + (y+5)^2 = 9$

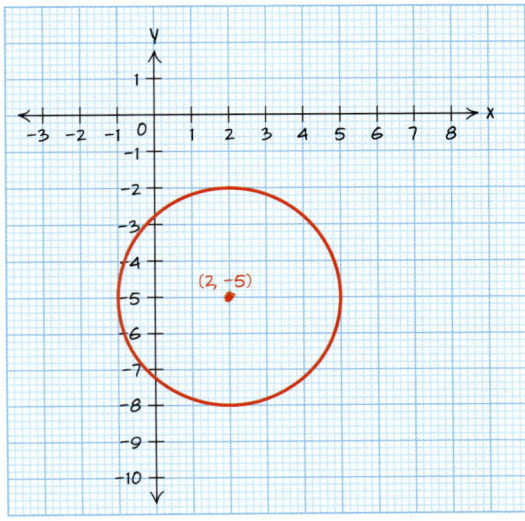

PP for Example 8.

$(x-4)^2 + (y-2)^2 = 16$

Section 9.2

PP for Example 1.

a. $\dfrac{x^2}{4} + \dfrac{y^2}{36} = 1$

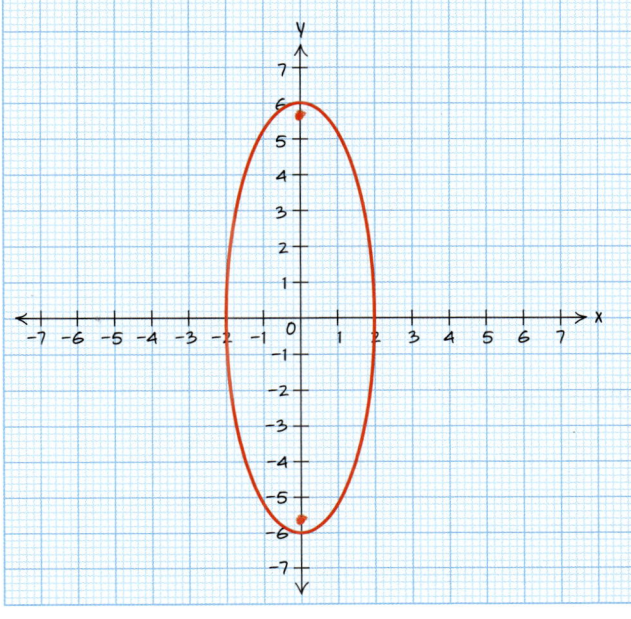

b. $\dfrac{x^2}{100} + \dfrac{y^2}{16} = 1$

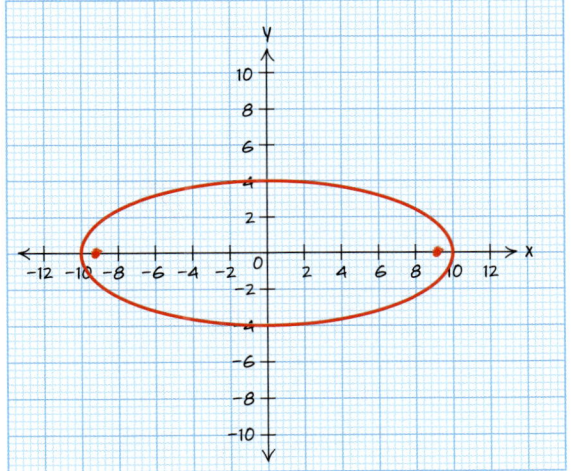

PP for Example 2.

$$\frac{x^2}{4} + \frac{y^2}{16} = 1$$

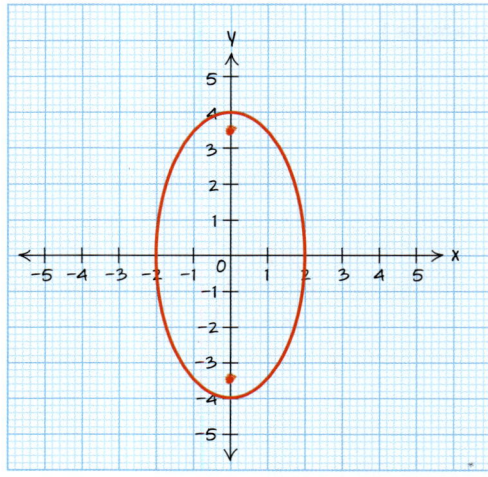

PP for Example 3.

$\frac{x^2}{4} + \frac{y^2}{9} = 1$. Foci: $(0, 2.24)$ and $(0, -2.24)$.

PP for Example 4.

The equation for the orbit of Mars is

$$\frac{x^2}{20022.25} + \frac{y^2}{19840} = 1$$

PP for Example 5.

a. $\frac{x^2}{36} - \frac{y^2}{64} = 1$

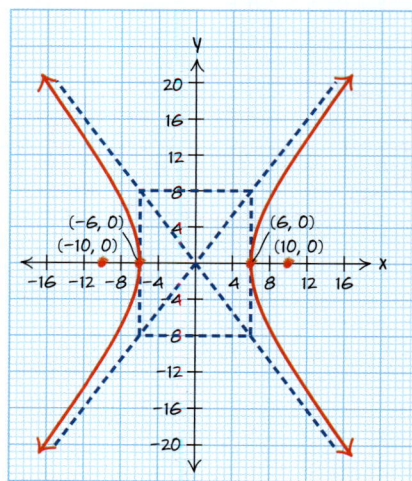

b. $16y^2 - 9x^2 = 144$

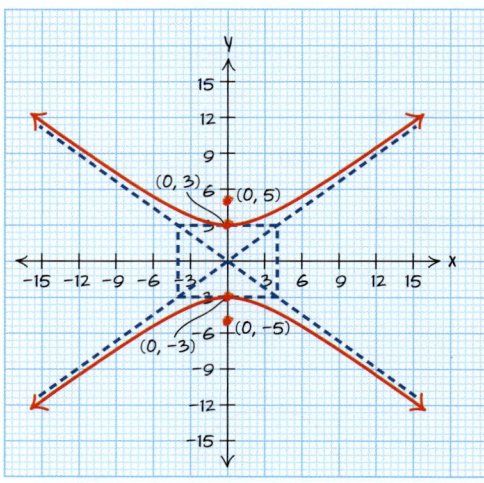

PP for Example 6.

$\frac{x^2}{36} - \frac{y^2}{9} = 1$. Foci: $(6.71, 0)$ and $(-6.71, 0)$.

PP for Example 7.

a. $9x^2 + 9y^2 + 2x + 7y - 4 = 0$

The coefficients of the two squared terms are the same, so this is the equation of a circle.

b. $16x^2 - 4y^2 - 64 = 0$

The coefficients of the two squared terms are different and have different signs. This is the equation of a hyperbola.

c. $4x^2 + y^2 - 4 = 0$

The coefficients of the two squared terms are different and are both positive. This is the equation of an ellipse.

Section 9.3

PP for Example 1.

a. 12, 17, 22, 27, and 32.   b. $\frac{1}{3}, \frac{1}{9}, \frac{1}{27}$, and $\frac{1}{81}$.

PP for Example 2.

a. -11 is the fifth term of the sequence.

b. 204 is the 14th term in the sequence.

PP for Example 3.

a. $a_n = 3 + (n - 1) \cdot 5$

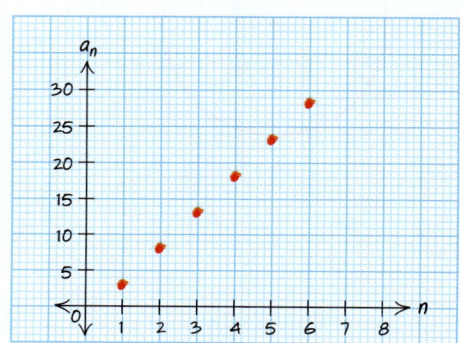

b. $f(n) = (-1)^n$

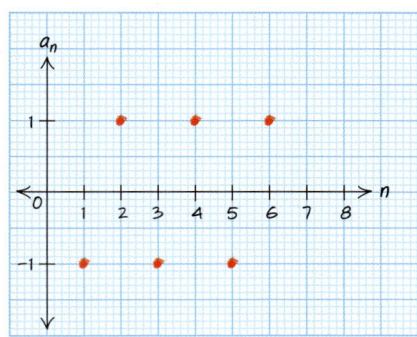

PP for Example 4.

a. The common difference $d = -6$ is constant; therefore, the sequence is arithmetic.

b. The common difference $d = 2$ is constant; therefore, the sequence is arithmetic.

c. The differences are not equal, so the sequence is not arithmetic.

PP for Example 5.

a. Common difference: $d = -2$. General term: $a_n = -2n + 11$.

b. The differences are not equal, so the sequence is not arithmetic.

c. Common difference: $d = \frac{1}{3}$. General term: $a_n = \frac{1}{3}n - 2$.

PP for Example 6.

a. $a_n = 1200n + 42{,}800$

b. $a_{14} = 59{,}600$. In the 14th year, the salary will be $59,600.

c. $21 = n$. The salary will reach $68,000 in the 21st year.

PP for Example 7.

a. $a_n = \frac{2}{5}n - 4$ b. $a_n = 3n - 8$

Section 9.4

PP for Example 1.

a. The sequence is geometric and the common ratio is $r = 0.25$.

b. The sequence is arithmetic.

PP for Example 2.

a. The general term of this geometric sequence is $a_n = 25\left(\frac{6}{5}\right)^{(n-1)}$.

b. The general term of this geometric sequence is $a_n = 200\left(-\frac{2}{5}\right)^{(n-1)}$.

PP for Example 3.

a. $a_n = 40{,}000(1.02)^{(n-1)}$.

b. $a_5 \approx 43{,}297.29$. In the fifth year, the salary will be $43,297.29.

c. $n \approx 47.27$. The salary will reach $100,000 in the 48th year. We must round n to a whole number, so choosing 48 implies that the salary will be at least $100,000.

PP for Example 4.

a. $a_n = 10(-3)^{(n-1)}$ b. $a_n = 320(0.2)^{(n-1)}$

Section 9.5

PP for Example 2.

a. $\sum_{n=1}^{4}(4n - 20) = -16 + (-12) + (-8) + (-4)$

$\sum_{n=1}^{4}(4n - 20) = -40$

b. $\sum_{n=1}^{5}(2(-3)^n) = -6 + 18 + (-54) + 162 + (-486)$

$\sum_{n=1}^{5}(2(-3)^n) = -366$

PP for Example 3.

$a_n = 10 + 3n$, $S_8 = 188$. This amphitheater has a total of 188 seats.

PP for Example 4.

$S_{44} = 2640$

PP for Example 5.

$S_9 \approx 105{,}152.65$. This family used about 105,152.65 gallons of water during the first 9 months of the year.

PP for Example 6.

a. $S_{12} = 13{,}650$ b. $S_{15} \approx 399.9996$

Answers to Selected Exercises

APPENDIX E

CHAPTER 1
Section 1.1

1. $x = 15$
3. $t = 10$
5. $x = 10$
7. $x = 4$
9. $x = 5000$

11. **a.** After 1 hour of training, a new employee can produce 30 candies per hour.
    b. After 4 hours of training, a new employee can produce 60 candies per hour.
    c. A new employee can produce 150 candies per hour after 13 hours of training.

13. **a.** In 1992 there were approximately 4178 homicides of 15- to 19-year-olds in the United States.
    b. In 2002 there were approximately 1019 homicides of 15- to 19-year-olds in the United States.
    c. In 1982 there were approximately 7337 homicides of 15- to 19-year olds in the United States.

15. **a.** If you sell 100 printed T-shirts you will lose $150.
    b. If you sell 400 printed T-shirts you will make $300 profit.
    c. To make $1000 profit you must sell 867 printed T-shirts.

17. **a.** It costs $52.50 to take a 25-mile taxi ride in NYC.
    b. For $100 you can take about a 48-mile taxi ride in NYC.

19. $P = 4200$ looks correct. This would mean that 4,200,000 people live in Kentucky. $P = 3.5$ is too small and $P = -210$ is not possible.

21. $T = -50$ looks correct. This would mean that the temperature at the South Pole is -50 degrees Fahrenheit. Both $T = 75$ and $T = 82$ are too warm.

23. **a.** On sales of $2000 you will make $80 in commissions.
    b. On sales of $50,000 you will make $3920 in commissions.
    c. To make $500 per week, you will need $7250 in sales each week.

25. **a.** $B = 29.95 + 0.55m$
    b. If you drive the 15-foot truck 75 miles it will cost you $71.20.
    c. For a total of $100 you can rent the 15-foot truck from Budget and drive it 127 miles.

27. **a.** $P = 250 + 0.07s$
    b. If you have sales of $2000 in a week, your pay will be $390.
    c. To earn $650 per week, you must have $5714.29 in sales each week.

29. **a.** Let C be the total cost (in dollars) for a trip to Las Vegas, and let d be the number of days you stay. $C = 125 + 100d$.
    b. A 3-day trip to Las Vegas will cost $425.
    c. If you have $700 and gamble half of it, you can stay in Las Vegas for only 2 days.

31. **a.** Let C be the total cost (in dollars) for shooting a wedding and let p be the number of proofs edited and printed. Then $C = 5.29p + 400$.
    b. If the photographer edits and prints 100 proofs the cost will be $929.
    c. With a budget of $1250, the photographer can edit and print 160 proofs.

33. **a.** Let C be the total cost (in dollars) for selling s snow cones for a month: $C = 7000 + 0.45s$.
    b. The monthly cost for selling 3000 snow cones is $8350.
    c. For a $10,600 budget the vendor can sell up to 8000 snow cones.

35. **a.** Let C be the total cost (in dollars), for the Squeaky Clean Window Company to clean windows for a day, when w windows are cleaned.
$$C = 1.5w + 230$$
    b. If the Squeaky Clean Window Company cleans 60 windows in a day it will cost the company $320.
    c. To stay within a budget of $450 the Squeaky Clean Window Company can clean up to 146 windows.

37. Maria's work is correct. Javier needs a decimal to correctly represent 55 cents per bottle in terms of dollars per bottle.

39. **a.** Let C be the total cost (in dollars) for pest management from Enviro-Safe Pest Management when m monthly treatments are done,
$$C = 150 + 38m$$
    b. If your house is treated for 18 months after the initial treatment it will cost $834.

41. **a.** Let C be the total monthly cost (in dollars) for a manufacturer to produce g sets of golf clubs.
$$C = 23{,}250 + 145g$$
    b. It costs the manufacturer $37,750 to produce 100 sets of golf clubs.
    c. $g \approx -22.4$. This is model breakdown. Their costs can never be lower than their fixed costs of $23,250.
    d. To break even selling 100 sets of golf clubs per month the manufacturer must sell each set for $377.50.

43. a. $C = 1500 + 1.5n$ for $n \leq 500$.

b. It costs Rockon $1875 to make 250 CDs.

c. With a budget of $2000, Rockon can order 333 CDs.

d. With a budget of $3000, Rockon can order 1000 CDs. This is a model breakdown. They can only order up to 500 CDs.

45. $x = 10$ 47. $d = 20$ 49. $m = 8$

51. $x = -\dfrac{20}{11}$ 53. $d = \dfrac{301}{10}$ 55. $d \approx -0.45$

57. $c = \dfrac{113}{3}$ 59. $d \approx -7.27$ 61. $z = \dfrac{17}{6}$

63. $v = \dfrac{42}{31}$ 65. $t = \dfrac{34}{15}$ 67. $a = \dfrac{F}{m}$

69. $F = \dfrac{J}{t}$ 71. $\alpha = \dfrac{\omega - \omega_0}{t}$ 73. $I = \dfrac{2k}{\omega^2}$

75. $m = \dfrac{2k}{v^2}$ 77. $y = \dfrac{c - ax}{b}$ 79. $x = \dfrac{y - 5}{a}$

81. $c = \dfrac{b - 3d}{2}$ 83. $y = \dfrac{z - 5x^2}{3}$

85. Yes, rounding the outside temperature to 73 degrees Fahrenheit is appropriate because a difference of 0.4 degree Fahrenheit would not be noticed.

87. The discounted price should be rounded to $236.57 because our monetary units extend to the hundredth place value.

89. The number of cars should be rounded to 313. To make the profit of $400 the company needs to wash just slightly more than 312 cars. Therefore, the next possible whole number greater than that is 313.

Section 1.2

1. a. The total prize money at Wimbledon in 2002 was 8.5 million British pounds.

b. In 2003 the total prize money at Wimbledon was 8.5 million British pounds.

c. (0, 7) In the year 2000 the winnings were 7 million British pounds

d. Domain: [2, 11]; Range: [8.5, 14]

3. (A) and (E) 5. (C) and (F)

7. 50 9. 0.01

11. 50 13. 10,000

15. a. Dependent variable: C = cost in dollars for producing chocolate-dipped key lime pie bars.

Independent variable: b = the number of bars produced.

b. Graphic title: Cost of Key Lime Bars on a Stick

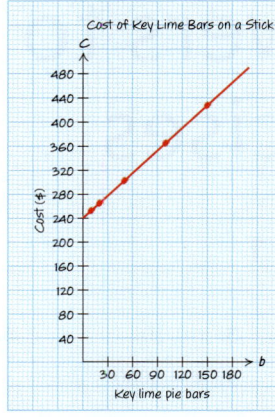

c. When $b = 200$ then $C = \$485$

The cost of producing 200 key lime bars on a stick is about $485.

d. Domain $0 \leq b \leq 200$; Range $240 \leq C \leq 485$

17. a. Dependent variable: G = gross profit for Quiksilver Inc. in millions of dollars.

Independent variable: t = years since 2000.

b.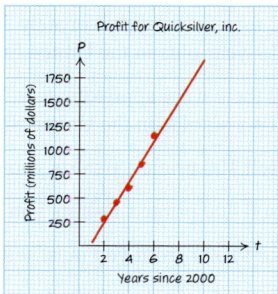

c. When $t = 10$ then $G = 1850$. The gross profit for Quiksilver Inc. in 2010 will be $1850 million.

d. Domain: $1 \leq t \leq 10$; Range: $50 \leq G \leq 1850$

19. Maria's model fits the data better. There is a smaller variance between the data points and the "Line of Best Fit."

21. $0 \leq t \leq 10$ is not a reasonable domain because the model predicts a negative value of cases per 100,000 population for t values of about 8.2. Domain: [0, 7]; Range: [0.6, 4.1]

23. a. Dependent variable: P = population of the United States in millions

Independent variable: t = years since 1995

b.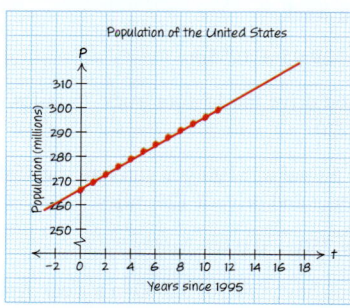

c. When $t = 14$ then $P \approx 308$. The population of the United States in 2009 will be around 308 million people.

d. Domain: $[0, 14]$;   Range: $[266, 308]$

e. $(0, 266)$

f. In 1995 there were approximately 266 million people in the United States.

25. a. W = Number of deaths of women induced by illegal drugs in the United States
    t = Years since 2000.

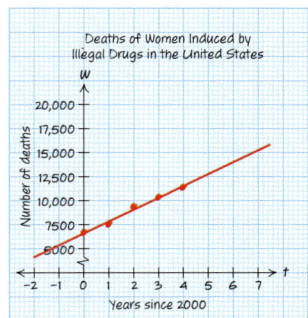

b. When $t = 7$ then $W \approx 15{,}000$. The number of drug-induced deaths of females in the United States in 2007 was around 15,000.

c. The number of drug-induced deaths of females in the United States reached 4000 in 1998.

d. Domain: $-1 \le t \le 7$;   Range: $8200 \le W \le 15{,}000$.

27. a. T = The number of years someone will live. a = Age in years

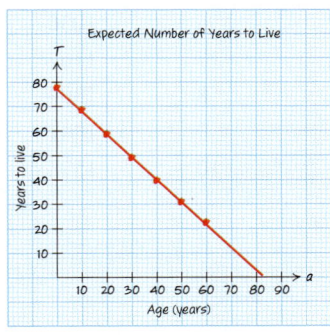

b. When $a = 45$ then $T = 35$.
A 45-year-old person would live around 35 years.

c. Domain: $[0, 80]$;   Range: $[3, 77]$

d. The graphical model predicts a 90-year-old person to live negative years. This is model breakdown.

e. The vertical intercept is $(0, 77)$.

f. At birth, a person will live about 77 years.

29. a. $(0, 2)$ b. $(-3, 0)$ c. $x = 3$ d. $y = 2.6$

31. a. $(0, 5)$ b. $(4, 0)$ c. Input value $= -10$
    d. Input value $= 15$ e. Output value $= -8$

33. a. $(0, -8)$ b. $(30, 0)$ c. $x = 11$
    d. $x = -9$ e. $y = -3$

35. a. False, $(0, -2)$ b. True
    c. False, $x = -4.5$ d. False, $x = -1.5$
    e. True

37. a. $(-15, 0)$ b. $(0, -10)$ c. $x = 15$ d. $x = 7.5$
    e. $y = -3$

39. a. $(0, 1)$ b. $(-2, 0)$ c. Input value $= 1$
    d. Input value $= -4$ e. Output value $= 2.5$

Section 1.3

1. a. Slope $= \dfrac{1}{2}$ b. Increasing   c. $(0, -1)$ d. $(2, 0)$

3. a. Slope $= \dfrac{5}{2}$ b. Increasing   c. $(0, -7.5)$ d. $(3, 0)$

5. a. Slope $= -5$ b. Decreasing
   c. $(0, 5)$ d. $(1, 0)$

7. $y = 2x + 3$

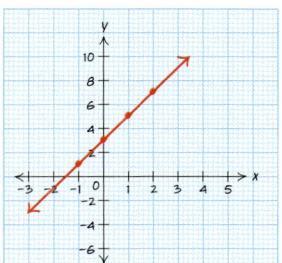

9. $x = 5y + 4$

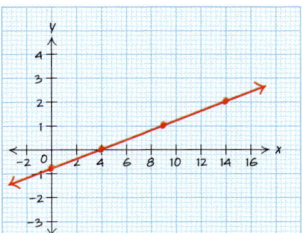

11. $y = x^2 + 2$

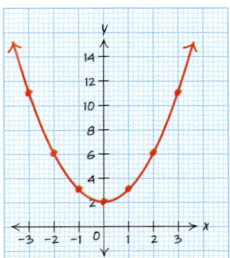

13. $y = \dfrac{2}{3}x + 6$

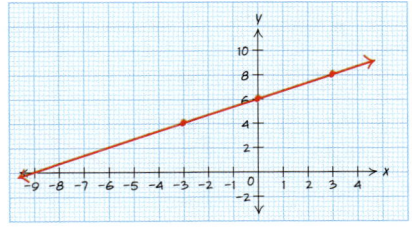

15. $x = \dfrac{2}{3}y - 4$

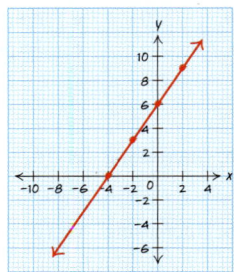

17. $y = 0.5x - 3$

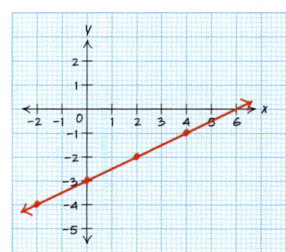

19. $x = -1.5y + 7$

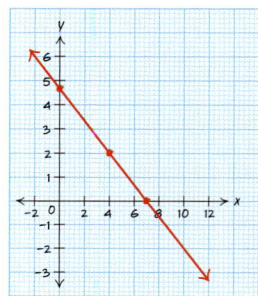

21. $x = -2x^2 + 15$

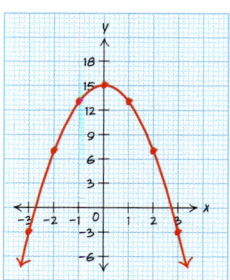

23. $B = 0.55m + 29.95$

a.

m	B
10	35.45
20	40.95
30	46.45
40	51.95

b.

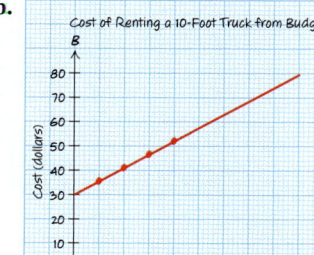

25. a. W = The sales clerk's weekly salary in dollars.

s = Total dollars of merchandise the sales clerk sells during the week.

$$W = 0.04s + 100$$

b.

s	W
100	104
500	120
1000	140
1500	160

c.

27. a. T = Total tuition and fees Western Washington University charges its resident undergrad students, in dollars.

u = Total units taken by the student, up to 10 units.

$$T = 137u + 208$$

b.

u	T
5	893
6	1030
7	1167
8	1304
9	1441
10	1578

c.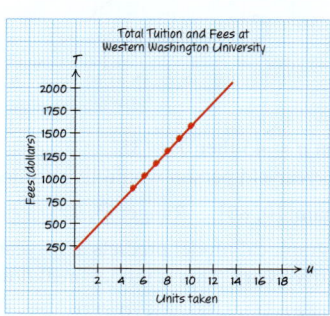

29. $P = -3t + 50$

a.

t	P
1	47
2	44
3	41
4	38

b.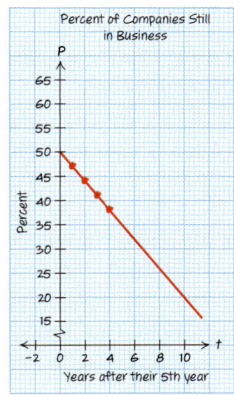

31. The slope is 2.
33. The slope is -7.
35. The slope is 4.
37. Yes, the points given in the table all lie on a line.
39. No, the points given in the table do not all lie on a line.
41. The slope is 3, y-intercept is $(0, 5)$
43. The slope is -4, y-intercept is $(0, 8)$
45. The slope is $\frac{1}{2}$, y-intercept is $(0, 5)$
47. The slope is 0.4, y-intercept is $(0, -7.2)$
49. The slope is $\frac{4}{3}$, y-intercept is $\left(0, \frac{7}{5}\right)$
51. No, the y-intercept is $(0, -9)$
53. The slope is -2, y-intercept is $(0, 20)$
55. The slope is 2, y-intercept is $(0, -10)$
57. The slope is $-\frac{3}{5}$, y-intercept is $\left(0, \frac{12}{5}\right)$
59. No, the slope is $-\frac{3}{2}$, y-intercept is $(0, 3)$
61. The slope is 0.55. The cost of renting a Budget truck increases by $0.55 per mile.
63. The slope is 0.04. For every dollar sold, the sales clerk earns $0.04.
65. The slope is 137. Western Washington University charges $137 per unit taken by resident undergraduate students.
67. The slope is -3. The percent of companies still in business decreases by 3 percentage points each year.
69.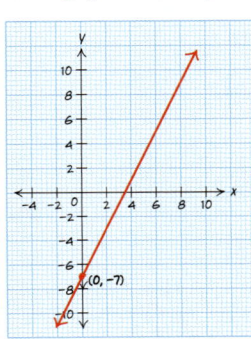

71.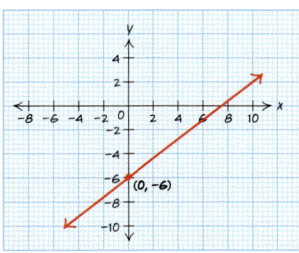

73.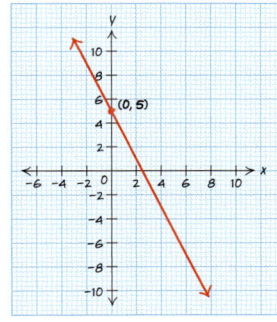

75.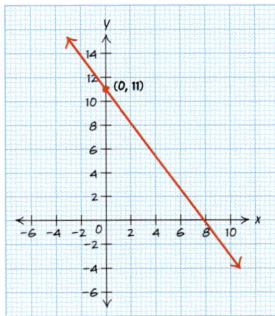

77. The graph has the y-intercept at $(0, 1.5)$, but it should be at $(0, -3)$.
79. The graph uses the slope $\frac{2}{3}$, but it should use the slope $-\frac{2}{3}$.
81.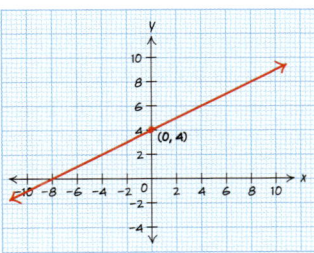

83.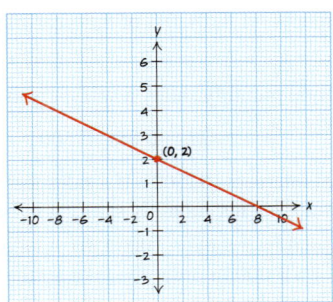

85.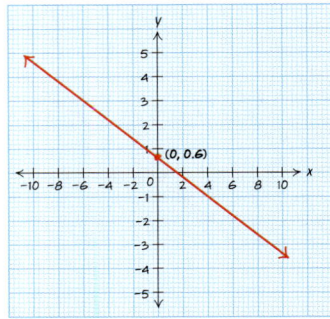

87.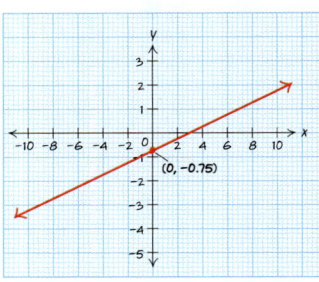

89.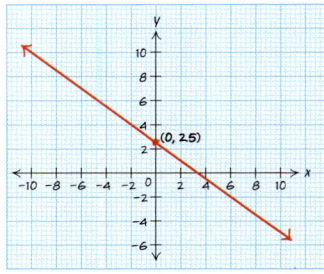

Section 1.4

1. $5x - y = -8$ **3.** $4x + y = 15$ **5.** $2x - 3y = 24$

7. $5x - 10y = -4$ **9.** $12x + 15y = -5$

11. $B = 0.55m + 29.95$

  a. Vertical intercept $(0, 29.95)$. It will cost you $29.95 to rent a 10-ft truck from Budget and drive it 0 miles.

  b. Horizontal intercept $(-54.45, 0)$. To rent a 10-ft truck from Budget for $0, you would have to drive it -55 miles. This is a model breakdown.

13. $P = 5a + 7$

  a. P-intercept $(0, 7)$. 7% of 10-year-old girls are sexually active.

  b. a-intercept $(-1.4, 0)$. 0% of 8.6-year-old girls are sexually active.

15. a. M = Total monthly salary a salesperson earns in dollars.

$$M = 0.05s + 500$$

  b. Vertical intercept $(0, 500)$. If a salesperson sells $0 during the month, his/her monthly salary will be $500.

  c. Horizontal intercept $(-10000, 0)$. A salesperson will need to sell $-\$10000$ in order to make $0 for the month. This is model breakdown.

17. $P = 9.5t + 1277$

  a. P-intercept $(0, 1277)$. The population of Maine in 2000 was 1277 thousand people.

  b. t-intercept $(-134.4, 0)$. The population of Maine was 0 people in 1866. This is model breakdown.

19. $M = 44t + 2798$

  a. M-intercept $(0, 2798)$. The number of Florida residents enrolled in Medicare in 2000 was 2798 thousand people.

  b. t-intercept $(-63.6, 0)$. 0 residents of Florida were enrolled in Medicare in 1936. This is model breakdown.

21. $2x + 4y = 8$

$(0, 2), (4, 0)$

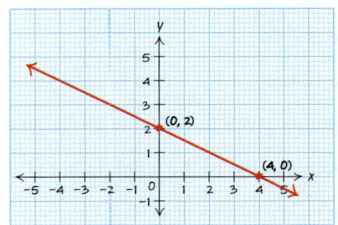

23. $3x - 5y = 15$

$(0, -3), (5, 0)$

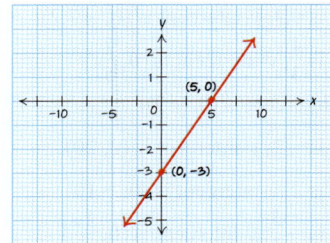

25. $4x + 6y = 30$

$(0, 5), (7.5, 0)$

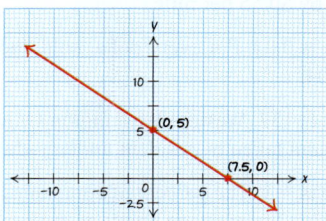

27. $-3x + 2y = 16$

$(0, 8), \left(-5\frac{1}{3}, 0\right)$

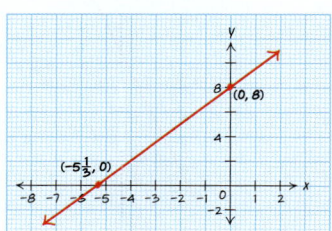

APPENDIX E Answers to Selected Exercises   E-7

29. $-2x - 4y = -20$
(0, 5), (10, 0)
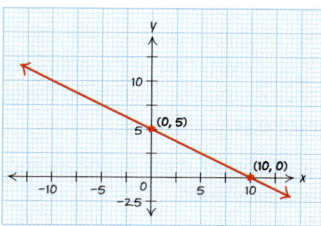

31. $x = 4$
(4, 0)
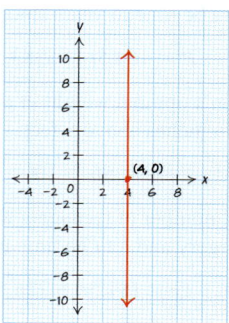

33. $y = 3$
(0, 3)
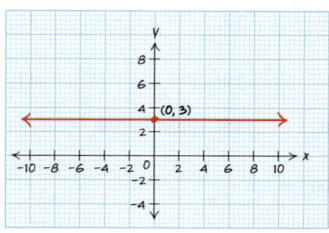

35. The graph of the line goes through the point (0, 0), but if you plug in $x = 0$ and $y = 0$ into the equation you get a false statement.

37. This is the graph of the line $y = -1$.

39.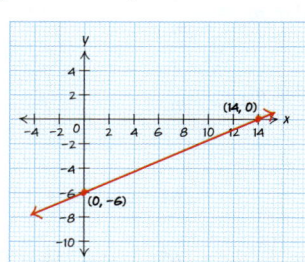

41.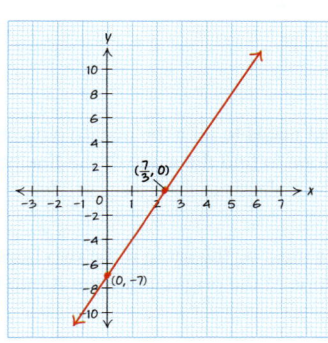

43.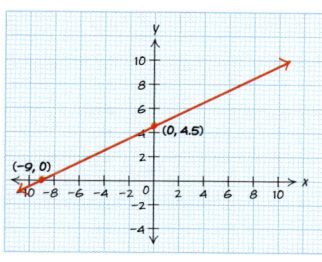

45.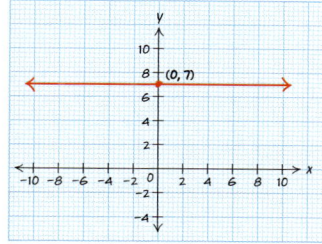

47.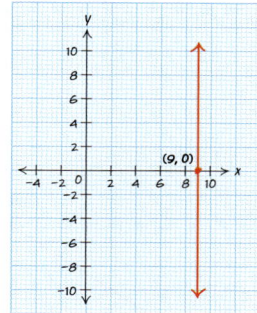

49.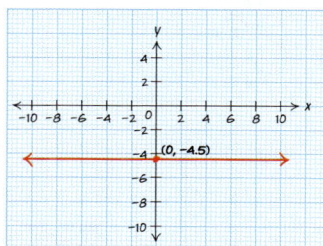

51.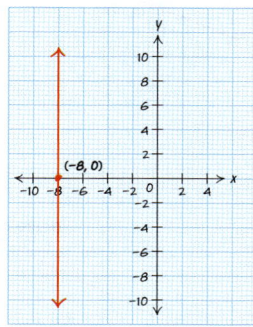

53.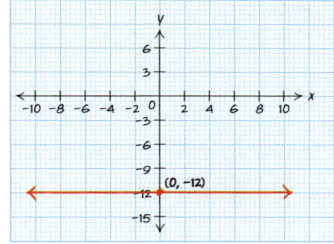

55.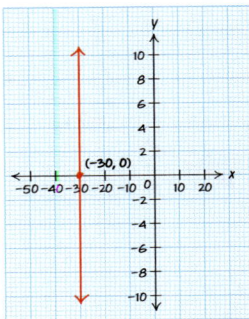

57.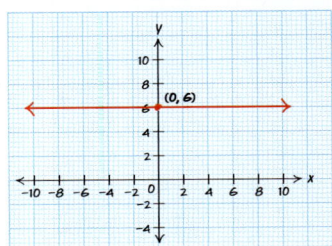

59.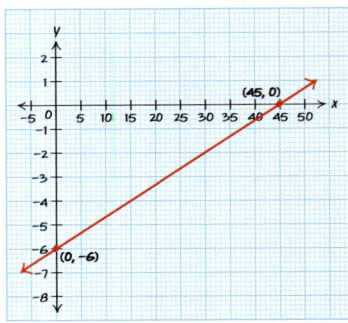

61.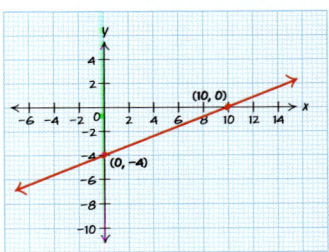

63.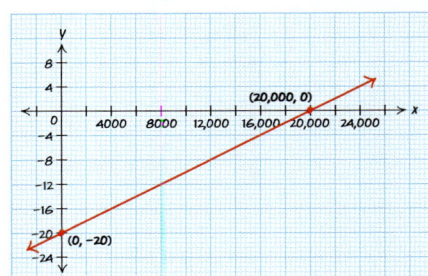

Section 1.5

1. $y = \dfrac{1}{2}x - 1$ **3.** $y = \dfrac{5}{2}x - \dfrac{15}{2}$ **5.** $y = -5x + 5$

7. $y = \dfrac{3}{10}x + 3$ **9.** $y = -\dfrac{5}{4}x + \dfrac{15}{2}$ **11.** $y = 4$

13. $x = 3$

15. Sherry's work is correct. One of the points that Maritza used did not clearly cross the intersection on the graph paper. (0, 1.5) was incorrect.

17. Let C be the total cost of shirts in dollars and let n be the total number of shirts purchased.
$$C = \dfrac{17}{2}n + 25$$

19. Let P be the total population of Washington State in thousands and let t be time in years since 2000.
$$P = 104.5t + 5383.5$$

21. Let W be a woman's optimal weight in pounds and let h be a woman's height in inches above 5 feet.
$$W = 5h + 100$$

23. Let G be the total number of teenagers who underwent gastric bypass surgery and let t be time in years since 2000.
$$G = 181.5t + 226.5$$

25. $y = -2x + 13$ **27.** $y = \dfrac{2}{3}x + 12$ **29.** $y = 3x$

31. $y = -\dfrac{1}{2}x + \dfrac{19}{2}$ **33.** $y = 4x + 11$ **35.** $x = 7$

37. $y = 8$ **39.** $y = 2x + 7$ **41.** $y = -7x + 15$

43. $y = 4x - 26$ **45.** Parallel, $m_1 = m_2$

47. Perpendicular, $m_1 = -\dfrac{1}{m_2}$

49. Perpendicular, $m_1 = -\dfrac{1}{m_2}$

51. Parallel, $m_1 = m_2$

53. Perpendicular, $m_1 = -\dfrac{1}{m_2}$

55. Perpendicular, $m_1 = -\dfrac{1}{m_2}$

57. Neither

59. $y = 4x$ **61.** $y = \dfrac{2}{3}x + 12$

63. $y = -\dfrac{1}{2}x + \dfrac{15}{2}$ **65.** $y = 0.25x - 0.25$

67. $y = -5x + 13$

69. $x = 4$

71. The student needs to take the negative reciprocal of the original slope to find the slope of the line perpendicular. $m = -\dfrac{3}{2}$

73. The slopes must be the same if the two lines are parallel. To find the slope of the original line put it into slope-intercept form. $m = -\dfrac{2}{5}$

75. The point (2, 6) is not the y-intercept, so b is not 6. The correct equation is $y = 4x - 2$.

77. a. Slope = 75.8. The population of Washington state increases 75.8 thousand each year.

b. The vertical intercept is (0, 5906.2). In 2000 the population of Washington state was 5906.2 thousand.

c. The horizontal intercept is (−77.9, 0). In 1922, the population of Washington state was 0. This is model breakdown.

79. a. Slope = 4.02. The annual amount of public expenditure on medical research in the United States increases $4.02 billion each year.

b. The vertical intercept is (0, 25.8). The annual amount of public expenditure on medical research in the United States in 2000 was $25.8 billion.

c. The horizontal intercept is (−6.42, 0). The annual amount of public expenditure on medical research in the United States was $0 in 1994. This is model breakdown.

81. a. Slope = 500. The monthly profit for a small used car lot increases by $500 for every car that is sold.

b. The vertical intercept is (0, −6000). If the small used car lot sells 0 cars, it will lose $6000.

c. The horizontal intercept is (12, 0). The small used car lot will need to sell 12 cars in order to break even in their monthly profit.

83. a. Let P the total profit in dollars Dan earns teaching surfing lessons and let s be the number of 1-hour surf lessons Dan gives. This can be represented by the equation
$$P = 30s - 700.$$

b. Slope = 30. Dan earns $30 for every 1-hour surf lesson he gives.

c. The vertical intercept is (0, −700). If Dan doesn't give any surf lessons, he will lose $700 in profit.

d. The horizontal intercept is (23.33, 0). Dan will need to give 24 one-hour surf lessons in order to break even in profit.

85. a. Let T be the total amount of money raised by the PTA for a new track at Mission Meadows Elementary School in dollars and let a be the total pounds of aluminum cans the PTA recycles. $T = 1.24a + 2000$

b. Slope = 1.24. The PTA at Mission Meadows Elementary School earns $1.24 for every pound of aluminum cans they recycle.

c. The vertical intercept is (0, 2000). The PTA at Mission Meadows Elementary School will have $2000 to put toward a new track if they recycle 0 pounds of aluminum cans.

d. The horizontal intercept is (−1612.9, 0). The PTA at Mission Meadows Elementary School will need to recycle −1612.9 pounds of aluminum cans in order to earn $0 for a new track. This is model breakdown.

87. a. Let P be the total percentage of Americans who have been diagnosed with diabetes and let t be time in years since 2000.
$$P = 0.15t + 4.5$$

b. When $t = 10$, then $P = 6$. This implies 6% of Americans will be diagnosed with diabetes in 2010.

c. Slope = 0.15. The percentage of Americans diagnosed with diabetes increases 0.15 percentage points each year.

89. Let C be the cost and n be the number of shirts.

a. When $n = 50$ then $C = 450$. It will cost $450 for 50 shirts.

b. Slope = 8.5. The cost of the shirts increases by $8.50 per shirt.

c. The vertical intercept is (0, 25). It costs $25 if 0 shirts are purchased. This is model breakdown.

d. The horizontal intercept is (−2.94, 0). To pay $0, you would have to purchase −3 shirts. This is model breakdown.

91. a. If $t = 30$ then $P = 8518.5$. In 2030, the population of Washington state will be about 8518.5 thousand.

b. Slope = 104.5. The population of Washington state increases 104.5 thousand each year.

c. The vertical intercept is (0, 5383.5). In 2000 the population of Washington state was 5383.5 thousand.

d. The horizontal intercept is (−51.52, 0). In 1949, the population of Washington state was 0. This is model breakdown.

93. $W = 5h + 100$ where W is the optimal weight of a woman in pounds and h is the number of inches past 5 feet.

a. If $h = 12$ then $W = 160$. The optimal weight of a woman who is 6 feet tall is 160 pounds.

b. Slope = 5. For every inch taller than 5 feet, a woman's optimal weight increases 5 pounds.

c. The vertical intercept is (0, 100). A woman's optimal weight at 5 feet tall is 100 pounds.

d. The horizontal intercept is (−20, 0). At 3 feet 5 inches, a woman's optimal weight is 0 pounds. This is model breakdown.

95. Let G be the number of gastric bypasses performed and t be the number of years past 2000.

a. If $t = 5$ then $G = 1134$. The number of teenagers who had gastric bypass surgery in 2005 was 1134.

b. Slope = 181.5. The number of teenagers who undergo gastric bypass surgery increases by about 182 each year.

c. The vertical intercept is (0, 226.5). In 2000, 227 teenagers underwent gastric bypass surgery.

d. The horizontal intercept is (−1.25, 0). In 1998, 0 teenagers underwent gastric bypass surgery. This is model breakdown.

Section 1.6

1. b **3.** c **5.** c

7. a **9.** x-scl = 1 y-scl = 5

11. x-scl = 1 y-scl = 0.1

13. x-min $= -45$ x-max $= -5$ x-scl $= 5$
    y-min $= -20$ y-max $= 35$ y-scl $= 5$

15. **a.** $R =$ Revenue in thousands of dollars for Quick Tire Repair Inc., $t =$ Years from 2000. $R = 4t + 584$
    b. Domain: $[2, 14]$; Range: $[592, 640]$
    c. Quick Tire Repair Inc. will have $700 thousand revenue in 2029.
    d. Slope $= 4$. Quick Tire Repair Inc's. revenue is increasing about $4 thousand per year.

17. $y = 2.5x + 8$ 19. $y = -0.75x + 20$

21. **a.** Let $M =$ Millions of metric tons of beef and pork, $t =$ Years since 2000. $M = 2.3t + 132.1$
    b. In 2006 there were about 145.9 million metric tons of beef and pork produced world wide.
    c. In 2019 the world wide production of beef and pork will be about 175 million metric tons.
    d. Domain: $[-3, 10]$; Range: $[125.2, 155.1]$
    e. Slope $= 2.3$. There is an increase of 2.3 million metric tons per year in the production of beef and pork world wide.

23. The slope is incorrect; the numerator and denominator are reversed. The slope should be
    $$m = \frac{365 - 265}{100 - 20} = 1.25 \quad \$1.25 \text{ per bar}$$

25. The slope is incorrect. The order of subtraction is reversed in the numerator. The slope should be
    $$m = \frac{1105 - 90}{3000 - 100} = 0.35 \quad \$0.35 \text{ per pen}$$

27. **a.** Let C be the number of reported chlamydia cases in thousands, and t the number of years since 2000. $C = 58.3t + 702$
    b. Slope $= 58.3$. The reported number of chlamydia cases increases by 58.3 thousand cases per year.
    c. There 1,051,800 estimated cases in 2006 chlamydia.

29. **a.** Let E be the total egg production in the United States (in billions), t the years since 2000. $E = 1.1t + 84.7$
    b. In 2008 the total egg production in the United States was 93.5 billion.
    c. Domain: $[-4, 10]$; Range: $[80.3, 95.7]$
    d. Slope $= 1.1$ billion/year. The egg production is increasing by 1.1 billion eggs per year in the United States.

31. Let U, S, and F be the number of Internet users in the United Kingdom, Spain, and Finland in millions, respectively, and let t be time in years since 1990.
    $$U = 5.5t - 37$$
    $$S = 2.27t - 17.6$$
    $$F = 0.25t - 0.6$$

33. The United Kingdom has the greatest growth. United Kingdom (5.5) > Spain (2.27) > Finland (0.25).

35. **a.** Let R be total revenue for FedEx in millions of dollars and let t be time in years since 2000.
    $$R = 2433.5t + 17195.5$$
    b. Domain: $[-3, 10]$; Range: $[9895, 41531]$
    c. Slope $= 2433.5$. The total revenue for FedEx increases $2433.5 million each year.
    d. If $t = 10$ then $R = 41530.5$. In 2010, the total revenue for FedEx will be $41530.5 million.

37. **a.** Let M be the total consumption of milk in the United States in millions of gallons, and let t be time in years since 2000.
    $$M = -36.5t + 9301.5$$
    b. Domain: $[-3, 10]$; Range: $[8936.5, 9411]$
    c. The vertical intercept is $(0, 9301.5)$. In 2000, the total consumption of milk in the United States was 9301.5 million gallons of milk.
    d. In 2008, Americans consumed approximately 9000 million gallons of milk.
    e. The horizontal intercept is $(254.8, 0)$. In 2255, Americans will consume 0 gallons of milk. This is model breakdown.

39. **a.** Let H be the total amount spent by individuals on health care expenses in the United States in billions of dollars, and let t be time in years since 2000.
    $$H = 17t + 147$$
    b. Domain: $[3, 14]$; Range: $[198, 385]$
    c. Slope $= 17$. The amount of money Americans spend on health care expenses increases $17 billion each year.
    d. If $t = 15$ then $H = 402$. In 2015, Americans will spend $402 billion on health care expenses.
    e. The amount that Americans will spend on health care expenses will reach $500 billion in 2021.

41. $y = 6x - 20$ 43. $y = -3x + 8$
45. $y = \frac{1}{3}x - 5$ 47. $y = -0.19x + 2.66$
49. $y = 15x + 342$ 51. $y = -41x - 3468$

Section 1.7

1. This is not a function because two children the same age may not be in the same grade.

   Input: $a =$ years old.

   Output: $G =$ grade level of student.

3. This is not a function because two children the same age may not be the same height.

   Input: $a =$ years old.

   Output: $H =$ heights in inches of children attending Mission Meadows Elementary School.

5. This is a function.

Input: t = years after starting the investment.

Output: I = interest earned from an investment in dollars.

7. This is not a function because each year there will be several songs at the top of the pop charts.

Input: y = year.

Output: S = song at the top of the pop charts.

9. This is a function.

Input: t = year.

Output: T = amount of taxes you paid in dollars.

11. This is a function.

13. This is not a function, because the input of 1 hour of poker play is related to two different amounts of winnings.

15. This is a function.

17. This is not a function, because the input of Monday is related to two different amounts of money spent on lunch.

19. Domain = {Jan., Feb., May, June}

Range = {5689.35, 7856.12, 2689.15, 1005.36}

21. Domain = {1, 2, 3, 4, 5}

Range = {−650, −100, −150, 60, 125, 200, 300}

23. The death rate from HIV for 20-year-olds is 0.5 per 100,000.

25. In 2002 there were 14304 small business loans made to minority-owned small businesses.

27. The amount spent on lunch on Tuesday was $5.95 and $6.33.

29. Yes, this is a function. Each input has one output.

31. No, this is not a function. The y^2 means some inputs will have more than one output.

33. Yes, this is a function. Each input has one output.

35. Yes, this is a function. Passes the Vertical Line Test.

37. Yes, this is a function. Passes the Vertical Line Test.

39. No, this is not a function. Fails the Vertical Line Test.

41. a. On the first day of starting a diet, a person's weight will be 86.5 kg.

b. 10 days after starting a diet, a person's weight will be 82 kg.

c. 30 days after starting a diet, a person's weight will be 75 kg.

d. 100 days after starting a diet, a person's weight will be 88 kg.

43. a. Eight ounces of chocolate were consumed in the Clark household on the 5th day of the month.

b. Twenty ounces of chocolate were consumed in the Clark household on the 15th day of the month.

c. Twenty eight ounces of chocolate were consumed in the Clark household on the 30th day of the month.

45. a. $P(\text{Ohio}) = 11.48$

b. $P(\text{Texas}) = 23.507783$

c. $P(\text{Wyoming}) = 0.515004$

47. a. $B(t) = 38t + 748$

b. $B(9) = 1090$. In 2009 the average Social Security benefit for retired workers was $1090.

c. $t = 33$. The average monthly Social Security benefit will be $2000 in 2033.

d. Domain: $[-1, 10]$;   Range: $[710, 1128]$

e. $(0, 748)$. In 2000 the average monthly Social Security benefit was $748.

49. a. $M(t) = 0.67t + 39.16$

b. Domain: $[-3, 10]$;   Range: $[37.15, 45.86]$

c. $(0, 39.16)$. In 2000 there were 39.16 million people enrolled in Medicare.

d. $M(10) = 45.86$. In 2010 there were 45.86 million people enrolled in Medicare.

e. $t \approx 16.18$. In 2016 there will be 50 million people enrolled in Medicare.

51. a. $f(5) = 3$ **b.** $f(-10) = -27$ **c.** $x = 3$

d. Domain: All real numbers

Range: All real numbers

53. a. $h(15) = \dfrac{31}{3}$ **b.** $h(-9) = -\dfrac{17}{3}$ **c.** $x = \dfrac{11}{2}$

d. Domain: All real numbers

Range: All real numbers

55. a. $g(2) = -18$ **b.** $g(-11) = -18$

c. Domain: All real numbers

Range: $\{-18\}$

57. Ashlyn set $x = 20$ and solved for $f(20)$. The correct solution is $x = 3$.

59. Alicyn should not have restricted the domain and range. The correct solution is:

Domain: All real numbers

Range: All real numbers

61. a. $f(2) = 1.6$

b. $f(-14) = -49.6$

c. $x = -1.625$

d. Domain: All real numbers

Range: All real numbers

63. a. $h(105) = 1970$

b. $x = -\dfrac{320}{7}$

c. Domain: All real numbers

Range: All real numbers

65. **a.** $f(3) = 2$
    b. $f(-2) = -8$
    c. $x = -1$
    d. Domain: All real numbers
       Range: All real numbers
    e. Vertical intercept $(0, -4)$
       Horizontal intercept $(2, 0)$

67. **a.** $f(25) = 50$
    b. $f(100) = 275$
    c. $x = 75$
    d. Domain: All real numbers
       Range: All real numbers
    e. Vertical intercept $(0, -25)$
       Horizontal intercept $(8, 0)$

69. **a.** $h(2) = 3$
    b. $h(5) = -4$
    c. $x = -0.2$ and $x = 4.2$
    d. Domain: All real numbers
       Range: $(-\infty, 3]$
    e. Vertical intercept $(0, 0)$
       Horizontal intercepts $(0, 0)$ and $(4, 0)$

Chapter 1 Review Exercises

1. $x = \dfrac{61}{5}$
2. $x = 17$
3. $t = 22.9$
4. $x = -\dfrac{4}{9}$
5. **a.** $h = 9.75$ in. The height after 1 week (7days) is 9.75 inches.
   b. 10 days. The course grass attendants should cut the grass in the rough 10 days before the tournament.
6. Let C be the cost in dollars of the satellite phone service and m be the number of minutes used.
   a. $C = 10m + 10$
   b. $C = 40$. A 3-minute call cost \$40.
   c. $m = 29$. The satellite call can last 29 minutes for \$300.
7. Let C be the cost in dollars of renting a Bobcat for h hours.
   a. $C = 40h + 15$
   b. $C = 95$. Renting a Bobcat for 2 hours will cost \$95.
   c. $C = 975$. A three-day rental of a Bobcat will cost \$975.
8. $C =$ The cost to produce putters, n is the number of putters.
   a. $C = 24.44n + 477.80$
   b. Vertical intercept is $(0, 477.80)$. It cost \$477.80 to produce 0 putters.
   c. $C = \$2921.80$. The cost to produce 100 putters is \$2921.80.
   d. Domain: $[0, 100]$; Range: $[477.80, 2921.80]$
   e. Slope $= 24.44$. For every putter produced the cost increases by \$24.44.

9. $P =$ profit for Costco Wholesale Corporation in millions of dollars, t the years since 2000.
   a. $P = 634.2t + 3510.6$
   b. 2008. The gross profit for Costco Wholesale will be 9 billion dollars in 2008.
   c. 9218.4 million. The gross profit for Costco Wholesale in 2009 will be 9.2184 billion dollars.
   d. Domain: $[0, 9]$; Range: $[3510.6, 9218.4]$
   e. Slope 634.2 million per year. The gross profit increases by 634.2 million for each year of operation.

10. **a.** Let P be the percent of full-time workers in private industry who filed an injury case t years since 2000.
$$P(t) = -0.167t + 5.2$$
   b. Slope $= -0.167$. The percent of full-time workers in private industry who filed an injury case is decreasing by about 0.167 percentage points per year.
   c. $P(3) = 4.699$. In 2003 about 4.7 percent of full-time workers in private industry filed an injury case.
   d. Vertical intercept $(0, 5.2)$. In 2000 about 5.2% of full-time workers in private industry filed an injury case.
   e. Horizontal intercept $(31.14, 0)$. In about 2031 there will be no injury cases filed by full-time workers in private industry. This is probably model breakdown since there will most likely always be some injury cases.

11. **a.** Let S be the number of kindergarten through twelfth grade students in millions in Texas public schools t years since 2000.
$$S(t) = 0.08t + 4$$
   b. Slope $= 0.08$. The number of kindergarten through twelfth grade students in Texas public schools is increasing by about 0.08 million (80 thousand) per year.
   c. $S(11) = 4.88$. In 2011 Texas can expect to have about 4.88 million kindergarten through twelfth grade students in public schools.

12. **a.** 7.1 million candles will be produced in 2015.
    b. 5.98 million candles will be produced in 2013.
    c. 10 million candles will be produced in 2020.

13. $T = \dfrac{v}{b}$

14. $m = \dfrac{F}{a}$

15. $x = \dfrac{c - by}{a}$

16. $y = \dfrac{b - 2x}{-a} = \dfrac{2x - b}{a}$

17.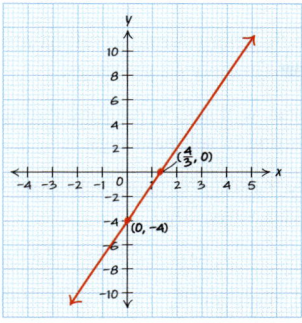

18.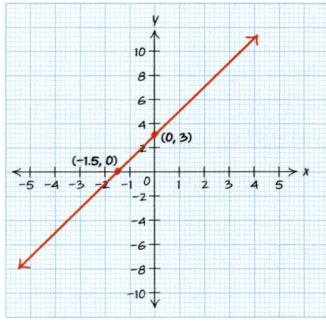

19.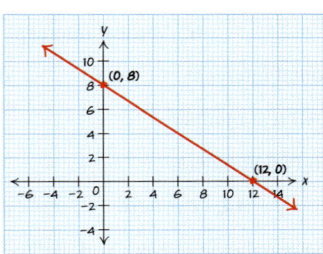

20.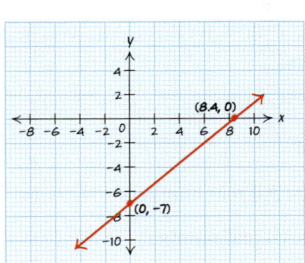

21.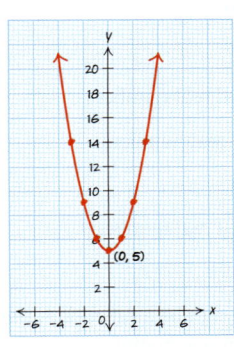

22.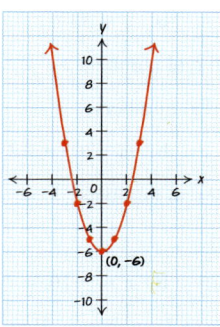

23.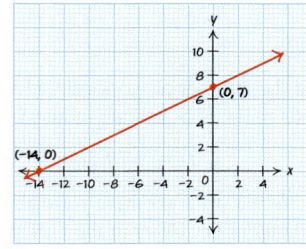

24.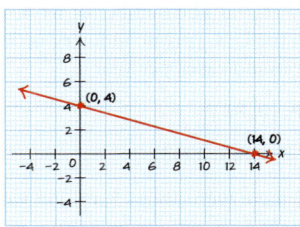

25. $y = 4x - 1$ 26. $y = -2x + 17$ 27. $y = 3x - 2$

28. $y = -0.5x + 17$ 29. $y = \dfrac{1}{4}x + \dfrac{25}{4}$ 30. $y = -\dfrac{5}{2}x + 4$

31. a. $(0, 4)$ b. $(1.3, 0)$

   c. slope $= -3$ d. $f(4) = -8$

   e. $f(x) = 16$ when $x = -4$

   f. $f(x) = -3x + 4$

32. a. $(0, -5)$ b. $(3.3, 0)$ c. slope $= \dfrac{3}{2}$

   d. $h(-2) = -8$ e. $h(x) = -2$ when $x = 2$

   f. $h(x) = \dfrac{3}{2}x - 5$

33. a. slope $= \dfrac{2}{3}$ b. $(0, 6)$ c. $y = \dfrac{2}{3}x + 6$

34. a. slope $= -0.45$ b. $(0, 3)$

   c. $y = -0.45x + 3$

35. The electricity cost for using holiday lights 6 hours a day is $13.

36. a. $P(10) = 10279$. In 2010 the population of Michigan will be about 10,279 thousand.

   b. The population of Michigan will reach 11,000 thousand in about 2034. This may be model breakdown if the current population growth does not continue.

37. a. $f(10) = 12$ b. $x = 4$

   c. Domain: All real numbers

   Range: All real numbers

38. **a.** $h(12) = 6$ **b.** $x = 33$

c. Domain: All real numbers

Range: All real numbers

39. **a.** $g(3) = 12$ **b.** $g(-20) = 12$

c. Domain: All real numbers

Range: {12}

40. **a.** $f(5) = 10.75$ **b.** $x = -10.4$

c. Domain: All real numbers;   Range: All real numbers

Chapter 1 Test

1. **a.** Let C be the number of work-related injury cases in thousands in the U.S. private industry t years since 2000.

$$C(t) = -79.3t + 4333$$

b. $C(9) = 3619.3$. In 2009 there were about 3619.3 thousand work-related injury cases in the U.S. private industry.

c. Slope $= -79.3$ The number of work-related injury cases in the U.S. private industry is decreasing about 79.3 thousand cases per year.

d. Domain: [0, 8]

Range: [3698.6, 4333]

e. In 1992 there were about 5 million work-related injury cases in the U.S. private industry.

2.

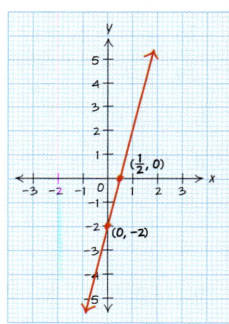

3.

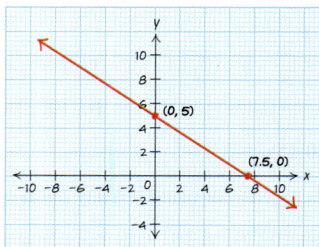

4.

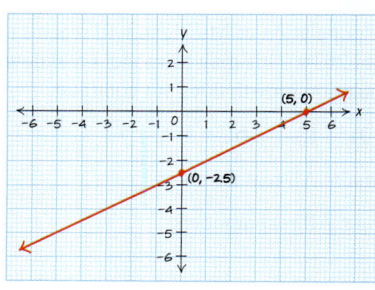

5. $x = \dfrac{c + by}{a}$ **6.** $y = 0.2x + 8.8$

7. $y = 2x - 2$

8. **a.** Six months after starting to sell paintings John Clark will sell approximately 42 paintings.

b. John Clark will sell 50 painting during the 10th month after starting to sell paintings.

9. **a.** Let R be the total revenue in millions of dollars for Apple Inc. t years since 2000.

$$R(t) = 5518t - 13793$$

b. Vertical intercept $(0, -13793)$. In 2000 Apple Inc. had total revenue of $-\$13{,}793$ million. This is model breakdown since Apple could not have a negative revenue.

c. Domain: [3, 9];   Range: [2761, 35869]

d. $R(10) = 41387$. In 2010 Apple Inc. had a total revenue of about $\$41{,}387$ million.

e. Apple Inc.'s total revenue was about $30 billion in 2008.

f. Slope $= 5518$. Apple Inc.'s total revenue increases about $5518 million dollars per year.

10. **a.** Slope $= -\dfrac{4}{5}$ **b.** y-intercept $= (0, 12)$

c. x-intercept $= (15, 0)$ **d.** $y = -\dfrac{4}{5}x + 12$

11. **a.** Slope $= 2$ **b.** Vertical intercept $= (0, 8)$

c. Horizontal intercept $= (-4, 0)$ **d.** $y = 2x + 8$

12. Domain: All real numbers

Range: All real numbers

13. $h = \dfrac{W}{t^2}$ **14.** $x = 1.76$

15. In 2010 the population of New York will be about 19.4 million.

16. **a.** To earn a 70% on the exam you need to study about 19 hours.

b. To earn a 100% on this exam, you need to study about 25 hours.

17. **a.** $f(4) = 25$

b. $x = -4$

c. Domain: All real numbers

Range: All real numbers

18. **a.** $h(35) = 26$

b. $x = -\dfrac{7}{2}$

c. Domain: All real numbers

Range: All real numbers

19. **a.** $g(6) = -9$

b. Domain: All real numbers

Range: {−9}

20. A relation may not be a function if the input value was paired with more than one output value.

CHAPTER 2
Section 2.1

1. (3, 2). This is a consistent system.
3. (−7, 3). This is a consistent system.
5. No solution. This is an inconsistent system. The two lines are parallel. Therefore, they will never intersect
7. a. At 15%, the percentage of people under 18 years was the same as the percentage of people 65 years and over with family income below the poverty level in 1974.
   b. At 13%, the percentage of people 65 years and over with family income below the poverty level was first equal to the percentage of people 18–64 years old in 1993.
9. (3, 10). Consistent; independent.
11. (−6, −8). Consistent; independent.
13. No solutions inconsistent.
15. Consistent; dependent. Infinite number of solutions.
17. a. Let $U(t)$ and $N(t)$ be the total amount of manufacturing output per hour in dollars by the United States and Norway, respectively, and let t be time in years since 2000.
    $$U(t) = 9.26t + 147.7 \quad N(t) = 5.3t + 105.9$$
    b.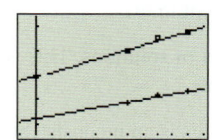
    c. The amount of manufacturing output per hour by the United States was equal to the amount of manufacturing output per hour by Norway in 1989.
19. a. Let $M(t)$ and $F(t)$ be the total percentages of white male and female Americans 25 years old or older who are college graduates, respectively, and let t be time in years since 1980.
    $$M(t) = 0.36t + 21.3 \quad F(t) = 0.58t + 12.3$$
    b.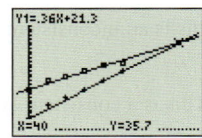
    c. The percentage of white American males with college degrees will be the same as the percentage of white American women with college degrees in 2020.
21. a.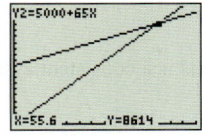
    b. Hope's Pottery will break even after producing and selling 56 vases. At this point, she will have expenses of $8640 and revenue of $8680. Before this point, she will lose money.

23.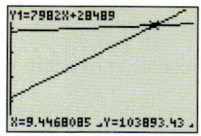
   a. *La Opinion* will have approximately the same circulation as the *Long Beach Press Telegram* in 1999, with a circulation of about 104,289.
   b. *La Opinion*'s model has a slope of 7982, and the *Long Beach Press Telegram*'s model has a slope of 726. This means that the circulation of *La Opinion* is increasing much faster than that of the *Long Beach Press Telegram*.
25. Salary option number 2 will never surpass salary option number 1 because these options are an inconsistent system. Salary option number 1 will always give you more money in the end.
27. (−3, −3). $x = −3, y = −3$. This is a consistent system with independent lines.
29. No solution. This is an inconsistent system. The two lines are parallel.
31. Infinite number of solutions. This is a consistent system with dependent lines.
33. (1.4, 10.3). $t = 1.4, R = 10.3$. This is a consistent system with independent lines.
35. (2, −3). $x = 2, y = −3$. This is a consistent system with independent lines.
37. The two lines are parallel. The system is inconsistent. No solution to the system.
39. A consistent system has exactly one solution if the lines are independent, but it has an infinite number of solutions if the lines are dependent.
41. Many possible answers. Should be graph with two lines intersecting at the point (2, 4).
43. Many possible answers.
45. (2, 11). $x = 2, y = 11$. This is a consistent system with independent lines.
47. No solution. This is an inconsistent system. The two lines are parallel.
49. Infinite number of solutions. This is a consistent system with dependent lines.
51. No solution. This is an inconsistent system. The two lines are parallel.
53. (9, 35). $t = 9, p = 35$. This is a consistent system with independent lines.
55. No solution. This is an inconsistent system. The two lines are parallel.
57. Infinite number of solutions. This is a consistent system with dependent lines.
59. (15, 10). $x = 15, y = 10$. This is a consistent system with independent lines.

Section 2.2

1. (4, 11). This is a consistent system with independent lines.

3. (4, 0). $t = 4$, $P = 0$. This is a consistent system with independent lines.

5. (6, 7). This is a consistent system with independent lines.

7. (12, 20). $k = 12$, $H = 20$. This is a consistent system with independent lines.

9. (5, 4). This is a consistent system with independent lines.

11. $v \approx 55.6$. Hope's Pottery will need to produce and sell 56 vases to break even.

13. The break-even point for Optimum Traveling Detail is 52 automotive details.

15. $t \approx 25.2$. In 2026, the number of associate's degrees conferred will be the same as the number of master's degrees.

17. **a.** Let $P(t)$ be the student-teacher ratio for public schools in the United States, let $R(t)$ be the total student-teacher ratio for private schools, and let t be the years since 2000.
$$P(t) = -0.05t + 16 \quad R(t) = -0.175t + 14.5$$
b. $t = -12$. In 1988, the student-teacher ratio was the same for public and private schools.

19. **a.** Let $H(t)$ be the percentage of Hispanics 25 years old or older who have a college degree, and let t be time in years since 2000. $H(t) = 0.3t + 10.6$
b. The percentage rate of Hispanics who have a college degree is increasing at a rate of 0.3% each year.
c. Let $A(t)$ be the percentage of African Americans 25 years old or older who have a college degree, and let t be time in years since 2000. $A(t) = 0.33t + 16.5$
d. The percentage rate of African Americans who have a college degree is increasing at a rate of 0.33% each year.
e. $t \approx -196.7$. In 1803, the percentages of Hispanics and African Americans who have a college degree will be the same. This is model breakdown.

21. (3, 1). This is a consistent system with independent lines.

23. $(-2, 0)$. $k = -2$, $H = 0$. This is a consistent system with independent lines.

25. (3, 7). $a = 3$, $G = 7$. This is a consistent system with independent lines.

27. $\left(2, \dfrac{19}{3}\right)$. $t = 2$, $P = \dfrac{19}{3}$. This is a consistent system with independent lines.

29. $x = 5$, $y = 1.5$. This is a consistent system with independent lines.

31. Sales of $40,000 worth of high-end fashion will result in the two options having the same monthly salaries.

33. **a.** Let A be the amount of money Damian invests in the account paying 5% simple interest, and let B be the amount of money Damian invests in the account paying 7.2% simple interest.
Total money: $A + B = 150,000$
Total interest: $0.05A + 0.072B = 9600$
b. Damian needs to invest $54,545.45 in the account paying 5% interest and $95,454.55 in the account paying 7.2% interest.

35. **a.** Let A be the amount of money Henry invests in the account paying 5% simple interest, and let B be the amount of money Henry invests in the account paying 8% simple interest.
Total money: $A + B = 1,375,000$
Total interest: $0.05A + 0.08B = 87,500$
b. Henry needs to invest $750,000 in the account paying 5% interest and $625,000 in the account paying 8% interest.

37. Joan needs to invest $93,750 in the account paying 5% interest, and $81,250 in the account paying 9% interest.

39. Truong invested $11,000 in the bond with a return of 9% and $29,000 in the stock with a return of 11%.

41. A total of 40 people would result in the same price from either of the two tour companies.

43. Tom did the work correctly. Matt wrote that the system is consistent, which is false. The system is inconsistent because the variables were both eliminated and the remaining statement was false.

45. (0.27, 1.57). $d = 0.27$, $r = 1.57$. This is a consistent system with independent lines.

47. Infinite number of solutions. This is a consistent system with dependent lines.

49. No solution. This is an inconsistent system. The two lines are parallel.

51. $\left(\dfrac{25}{7}, -\dfrac{82}{7}\right)$. This is a consistent system with independent lines.

53. Infinite number of solutions. This is a consistent system with dependent lines.

55. Infinite number of solutions. This is a consistent system with dependent lines.

57. No solution. This is an inconsistent system. The two lines are parallel.

59. $g = 2$, $h = 7$. This is a consistent system with independent lines.

61. $x = -6$, $y = -7$. This is a consistent system with independent lines.

Section 2.3

1. $x = 3$, $y = 2$. This is a consistent system with independent lines.

3. $(-3, -5)$. $g = -3$, $h = -5$. This is a consistent system with independent lines.

5. $x = 5$, $y = 2$. This is a consistent system with independent lines.

7. (4, 3). $k = 4$, $H = 3$. This is a consistent system with independent lines.

9. (4, 1). $t = 4, k = 1$. This is a consistent system with independent lines.

11. The chemistry student will need 15.556 ml of the 5% HCl solution and 4.444 ml of the 50% HCl solution to make 20 ml of a 15% HCl solution.

13. Kristy will need 15.625 ml of the 2% NaCl solution and 9.375 ml of the 10% NaCl solution to make 25 ml of a 5% NaCl solution.

15. The store manager must use 33.33 lb of the 20% premade mix with 66.67 lb of the 5% premade mix to make a 100-lb of mix with 10% peanuts.

17. $(6, -6)$. $w = 6, s = -6$. This is a consistent system with independent lines.

19. Infinite number of solutions. This is a consistent system with dependent lines.

21. No solution. This is an inconsistent system. The two lines are parallel.

23. $(2.3, -3.27)$. $c \approx 2.3, W \approx -3.27$. This is a consistent system with independent lines.

25. Infinite number of solutions. This is a consistent system with dependent lines.

27. a. Let A be Option 1 for admission and ride tickets at a local fair in dollars, let B be Option 2 for admission and ride tickets at a local fair in dollars, and let r be the total number of rides purchased. $A = 0.50r + 22; B = 0.75r + 15$. Therefore, 28 rides will result in the same cost for both options for admission and ride tickets at a local fair.

    b. Option 1 is a better deal if you are planning on riding a larger number of rides at the fair.

29. Each plank costs $1.26, and each 4 × 4 cost $4.24.

31. Ana worked a total of 140 regular hours and 25 overtime hours.

33. Substitution, because H is already isolated.

35. Substitution, because T is already isolated.

37. In solving an inconsistent system using the elimination or substitution method, both variables will be eliminated, and the remaining statement will be false.

39. Use a table or graph the two equations and see where the lines intersect.

41. $d = 15, g = -45$. This is a consistent system with independent lines.

43. $x = 0, y = \frac{2}{11}$. This is a consistent system with independent lines.

45. $x = 3.7, y = -4$. This is a consistent system with independent lines.

47. $d = 19.6, W = 14.4$. This is a consistent system with independent lines.

49. Infinite number of solutions. This is a consistent system with dependent lines.

51. Frank did not multiply the 30 by 2.

53. $x \approx -0.58, y \approx 2.1394$. This is a consistent system with independent lines.

55. $\left(0, -\frac{8}{7}\right)$. $c = 0, d = -\frac{8}{7}$. This is a consistent system with independent lines.

57. $x = 3, y = -6$. This is a consistent system with independent lines.

59. $x = 7.5, y = 3$. This is a consistent system with independent lines.

61. Infinite number of solutions. This is a consistent system with dependent lines.

63. No solution. This is an inconsistent system. The two lines are parallel.

65. $t = -0.7, y = 0.87$. This is a consistent system with independent lines.

Section 2.4

1. $x > 6$
3. $P \geq 8$
5. $V < -8$
7. $t > -9$
9. $K \geq -56$

11. Michelle's work is correct. Amy reversed the inequality symbol for subtracting 8 from both sides, which is incorrect.

13. a. Let $F(t)$ and $P(t)$ be the number of full- and part-time faculty in U.S. higher education institutions and let t be time in years since 1990.

    $$F(t) = 23.65t + 328.45 \quad P(t) = 29.1t + 173.3$$

    b. The number of part-time faculty will be greater than the number of full-time faculty after 2018.

15. a. Let $A(t)$ and $E(t)$ be the total population of Africa and of Europe, respectively, in millions, and let t be time in years since 1900.

    $$A(t) = 13.25t + 361 \quad E(t) = 2.43t + 663.1$$

    b. Africa has a greater population than Europe after 1997.

17. $<$ 19. $\geq$ 21. $>$ 23. $<$

25. The cabinet manufacturer must sell at least 118 cabinets to break even or make a profit.

27. Optimum Traveling Detail must detail at least 52 cars to break even or make a profit.

29. a. Let $F(m)$ be the cost in dollars for the Freedom plan from Uptown Wireless when used for m minutes.

    $$F(m) = 69.99t + 0.3m$$

    Let $N(m)$ be the cost in dollars for the No-Strings plan from Go Wireless when used for m minutes.

    $$N(m) = 69.99t + 0.35m$$

    Let $Y(m)$ be the cost in dollars for the Your Free plan from U-R Mobile when m minutes are used.

    $$Y(m) = 79 + 0.25m$$

    b. U-R Mobile will have the cheapest plan if used for more than 180 minutes.

31. Before 1992, the percentage of births to unmarried women in the United States was greater than the percentage in the United Kingdom.

33. Sales need to be at least $2976.19.

35. $Y1 < Y2$ when $x < 2$

37. $x < -7$ **39.** $x > 5$ **41.** $x \geq 4$

43. $x \leq 3.5$ **45.** $x < -3.33333$ **47.** $x > -5$

49. $x \leq -3$ **51.** $x > 2$ **53.** $x < 12$

55. $x \leq -\dfrac{16}{3}$ **57.** $x \geq -\dfrac{9}{2}$ **59.** $d > \dfrac{6}{7}$

61. $v > -8.49$ **63.** $k \leq 2.3$ **65.** $w \leq \dfrac{415}{74}$

67. $x < -11.59$

Section 2.5

1. $x = \pm 12$ **3.** $h = 8$ or $h = -22$

5. $b = 20$ or $b = 4$ **7.** No solution.

9. $r = 7$ or $r = -29$ **11.** No solution.

13. No solution. **15.** $x = \dfrac{41}{3}$ or $x = \dfrac{31}{3}$

17. Ricardo should travel at an average speed of 45 mph or 75 mph.

19. Ricardo will be 50 miles away from Fresno after about 10.83 hours and again after about 12.5 hours.

21. $-6 < x < 6$ **23.** $x \leq -3$ or $x \geq 3$

25. $x < -5$ or $x > -2$ **27.** $-3 \leq x \leq 8$

29. $x < -6$ or $x > 5$

31. a. $-4 < x < 4$
   b. $x < -4$ or $x > 4$

33. a. $0 \leq x \leq 4$
   b. $x \leq 0$ or $x \geq 4$

35. a. $-5 < x < 5$
   b. $x < -5$ or $x > 5$

37. $-5 < x < 5$

39. $-1 \leq h \leq 7$

41. $-5 < b < 7$

43. The acceptable lengths of this coupler are $2.999 \leq L \leq 3.001$ centimeters.

45. The acceptable pressures for this experiment are $19.5 \leq P \leq 20.5$ psi.

47. $-3 < x$ or $x > 3$
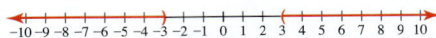

49. $2 \leq p$ or $p \geq 6$

51. $y < -9$ or $y > -3$
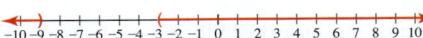

53. $h < 64$ or $h > 74$. Men's heights that are less than 64 inches or greater than 74 inches are considered unusual.

55. a. IQ scores that are below 68 or above 132 would be considered unusual.
   b. Yes. Albert Einstein's IQ was unusual. His IQ was greater than 172.

57. $-10 < r < -2$
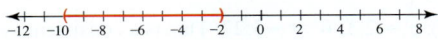

59. $x > 8$
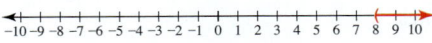

61. No solution.

63. $m < -9$ or $m > 7$
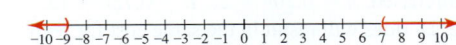

65. $x < -24$
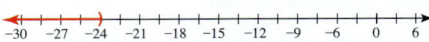

67. All real numbers.

69. $2 \leq x \leq 6$
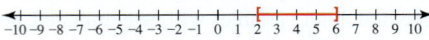

71. $x < \dfrac{1}{2}$ or $x > \dfrac{7}{2}$

73. No solution.

75. $x \leq \dfrac{4}{3}$

77. $-16 \leq x \leq 4$

Section 2.6

1. A 61" tall person weighing 105 pounds falls into the normal range.

3. A 5'9" (69") tall person is considered overweight if the person weighs above 172 pounds.

5. A person who weighs 150 pounds must be between 5'5" and 6'3" to fall in the normal range.

7. A person who is 67" tall has a normal weight range of 120 pounds to 160 pounds.

9. They can build at most 370 mountain bikes per month.

11. Yes, they can build 400 cruisers and 100 mountain bikes per month. This is at the 500 bike per month limit and under the cost limitations for a month.

13. **a.** Let C be the total number of cruisers Bicycles Galore produces monthly and let M be the total number of mountain bikes Bicycles Galore produces monthly.

    $$C + M \leq 500 \quad 65C + 120M \geq 40{,}000$$

    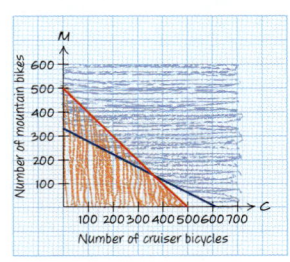

    b. Bicycles Galore will be able to meet a demand by the board of directors for $40,000 in profit per month. Any combination of mountain bikes and cruisers in the overlapping shaded region will meet the board's demand.

    c. They must make 137 mountain bikes and 363 cruisers to make $40,000 profit. If they make more mountain bikes they will make more profit.

15.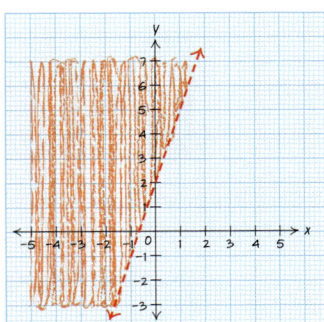

17.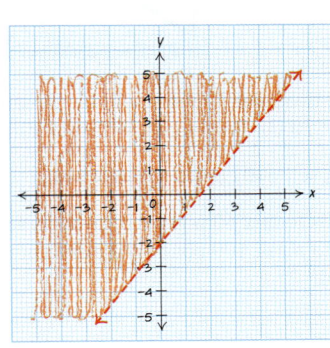

19.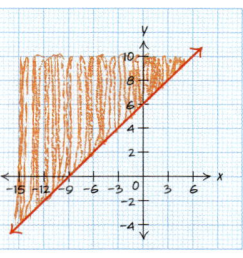

21.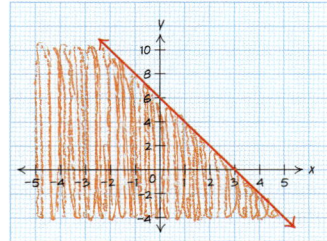

23.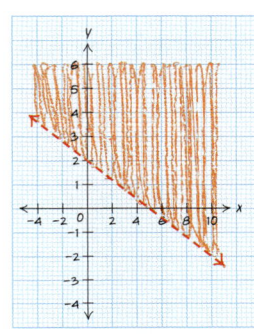

25.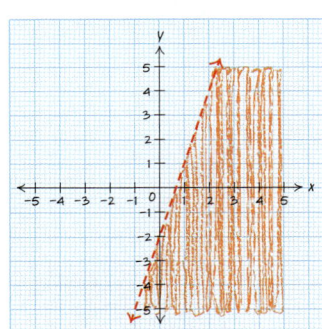

27. $y \leq 2x - 3$

29. $y > -\dfrac{3}{2}x + 6$

31. $y > -\dfrac{1}{2}x + 9$

33. $y \leq -x + 1$

35. Let A be the amount in dollars Juanita invests in the account paying 4.5% interest, and let B be the amount in dollars invested in the account paying 3.75% interest.

    $$A + B \leq 750{,}000 \quad 0.045A + 0.0375B \geq 30{,}000$$

    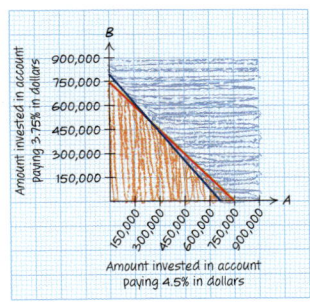

37. a. Let P be the number of power bars, and let D be the number of sports drinks that need to be consumed by the athletes.

$$240P + 300D \geq 2000 \quad 30P + 70D \geq 350$$

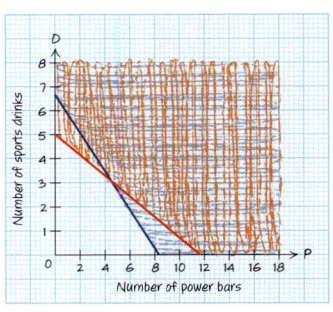

b. The point of intersection gives the least amount of each. Each athlete can carry a minimum of 3 drinks and 5 power bars. This combination will meet the minimum calorie and carb requirement.

c. If a racer carries 4 drinks, he or she must also carry 4 power bars.

39.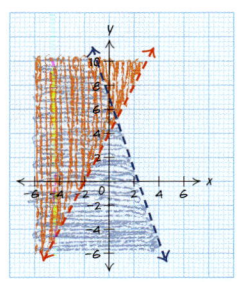

41.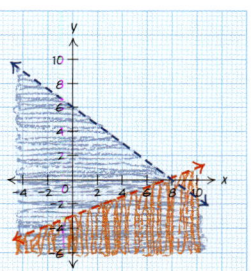

43. No solution.

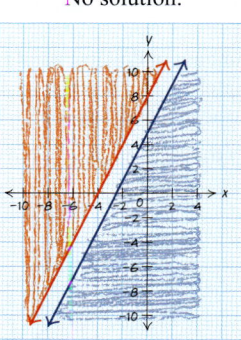

45.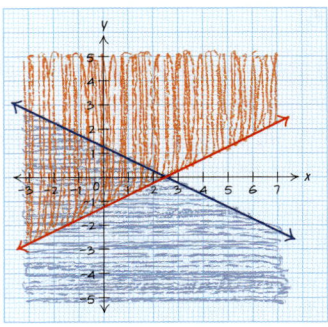

47.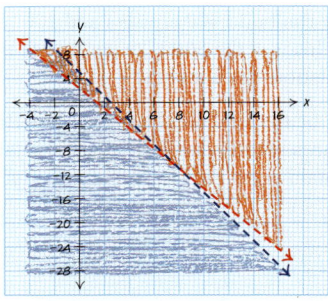

49.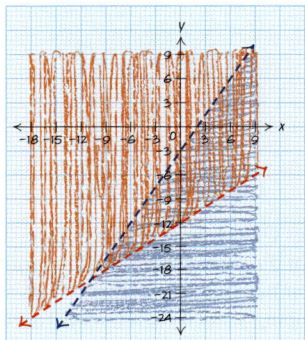

Chapter 2 Review Exercises

1. a.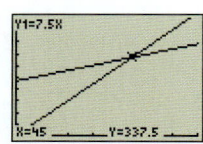

b. Frank's Shoe Repair will break even when they repair 45 pairs of shoes a month.

2. a.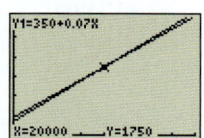

b. These two salary options will be the same when the sales reach $20,000 per month, with a monthly salary of $1750.

3. You must sell 640 copies to break even.

4. John needs to mix 60 pounds of peanuts and 40 pounds of cashews to make 100 pounds of mixed nuts that cost $3.00 per pound.

APPENDIX E Answers to Selected Exercises   E-21

5. **a.** Let $R(a)$ be the residual value in dollars of a Mazda RX-8 that is a years old.

$$R(a) = -2987.5a + 27{,}200$$

Let $C(a)$ be the residual value in dollars of a Mini Cooper S that is a years old.

$$C(a) = -2112.5a + 20{,}900$$

b. The two cars will have the same residual value of approximately $6200.00 when they are both about 7 years old.

6. **a.** Let $G(t)$ be the percentage of births to unmarried women in Germany t years since 2000.

$$G(t) = t + 24.0$$

b. Slope = 1. The percentage of births in Germany to unmarried women is increasing by 1 percentage point per year.

c. Let $U(t)$ be the percentage of births to unmarried women in the United Kingdom t years since 2000.

$$U(t) = 0.7t + 39.4$$

d. The percentage of births to unmarried women in the United Kingdom will be greater than the percentage in Germany after 2051. This is probably model breakdown.

7. Consistent system with independent lines; (10, 1)

8. Consistent system with independent lines. $\left(-\dfrac{57}{16}, -\dfrac{45}{16}\right)$

9. Consistent system with independent lines.

$$d = \dfrac{-91}{22} \quad w = \dfrac{-14}{11}$$

10. Inconsistent system, no solution.

11. Consistent system with dependent lines. Infinite number of solutions.

12. The foundation should invest $1,750,000 in the account that pays 7% simple interest and $1,250,000 in the account that pays 11% simple interest.

13. Brian should use 105 gallons of the 15% solution and 45 gallons of the 5% solutions to make the 150 gallons of the 12% solution of test chemical AX-14.

14. Consistent system with independent lines; $(-2, -5)$.

15. Consistent system with independent lines; (2, 1). $w = 2, t = 1$.

16. Consistent system with independent lines; $(-3.38, 2.69)$.

$$d \approx -3.38, C \approx 2.69$$

17. Consistent system with independent lines; (15, 12.5). $m = 15, n = 12.5$.

18. Consistent system with independent lines; $(-0.7, -5.07)$.

19. Consistent system with dependent lines; Infinite number of solutions.

20. Consistent system with independent lines; $(w \approx -0.587, t \approx 2.03)$.

21. Inconsistent system. No solution.

22. Consistent system with independent lines; $\left(0, \dfrac{1}{3}\right)$.

23. Consistent system with independent lines; $(100, -27)$. $w = 100, z = -27$.

24. Consistent system with independent lines; $(2.765, -6.144)$.

25. $t \approx 21.86$. The percentage of births that are to teenage mothers in Louisiana will be less than the percentage in Delaware after 2021. This is probably model breakdown.

26. $x > 12$

27. $x > 7$

28. $x \geq \dfrac{55}{3}$

29. $x \leq \dfrac{11}{12}$

30. $t > -\dfrac{16}{27}$

31. $d < -5$

32. $V > -2.54$

33. $k \leq 259.76$

34. $x = 27$ or $x = -13$

35. $x = 23.5$ or $x = -26.5$

36. $x = 11$ or $x = -49$

37. $x = 24$ or $x = -10$

38. $-8.5 < x < 8.5$

39. $x < -6.5$ or $x > 6.5$

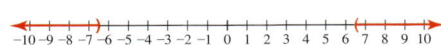

40. $x \leq -15$ or $x \geq 9$

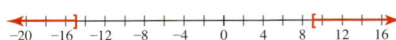

41. $4 \leq x \leq 10$

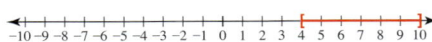

42. $x < -15$ or $x > 5$

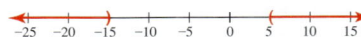

43. $x \leq -6$ or $x \geq -4$

44.

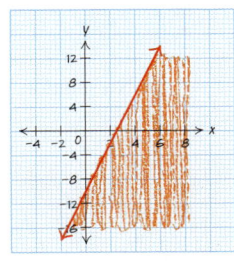

45.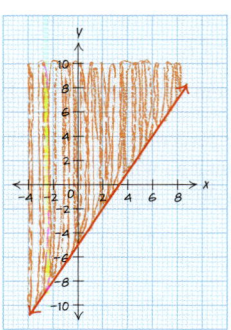

46.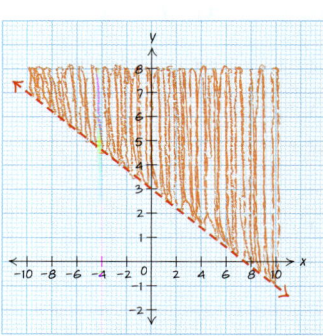

47.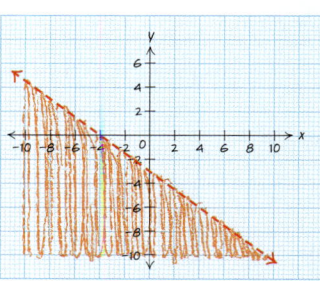

48.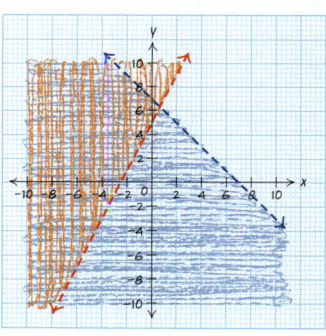

49. No solution.

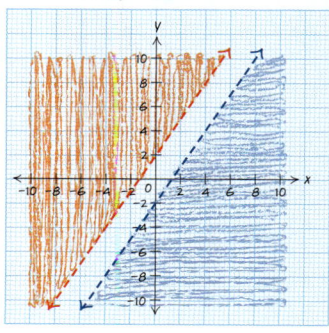

50.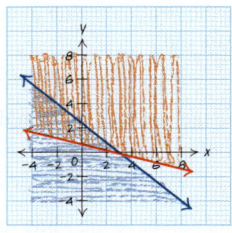

51.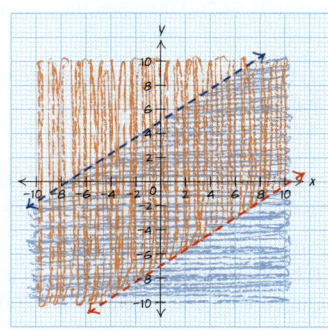

52. Let R be regular tickets, and let S be student tickets.

$$S + R \leq 12{,}000 \qquad 9S + 15R \geq 140{,}000$$

53. $y \leq \dfrac{1}{3}x - 6$

54. $y > -\dfrac{1}{2}x + 6$

Chapter 2 Test

1. a. Let $C(t)$ be the average hourly earnings, in dollars per hour, of production workers in manufacturing industries in California.

$$C(t) = 0.31t + 14.11$$

Let $M(t)$ be the average hourly earnings, in dollars per hour, of production workers in manufacturing industries in Massachusetts.

$$M(t) = 0.60t + 14.66$$

b. $t \approx -1.9$. In about 1998, the average hourly earnings for production workers in manufacturing industries in California were approximately the same as those in Massachusetts at about $13.52 per hour.

2. a. Let A be the amount in dollars invested in the account paying 12%, and let B be the amount in dollars invested in the account paying 7%.

$$A + B = 500{,}000 \quad 0.12A + 0.07B = 44{,}500$$

b. Christine should invest $190,000 in the account paying 12% interest and $310,000 in the account paying 7% interest.

3. Let M be the number of hours per week Georgia needs to study for her math class, and let h be the number of hours per week Georgia needs to study for her history class.

$$M \geq 12 \qquad M + h \leq 20$$

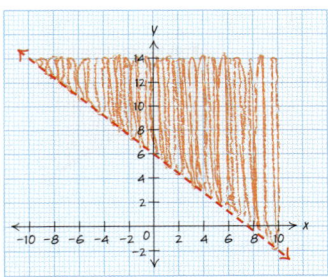

20.

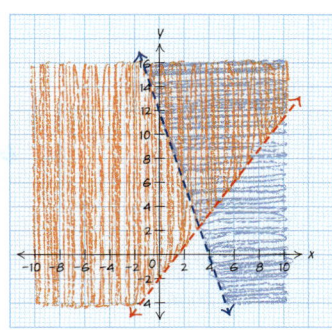

4. Wendy should use 1.6 liters of the 5% HCl solution and 0.4 liter of the 20% HCl solution to make 2 liters of 8% HCl solution.

5. $x > \dfrac{15}{7}$

6. $m \leq -1.8$

7. $a \leq -4.1$

8. Consistent system with independent lines; $(12, -2)$.

9. Consistent system with dependent lines; infinite number of solutions.

10. Inconsistent system. No solution.

11. Consistent system with independent lines; $(-4.2, 3.5)$.

12. $x \leq 8$

13. Scott needs to sell 49 hammock stands to break even.

14.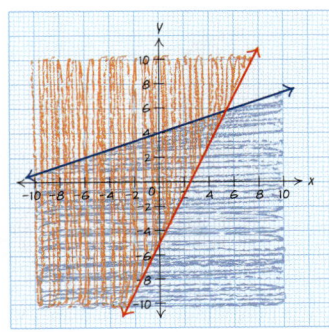

15. The revenue for cellular phone providers will be greater than that of local telephone providers after 2026.

16. $x < -4$

17. $x = 9$ or $x = -19$

18. $-9 < x < 17$

19.

Cumulative Review for Chapters 1–2

1. **a.** Let P be the population of the United States in millions, and let t be the years since 2000.

   $P = 2.7t + 282.1$

   b. The population of the United States will be 314.5 million in 2012.

   c. Domain: $[0, 13]$; range: $[282.1, 317.2]$

   d. Vertical intercept: $(0, 282.1)$. The population of the United States was about 282.1 million in 2000.

   e. $(15.9, 325)$. The United States. population will reach 325 million in about 2016.

2. $y = 3x - 4$

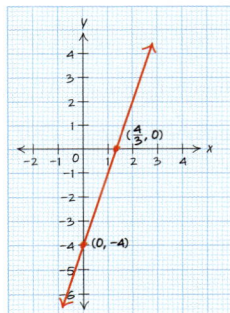

3. $y = -\dfrac{1}{3}x + 2$

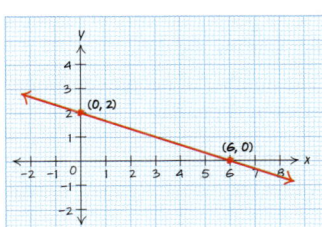

4. $5x - 6y = 9$

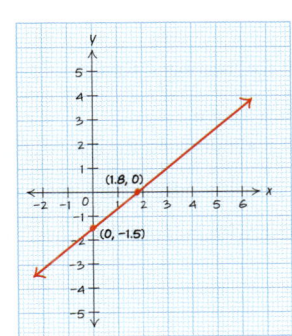

5. $c = \dfrac{a-b}{ab}$

6. $m = \dfrac{2(P-3)}{n^2}$

7. $y = 2x + 10$

8. $y = -\dfrac{3}{4}x + \dfrac{13}{2}$

9. a. Steve plans to run 11 miles in week 4.
   b. Steve plans to run 20 miles in week 10.

10. a. Let $R(t)$ be the total revenue (in billions of dollars) for Cisco Systems, Inc., and let t represent the years since 2000.
    $R(t) = 4.9t + 0.3$
    b. Vertical intercept: $(0, 0.3)$. The total revenue for Cisco Systems, Inc. in 2000 was about 300 million.
    c. Domain: $[3, 8]$ (due to the economic downturn) Range: $[15, 39.5]$
    d. The model predicts Cisco Systems' revenue to be $49.3 billion in 2010.
    e. The model predicts Cisco Systems' revenue to be $64 billion in 2013.
    f. The slope is 4.9, which means that Cisco Systems' revenue is increasing at a rate of $4.9 billion per year.

11. a. $m = \dfrac{2}{3}$ b. $(0, 4)$ c. $(-6, 0)$ d. $y = \dfrac{2}{3}x + 4$

12. a. $m = \dfrac{15}{4}$ b. $(0, -5)$ c. $\left(\dfrac{4}{3}, 0\right)$ d. $y = \dfrac{15}{4}x - 5$

13. Domain: All real numbers; Range: All real numbers

14. Domain: All real numbers; Range: $\{20\}$

15. $a = -12$

16. $x = 1.76$

17. $x = 64$

18. $p = \dfrac{248}{35}$

19. On the sixth day of the month, a person sleeps 7 hours.

20. a. $P(14) = 58$. 58 people out of 100 remain on the diet plan 14 days after starting the diet.
    b. $m = -3$. The number of people remaining on the diet plan decreases by 3 per day.

21. a. $f(3) = -18$ b. $x = 11$
    c. Domain: All real numbers; Range: All real numbers

22. a. $h(32) = -9$ b. $x = 15$
    c. Domain: All real numbers; Range: All real numbers

23. a. $g(10) = 7$
    b. Domain: All real numbers; Range: $\{7\}$

24. a. $f(6) = 8$ b. $x = -6$
    c. Domain: All real numbers; Range: All real numbers

25. Consistent system with independent lines; $x = 3, y = 4$.

26. Consistent system with independent lines; $(4, 22)$. $d = 4, c = 22$.

27. Consistent system with dependent lines; infinite number of solutions.

28. Consistent system with independent lines; $\left(\dfrac{1}{2}, \dfrac{1}{3}\right)$
    $a = \dfrac{1}{2}, b = \dfrac{1}{3}$.

29. Consistent system with independent lines; $\left(\dfrac{13}{15}, -\dfrac{11}{15}\right)$
    $x = \dfrac{13}{15}, y = -\dfrac{11}{15}$

30. Inconsistent system. No solution.

31. a. $x + y = 1{,}200{,}000 \quad 0.05x + 0.035y = 53{,}250$
    b. $750,000 at 5% and $450,000 at 3.5%.

32. Hamid should mix $2\dfrac{2}{3}$ liters of 6% HCl solution with $1\dfrac{1}{3}$ liters of 24% HCl solution to get 4 liters of 12% HCl solution.

33. $x < 4$

34. $m \geq -\dfrac{68}{5}$ or $m \geq -13.6$

35. $a \geq \dfrac{115}{57}$

36. $x \leq 3$

37. Sandra must have $2750 or more per week in sales to earn at least $565 per week.

38.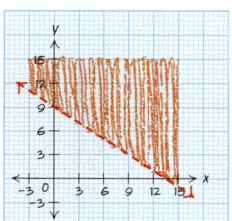

39.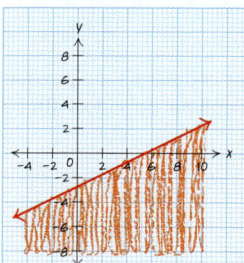

40.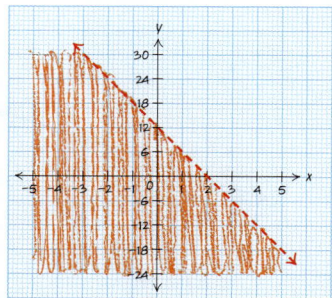

41. $t \geq 16.25$. The percentage of Canadian TV subscribers who use satellite services will be greater than those who use cable services after 2016.

42. $x < 4$

43. $x = 10$ or $x = -14$
44. $a = 24$ or $a = -8$
45. No solution.
46. $r = -\dfrac{1}{3}$ or $r = -3$
47. $-5 < x < 9$

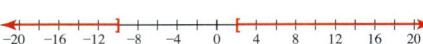

48. $-9 \leq x \leq 6$

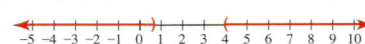

49. $x \leq -10$ or $x \geq 2$

50. $x < \dfrac{2}{3}$ or $x > 4$

51.
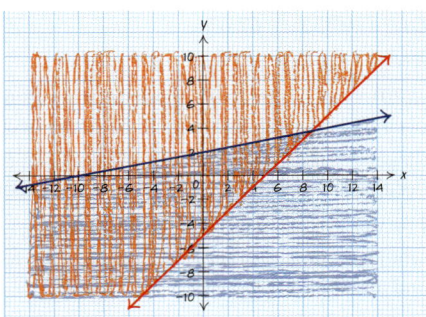

52.
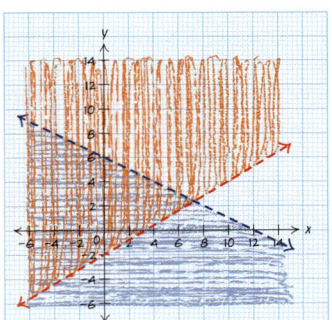

CHAPTER 3
Section 3.1

1. 113
3. $2^9 = 512$
5. w^{10}
7. $7^3 = 343$
9. z^4
11. $3^3 = 27$
13. s^2
15. $\dfrac{4r^5}{t^4}$
17. $\dfrac{-9a^3}{b^7}$
19. $3b^2$
21. $\dfrac{y^9}{3x^8}$
23. $\dfrac{-4y^5}{x^6}$
25. 189. Fred added the bases and added the exponents.
27. $\dfrac{-7x^2}{y^3}$. Bob treated -7 as a negative exponent and incorrectly moved it to the denominator.
29. m^7. Karen multiplied the exponents instead of adding.
31. $8a^3c^2$
33. $15x^6y^4$
35. $\dfrac{16}{81}a^{12}b^{28}c^4$
37. $\dfrac{1}{4x^4y^6}$
39. $a^2 + 10ab + 25b^2$
41. 1
43. $\dfrac{25}{9}$
45. $\dfrac{81y^8}{x^{12}}$
47. a. $7x^2y$
    b. $7x^2y$
    c. Yes; because the bases are the same, you can subtract exponents.
    d. $49x^4y^2$
49. a. No, because you cannot apply the exponent over addition.
    b. $4x^2 + 24x + 36$
51. $x^2 + 18xy + 81y^2$
53. $\dfrac{6}{7x^2y^4}$
55. $\dfrac{4h^6}{25g^6}$
57. $\dfrac{9}{8x^5}$
59. $\dfrac{75}{2c^3}$
61. a. The power rule.
    b. The product rule.
    c. Negative exponents.
63. $x^{\frac{1}{2}}$
65. $m^{\frac{1}{5}}$
67. $c^{\frac{3}{2}}$
69. $t^{\frac{2}{3}}$
71. $(5xy)^{\frac{1}{2}}$
73. $4^{\frac{1}{7}}m^{\frac{3}{7}}n^{\frac{6}{7}}p^{\frac{2}{7}}$
75. $\sqrt[3]{r}$
77. $\sqrt[3]{n^2}$
79. $\sqrt[3]{xy}$
81. $\sqrt[5]{rs^2}$
83. $\sqrt{xy^3z}$
85. 3
87. 10
89. -4
91. $\dfrac{1}{5}$
93. $2xy^3$
95. $8m^3n^6$
97. $\dfrac{11b^4}{a^3}$
99. $\dfrac{7x^2}{5y^3}$
101. $\dfrac{4mn^4}{p}$
103. $\dfrac{a}{10b^2c}$

Section 3.2

1. Two terms; $5x$ variable term with coefficient 5; 9 constant term.
3. Three terms; $-3x^2$ and $2x$ variable terms with coefficients -3 and 2; -8 constant term.
5. Three terms; all variable terms with coefficients 12, 3, and -8.
7. Yes, it is a polynomial.
9. No, it is not a polynomial because the first term has a negative exponent on the variable.
11. Yes, it is a polynomial.
13. No, it is not a polynomial because the first term has a rational exponent on the variable.
15. Yes, it is a polynomial.

17. $2x^5$ has degree 5; -7 has degree 0; the polynomial has degree 5.

19. $5p^2$ has degree 2; $4p$ has degree 1; -87 has degree 0; the polynomial has degree 2.

21. $4m$ has degree 1; 8 has degree 0; the polynomial has degree 1.

23. $2a^2b^3$ has degree 5; $3ab^2$ has degree 3; $-8b$ has degree 1; the polynomial has degree 5.

25. $\frac{2}{3}gh$ has degree 2; $\frac{1}{4}g^3h^5$ has degree 8; $-\frac{2}{9}g$ has degree 1; 7 has degree 0; the polynomial has degree 8.

27. $7x + 14$

29. $4x + 4$

31. $8x^2 + 4x - 20$

33. $2x^3z^2 - 4x^2z + 17xz + 3z$

35. $-4x + 16$. Chie distributed the subtraction sign and then turned the addition problem into a multiplication problem.

37. $25x^2 - 80x + 64$. Gordon distributed the exponent across the two terms in the binomial.

39. $15a^2b + 20ab^2 - 24ab - 32b^2$

41. $6x^3 + 5x^2 - 11x + 35$

43. $9x^2 + 42xy + 49y^2$

45. $51x^2 + 31x - 5$

47. $21xy^2 + 43xy - 42x$

49. a. $T(0) = 712.3$, so in 2000, the average American ate 712.3 pounds of fruits and vegetables.
b. $F(0) = 289.6$, so in 2000, the average American ate 289.6 pounds of fruit.
c. $T(0) - F(0) = 422.7$, so in 2000, the average American ate 422.7 pounds of vegetables.
d. Let $V(y)$ be the average number of pounds of vegetables each American eats per year.
$$V(y) = -y + 422.7$$
e. $V(0) = 422.7$, so in 2000, the average American ate 422.7 pounds of vegetables.
f. $V(10) = 412.7$, so in 2010, the average American will eat approximately 412.7 pounds of vegetables. $V(15) = 407.7$, so in 2015, the average American will eat approximately 407.7 pounds of vegetables. $V(20) = 402.7$, so in 2020, the average American will eat approximately 402.7 pounds of vegetables.

51. a. Let $O(t)$ be the per capita consumption of milk products in gallons per person other than whole milk in the United States t years since 2000.
$$O(t) = -0.08t + 14.33$$
b. $W(5) = 7.04$, so 7.04 gallons of whole milk were consumed per person in the United States in 2005.
c. $O(5) = 13.93$, so 13.93 gallons of milk products other than whole milk were consumed per person in the United States in 2005. $O(10) = 13.53$, so 13.53 gallons of milk products other than whole milk were consumed per person in the United States in 2010.
d. Slope $= -0.21$. The amount of whole milk consumed per person in the United States is decreasing by approximately 0.21 gallon per year.
e. M-intercept: $(0, 22.42)$. In 2000, 22.42 gallons of milk products were consumed per person in the United States.

53. $U(t) - M(t)$ is the number of women in the United States t years since 1900.

55. $ST(d)$ is the miles driven on day d of a cross-country trip.

57. $\frac{D}{U}(t)$ is the average amount of national debt per person in the United States in dollars t years since 1900.

59. $U(t) \cdot D(t)$ is the total amount of personal debt in the United States t years since 1900.

61. a. $M(t) + F(t) = 100$ because the percentage must add up to 100.
b. $F(t) = 0.24t + 11.36$
c. $F(5) = 12.56$. In 2005, approximately 12.56% of jail inmates in United States federal and state prisons were female.
d. Slope $= 0.24$. The percentage of jail inmates who are female in U.S. federal and state prisons is increasing by approximately 0.24 percentage points per year.
e. M-intercept $= (0, 88.64)$. The percentage of jail inmates who are male in U.S. federal and state prisons was 88.64% in 2000.

63. a. $O(t) = 0.433t + 62.67$
b. $R(t) = -0.433t + 37.33$
c. $R(20) = 28.67$. In 2010, approximately 28.67% of occupied housing units should be renter occupied.
d. Slope $= -0.433$. The number of housing units that are renter occupied is decreasing by 0.433 percentage point each year.

65. a. Let $N(w)$ be the total area in square inches of a Norman window with a height of 75 inches for the rectangle part and a width of w inches.
$$N(w) = \frac{\pi}{8}w^2 + 75w$$
b. $N(36) = 3208.94$. A Norman window with a height of 75 inches for the rectangular part and a width of 36 inches will have a total area of 3208.94 square inches.

67. a. $-3x + 17$ b. $9x - 1$
c. $-18x^2 - 21x + 72$ d. $\dfrac{3x + 8}{-6x + 9}$

69. a. $\dfrac{5}{2}x - \dfrac{2}{5}$ b. $\dfrac{3}{2}x - \dfrac{4}{5}$
c. $x^2 + \dfrac{1}{10}x - \dfrac{3}{25}$ d. $\dfrac{\frac{1}{2}x + \frac{1}{5}}{2x - \frac{3}{5}}$

71. a. $x^2 + 7x + 15$ b. $x^2 + x + 5$
c. $3x^3 + 17x^2 + 50x + 50$ d. $\dfrac{3x + 5}{x^2 + 4x + 10}$

73. $f(x) + g(x) = 10x - 2$. Jose set the expression equal to zero and solved for x.

75. a. $R(10) = 153.9$. In 2010, IBM had revenue of $153.9 billion for goods sold. $C(10) = 97.1$. In 2010, IBM had a cost of $97.1 billion for selling goods.
b. Let $P(t)$ be the profit of IBM in billions of dollars t years since 2000.
$$P(t) = 0.6t^2 - 4.8t + 44.8$$

c. $P(10) = 56.8$. According to the model, IBM's profit in 2010 was approximately $56.8 billion.

77. a. $R(11) = 11{,}343.2$. The revenue earned by Pearson Publishing Co. in 2011 is approximately 11,343.2 million British pounds. $P(11) = 6160.8$. The profit earned by Pearson Publishing Co. in 2011 is approximately 6160.8 million British pounds.

b. Let $C(t)$ be the publishing costs at Pearson Publishing Co. in millions of British pounds t years since 2000.
$$C(t) = 76.2t^2 - 645t + 3057.2$$

c. $C(11) = 5182.4$. The costs at Pearson Publishing Co. for 2011 are approximately 5182.4 million British pounds.

79. a. $S(100) = 60$. Car stereo producers are willing to supply 60 thousand car stereos if the price is set at $100.

b. Let $R(p)$ be the revenue in thousands of dollars from selling the supplied car stereos at a price of p dollars.
$$R(p) = 0.009p^3 - 0.5p^2 + 20p$$

c. If car stereos are sold for $90 each, the projected revenue will be about $4,311,000.00.

81. a. $S(25) = 33.75$. Car manufacturers are willing to supply about 33,750 minivans if they can be sold for $25,000 each.

b. Let $R(p)$ be the revenue in millions of dollars from selling minivans at a price of p thousand dollars.
$$R(p) = 0.11p^3 - 3.2p^2 + 45p$$

c. $R(31) = 1596.81$. If minivans sell for $31,000 each, the projected revenue will be about $1.6 billion.

83. a. $I(5) = 114{,}111$. There were about 114,111 immigrants admitted to the United States as permanent residents under refugee acts in 1995.

b. $E(5) = 47{,}414$. There were about 47,414 European immigrants admitted to the United States as permanent residents under refugee acts in 1995.

c. Let $N(t)$ be the number of non-European immigrants admitted to the United States as permanent residents under refugee acts t years since 1990.
$$N(t) = -11{,}422t^2 + 138{,}443t - 339{,}968$$

d. $N(8) = 36{,}568$. There were about 36,568 non-European immigrants admitted to the United States as permanent residents under refugee acts in 1998.

85. a. Let $U(t)$ be the number of U.S. residents in millions who are Caucasian and under 18 years old t years since 1990.
$$U(t) = -0.014t^2 + 0.524t + 51.579$$

b. $U(12) = 55.851$. In 2002, there were approximately 56 million Caucasian U.S. residents who were under 18 years old.

87. a. 2   **b.** 44

c. -483 **d.** $-\dfrac{23}{21}$

89. a. $\dfrac{23}{5} = 4.6$ **b.** $\dfrac{11}{5} = 2.2$

c. $\dfrac{102}{25} = 4.08$ **d.** $\dfrac{6}{17} \approx 0.35$

91. a. 75   **b.** 35

c. 1100   **d.** $\dfrac{4}{11}$

93. c   **95.** a

97. a. $9x + 8$ **b.** 26

c. $20x^2 + 33x + 7$ **d.** 1551

99. a. $4x^2 - 1$ **b.** 15

c. $4x^3 + 9x^2 + 3x + 2$ **d.** 76

Section 3.3

1. a. $f(g(x)) = 12x + 22$ **b.** $g(f(x)) = 12x + 33$

3. a. $f \circ g(x) = -18x + 35$ **b.** $g \circ f(x) = -18x - 39$

5. a. Let $K(w)$ be the total weekly cost in dollars for week w of production.
$$K(w) = C(T(w)) = 875w + 10{,}250$$

b. $T(5) = 5500$. There were 5500 toys produced during week 5.

c. $C(5000) = 13{,}750$. The total weekly cost from production of 5000 toys per week is $13,750.

d. $K(7) = 16{,}375$. The total weekly cost of production for week 7 was $16,375.

e. $K(w) = 18{,}500$ when $w \approx 9.4$. The weekly cost will reach and surpass $18,500 in week 10 of production.

7. a. $V(10) = 4$. A West Tech employee will get 4 weeks of vacation per year after working with the company for 10 years.

b. $C(4) = 6575$. West Tech's cost for a 10-year employee's vacation is $6575.

c. Let $K(y)$ be the cost for vacation taken by an employee who has been with the company for y years.
$$K(y) = C(V(y)) = 375y + 2825$$

d. $K(20) = 10{,}325$. West Tech's cost for a 20-year employee's vacation is $10,325.

e. $K(30) = 14{,}075$. West Tech's cost for a 30-year employee's vacation is $14,075.

9. a. $A(4) = 6600$. On the fourth day of the Renaissance fair, there were 6600 people in attendance.

b. $P(6600) = 10700$. On the fourth day of the Renaissance fair, a profit of $10,700 was made.

c. Let $T(d)$ be the profit in dollars made at the Renaissance fair d days after the fair opens.
$$T(d) = P(A(d)) = -800d^2 + 7000d - 4500$$

d. $T(3) = 9300$. The Renaissance fair made $9300 profit on the third day of the fair.

e. $T(1) + T(2) + T(3) + T(4) + T(5) + T(6) + T(7) = 52{,}500$. If the Renaissance fair lasts a total of 7 days, the total profit would be $52,500.

11. $P(B(t)) =$ The profit in thousands of dollars for Pride Bike Co. in year t.

13. $H(T(d)) =$ The number of homeless people who seek space at the homeless shelter on day d of the week.

15. a. $8x + 29$ **b.** $8x + 16$
17. a. $-56x + 116$ **b.** $-56x - 73$
19. a. $x - \dfrac{1}{10}$ **b.** $x - \dfrac{1}{5}$
21. a. $2.4276x + 6.7732$ **b.** $2.4276x + 14.9152$
23. a. $3x^2 + 12x + 35$ **b.** $9x^2 + 42x + 55$
25. a. $4x^2 + x + 3$ **b.** $4x^2 + 17x + 19$
27. $f(g(x)) = 21x + 14$. Lilybell did $f(x)g(x)$.
29. $f(g(x)) = 21x + 14$. Colter did $g(f(x))$.
31. a. -55 **b.** -129
33. a. $\dfrac{19}{10} = 1.9$ **b.** $\dfrac{9}{5} = 1.8$
35. a. -0.5096 **b.** 7.6324
37. a. 170 **b.** 490
39. a. 21 **b.** 69
41. $N(t) = T(t) - K(t) =$ Number of people over 12 years old who go on the tour in year t.
43. $A(T(t) - K(t)) =$ Cost for all people over 12 years old who traveled on the tour in year t.
45. $C(K(t)) - A(T(t) - K(t)) =$ Difference between the cost for all children under 12 years old and the cost for all people over 12 years old who traveled on the tour in year t.
47. a. $f(x)g(x) = 27x^2 + 69x + 14$
   b. $f(x) + g(x) = 12x + 9$
   c. $f(g(x)) = 27x + 65$
49. a. $f(3) + g(3) = -15$
   b. $f(4)g(4) = -420$
   c. $f(g(7)) = -134$
51. a. $f(g(x)) = \dfrac{1}{6}x + \dfrac{5}{6}$
   b. $f(42)g(42) = 318.75$
   c. $f(x) + g(x) = \dfrac{11}{12}x + \dfrac{13}{12}$
53. a. $f(g(x)) = 5.25x + 18$
   b. $g(f(6)) = 56.25$
   c. $g(x) - f(x) = 2x + 4.5$

Section 3.4

1. $2(3x + 4)$ **3.** $h(4h + 7)$
5. $x(25x^2 + x - 2)$ **7.** $2ab(2a + 3)$
9. $2(2x^2 + 3x - 1)$ **11.** $3xyz(5x + 2z - y)$
13. $(x + 5)(3x + 2)$ **15.** $(5w + 4)(7 - 2w)$
17. $(x - 8)(x + 7)$ **19.** $(r + 4)(7r + 2)$
21. $(8x + 9)(2x + 3y)$ **23.** $(2m - 7)(4m - 5n)$
25. $(x - 7)(x + 3)$ **27.** $(x + 5)(x + 1)$
29. $(w - 9)(w + 2)$ **31.** $(t - 4)(t - 7)$
33. $(2x + 3)(x + 5)$ **35.** $(7m - 4)(m - 3)$
37. $(4x + 5)(x - 9)$ **39.** Prime.
41. Dusty grouped the last two terms without first rewriting with addition. This caused a positive 7 to be factored out of the 21 when in fact a -7 needed to be factored out. Then the student continued to factor as if there were common factors and, when none were available, wrote all the binomial groupings as factors. Here is the correct factorization:
$(x - 3)(2x - 7)$
43. $2(x^2 + 8x + 1)$ **45.** $(3w + 4)(2w + 7)$
47. $(5t + 2)(2t - 9)$ **49.** Prime.
51. Prime.   **53.** Prime.
55. $2(m + 3)(m + 4)$ **57.** $6(x - 4)(x - 7)$
59. $5(4x - 3)(2x + 3)$ **61.** $x(x + 3)(x - 7)$
63. $5x(2x + 3)(x + 2)$ **65.** $y(x + 9)(x + 4)$
67. $(2x + 5y)(x + 3y)$ **69.** $(3a + b)(2a - 5b)$
71. $-2m(3n + 4)(n + 2)$ **73.** $(6x - 5)(4x - 3)$
75. $4(3x + 2)(2x - 5)$ **77.** $(5n - 8)(4n - 3m)$
79. $4g(3f + 7)$ **81.** Prime.
83. $7(d + 5)(d - 2)$ **85.** $5(2x^2 - 4x + 3)$

Section 3.5

1. $(x + 4)^2$ **3.** $(3g + 2)^2$
5. $(2t - 7)^2$ **7.** $(5x + 3y)^2$
9. $(x + 6)(x - 6)$ **11.** $(3k + 4)(3k - 4)$
13. $(m - 4)(m^2 + 4m + 16)$ **15.** $(x + 5)(x^2 - 5x + 25)$
17. $(2x + 3)(4x^2 - 6x + 9)$ **19.** $3(g - 2)(g^2 + 2g + 4)$
21. $2(5x + 3)(5x - 3)$
23. Prime. The sum of two squares does not factor.
25. Prime. The polynomial is not a perfect square trinomial.
27. Prime.
29. $(r + 2)(r - 2)(r^2 - 2r + 4)(r^2 + 2r + 4)$
31. $(4b^2 + 25c^2)(2b + 5c)(2b - 5c)$
33. $(h^2 + 5)(h^2 - 2)$ **35.** $(g^4 + 3)^2$
37. $(2t^6 + 3)(t^6 + 5)$ **39.** $2(4x^8 + 6x^4 + 9)$
41. $\left(\sqrt{H} + 3\right)^2$ **43.** $\left(2t^{\frac{1}{2}} - 5\right)^2$
45. $7(g^2 + 9h^2)(g + 3h)(g - 3h)$
47. $5wxz(w^2x + 8z)(w^2x - 3z)$
49. $ab^2(a - b)(24a + 35b)$
51. $(a - b^3)(a + b^3)(a^2 + ab^3 + b^6)(a^2 - ab^3 + b^6)$
   $(a^4 - a^2b^6 + b^{12})(a^2 + b^6)$
53. $(x - 1)(x + 1)(x^2 + 1)(x^4 + 1)(x^8 + 1)$
55. $(5x - 3)(4x + 7)$
57. $(3m - 4)^2$
59. Prime.

61. $(5x - 4)(25x^2 + 20x + 16)$

63. $(3r^{\frac{1}{2}} - 4)(2r^{\frac{1}{2}} + 5)$

65. $-2x(5x + 2)(3x - 7)$

Chapter 3 Review Exercises

1. $\dfrac{8x^{15}y^8z}{25}$ **2.** $15x^4y^7$ **3.** $\dfrac{15a^4c^5}{b^3}$

4. 1 **5.** $2x^2y^4$ **6.** $\dfrac{b^{10}c^4}{a^6}$

7. 45 **8.** $\dfrac{1}{ab^8c^2}$ **9.** $\dfrac{a^4}{9b^{10}c^4}$

10. $\dfrac{9x^3y^3z^3}{10}$

11. a. $\sqrt[4]{xy^3}$ **b.** $\sqrt{3a}$

12. a. $4^{\frac{1}{5}}x^{\frac{2}{5}}$ **b.** $7^{\frac{1}{2}}a^{\frac{1}{2}}b^{\frac{1}{2}}$

13. Let $P(t)$ be the monthly profit in dollars for Hope's Pottery when v vases are produced and sold in the month.
$$P(t) = 90v - 5000$$

14. a. Let $P(t)$ be the total prescription drug sales in the United States in billions of dollars t years since 2000.
$$P(t) = 17t + 149.4$$

  b. Let $M(t)$ be the amount of mail-order prescription drug sales in the United States in billions of dollars t years since 2000.
$$M(t) = 4.75t + 21.7$$

  c. $P(10) = 319.4$. In 2010, the total prescription drug sales in the United States were about \$319.4 billion.

  d. $M(10) = 69.2$. In 2010, the mail-order prescription drug sales in the United States were about \$69.2 billion.

  e. Let $N(t)$ be the non-mail-order prescription drug sales in the United States in billions of dollars t years since 2000.
$$N(t) = 12.25t + 127.7$$

  f. $N(9) = 237.95$. In 2009, the non-mail-order prescription drug sales in the United States were about \$237.95 billion.
$N(15) = 311.45$. In 2015, the non-mail-order prescription drug sales in the United States will be about \$311.45 billion.

15. a. $J(9) = 778.31$. In 1999, the federal funding for research and development at Johns Hopkins University was about \$778.31 million.

$W(t) = 382.79$. In 1999, the federal funding for research and development at the University of Washington was about \$382.79 million.

  b. Let $D(t)$ be the difference between the federal funding at Johns Hopkins University and the University of Washington in millions of dollars t years since 1990.
$$D(t) = 13.38t^3 - 267.06t^2 + 1749.66t - 3473.58$$

  c. $D(10) = 697.02$. In 2000, Johns Hopkins University had about \$697.02 million more federal funding for research and development than the University of Washington.

16. $5x - 6$ **17.** $2x^2y - 2xy - 13y^2$

18. $6x^2 - 13x - 63$ **19.** $15x^2 - xy - 28y^2$

20. $25x^2 - 30x + 9$

21. Three terms.

Term	$8a^3b^2$	$-7a^2b$	$19ab$
Type	Variable	Variable	Variable
Coefficient	8	-7	19

22. Three terms.

Term	$4m^7np^3$	$24m^5n^4p^3$	14
Type	Variable	Variable	Constant
Coefficient	4	24	

23. Two terms.

Term	$5t$	8
Type	Variable	Constant
Coefficient	5	

24. Degree of polynomial $= 5$

Term	$8a^3b^2$	$-7a^2b$	$19ab$
Degree of term	5	3	2

25. Degree of polynomial $= 12$

Term	$4m^7np^3$	$24m^5n^4p^3$	14
Degree of term	11	12	0

26. Degree of polynomial $= 1$

Term	$5t$	8
Degree of term	1	0

27. Polynomial.

28. Not a polynomial as it contains a square root.

29. Polynomial.

30. Not a polynomial as it contains a negative exponent.

31. $U(t) - C(t)$. The number of adults 20 years old or older in the United States t years since 1900.

32. $(R - C)(l)$. The profit in thousands of dollars for Luxury Limousines, Inc. to produce and sell l limousines per year.

33. $ES(t)$. The total number of sick days taken by Disneyland employees in year t.

34. $W(E(t))$. The annual workers compensation insurance costs in dollars at Disneyland in year t.

35. $(M + R)(t)$. The total amount spent on cancer treatments and research in the United States in year t.

36. a. $2x + 11$ **b.** $10x - 5$

  c. $-24x^2 + 36x + 24$

37. **a.** $-2x + 68$ **b.** $-32x$
    c. $-255x^2 - 68x + 1156$ **d.** $\dfrac{15x + 34}{-17x + 34}$

38. **a.** $4.4x - 6.4$ **b.** $-3.6x + 7.6$
    c. $1.6x^2 - 0.4x - 4.2$

39. **a.** $\dfrac{16}{7}$ **b.** $\dfrac{2}{7}$ **c.** $\dfrac{9}{7}$

40. **a.** Let $L(t)$ be the number of people in California's labor force in millions t years since 1990.
    $$L(t) = 0.37t + 13.2$$
    b. Let $U(t)$ be the number of people in California who are considered unemployed in millions t years since 1990.
    $$U(t) = -0.07t + 1.56$$
    c. In 2005, there were about 18.75 million people in California's labor force.
    d. In 2005, there were about 510,000 people in California who are considered unemployed. This might be too low and might be model breakdown.
    e. Let $P(t)$ be the percentage as a decimal of California's labor force that is considered unemployed t years since 1990.
    $$P(t) = \dfrac{U(t)}{L(t)} = \dfrac{-0.07t + 1.56}{0.37t + 13.2}$$
    f. In 1998, 6.2% of California's labor force was considered unemployed.
    In 1999, 5.6% of California's labor force was considered unemployed.
    In 2000, 5.1% of California's labor force was considered unemployed.
    In 2005, 2.7% of California's labor force was considered unemployed. This might be too low and might be model breakdown.

41. **a.** $-24x + 51$ **b.** $-24x - 4$
42. **a.** $-255x + 544$ **b.** $-255x - 544$
43. **a.** $\dfrac{8}{5}x - \dfrac{11}{5}$ **b.** $\dfrac{8}{5}x - \dfrac{23}{5}$
44. **a.** $\dfrac{6}{7}$ **b.** $\dfrac{-31}{7}$
45. **a.** -6.1 **b.** -7.5
    c. -4.76 **d.** -1.58
    e. 6.15
46. **a.** -3.58 **b.** 0.58
    c. 3.12 **d.** -3.305
47. $(x + 7)(x - 5)$ 48. $(x + 3)^2$
49. $(3x - 8)(2x + 7)$ 50. $(5b + 1)(b - 3)$
51. $(2p + 5)(3p + 4)$ 52. $7(2k - 1)(k - 1)$
53. $4x(5x + 2)(x - 3)$ 54. $(a - 1)(a + 6b)$
55. $(3m + 10)(3m - 10)$ 56. Prime.
57. $(3x + 5)^2$
58. $(3x - 1)(9x^2 + 3x + 1)$
59. $(5w + 2x)(25w^2 - 10wx + 4x^2)$
60. $(t + 2)(t - 2)(t^2 - 2t + 4)(t^2 + 2t + 4)$
61. $(5m^3 - 7)(2m^3 - 3)$
62. $5(4h + 5)(4h - 5)$
63. $2b(3a + 7)^2$
64. $3(r^2 + 4)(r + 2)(r - 2)$

Chapter 3 Test

1. $\dfrac{32b^{26}}{9c^2}$ 2. $\dfrac{1}{2b^5c^2}$ 3. $\dfrac{5x}{7y^5}$

4. **a.** $T(9) = 16.74$. In 1999, New Zealand spent about 16.74 million New Zealand dollars on the treatment of diabetes.
   b. $R(t) = 10$ when $t \approx 7.3$. New Zealand spent about 10 million New Zealand dollars on diabetes research in 1997.
   c. Let $D(t)$ be the total New Zealand spent on diabetes research and treatment in millions of New Zealand dollars t years since 1990.
   $$D(t) = R(t) + T(t) = 2.4t + 7.36$$
   d. $D(10) = 31.36$. New Zealand spent about 31.36 million New Zealand dollars on research and treatment of diabetes in 2000.

5. **a.** $f(x) - g(x) = 2x + 24$
   b. $f(x)g(x) = 8x^2 + 6x - 119$
   c. $f(g(x)) = 8x - 11$

6. **a.** $(f + g)(4) = 14.15$
   b. $fg(-2) = 36.075$
   c. $f \circ g(6) = 20.485$
   d. $g(f(0)) = -2.82$

7. **a.** $S(10) = 544.6$. In 2000, there were approximately 544.6 thousand science/engineering graduate students in doctoral programs.
   b. Let $M(t)$ be the number of male science/engineering graduate students in doctoral programs in thousands t years since 2000.
   $$M(t) = 0.8t^2 + 0.7t + 225.7$$
   c. $M(10) = 312.7$. In 2010, there were approximately 312.7 thousand male science/engineering graduate students in doctoral programs.

8. **a.** $P(6) = 51.8$. In 2006 about 51.8% of murders were committed by using a handgun.
   b. Let $H(t)$ be the number of murders in thousands that were committed by using a handgun t years since 2000.
   $$H(t) = M(t)\dfrac{P(t)}{100} = -0.00006t^3 - 0.0013t^5 + 0.0095t^4$$
   $$- 0.075t^3 + 0.58t^2 - 1.43t + 6.78$$
   c. $H(7) = 7.6$. In 2007, about 7.6 thousand murders were committed by using a handgun.

9. $(T - B)(t)$ is the number of girls attending the tennis camp in year t.

10. $M(B(t))$ is the number of boys' matches at the camp in year t.

11. $C(T(t))$ is the total cost for the camp in year t.

12. $A(t) \cdot M(T(t))$ is the time in minutes it takes for all the matches to be played in year t.

13. $x^2 + 2x - 16$ 14. $9x^2y - 4xy - 2y^2 + 8$

15. $6x^2 + 10x - 56$
16. $6a^2 - 23ab + 20b^2$
17. Three terms; $7x^2$ and $8x$ variable terms with coefficients 7 and 8; -10 constant term.
18. Three terms; $14m^3n^6$ and $-8m^2n$ variable terms with coefficients 14 and -8; 205 constant term.
19. $7x^2$ has degree 2; $8x$ has degree 1; -10 has degree 0; the polynomial has degree 2.
20. $14m^3n^6$ has degree 9; $-8m^2n$ has degree 3; 205 has degree 0; the polynomial has degree 9.
21. No, it is not a polynomial because the second term has a negative exponent on the variable.
22. Yes, it is a polynomial.
23. $(x - 9)(x - 5)$
24. $2p(p + 7)(p - 3)$
25. $(2m - 9)(6m + 11n)$
26. $(6b + 7)(6b - 7)$
27. $(t - 10)^2$
28. $7x(x + 2)(x - 2)$
29. $(2m + 5n)(4m^2 - 10mn + 25n^2)$
30. Prime.
31. $(3m^4 - 4)(2m^4 - 7)$
32. $(a - 5b)(a^2 + 5ab + 25b^2)$

CHAPTER 4
Section 4.1

1. This function is a quadratic function.
3. This function is a linear function.
5. This function is a quadratic function.
7. Other.
9. This function is a quadratic function.
11. This function is a quadratic function.
13. This function is a quadratic function.
15. a. $(-5, 1)$ b. $x < -5$ c. $x > -5$
    d. $(-7.5, 0)(-2.5, 0)$ e. $(0, -3)$
17. a. $(4, 25)$ b. $x < 4$ c. $x > 4$
    d. $(-1, 0)(9, 0)$ e. $(0, 9)$ f. 21
19. a. $(-1, 4)$ b. $x > -1$ c. $x < -1$
    d. None. e. $(0, 5)$ f. 9.5
    g. $x = 3$ and $x = -5$
21. Horizontal intercepts: $(-4, 0), (3, 0)$
    Vertical intercept: $(0, 6)$
23. Horizontal intercepts: $(-10, 0), (15, 0)$
    Vertical intercept: $(0, -15)$
25. a. D = Average number of days above 70°F in San Diego, California, during month m.
    m = Month of the year (i.e., $m = 1$ represents January)

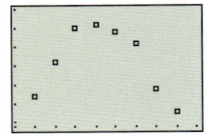

b. A quadratic would fit best. The data looks like a downward facing parabola.
c. The vertex is at about $(8, 31)$ and is a maximum point. Thus, August has the most number of days on average above 70°F in San Diego, California.
d. Using the shape of the distribution, there would be about 7 days above 70°F in San Diego, California, during the month of April.

27. a. 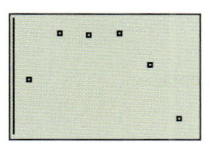 b.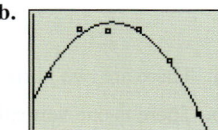

c. This model fits the data reasonably well; the point at the top is low for a vertex, but otherwise the data is shaped like a quadratic.
d. The vertex for the model is at about $(4, 11{,}238)$ and is a maximum point for this function.
e. This vertex represents that in 1994, the number of cable television systems in the United States reached a maximum of 11,238.

29. a.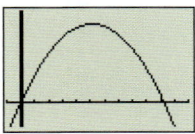

b. The vertex of this parabola is at about $(5, 128.5)$, which means that the tennis ball will reach its highest point in the air at 128.5 meters after about 5 seconds.
c. $H(2) = 81.4$. About 2 seconds after being hit, the tennis ball will reach a height of 81.4 meters.
d. According to the graph, the tennis ball will hit the ground at about 10.224 seconds.

31. a. According to the graph, the revenue from selling T-shirts for $10 each would be about $1000.
b. According to the graph, the maximum monthly revenue would be about $1100.
c. The bookstore should charge $15 for each T-shirt to maximize the monthly revenue.
d. The revenue may go down after the vertex in this situation because people are less likely to purchase these T-shirts for more than $15 each, according to this graph.

33. A quadratic model would best fit these data. The vertex is $(-5, 8)$, which is a maximum point.
35. A quadratic model would best fit these data. The vertex is $(5, -5)$, which is a minimum point.
37. A linear model would best fit these data.
39. A linear model would best fit these data.

41.

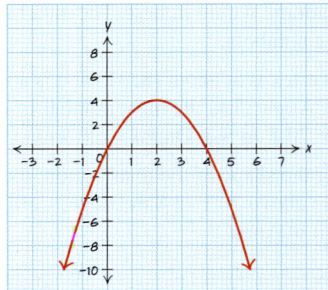

   a. There are two horizontal intercepts.
   b. There is one vertical intercept.

43.

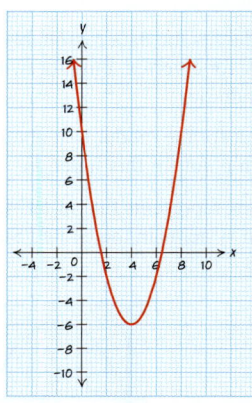

   a. There are two horizontal intercepts.
   b. There is one vertical intercept.

45. The vertex has to touch the x-axis.

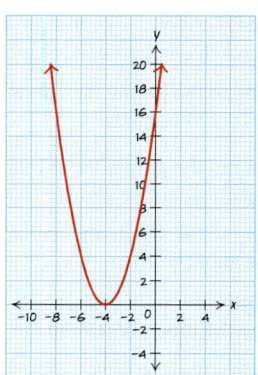

47. a. Above the horizontal axis
   b. Below the x-axis

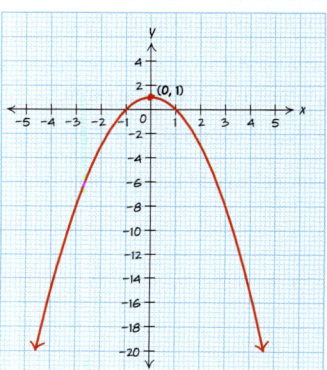

Section 4.2

1. h; positive.
3. k; negative.
5. $|a| > 1$
7. a; positive.
9. $h = 0, k = 0$
11. **a.** $(-2.5, -3)$ **b.** $x = -2.5$ **c.** $(-5.5, 4)$
13. **a.** $(5.5, -6)$ **b.** $x = 5.5$ **c.** $(2, 5)$
15. **a.** $(-4, -8)$ **b.** $x = -4$ **c.** Positive.
    d. $h = -4, k = -8$ **e.** $(2, 15)$
17. **a.** $(2, 10)$ **b.** $x = 2$ **c.** Negative.
    d. $h = 2, k = 10$ **e.** $(5, -6)$

19.

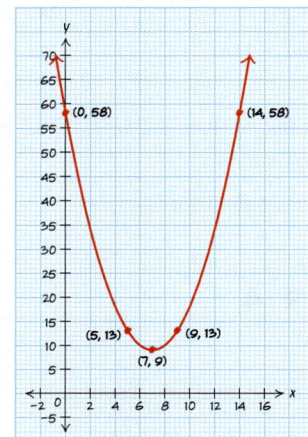

21.

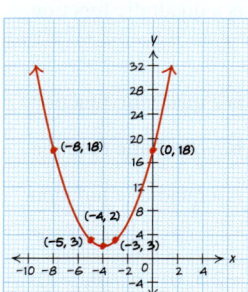

23.

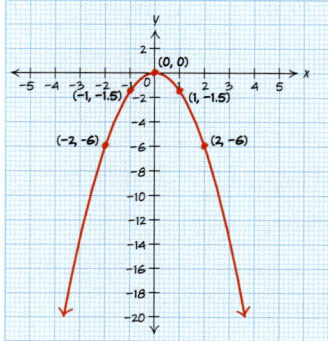

25.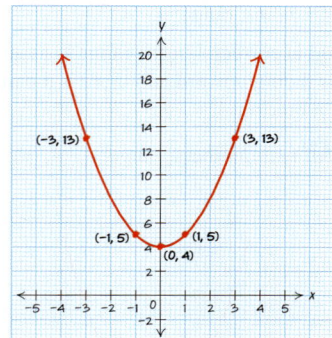

27.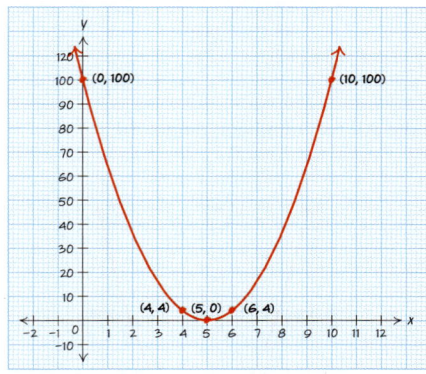

29. a. $P(90) = 6796.55$, so in 1990, the poverty threshold was about $6,796.55 for individuals under 65 years old in the United States.

b. $P(80) = 4024.55$, so in 1980, the poverty threshold was about $4,024.55 for individuals under 65 years old in the United States.

c.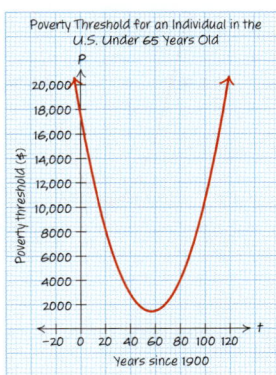

d. According to this model, the poverty threshold for individuals under 65 years old in the United States reached a minimum of about $1406 in 1957, which seems to make sense.

e. According to the graph, the poverty threshold for individuals under 65 years old in the United States was $3000 in 1975 and 1939; 1939 shows model breakdown.

f. Range: [1406, 6796.55]

31. a.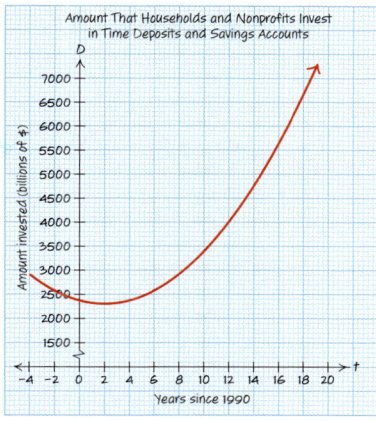

b. In 1996, households and nonprofit organizations invested about $2453 billion in time and savings accounts.

c. In about 1993, these time and savings accounts were at their lowest levels with about $2300 billion invested.

d. According to the graph, households and nonprofit organizations invested $3000 billion in about 1987 and again in 1999.

e. Range: [2300, 6125]

33. a. $h(2) = 192$, which means that 2 seconds after being dropped from the roof of a building, the object dropped will be at a height of about 192 feet.

b.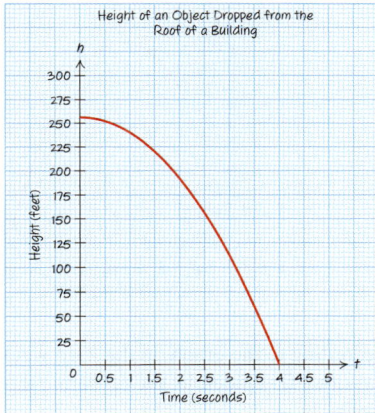

c. According to the graph, the object will hit the ground about 4 seconds after being dropped from the roof of the building.

d. Domain: [0, 4]; range: [0, 256]

35. a. $P(100) = 2.74$, so the monthly profit made from selling round-trip airline tickets from New York City to Orlando, Florida, when the tickets are sold for $100 each is $2.74 million.

b. Selling round-trip airline tickets from New York City to Orlando, Florida, for $120 each will produce a maximum profit of $2.75 million.

c.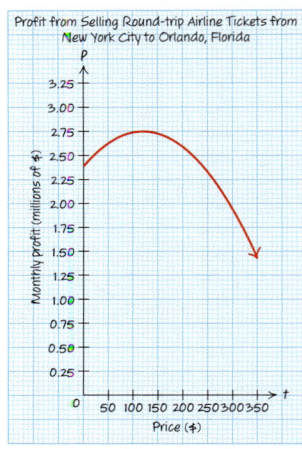

d. According to the graph, the company will have a monthly profit of about $2 million when selling round-trip airline tickets from New York City to Orlando, Florida, for $295 each and −$50; −$50 is model breakdown.

e. Domain: [60, 300]; range: [1.95, 2.75]

37. a. $E(4) = 7.784$, so the epidemic threshold for the fourth week of 2006 was 7.784%

b.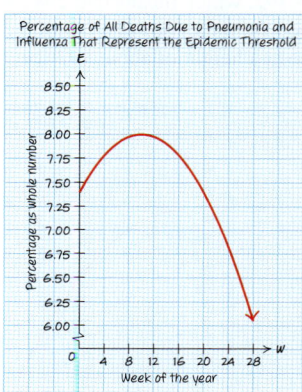

c. According to the graph, in the 23rd week of 2006, there was not a flu epidemic; the epidemic threshold was about 7%.

d. The vertex for this model is (10, 8), which means that in week 10 of 2006, the epidemic threshold reached a maximum of 8%.

e. Domain: [1, 26]; range: [6.5, 8]

39. Domain: All real numbers; range: $(-\infty, 15]$

41. Domain: All real numbers; range: $[-58, \infty)$

43. Domain: All real numbers; range: $[0, \infty)$

45. Domain: All real numbers; range: $(-\infty, 4]$

47. Domain: All real numbers; range: $y \geq -9$

The domain should be all real numbers. The range is using a greater than sign, but it should use a greater than or equal to sign.

49. Domain: $(-\infty, \infty)$; range: $(-\infty, 7]$

The domain is correct, but the range is assuming that the parabola is facing upward and there is a parenthesis on the 7 instead of a bracket.

51. Domain: All real numbers; range: $[-4, \infty)$

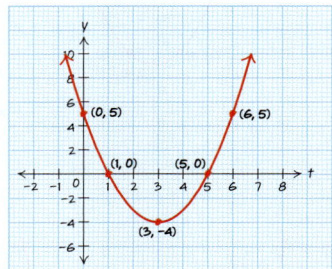

53. Domain: All real numbers; range: $[0, \infty)$

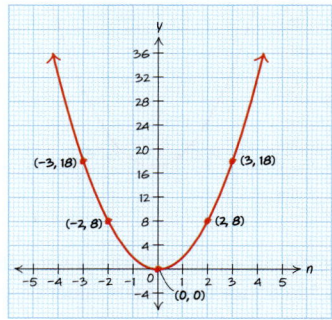

55. Domain: All real numbers; range: $[2, \infty)$

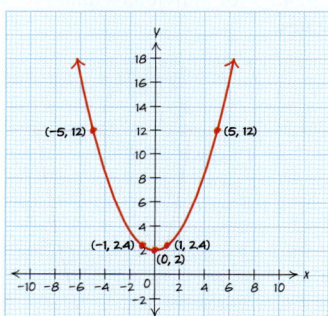

57. Domain: All real numbers; range: $(-\infty, 0]$

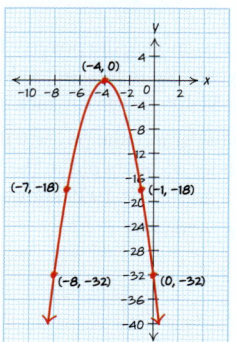

59. Domain: All real numbers; range: $[2, \infty)$

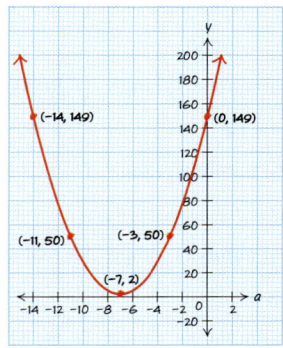

61. Domain: All real numbers; range: $[-8, \infty)$

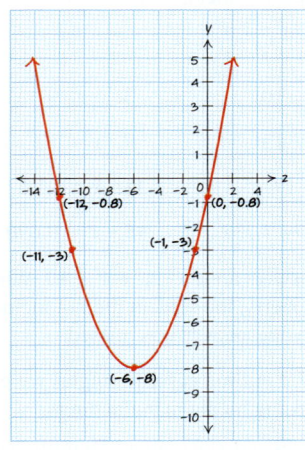

63. Domain: All real numbers; range: $(-\infty, -15]$

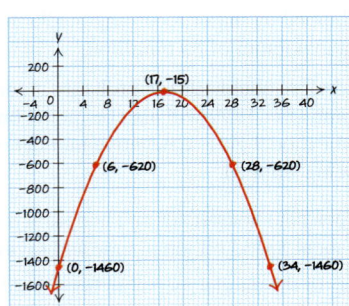

65. Domain: All real numbers; range: $(-\infty, 25]$

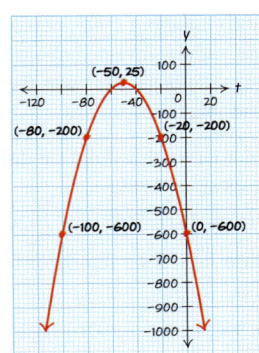

67. Domain: All real numbers; range: $[-2500, \infty)$

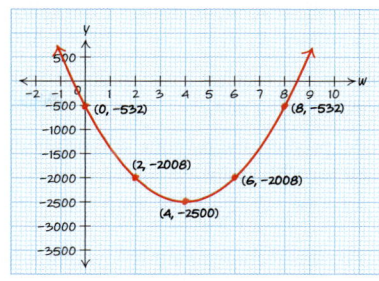

69. a. Vertex: $(0, 100)$
  b. The parabola is narrow.
  c. The parabola faces upward.
  d. X Min: -10 X Max: 10
   Y Min: 0 Y Max: 600

71. a. Vertex: $(-30, -50)$
  b. The parabola is neither wide nor narrow.
  c. The parabola faces upward.
  d. X Min: -45 X Max: -10
   Y Min: -60 Y Max: 70

73. a. Vertex: $(-20, 50)$
  b. The parabola is wide.
  c. The parabola faces upward.
  d. X Min: -100 X Max: 60
   Y Min: 45 Y Max: 60

75. a. Vertex: $(1000, 1000)$
  b. The parabola is wide.
  c. The parabola faces upward.
  d. X Min: 0 X Max: 2000
   Y Min: 900 Y Max: 1500

77. a. Vertex: $(-25{,}000, -10{,}000)$
  b. The parabola is narrow.
  c. The parabola faces downward.
  d. X Min: -25010 X Max: -24990
   Y Min: -11000 Y Max: -9900

Section 4.3

1. $f(x) = 2.0(x + 5)^2 - 15$
3. $f(x) = 0.4(x - 4)^2 + 3.2$
5. $f(x) = 2(x + 1.5)^2 - 20$
7. $f(x) = 2.7(x - 5)^2 - 8$
9. $f(x) = -2(x - 7)^2 + 15$
11. $f(x) = 4(x + 2)^2 - 10$
13. Change k to shift the graph up.
15. Change h to shift the graph right.
17. Change a to make the graph wider.

19. Change both h and k to shift the graph left and up.

21. **a.** Let $R(p)$ be the revenue in dollars from selling gloves at a price of p dollars. $R(p) = -(p - 20)^2 + 500$
    b. The total revenue reaches a maximum of $500 from selling the gloves for $20 a pair.
    c. This model predicts the revenue will be about $496 if the gloves sell for $22 a pair.
    d. Domain: [6, 35]; range: [275, 500]

23. **a.** Let $T(m)$ be the average monthly low temperatures (in degrees Fahrenheit) in Anchorage, Alaska, during month m (i.e., $m = 1$ represents January). $T(m) = -2.56(m - 7)^2 + 51$
    b. The highest average low temperature for Anchorage, Alaska, is about 51°F in July.
    c. This model predicts the average low temperature in March to be 10°F.
    d. Domain: [3, 11]; range: [10, 51]

25. **a.** Let $F(t)$ be the number of Hispanic families in thousands in the United States below the poverty level t years since 1990.
    $$F(t) = 46(t - 9.5)^2 + 1520$$
    b. In 2005, there were approximately 2,911,500 Hispanic families in the United States under the poverty level.
    c. In about 1999 or 2000, the number of Hispanic families in the United States under the poverty level reached a minimum of about 1,520,000. This is probably true only for this close time period.
    d. Domain: [6, 15]; range: [1520, 2911.5]

27. **a.** Let $H(t)$ be the number of hours per year the average person spent using the Internet t years since 1990.
    $$H(t) = 5.7(t - 5)^2 + 5$$
    b. If you use this model for years before 1995, the model shows more and more hours spent on the Internet as you go back in time. This is model breakdown because the Internet was just beginning to take off in the early 1990s.
    c. Domain: [5, 12]; range: [5, 284]
    d. $H(11) = 210.2$. In 2001, the average person spent 210 hours using the Internet.
    e. According to this model, people will spend 365 hours per year on the Internet in 2003.

29. **a.** Let $H(d)$ be the height of the baseball (in feet) from the ground, and let d be the horizontal distance of the ball (in feet) from home plate.
    $$H(d) = -0.0026(d - 230)^2 + 140$$
    b. Domain: [0, 400]; range: [2.46, 140]
    c. This vertex of this model is (230, 140), which means that at a distance of 230 feet after being hit, the baseball will be at a maximum height of about 140 feet from the ground.
    d. $H(450) = 14.16$, If the center wall of the stadium is 450 feet from home plate and is 10 feet tall, the ball will make it over the wall.

31. $y = -4(x + 5)^2 + 8$
    Domain: All real numbers; range: $(-\infty, 8]$

33. $y = (x - 5)^2 - 4$
    Domain: All real numbers; range: $[-4, \infty)$

35. $y = 12(x + 55)^2 - 9000$
    Domain: All real numbers; range: $[-9000, \infty)$

37. $y = -(x - 6)^2 + 14$
    Domain: All real numbers; range: $(-\infty, 14]$

39. $y = 3.2x - 15$
    Domain: All real numbers; range: All real numbers

41. $y = -1.5x^2 + 8$
    Domain: All real numbers; range: $(-\infty, 8]$

43. $y = -0.4(x + 5)^2$
    Domain: All real numbers; range: $(-\infty, 0]$

Section 4.4

1. $x = 10 \quad x = -10$
3. $x = 9 \quad x = -9$
5. $x = 13 \quad x = -13$
7. $x = 12 \quad x = -2$
9. $x = 3 \quad x = -17$
11. $x = 21.6 \quad x = -7.6$
13. $x \approx 4.268 \quad x \approx 7.732$
15. $x \approx 0.690 \quad x \approx -8.690$

17. The correct answers are $x = 2$, $x = -10$. When applying the square root property, the student did not include the negative square root, so one answer was missed.

19. **a.** In 1998, 27.2 million cassette singles were shipped by major recording media manufacturers.
    b. In about 1989 and 1997, about 50 million cassette singles were shipped by major recording media manufacturers.
    c. Vertex = (13, 85.6). In 1993, the number of cassette singles shipped by major recording media manufacturers reached a maximum of 85.6 million.
    d. Anytime before about 1988 is probably model breakdown, and anytime after 1999 will also be model breakdown because the model gives a negative number of cassette singles being shipped.

21. **a.** In 2000, personal households and nonprofit organizations invested about $3133 billion in time and savings accounts.
    b. According to this model, personal households and nonprofit organizations invested about $7500 billion in time and savings deposits in the years 1976 and 2010. 1976 might be model breakdown.
    c. Vertex: (3, 2300). In about 1993, personal households and nonprofit organizations invested about $2300 billion in time and savings deposits. This was the minimum amount invested around this time.

23. **a.** Let $P(t)$ be the poverty threshold (in dollars) for a family of four in t years since 1900.
    $$P(t) = 17(t - 67)^2 + 2920$$
    b. The poverty threshold for a family of four in 1990 was about $11,913.
    c. Domain: [60, 95]; range: [2920, 16,248]
    d. The poverty threshold for a family of four was $5000 in about 1956 and 1978; 1956 would be model breakdown.

25. a. After 5 minutes, there would be about 1125 gallons of water remaining in the tank.

b. After about 10 minutes, there would be only 500 gallons of water remaining in the tank.

c. After about 18 minutes, there would be only 20 gallons of water remaining in the tank.

27. $x = 1 \quad x = -7$

29. $x \approx 16.185 \quad x \approx -0.185$

31. No real solution.

33. $x \approx 0.352 \quad x \approx -11.352$

35. $x = 2 \quad x = 7$

37. No real solution.

39. $x = 1 \quad x = -5$

41. $x = 0 \quad x = -\dfrac{7}{5}$

43. $x = 2 \quad x = -\dfrac{10}{7}$

45. A 6.1% interest rate would turn a deposit of $800 into $900 in 2 years.

47. A 3.68% interest rate would turn a deposit of $2000 into $2150 in 2 years.

49. The guy wire is approximately 2236.068 feet long.

51. Abida should put the base of the ladder 15 feet away from the house.

53. $f(x) = (x + 3)^2 - 1$

55. $g(t) = (t - 4)^2 - 36$

57. $f(x) = \left(x - \dfrac{7}{2}\right)^2 - \dfrac{9}{4}$

59. $h(x) = 3(x + 2)^2 + 12$

61. $f(t) = 2(t - 4)^2 - 44$

63. $f(x) = 4\left(x + \dfrac{5}{8}\right)^2 - \dfrac{345}{16}$

65. $g(x) = 0.5(x + 7)^2 - 54.5$

67. $f(x) = 0.2(x - 17.5)^2 - 71.25$

69. $c(p) = \dfrac{2}{7}\left(p - \dfrac{35}{4}\right)^2 - \dfrac{1249}{56}$

71. Domain: All real numbers; range: $[-16, \infty)$

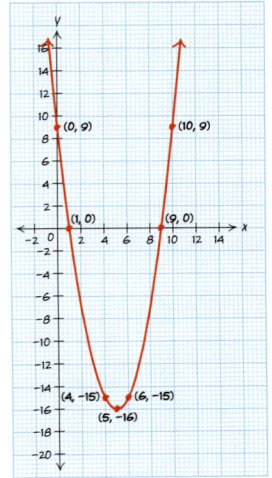

73. Domain: All Real Numbers; Range: $[-5, \infty)$

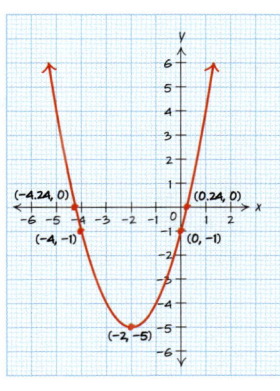

75. Domain: All real numbers; range: $[10, \infty)$

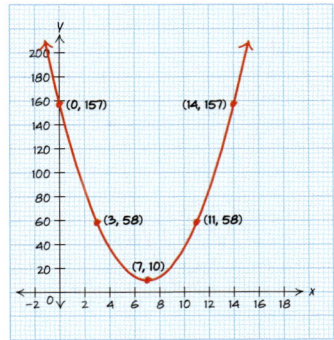

77. Domain: All real numbers; range: $(-\infty, 18]$

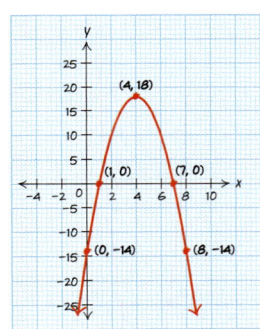

79. Domain: All real numbers; range: $(-\infty, 32]$

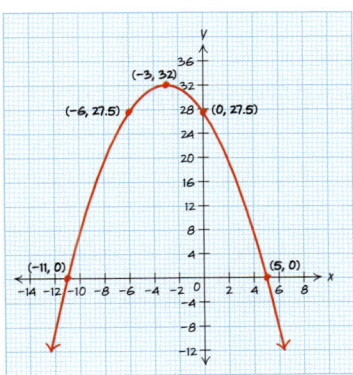

81. Domain: All real numbers; range: $[-12, \infty)$

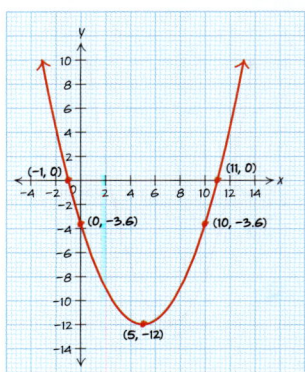

83. Domain: All real numbers; range: $[0.2, \infty)$

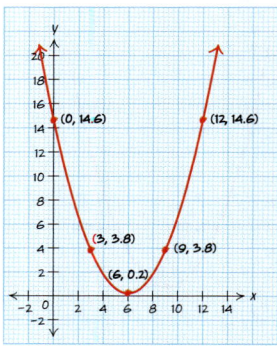

85. Domain: All real numbers; range: $[-16, \infty)$

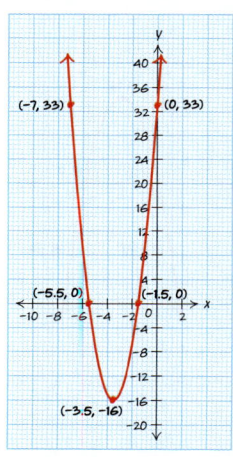

87. Domain: All real numbers; range: $[-1, \infty)$

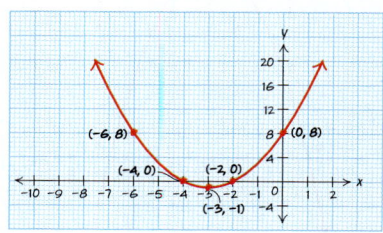

89. Domain: All real numbers; range: $[-36, \infty)$

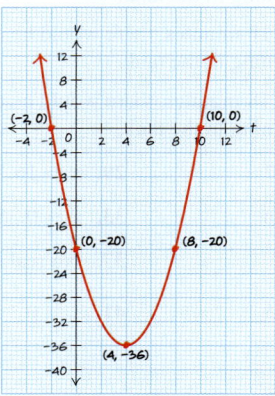

91. Domain: All real numbers; range: $[-2.25, \infty)$

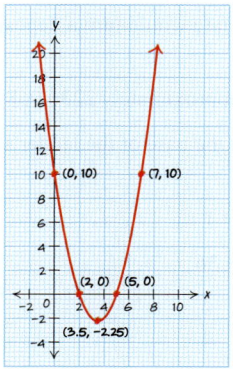

93. Domain: All real numbers; range: $[12, \infty)$

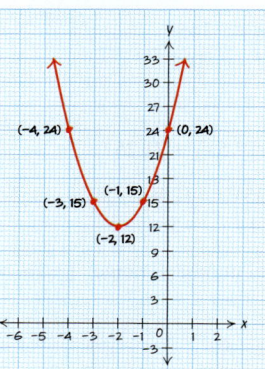

95. Domain: All real numbers; range: $[-44, \infty)$

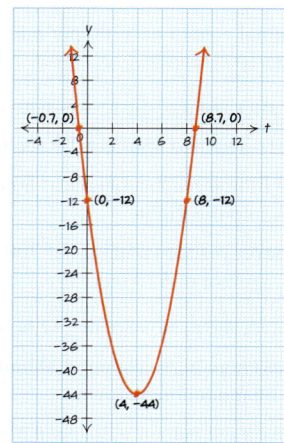

97. Domain: All real numbers; range: $[-21.5625, \infty)$

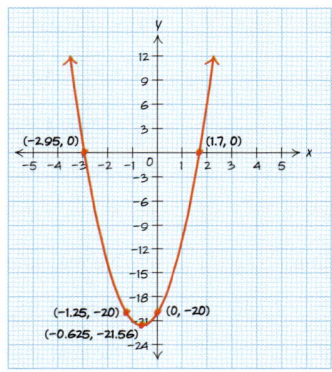

99. Domain: All real numbers; range: $[-54.5, \infty)$

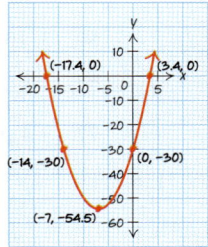

Section 4.5

1. $x = -3, x = 2$
3. $w = -7, w = -\dfrac{10}{3}$
5. $x = 0, x = 4, x = -5$
7. $x = 0, x = 9, x = -7$
9. $w = 0, w = 6, w = \dfrac{8}{3}, w = -\dfrac{5}{7}$
10. $m = 0, m = -3.8, m = 7.5, m = 4.5, m = -8$

11. a. The height of the professor $\frac{1}{2}$ second after jumping off the waterfall was about 18 feet.

   b. It took about 1.25 seconds for the professor to hit the pool of water below.

   c. It took about 1 second after jumping for the professor to reach a height of 8 feet above the pool of water.

13. a. The company needs to make either 2 or 18 parts to make $144 profit per part.

   b. The company needs to make either 5 or 15 parts to make $300 profit per part.

   c. The company needs to make 10 parts to make $400 profit per part.

15. a. The company needs to manufacture either 5 or 13 machines to have an average cost of $390 per machine.

   b. The company needs to manufacture either 8 or 10 machines to have an average cost of $480 per machine.

17. $x = -3, x = 7$
19. $h = -3, h = -9$
21. $x = -6, x = 6$
23. $x = -4, x = 4$
25. $x = -10, x = 10$
27. $x = 0, x = -4$
29. $x = 0, x = \dfrac{5}{3}$
31. $x = -4, x = -5$

33. The two factors must be equal to zero. Set the equation equal to zero and factor. $x = -1, x = -4$

35. $t = 4, t = 7$
37. $m = \dfrac{4}{7}, m = 3$
39. $x = -\dfrac{5}{4}, x = 9$
41. $t = -\dfrac{2}{5}, t = \dfrac{9}{2}$
43. $x = -5, x = 5$
45. $p = -\dfrac{5}{4}, p = \dfrac{8}{7}$
47. $h = -1.5, h = 7$
49. $x = 0, x = -3, x = 7$
51. $w = 0, w = 5, w = 2$
53. $x = 0, x = -1.5, x = -2$

55. a. The profit for this school fundraiser is $4125 total if the profit per candy bar is $0.75.

   b. Let $P(n)$ represent the total profit made at the school's sports program fundraiser, and let n represent the number of each $0.50 increase on the total price of the candy bar.
   $$P(n) = (0.75 + 0.50n)(5500 - 1000n)$$

   c. $P(2) = 6125$. If the price per candy bar is increased by $1, the total profit would be $6125.

   d. $P(4) = 4125$. If the price per candy bar is increased by $2, the total profit would be $4125.

   e. $P(2) = 6125$. Increasing the price of each candy bar by $1 would bring the school the highest profit.

57. a. The store's weekly revenue from selling boxes of golf balls for $15 would be $1500.

   b. The store's weekly revenue from selling boxes of golf balls for $17 would be about $1020.

   c. Let $R(x)$ represent the store's weekly revenue from selling boxes of golf balls if the store lowers the price by x dollars.
   $$R(x) = (100 + 20x)(15 - x)$$

   d. $R(7) = 1920$. The stores weekly revenue from selling boxes of golf balls would be $1920 if each box was sold for $7.

   e. Selling the boxes of golf balls for $10 each would maximize the store's weekly revenue.

59. The enclosure was either 30 feet by 80 feet or 40 feet by 60 feet.

61. The width of the frame is 3.5 inches.

63. $f(x) = x^2 + x - 6$

65. $f(x) = -x^2 + 9$

67. $f(x) = -2x^2 + 8$

69. $f(x) = \dfrac{10}{9}(x + 2)^2 + 5$

71. $f(x) = x^2 + 4x - 21$

73. $f(x) = x^2 - 6x + 8$

75. $f(x) = 3x^2 - 14x + 8$

  or $f(x) = x^2 - \dfrac{14}{3}x + \dfrac{8}{3}$

77. $f(x) = 5x^2 - 30x + 40$

79. $f(x) = -3x^2 - 5.25x + 1.5$

81. $f(x) = -4x^2 + \dfrac{50}{3}x + 6$

83. $x = -5, x = 5$

85. $n = 0, n = -3.5, n = 1$

87. $h = \dfrac{7}{3}$

89. $p = 5, p = 2$

91. $b = -3, b = 0$

93. No real solution.

95. $m = 2$

97. $y = -5 + \sqrt{10} \approx -1.84, \quad y = -5 - \sqrt{10} \approx -8.16$

Section 4.6

1. $x = -5, x = -3$

3. $x = 5, x = 2$

5. $x = -4, x = 1$

7. $x = -0.5, x = 5$

9. $x \approx -1.07, x \approx 2.82$

11. a. $L(55) = 22.63$. In 1955, about 22.63 thousand tons of lead were used in paint in the United States

    b. In about 1965, 5500 tons of lead were used in paints in the United States. 1984 would be considered model breakdown.

    c. In 1945, about 51 thousand tons of lead were used in paints in the United States. 2005 would be model breakdown.

13. a. The current in the wire after 0.5 second would be about 3.75 amperes.

    b. After about 3 seconds, the current would reach 20 amperes.

    c. After about 5 seconds, the current would reach 60 amperes.

15. a. The marginal cost of producing the 201st pair of shoes would be $27.

    b. For the marginal cost to be $20 per pair of shoes, you would be producing the 163rd pair of shoes.

17. a. The marginal cost of producing the 401st bike would be $158.

    b. For the marginal cost to be $700 per bike, you would be producing the 1073rd bike.

19. a. One second after launch, the rocket was 186 feet high.

    b. The rocket first reached 450 feet 2.9 seconds into the flight.

    c. The rocket will first reach a height of 600 feet about 4.95 seconds into the flight.

    d. The rocket never reached a height of 700 feet. The rocket reached a maximum of 627 feet.

21. a. In 2000, an average person spent 141 hours on the Internet.

    b. In 1999, an average person spent 100 hours on the Internet. 1989 would be model breakdown.

    c. In 2004, an average person will spend about 1 hour per day on the Internet. 1984 would be model breakdown.

23. $a = 3, a = -5$

25. $t \approx -1.67, t \approx 1.67$

27. $x = 0, x = 5$

29. $x = 0, x \approx 0.55$

31. $x = 18.4$

33. $x = 0, x = -\dfrac{3}{5}$

35. $x = 13, x = 5$

37. $x = 2, x = 0.5$

39. $w \approx 5.93, w \approx 2.07$

41. $f \approx 3.22, f \approx -4.62$

43. $d = -7, d = 5$

45. $b = -4, b = 7$

47. No real solution.

49. No real solution.

51. $f \approx 14.46, f \approx -10.46$

53. $p \approx -3.95, p \approx 2.28$

55. No real solution.

57. No real solution.

59. $x = 12, x = 0, x = -7$

61. $x = 1$

63. $d \approx -6.43$

65. No real solution.

67. $x \approx 6.62, x \approx -3.62$

69. $(3, 10), (-4, 10)$

71. No solution.

73. $(-5, -1)$

75. $(-3, -25)$

77. $(-4, 12), (3, 12)$

79. $\approx (-3.71, 0.295), (2.71, 6.71)$

81. $(-0.41, -10.07), (2.41, 4.07)$

83. No solution.

85. $(-4, 19), (1.71, 8.39)$

87. $(2.5, 7.25), (3, 8)$

89. $(-2.54, -1.04), (1.21, -0.72)$

91. $(-3, -29)$

93. No solution.

95. No solution.

97. $(-1.41, 0.17), (1.41, 5.83)$

Section 4.7

1. $y = -2x^2 - 6x - 7$: C, downward, y-int: $(0, -7)$.

   $y = -0.5x^2 - x + 7$: A, downward, y-int: $(0, 7)$.

   $y = x^2 + 4x + 4$: B, upward, y-int: $(0, 4)$.

   $y = x^2 - 5x - 6$: None, upward, y-int: $(0, -6)$.

3. $y = -2(x - 3)^2 - 9$: None, downward.

   $y = 2(x - 3)^2 - 9$: A, upward, vertex: $(3, -9)$.

   $y = 2(x + 4)^2 + 1$: C, upward, vertex: $(-4, 1)$.

   $y = 2(x + 4)^2 - 9$: B, upward, vertex: $(-4, -9)$.

5. Vertex: $(-3, -1)$

   Vertical intercept: $(0, 8)$

   Horizontal intercepts: $(-4, 0), (-2, 0)$

7. Vertex: $(3, -12)$

   Vertical intercept: $(0, 15)$

   Horizontal intercepts: $(1, 0), (5, 0)$

9. Vertex: $(8, -20)$

   Vertical intercept: $(0, 236)$

   Horizontal intercepts: $(5.764, 0), (10.236, 0)$

11. Vertex: $(-1.2, 2.8)$

    Vertical intercept: $(0, 10)$

    Horizontal intercepts: None

13. Vertex: $(2, -2)$

    Vertical intercept: $(0, 4)$

    Horizontal intercepts: $(0.845, 0), (3.155, 0)$

15. a. The cost to produce 30 uniforms would be about $2975.

    b. The vertex is $(50, 2875)$, which means the minimum cost would be $2875, producing 50 uniforms.

    c. According to the model, the school can never produce uniforms for $1600, which would be model breakdown.

17. a. The revenue from selling 5 thousand digital cameras would be $375 thousand.

    b. The company must sell 10 thousand cameras or 20 thousand cameras to generate revenue of $600,000.

    c. The company must sell 15 thousand cameras to maximize its revenue.

19. a. According to this model, the annual net sales for Home Depot in 1995 was about $14,471 million.

    b. Home Depot's net sales were $30,000 million in about 1998.

    c. The vertex is $(-0.728, 2118.9)$, which means that in 1989 Home Depot's net sales were at a low point of $2118.90 million.

21. a. According to this model, the net income for Quiksilver in 1999 was about $30.4 million.

    b. According to this model, Quiksilver's net income reached $18 million in 1997 and again in 2004.

    c. The vertex is $(10.77, 35.08)$, which means that in about 2001, Quiksilver reached a maximum net sales of $35.08 million.

23. a. The ball is at a height of 4 feet when it is hit.

    b. The ball reached a height of 20 feet about 0.5 second and again 2 seconds after being hit.

    c. The ball reached its maximum height 1.25 seconds after being hit.

    d. The ball reached a maximum height of 29 feet.

    e. If the ball does not get caught, the ball will hit the ground 2.596 seconds after being hit.

25. a. The ball is at a height of 4.2 feet when it is hit.

    b. The ball reached a height of 40 feet at 0.744 second and again 3.0055 seconds after being hit.

    c. The ball's maximum height is 60.45 feet.

    d. If the ball does not get caught, the ball will hit the ground 3.819 seconds after being hit.

27. a. $H(6) = 58.4$. In June, the average high temperature in Melbourne, Australia, is 58.4°F.

    b. The vertex is $(7.22, 57.056)$, which means that in July, the average high temperature in Melbourne, Australia, reaches its minimum of 57.056°F.

    c. During September and June, the average temperature of Melbourne, Australia, is about 60°F.

29. Domain: All real numbers; range: $[-16, \infty)$

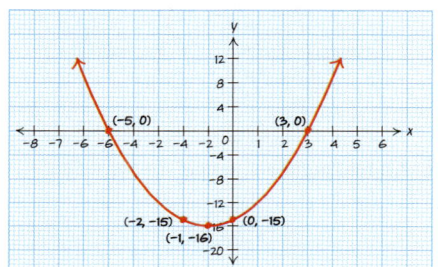

31. Domain: All real numbers; range: $(-\infty, 6.25]$

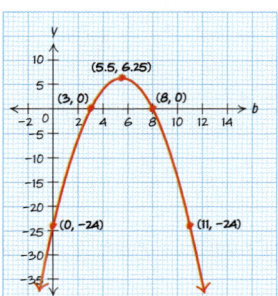

33. Domain: All real numbers; range: $[-264.5, \infty)$

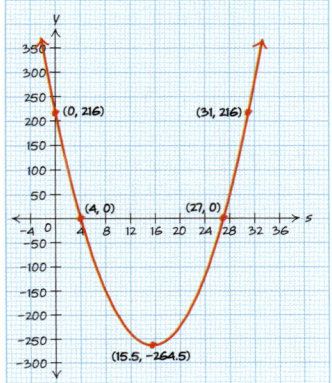

35. Domain: All real numbers; range: $[5, \infty)$

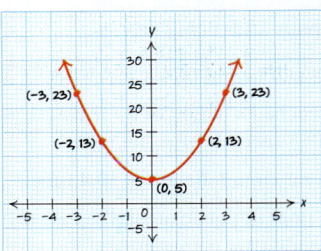

37. Domain: All real numbers; range: $(-\infty, -3]$

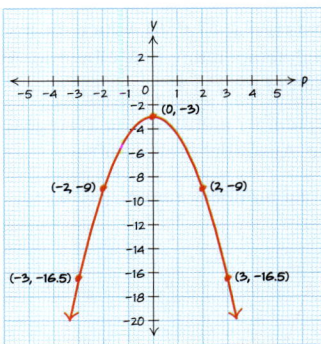

39. Domain: All real numbers; range: $(-\infty, -5]$

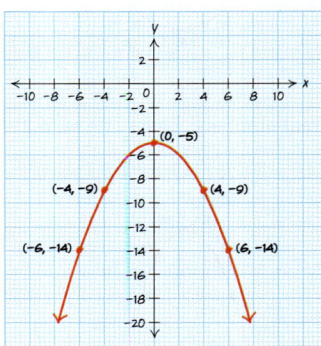

41. Domain: All real numbers; range: $(-\infty, 2.8125]$

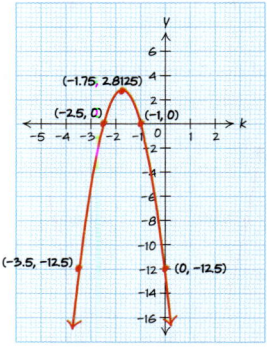

43. Domain: All real numbers; range: $[-52.9, \infty)$

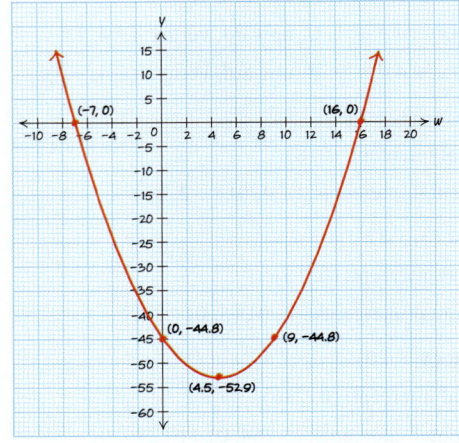

45. Domain: All real numbers; range: $[-3.1, \infty)$

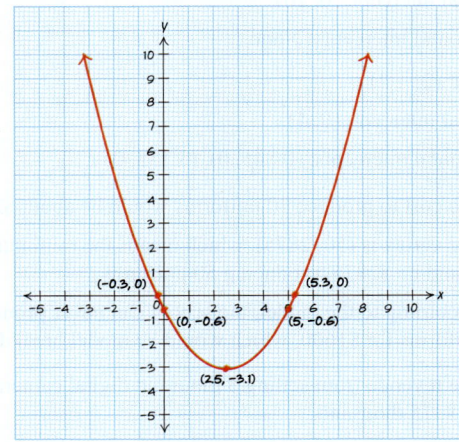

47. Domain: All real numbers; range: $(-\infty, 86.8]$

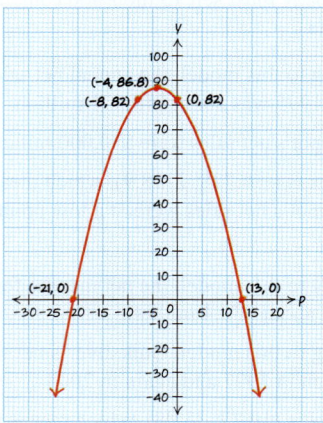

49. Domain: All real numbers; range: $(-\infty, 17]$

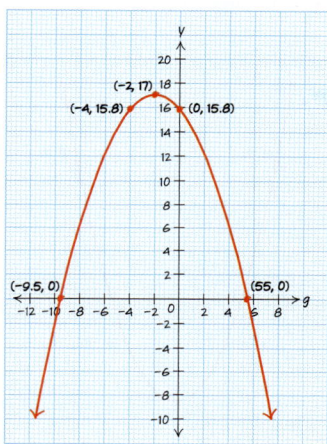

51.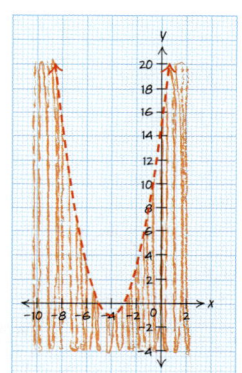

53.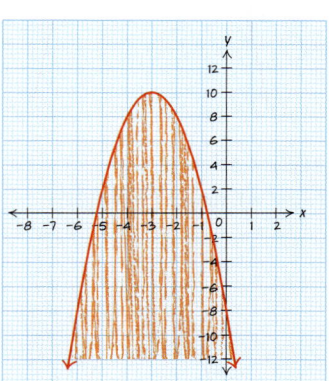

55.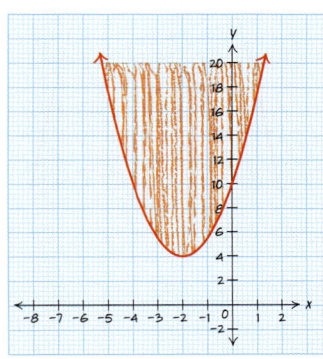

57.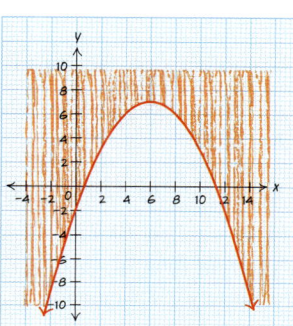

59.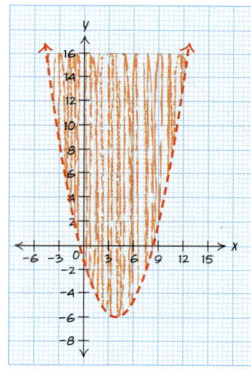

61.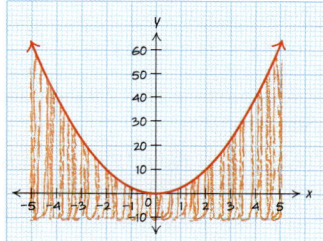

63.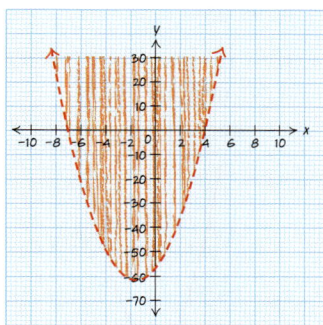

65. $y < x^2 + x - 12$ **67.** $y \geq -0.5(x + 5)(x + 1)$

Chapter 4 Review Exercises

1. a. Vertex = $(-1, 9)$ **b.** $x < -1$
   c. $x > -1$ **d.** $(-4, 0), (2, 0)$ **e.** $(0, 8)$

2. a. $x = -1$
   b. a would be negative, since the graph faces downward.
   c. $h = -1$, $k = 9$ **d.** $f(1) = 5$ **e.** $x = -5.5, x = 3.5$

3. **a.** $x = 1$
   b. a would be positive, since the graph faces upward.
   c. $h = 1, k = -4$ **d.** $f(-1.5) = 2.25$
   e. $x = -1.6, x = 3.6$

4. **a.** Vertex: $(1, -4)$ **b.** $x > 1$ **c.** $x < 1$
   d. $(-1, 0), (3, 0)$ **e.** $(0, 23)$

5. **a.** $P(8) = 637.87$. In 1998, the population of North Dakota was approximately 638 thousand.
   b.

   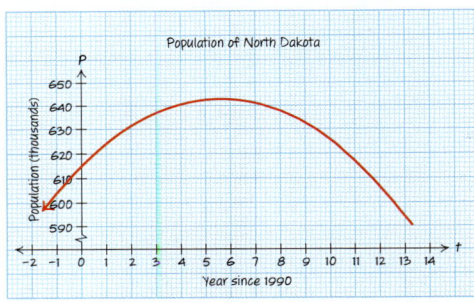

   c. Vertex = $(5.6, 643)$. In about 1996, the population of North Dakota reached a maximum of about 643 thousand people.
   d. The population of North Dakota was about 640,000 in about 1994 and about 1998.
   e. Range: $606.55 \leq P(t) \leq 643$

6. **a.** There were approximately 17.9 thousand murders in the United States in 1996.
   b. Vertex: $(2.5, 23)$. In about 1993, the most murders occurred in the United States at 23 thousand.
   c. In 1997 and 1988, there were about 14,500 murders in the United States.
   d. Range: $[10.295, 23]$

7. $h(x) = (x + 3)^2 - 16$
   Domain: $(-\infty, \infty)$; range: $[-16, \infty)$

   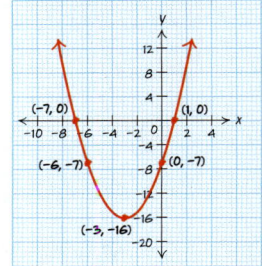

8. $f(x) = 1.25(x - 4)^2 - 20$
   Domain: $(-\infty, \infty)$; range: $[4, \infty)$

   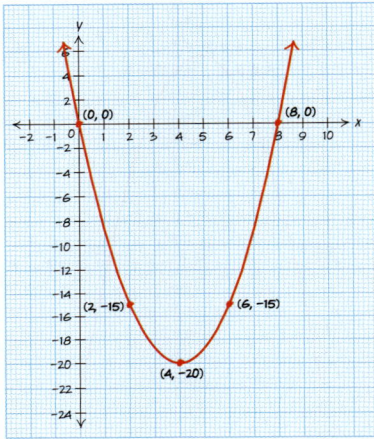

9. $g(x) = -\dfrac{1}{4}(x - 8)^2 + 9$
   Domain: $(-\infty, \infty)$; range: $(-\infty, 9]$

   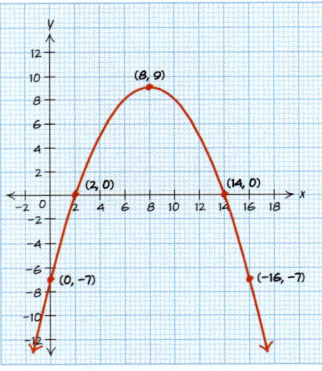

10. $h(x) = -2(x + 4.5)^2 + 12.5$
    Domain: $(-\infty, \infty)$; range: $(-\infty, 12.5]$

    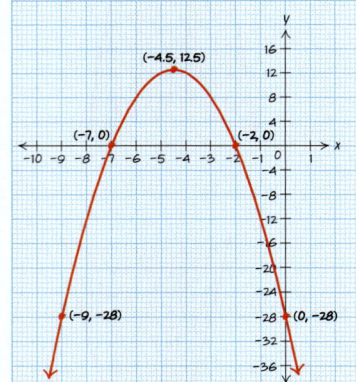

11. $g(x) = -0.5(x+7)^2 - 3$
    Domain: $(-\infty, \infty)$; range: $(-\infty, -3]$

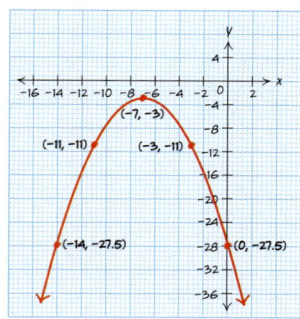

12. $f(x) = 3(x-8)^2 + 4$
    Domain: $(-\infty, \infty)$; range: $[4, \infty)$

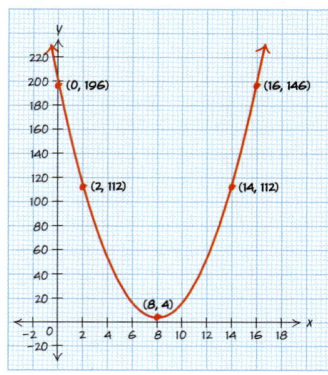

13. a. Let $J(t)$ be the number of juveniles in thousands arrested for possession of drugs t years since 1990.
    $$J(t) = -3095(t-7)^2 + 124{,}683$$
    b. Domain: $[1, 11]$; range $[13{,}263, 124{,}683]$
    c. In 2000, there were approximately 96,828,000 juveniles arrested for drug possession.
    d. The juvenile arrests were down to 50,000,000 again in about 2002.

14. a. Let $R(t)$ be the obligations of the U.S. Department of Commerce for research and development in millions of dollars t years since 1990. $R(t) = 17(t-8.1)^2 + 982$
    b. Domain: $[4, 12]$; range: $[982, 1267.77]$
    c. $R(8) = 982.17$. In 1998, the research and development obligations of the U.S. Department of Commerce was approximately $982 million.
    d. The research and development obligations for the U.S. Department of Commerce will reach $1.5 billion in about 1993 and in 2004.

15. a. In 1995, the median sales price of a new home in the western United States was $142,310.
    b. In 2003, the median sales price of a new home in the western United States reached $250,000.
    c. Vertex: (2.72, 136.73). In about 2003, the median sales price of a new home in the western United States reached a low of $136,730.

16. a. In 2004, the median asking price of houses in Memphis, Tennessee, was $158,360.
    b. In 2005 and again in 2009, the median asking price of houses in Memphis, Tennessee, reached $165,000.
    c. Vertex: (6.95, 170.84). In about 2007, the median asking price of houses in Memphis, Tennessee, reached a high of $170,840.

17. $t = \pm 13$
18. $m = \pm 7$
19. $x = -4, x = -10$
20. $x \approx 11.66, x \approx 0.34$
21. No real solutions.
22. No real solutions.
23. $c \approx -3.76, x \approx -8.23$
24. $p \approx 4.87, p \approx 3.13$
25. $x \approx -27.11, x \approx 1.11$
26. $x = 13, x = -1$
27. $x = 6, x = -11$
28. No real solutions.
29. $f(x) = (x+4)^2 - 5$
30. $g(x) = -6\left(x - \dfrac{5}{3}\right)^2 - \dfrac{4}{3}$
31. $t = 10, t = 2$
32. $p = -9, p = 3$
33. $x = 0, x = \dfrac{4}{3}$
34. $x = 1, x = -\dfrac{2}{3}$
35. $m = \pm 8$
36. $m = \pm \dfrac{5}{2}$
37. $x = \dfrac{4}{3}$
38. $x = -6, x = 10$
39. $x = 0, x = -5, x = -2$
40. $h = 0, h = -2, h = 7$
41. $y = (x+1)^2 - 9$
42. $y = -3(x+1)^2 + 12$
43. $x = -2, x = 6$
44. $t = -9, t = \dfrac{7}{2}$
45. No real solutions.
46. $x \approx 2.19, x \approx -5.19$
47. $x \approx 2.10, x \approx -1.32$
48. $n = \dfrac{57}{13}, n \approx -3$
49. $x \approx -0.28, x \approx -2.39$
50. $a \approx -3.87, a \approx +3.87$
51. Domain: All real numbers; range: $(-\infty, 2.5]$

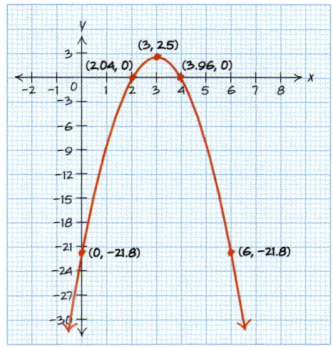

52. Domain: All real numbers; range: $[7, \infty)$

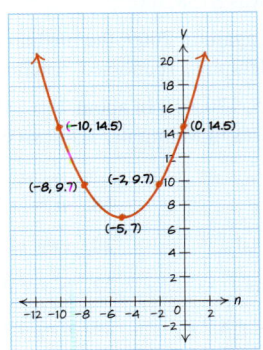

53. Domain: All real numbers; range: $[-15, \infty)$

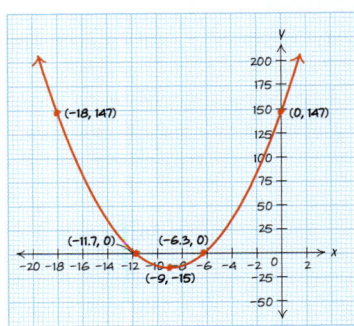

54. Domain: All real numbers; range: $(-\infty, 28]$

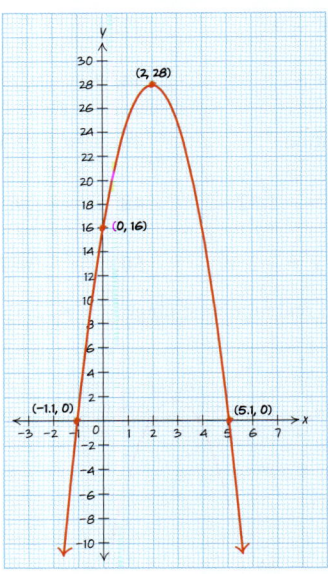

55. $(-3, -24), (4, 11)$ **56.** $(-2, 11), (6, -21)$

57. $(1, 13), (4.5, 118)$ **58.** $(-1.46, -12.32), (5.46, 22.32)$

59. $(3, 9), (5, 9)$

60. a. The largest possible rectangular area is 1250 ft².

  b. The largest possible area requires a length of 50 feet and a width of 25 feet.

61. $y < 2x^2 + 24x - 10$

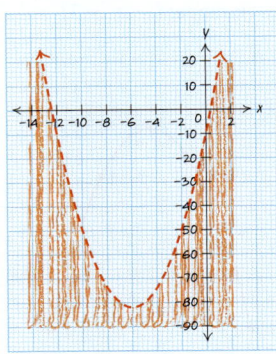

62. $y \geq 0.5x^2 - 4x - 6$

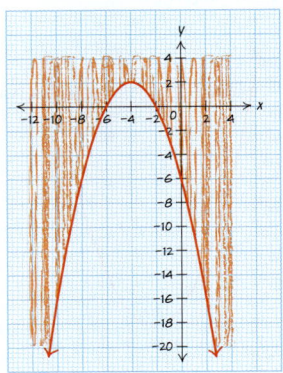

63. $y \leq 2.5(x - 4)^2 - 18$

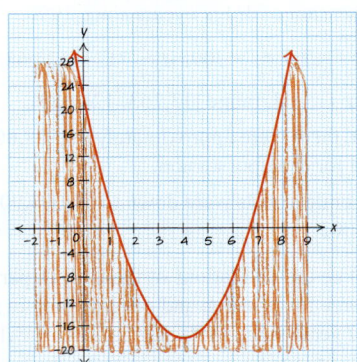

64. $y > -0.8(x + 1)^2 - 3$

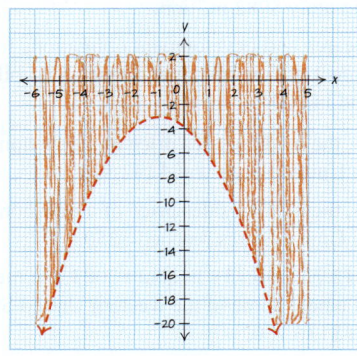

65. $(-4, 3), (2, 6)$

66. $(-4, 0), (2, 9)$

Chapter 4 Test

1. a. $V(8) \approx 1578$. In 2008, there were approximately 1578 thousand violent crimes in the United States.

b.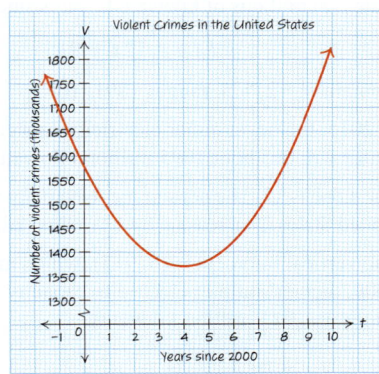

c. Vertex $= (4, 1370)$. In 2004, the number of violent crimes in the United States was at a minimum of 1370 thousand.

d. According to this graph, there were 1.5 million violent crimes in 2001 and again in 2007.

2. $(-6, 4)$, and $(4, 8)$

3. $x = -\dfrac{7}{3}, x = 4$

4. $x = \dfrac{7}{4}, x = \dfrac{5}{2}$

5. a. Let $O(t)$ be the total outlays in billions of dollars for national defense and veterans benefits by the United States t years since 1990. $O(t) = 1.9(t - 6)^2 + 308$

b. Domain: $[1, 11]$; range: $[308, 355.5]$

c. In 1998, the total outlays for national defense and veterans benefits by the United States were about $315.6 billion.

d. Total outlays for national defense and veterans benefits by the United States will reach half a trillion dollars in about 2006.

6. a. Let $R(t)$ be the revenue for the U.S. commercial space industry from the satellite manufacturing in billions of dollars t years since 1990.

$$R(t) = -1.225(t - 8)^2 + 11.8$$

b. Domain: $[6,10]$; range: $[6.9, 11.8]$. In this case, we did not go back from the data because it resulted in a very small revenue in 1995, and that seemed like model breakdown.

c. $R(10) = 6.9$. In 2000, the revenue from satellite manufacturing was about $6.9 billion.

d. Vertex $= (8, 11.8)$. In 1998, the revenue from satellite manufacturing reached a maximum of $11.8 billion.

7. Domain: All real numbers; range: $(-\infty, 3.5]$

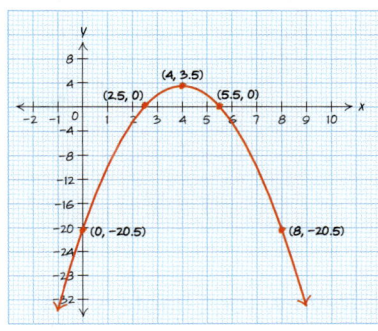

8. Domain: All real numbers; range: $[10, \infty)$

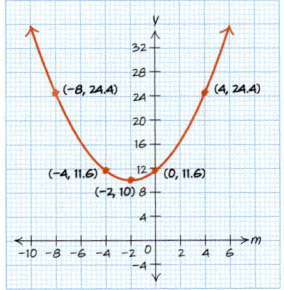

9. Domain: All real numbers; range: $(-\infty, -17]$

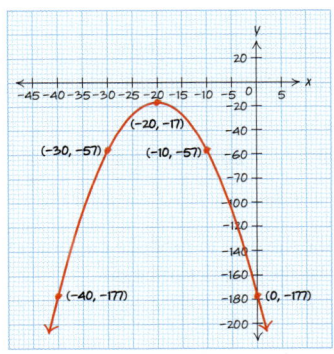

10. $(1, -2)$ and $\left(-\dfrac{11}{9}, -\dfrac{982}{81}\right)$

11. a. In 2000, there were about 1230.8 thousand privately owned single-unit houses started.

b. There were 1000 thousand privately owned single-unit houses started in about 2001 and in about 1996.

12. The ladder is about 19.6 feet up the house.

13. $(-7, 4), (-4, 8)$

14. a. $(-3, -4)$ **b.** $x > -3$ **c.** $x < -3$
d. $x = -3$ **e.** $(0, 5)$ **f.** $(-5, 0), (-1, 0)$
g. $f(x) = x^2 + 6x + 5$

15. $(-8.449, 0), (-3.551, 0)$

16. $x = 3, x = 15$ **17.** $x = \pm 2.5$ **18.** $f(x) = 3(x + 4)^2 - 78$

19.

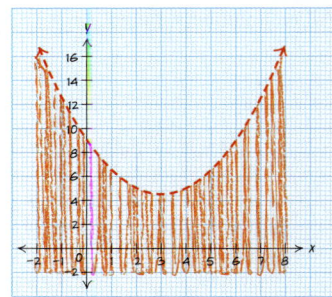

Cumulative Review for Chapters 1–4

1. $m = 4$
2. $t = 1, t = 10$
3. $x = 9, x = 5$
4. $a = \dfrac{74}{3}$
5. $x = 0, x = \dfrac{8}{3}$
6. $m = \dfrac{108}{17} \approx 6.35, m = -5$
7. $c = 12$
8. $n = 0, n = -18, n = 12$
9. $x = -1, x = 5$
10. $h \approx -8.68$
11. $m = \dfrac{n+5}{4n}$
12. $b = \dfrac{\tfrac{5}{2}(c-7a)}{a^2} = \dfrac{5c}{2a^2} - \dfrac{35}{2a}$
13. $x = 5, y = 4$
14. $(-5, -22)$ and $(2, -1)$
15. Inconsistent system; no solution.
16. $m = 1, n = -1$
17. $(2.13, 2.08)$ and $(-2.04, -11.60)$
18. Consistent system with dependent lines; infinite number of solutions.
19. $y = \dfrac{9}{2}x - 10$
20. $y = -\dfrac{1}{8}x + 6$
21. $y = -0.4x + 8.8$
22. **a.** Let $C(n)$ be the cost to produce custom printed tote bags, and let n be the number of bags.

$$C(n) = 3.55n + 50$$

   b. 200 bags will cost $760.

   c. Slope $= 3.55$. The cost to produce custom printed tote bags is increased by \$3.55 per bag.

23. **a.** Let $P(t)$ be the population of Pittsburgh, Pennsylvania, in millions t years since 2000.

$$P(t) = -0.012t + 2.434$$

   b. Domain: $[-2, 12]$; range: $[2.29, 2.458]$

   c. The population of Pittsburgh, Pennsylvania, will be about 2.29 million in 2012.

   d. The slope is -0.012. The population of Pittsburgh, Pennsylvania, is decreasing by 12,000 people per year.

   e. In 2015, the population of Pittsburgh, Pennsylvania, will be 2.25 million.

24. **a.** Let $R(p)$ be the revenue from selling four-person tents (in dollars) if they are priced at p dollars each.

$$R(p) = -0.989(p - 70)^2 + 3250$$

   b. The vertex is $(70, 3250)$. When the price is \$70 per four-person tent, the revenue generated is at a maximum of \$3250.

   c. When the price is \$60 per four-person tent, the revenue generated is \$3151.10.

   d. Domain: $[20, 100]$; range: $[777.5, 3250]$

25. **a.** Let $E(t)$ be the net summer electricity capacity in the United States (in millions of kilowatts) t years since 2000.

$$E(t) = 12t + 915$$

   b. Domain: $[-2, 12]$; range: $[891, 1059]$

   c. In 2015, the net summer electricity capacity in the United States will be about 1095 million kilowatts.

   d. The slope is 12. The net summer electricity capacity in the United States is increasing by 12 million kilowatts per year.

   e. The vertical intercept is $(0, 915)$, which means in the year 2000, the net summer electricity capacity in the United States was 915 million kilowatts.

26.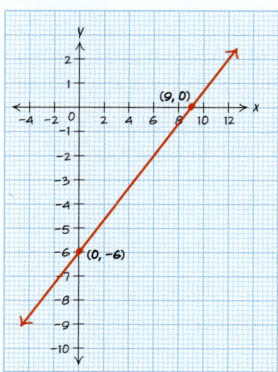

27.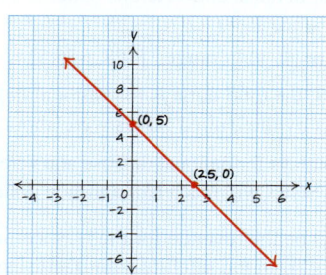

28.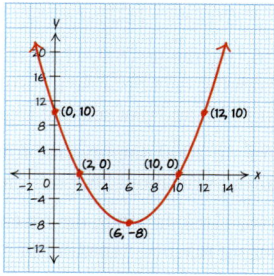

29.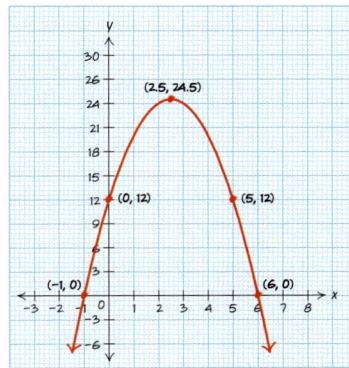

30.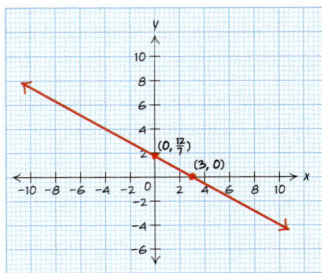

31.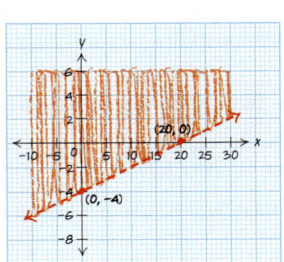

32.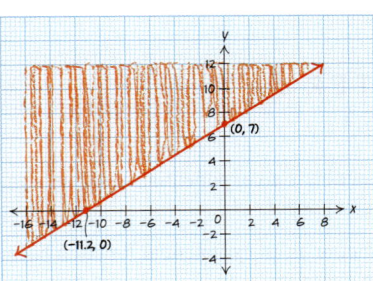

33.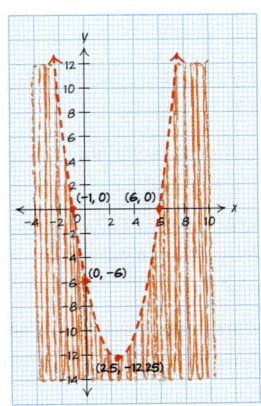

34. a. $-5x + 26$ b. $14x^2 + 20x - 144$ c. $14x - 28$

35. a. $-2x^2 - 2x + 5$ b. $6x^3 + 7x^2 - 47x + 36$
    c. $6x^2 + 15x - 31$ d. $18x^2 - 33x + 3$

36. a. 20.6 b. -104
    c. 1.4 d. -6.72

37. a. -12 b. 90
    c. 32 d. -2

38. a. $f(21) = -23$ b. $x = -6.5$
    c. Domain: All real numbers; range: All real numbers

39. a. $g(9) = 20$
    b. Domain: All real numbers; range: $\{20\}$

40. a. $h(-8) = -12$ b. $x = 13.75$
    c. Domain: All real numbers; range: All real numbers

41. During the first 8 minutes of a jog-a-thon, an eighth grader has run 3.5 laps.

42. a. $(0, 9)$ b. $(15, 0)$ c. $m = -\dfrac{3}{5}$
    d. $y = -\dfrac{3}{5}x + 9$

43. a. $f(8) = 4$ b. $x = -4$
    c. Domain: All real numbers; range: All real numbers

44. a. $m = -\dfrac{2}{9}$ b. $(0, 2)$ c. $(9, 0)$
    d. $y = \dfrac{-2}{9}x + 2$

45. a. Let x be the amount in dollars Greg is investing at 3%, and let y be the amount in dollars Greg is investing at 2.5%.
    $$x + y = 900{,}000$$
    $$0.03x + 0.025y = 25{,}800$$
    b. \$240,000 should be invested at 2.5%, and \$660,000 should be invested at 3%.

46. They need to mix 80 pounds of the \$11.99 per pound mild roast coffee with 120 pounds of the \$8.99 per pound bold roast coffee to make 200 pounds of a \$10.19 per pound blend.

47. Deeva needs to make at least \$2750 per week in sales to earn at least \$410 per week.

48. There are four terms. $24x^3y$: variable term, coefficient 24
    $-7x^2y^2$: variable term, coefficient -7
    $12y^3$: variable term, coefficient 12
    -10: constant term

49. There are four terms. $-6x^5$: variable term, coefficient -6
    $4x^2$: variable term, coefficient 4
    $-2x$: variable term, coefficient -2
    4: constant term

50. $24x^3y$ has degree 4.

$-7x^2y^2$ has degree 4.

$12y^3$ has degree 3.

-10 has degree 0.

The degree of the polynomial is 4.

51. $-6x^5$ has degree 5.

$4x^2$ has degree 2.

$-2x$ has degree 1.

4 has degree 0.

The degree of the polynomial is 5.

52. Polynomial

53. Not a polynomial. The variable y has a negative exponent.

54. $x < -2$

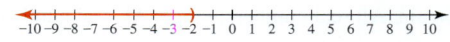

55. $x \geq -2$

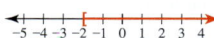

56. $3 < x < 21$

57. $-2 \leq x \leq \dfrac{16}{3}$

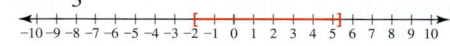

58. $x \leq 6$ or $x \geq 12$

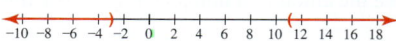

59. $x < -3$ or $x > 11$

60. $y1 < y2$ for $x < 8$

61. $m = -5$ or $m = 10$

62. $a = 7$ or $a = 11$

63.

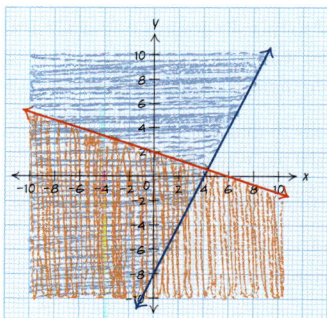

64.

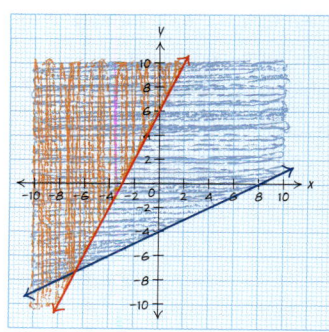

65. $\dfrac{81m^{16}n^{20}}{625}$

66. $-40x^4y^9$

67. $\dfrac{3g^2h^4}{2}$

68. $-\dfrac{7x^3y^6}{10}$

69. $-\dfrac{3b^4}{2a}$

70. $\dfrac{6m^2n^2}{5}$

71. a. $S(6) = 65.56$. In 2006, Sweden's GDP per employed person was $65,560.

  b. Sweden's GDP per employed person will be greater than that of the Netherlands after 2008.

72. a. In 2007, there were about 8780 loans to African American owned small businesses.

  b. Let $T(t)$ be the total number of loans (in thousands) given to African American and Hispanic American owned small businesses t years past 2000. $T(t) = 2.59t + 2.27$

  c. There will be about 28,170 loans given to African American and Hispanic American owned small businesses in 2010.

73. $(a + 3)(a + 8)$

74. $3n(n - 7)(n - 2)$

75. $(2x - 5y)(8x - 7)$

76. $(5k + 7)(5k - 7)$

77. $(4t + 3)(3t - 7)$

78. $5(z - 2)(z^2 + 2z + 4)$

79. a. The population of this city will be about 78 thousand in 2015.

  b. There are about 47 thousand burglaries in this city in 2015.

  c. Let $K(t)$ be the number of burglaries in this city t years since 2000. $K(t) = 1.2t + 29$

  d. In 2020, there will be about 53 thousand burglaries in this city.

80. $f(x) = (x + 5)^2 - 39$

CHAPTER 5
Section 5.1

1. This function is an exponential function because the variable is in the exponent.

3. This function is an exponential function because the variable is in the exponent.

5. This function is a quadratic function because the degree of the polynomial is 2.

7. This function is a quadratic function because the degree of the polynomial is 2.

9. Other, because there is a variable in the numerator and in the denominator.

11. Let $L(h)$ be the number of *Lactobacillus acidophilus* bacteria present after h hours have passed.

  a. $L(h) = 30(2)^h$

  b. After 12 hours, there are 122,880 bacteria present.

  c. After 24 hours, there are 503,316,480 bacteria present.

  d. After just 15 hours, there are about 1 million bacteria present.

13. Let $E(h)$ be the number of *E. coli* bacteria present after h hours have passed.

   a. $E(h) = 3(2)^{2h}$

   b. After 5 hours, there are approximately 3072 bacteria present.

   c. After about 9.17 hours, there are 1 million bacteria present.

15. Let $S(n)$ be the number of *Streptococcus lactis* bacteria present after n 26-minute time intervals have passed.

   a. $S(n) = 8(2)^n$

   b. $S(15) = 262144$. After 6.5 hours, there are 262,144 bacteria present.

17. Let $B(h)$ be the number of bacteria present after h hours have passed.

   a. $B(h) = 8(3)^{2h}$

   b. After 10 hours, there are approximately 2.7894×10^{10} bacteria present.

19. Let $T(h)$ be the number of *Treponema pallidum* bacteria present after h hours have passed.

   a. $T(h) = 12(2)^{\frac{h}{33}}$

   b. $T(168) = 409$. After 1 week, there are approximately 409 bacteria present.

21. Let $F(n)$ be the number of square meters burnt after n 6-minute time intervals have passed.

   a. $F(n) = 2^{n+1}$ or $F(n) = 2(2)^n$

   b. After $\frac{1}{2}$ hour, there are 64 square meters burnt.

23. Let $H(t)$ be the number of Hispanic centenarians in the year t years since 1990.

   a. $H(t) = 2072(2)^{\frac{t}{7.5}}$

   b. In 2050, there will be approximately 530,432 Hispanic centenarians in the United States.

25. Let $I(t)$ be the population of India (in millions) in the year t years since 1971.

   a. $I(t) = 560(2)^{\frac{t}{33}}$

   b. In 2015, there will be approximately 1411 million people in India.

   c. Midway through the year 2031, there will be approximately 2 billion people in India.

27. Internet search: approximately 65 years.

29. a. $50,000

   b. Your opinion.

   c. Let $S(w)$ be salary option 2 (in dollars) for week w of the job: $S(w) = 0.01(2)^{w-1}$

   d. The total salary from option 2 for 25 weeks of work is $335,544.31.

   e. Option 2 is the best deal for 25 weeks of work.

   f. Option 1 offers $40,000 for 20 weeks of work, and option 2 offers $10,485.75 for 20 weeks of work. Therefore, option 1 is the best deal for 20 weeks of work.

31. Let $P(t)$ be the percentage of lead-210 left in a body t years after a person has died.

   a. $P(t) = 100\left(\frac{1}{2}\right)^{\frac{t}{22}}$

   b. Approximately 60 years after death, there will be about 15.10% of lead-210 left in the body.

   c. Approximately 95 years after death, there will be about 5% of lead-210 left in the body.

33. Let $P(d)$ be the percentage of radon-222 left after d days.

   a. $P(d) = 100\left(\frac{1}{2}\right)^{\frac{d}{3.825}}$

   b. After approximately 30 days, there will be about 0.44% of radon-222 left.

   c. After approximately 16.5 days, there will be about 5% of radon-222 left.

35. a. The graph is decreasing.

   b. This is an example of exponential decay.

   c. $f(10) = 22$

   d. $x = 6$ for $f(x) = 40$

   e. $f(0) = 100$

37. a. The graph is increasing.

   b. This is an example of exponential growth.

   c. $h(5) = 20$

   d. $x = 15$ for $h(x) = 100$

   e. $h(0) = 10$

39. a. The graph is decreasing.

   b. This is an example of exponential decay.

   c. $g(7.5) = 20$

   d. $x = 4$ for $g(x) = 30$

   e. $g(0) = 60$

41. $f(x) = 25(4)^x$

43. $f(x) = -35(7)^x$

45. $f(x) = 2000(0.2)^x$

47. $f(x) = 32(1.5)^x$

49. $f(x) = 6400(0.125)^x$

51. The base $b = 3$, and the coefficient $a = 5$.

53. The base $b = \frac{1}{5}$, and the coefficient $a = 200$.

55. The base $b = 2$, and the coefficient $a = -4$.

57. Let $E(h)$ be the number of *E. coli* bacteria present after h hours have passed: $E(h) = 3(2)^{2h}$.

   a. The base $b = 2$.

   b. The coefficient $a = 3$.

   c. The coefficient $a = 3$ represents the number of *E. coli* bacteria present at time $t = 0$ (at the start).

59. Let $M(h)$ be the number of *Treponema pallidum* bacteria present after h hours have passed: $T(h) = 12(2)^{\frac{h}{33}}$

   a. The base $b = 2$.

   b. The coefficient $a = 12$.

   c. The coefficient $a = 12$ represents the number of *Treponema pallidum* bacteria present at time $t = 0$ (at the start).

61. $f(x) = 3(4)^{\frac{x}{5}}$

63. $f(x) = -7(3)^{\frac{x}{10}}$

65. $f(x) = 1701\left(\frac{1}{3}\right)^{\frac{x}{6}}$
67. $f(x) = 80(4)^{x-2}$
69. $f(x) = 2(6)^{x-5}$
71. $f(x) = 3584(0.25)^{x-4}$

73. The function is increasing because the base is greater than 1.

75. The function is decreasing because the base is less than 1.

77. The function is increasing because the base is greater than 1.

79. The function is decreasing because the base is less than 1.

Section 5.2

1. Exponential equation.
3. Power equation.
5. Power equation.
7. Exponential equation.
9. Power equation.
11. $x = 3$
13. $c = 5$
15. $t = -2$
17. $x = -4$
19. $d = 5$
21. $w = 2$
23. $x = 4$
25. $x = -4$
27. $x = -5$
29. $x = 3$
31. $x = 0$
33. $m = 0$
35. $t = 0$
37. $x = -2$. Tom evaluated $\frac{1}{16}$ as 4^2 instead of $\frac{1}{16} = 4^{-2}$. So he missed the negative sign in the exponent.
39. a. There are 25,600 *Lactobacillus acidophilus* bacteria present after 8 hours have passed.
    b. There are 800 *Lactobacillus acidophilus* bacteria present after 3 hours have passed.
41. a. After 8 hours have passed, 32,805 people have heard the rumor.
    b. It takes 4 hours for the rumor to have spread to 405 people.
43. $x = 3$
45. $c = 3$
47. $c = 3$
49. $t = 2$
51. $m = 2$
53. $x = 2$
55. $x = 4$
57. $x = 4$
59. $w = 3$
61. $x = 2$
63. $x = \pm 5$
65. $w = \pm 5$
67. $x = -6$
69. Warrick forgot to put the $\pm$ symbol on the final answer. Both positive 6 and negative 6 are correct answers.
    $x = \pm 6$ is the correct answer.
71. Frank divided 216 by 3 instead of raising 216 to the power $\frac{1}{3}$.
    $r = 6$ is the correct answer.
73. a. An NBA basketball with radius 12 cm has a volume of about 7238.2 cm³.
    b. An International Basketball Federation basketball with a volume of 8011 cm³ has a radius of about 12.4 cm.
75. The length of the side is 4.5 in.
77. $x = 2$
79. $x \approx \pm 2.6$
81. $c \approx \pm 1.2$
83. $x = 5$
85. $b = \pm 6$
87. Exponential $k = 4$
89. Power $r \approx 1.6$
91. Exponential $x = 3$
93. Power $g = 2$
95. Power $h \approx \pm 2.2$
97. Exponential $x = 3$
99. Power $x = \pm 3$
101. Exponential $x = 7$

Section 5.3

1. a. a is positive because the graph is above the x-axis.
   b. $b > 1$ because we have exponential growth with $a > 0$.
   c. The graph is increasing.
   d. Domain: All real numbers $(-\infty, \infty)$
   e. Range: $(0, \infty)$
3. a. a is positive because the graph is above the x-axis.
   b. $b < 1$ because we have exponential decay with $a > 0$.
   c. The graph is decreasing.
   d. Domain: All real numbers
   e. Range: $(0, \infty)$
5. a. a is negative because the graph is below the x-axis.
   b. $b < 1$ because we have exponential growth with $a < 0$.
   c. The graph is increasing.
   d. Domain: All real numbers $(-\infty, \infty)$
   e. Range: $(-\infty, 0)$
7. a. a is negative because the graph is below the x-axis.
   b. $b > 1$ because we have exponential decay with $a < 0$.
   c. The graph is decreasing.
   d. Domain: All real numbers
   e. Range: $(-\infty, 0)$
9. a. a is negative because the graph is below the x-axis.
   b. $b > 1$ because we have exponential decay with $a < 0$.
   c. The graph is decreasing.
   d. Domain: All real numbers $(-\infty, \infty)$
   e. Range: $(-\infty, 0)$
11. $f(x) = 7(2)^x$
    $a = 7$, so the y-intercept is $(0, 7)$.
    $b = 2$, $b > 1$, and $a > 0$, so the graph is increasing.
    Domain: $(-\infty, \infty)$, range: $(0, \infty)$

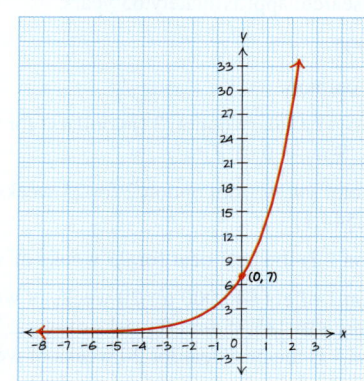

13. $g(x) = 3(1.2)^x$

$a = 3$, so the y-intercept is $(0, 3)$.

$b = 1.2$, $b > 1$, and $a > 0$, so the graph is increasing.

Domain: $(-\infty, \infty)$, range: $(0, \infty)$

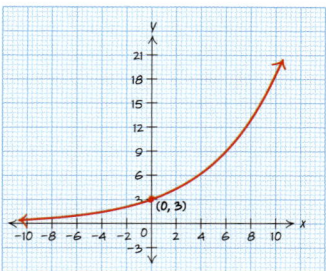

15. $h(x) = 12(1.4)^x$

$a = 12$, so the y-intercept is $(0, 12)$.

$b = 1.4$, $b > 1$, and $a > 0$, so the graph is increasing.

Domain: $(-\infty, \infty)$, range: $(0, \infty)$

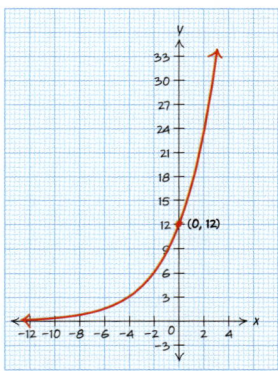

17. $p(t) = 140\left(\dfrac{1}{2}\right)^t$

$a = 140$, so the t-intercept is $(0, 140)$.

$b = \dfrac{1}{2}$, $b < 1$, and $a > 0$, so the graph is decreasing.

Domain: $(-\infty, \infty)$, range: $(0, \infty)$

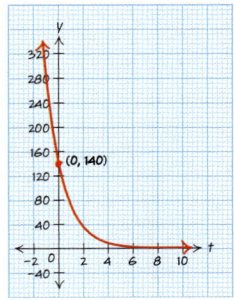

19. $f(x) = 250\left(\dfrac{1}{4}\right)^x$

$a = 250$, so the x-intercept is $(0, 250)$.

$b = \dfrac{1}{4}$, $b < 1$, and $a > 0$, so the graph is decreasing.

Domain: $(-\infty, \infty)$, range: $(0, \infty)$

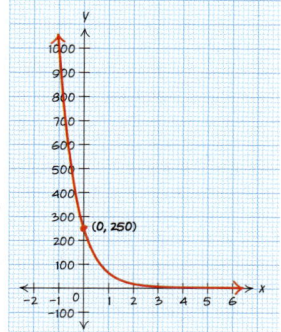

21. a.

b. The B-intercept is $(0, 4)$. At the start, there were 4 E. coli bacteria present in the sample.

23. a.

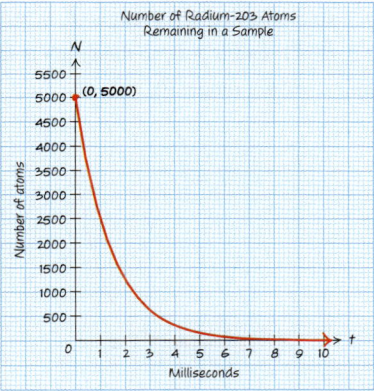

b. The N-intercept is $(0, 5000)$. At the start, there were 5000 radium-203 atoms present in the sample.

25. $f(x) = -2(1.4)^x$

$a = -2$, so the y-intercept is $(0, -2)$.

$b = 1.4$, $b > 1$, and $a < 0$, so the graph is decreasing.

Domain: $(-\infty, \infty)$, range: $(-\infty, 0)$

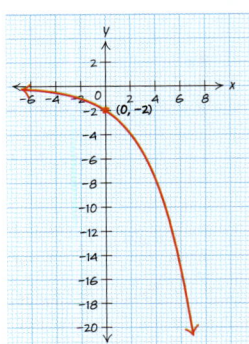

27. $g(t) = -3(0.7)^t$

$a = -3$, so the vertical intercept is $(0, -3)$.

$b = 0.7$, $b < 1$, and $a < 0$, so the graph is increasing.

Domain: $(-\infty, \infty)$, range: $(-\infty, 0)$

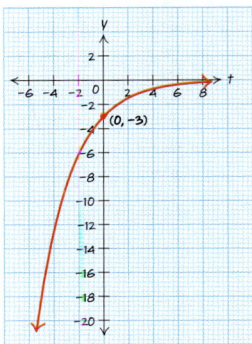

29. $h(m) = 0.5(2.5)^m$

$a = 0.5$, so the vertical intercept is $(0, 0.5)$.

$b = 2.5$, $b > 1$, and $a > 0$, so the graph is increasing.

Domain: $(-\infty, \infty)$, range: $(0, \infty)$

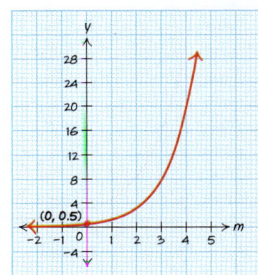

31. $j(w) = -0.5(4)^w$

$a = -0.5$, so the vertical intercept is $(0, -0.5)$.

$b = 4$, $b > 1$, and $a < 0$, so the graph is decreasing.

Domain: $(-\infty, \infty)$, range: $(-\infty, 0)$

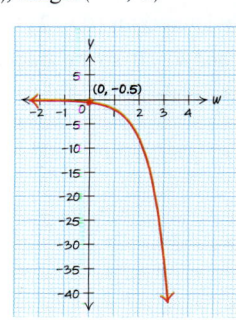

33. $h(t) = -0.4(1.5)^t$

$a = -0.4$, so the vertical intercept is $(0, -0.4)$.

$b = 1.5$, $b > 1$, and $a < 0$, so the graph is decreasing.

Domain: $(-\infty, \infty)$, range: $(-\infty, 0)$

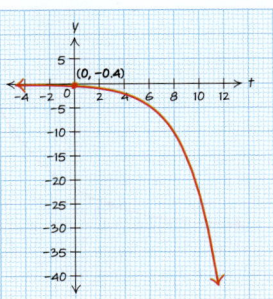

35. $c(n) = 550\left(\dfrac{3}{4}\right)^n$

$a = 550$, so the vertical intercept is $(0, 550)$.

$b = \dfrac{3}{4}$, $b < 1$, and $a > 0$, so the graph is decreasing.

Domain: $(-\infty, \infty)$, range: $(0, \infty)$

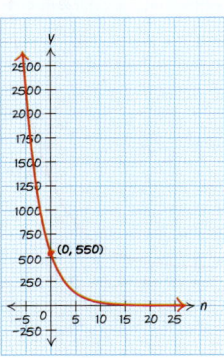

37. $f(x) = -500\left(\dfrac{1}{5}\right)^x$

$a = -500$, so the vertical intercept is $(0, -500)$.

$b = \dfrac{1}{5}$, $b < 1$, and $a < 0$, so the graph is increasing.

Domain: $(-\infty, \infty)$, range: $(-\infty, 0)$

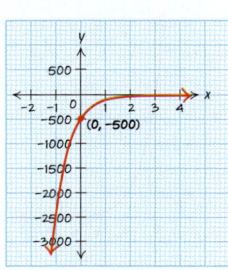

39. $f(x) = -700(0.95)^x$

$a = -700$, so the vertical intercept is $(0, -700)$.

$b = 0.95$, $b < 1$, and $a < 0$, so the graph is increasing.

Domain: $(-\infty, \infty)$, range: $(-\infty, 0)$

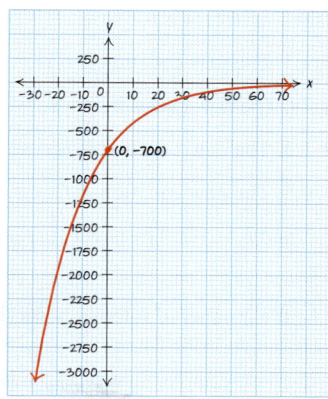

41. The student wrote that the domain started at zero and the range started at 7.

Domain: $(-\infty, \infty)$, range: $(0, \infty)$

43. The student wrote the domain correctly, but the range should be from zero to infinity.

Domain: $(-\infty, \infty)$, range: $(0, \infty)$

45. a. $y = 10$ **b.** Domain: $(-\infty, \infty)$
c. Range: $(10, \infty)$

47. a. $y = -5$ **b.** Domain: $(-\infty, \infty)$
c. Range: $(-\infty, -5)$

49. a. $y = 3$ **b.** Domain: $(-\infty, \infty)$ **c.** Range: $(3, \infty)$

51. a. $y = 0$ **b.** Domain: $(-\infty, \infty)$ **c.** Range: $(0, \infty)$

53. a. $y = -6$ **b.** Domain: $(-\infty, \infty)$ **c.** Range: $(-\infty, -6)$

55. Horizontal asymptote: $y = 30$

Domain: $(-\infty, \infty)$, range: $(30, \infty)$

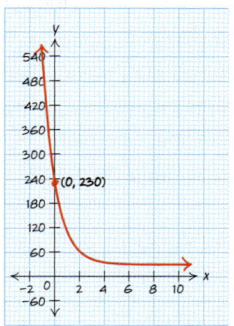

57. Horizontal asymptote: $y = 3$

Domain: $(-\infty, \infty)$, range: $(3, \infty)$

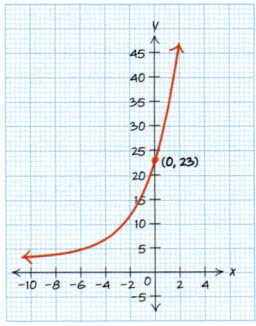

59. Horizontal asymptote: $y = -12$

Domain: $(-\infty, \infty)$, range: $(-\infty, -12)$

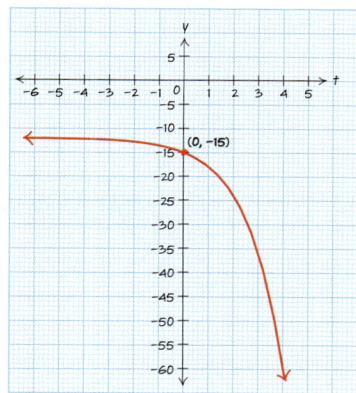

61. Horizontal asymptote: $y = -50$

Domain: $(-\infty, \infty)$, range: $(-\infty, -50)$

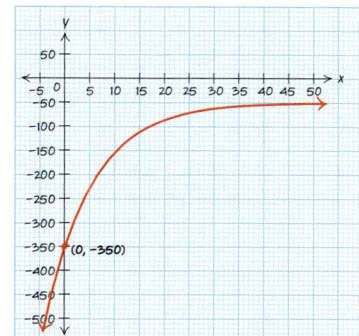

63. Horizontal asymptote: $y = 25$

Domain: $(-\infty, \infty)$, range: $(25, \infty)$

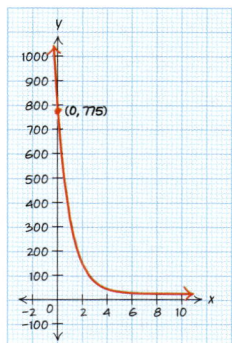

65. Domain: $(-\infty, \infty)$, range: $(-\infty, \infty)$

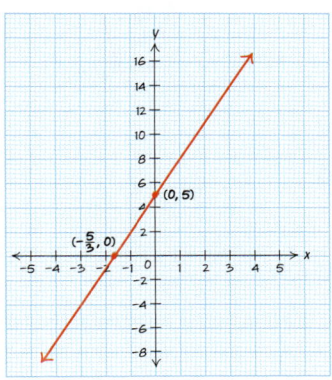

67. Domain: $(-\infty, \infty)$, range: $[3, \infty)$

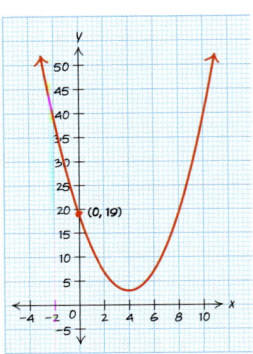

69. Domain: $(-\infty, \infty)$, range: $(0, \infty)$

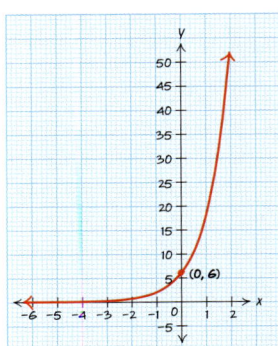

71. Domain: $(-\infty, \infty)$, range: $(-\infty, \infty)$

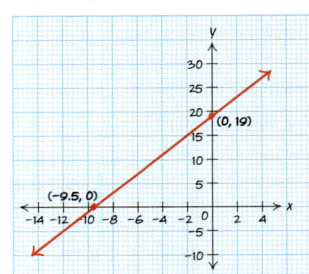

73. Domain: $(-\infty, \infty)$, range: $[-75, \infty)$

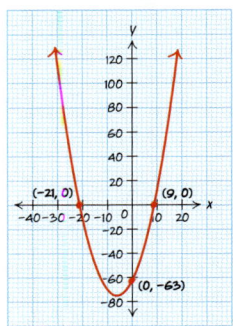

Section 5.4

1. $f(x) = 25(1.5)^x$

3. $f(x) = 900(0.9)^x$

5. *b*, because the function is increasing too rapidly.

7. *a*, because the *y*-intercept is too low.

9. Adjust $a = 3.5$. The new equation is $f(x) = 3.5(1.2)^x$.

11. Adjust $a = 78$. The new equation is $f(x) = 78(0.45)^x$.

13. Adjust $b = 1.3$. The new equation is $f(x) = 8(1.3)^x$.

15. Adjust $b = 0.5$. The new equation is $f(x) = 120(0.5)^x$.

17. Adjust $a = 5$ and $b = 1.2$. The new equation is $f(x) = 5(1.2)^x$.

19. a. Let $I(t)$ be the number of Internet hosts (in millions) in the year *t* years since 1990. Then $I(t) = 0.184(1.92)^t$.
  b. In 1997, there were approximately 17.7 million Internet hosts.
  c. Domain: $[0, 8]$, range: $[0.18, 33.98]$
  d. Partway through the year 1998, the number of Internet hosts reached 35 million.

21. a. Let $N(t)$ be the number of nuclear warheads in the U.S. arsenal in the year *t* years since 1940. Then $N(t) = 0.78(1.70)^t$.
  b. In 1970, there were approximately 157 nuclear warheads stockpiled in the U.S. arsenal.
  c. Domain: $[5, 20]$, range: $[11, 31701]$
  d. By the end of 1960, the number of stockpiled nuclear warheads in the U.S. arsenal surpassed 50,000.

23. a. Let $V(t)$ be the volume of water left in the cylinder *t* seconds after the start of the experiment. Then $V(t) = 15.86(0.994)^t$.
  b. Six minutes after the start of the experiment, approximately 1.82 liters of water remain in the cylinder.
  c. Domain: $[0, 500]$, range: $[0.78, 15.86]$
  d. It took approximately 459 seconds for there to be only 1 liter of water remaining in the cylinder.

25. a. Let $R(d)$ be the river gauge height, in feet above normal, *d* days after the rainfall event. Then $R(d) = 14(0.553)^d$.
  b. Ten days after the rainfall event, the gauge height will read 0.037 foot above normal.
  c. Domain: $[0, 12]$, range: $[0.01, 14]$

27. $f(x) = 4(1.32)^x$
  Domain: $(-\infty, \infty)$, range: $(0, \infty)$

29. $f(x) = 94(0.63)^x$
  Domain: $(-\infty, \infty)$, range: $(0, \infty)$

31. $f(x) = 4.2(1.27)^x$
  Domain: $(-\infty, \infty)$, range: $(0, \infty)$

33. $f(x) = 462(0.54)^t$
  Domain: $(-\infty, \infty)$, range: $(0, \infty)$

35. The data plot shows an increasing curve with positive values, so we know that $a > 0$ and $b > 1$. The student's model, therefore, is incorrect. The student miscalculated the value of *b* by multiplying both sides of $3.58 \approx b^7$ by $\frac{1}{7}$ instead of raising each side of the equation to the $\frac{1}{7}$ th power. As a result the value of *a* is incorrect as well.

37. $f(x) = -17(1.22)^x$
  Domain: $(-\infty, \infty)$, range: $(-\infty, 0)$

CHAPTER 5 Exponential Functions

39. $f(x) = -56(0.45)^x$
Domain: $(-\infty, \infty)$, range: $(-\infty, 0)$

41. $f(x) = -247(1.24)^x$
Domain: $(-\infty, \infty)$, range: $(-\infty, 0)$

43. $f(x) = -178(0.63)^x$
Domain: $(-\infty, \infty)$, range: $(-\infty, 0)$

45. Exponential: Text messaging continues to grow rapidly.

47. Exponential: Music downloads continue to grow rapidly.

49. Linear: Steady growth of New Hampshire's population.

51. a. Let $V(m)$ be the number of visitors to MySpace.com (in millions) m months after the site started. Then $V(m) = 0.15(1.21)^m$.
   b. Domain: $[9, 36]$, range: $[0.8, 143.3]$
   c. MySpace.com reached 50 million visitors after about 30 months.

53. a. Let $P(t)$ be the population of Georgia (in millions) in the year t years since 2000. Then $P(t) = 0.195t + 8.13$.
   b. Domain: $[0, 15]$, range: $[8.13, 11.06]$
   c. The population of Georgia will reach 11 million in the year 2014.

55. $f(x) = 3(x + 4)^2 - 10$

57. $f(x) = 79.9(0.70)^x$

59. $f(x) = 0.2x + 0.3$

Section 5.5

1. 3% growth   **3.** 25% growth
5. 250% growth   **7.** 5% decay
9. 64% decay

11. a. In 1990, the population of white-tailed deer was approximately 2.57 million.
   b. The population of white-tailed deer grows at a rate of 8% per year.
   c. This population might stop growing at the rate of 8% per year because of limited food or space.

13. a. In 2015, the population of the Virgin Islands will be approximately 108.14 thousand.
   b. The population of the Virgin Islands is decreasing at a rate of 0.061% per year.

15. Let $P(t)$ be the population after t years have passed. Then $P(t) = 40(1.03)^t$.

17. Let $P(t)$ be the population after t years have passed. Then $P(t) = 200(0.98)^t$.

19. a. Let $H(t)$ be the population of humpback whales t years after 1981. Then $H(t) = 350(1.14)^t$.
   b. Domain: $[-2, 15]$, range: $[269, 2498]$
   c. In 1990, the humpback whale population was approximately 1138.

21. a. Let $D(t)$ be the population of Denmark (in millions) t years after 2008. Then $D(t) = 5.5(1.00295)^t$.
   b. In 2020, the population of Denmark will be approximately 5.70 million.

23. Compounding interest formula: $A = P\left(1 + \dfrac{r}{n}\right)^{nt}$

25. Compounding interest formula: $A = P\left(1 + \dfrac{r}{n}\right)^{nt}$

27. Compounding interest formula: $A = P\left(1 + \dfrac{r}{n}\right)^{nt}$

29. Continuously compounding interest formula: $A = Pe^{rt}$

31. The missing variable is A.
$A = ?$
$P = 20{,}000$
$r = 0.03$
$n = 365$
$t = 5$

33. The missing variable is t.
$A = 8000$
$P = 4000$
$r = 0.0275$
$n = 12$
$t = ?$

35. The missing variable is r.
$A = 1000$
$P = 500$
$r = ?$
$n = 365$
$t = 10$

37. $\$16{,}470.09$ **39.** $\$16{,}486.65$
41. $\$1{,}928{,}712.77$ **43.** $\$1{,}928{,}699.39$
45. $\$400{,}639.19$ **47.** $\$405{,}520.00$

49. The second option, with an interest rate of 3.95% compounded continuously, will have the larger balance after 5 years.

Chapter 5 Review Exercises

1. a. Let $P(t)$ be the population of Africa (in millions) in the year t years since 1996. Then $P(t) = 731.5(2)^{\frac{t}{28}}$.
   b. In 2005, the population of Africa was approximately 914.06 million.

2. a. Let $F(d)$ be the number of people with flu symptoms after d days. Then $F(d) = (3)^d$.
   b. After 2 weeks, approximately 4,782,969 people have flu symptoms. This is possibly model breakdown.

3. a. Let $P(d)$ be the percentage of polonium-210 left in a sample after d days. Then $P(d) = 100\left(\dfrac{1}{2}\right)^{\frac{d}{138}}$.
   b. After 300 days, there will be about 22.16% of the polonium-210 sample left.

4. a. Let $P(t)$ be the percentage of thorium-228 left in a sample after t years. Then $P(t) = 100\left(\dfrac{1}{2}\right)^{\frac{t}{1.9}}$.
   b. After 50 years, there will be about 0.00000197% of the polonium-210 sample left. This amount approaches 0%.

5. $f(x) = 5000(0.8)^x$

6. $f(x) = 0.2(7.6)^x$

7. $x = 3$
8. $x = 5$
9. $x = 4$
10. $x = 7$
11. $x = 3.5$
12. $x = \pm 5$
13. $x = 6$
14. $x = \pm 6$
15. $x = -5$
16. $x = -4$
17. $x = -3$
18. $x = -5$

19. **a.** a is positive because the y-intercept is above the origin.
   b. b is greater than 1, because $a > 0$ and the function is increasing.
   c. Increasing.
   d. Domain: $(-\infty, \infty)$, range: $(0, \infty)$

20. **a.** a is negative because the y-intercept is below the origin.
   b. b is greater than 1, because $a < 0$ and the function is decreasing.
   c. Decreasing.
   d. Domain: $(-\infty, \infty)$, range: $(-\infty, 0)$

21. **a.** $a > c$, because the y-intercept for $f(x)$ is greater than the y-intercept of $g(x)$.
   b. $b < d$, because $f(x)$ is increasing at a slower rate than $g(x)$.

22. **a.** Decreasing.
   b. This is an example of exponential decay.
   c. $f(20) = 5$
   d. $x = 10$
   e. $(0, 20)$

23. **a.** Increasing, because $b > 1$ and $a > 0$.
   b. This is an example of exponential growth.
   c. $(0, 2)$

24. Since $a > 0$ and $b > 1$, the function is increasing.

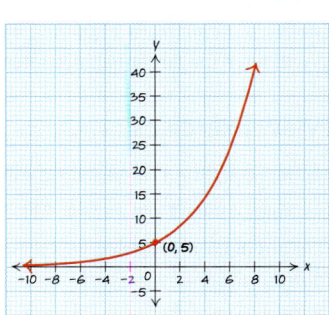

25. Since $a < 0$ and $b < 1$, the function is increasing.

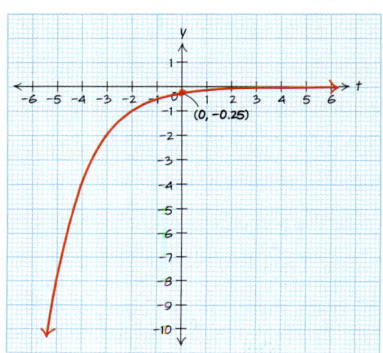

26. Since $a > 0$ and $b < 1$, the function is decreasing.

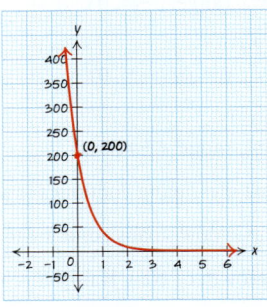

27. Since $a < 0$ and $b < 1$, the function is increasing.

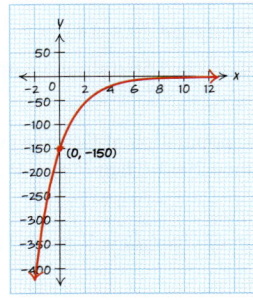

28. Since $a > 0$ and $b < 1$, the function is decreasing.

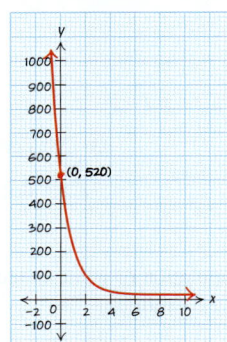

29. Since $a < 0$ and $b > 1$, the function is decreasing.

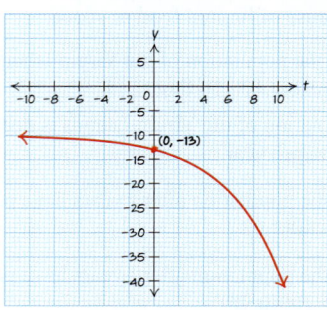

30. Domain: $(-\infty, \infty)$, range: $(0, \infty)$
31. Domain: $(-\infty, \infty)$, range: $(-\infty, 0)$
32. Domain: $(-\infty, \infty)$, range: $(20, \infty)$
33. Domain: $(-\infty, \infty)$, range: $(-\infty, -10)$

34. a. Let $W(t)$ be the number of nonstrategic warheads in the U.S. arsenal in the year t years since 1950. Then $W(t) = 52.3(1.74)^t$.
  b. Domain: [0, 11], range: [52, 23,150]
  c. The model predicts that there were approximately 212,198 nonstrategic warheads in the U.S. arsenal in the year 1965. This could be model breakdown.

35. a. Let $C(s)$ be the number of counts from a Geiger counter from the decay of barium-137 after s seconds. Then $C(s) = 3098(0.996)^s$.
  b. After 10 minutes, the Geiger counter ticks off about 280 counts per minute.
  c. Domain: [0, 1800], range: [2, 3098]

36. a. In 2005, the population of Franklin's gulls was approximately 551.89 thousand.
  b. The population of Franklin's gulls decays at a rate of 5.95% per year.
  c. This population might stop declining at the rate of 5.95% per year because of environmental intervention.

37. a. Let $E(t)$ be the population of the European Union (in millions) t years after 2008. Then $E(t) = 491(1.0012)^t$.
  b. In 2020, the population of the European Union will be approximately 498.12 million.

38. The amount of money in the account after 10 years is $12,735.78.

39. The amount of money in the account after 25 years is $837,289.75.

40. The amount of money in the account after 15 years is $91,102.95.

Chapter 5 Test

1. a. Let $B(h)$ be the number of bacteria present (in millions) h hours after 12 noon have passed. Then $B(h) = 5(2)^h$.
  b. By 6:00 P.M. there are 320 million bacteria on the bathroom door handle.
  c. $B(h) = 5(2)^{4h}$

2. a. Let $P(t)$ be the percentage of thallium-210 left in a sample after m minutes. Then $P(m) = 100\left(\frac{1}{2}\right)^{\frac{m}{1.32}}$.
  b. After 15 minutes, there is 0.038% of the thallium-210 sample left, which means that there is approximately 0% left.

3. The investment will be worth $1521.03 after 12 years.

4. The investment will be worth $326,646.80 after 35 years.

5. a. a is negative because the graph is below the x-axis.
  b. $b < 1$ because we have exponential growth with $a < 0$.
  c. Domain: $(-\infty, \infty)$, range: $(-\infty, 0)$

6. a. The graph is increasing.
  b. $f(1) = -4$
  c. $x = -2$
  d. $(0, -8)$.

7. a. Let $C(t)$ be the number of CD singles shipped (in millions) in the mid-1990s in the year t years since 1990. Then $C(t) = 1.56(1.71)^t$.
  b. Domain: [0, 10], range: [1.56, 333.49]
  c. $C(10) = 1.56(1.71)^{10} \approx 333.49$. In 2000, there were approximately 333.49 million CD singles shipped.

8. a. Let $P(t)$ be the number of professionals in developing countries (in millions) in the year t years since 2009. Then $P(t) = 4(1.065)^t$.
  b. By the year 2013, there will be approximately 5.15 million professionals in developing countries.

9. Since $a < 0$ and $b > 1$, the function is decreasing. Domain: $(-\infty, \infty)$, range: $(-\infty, 0)$

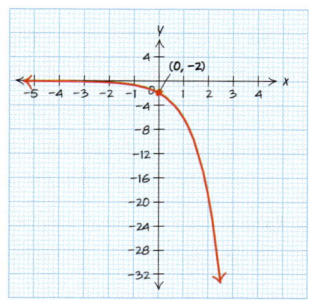

10. Since $a > 0$ and $b < 1$, the function is decreasing. Domain: $(-\infty, \infty)$, range: $(0, \infty)$

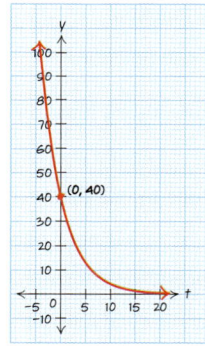

11. Since $a > 0$ and $b < 1$, the function is decreasing. Domain: $(-\infty, \infty)$, range: $(50, \infty)$

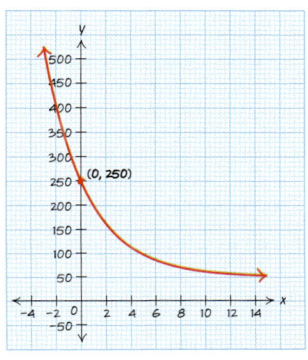

12. Since $a < 0$ and $b < 1$, the function is increasing.
Domain: $(-\infty, \infty)$, range: $(-\infty, -20)$

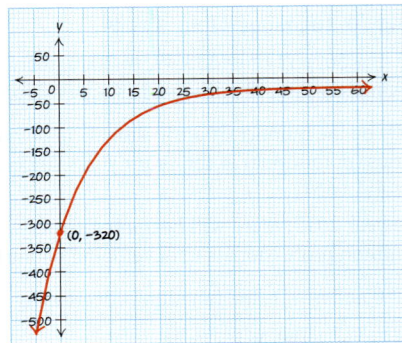

13. $x = 5$ **14.** $x = 3$ **15.** $x = 4$
16. $x = 6$ **17.** $x = -4$ **18.** $x = \pm 3$
19. $x = \pm 5.2$

CHAPTER 6
Section 6.1

1. This is a one-to-one function.
3. This is a one-to-one function.
5. This is not a one-to-one function.
7. This is a one-to-one function.
9. This is not a one-to-one function.
11. $f(x) = \frac{1}{2}x + 3$ is a one-to-one function.
13. $h(x) = 2.5x^2 + 3x - 9$ is not a one-to-one function.
15. $f(x) = 2x^3 + 4$ is a one-to-one function.
17. $f(x) = 4x^3 + 2x^2 - 5x - 4$ is not a one-to-one function.
19. $f(x) = 2x^4$ is not a one-to-one function.
21. $g(x) = 3(1.2)^x$ is a one-to-one function.
23. $f(x) = 100(0.4)^x + 20$ is a one-to-one function.
25. $f(x) = 20$ is not a one-to-one function.
27. a. $t(N) = -0.003N + 15.23$
b. Domain: $[1019, 5125.7]$, range: $[-1, 12]$
c. In 1999, there were about 2000 homicides of 15 to 19-year-olds in the United States.
29. a. Range: $-2500 \le P \le 25,000$
b. $b(P) = 0.18P + 454.55$
c. There is a \$3000 profit from selling 1000 books.
d. You would need to sell 1355 books to make a \$5000 profit.
e. Domain: $-2500 \le P \le 25,000$, range: $0 \le b \le 5000$
31. a. $t(P) = 0.389P - 97.191$
b. The U.S. population reached 260 million in about 1994.
c. For the inverse function $t(P) = 0.389P - 97.191$, the input variable P represents the U.S. population in millions, and the output variable t represents the number of years since 1990.

33. a. $c(M) = 0.667M - 6666.667$
b. They would have to sell about 20,014 CDs for "Math Dude" to earn \$40,000.
c. Domain: $25,000 \le M \le 85,000$, range: $10,000 \le c \le 50,000$

35. $f^{-1}(x) = \frac{1}{3}x - \frac{5}{3}$ **37.** $g^{-1}(t) = -\frac{1}{4}t + 2$

39. $h^{-1}(x) = \frac{3}{2}x + \frac{27}{2}$ **41.** $f^{-1}(x) = 5x - 3$

43. $h^{-1}(x) = 2.5x + 4$ **45.** $P^{-1}(t) = -0.4t - 3$

47. Yes, they are inverses. **49.** Yes, they are inverses.
51. No, they are not inverses.

53.

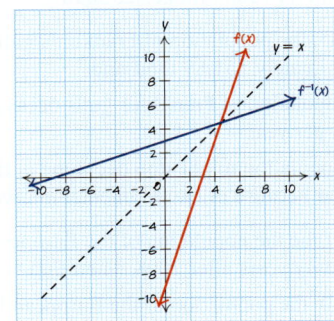

55.

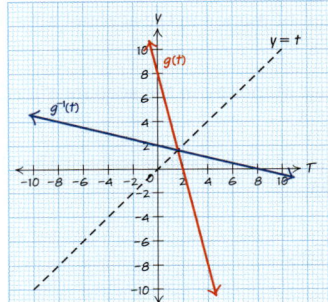

57.

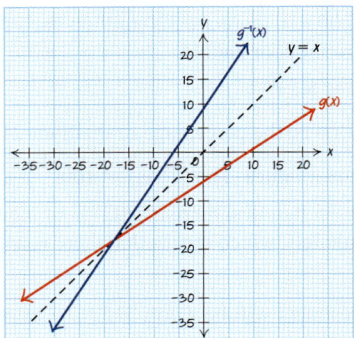

59.

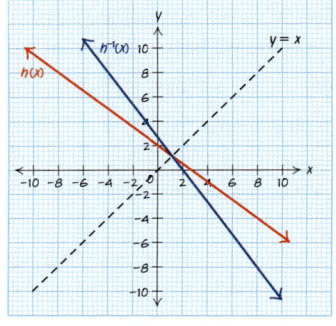

APPENDIX E Answers to Selected Exercises   E-61

61. a.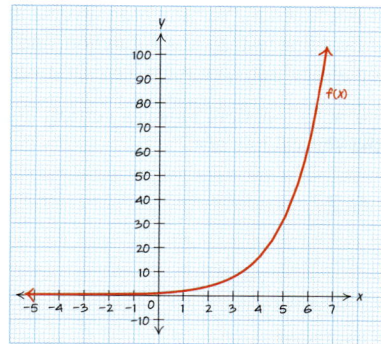

61. b.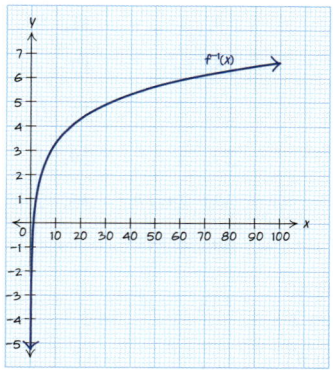

63. a.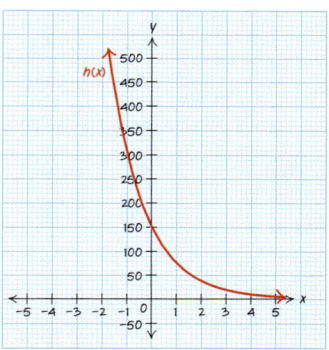

63. b.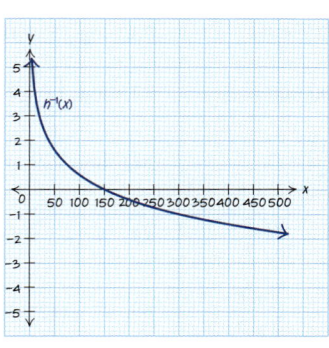

65. a.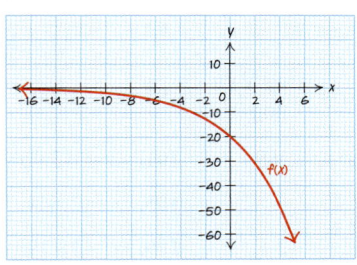

65. b.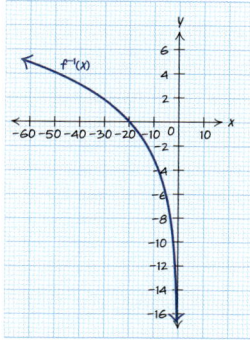

67. a.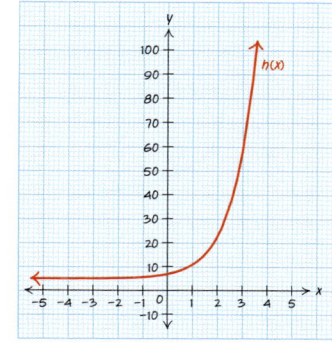

67. b.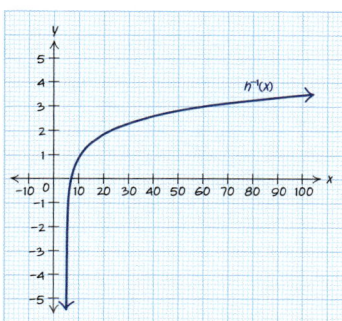

69. a.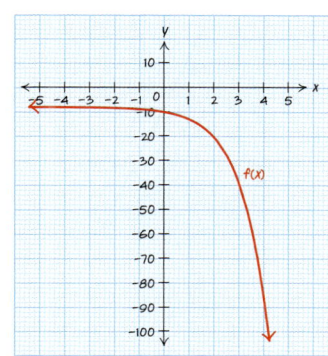

69. b.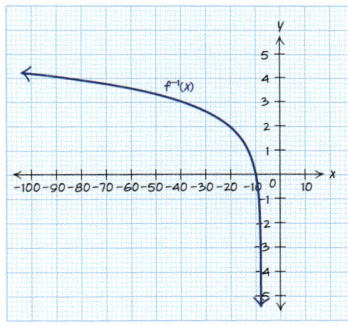

71. a.
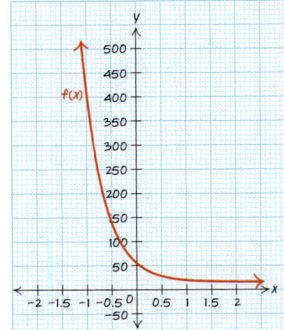

71. b.
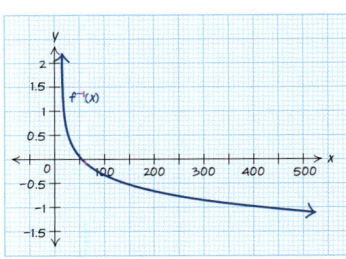

73.
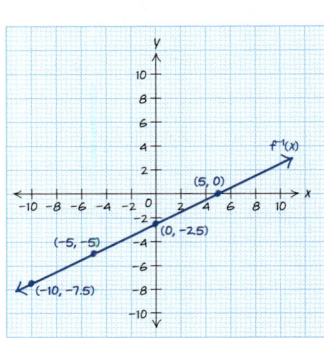

75.
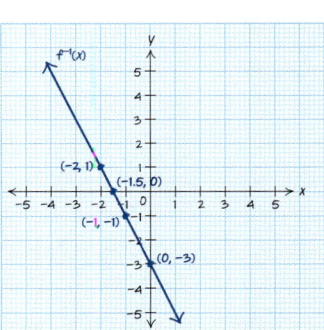

77.
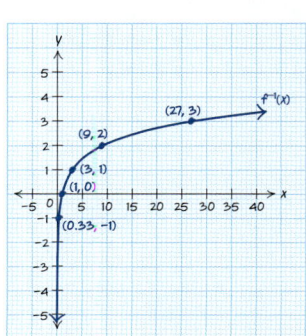

79. Yes, it is a function. Yes, it has an inverse. It passes the vertical line test, so it is a function, and it passes the horizontal line test, so it must have an inverse.

81. No, it is not a function. This is a vertical line, and it will not pass the vertical line test.

Section 6.2

1. The base is 5.
3. The base is 10.
5. The base is e.
7. 3
9. 3
11. 8
13. 3
15. -1
17. -1
19. -2
21. -2
23. 0
25. 4
27. 3
29. 1
31. 1
33. 2.097
35. 2.439
37. 3.807
39. 4.787
41. a. $\dfrac{\log 40}{\log 3} \approx 3.358$ b. $\dfrac{\ln 40}{\ln 3} \approx 3.358$
43. a. $\dfrac{\log 63}{\log 5} \approx 2.574$ b. $\dfrac{\ln 63}{\ln 5} \approx 2.574$
45. 1.654
47. -1.080
49. 2.338
51. -0.572
53. 4.227
55. $10^3 = 1000$
57. $10^{-2} = 0.01$
59. $2^3 = 8$
61. $3^4 = 81$
63. $e^5 = e^5$
65. $3^{-2} = \dfrac{1}{9}$
67. $5^{-2} = \dfrac{1}{25}$
69. $\log_2 1024 = 10$
71. $\log_5 625 = 4$
73. $\log_{25} 5 = 0.5$
75. $\log 100{,}000 = 5$
77. $\log_{\frac{1}{5}}\left(\dfrac{1}{625}\right) = 4$
79. $\log_{\frac{1}{2}}\left(\dfrac{1}{8}\right) = 3$
81. $\log_3 729 = 2x$
83. $f^{-1}(x) = \dfrac{\log x}{\log 7}$
85. $h^{-1}(c) = \log c$
87. $m^{-1}(r) = \dfrac{\log r}{\log 5}$
89. $f^{-1}(x) = 3^x$
91. $g^{-1}(x) = e^x$
93. $f^{-1}(x) = \dfrac{\log\left(\dfrac{x}{3}\right)}{\log 4}$
95. $h^{-1}(x) = \dfrac{\log(2x)}{\log 9}$
97. $x = 1000$
99. $t = e^2 \approx 7.389$
101. $w \approx 13.967$
103. $m = 4$
105. $x \approx 1.644$
107. $t \approx \dfrac{25}{2} = 12.5$
109. $x \approx 18.199$

Section 6.3

1. a. $f(50) = 2$ b. $x = 125$
3. a. $h(250) = 4$ b. $x = 20$
5. a. $g(40) = -3$ b. $x = 13$
7. a. $h(3) = -11$ b. $x = 5$
9. The base is greater than 1, because the graph is increasing.
11. The base is less than 1, because the graph is decreasing.

13.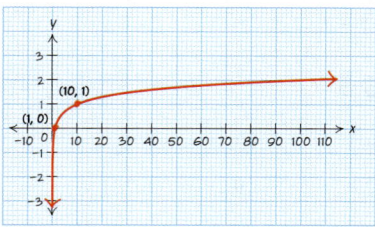

15.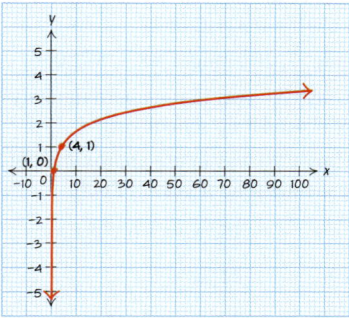

17.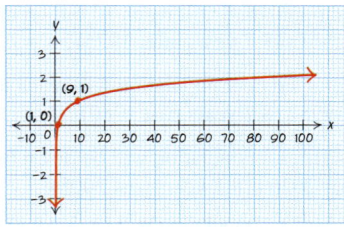

19.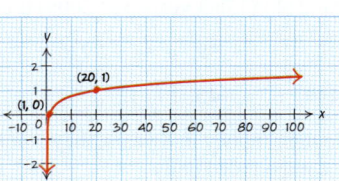

21.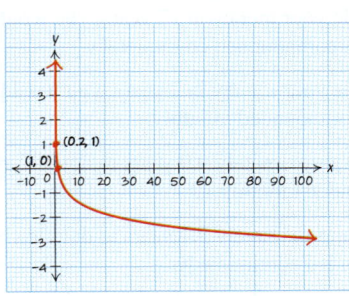

23.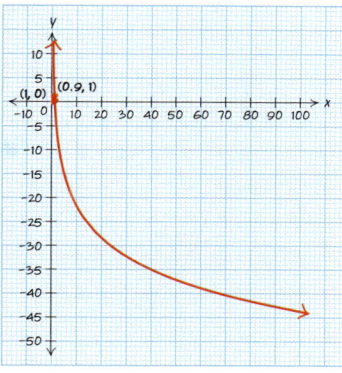

25.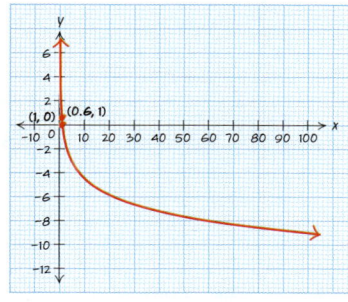

27. Domain: $(0, \infty)$, range: $(-\infty, \infty)$
29. Domain: $(0, \infty)$, range: $(-\infty, \infty)$
31. Domain: $(0, \infty)$, range: $(-\infty, \infty)$
33. Domain: $(0, \infty)$, range: $(-\infty, \infty)$
35. Domain: $(0, \infty)$, range: $(-\infty, \infty)$
37. Domain: $(0, \infty)$, range: $(-\infty, \infty)$
39. $f^{-1}(x) = \dfrac{1}{2}x - 6$
41. $f^{-1}(x) = \dfrac{\log x}{\log 4}$
43. $g^{-1}(x) = 9^x$

45.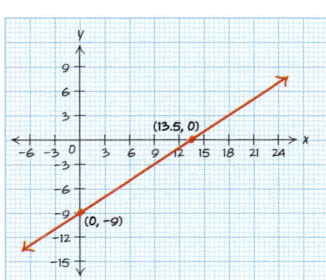

47.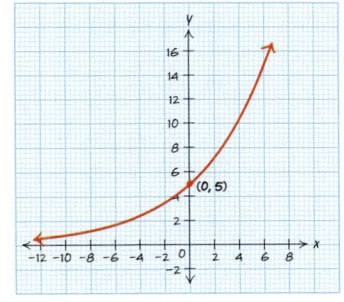

49.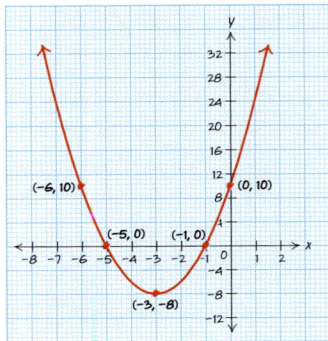

51.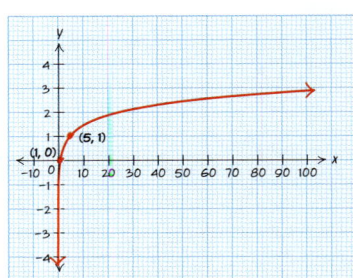

53. 2.477 **55.** 3.091 **57.** 4.005
59. 4.744 **61.** −0.446 **63.** $x = 512$
65. $x = 10{,}000{,}000$ **67.** $x = \dfrac{1}{16}$

Section 6.4

1. $\log 5 + \log x$ **3.** $\ln x + \ln y$
5. $\ln 4 + \ln a + 3 \ln b$ **7.** $\ln 2 + 2 \ln h + 3 \ln k$
9. $\log_3\left(\dfrac{1}{2}\right) + \log_3 a + 2 \log_3 b + 3 \log_3 c$
11. $\log x - \log y$ **13.** $\log 12 - \log m$
15. $\log 2 + 2 \log x - \log y$
17. $\ln 3 + 4 \ln x + 3 \ln y - \ln z$
19. $\dfrac{1}{2} \log 5 + \dfrac{1}{2} \log x$
21. $\dfrac{1}{2} \ln 7 + \dfrac{1}{2} \ln a + \dfrac{1}{2} \ln b$
23. $\dfrac{1}{2} \log_9 2 + \log_9 w + \dfrac{3}{2} \log_9 z$
25. $\dfrac{2}{5} \ln m + \dfrac{3}{5} \ln p$
27. $\dfrac{1}{2} \log_{15} 3 + 2 \log_{15} x + \dfrac{3}{2} \log_{15} y - 5 \log_{15} z$
29. $\ln (xy)$ **31.** $\log (a^2 b^3)$ **33.** $\ln\left(\dfrac{x}{y}\right)$
35. $\log(5x^2 yz^3)$ **37.** $\log_3\left(\dfrac{ab^2}{c^5}\right)$
39. $\log_5(7\sqrt{xy})$ **41.** $\ln\left(\dfrac{\sqrt{7ab^3}}{c^4}\right)$
43. $\ln\left(\dfrac{16{,}807 a^2 b^4}{c^3 d^2}\right)$ **45.** $\log_3\left(\dfrac{a^2 b^2}{243 c^3}\right)$
47. $\log_5(7x^3 y^3)$ **49.** $\log\left(\dfrac{\sqrt{15xy^5}}{z^4}\right)$

51. $\log(x^2 + 4x - 12)$ **53.** $\ln(x^2 - 10x + 21)$
55. There is no expansion for the expression $\log(x + 5)$. We can expand when there is multiplication or division inside the logarithm, not for addition or subtraction inside the logarithm.
57. The student incorrectly multiplied two separate log functions instead of multiplying just inside one log function. The expression $\log x + \log y$ comes together as $\log(xy)$.
59. $\log_b 32 = 7$ **61.** $\log_b 8 = 1$
63. $\log_b 15 = -6$ **65.** $\log_b 50{,}000 = 10$
67. $\log_b 0.25 = 6$

Section 6.5

1. $w \approx 4.395$ **3.** $x \approx 3.679$ **5.** $x \approx -0.410$
7. $x \approx 2.227$ **9.** $x \approx -6.221$

11. a. Let $C(t)$ be the number of hours the Cray C90 has been used per academic year t years since 1980. Then $C(t) = 0.311(2.057)^t$.
 b. Domain: $[7, 20]$, range: $[48.5, 572{,}066]$
 c. The Cray C90 was in use approximately 65,727 hours in the 1997 academic school year.
 d. The Cray C90 was in use approximately 500,000 hours in the academic school year of 2000.

13. $x = 8$ **15.** $x \approx -6.208$ **17.** $x \approx 9.002$
19. $x \approx 2.319$ **21.** $x = 7$ **23.** $t = 6$
25. $h \approx 11.000$ **27.** $m \approx 7.600$ **29.** $n = 3$

31. The world population will reach 10 billion midway through the year 2023 (48.7 years after 1975).
33. The white-tailed deer population will reach 5 million late in the year 1998 (18.7 years past 1980).
35. a. South Africa's population will be down to 40 million by the year 2026 (18.1 years past 2008).
 b. South Africa's population will be down to 21.9 million by the year 2146 (138.3 years past 2008) if the trend continues.

37. $x = 2$ **39.** $m \approx -2.926$ **41.** $x = \pm 2$
43. $x \approx \pm 2.449$ **45.** $t = 12$ **47.** $x \approx 2.268$

49. To double their money in 8 years, they would need to invest at 8.67% compounded daily.
51. To double your money in 12 years, you would need to invest at 5.78% compounded continuously.
53. To double your money in 7 years, you would need to invest at 9.94% compounded monthly.
55. It takes approximately 9.9 years to double your money at 7% compounded monthly.
57. It takes approximately 7.7 years to double your money at 9% compounded continuously.
59. It takes approximately 54.9 years to triple your money at 2% compounded daily.
61. It takes approximately 43.9 years to triple your money at 2.5% compounded continuously.

63. $f^{-1}(x) = \dfrac{\log\left(\dfrac{x}{5}\right)}{\log(3)}$ **65.** $h^{-1}(x) = \dfrac{\ln\left(\dfrac{x}{-2.4}\right)}{\ln 4.7}$

67. $g^{-1}(x) = \ln\left(\dfrac{x}{-3.4}\right)$ **69.** $n \approx 1.088,\ n \approx -1.838$

71. $d \approx 2.091$ **73.** $x \approx 10.514$

75. $n \approx 5.236,\ n \approx 0.764$ **77.** $(3.098, 75.145)$

79. $(-4.130, -12.566)$

81. $(7.214, 8.013)$ and $(1.859, 54.111)$

Section 6.6

1. $x = 10{,}000$ **3.** $x \approx 148.413$ **5.** $x = 16$

7. $x = 19.6$ **9.** $x \approx 4.1297$

11. An earthquake that has an intensity of 2000 cm has a magnitude of about 7.3.

13. An earthquake that has an intensity of 500,000 cm has a magnitude of about 9.7.

15. The intensity of this 8.0 magnitude earthquake was 10,000 cm.

17. The intensity of this 6.6 magnitude earthquake was 398.11 cm.

19. Scope mouthwash with a hydrogen ion concentration of 1.0×10^{-7} M has a pH of 7.

21. Car battery acid with a pH of 1 has a hydrogen ion concentration of 1.0×10^{-1} M.

23. Oven cleaner with a pH of 14 has a hydrogen ion concentration of 1.0×10^{-14} M.

25. Blood plasma with a pH level between 7.35 and 7.45 has a hydrogen ion concentration between 3.55×10^{-8} M and 4.47×10^{-8} M.

27. The chain saw with an intensity of 10^{-1} watt/m² has a decibel level of 110 dB.

29. Raindrops with an intensity of 10^{-8} watt/m² has a decibel level of 40 dB.

31. A jack hammer with a decibel level of 120 dB has an intensity of 1 watt/m².

33. $x = 50$ **35.** $h = -8192$ **37.** $t \approx 0.045$

39. $r \approx 3.389$ **41.** $p \approx -0.342$ **43.** $x \approx 14.434$

45. $x \approx 4.405$

47. Matt made the base 10 on the second step when it should have been e.

49. Frank added 3 and 5 on the second step when the student should have subtracted 5 from both sides of the equation.

51. $x \approx 6.325$ **53.** $x \approx 6.347$

55. $x = -13.5,\ x = -5.5$ **57.** $x = 1$

59. $x \approx 6.249,\ x \approx -7.916$ **61.** $x \approx 0.0003$

63. $x = 3$ **65.** $x = -2.5,\ x = 1$

67. $t = 0,\ t = -5,\ t = -1$ **69.** $d \approx 0.667$

Chapter 6 Review Exercises

1. $f^{-1}(x) = \dfrac{5}{7}x + 5$ **2.** $g^{-1}(x) = -\dfrac{1}{2}t + 3$

3. $h^{-1}(x) = \dfrac{\log x}{\log 5}$ **4.** $f^{-1}(x) = \dfrac{\log\left(\dfrac{x}{3.5}\right)}{\log 6}$

5. $g^{-1}(x) = \ln\left(\dfrac{x}{2}\right)$ **6.** $f^{-1}(x) = 5^x$

7. $h^{-1}(t) = 0.2^t$

8. a. $e(D) = 0.0025D - 0.125$

   b. The AEM Toy Company can manufacture 4050 dolls when they have 10 employees working.

   c. The AEM Toy Company must have 3 employees working on a particular day to manufacture 1000 dolls.

   d. Domain: $2050 \le D \le 12{,}050$, range: $5 \le e \le 30$

9. a. $d(P) = 0.222P + 66.667$

   b. The AEM Toy Company must sell 2731 dolls to make a $12,000 profit.

   d. Domain: $1950 \le P \le 17{,}700$, range: $500 \le d \le 4000$

10. This function is a one-to-one function.

11. This function is not a one-to-one function.

12. $\log_4 64 = 3$ **13.** $\log_3 1 = 0$

14. $\log_8 8^5 = 5$ **15.** $\log_2\left(\dfrac{1}{8}\right) = -3$

16. $f(x) = \log_{2.5}(x)$

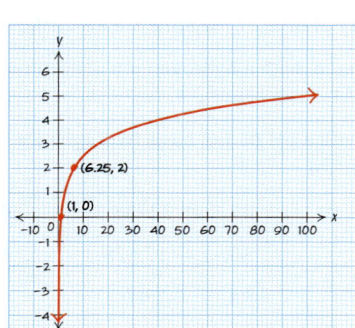

17. $f(x) = \log_{0.4}(x)$

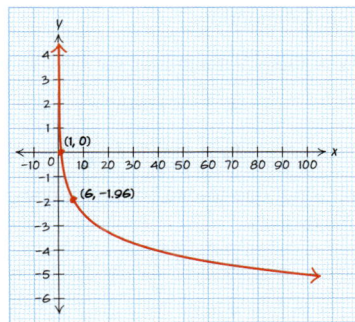

18. Domain: $(0, \infty)$, range: $(-\infty, \infty)$

19. Domain: $(0, \infty)$, range: $(-\infty, \infty)$

20. $\log 7 + \log x + \log y$ 21. $\log_5 3 + 3\log_5 a + 4\log_5 b$

22. $\ln 2 + 3\ln x + 4\ln y$ 23. $\frac{1}{5}\log_7 3 + \frac{3}{5}\log_7 x + \frac{1}{5}\log_7 y$

24. $\frac{1}{2}\log_3 5 + \frac{1}{2}\log_3 a + \log_3 b + \frac{3}{2}\log_3 c$

25. $\log 3 + 3\log x + 5\log y - 4\log z$

26. $\log 4 + 5\log a + \log b - 3\log c$

27. $\log(4x^3 y^2 z)$ 28. $\log\left(\frac{2x^4 y^5}{z^2}\right)$

29. $\log_3\left(\frac{ab^3}{c^2}\right)$ 30. $\ln\left(\frac{7x^2}{y^5 z^2}\right)$

31. $\log(x^2 - 5x - 14)$

32. **a.** In 2005 the population of Franklin's gulls was approximately 551.89 thousand.
   b. According to the model, the population of Franklin's gulls reached 500,000 midway through the year 2006.
   c. This population might stop declining at this rate, owing to environmental intervention.

33. It takes approximately 11.6 years to double your money at 6% compounded monthly.

34. It takes approximately 8.2 years to double your money at 8.5% compounded continuously.

35. $w \approx 1.822$ 36. $t = -5$ 37. $x \approx 2.656$

38. $x \approx -0.021$ 39. $x \approx 1.841$ 40. $x = -2$

41. $x \approx 0.466$ 42. $x \approx \pm 1.414$

43. The intensity of this 5.7 magnitude earthquake was 50.12 cm.

44. Cranberry juice, with a hydrogen ion concentration of 3.7×10^{-3} M, has a pH of 2.4.

45. Shampoo, with a pH of 5.6, has a hydrogen ion concentration of 2.5×10^{-6} M.

46. $x = 0.0001$ 47. $t = 32768$ 48. $x = 9998$

49. $t \approx 0.003$ 50. $x \approx 17.678$ 51. $x \approx 7.266$

52. $x = -3$ 53. $x \approx 3.976$

Chapter 6 Test

1. It takes approximately 19.8 years to double your money at 3.5% compounded monthly.

2. The intensity of this 7.0 magnitude earthquake was 1000 cm.

3. $x = -0.375$ 4. $x \approx 11.462$

5. $x \approx 2.524$ 6. $x \approx 0.413$

7. No, the graph is not one-to-one because it does not pass the horizontal line test.

8. $\log(16a^{17} b^{15})$ 9. $\log\left(\frac{5xy^3}{z^4}\right)$

10. $\ln 4 + 3\ln a + \ln b + 4\ln c$

11. $\log 3 + \log x + 5\log y - \frac{1}{2}\log z$

12.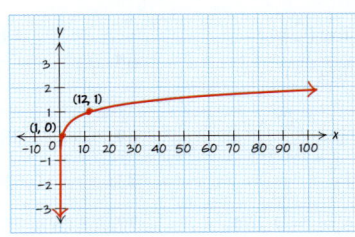

13. Domain: $(0, \infty)$, range: $(-\infty, \infty)$

14. **a.** Let $C(t)$ be the number of CD singles shipped (in millions) in the year t years since 1990. Then $C(t) = 1.56(1.71)^t$.
   b. Domain: $[0, 10]$, range: $[1.56, 333.49]$
   c. In 2000, there were approximately 333.49 million CD singles shipped.
   d. In 1992, there were approximately 5 million CD singles shipped.

15. $f^{-1}(x) = \frac{1}{5}x - \frac{2}{5}$ 16. $g^{-1}(x) = 15^x$

17. $f^{-1}(x) = \frac{\log x}{\log 5}$ 18. $h^{-1}(x) = \frac{\log\left(\frac{x}{2}\right)}{\log 7}$

19. A solution with a pH of 2.58 has a hydrogen ion concentration of 2.63×10^{-3} M.

20. $\log_3 200 \approx 4.8227$

Cumulative Review for Chapters 1–6

1. $x = \frac{\ln\left(\frac{16}{3}\right)}{\ln(4.2)} \approx 1.166$ 2. $x = 6$

3. $t = 1, t = 9$ 4. $c = \frac{e^5 + 7}{2}$ 5. $b = -1$

6. $n \approx 0.984, n \approx -1.544$ 7. $h = 0, h = -2 \pm 2\sqrt{3}$

8. $x = \frac{\frac{\ln 45}{\ln 2} + 7}{3} \approx 4.164$ 9. $g \approx -4.046$

10. $x = \frac{-1 \pm \sqrt{57}}{2}$ 11. $x = \frac{5 + \sqrt{481}}{2} \approx 13.466$

12. $x = -5 + \ln 89$ 13. $b = 13, b = -27$

14. $t = 9, t = 3$ 15. $x = 4$ 16. $x = \pm 2$

17. $(7, 15)$ 18. $(3.2, 4.2)$

19. Infinite number of solutions.

20. $(-6, 18)$ and $(4, 48)$ 21. $(3, 6)$ and $\left(\frac{1}{3}, -\frac{98}{9}\right)$

22. No solution.

23. Let $P(t)$ be the gross profit for UTstarcom, Inc. (in millions) t years since 1990.
   a. $P(t) = 0.48(1.74)^t$
   b. Domain: $[4, 15]$, range: $[4.4, 1947.5]$
   c. In 2005, the gross profit for UTstarcom, Inc reached $2 billion.

24. Let $C(t)$ be the number of cocaine-related emergency department episodes (for people over the age of 35) t years since 1990.
    a. $C(t) = 7336.3t + 23{,}245.7$
    b. In the year 2003, the number of cocaine-related emergency department episodes (for people over the age of 35) was 118,618.
    c. In 2006, the number of cocaine-related emergency department episodes (for people over the age of 35) reached 140,000.
    d. The slope is 7336.3. The number of cocaine-related emergency department episodes (for people over the age of 35) increases by 7336.3 cases each year.
    e. Domain: $[-1, 15]$, range: $[15{,}909, 133{,}290]$

25. a. $f(9) = 29$ b. $f\left(\dfrac{5}{4}\right) = -2$
    c. Domain: $(-\infty, \infty)$, range: $(-\infty, \infty)$

26. a. $f(x) + g(x) = 1.5x + 9.5$
    b. $f(g(x)) = -7x + 23.25$
    c. $f(x)g(x) = -7x^2 + 11.25x + 22$

27. a. $f(3) - g(3) = 12$ b. $f(g(5)) = -123$
    c. $f(4)g(4) = -21$

28. a. $f(x) - g(x) = 3x^2 + 9x - 9$
    b. $f(g(x)) = 48x^2 - 148x + 110$
    c. $f(x)g(x) = 12x^3 - x^2 - 43x + 14$

29. a. $f(5) - g(5) = -101$ b. $f(g(2)) = -1173$
    c. $f(3)g(3) = -414$

30. $L(V(t))$ is the amount of liter in tons left in Yellowstone National Park in year t.

31. a. In 1995, 20.5% of the U.S. triathlon members were women.
    b. In the year 2017, the percentage of U.S. triathlon members that are women will be greater than that of men.

32. $x > \dfrac{8}{3}$ 33. $x \geq 3.2$

34. $x < 2$ or $x > 12$ 35. $-\dfrac{13}{3} \leq x \leq 1$

36. a. Slope: $m = -\dfrac{5}{2}$ b. x-intercept: $(2, 0)$
    c. y-intercept: $(0, 5)$ d. $y = -\dfrac{5}{2}x + 5$

37. $y = \dfrac{2}{3}x - 12$ 38. $y < \dfrac{2}{3}x - 9$

39. Let $P(t)$ be the population of Detroit, Michigan, in thousands t years since 2000.
    a. $P(t) = -31t + 4673$
    b. In 2015, the population of Detroit, Michigan, will be about 4208 thousand.
    c. The slope is -31. The population of Detroit, Michigan, is decreasing by 31 thousand each year.

40. Sarah needs $13\tfrac{1}{3}$ ml of 15% HCl solution and $16\tfrac{2}{3}$ ml of 60% HCl solution to make 30 ml of 40% HCl solution.

41. Let $W(t)$ be the number of women triathletes in the United States t years since 1990.
    a. $W(t) = 106.5t^2 + 110$
    b. Domain: $[0, 20]$, range: $[1100, 43{,}700]$
    c. In 2005, there were 25,063 women triathletes in the United States.
    d. The vertex is $(0, 1100)$. In 1990, there were 1100 women triathletes in the United States. This was the lowest number during this time.
    e. In 2009, there were about 40,000 women triathletes in the United States.

42.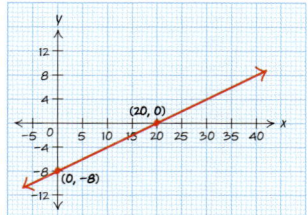

43.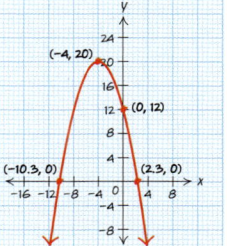

44.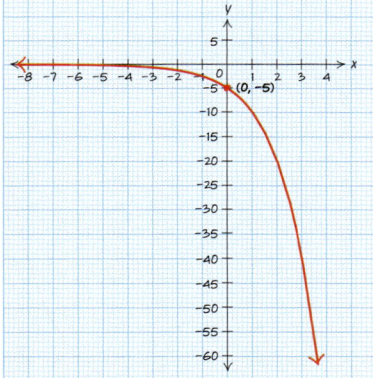

45.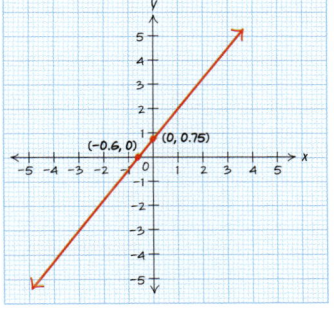

46.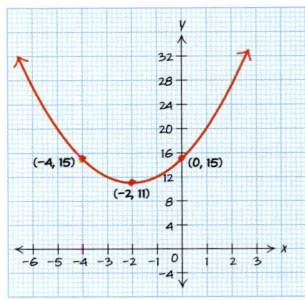

47.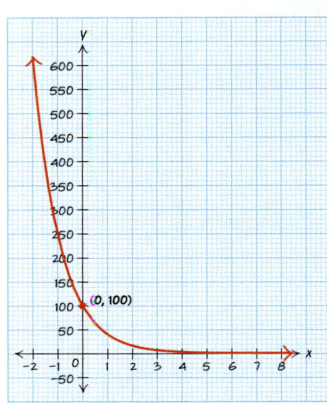

48.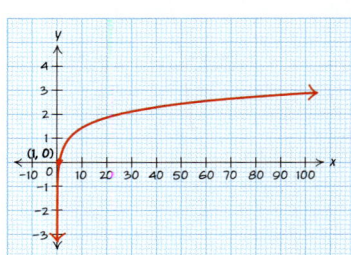

49.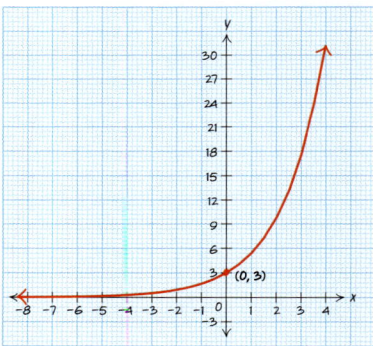

50.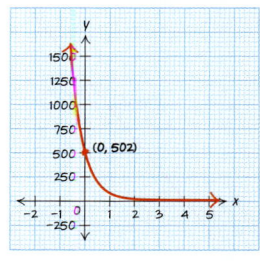

51.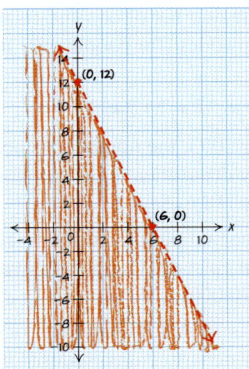

52.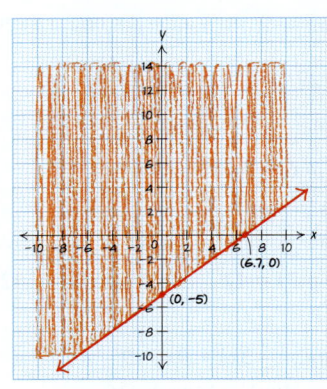

53.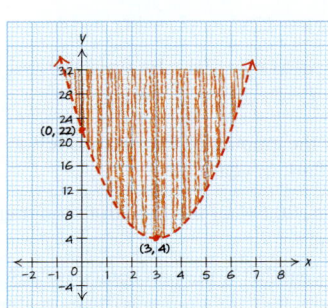

54. $11d - 6$
55. $14x^2 + 19x - 40$
56. $2x^2 - 5x + 8$
57. $9t^2 - 24t + 16$
58. $14p^2 - p - 24$

59.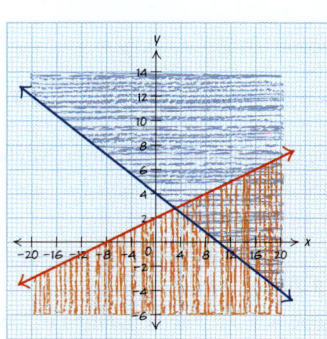

60. $(2x + 3)(x - 7)$
61. $(5r + 8t)(5r - 8t)$
62. $7g(3g - 8)(2g + 5)$
63. $(t + 7)(t^2 - 7t + 49)$
64. $(2x - 3)^2$

65. Second-degree polynomial
66. $f(x) = 3(x + 4)^2 - 58$
67. a. The average price for baseball tickets in 2010 is $28.00.
    b. The vertex is $(-10.7, 5.67)$. In 1979, the average price for baseball tickets was $5.67. This was a minimum for this model.
    c. In 2017, the average price for baseball tickets will be $40.77.
68. Let $T(m)$ be the number of thorium-233 atoms after m minutes.
    a. $T(m) = 5000\left(\frac{1}{2}\right)^{\frac{m}{22}}$
    b. After about 51 minutes, there will be only 1000 of the thorium-233 atoms left.
    c. By the end of the experiment, there will be only 17 of the thorium-233 atoms left.
69. $f^{-1}(x) = \frac{1}{3}x - 4$
70. $g^{-1}(x) = \frac{\log\left(\frac{x}{4}\right)}{\log 3}$
71. $H^+ = 1 \times 10^{-6}$ M
72. Let $P(t)$ be the Muslim population in Israel (in millions) t years since 2008.
    a. $P(t) = 1.24(1.028)^t$
    b. The Muslim population in Israel in 2012 will be 1.38 million.
    c. The Muslim population in Israel will reach 2 million by the year 2025.
73. $60m^6n^5$
74. $\frac{81g^{12}}{h^{16}}$
75. 8
76. $\frac{12x^2z^7}{5y^7}$
77. $\frac{16a}{5b^{13}}$
78. $\frac{3m}{2n}$
79. $a > 0$. The graph is above the x-axis.
80. $b > 1$. The graph is increasing and $a > 0$.
81. Domain: $(-\infty, \infty)$, range: $(0, \infty)$
82. $a > 0$. The graph is facing upward.
83. $(h, k) = (-1, -2)$ (h, h) is the vertex of the parabola.
84. $f(2.5) = 18$
85. $f(2) = 12$
    $f(-4) = 12$
86. Domain: $(-\infty, \infty)$, range: $[-2, \infty)$
87. $m > 0$. The graph is increasing.
88. $b = -1$. The y-intercept is $(0, -1)$.
89. $m = \frac{2}{3}$
90. $f(3) = 1$
91. Domain: $(-\infty, \infty)$, range: $(-\infty, \infty)$

CHAPTER 7

Section 7.1

1. a. Let $C(p)$ be the per player cost in dollars for a charity poker tournament if p players participate. $C(p) = \frac{600}{p}$
   b. The per person cost for 75 players to participate is $8 per person.
   c. Domain: $2 \leq p \leq 100$; range: $6 \leq C \leq 300$

3. a. The total cost for 100 people to attend the event is $820.
   b. The per person cost for 100 people to attend the event is $8.20 per person.
   c. Let $F(p)$ be the per person cost in dollars if p people attend the charity event. Then
   $$F(p) = \frac{4.55p + 365.00}{p}$$
   d. The per person cost for 150 people to attend the event is $6.98 per person.
   e. Domain: $1 \leq p \leq 250$; range: $6.01 \leq F \leq 369.55$

5. a. Let $C(n)$ be the cost in dollars for a room at this particular hotel if you stay for n nights. $C(n) = 150n$
   b. The total cost for a 7-night stay at this hotel is $1050.
   c. With a budget of $800, you can stay 5 nights at this hotel.

7. a. Let $V(t)$ be the volume in liters it takes to store the helium if the temperature is t Kelvin. $V(t) = 0.0020525t$
   b. It takes 0.513 liter to store 0.50 mole of helium at 250 K.
   c. If you have 0.75 liter to store the helium, the temperature needs to be 365.4 K.

9. a. Let $I(d)$ be the illumination (in foot-candles) of a light source d feet from the source. $I(d) = \frac{2550}{d^2}$
   b. The illumination of this light at a distance of 5 feet from the source is 102 foot-candles.
   c. The illumination of this light at a distance of 30 feet from the source is 2.83 foot-candles.
   d. An illumination of 50 foot-candles occurs at a distance of approximately 7.1 feet from the source.

11. A 220-pound man weighs only 97.78 pounds when he is 2000 miles above the earth.

13. The current is 1.8 amps when the resistance is 130 ohms.

15. a. $P(2) = 15$. The pressure of the gas is 15 pounds per square inch when the volume of the balloon is 2 cubic inches.
    b. $P(120) = 3$. The pressure of the gas is 3 pounds per square inch when the volume of the balloon is 10 cubic inches.

17. a. $y = 2.35x^3$
    b. $y = 1203.2$

19. a. $y = \frac{10{,}935}{x^3}$
    b. $y = 87.48$

21. a. $M = \frac{1725}{5\sqrt{t}}$
    b. $M = 69$

23. a. The average benefit for a person participating in the food stamp program in 1995 was $853.29.

  b. The average benefit for a person participating in the food stamp program was approximately $800 in 1992 and again in 2000.

25. a. Let $N(t)$ be the average amount of national debt per person in dollars in the U.S. t years since 1970.
$$N(t) = \frac{8215.1t^2 - 23{,}035.4t + 413{,}525.6}{0.226t^2 + 1.885t + 204.72}$$

  b. The average amount of national debt per person in 2000 was $15,314.21.

  c. The average amount of national debt per person was $10,000 midway through 1990.

27. Domain: x is all real numbers except $x = 0$.

29. Domain: x is all real numbers except $x = 3$.

31. Domain: b is all real numbers except $b = -11$.

33. Domain: t is all real numbers except $t = -\frac{5}{4}$.

35. Domain: a is all real numbers except $a = -\frac{7}{2}$, $a = 3$.

37. Domain: n is all real numbers except $n = -4$, $n = -10$.

39. Domain: m is all real numbers except $m = -3$ and $m = -4$.

41. Domain: x is all real numbers.

43. Domain: t is all real numbers except $t = -7$ and $t = 7$.

45. Domain: r is all real numbers except $r = -3$ and $r = -2$.

47. Domain: t is all real numbers except $t = 5$ and $t = 2$.

49. Domain: a is all real numbers.

51. a. Domain: x is all real numbers except $x = -5$ and $x = 7$.

  b. $f(0) = 0$

  c. $f(x) = 5$ when $x \approx -4.5$ and $x \approx 7.5$

53. a. Domain: x is all real numbers except $x = -3$ and $x = 4$.

  b. $h(1) \approx -1$

  c. $h(-4) \approx 1.9$

  d. $h(x) = -2$ when $x \approx -1.9$ and $x \approx 2.8$

55. Domain: x is all real numbers except $x = -3$ and $x = -5$.

57. Domain: x is all real numbers except $x = -4$ and $x = 6$.

59. Domain: x is all real numbers except $x = -3$ and $x = -5$.

61. Domain: x is all real numbers except $x = -3$. The value represents a vertical asymptote because the factor $(x + 3)$ does not simplify with the numerator.

63. Domain: x is all real numbers except $x = -3$ and $x = -1$. The values represent vertical asymptotes because the factors $(x + 3)$ and $(x + 1)$ do not simplify with the numerator.

65. Domain: x is all real numbers except $x = -5$ and $x = -2$. The value $x = -5$ represents a vertical asymptote because the factor $(x + 5)$ does not simplify with the numerator. The value $x = -2$ represents a hole because the factor $(x + 2)$ simplifies with the numerator.

67. Domain: x is all real numbers except $x = \frac{4}{3}$ and $x = -\frac{5}{2}$. The values represent vertical asymptotes because the factors $(3x - 4)$ and $(2x + 5)$ do not simplify with the numerator.

69. Domain: x is all real numbers except $x = -1$ and $x = -2$. The value $x = -1$ represents a vertical asymptote because the factor $(x + 1)$ does not simplify with the numerator. The value $x = -2$ represents a hole because the factor $(x + 2)$ simplifies with the numerator.

71. Domain: x is all real numbers except $x = 0$, $x = -2$, and $x = 4$. The values $x = -2$ and $x = 4$ represent vertical asymptotes because the factors $(x + 2)$ and $(x - 4)$ do not simplify with the numerator. The value $x = 0$ represents a hole because the factor (x) simplifies with the numerator.

73. Domain: x is all real numbers except $x = 4$ and $x = -4$. The value $x = 4$ represents a vertical asymptote because the factor $(x - 4)$ does not simplify with the numerator. The value $x = -4$ represents a hole because the factor $(x + 4)$ simplifies with the numerator.

75. Domain: x is all real numbers. There are no vertical asymptotes or holes.

Section 7.2

1. $\dfrac{10x^2}{7}$ **3.** $\dfrac{12}{x+5}$ **5.** $\dfrac{1}{2}$

7. $-\dfrac{1}{2}$ **9.** $-\dfrac{4}{3}$ **11.** $\dfrac{x+2}{7-x}$

13. $\dfrac{x+5}{x-2}$ **15.** $\dfrac{x+3}{x+2}$ **17.** $\dfrac{t+5}{t-6}$

19. $\dfrac{w-4}{w-3}$ **21.** $\dfrac{5(x+7)}{x+17}$ **23.** $\dfrac{m+3}{2m+1}$

25. $\dfrac{2x+5}{x-7}$ **27.** $2x^2 + x - \dfrac{4}{5}$ **29.** $\dfrac{3x}{4} - 2 + \dfrac{1}{4x}$

31. $5a^2b + 6ab^2 - 7b$

33. $2g^3h - 6g^2 + \dfrac{4}{h}$

35. Tom should have factored the numerator. $x + 2$

37. Amy did not distribute the negative through the binomial on the second step. The binomial $-6x^2 - 15x$ should have been $-6x^2 + 15x$. $3x + 7$

39. The length is $7x - 24$.

41. The length is $5x^2 + 7x + 3$.

43. $6x + 5$ **45.** $x^2 + 5x - 4$ **47.** $3x + 5$

49. $4b^2 - 3b + 5$ **51.** $5n^2 + 4n - 8$ **53.** $x + 4$

55. $2m + 7$ **57.** $4x + 5$

59. $3x + 5 + \dfrac{5}{x+4}$ **61.** $6x + 7$ **63.** $2x^2 + 3x + 5$

65. $x^2 + 2x + 5$ **67.** $x^2 - 3x + 4$ **69.** $t^2 - 12 + \dfrac{52}{t+3}$

71. $a^2 + 2a + 3$ **73.** $t^2 + 6t + 12$

75. $4a^2 - 8a + 18 + \dfrac{4}{a+2}$

77. **a.** 13 **b.** $5 \cdot 13$

79. **a.** $4x + 3$ **b.** $(x - 7)(4x + 3)$

81. **a.** $4x - 7$ **b.** $(x + 3)(4x - 7)$

83. **a.** $2x^2 - 9x - 18$ **b.** $(x + 5)(x - 6)(2x + 3)$

85. **a.** $3t^2 - 17t + 20$ **b.** $(t^2 + 9)(3t - 5)(t - 4)$

Section 7.3

1. xy^2

3. $\dfrac{350}{3a^3b^2c^2}$

5. $\dfrac{x + 3}{x - 5}$

7. $-\dfrac{2x + 3}{x + 3}$

9. $\dfrac{2(x + 5)}{x - 7}$

11. $-\dfrac{1}{2}$

13. $\dfrac{(x + 7)^2}{(x - 3)(x - 9)}$

15. $-\dfrac{k + 6}{k + 9}$

17. $\dfrac{m + 3}{m + 2}$

19. $\dfrac{(x - 11)(x + 9)}{(x - 4)^2}$

21. Frank missed the factor of (-1) when dividing out $(7 - x)$ and $(x - 7)$.
$-\dfrac{3x + 5}{x + 2}$

23. $\dfrac{xy}{z^2}$

25. $\dfrac{(x + 5)(x - 7)}{(x - 3)(x - 5)}$

27. $\dfrac{5(x + 7)}{3(x - 4)}$

29. $-\dfrac{x + 9}{x + 7}$

31. $-\dfrac{10(b - 4)}{3(b + 4)}$

33. $\dfrac{x + 3}{x - 8}$

35. $\dfrac{w - 4}{w - 8}$

37. $\dfrac{(c - 5)(3c - 5)}{(2c + 3)(c + 4)}$

39. 1

41. $\dfrac{(x - 5)(x + 8)(x + 1)}{(x + 2)^2(x - 2)}$

43. $\dfrac{a - 5}{a + 4}$

45. $\dfrac{6h(h - 3)}{(h + 4)(h + 8)}$

47. $\dfrac{3x - 2}{4x - 3}$

49. $\dfrac{5t + 3}{8t + 5}$

51. $x + 3$

53. $\dfrac{(x - 3)^2}{5x}$

Section 7.4

1. LCD = $84x^2y$

   $\dfrac{49}{84x^2y}$ and $\dfrac{9x}{84x^2y}$

3. LCD = $900n^2p$

   $\dfrac{35m}{900n^2p}$ and $\dfrac{24n^3}{900n^2p}$

5. LCD = $(x + 8)(x + 7)$

   $\dfrac{(x + 2)(x + 7)}{(x + 8)(x + 7)}$ and $\dfrac{(x + 8)(x - 4)}{(x + 8)(x + 7)}$

7. LCD = $(x + 6)(x + 5)$

   $\dfrac{(x - 5)(x + 5)}{(x + 6)(x + 5)}$ and $\dfrac{(x + 6)(x - 5)}{(x + 6)(x + 5)}$

9. LCD = $(x - 5)$

   $\dfrac{-2x}{(x - 5)}$ and $\dfrac{3x}{(x - 5)}$

11. LCD = $(x + 1)$

    $\dfrac{2x(x + 1)}{(x + 1)}$ and $\dfrac{5}{(x + 1)}$

13. LCD = $(h + 3)(h + 7)(h + 5)$

    $\dfrac{(h + 2)(h + 5)}{(h + 3)(h + 7)(h + 5)}$ and $\dfrac{(h - 4)(h + 7)}{(h + 3)(h + 7)(h + 5)}$

15. LCD = $(n - 7)(n - 2)(n - 8)$

    $\dfrac{(n + 1)(n - 8)}{(n - 7)(n - 2)(n - 8)}$ and $\dfrac{(n + 2)(n - 7)}{(n - 7)(n - 2)(n - 8)}$

17. LCD = $(x + 3)(x + 2)(x - 4)$

    $\dfrac{(x + 1)(x - 4)}{(x + 3)(x + 2)(x - 4)}$ and $\dfrac{(x - 3)(x + 3)}{(x + 3)(x + 2)(x - 4)}$

19. LCD = $(t - 7)(t + 5)(t - 4)$

    $\dfrac{(t + 3)(t - 4)}{(t - 7)(t + 5)(t - 4)}$ and $\dfrac{(t - 7)(t - 7)}{(t - 7)(t + 5)(t - 4)}$

21. $\dfrac{2(x + 5)}{x + 7}$

23. -3

25. $\dfrac{2d^2 + 13d + 29}{(d + 3)(d + 7)}$

27. $\dfrac{2(c + 1)}{(c + 3)(c + 7)}$

29. $\dfrac{11}{n - 2}$

31. $\dfrac{15}{x - 9}$

33. $\dfrac{5x + 37}{x + 7}$

35. $\dfrac{2x^2y^2 + 7}{xy^2}$

37. Marie added the numerators together and then added the denominators together, but she needed to get a common denominator and add only the numerators.

    $\dfrac{2x^2 + 3x - 11}{(x - 7)(x + 2)}$

39. Matt cross-canceled the denominator in the first expression with the numerator in the second expression. He should cross-cancel only when multiplying two fractions.

    $\dfrac{2x^2 - 9x + 53}{(x - 7)(x + 4)}$

41. **a.** $\dfrac{125}{d^2}$

    b. If both lights are placed a distance of 5 feet away, the illumination is 5 foot-candles.

43. **a.** $\dfrac{5.5p + 850}{p}$

    b. If 100 people attend this event, the average cost will be $14 per person.

45. $\dfrac{2}{r + 5}$

47. $\dfrac{2(x^2 - 6x - 11)}{(x - 8)(x - 7)(x - 2)}$

49. $\dfrac{-2(w + 5)(w + 1)}{(w + 3)(w + 4)(w - 6)}$

51. $\dfrac{2(3h - 26)}{(h - 8)(h - 7)(h - 2)}$

53. $\dfrac{2t^2 + 18t + 55}{(t + 4)(t + 9)(t - 3)}$

55. $\dfrac{2(3k^2 - 2k - 7)}{(4k - 5)(2k + 1)(k + 3)}$

57. $\dfrac{r - 3}{r + 7}$

59. $\dfrac{d + 5}{d + 3}$

61. $\dfrac{3x + 6}{2x - 5}$

63. $\dfrac{7a^2 + 4}{5a^2 + 2a}$

65. $\dfrac{3(r^2 + 25)}{2r(r - 5)}$

67. $\dfrac{3x^3y^2 + 5x^4}{7x^4y - 6y^2}$

69. $\dfrac{3x^2 + 45y^2}{60y^2 + 5xy}$

71. $\dfrac{3(x + 1)}{2(x + 5)(2x^2 + 2x + 1)}$

73. $\dfrac{(x - 3)(7x + 58)}{x + 8}$

75. $\dfrac{(x - 3)(-5x^2 - 13x + 4)}{(x + 3)(7x^2 + 18x - 12)}$

77. **a.** $R = \dfrac{R_1 R_2 R_3}{R_2 R_3 + R_1 R_3 + R_1 R_2}$ **b.** $R \approx 1.636$ ohms

79. **a.** $f = \dfrac{cd}{c + d}$ **b.** $f \approx 0.65$ feet

81. $\dfrac{a + 2}{(a + 3)(a - 3)(a - 1)}$

83. $\dfrac{-2(7m + 37)}{(m + 1)(m + 7)}$

85. $-\dfrac{(x + 7)}{(x + 2)}$

87. $\dfrac{6}{x + 3}$

89. $\dfrac{5x^2 + 8x + 28}{(x + 3)(x + 4)(x - 4)}$

Section 7.5

1. $x = 3$
3. $x = -\dfrac{11}{10}$
5. $t = -1, t = 4$
7. $x = -3, x = -1$

9. **a.** There must be 30 players participating for the cost to be $20 per player.
   b. There must be 86 players participating for the cost to be about $7 per player.

11. **a.** Let $C(p)$ be the per person cost if p people attend the charity event. Then $C(p) = \dfrac{3400}{p}$.
    b. To keep the cost down to $40 per person, 85 people need to attend the charity dinner.

13. $x = 5$
15. $m = \dfrac{13}{30}$
17. $x = 2$
19. $h = 1$

21. **a.** Approximately 8.61% of the housing units in Colorado were vacant in 1995.
    b. There was an 11.5% vacancy rate in Colorado housing units in 1991 and again in mid-2000.

23. **a.** Let $N(t)$ be the average amount California spent in dollars per person t years since 1900.
    $N(t) = \dfrac{55.125t^2 - 6435.607t + 186,914.286}{0.464t - 12.47}$

b. The per capita spending in California in 1995 was $2310.49.
c. The per capita spending in California reached $3000 in 2002. 1939 represents model breakdown.

25. **a.** The per capita spending in Connecticut in 2008 was $5528.64.
    b. The per capita spending in Connecticut reached $4000 in 2003.

27. The resistance is 266.7 ohms when the current is 0.9 amps.

29. **a.** $y = \dfrac{1375}{x^3}$ **b.** $x \approx 7.005$

31. **a.** $W = \dfrac{10{,}935}{p^4}$ **b.** $W = 1.0935$ **c.** $p \approx 2.287$

33. $x = 9$
35. $x = 5$
37. $w = -3$

39. $x = 1$
41. $r = \dfrac{23 \pm \sqrt{313}}{4}$

43. $x = 7, x = -4$
45. $a = -\dfrac{122}{19}, a = 4$

47. $v = 50$
49. $d = -4.25$

51. $x = -10 \pm 3\sqrt{11}$
53. $x = 63$
55. $k = -4, k = 8$
57. $t = 15$
59. $w = -9, w = 1$

61. It would take about 6.9 hours for Gina and Karen working together to solve all the problems in the chapter.

63. It would take 3 hours for Rosemary and Will working together to mow the lawn.

65. $\dfrac{x^2 + 9x + 22}{(x + 2)(x + 5)}$
67. $\dfrac{10}{m + 6}$
69. $x = \dfrac{34}{7}$

71. $c = \dfrac{-7 \pm \sqrt{57}}{2}$
73. $x = -7.5$

75. Quadratic equation
$x \approx 1.553$ and $x \approx -3.220$

77. Linear equation
$x = \dfrac{-75}{8} = -9.375$

79. Logarithmic equation
$x \approx 51.598$

81. Rational equation
$x \approx -0.446$ and $x \approx 13.446$

83. Logarithmic equation
$x \approx 6.217$

Chapter 7 Review Exercises

1. **a.** $I(d) = \dfrac{36{,}000}{d^2}$
   b. The illumination of this light at a distance of 40 feet from the source is 22.5 foot-candles.
   c. An illumination of 50 foot-candles occurs at a distance of approximately 26.8 feet from the source.

2. **a.** Let $C(m)$ be the cost in dollars for an international call lasting m minutes. $C(m) = 0.27m$
   b. The total cost for a 30-minute international phone call is $8.10.
   c. You could make a 74-minute call for $20 with this calling card.

3. **a.** $y = 4x^2$ **b.** $y = 16$

4. **a.** $y = \dfrac{224}{x^5}$ **b.** $y = 0.21875$

5. **a.** Let $E(p)$ be the entrance fee in dollars for the math competition when p people compete. $E(p) = \dfrac{7.50p + 500}{p - 10}$
   b. If 100 people attend, then each nonscholarship person will pay $13.89.
   c. For each nonscholarship person paying $10.00, there needs to be 240 people participating.

6. **a.** Americans ate about 27 pounds of cheese per capita in the year 1995.
   b. According to the model, Americans ate about 30 pounds of cheese per capita in the year 2000 and will do so again in 2024.

7. Domain: All real numbers except $x = 9$

8. Domain: All real numbers except $x = -5$ and $x = 3$

9. Domain: All real numbers except $x = -\dfrac{3}{2}$ and $x = 5$

10. Domain: All real numbers

11. Domain: All real numbers except $x = -6$ and $x = 4$

12. Domain: All real numbers except $x = -3$ and $x = 6$

13. $6x^5 + 4x^2 - \dfrac{15}{2x}$ 14. $\dfrac{-1}{m + 5}$ 15. $\dfrac{4}{a + 4}$

16. $\dfrac{3p + 2}{p + 9}$ 17. $4h - 3$

18. $b + 7 + \dfrac{33}{b - 4}$ 19. $x^2 + 6x - 5$ 20. $c^2 + 3c - 4$

21. $2t + 5$ 22. $5r - 3 + \dfrac{2}{r + 4}$

23. $4x^2 + 7x - 2$ 24. $5y^2 + 15y + 25$

25. $\dfrac{x + 3}{x - 2}$ 26. $\dfrac{4(m + 9)}{(m + 3)(m - 7)}$

27. $\dfrac{x + 2}{x + 7}$ 28. $-\dfrac{h - 4}{h + 2}$

29. $\dfrac{(d + 3)(d - 1)}{(d + 2)(d - 3)}$ 30. $\dfrac{x + 4}{x - 5}$

31. $\dfrac{6}{(n + 4)(n + 7)}$ 32. $\dfrac{v + 7}{v - 8}$ 33. $\dfrac{4x + 15}{x + 6}$

34. $\dfrac{-4h + 41}{(h - 5)(h + 2)}$ 35. $\dfrac{x^2 + x + 16}{(x + 3)(x + 5)(x - 7)}$

36. $\dfrac{2a^2 + 8a + 27}{(a + 1)(a - 4)(a + 3)}$

37. $\dfrac{t^2 + 12t + 5}{(3t + 2)(t - 5)(t + 4)}$

38. $\dfrac{3x^2 + 6x - 11}{(x - 7)(x + 2)(x + 3)}$

39. $\dfrac{3a + 4}{4a - 7}$ 40. $\dfrac{20n - 7}{8n + 24}$ 41. $\dfrac{2(x + 5)(x - 2)}{5(x + 3)}$

42. $\dfrac{(2t + 19)(t - 2)}{(4t + 13)(t + 2)}$ 43. $m = 3$ 44. $c = -\dfrac{41}{3}$

45. $x = 0, x = 21$ 46. $b = \dfrac{1}{12}$

47. $x = \dfrac{-19 \pm \sqrt{2065}}{12}$ 48. $h = -\dfrac{15}{4}, h = 1$

49. $x = -7$ 50. $w = 3$ 51. $x = 3$

52. $k = -\dfrac{23}{3}, k = 2$ 53. $x = -\dfrac{10}{3}, x = -2$

54. $b = \dfrac{3 \pm \sqrt{29}}{2}$

55. It would take about 8.9 hours for Sam and Craig working together to put up the fence.

56. It would take 2.4 years for Mark and Cindy working together to write the book.

Chapter 7 Test

1. **a.** Let $I(d)$ be the illumination (in foot-candles) of a light source d feet from the source. Then $I(d) = \dfrac{4320}{d^2}$
   b. The illumination of this light at a distance of 7 feet from the source is 88.2 foot-candles.
   c. An illumination of 20 foot-candles occurs at a distance of approximately 14.7 feet from the source.

2. **a.** Let $L(d)$ be the length of a wall (in feet) if you know the length of the drawing d (in inches) on the blueprint. Then $L(d) = 2.5d$.
   b. A 2.5-inch drawing represents a wall 12.5 feet long.
   c. A 20-foot wall is represented by an 8-inch drawing.

3. $\dfrac{2x(x + 3)}{(x + 7)(x - 4)}$ 4. $\dfrac{x + 4}{x + 2}$ 5. $\dfrac{2}{(w + 3)(w - 4)}$

6. $\dfrac{2(m + 1)}{(m + 4)}$ 7. $\dfrac{8(x - 1)}{2x - 7}$ 8. $\dfrac{14(x + 1)}{(x - 3)(x + 4)}$

9. $\dfrac{2(3x^2 + 18x + 1)}{(x + 3)(x + 2)(x + 7)}$ 10. $\dfrac{12d + 5}{24(d - 2)}$

11. $\dfrac{(3x + 5)(x - 1)}{x - 9}$

12. **a.** Domain: x is all real numbers except $x = -9$ and $x = 3$.
    b. Domain: x is all real numbers except $x = -2$ and $x = 4$.

13. $x^2 + 4x - 5$ 14. $x^3 + 3x + 4$

15. $x = 17$ 16. $x = 19$ 17. $x = \pm 3$

18. $x = 1$ ~~$x = 2$~~

19. $x = \dfrac{49}{5}$ ~~$x = 2$~~

CHAPTER 8
Section 8.1

1. a.

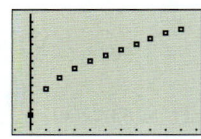

A radical function seems appropriate here because the scatterplot increases rapidly and curves slightly.

b.

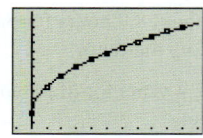

The graph fits the data well and seems to follow the pattern.

c. The cost to produce 550,000 heaters is approximately $8.57 million.

d. The cost to produce 1,500,000 heaters is approximately $12.07 million.

e. Domain: [0, 15], range: [3.2, 12.07]

3. a. The graph fits the data well and seems to follow the pattern.

b. A basilisk lizard with body mass of 50 grams will have a leg length of ≈0.061 meter.

c. A basilisk lizard with body mass of 100 grams will have a leg length of ≈0.077 meter.

d. A basilisk lizard with leg length of 0.1 meter will have a body mass of ≈222.6 grams.

5. a. The airspeed of a butterfly with a thoracic mass of 0.05 gram is 2.92 meters per second.

b. The airspeed of a butterfly with a thoracic mass of 0.12 gram is 4.93 meters per second.

c. Range: [0.28, 12.95]

7. a. $f(81) = 22.5$ **b.** $f(25) = 12.5$
9. a. $f(45) \approx 12.847$ **b.** $f(32) = 12$
11. a. $f(28) = 6$ **b.** $f(113) = 11$
13. a. $f(50) \approx 6.325$ **b.** $f(59) = 7$
15. a. $f(-60) \approx -1.694$ **b.** $f(-2207) = -3$
17. a. $f(5) \approx 3.344$ **b.** $f(27.473) \approx 12$
19. a. $f(5) = 3$ **b.** $f(-3) = 1$
21. a. $f(4) = 2$ **b.** $f(-7.5) = -1.5$

20. a. Let $U(t)$ be the unemployment rate for the state of Florida t years since 2000. Then $U(t) = \dfrac{80.5t + 287.5}{141.5t + 7826.8}$

b. In 2003, the unemployment rate for the state of Florida was about 6.4%.

c. The unemployment rate for the state of Florida reached 8% in 2005.

23. a. This is an odd root.   **b.** Domain: $(-\infty, \infty)$
  c. Range: $(-\infty, \infty)$
25. a. This is an even root.   **b.** Domain: $[-4, \infty)$
  c. Range: $[0, \infty)$
27. a. This is an even root.   **b.** Domain: $(-\infty, -5]$
  c. Range: $[0, \infty)$
29. a. This is an odd root.   **b.** Domain: $(-\infty, \infty)$
  c. Range: $(-\infty, \infty)$
31. a. Even   **b.** Domain: $x \leq -2$
  c. Range: $(-\infty, 0]$
33. Domain: All real numbers, range: All real numbers
35. Domain: $x \geq 0$, range: $[0, \infty)$
37. Domain: All real numbers, range: All real numbers
39. Domain: $x \geq 0$, range: $[0, \infty)$
41. Domain: $x \geq -12$, range: $[0, \infty)$
43. Domain: $x \geq -7$, range: $[0, \infty)$
45. Domain: All real numbers, range: All real numbers
47. Domain: $x \geq -2$, range: $[0, \infty)$
49. Domain: $x \geq 23$, range: $[0, \infty)$
51. Domain: $x \leq 3$, range: $[0, \infty)$
53. Domain: $x \leq -10$, range: $[0, \infty)$
55. Domain: All real numbers, range: All real numbers
57. Domain: $x \geq -\dfrac{5}{3}$, range: $[0, \infty)$
59. Domain: $x \leq \dfrac{9}{4}$, range: $[0, \infty)$
61. $f(x) = \sqrt{x+7}$

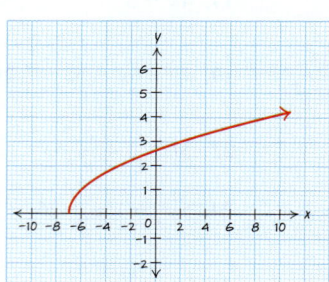

63. $9(x) = \sqrt{x-8}$

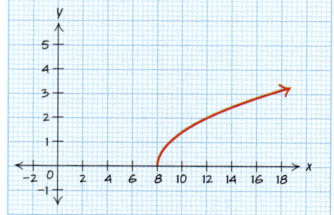

65. $f(x) = -\sqrt{x+3}$

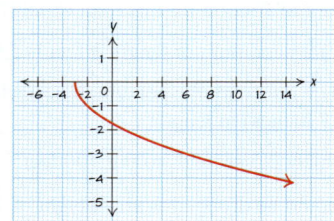

67. $f(x) = -\sqrt{x-6}$

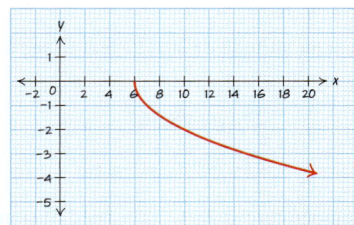

69. $f(x) = \sqrt[3]{x+2}$

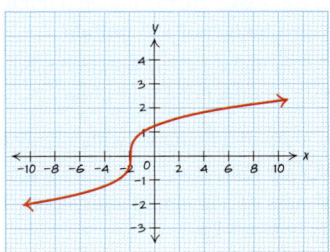

71. $h(x) = \sqrt[3]{x-11}$

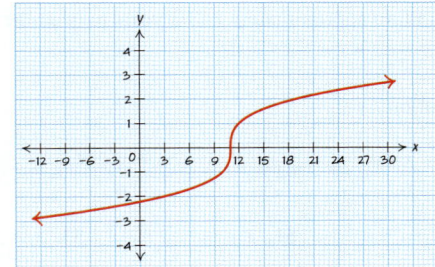

73. $f(x) = \sqrt{9-x}$

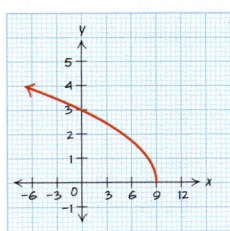

75. $g(x) = -\sqrt{x-3}$

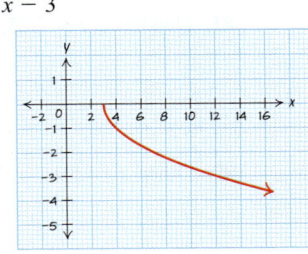

77. $h(x) = \sqrt[5]{2x}$

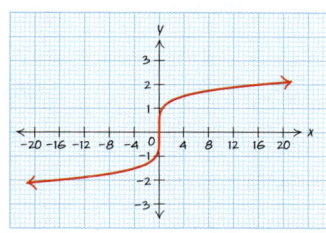

79. $g(x) = 2\sqrt{x+3}$

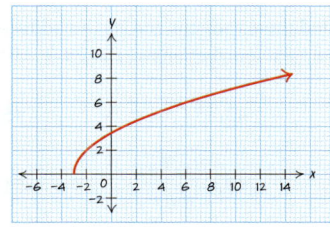

Section 8.2

1. 10   **3.** 11
5. 4   **7.** -2
9. Not a real number.
11. $5\sqrt{2}$ **13.** $6\sqrt{5}$ **15.** $2\sqrt[3]{5}$
17. $-2\sqrt[3]{2}$ **19.** $7x$ **21.** $5y^2$
23. Mark simplified the square root of 10 to 5. $xy^4\sqrt{10}$
25. $14m^2n$ **27.** $6a\sqrt{a}$ **29.** $36a^2b^4c^7\sqrt{ac}$
31. $2xy^2$ **33.** $6xyz^2\sqrt[3]{18y^2z^2}$ **35.** $2np\sqrt[4]{60m^3p}$
37. $-2cd^2$ **39.** $2mn^2$ **41.** $4\sqrt{5x}$
43. $2\sqrt{t}$ **45.** $14\sqrt[3]{5b}$ **47.** $13n\sqrt{6}$
49. $-3t\sqrt{11r}$ **51.** $x\sqrt{y}$ **53.** $6\sqrt[3]{a}$
55. Matt added the radicands. The correct answer is $11\sqrt{5}$.
57. Amy added terms that are not like terms. These can not be added. The correct answer is just $4\sqrt{7} + 10\sqrt[3]{7}$.
59. $17ab^2\sqrt{2ac}$
61. $35x^2y^3\sqrt[3]{z^2} - 54x^2y^2\sqrt[3]{z^2}$
63. $-7\sqrt{13x} + 10x\sqrt{2}$ **65.** $17\sqrt{2} - 28\sqrt{3}$
67. $7xyz^2\sqrt{z} + 7xz^2\sqrt{yz}$ **69.** $12\sqrt{2x} + 4\sqrt[3]{2x}$
71. $-4ab^2\sqrt[5]{2c^3}$ **73.** $23xy^2\sqrt[4]{x}$
75. $-3\sqrt[3]{7xy} + 3\sqrt[5]{7xy}$ **77.** $12\sqrt{2}$ inches
79. Yes: $\sqrt[6]{64} = 2$ **81.** 64

Section 8.3

1. $\sqrt{21}$ **3.** $2\sqrt{15}$ **5.** $5x\sqrt{2}$
7. $6\sqrt{mn}$ **9.** $2x$ **11.** $7m\sqrt{2n}$

13. $6a^2b^3\sqrt{5a}$
15. $35xy\sqrt{6x}$
17. $35m^3n^6\sqrt[4]{mn}$
19. $14x^3y\sqrt[3]{20y^2}$
21. $36x^4y^6\sqrt[5]{686x^2}$
23. $12 + 4\sqrt{2} + 3\sqrt{5} + \sqrt{10}$
25. $15 - 5\sqrt{15} + 3\sqrt{6} - 3\sqrt{10}$
27. $18 + 3\sqrt{6x} + 6\sqrt{2x} + 2x\sqrt{3}$
29. $3\sqrt{2} - 3\sqrt{c} - \sqrt{6c} + c\sqrt{3}$
31. $9 + 4\sqrt{5}$
33. $16 + 8\sqrt{3m} + 3m$
35. 2
37. $64 - 175ab$
39. Tom distributed the exponent across the addition. This must be FOILed out. The correct answer is $28 + 10\sqrt{3}$.
41. 2
43. 3
45. $\dfrac{x^2\sqrt{7}}{5}$
47. $\dfrac{x}{3y}$
49. $\dfrac{7\sqrt{5n}}{5n}$
51. $\dfrac{d\sqrt{21}}{3}$
53. $\dfrac{5\sqrt{14xy}}{21y}$
55. $\dfrac{\sqrt[3]{245x^2y}}{7xy}$
57. $\dfrac{\sqrt[3]{36xy^2}}{14xy}$
59. $\dfrac{\sqrt[4]{10a^2b^3}}{2ab^2}$
61. $\dfrac{4\sqrt[4]{36y^2z^2}}{3yz}$
63. Hung multiplied by the reciprocal, but Hung was supposed to simplify first and then multiply the numerator and the denominator by $\sqrt{2}$.

$4\sqrt{14y}$

65. $\dfrac{2\sqrt{3} + 3\sqrt{2}}{3}$
67. $\dfrac{2\sqrt{15} + 3\sqrt{5}}{120}$
69. $10 - 5\sqrt{3}$
71. $\dfrac{10 + 2\sqrt{7} + 5\sqrt{3} + \sqrt{21}}{18}$
73. $\dfrac{27 - 9\sqrt{2x}}{9 - 2x}$
75. $\dfrac{2 + 6\sqrt{3x} + \sqrt{7x} + 3x\sqrt{21}}{1 - 9x}$
77. $\dfrac{6 + 2\sqrt{mn} + 15\sqrt{m} + 5m\sqrt{n}}{9 - mn}$
79. $\dfrac{4 - 2\sqrt{30x} + 8\sqrt{5x} - 20x\sqrt{6}}{2 - 15x}$
81. $7\sqrt{2} - 7\sqrt{5x}$
83. $8 + 2\sqrt{15}$
85. $\dfrac{24 - 8\sqrt{5x}}{9 - 5x}$
87. $28 + 4\sqrt{6} + 7\sqrt{2x} + 3\sqrt{3x}$
89. $\dfrac{2\sqrt[3]{12mn^2}}{55mn}$

Section 8.4

1. $x = 23$
3. $x = \dfrac{221}{3}$
5. $t = -\dfrac{45}{2}$
7. $x = \dfrac{11}{2}$

9. The speed of the car is approximately 67 mph.
11. The coefficient of friction for this road is 0.877.
13. The period of a pendulum of length 3 feet is 1.92 seconds.
15. For a period of 1 second, the pendulum needs to be about 0.81 foot (9.7 inches) in length.
17. It takes an object 2.5 seconds to fall 100 feet.
19. An object should be dropped from a height of 1600 feet if you want it to fall for 10 seconds.
21. A person can see a distance of 3 miles if their eye is 6 feet off the ground.
23. A person is at an altitude of 20,417 feet to see a distance of 175 miles.
25. $x = 32$
27. $g \approx 8.567$
29. $x = 9$
31. $z = -\dfrac{1}{9}$
33. When the temperature is 28°C, the speed of sound is 347.89 meters per second.
35. a. Jim Bob's can make 26.552 hundred thousand space heaters with a budget of $15 million.
    b. Jim Bob's can make 14.767 hundred thousand space heaters with a budget of $12 million.
37. a. The body mass of a mammal with a body length of 1 meter is 27.8 kg.
    b. The body mass of a mammal with a body length of 2.4 meters is 384.7 kg.
39. The average distance from the sun to Uranus is approximately 19.2 AU.
41. It takes Pluto approximately 248.3 years to orbit the sun.
43. $x = 16$
45. No solution.
47. $x = \dfrac{1 - \sqrt{241}}{6} \approx -2.42$ $x = \dfrac{1 + \sqrt{241}}{6} \approx 2.75$
49. $w = 2$
51. $x = 4$
53. $x = \dfrac{2}{3}$
55. No solution.
57. $x = -1, x = 3$
59. When Marie squared both sides, the operation was done incorrectly on the left hand side.

$x = 12$

61. $x = 105.5$
63. $x = 5$
65. $x = 24.6$
67. $x = 32$
69. $x = 1$
71. $2\sqrt[3]{5x^2}$
73. $36 - 2x$
75. $x = 12$
77. $2\sqrt[6]{4ab^2} + 27b\sqrt{2a}$
79. $a = 4, a = 8$
81. Quadratic. $t = 7.25, t = -8$
83. Linear. $g = -\dfrac{20}{3}$
85. Other. $x = 0, x = -6, x = 2$
87. Radical. $n = 5$
89. Other. $x = \pm 9$

Section 8.5

1. $8i$
3. $9i$
5. -6
7. $2i\sqrt{5} \approx 4.472i$
9. $10i\sqrt{2} \approx 14.142i$
11. $12 - 10i$
13. $8i$
15. $3 + 4i$
17. $\approx 4.098i$
19. $-12 - 12i\sqrt{7} \approx -12 - 31.749i$
21. Real part $= 2$ Imaginary part $= 5$
23. Real part $= 2.3$ Imaginary part $= -4.9$
25. Real part $= 3$ Imaginary part $= 0$
27. Real part $= 0$ Imaginary part $= 7$
29. $8 + 9i$
31. $7.9 - 1.7i$
33. $3 - 4i$
35. $-5 + 7i$
37. $-4 - 3i$
39. $6.5 - 9.2i$
41. 5
43. $-13 + 26i$
45. $-14 - 22i$
47. $60.75 - 54i$
49. $-60 + 32i$
51. 58
53. $13 + 84i$
55. $4 + 3i$
57. $\dfrac{7}{2} - 2i$
59. $\dfrac{37}{72} - \dfrac{1}{3}i \approx 0.514 - 0.333i$
61. $\dfrac{16}{13} - \dfrac{11}{13}i$
63. $\dfrac{17}{29} + \dfrac{1}{29}i$
65. $-\dfrac{29}{85} + \dfrac{82}{85}i$
67. $\dfrac{1082}{221} + \dfrac{197}{221}i$
69. $x = \pm 3i$
71. $t = \pm 4i$
73. $x = 8 \pm \dfrac{2\sqrt{6}}{3}i \approx 8 + 1.633i$
75. $x = -\dfrac{5}{4} \pm \dfrac{\sqrt{39}}{4}i \approx -1.25 + 1.561i$
77. $x = 0, x = 1, x = -\dfrac{7}{5}$
79. $x = 0, x = 7 \pm 4i$
81. $x = 7 \pm \dfrac{\sqrt{2}}{2}i$

Chapter 8 Review Exercises

1. a. $P(3) \approx 13.76$. In the third month of the year, Big Jim's Mart had a profit of about $13.76 thousand.
   b. $P(8) \approx 15.96$. In August, Big Jim's Mart had a profit of about $15.96 thousand.
2. $W(4) \approx 51.39$. A four-year-old alpaca weighs about 51.39 kilograms.
3. Domain: $x \geq 0$, range: $y \leq 0$
4. Domain: $x \geq 0$, range: $y \geq 0$
5. Domain: $x \geq -\dfrac{5}{3}$, range: $y \geq 0$
6. Domain: $x \geq \dfrac{7}{2}$, range: $y \geq 0$
7. Domain: $x \leq 4$, range: $y \geq 0$
8. Domain: $x \leq 7$, range: $y \geq 0$
9. Domain: All real numbers, range: All real numbers
10. Domain: All real numbers, range: All real numbers
11. Domain: $x \geq 0$, range: $y \geq 0$
12. Domain: $x \geq 0$, range: $y \leq 0$
13. Domain: All real numbers, range: All real numbers
14. Domain: All real numbers, range: All real numbers
15. Domain: $x \geq 5$, range: $y \leq 0$
16. Domain: $x \leq 18$, range: $y \leq 0$

17.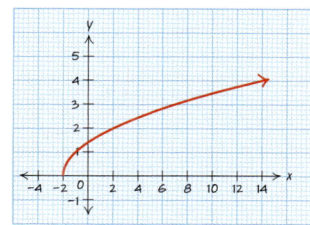

18.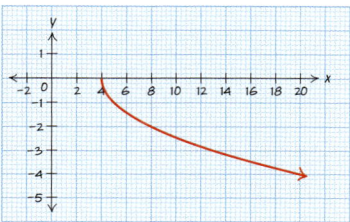

19.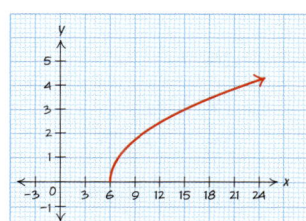

20.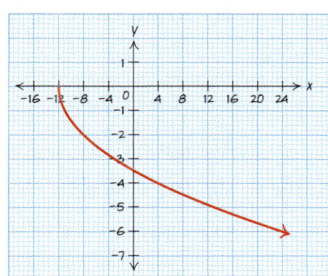

21.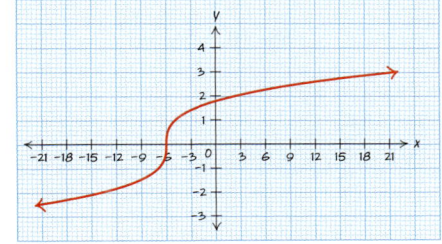

22.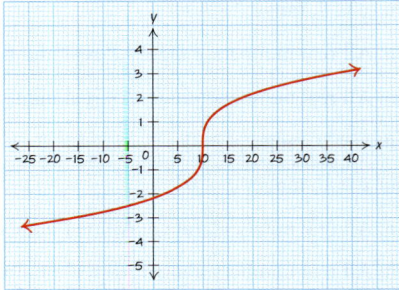

23.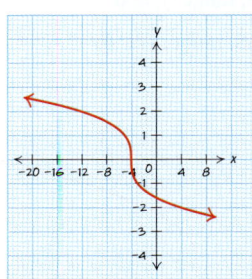

24.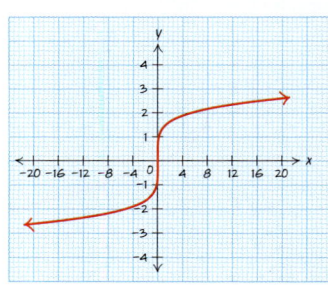

25. 11
26. 7
27. $2\sqrt{6x}$
28. $6\sqrt{2m}$
29. $12xy^2$
30. $10a\sqrt{ab}$
31. $30m^3n^5\sqrt{5n}$
32. $4c^5d^4\sqrt{55c}$
33. $-2xy^2\sqrt[3]{y}$
34. $4xy\sqrt[3]{y^2}$
35. $b^2\sqrt[4]{56a^2b^2}$
36. $n^2\sqrt[4]{25m^3n^3}$
37. $2ab^2c^6$
38. $ac\sqrt[5]{-40a^2b^3}$
39. $7\sqrt{3x}$
40. $-3x\sqrt{5y}$
41. $26xy\sqrt{3xy}+4\sqrt{3y}$
42. $12\sqrt{3a}+7\sqrt[3]{3a}$
43. $-2a\sqrt[3]{b^2}$
44. $21ac^2\sqrt[5]{2a^2b^3}$
45. $10\sqrt{3}$
46. $3x\sqrt{2y}$
47. $12a^2b\sqrt{10}$
48. $3ab\sqrt[3]{2b}$
49. $18x^3y^5z^2\sqrt[3]{10x^2}$
50. $101+36\sqrt{5}$
51. $-99+\sqrt{7}$
52. -141
53. $3\sqrt{2}$
54. $2\sqrt{2}$
55. $\dfrac{\sqrt{15}}{3}$
56. $\dfrac{\sqrt{21bc}}{3c}$
57. $\dfrac{5\sqrt[3]{4xy^2}}{2y}$
58. $10-5\sqrt{3}$
59. $\dfrac{15+26\sqrt{x}+8x}{9-16x}$
60. $\dfrac{8-4\sqrt{6}-2\sqrt{7}+\sqrt{42}}{-2}$

61. $T(2) \approx 1.57$. A 2-foot pendulum will have a period of about 1.57 seconds.

62. For a pendulum to have a period of 2.5 seconds, it would have to be about 5 feet long.

63. Big Jim's Mart will have a profit of about $17,000 in the eleventh month of the year.

64. A 5-year-old alpaca will weigh about 60 kilograms.

65. $x=11$
66. $x=\dfrac{11}{3}$
67. $x=16$
68. No solution.
69. $x=4.5$
70. $x=9$
71. No solution.
72. $x=20$
73. $x=13.5$
74. $x=2.8$
75. $5i$
76. $4i\sqrt{2}$
77. $7i$
78. $(6+5\sqrt{2})+(-2\sqrt{10}+3\sqrt{5})i$
79. $7+10i$
80. $5.9-2.4i$
81. $1+4i$
82. $2.9+7.1i$
83. $34-13i$
84. $-23+52i$
85. 13
86. $46.23-5.99i$
87. $4+\dfrac{7}{3}i$
88. $4+3i$
89. $1.4-0.8i$
90. $\dfrac{14}{13}-\dfrac{21}{13}i$
91. $-\dfrac{13}{25}+\dfrac{41}{25}i$
92. $-\dfrac{19}{25}-\dfrac{42}{25}i$
93. $0.777+0.008i$
94. $-0.901+0.984i$
95. $x=\pm 5i$
96. $m=\pm 4i$
97. $a=7\pm i\sqrt{3}$
98. $g=9\pm 2i$
99. $x=-2\pm\dfrac{\sqrt{3}}{3}i$
100. $b=-2\pm i\sqrt{5}$
101. $t=0, t=3\pm 2i$
102. $h=0, h=-4\pm 7i$

Chapter 8 Test

1. $v(20)\approx 344.1$. The speed of sound in air is about 344.1 meters per second when the temperature is 20°C.

2. $T(h)\approx 90$. When $h\approx 137174.2$, add 1500 to allow time to pull the cord. The plane would have to be at an altitude of about 138,674 feet. This is not reasonable, so it is not possible for the daredevil to free-fall for 1.5 minutes.

3. Domain: $x\geq -7$, range: $(-\infty, 0]$

4. Domain: All real numbers, range: All real numbers

5. Domain: $x\leq 3$, range: $[0, \infty)$

6.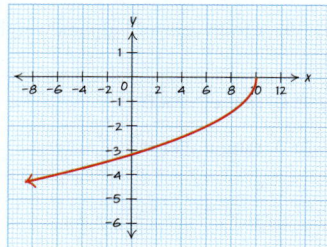

7.

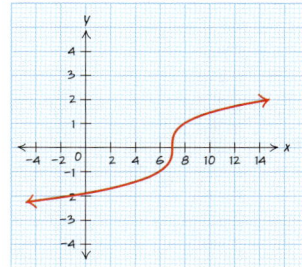

8. $\dfrac{\sqrt{30x}}{5x}$ **9.** $\dfrac{2\sqrt{5ab}}{5a}$ **10.** $\dfrac{5m\sqrt[3]{9m^2n}}{3mn}$

11. $-6 + 3\sqrt{5} + 2\sqrt{2} - \sqrt{10}$

12. $6y\sqrt{x}$ **13.** $14ab\sqrt{30b}$ **14.** $-2np^2\sqrt[5]{m^3n^2}$

15. $x = -5 + i\sqrt{0.5}, x = -5 - i\sqrt{0.5}$

16. $x \approx -1 + 0.86i, x \approx -1 - 0.86i$

17. To see 20 miles to the horizon, a person's eye would have to be about 266.67 feet off the ground.

18. $n\sqrt{6m}$ **19.** $16ab^2\sqrt{7ab} + 4b\sqrt{7a}$

20. $6a\sqrt{5bc}$ **21.** $2ab\sqrt[3]{9a}$ **22.** $30x^4yz^4\sqrt[3]{4x^2y}$

23. $-14 - 16\sqrt{3}$ **24.** $x = \dfrac{12}{5}$ **25.** $x = 49$

26. $x = 23$ **27.** $x = 59$ **28.** $4.1 - 1.4i$

29. $1 + 18i$ **30.** $31 - 29i$

31. $-25.875 - 19.875i$ **32.** $\dfrac{32}{41} - \dfrac{40}{41}i$

33. $\dfrac{4}{85} + \dfrac{33}{85}i$

34. a. $v(20) \approx 1006.5$. The speed of sound in helium at 20°C is about 1006.5 meters per second.
 b. According to part a, the speed of sound in air is much slower than the speed of sound in helium.

Cumulative Review for Chapters 1–8

1. $x = 4$ **2.** $h = -14$ **3.** $n = \pm 8$

4. $a = \dfrac{62}{3}$ **5.** $t = -\dfrac{7}{3}, t = 8$ **6.** $x = \dfrac{e^3 + 9}{4}$

7. $x = 59$ **8.** $x = -7$ **9.** $c = -6$

10. $x = -7$ **11.** $x = 35 - 6\sqrt{31}$

12. $x = 12, x = -19$ **13.** $g = 7, g = -\dfrac{111}{17}$

14. $w = -18$ **15.** $b = \dfrac{64}{3}$ **16.** $x = 258.25$

17. $r = \dfrac{3}{2} \pm \dfrac{3\sqrt{3}}{2}i$ **18.** $t = 6.5, t = -1.5$

19. $x = \pm\sqrt{10}$ **20.** $d = 0, d = 9, d = -7$

21. No solution. **22.** $(2, 2)$

23. $(1, -2)$ and $\left(-\dfrac{31}{12}, -\dfrac{139}{48}\right)$

24. $(4, 32)$ and $\left(-\dfrac{14}{3}, \dfrac{2}{9}\right)$

25. Let $M(t)$ be the number of trademark applications for items with stars-and-stripes motif t years since 2000.
 a. $M(t) = -37(t-2)^2 + 380$. *Note:* The vertex needed adjustment to better fit the data.
 b. Domain: $[-1, 5]$, range: $[47, 380]$
 c. In 2005, there were approximately 47 trademark applications for items with a stars-and-stripes motif.
 d. The vertex is $(2, 380)$. This means that in the year 2002, there were 380 trademark applications for items with a stars-and-stripes motif. This was the highest number of trademark applications.

26. Let $N(t)$ be the net sales (in millions of dollars) for Odyssey Health Care, Inc. t years since 2000.
 a. $N(t) = 85.3(1.476)^t$
 b. Domain: $[-1, 5]$, range: $[57.8, 597.6]$
 c. In 1998 Odyssey Health Care, Inc. had about $39.2 million in net sales.
 d. According to the model, the net sales for Odyssey Health Care, Inc. reached $600 million in the year 2005.

27. Let $H(t)$ be the average number of hours per year Americans spent listening to recorded music t years since 2000.
 a. $H(t) = -9t + 246.5$
 b. In 1995, Americans spent an average of 291.5 hours per year listening to recorded music.
 c. In 2010, Americans will spend an average of 150 hours per year listening to recorded music.
 d. The slope is -9. This means that the number of hours per year that Americans spend listening to recorded music is decreasing by 9 hours per year.
 e. Domain: $[-5, 10]$, range: $[156.5, 291.5]$

28. Domain: x is all real numbers except $x = -2$ and $x = -4$.

29.

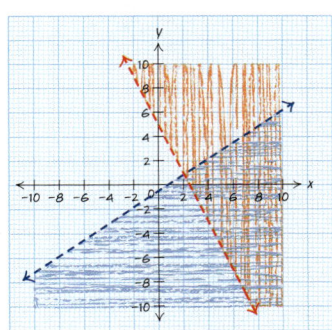

30. $x < 5$ **31.** $n \geq 18$

32. $w \geq 8$ or $w \leq 5$ **33.** $d < -10$ or $d > 0$

34. No solution.

35. a. $V(15) \approx 106.06$. In 2005, U.S. amusement parks had approximately 106.06 million visitors.
 b. $(5.04, 78.76)$. In about 1995, U.S. amusement parks had their lowest number of visitors with about 78.76 million visitors.
 c. U.S. amusement parks had about 90 million visitors in 1989 and again in 2001.

36. a. Let $B(m)$ be the number of salmonella bacteria on the counter m minutes after 2:00 PM.

$$B(m) = 4000(2)^{\frac{m}{10}}$$

b. After about 80 minutes, there will be about 1 million salmonella bacteria on the counter.

c. $B(240) = 67{,}108{,}864{,}000$. If the spill stays unnoticed until dinner at 6:00 pm, there will be 67,108,864,000 bacteria present. This may be model breakdown.

37. $f^{-1}(x) = 2x + 10$ **38.** $g^{-1}(x) = \dfrac{\log\left(-\dfrac{x}{3}\right)}{\log 5}$

39. a. $f(9) = -7$
b. Domain: All real numbers, range: $\{-7\}$

40. a. $f(x) + g(x) = 3.6x + 12.2$
b. $f(g(x)) = -7x + 25$
c. $f(x)g(x) = -7x^2 + 3.4x + 28.8$

41. a. $f(6) - g(6) = 45$
b. $f(g(7)) = -26.5$
c. $f(3)g(3) = -9.5$

42. a. $f(x) + g(x) = 2x^2 + 9x - 13$
b. $f(g(x)) = 8x^2 - 26x + 7$
c. $f(x)g(x) = 4x^3 + 4x^2 - 51x + 40$

43. An earthquake with a magnitude of 7.8 has an intensity of about 6309.6 cm.

44. $y = \dfrac{5}{2}x + 27$

45. $y > -x - 3$

46.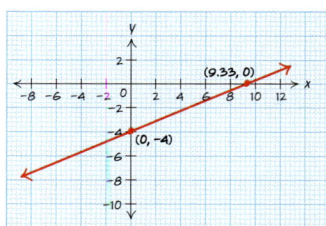

47.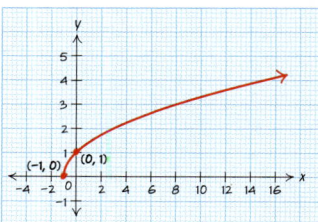

48.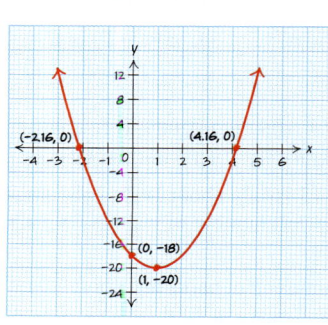

49.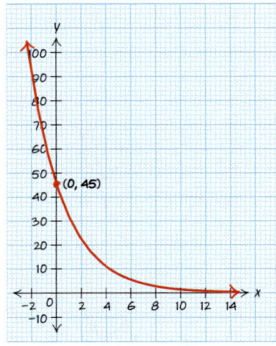

50.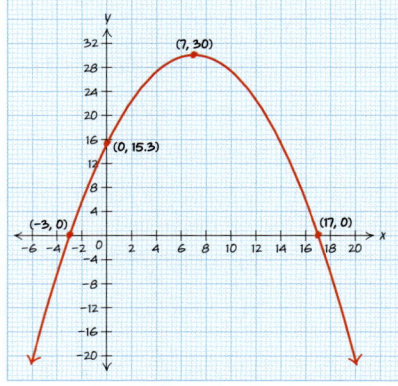

51.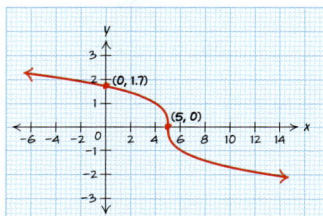

52.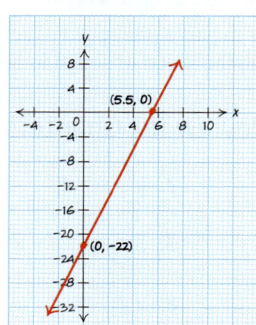

53.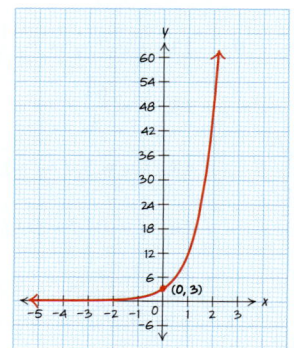

54.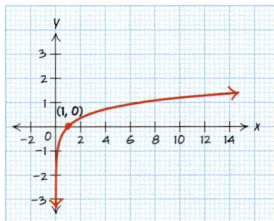

55.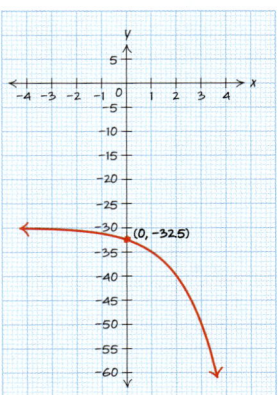

56.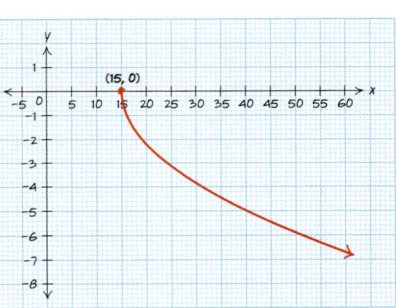

57.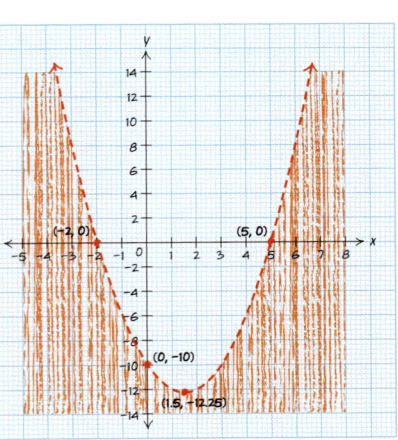

58.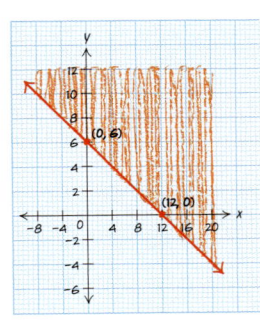

59.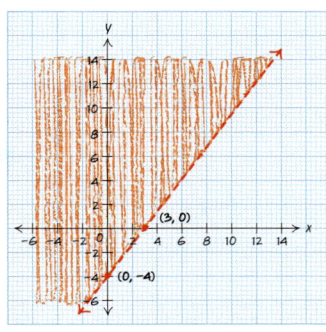

60.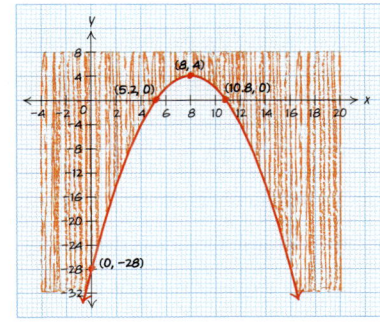

61. a. Let $P(v)$ be the pressure in pounds per square inch in the syringe when the volume of the syringe is v cubic inches.

$$P(v) = \frac{8}{v}$$

b. $P(1) = 8$. If the volume in the syringe is 1 cubic inch, the pressure is 8 pounds per square inch.

62. $19\sqrt{x} - 10$ **63.** $-2x^2 + 11x - 19$

64. $\dfrac{(x+2)(x+5)}{(x-8)(x+3)}$ **65.** $6c + 13\sqrt{2c} + 12$

66. $8b^2 + 14b - 30$ **67.** $9x^2 - 48x + 64$

68. $\dfrac{(4h+7)(h-4)}{(h+3)(h-6)}$ **69.** $\dfrac{2x^2 - 9x + 12}{(x+4)(x-6)}$

70. a. $G(10) = 1882.7$. The National Endowment for the Arts gave about 1883 grants in 2000.

b. $F(10) = 84010$. The National Endowment for the Arts gave about $84,010 thousand in grants in 2000.

c. The average grant given by the National Endowment for the Arts in 2000 was about $44.62 thousand.

d. $\dfrac{F(t)}{G(t)} = \dfrac{3060t^2 - 55290t + 330910}{210.9t - 226.3}$ gives the average size of a grant in thousands of dollars, given by the National Endowment for the Arts t years since 1990.

e. The National Endowment for the Arts average grants in 1998 and 2002 averaged $50,000 each.

71. $8(r+4)(r-4)$ **72.** $(c+5)^2$

73. $2(t+3)(4t-5)$ **74.** $(t-4)(t^2 + 4t + 16)$

75. $x(7x-4)(3x-8)$ **76.** Degree $= 2$

77. $f(x) = -2(x-5)^2 + 6$

78. **a.** $W(10) \approx 20$. The average weight of girls who are 10 months old is about 20 pounds.

  b. The average weight of girls who are about 19 months old is 25 pounds.

79. $x = 1$

80. Domain: $x \leq 10$, range: $y \geq 0$

81. a is negative, since the vertical intercept of the graph is negative.

82. b is less than 1, since the graph is increasing and a is negative.

83. Domain: All real numbers, range: $(-\infty, 0)$

84. a is negative because the parabola is facing downward.

85. $h = -1.5$, $k = 18$ because the vertex of the parabola is at the point $(-1.5, 18)$

86. $f(-3) = 11$ 87. $x = 1, x = -4$

88. Domain: All real numbers, range: $(-\infty, 18]$

89. m is negative because the line is decreasing.

90. Because the line crosses the y-axis at (0.4). $b = 4$

91. $m = -\dfrac{2}{5}$ 92. $x = 7.5$

93. Domain: All real numbers, range: All real numbers

94. $10x^2y^4z$ 95. $2ab^2c^2\sqrt[3]{4ab^2}$ 96. $\dfrac{\sqrt{10y}}{5y}$

97. $\dfrac{20 + 5\sqrt{7} + 4\sqrt{3} + \sqrt{21}}{9}$

98. $11 - 5i$ 99. $-3 + 8i$ 100. 40

101. $50 - 10i$ 102. $\dfrac{3}{7} - \dfrac{2}{7}i$ 103. $\dfrac{46}{73} + \dfrac{1}{73}i$

104. $\dfrac{2}{97} + \dfrac{45}{97}i$

105. $x = 2 \pm 4i$

106. $x = 0, x = 5 \pm 3i$

107. $x = \pm 3i$

108. $x = 3 \pm 5i$

CHAPTER 9
Section 9.1

1. Opens up, vertex $(-4, -57)$, axis of symmetry $x = -4$

3. Opens right, vertex $(-19, -2)$, axis of symmetry $y = -2$

5. Opens left, vertex $(4, 0)$, axis of symmetry $y = 0$

7. Opens down, vertex $(12.5, 28.25)$, axis of symmetry $x = 12.5$

9. Opens right, vertex $(0, 0)$, axis of symmetry $y = 0$

11. $x = y^2 + 10y + 21$

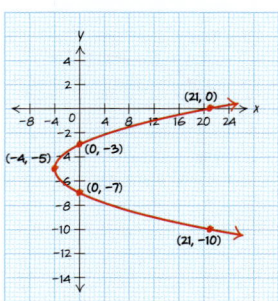

13. $x = -2y^2 + 15y + 8$

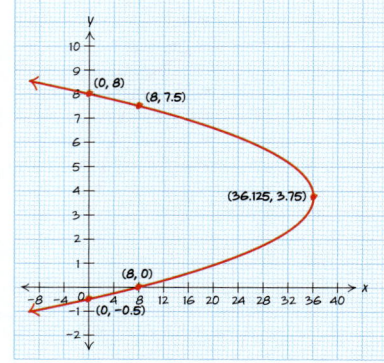

15. $x = \dfrac{1}{2}(y - 4)^2 + 5$

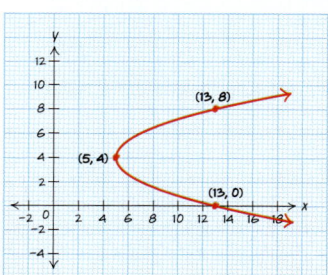

17. $x = (y + 2)^2 - 6$

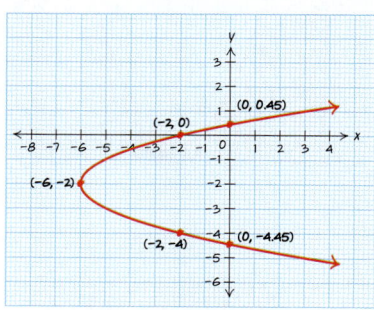

19. $x = -2(y + 5)^2 + 4$

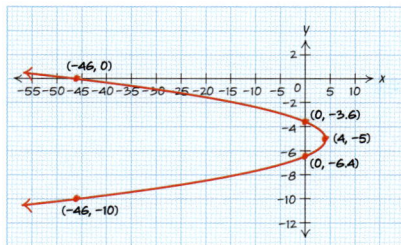

21. $x = \dfrac{y^2}{36}$

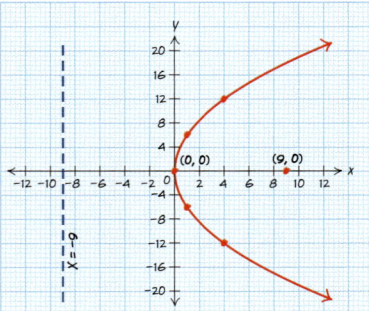

23. $y = \dfrac{1}{15}x^2$

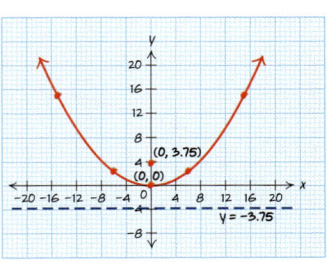

25. $28x = y^2$

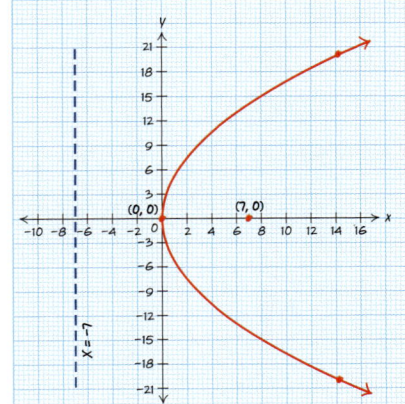

27. $y = \dfrac{x^2}{4}$ 29. $x = \dfrac{y^2}{24}$ 31. $x = -\dfrac{y^2}{40}$

33. The receiver should be $1\tfrac{1}{8}$ ft above the center of the satellite along the axis of symmetry.

35. The arch that is 40 feet from the center of the bridge is 42 feet high.

37. The cables that are 1000 feet from the center of the bridge are 192 feet high.

39. $x^2 + y^2 = 36$

41. $(x - 4)^2 + (y - 9)^2 = 16$

43. $(x + 2)^2 + (y - 3)^2 = 144$

45. $(x + 1)^2 + (y + 4)^2 = 0.25$

47. $x^2 + y^2 = 9$; center $= (0, 0)$, radius $= 3$.

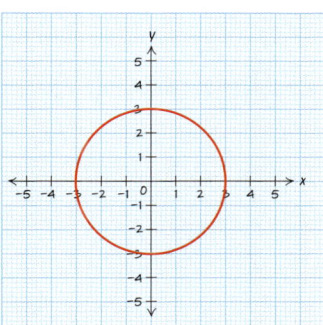

49. $(x - 5)^2 + (y - 4)^2 = 25$; center $= (5, 4)$, radius $= 5$.

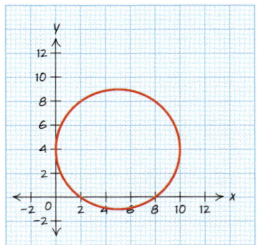

51. $(x + 4)^2 + (y + 1)^2 = 1$; center $= (-4, -1)$, radius $= 1$.

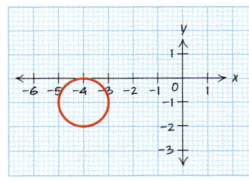

53. $(x - 8)^2 + (y + 2)^2 = 36$; center $= (8, -2)$, radius $= 6$.

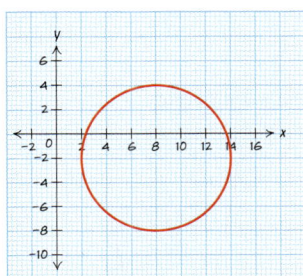

55. $(x + 1)^2 + (y - 3)^2 = 1$; center $= (-1, 3)$, radius $= 1$.

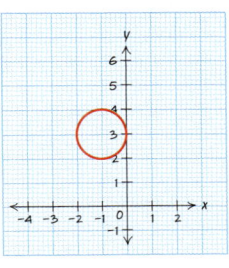

57. $(x + 3)^2 + (y - 6)^2 = 36$; center $= (-3, 6)$, radius $= 6$.

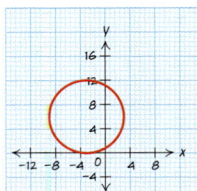

59. $\left(x + \frac{1}{2}\right)^2 + \left(y - \frac{1}{2}\right)^2 = 9$; center $= \left(-\frac{1}{2}, \frac{1}{2}\right)$, radius $= 3$.

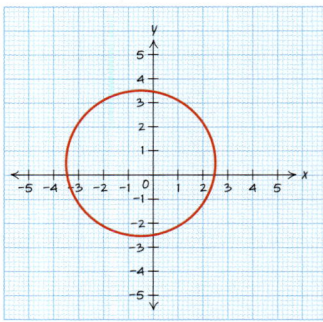

61. $x^2 + (y - 2)^2 = 4$

63. $(x + 4)^2 + (y - 2)^2 = 9$

65. $(x - 10)^2 + (y - 10)^2 = 25$

67. $x^2 + y^2 = 9$
This is circle with center $= (0, 0)$ and radius $= 3$.

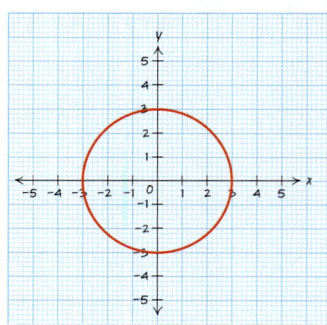

69. $y = x^2 - 4x - 12$
This is a parabola with vertex $= (2, -16)$.

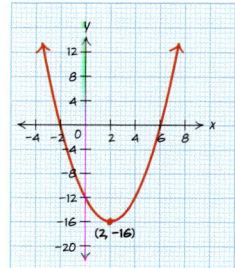

71. $x = y^2 + 8y + 15$
This is a parabola with vertex $= (-1, -4)$.

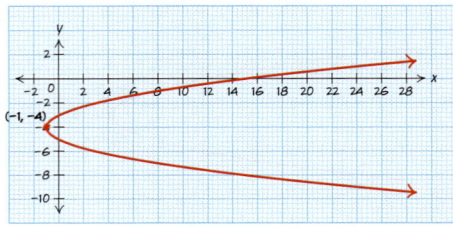

73. $x^2 + (y - 3)^2 = 49$
This is a circle with center $= (0, 3)$ and radius $= 7$.

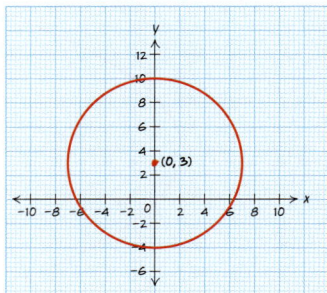

75. $x = \frac{1}{32} y^2$
This is a parabola with vertex $= (0, 0)$.

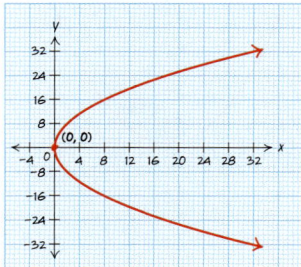

Section 9.2

1. $\dfrac{x^2}{4} + \dfrac{y^2}{16} = 1$

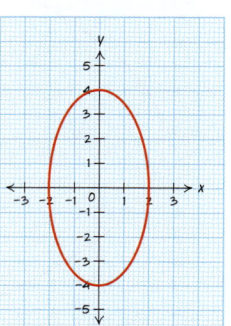

3. $\dfrac{x^2}{25} + \dfrac{y^2}{9} = 1$

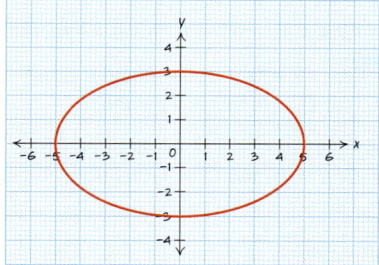

5. $x^2 + \dfrac{y^2}{9} = 1$

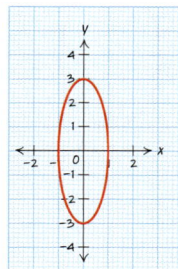

7. $\dfrac{x^2}{10} + \dfrac{y^2}{49} = 1$

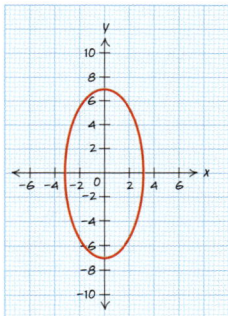

9. $\dfrac{x^2}{9} + \dfrac{y^2}{4} = 1$

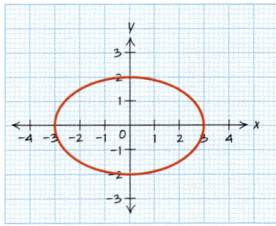

11. $\dfrac{x^2}{36} + y^2 = 1$

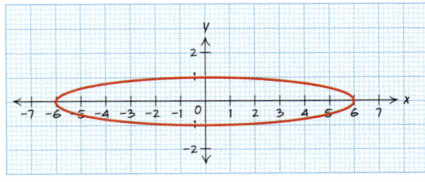

13. $\dfrac{x^2}{9} + \dfrac{y^2}{20} = 1$

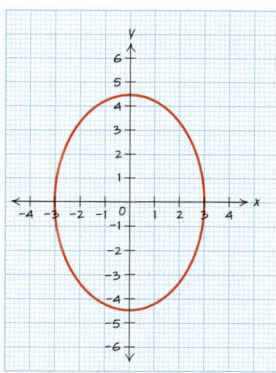

15. $\dfrac{x^2}{64} + \dfrac{y^2}{16} = 1$

Foci: $(-6.93, 0)$ and $(6.93, 0)$.

17. $\dfrac{x^2}{4} + \dfrac{y^2}{9} = 1$

Foci: $(0, -2.24)$ and $(0, 2.24)$.

19. $\dfrac{x^2}{225} + \dfrac{y^2}{100} = 1$

Foci: $(-11.18, 0)$ and $(11.18, 0)$.

21. $\dfrac{x^2}{13{,}479{,}912.25} + \dfrac{y^2}{12{,}649{,}080} = 1$

23. $\dfrac{x^2}{784{,}996} + \dfrac{y^2}{782{,}880} = 1$. Aphelion ≈ 932 million miles.

25. Width: 135 cm, depth: 30.75 cm.

27. Let w be half the width of the arch, and let h be the height of the arch.
$$\dfrac{w^2}{3600} + \dfrac{h^2}{1225} = 1$$

29. $\dfrac{x^2}{4} - \dfrac{y^2}{9} = 1$

Vertices $= (2, 0), (-2, 0)$

Foci $= (\sqrt{13}, 0), (-\sqrt{13}, 0) \approx (3.6, 0), (-3.6, 0)$

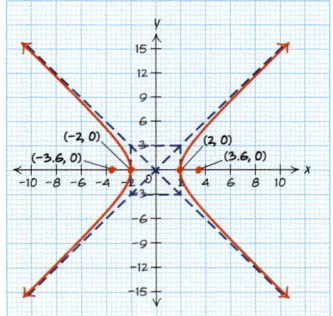

31. $\dfrac{y^2}{36} - \dfrac{x^2}{4} = 1$

Vertices $= (0, 6), (0, -6)$

Foci $= (0, \sqrt{40}), (0, -\sqrt{40}) \approx (0, 6.3), (0, -6.3)$

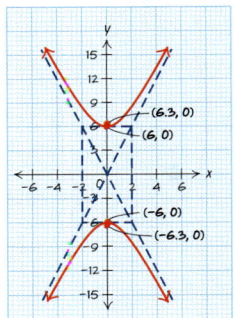

33. $\dfrac{x^2}{25} - y^2 = 1$

Vertices $= (5, 0), (-5, 0)$

Foci $= (\sqrt{26}, 0), (-\sqrt{26}, 0) \approx (5.1, 0), (-5.1, 0)$

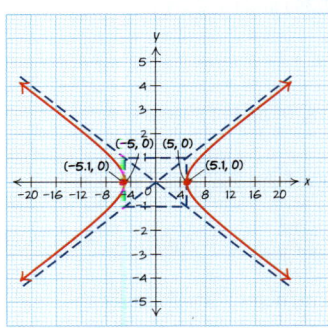

35. $\dfrac{y^2}{9} - x^2 = 1$

Vertices $= (0, 3), (0, -3)$

Foci $= (0, \sqrt{10}), (0, -\sqrt{10}) \approx (0, 3.2), (0, -3.2)$

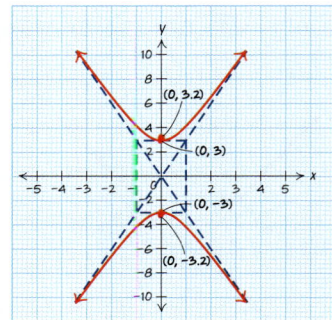

37. $\dfrac{x^2}{16} - \dfrac{y^2}{9} = 1$

Foci $= (5, 0), (-5, 0)$

39. $\dfrac{y^2}{4} - \dfrac{x^2}{1} = 1$

Foci $= (0, \sqrt{5}), (0, -\sqrt{5})$

41. $3x^2 + 3y^2 = 27 \qquad A = 3$ and $C = 3$

Since $A = C$, this is a circle.

43. $4x^2 - 9y^2 = 36 \qquad A = 4$ and $C = -9$

Since $A \ne C$, $A < 0$, or $C < 0$, this is a hyperbola.

45. $6y^2 + 2y + x - 9 = 0 \qquad A = 0$ and $C = 6$

Since $A = 0$ or $C = 0$, this is a parabola.

47. $64x^2 + 36y^2 = 576 \qquad A = 64$ and $C = 36$

Since $A \ne C$, $A > 0$, and $C > 0$, this is an ellipse.

49. Ellipse; $\dfrac{x^2}{16} + \dfrac{y^2}{36} = 1$

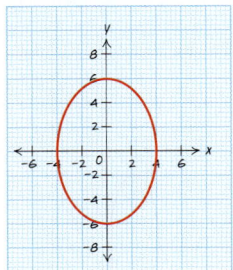

51. Hyperbola; $\dfrac{x^2}{49} - \dfrac{y^2}{100} = 1$

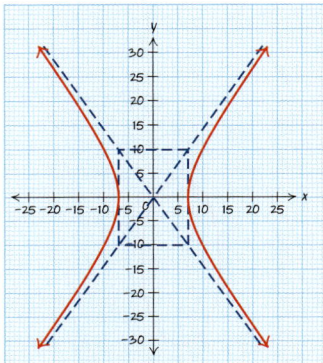

53. Ellipse; $\dfrac{x^2}{2} + \dfrac{y^2}{4} = 1$

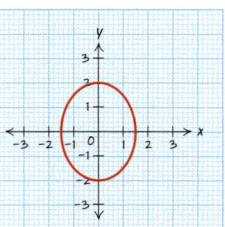

55. Circle; $x^2 + y^2 = \dfrac{1}{4}$

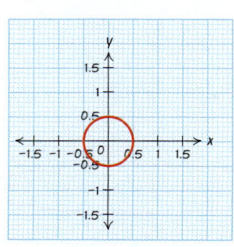

57. Circle; $(x + 1.5)^2 + (y - 2.5)^2 = 11$

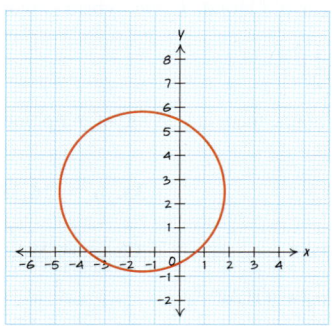

Section 9.3

1. $a_1 = 4, a_2 = 10, a_3 = 16, a_4 = 22$

3. $f(1) = 2, f(2) = 8, f(3) = 18, f(4) = 32$

5. $a_1 = -4, a_2 = 4, a_3 = -4, a_4 = 4, a_5 = -4$

7. $-17, -11, 7, 61, 223$.

9. $\dfrac{1}{2}, \dfrac{8}{9}, \dfrac{6}{5}, \dfrac{16}{11}, \dfrac{5}{3}$

11. $\dfrac{1}{4}, \dfrac{1}{16}, \dfrac{1}{64}, \dfrac{1}{256}, \dfrac{1}{1024}$

13. $-1, \dfrac{1}{4}, -\dfrac{1}{9}, \dfrac{1}{16}, -\dfrac{1}{25}$

15. $n = 20$ **17.** $n = 14$ **19.** $n = 11$

21. $n = 12$ **23.** $n = 162$

25. $a_n = 4n - 5$

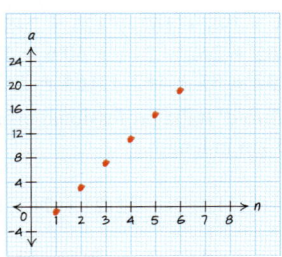

27. $a_n = 4 + (n - 1)6$

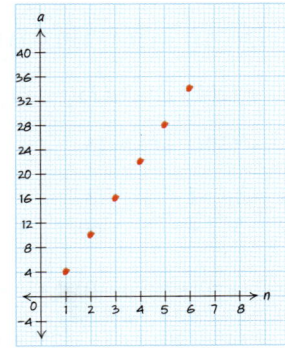

29. $f(n) = 3^n - 20$

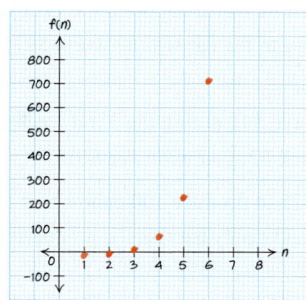

31. $a_n = 4n(-1)^n$

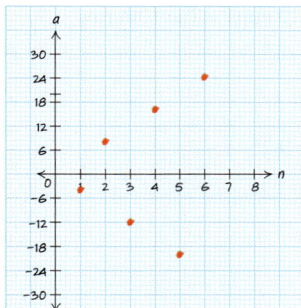

33. $a_n = \dfrac{4n}{n + 7}$

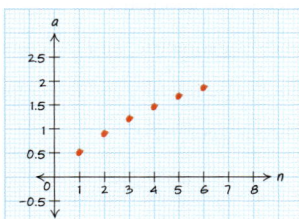

35. Arithmetic, common difference $d = 4$.

37. Arithmetic, common difference $d = 2.35$.

39. Not arithmetic.

41. Arithmetic, common difference $d = -3.45$.

43. Arithmetic, common difference $d = -\dfrac{1}{2}$.

45. Arithmetic, common difference $d = 6$.
$$a_n = -2 + 6n$$

47. Arithmetic, common difference $d = 1.25$.
$$a_n = 2.75 + 1.25n$$

49. Not arithmetic.

51. Arithmetic, common difference $d = -\dfrac{2}{3}$.
$$a_n = \dfrac{14}{3} - \dfrac{2}{3}n$$

53. a. $a_n = 35{,}100 + 900n$

   b. Your salary will be $45,900 in the 12th year of employment.

   c. In your 28th year of employment, your salary will be $60,300.

55. **a.** $a_n = 52{,}500 - 4500n$
    b. The car will be worth $3000 in the 11th year.
    c. In the 7th year, the car will be worth $21,000.
57. **a.** $a_n = 9 + n$
    b. You will save $41 during the 32nd week.
    c. During week 52, you will save $61.
59. **a.** It costs $415 to produce 100 candy bars.
    b. One thousand candy bars can be produced for a cost of $550.
    c. Common difference $d = 0.15$. The cost of production increases by $0.15 per candy bar produced.
61. **a.** $a_n = n$
    b. There are 10 cups in the 10th row.
63. $a_n = -2n + 17$
65. $a_n = 3.5n + 3.5$
67. $a_n = 4.75n + 9.25$
69. $a_n = 0.2n - 37.2$

Section 9.4

1. Yes, the sequence is geometric, and the common ratio is 1.5.
3. Yes, the sequence is geometric, and the common ratio is $\frac{1}{3}$.
5. No, the sequence is not geometric.
7. Yes, the sequence is geometric, and the common ratio is 2.
9. Yes, the sequence is geometric, and the common ratio is -3.
11. Yes, the sequence is geometric.
    $$a_n = 3(2)^{n-1}$$
13. Yes, the sequence is geometric.
    $$a_n = 2(-4)^{n-1}$$
15. Yes, the sequence is geometric.
    $$a_n = 625(0.2)^{n-1}$$
17. No, the sequence is not geometric.
19. Yes, the sequence is geometric.
    $$a_n = -4(0.5)^{n-1}$$
21. **a.** $a_n = 56{,}000(1.015)^{n-1}$
    b. $a_7 = 61{,}232.82$. Your salary will be $61,232.82 in the 7th year of employment.
    c. During your 25th year of employment, you will earn a salary of $80,052.
23. **a.** $a_n = 8\left(\frac{1}{2}\right)^{n-1}$
    b. $a_3 = 2$. The ball bounces to a height of 2 ft after the 3rd bounce.
25. **a.** $a_n = 5(2)^{n-1}$
    b. $a_7 = 320$. On day 7, there were 320 people with the flu.
27. **a.** $a_n = 21{,}000(0.93)^{n-1}$
    b. A four-year-old Honda Civic hybrid is worth $16,891.50.
29. **a.** $a_n = 10\left(\frac{1}{2}\right)^{\frac{n}{2}}$
    b. After six hours, there is 1.25 mg of morphine left in the patient.
31. $a_n = 2(5)^{n-1}$
33. $a_n = 16{,}384\left(\frac{1}{4}\right)^{n-1}$
35. $a_n = 3(-2)^{n-1}$
37. $a_n = 256\left(-\frac{1}{2}\right)^{n-1}$
39. Geometric, $a_n = 11(2)^{n-1}$
41. Arithmetic, $a_n = 8 + 6n$
43. Geometric, $a_n = 7776\left(\frac{1}{2}\right)^{n-1}$
45. Arithmetic, $a_n = -41.25 + 3.25n$
47. **a.** $a_n = 180n$
    b. $a_{10} = 1800$. The sum of the interior angles of a 12-sided polygon is 1800°.
49. **a.** $a_n = 200{,}000 - 8{,}000n$
    b. The machine will be worth $120,000 when it is 10 years old.
51. **a.** $a_n = 10 + 5n$
    b. You should be jogging 35 minutes during the 5th week after surgery.

Section 9.5

1. **a.** $20.5 + 156 + 198 + 162 + 125 + 248 + 244 + 209 + 165 + 262 + 214 + 60.6 + 176 + 172 + 161 + 237 + 83 + 182 + 164 + 203 + 14.4$
   b. The 2009 Giro d' Italia professional bike race was 3456.5 km in length.
3. $\sum_{n=1}^{5}(2n + 7) = (2(1) + 7) + (2(2) + 7) + (2(3) + 7) + (2(4) + 7) + (2(5) + 7) = 65$
5. $\sum_{n=1}^{4}(4^n) = 4 + 4^2 + 4^3 + 4^4 = 340$
7. $\sum_{n=1}^{6}(5n(-1)^n) = (5(-1)) + (10(-1)^2) + (15(-1)^3) + (20(-1)^4) + (25(-1)^5) + (30(-1)^6) = 15$
9. $\sum_{n=1}^{5}\left(\frac{1}{n}\right) = 1 + \frac{1}{2} + \frac{1}{3} + \frac{1}{4} + \frac{1}{5} = \frac{137}{60}$
11. $\sum_{n=1}^{8}(n^2 + 5n) = (1 + 5) + (2^2 + 10) + (3^2 + 15) + (4^2 + 20) + (5^2 + 25) + (6^2 + 30) + (7^2 + 35) + (8^2 + 40) = 384$
13. $\sum_{n=1}^{8}(3(-2)^n) = (3(-2)) + (3(-2)^2) + (3(-2)^3) + (3(-2)^4) + (3(-2)^5) + (3(-2)^6) + (3(-2)^7) + (3(-2)^8) = 510$

15. **a.** $a_n = 41{,}200 + 800n$
    b. Your salary will be $53,200 in the 15th year of employment.
    c. You will have earned a total of $714,000 over a 15-year period of employment.
17. **a.** $a_n = 6 + 2n$
    b. You will save $22 during the 8th week.
    c. You will have saved a total of $858 during the first 26 weeks.
19. **a.** $a_n = 14 + n$
    b. A total of 246 blocks are needed to build a wall with 12 rows.
21. $S_{30} = 1890$ 23. $S_{15} = 825$
25. $S_{40} = -7460$ 27. $S_{52} = 3629.6$
29. $S_{27} = 234$
31. **a.** $a_n = 42{,}000(1.02)^{n-1}$
    b. Your salary will be $55,418.11 during the 15th year of employment.
    c. You will have earned a total of $726,323.51 over a 15-year career.
33. **a.** $a_n = 10(1.10)^{n-1}$
    b. You will save $19.49 during the 8th week.
    c. You will have saved a total of $1091.82 during the first 26 weeks
35. **a.** $a_n = 3(2)^{n-1}$
    b. There are 48 people diagnosed with the flu on the 5th day.
    c. There were 49,149 people diagnosed with the flu during the first two weeks.
37. $S_9 = 29{,}523$ 39. $S_6 = 274{,}512$
41. $S_{11} = -621{,}937{,}812$ 43. $S_{12} \approx 449.999$
45. $S_{10} \approx -666.597$
47. Geometric.   49. Geometric.
    $S_{10} = 10{,}230$ $S_6 = 1{,}092$
51. Arithmetic.   53. Geometric.
    $S_{23} = 2{,}783$ $S_{12} \approx 7{,}389.524$
55. Arithmetic.
    $S_{10} = 272.5$

Chapter 9 Review Exercises

1. $x = y^2 + 6y - 16$

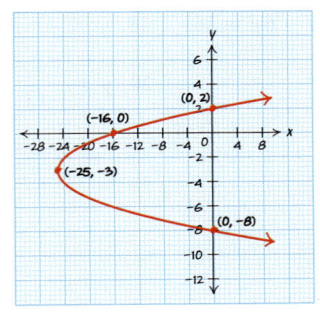

2. $y = 2x^2 - 24x - 30$

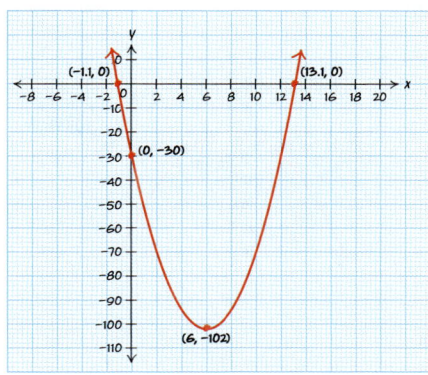

3. $x = -1.5(y + 2)^2 + 6$

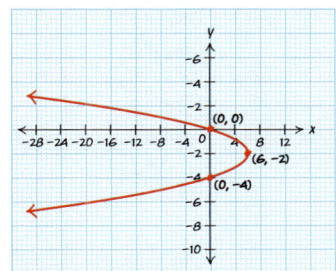

4. $x = -\dfrac{1}{2}y^2 - 3y - 8$

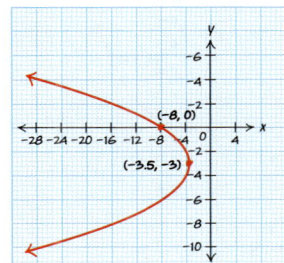

5. $x = \dfrac{y^2}{48}$

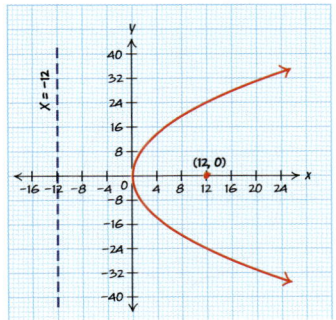

6. $y = \dfrac{x^2}{-20}$

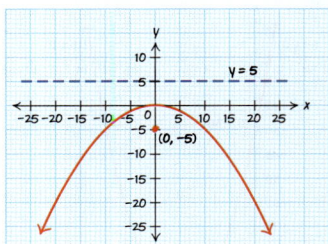

7. $x = \dfrac{y^2}{-8}$

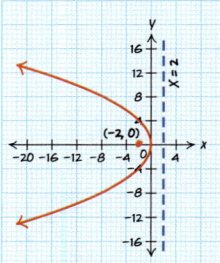

8. $y = \dfrac{x^2}{14}$

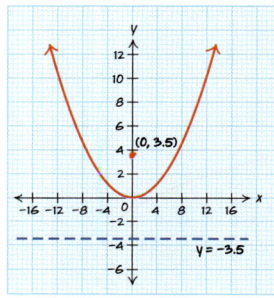

9. $(x-2)^2 + (y-8)^2 = 9$
10. $(x-5)^2 + (y+2)^2 = 6.25$
11. $x^2 + y^2 = 16$; center = $(0, 0)$, radius = 4.

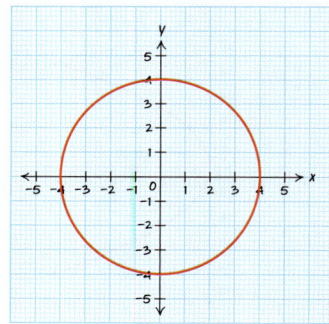

12. $(x-5)^2 + (y-2)^2 = 4$; center = $(5, 2)$, radius = 2.

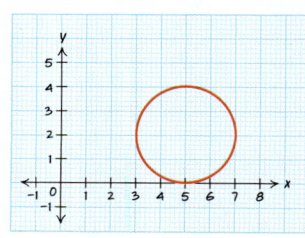

13. $(x+4)^2 + (y-6)^2 = 1$; center = $(-4, 6)$, radius = 1.

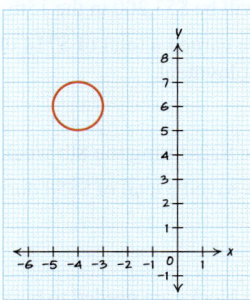

14. $(x+3)^2 + (y+7)^2 = 0.25$; center = $(-3, -7)$, radius = 0.5.

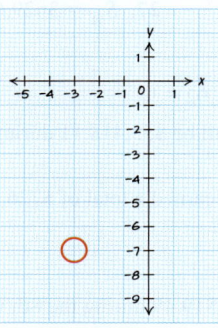

15. $(x+6)^2 + (y-5)^2 = 100$; center = $(-6, 5)$, radius = 10.

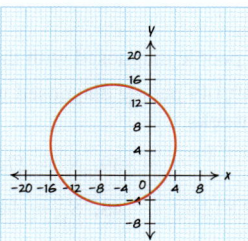

16. $(x-4)^2 + (y+2)^2 = 25$; center = $(4, -2)$, radius = 5.

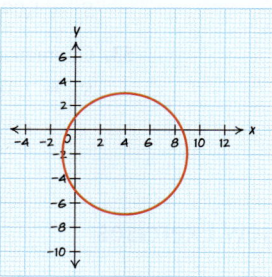

17. $(x-2)^2 + (y-6)^2 = 16$
18. $(x+4)^2 + (y-2)^2 = 36$
19. $x = \dfrac{y^2}{20}$
20. $y = \dfrac{x^2}{-12}$

21. $\dfrac{x^2}{16} + \dfrac{y^2}{25} = 1$

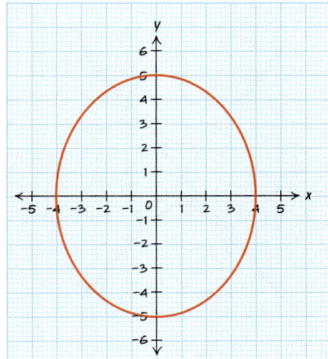

22. $\dfrac{x^2}{0.25} + \dfrac{y^2}{1} = 1$

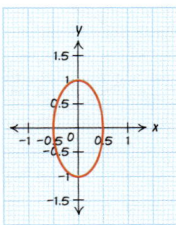

23. $\dfrac{x^2}{4} + \dfrac{y^2}{100} = 1$

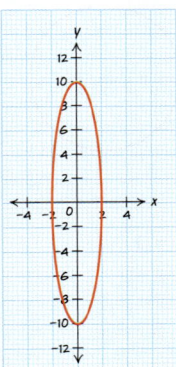

24. $\dfrac{x^2}{4} + \dfrac{y^2}{20} = 1$

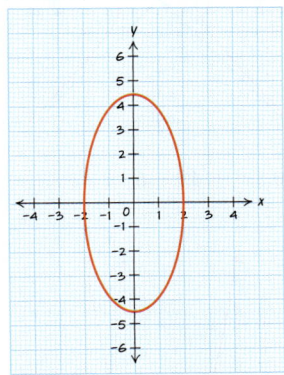

25. $\dfrac{x^2}{16} - \dfrac{y^2}{25} = 1$

Vertices $= (4, 0), (-4, 0)$

Foci $= \left(\sqrt{41}, 0\right), \left(-\sqrt{41}, 0\right) \approx (6.4, 0), (-6.4, 0)$

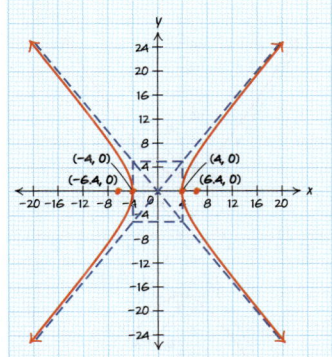

26. $\dfrac{y^2}{100} - \dfrac{x^2}{4} = 1$

Vertices $= (0, 10), (0, -10)$

Foci $= \left(0, \sqrt{104}\right), \left(0, -\sqrt{104}\right) \approx (0, 10.2), (0, -10.2)$

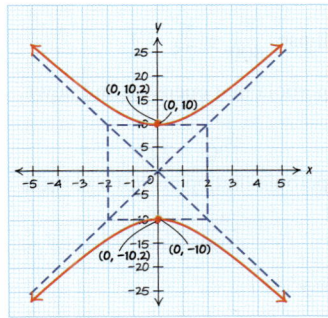

27. $\dfrac{x^2}{16} - y^2 = 1$

Vertices $= (4, 0), (-4, 0)$

Foci $= \left(\sqrt{17}, 0\right), \left(-\sqrt{17}, 0\right) \approx (4.1, 0), (-4.1, 0)$

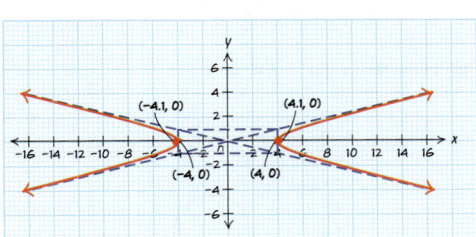

28. $\dfrac{y^2}{25} - \dfrac{x^2}{9} = 1$

Vertices = $(0, 5), (0, -5)$

Foci = $(0, \sqrt{34}), (0, -\sqrt{34}) \approx (0, 5.8), (0, -5.8)$

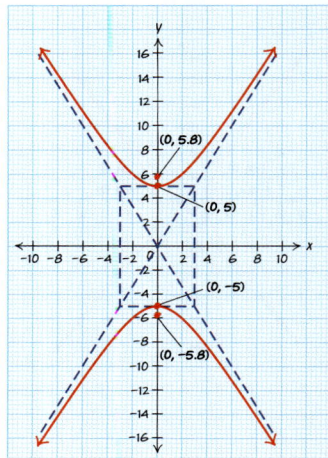

29. This is an ellipse. $\dfrac{x^2}{100} + \dfrac{y^2}{225} = 1$

30. This is an ellipse. $\dfrac{x^2}{400} + \dfrac{y^2}{25} = 1$

31. This is a hyperbola that opens up and down.

$$\dfrac{y^2}{100} - \dfrac{x^2}{225} = 1$$

32. This is a hyperbola that opens left and right.

$$\dfrac{x^2}{100} - \dfrac{y^2}{64} = 1$$

33. Ellipse; $\dfrac{x^2}{9} + y^2 = 1$

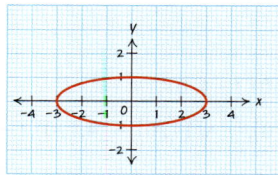

34. Circle; $(x - 9)^2 + (y - 6)^2 = 9$

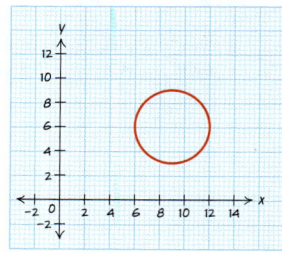

35. Hyperbola; $\dfrac{y^2}{25} - x^2 = 1$

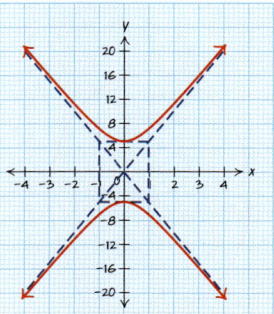

36. Parabola; $x = \dfrac{2}{3}y^2 - 2y + 7$

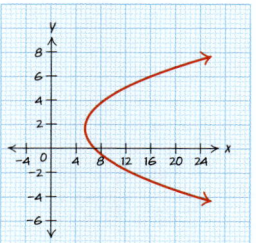

37. Circle; $(x + 3)^2 + y^2 = 9$

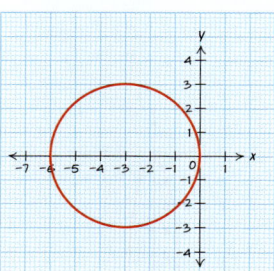

38. Hyperbola; $\dfrac{x^2}{36} - \dfrac{y^2}{9} = 1$

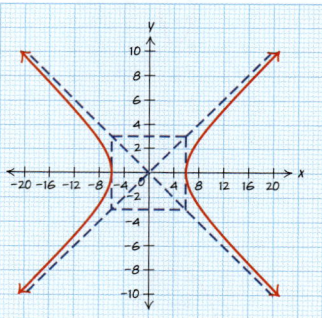

39. Parabola; $y = 0.5x^2 + 4x - 6$

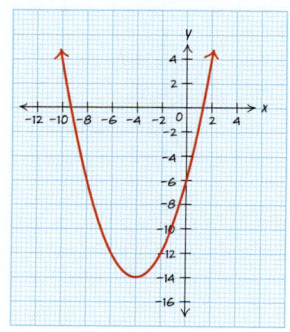

40. Ellipse; $\dfrac{x^2}{49} + y^2 = 1$

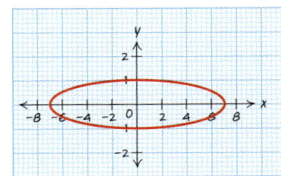

41. $-13, -6, 1, 8, 15$; $a_n = 7n - 20$

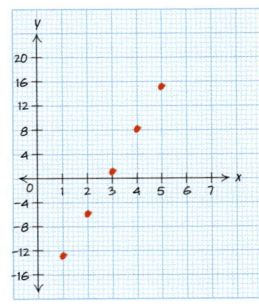

42. $6, 9, 13.5, 20.25, 30.375$; $a_n = 4(1.5)^n$

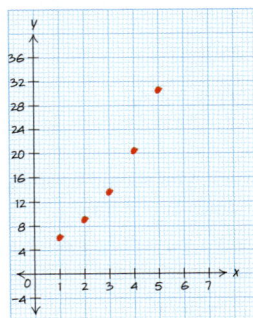

43. $-\dfrac{1}{8}, \dfrac{1}{16}, -\dfrac{1}{24}, \dfrac{1}{32}, -\dfrac{1}{40}$; $a_n = \dfrac{(-1)^n}{8n}$

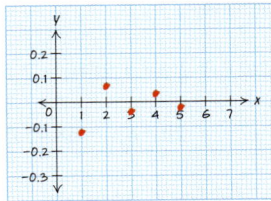

44. $0.22, 0.4, 0.55, 0.67, 0.77$; $a_n = \dfrac{2n}{n+8}$

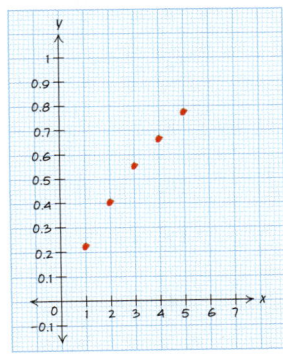

45. $n = 15$
46. $n = 9$
47. $n = 22$
48. $n = 66$
49. Arithmetic, common difference $d = -3$.
$$a_n = 85 - 3n$$
50. Arithmetic, common difference $d = 2.5$.
$$a_n = -6.5 + 2.5n$$
51. Not arithmetic.
52. Arithmetic, common difference $d = -\dfrac{5}{2}$.
$$a_n = \dfrac{17}{2} - \dfrac{5}{2}n$$
53. a. $a_n = 3800 + 200n$
   b. In 2015, the attendance on Easter Sunday at North Coast Church will be 6800.
   c. The attendance on Easter Sunday at North Coast Church will be 6000 in the year 2011.
54. a. $a_n = 29{,}500 - 2500n$
   b. The car will be worth \$14,500 in the sixth year.
   c. The car will be 11 years old when it is worth \$2000.
55. $a_n = 6.5n + 33.5$
56. $a_n = -8n + 20$
57. Geometric, $a_n = 3(4)^{n-1}$
58. Geometric, $a_n = 36\left(\dfrac{1}{3}\right)^{n-1}$
59. Geometric, $a_n = 0.5(-2)^{n-1}$
60. Not geometric.
61. a. $a_n = 4000(1.02)^n$
   b. $a_8 = 4686.64$. After eight years, the product will cost \$4686.64.
62. a. $a_n = 42{,}000(1.0125)^{n-1}$
   b. During your fifth year of employment, your salary will be \$44,139.70.
   c. During your 16th year of employment, your salary will be \$50,603.
63. $a_n = 32(1.5)^{n-1}$
64. $a_n = 81\left(\dfrac{1}{3}\right)^{n-1}$
65. $\displaystyle\sum_{n=1}^{7}(4n-20) = (4(1)-20) + (4(2)-20) + (4(3)-20)$
$\phantom{\displaystyle\sum_{n=1}^{7}(4n-20) =} + (4(4)-20) + (4(5)-20)$
$\phantom{\displaystyle\sum_{n=1}^{7}(4n-20) =} + (4(6)-20) + (4(7)-20)$
$\phantom{\displaystyle\sum_{n=1}^{7}(4n-20) } = -28$
66. $\displaystyle\sum_{n=1}^{6}(5(-2)^n) = (5(-2)) + (5(-2)^2) + (5(-2)^3)$
$\phantom{\displaystyle\sum_{n=1}^{6}(5(-2)^n) =} + (5(-2)^4) + (5(-2)^5) + (5(-2)^6)$
$\phantom{\displaystyle\sum_{n=1}^{6}(5(-2)^n) } = 210$
67. a. $a_n = 37{,}100 + 900n$
   b. Your salary will be \$50,600 in the 15th year of employment.
   c. You will have earned a total of \$664,500 over a 15-year period of employment.

68. a. $a_n = 2 + 3n$
  b. On the 5th day, you will pick 17 apricots.
  c. You will pick a total of 343 ripe apricots during the first two weeks.

69. During a 15-year career, you will have earned a total of $688,226.05.

70. a. $a_n = 10{,}000(1.2)^{n-1}$
  b. During the seven-day event, 129,159 people attended the car show.

71. $S_{14} = 518$ **72.** $S_{23} = 6267.5$

73. $S_{11} = 10{,}106.52459$ **74.** $S_{20} = -15.99998$

Chapter 9 Test

1. $(x + 7)^2 + (y - 4)^2 = 64$

2. Focus = $(9, 0)$, directrix : $x = -9$.

3. $\dfrac{x^2}{100} - \dfrac{y^2}{36} = 1$

This is a hyperbola in standard form.

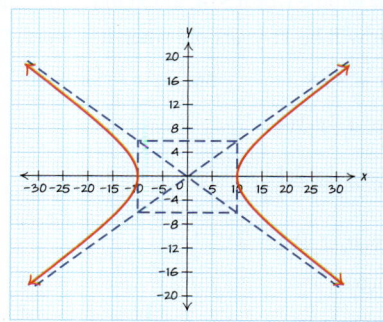

4. $x = -y^2 + 12$
This is a parabola.

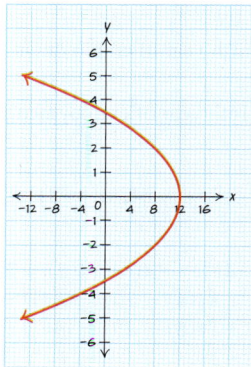

5. $\dfrac{x^2}{4} + \dfrac{y^2}{49} = 1$
This is an ellipse.

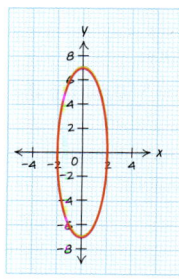

6. $(x - 3)^2 + (y - 7)^2 = 81$
This is a circle in standard form.

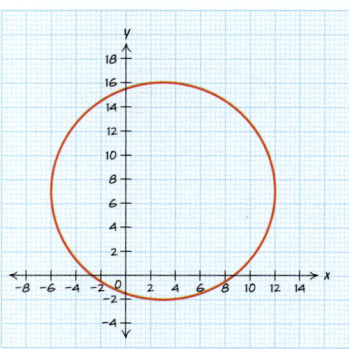

7. $x = \dfrac{y^2}{-8}$ **8.** $\dfrac{x^2}{4} + \dfrac{y^2}{36} = 1$

9. An arithmetic sequence has a common difference; therefore, the difference between successive terms is constant. A geometric sequence has a common ratio; therefore, the ratio between successive terms is constant.

10. $-12, 36, -108, 324, -972$.

11. $2, 3, \dfrac{18}{5}, 4, \dfrac{30}{7}$ **12.** $n = 23$ **13.** $n = 316$

14. Geometric, $a_n = 250\left(\dfrac{1}{5}\right)^{n-1}$

15. Arithmetic, $a_n = 31.5 - 7.5n$

16. a. $a_n = 2 + 3n$
  b. By the third week, you should be stretching your injured knee for 11 minutes at one time.
  c. In the tenth week, you will be stretching your injured knee for half an hour at one time.

17. a. $a_n = 60{,}000(1.025)^{n-1}$
  b. Your salary will be $95,919.01 in the 20th year.
  c. You will have earned a total of $1,532,679.46 in salary over a 20-year career.

18. $\displaystyle\sum_{n=1}^{6}(2n^2) = 2 + 2(2)^2 + 2(3)^2 + 2(4)^2 + 2(5)^2 + 2(6)^2 = 182$

19. $S_{15} = 7110$ **20.** $S_{12} = 38{,}453.56699$

Cumulative Review for Chapters 1–9

1. $a = 6$ **2.** $t = 3, t = -9$

3. $x = \dfrac{23}{2}$ **4.** $m = \dfrac{31}{3}$

5. $h = 14$ **6.** $n = 3$

7. $x = \dfrac{e^4 + 5}{2} \approx 29.799$ **8.** $b = -\dfrac{11}{2}, b = 5$

9. $h = 8, h = -\dfrac{24}{5}$ **10.** $x = -1$

11. $t = 4$ **12.** $n = 9$

13. $g = 4.5, g = -5.25$ **14.** $a = -\dfrac{33}{7}$

15. $x = \dfrac{41}{2}$ **16.** $m = -\dfrac{236}{3}$

17. $r = 0, r = -8, r = 5$ 18. $h = \frac{1}{3}, h = -\frac{5}{3}$

19. $x = -2$ 20. $t \approx \pm 3.98$

21. No solution.   22. $(0.5, 4)$

23. $\left(-\frac{3}{8}, \frac{125}{16}\right)$ and $(2, 9)$ 24. $\left(-\frac{7}{3}, -\frac{71}{9}\right)$ and $(5, -25)$

25. Let $P(n)$ be the profit in dollars from manufacturing and selling n custom MP3 players.

   a. $P(n) = \frac{-1}{250}(n - 1200)^2 + 4800$

   b. Domain: $[150, 2000]$, range: $[390, 4800]$.

   c. The profit from manufacturing and selling 1000 custom MP3 players is $4640.

   d. Vertex: $(1200, 4800)$. The profit from manufacturing and selling 1200 custom MP3 players is $4800. This is the maximum profit.

26. Let $R(t)$ be the primary revenue for the Wikimedia Foundation (in millions of dollars) t years since 2000.

   a. $R(t) = 0.003(2.62)^t$

   b. Domain: $[2, 10]$, range: $[0.021, 45.723]$

   c. The primary revenue for the Wikimedia Foundation was approximately $45.7 million in 2010.

   d. According to this model, the Wikimedia Foundation will reach $50 million in annual primary revenue in the year 2010.

27. Let $N(t)$ be the total circulation of U.S. Sunday newspapers (in millions) t years since 2000.

   a. $N(t) = -2.05t + 65.55$

   b. In 2010, the total circulation of U.S. Sunday newspapers was 45.05 million.

   c. The model predicts that the total circulation of U.S. Sunday newspapers will be only 25 million by the year 2020.

   d. Slope $= -2.05$. Each year, the total U.S. Sunday newspaper circulation decreases by 2.05 million.

   e. Domain: $[-5, 20]$, range: $[24.55, 75.8]$.

28. Domain: All real numbers except for -9 and 2.

29.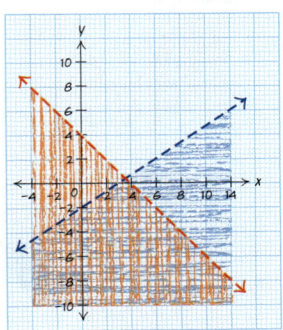

30. $a > -7$ 31. $n \geq \frac{79}{3}$

32. $h \leq \frac{1}{2}$ or $h \geq 4$ 33. All real numbers.

34. $x \leq 0.9$ or $x \geq 1.9$

35. a. In 2008, there were 398.7 thousand apprenticeship training registrations for all major trade groups in Canada.

   b. $(5.1, 165.98)$. In 1995, there were 165.98 thousand apprenticeship training registrations for all major trade groups in Canada. This is the least number of registrations during this time.

   c. In 1985 and again in 2005, there were 300 thousand apprenticeship training registrations for all major trade groups in Canada.

36. Let $B(h)$ be the number of bacteria present after h hours.

   a. $B(h) = 2000(2)^{3h}$

   b. After approximately 3 hours, there will be 1 million bacteria.

   c. There will 65,536,000 bacteria present after 5 hours.

37. $f^{-1}(x) = \frac{3}{2}x + 12$ 38. $g^{-1}(x) = \frac{\log\left(\frac{x}{4}\right)}{\log(3)}$

39. a. $f(3) = 2.5$

   b. Domain: $(-\infty, \infty)$, range: $\{2.5\}$.

40. a. $f(x) + g(x) = 5.4x - 4.3$

   b. $f(g(x)) = 6.8x - 1.6$

   c. $f(x)g(x) = 6.8x^2 - 18.4x - 18.9$

41. a. $f(2) - g(2) = 10.5$

   b. $(f \circ g)(7) = -69.25$

   c. $f(2)g(2) = 161.5$

42. a. $f(x) + g(x) = 3x^2 + 7x - 10$

   b. $f(g(x)) = 27x^2 - 60x + 26$

   c. $f(x)g(x) = 9x^3 - 34x + 24$

43. The intensity of this earthquake is 316.228 cm.

44. $y = -\frac{7}{3}x - \frac{29}{3}$ 45. $y > \frac{3}{5}x - 6$

46. $y = \frac{1}{4}x - 6$

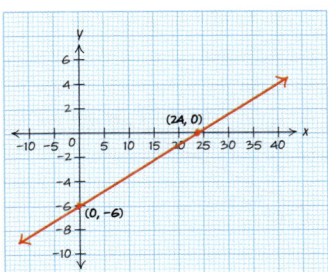

47. $y = -3x^2 - 24x + 2$

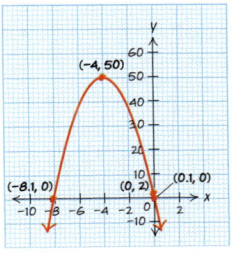

48. $y = -1.5(2.5)^x$

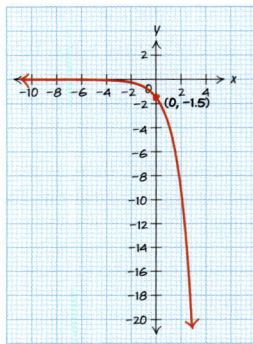

49. $y = \sqrt{x + 17}$

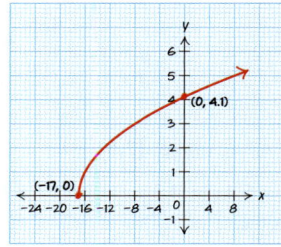

50. $y = \frac{1}{3}(x + 4)^2 - 10$

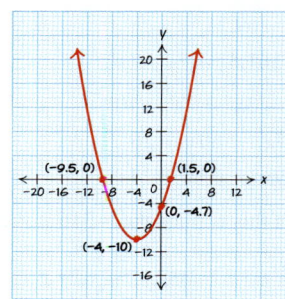

51. $5x - 30y = 180$

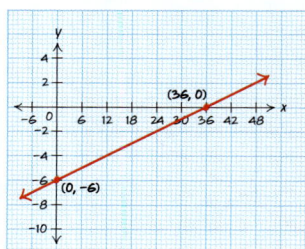

52. $y = 800(0.2)^x$

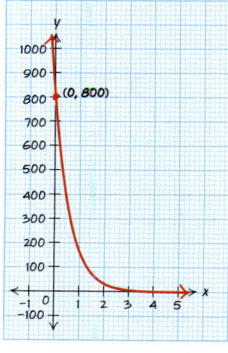

53. $y = \log_9 x$

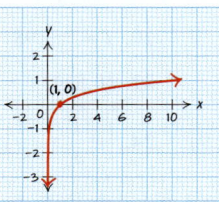

54. $y = \sqrt[3]{x + 9}$

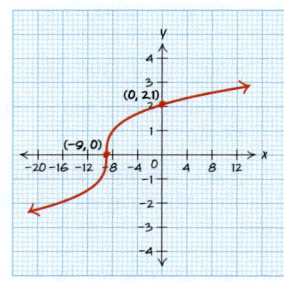

55. $y = 3(1.6)^x + 20$

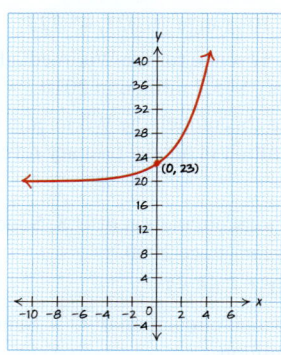

56. $y < x^2 - 5x - 14$

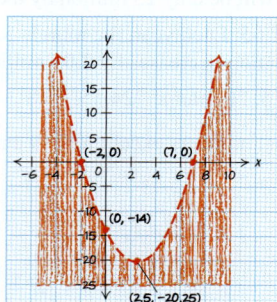

57. $y \geq 2x + 3$

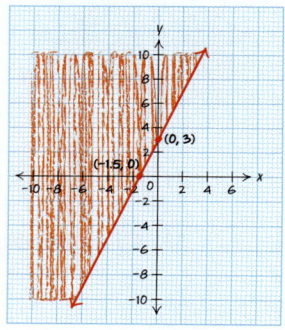

58. $5x - 6y < 6$

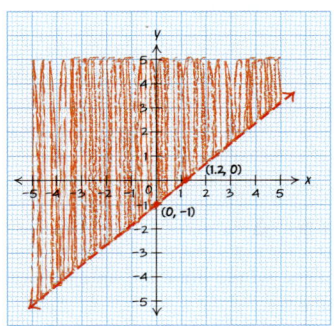

59. $y \geq -3(x - 1)^2 + 8$

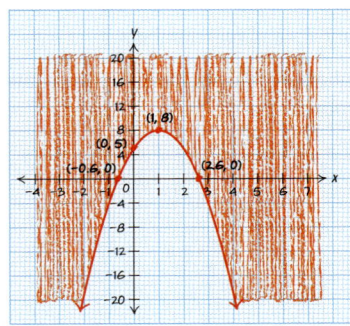

60. a. Let P be the profit in millions of dollars, and let t be the time in seconds that it takes to produce a product.

$$P = \frac{14.4}{t}$$

b. The profit is $1.44 million if it takes 10 seconds to produce each product.

61. $22\sqrt{3a} - 10\sqrt{a}$

62. $9m^3 + 11m^2 - 2m + 9$

63. $\frac{(x - 4)(x + 2)}{(x - 3)}$

64. $-14 + 12\sqrt{3w} + 6w$

65. $4a^2 + 27a + 35$

66. $25x^2 - 60x + 36$

67. $\frac{(3h - 4)(h + 7)}{(h - 2)(h + 2)}$

68. $\frac{5m^2 + 39m - 12}{(m + 7)(m - 3)}$

69. a. There will be 5260 graduating students at this college in 2012.

b. In 2012, 50% of the graduating students at this college will attend graduation ceremonies.

c. There will be 2630 graduating students attending graduation ceremonies at this college in 2012.

d. Let $A(t)$ be the number of graduating students attending graduation ceremonies at this college t years since 2000. $A(t) = 2924 + 31.3t - 4.65t^2$.

e. There will be approximately 2347 graduating students attending graduation ceremonies at this college in 2015.

70. $(2x + 7)(x - 4)$

71. $5(2a + 3b)(2a - 3b)$

72. $(h + 5)(h^2 - 5h + 25)$

73. $3m(7m + 2)(m - 8)$

74. $5(3x - 8)(4x - 7)$

75. Degree 3.

76. $f(x) = 4(x - 6)^2 - 18$

77. a. The profit for Beach Shack Rentals one year after it opened was $16.89 thousand.

b. After approximately 38 months, the profit for Beach Shack Rentals will be $25,000.

78. $x = -6$

79. Domain: $(-\infty, 10]$; range: $[0, \infty)$

80. The value of "a" is negative. The y-intercept is below the x-axis.

81. The value of "b" is less than 1. This is an increasing function with $a < 0$.

82. Domain: $(-\infty, \infty)$, range: $(-\infty, 0)$

83. The value of a is positive because the parabola opens upward.

84. The vertex of the parabola is $(-2, -16)$. Therefore $h = -2$ and $k = -16$.

85. $f(-4) = -12$

86. $x = -6$ and $x = 2$ when $f(x) = 0$.

87. Domain: $(-\infty, \infty)$, range: $[-16, \infty)$

88. The value of m is positive because the slope of the line is positive.

89. $b = -3$

90. $m = \frac{2}{5}$

91. $x = -5$ when $f(x) = -5$.

92. Domain: $(-\infty, \infty)$, range: $(-\infty, \infty)$

93. $6a^2b^3c$

94. $3xy\sqrt[3]{2y^2}$

95. $\frac{\sqrt{14ab}}{2b}$

96. $12 + 6\sqrt{3} + 2\sqrt{5} + \sqrt{15}$

97. $9 - 3i$

98. $-4 + 8i$

99. 97

100. $94 - 2i$

101. $\frac{5}{4} - \frac{3}{2}i$

102. $\frac{46}{29} + \frac{28}{29}i$

103. $\frac{6}{5} - \frac{8}{5}i$

104. $x = -2 \pm 5i$

105. $x = -6 \pm 2i, x = 0$

106. $x = \pm i\sqrt{70}$

107. $x = 6 \pm 2i$

108. $(x - 3)^2 + (y + 9)^2 = 144$

109. Focus $= (-11, 0)$, directrix: $x = 11$

110. Hyperbola. $\frac{x^2}{144} - \frac{y^2}{64} = 1$

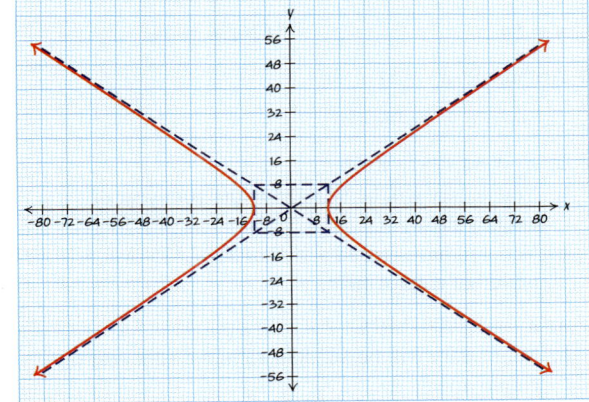

111. Ellipse. $\dfrac{x^2}{4} + \dfrac{y^2}{121} = 1$

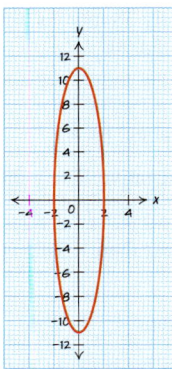

112. Parabola. $x = -y^2 - 2y + 10$

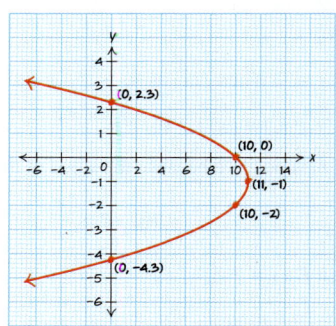

113. Circle. $(x + 8)^2 + (y - 3)^2 = 36$

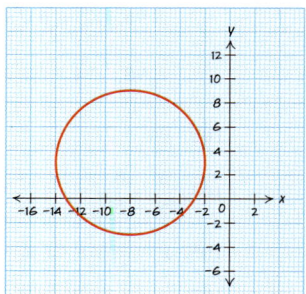

114. $y^2 = -24x$

115. $\dfrac{x^2}{144} + \dfrac{y^2}{25} = 1$

116. $a_n = 23 + 11n$

117. $-10, 20, -40, 80, -160$

118. $-\dfrac{3}{2}, -\dfrac{24}{11}, -\dfrac{18}{7}, -\dfrac{48}{17}, -3$

119. $n = 84$

120. $n = 44$

121. Arithmetic, $a_n = 51 - 3n$

122. Geometric, $a_n = 96(4)^{n-1}$

123. a. $a_n = 54{,}175 + 825n$
   b. Your salary will be $62,425 in the 10th year of employment.
   c. You will have earned a total of $587,125 over a 10-year period of employment.

124. a. $a_n = 55{,}000(1.015)^{n-1}$
   b. Your salary will be $62,886.45 in the 10th year of employment.
   c. You will have earned a total of $588,649.69 over a 10-year period of employment.

125. $\sum\limits_{n=1}^{6}(2n^3) = 2 + 2(2)^3 + 2(3)^3 + 2(4)^3 + 2(5)^3 + 2(6)^3 = 882$

126. $S_{15} = 495$

127. $S_{12} \approx 28{,}987.51066$

APPENDIX A

1. Natural numbers, whole numbers, integers, rational numbers, and real numbers.

3. Whole numbers, integers, rational numbers, and real numbers.

5. Irrational numbers and real numbers.

7. Rational numbers and real numbers.

9. Natural numbers, whole numbers, integers, rational numbers, and real numbers.

11. – 19.

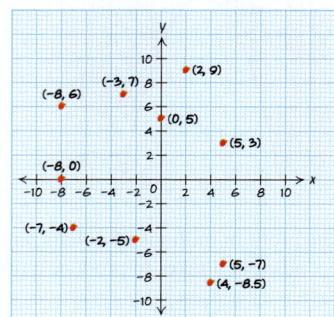

21. -18 23. -14 25. -2
27. -16 29. 20 31. -6
33. 4 35. 24 37. -17
39. 3 41. $\dfrac{7}{34}$ 43. $\dfrac{3}{2}$
45. $\dfrac{6}{35}$ 47. $\dfrac{41}{75}$ 49. $\dfrac{15}{28}$
51. $\dfrac{7}{40}$ 53. $\dfrac{4}{65}$ 55. -11
57. 110 59. 159 61. $\dfrac{25}{9}$
63. -2214 65. 8 67. $\dfrac{34}{5}$
69. -22 71. $x = 4$ 73. $x = \dfrac{27}{2}$
75. $m = 31$ 77. $h = 79$ 79. $k = 53$
81. $b = \dfrac{55}{4}$ 83. $w = \dfrac{225}{14}$ 85. $m = 18$
87. $x = -\dfrac{37}{2}$ 89. $274{,}000{,}000$
91. $-4{,}630{,}000{,}000{,}000$
93. 0.0000128
95. -0.00000000000096

97. 1.4×10^{10} **99.** -5.47×10^8
101. 7.8×10^{-8} **103.** -7.51×10^{-12} **105.** $[-6, 4]$
107. $(-5, 10]$ **109.** $[-8, \infty)$ **111.** $(4, \infty)$

APPENDIX B

1. a. The point $(x, y, z) = (0, 5, 1)$ is not a solution of the system.
  b. The point $(x, y, z) = (0, 5, -3)$ is not a solution of the system.
3. a. The point $(x, y, z) = (-5, 0, 1)$ is not a solution of the system.
  b. The point $(x, y, z) = (10, 0, -3)$ is a solution of the system.
5. a. $x = 16, y = 1, z = 0$ is not a solution of the system.
  b. $x = 0, y = 7, z = -8$ is a solution of the system.
7. $(x, y, z) = (4, 0, 1)$ **9.** $(x, y, z) = (-2, 2, 3)$
11. $(x, y, z) = (1, -1, 2)$ **13.** $(x, y, z) = (1, -2, 3)$

15. a. This is a 2×2 matrix.
  b. -1 is in the $(1, 2)$ position.
17. a. This is a 2×3 matrix.
  b. -2 is in the $(1, 3)$ position.
19. a. This is a 4×3 matrix.
  b. -3 is in the $(4, 1)$ position.

21. $\begin{bmatrix} 2 & -1 & -5 \\ 4 & 6 & -1 \end{bmatrix}$ **23.** $\begin{bmatrix} 1 & -1 & 0 \\ -3 & 4 & -1 \end{bmatrix}$

25. $\begin{bmatrix} -2 & 3 & -1 & 0 \\ 1 & 1 & -1 & 5 \\ 1 & -1 & 1 & 2 \end{bmatrix}$ **27.** $\begin{bmatrix} -1 & 1 & 1 & 3 \\ 0 & 3 & -2 & 1 \\ -1 & 0 & 2 & 0 \end{bmatrix}$

29. $(x, y) = \left(\dfrac{7}{3}, -2\right)$

31. $(x, y) = (-2, -5)$

33. $(x, y, z) = \left(\dfrac{9}{2}, \dfrac{5}{2}, -2\right)$

35. $(x, y, z) = \left(\dfrac{23}{4}, -\dfrac{1}{2}, -5\right)$

37. $\begin{bmatrix} 1 & -1 & -5 \\ 0 & -1 & -3 \end{bmatrix}$

39. $\begin{bmatrix} 4 & 3 & 7 \\ 0 & 8 & 17 \end{bmatrix}$

41. $\begin{bmatrix} 1 & 1 & 1 & 2 \\ 0 & 1 & 0 & -1 \\ 0 & 0 & 2 & 4 \end{bmatrix}$

43. $\begin{bmatrix} -12 & 0 & -8 & -36 \\ 0 & -6 & 4 & 24 \\ 0 & 0 & 1 & 3 \end{bmatrix}$

45. $(x, y) = (-2, 5)$
47. $(x, y) = (-3, 6)$
49. $(x, y, z) = (3, 0, -1)$
51. $(x, y, z) = (-1, 0, -3)$
53. $(x, y, z) = (4, 6, -1)$
55. $f(x) = -x^2 + 4x - 1$
57. $f(x) = x^2 + 5x + 4$
59. $f(x) = 2x^2 + 5x - 3$
61. $f(x) = -3x^2 + 10x + 8$

Index

A

Absolute value, 175–176, 207
Absolute value equations, 176, 207
   solving, 176–178, 207
Absolute value inequalities
   greater than/greater than equal to, 182–185, 207, 208
   less than/less than equal to, 179–182, 207, 208
   solving, 180–182, 183–185, 207–208
AC method of factoring, 262–267, 273, 278, 283–284, 351, 569
Addition
   complex numbers, 657, 668
   fractions, A-5
   functions, 236–240, 282, 283
   integers, A-3–A-4
   radical expressions, 632–633, 665
   rational expressions, 584–586, 604
Addition property of equality, 216, 407, 545, 676
Algebraic solutions
   combining functions, 239–240, 242–243, 254
   linear inequalities, 164–166
   quadratic equations, 370–371, 395, 397
   radical models, 645–647
Allometry, 613
Aphelion, 708
Appollonius, 684
Arithmetic sequences, 722, 725, 752–753
   common difference, 722–723, 730, 752
   general term, finding, 723–725, 726, 752
   general term formula, 723–725, 752
   graphing, 720–721
   summing, 741–743, 753, 754
Arithmetic series, 741–743, 753, 754
Associative property, 223
Asymptotes, 440–441, 443, 506, 508, 560–563, 711
Augmented matrices, B-5–B-6
Axes
   ellipse, 704
   graph, 688
Axis of symmetry, 306–307, 390, 687

B

Back-solving, B-6–B-7
Bacterial growth, 414–415, 418–420
Bases
   exponential functions, 415, 472, 506, 508, 518
   exponents, 222, 227, 229
   logarithmic functions, 497–498, 499–500, 506–509, 518, 537

Best-fit line, 18–21, 113
   graphing calculators, 83, 84, 86, 326–327, 329
   quadratic models, 298, 323–324, 326–327, 329
Binomials, 235, 273–275
Bombelli, 655
Brackets, 179–180, 183
Break (graph), 17
Break-even point, 3, 4

C

Canceling out, 567
Cardano, Geronimo, 655
Cell phone cost problems, 72, 214
Center (circle), 696
Change of base formula (logarithmic functions), 499–500, 518, 537
Circles, 696
   equation for, 696–697, 698–700, 714, 748, 750, 751
   geometric definition, 696, 748
   graphing, 697–698, 750
   vertical line test, 685
Coefficients, 39, 234–235, 295
Comet orbits, 711, 759
Common denominator. See Least common denominator (LCD)
Common difference (arithmetic sequences), 722–723, 730, 752
Common logarithms, 498, 500, 537
Common ratio (geometric sequences), 729–730, 731–732
Commutative property, 223, 254
Completing the square, 339–343, 366, 367–368, 392, 545, 676
   converting to vertex form, 343–344, 392, 393
   equations for circles, 698–699
Complex conjugates, 659–660
Complex fractions, 587
Complex numbers, 656–657
   addition/subtraction, 657, 668
   calculating, C-3–C-4
   conjugates, 659–660, 661–662, 668
   division, 660–661, 668
   multiplication, 658–660, 668
   rationalizing the denominator, 660
   simplifying, 658–660
   unreal solutions, 642
Complex numbers equations, 661–662
Composition of functions, 251–256, 283
   inverses, 490, 492
Compound inequalities, 179
Compound interest, 466–469, 475–476, 523

Conic sections, 684–685
   circles, 684–685, 696–700, 714, 748, 750
   ellipses, 683, 684–685, 703–709, 714, 750, 751
   general equation, 713–714, 751
   hyperbolas, 684–685, 709–713, 714, 750, 751–752
   parabolas, 683, 684–696, 714, 748
Conjugates, 641
   complex numbers, 659–660
   quadratic formula, 641, 661–662, 668
   radical expressions, 641–642, 659, 660, 666
Consistent systems, 137, 138–139, 149–150, 203
Constant terms, 46, 234, 441–442
   exponential functions, 439–443, 446, 472, 473, 518
   logarithmic functions, 497–498, 499–500, 518
   parabolas, 294–295, 304–306, 375, 390, 687
   quadratic functions, 294–295, 304–306, 375, 390
Cost, 3–4
Crichton, Michael, 414
Cube roots. See also nth roots; Radical equations; Radical expressions
   exponents and, 230–231
   perfect cubes, 630
   rationalizing the denominator, 639–640
Cubes
   perfect, 275, 630
   sum/difference of, 274–275, 276, 278, 279, 285

D

Data, 16
   collection, 15, 125, 126, 214, 404–405, 481, 610, 674–675
   comparing, 129–130
   graphing calculators, 81
   linearly related, 18, 43–45, 125, 126
   manipulation/conversion, 17–18
   in tables, 15–16
Decay problems
   cooling, 480
   half-life, 420–422
   light intensity, 422–423, 556–557, 609
   population, 475
Decay rate, 464, 475. See also Exponential decay
Decay. See Exponential decay
Decreasing pattern, 420
Degree
   polynomial, 236, 282
   term, 236, 282, 339

I-1

Dependent lines, 137, 138–139, 203
   substitution method, 149–150, 204
Dependent variables, 16, 113. See also
   Ranges
Depreciation, 65, 725
Depreciation problems, 65–66, 71, 725, 753
Difference of two cubes, 274–275, 276,
   278, 279, 285
Difference of two squares, 273–274, 276,
   278, 284–285
Direct variation, 554, 555, 601
Directrix, parabola, 691
Discriminants, 371
Distance formula, 691, 696
Distributive property, 216, 407, 545, 676
   in polynomials, 242, 282
   in subtraction, 5
Dividends, 571
Division
   by zero, 59
   complex numbers, 660–661, 668
   fractions, A-5–A-6
   functions, 240–243, 282, 283
   graphing calculator, 521
   integers, A-4–A-5
   long division, 570–573, 602
   polynomials, 570–574, 602–603
   radical expressions, 637–641, 666
   rational expressions, 578–580, 602–603
   synthetic division, 573–574, 602, 603
Divisors, 571
Domain values. See Independent variables
Domains, 24–27, 113
   exponential functions/models, 443–445,
      454–455
   in functions, 104–105, 119, 312–315,
      327, 329–330, 390
   inverse functions, 486
   logarithmic functions, 501, 509–510, 538
   radical functions, 614, 617–621
   rational functions, 553–554, 560–563,
      601, 602
   relations, 93–94
   sequences, 718

E

e, 468–469
   natural logarithms, 498, 499
Earthquake intensity problems, 529–530
Elimination method, linear equations, 144,
   155–160, 205, 216, 407, 545, 676
Ellipses, 683, 703–704, 708–709, 711
   equation for, 704–708, 714, 750, 751
   geometric definition, 704–705, 750
   graphing, 705–707, 751
   vertical line test, 685
Encryption, 221, 290
Endpoints (interval), 180
Engineering problems
   flashlight design, 695–696
   satellite dish design, 696
   tolerances, 182
   turn radius, 673

Equations, systems of, 130, 203. See also
   Specific types of systems of equations
Equations. See Specific types of equations
Equation-solving tools, 216, 407, 545,
   676–677, A-8–A-10
Euler, Leonhard, 655
Exercise problems, 556
Exponential decay, 420–423, 472, 475, 551
Exponential equations, 436
   compound interest, 466–469, 475–476,
      523–525
   finding, 451–452
   systems of, 451–452
Exponential equations, solving
   inspection, 431–432, 472–473
   trial and error, 431, 432–434
   using logarithms, 517–522, 532,
      539–540, 545, 676, 734
Exponential form, 500–501
Exponential functions, 415, 439, 446, 472,
   489, 496–497, 502–503. See also
   Geometric sequences
   domains/ranges, 443–445, 454–455,
      730–731
   exponential decay, 420–423
   exponential growth, 413–420, 421
   graphing, 439–443, 446–447, 473–474,
      506, 508
Exponential functions, inverse. See Logarithmic functions
Exponential growth, 413–420, 421,
   462–465, 472, 475
Exponential models
   domains/ranges, 454–455
   exponential decay, 420–423, 466, 475
   exponential growth, 415–420, 434,
      463–465
   solving, 519–521, 522–523
Exponential models, finding, 452–454,
   474–475, 519–520, 546, 677
   checking, 417
   exponential decay, 420–423, 466
   exponential growth, 415–420, 434,
      463–465, 474–475
   pattern recognition, 415–425, 472
   trial and error, 434
   vertical intercepts, 455–457
Exponents, 222. See also Logarithmic
   functions
   in like terms, 237
   negative, 227–228, 229–230, 281, 431, 517
   order of operations, 226
   power rule, 224, 281, 431, 517
   powers of products and quotients rules,
      224–225, 281, 431, 517, 630–631
   product rule, 222–223, 281, 431, 517
   quotient rule, 223–224, 281, 431, 517,
      570
   rational, 230–232, 281, 431, 517, 631
   zero, 228–229, 281, 431, 517
Extraneous solutions, 650, 662
Eyeball best-fit line, 18–21, 113
   graphing calculators, 83, 84, 86,
      326–327, 329

   quadratic models, 298, 323–324,
      326–327, 329

F

Factoring polynomials, 260
   AC method, 262–267, 273, 278,
      283–284, 351, 569
   by grouping, 261–262, 278, 367
   difference of two squares, 273–274, 276,
      278, 284–285
   difference/sum of two cubes, 274–275,
      276, 278, 279, 285
   GCF, 260–261, 265, 278
   multistep, 275–277
   pattern recognition, 272–277, 278–279,
      284–285
   perfect square trinomials, 272–273, 278,
      283
   prime polynomials and, 269–270
   quadratic form, 277, 278, 279, 285
   quadratics, solving, 349–354, 366, 367,
      368, 394–395, 407, 545, 676
   quadratics and, 262–267
   rational expressions, 567–570
   trial and error method, 267–269, 278
Factorization. See Factoring polynomials
Factors, 260
   GCF, 260–261, 265, 278
   zero factor, 350, 354–355
Fibonacci sequence, 759
Finite sequences, 718, 752
Finite series, 739
First-degree term, 339
Fixed costs, 3–4
Foci
   ellipse, 703–704, 708
   parabola, 691
Foot-candles, 556
Formulas, 9. See also Quadratic formula
   change of base formula, 499–500,
      518, 537
   compound interest, 466
   distance formula, 691, 696
   general term, arithmetic sequence,
      723–725, 752
   general term, geometric sequence, 730,
      732–733, 753
   point-slope formula, 64, 67–68, 116, 723,
      726
Fractions, A-5–A-6. See also Rational functions
   complex, 587
   general form of a line, 54
   graphing calculators, 40
   LCD, 582, A-5
   in linear equations, 7–9
   rational exponents, 230–232, 281, 431
   rationalizing the denominator, 637–642,
      660, 666
   reciprocals, 227
   simplifying, 568, 637–642
Function notation, 99–100, 103
   combining functions, 237, 239

composition, 251
finding models, 100–104
Functions, 94, 119, 488. *See also* Polynomials; *Specific types of functions*
   horizontal line test, 488–490
   relations vs., 94–97
   undefined, 59, 104–105, 509, 533
   vertical line test, 96–99, 119, 684–685
Functions, combining, 236–239
   addition/subtraction, 236–240, 282, 283
   algebraically, 239–240, 242–243, 254
   composition, 251–256, 283, 490, 492
   function notation and, 239, 240–241
   multiplication/division, 240–243, 282, 283
   substitution, 251–253
Fundamental rectangle, 711

G

Gauss, Carl Friedrich, 655
General form (lines), 54–55, 115
   intercepts and, 57, 115–116
General terms, 718, 726
   arithmetic sequence formula, 723–725, 752
   finding, 723–725, 726, 730, 732–733, 735–736
   geometric sequence formula, 730, 732–733, 753
Geometric growth. *See* Exponential growth
Geometric sequences, 729–730, 734, 753
   common ratio, 729–730, 731–732
   domain, 730–731, 753
   general term, finding, 730, 732–733, 735–736
   general term formula, 730, 732–733, 753
   graphing, 731
   summing, 743–746, 754
Geometric series, 743–746, 754
Geometry problems, 238
Graphical models
   linear, 18–27, 113
   model breakdown, 24, 55, 325, 329–330
   quadratic, 298–299
Graphical solutions
   linear equations, 130–134, 143–144, 203
   linear inequalities, 170–171, 180–182, 184–185, 195–199, 206, 208–209, 216
   quadratic equations, 295, 369, 396, 407
   radical models, 613–615
Graphing
   arithmetic sequences, 720–721
   circles, 697–698, 750
   ellipses, 705–707, 751
   exponential functions, 439–443, 446–447, 473–474, 506, 508
   geometric sequences, 731
   hyperbolas, 711–712, 751–752
   inverse functions, 491–492, 502
   logarithmic functions, 505–509
   parabolas, 688–690, 693–695, 748, 749
   radical functions, 617
   rational functions, 560–563
   scatterplots, 16–18, 113–114

Graphing, equations
   horizontal/vertical lines, 59–61
   intercepts, using, 57–58, 345–346
   plotting points, 35–39
   slope test, 43–45
   slope-intercept form, using, 48–49
   systems of, 130–134
Graphing, inequalities
   linear, 170–171, 189–195, 199, 208–209
   quadratic, 383–384, 397, 398
Graphing, quadratic functions, 295–299, 389
   standard form, 375–382, 397–398
   vertex form, 304–315, 345–346, 392–393
Graphing calculators
   absolute value, 176, C-3
   ANS feature, 744
   basic keys, C-1–C-2
   best-fit line, 83, 84, 86, 323–324, 326–327, 329
   calculations, C-2
   checking solutions, 132, 134–135
   complex numbers, C-3–C-4
   compound interest, 467
   converting decimals to fractions, C-3
   data entry, 81
   delete/undelete, 81
   display distortion in, 69, 97, 132, 492, 562
   division, 521
   division by zero, 59
   e, 469
   equations, 131–132, C-4
   error messages, 82, 562, C-8–C-9
   exponential functions, 415, 419
   exponents, 225, 292, 294, 614, 744
   fractions, 40
   functions, basic, C-6
   inequalities, 169–171, 199, C-8
   intersect feature, C-11
   linear equations, 36
   linear models, 83–87
   LN button, 499
   LOG button, 499
   logarithmic functions, 498, 499, 518
   minimum/maximum feature, C-10
   minus vs. negative button, 436
   negative numbers, 436
   nonfunctions, 99
   parentheses, 40, 419, 467, 521, 559, 614
   quadratic models, 298–299
   radical functions, 617
   rational functions, 559, 561–562
   regression equations, C-11–C-12
   scatterplots, 80–82, C-6–C-8
   scientific notation, 531
   slope, 39–40, 43
   slope-intercept form and, 46
   square/cube roots, 231, 435, 614
   STAT menu, 81–82, C-7–C-8
   stat plots, 134–135
   TABLE feature, 37, 38, 132, 135–136, 169–170, 468, 559, C-4–C-5

TRACE button, 132, 134, 299, C-6
vertex of a quadratic model, 299
window, setting, 69, 97, 132, 492, 562, C-5–C-6
zero feature, C-10
ZOOM menu, 69, 97, 492, C-9
Graphs. *See also* Scatterplots
   asymptotes, 440–441, 443, 506, 508, 560–563
   axes, 688
   best-fit lines, 18–21, 83, 84, 86, 113, 298, 323–329
   data manipulation in, 17–18
   domains, 24–27
   holes, 560, 562
   intercepts, 21–24, 26–27
   ranges, 24–27
   rectangular coordinate system, A-2–A-3
   scale in, 69, 97, 132, 308, 505
   vertical line test, 96–99, 119
Greatest common factor (GCF), 260–261, 265, 278
Greatest value. *See* Vertices
Gregory, James, 691
Growth problems
   age/disease incidence, 519–520
   bacteria, 418–419, 434, 472
   business growth, 474–475
   deer, 734
   gross domestic product, 465
   growth rate comparison, 614–615
   insects, 463–464
   loan sharks, 417, 435
   open source software, 452–454, 455
   population, 465–466, 523
   rabbits, 480
   rumors, 416–417
   sea otters, 464–465, 522
   tumors, 544
   weeds, 455–456
Growth rate, 462, 464, 475. *See also* Exponential growth
Growth. *See* Exponential growth

H

Half-life, 420, 421
Hedge fund, 51
Higher roots. *See* nth roots
Hole (rational functions), 560, 562
Horizontal asymptotes, 440–441, 443, 506, 508
Horizontal intercepts, 21–24, 113, 297
Horizontal line test, 488–490
Horizontal lines, 59–61, 490
   domain/range and, 105
Hyperbolas, 709–710, 711
   direction, 710–711
   equation for, 710, 713, 714, 750–751
   fundamental rectangle, 711
   geometric definition, 710, 750
   graphing, 711–712, 751–752
   vertical line test, 97, 685

I

i. See Imaginary numbers
Illumination, 556
Imaginary numbers, 655–656, 668. *See also* Complex numbers
    calculating, 658
Imaginary unit, 655, 668
Inconsistent systems, 137, 138, 149–151, 203, 204, 205
Increasing pattern, 420, 441
Independent lines, 137, 138, 150, 203
Independent variables, 16, 113. *See also* Domains
Index (roots), 612, 617–622
Infinite sequences, 718, 752
Infinity, 105
Input variables. *See* Domains; Independent variables
Integers, 54, 656, A-1–A-2, A-3–A-5
Intensity (earthquake), 529
Intercepts, 21–24, 26–27, 455
    constant terms, 46, 441–442
    finding, 55–57, 115–116, 297, 375
    in graphing, 57–58, 345–346
    interpreting, 55–57, 71–72
Interest
    compound, 466–469, 475–476, 523
    compounded continuously, 475, 523
    simple, 148
Interval notation, 179–180, 183, A-12–A-14
Intervals (absolute value), 179–180, 183
Inverse functions, 484–488, 502–503, 536–537
    composition of, 490, 492
    domains/ranges, 486
    graphing, 491–492, 502
    horizontal line test, 488–490
Inverse notation, 486
Inverse variation, 554–555, 556–558, 601
Investment problems, 148–149
    comparisons, 197–198
    compound interest, 467–468, 469, 475–476, 523–525
Irrational numbers, 468–469, 656, A-1–A-2

K

Kepler, Johannes, 684
Kepler's Laws of Planetary Motion, 673, 684, 708

L

Least common denominator (LCD), 581–584, 587, A-5
    general form of a line, 54–55
Least value. *See* Vertices
Like radical expressions, 632
Like terms, 237, 282
Linear equations, 46. *See also* Systems of linear equations
    absolute value in, 176–178
    coefficients, 39
    constant terms, 46
    function notation, 99–100
    general form, 54–55, 57, 115
    literal equations, 9–10
    parallel, 69–70, 117
    perpendicular, 69–70, 117
    point-slope formula, 64, 116
    slope, 39–43, 46–48, 114–115
    slope-intercept form, 41, 46–47, 64, 103, 114, 116
    symbols, 2
    systems of, 130
    variables, 6–7, 16
Linear equations, finding
    point-slope formula, 64, 67–68, 116
    slope-intercept form, 64–66, 68–69, 116
Linear equations, graphing
    horizontal/vertical lines, 59–61
    intercepts, 55–58
    plotting points, 35–39
    slope, 41–42
    slope-intercept form, 48–49
Linear equations, solving, 2–10, 112–113, 532
    addition property of equality, 216, 407
    distributive property, 5, 216, 407
    fractions, 7–9
    isolating variables, 3
    multiplication property of equality, 216, 407, 545
    reasonable solutions, 6, 7
    toolbox summary, 216
    variables, defining, 6–7
Linear functions, 292–293. *See also* Arithmetic sequences; Functions; Polynomials
    domains/ranges, 104–105, 119
    inverse, 488, 489, 490, 496
Linear inequalities, 163–164, 179
    graphing, 170–171, 189–195, 199, 208–209
    intervals, 179–180, 183
Linear inequalities, solving, 206
    absolute value and, 180–182, 183–185
    algebraically, 164–166
    graphically, 180–182, 184–185, 206, 208–209, 216
    negative numbers and, 163–164
    numerically, 206, 216
Linear models, 133–134, 145–149
    eyeball best-fit line, 18–21, 83, 84, 86, 113
    finding, 83–88, 117–118, 216, 407, 546, 677
    finding with function notation, 100–104
    predictions, 21, 87, 102
    scatterplots and, 16–21
    systems of inequalities, 167–169
    using graphing calculators, 83–88
Literal equations, 9–10
Logarithm form, 500–501
Logarithmic equations, 528–533, 537–538, 540–541, 545, 676
Logarithmic functions, 497, 502–503, 537
    change of base formula, 499–500, 518, 537
    domains/ranges, 501, 509–510, 538
    properties, 497–498, 512–514, 517, 539
    undefined, 509, 533
Logarithmic functions, graphing, 505–506, 538
    base and, 506–509
    radical functions vs., 614
Logarithmic functions, inverse. *See* Exponential functions
Logarithmic functions, solving, 497–498, 503, 537–538
    using properties, 498, 512–516, 517–523
Logarithmic models, 529–531
Logarithms. *See* Logarithmic functions
Long division, polynomials, 570–573, 602
    synthetic division, 573–574, 602, 603
Lumen, 422

M

Magnitude (earthquake), 483, 529–530
Major axis, ellipse, 704
Matrices, B-4–B-5
    augmented, B-5–B-6
    back-solving, B-6–B-7
    linear equations, B-1–B-12
    quadratic equations, B-12–B-14
    row operations, B-7–B-9
Maximum points. *See* Vertices
Mean distance (planet to sun), 708
Menaechmus, 684
Minimum points. *See* Vertices
Minor axis, ellipse, 704
Mixture problems, 156–157
Model breakdown, 24, 55
    in applications, 104, 391, 454
    in exponential models, 454
    quadratic models, 325, 329–330, 371
    reasonable solutions, 6, 7
Models, finding, 216, 407, 546, 677. *See also* Best-fit line
    exponential models, 415–425, 434, 452–457, 463–466, 472, 474–475, 519–520, 546, 677
    linear models, 83–88, 117–118, 216, 407, 546, 677
    quadratic models, 321–330, 391–392, 407, 546, 677
    regression, C-11–C-12
Molarity, 530
Monomials, 235
Multiplication
    complex numbers, 658–660, 668
    fractions, A-5–A-6
    functions, 240–243, 282, 283
    integers, A-4
    radical expressions, 635–636, 666
    rational expressions, 577–578, 603
Multiplication property of equality, 216, 407, 545

N

Natural logarithms, 498, 500, 537
Natural numbers, 656, 718, A-1–A-2
Negative exponents, 227–228, 229–230, 281, 431, 517
Negative numbers, 163–164, 436
Negative radicands, 665
   imaginary numbers, 655–656, 668
   nonreal solutions, 335, 371, 392, 395, 629–630, 649–650
Negative roots, 630, 665
Newton, Isaac, 691
Nonnegative numbers, 618
Normal range problems, 185
nth roots, 612, 629, 630, 664. *See also* Radical equations; Radical expressions; Square roots
   graphing, 617–618
   negative, 630, 665
   rationalizing the denominator, 639–641, 666
Numerical solutions
   linear equations, 130, 135–136, 143–144, 203, 216
   linear inequalities, 169–170, 206, 216
   quadratic equations, 369, 395, 407

O

Ohm's Law, 599
One-to-one functions, 488–490, 536
Order of operations, 226, 436, 521, 733, A-6–A-7
Ordered lists, 717
Ordered pairs, 94, A-3
Origin, A-2–A-3
Output variables. *See* Dependent variables; Ranges

P

π, 468
Pappus of Alexandria, 691
Parabolas, 289, 293–294, 335, 683–686, 691, 695–696, 711. *See also* Quadratic functions
   direction in, 305–306, 390, 686–687
   equations for, 688, 691–692, 695, 714, 748, 749
   geometric definition, 691, 748
   graphing, 688–690, 693–695, 748, 749
   horizontal line test, 489
   symmetry, 306–307, 687
   vertical line test, 96–97, 685
   width in, 304–305, 306, 390
Parallel lines, 69, 70, 117
   inconsistent systems, 137, 138, 149–151, 203, 204, 205
Parentheses
   graphing calculators, 40, 419, 467, 521, 559, 614
   interval notation, 179–180, 183

Pattern recognition factoring, 272–277, 278–279
PEMDAS, 226, 436, 521, 733, A-6–A-7
Percentage points, 48
Percentages, 48
Perfect cubes, 275, 630
Perfect powers, 630
Perfect square trinomials, 272–273, 278, 283
Perfect squares, 273, 630
Perihelion, 708
Perpendicular lines, 69, 70, 117
PH, 530
pH problems, 530–531
Physics problems
   airspeed/mass, 613–614
   comet orbits, 759
   distance to horizon, 664, 667
   force/acceleration, 557, 593
   light decay, 422–423, 556–557, 609
   pendulum, 645–646
   planetary orbits, 709
   planetary periods, 673–674
   speed of sound, 646–647
   speed/acceleration, 405
Planetary orbits, 708–709
Point-slope formula, 64
   arithmetic sequences, 723, 726
   finding an equation, 64, 67–68, 116
Polynomials, 234–236, 282, 339
   distributive property, 242, 282
   division, 570–574, 602–603
   LCD in, 581–584, 587
   prime, 269–270
Polynomials, solving. *See* Factoring polynomials
Positive numbers, 618
Power equations, 435–436, 472–473
Power property for logarithms, 514, 517, 539
Power rule for exponents, 224, 281, 431, 517
Powers of products and quotients for exponents, 224–225, 281, 431, 517
   radical expressions and, 630–631
Powers of products and quotients rules for exponents, 224–225, 281, 431, 517, 630–631
Prime numbers, 269
Prime polynomials, 269–270
Principal nth root, 630
Principal root, 630, 665
Processes, 251
Product property
   for logarithms, 512–513, 517, 539
   of radicals, 630–631, 665
   of zero, 350
Product rule for exponents, 222–223, 281, 431, 517
Production problems
   appliances, 196–197
   bicycles, 190–191, 195–196
   cameras, 241–242

   fruit, 404
   orange juice, 252–253
   planes, 353
   SUVs, 242
Profit, 3, 4, 237
Properties. *See also* Square root property
   addition property of equality, 216, 407, 545, 676
   associative property, 223
   commutative property, 223, 254
   distributive property, 5, 216, 242, 282, 407, 545, 676
   multiplication property of equality, 216, 407, 545
   power property for logarithms, 514, 517, 539
   product property for logarithms, 512–513, 517, 539
   product property of radicals, 630–631, 665
   product property of zero, 350
   quotient property for logarithms, 512–513, 517, 539
   quotient property of radicals, 637
   zero factor property, 350, 354–355
Pythagorean Theorem, 705

Q

Quadrants, A-2–A-3
Quadratic, 293
Quadratic equations, 335. *See also* Systems of quadratic equations
   finding, 354–358, 394, 395
Quadratic equations, solving
   completing the square, 339–343, 366, 367–368, 392–393, 407, 545, 676
   factoring, 349–354, 366, 367, 368, 394–395, 407, 545, 676
   nonreal solutions, 335, 371, 392, 395
   quadratic formula, 362–365, 366–367, 368–369, 371, 395–396, 407, 545, 661–662, 668, 676
   square root property, 335–339, 366, 368, 392–393, 407, 545, 676
   standard form, 349–358
   vertex form, 335–343
Quadratic expressions, 335
Quadratic form, trinomials, 277, 285
Quadratic formula, 362–363
   complex numbers and, 661–662, 668
   conjugates, 641, 661–662, 668
   using, 363–365, 366–367, 368–369, 395–396, 407, 545, 676
Quadratic functions, 262, 292–295, 293, 335. *See also* Functions; Parabolas
   domains/ranges, 312–315, 325–330, 390
   graphing, 289, 295–299, 304–315, 345–346, 375–382, 389–391, 397–398
   inverse, 489, 496
   standard form, 293, 294–295, 349–358, 375–382, 389, 397–398
   vertex form, 293, 294, 304–315, 343–346, 375, 389, 390–391, 392–393

Quadratic inequalities, 383–384, 397, 398
Quadratic models
  best-fit lines, 298, 323–324, 326–327, 329
  domain/range, 390, 391
  finding, 321–330, 391–392, 407, 546, 677
  model breakdown, 325, 329–330, 371
  predictions, 325–330, 391
  using graphing calculators, 298–299, 321–330
Quotient property
  for logarithms, 512–513, 517, 539
  of radicals, 637
Quotient rule for exponents, 223–224, 281, 431, 517, 570

R

Radical, 230–231, 281, 612, 664
Radical equations
  extraneous solutions, 649–650, 662
  nth roots, 650–652, 666
  square roots, 644–650, 666–667, 677
Radical expressions, 612, 630
  addition/subtraction, 632–633, 665
  conjugates, 641–642, 659, 660, 666
  division, 637–641, 666
  like, 632
  multiplication, 635–636, 666
  properties, 630–631, 637, 665
  rationalizing the denominator, 638–642, 660
Radical expressions, simplifying, 630–632, 637–642, 665
  complex numbers, 655–657
Radical functions, 611, 616–617, 664
  domain/range, 614, 617–621, 664–665
Radical functions, graphing, 617
  even vs. odd roots, 619–624
  logarithmic function vs., 614
Radical models, 612–615, 645–647
Radicands, 612, 665. See also Negative radicands
Radius, 696
Range values. See Dependent variables
Ranges, 24–27, 113
  exponential functions/models, 443–445, 454–455
  in functions, 104–105, 119, 312–315, 327, 329–330, 390
  inverse functions, 486
  logarithmic functions, 501, 509–510, 538
  radical functions, 617–621
  rational expressions, 553–554
  relations, 93–94
Rate of change. See Slope
Rational equations, 592–597, 604–605, 676
Rational exponents, 230–232, 281, 431, 517, 614
Rational expressions, 553, 601
  addition, 584–586, 604
  complex fractions, 587–588

division, 578–580, 602–603
  LCD in, 581–584, 587
  multiplication, 577–578, 603
  simplifying/reducing, 567–570, 577, 587–588, 602–603
  subtraction, 586–587, 604
Rational functions, 552–553, 601
  domain/range, 553–554, 560–563, 601, 602
  graphing, 560–563
Rational models, 555–560, 593, 594–595, 597
Rational numbers, 656, A-1–A-2. See also Fractions
Rationalizing the denominator, 638–641, 666
  using conjugates, 641–642, 660, 666
Ratios, 553
  geometric sequences, 729
Real numbers, 656, A-1–A-2
Rectangular coordinate system, A-2–A-3
Regression, C-11–C-12
Relations, 93–94, 119
  functions vs., 94–97
Rental cost problems, 2–3, 37–39, 485, 553–554, 593
  comparisons, 131–132, 166–167
Repair cost problems, 255, 555
  flooring, 125–126
Revenue, 3
Revenue/cost/profit problems, 4–6, 47–48, 56, 112–113
  amusement park, 381–382
  business trends, 85–88, 113–114, 118, 241, 251–253, 382, 474–475
  movie theater, 353–354
  photographer, 352–353
  radio play time, 536
  restaurant, 354
Richter scale, 529
Rise (change in y), 40, 41
Rounding, 3
Rules for exponents, 222–224, 281, 431, 517, 570, 630–631
Run (change in x), 40, 41

S

Salary growth rate, 462
Salary structure problems, 147–148, 725, 734, 742, 744
Satellite dish design, 691, 696
Scatter diagrams. See Scatterplots
Scattergrams. See Scatterplots
Scatterplots
  best-fit lines, 18–21, 83, 84, 86, 113, 298, 323–324, 326–327, 329
  creating, 16–18, 80–82, 113–114
Scientific notation, 419, 531, A-10–A-12
Second-degree term, 339
Sequences, 683, 717–720, 752, 759
  arithmetic, 722–726, 752–753
  domain, 718, 730–731

geometric, 729–736, 753
  graphing, 720–722, 731
Series, 683, 738–740, 753
  arithmetic, 741–743, 753, 754
  geometric, 743–746, 754
  summation notation, 740–741, 753
Set of numbers, 718
Simple interest, 148
Slope, 39–41, 46, 555
  arithmetic sequence, 722
  finding, 41–43, 114–115
  interpreting, 47–48, 71–72
  undefined, 70
Slope-intercept form (line), 41, 46–47, 114
  arithmetic sequence, 723, 726
  finding an equation, 64–66, 68–69, 116
  function notation vs., 99, 103
  graphing, 48–49
Solution set, 189–190
Solutions
  reasonable, 6, 7
  of systems of two linear equations, 130
Square root property, 335–336
  solving equations, 336–339, 366, 368, 392, 407, 532, 545, 676
  solving power equations, 435–436
Square roots, 629. See also Radical equations; Radical expressions
  exponents and, 230–232, 281, 431, 517
  perfect squares, 630
  rationalizing the denominator, 638–639, 666
Square roots, negative numbers
  imaginary numbers, 655–656, 668
  nonreal solutions, 335, 371, 392, 395, 629–630, 649–650
Squares
  difference of, 273–274, 276, 278, 284–285
  perfect, 273, 630
  perfect square trinomials, 272–273, 278, 283
  sum of, 273, 274, 285
Standard earthquake, 529
Standard equations
  circles, 696–697, 698–699, 748
  ellipses, 704, 706–707, 750
  hyperbolas, 710, 750–751
  parabolas, 692, 748
Standard form (quadratic function), 293, 294–295, 389
  converting, 343–344, 375, 392, 393
  graphing, 375–382, 397–398
  solving equations, 349–358
Statplots. See Scatterplots
Substitution method
  dependent lines, 149–150, 204
  linear equations, 144–151, 204–205, 216, 407, 545, 676
  linear functions, 251–253
Subtraction
  complex numbers, 657, 668
  distributive property, 5

fractions, A-5
functions, 236–240, 282, 283
integers, A-4
radical expressions, 632–633, 665
rational expressions, 586–587, 604
Sum of two cubes, 274–275, 276, 278, 279, 285
Sum of two squares, 273, 274, 285
Summation notation, 740–741, 753
Symbols. *See also* Function notation
   a_n, 718
   approximation, 2
   brackets (intervals), 179–180, 183
   composition of functions, 251
   dashed line (inequality), 190, 383
   e, 468–469, 498, 499
   equals sign, 2
   i, 655
   infinity, 105
   inverse notation, 486
   less than/greater than, 164
   parentheses (intervals), 179–180, 183
   radicals vs. exponents, 230–231, 281, 612
   solid line (inequality), 192
   summation notation, 740–741, 753
Symmetric points, 307, 322
Symmetry, 306
   in inverse functions, 491–492, 505–506
   in parabolas, 306–307, 687
Synthetic division, 573–574, 602, 603
Systems of equations, 130, 203
Systems of exponential equations, 451–452
Systems of linear equations
   three unknowns, B-1
   two unknowns, 130, 136–139, 203
Systems of linear equations, solving
   elimination method, 144, 155–160, 205, 216, 407, B-2–B-4
   graphically, 130–134, 143–144, 203
   matrices, B-4–B-12
   numerically, 130, 135–136, 143–144, 203, 216
   substitution method, 144–151, 204–205, 216, 407
   three unknowns, B-1–B-14
   toolbox summary, 216
Systems of linear inequalities, graphing, 189–195, 208–209
Systems of linear inequalities, solving, 166–169
   absolute value and, 180–182, 183–185
   graphically, 170–171, 180–182, 184–185, 195–199, 208–209, 216
   numerically, 169–170, 216
Systems of quadratic equations
   algebraically, 370–371, 395, 397
   graphically, 295, 369, 396, 407

matrices, B-12–B-14
numerically, 369, 395, 407

T

Tartaglia, Niccolo, 655
Terms (polynomials), 234–235, 282. *See also* Constant terms
   degree, 236, 282, 339
   like, 237, 282
Terms (sequences), 718, 720
Tolerance (machine parts), 182
Torricelli's law, 348
Travel problems, 178–179
Trend problems. *See also* Decay problems; Growth problems
   air pressure, 48, 56, 486
   cell phone subscribers, 313–314, 328–330
   charitable donations, 66, 71–72
   collections, 752–753
   college enrollment, 55–56, 169
   college enrollment/costs, 239
   disease rate/costs comparison, 146
   disease rates, 72–73
   employment/costs, 241, 251–252
   food consumption comparisons, 133–134, 145
   male/female athletes, 167–169
   Medicare enrollment, 559
   Medicare payments, 594–595
   mortality rates, 16–18, 21–22, 24–25, 55–56, 119, 133, 739–740
   population, 18, 21, 25, 80–82, 83–85, 101–102
   population/national debt, 240–241
   population/pollution, 253, 255–256
   public libraries, 325–327, 365
   revenue/cost/profit, 85–88, 113–114, 118, 382
   series, 742
   smoking rates, 47–48
   solar flares, 391–392
   store growth/contraction, 25–27, 67, 72
   street improvement costs, 283
   student/teacher ratio, 558–559
   temperature, 321–323
   TV cable systems, 323
   TV opinions, 206
   TV watching, 312–313
   unemployment rate, 365
   water use, 744
Trends, 1, 15–16, 129
Trial and error factoring, 267–269, 278
Trinomials, 235
   factoring, 272–273, 277
   quadratic form, 277
Turn radius, 611, 673

U

Undefined functions, 59, 104–105, 509, 533, 553, 560
Units, 7, 239
Upper triangular form, B-6, B-7–B-9

V

Variable costs, 3–4
Variables, 234. *See also* Domains; Ranges
   combining functions and, 237, 240
   defining, 6–7
   dependent/independent, 16
   isolating, 3
Variation, 554–555, 601
   models, 555–560
Vertex form (quadratic function), 293, 294, 389
   converting, 343–344, 375, 392, 393
   graphing, 304–315, 345–346, 390–391, 392–393
   solving equations, 335–343
Vertical asymptotes, 440, 506, 508, 560–563
Vertical intercepts, 21–24, 26–27, 375, 455
   constant terms, 46, 441–442
Vertical line test, 96–99, 119, 684–685
Vertical lines, 59–61
   domain/range and, 105
Vertices, 293
   ellipses, 704
   parabolas, 305, 375–376

W

Whole numbers, 656, A-1–A-2
Work problems, 597, 609–610

X

x values. *See* Domains; Independent variables
x-axis, 688, A-2–A-3
x-intercepts. *See* Horizontal intercepts

Y

y values. *See* Dependent variables; Ranges
y-axis, 688, A-2–A-3
y-intercepts. *See* Vertical intercepts

Z

Zero
   division by, 59
   exponent, 228–229, 281, 431, 517
Zero factor property, 350, 354–355

Photo Credits

VII: © 2010 Jose Luis Pelaez Inc/Blend Images/Jupiter Images; VII: © 2010 Getty Images/Jupiter Images; VIII: © 2010 Fuse/Jupiterimages Corporation; VIII: Patryce Bak/Getty Images; IX: © Matthew Watkinson/Alamy; X: (t) © Istockphoto.com/vanbeets; X: (b) © 2010 Kenneth C. Zirkel/Jupiter Images; XI: (t) © Andrew Charlesworth/Alamy; XI: (b) © Istockphoto.com/AccescodeHFM; XXII: istockphoto.com/DOUGBERRY; XXII: Image copyright seanelliottphotography 2010. Used under license from Shutterstock.com ; XXII: © Ian Leonard/Alamy; XXI: Image copyright © ZTS. Used under license from Shutterstock.com; XXII: Courtesy of White Loop Ltd; XXII: © Istockphoto.com/Chris Schmidt; XXI: © Istockphoto.com/clu; XXII: © Istockphoto/Carmen Martínez Banús

Chapter 1

1: © 2010 Jose Luis Pelaez Inc/Blend Images/Jupiter Images; 3: © iStockphoto/robert van beets; 4: © Image copyright Reinhold Foeger, 2009. Used under license from Shutterstock.com; 11: AP Photo/The Plain Dealer, Marvin Fong; 12: © iStockphoto.com/Elena Korenbaum; 14: © iStockphoto.com/malerapaso; 15: © iStockphoto.com/tillsonburg; 29: Courtesy of Debbie Hickman; 87: © Mark Clark; 90: Courtesy of Debbie Hickman; 108: © Image copyright Valentyn Volkov, 2009. Used under license from Shutterstock.com

Chapter 2

129: © 2010 Getty Images/Jupiter Images; 141: © Corbis 146: © Image copyright Jennifer King, 2009. Used under license from Shutterstock.com;147: AP Photo/The Vindicator, Bruce W. Palmer; 161: © iStockphoto.com/Damir Cudic; 178: 2010 © staceyb. Image from BigStockPhoto.com; 179: 2010 © staceyb. Image from BigStockPhoto.com; 186: 2010 Google-Map data © 2010 Google, INEGI; 190: © Image copyright Vadim Ponomarenko, 2009. Used under license from Shutterstock.com; 190: © Image copyright Golf Money, 2009. Used under license from Shutterstock.com 202: Javier Soriano/AFP/Getty Images; 212: © Image copyright Michelle Donahue Hillison, 2009. Used under license from Shutterstock.com

Chapter 3

221: © 2010 Fuse/Jupiterimages Corporation; 238: © Image copyright William Attard McCarthy, 2009. Used under license from Shutterstock.com; 242: © iStockphoto.com/Aldo Murillo; 248: © Image copyright Galina Barskaya, 2009. Used under license from Shutterstock.com; 252: © AGStockUSA/Alamy; 254: © iStockphoto.com/Daniel Stein; 255: lisafx/Big Stock Photo; 258: © Richard McGuirk/iStockPhoto

Chapter 4

291: Patryce Bak/Getty Images; 301: (bl) © iStockphoto.com/Kirby Hamilton; 301: (br) AP Photo/Paul Sakuma, file; 301: (br) Jay Laprete/Bloomberg via Getty Images; 302: Image copyright © JustASC. Used under license from Shutterstock.com; 312: Copyright © Bill Aron/Photo Edit Inc.; 314 © BananaStock/Alamy; 318: © BananaStock/Alamy; 325: © Blend Images/Alamy; 332: © iStockphoto.com/Emre YILDIZ; 332: Image copyright Gelpi, 2009. Used under license from Shutterstock.com; 333: © Image copyright Feverpitch, 2009. Used under license from Shutterstock.com; 334: AP Photo/The Vindicator, Bruce W. Palmer; 348: © 2010 Getty Images/Jupiter Images; 348: © Stacy Barnett/Fotolia; 352: AP Photo/Matt Sayles; 353: Erik Dreyer/Taxi/Getty Images; 358: © Christine Clark; 365: © BananaStock/Alamy; 386: © EuToch/Fotolia; 399: rusak/Big Stock Photo; 408: (bl) © iStockphoto.com/Don Nichols; 408: (cr) Image copyright © scoutingstock 2010. Used under license from Shutterstock.com

Chapter 5

413: © Matthew Watkinson/Alamy; 414: © Ted Horowitz/CORBIS; 420: © A. Huber/U. Starke/Corbis; 422: © istockphoto.com/Bluberries ; 427: © Horacio Villalobos/Corbis; 428: © Tony Freeman/Photo Edit Inc.; 438: (tr) Image copyright grzym, 2010. Used under license from Shutterstock.com; 438: (tr) Image copyright Vahe Katrjyan 2010. Used under license from Shutterstock.com; 463: © Daniel L. Geiger/SNAP/Alamy; 464: © Ron Niebrugge/Alamy; 470: (l) © Stephen Frink Collection/Alamy 470: (r) © Elvele Images Ltd/Alamy; 478: Image copyright © zixian. Used under license from Shutterstock.com

Chapter 6

483: © Istockphoto.com/vanbeets; 494: (bcl) Courtesy of US Census Bureau; 494: (cl) © iStockphoto.com/Janine Lamontagne; 522: © iStockphoto.com/Ernest Bielfeldt; 529: © ImageState/Alamy; 544: © iStockphoto.com/BanksPhotos

Chapter 7

551: © 2010 Kenneth C. Zirkel/Jupiter Images; 563: © Mark Clark

Chapter 8

611: ©Andrew Charlesworth/Alamy; 613: Istockphoto.com/Liliboas ; 625: © Buddy Mays/CORBIS; 664: Bill Ross/CORBIS; 669: Image copyright Michael Woodruff 2010. Used under license from Shutterstock.com; 671: Istockphoto.com/2happy

Chapter 9

683: © Istockphoto.com/AccescodeHFM; 691: istockphoto.com/bkindler; 701: (tr) istockphoto.com/OvertheHill; 701: (mr) Image copyright adv 2010. Used under license from Shutterstock.com; 701: (br) © Sean Elliott Photography 2010. Used under license from Shutterstock.com; 702: Image copyright sozdatel 2010. Used under license from Shutterstock.com; 715: (cr) © Elizabeth Whiting & Associates/Alamy; 715: (br) Image copyright yampi 2010. Used under license from Shutterstock.com; 716: (tl) © istockphoto/Peter Garbet; 716: (tcl) Image copyright David Hughes 2010. Used under license from Shutterstock.com; 728: istockphoto.com/foxtalbot; 737: © Ted Foxx/Alamy; 737: Image copyright 6th Gear Advertising 2010. Used under license from Shutterstock.com; 742: Image copyright Chad McDermott 2010. Used under license from Shutterstock.com

Unit Conversions

Capacity

U.S. system
8 fluid ounces = 1 cup
2 cups = 1 pint
2 pints = 1 quart
4 quarts = 1 gallon

Metric system
1 liter = 1.06 quarts
1 liter = 0.26 gallons
3.79 liters = 1 gallon
0.95 liters = 1 quart
29.57 milliliters = 1 fluid ounce
1,000 milliliters = 1 liter

Metric to U.S. system
1 liter = 1.06 quarts
1 liter = 0.26 gallons
3.79 liters = 1 gallon
0.95 liters = 1 quart
29.57 milliliters = 1 fluid ounce
1,000 milliliters = 1 liter

Length

U.S. system
12 in = 1 ft
3 ft = 1 yd
5,280 ft = 1 mile

Metric system
1 cm = 10 mm
100 cm = 1 m
1 km = 1000 m

Metric to U.S. system
1 meter = 1.09 yards
1 meter = 3.28 feet
1 kilometer = 0.62 miles
2.54 centimeters = 1 inch
0.30 meters = 1 foot
1.61 kilometers = 1 mile
1 AU = 1.496×10^8 km

Area

$9 \text{ ft}^2 = 1 \text{ yd}^2$
$43,560 \text{ ft}^2 = 1 \text{ acre}$

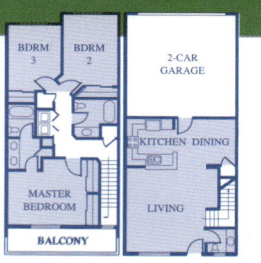

Temperature

$$F = \frac{9}{5}C + 32$$

$$C = \frac{5}{9}(F - 32)$$

Weight

1 kilogram = 2.20 pounds
1 gram = 0.04 ounces
0.45 kilograms = 1 pound
28.35 grams = 1 ounce
1 ton = 2000 pounds